Botany for Degree Students

Fungi

Botany for Degree Students

Fungi

B R VASHISHTA
Formerly Reader, Biology Study Group
Department of Botany
Punjab University, Chandigarh
Head of the Department of Botany
Multanimal Modi College, Modi Nagar and
Punjab University College, Hoshiarpur

A K SINHA
M.Sc., Ph.D., F.P.S.I., F.S.B.
Formerly Reader and Head
Department of Botany
Feroze Gandhi Postgraduate College
(Kanpur University), Raebareli (U.P.)

ANIL KUMAR
M.Sc., Ph.D.
Department of Botany
Feroze Gandhi Postgraduate College
(Kanpur University), Raebareli (U.P.)

S. CHAND
PUBLISHING

S Chand And Company Limited
(ISO 9001 Certified Company)

S. CHAND
PUBLISHING

S Chand And Company Limited

(ISO 9001 Certified Company)

Head Office: D-92, Sector–2, Noida – 201301, U.P. (India), Ph. 91-120-4682700

Registered Office: A-27, 2nd Floor, Mohan Co-operative Industrial Estate, New Delhi – 110 044
Phone: 011-49731800

www.**schandpublishing.com**; e-mail: **info@schandpublishing.com**

Marketing Offices:

Chennai	:	Ph: 23632120; chennai@schandpublishing.com
Guwahati	:	Ph: 2738811, 2735640; guwahati@schandpublishing.com
Hyderabad	:	Ph: 40186018; hyderabad@schandpublishing.com
Jalandhar	:	Ph: 4645630; jalandhar@schandpublishing.com
Kolkata	:	Ph: 23357458, 23353914; kolkata@schandpublishing.com
Lucknow	:	Ph: 4003633; lucknow@schandpublishing.com
Mumbai	:	Ph: 25000297; mumbai@schandpublishing.com
Patna	:	Ph: 4011400; patna@schandpublishing.com

First Edition 1962
Subsequent Editions and Reprints 1995, 96, 98, 99, 2000, 2002, 2003, 2005, 2007, 2008, 2010, 2011, 2014, 2016
Revised Edition 2016
Reprint 2018 (Twice), 2019, 2020 (Twice), 2021, 2022, 2025 (Twice)

Reprint 2026

ISBN: 978-93-525-3300-8 **Product Code:** H6BFU68BOTN10ENZX16O

PRINTED IN INDIA

By Shri Shyamlal Printing Press Private Limited, Plot 20/4, Site-IV, Industrial Area Sahibabad, Ghaziabad – 201 010 and Published by S Chand And Company Limited, A-27, 2nd Floor, Mohan Co-operative Industrial Estate, New Delhi – 110 044.

Preface to the Revised Edition

It is a matter of great pride for us to present the new edition of ***Botany for Degree Students—Fungi***. Every effort has been made to maintain the sanctity of the book by ensuring that the original format remains unchanged while revising the book.

Though the Fungi is the most fascinating group of plants, many of its members are being considered as animals. Due to its unique features, several new findings have come up in the field. In this new edition, a chapter on "Plant Parasitic Nematodes," has been incorporated which discusses morphology and anatomy, life cycle, common symptoms of nematodes causing diseases such as root knot nematodes, cyst nematodes, seed gall nematodes, etc. Additionally, a topic on "Life History and Contributions of Many Important Scientists" in the field of fungi and plant pathology has been included in the chapter "Fungi: General Topics and Discussion".

Student's suggestions have been considered while revising the book and it is hoped that it will be appreciated by the readers.

Our appreciation and thanks are due to many people for their help and encouragement. We are especially grateful to our undergraduate and postgraduate students whose suggestions helped in updating and enhancing the contents of the book. We are also thankful to the management and editorial team of S. Chand and Company Pvt. Ltd. for their help and support in publication of this revised edition.

We hope this edition will fulfil the expectations of its readers. Any comments and suggestions for the improvement of the book will be appreciated.

AUTHORS

Preface to the First Edition

The present text deals with an interesting and a fascinating group of primitive plants that have been placed by the taxonomists under the subdivision 'Fungi'. With the advancement of our knowledge of the subject, a large number of texts have appeared which are useful in their own way, but do not fully meet the needs of Indian students.The present treatise has been produced in a simple and a lucid style, so that even a mediocre student of Botany may very well understand the life phenomenon of these plants of 'fungal organisation'. A simple, clear,well illustrated and a concise account of the structure, reproduction, functions and relationships of important members of each class of fungi has been given. An introductory chapter in the beginning of the text is meant to acquaint the student with a general outline of the various aspects of the life history of the fungi. It also deals with the classification of this subdivision.

The subject matter of the book has been divided into four sections. Section I introduces the student with the Myxomycetes and the True Fungi. Life histories of important members of each class have been dealt with in a simple way and all the relevant details have been given. Graphic representations of the life cycles have been added at appropriate places so as to enable the reader to form a clear idea of the phenomenon of generations. A general account of the 'Fungi Imperfecti'has also been given at the end of this section. Section II includes a general consideration of the bacteria and thier importance to man. A preliminary account of the viruses has also been added in this section. Virology is a new and recently developed branch of mycology which deals with the little known and as yet imperfectly explored but extremely important organisms of debatable nature. A concise but a clear account of these organisms is given in this section, so as to make the students aware of the nature and other important aspects of the viruses. Section III is devoted to a brief account of the Lichens. Section IV deals with the plant diseases. It starts with an introductory chapter that deals with this important subject and gives an overall picture of the symptoms and control measures of the plant diseases in general.The concluding chapter deals with some of the plant diseases caused by the parasitic fungi, the bacteria and the viruses. Revision questions have been given at the end of each chapter.

No claim of originality is made in the preparation of this book. It is simply a compilation work done in a manner so as to meet the needs of students of Botany of the Indian universities. A large number of standard books on the subjects and research journals have been consulted.Grateful thanks are due to the authors, editors and publishers of these books and journals. I thank my colleagues in the department, Dr. V. P. Dubey and Dr. M.R. Sharma of Meerut College, Meerut,for their suggestion in the preparation of the manuscript. My sincere thanks go to Dr. A. B. Gupta, Head of Botany Department, Christ Church College, Kanpur and Prof. D. N. Srivastava, Head of Botany Department, Bareilly College, Bareilly, for the encouragement and inspiration received from them. Finally I wish to express my thanks to my publishers for bringing out the book in a nice form.

Suggestions towards improvement, if any, will be highly appreciated by me.

B.R.VASHISHTA

CONTENTS

SECTION I - DIVISION MYCOTA

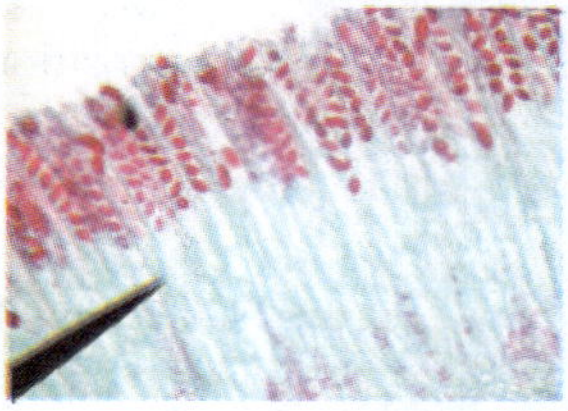

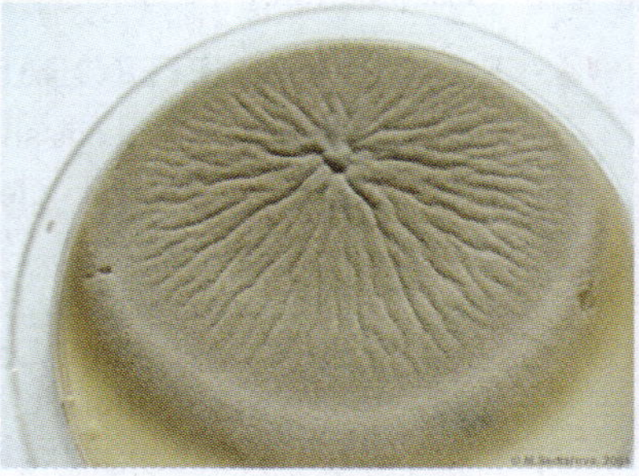

SECTION II - SCHIZOMYCOTA, VIRUSES AND MYCOPLASMA

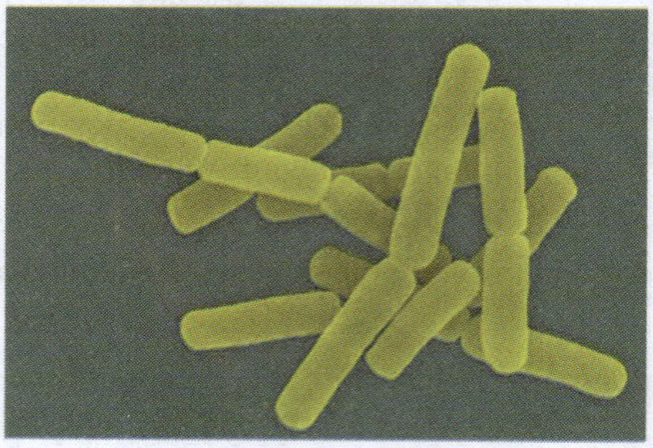

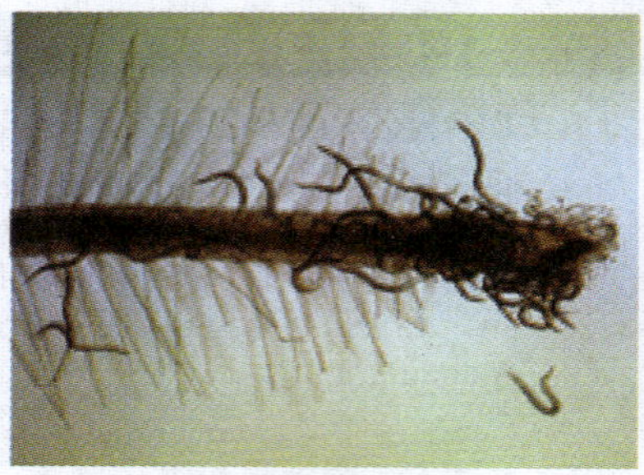

SECTION I

DIVISION MYCOTA

Mushroom – an edible fungus

CHAPTER 1

THE FUNGI General Characters and Classification

- Somatic or Vegetative Phase
- Reproductive Phase – Reproduction in fungi
- Whittaker's five kingdom System & Beyond
- Classification of fungi
  - A. The lower fungi
  - B. The higher fungi
  - C. Recent development in fungal taxonomy
- Brief History of Mycology
- Origin and phylogeny of fungi

FUNGI is the plural of the word FUNGUS which is derived from the latin word FUNGOUR which means **to flourish**. The word was primarily used with reference to MUSHROOMS which develop overnight. In usage, the meaning of the word has been expanded to include thallus like achlorophyllous plants such as the molds and other similar organisms related to mushrooms. The fungi, thus are a large group of simple thallus like plants which lack chlorophyll. The older botanists employed the term in a wider sense to include bacteria, slimemolds as well as true fungi. About 5100 genera and more than 50,000 species of fungi are known today. Familiar examples of the fungi are the yeasts, molds, toad stools mushrooms, polypores, puffballs, rusts and smuts.

The branch of botany that deals with fungi is called **mycology** (Gr. *mykes* means mushroom and

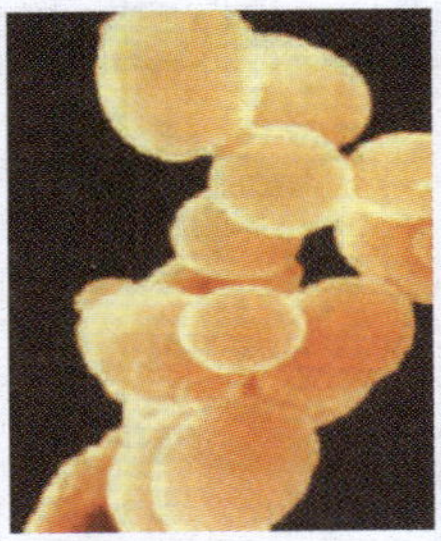

Yeasts

Molds

Stools

Mushrooms

Polypores

Puffballs

Rusts

Smuts of barley

logos means study of). Etymologically, mycology is the study of mushrooms. It deals with life histories, relationships and evolutionary tendencies of fungi. The scientist who is concerned with fungi is called a **mycologist**. He studies structure, reproduction, physiology and taxonomy of fungi.

In common language, the fungi may be defined as non-green, nucleated thallophytes. However, mycologists have defined fungi more scientifically. According to Alexopoulos (1962), the fungi include nucleated spore bearing achlorophyllous organisms that generally reproduce sexually and whose filamentous branched somatic structures are typically surrounded by cellwalls containing cellulose or chitin or both. A more technical definition of fungi was later given by Bessey (1968) which says that fungi are chlorophyll-less non-vascular plants whose reproductive or vegetative structures do not permit them to be assigned to positions among recognised groups of higher plants or algae.

On the basis of the above mentioned, the characteristics of fungi may be summarised as follows :

Habitat : Ubiquitous i.e found everywhere as saprophytes, parasites or hyper parasites and/or symbionts.

Thallus : On or in the substratum and plasmodial amoeboid; or unicellular or multicellular, filamentous; if filamentous septate or aseptate

Cell wall : Well defined; made up of chitin or of cellulose

Nuclear status : Eukaryotic; Uni–bi–or multi nucleate; homo–or heterokaryotic.

Nutrition : Heterotrophic

Reproduction : Asexual, sexual/or parasexual through spores.

Distribution : Cosmopolitan.

Habitat. The fungi grow in diverse habitats. In fact they are found in almost every available habitat on earth where organic material (living or dead) is present. They are, thus, universal in their distribution. Great many of them are terrestrial. They occur in soil which abounds in dead, decaying organic materials. The terrestrial fungi thrive best in humus soil. They are considered more advanced. They produce non-motile reproductive cells which are dispersed passively by wind, water or animals. Some fungi attack living organisms. They live in tissues of plants and animals. Some fungi are aquatic. The aquatic fungi are considered primitive. They live on decaying organic matter and living organisms found in fresh water and produce flagellate (motile) reproductive cells which swim to new localities. Many fungi grow on our foodstuffs such as bread, jams, pickles, fruits and vegetables. Some fungi are found in drinking water. Thus, the fungi contaminate both our food and drinking water. They are present all the time in air that we breathe. Majority prefer to grow in darkness and dim light in moist habitats.

Habits or modes of Life. The fungi are plants of very varied habits. They lack chlorophyll and like animals are unable to manufacture their own food from carbondioxide and water. In their mode of nutrition, the fungi thus differ from all green plants. They obtain food readymade from an external source. All fungi are, therefore, **heterotrophs.** In their mode of nutrition, they are **heterotrophic.** However, like all other plants, they cannot ingest solid food but absorb it directly through cell membranes either by living as **saprophytes** (*saprobes*) or **parasites.** Thus, according to their mode of nutrition, the fungi are classified into two cateogries, the **saprophytes** or **saprobes** and the **parasites.** The saprophytes grow where dead organic matter abounds in the substratum. This mode of life is called **saprophytic** (saprobic). The parasites live in or on the living bodies of other organisms (plants and animals) and obtain food from them. This mode of life is called **parasitic.** The best examples of parasitic fungi are the *rusts* and *smuts*. The organism on which the parasite feeds is called the **host** or **suscept**. The presence of the parasite may cause an abnormal condition of the host which is called **disease.** The parasitic fungi are harmful organisms. The thallus of a parasitic fungus may grow on the external surface of the host but usually it is hidden from view. The former are called the **ectoparasites** (Vine molds) and the latter **endoparasites** (*Pythium debaryanum*, *Ustilago* and *rusts*). In the endoparasites, the unseen but vitally active thallus grows in the tissues of the host plant.

There is, however, not always a sharp distinction between parasites and saprophytes. Among the parasites, one can distinguish the following three degrees of parasitism :—

1. Obligate parasites. They can grow only upon suitable living host tissues. The best examples of obligate parasites are the *downy* and *powdery* mildews. So far all attempts to grow them on dead artificial media have failed.

Downy mildews

Powdery mildews

2. Facultative saprophytes. Normally they live as parasites and attain their best development as such. However, they are able to grow apparently indefinitely as saprophytes under emergent circumstances. To this category belong *Taphrina deformans*, the *leaf curl fungi* and some smuts. *Phytophthora infestans* is a parasite on potato. It causes *potato blight diseases*. In the laboratory, it can be grown as a saprophyte on cultures of oat meal agar.

Phytophthora infestans on potato

Taphrina deformans on peach leaves

3. Facultative parasites. Normally they are saprophytes in their habit but may now and then become parasites. Some *Fusarium* species are good examples of this type. They live as saprophytes in the soil. When suitable host plants are sown in such a soil, they attack them and start living as parasites. *Botrytis cinerea* is another good example of a facultative parasite.

Orange fruit infected with Penicillium

The fungi which live strictly as saprophytes are called the **obligate saprophytes**. They are incapable of infecting plants or animals. The common examples of obligate saprophytes are *Mucor*, many field toadstools and most of the species of *Penicillium*.

Fusarium ear rot of maize

Mucor

Botrytis cinerea

According to the damage done to the host plant, the parasitic fungi may be divided into two sorts, **destructive parasites** and **balanced parasites.** The destructive parasites cause the worst diseases of our field crops, garden crops and forest trees. They kill the host cells or tissues or produce substances toxic to the host. The balanced parasites are so adapted in their demands upon the host that the latter is able to supply the needs of the parasite and continues to live itself.

Lichens Exhibiting Symbiosis.

In parasitism, advantages are all on one side *i.e.* to the parastie. The disadvantages to the host may vary from a trifling inconvenience to complete destruction. There is, however, another type of association in which the fungus enters into a partnership with another organism. In this case, the advantages are not all on one side. Both the partners are mutually benefited. This type of association is known as **symbiosis.** The best example of symbiosis is furnished by lichens. The thallus of a lichen consists of a fungus living symbiotically with an alga. In this partnership, each member profits by the presence of the other.

Pine Root with Mycorrhiza.

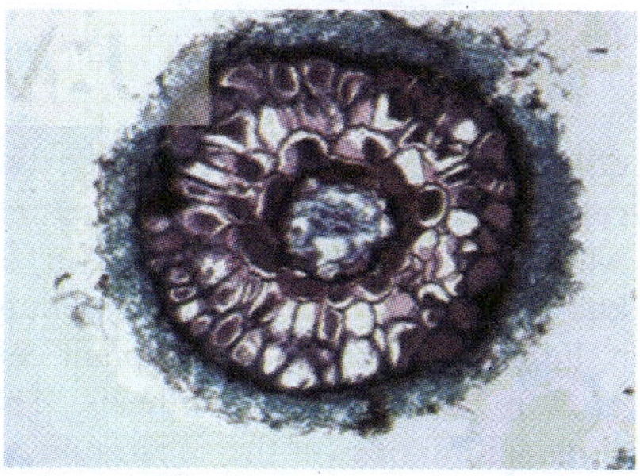

Ectotrophic Mycorrhiza on Pine Root.

Another example of symbiosis is furnished by **mycorrhiza**. In this case, a fungus establishes a relationship with the roots of certain higher plants particularly the forest trees. For example, in *Pinus* the fungal hyphae form a weft around the pine roots. The root hairs are absent in this region of the root. It is suggested that the fungal hyphae perform the work of the root hairs. They absorb water and minerals in solution from the soil and transmit to the roots of the tree. The tree provides food to the fungus. This type of mutually beneficial association between a fungus and the roots of a higher plant is called **mycorrhiza.** It is **ectotrophic** when the fungal hyphae live on the surface and **endotrophic** when they penetrate the root and grow inside the root tissues.

The life cycle of a fungus consists of two phases. the **somatic** or **vegetative phase** and the **reproductive phase.**

SOMATIC OR VEGETATIVE PHASE

1. Thallus. The vegetative phase of a fungus is a **thallus**. It may be **unicellular or filamentous**.

(*a*) Unicellular Thallus (Fig. 1.1). In some of the lower fungi such as the **chytrids,** the thallus is more or less a spherical, single-celled structure (A). At the time of reproduction, it becomes a reproductive unit. The latter produces the asexual or sexual cells. Such fungi are called **holocarpic.** In them, the vegetative and reproductive stages do not occur together in the same thallus. *Plasmodiophora* has a vegetative phase consisting of a naked, multi-nucleate, amoeboid mass of protoplasm (D). It is termed **plasmodium.** The protoplast of the diploid plasmodium cleaves to form the resting spores. The yeasts, which are related to the filamentous forms, also have a unicellular thallus (B). In the unicellular holocarpic forms (*Synchytrium*, Fig. 4-4D), the mycelium is absent.

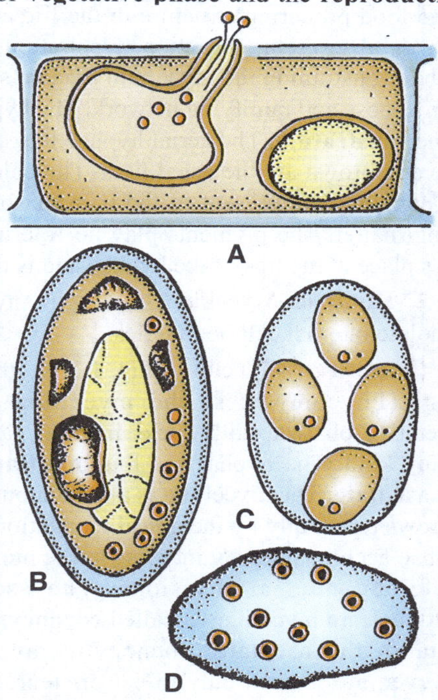

Fig. 1.1. (A—D). *Fungi.* Unicellular Holocarpic Thalli. A, *Olpidium endogenum* (holocarpic chytrid), two organisms within the algal host; B, **Yeast,** unicellular thallus; C, Reproductive phase of B; D, Diploid plasmodium of *Plasmodiophora*.

(b) Filamentous Thallus (Fig. 1.2). The vast majority of the fungi have a **filamentous** thallus. It originates through the germination of a spore. The spore germinates as it lands on a suitable substratum where other conditions of life are also favourable. In some species, the spore, on germination, produces only a short, tubular structure of limited growth. It constitutes the thallus and is technically called a **hypha.** The spores of most of the fungi, however, give rise to a fluffy thallus consisting of a cottony mass of fine, branched filaments. These long, fine filaments are called the **hyphae** (sing. hypha). Some of these hyphae, at a certain stage of maturity, extend into the air and bear the reproductive bodies. The rest spread over or within the substratum and continue the normal activities. Such fungi are called **eucarpic.** Collectively the hyphae comprise the vegetative body (thallus) of a fungus which is called the **mycelium.** The hypha is thus a structural unit of the **mycelium.** It consists of a thin, transparent wall filled or lined with a layer of cytoplasm. The medium upon which the mycelium grows is known as **substratum**. The mycelium is the food procuring structure in the life cycle. It carries on the general activities of a plant cell such as absorption, digestion, respiration, excretion and growth but not photosynthesis. The hyphae constituting the mycelium branch, spread in all directions within or over the substratum to form a loose and ramifying network. The hyphae are usually colourless particularly those embedded in the **substratum.** The aerial hyphae in some fungi become coloured. Black, orange, yellow, red, blue and brown are the usual tints. The colour is usually confined to the hyphal wall. Even when the pigments are present in the protoplasmic contents, they do not form an integral part of the living matter. The pigments play no role in the physiology of the fungus. The growth in length takes place at the tips of the hyphae and is thus termed **apical.**

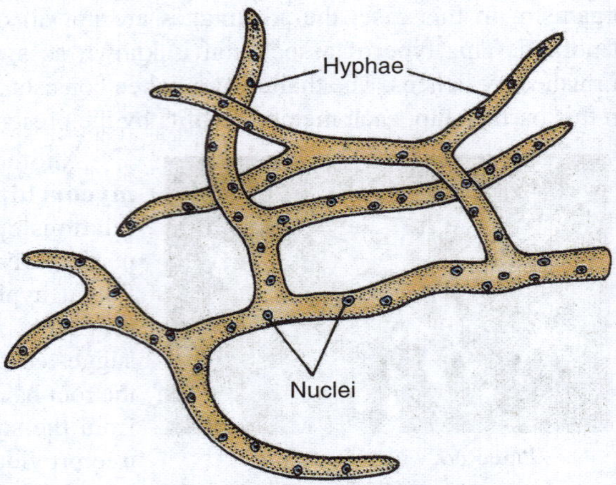

Fig.1.2 *Fungi.* Filamentous Thallus. Aseptate coenocytic mycelium.

2. Kinds of Mycelia. The fungus mycelium, in fact, is a multinucleate structure lacking complete internal cell bounderies.

(i) Aseptate Mycelium (Fig. 1.2). In the algal fungi (Phycomycetes) the mycelium in the vegetative phase usually lacks internal partitions of any kind. The hyphae are thus **multinucleate** and **aseptate.** The mycelium is a continuous mass. It grows terminally by the apical elongation of the hyphae accompained by increase in the number of nuclei by nuclear divisions. The aseptate, multinucleate mycelium is called **coenocytic.** The septa in the aseptate, coenocytic mycelium, however, are formed only to cut off reproductive structures, or to seal off a damaged portion. Such septa are solid plates. They lack pores.

(ii) Septate Mycelium (Fig. 1.3). The members of other classes of fungi (Ascomycetes and Basidiomycetes) develop internal cross walls

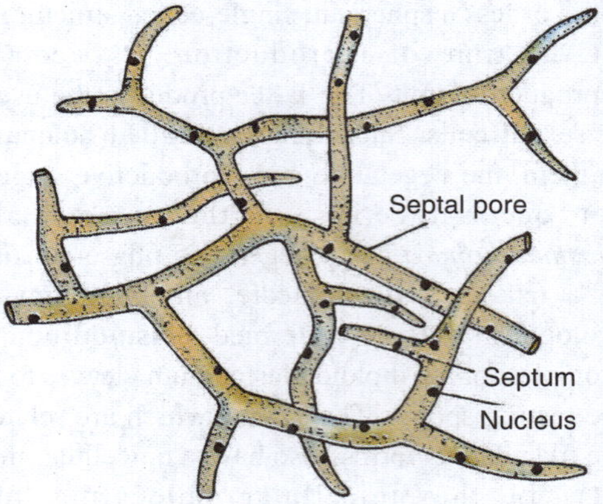

Fig. 1.3. *Fungi.* Filamentous Thallus. Septate mycelium with uninucleate cells.

called the **septa** which divide the hyphae into segments. The septa appear at regular intervals behind the hyphal tip. The segments may be uninucleate (Fig. 1.3) or multinucleate (Fig. 1.4). In a **septate** mycelium the septa between the cells are transverse. Longitudinal or oblique septa are rare. The formation of septum is always preceded by the division of nucleus. If the cell has more than one nuclei there is simultaneous division of all the nuclei before wall formation. It must, however, be borne in mind that the septa in the septate mycelium of fungi are incomplete. Each has a central pore and rarely more than one pore. Through these the cytoplasm, cytoplasmic organelles and even nuclei can regularly pass from one cell to other. It is evident, therefore, that the distinction between coenocytes and septate hyphae is not so profound as it was formerly thought to be. The presence of septa gives mechanical support to the hyphae. Complete partitions do not occur in the vegetative phase of fungi. In some cases, the septa possess more than one pore and rarely none at all.

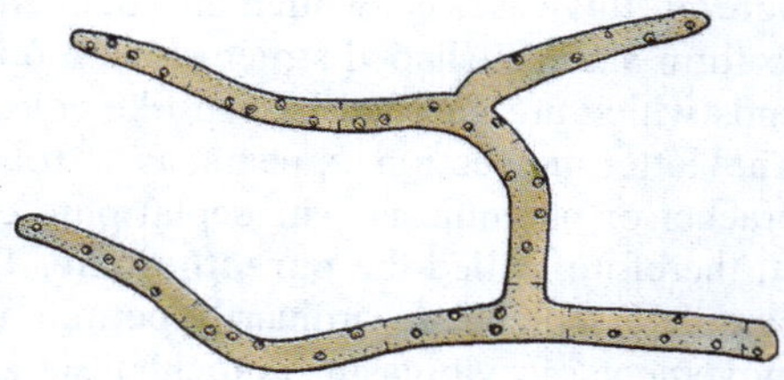

Fig. 1.4. *Fungi*. Filamentous Thallus. Septate mycelium with multinucleate cells.

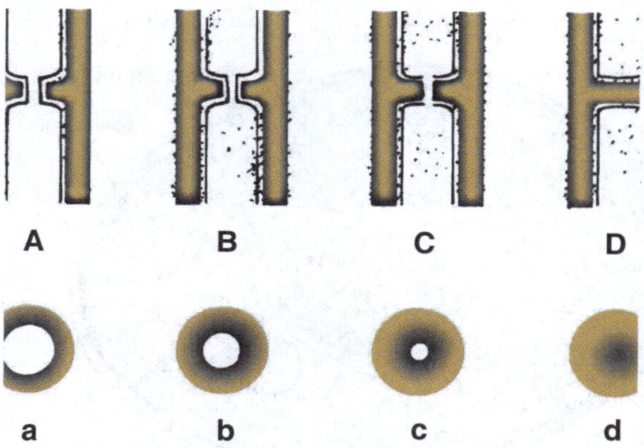

Fig. 1.5 (A-D). *Fungi*. Stages in the development of a transverse septum.

Origin of Septa (Fig. 1.5). The septa in all the fungi (septate Phycomycetes, Ascomycetes, Basidiomycetes and Imperfect Fungi) have a similar origin. They arise at regular intervals behind the hyphal tip. The septum originates at the periphery on the inside of tubular hyphal wall as a ring (annular) or rim of wall material. The annular growth grows slowly inwards towards the centre (centripetally) like an iris diaphragm increasing in width and decreasing the diameter of pore. A cross wall or septum is finally formed which is seldom complete. The centripetal growth of the septum in fungi is rapid. According to Buller (1933), the aseptate *Rhizopus* takes 20-25 minutes to completely seal off a damaged portion. The complex dolipore septum in *Rhizoctonia* (Basidiomycetes) is completed in 10 minutes. *Ciboria* takes only 6 minutes.

Septal Pore (Fig. 1.6). Usually a small hole or a pore remains in the centre of the septum (A). The perforated septum gives rigidity to the tubular hypha maintaining, at the same time, the protoplasmic continuity from cell to cell. In all the septate fungi with the exception of the Basidiomycetes, the septum has a simple central pore in the middle of the cross wall (Fig. 1.6). The septum in some may be slightly swollen where the pore is formed. In the Basidiomycetes (Fig. 1.7), there is further elaboration of the septum to produce a more complex pore named a *dolipore* (Latin *dolium* meaning a large jar). The rim of the central

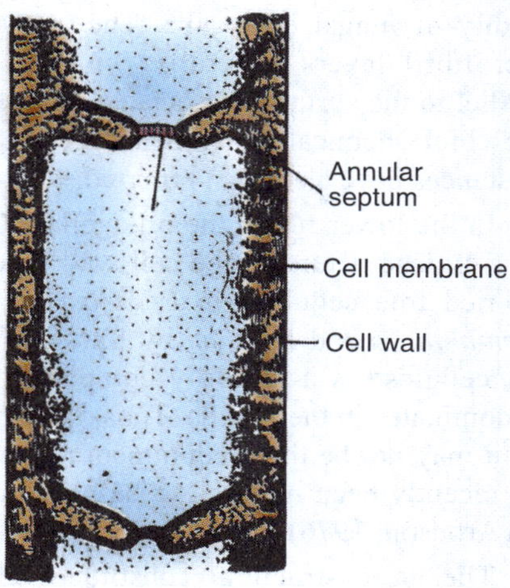

Fig. 1.6. *Fungi*. Annular septum with a simple pore.

pore, in this case, is swollen and thickened to form a barrel-shaped structure with open ends which are guarded by cap-like covers. The latter in section appears as a round bracket or parenthesis. The septal pore cap is, therefore, called the **parenthesome**. The complete septal pore ordinarily permits the movement of cytoplasm, mitochondria and nuclei from cell to cell but under certain conditions the parenthesome may close the passage. The dolipore parenthesome is the hallmark of the Basidiomycetes.

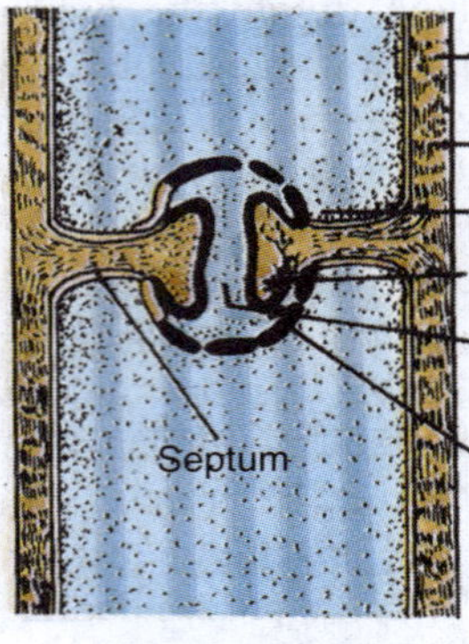

Fig. 1.7. *Fungi.* Annular septum of a basidiomycete with a doplipore septum.

3. Structure of Fungal cell (Fig. 1.8). With the exception of slime molds, the fungal cell usually consists of a strong, rigid cell wall enclosing the protoplast.

(*a*) The Cell Wall. The composition of cell wall is variable among the different groups of fungi or between the different species of the same group. In the majority of fungi, the wall lacks cellulose but contains a form of chitin known as the **fungus cellulose** which is strictly not identical with insect chitin. The suggested formula for fungus chitin is $(C_{22}H_{54}N_4O_{21})_n$. Electron microscope studies reveal that chitin occurs as elongated variously oriented microfibrillar units. These are laid down in layers and form the basis of the structural rigidity of fungal cell walls. The microfibril layers generally run parallel to the surface. Associated with the microfibrillar components is the nonfibrillar material. The chief chemical constituents are various polysaccharides, but proteins, lipids besides other substances have also been reported.

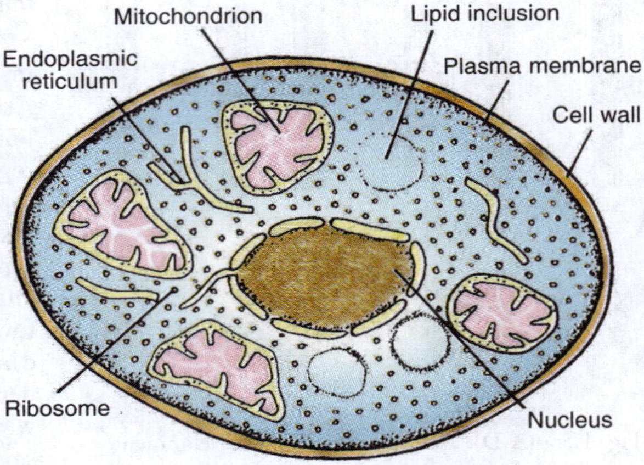

Fig. 1.8. *Fungi.* Fine structure of **Torula Yeast** cell based on an electron micrograph.

In the lower fungi, the biflagellate Oomycetes are said to be distinct from all other fungi in the **cellulose** nature of the cell wall (Frey, 1950 and Parker *et al,* 1963). De Bary (1866, 1887) reported true cellulose in *Peronospora* and *Saprolegnia.* Precise analysis of the cell wall of *Phytophthora* and *Pythium* by Bartnicki-Garcia (1966), Mitchell and Sabar (1968) has revealed that cellulose is a minority component or even absent altogether. On the other hand, glucan predominates in their walls. Thus, the Oomycetes may be said to have cellulose in their cell walls but it may not be the predominant material. Chitin which had long been considered to be absent has recently been reported to be present even in the cell walls of some Oomycetes (Lin, Sicher and Aronson, 1976)

The basic structural constituent of the cell wall in the Zygomycetes and higher fungi (Ascomycetes and Basidiomycetes) is chitin. It is a polysaccharide based on the nitrogen containing sugar (glucosamine). It is probable that more or less closely associated with chitin in the cell wall are pectic materials, protein, lipids, cellulose, callose and minerals. The clear evidence of such an association is, however, lacking. Burnet (1968) is of the opinion that insoluble *B* glucan forms the predominant structural material of the wall of Ascomycetes and Basi-diomycetes. In addition chitin may as well be present in appreciable amounts. In the yeasts and a few other Hemiascomycetideae chitin is absent. Their walls are mainly composed of microfibrils of mannan and *B* glucan (Hawker, 1966). Mannan is a polymer of hexose sugar mannose whereas glucan is polymer of glucose. Some investigators have reported the occurrence of lignin in several fungi. It is doubtful whether this substance is chemically the same as the lignin of higher plants. It is obvious that our present knowledge of the chemical composition of the cell wall of fungi is incomplete. Like the cellulose wall, the chitin wall of most fungi is permeable both to water and substances in true solution.

(*b*) The Protoplast. The living substance of the cell within the cell wall is the protoplast. It lacks the chloroplasts but is differentiated into the other usual cell parts such as plasma or cell membrane, vacuolated cytoplasm, cell organelles and one or more nuclei.

Cell Membrane. It is a delicate, extremely thin, living membrane which closely invests the protoplast. The cell or plasma membrane is pressed against the cell or hyphal wall except for occasional invaginations in some regions. The invagination is either in the form of an infolded convoluted pocket or a pouch enclosing granular or vesicular material. Moore and Mc Lear (1961) named it **lomasome.** Actually the plasma membrane is the surface layer of the protoplast altered to perform special functions. It is differentially **permeable** and shows a typical tripartite structure under the electron microscope. There is an electron dense layer on either side of the less dense central region.

Cytoplasm. Within the plasma membrane is the colourless cytoplasm in which sap-filled vacuoles may occur. In young hyphae and hyphal tips, the cytoplasm appears rather uniform and homogeneous. Immersed in the cytoplasm are structures known as the **organelles** and **inclusions.** The organelles are living structures, each with a specific function. The inclusions are dead, have no specific function and thus are not essential to cell survival. Amongst the cell organelles are included the endoplasmic reticulum, mitochondria, ribosomes, golgi apparatus and vacuoles. Lomasomes which are membranous structures lying between the cell wall and plasma membrane are common. Examples of inclusions are the stored foods (glycogen, and oil drops), pigments and secretory granules.

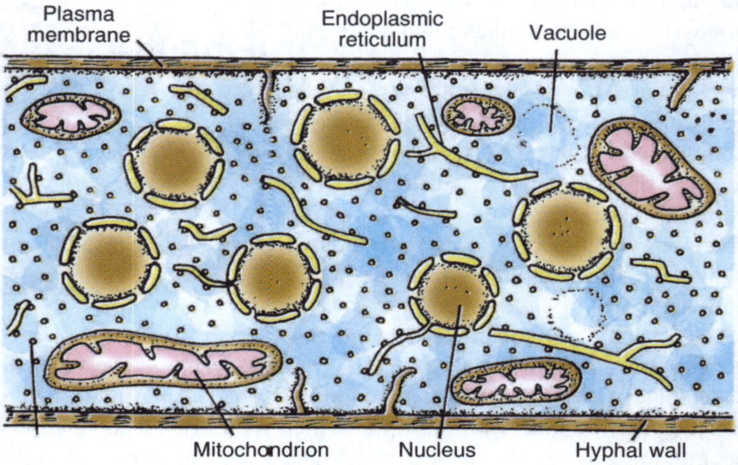

Fig. 1.9. *Fungi.* Fine structure of a hypha near the growing tip of *Mucor* based on an electron micrograph.

(*i*) *Endoplasmic Reticulum.* The presence of endoplasmic reticulum in the fungal cytoplasm has been demonstrated by the use of electron micro-scope. It is composed of a system of membranes or microtubular structures usually beset with small granules which by some scientists are likened to the ribosomes. In many fungi, the endoplasmic reticulum is highly vesicular. Usually it is loose and more irregular than in the cells of green plants.

(*ii*) *Mitochondria.* The cytoplasm contains small, usually spherical bodies known as the **mitochondria.** Each mitochondrion is enveloped by a double membrane. The inner membrane is infolded to form the **cristae** which are in the form of parallel flat plates or irregular tubules. The cristae contain the same fluid that fills the space between the two membranes. The mitochondria function as the power house of the cell. There is no fundamental difference between the mitochondria of fungi and those of green plants. However, Hawker (1965) holds that the cristae of fungal mitochondria are fewer, flatter and more irregular than those of the green plants.

(*iii*) *Golgi Apparatus* (*Dictyosomes*). With the exception of Oomycetes there is less certainty of the occurrence of structures similar to those of the **golgi apparatus (dictyosomes)** in fungi. Moore and Muhlethaler (1963) reported a golgi apparatus consisting of three flattened sacs surrounded by many bubble-like structures in *Saccharomyces* cells.

(*iv*) *Vacuoles.* The cytoplasm of young hyphae or fungal cells and hyphal tips lacks vacuoles. They appear further back or in the old cells. With age, they enlarge and show a tendency to coalesce and ultimately reduce the cytoplasm to thin lining layer immediately within the cell wall.

(*v*) *Inclusions.* The cytoplasm contains various kinds of inclusions. Examples of stored foods are lipid globules, granules of glycogen, oils and the carbohydrate trehalose, proteinaceous material and volutin. The glycogen may occur in vacuoles. There are no starch grains. Of the pigments, the fungi lack chlorophyll. Carotenoids are often conspicuous by their presence and may occur throughout the cytoplasm or concentrated in the lipid granules or distributed in the cell walls. The cytoplasm, in addition, secretes several kinds of ferments, enzymes and organic acids.

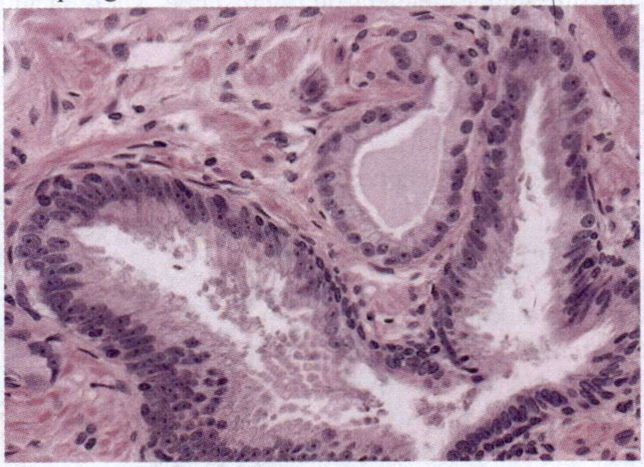

Nucleolus.

Nucleus. The cytoplasm in the individual cells contains one, two or more globose or ellipsoid nuclei which in the somatic portion are small and usually range from 1-2 or 3μ in dimaeter. They cannot be seen without special techniques. Structurally the nucleus consists of: (*i*) a central, dense body with a clear area around it. (*ii*) Chromatin strands, and (*iii*) the whole structure surrounded by a definite nuclear membrane. The central body takes heavy iron haematoxylin stain and is usually Feulgen-negative. In electron micrographs, it appears as an amorphous or granular mass. Mycologists usually designate it as the **nucleolus.** Bakerspigel (1960) stated that it contains **RNA.** During nuclear division, the chromatin strands become organised into chromosomes which are extremely small and difficult to count. Under the electron microscope, the nuclear membrane is seen to consist of inner and outer layers of electron dense material and the middle one of eletron transparent substance. The nuclear membrane has pores. At certain points, the nuclear membrane is continuous with the endoplasmic reticulum.

Somatic Nuclear Division. The older mycologists believed that the somatic nuclear division in most fungi (except yeasts) differs markedly from normal mitosis. Moore (1964) termed it **karyochorisis** in contradiction to normal **mitosis** or **karyokinesis.** Lu (1974), however, stated that in fungi mitosis occurs more or less normally and the fungal nucleus passes through the stages characteristic of mitosis namely, interphase, prophase, metaphase, anaphase and telophase. The cell division is closed and thus termed **intranuclear.** It means the nuclear membrane remains more or less intact and does not disorganise during prophase. Centrioles appear in close association with the nuclear membrane. They apear in a pair and each migrates to the opposite poles of the nucleus to control the formation of the spindle.

Fungal Flagella (Fig. 1.10). The Higher Fungi (Ascomycetes, Basidiomycetes and Deuteromycetes) lack motile cells in the life cycle. With the exception of Zygomycetes, they are characteristic of the lower Fungi (Phycomycetes). The motile cells (zoospore and gametes) in the Lower Fungi are furnished with one or two fine, protoplasmic, whip-like threads known as the *flagella*. These are thin, hairlike emergences of the cell cytoplasm which function as propelling or locomotory structures of the cell. There is either a single anterior or posterior flagellum or they occur in a pair on the cell.

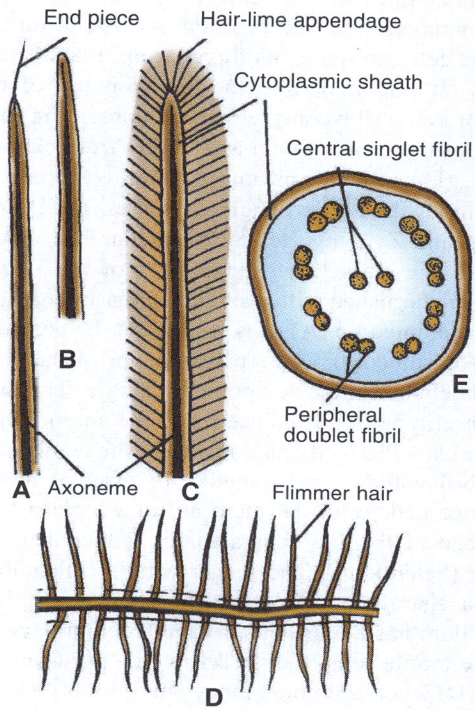

Fig. 1.10 (A-E). *Fungal Flagella.* A,L.S. upper portion of an acronematic or whiplash flagellum with a drawn out pointed tip or end piece; B, L.S. whiplash flagellum with blunt tip; C, L.S. Pantonematic or tinsel flagellum; E, T.S. eucaryotic flagellum showing 9+2 arrangement of fibrils; D, Diagrammatic representation of a part of the tinsel flagellum from an electron micrograph (Based on

(*a*) *Structure of the Flagellum* (E). Forming the core of the flagellum is a central or axial filament known as the **axoneme** (A-C). The axoneme is surrounded by a double cytoplasmic sheath or membrane which is an extension of the plasma membrane of the motile cell. In a cross section, the axoneme is seen to be composed of eleven fibrils (E). Of these, two are central and nine peripheral. The two central single fibrils lie side by side and form an elastic axial thread which is sometimes enclosed by a subsidiary sheath of its own. Of the nine peripheral fibrils, each consists of two thin fibrils. The nine peripheral doublet fibrils surround the central pair of singlet fibrils. The entire growth of 9+2 fibrils is enclosed by a double cytoplasmic sheath. This 9+2 arrangement of fibrils is the basic structure of a eucaryotic flagellum found in the motile cells of all organisms except the bacteria.

(*b*) *Kinds of flagella.* The fungal flagella are reported to be of the following three kinds:—

(*i*) Whiplash flagellum with an end piece (Fig. 1.10 A). It has a smooth surface and narrows abruptly towards the tip to form a distinct **end piece.** The end piece is a thin extension of variable length. It is flexible. According to Manton, the end piece is the naked portion of the axoneme. It is composed of eleven fibrils. Koch (1956) and Kole (1957) stated that the end piece consists of the two central fibrils only.

(*ii*) Whiplash flagellum with a blunt tip (Fig. 1.10 B). The blunt whiplash flagellum has smooth surface but lacks the narrow tip (end piece).

(*iii*) Tinsel flagellum (Fig. 1.10 C and D). It does not obviously end in a narrow tip like the blunt whiplash type flagellum but bears fine lateral hair-like appendages known as the **flimmer hairs** or mastigoneme (C). The flimmer hairs on the tinsel flagellum arise from the axoneme (D). The tinsel flagellum is also known by other names such as **pantonematic, flimmer** or **flimmergeisel flagellum.** The other names of the whiplash flagellum are **acronematic** or **peitchgeisal flagellum**.

Flagellum in the Lower Fungi (Fig. 1.11). The number, kind and position of flagella on the motile cell constitutes its **flagellation.** Flagellation plays an important role in the taxonomy of the Lower Fungi (Phycomycetes). It is constant in each class of the Lower Fungi and differs from class to class. Thus, it forms an important feature in classifying the Lower Fungi into classes. The Higher Fungi and Zygomycetes lack flagella. The motile cell in the class Chytridiomycetes of the Lower Fungi is furnished with a single whiplash flagellum with a pointed tip (A). It is inserted at the posterior end (opisthocont). The representatives of the class Hyphochytridiomycetes posses a single flagellum of tinsel type (B). It is inserted at the anterior end. In the class Plasmodiophoromycetes, the motile cells are biflagellate. The flagella are inserted at the anterior end. Both the them are of whiplash type but one of them is several times longer than the other (heterokont). The longer whiplash flagellum has a sharply pointed tip. The shorter whiplash flagellum has a blunt tip. Biflagellate motile cells are also characteristic of the class Oomycetes. In the motile cells of this class with a pyriform shape, the two flagella are inserted at the anterior end (D) whereas in the kidney-shaped motile cells, the flagella arise from the depression. One of these is of tinsel type and the other whiplash with a sharply pointed tip. In motion, the tinsel flagellum is directed forward and the whiplash is directed backward.

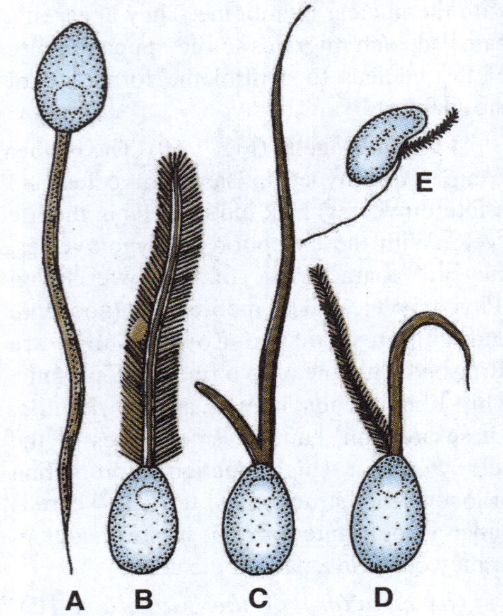

Fig. 1.11. (A-D). Flagellation in the *Lower Fungi.*

Aggregation and Modification of the Hyphae. In majority of fungi, hyphae are interwoven loosely to form *mycelium*. In some fungi, the whole mycelium or parts thereof aggregate in different ways or undergo various modifications which are used to perform various somatic and reproductive functions. Different types of aggregations or modifications of hyphae mycelium are as follows :

***(i)* Plectenchyma** (Fig. 1.12). A false tissue formed by aggregation of hyphae is known as Plectenchyma. All fungal tissues come under this general term. Plectenchyma is of two types :

(a) **Prosenchyma** (B). It is rather a loosely woven tissue of hyphae. The hyphae composing it do not loose their identity. They run more or less parallel to one another and are composed of elongated cells.

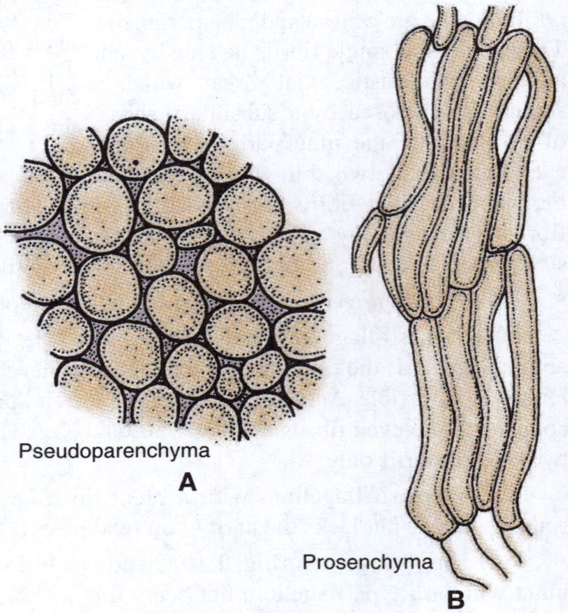

Fig. 1.12. (A-B). *Fungi.* Kinds of fungal tissues.
A. Pseudoparenchyma; B. Prosenchyma (Diagrammatic)

(*b*) **Pseudoparenchyma** (A). In the fructifications of higher fungi, the hyphae become woven and intertwined into a compact mass.

The walls of the hyphae in the mass get fused and they lose their individuality. As a result, the hyphal mass appears to be continuous structure consisting of isodiametric or oval cells (A). It bears a striking superficial resemblance to the parenchyma tissue of the higher plants and is called **pseudoparenchyma.**

(*ii*) **Rhizomorph** (Fig. 1.14 A, B). A thick strand or root like aggregation of somatic hyphae is called **Rhizomorph.** The hyphae loose their identity and individuality and the whole mass behaves as an organised unit. It is believed that rhizomorph has a higher infection capacity than individual hyphae. Examples of Rhizomorphs are found in *Armillariella mellea*.

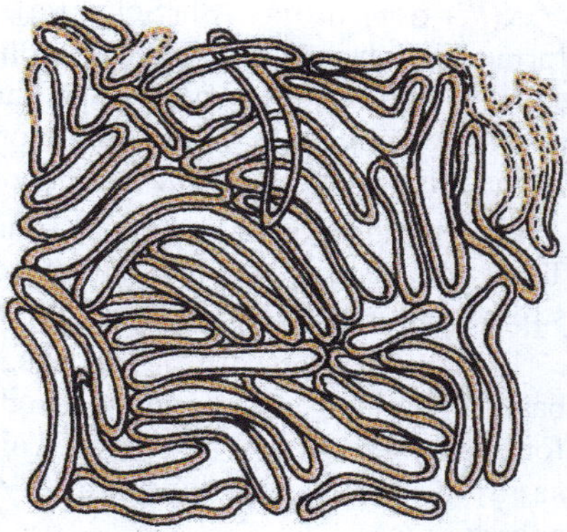

Fig. 1.13. *Fungi.* Prosenchymatous tissue in the pith of *Claviceps purpurea*; the hyphal elements are recognisable (Diagrammatic).

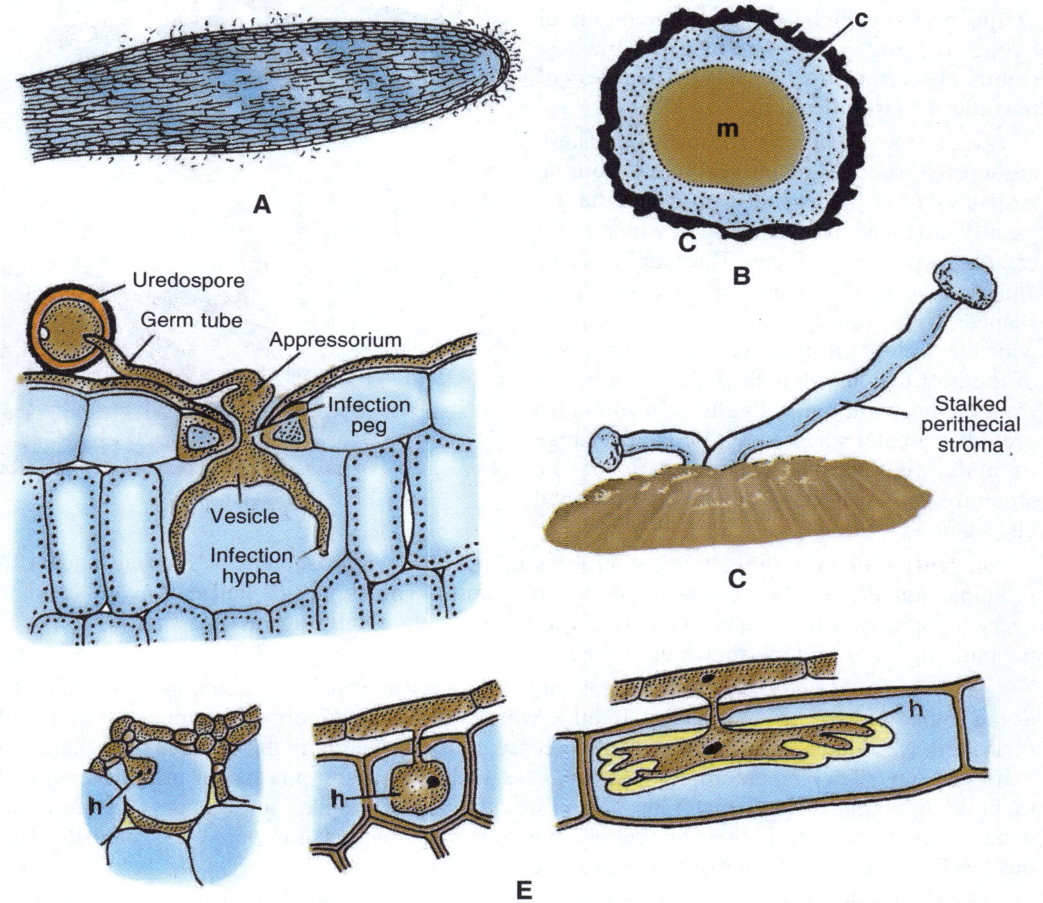

Fig. 1.14. Aggregation and Modifications of hyphae. A-B, Rhizomorph; C, Sclerotia; D, Appressorium; E, Haustorium.

***(iii)* Sclerotium** (Pl. Sclerotia). A sclerotium is a compact globose or elongated structure formed by the aggregation and adhesion of hyphae. It may survive for long periods of time sometimes for several years and thus represent the resting stage of the fungus. The sclerotia usually germinate to form hyphae or may form reproductive structures. Sclerotia are commonly formed in *Claviceps purpurea, Rhizoctonia solani* and *Macrophomina phaseoli* (Fig. 1.14 C).

***(iv)* Stromata.** Any fungal tissue that forms reproductive structures are called **Stromata**. These are compact somatic structures like mattresses. Generally, sclerotia on germination form stromata in which reproductive structures develop.

***(v)* Pseudosclerotium** (pl. pseudosclerotia). These sclerotia like bodies are formed at the base of various fruit bodies of higher fungi. In *Polyporus basilapiloides*, the pseudosclerotia formed below the soil surface are composed of sand particles surrounded by hyphal aggregations.

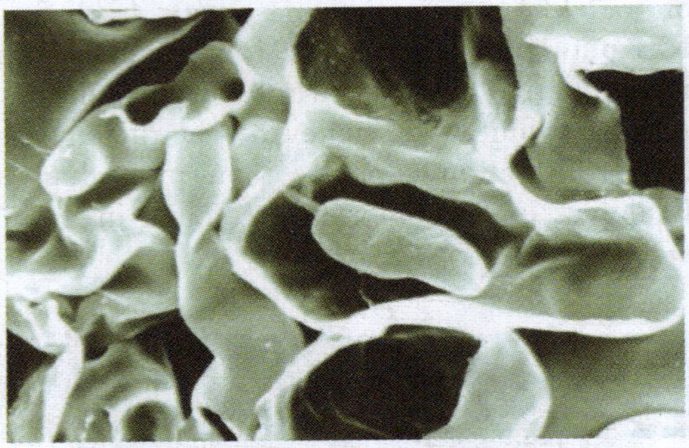

Rust SEM Intracellular Haustorium

***(vi)* Appressorium** (pl. Appressoria). These are common in parasitic fungi mostly ectoparasites. An appressorium is a terminal simple or lobed swollen structure of germtubes or infection hyphae. It adheres to the surface of the host and helps in the penetration of hyphae of the pathogen. Appressoria are commonly formed by the parasitic members of the order **Erysiphales** (Fig. 1.14 D).

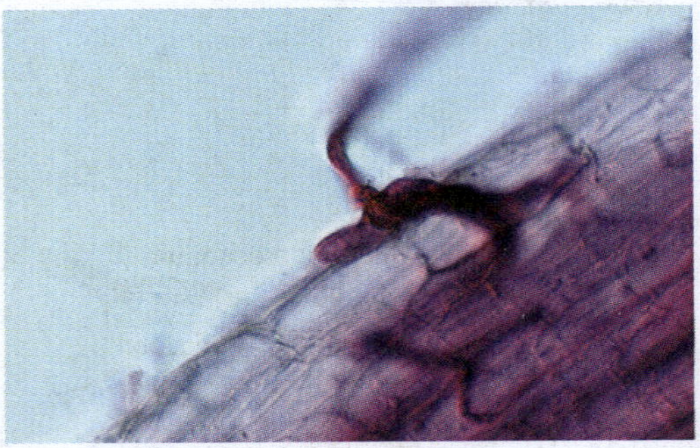

Appressorium

***(viii)* Haustorium** (pl. Haustoria). These are mostly produced as intracellular absorbing structures of obligate parasites. Haustoria are usually produced in those fungi in which inter-cellular mycelium are found. They vary in shape and may be knob shaped or branched finger shaped. They secrete certain specific enzymes which hydrolyse the proteins and carbohydrates of the host cell and thus they absorb nutrients from the host without killing it. Haustoria also provide a greater surface area for the exchange of materials (Fig. 1.14 E). Details of the structure of haustorium are being described elsewhere.

4. Nutrition in Fungi. It is the process of procuring nourishment or nutrient and is the principal function of the vegetative phase (mycelium). The supply of nutrients is essential for every living organism because the nutrients serve as structural units and also as energy sources for the building up of protoplasm which is the basis of life.

***(a)* Nutritional requirements.** The fungi utilise both organic compounds and inorganic materials as the source of their nutrient supply. In other words, organic and inorganic compounds constitute their food. No fungus is able to make any increase in its dry weight in the absence of organic food materials, why? Lacking chlorophyll the fungi are unable to photosynthesize or use carbon dioxide to build up organic food materials. They are, thus heterotrophic for carbon (organic) food compounds which they in their natural habitats obtain by living as saprophytes or parasites from dead or living plants, animals or micro-organisms or their wastes.

***(b)* Essential Elements.** The constituent elements of the organic and inorganic substances which fungi make use of are C, O, H, N, P, K, Mg, S, P, Mn, Cu, Mo, Fe and Zn. Calcium is required by some fungi but not all. These elements which fungi require as food are termed the essential elements. Some of these elements. the fungi need in extremely small trace amounts and the others in comparatively larger amounts. The former are called the **trace** or **micro elements** and the latter **macro elements.** The fungal growth is adversely affected or the fungus fails to grow if any one of the essential elements is lacking in the culture medium. Examples of the macro elements are C. N, O, H, S, P, K and Mg. The macro elements are body builders and provide energy for metabolic processes.

(c) Sources of Macroelements. The organic substances ususally utilized by fungi are very varied in nature. Yeasts, for example, can use acetates as sources of **carbon** but for most fungi, the chief sources of carbon are the carbohydrates. The carbohydrates are needed for building up the body and also as a source of energy. In a typical fungus, 50% of the dry weight is carbon. Of the carbohydrate source of carbon, most fungi use simple sugars. Glucose, for instance, is suitable for almost all fungi. Next in preference are the fructose. Less commonly used are the hexose sugars and some pentoses. Xylose among the pentoses has been reported to be superior to glucose for some fungi. Mannitol is equivalent to glucose for many fungi. Maltose which occurs in nature as a byproduct of starch hydrolysis is utilized by many fungi. Sucrose is also a good source of carbon for some. From among the polysaccharides, starch and cellulose are utilised by a fewer fungi which can synthesize the appropriate hydrolytic enzymes. Some fungi are able to make good growth on fats as the only source of carbon. Organic acids are generally poorer sources of carbon for most fungi. Basidiomycetes include most of the lignin-utilizing fungi. Proteins, lipids, some organic acids and higher alcohols are utilised by some fungi as a sole energy source. Growth, however, is always better on a substance containing a suitable carbohydrate.

Besides carbon, fungi require **nitrogen.** To obtain nitrogen, they utilise both organic and inorganic materials as the source. The chief organic sources of nitrogen are protein, peptide or an amino acid. Certain groups of fungi show specializations in respect of certain nitrogen sources. For example, the Saprolegniaceae and Blastocladiales include a number of species which grow only with organic nitrogen such as amino acid. In nature, fungi decompose proteins and other materials to obtain their supply of nitrogen. In pure cultures amino acids, peptides, or peptones, gelatin, casein and egg albumin can serve as sources of organic nitrogen for building up protoplasm. Urea is also considered as a utilisable nitrogen source for some fungi.

Many fungi, however, obtain nitrogen from inorganic sources. A number of fungi are known which use both nitrate and ammonium salts. Robbin (1939) and Lindberg (1944) reported that *Absidia sp., Mucor hiemalis, Lenzites trabea and Marasmius sp.* use ammonia but are incapable of utilizing nitrate salts. Fewer fungi are able to utilize nitrate salts. Nitrites can be toxic. Organic sources of nitrogen can also serve as sources of carbon. There is not much evidence to support nitrogen fixation in saprophytic fungi. Metcalfe and Chayen (1954) reported that soil inhabiting *Rhodotorula* and yeast-like *Pullularia pullans* fix atmospheric nitrogen. It is certain, however, that nitrogen fixation is not a widespread ability in fungi.

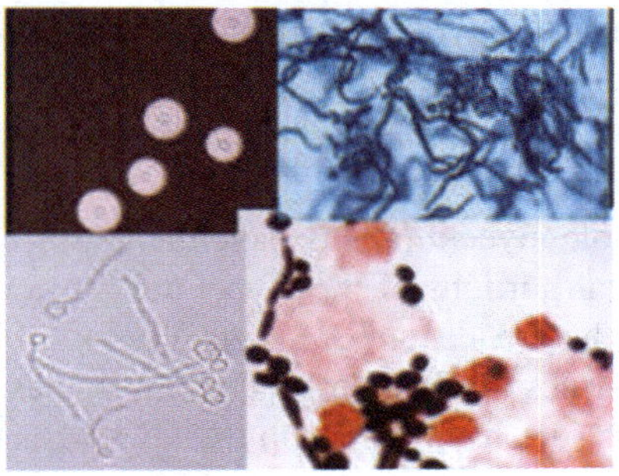

Rhodotorula

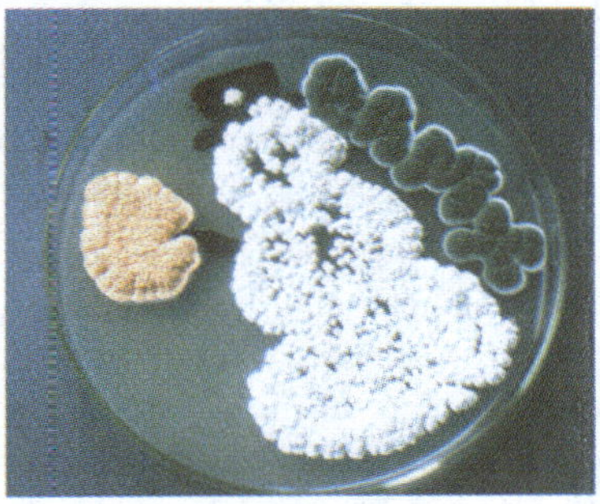

Pullularia pullans.

Hydrogen and oxygen are supplied in the form of water which is the major constituent of fungus mycelium forming about 85-90% of the entire weight.

The chief among the inorganic nutrients which the fungi require in fairly large amounts for their mineral nutrition are **sulphur, phosphorus, potassium** and **magnesium.** These **macronutrients** the fungi obtain from simple inorganic salts or sources such as sulphates for sulphur, and phosphates for phosphorus. These must be supplied in any culture medium. Calcium is not known to be needed by the fungi in general. Some, however, require it as **a micronutrient.** Some fungi are reported to require only minute traces of iron, zinc, copper, manganese, boron, and cobalt and molybdenum. These trace elements or micronutrients are considered essential for growth. The form in which the major and the minor metallic element requirements are utilised is the **anion.** Fungi store excess food in the form of glycogen or lipids.

The fungi, like all other organisms, require minute amounts of specific, relatively complex organic compounds for growth. These are the **vitamins** or **growth factors.** Many fungi synthesize their own supply of appropriate growth factors from a simple nutrient solution of defined composition. Such fungi are thus autotrophic for vitamins and are called **auxo-autotrophic.** They need no exogenous supply. There are others which depend in whole or in part on an external source, because they are unable to synthesize one or more of the essential growth substances. The fungi heterotophic for their needs of growth factors are termed **auxo-trophic.** There are marked difference between the vitamin demands of the different species of a genus or even the strains of a single species. The important fungal vitamins, which may function in **enzyme systems** include thiamine (B_1), biotin, pyredoxine (B_6) and riboflavin (B_2). A few fungi also need nicotinic acid and pantothenic acid. The vast majority, however, require thiamine (B_1). The growth factors are catalytic in their actions. To sum up, the basic nutritional needs of fungi are:—

(*i*) A suitable organic compound as a source of carbon and energy.

(*ii*) A suitable source of nitrogen.

(*iii*) Inorganic ions of sulphur, phosphorus, potassium and magnesium in significant amounts.

(*iv*) Inorganic ions of iron, zinc, copper, manganese and molybdenum only in minute traces.

(*v*) Certain vitamins or organic growth factors in trace amounts.

Besides the nutritional requirements listed above the growth of fungi is influenced by certain habitat factors such as temperature, oxygen supply, moisture, pH value and by-products of metabolism.

(d) Modes of nutrition. The fungi lack chlorophyll. They are, therefore, unable to synthesize carbohydrate food from inorganic materials and get it readymade from sources external to themselves. These heteromorphs according to their method of obtaining food are divided into two categories, namely, the **saprophytes** or **saprobes** and **parasites.** The former absorb the required organic nutrients directly through the cell membrane from the substratum which abounds in dead organic matter of both animal and plant origin. The saprophytes cannot ingest solid food. Yeast and molds are the common examples of saprophytic fungi. The parasite lives in or on the living body of a plant or animal and absorbs organic molecules as nutrients through the cell walls from the tissues of the host. Rusts and smuts are the common parasites.

(e) Mechanism of nutrition (Fig. 1.15). The whole mycelium may have the power to absorb these nutrients or this task may be assigned to special portions of the mycelium. In saprophytic fungi, the hyphae (*Mucor mucedo*) or rhizodial hyphae (*Rhizopus stolonifer*) come in intimate contact with the nutrients in the substratum (A) and absorb soluble smaller molecules such as sugars and amino acids. Insoluble complex substances such as proteins, lipids and polysaccharides are first broken into soluble monomeres (digested) by secreting extra-cellular enzymes and then

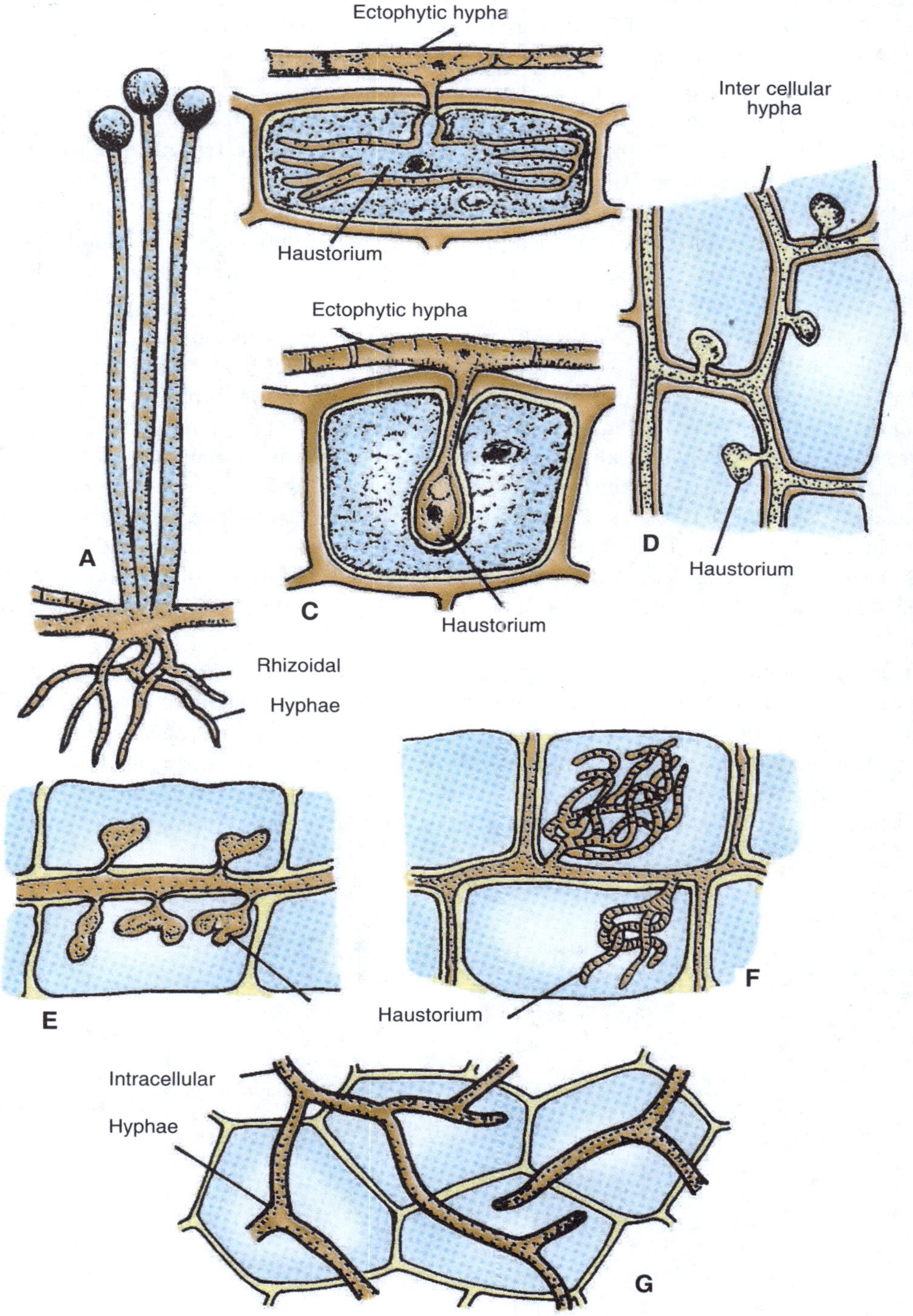

Fig. 1.15 (A–G). *Fungi*. Modes of Nutrition. Explanation in the text.

absorbed. The fungal hyphae secrete enzymes which convert insoluble complex food materials in the substratum to soluble ones. The latter are then absorbed by direct diffusion either through the hyphal walls of the hyphae that penetrate the substratum or by the rhizoidal hyphae.

The mycelium of the parasites is rarely ectophytic but frequently it grows inside the host. The hyphae either ramify in the intercellular space between the host cells (D) or penetrate into the host cells (G). The former are called **intercellular hyphae** and the latter **intracellular hyphae.** The intercellular hyphae obtain nutrition through the cell walls or membranes of the host cells. This they do by secreting an enzyme upon the plasma membrane of the host cell. It makes the membrane more permeable to the contained solutes. The latter diffuse out and are absorbed by the hyphal walls. The hyphal walls of the intracellular hyphae come in direct contact with the host protoplasm (G) and obtain food by direct diffusion.

The intercellular hyphae of some highly specialised (obligate) plant parasites give out slender, lateral outgrowths. The hyphal outgrowth punctures the host cell wall making a minute pore through which it enters the host cell. Within the host cell, it enlarges to form a globose (D), lobed (B), or branched (F) absorptive organ. This type of feeding organ of the parasitic fungi is called a **haustorium.** It is markedly specialised in structure to absorb nutrition from the host tissues. The haustorium is intracellualr and thus robs the host of its food without killing it. Haustoria are characteristic of obligate parasites. They vary in shape and size in different fungi. In *Albugo*, the haustorium is a button-like (D) or spherical structure. *Peronospora parasitica* has sac-like haustoria (E) in the leaf cells of *Capsella*. *Peronospora calotheca* produces branched filamentous haustorium in the stem cells of *Galium* (F). *Erysiphe graminis* forms an elongated branched haustorium inside the host cell (B).

Each haustorium (Fig. 1.16) usually consists of two parts, a constricted region which is in the form of a narrow penetration tube and the expanded or branched region on the host cell. The penetration tube is usually clasped by a 'collar' of host wall material. The enlarged or expanded region of the haustorium causes invagination of the cytoplasmic membrane of the host cell. The latter remains closely appressed to the wall of the haustorium. There is a zone of apposition enclosing the haustorium between the fungal wall and the unbroken cytoplasmic membrane of the

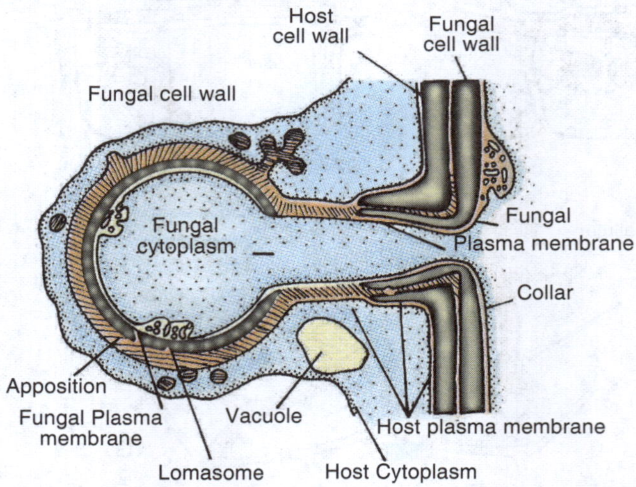

Fig. 1.16 Fungi. A section through an haustorium of *Albugo candida* based on an electron micrograph. (After Berlin and Bowen)

host. Its origin is in dispute. The secretion from the haustorium upon the plasma membrane of the host makes it permeable to solutes contained in the sap cavity. They diffuse out and are then absorbed by the haustorium. The parastic fungi do not produce haustoria in artificial cultures. The haustoria are also not produced by fungi which live as parasites on animals.

The fungi, as mentioned above, are unable to synthesize sugars from carbon dioxide and water. They, however, can synthesize from soluble sugars the more complex carbohydrates which are the chief components of their cell walls. They are also able to synthesize proteins and eventually protoplasm if supplied with carbohydrates and simple nitrogen compounds such as ammonium salts. Besides ammonium salts, they can absorb and utilize many complex but soluble organic nitrogenous compounds.

Many fungi obtain nutrition by living in mutually beneficial associations with other plants. The association is not causal but permanent and is established during long process of evolution. The two best known examples of mutualisitc associations of fungi with other plants are *Symbiosis* and *Mycorrhiza*.

(a) Symbiosis. The common example of symbiosis is an association of a fungus and an alga in a lichen thallus. The two organisms in this association are so intertwined as to form a single composite thallus plant which differs from either of the partners in form and habit. The duty of alga in this partnership is to synthesize food with the help of green chloroplasts and share it with its fungal partner. The fungus absorbs minerals in solution and water from the substratum and passes them on to the alga. The fungal hyphae which form the bulk of the lichen thallus provide shelter to the alga, in addition.

(b) Mycorrhiza. (pl. Mycorrhizae or mycorrhizas). It is defined as the symbiotic association between the hyphae of certain fungi and roots of plants. It is of three types : (*i*) ectomycorrhiza, (*ii*) endomycorrhiza, and (*iii*) ectoendomycorrhiza.

(*i*) *Ectomycorrhiza*. The fungal hyphae in this case form a complete envelope around the root tip and also penetrate and extend into the first few cortical layers to form an intercellular network of hyphae known as the **Hartignet**. The hyphal strands that extend into the substrate from the envelope absorb water and nutrients from the soil and pass them on to the roots of the plant through the Hartig's net. The presence of the fungus thus increases root absorption. In return the fungus receives food and shelter.

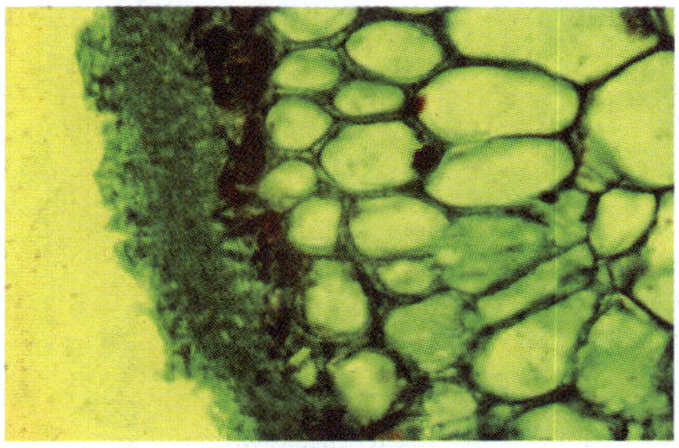

Ectomycorrhiza c.s Pine Root

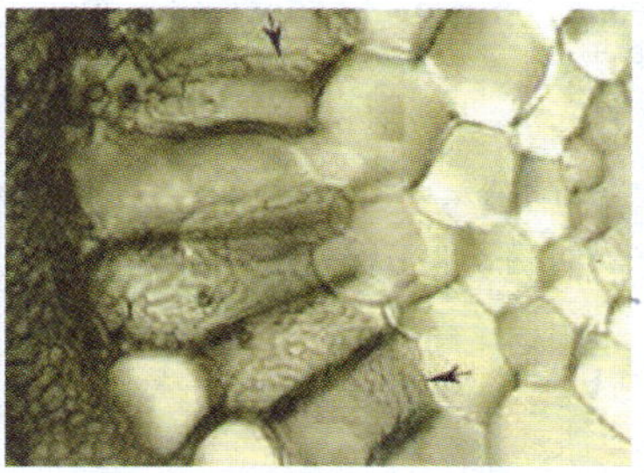

Hartig Net of Populus

(*ii*) *Endomycorrhiza*. The fungal hyphae, in this case, penetrate root hairs, epidermis and reach the cortex where they grow intracellularly forming fungal knots in the cortical cells. A portion of the mycelium lives in the soil but it forms no dense hyphal growth (envelope on the surface of the root).

(*iii*) *Ectoendomycorrhiza*. It is a combination of the first two. The fungal hyphae form a sheath at the surface of the root. Within the root, they grow intercellularly and intracellulary.

(Discussed in detail in chapter 17).

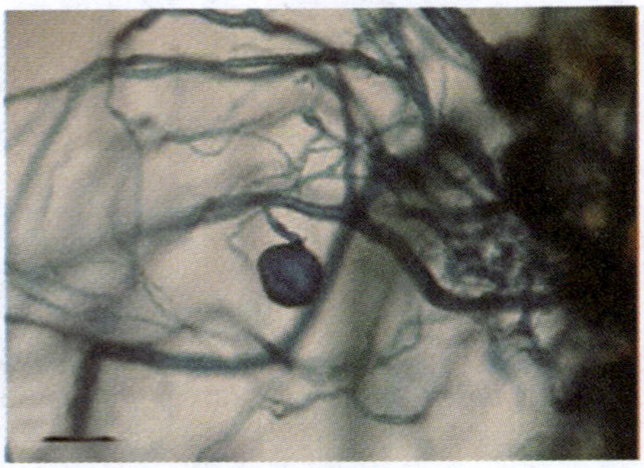

Endomycorrhiza

Ectoendomycorrhiza

REPRODUCTIVE PHASE

When the mycelium has reached a certain stage of maturity and has accumulated reserve food, it starts reproducing. Reproduction involves the production of new individuals resembling the parent or parents.In fungi, reproduction is of three kinds, **vegetative, asexual** and **sexual.** Many mycologists, however, include all methods of reproduction (vegetative and spore formation) which do not involve union of nuclei or sex cells or sex organs under **asexual** or **somatic** reproduction.

ASEXUAL REPRODUCTION IN FUNGI

Most of the fungi prefer asexual mode of reproduction which lacks meiosis and nuclear fusion during the process. It takes place through Fragmentation, Budding, Fission and spore formations. Fragmentation is very common in fungi where fragmented mycilia can develops into a new individual (Fig. 1.17). Members of *Saccharomycetes* (yeast) reproduce via budding. The parent cell develops simply bud like protuberances which detach from the parent cell and develops a new individual. Yeast also multiply through fission in which each cell divides into two by forming a cell wall in between and each cell enlarges to form a new individual. (Fig. 1.18) The process is common in bacteria also. Most common type of asexual reproduction in fungi are through spore formation, Fungal spores are minute, globose to oval in structure having different colour, shape and size. Broadly fungal spores are divided into two group conidia and the sporangiospores (Fig. 1.19).

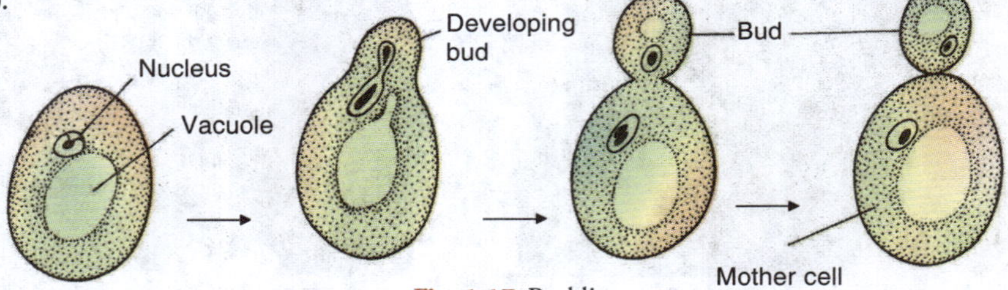

Fig. 1.17: Budding

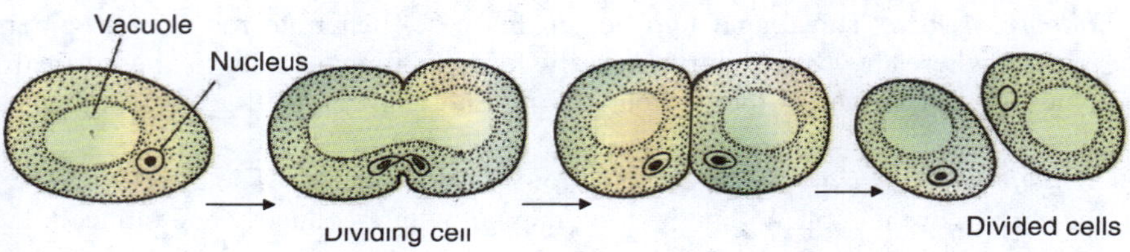

Fig. 1.18: Fissian

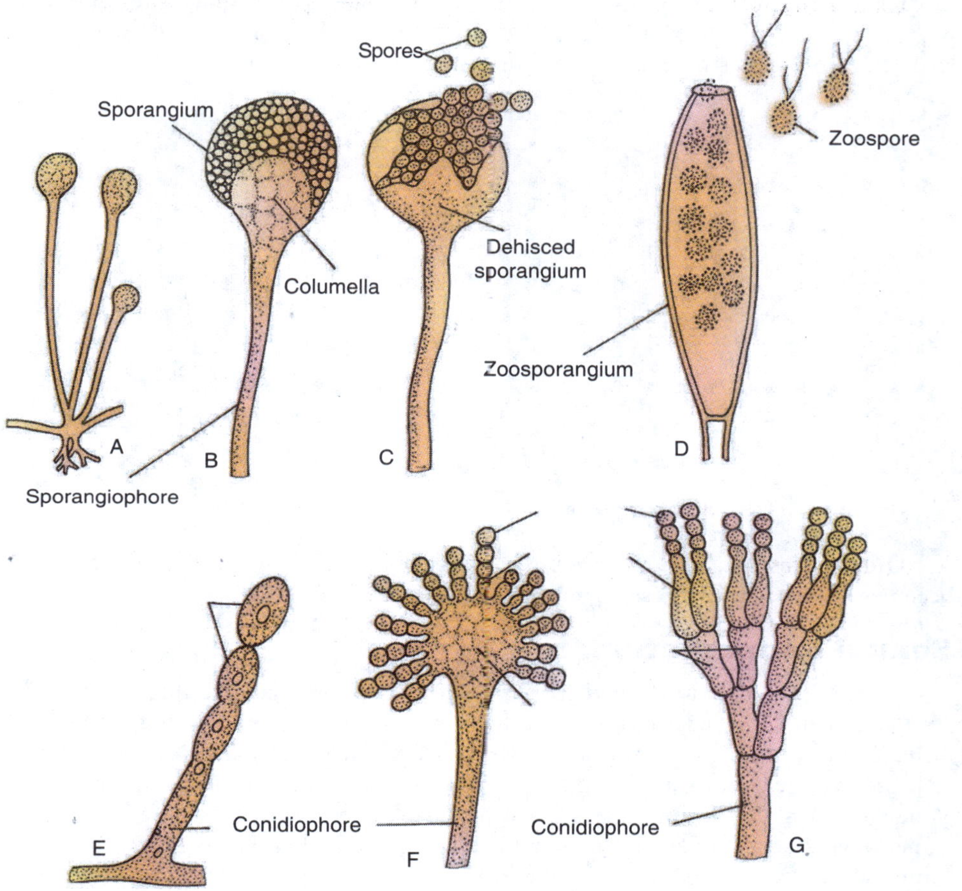

Fig.1.19: Common types of asexual reproduction A – C Spores; D. Zoospores; E – G Conidia

Sporangiospores

The spores which are produced inside the sporangia are called sporangiospores. Since these spores are produced inside the sporangia also termed as Endospores. Sporngiospores are of two types:

(i) *Aplanospores*: When spores are non motile like produced in sporangium of *Rhizopus* and *Mucor they* are termed as aplanospores.

(ii) *Zoospores*: When spores produced inside the sporangium are flagellate and motile they are termed as Zoospores. These zoospores may be uniflagellate or biflagellate. E.g. Uniflagellate zoospores in *Synchytrium endobioticum* and biflagellate in *Saprolegnia*, *Albugo*, *Phytophthora* etc.

Conidia (Sing : Conidium)

These spores are produced exogenously either singly or in the form of chain on the hyphal tip or lateral branch of the hyphae. The specialized branches having conidia or conidial chain are called conidiophores. Conidia germinate directly after falling on the suitable host or surface by giving out gem tube. Eg. Conidial chain in *Albugo*, conidia of Fungi imperfecti.

Chlamydospores: (Fig.1.20) Such spores are formed by rounding off of the contents of one or more hyphal segments. They are difficult to detach from fungal mycelium, thick walled oval to rouh structures remain dormant for long time some times upto many years.

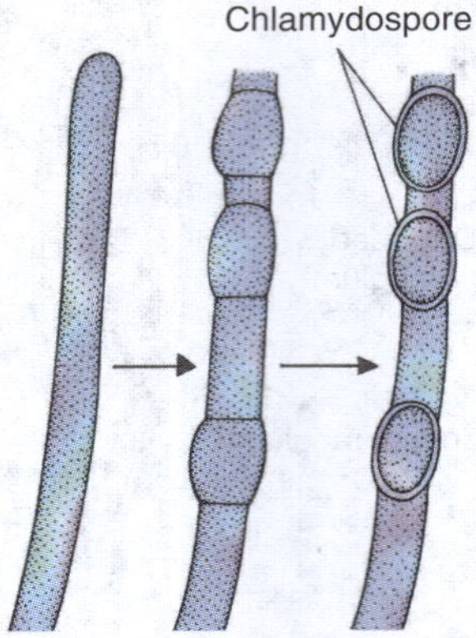

Fig.1.20: Chlamydospores

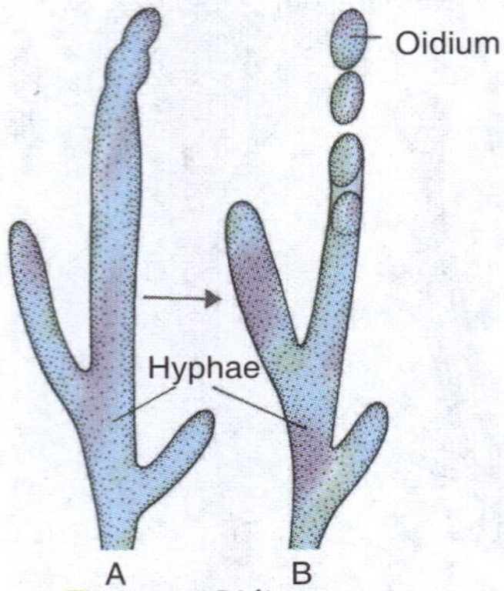

Fig.1.21: Oidiospores

Oidiospores: (Fig.1.21) Spores produced under artificial laboratory conditions. Fungal hypha becomes beaded because of segmentation of hypha .e.g *Penicilium sp*.

Sexual Reproduction:

Sexual reproduction in fungi like any other organisms, involve three steps : Plasmogamy (fusion of cell), karyogamy (fusion of nucleus) and meiosis. Two haploid cells of genetically different, strain fuses to form a dikaryotic cell then karyogamy takes place to develop diploid state of life cycle. Meiosis finally restores the haploid state of the fungus during sexual reproduction. In aquatic fungi fertilization takes place by fusion between free living gametes or free living gametes with gametangium is common. In other form of fungi where gametangia either differ in form and function (oogonia and antheridia) or in size only or even none at all. In most of the fungi anastomosis (fusion of fungal tips) is common for fertilization. Meiosis, thereafter, confined to specialized cell structure, frequently followed by spore production from it or in apomictic forms its equivalent without meiosis. Two such specialized cell structures are the Ascus and Basidium producing somewhat oval shaped Ascospores and Basidiospores respectively. In Ascomycetes the four nuclei derived from meiosis in ascus undergo further mitotic division to produce eight ascospores endogenously. These ascospores are subsequently released from the ascus. similarly in Basidiomycetes nuclear fusion is followed by meiosis but here the basidiospores generally develop immediately after meiosis at the tip of four tapering sterigmata. Thus four Basidiospores develop exogenously from one basidium. Basidiospores are said to be discharged violently from Basidium.

LIFE CYCLE IN FUNGI

Raper (1954) has originally described seven type of life cycle in fungi that has been reduced to five basic life cycles (Burnett,1976). They are

1. Asexual Life Cycle : Where sexual reproduction is apparently lacking entirely. As described in many fungi belonging to deuteromycetes.

2. Haploid Life Cycle: Where diploid phase is of minimal duration. Meiosis immediately follows nuclear fusion and the mitotic products are dispersed. Such life cycles are found in many Phycomycetes and few Ascomycetes.

3. Haploid–dikaryotic Life Cycle : Such life cycles are common in many Ascomycetes and basidiomycetes. Binucleate ascogenous hyphae are developed just prior to ascus development, and such a dikaryon cannot apparently exist independently of the haploid phase. At the other extreme is the condition where the meiospore fuse to reform a dikaryon so that the fungus is dikaryotic throughout its life cycle , except for the movement of fertilization and during subsequent meiosis(eg saccharomycetales and more common in Ustilagiales). In many basidiomycetes mycelium derived from germinating meiospores may persist in haploid condition as monokaryon but once a dikaryon is formed by hyphal fusion for that it may comprise the most long lived phase of life cycle. eg various fairy ring fungi.

4. Haploid-diploid Life Cycle : where both the haploid and diploid phases alternate regularly. An unusual life cycle mostly restricted to a few aquatic Oomycetes.

5. Diploid Life Cycle: where haploid phase is very short, restricted only to gametes or gametangial contact. Most of the life cycle is diploid.. Most of the oomycetes follow this - life cycle.

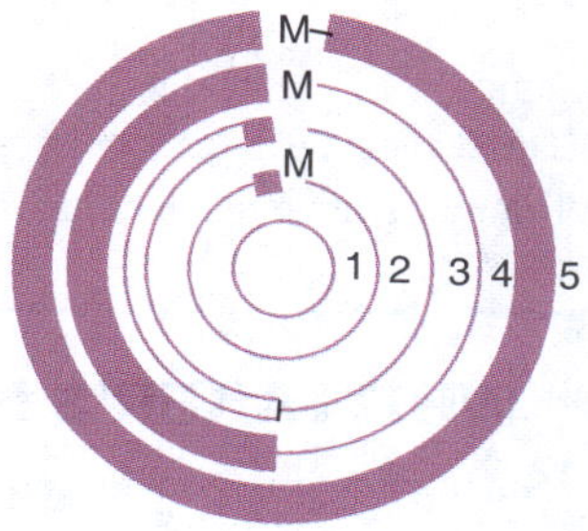

Fig. 1.22. A diagram to illustrate the five basic life cycles in fungi. Each circle represents a life cycle and should be followed clockwise; *M* represents meiosis, ——— a haploid phase, ═══ a dikaryotic phase and ▬▬ a dipoloid phase. The life cycles shown are, 1, asexual; 2, haploid; 3, haploid-dikaryotic; 4, haploid–diploid; 5. diploid. (Based on Raper, 1954)

From the general account of the fungi given in the preceding pages, it is evident that they are sharply set off from the algae in general, in the absence of chlorophyll, mode of nutrition and accumulation of carbohydrate reserve as glycogen instead of starch. The detailed account of these differences is given in Table I.

Inspite of the differences between the algae and fungi enumerated in Table I; the two are similar in their methods of reproduction, unicellular nature of sex organs, thallus-like plant body, lack of tissue differentiation of any kind and no embryo formation after gametic union.

TABLE — I

Algae	*Fungi*
1. The algae prossess **chlorophyll** which in some of them is masked by other pigments such as brown, red, etc.	1. The fungi lack chlorophyll.
2. With the help of chlorophyll, the algae are able to synthesize their own carbohydrate food and thus are called **autotrophic** or **autophytic** in their mode of nutrition.	2. Lacking chlorophyll, the fungi are unable to synthesize their own organic food. They get it readymade from external sources either by living as **parasites** or **saprophytes**. For this reason, the fungi are referred to as **heterotrophic** or **heterophytic** in their mode of nutrition.
3. The algal thallus is composed of true parenchyma cells. The unit of structure of an algal thallus thus is a cell.	3. The fungal thallus which is termed the mycelium is not formed by the division and differentiation of cells but is composed of a false tissue called **pseudoparenchyma**. The latter is an interwoven mass of fine, tubular structures called the **hyphae**. Thus the unit of structure of a fungal thallus or mycelium is a **hypha** and not a cell.

4. The cell wall in algae typically consists of cellulose.	4. The cell or hyphal wall generally consists of fungus cellulose often called **chitin**.
5. The food reserve is chiefly in the form of starch.	5. The reserve food invariably accumulates in the form of **glycogen** and not starch.
6. The algae usually grow in habitats where light is available.	6. Fungi generally grow in darkness and dim light.
7. The algae grow in water or damp soil and sometimes as epiphytes rarely as endophytes.	7. The fungi grow in a wide variety of habitats; as parasites in the tissues of plants and animals causing disease; as saprophytes in decaying organic matter in soil or water.
8. The sexual apparatus increases in complexity from the simple to the higher forms.	8. There is gradual and progressive simplification and ultimate elimination of the sexual apparatus from the lower to higher forms.

WHITTAKER'S FIVE KINGDOM SYSTEM AND BEYOND

Since ages it is considered that Plants and animals are fundamentally two different categories of living organisms. We recognize these two groups of living organisms on the basis of sharp differences on their mode of nutrition and mobility. Plant are non-motile, autotrophs and apparently drawing their nutrition from the soil or sometimes from other plants where as animals on the other hand are motile and heterotrophic and food ingesting organisms. This is the simplest and traditional system of classification or two kingdom system: **Plantae** (plants) and **Animalia** (animals).

But the problem cropped up with the development of knowledge in the diversity of living systems and confusion to place microorganisms in either of the kingdom when so many organisms were found to be having characteristics of both plants as well as of animals, e.g. Fungi are a unique group of living organisms having characters of both plants and animals being heterotrophic and non-motile. Simply it can be said that fungi could be assigned with plants as it is non-motile and draw their nutrition from substratum. However like animals fungi are heterotrophs, depend upon other plants or sometimes even on animals for their nutrition. With these characters scientists of that got confused and it has become difficult to place fungi either with plants or with animals. Placement of microorganisms specially the unicellular one was also a complicated question. Discovery of certain motile and photosynthetic organisms (*Euglena*) and in certain cases species are closely related but differ in their mode of nutrition or mobility has exaggerated the problem further. Existence of viruses, bacteria, fungi, blue-green algae and protozoan has compelled the scientist of that time to think beyond two kingdom system of classification. Haeckel (1866), a German Zoologist constituted a separate group called **Prostista** for such microorganisms. The new group included all the unicellular organisms or organism without tissue system (Algae, fungi, bacteria etc.) and also for all those organisms which posses the character of both plants and animals. This was the beginning of **three kingdom system**: the **Plantae**, the **Animalia** and the **Protista**.

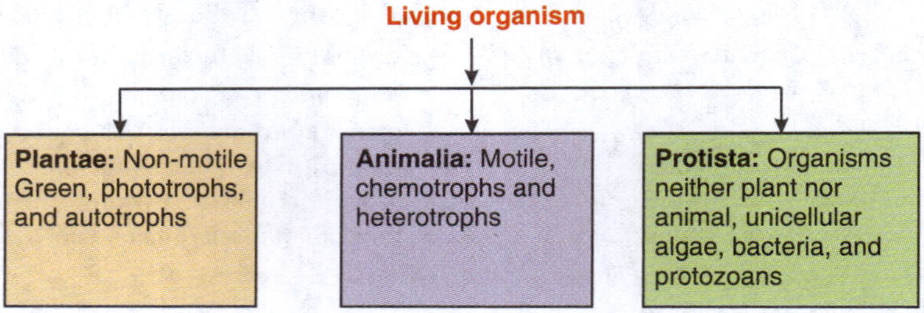

Evolution of three kingdom system has partially solved the problem of placing microorganisms in a certain better way. Several attributes were taken into account like morphological and anatomical complexities, mode of nutrition, mobility and division of labour in the tissue system. A large sum of knowledge were accumulated during the era of electron microscopy with respect to the cellular and molecular organization of organisms and placed the whole organisms in two fundamentally different groups as **Eukayotes** (organisms possessing true nucleus) and **Prokaryotes** (organisms without true nucleus).

Antony van Leeuwenhoek (1677) was the first one to describe microorganisms through his own constructed microscope having a magnification of 50 X to 300X, He called them as dierkens means animalcules (Royal Society of London).

Staneir and van Neil (1962) differentiated organisms in two groups where one group of organism that do not possess membrane bound nucleus in their cells are called **Prokaryotes** (bacteria and cyanobacteria) and the rest which possess membrane bound nucleus or true nucleus in their cell are called **Eukaryotes.**

This has provided the basis for further refinement of three kingdom system of classifying organisms. Knoll and Ruska (1932) divided the kingdom Protista into two groups: the Lower Protista for prokaryotic microorganisms and the **Higher Protista** for eukaryotic microorganisms. **Monera** a separate name proposed for Lower Protista whch includes bacteria and bluegreen algae (Copeland, 1956; Barkely, 1968). The higher Protista was retained as such. Thus system of classification transformed the concept of three kingdom system to **Four kingdom system**. Plantae, Animalia, Protista and Monera.

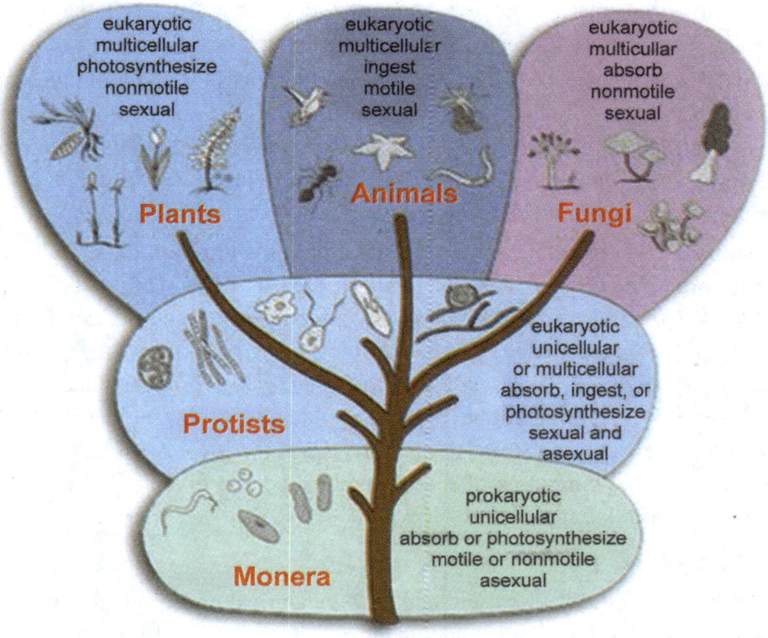

Whittaker's Five Kingdom System

Although the four kingdom system provided a greater clarity in classifying fundamentally different living organisms, especially the microbial organisms, but still several doubts prevailed related to nutritional characteristics, like fungi has unique mode of nutrition i.e. completely absorptive not found in any other organisms. This nutritional feature of fungi has prompted scientists of that time to rethink over four kingdom system of classification. It was R.H. Whittaker in the year 1969 proposed **five kingdom system** of classification by introducing Mycota as a separate kingdom for absorptive fungi.

KINGDOM:

1. MONERA (prokaryotic organisms)
2. PROTISTA (primitive eukaryotic organisms)
3. FUNGI (absorptive)
4. PLANTAE (advanced eukaryotic plants, Photosynthetic)
5. ANIMALIA (all multicellular animals, ingestive)

The Monerans which represents the earliest group of organisms are thought to have given rise to Protista. Protists are unicellular eukaryotes (protozoa and unicellular algae) from which multicellular eukaryotes evolved along three separate lines having different mode of nutrition. All were given rank of kingdom namely, the fungi (absorptive in nature) Plantae (photosynthetic in nature) and Animalia (ingestive in nature). It is found that Fungi were the first to appear from Protista. Later, about a billion years ago some protists must have evolved into primitive multicellular animals. Still later, probably about 350 millions years ago, some protists must have evolved into higher forms of plants.

Merits and Demerits of Five Kingdom Classification

Whittaker's five-kingdom classification system with several merits has also inherited several drawbacks. But still the most accepted system of modern classification mainly because of the phylogenetic placing of different groups of living organisms.

Separation of prokaryotes into an independent kingdom, Grouping of all unicellular eukaryotes under the kingdom Protista, Elevation of the group fungi to the status of a kingdom are some of the highlights of Whittaker's five kingdom system of classification. The kingdoms Plantae and Animalia are now more homogenous groups than they were in the two kingdom classification as it shows the phylogeny of different life styles.

Though Monera and Protista are two separate kingdoms but both include autotrophic as well as heterotrophic organisms, cells with cell wall as well as cells without cell wall. Unicellular green algae like *Volvox* and *Chlamydomonas* have not been included under Protista because of their resemblance to other green algae. Slime moulds differ totally from other members of Protista in their general organization. Viruses have not been given proper place in this system of classification. None of the three higher kingdoms include a single ancestor of all its forms. Multicellular lines have originated from protistans several times (polyphyletic). This classification of fungi where all the five kingdoms of living organisms were given equal taxonomic status has wide acceptance and popular appeal.

The Fungal Domain:

During the last two decades new molecular, biochemical and phylogenetic studies on fungi yielded a wealth of information about fundamental similarities and differences between organisms, and these new findings have been used to classify the organisms into three groups, called 'domains, of which two belong to prokaryotes and the third one belong to eukaryotes. **Fungi** are now recognized as one of the five eukaryotic kingdoms, the other being **Animalia, Plantae, Chromista** (corresponding roughly to algae and also known as **Stramenopila**) and **Protozoa,** which includes a wide variety of mainly phagotrophic unicelular organisms. (detailed classification can be seen further in the classification of fungi segment of the book).

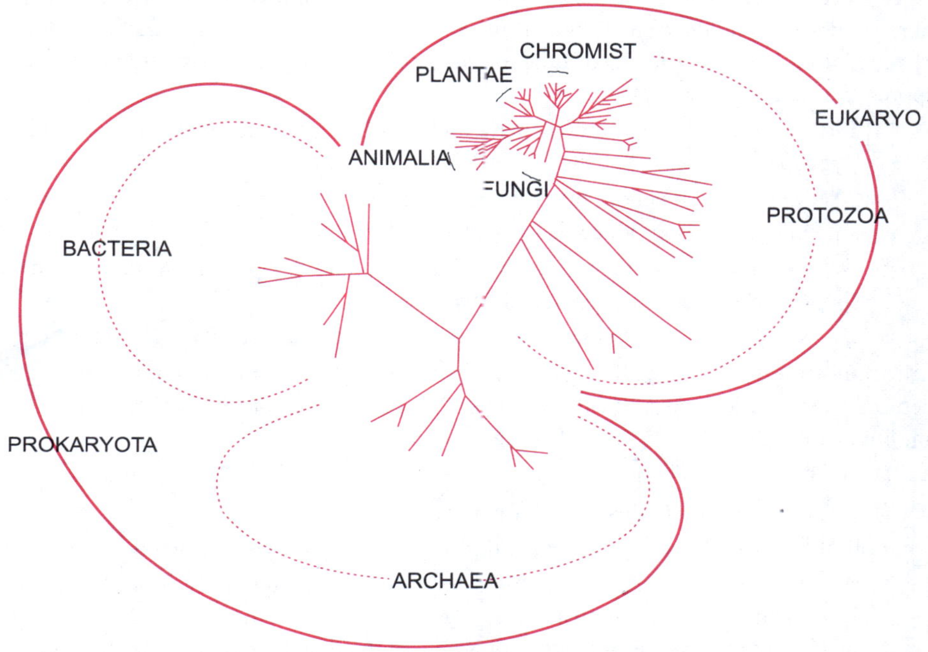

Figure 1.23. A phylogenetic tree showing the relationships between the two Prokaryote and five Eukaryote kingdoms. After Hawksworth, D.L. *et al.* (1995) Ainsworth & Bisby's Dictionary of the Fungi, 8th edn. CAB International, Wallingford.

CLASSIFICATION OF FUNGI

The classification of fungi is still in a state of flux. A stable and ideal system of classification is yet to be proposed. Many systems have been proposed from time to time by mycologists. With the advancement of knowledge newer systems are proposed while the older ones are discarded. The classification of fungi is mostly based on the features noted below :

(*a*) Nature of somatic phase **whether** unicellular or a mycelium, if latter septate or aseptate.

(*b*) Kinds of asexual spores (mitospores)–sporangiospores or conidia; if former motile or non-motile; number, form and arrangement of flagella in the motile spores.

(*c*) Kinds of sporangia.

(*d*) Nature of the life cycle–whether asexual, haplontic, diplontic, haplodiplonitc or haploid dikaryotic.

(*e*) Presence or absence of the perfect or sexual stage.

It was probably **Bauhin** who for the first time tried to classify fungi in his book **Pinax Theatri Botanici** (1623) wherein he described 100 species of fungi. Later, fungi were divided into six classes by **Tournefort (1694)** in his book **Elements de Botanique** while **Carolus Linnaeus** (1753) placed all fungi in his 24th class **Cryptogamia** in his famous book **Species plantarum.** The first detailed classification of fungi, however, was proposed by **Elias Magnus Fries** in his three volume book **Systema Mycologicum** published during 1821-1832. He divided fungi into four classes: (*i*) Coniomycetes (*ii*) Hyphomycetes (*iii*) Gastromycetes and (*iv*) Hymenomycetes. **Saccardo (1882)** included Bacteria and slime molds in fungi in his famous book **Sylloge Fungorum** and divided fungi into six classes : (*i*) Schizomycetes (*ii*) Myxomycetes (*iii*) Phycomycetes (*iv*) Ascomycetes (*v*) Basidiomycetes and (*vi*) Deuteromycetes.

Gwynne-Vaughan and Barnes (1926) rejected the contention of Saccardo and excluded bacteria and slime molds from fungi. Slime molds were treated as **Forms resembling Fungi**. They recognised only four classes of fungi on the basis of structure of mycelium and characteristics of spores. These four classes are :

(*i*) **Phycomycetes.** Mycelium aseptate and multinucleate, sexual spores are Oospores and Zygospores.

(*ii*) **Ascomycetes.** Mycelium fairly branched and septate; Septa simple; Characteristic Spores are endogenously produced ascospores.

(*iii*) **Basidiomycetes.** Mycelium branched and septate, Septa with Dolipores, Characteristic spores are exogenously produced basidiospores.

(*iv*) **Deuteromycetes.** Mycelium branched and septate, sexual spores absent.

Gaumann and Dodge (1928), who also did not include slime molds in Fungi, proposed a system of classification based on the life-cycle and development of fructification. According to them, Fungi were divided into the following four classes :

(*i*) **Archimycetes.** Thallus naked.

(*ii*) **Phycomycetes.** Thallus with definite cell wall.

In both the above classes zygote is the diploid stage and meiosis occurs at its germination.

(*iii*) **Ascomycetes.** Thallus more highly developed, plasmogamy results in a distinct mycelium; Karyogamy delayed; meiosis takes place in a distinct structure leading to the formation of sporangial structures called asci producing endogenously the **ascospores.**

(*iv*) **Basidiomycetes.** Thallus more highly developed consisting of dikaryotic mycelium in the somatic phase; meiosis shifted to a much later stage in a structure called conidiophores or **basidia** exogenously producing basidiospores.

Bessey (1950) agreed with De Bary (1887) and placed the slime molds under the name **Mycetozoa**. His system of classification was largely based on the reproductive characters. True fungi were divided as follows :

1. *Lower Fungi*. Fruit bodies not formed. Only one class **Phycomyceteae.**
2. *Higher Fungi*. Fruit bodies are formed, divided into three classes :
   (*i*) **Ascomyceteae** Fruit body asci; ascospores endogenously formed.
   (*ii*) **Basidiomyceteae** Fruit body basidia; basidiospores formed exogenously.
   (*iii*) **Deuteromycetae.** Sexual stage absent.

The practice of dividing Fungi into four classes continued till 1958 when **Sparrow** (1958) created nine classes emphasising the importance of flagellation of the zoospores in the classification. He considered that flagellation should be treated as a good taxonomic and phylogenetic criterion. These nine classes are : (*i*) Chytridiomycetes (*ii*) Hyphochytridiomycetes (*iii*) Plasmodiophoromycetes (*iv*) Oomycetes (*v*) Zygomycetes (*vi*) Trichomycetes (*vii*) Ascomycetes (*viii*) Basidiomycetes (*ix*) Deuteromycetes.

Again in 1961 **Martin** proposed a system of classification based on the septation of mycelium and spore characters. The classification proposed by him is as follows :

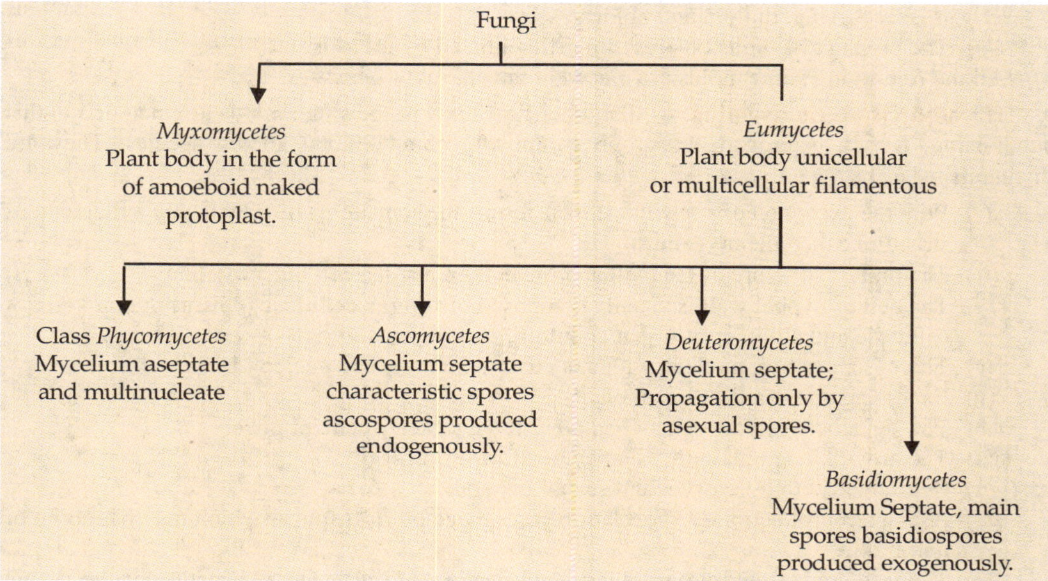

It would be clear from the above that modern mycologists are unanimous on the exclusion of bacteria from the fungi on the basis of lack of a true nucleus in the bacterial cell. There is a difference of opinion as regards the slime molds. Many mycologists favour their inclusion among the fungi. The Committee on the International Rules of Botanical Nomenclature has made the following recommendations :–

(*a*) The names of divisions of fungi should end in–**mycota.**
(*b*) The sub-divisions should end in–**mycotina.**
(*c*) The names of classes should have a suffix–**mycetes.**
(*d*) The sub-classes should end with a suffix–**mycetideae.**
(*e*) The names of orders should end in–**ales.**
(*f*) The families should end with a suffix–**aceae.**

Division Mycota. The mycologists place division **Mycota** in the kingdom **Plantae**. This division includes non-green (achlorophyllous) nucleated thallophytes (fungi) which are characterised by the following features :–

(*a*) Obtain nutrition either by living as **saprophytes** or as **parasites**. Lacking chlorophyll, the mode of nutrition is thus heterotrophic and absorptive except slime molds in which it is phagotrophic.
(*b*) Somatic phase may be **filamentous, unicellular** or a **plasmodium**.
(*c*) Presence of a **true nucleus** with a **nuclear membrane** and **nucleoli** in the cell.
(*d*) Typically the cell wall is chitinised but some Oomycetes have a cellulose cell wall.
(*e*) Reserve food accumulates in the form of **glycogen** and not starch. In addition there are oil drops.
(*f*) Reproduce both by asexual and sexual methods.

The division Mycota includes two sub-divisions, Myxomycotina and Eumycotina.

I. Subdivision Myxomycotina. This subdivision includes the **slime molds** which possess the following distinctive features :–

(*i*) The thallus is a non-green, multinucleate mass of protoplasm called the **plasmodium.**
(*ii*) The plasmodium lacks a definite cell wall, and thus is amoeboid in shape.
(*iii*) It is free living, diploid and holocarpic.
(*iv*) The propagative units (spores) are differentiated by **meiosis**.

All the free-living slime molds are placed in the class *Myxomycetes.*

II. Subdivision Eumycotina. All the other fungi except the slime molds are included in this subdivision. It is a heterogeneous group comprising about 80,000 known species. They are distinguished as below :–

(*i*) With the exception of a few unicellular forms, the somatic phase is typically a filamentous structure called the **mycelium.**
(*ii*) The unit of structure of the thallus or mycelium is not a cell but a **hypha.**
(*iii*) The cell or hyphal wall is usually composed of **fungus cellulose** (**chitin**) or a mixture of cellulose and chitin or rarely of cellulose.
(*iv*) There is a **definite nucleus** with a nuclear membrane and nucleoli in the cell.
(*v*) Excepting a few, all are **eucarpic.**
(*vi*) The septa between the cells when present have each a **central pore.**
(*vii*) The growth in length in the filamentous forms is **apical.**
(*viii*) Reproduce both by asexual and sexual methods.
(*ix*) Asexual reproduction or sporulation takes place by **mitospores** which may be motile or non-motile.
(*x*) There is gradual and progressive simplification and ultimate elimination of the sexual apparatus from the lower to the higher forms.

The older mycologists divide the subdivision **Eumycotina**, which includes the true fungi, into four classes, namely, *Phycomycetes, Ascomycetes, Basidiomycetes* and *Deuteromycetes*. This scheme of classification of true fungi is still in vogue in many countries. The modern mycologists, however, consider that the class Phycomycetes, which includes all the lower fungi, is not a natural group of closely related forms. Alexopoulos (1962), therefore, took a logical step and split the lower fungi included in the class Phycomycetes into six classes.

An outline scheme of classification of **Eumycotina** (ture fungi) followed in this edition of the text, is set out below. It is customary to divide the true fungi into two main groups, the **lower** and the **higher Fungi.**

A. THE LOWER FUNGI (PHYCOMYCETES)

They have comparatively a simple thallus which in some is unicellular and in others filamentous (mycelium). The septa usually remain suppressed in the actively growing vegetative mycelium. They reproduce asexually by sporangiospores (generally motile, rarely non-motile), rarely by conidia. The early mycologists placed the lower fungi in a single **class** *Phycomycetes* which is split in this text into the following **six classes** on the basis of presence or absence of motile cells in the life cycle and the number, form and position of flagella on motile cells.

Motile Cells Produced

1. *Class Chytridiomycetes.* Included in this class are the lower fungi in which motile cells have a **single** flagellum of **whiplash type** inserted at the **posterior end** (**opisthocont**). The members of this class are called the **chytrids.**

2. *Class Hyphochytridiomycetes.* The motile cells (zoospores) possess a **single** flagellum of **tinsel type** which is inserted at the **anterior end.**

3. *Class Plasmodiophoromycetes.* The motile cells (zoospores) are **biflagellate**. Both the flagella are of **whiplash type** but one of these is longer than the other. The longer one has a sharply pointed end and the shorter one has a blunt end. The club root organisms are the examples of this class.

4. *Class Oomycetes.* The motile cells are **biflagellate.** The two flagella are usually of nearly equal length. One of these points forwards and the other trails behind. The former is of **tinsel type** and latter of **whiplash type.** The common representatives of this class are the water molds, downy mildews and blights.

Motile Cells Lacking

5. *Class Zygomycetes.* Motile cells are absent. Asexual reproduction takes place by sporangiospores which are encapsulated and wind disseminated. The common examples are the pin molds and the bread molds.

6. *Class Trichomycetes.* The motile cells are lacking. The asexual spores are produced exogenously at the tips of special hyphae (conidiophores) and are called the **conidia.**

B. THE HIGHER FUNGI

The somatic phase consists mostly of a mycelium which is usually septate. The septa have each a central pore and thus are incomplete. No motile cells are produced in the life cycle. The higher fungi comprise the following three classes :

7. *Class Ascomycetes.* The characteristic spores of the sexual or perfect stage are **endogenous** in origin and are called the **ascospores**. They are produced within special, sac-like structures called the asci. Plasmogamy takes place by the fusion of gametangia or somatic hyphae. Somatic phase is usually a **septate mycelium** which multiplies asexually by **conidia,** rarely it is unicellular (as in the yeasts) and is propagated by budding or rarely by fission.

8. *Class Basidiomycetes.* The characteristic spores of the sexual or perfect stage are **exogenous** in origin and are called the **basidiospores.** They are borne externally usually on club-shaped structures called **basidia.** The somatic phase consists of a simple or complex mycelium with or without clamp connections. Sexual organs are lacking. Plasmogamy, however, takes place by **hyphal fusion.**

9. *Form-class Deuteromycetes* or *Fungi Imperfecti.* Included in this class are the higher fungi in which the sexual or perfect stage is unknown. The somatic phase consists of a **septate mycelium** which multiplies asexually by **conidia.** The conidia are produced on conidiophores which may occur isolated or may be aggregated to form complex structures such as **synnema, pycnidia** and **acervuli.**

The classification of Fungi outlined above has been followed in this text. However, for the benefit of the students a brief account of the classification proposed by Alexopoulos and Mims (1979) is given here. They place the fungi including the slime molds in the Kingdom **Mycetae** of the superkingdom **Eukaryota** which, in addition, includes four other kingdoms. They divide the Kingdom Mycetae into three divisions namely (*i*) Gymnomycota (*ii*) Mastigomycota, and (*iii*) Amastigomycota. The division is subdivided into subdivisions, classes, sub-classes and orders.

Division I. Gymnomycota. It includes phagotrophic organisms devoid of cell walls. This division comprises two subdivisions. These are Acrasiogymnomycotina and Plasmodiogymnomycotina.

Subdivision 1. Acrasiogymnomycotina. It includes a single class Acrasiomycetes.

Class 1. Acrasiomycetes. Lacks flagellated cells except one species. The class comprises two sub-classes Acrasiomycetidae and Dictyosteliomycetidae.

Subdivision 2. Plasmodiogymnomycotina. It is split into two classes :

Class 1. Protosteliomycetes.

Class 2. Myxomycetes. It includes the true slime mold and comprises three sub-classes namely,

Sub-class 1. Ceratiomyxomycetidae. It includes one order Ceratiomyxales.

Sub-class 2. Myxogasteomycetidae. It comprises four orders namely, Liceales, Echinosteleales, Trichlales & Physarales.

Subclass 3. Stemonitomycetidae with one order Stemonitales.

Division II. Mastigomycota. Includes fungi with absorptive nutrition, unicellular or filamentous, mycelium coenocytic. It comprises two subdivisions :

Subdivision 1. Haplomastigomycotina. Includes fungi with uni-orbi-flagellate zoospores. It comprises 3 classes :

Class 1. Chytridiomycetes. Fungi producing zoospores furnished with a single whiplash flagellum inserted at the posterior end.

Class 2. Hyphochytridiomycetes. Motile cells with a single tinsel flagellum inserted at the anterior end.

Class 3. Plasmodiophoromycetes. Parasitic fungi producing biflagellate motile cells, with both the flagella of whiplash type inserted at the anterior end.

Subdivision 2. Diplomastigomycotina. Sexual reproduction oogamous, zoospores biflagellate, meiosis gametangial. This subdivision includes a single class :–

Class 1. Oomycetes. It comprises four orders namely, Lagenidiales, Saprolegniales, Leptomitales and Peronosporales.

Division III. Amastigomycota. Fungi with absorptive nutrition, motile cells lacking, mycelium aseptate or septate. This division includes four subdivisions, namely Zygomycotina, Ascomycotina, Basidiomycotina and Deuteromycotina.

Subdivision 1. Zygomycotina with two classes :

Class 1. Zygomycetes. It includes six Orders.

Class 2. Trichomycetes. It comprises five orders.

Subdivision 2. Ascomycotina. Fungi usually with a septate mycelium producing haploid ascospores in sac-like cells called Asci. This subdivision is represented by a single class :–

Class Ascomycetes. It is subdivided into five subclasses as follows :–

Subclass 1. Hemiascomycetidae comprising three orders

Subclass 2. Plectomycetidae with five orders.

Subclass 3. Hymenoascomycetidae with ten orders.

Subclass 4. Laboulbeniomycetidae including two orders

Subclass 5. Loculoascomycetidae with five orders.

Subdivision 3. Basidiomycotina. Septate mycelium, produces basidiopores (meiospores) exogenously on various types of basidia. The subdivision includes a single class.

Class 1. Basidiomycetes. It is split into three subclasses :

Subclass 1. Holobasidiomycetidae.

Subclass 2. Phragmobasidiomycetidae

Subclass 3. Teliomycetidae.

Subdivision 4. Deuteromycotina. It includes imperfect fungi in which sexual stage is unknown. It comprises a single form class.

Form Class 1. Deuteromycetes with three form subclasses namely Blastomycetidae, Coelomycetidae, and Hyphomycetidae.

Lichens. Ascomycetous, Basidiomycetous and Deuteromycetous mycobionts, symbiotically associated with green or blue-green algae (phycobionts), form various types of thalli.

A more natural system of classification of fungi was proposed by Ainsworth (1973) which has been accepted by many mycologists of today like Webster (1980), Bilgrami (1985) and Dube (1987). Ainsworth placed all the fungi in the kingdom **Mycota.** His classification is as follows :

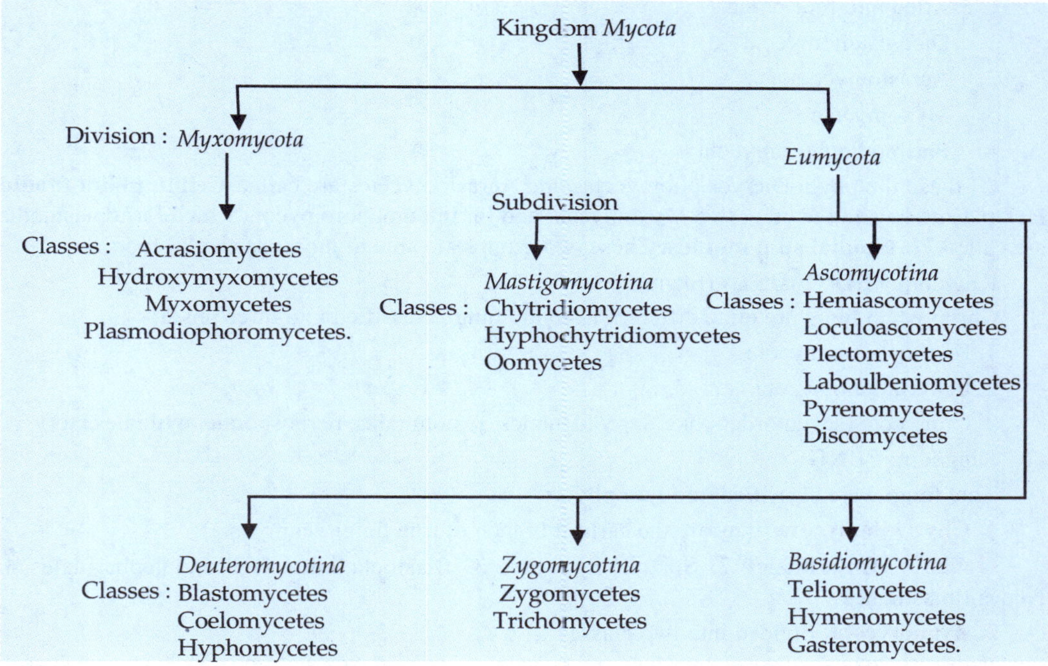

In the 7th. edition of Ainsworth and Bisby's dictionary of fungi (1983) there were 8 classes instead of 4 in the division Myxomycota. Sub-Division Ascomycotina has 37 orders arranged alphabetically but no class. Under sub-division Basidiomycotina class Teliomycetes was replaced by Urediniomycetes and Ustilaginomycetes. Class Blastomycetes was dropped from sub-division Deuteromycotina

However, None of the taxonomic schemes were universally accepted by mycologists. Still Mycologists are trying to arrive on a common plate form as for as fungal taxonomy is concern. The traditional system of fungal classification which were based on morphological features of fungi have now been replaced by the system based on phylogenetic relationships. New International Code of Botanical classification (McNeill et. al. 2006) has 7 ranks, use phylum and sub-phylum instead of Division and Sub-Division. Besides Kingdom and Sub-kingdom other ranks are to end with-

Phylum....................mycota (except Microsporidia)

Sub-phylum .... mycotina

Order .... ales

Class ....mycetes

Sub-class.................mycetidae

In the new approach of fungal taxonomy Mycologists placed organisms studied by them under three kingdoms based on their morphological characters as well as their phylogenetic relationships, the **Protozoa**, the **Chromista** and the **Fungi** under domain Eukaryota. The classification mentioned here is based on 8th edition of Ainsworth & Bisby's Dictionary of the Fungi. (Hawsworth, Kirk, Pegler & Sutton, 1995)

Kingdom: **PROTOZOA** (Mycetozoa)

Classified into four phyla

1. Dictyosteliomycota
2. Acrasiomycota
3. Myxomycota
4. Plasmodiophoromycota

Of these the phyla Dictyosteliomycetes and Acrasiomycetes are called **Cellular slim mould** lack plasmodia and the other two Myxomycota and Plasmodiophoromycota which form plasmodia are called **Plasmodial slim moulds.** These two groups are able to ingest particular food.

Kingdom: **CHROMISTA** (Staminipila)

Chromistian fungi are quite different from the fungi, classified into three phyla:

1. Hyphochytriomycota
2. Labyrinthulomycota
3. Oomycota (seven orders like Saprolegniales, leptomitales, pernosporale, pythiales etc.)

Kingdom: **FUNGI**

True fungi were classified into four phyla

1. Chytridiomycota (they are the earliest branch of true fungi : eumycota):

Order 1. Chytridiales 2. Spizellomycetales 3. Blastocladiales 4. Monoblepharidales 4. Neocallimastigales

2. Zygomycota: divided into two classes

Class 1. Zygomycetes – seven orders are recognized. Some of them are Mucorales, Entomopthorales, Kickxellales, zoopagales and Glomales.

Class 2. Trichomycetes – they are obligate parasite on arthropods.

3. Ascomycota: Forty six orders are recognized under this phylum which also includes lichenized fungi. Some of the orders are saccharomycetales, Erysiphales, Eurotiales, Lecanorales, Peltigerales, pezizales and xylariales etc.

4. Basidiomycota: The phylum is divided into three classes.

Class 1. Basidiomycetes with thirty two orders. Like Agaricales, Ceretobasidiales, Nidulariales, Tremellales etc.

Class 2. Ustilagomycetes – The Smut with seven orders like Ustilaginales and Sporidiales.

Class 3. Teliomycetes with two order Uredinales (The Rust), and Septobasidiales (parasites on Scale insects.)

5. Mitosporic Fungi (informal group): previously known as Fungi Imprfecti. Classified on the basis of asexual morphology into three classes

Class 1. Coelomycetes – conidia are formed in a cavity composed of fungus or mixed host and fungal tissues.

Class 2. Hyphomycetes – conidia not within cavity, also include yeasts.

Class 3. Agonomycetes – known as mycelia Sterilia. Lacking spores. Not a mitosporic fungi but included with this group just for convenience.

RECENT DEVELOPMENTS IN FUNGAL TAXONOMY

Some revolutionary changes have been witnessed in the taxonomy of fungi. Following the establishment of communications platform (Deep hypha), the AFTOL (Assembling Fungal Tree of Life)) project based on molecular studies, gene sequencing, multigene phylogenies and data analysis come up with a scheme of fungal classification (Hibbett et. al., 2007) that has been shaped by 67 mycologists. These authors emphasized the broad support and input called on worldwide mycologists to adopt this unified system of fungal taxonomy.

Hibbett's classification represents a set of three diagram and taxa of uncertain position listed as "incertae sedis". The classification has 1 kingdom, 1 subkingdom, 7 phyla, 10 sub-phyla, 12 subclasses and 179 orders. Summary of this classification is discussed below:

Kingdom **FUNGI**

Phylum: 1. **Chytridiomycota** (Archemycota)

2. **Neocallimastigomycota**

3. **Blastocladiomycota**

4. **Microsporidia**

5. **Glomeromycota** — sub-phyla →
- Mucoromycotina
- Etomophoromycotina
- Zoopagomycotina
- Kickxellomycotina

Sub-Kingdom : **Dikaraya**

Phylum: 1. **Ascomycota:** — sub-phyla →
- Taphrinomycota
- Saccharomycotina
- Pezizomycotina

2. Phylum .... sub-phyla — sub-phyla →
- Pucciniomycotina
- Ustilaginomycotina
- Agaricomycotina

Basidiomycota *incertae sedis* not placed in any sub-phylum are class-Wallemiomycetes and Entorrhizomycetes.

In this system of classification to minimize the confusion between taxon name and informal term the Basidiomycetes, Urediniomycetes and Ustilaginomycetes were renamed and elevated to the rank of sub-phylum: the Agaricomycotina, Pucciniomycotina and Ustilaginomycotina, respectively. The most significant dramatic changes in the classification concern with *Basal Fungal Lineages* which include taxa that have traditionally been placed under zygomycota and Chytridiomycota. Based on molecular and cytolcgical analysis of rRNA, tef1 and rbp1 these group have been found to be of polyphyletic origin. Phylum Chytridiomycota have been retained., Blastocladiales, a traditional member of Chytridiomycota elevated to the rank of phylum Blastocladiomycota (Jams et. al., 2007). Similarly Neocallimasigales also elevated to Phylum level Neocallimastigomycota. Mostly there are four major groups of true fungi recognized by mycologists are – Ascomycota, Basidiomycota, Chytridiomycota and Zygomycota (Alexopolous et. al. 1996, Webster and Weber, 2007). Recent studies have provided the support for recognition of an additional phyla the Glomeromycota which was traditionally placed under Zygomycota that form an association with roots of most plants. A group of parasitic organisms called Microsporidia that live inside the cells of animals are also now considered to belong to fungal Kingdom. Based on rRNA analysis a new clad containing Glomeromycota and Dikrya was classified as SYMBIOMYCOTA but the taxon is not included in the classification as its placement is not clear yet.

Weber (2009) in his 'Recent development in the molecular Taxonomy of fungi' tried to summarized current taxonomic concepts of phylogenetic and taxonomic placements of fungi as:

I. Non-Fungal Organisms studied by Mycologists-

A. Slime Moulds

B. *Plasmodiophora* and related species

C. Staminipila

D. *Haptoglossa* (species infect rotifiers or nematodes like *Plasmodiophora* and walled sporidium enters the living host, zoospores formed appears to be anisokonts.)

II. The Basal Fungi–

A. Microsporidia

B. Chytrids

C. Zygomycetes-Type fungi

D. Glomeomycota

III. Dikarya (further classified as represented in Hibbett's classification)

However these classifications are not going to be the final one accepted by most, the process of modification addition deletion will keep on in refining the fungal classification system. Scientists are using additional tools, new methods to study fungi. e.g. jones *et.al.*, 2011 described Cryptomycota, a potentially new phylum of organism within kingdom Fungi.

AINSWORTH AND BISBY'S DICTIONARY OF FUNGI

The dictionary aims to provide entry point into sum total of accumulation of knowledge in the field of systemic mycology for all those who work with fungi be it university students or a mycologist. Now the dictionary has become popular for gaining knowledge in fungi their classification and updated information for wider mycological audience. All organisms traditionally studied by mycologists are covered in the dictionary including lichens, mushrooms, slim moulds, water moulds and yeast in the later editions. First edition of the dictionary appeared in 1943 and by now 10 editions have come up with recent informations in the fungal taxonomy. The classification based on 8th. edition of Ainsworth and Bisby's dictionary is presented in the book above in classification of fungi chapter. 9th edition published in 2001 (Kirk et. al. 2001) has some significant changes in the classification of Ascomycota and Basidiomyota using traditional knowledge considered with state of art in phylogenetic analysis. Few important changes in this edition are reduction in the number of order in Basidiomycota from 32 to 16, reclassification of basidiomycetous yeast into three groups and for the first time placed anamorphs (Fungi lacking sexual spores) in a more cohesive group using data generated in molecular phylogeny.

In the 10th. Edition that appeared in 2008 (Kirk et. al. 2008) most of the revisions are related to higher ranks within fungi, largely based on the results obtained from AFTOL (Assembling Fungal Tree of Life) project. Phylogenetic informations gained from multigene sequence analysis during first phase of AFTOL project that has revolutionised the understanding of how the kingdom should be classified. The dictionary has three major categories viz:

I. A dictionary of the Fungi

II. A dictionary of the Chromistian / Stramenopile Fungi – like organism

III. A Dictionary of the Protozoan Fungi-like organism

The latest edition of Dictionary of fungi also throws some light on the evolutionary origin of fungi. Information gathered in the field shows that the sister group of kingom fungi is closer to *Animalia* than to *Plantae.* But this statement also hides the fact that Chromistan fungus-like organisms, of which *Phytophthora infestans* is perhaps the best example, are very distantly related to fungi, being instead more allied with the brown seaweeds, among other a clear indication that mycelia way of life evolved at least two separate occasions. Surprisingly, however, the protozoan fungus like organisms are closer to the Fungi, being classified in the *Amoebozoa* with other protozoan amoebae. Fungi together with *Animalia* and a few other Protozoan groups constitute the *Opisthokonta*; and this group and Amoebozoa form the first major branch at the base of the *Eukaryota*.

(Bisby passed away in 1958, and Ainsworth passed away in 1998)

Cladogram of Fungi:

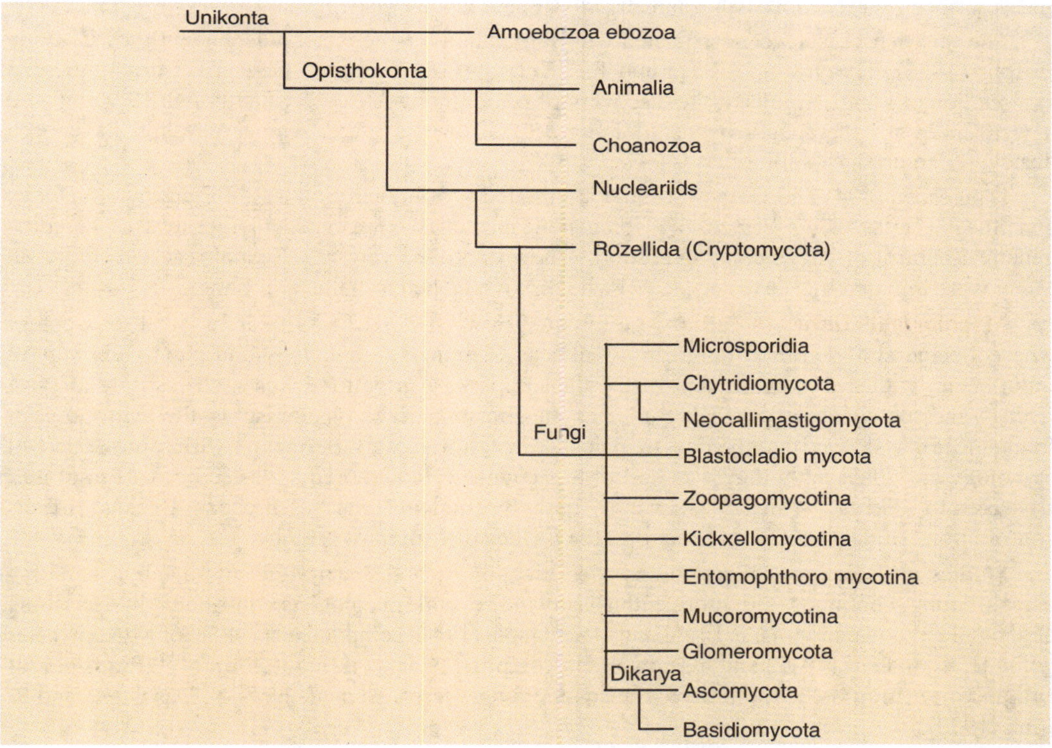

BRIEF HISTORY OF MYCOLOGY

Beautifully coloured umbrella shaped mushrooms and toad stools growing on soils forming 'fairy rings' attracted Man from the very beginning. One finds references to fungi in Roman and Greek classics. Records of plant diseases could also be found in Vedas and the Bible. Herbals or books containing descriptions of fungi were published by several authors like Clusius (1601) J. F. van Steerbeck (1675), Robert Hooke & others. The most famous work of their period was published by P. A. Micheli (1679-1737) who included scientific descriptions and illustrations of many fungi known till then. His famous book *Nova plantarum genera* was published in 1729. He experimentally proved that fungi originated from spores thus putting an end to the theory of fungal origin from decaying materials. He was much ahead of his time and thus has the honour of being referred to as **Father of Mycology.**

'Fairy Ring' on Soils

The number of species and genera discovered and described increased rapidly due to the diligently pursued researches by several renowned mycologists of that period. They included Persoon (1801), Fries (1821-32), DeBary (1831-1888), Saccardo (1845-1920), Sydow (1851-1924) and others. Soon it was proved beyond doubt that fungi caused diseases in plants, mycologists became interested in studying diseases of plants, animals and human beings. Cereal rusts, smuts warts etc. were observed and described by French, Italian, English and Russian mycologists.

The fact that yeasts were associated with fermentation announced independently by C. C. de la Tour (1836), T. Schwann (1837) and F.T. Kutzing (1837) opened new vistas in commercial exploitation of Fungi. Similarly, the discovery of penicillin by Alexander Fleming in 1928 initiated new avenues for therapautic value of Fungi. Since then, a large number of fungi have been discovered to produce antibiotics.

Dangeard (1894) and Harper (1896-97) initiated Cytological Studies on Fungi while Dodge and Shear (1898) using *Neurospora* for genetical studies contributed to the laws of heredity, genetical control of enzyme and evolution of new improved varieties. Sexual reaction studies in Fungi were initiated by Blakeslee (1904) who discovered heterothallism in Fungi.

Mycological studies in India were initiated by Lt. Col. K.R. Kirtikar in late 19th Century who collected and identified fungi. In a real sense, studies on mycology and plant pathology in India began with the establishment of the then Imperial Agricultural Research Institute at **Pusa** (Bihar) and the arrival of Sir E.J. Butler as the first imperial Mycologists to the then Govt. of India in 1905. It was mainly due to painstaking efforts of Butler that a firm foundation of mycology and Plant pathology was laid in this country. He authored a classic book **"Fungi and Diseases in Plants"** which is still used both by students and researchers. Because of his contributions Butler is aptly referred to as the **Father of Indian Mycology.**

Studies on isolation, collection and identification of both macro and micro fungi including parasitic fungi continued with more enthusiasm and several mycologists contributed in describing new fungi unknown to India. P. Bruhl and J. SenGupta (1927) compiled a checklist of Myxomycetes while H. Chaudhari (1947) and his coworkers compiled their work in the form of **"Handbook of Indian water moulds".** Additions to the aquatic fungi were also made by S. N. Das Gupta and R. John (1953).

Parasitic fungi causing diseases of important crops received greater attention. B.N. Uppal and J.H. Weston (1936) made comprehensive studies of *Sclerospora*–the cause of Green ear disease of Bajra. Species of *Pythium* and *Phytophthora* causing diseases of rubber plants and palms received attention from S.L. Ajrekar and K.D. Rajulu (1931). In 1931 itself, M. Mitra recorded a new bunt or Karnal Bunt of wheat while a list of Indian Species and varieties of *Aspergillus* was published by U.N. Mohanty (1948).

B.B. Mundkur and M.J. Thirumalachar in 1952 published a consolidated list of Indian Ustilaginales. Earlier in 1947, Mundkur founded the Indian Phytopathological Society with S.R.

Bose as its first president. K.C. Mehta and Coworkers (1940) studied the problem of the recurrence of the wheat rusts in plains of India.

Indian Mycological researches got an impetus with the establishment of the universities of Bombay, Madras and Calcutta. The then mycologists were purely plant pathologists who gave more importance to the listing and description of parasitic fungi causing diseases in plants. In addition to mentioned above, the prominent names include K.D. Bagchee, T.S. Ramakrishnan, J.H. Mitter, K.J. Narsimhan, S.N. Das Gupta, R.N. Tandon, R. Prasad, T.S. Sadasivan, C.V. Subramaniam and others. The mycologists who studied slime molds and soil fungi were B. S. Mehrotra (1952-1970) on Mucorales; B. Bakshi (1954) on Fungi from Himalayas, V. Agniothrudu (1954-68) slime molds and soil fungi from North and North East India; K. S. Bhargava (1968-80) on taxonomy of soil fungi; J. N. Rai (1961-83) on taxonomy of soil fungi from usar soils and Mangroves.

Ecological, physiological and Biochemical studies on fungi were also undertaken at different centres in the country. T. S. Sadasivan at Madras University initiated studies on ecology and physiology of Soil Fungi; S.N. Das Gupta at Lucknow University established a centre of studies on histopathology and deficiency disorders, A. Mahadevan at Madras University made significant contributions on physiology and biochemistry of fungi; at the University of Allahabad, R.K. Saksena initiated detailed studies on storage diseases of fruits and vegetables which were ably continued by R.N. Tandon. K.S. Bilgrami at Bhagalpur contributed on physiology and biochemistry of fungi especially the aflatoxins. Studies on phyllosphere fungi were initiated by S. Sinha at Agra. J.N. Rai at Lucknow conducted Ecological studies on soil fungi especially those from usar soils. Other prominent mycologists who are engaged in studies on taxonomy, ecology and biochemistry of fungi include R.S. Dwivedi, B. Rai, K.G. Mukergi, P.D. Sharma, H.C. Dube, S. Chandra, Kamal & A K Sinha.

Butler and Bisby (1931) published the first systematic account of Indian Fungi. Later several supplements were brought out by Mundkur (1938), Ramkrishnan and Subramanian (1952), Vasudeva (1977), and Bilgrami, Jamaluddin and Rizvi (1979).

Origin and phylogeny of Fungi. The origin and evolutionary relationships of fungi are not definitely known. They are a matter of speculation and problem for future research for mycologists. Concerning their origin, the first and the traditional hypothesis regards algae as the ancestral stock. Some mycologists consider protozoa as the ancestral stock.

The supporters of the **algal hypothesis** hold that the fungi are degenerate algae which have lost their chlorophyll and thus have adapted themselves of a heterotrophic mode of nutrition. It is not difficult to believe it because certain flagellates under different conditions develop chlorophyll and prepare their own food or lose chlorophyll and live as saprobes.

The proponents of algal hypothesis are divided into two camps. Some suggest a **monophyletic** origin. They hold that with the exception of slime molds and some simplest fungi, all others, namely, Phycomycetes, Ascomycetes and Basidiomycetes represent one main evolutionary line of fungi evolved from an algal ancestor. To view that all fungi have a common ancestry and arose in a monophyletic series appears untenable to most of the mycologists. They support the **polyphyletic** origin from the various groups of algae. On the basis of multiple origin, the fungi are considered a heterogeneous aggregation, the Phycomycetes having evolved from one class of algae and Ascomycetes from another. The basidiomycetes are thought to have been evolved from the Ascomycetes.

The supporters of the algal origin of the Phycomycetes further belong to two opposite schools of thought. One favours algal origin from a siphonaceous ancestor belonging to the green algae and the other from a Xanthophycean ancestor such as *Vaucheria*. There are differences in metabolism and type of flagellation of the motile cells in the Phycomycetes and green algae (Table II).

TABLE – II

Green Algae	Phycomycetes
1. The reserve food in the green algae including even the saprophytic and parasitic species accumulates in the form of starch.	The reserve food in the Phycomycetes accumulates in the form of glycogen and not starch.
2. The motile cells in the green algae are usually biflagellate. The flagella are of equal length and of whiplash type. They are inserted at the anterior end.	The motile cells in some Phycomycetes are uniflagellate and in others biflagellate. The single flagellum when inserted at the anterior end is of tinsel type and at the posterior end of it is whiplash type. Of the two flagella in the biflagellate cells, one is of tinsel type and the second of whiplash type.

In view of the above-mentioned differences in metabolism and type of flagellation in the two groups, the opponents of the chlorophycean origin of Phycomycetes consider this hypothesis untenable.

The advocates of the Xanthophycean origin of the Phycomycetes base their hypothesis on the resemblances of structure and reproduction between the Oogamous Phycomycetes (Oomycetes) and Oogamous yellow-green algae (Xanthophyceae) such as *Vaucheria.* In both the somatic phase is a coenocytic, aseptate filamentous thallus. They have more or less an identical life cycle and oogamous sexual reproduction. There is similarity in the chemical composition of cell wall in both. This viewpoint of phycomycetean origin of fungi from Vaucheria-like ancestors was propounded by De Bary in 1881. The name Phycomycetes of algal fungi emphasises this viewpoint. De Bary envisaged the origin of *Saprolegnia*-like oomycete from a vaucherial ancestor by loss of chlorophyll and change in the mode of nutrition. It further gave rise to the chytrids and to the Zygomycetes by retrogression.

Some mycologists led by Bessey (1942) advocated algal origin of fungi from the unicellular coccoid Xanthophyceae. This hypothesis is based on the similarity in structure, anterior position of flagella, presence of cellulose in the cell wall and accumulation of reserve food in the form of glycogen in both the Phycomycetes and the **coccoid** ancestor. The coccoid ancestor gave rise to the Phycomycetes along two divergent lines, one with anteriorly uniflagellate swarmers and the other with biflagellate swarmers. The opponents of this hypothesis point out to the occurrence of posteriorly uniflagellate swarmers and absence of cellulose in the cells of some uniflagellate Phycomycetes. They argue that all the uniflagellate and biflagellate Phycomycetes have not evolved from a common ancestor. They are **polyphyletic** in their origin, some having evolved from an algal ancestor and the others from a protozoan (flagellate) ancestor.

Fischer and Dangeard advocate the protozoan (flagellate) ancestry of the Phycomycetes. To the modern mycologists, this hypothesis seems more probable. It is based on similar metabolism and type of flagellation in both the Phycomycetes and protozoans. On the basis of this theory, the Phycomycetes have evolved from the flagellates via the chytrids. The uniflagellate forms arose from the uniflagellate protozoa and biflagellate forms evolved from the biflagellate protozoa. These non-mycelial forms, in turn, gave rise to the more advanced mycelial forms by progressive evolution.

Hawker (1967) suggests a polyphyletic origin of the lower fungi (Phycomycetes) from the aquatic flagellates along parallel lines. According to him, Chytridiales, Hyphochytridiomycetes and Plasmodiophoromycetes have originated from flagellates having appropriate types of flagella. The Oomycetes which resemble certain filamentous algae in the composition (cellulose nature) of cell wall, form of sex organs, life cycle, and form of endoplasm and mitochondria may have evolved from an algal ancestor by the loss of chlorophyll. Hawker suggests common ancestry for the chytrids and Zygomycetes at the flagellate level. The floridian origin of higher fungi based on the similarities between the reproductive organs does not survive critical examination.

To hold that different classes of fungi have an independent (polyphyletic) origin means there is no relationship between them. In that case, the classes of fungi should be raised to the status of divisions of phyla. This view suffers from a drawback as how to explain the close similarities which exist between the different classes of fungi such as :–

1. Similarity between the antheridia and oogonia of Phycomycetes and the sex organs of Ascomycetes.
2. Similar origin, physiology and phylogeny of the ascus and the basidium.
3. Similar nature, origin and development of conidia.
4. Similarity of the somatic phase (mycelium) differing, of course, in the magnitude of specialisation and differentiation.

The above-mentioned similarities indicate that the higher fungi have a common origin. The tendency among the Zygomycetes towards evolution of the conidium and protection of the zygote indicate an affinity with the Ascomycetes. Gaumann thus advocates that the Ascomycetes have evolved from the Zygomycetes. Similar origin, physiology and phylogeny of the ascus and the basidium support the origin of the Basidiomycetes from the Ascomycetes. Saivle (1955) postulated a highly plausible theory of the origin of the Ascomycetes and Basidiomycetes from a Taphrinalike ancestor.

Burnett (1968) disapproves speculation of the phylogenetic trends based on the similarities between the reproductive parts and on the study of comparative morphology. According to him speculation based on "this type of study can hardly be proved. Its plausibility cannot be subjected to scientific testing".

QUESTIONS

ESSAY TYPE

1. Give the scheme of classification of fungi as proposed by Alexopoulos and mention the chief characteristics of each group. *(Kanpur, 1996, 1998)*
2. Give detailed classification of Fungi and provide important characters of the classes. *(Awadh, 1992; Himachal Pradesh, 1995)*
3. Give a brief account of the general classification as given by Alexopoulos. *(Awadh, 1995)*
4. Give an account of Fungi as studied by you giving characteristic features of different classes. *(Allahabad, 1999)*
5. Describe in detail, giving examples, the various modes of nutrition met with in Fungi.
6. Give an outline classification of fungi as proposed by Ainsworth. Enumerate main features of different classes of fungi. *(Gorakhpur, 2003 ; Meerut, 1994)*
7. Give a suitable classification of fungi upto the class level placing different genera in your syllabus in their respective classes. *(Kumaon, 1997)*
8. Describe the classification of fungi by Alexopoulos and Mims. Give characters upto class level. *(Rohilkhand , 1993)*
9. Describe the structure, reproduction and diagnostic features of fungi. *(Bihar, 1993)*
10. Describe the classification of fungi as proposed by Alexopoulos.*(Gorakhpur, 1998)*
11. What are fungi ? Describe their asexual reproduction. *(Kumaon, 2004)*
12. Give an account of classification of fungi as studied by you giving characteristic features of subdivisions and divisions. *(Allahabad, 2004)*
13. Name the different classes of fungi given by Alexopoulos. Write the characters of any four of them. *(Rohilkhand, 2003)*
14. Discuss the salient features of important groups of fungi. *(Meerut, 1997)*
15. (*a*) Describe any recent system of classification of Fungi giving its important characteristic features. *(Kanpur M.Sc., 2003)*

    (*b*) Describe taxonomic status of fungi as suggested by Alexopoulos and Mims. *(Rohilkhand M.Sc., 2006)*

    (*c*) Write in detail the classification of fungi given by Alexopoulos. *(Kanpur, 2006)*

SHORT ANSWER TYPE

16. What is meant by facultative saprophytes?
17. Who is credited with the discovery of heterothallism?
18. Give three modifications of fungal mycelium and mention their uses, if any.
19. What do you know about the contributions of
    (*a*) S.N. Das Gupta (*b*) C.V. Subramanian
    (*c*) J.N. Rai (*d*) Micheli
    (*e*) Butler
    (*f*) Alexopoulos (*Kanpur M.Sc., 2006*)
20. Define Fungi.
21. Compare Algae with Fungi.
22. (*a*) Differentiate between Green algae and Phycomycetes.
    (*b*) Parasite and Saprophyte (*Kanpur, 2003*)
23. Give salient features of parasites.
24. Discuss the mechanism of nutrition in Fungi. (*GND Univ., 1991, 1992*)
25. Discuss the salient features of Fungi. How do they differ from Algae?(*GND Univ., 1992*)
26. (*a*) Enumerate the various classes of Fungi. Give two distinctive features of each class.
    (*b*) Describe the structure of fungal flagellum.
    (*c*) What is meant by the term obligate parasites? (*Himachal Pradesh, 1993*)
27. Give an outline of any one system of classification of fungi. (*Himachal Pradesh, 1996*)
28. Describe the role of carbon or nitrogen in the nutrition of fungi. (*Kanpur, M.Sc. I, 1986*)
29. What are fungi? Give an account of asexual reproduction in fungi. (*Kumaon, 1999; Lucknow, 1994, 1996, Poorvanchal, 2001*)
30. Enumerate the salient features of fungi as a group.
31. Comment briefly on the origin of fungi.
32. Give a brief account of History of Mycology in India.
33. Write short notes on
    (*a*) Martin's classification of fungi (*Kanpur, 1997*)
    (*b*) Classification of Fungi (*Awadh, 1996*)
    (*c*) Nutrition in fungi (*Poorvanchal 2001; Lucknow, 2004 ; Gorakhpur, 1992, 95, 2003*)
    (*d*) Eukaryotic Cell Structure (*Lucknow, 2003*)
    (*e*) Saprophyte fungi (*Kumaon, 2004*)
    (*f*) Symbiosis
    (*g*) Sclerotium (*Himachal Pradesh, 1995*)
    (*h*) Rhizomorph (*Kanpur, 2003*)
    (*i*) Somatic structure of fungi (*Kumaon, 1999*)
    (*j*) Facultative parasite (*Allahabad, 2000*)
    (*k*) Mycorrhiza (*Allahabad, 2000, Poorvanchal, 2001*)
    (*l*) Plectenchyma (*Allahabad, 2001*)
    (*m*) Obligate parasite (*Allahabad*, 2001)
    (*n*) Classification of fungi upto the rank of classes given by Ainsworth (*Poorvanchal, 2002*)
    (*o*) Distinguishing Characters of fungi (*Poorvanchal, 2002*)
    (*p*) Ultra Structure of fungal cell (*Meerut, 1994*)
34. Write notes on
    (*a*) Classification of fungi as given by Alexopoulos (1962) (*Gorakhpur, 1995*)
    (*b*) Methods of plasmogamy in fungi. (*Gorakhpur, 1995*)
    (*c*) Habit and Habitat of fungi. (*Poorvanchal, 2001 ; Gorakhpur, 1996*)
    (*d*) Thallus organisation in fungi. (*Lucknow, 2003; Gorakhpur, 2003*)
    (*e*) Vegetative reproduction in fungi (*Kumaon, 2004*)
35. (*a*) On the basis of nutrition describe various categories of fungi giving suitable examples. (*Kumaon, 1998*)
    (*b*) Give an account of nutrition in fungi. (*Kanpur M.Sc., 2002*)
36. Discuss the status of fungi in the five Kingdom classification system. (*Lucknow, 2002*)
37. Name the father of Modern Mycology and Plant Pathology. (*Poorvanchal, 2003*)
38. Differentiate between
    (*a*) Simple septum and Dolipore septum (*Allahabad, 2004*)
    (*b*) Saprophyte and Parasites
    (c) Autotroph and Heterotroph
    (d) Coenocytic and Septate mycelia
39. Write the classification of fungi proposed by Sparrow. (*Kanpur, 2002*)
40. What is plasmogamy ? (*Poorvanchal, 2002*)

OBJECTIVE TYPE

41. Fill in the blanks :

(*i*) The word fungus is derived from the latin word...

(*ii*) The science of study of Fungi is known as..

(*iii*) The fungi which grow on dead or decaying organic matter are called.......................................

(*iv*) Obligate parasites are those fungi which derive their nutrition only from....................................

(*v*) The nutrition in fungi is ....

(*vi*) Fungi lack ....

(*vii*) The mass of fungal hyphae is known as ....

(*viii*) A false tissue formed by aggregation of hyphae is known as ....

(*ix*) Fungi which grow on dung are called ....

(*x*) Monoecious fungus is that which bears male and female sex organs on the .... thallus.

(*xi*) The cell wall of the fungi is made up of ....

(*xii*) Father of modern Mycology is ....

(*xiii*) The absorptive organs produced inside the parasitic fungal mycelia are known as ....

(*xiv*) The classification of fungi is based both on mycelial stages and ....

(*xv*) The group of fungi which lack sexual reproduction are placed in the class ....

42. Select the correct answer :

(*i*) One important character of rhizomorph is that it
(*a*) grows actively even under adverse conditions.
(*b*) is produced only by lower fungi.
(*c*) grows only to a short distance.
(*d*) is resistant to adverse conditions.

(*ii*) Imperfect stage of the fungus refers to
(*a*) a poorly developed mycelium.
(*b*) poorly developed sex organs.
(*c*) conidial stage.
(*d*) immature fruting body.

(*iii*) The branch of Botany that studies fungi is known as
(*a*) morphology (*b*) mycology
(*c*) microbiology (*d*) phycology.

(*iv*) Haustoria are produced
(*a*) by fungi parasitising animal tissues.
(*b*) by hyphae of obligate parasites.
(*c*) in artificial cultures.
(*d*) after the hyphae become intracellular.

(*v*) The common food stored in fungi is
(*a*) starch (*b*) glycogen
(*c*) oil (*d*) fat.

(*vi*) A sclerotium refers to a modified mycelium which is
(*a*) an underground structure
(*b*) a hard resting body
(*c*) mainly food storing organ
(*d*) easily carried off by wind.

(*vii*) Somatogamy is the
(*a*) fusion of gametes
(*b*) fusion of vegetative cells
(*c*) contact between two gametangia
(*d*) copulation between two gametangia.

(*viii*) Stroma is
(*a*) compact somatic hyphae with fruit bodies
(*b*) loosely interwoven hyphae
(*c*) a small hyphae branch
(*d*) a group of spores.

(*ix*) All fungi are
(*a*) autotrophs (*b*) bneterotrophs
(*c*) saprophytes (*d*) parasites.

(*x*) Non-motile spores produced internally are known as
(*a*) zoospores (*b*) chlamydospores
(*c*) aplanospores (*d*) oidiospores

(*xi*) The unicellular true fungi in which the complete cell is converted to reproductive structures are called
(*a*) eucarpic (*b*) holocarpic
(*c*) zoocarpic
(*d*) holocarpic and Zoocarpic

(*xii*) The formula of Chtin is
(*a*) $(C_{20}H_{54}N_2O_{21})_n$ (*b*) $(C_{22}H_{54}N_4O_2)_n$
(*c*) $(C_{22}H_{24}N_4O_{21})_n$ (*d*) none

(*xiii*) The parasitic fungi capable of growing as saprophytes are called
(*a*) facultative parasite
(*b*) facultative saprophyte
(*c*) obligate parasite
(*d*) obligate saprophyte

(*xiv*) Perfect stage of fungus means
(*a*) the fungus reproduces asexually
(*b*) the fungus reproduces sexually
(*c*) the fungus is perfectly healthy
(*d*) it forms resting spores

(*xv*) The artificial class of fungi is called
(*a*) deuteromyceles (*b*) phycomycetes
(*c*) ascomycetes (*d*) basidiomycetes

(*xvi*) An obligate parasite is
(*a*) *Mucor* (*b*) *Albugo*
(*c*) *Fusarium* (*d*) *Saccharomyces*

Stemonitis : A slime mold

CHAPTER 2

SUBDIVISION – MYXOMYCOTINA The Slime Molds

- Class : Myxomycetes (The True Slime Molds)
- Order : Physarales
- Family : Physaraceae; Genus : *Physarum*
- A. Somatic Phase
- B. Reproductive Phase

General Account. This subdivision includes organisms which are phagotrophic in their mode of nutrition. The somatic phase is a multinucleate protoplasmic mass. The reproductive phase consists of a sporophore which produces one to several encapsulated spores. Popularly these organisms are called the **slime molds** because they contain and secrete slime. The slime molds are of two types, **acellular** and **cellular**. The somatic phase in the former is called the **plasmodium**. It is a free living, multinucleate, apparently naked protoplasmic mass. It is holocarpic. The somatic phase in the cellular slime molds is termed the **pseudoplasmodium**. The latter is an aggregation of amoeba-like organisms termed the **myxamoebae**. The latter do not fuse in the aggregation. They retain their identity but function as a unit until the time of sporulation. Thus, the unit of structure in the pseudoplasmodium is a nucleate, naked **myxamoeba**.

Are slime molds true fungi ? No, they differ from the true fungi in their phagotrophic mode of

Slime Molds

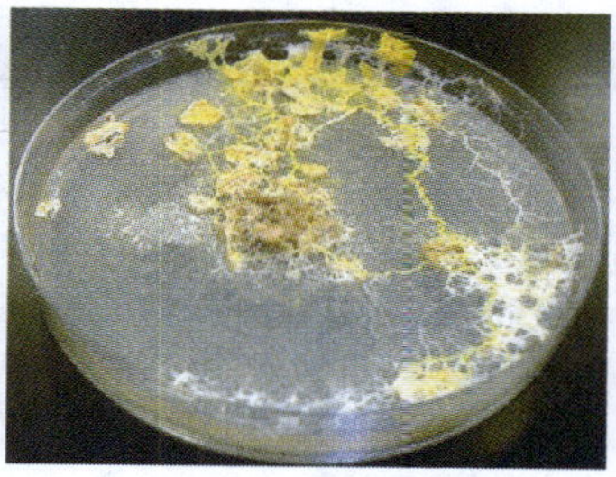

Acellular Slime Molds

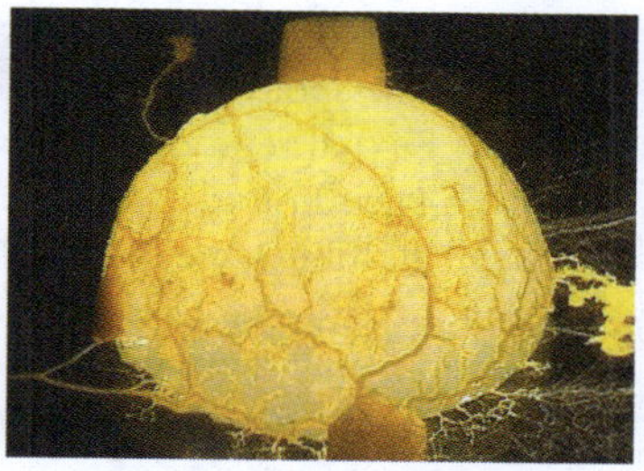

Cellular Slime Molds

nutrition and the nature of the somatic phase which is a plasmodium or pseudoplasmodium. The true fungi are absortive in their mode of nutrition and the unit of thallus structure is a hypha. For these reasons the slime molds have been separated from the true Fungi (Eumycotina) and placed in the subdivision Myxomycotina.

Classification. This subdivision includes three classes namely *Acrasiomycetes* (the cellular slime molds), *Protostaliomycetes* and *Myxomycetes*. Of these, class *Myxomycetes* is widely distributed and well investigated.

CLASS : MYXOMYCETES (The True Slime Molds)

Introduction. It is a group of organisms of great scientific interest. Some of these are of remarkable beauty, have delicate structure and brilliant colours. They exist in non-green slimy masses of protoplasm sending out pseudopodia. This has earned for them the name slime molds or slime fungi. They are found in cold, moist shady places in dead organic matter such as decaying logs and fallen dead leaves in the woods. The class includes the true plasmodial acellular slime molds most of which produce pigmented spores in small, delicate sporangia. The chief characteristics of this class are :

1. The somatic phase is represented by a multinucleate apparently naked acellular slimy protoplasmic mass called the plasmodium.
2. The plasmodium is the product of syngamy, hence a diploid structure.
3. The diploid plasmodium is holocarpic, free living and active. It contains and secretes slime.
4. Normally at the fruiting time, the entire plasmodium is organised into one or more plant-like reproductive structures, the sporangia (sporophores) or under conditions of stress and strains it becomes converted into an irregular hard structure, the **sclerotium**.
5. The reproductive phase is stationary.
6. With the exception of three species which are exosporous, all the others are endosporous and produce spores within sporangia.
7. The sporangium generally develops a tough non-cellular layer or wall called the peridium which is often studded with tiny crystals of calcium salts.
8. Within the peridium is usually an intricate network of fine tube-like structures constituting the *capillitium*.
9. The numerous spores are differentiated from the diploid protoplast of the sporangium by meiosis.
10. The encapsulated haploid spores or meiospores are close packed between the fine tubes of the capillitium but are free from them when mature.
11. The spore wall is differentiated into two layers, the outer of which is sculptured or spiny.
12. On germination, the haploid spores or meiospore give rise either to myxamoebae or biflagellate swarm cells which function as gametes.

13. The swarm cells or myxamoebae do not produce slime.
14. The sporangium in *Physarum* thus functions as on organ of sexual reproduction.
15. Sexual reproduction is of isogamous type.
16. The diploid zygote, by repeated mitoses but no cytokinesis, directly gives rise to multinucleate plasmodium.

Somatic phase. The creeping multinucleate apparently naked **plasmodium** which represents the somatic or assimilative phase in the life cycle of Myxomycetes, varies in structure in the different species. Alexopoulos (1960 b) described three types namely protoplasmodium, aphanoplasmodium and phaneroplasmodium.

1. Protoplasmodium. The plasmodium in some Myxomycetes (order Echinosteliales) is simple and of a primitive type. It is a uninucleate tiny mass of nearly homogeneous slimy protoplasm which form pseudopode but shows no distinction into veins. It is the smallest among the Myxomycetes and remains microscopic as long as it exists. The cytoplasmic stream is indistinct, slow and irregular. At the fruiting time it gets converted into a single sporangium (Fig. 2.1A).

2. Aphanoplasmodium. This type of plasmodium is characteristic of the order Stemonitomycetales. Early in its development the aphanoplasmodium looks very much like a protoplasmodium. During further growth it elongates and branches, finally resulting in a network of delicate strands. The aphanoplasmodium lacks the slimy sheath. The plasmodial protoplasm is less granular and thus transparent and not easily visible. The distinction into ectoplasm and endoplasm in not conspicuous. The cytoplasmic streaming is, however, rapid and confined by a fine membrane (Fig. 2.1 B).

3. Phaneroplasmodium. It is the most common type and characteristic of the order Physarales. The mature phaneroplasmodium is a massive structure. In its initial stages of development it is very much like the proloplasmodium. The multinucleate slimy protoplasm of the phaneroplasmodium is highly granular. It is differentiated into ectoplasm and endoplasm. At maturity it is divisible into an anterior fan-shaped perforated sheet of protoplasm and posterior zone consisting of a reticulate network of tubular veins or strands in which flows the rapid endoplasmic stream (Fig. 2.1 C).

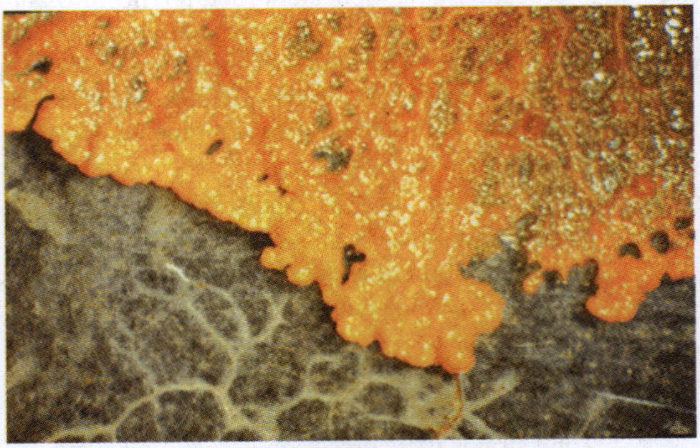

Highly Granular Protoplasm of a Typical Phaneroplasmodium.

Sclerotium formation. Under conditions of stress and strain, the phaneroplasmodium becomes converted into an irregular hardened mass of thick-walled cellular units. It is termed the **sclerotium**. The polynucleate thick-walled units constituting the sclerotium are termed **spherules**. The sclerotia and spherules remain dormant under conditions unfavourable for vegetative growth. With the return of conditions suitable for growth, the sclerodium germinates to give rise to a new plasmodium.

Reproductive phase. Normally on reaching a certain stage of maturity, the Myxomycete plasmodium passes into the reproductive stage. During this stage the entire plasmodium becomes converted into one or more fruit-like bodies, the *sporophores* or *spororigia* which bear the *spores*. This process is termed *sporulation*. With the exception of three species belonging to the order Ceratiomyxales which bear spores externally and are termed *exosporous*, all other Myxomycetes are *endosporous*. The latter bear spores within sporophores. The sporophores in the endosporous Myxomycetes chiefly are of three types namely sporangia, aethalia and plasmodiocarps.

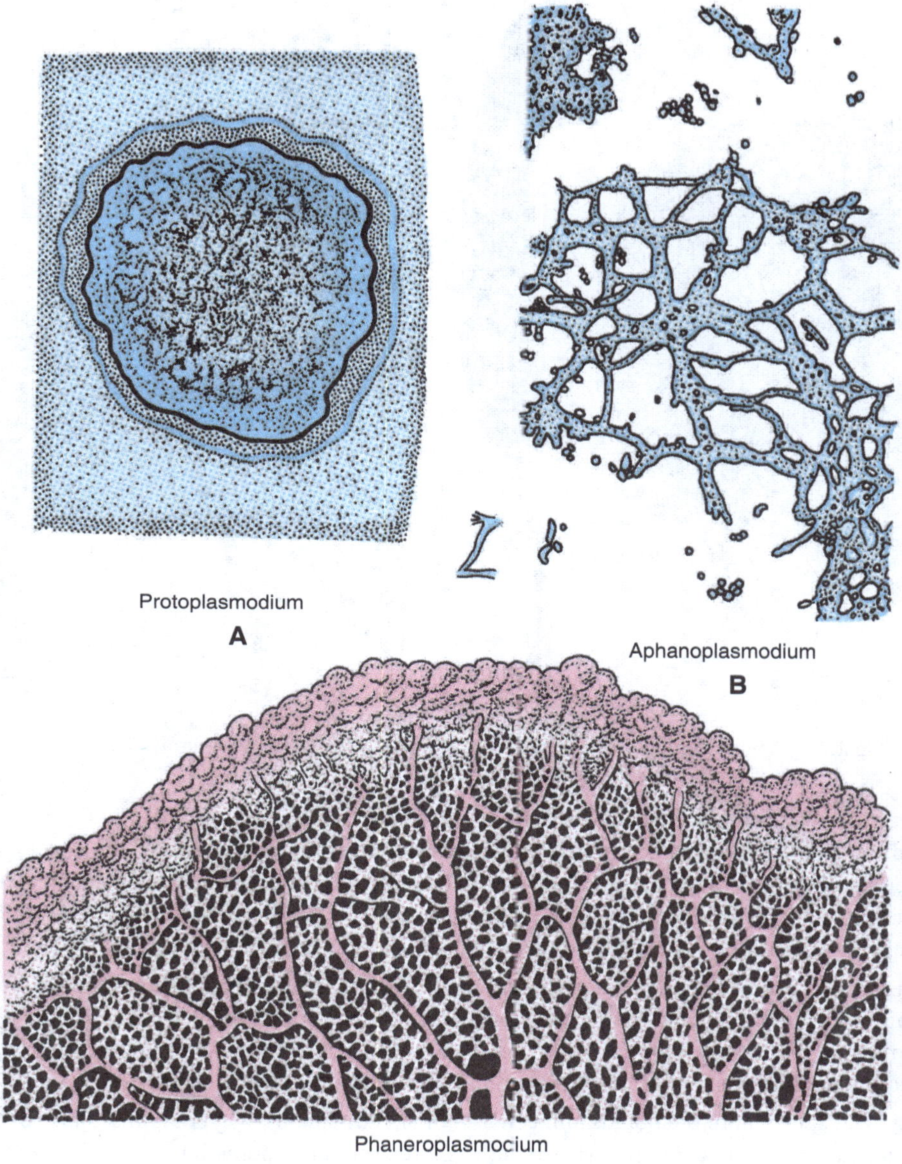

Fig. 2.1. Different types of Plasmodia in Slime molds

Sporangia. Most of the endosporous Myxomycetes (order Physarales) produce fruit bodies of this type. At the fruiting time the planeroplasmodium becomes converted into a group of several stalked, sometimes sessile sac-like structures, the *sporangia*. The sporangia in the group remain separate from one another. However, they arise in close proximity from a common thin transparent cellophane like base called the *hypothallus*. The latter is secreted and deposited during conversion of the plasmodium on the portion of the substratum formerly occupied by it. In some cases the hypothallus is absent (Fig. 2.2). Examples are *Arcyria*, *Physarum*, *Trichia*, *Didymium* and others.

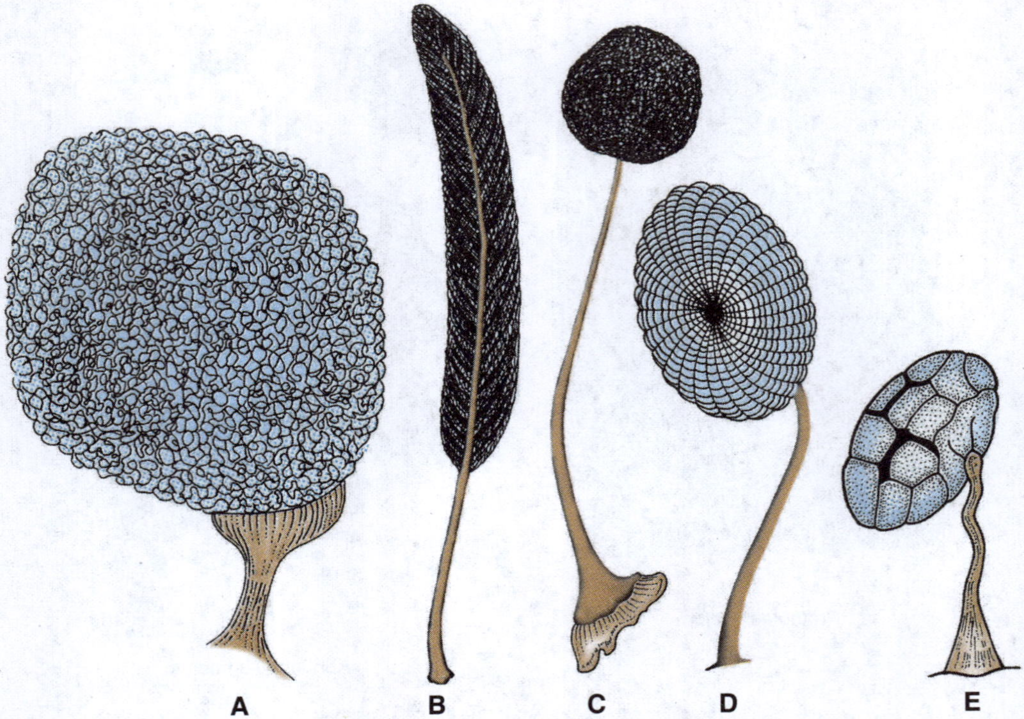

Fig.2.2. Different types of sporangia in slime molds : A, *Arcyria*; B, *Stemonitis*; C, *Comatricha;* D, *Didymium;* E, *Physarum*

Aethalia. Aethalium type of fruit body is characteristic of *Lycogala* and *Fuligo*. The plasmodium, at the fruiting time is fairly large structure. It becomes converted into a group of sac-like sporangia that do not separate from one another. The entire fructification which is termed aethalium, is enclosed in a common peridium and share a common hypothallus. The sporangial walls in the aggregation may be distinct, hardly visible or not present at all (Fig. 2.3A).

Plasmodiocarp. In forms like *Hemitrichia* the fruit body is very much like a sessile sporangium that retains the shape of the plasmodial venation. This type of sporophore is called a *plasmodiocarp*. It is formed by the concentration of the plasmodial protoplasm around some of the main veins followed by the development of peridium around each (Fig. 2.3B).

A typical sporophore or shorangium consists of six parts namely hypothallus, stalk, peridium, columella, capillitium and spores. All the sporophores produce spores but may or may not have all the other five parts. Generally the spores are globose in form and have a thick cell wall which is usually sculptured and rarely smooth. The mature spores are uninucleate and haploid. Meiosis occurs in the spores when young.

Classification. The class Myxomycetes is often divided into the following three subclasses :

(i) *Ceratiomyxomycetidae.* It contains a single order Ceratiomyxales represented by a single family Ceratiomyxaceae. *Ceratiomyxa* with three exosporous species is the only genus included in this family.

(ii) *Stemonitomycetidae.* It includes a single order Stemonitales which comprises only one family Stemonitaceae. It is endosporous. The somatic phase is of aphano-plasmodium type.

(iii) *Myxogastromycetidae*. It comprises four orders namely Physarales, Siceales, Echinosteliales and Trichiales. All are endosporous. The somatic phase is usually of protoplasmodial or phaneroplasmodial type. The class Myxomycetes thus contains six orders. Of these Physarales belonging to the subclass Myxogastromycetidae is the most important.

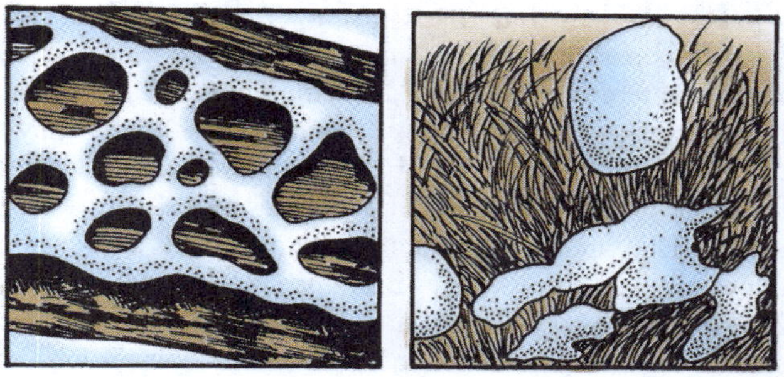

Fig. 2.3. Aethalia (A) and plasmodiocarp (B) in slime molds.

ORDER : PHYSARALES

The somatic phase of the true slime molds included in this order is of phaneroplasmodia type. Sporangium development is generally of myxogastroid type. Presence of abundant lime on the sporangium and purplish brown spores are the other important features . The order is represented by about 142 species and 12 genera placed in two families. *Physaraceae* and *Didymataceae*. Of these Physaraceae with the type genus *Physarum* is well known.

FAMILY : PHYSARACEAE; GENUS : Physarum

The genus *Physarum* with about 100 species is the largest and the best studied slime mold in the class Myxomycetes. The species *P. polycephalum* is usually taken as a type to illustrate structure and life history of true slime molds. Some other important and widely distributed species of this genus are *P. cinereum*, *P. compressum*, *P. viride* and *P. bivalve*. The other common genera of this family are *Fuligo* and *Craterium*.

P. polycephalum

P. cinereum

P. compressum

P. viride

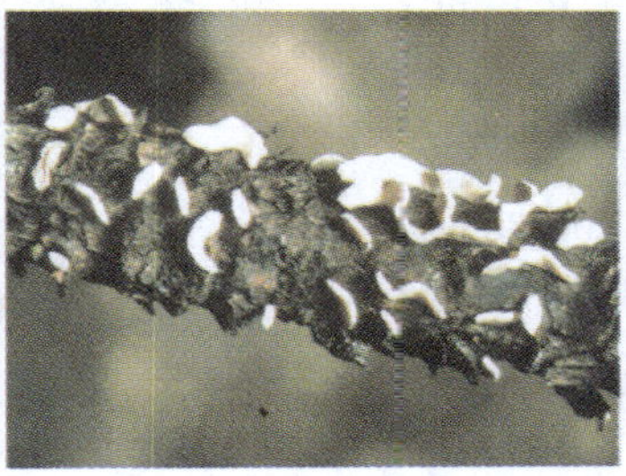

P. bivalve

Craterium

Occurrence and distribution. The slime molds feed on bacteria, protozoa, other small organisms and dead organic matter. The Physarales usually live as saprophytes. They can easily be encountered on dead and decaying undersides of fallen rotting leaves, twigs and logs of wood often under bark, decaying vegetable matter, and fleshy fungi, on damp soil and frequently on damp rocks in high humidity, cool temperature and away from the direct rays of sun. In fact they love shady, cool, damp and dingy places in the forests where there is plenty of decaying vegetable matter. They have been reported from all parts of world but are quite common and active in the humid tropical forests. They also appear in abundance following warm, rainy periods in temperate zone forests, especially in late spring when there is plenty of decaying vegetable matter. A few species have been reported to occur on the alpine region. A little after and during rainy season the slime molds are found on the leaves of grasses or other plants on the lawns.

The life cycle of a slime mold consists of two distinct stages or phases namely, (i) the vegetative growth (somatic) stage as represented by the diploid plasmodium, and (ii) the reproductive (or sporulative) stage.

A. SOMATIC PHASE

Zygote formation. On the onset of conditions favourable for growth the spiny thick-walled resting meiospores germinates releasing usually 1 to 4 biflagellate **swarm cells** or **myxamoebae** which function as gametes. Copulation between two swarm cells or myxamoebae is followed by karyogamy. The resultant fusion cell is the **zygote**. It is amoeboid in form and has a single diploid nucleus. The amoeboid zygote constantly puts forth slender processes, the pseudopodia and withdraws them. Thus, it has no definite shape. With the help of *pseudopodia* it creeps over the substratum engulfing food particles. Assimilation of food results in the synthesis of more protoplasm resulting in growth which is irreversible increase in size. Growth in zygote is accompanied by repeated and successive meiotic divisions of the diploid parent nucleus. There is thus repeated karyokinesis but no cytokinesis. The numerous nuclei come to lie in the cytoplasm of the same zygote. It becomes clear from the above account that as a result of growth, repeated karyokinesis but no cytokinesis, the amoeba-like zygote gradually changes into a multinucleate amoeboid mass of protoplasm with diploid nuclei. It is termed the plasmodium.

Plasmodium (Fig. 2.4). The somatic phase of a true slime mold is thus a thallus consisting of a free living irregularly shaped mass of slimy protoplasm with several diploid nuclei embedded in it. Technically it is called the **plasmodium**. The plasmodium is acellular and is the product of syngamy, hence a diploid structure. It has neither any definite shape nor definite size. The plasmodium constitutes the assimilative phase in the life cycle. It contains and secretes slime which protects it from dehydration. Probably correlated with the presence of slime in the growing plasmodium is its inability to undergo cytokinesis. The plasmodia are often very colourful. The colour ranges from red, violet, orange, brown, yellowish, black, slightly greenish to even colourless. The plasmodium of

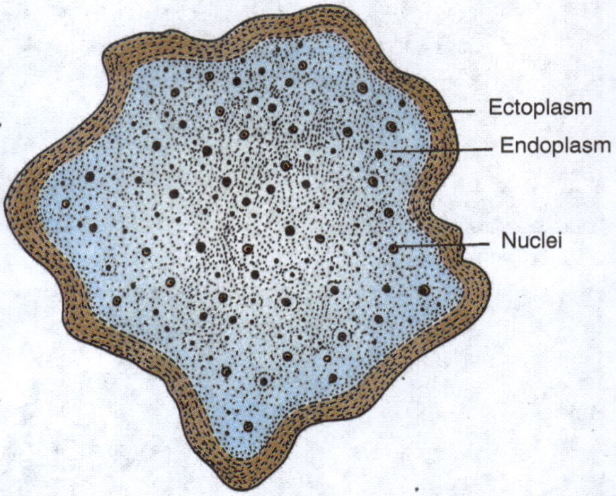

Fig. 2.4. *Slime molds.* Young diploid plasmodium showing gross structure. (Diagrammatic)

P. polycephalum is bright yellow depending upon the species. However, the plasmodium lacks chlorophyll.

Phaneroplasmodium. In some slime molds, under favourable conditions, the plasmodium increases in size, as it develops, by repeated synchronous (simultaneous) mitotic nuclear division of diploid nuclei and also by annexing zygotes and other plasmodia of the same species it comes across as it creeps over the substratum. Their nuclei, however, do not unite. Consequently the plasmodium, in some cases, may be organised into an extensive multinucleate amoeboid mass of protoplasm bounded by a plasma membrame (syncytium) attaining a size several square centimetres. It is known as the **macroplasmodium** or more appropriately the *phaneroplasmodium*. It is characteristic of *Physarum polycephalum*.

The mature plasmodium (phaneroplasmodium, Fig. 2.5) is a massive structure which becomes differentiated into an anterior fan-shaped sheet of granular protoplasm with a network of veins and a posterior zone which consists of a complex reticulate network of thick branched veins or strands, also called the channels (Fig. 2.5.) Each strand is a tubular structure of jellified protoplasm. It contains more fluid portion of the protoplasm (endoplasm) which circulates actively within the strands; carrying with it the numerous nuclei, vacuoles and other inclusions.

Cyclosis. The enodplasm in the strands, as mentioned above, shows streaming movements. It is in a constant state of rotation. It circulates actively within the strands carrying with it the numerous nuclei, vacuoles and other inclusions first in one direction for about 40-60 seconds. It then slows down, finally stops momentarily and then streams in the reverse. This unending, oscillatory endoplasmic streaming is termed **cyclosis**. The motive force which brings about cyclosis has been considered to be associated with metabolism. It is generated through the interaction of a contractile protein with ATP (Adenosin triphosphate). It has now been determined that **actin** (Hatano and Oosawa, 1966) and **myosin** (T'so *et.al*, 1957) are the two contractile proteins which occur in the plasmodium. Wholfarth (1962) showed that these are located in the cytoplasmic filaments and fibril structures. Daniel and Jarlfors (1972) opined that "location and structure of the microchannel/cortex system strongly suggests it as the site for localisation of contractile function implicated in cyclosis and motility".

Actin, actinin and myosin have now been extracted from the plasmodium of *Physarum*. These contractile proteins, according to Hatano *et al,* (1980) organise to form the subcellular structures responsible for the generation of motive force involved in cyclosis. With the cyclic contraction of plasmodial gel (ectoderm) induced by contraction of actomyosin systems, F-actin filaments in the gel layer form bundles. The inner sol (endoplasm) is squeezed out into the plasmodial strands. According to Kumiya and Kurdo (1958) this causes the flow of sol in the strands. On relaxation of the gel layer, the fibrils disappear. Thus, actin which exists as cytoplasmic fibrils or actin filaments are not perpetual structures. These appear and disappear according to the contraction and relaxation of the gel layer. Actinin is considered to play a role in regulating sol-gel transformation.

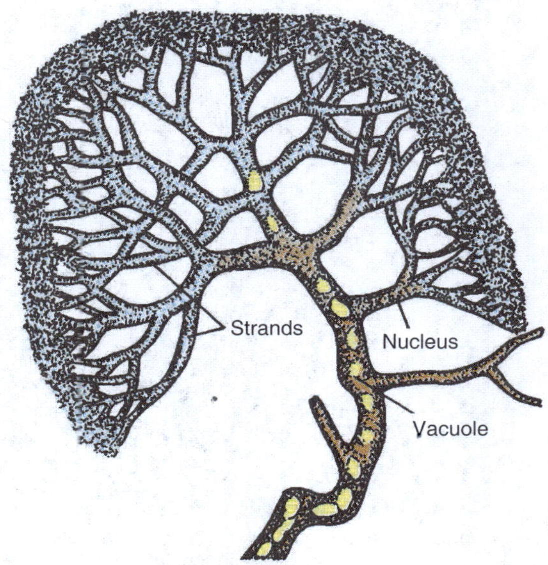

Fig. 2.5. *Slime molds.* Older fan-shaped plasmodium of *Physarum polycephalum.*

Structure of Plasmodium. (*i*) *Gross structure*. Careful observation has revealed that the apparently naked plasmodium is bounded by a thin flexible non-cellular slimy layer distinct from the protoplasm. It is very thin at the advancing front end where foodparticles are engulfed in an amoeboid manner. Within the *slime* layer or sheath, the plasmodial protoplasm at its air interface is differentrated into a distinct plasma membrane. It is an integral part of the protoplasm and completely invests it. Within the plasma membrane, the plasmodial protoplasm is differentiated into two zones, the outer and inner. The outer zone is gelatinous. It consists of firmer cytoplasm of less liquid consistency. It is devoid of nuclei and other cell organelles. The inner zone which is called the endoplasm contains cytoplasm of more fluid consistency. It is in the plasmosol state. It contains numerous tiny nuclei, small vacuoles, contractile vacuolas, pigment granules and other inclusions. In fact the endoplasm is coarsely granular. The two phases of the protoplasm blend gradually.

(*ii*) *Fine structure* (Fig. 2.6). The plasmodial protoplasm at its air interface is differentiated into a distinct plasma membrane.In *Physarum* it undergoes extensive invaginations. Coating the plasma membrane is a non-cellular thin slime layer that contains microfibrils. Probably the slime

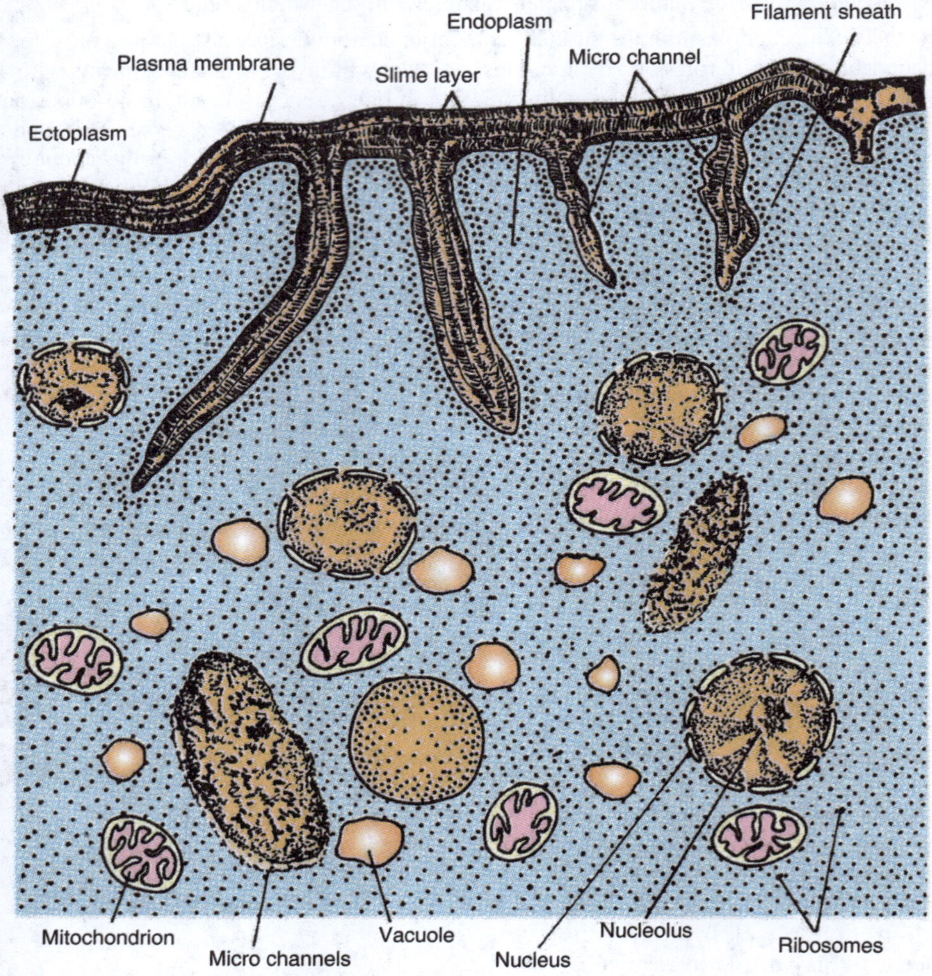

Fig. 2.6. *Physarum polycephalum*. A diagrammatic representation of the fine structure of plasmodium.

sheath consists of mucopolysaccharide protein complex. Henney and Asgari (1975) reported that slime is composed of protein consisting of 17 different aminoacids. The other component is a polysaccharide consisting of hexose-galactose. Within the plasma membrane, the plasmodial protoplasm is differentiated into two zones, the outer and the inner. The outer zone consists of finer cytoplasm of less liquid consistency. It is in the plasmogel state and is devoid of nuclei and other larger cell organelles and large vacuoles. This layer or zone of cytoplasm is termed the *ectoplasm* or *cortex*. Occasionally the ectoplasm exhibits fine fibrils. The inner zone which is called the *endoplasm* contains cytoplasm of more fluid consistency which is in the plasmosol state. The endoplasm shows a complex network of microchannels and contains numerous discrete vacuoles. The obvious cell organelles which the endoplasm contains are the nuclei, mitochondria, endoplasmic reticulum and the numerous ribosomes.

Microchannels. These are fine tubular structures, approximately 1μ in diameter coursing through the endoplasm. They arise as invaginations of the plasma membrane from the exterior into the cytoplasm. Each microtubule contains a fibrous material similar to and contiguous with the slime layer coating the plasma membrane. In the cytoplasm adjacent to each microtubule is a layer of fine filaments constituting the *filament sheath*. The latter surrounds each microchannel and in the area of invagination at the plasmodial surface it appears as a continuation of the ectoplasm. The filaments in the sheath run parallel to each other and to the microchannel.

Vacuoles. These are membrane bound, spherical structures of varying sizes found dispersed in the endoplasm. They lack cytoplasm and are devoid of filament sheath.

Nuclei. The diploid nuclei have each a distinct nucleolus. The nuclear membrane has pores.

Mitochondria. The mitochondria are round or elongate in form. The cristae form ribbon-like configurations.

Locomotion. Plasmodial movement is termed *locomotion*. Under nongrowth conditions, it is rapid whereas under conditions of rapid growth it is very slow. The plasmodium moves with the help of pseudopodia which are blunt finger-like processes or lobes of its body. It puts out one or more pseudopodia in one direction. The entire protoplast of the body then flows in the direction of the pseudopodium thus formed. In this way the plasmodium creeps slowly over the surface of the substratum. The contractile proteins which occur in the microchannel-cortex system form the mechanico-chemical basis for motility. Actinin is suggested to play a vital role in regulating sol-gel transformation. It moves towards moisture and shuns strong light during the somatic phase, which is thus found in the dark.

Nutrition. The decaying vegetable matter constitutes the chief food of the slime mold plasmodium. It is absorbed saprophytically through the ectoplasm. The plasmodium also feeds on small organisms such as bacteria, protozoa or fungi, other smaller micro-organisms and bits of non-living solid, nutritious decayed organic matter it comes in contact with. It is thus **phagotrophic** in its nutrition. The food is engulfed by the plasmodium flowing around it. The ingested food is digested within food vacuoles by the enzymes and is used to build new protoplasm. The waste matters are rejected and simply left behind with the slime trail as the plasmodium moves ahead. The plasmodium also contains contractile vacuoles which probably have an excretory function. The somatic or vegetative stage of slime molds thus resembles amoeba of the animal kingdom.

To sum up, the diploid plasmodium, which represents the acellular somatic phase in the life cycle, has the capacity for engulfment of particulate matter, for migration over the substratum and for somatic fusion (plasmodial coalescence). The duration of this phase in the life cycle depends on certain external conditions such as low light intensity, abundance of moisture and plenty of available food. These conditions prolong the growth of the somatic or assimilative phase of the slime mold plasmodium.

After a period of food intake and crawling movements the plasmodium attains its maximum size and reaches maturity. In the normal course of events the reproductive phase ushers in at this stage and the plasmodium forms fruit bodies. However, under conditions of stress and strains the phaneroplasmodium of *Physarum* becomes converted into a hard, irregularly shaped structure known as the *sclerotium*.

Sclerotia formation (Fig. 2.7). Under certain conditions such as drought, starvation, cold and absence of light the phaneroplasmodium by differentiation and cleavage becomes transformed into an irregular hard structure consisting of thick-walled cellular units. It is the **sclerotium**. The sclerotium thus represents the only diploid cellular stage in the slime mold life cycle. It remains dormant for a long period under conditions unfavourable for growth. With return of suitable conditions, the sclerotium grows into a plasmodium again. During the early period of sclerotium development, there is condensation of the plasmodial protoplast through desiccation and extrusion of slime. This is followed by cleavage of the protoplast into polynucleate units or portions. Finally a thick, two-layered wall is formed around each unit or portion. The sclerotium thus consists of small cells named *macrocysts* or more appropriately called *spherules*. The latter in a sclerotium remain grouped together in a berry-like form and are covered by a common coat. The thick walls of the spherules separate them from the adjacent ones. The spherules in old sclerotia are lobed and not rounded. According to Jump (1954), the sclerotia are induced under gradual desiccation, low temperature (5°C), high osmotic solutions of certain heavy metals, starvation and low (2.0)*p*H.

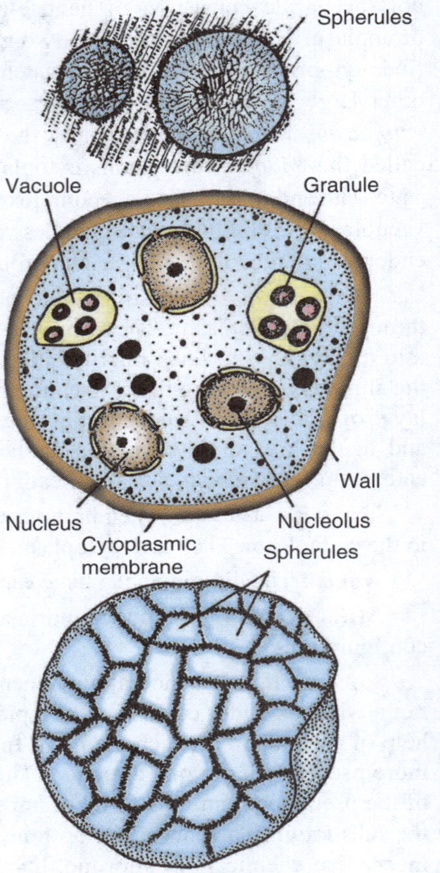

Fig. 2.7 (A-C). *Physarum polycephalum*. A. manitol induced sperule; B. spherule showing fine structure; C. sclerotium (After Chet and Kislev)

Spherules (Fig. 2.7. A-B). The manitol-induced spherules exist as separate units (A). They are round in form and vary in size from 6-12 μ in dia (Chet and Kislev, 1974). The spherule has a thick hard wall. Within the cytoplasmic membrane which closely invests the cytoplasm (B), the latter contains a few diploid nuclei and vacuoles. Each nucleus has a nucleolus. The vacuoles contain small dense hollow granules which vary in dia. from 0.1-2 μ. The spherule germinates under normal conditions. The thick hard wall cracks. The polynucleate spherule protoplast emerges through the split. It feeds and grows to form the new diploid plasmodium.

The sclerotia and spherules primarily serve as means of perennation. Secondarily they constitute the vegetative methods of asexual reproduction in the life cycle and serve to prolong the diploid stage in the life cycle of *Physarum*. The prolonged diploid stage is unusual among the true fungi. Some myxomycetes including *Physarum* overwinter in the sclerotial stage.

Alexopoulos (1966) reported that the aphano-plasmodia of some myxomycetes (stemomtales) do not form sclerotia. Instead they form cysts directly from the plasmodial veins.

B. REPRODUCTIVE PHASE

Normally when the phaneroplasmodium of *Physarum* attains its maximum size and reaches maturity, it passes into the reproductive phase. The Phaneroplasmodium at this stage contains about one hundred to many thousand diploid nuclei. During the reproductive phase, the entire plasmodium is converted into one or more fruiting bodies which bear the spores. The fruiting body is called the *sporophore* or *sporangium*. The spores are cleaved from the sporangial protoplast. This process is termed sporulation. Since both the somatic and reproductive phases do not coexist in the same plasmodium, the slime molds are said to be holocarpic. The conditions which promote fruit formation are not definitely known. Moisture, light, temperature, pH and exhaustion of food have been suggested to be ralated to fruiting by some workers. There are others who proved this claim to be untenable on the basis of their experimental results.

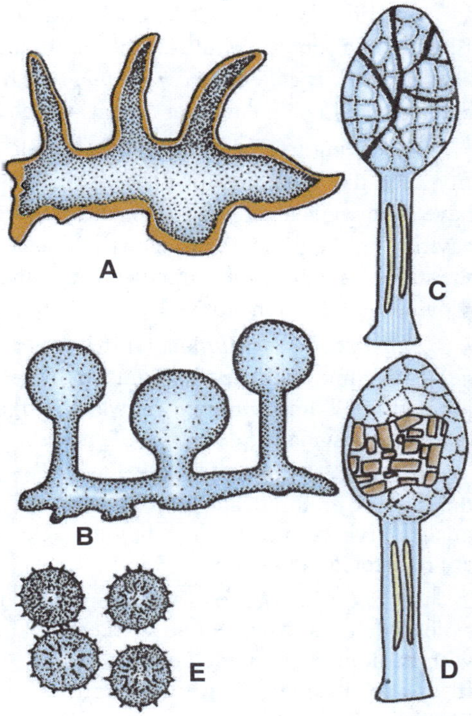

Fig. 2.8. (A-E). *Slime molds*. Formation of fruit bodies and differentiation of resting spores. Explanation in the text.

Sporulation. Under this heading we study (*i*) development of sporangium (*ii*) its structure and (*iii*) sporogenesis.

(*i*) *Development of Sporangium*. (Fig. 2.8) At the fruiting time which is influenced by conditions not definitely known, the phaneroplasmodium changes its behaviour. It comes out of darkness or diffuse light and moves to exposed sites which are dry. The amoeboid life ceases. The protoplast of the quiescent plasmodium then becomes concentrated in some places to form hemispherical mounds. The slime layer of the plasmodium dries or simply remains as a sheath on the substratum to constitute the *hypothallus*. Owing to the upward pressure of the protoplast from within with a pulsating movement, each mound elongates into a column like structure known as the *papilla* (A). As the papillae elongate their bases constrict to form the *stalk* of the sporangium and free ends swell up to form the body (B). A large number of nuclei along with cytoplasm stream into the swollen tip which is then separated from the stalk by a septum. The swollen multinucleate terminal body along with the stalk functions as the sporangium (C).

(*ii*) *Structure of Sporangium*. The sporangia in slime molds may be stalked or sessile. They are beautifully coloured. White, purple, orange and brown are the usual tints. They vary in form and may be spherical, ovoid, elongated etc. The stalk which is thin and slender, also varies in length. *P. polycephalum* has stalked, clustered sporangia. Each stalk may end in one or more dark-coloured more or less globular sporangia. In height the sporangium varies from 1-3 mm including the stalk. The sporangium in *Physarum* is enclosed in a tough, slimy, non-cellular layer, or wall called the *peridium*. It is hard, brittle and has a rigid wrinkled texture. According to Kislev and Chet (1973), the peridium surface is marked by evenly distributed small holes or cavities about 5-10 μ in dia. and is covered with numerous head-like granules up to 2 μ in dia. Within the peridium, in a mature sporangium, are the numerous tiny, rounded spores which are close packed in between fine tube-like structures constituting the *capillitium*. The spores are, however, free from them when mature.

Capillitium. The cleavage furorows in the dividing proloplast of the young sporangium are filled with a fungus cellulose material in the form of an intricate network of fine hyaline threads. This network is called the *capillitium* (C). It appears to consist of a network of calcareous nodules connected by hyaline tubular threads which adhere to the peridial wall.

According to Kislev and Chet (1973), the capillitium in *Physarum* is an intricate network of fine tube-like structures of differing diameters. The capillitial tubes may be simple or branched, have rigid walls and open at the peridial surface (wall). These capillitial openings look funnel-like cavities on the peridial surface (Kislev and Chet, 1973). Fine particles or granules are found on the surface and bead-like granules within the tubes. These structures probably of calcium carbonate provide rigidity to the tubes.

Functions of capillitium. (*i*) It serves as an endoskeleton providing mechanical support to the delicate, slime mold sporangium. (*ii*) It serves as a conducting system through which excretory products and lime migrate outward to the peridium and become deposited thereon and (*iii*) perhaps assists in the despersal of spores.

(*iii*) *Sporogenesis.* It is a process where by haploid spores (meiospores) are fashioned from the diploid multinuleate protoplast of the sporangium by *meiosis*. Meiosis is a special kind of nuclear division in which the diploid nuclei undergo two successive divisions but the chromosomes are replicated only once.

Location of Meiosis. The numerous diploid nuclei in the sporangium protoplast undergo synchronous division. Some workers hold that this division is *meiotic*. The resultant protoplast with haploid nuclei undergoes cleavage into uninucleate tiny daughter protoplasts. Each of these becomes rounded and secretes a thick wall around it to become a meiospore. Thus, according to this view, meiotic division in *Physarum* occurs in the *precleavage sporangium.*

The opposite view held by Stock *et al.* (1964), Aldrich (1967) and Pandall and Lynch (1974) advocates meiosis in *P. polycephalum* in the *post cleavage spores*. According to them, nuclear division in the precleavage sporangium is mitotic. Randall and Lynch (1974) observed nuclei in mature spores containing distinct synaptonemal complexes, a feature characteristic of meiotic prophase. Their observation thus confirms the view that meiosis in this slime mold occurs after spore cleavage and not prior to it. In fact the young cleaved spores are uninucleate and diploid. The diploid nucleus later undergoes division in which synaptonemal complexes characteristic of meiosis become apparent. Immediately the second division follows.

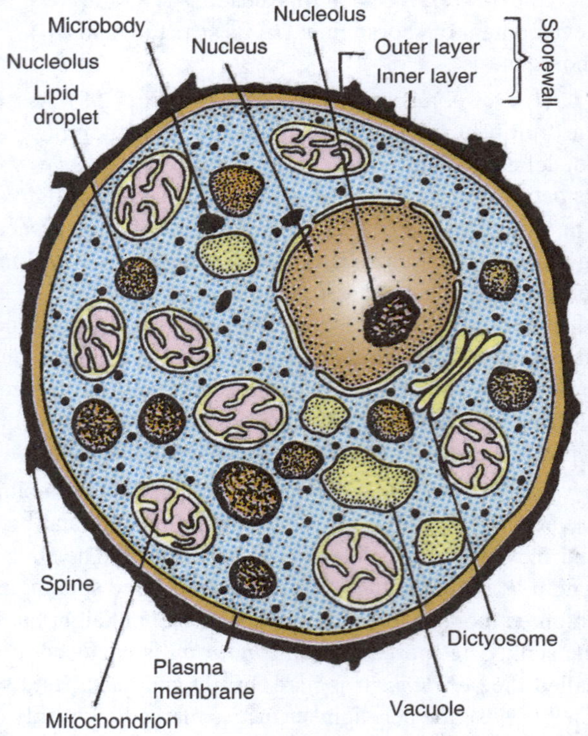

Fig. 2.9. *Physarum polycephalum.* Diagrammatic representation of the fine structure of a resting spore(meiospore) (Based on Randall and Lynch)

Each spore becomes quadrinucleate. The four nuclei represent the tetrad. Three of these disintigrate, consequently the spore becomes uninucleate and haploid. Technically the resultant uninucleate haploid spore is called a meiospore.

(*iv*) *Dispersal of spores*. Wind, rain and mites play an important role as agents of spore dispersal in slime molds. The peridium disintegrates exposing the capillitium and mass of mature spores it supports. The hygroscopic capillitium twists in various ways as it absorbs and loses water. These movements cause the spore mass to break and thus loosen and free the spores which are dispersed by air currents. At the same time the capillitium expands carrying the spores high up with it. From this position the dry, powdery spores are easily disseminated by wind. The wind disseminated spores is a plant characteristic. Only plants produce wind borne spores and sporangia. The fruiting stage of slime molds thus reminds one of a fungus.

(*v*) *Spore ultrastructure* (Fig. 2.9). The purplish brown meiospores of *P. polycephalum* are resting spores. They are small, round, uninucleate structures not more than 10μ in diameter with a spiny spore wall. The thick, beautiful sculptured spore wall is reported to contain cellulose which is typical of plant cells. McCormick, Blomquist and Rusch (1970) found it to contain mainly galactosamine and some protein and melanin. The spore wall is differentiated into two layers, the outer and the inner. According to Randall and Lynch (1974), the outer layer, which is thicker, is electron opaque with a granular composition. Externally it bears randomly distributed cone-shaped spines. The inner layer of the spore wall which is comparatively thinner, is electron transparent and has a fibrous appearance. Within the spore wall is the spore protoplast which at its periphery is differentiated into a distinct unit membrane, the *plasma membrane*. The latter occasionally invaginates to form pocket-shaped vesicles which extend into the spore cytoplasm. Gaither (1974) reported that the meiospores contain the usual complement of discrete organelles embedded in a dense cytoplasm. Numerous spheroidal mitochondria are dispersed throughout the cytoplasm. Randall and Lynch (1974) reported the occurrence of electron-dense granules and occasional presence of electron-dense helical filaments within mitochondrial cristae. The single nucleus has a distinct nucleolus. The nuclear membrane has pores. Near the periphery of mature spores in close proximity to vacuoles Randall and Lynch (1974) rarely observed dictyosomes consisting of stacks of 3 or 4 parallel cisternae. The other inclusions found in the cytoplasm are the lipid droplets, glycogen-like granules, vacuoles and microbodies.

(*iv*) *Germination of Meiospores* (Fig. 2.10). The meiospores (A) remain viable for a considerable period. They are exceptionally resistant to prolonged periods of desiccation owing to the thick spore wall and physico-chemical structure of spore protoplasm (A). They germinate on moist substrata such as dead leaves, logs, soil or substratum in favourable temperature and in rain water in nature. On germination the spore wall either cracks open (B) or a tiny pore appears in it. The contents usually escape in the form of spindle-shaped structures through it (C).

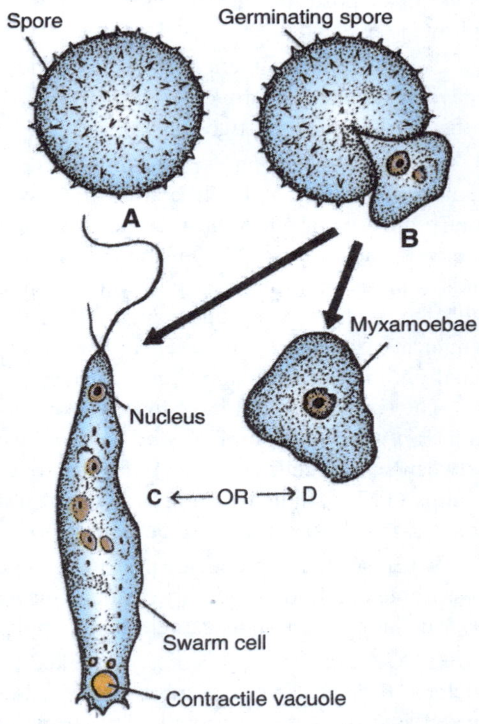

Fig. 2.10 (A-D). Slime molds. Germination of meiospore.

Each has two flagella inserted at its anterior end. One of them is shorter than the other but both are of whiplash type. The shorter flagellum is directed backward or appressed against the body. This flagellated structure is known as the **swarm cell**. The swarm cells greatly resemble colourless algal flagellates or protozoans (zooflagellates). The number of swarm cells produced by each resting spore varies from one to four. The anteriorly biflagellate swarm cells are naked and amoeboid. Each contains a single nucleus toward the anterior end and contractile vacuoles at the posterior end (C). Sometimes, instead of swarm cells, one to four amoeboid cells known as **myxamoebae** emerge when a spore germinates (D). Presence of moisture favours the flagellate form and drier conditions induce the amoeboid form. The swarm cells as such do not divide. The myxamoebae normally divide and increase in number.

(C) *Sexual Reproduction* (Fig. 2.11). The sexual stage intervenes between the resting meiospore stage and diploid plasmodium. The swarm cells or myxamoebae which function as gametes are produced by the germination of meiospores or resting spores differentiated by meiosis from the diploid protoplast of the *sporangium*. The sporangium in *Physarum* thus is an organ of sexual reproduction. The swarm cells (A) or myxamoebae (a) get their nutrition from the surrounding medium by absorption, and also by ingesting bacteria, fungal spores, yeast cells and small particles of organic matter at their sticky posterior end. The sexual process consists of the following events :

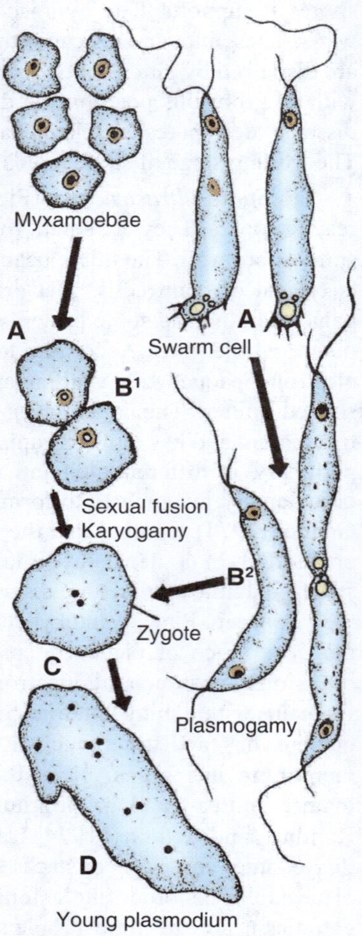

Fig.2.11 (A-D). Slime molds. Sexual reproduction

Sexual fusion. Depending upon the species, fusion between gametes (swarm cells or myxamoebae) takes place in either of the following two ways :

(*i*) In *P. polycephalum* swarm cells directly function as planogametes (A). After liberation they swim about with a rotary movement and finally come in contact at their sticky posterior ends in pairs (B). Fusion thus starts at their posterior ends. The cytoplasm fuses with the cytoplasm and nucleus with the nucleus. The young zygote thus formed is, at first, binucleate and flagellate (B2). It swims about for a while. Finally it retracts the flagella and changes into a myxamoeba (C). The two nuclei in it have fused to form a diploid nucleus. In *P. polycephalum* and *P. flavicomum* fusion takes place between swarm cells (gametes) of opposite mating strains. They are thus *heterothallic*. Clark and Collins (1976) studied mating systems of eleven species of myxomycetes. They reported that *P. cinereum* is also heterothallic but *P. compressum*, *P. gyrosum* and *P. pusillum* are *homothallic*.

(*ii*) In some other species myxamoebae emerge from the resting spores one each (*a*). Under favourable conditions each myxamoeba may divide repeatedly to form a number of daughter cells (a). The latter function as gametes and copulate in pairs (b) to form a zygote (c).

(*b*) *Germination of Zygote*. The naked zygote (C) formed in either case has a single diploid nucleus. It creeps over the substratum feeding on bacteria and organic matter synthesising more protoplasm resulting in growth. The growth in size is accompanied by repeated and successive divisions of the diploid parent nucleus. The divisions are mitotic. As a result of subsequent growth, repeated karyokinesis but no cytokinesis, the zygote gradually becomes changed into a large multinucleate amoeboid mass of protoplasm, called the **macro plasmodium** (D). The numerous nuclei embedded in its protoplasm are diploid in nature. Favourable temperature, abundant moisture and food favour its growth, movement and reproduction.

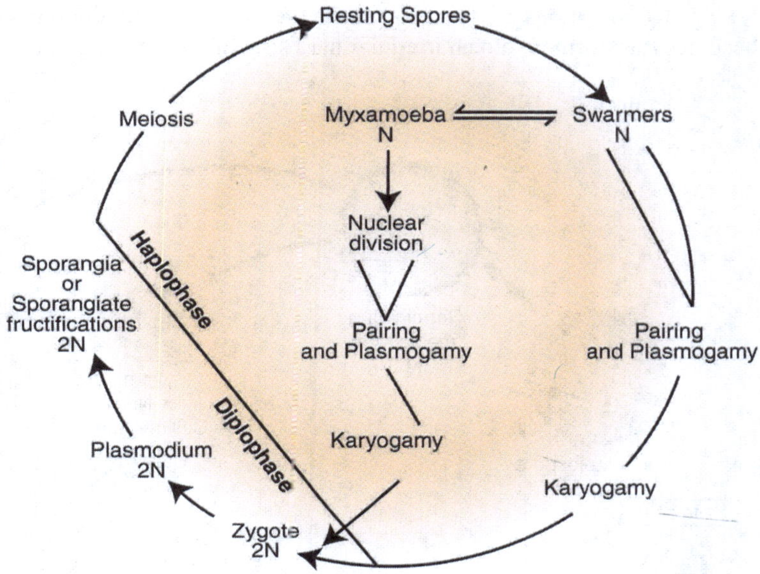

Fig. 2.12. Word diagram of general life cycle of a slime mold.

In many cases, the young diploid plasmodia may combine with zygotes or other plasmodia of the same species or a number of zygotes may coalesce to form a single larger plasmodium. In all these cases, the union involves the fusion of their cytoplasm only. There is no fusion between the nuclei.

Summary of the Life Cycle (Fig. 2.13). The life cycle of an endogenous slime mold such as *Physarum* starts with the germination of a meiospore under favourable conditions. One to four uninucleate haploid biflagallate *swarm cells* or *myxamoebae* are released through a slit or pore, the former under wet conditions and the latter under drier conditions. Fusion between two swarm cells or myxamoebae followed by karyogamy results in the formation of a zygote which is uninucleate, diploid and amoeboid in form. It is *acellular*. It creeps over the substratum feeding on bacteria and organic matter synthesizing more protoplasm resulting in growth. The growth in size is accompanied by repeated and successive divisions of the diploid parent nucleus. The divisions are mitotic. As a result of subsequent growth, repeated karyokinesis but no cytokinesis, the zygote gradually becomes changed into a multinucleate amoeboid mass of protoplasm, called the **plasmodium**. The numerous nuclei embedded in its protoplasm are diploid in nature. Favourable temperature, abundant moisture and food favour its growth, movement and reproduction. In many cases the young diploid plasmodia may combine with zygotes or other plasmodia of the same species or a number of zygotes may coalesce to form a single larger plasmodium. In all these cases, the union involves the fusion of their cytoplasm only. There is no fusion between the nuclei. Eventually the plasmodium in many cases becomes a massive structure called the *macroplasmodium* or *phaneroplasmodium*.

Normally the plasmodium after attaining a certain size and stage of maturity enters the reproductive phase. The amoeboid life ceases. The slime layer dries. The quiescent plasmodium thickens. The diploid protoplast concentrates at a few points forming a mound like structure. Th latter grows into a stalked *sporangium*. The diploid protoplasm or of the sporangium cleaves numerous young spores each of which has a diploid nucleus. The diploid nuclei of the y spores undergo meiosis to form meiospores. When mature the meiospores are released and disp by wind. On falling on a suitable substratum and under favourable conditions, the mei germinates to release swarm cells or myxamoebae which fuse in pairs to form the *zygote* in the diploid condition is re-established.

Under conditions of stress and strain the mature plasmodium by differentiation and cleavage becomes transformed into an irregular hard structure consisting of thick-walled cellular units. The

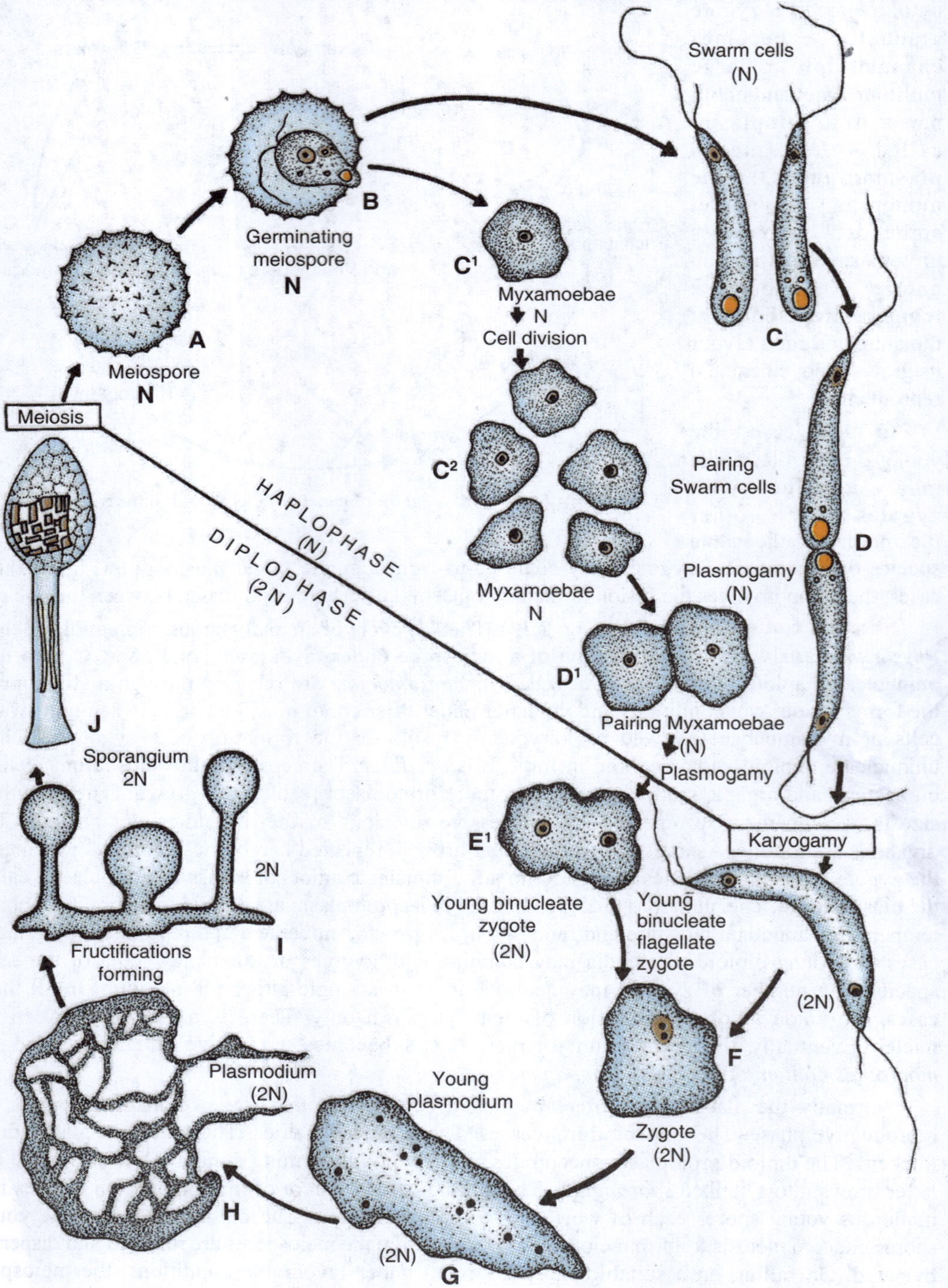

Fig. 2.13 (A-J). *Slime molds.* Pictorial life cycle of *Physarum polycephalum*.

former is called the sclerotium and the latter *spherules*. The sclerotium remains dormant when the conditions are unfavourable for growth. With the return of suitable conditions it grows into a new plasmodium.

Alternation of Generations. Strictly speaking, there is no alternation of two distinct generations in the life cycle of true slime molds. The diploid plasmodium is the *sporophyte*. Along with other diploid structures such as the zygote, sporangia and the young diploid spores, it constitutes the *sporophyte generation* or *diplophase*. There is, however, no gametophyte plant. Meiosis occurs in the young diploid spores. The mature haploid spores (meiospore) and the gametes (swarm cells or myxamoebae) they give rise to on germination represent the extremely reduced *gametophyte* or *haplophase* limited only to a few haploid cells. Truely speaking there is alternation of sporophyte generation with a few haploid cells in the life cycle. Such a life cycle is called *diploid* or *diplontic*. The sclerotia and spherules do not play any role in the phenomenon of alternation of generations. They primarily serve as means of perennation and serve to prolong the diplophase in the life cycle.

Economic Importance. Th slime molds are not of much economic value. During the rainy season and just after, they are most abundant on lawns and in gardens. They creep over ornamentals and render them unsightly. Sometimes they even smother the seedlings in beds. Like many bacteria and true fungi, the slime molds bring about decay and decomposition of organic materials on and in the soil.

Intricately Designed Plasmodium.

In recent years, the slime molds have become the subject of intensive laboratory studies on biological problems which challenge the biologists. Particular attention has been given to protoplasmic streaming, physiology of reproduction and morphogenesis.

The plasmodia of true slime molds are acellular, hence form an excellent material for the study of structure and physiology of protoplasm. They are considered ideal tools for experimental studies on physiology and development in lower organisms. Their remarkable beauty, attractive colours and delicately constructed fruiting bodies of intricate designs are of immense artistic value.

Origin and Phylogenetic relationships. The evolutionary origin of slime molds and their inter-relationships are obscure. Both are a matter of speculation. Some biologists hold that the slime molds form a connecting link between the plants and animals as they possess characteristics common to both. The somatic phase is animal-like but the fruiting bodies are more plant-like. The supporters of this view consider the slime molds to have evolved from the flagellate ancestors and consider them closely related to animals like *Amoeba* of the class **Sarcodina**, phylum **Protozoa**.

The formation of a fruiting body (sporangium) in the slime molds and the cellulose nature of the spore wall are plant characteristics. The absence of chlorophyll indicates their fungal or bacterial nature. The slime molds resemble one group of bacteria called Myxobacteriales in several ways. Thus the possibility of Myxobacteriales as a potential ancestral stock of slime molds is not excluded. Some mycologists are, however, of the opinion that the resemblance between the slime molds and Myxobacteriales is simply an instance of parallel evolution.

Furthermore the slime molds differ from the bacteria in their cellular organisation. In slime molds, the nuclei are bound by membranes and chromosomes are involved in mitosis. Besides, they are unlike the bacteria and true fungi in their plasmodial stage and phagotrophic nutrition.

Some mycologists, however, include the slime molds in the fungi and connect them to the Phycomycetes by such forms as Plasmodiophoromycetes. They consider that the slime molds (Myxomycetes) are related to the Plasmodiophoromycetes which include the endoparasitic slime molds. The two resemble in having plasmodium as the vegetative phase and in the type of motile cells. The recent discovery that the slime molds and Plasmodiophoromycetes are similar in having two unequal whiplash type flagella has tempted them to group the two together under the division Myxomycophyta. However, the differences between the Myxomycetes and Plasmodiophoromycetes are more fundamental. In their life cycle, general organisation, mode of life (being obligate endoparasites), haploid somatic phase and absence of fructifications, the Plasmodiophoromycetes differ from the slime molds.

The presence of naked biflagellate swarm spores in the life history of slime mold is suggestive of the fact that they have originated from biflagellated ancestors among the Phycomycetes (Oomycetes). Since both the flagella of Myxomycetes are of whiplash type though unequal in size indicates that the slime molds originated independently of the Phycomycetes in which one of the two flagella is of whiplash type and the other of tinsel type. Moreover, occurrence of a prolonged diploid phase is unusual among the true fungi. Martin, therefore, considers the Myxomycetes to have evolved independently of the true fungi and thus places the slime molds in a subdivision of their own which he calls Myxomycotina. They have given rise to no higher forms of life. Alexopoulos subscribes to this view.

Taxonomic Position. The Myxomycetes are a group of doubtful taxonomic position. They possess features common both to plants and animals. As mentioned above, their somatic phase (plasmodium) in its structure and physiology reminds one of ameoba-like animals. In fact De Bary (1887) was so much impressed by the animal-like characteristic of the slime molds that he called them **Mycetozoa** which means fungus animals. He was supported by Lister (1925), Bessey (1950), Kudo (1954) and many others. Bessey and Kudo actually placed the slime molds in the phylum Protozoa, class Sarcodina which also includes Amoebae. But the fruiting or reproductive stage constituted by the stationary sporophores or sporangia producing spores covered by cell walls and absence of chlorophyll are, however, suggestive of fungal nature.

Slime molds were first recognised as a natural group of fungi in 1829 by Elias Magnus Fries—a Swedish mycologist. He placed them within what we now call the Basidiomycetes. Four years later, another botanist elevated them to a rank coordinate with the other major groups of fungi. It was Thomas H. Macbride (1899), who in his monograph on the slime molds used the term Myxomycetes (Myxa = slime + myketes = mushroom, fungi) for this group meaning slime fungi. Later G. W. Martin (1961), an authority on the slime molds, supported the view of the fungal nature of the slime molds. He advanced convincing arguments in favour of this view. Some mycologists thus place the Myxomycetes as a class co-ordinate with the other four classes of true fungi (Eumycota). There are others who point out the differences between the true fungi and the slime molds. These are given in table III.

On the basis of the fundamental differences between the slime molds and the true fungi (Phycomycetes) majority of the mycologists, much as Martin, Alexopoulos and many others place slime molds in the subdivision **Myxomycotina** of the division Mycota which also includes the true fungi under subdivision **Eumycotina**. According to this view, the systematic position of slime mold *Physarum* would be as under :

Alexopoulos (1962)

Division	:	**Mycota**
Subdivision	:	**Myxomycotina**
Class	:	**Myxomycetes**
Order	:	**Physarales**
Family	:	**Physaraceae**
Genus	:	*Physarum*
Species	:	*polycephalum*

Alexopoulos and Mims (1979)

Superkingdom	:	***Eukaryota***
Kingdom	:	***Mycetae***
Division	:	***Gymnomycota***
Subdivision	:	***Plasmodiogymnomycotina***
Class	:	***Myxomycetes***
Sub-class	:	***Myxogastromycetedae***
Order	:	***Physarales***
Family	:	***Physaraceae***
Genus	:	*Physarum*
Species	:	*polycephalum*

TABLE – I

Myxomycetes (Slime molds)	*Phycomycetes* (Lower fungi)
1. The somatic phase is a multinucleate,, diploid holocarpic **plasmodium** which is the product of syngamy.	1. With the exception of a few, the somatic phase is a filamentous,, haploid structure called a **mycelium**. It is generally eucarpic and is the product of germination of spore.
2. The plasmodium is free living and capable of engulfing particulate matter and of executing movements over the substratum. It has the capacity for plasmodium coalescence (somatic fusion) in which fusion of nuclei is not involved.	2. The mycelium is anchored to the substratum and lacks all the attributes possessed by the plasmodium. It absorbs nutrition either by living as a parasite or as a saprophyte.
3. The diploid phase, as represented by the plasmodium,, is propagated by fission and prolonged by the formation of polynucleate, thick-walled cysts or a sclerotium-like structure.	3. The prolongation of the diploid phase by the formation of special structures is unusual among the Phycomycetes.
4. The plasmodium reproduces by the formation of uninucleate,, thick-walled resting spores which are produced within minute fruiting bodies called **sporangia**.	4. The mycelium reproduces asexually by the formation of uninucleate or multinuleate spores produced endogenously within sporangia and sometimes by conidia produced exogenously at the tips of hyphae.
5. The spores in the slime molds are differentiated by meiosis from the diploid protoplast of the sporangium.	5. The spores and conidia are differentiated from the haploid protoplast by mitosis.
6. The thick-walled resting spores which belong to the category of meiospores are uninucleate and nonmotile.	6. The sporangiospore and conidia which belong to the category of mitospores may be uninucleate or multinucleate. The sporangiospores may be motile or nonmotile but conidia are always nonmotile.

7. On germination, each resting spore produces 1-4 biflagellate swarm cells which fuse in pairs to form the zygote. The latter by growth gradually becomes changed into a multinucleate plasmodium.	7. On germination, the mitospore (sporangiospore or conidium) produces the mycelium.
8. The motile cells in the life cycle are biflagellate. Both the flagella are of whiplash type but unequal in length.	8. The motile cells may be uni- or biflagellate. In the latter case one flagellum is of whiplash type and the other tinsel.
9. The haploid phase in the slime mold is inconspicuous and restricted to the resting spores and swarm cells.	9. Reverse is the case in the Phycomycetes. The diploid phase, as represented by the zygospore, is inconspicuous.

QUESTIONS

ESSAY TYPE

1. What are the slime molds ? Give an account of their habit, habitat and somatic phase.
2. What do you know about the method of nutrition of slime molds? In what respects do they resemble the (*a*) animals and (*b*) plants?
3. Describe structure and reproduction in the Myxomycetes. How do they differ from the Phycomycetes (Lower fungi)? Do slime molds exhibit alternation of generations in their life cycle?
4. Give an account of generalised life cycle of a myxomycetous fungus.
5. Describe the structure and reproduction in *Physarum*.
6. What are myxomycetes? Give their characteristic features. (*GND Univ., 1991, 1992*)
7. What are the distinguishing features of Myxomycetes? Give a concise but illustrated account of reproduction in this group. (*Punjab, 1991*)
8. Discuss distinguishing features of Myxomycetes. Give a concise account of reproduction in this group. (*Kanpur, M.Sc., 1997*)
9. Discuss briefly the classification, phylogeny and affinities of Myxomycetes. (*Kanpur, M.Sc., 1994, 1996, 1989*)
10. Give an account of somatic structure and modes of reproduction in Myxomycetes. (*Kanpur, M.Sc. 1995*)
11. Describe the distinguishing features of Myxomycetes. Give its classification, phylogeny and affinities in brief. (*Kanpur, M.Sc; 2004*)

SHORT ANSWER TYPE

12. Write short notes on :
    (*a*) Sporangium of Myxomycetes. (*Punjab, 1993*)
    (*b*) *Physarum* (*Kumaon, 2001*)
    (*c*) *Stemonitis* (*Kanpur, M.Sc., 1999, 1998, Poorvanchal M.Sc., 2003; Kumaon, 2000, 2003*)
    (*d*) *Didymium*
    (e) *Slime Molds* (*Poorvanchal M.Sc., 2003*)
13. Why are slime molds not classified with true fungi? (*Kanpur M.Sc., 1990*)

OBJECTIVE TYPE

14. Fill in the blanks :

(*i*) Plasmodium is always ....

(*ii*) The somatic phase in a myxomycetous fungus is known as ....

(*iii*) The nutrition in plasmodium is always ....

(*iv*) The sporangium in *Physarum* is ....

(*v*) The spores in *Physarum* is ....

15. Select the correct answer :

(*i*) The Plasmodium is always:

(*a*) Haploid (*b*) Diploid
(*c*) Triploid (*d*) Tetraploid

(*ii*) The somatic phase in a myxomycetous fungus is called

(*a*) Hypha (*b*) Mycelium
(*c*) Plasmodium (*d*) None of the above

(*iii*) Myxomycetous fungi are also known as

(*a*) Slime molds (*b*) Bread molds
(*c*) Black molds (*d*) Blue molds

(*iv*) The nutrition in slime moulds is

(*a*) Autotrophic (*b*) Heterotrophic
(*c*) Holozoic (*d*) Holocarpic

(*v*) Meiosis in the life cycle of a slime molds occur in

(*a*) diploid spores (*b*) plasmodium
(*c*) the gametes (*d*) zoospores

CHAPTER 3

SUBDIVISION – EUMYCOTINA The Lower Fungi : Phycomycetes

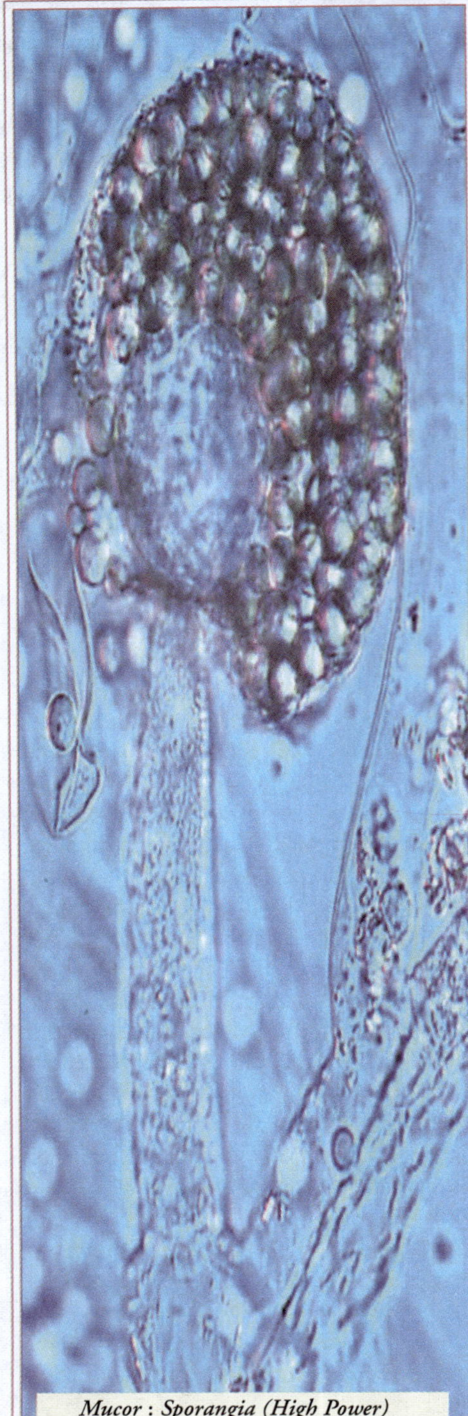

Mucor : *Sporangia (High Power)*

- General Features
- Evolution of Thallus
- Reproduction
- Classification of Lower Fungi

General Features. The lower fungi comprise the simplest and the primitive group of true fungi represented by about 1,500 species. The majority of these are aquatic (*Saprolegnia*), some are amphibious (*Phytophthora*) and some terrestrial (*Rhizopus*). They live either as strict parasites or saprophytes. Strictly parasitic species live on algae, ferns and seed plants. A number of them cause diseases of economic plants, such as the downy mildew of grape, brown rot of lemon and potato blight. All of these were first recognised as a coherent group in 1873 by the German mycologist Heinrich Aton De Bary. He placed them in the class **Phycomycetes** (Gr. *phykos* meaning alga, *mycetes* meaning mushrooms, fungi, hence alga-like fungi). Popularly they are called the **algal fungi** because they bear a striking resemblance to some of the algae (*Vaucheria*) in morphology of the thallus and sex

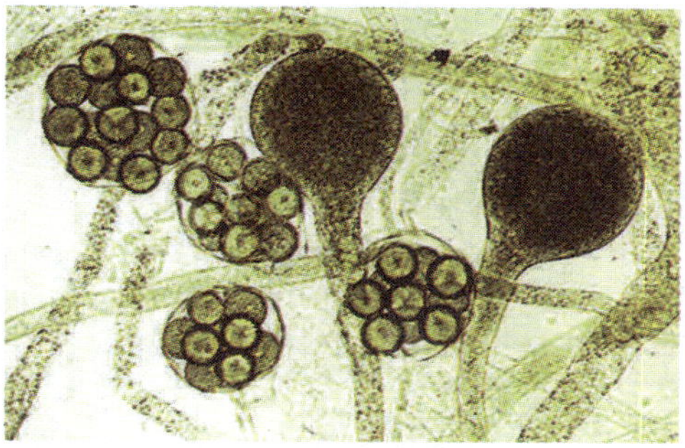

Saprolegnia

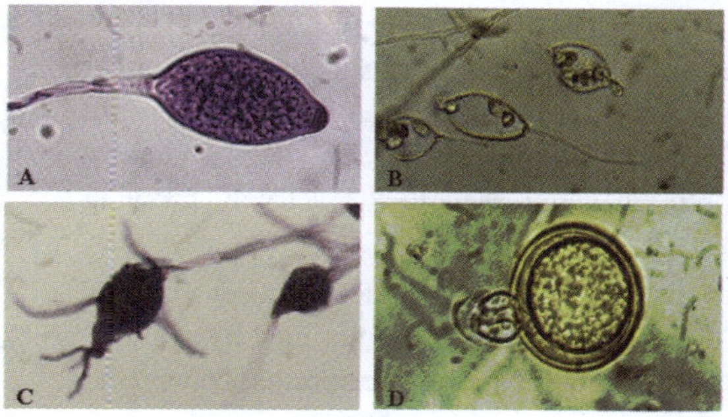

Phytophthora

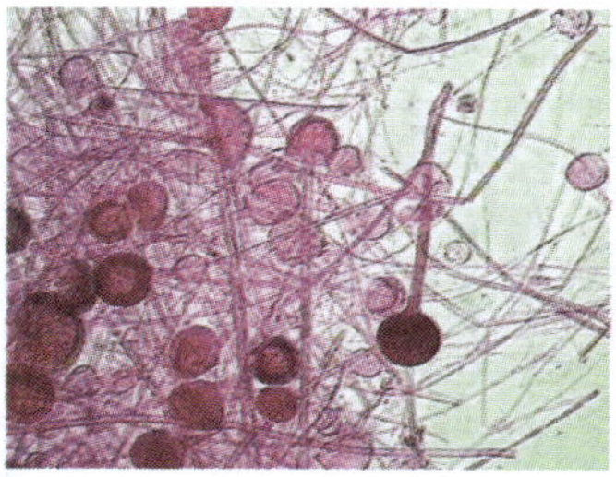

Rhizopus

Potato Blight

Downy mildew of Grape

Black Molds

Brown Rot of Lemon

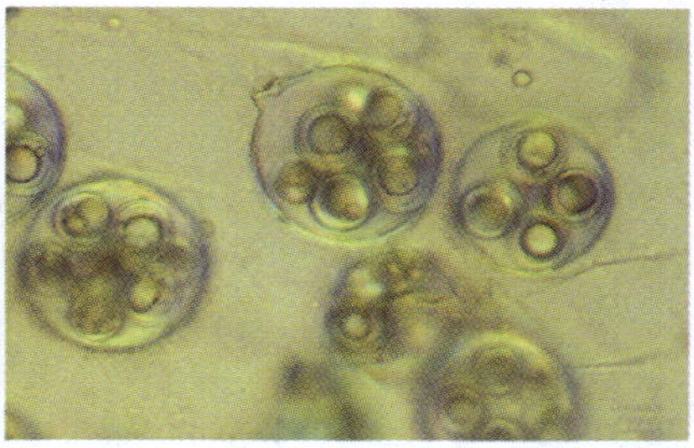

Water Molds

Downy Mildews

White Rusts.

organs. The lower Fungi or Phycomycetes are characterised by:

(*a*) The somatic phase consists of either a unicellular thallus or a non-septate coenocytic mycelium. The septa usually remain suppressed in the actively growing stage.

(*b*) Asexual reproduction takes place by sporangiospores and sometimes by conidia. The sporangiospores are motile (zoospores) in aquatic species and non-motile in the terrestrial species.

(c) The spore producing units are usually not grouped into a complex fructification.

(*d*) Sexual reproduction, when present, is either isogamous or heterogamous.

(*e*) Karyogamy generally follows plasmogamy almost immediately so that there is no **dikaryophase** in the life cycle.

The Lower Fungi (*Phycomycetes*) include the simplest members of true Fungi. The most primitive among them are the aquatic species. The common examples of Lower Fungi are the black molds, water molds, downy mildews and white rusts.

Evolution of Thallus (Fig. 3.1). In the simplest and primitive members, which are aquatic, there is no mycelium. The thallus in these forms is microscopic and unicellular. In *Olpidium endogenum,* it consists of a tiny mass of naked, uninucleate protoplast which enters the host cell, enlarges and lays down a cell wall of chitin (A_1). At the reproductive stage, the entire cell protoplast is involved in zoospore formation (A_2). In the more advanced unicellular forms the thallus is fixed to the substratum by a few thin, hair-like structures, the **rhizoids** (B_1). The coenocytic thallus is ***holocarpic.*** At maturity, it is converted either into a single reproductive organ (zoosporangium or gametangium) as in family *Olpidiaceae* (B_2) or into a cluster of such organs (family *Synchytriaceae*). Some more advanced aquatic members have a little

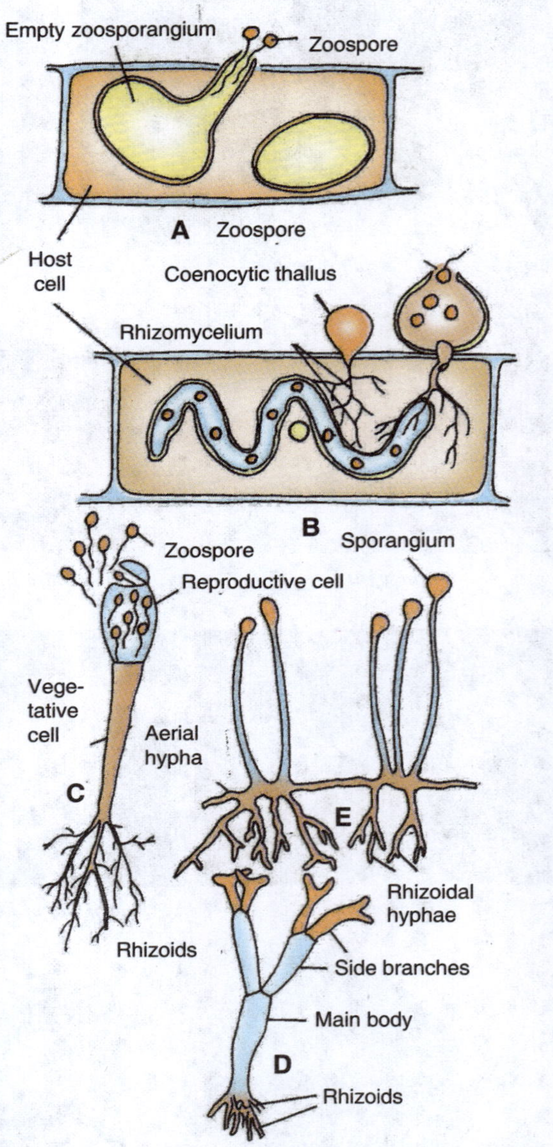

Fig. 3.1 (A-E). *Lower Fungi.* Evolution of thallus. A, two unicellular organisms of *Olpidium endogenum* inside the host cell of a green alga; B, two individuals of *Rhizophidium* (young and old) growing upon the host cell of *Spirogyra*; C, club-shaped thallus of *Blastocladiella variabilis* liberating zoospores from the upper reproductive cell; D, small mycelium of *Allomyces* consisting of dichotomously branched hyphae arising from the trunk-like body; E, eucarpic mycelium of *Rhizopus* differentiated into aerial portion and holdfast.

more complex vegetative body. It may consist of two multinucleate compartments. One of these is vegetative in function and the other reproductive (C). *Blastocladiella* is an example. The gametothallus of *Allomyces* which is a soil dwelling fungus, consists of branched, multinucleate hyphae forming a small mycelium. It compries, (*i*) a group of rhizoids which fixed the thallus to the substratum, (*ii*) a short, slender trunk-like body bearing, (*iii*) dichotomously branched hyphae (D), the latter bear the sex organs.

The vast majority of the Lower Fungi (Phycomycetes), aquatic as well as terrestrial, have a well developed, filamentous thallus, celled the **mycelium.** It is differentiated into distinct, sterile and fertile portions and thus is **eucarpic** (E). The eucarpic mycelium is of two types. In one type, the mycelium is differentiated into an aerial portion and holdfast (*Rhizopus*). The holdfast is branched and tapering (E). It penetrates the substratum. The aerial portion consists of usually stout, cylindrical, branched and coenocytic aseptate hyphae. In the second type the mycelium (as in Peronosporales) consists of slender, cylindrical, coenocytic hyphae which grow hidden from view within the tissues of the host (Fig. 1.14D) or in dead organic matter in the soil or water.

The hyphae in the mycelium of majority of the species, as mentioned previously, are **coenocytic** and **aseptate.** The septa usually remain suppressed in the actively growing mycelium. They are, however, not totally absent. They appear in the older hyphae of some species. In a few species they have been reported to be formed regularly by the young hyphae. However, these septa do not form complete partitions. In the majority of cases, they have a central pore. The septa also appear in connection with the formation of the reporoductive organs and also to seal a damaged portion of hypha. In the latter cases the septa are complete partitions. They are solid, plate-like structures.

The filamentous Lower Fungi (Phycomycetes) can be arranged in a series illustrating a progressive transition from an aquatic to a terrestrial mode of life through the amphibious way of life. The series starts with *Saprolegnia* at the lower end. It is a saprophytic water mold. It grows in calm fresh water and also in well-irrigated soils. *Pythium* is a little more soil-minded. More often it grows in damp soil than in water. However, they both produce zoospores. This shows that basically they are **aquatic types.**

Phytophthora is a distinctly **amphibious type.** Under normal conditions, the fungal mycelium bears sporangia which develop zoospores. But under adverse circumstances which do not permit dispersal by zoospores such as lack of free water or too high a temperature, the sporangia get detached and behave like conidia. The conidia are dispersed by wind. On germination, the conidium directly produces a **germ tube** which grows into a new fungus mycelium. Similarly, a species of *Plasmopara* which is a causative agent of *downy mildew disease* of grapes is *amphibious.* It can adjust itself to both terrestrial and aquatic habitats. In the former case the sporangia behave as conidia and in the latter case they produce zoospores. On the other hand, a species of *Peronospora* (*P. destructor*) which causes *downy mildew disease* in onions is fully terrestrial. It produces sporangia which invariably function as conidia and grow by putting out germ tubes.

Mucorales which are the most advanced members of the Lower Fungi (Phycomycetes) are strictly terrestrial. They bear sporangia in which are produced non-motile, encapsulated, wind-disseminated sporangiospores—an adaptation to a terrestrial mode of life. The sporangiospore has a wall which makes it more resistant to drying and other unfavourable conditions (high or low temperature) than the zoospores. It can survive for some period while being blown about in the air till the time it falls on a suitable soil where it can germinate to give rise to a new plant.

Reproduction. The Lower Fungi (Phycomycetes) reproduce asexually as well as sexually.

1. Asexual Reproduction (Figs. 3.2 and 3.3). Besides multiplying by the vegetative method of fragmentation, the lower fungi reproduce asexually by the formation of following types of spores :—

(a) Sporangiospores. These are asexual spores produced endogenously in special sac-like asexual reproductive organs called the **sporangia.** The sporangia vary widely in form in the lower fungi. These may be flask-shaped (A), cylindrical, irregularly lobed (D), club-shaped (C), elongate or even globose in form (B). The protoplast of the sporangium undergoes cleavage to form sporangiospores in large numbers. From a mature sporangium, the sporangiospores may escape through the exit papillae (B) or a pore on the sporangial wall or at the end of a discharge tube or through an apical opening formed by the separation of a minute circular cap called the lid (A). In the land forms, the sporangiospores are released by the rupture of the sporangial wall (Fig. 3.3B). The released spores may be motile in the

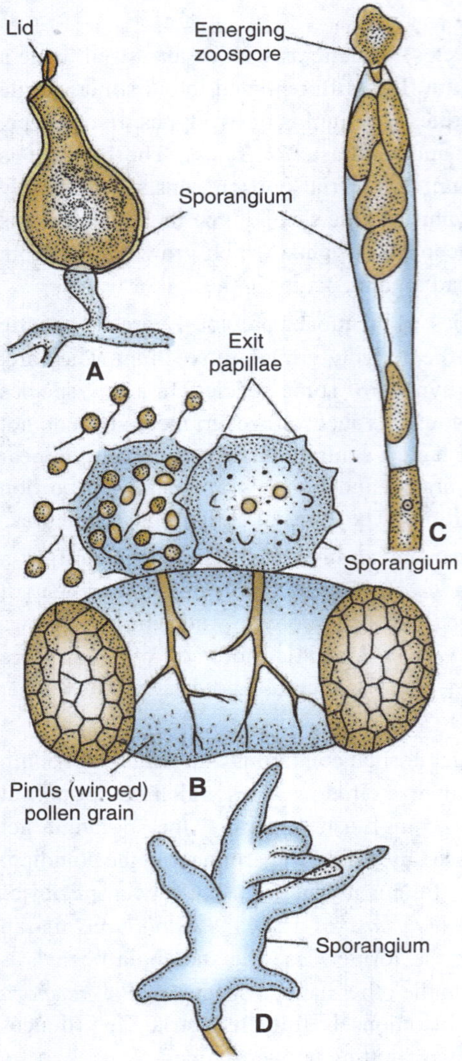

Fig. 3.2 (A-D). *Lower Fungi.* Forms of sporangia. A, flask-shaped sporangium of *Nowakowskiella ramosa*; B, Zoosporangium with exit papillae of *Rhizophidium halophilum* on pollen grain of *Pinus*; C, club-shaped zoosporangium of *Monoblepharis*; D, irregularly-shaped sporangium of *Pythium aphanidermatum* (A, after Karling, B, after Vebelmesser, C, after Woronin and D, after Middleton).

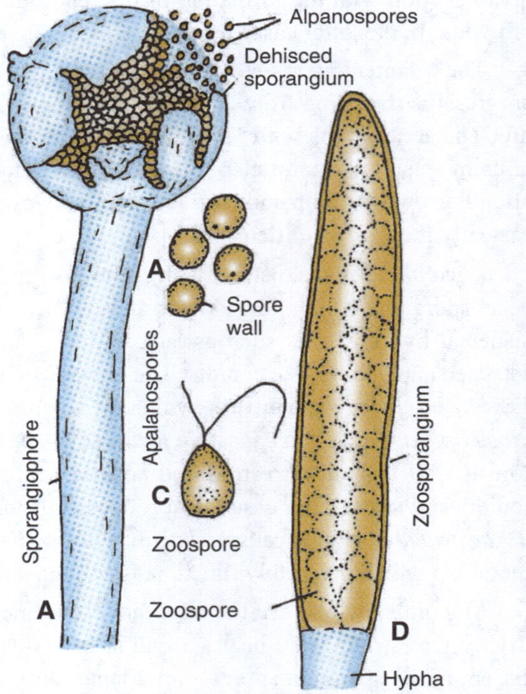

Fig. 3.3 (A-D). *Lower Fungi.* Sporangiospores. A. aplanospores produced endogenously in the terminal, globose sporangium of *Mucor* B. Spores; C. Biflagellate zoospore produced within the cylindrical zoosporangium of *Saprolegnia*; D. Cylindrical zoosporangium.

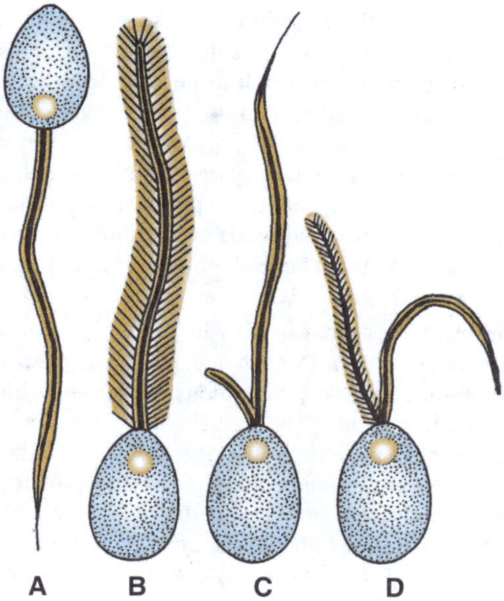

Fig. 3.4 (A-D). *Lower Fungi.* Kinds of zoospores.

aquatic species and nonmotile in the land forms. The former are called zoospores (Fig. 3.3C) and the latter aplanospores (Fig. 3.3A). In the holocarpic species, the entire thallus functions as the sporangium (Fig. 3.1 A_2). The eucarpic species have the sporangia borne on special reproductive hyphae known as the sporangiophores which may be simple (Fig. 3.1E) or branched.

(*i*) Zoospores (Fig. 3.4). The zoospores may be uniflagellate (Chytridiomycetes and Hyphochytridiomycetes) or biflagellate (Plasmodiophoromycetes and Oomycetes). The Chytridiomycetes which include the simplest and the most primitive forms of lower fungi produce uniflagellate zoospores (A) with a whiplash flagellum inserted at the posterior end. In the Hyphochytridiomycetes the single flagellum is of tinsel type. It is borne at the anterior end (B). A small group of endoparasitic slime molds which belong to the class Plasmodiophoromycetes produce biflagellate zoospores which bear two whiplash flagella of unequal size at the anterior end (C). The smaller flagellum has a blunt tip and the longer has a pointed tip.

The more advanced members of the lower fungi which belong to the class Oomycetes produce biflagellate zoospores with the two flagella inserted either at the anterior end (D) or laterally (Fig. 3.6C). One of these is of tinsel type and the other whiplash. The anteriorly flagellate zoospore is usually pear shaped while the laterally flagellate zoospore is reniform. The phenomenon of the presence of two types of zoospores is called *Diplanetism.*

(*ii*) Aplanospores (Fig. 3.5). Asexual reproduction in the advanced lower fungi, which are strictly terrestrial (Mucorales), is by means of non-motile, encapsulated, wind-disseminated spores known as the **aplanospores.** They are produced in globose sporangia borne on simple or branched sporangiophores. The primitive genera of the Mucorales (*Mucor* and *Rhizopus*) have large

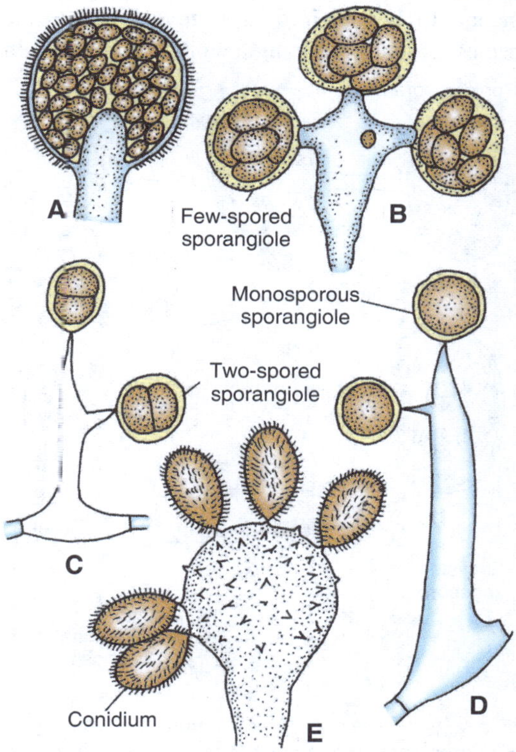

Fig. 3.5 (A-E). *Lower Fungi.* Stages showing transition from a multispored sporangium to a conidium in Mucorales. A, multispored sporangium of *Blakeslea trispora*; B, sporangiole of B. *trispora* with a few spores; C-D. two spored and one spored sporangiole of *Haplosporangium bisporale*; E, swollen tip of conidiphore of *Cunninghamella echinulata* bearing conidia (After Thaseter).

multispored, columellate sporangia borne singly and terminally on unbranched sporangiophores (A). The sporangiophores in other genera of the Mucorales, which are considered more advanced, are branched, each ultimate branch bearing one or more sporangia which in general are of three types, (i) large sporangia with many spores as in *Mucor*, (ii) small sporangia or sporangiole with few spores (B-C) even two or three and (iii) single spored sporangia (D). A species may bear all the three types showing thereby that the monosporous sporangiole have arisen from multispored sporangia by reduction in size and number of spores from many to one. One spored sporangiole is regarded as a conidium particularly so when the spore and sporangial walls are completely fused as in *Cunninghamella echinulata* (E). The latter produces numerous conidia on the inflated tip of the conidiophore.

(*b*) Conidia (Fig. 3.6). The transition from sporangia to conidia in the Mucorales cited in the paragraph above is paralleled by the *Pythium—Peronospora series of Oomycetes.* Certain of the more advanced families (Peronosporaceae and Albuginaceae) have deciduous sporangia which get detached. They are dispersed, in toto, mainly by air currents. The detachable sporangia are often called the conidia and the hyphae bearing them **conidiophores.** Some mycologists call them **conidiosporangia** (A). On germination, in a drop of water, they may function as **zoosporangia** (B). Their contents produce one or more zoospores. When the environment is dry they function as **conidia** (b), each germinating directly by putting out a short, slender hypha, the germ tube (C). The latter develops into a new mycelium. On the other hand, *Peronospora destructa*, a terrestrial species, produces sporangia which invariably behave as conidia.

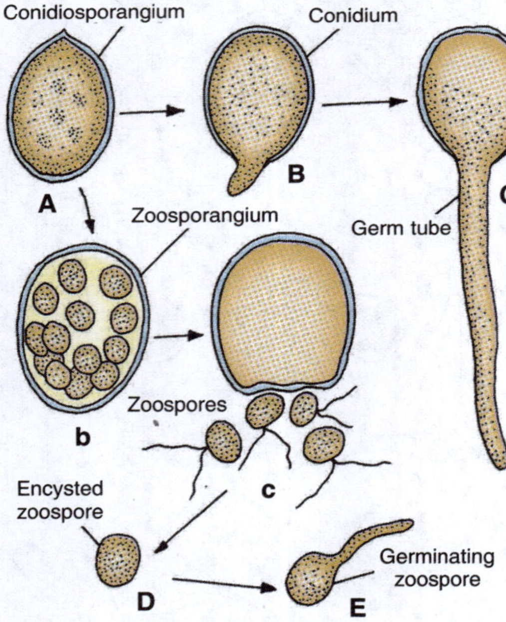

Fig. 3.6 (A-E). *Lower Fungi.* A-E, indirect germination of conidiosporangium of *Bremia lactuca* and germination of encysted zoospore; B-C, direct germination of conidiosporangium by putting out a germ tube (After Milbrath)

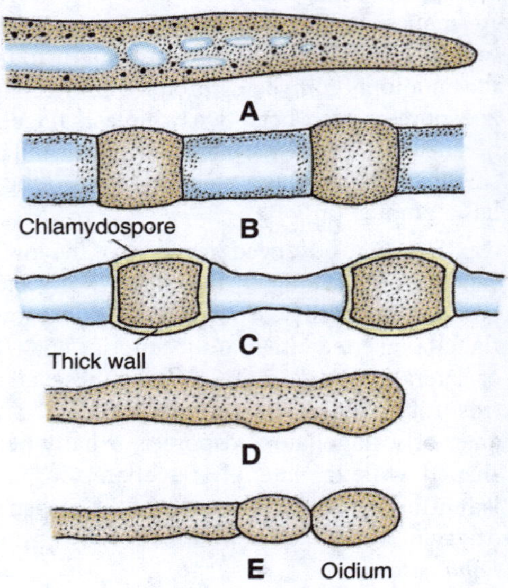

Fig. 3.7 (A-E). *Lower Fungi.* A-C, stages in chlamydospore formation; D-E, oidia formation of *Mucor.*

(*c*) Chlamydospores (Fig. 3.7 A-C). Some members of Lower Fungi (Phycomycetes) produce thick-walled resting cells, **the chlamydospores.** They are intercalary in position. Under conditions unfavourable for growth, the protoplasmic contents of coenocytic hyphae (A) accumulate at intervals (B). The intervening portions remain empty. Each localised accumulation is delimited by septa (B). Later these thin-walled segments of the hyphae secrete thick walls around them to become **chlamydospores** (C). The chlamydospores are rich in reserve food and are very resistant to desiccation.

(d) Oidia (Fig. 3.7 D-E). The Mucorales frequently produce rounded or oval thin-walled cells when grown in nutrients. The hyphae undergo segmentation and produce yeast-like cells called **oidia** (E). Each oidium, on germination, produces a new mycelium under suitable conditions.

2. Sexual Reproduction. There is a great diversity in the sexual process of the lower fungi (Phycomycetes). It ranges from isogamy to oogamy through anisogamy. Most of the speices are **homothallic.** In both, kinds of gametes are produced on the same thallus. Many species are **heterothallic.** The single thallus in the latter produces only one kind of gametes either male or female so that there are separate male and female plants. The heterothallic speices are eucarpic. The first step in sexual reproduction is the bringing together of male and female protoplasts or nuclei in the same cell. This process is known as **plasmogamy.** In the lower fungi, it is accomplished in three ways, namely, **planogametic copulation, gametangial contact** and **gametangial copulation.**

(*a*) Planogametic Copulation. It is characteristic of the primitive planogametic lower fungi (Chytridiomycetes and Plasmodiophoromycetes). In this kind of gametic union, one or both the gametes of a uniting pair are motile (flagellate) and are called the **planogametes.** The planogametes in the Chytridiomycetes are uniflagellate. The single flagellum is of whiplash type. It is inserted at the posterior end **(opisthocont)**. The Plasmodiophoromycetes produce biflagellate gametes. Both the flagella are of whiplash type but of unequal length. They are inserted at the anterior end. The shorter one has blunt tip and the longer one has a pointed tip.

(*i*) Isoplanogametic Copulation (Fig. 3.8 A-C). The gametic union in the primitive chytrids (*Olpidium*) and Plasmodiophoromycetes (*Plasmodiophora*) takes place between two morphologically similar motile gametes known as the isoplanogametes (A). Sexual fusion between isoplanogametes is described as **isoplanogametic copulation** (B-C).

(*ii*) Anisoplanogametic Copulation (Fig. 3.9). Members of the order Blastocladiales belonging to the genus *Allomyces* produce two types of planogametes which are morophologically

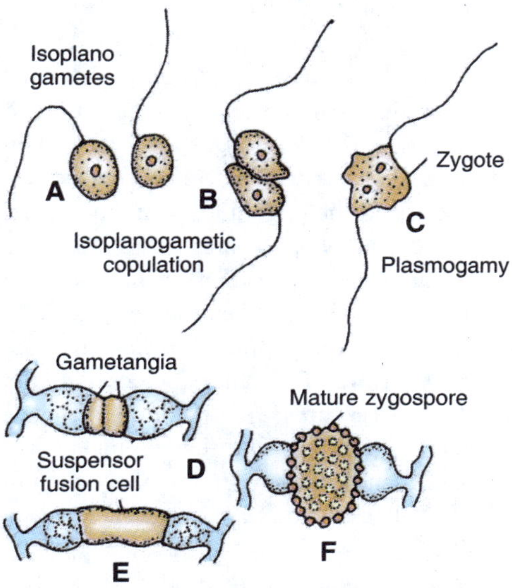

Fig. 3.8 (A-E). *Lower Fungi*. Isogamous sexual reproduction. A-C, plasmogamy by isoplanogametic copulation in *Olpidium*; D-F, plasmogamy by isogametangial copulation in *Rhizopus* (A-C, after Kusano).

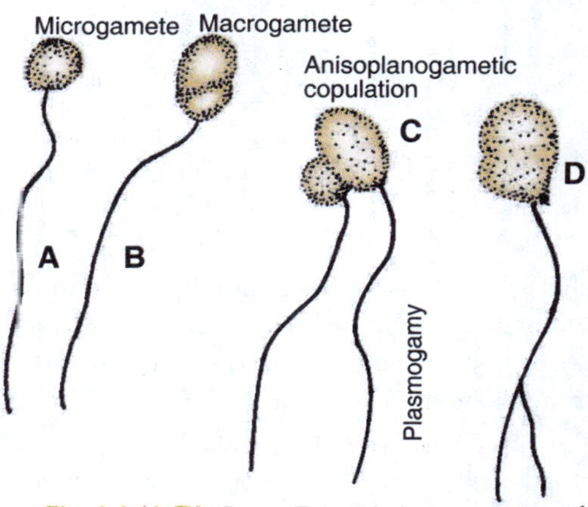

Fig. 3.9 (A-D). *Lower Fungi*. Anisogamous sexual reproduction in *Allomyces*. Plasmogamy by anisoplanogametic copulation (After Emerson).

similar but differ in size. These are termed **anisoplanogametes** (A, B). The gametic union between anisoplanogametes (C) is known as **anisoplanogametic copulation** (D). It is illustrated by *Allomyces*.

(iii) **Ooplanogametic Copulation** (Fig. 3.10). *Monoblepharis* illustrates the most advanced type of planogametic copulation. It produces small opisthocont male gametes in large numbers in the sex organ called **antheridium** (A). These are known as the **sperms.** The oogonium produces a single large immobile female gemete, the **egg** or **ovum** (A). The sperm enters the oogonium through the apical pore to fertilize the egg (B). The gametic union between a motile male gemete (sperm) and an immobile female gamete (egg or ovum) is called **ooplanogametic copulation.** It is significant to note that Monoblepharidales are the only oogamous fungi with a motile male gamete (sperm).

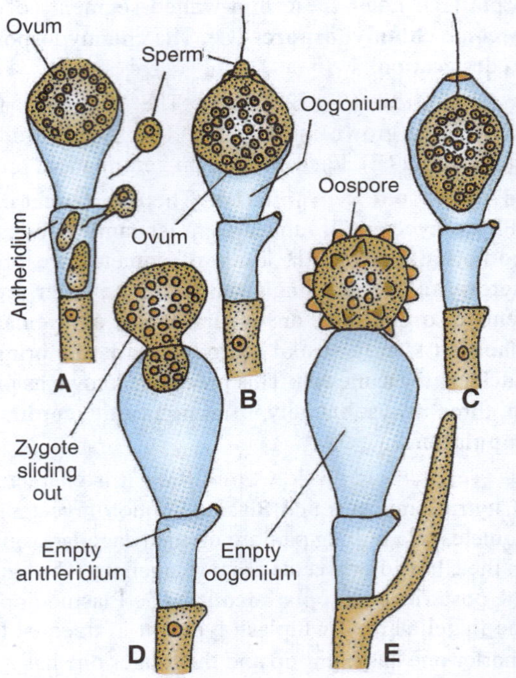

Fig. 3.10 (A-E). *Lower Fungi.* Oogamous sexual reproduction in *Monoblepharis*. Plasmogamy by ooplanogametic copulation (After Laibach).

(b) **Gametangial Contact** (Fig. 3.11). A group of aplanogametic lower fungi (Oomycetes) produce non-motile gametes (aplanogametes) that are always oogamous. The aplanogametes are never released from the gametangia to the outside. The gametangia in which the gemetes are produced are hightly differentiated. The male gametangium is called the **antheridium** and the female **oogonium.** No sperms are organised. The oogonium may contain one to many eggs. At the time of sexual act, the antheridium and the oogonium come in contact (A). They do not actually fuse and do not lose their identity. The antheridium in this position puts out a slender, tubular outgrowth at the point of contact with the oogonium. It is known as the **fertilization tube.** The latter pierces the thin spot in the oogonial wall and dips deep into the oogonium. It then ruptures at its tip to introduce the male nucleus (or nuclei) into the oogonium (B) where it fuses with the female nucleus (or nuclei) to accomplish fertilisation. In rare cases, the male nucleus may enter the oogonium through a pore formed by the dissolution of the intervening walls at the point of contact.

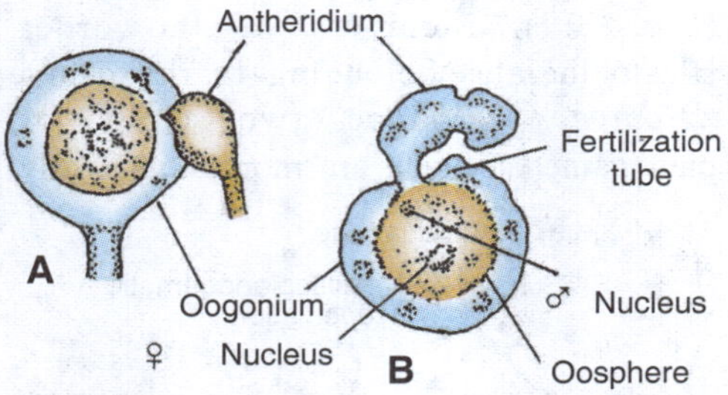

Fig. 3.11 (A-E). *Lower Fungi.* Oogamous sexual reproduction in *Pythium* by gametangial contact (After Mijake).

(c) **Gametangial Copulation.** In another group of aplanogametic lower fungi which are strictly terrestrial (Mucorales), gametic union is brought about by the fusion between the gametangia. The uniting gametangia lose their identity in the sexual act. This kind of gametic union is called **gemetangial copulation.** The uniting gametangia are usually morphologically similar (isogamous). Rarely they are dissimilar and unequal in size (anisogamous).

(i) Isogametangial Copulation (Fig. 3.8 D-F). Typical examples of this method of gametic union are *Rhizopus* and *Mucor*. Some species of these genera are homothallic and certain other are heterothallic. In the latter case sexual reproduction takes place only when two hyphae of opposite strains come in contact. Morphologically these are similar. At this point of contact, each of them produces a short, lateral copulating branch, the **progametangium.** The tips of two progametangia enlarge. Later a septum appears near the swollen tip of each progametangium dividing it into a terminal **gametangium** and a basal **suspensor** (D). The intervening walls between the two gametangia finally dissolve forming a common fusion cell (E). The multinucleate protoplasts (coenogametes) of the two copulating gametangia mingle in the fusion cell. The nuclei of opposite strains in the heterothallic species fuse in pairs to form diploid nuclei. Within the parent gametangial wall, the young zygospore secretes around it a cell wall consisting of two layers. The outer spore wall layer is thick, dark and sculptured. As the zygospore increases in size, the parent gametangial wall ruptures exposing the dark, rugged exospore of the zygospore (F).

(ii) Anisogametangial Copulation (Fig. 3.12, A-C). In another genus of the Mucorales called *Zygorhynchus*, the copulating gametangia and their suspensors are unequal in size (B). Fusion between unequal gametangia is termed **anisogametangial copulation.**

(iii) Anisogametangial Contact (Fig. 3.12, D-G). *Endogone lactiflua* exhibits distinct anisogamy. Of the two progametangial hyphae which come in contact, one is larger in diameter thn the other (D). Both are multi-nucleate. Multinucleate end cells known as gametangia are delimated at their tips. All the nuclei except one degenerate in the two contacting gametangia (E) which are unequal. They do not actually fause or lose their identity during the sexual act. The surviving nucleus of the smaller male gametangium migrates into the larger female gametangium through a pore formed by dissolution of the intervening gametangial walls at the points of contact (F). Later both the nuclei, male as well as female, migrate into an outgrowth of the female gametangium (G). There they fuse to accomplish fertilisation.

In Phycomycetes, plasmogamy is immediately followed with karyogamy.

Germination of zygospore in the Lower Fungi (Fig. 3.13). On the return of conditions favourable for growth, the zygospore germinates in either of the following ways :—

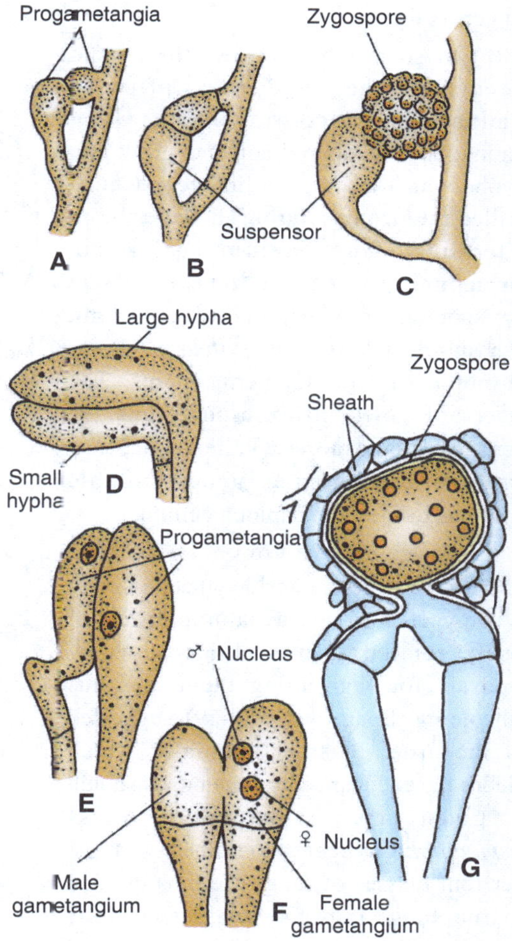

Fig. 3.12 (A-G). *Lower Fungi.* Anisogamous sexual reproduction in Mucorales. A-C, by anisogametangial copulation in *Zygorhynchus moelleri*; D-G, by anisogamous gametangial contact in *Endogone lactiflua* (After Hawker.)

(*a*) The diploid contents of the zygospore (A) undergo meiosis but meiospores are not formed. The zygospore containing the haploid protoplast then directly germinates to give rise to new **haplomycelium** (B). *Monoblepharis* is an example. Rarely as in *Allomyces,* the diploid zygospore (C) germinates by putting out a germ tube (D) which grows and branches to form the **diplomycelium.** No meiosis takes place prior to or during germination of the zygospore.

(*b*) Prior to germination, the diploid unclei undergo zygotic meiosis. The resultant haploid nuclei are fashioned into **meiospores**. The latter are extruded either by the rupture of zygospore wall into a sessile vesicle (E, *Albugo candida*) or the zygospore forms a short exit tube which terminates in a vesicle (F, *Pythium perioplocum*). With the bursting of the vesicle wall, the meiospores are liberated. They swim about for a while, retract their flagella and encyst. The encysted meiospore germinates by a germ tube.

(*c*) In some other species, the diploid contents of the zygospore, prior to germination, undergo meiosis. Upon germination, the haploid contents grow into a hypha which grows to a limited extent. It is called the **promycelium.** The latter bears a terminal **sporangium** or **germ sporangium.** The haploid contents of the germ sporangium undergo cleavage to form either motile (*Plasmopara viticola,* G), or non-motile (*Mucor,* H) germ spores. The former are called **meiozoospores** and the latter **meioaplanospores.** On liberation, each germ spore germinates under suitable conditions to form the haplomycelium.

Classification of Lower Fungi. The modern mycologists consider that the class Phycomycetes is not a natural group of closely related forms. Numerous new proposals for regrouping them are thus forthcoming. Sparrow (1958-59) suggested that the order Plasmodiophorales which includes the endoparasitic slime molds should be given the rank of a class *Plasmodiophoromycetes* coordinate with the other four classes of fungi. He thus divided the true fungi into five classes namely, *Plasmodiophoromycetes, Phycomycetes, Ascomycetes, Basidiomycetes* and *Deuteromycetes*. The rest of the Phycomycetes, he classified, on the basis of

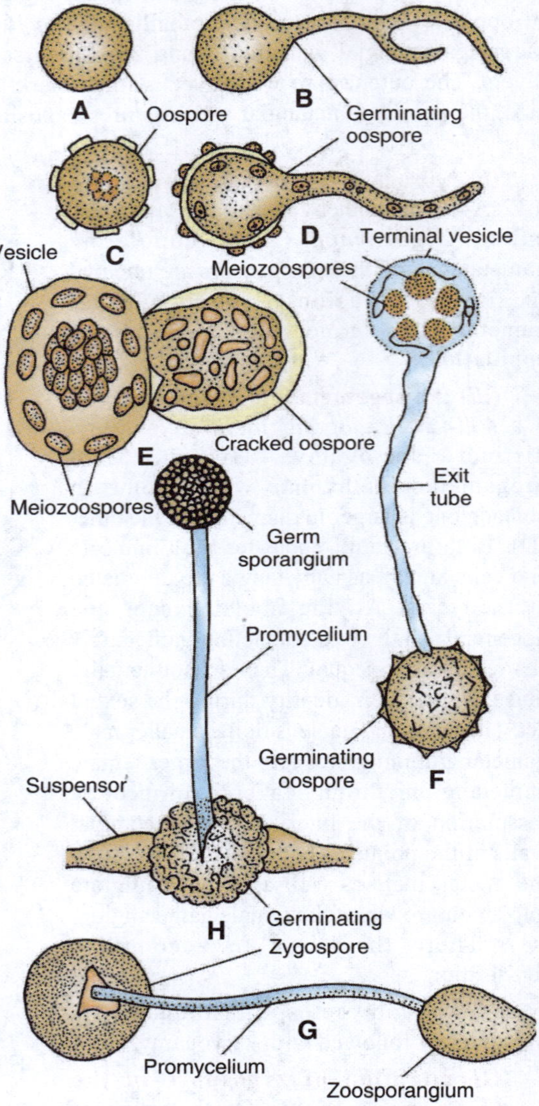

Fig. 3.13 (A-H). *Lower Fungi.* Germination of Zygospore. Explanation in the text. (A-B, after Smith, C-D, after Woronin, E, after de Bary, F after Drechsler and G after Avans.)

presence or absence of motile cells in the life cycle and the kind, number and arrangement of flagella on the motile cells, as follows :—

Phycomycetes (Lower Fungi)

- Sporangiospores **motile**
  - Motile sporangiospores **uniflagellate**
    - Sub-class 1. **Uniflagellatae**
      - Order 1. Chytridiales
      - Order 2. Blastocladiales
      - Order 3. Monoblepharidales
  - Motile sporangiospores **biflagellate**
    - Sub-class 2. **Biflagellatae**
      - Order 1. Saprolegniales
      - Order 2. Leptomitales
      - Order 3. Lagenidiales
      - Order 4. Peronosporales
- Sporangiospores **non-motile**
  - Sub-class 3. **Aflagellatae** or **Aplanatae**
    - Order 1. Mucorales
    - Order 2. Entomophthorales

Alexopoulos, as mentioned before, took a more logical step to split the Phycomycetes directly into six classes, namely, *Chytridiomycetes Hyphochytridiomycetes, Oomycetes, Plasmodiophoromycetes, Zygomycetes* and *Trichomycetes.* Details may be referred to in Chapter 1.

Hawker (1967) supported Sparrow's suggestion and presented his own scheme of classification of Lower Fungi (Phycomycetes) with slight modifications. He divided the Lower Fungi into three groups as follows :—

Group A. Uniflagellatae. It includes three classes :—

1. *Class Chytridiomycetes* (motile cells posteriorly uniflagellate) with three orders, *Chytridiales, Blastocladiales* and *Monoblepharidales.*

2. *Class Hyphochytridiomycetes* (motile cells anteriorly uniflagellate) including a single order *Hyphochytridiales*.

3. *Class Plasmodiophoromycetes* (motile cells with anteriorly inserted one long flagellum and another very short flagellum) (both whiplash type) including a single order *Plasmodiophorales.*

Group B. Biflagellatae (motile cells biflagellate, one flagellum of whiplash type and the other tinsel type) including one class.

4. *Class Oomycetes* with four orders, namely, *Saprolegniales, Leptomitales, Lagenidiales* and *Peronosporales.*

Group C. Aplanatae (motile cells absent) including two classes :

5. *Class Zygomycetes* with three orders, Mucorales, *Entomophthorales*, and *Zoopagales.*

6. *Class Trichomycetes with* **five orders,** *Amoebidiales, Eccrinales, Harpellales, Genistellales* and *Asellariales.*

Ainsworth (1973) classified the lower fungi into two subdivisions namely *Mastigomycotina* and *Zygomycotina.* Subdivision Mastigomycotina, characterised by the **presence** of motile stages in the life cycle, has been classified into four classes : Chytridiomycetes, Hyphochytridiomycetes, Plasmodiophoromycetes and Oomycetes. Subdivision Zygomycotina, characterised by the complete **absence** of motile stages, has been classified into two classes—Zygomycetes and Trichomycetes.

In this book the system of classification of lower fungi—Phycomycetes as proposed by Alexopoulos (1979) has been followed.

QUESTIONS

ESSAY TYPE

1. List the diagnostic features of the class Phycomycetes.
2. Give an account of somatic structures in the Phycomycetes.
3. Describe the various methods of asexual and sexual reproduction met with in the Phycomycetes.
4. Give a broad outline of the classification of the Phycomycetes.

SHORT ANSWER TYPE

5. Write short notes on
   (*a*) Planogametic fusion.
   (*b*) Phycomycetes.
6. Differentiate between
   (*a*) Gametangial Copulation and Gametangial contact. (*Allahabad University, 1993, 1997, 1999*)
   (*b*) Isoplanogametic fusion and Anisoplanogametic fusion. (*Allahabad, 1993*)
   (*c*) Myxomycotina and Eumycotina. (*Kanpur M.Sc., 1994, 1998*)

OBJECTIVE TYPE

7. **Fill in the blanks :**
   (*i*) Aseptate multinucleate mycelium is characteristic of the class..................................
   (*ii*) Aseptate multinucleate mycelium is also known as .... mycelium.
   (*iii*) Diplanetism is defined as the phenomenon of presence of .... types of zoospores.
   (*iv*) Motile asexual reproductive bodies are known as ....
   (*v*) In the planogametic copulation the gametes are ....
8. **Select the correct answer :**
   (*i*) Motile asexual reproductive bodies are called
   (*a*) Chlamydospores (*b*) Zoospores.
   (*c*) Aplanospores (*d*) Zygospores.
   (*ii*) The characteristic feature of the class Phycomycetes is the presence of
   (*a*) Dikaryotic mycelium
   (*b*) Septate multinucleate mycelium.
   (*c*) Aseptate multinucleate mycelium.
   (*d*) Septate uninucleate mycelium.
   (*iii*) The type of sexual reporduction in which both the gametes are motile and fuse outside the gametangia in presence of water is known as
   (*a*) Planogametic copulation
   (*b*) Gametangial contact
   (*c*) Oogamy
   (*d*) Gametangial Copulation.
   (*iv*) Which of the following is diploid?
   (*a*) Chlamydospores (*b*) Zoospores
   (*c*) Aplanospores (*d*) Zygospores.
   (*v*) The laterally biflagellate Zoospores are usually
   (*a*) pear shaped (*b*) kidney shaped
   (*c*) lemon shaped (*d*) irregular in shape.
   (*vi*) Which of the following groups contain typically coenocytic forms
   (*a*) Phycomycetes (*b*) Lichens
   (*c*) Deuteromycetes (*d*) Basidiomycetes.
   (*vii*) The fungal flagella are made up of fibrils arranged in
   (*a*) 10 + 2 (*b*) 7 + 4
   (*c*) 9 + 2 (*d*) 5+ 6
   (*viii*) Algal fungi are placed in
   (*a*) Ascomycetes (*b*) Basidiomycetes
   (*c*) Phycomycetes (*d*) Deuteromycetes

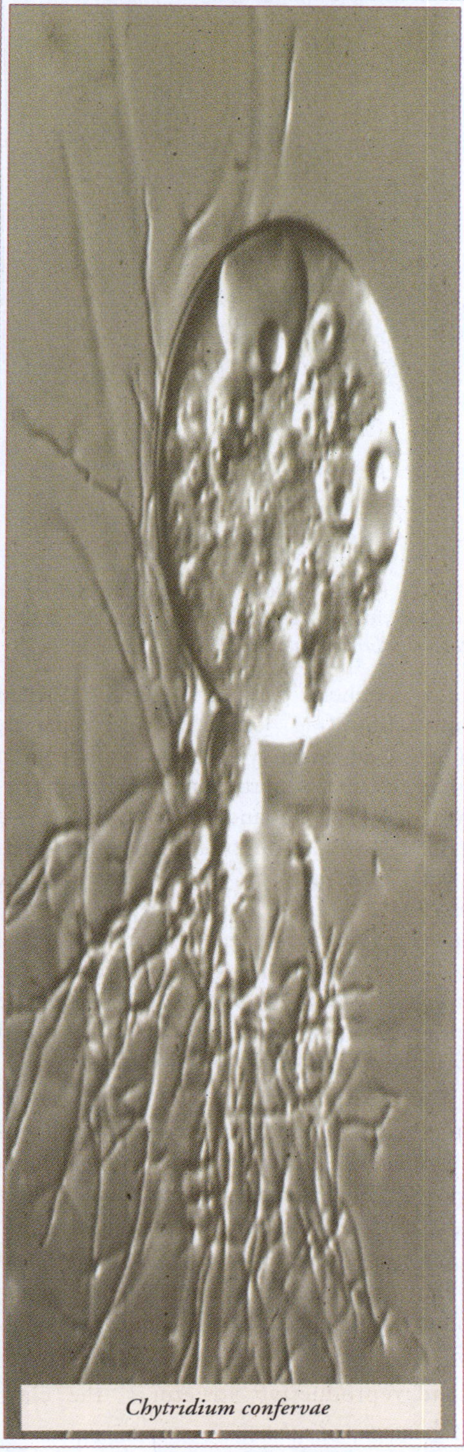

Chytridium confervae

CHAPTER 4

SUBDIVISION – EUMYCOTINA
Class : Chytridiomycetes

- Gereral Characteristics
- Classification
- Orcer (1): Chytridiales
- Family : Synchytriaceae
- *Syrchytrium endobioticum*
- Orcer (2) : Blastocladiales
- Family : Blastocladiaceae
- *Allcmyces*
- Orcer (3): Monoblepharidales
- Family : Monoblepharidaceae
- *Monoblepharis*

General characteristics. The chief characteristic feature of this class is the production of uniflagellate reproductive cells (zoospores and planogametes). The single flagellum is of a **whiplash** type and is inserted posteriorly. The zoospore with a posteriorly inserted flagellum is called **opisthocont.** The flagellum is attached to the **blepharoplast** within the cell. The motile cells of some species possess a nuclear cap which consists of RNA. It shields the nucleus at the anterior end of the cell. Majority of the members occur in water. Some are found in the soil. Many are parasitic on higher plants of economic value. The primitive members of this class are microscopic. The vegetative body in them is an acellular, coenocytic thallus (Fig. 4.1 A) which is **holocarpic.** It has no cell wall in the earlier stages (*Olpidium*). In the more advanced species, the unicellular thallus is drawn out at one point into fine, branching hairs (Fig. 4.1 B), the so-called **rhizoids** which aid in anchorage and intake of nutrients (*Rhizophidium*). In slightly more complex members, there is a much branched **rhizomycelium** (Fig. 4.1 C). They are eucarpic. In still more advanced types, a scanty mycelium consisting of a few short filamentous hyphae is in evidence. The advanced members have a mycelium consisting of typical hyphae woven into a eucarpic mycelial meshwork (*Monoblepharis*). The hyphe are coenocytic. Chitin is the chief constituent of the cell of hyphal wall. Besides, there is *B* glucan. The septa ordinarily remain suppressed during the vegetative phase but appear to delimit reproductive organs and are solid plates.

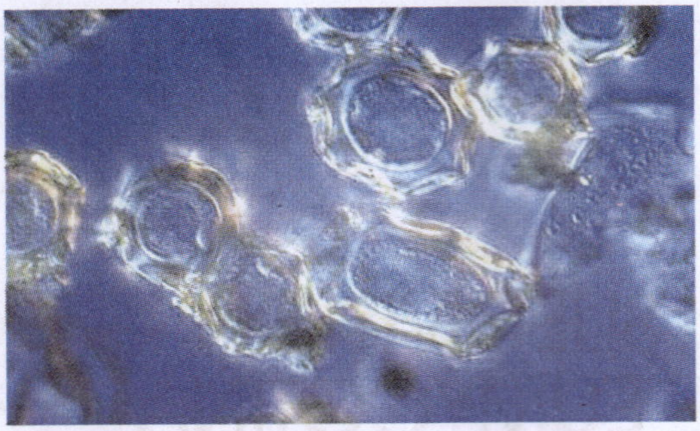

Olpidium

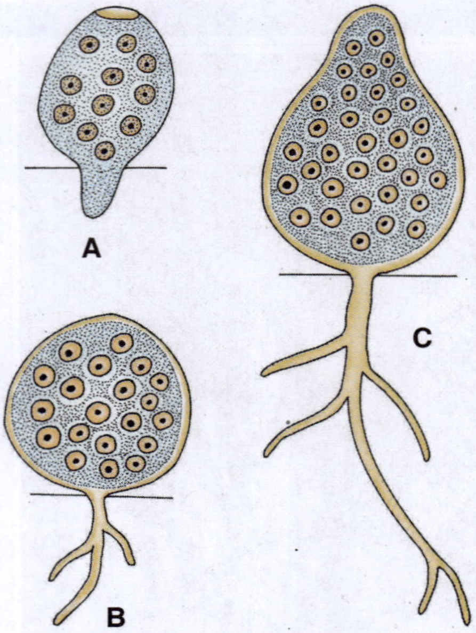

Fig. 4.1 (A-C). Evolution of Rhizomycelium in Chytrids. A, *Phlyctidium breviceps*; B, *Rhizophidium cyclotellae*; C, *Chytridium sp*.

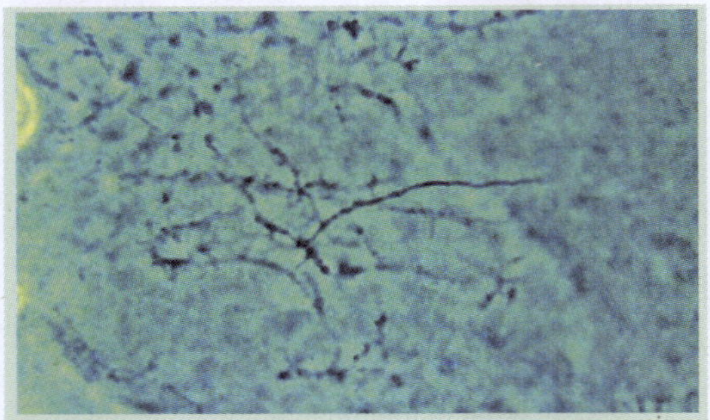

Rhizophidium

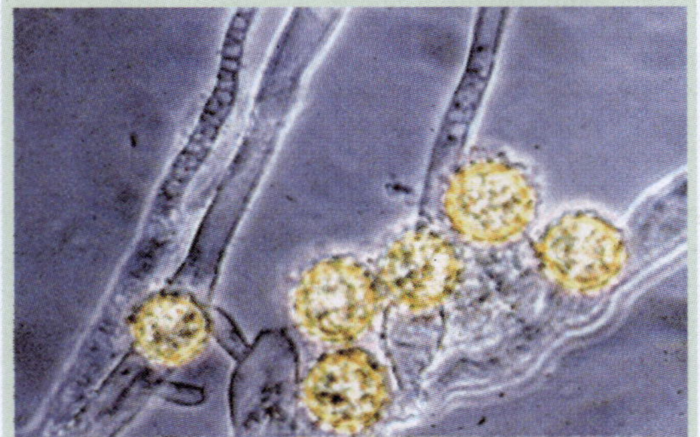

Monoblepharis

The asexual reproductive organs are the **sporangia,** each of which produces numerous tiny, uninucelate and uniflagellate opisthocont zoospores. The liberated zoospores swim for a time. Later each retracts its flagellum and undergoes encystment. After a short period of rest, the encysted spore germinates. The sexual reproduction may be isogamous or anisogamous. In some it is typically oogamous (Monoblepharidales).

Classification. On the basis of vegetative and reproductive structures, the class **Chytridiomycetes** is divided into the following four orders :—

1. Chytridiales. It is by far the largest order and includes the most primitive members of the class *Chytridiomycetes.* The somatic phase is a microscopic, holocarpic or eucarpic, single-celled structure which in some species is drawn out at one point into fine branching extensions constituting the **rhizomycelium.** True mycelium is lacking. The zoospore has a refractive oil globule but may lack a nuclear cap. It swims with a hopping movement or creeps over solid substratum like an amoeba.

2. Harpochytriales. It is a small order represented by two genera *Harpochytrium* and *Oedogoniomyces.* They are placed in the family Harpochytriaceae.

3. Blastocladiales. The vegetative body is a typical mycelium. Sexual reproduction is or aniso-planogamous. The sporangia are thick-walled and resistant. The zoospores has no refractive oil globule but has a nuclear cap. It swims evenly.

4. Monoblepharidales. The vegetative body is a typical mycelium as in the Blastocladiales. Sexual reproduction is heterogamous and takes place by the fusion of a motile male gamete and a non-motile female gamete. Resistant sporangia are lacking. The zoospores are similar to those of the Blastocladiales.

Sparrow (1973) divided the class into the following four orders on the basis of vegetative characters, zoospore characteristics, mode of plasmogamy and resting spores. The classification and key to the orders of the class Chytridiomycetes are given below :

Class : *Chytridiomycetes*

- Thallus uniaxial or surface only i.e. epiphytic or epizoic, Order : *Harpochytriales*
- Thallus variously constructed, penetrating the substratum by means of rhizoids or hyphae or enterely within it.
  - True mycelium absent, rhizomycelium may be present in some genera, zoospores with a conspicuous replactic globule, germination monopolar Order : *Chytridiales*
  - True mycelium present, globule lacking in zoospores, germination bipolar
    - Asexual resistant sporangia present, fusion of anisogametes Order : *Blastocladiales*
    - Asexual resistant sporangia not formed, sexual reproduction oogamous. Order : *Monoblepharidales*

Three of the four orders viz. Chytridiales, Blastocladiales and Monoblepharidales are being described in some detail.

ORDER (1) : CHYTRIDIALES

General Characters. This order includes 65 genera and 300 species. They are the most primitive and simplest members of true fungi. They are often called the **chytrids.** The chytrids include both aquatic and soil inhabiting species. The latter are restricted to the wet soil. A few species are marine. Many chytrids are parasitic and others saprophytic. The fresh water species parasitize algae (Fig. 4.2A) or water moulds. The common examples of parasitic species which attack vascular seed plants of economic value are *Synchytrium endobioticum, Physoderma zeaemaydis* and *Urophylyctis alfalfae.* Some

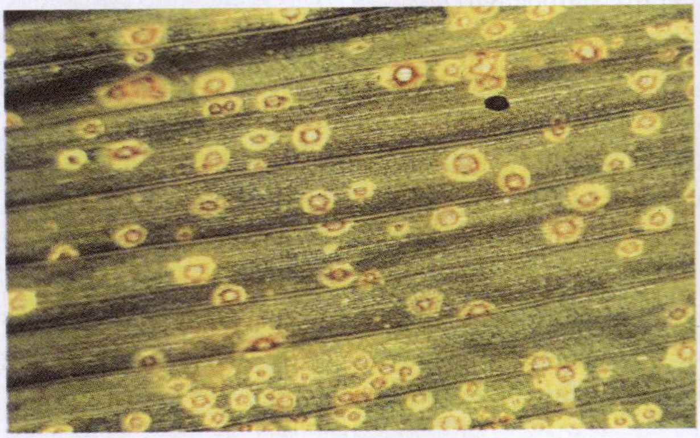

Physoderma zeaemaydis.

Urophylyctis alfalfae.

of the parasitic species are **endobiotic** and **holocarpic** (*Synchytrium*). During the early stages of their development they often lack a cell wall. They live within the host cell and in some of them

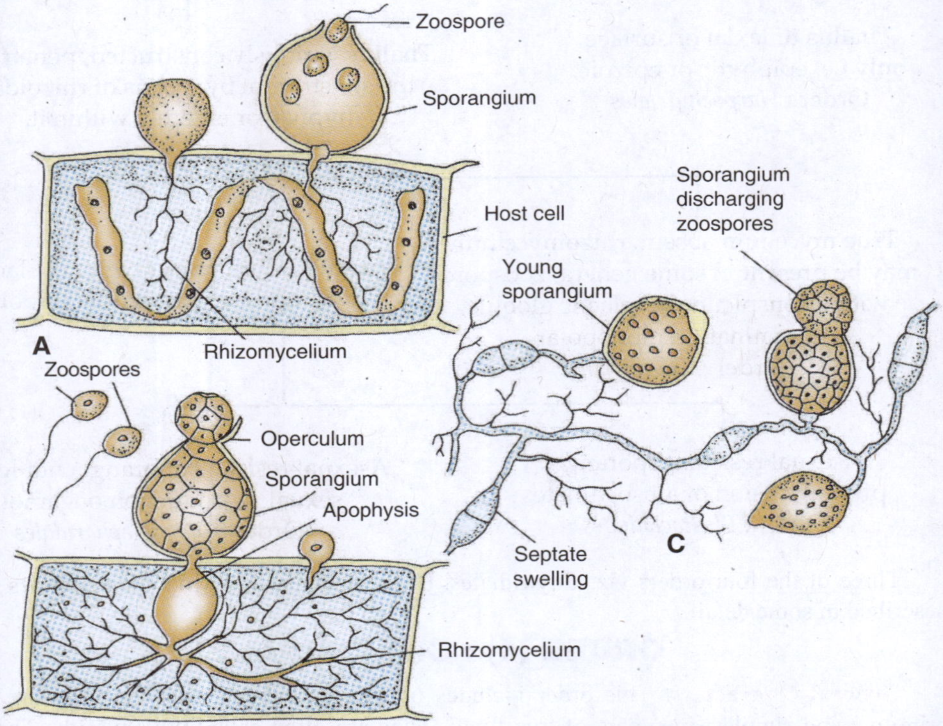

Fig. 4.2 (A-C). Epibiotic chytrids. A, *Rhizophidium* on algal host; B, *Chytridium lagenaria* on algal host (monocentric); C, *Cladochytrium tenue* (Polycentric).

reproductive structures are produced within the host cell. The more advanced forms are **epibiotic** and **eucarpic.** In these, the reproductive structures are developed outside the host. The rhizomycelium is embedded in the host tissue. The endobiotic species are considered primitive. The thallus in all the chytrids is a single cell (Fig. 4.3 D). The cell wall consists of fungal chitin or both fungal chitin and cellulose. In the eucarpic spce_es, the thallus develops a system of rhizoid-like fine hyphal branches constituting **rhizomycelium** (Fig. 4.2 B). It is an intergral part of the thallus. In some species, it bears a single reproductive structure (sporangium or resting spore) and in others more than one. The thallus in the former is called **monocentric** (Fig. 4.2 B) and in the latter **polycentric** (Fig. 4.2 C). The thallus of monocentric species superficially resembles *Botrydium,* an alga. Sometimes hypha-like branches develop on the thallus.

Asexual reproduction takes place by means of posteriorly uniflagellate zoospores produced in globose, cylindrical, flask-shaped or irregularly lobed sporangia. The single flagellum is of whiplash type. On finding a suitable substratum, the zoospore retracts its flagellum and germinates by putting out a small peg-like germtube. The germination is thus **monopolar.** Sexual reproduction in the primitive chytrids is isogamous. The isogametes morphologically resemble the zoospores. More advanced members of the order exhibit a tendency towards heterogamy.

Classification of Chytridiales. Brown (1856) introduced operculation as a criterion for classification of the chytridiales. However, it was Sparrow (1935, 1973) who on the basis of the liberation of zoospores from the sporangium (operculation) established the concept of two groups, one **operculatae** and the other **inoperculatae** in the chytrids. In the former zoospores escape through an opening formed by the separation of a minute, circular cap-like **lid** or **operculum** at the end of a discharge tube. The zoospores in the inoperculate species escape through a pore in the wall of the sporangium or a discharge tube. The operculate species are considered more advanced. On this basis, the order Chytridiales has been divided into nine families. These are : Olpidiaceae, Achlyogetonaceae, Synchytriaceae, Phlyctidiaceae, Rhizidiaceae, Physodermataceae Cladochytriaceae, Chytridiaceae and Megachytriaceae.

Barr (1980) proposed reclassification of the Chytridiales based on fundamental differences in the ultrastructure of zoospores. He splits it into two orders *Chytridiales* (amended) and a new order *Shizellomycetales.* Both the orders are divided into nine families on the basis of thallus development. These are: Olpidiaceae, Achlyogetonaceae, Synchytriaceae, Phlyctidiaceae, Rhizidiaceae, Cladochytriaceae, Physodermaceae, Chytridiaceae and Megachytriaceae. Genera in the families are recognized on zoospore ultrastructure. The importance of operculation has been deemphasized at the species level of classification. Of the families included in the order Chytridiales (amended), *Synchytriaceae* is described here.

FAMILY : SYNCHYTRIACEAE

General Features. The family includes parasitic chytrids possessing a microscopic, unicellular, holocarpic thallus. At the time of reproduction, the latter divides into a number of compartments, each functioning as a **sporangium** or **gametangium**. All the reproductive structures are invested by a common membrane to form a **sorus.** The sporangia are inoperculate (without a lid) which means they discharge their contents through an aperture in the sporangial wall or throuth a discharge tube. The best known genus of this family is *Synchytrium.*

Synchytrium taraxaci on the host

According to Karling (1964), the genus has six subgenera and 200 species. The best known and commonest is *S. endobioticum* which causes black wart disease of potato. While *S. taraxaci* has been reported to cause orange galls and hypertrophy in the leaves of dandelion (*Taraxacum officinale*), *S. trichosanthoides* is parasitic on many cucurbits. A detailed account of *Synchytrium endobioticum*—the cause of wart disease of potato is presented here.

SYNCHYTRIUM ENDOBIOTICUM

Synchytrium endobioticum causes the black wart or wart disease of potato (*Solanum tuberosum*). According to Karling (1964), it has a broad spectrum of Solanaceous sp. in its host range. The potato wart disease is widely distributed in the potato growing regions of the world. It is prevalent in areas with a cool moist climate. In India, it has been reported from the Darjeeling district (West Bengal) and areas of Nepal contiguous to the former. The fungal parasite cannot survive in hot places.

Black wart of Potato.

Symptoms. Usually the disease affects the underground parts of the host. Diseased potato tubers appear as dark brown or black cauliflower like outgrowths. Galls or tumours may be formed on aerial parts as well. On shoots, the disease appears in the form of outgrowths on green twisted leafy structures. Due to the presence of the fungus, the host cells are stimulated to divide leading to increase in the number of host cells. Most of the host cells contain resting sporangia (Fig. 4.3).

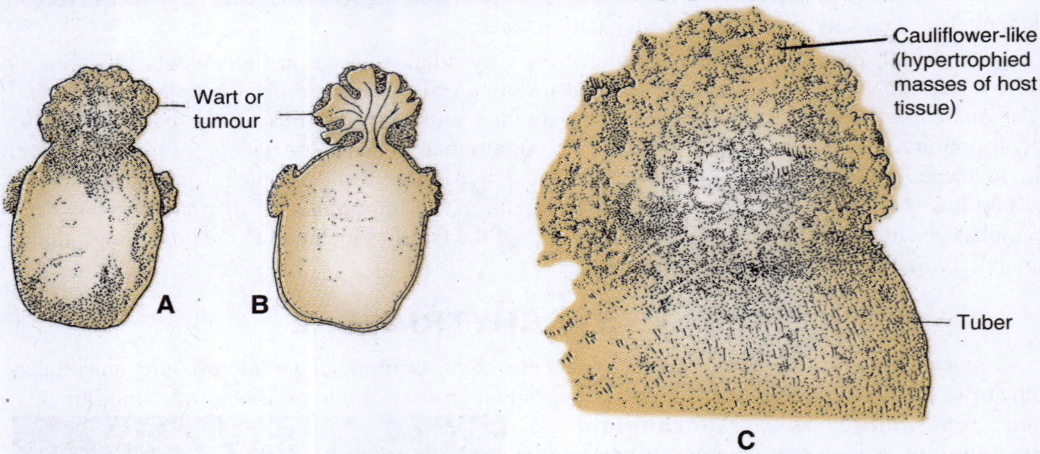

Fig.4.3 (A-C). A-B, Potato tubers infected by *S. endobioticum* showing black warts; C, portion of A slightly enlarged.

Thallus structure. *Synchytrium endobioticum* is a holocarpic endoparasite. It is a holocarpic fungus because the thallus is naked and unicellular. The thallus forms a mass of naked, uninucleate amoeboid mass of protoplasm. Soon, a double layered chitinous wall develops around the thallus. The fungus is an endoparasite because it is endobiotic and occurs as a parasite in the epidermal cells of the host.

Life cycle. The life cycle of *S.endobioticum* has been studied by Curtis (1921) and Kohler (1923, 1931).

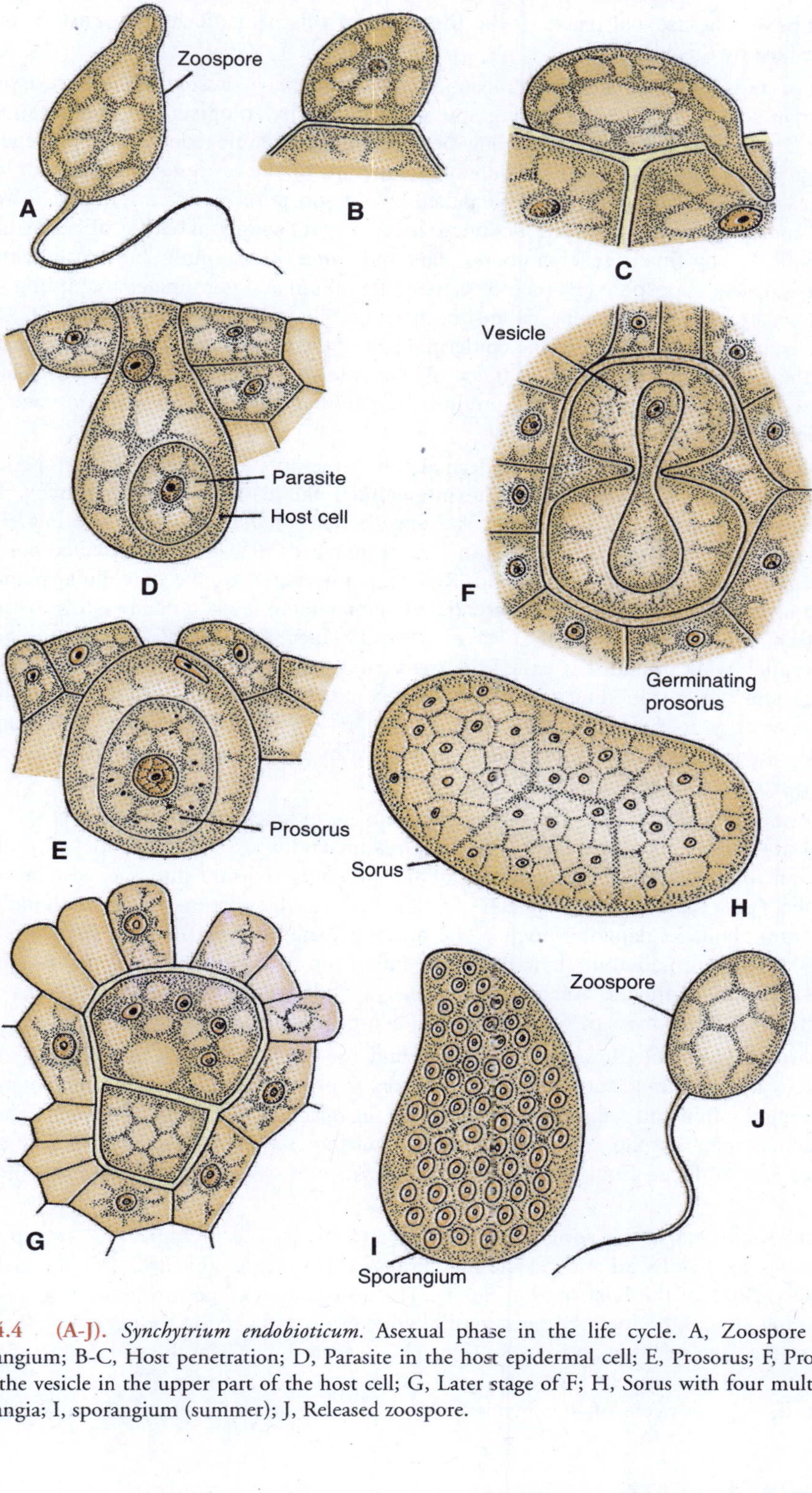

Fig. 4.4 (A-J). *Synchytrium endobioticum.* Asexual phase in the life cycle. A, Zoospore from a resting sporangium; B-C, Host penetration; D, Parasite in the host epidermal cell; E, Prosorus; F, Prosorus migrating into the vesicle in the upper part of the host cell; G, Later stage of F; H, Sorus with four multinucleate young sporangia; I, sporangium (summer); J, Released zoospore. (After Curtis)

Asexual Phase. The asexual phase of the life cycle of this parasitic fungus starts with the infection of the host by the parasite.

(*i*) Infection (Fig. 4.4). The causal organism (*S. endobioticum*) is present in the wart tissue of potato tuber or in soil in the form of resting sporangia. Some mycologists call them resting or winter spores. They remain viable for a long period. Under suitable conditions (presence of potato seedlings in the field, moist soil and suitable temperature), the resting sporangium or spore germinates releasing posteriorly uniflagellate, naked haloid zoospores which are liberated by the rupture of the enclosing membrane. The liberated zoospore (A) swims in a film of water in the contaminated soil. It may finally reach a potato plant and come to rest on the host surface (tuber or stolon). The quiescent zoospore (B) then retracts its flagellum and germinates by putting out a small, thin, naked, peg-like germ tube called the **infection peg** or **penetration tube** (C). The latter pierces the cuticle and the wall of the epidermal cell of the tuber or bud tissue in the stolons. Subsequently the uninucleate protoplast of the zoospore enters the host as a naked mass. The old view that the zoospore dissolves a minute pore in the epidermal wall through which it penetrates is now discarded.

(*ii*) Prosorus Stage. Once within the epidermal cell of the host it promptly sinks to its lower part (D) where it develops into a uninucleate intracellular thallus with a comparatively large nucleus (D) at the expense of the food material absorbed from the host cell. The unicellular parasite rounds off and secretes a thin wall around it. It increases in size as the infected host cell is induced to enlarge and become pear-shaped. Reaching a certain size, the unicellular pathogen thallus secretes a wall around it which is differentiated into an outer, thick, golden yellow **exospore** and an inner, thin, hyaline **endospore** (E). The nucleus also increases in size. The mature thallus with a heavy golden wall around it is called the **prosorus** (pl. prosori). Meanwhile the adjacent epidermal cells and the surrounding cortical cells become stimulated to activity. They divide rapidly and repeatedly to form a minute gall or tumor or a wart-like tissue. The infected hypertrophied cell containing the prosorus in its lower part by now is dead. It is in the centre of a rosette of more or less hardened epidermal cells.

(*iii*) Germination of Prosorus. The mature prosorus germinates within the dead host cell. A pore is formed in the thick exospore layer. The thin, hyaline endospore layer extrudes throuth the pore in the form of vesicle (F). The contents of the prosorus migrate into the vesicle which extends into the upper half of the host cell. During migration, prosorus nucleus divides (G) repeatedly to form about 32 daughter nuclei. The multinucleate protoplast of the vesicle is now partitioned by newly formed thin, hyaline walls into four to nine multinucleate polygonal compartments (H) or segments. Each segment has a wall of its own and functions as a **zoosporangium.** The whole mass of 4 to 9 sporangia compose a **sorus.**

(*iv*) Sporangia. The nuclei in each sporangium undergo further division to form 200 to 300 daughter nuclei. The multinucleate protoplast is finally organised into uninucleate daughter protoplasts by aggregation and rounding off of the cytoplasm round the nuclei (H). Each daughter protoplast becomes metamorphosed into a posteriorly uniflagellate zoospore. In this way about 1500 zoospores are produced from the single, original zoospore which initiated the infection of the host.

The mature sporangia in the sorus absorb water and swell. The pressure thus set up from within ruptures the host cell wall and vesicle membrane. The zoosporangia are pushed out of the sorus on to the surface of the host (potato tuber). The zoospores escape through an opening (a slit) in the sporangial wall, or upon the rupture of hyaline projections in the sporangial wall which are termed **papillae.** The released zoospores (J) swim in a film of water in the soil. Some of these may reinfect the host and repeat the sequence of events outlined above. This completes the **asexual** phase in the life cycle of *Synchytrium.* From the account of asexual phase of the life

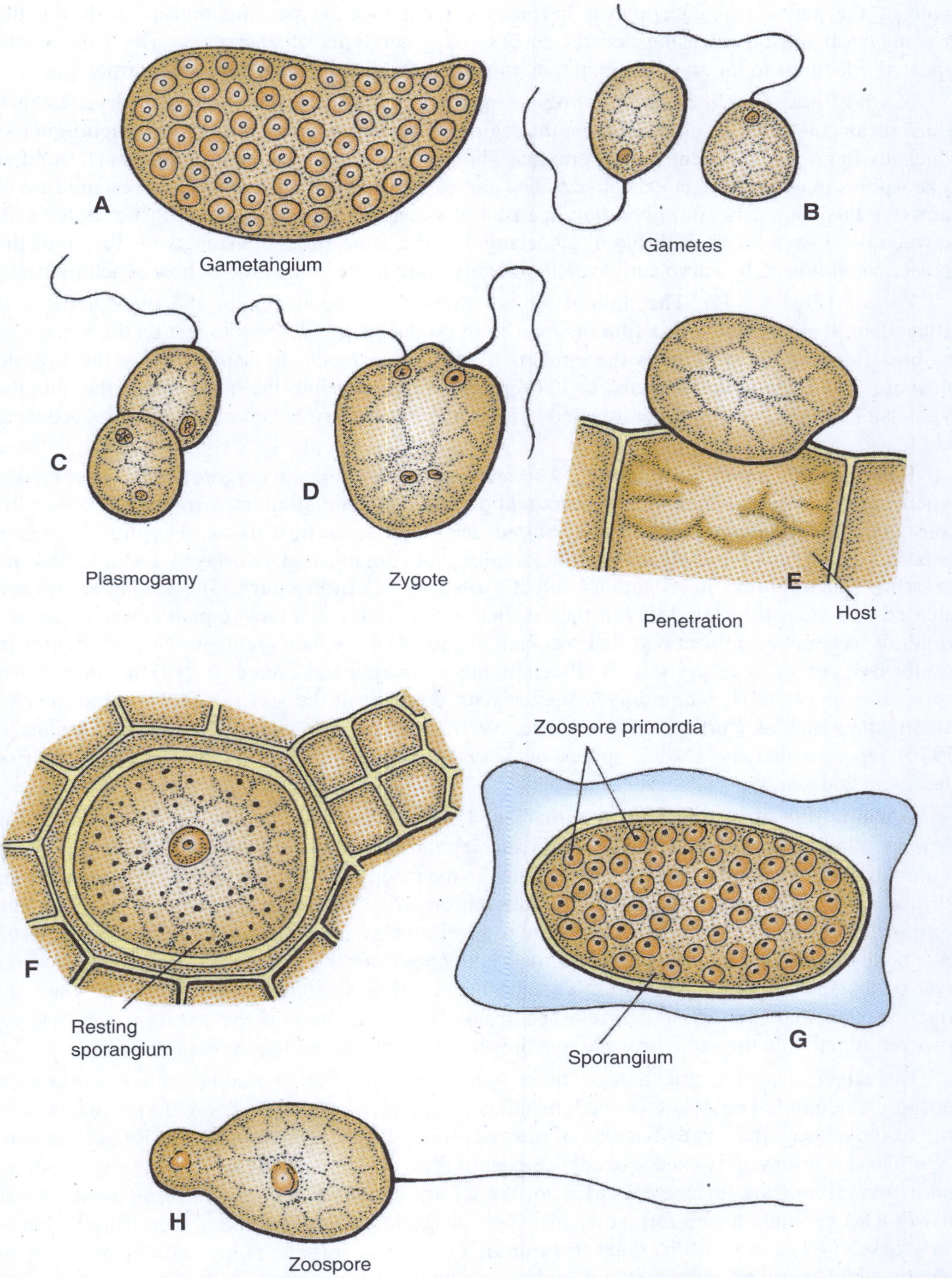

Fig. 4.5 (A-H). *Synchytrium endobioticum*. Sexual phase of the life cycle. A, Gametangium; B, two gametes released from the gametangium; C, Fusing gametes; D, Zygote; E, Zygote penetrating the host epidermal cell; F, Resting sporangium; G, Germinating resting sporangium with zoospore primordia; H, Released Zoospore.

(After Curtis)

cycle of the parasite given above it becomes evident that the parasite multiplies during the growing crop season releasing several successive generations of zoospores. The latter cause repeated infections to the standing crop thus spreading the disease in an epidemic form.

Sexual Phase (Fig. 4.5) : Gametangia. Under conditions of scarcity of water (dry weather), which means the end of growing season, the segments of the prosorus function as **gametangia** (A) which are in no way different from sporangia. The gametangia produce **planogametes** (B) similar to zoospores in every respect except size. The gametes are smaller than the zoospores and fuse in pairs (C). Fusion occurs after liberation in a film of water on the surface of the host or in the soil, between planogametes from different gametangia in the same sorus. Plasmogamy (fusion of the gametes) is followed by karyogamy which probably takes place at the time of host penetration.

Zygote (Fig. 4.5 D). The diploid zygote formed by the fusion of the planogametes is biflagellate. It swims about in a film of water in the soil. Finally it comes to rest on the surface of the host (tubers) and penetrates the epidermal cell (E) in much the same way as the haploid zoospore. The flagella are retracted and absorbed before entry into the host. Within the host the zygote sinks to the bottom of the infected hypertrophying epidermal cell of the host and grows in size.

Resting Sporangium (Fig. 4.5 F). The presence of the parasite (zygote) in the host causes hypertrophy and hyperplasia of the surrounding cells. They are stimulated to divide repeatedly. Consequently the infected cell is soon buried deep within the host tissue. The diploid zygote (parasite) enlarges and is enclosed in a thick, reticulately ornamented, two-layered wall to become a **resting sporangium.** Some authors call it a **resting** or **winter spore.** The resting spores are released into the soil by the decay of the infected tubers. In the soil they remain dormant through winter. It has, however, been reported that resting spores can remain viable in the warty tubers in soil for five years and air dry soil for fifteen monthes. Sharma and Commack (1976) reported that the resting spores of *S. endobioticum* may remain dormant in the soil for 25-30 years. Several investigators such as Curtis (1921), Glynne (1926), Karling (1967) and Sharma and Commack (1976) reported that the resting spores of *S. endobioticum* function as sporangia giving rise directly to zoospores.

Germination of resting Sporangium (Fig. 4.5 G). At the onset of conditions favourable for growth (following spring when the host is available), the resting sporangium germinates. A number of granule-like structures make their appearance in the cytoplasm. These are **zoospore primordia.** Further, each zoospore primordium extrudes chromatin. This is suggestive of meiosis. The multinucleate protoplast undergoes cleavage to form numerous uninucleate, dauthter protoplasts. Each daughter protoplast gets metamorphosed into a posteriorly uniflagellate zoospore. The outer layer of the sporangial wall ruptures. The zoospores (H) escape throuth the exit tube. They are larger in size than the zoospores released from the sporangium of the asexual cycle. They, however, function in the same way and may infect the host to repeat the asexual phase.

The effect of zygote infection on tuber is more serious and destructive as compared with zoospore infection. The zygote infection induces excessive rate of division (**hyperplasia**) not only of the adjacent host cells but also of the cells some distance away from the site of infection. Hyperplasia is followed by excessive enlargement of the resultant cells **(hypertrophy)**. Hypertrophy and hyperplasia of the surface cell layers of the infected potato tuber and the resultant abnormal growth activity leads to the formation of large, unsightly and useless masses of crinkled black warty tissue (4.3) known as the **wart** or **tumour.** The wart is soft and pulpy and thus invaded by bacteria which cause rot. The warty tubers become useless and a source of infection.

Control. To reduce spread of the pathogen, quarantine regulations are prasctised. In India, the entry of potatoes from Darjeeling Hills, where wart disease is known to occur, has been prohibited. A number of varieties which are immune to the disease have been produced in recent years. Their use has considerably prevented spread of disease. The application of lime and use of fungicides in the soil before planting have given some useful results.

Economic importance. The chytrids are of little direct economic importance. Some parasitic species cause serious disease of crop plants. The examples are the *potato wart disease, a leaf and fruit disease* of cranberries, the *crown wart of alfalfa* and the brown spot disease of corn. The Wart Disease of potatoes may cause a very great loss of tubers first in the ground and later in storage. Some species of *Allomyces* and *Blastocladiella* have been found to be ideal tools for

Brown Spot Disease of Corn.

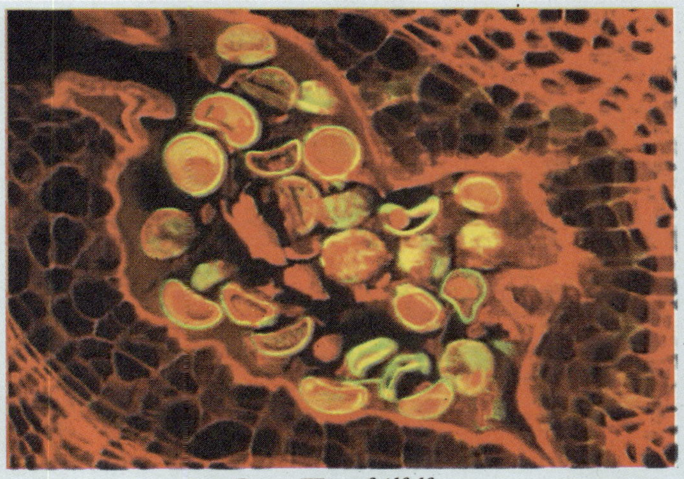

Crown Wart of Alfalfa.

experimental studies and research in the study of morphogenesis. Some chytrids are destructive parasites of algae which form a link in the food chain of aquatic animals. A few attack aquatic fungi.

Origin and Phylogeny of Chytrids. Some mycologists consider chytrids to be the simplest and the primitive members of the lower fungi. According to others, they are degenerated forms which were once more complex. Hawker (1967) who susbcribes to the first view derives their origin from the aquatic flagellates having the appropriate type of flagellation (Euglenoid-Pyrrophyta complex). Of considerable importance in this respect are those species of Euglenophyta and Pyrrophyta which live saprophytically on decaying organic matter. It has been suggested that the chytrids evolved from the ancient Euglenoid-Pyrrophyta complex through natural selection which favoured an organism that could survive as a decomposer within the ancient ecosystems. This assumption is based on certain similarities, (*i*) Both the

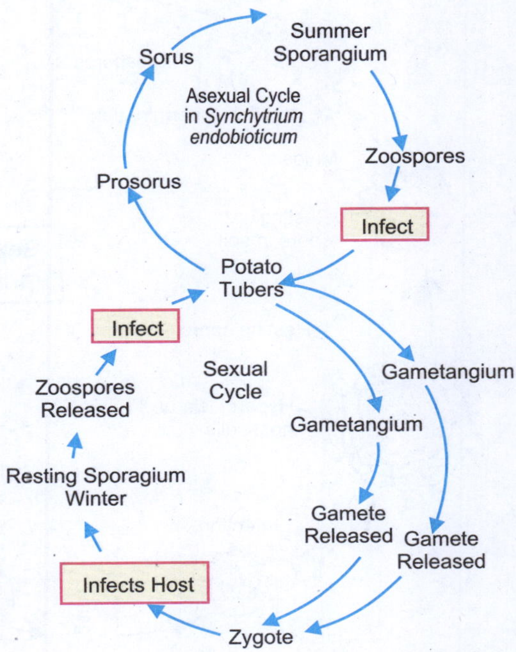

Fig. 4.6. Word diagram of the life cycle of *Synchytrium endobioticum*.

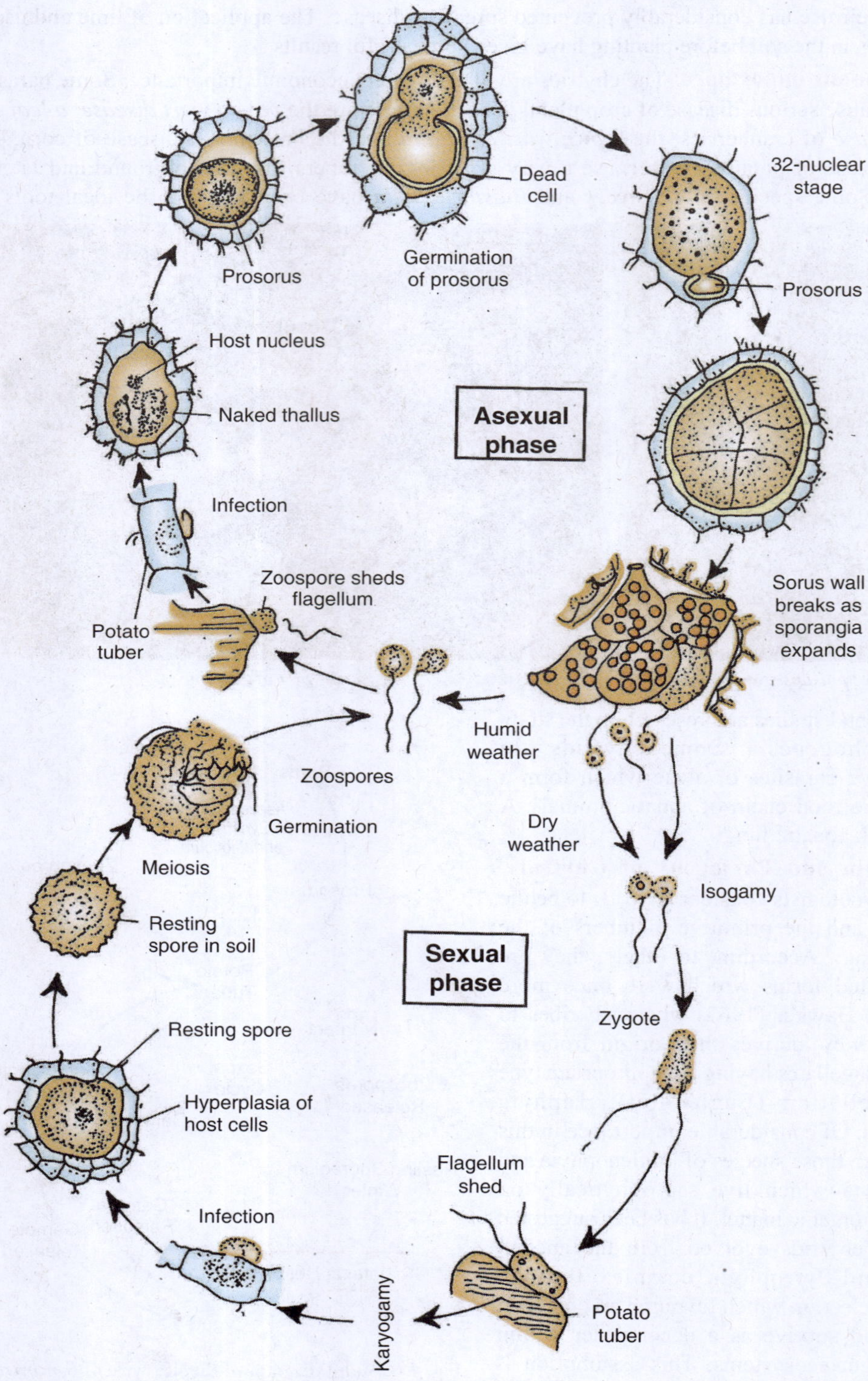

Fig. 4.7. Pictorial life-cycle of *Synchytrium endobioticum*.

chytrids and euglenoids resemble biochemically, (*ii*) The tendency of some chytrids to form thread-like rhizoids is also exhibited by some Pyrrophyta algae, (*iii*) Chytrids resemble both euglenoids and pyrrophytes in certain details of cell division. Several mycologists hold that the Oomycetes have evolved from the flagellates via the chytrids.

The chytrids in their life cycle and range of somatic phase resemble the Hyphochytridiales. Owing to this close resemblance the latter were formerly included in the order Chytridiales. The fundamental difference, however, is in the form of zoospores which in the Hyphochytridiales are anteriorly uniflagellate. The single flagellum is of tinsel type. In contrast, the zoospore of Chytridiales has an opisthocont flagellum of whiplash type. Some mycologists hold that during chytrid evolution, some members of this group at some ancient period, served as progenitors of two other groups of Fungi, namely, the true slime molds (Myxomycetes) and the bread molds (Zygomycetes). In the range of somatic phase, absence of cell walls and mode of nutrition, the chytrids bear resemblance to the slime molds. The presence of flagellated reproductive cells in the slime molds is considered as further evidence of a possible ancient chytrid ancestry. The bread molds resemble the chytrids in their use of chitin (instead of cellulose) as a major cell wall molecule and synthesis of lysine via the same biochemical pathway. The Zygomycetes, however, differ from both the chytrids and slime molds in the absence of flagellated cells of any kind.

Taxonomic Position. The systematic position of the chytrids is not very certain. The earlier mycologists placed the chytrids in the order Chytridiales of sub-class Oomycetes and class Phycomycetes. Gaumann and Dodge (1928) and Gaumann (1952) placed the primitive chytrids in a distinct class Archimycetes. Alexopoulos separated the Chytridiales, Blastocladiales and Monoblepharidales from the class Phycomycetes and constituted a distinct class Chytridiomycetes in coordination with the other classes of the Fungi. *Synchytrium endobioticum* was first described by Schelbersky (1895) but he placed it in the genus *Chrysophlyctis*. It was Percival who transferred the species to the genus *Synchytrium*. According to Alexopoulos (1962), the systematic position of *S. endobioticum* is :

Division	:	**Mycota**
Sub-Division	:	**Eumycotina**
Class	:	**Chytridiomycetes**
Order	:	**Chytridiales**
Family	:	**Synchytriaceae**
Genus	:	*Synchytrium*
Species	:	*enodobioticum*

Alexopoulos and Mims (1979) place *Synchytrium* in the Superkingdom Eukaryonta, Kingdom Myceteae, Division Mastigomycota, Subdivision Haplomastigomycotina, Class Chytridiomycetes, Order Chytridiales, Family Synchytriaceae.

Superkingdom	:	**Eukaryonta**
Kingdom	:	**Mycetae**
Division	:	**Mastigomycota**
Sub-division	:	**Haplomastigomycotina**
Class	:	**Chytridiomycetes**
Order	:	**Chytridiales**
Family	:	**Synchytriaceae**
Genus	:	*Synchytrium*
Species	:	*endobioticum*

ORDER (2) : BLASTOCLADIALES

General Features and Classification. The saprophytic members of this order occur chiefly in water which is often rather stagnant. Some saprophytic species are soil dwwllers. This occurs in the wet soil. Some species are parasitic. They attack fungi and invertebrates such as insects, worms and rotifers.

The somatic phase (Fig. 4.8) is more complex and variable than the Chytridiales. Often the thallus is large enough to be seen with the naked eye. The tendency to increased complexity of vegetative structure is evident. In the simplest forms, the soma shows only slight advance on the chytrids. For example in *Blastocladiella emersonii,* the simple eucarpic thallus consists of an insignificant basal cell and a tuft of attaching rhizoids, the former bearing a thin (A) or thick-walled terminal sporangium (B). The basal cell in *B. variabilis* is more prominent and cylindrical in form (C). There is no mycelium. *Blastocladia pringsheimii* has a thallus differentiated into rhizoids and a stout trunk like hypha arising from the former to project in the water (D). The plantule bears the reproductive structures. In *Allomyces* which is a soil dweller, thallus is hyphal. It is filamentous and consists of richly branched, multinucleate hyphae (Fig. 4.9 A). The hyphae carry on the metabolism of the fungus and reproductive structures are borne on the tips of dichotomies. The wall is of fungus chitin. The Blastocladiales are more advanced than the Chytrids in the physiology of their nutrition.

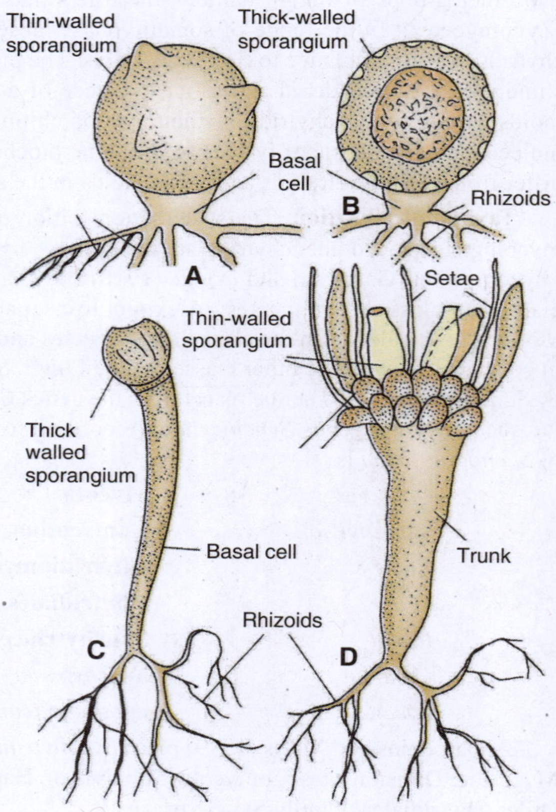

Fig. 4.8 (A-D). *Blastocladiales*. Thalli of *Blastocladiella emersonii* bearing thin-walled sporangium (A) and thick-walled sporangium (B); C, club-shaped sporothallus of *B. variabilis* with a resting sporangium; D, sporothallus of *Blastocladia pringsheimii* consisting of rhizoids, and a trunk bearing setae, thin and thick-walled sporangia (A, B and D after Hawker, C, after Harder and Sorgel).

Asexual reproduction is by uniflagellate zoospores. The single whiplash flagellum is inserted posteriorly (opisthocont). The zoospores are produced in two kinds of sporangia, (*i*) the thin-walled sporangia and (*ii*) thick-walled resting or resistant sporangia. The latter usually have thick pitted walls. The former are called **mitosporangia** and the latter **meiosporangia.** The two kinds of sporangia may occur on different plants (*Blastocladiella*) but in most species (*Blastocladia* and *Allomyces*) both types occur regularly on the same plant.

Sexual reproduction is not universal in the order. Where found it takes place by the primitive method of planogametic copulation as in the Chytrids. The fusing gametes are uniflagellate, isoplanogametes in some species and anisoplanogametes in others. The motile cells (zoospores and gametes, possess a centrally located nuclear cap composed of mass of ribonucleic acid. The order comprises 5 genera with about 40 species. It is divided into 3 families, namely, Coelomomycetaceae, Catenariaceae and Blastocladiaceae. The last one is discussed here as it is the most typical and well investigated.

FAMILY : BLASTOCLADIACEAE

The chief characteristics of the family are : Members with a well developed mycelium, simple or branched attached to the substratum by simple or branched rhizoids; the hyphal walls are chitinous and the hyphae multinucleate coenocytic and nonseptate but with a pseudosepta in the form of the thickened rings, species with sexual reproduction have two types of thalli (sporothallic and gametothallic); diplobiontic life cycle and sexual reproduction by planogametic copulation between isogametes or anisogametes. The family is represented by 5 genera, namely, *Allomyces, Blostocladia, Blastocladiella, Blastocladiopsis* and *Microallomyces.* Of these *Allomyces* is the best known and is described.

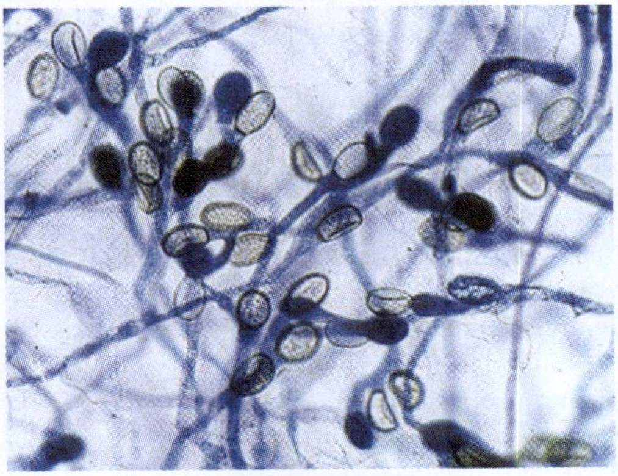

Allomyces

Blastocladiella

ALLOMYCES

Occurrence. The genus *Allomyces* was first discovered in India by E.J. Butler (1911). Subsequently it was found to be worldwide in distribution. It has been reported chiefly in tropical or warm countries. The genus comprises 5 species. Of these, *A. arbuscula, A. macrogynus* and *A. javanicus* are important. All the species are saprophytic, soil dwellers and are found in the wet soil.

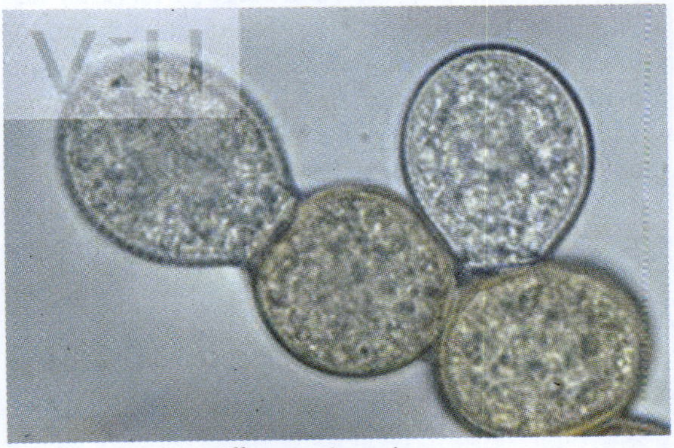

Allomyces javanicus.

Thallus Structure (Fig. 4.9 A). The thallus in all the species is hyphal and is called the mycelium.

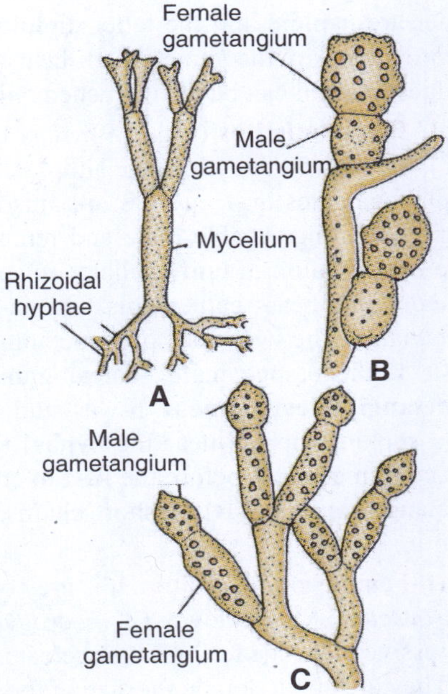

Fig. 4.9 (A-C). *Allomyces.* A, Gametothallus, B, part of gametothallus of *A. arbuscula* bearing gametangia, C, upper part of the gametothallus of *A. macrogynus* bearing gemetangia (A and C adapted from Alexopoulos and B from Hawker).

It is filamentous and attached to the substratum by a tuft of delicate but well formed, branched rhizoidal hyphae constituting the *rhizoidal system.* From the latter arises a single stout but slender hypha forming the lower trunk-like portion which subsequently undergoes several successive dichotomous branchings to form the main, more or less, dichotomously branched part of the mycelium on which the reproductive organs are borne terminally. At the time of dichotomy, the growing multinucleate hyphal tip divides into two new hyphal tips. Each of the new hyphal tips produces one of the two branches which develop nearly synchronously in the young mycelia. Septa are formed in both the hyphal branches nearly synchronously just at or above the dichotomy. In *Allomyces* they are not formed in conjuction with cell division. The hyphae are thus multinucleate and coenocytic. Meyer and Fuller (1985) reported that ultrastructurally the septum consists of three layers with the electron opaque layer between two outer electron translucent layers. In *Allomyces* each septum arises as 5 or 6 separate ingrowths of the wall material at each dichotomy. These protrusions broaden and fuse laterally and then grow centripetally leaving a central pore. The lateral fusion between the protrusions is not always complete. Small pores may be left here in the rim of the mature septum. The septa in *Allomyces* are often referred to as **false septa** because they do not completely separate the protoplasm of adjacent somatic segments.

Hyphal structure. The hyphal wall consists chiefly of Chitin, B glucan and ash. Within it, is the plasma membrane which closely investes the hyphal cytoplasm. Embedded in the cytoplasm are the numerous nuclei. Besides the nuclei there are mitochondria and roughly unique organelles termed the *concentric granules.* They are found in abundance in each hyphal segment. The concentric granules move about in a slow random fashion separate from the normal cytoplasmic streaming. According to Robert (1987) concentric granules act as plugs for septal pores and thus prevent protoplasmic loss of following damage.

The vegetative thalli in *Allomyces* are of two types, **gametothalli** and **sporothalli.** The former are haploid and the latter diploid. In the vegetative phase the two are indistinguishable. Towards maturity the gametothalli bear gametangia and sporothalli produce sporangia. Thus, the two tpes of thalli can be distinguished only when they begin to form the reproductive organs.

1. Gametothallus (Fig. 4.9). It is concerned with sexual reproduction and is homothallic (B,C). Sexual reproduction like simple chytrids is by planogametic copulation but unlike chytrid the planogametes in *Allomyces* are anisogometes and the two types of gametes are produced in distinctly distinguishable male and female gametangia. Anisogamous sexual reproduction takes place by the union of uniflagellate, anisoplanogametes. Reaching a certain stage of maturity, the gametothallus bears gametangia terminally on the side branches in close proximity in pairs or short chains. The two kinds of gametangia are more or less globose and usually occur in the ratio of 1 : 1, the orange male gametangium alternating with the colourless female gametangium **Gametangial development.** It was studied by Morrison (1977 *a*) in *A. macrogynus.* He reported that a septum appears near the hyphal tip to delimit the male gametangium. A second septum appears immediately before the first to cut off the female gemetangium into more or less globose gametangia in pairs (B) or short chains. The male and female gametangia alternate with each other.

(*i*) Female gametangium. It is grey or hyaline and may be borne above the male gemetangium (*A. arbuscula,* A) or below it (*A. macrogynus,* C). The multinucleate gametangial contents undergo progressive cleavage to form uninucleate, uniflagellate, female gametes, 10 μm in size (Fig. 4.10 C_2). They are about double the size of the male gametes.

(*ii*) Male gametangium. It is smaller in size than the female gametangium and is orrange or red in colour. The colour is due to the presence of carotenoid pigment in the cytoplasm. The uninucleate male gametes formed by progressive cleavage of the male protoplast are pigmented. They are about half the size of the female gametes and are uninucleate and uniflagellate (Fig. 4.10C).

Gametes. (Fig. 4.10C). Both types of gametes (male and female) are motile, uninucleate and posteriorly unflagellale. The single flagellum is of whiplash type. The gamete nucleus has a prominent nucleolus. The male gamete is smaller than female but has the same structure.

***(iii)* Gamete Discharge** (Fig. 4.10 B). Following gametogenesis (differentiation of gametes) within the respective gametangia, the motile gametes of both kinds are discharged singly one after the other into the water at about the same time through the exit papilla one or more in number located at the various positions in the gametangial cell wall. After discharge the orange or red male gametes swim away quickly from the male gametangia. The female gametes in *A. macrogynus* emerge from the gametangia. Some disperse and swim away but remain together in groups, small or large. From the clusters, they disperse with time. The separated female gametes do not swim for an extended period. Soon they settle down at the bottom. The female gemetes thus are not very active in *A. macrogynus.*

***(iv)* Plasmogamy.** The discharged female gametes produce a sexual hormone called *sirenin.* It serves as a chemoattractant causing directed movements of the male gametes towards the female gametes. According to Pommerville (1978), the contact of the male with the female gamete is still by a random hit. Sirenin, however, greatly increases the chances that a hit will not go amiss. The male and the female gametes copulate in pairs (D). Pairing of the anisoplanogametes is soon followed by the fusion of their membranes and mingling of their cytoplasm. It is termed **plasmogamy.** Plasmogamy thus results in the formation of a binucleate fusion cell (E).

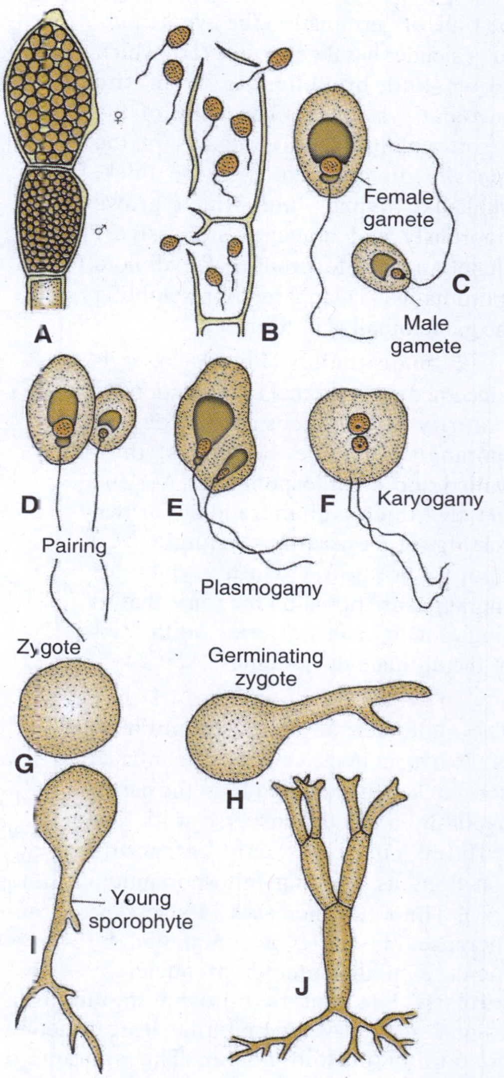

Fig. 4.10 (A-J). *Allomyces.* Gametangia (A) and liberation of gametes (B) of *A. arbuscula*; (C_{1-2}) male and female gemetes of *A. javanicus*; D, pairing of male and female gemetes; E, fusion of gametes; F, motile zygote showing karyogamy; G, zygote; H, germinating zygote; I, young sporophyte; J, sporothallus with a rhizoidal system (B after Smith, C-E, after Kniep, G-H, after Alexopoulos and I-J after Hawker).

***(v)* Karyogamy.** Soon after plasmogamy the two nuclei (male and female) fuse to accomplish karyogamy.

***(vi)* Zygote.** The resultant fusion cell with a diploid nucleus is called **zygote.** It is motile and biflagellate. It moves sluggishly for 5 to 10 minutes (F). It then comes to rest, retracts the flagella, rounds up and secretes a wall around it (G).

Germination of Zygote (Fig. 4.10 H-I). The zygote germinates immediately. There is thus no resting sexual spore in the life cycle. The function is taken over by the resting sporangia. At the time of germination the zygote puts out a slender basal germ tube (H), which grows and branches to form the rhizoidal system (I). The body of the zygote enlarges and elongates in the opposite direction to form the thick, trunk-like hyphal tube which grows vigorously and branches successively dichotomously to produce the diploid sporothallus (J), apparently resembling the gametothallus in form.

2. Sporothallus (Fig. 4.11). It is concerned with asexual reproduction. At maturity it produces two types of sporangia : (*i*) the colourless, thin-walled zoo or mitosporangia (A_3) and (*ii*) the thick-walled reddish brown **resting** or **resistant sporangia** (A_1) (meiosporangia). Both types of sporangia are borne on the same thallus single or in catenate series on the tips of the ultimate dichotomies (A_2).

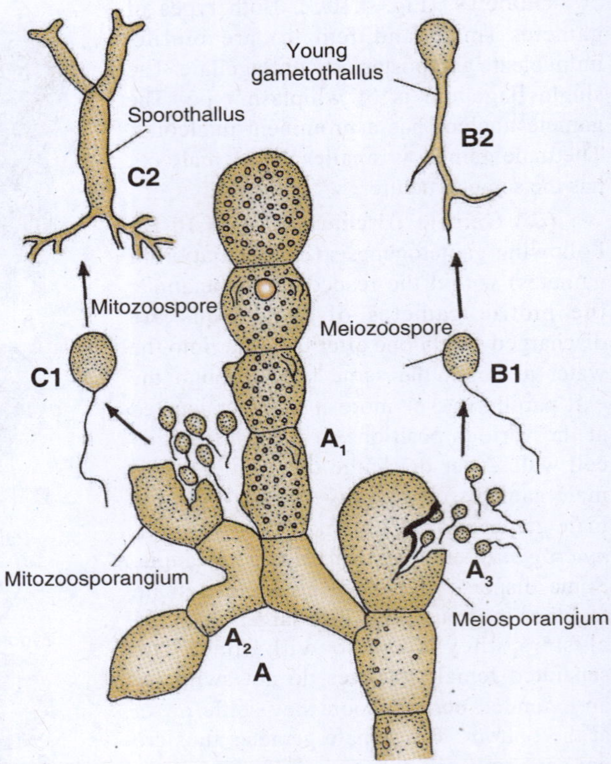

Fig. 4.11 (A-C). *Allomyces*. A, part of mature sporothallus bearing sporangia single (A_1 and A_3) or in catenate series (A_2); A_1, meiosporangium dehiscing to liberate meiozoospores; A_3, mitosporangium dehiscing to liberate mitozoospores; B_1, meiozoospore; B_2, young gametothallus; C_1, mitozoospores; C_2, sporothallus.

(*i*) Mitosporangium (Fig. 4.11 A_3). The multinucleate tip of the ulitimate of the hyphal branch swells. The inflated portion is then separated from the parent hypha by a basal transverse wall. The inflated multinucleate structure functions as a young mitosporangium (A_3). The latter increases in size. The increase in size is accompanied by increase in the number of nuclei by **mitosis.** The mature mitosporangium is a colourless, thin-walled multinucleate structure containing diploid nuclei. The sporangial contents are then divided by progressive cleavage into uninucleate daughter protoplasts by inward furrowing of the plasma membrane. Each uninucleate daughter protoplast metamorphoses into a diploid, uniflagellate, opisthocont zoospore. It is known as the **mitozoospore** (C_1). The diploid mitozoospores escape through pores, one or more in number developed in the sporangial wall (A^3). The discharged mitozoospore swims about for a while, comes to rest, retracts the flagellum, rounds up and secretes a wall. The encysted zoospore germinates immediately.

Zoospore Structure (Fig. 4.12). It is a

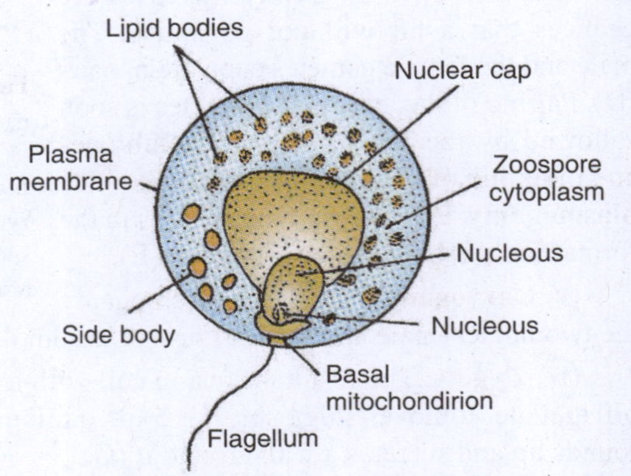

Fig. 4.12. *Allomyces macrogynus* light microscopic view of mitozoospore (After Fuller and Olson).

naked, uninucleate tiny mass of protoplast furnished with a single whiplash type of flagellum inserted at the posterior end. It has no cell wall and thus is termed naked. Within the plasma membrane, the cytoplasm contains a posteriorly located nucleus containing a distinct nucleolus. A single large basal mitochondrion forms a cap-like structure around the posterior region of the nucleus. The flagellum within the plasma membrane intergrates with a basal granule known as the functional Kinctosome. The latter is situated near the posterior end of the nucleus enclosed by the cup-like basal mitochondrion. A large, somewhat, horse shoe-shaped membrane bound organelle, **the nuclear cap** is located in the anterior region of the zoospore. It surrounds the major portion of the posteriorly located nucleus. The zoospore cytoplasm lacks the ribosomes and endoplasmic reticulum. The former, however, are abundant in the nuclear cap cytoplasm. Embedded in the cytoplasm of the anterior region of the zoospore are numerous lipid bodies, mitochondria and membrane bound vesicles. Mitochondria mostly occur along the nuclear cap membrane. The double-membraned envelope of the nuclear cap is perforated at intervals like the nuclear membrane. During encystment of zoospore the nuclear cap membrane breaks. The ribosomes become dispersed throughout the zoospore cytoplasm. The mitochondria become more numerous and are also found at the periphery of the encysted zoospore.

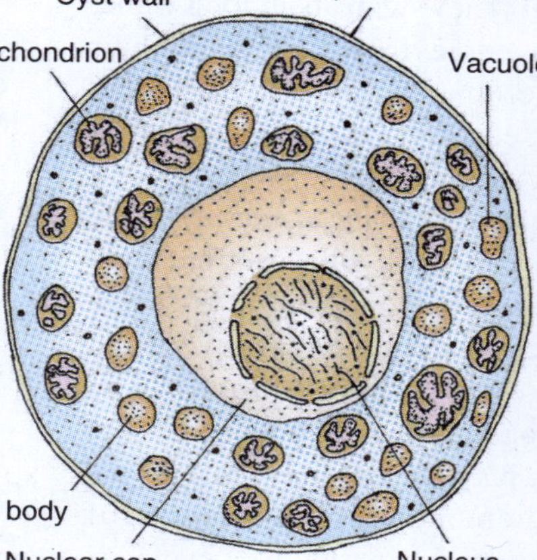

Fig. 4.13. *Allomyces macrogynus.* Encysted zoospore.

Germination of encysted mitozoospore (Fig. 4.11 C_1-C_2). The encysted zoospore (Fig. 4.13) germinates immediately. It contains stored mRNA which is used up during germination. In the germinating cyst, the nuclear cap loses its identity and is no longer visible as such. The ribosomes become dispersed in the cyst. At the time of germination, the cyst puts out a slender basal germ tube. The ribosomes become dispersed in the emerging germ tube which grows and branches to form the branched **rhizoidal system.** The main body of the cyst then elongates and grows in the opposite direction to form a tihcker germ tube. The later as it grows, branches repeatedly to produce the dichotomously branched part of the diploid **sporothallus** (Fig. 4.11 C_2). The diploid mitozoospores thus serve to reduplicate the sporophyte generation in the life cycle. They play no role in the phenomenon of alternation of generations.

***(ii)* Meiosporangium** (Fig. 4.11 A_1). The early development of meiosporangia is similar to that of the mitosporangia. There is, however, formation of thick, pitted wall internal to the original sporangial wall in addition. The mature meiosporangia are thus thick-walled and pitted, reddish brown in colour, the multinucleate diploid sporangial contents undergo a resting period of 2-6 weeks or even more prior to the cleavage of the protoplast into zoospores, hence the name resting or **resistant sporangia.** After the resting period the meiosporangia germinate. There is increase in the number of nuclei. The nuclear division, however, is **meiotic.** Meiotic Prophase I occurs during resting sporangium development. The two meiotic divisons take place after induction of resting sporangium germination [Emerson and Wilson, 1954 and Olson, 1974]. Meiosis is followed by the cleavage of the sporangial contents into uninucleate, uniflagellate, opisthocont zoospores which are haploid and slightly smaller than the diploid mitozoospores. They are known as the **meiozoospores** or **gonozoospores.** Emerson and Wilson (1954) reported the basic or haploid chromosome number of 14, and strains with 28, 50 (probably 56) for *A. macrogynus* indicating diploid and tetraploid series. The wall of the mature meiosporangium cracks (A_1) and the meiozoospores escape in the water. They swim about for a while, then come to rest, retract their flagellum and round off. Each quiescent meiozoospore secretes a wall around it and then germinates immediately in the same way as the mitozoospore but produces an alternate haploid plant in life cycle called the **gametothallus** (B_2) and not the sporothallus. The meiozoospores or gonozoospores thus play an important role in the phenomenon of alternation of generations in the life cycle of *Allomyces*.

Life Cycle of Allomyces (Fig. 4.14). Allomyces in its life cycle exhibits distinct alternation of generations—a significant feature which sets these fungi (Blastocladiales) apart from all other fungi, there occur two distinct but identical individuals in a single life cycle of *Allomyces*. One of these is the haploid **gametothallus** and the other diploid **sporothallus.**

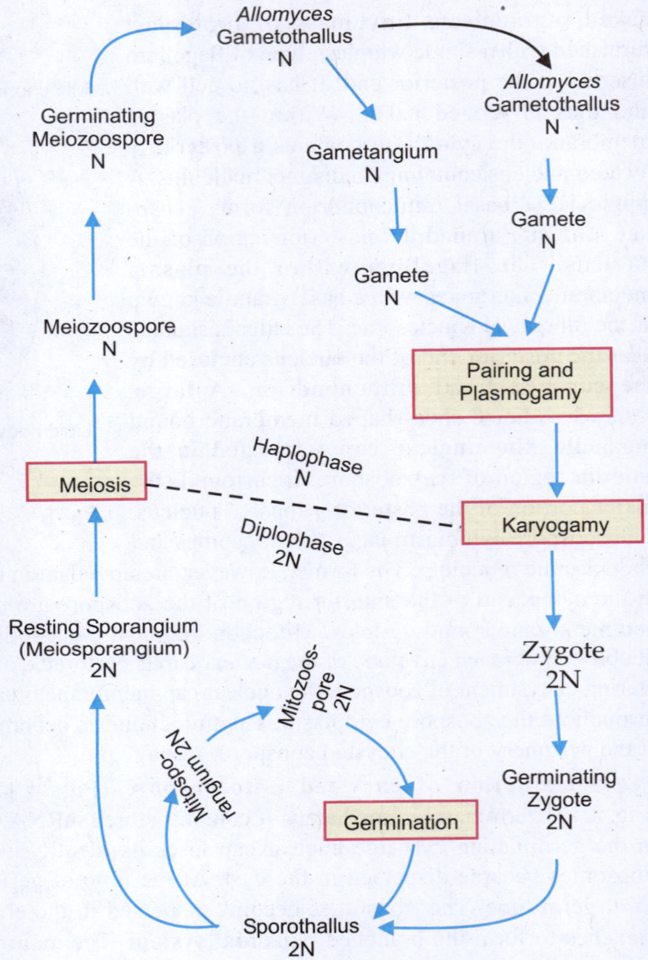

Fig. 4.14. *Allomyces.* Word diagram of the life cycle.

1. Gametothallus. It is **homothallic** and at a certain stage of maturity bears **gametangia** which are distinguished as male and female. Both produce gametes which are motile, uniflagellate and uninucleate.

The planogametes are unequal in size. The male gamete is about half the size of the female. The gametothallus, male and female gametangia and gametes which are all haploid structures constitue the **gametophyte generation** or **haplophase.** Meiozoospores are the pioneer structures of this phase and the anisoplanogametes, the last. The anisoplanogametes fuse in pairs to form the diploid zygote.

2. Sporothallus. The diploid zygote germinates immediately to produce an alternate plant in the life cycle. It is the diploid **sporothallus.** At maturity the sporothallus produces two typtes of sporangia and not the gametangia. These are thin-walled **mitosporangia** and thick-walled, pitted resting or resistant **meiosporangia**. The mitosporangia produce diploid zoosores. Since the nuclear division involved in their differentiation is mitotic, it is appropriate to call them **mitozoospores.** The latter, on germination, give rise to the diploid sporothalli. The mitozoospores thus serve to reduplicate the diploid generation or diplophase in the life cycle and thus play no role in the phenomenon of alternation of generations.

The nuclei in the meiosporangium prior to its germination undergos **meiosis**. The meiosporangial protoplast containing the haploid nuclei undergoes cleavage to form haploid zoospores known as the **meiozoospores** or **gonozoopores** in contradiction to the mitozoospores. The meiozoospore, on germination, produces an alternate plant in the life cycle called the gametothallus. The zygote, sporothallus, mitosporangia with meiozoospores and meiosporangia, which are all diploid, constitute the **sporophyte generation** or **diplophase.** The diploid zygote is the pioneer structure of this phase and meiosporangia the last. The meiozoospores produced in the meiosporangia are the pioneer structures of the future haplophase.

The gametothallus bearing the gametangia and the sporothallus bearing the sporangia occur regularly one after the other in the single sexual cycle of *Allomyces.* This phenomenon is called **alternation of generations.** Since the two alternating individuals (gametothallus and sporothallus) in the life cycle are morphologically similar, the alternation of generations in *Allomyces* is termed **isomorphic.** The life cycle of *Allomyces* in which there is alternation of two vegetative individuals, the gametophyte or sporophyte with sporogenic meiosis is called **Haploid-diploid** or **diplobiontic.**

On the basis of his extensive studies of the collections obtained from all over the world Emerson (1941) divided the genus **Allomyces** into there sub genera : **Euallomyces, Cystogenes** and **Brachyallomyces.** They differ from each other in exhibiting alternation of generation. While **Brachyallomyces** shows no alternation of generation because of the absence of gametophytic stage, **Euallomyces** shows isomorphic alternation of generation and **Cystogenes** shows heteromorphic alternation of generation having the gametophyte reduced to a small tiny cyst.

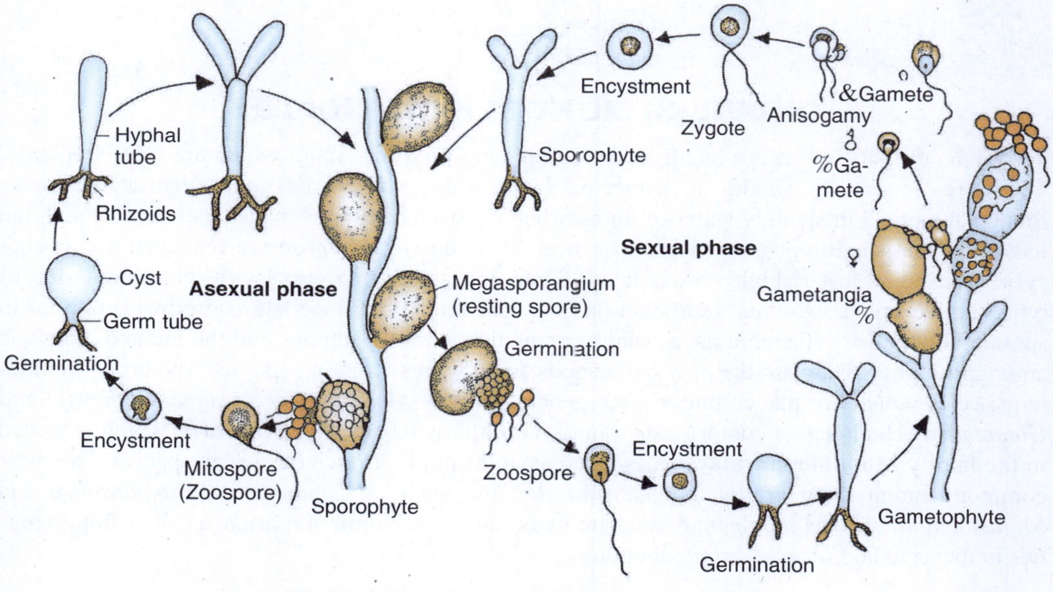

Fig. 4.15. Pictorial life cycle of *Allomyces arbuscula*

Affinities and taxonomic position of Blastocladiales. The presence of distinct alternation of generations and resistant sporangia and lack of sexual resting spores are features which set the Blastocladiales apart from all other fungi. In the greater development of the thallus (soma) composed of true hyphae, increase in size, visible to the naked eye and bearing reproductive organs and increased complexity of sexual reproduction, anisogamous gametes borne in separate distinguishable gametangia (male or female), they show distinct advance on the chytrids. However, the flagellate stages with one opisthocont flagellum, chitinous nature of cell wall and more or less tubular mycelium are features which are suggestive of a closer affinity with the Chytridiales from which they appear to have originated. Thus, Blastocladiales shows advancement over chytridiales in (*i*) Development of true hyphae in some genera, (*ii*) Differentiation of male and female gametangia and gametes and (*iii*) presence of distinct alternation of generations in some genera.

The taxonomic position *of Allomyces*, according to the consensus of opinion, is given below:

Division	:	**Mycota**	*Family*	:	**Blastocladiaceae**
Sub-division	:	**Eumycotina**	*Genus*	:	*Allomyces*
Class	:	**Chytridiomycetes**	*Species*	:	*arbuscula*
Order	:	**Blastocladiales**			

Alexopoulos and Mims (1979) place *Allomyces* in the Superkindgom Eukaryonta, Kingdom Myceteae. Division Mastigomycota, Subdivision Haplomastigomycotina, Class Chytridiomycetes, Order Blastocladiales, Family Blastocladiaceae.

Superkingdom	:	**Eukaryonta**
Kingdom	:	**Myceteae**
Division	:	**Mastigomycotina**
Sub-division	:	**Haplomastigomycotina**
Class	:	**Chytridiomycetes**
Order	:	**Blastocladiales**
Family	:	**Blastocladiaceae**
Genus	:	*Allomyces*
Species	:	*arbuscula*

ORDER (3) MONOBLEPHARIDALES

General features. It is a small order comprising saprobic fungi which are mostly aquatic. They are commonly found in winter on dead, water soaked but not decorticated twigs found submerged in shallow water of undisturbed ponds. The fungus makes its entry through the lenticels. The somatic phase consists of a true mycelium. The hyphae are branched and coenocytic. The protoplast is highly vacuolated. Septa appear in connection with the delimitation of reproductive organs. Asexual reproduction is by posteriorly uniflagellate zoospores produced in sporangia which are filamentous. Sexual reproduction is heterogamous and the method is unique among the fungi. They are the only oemaogous fungi with swimming sperms. The order includes ten species which are placed under three genera, namely *Monoblepharis, Monoblepharella* and *Gonapodya.* The last two comprise the family **Gonapodyaceae.** *Monoblepharis,* which is placed in the family **Monoblepharidaceae,** is taken as an example. It includes seven species. The most common among them are *M. polymorpha, M. hypogyna, M. sphaerica, M. macrandra* and *M. fasciculata.* The Monoblepharidales are of no direct economic importance. Their importance lies in their method of sexual reproduction.

FAMILY : MONOBLEPHARIDACEAE

Type : Monoblepharis (Fig. 4.16 A)

Occurrence. *Monoblepharis* is a saprophytic fungus commonly found in still water in the form of small pustules on decaying plant remains such as twigs submerged in the clear water of ponds and tanks. The growth of the fungus is favoured by low temperatures of about 3°C.

Mycelium (Fig. 4.16 A). The mycelium is filamentous and hypha-like. It is usually unbranched and attached to the substratum by tuft of rhizoidal hyphae. The hypha is rigid and coenocytic. The hyphal wall gives reactions of cellulose. According to Hawker (1967), the wall contains chitin. The protoplast is vacuolated and presents a foamy appearance—a feature characteristic of the entire order. The vacuolation is very regular. The cytoplasm forms a network. The nets enclose

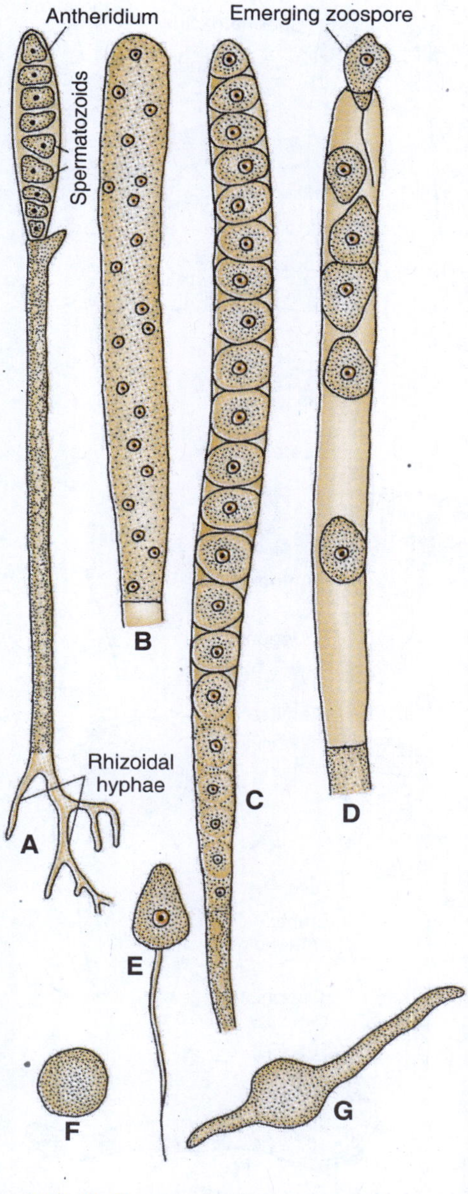

Fig. 4.16 (A-G). *Monoblepharis*. A, mycelium with a terminal antheridium and a small hyphal outgrowth below it indicating the initiation of the oogonium; B, young zoosporangium containing a multinucleate homogeneous protoplast; C, zoosporangium with partially differentiated zoospores; D, zoosporangium with emerging zoospores; E, liberated zoospore; F, Encysted zoospore; G, germination of encysted zoospore (A, diagrammatic, B after Laibach, C Hawker, D Woronin, and E after Saprrow).

vacuoles of regular size and form. The nuclei lie at equal distance from one another in a single row. The septa remain supressed during the vegetative phase.

Reproduction. The reproductive organs (sporangia and sex organs) are developed at the tips of hyphae on the same plant.

(*i*) Asxeual Reproduction (Fig. 4.16 B-E). It occurs within the range of 8° to 12° C and takes place by means of uniflagellate, opisthocont zoospores produced endogenously within elongated sporangia in large numbers. The sporangium is a narrow, cylindrical or club-shaped terminal cell delimited from the parent hyphae by a cross wall. The sporangia occur singly. At the time of sporangium development, the terminal portion of a hypha becomes slightly swollen—not in any way much larger than the diameter of a parent hypha. It is multinucleate and contains homogeneous protoplast. Soon the terminal slightly inflated portion is cut off by a basal wall to function as a sporangium (B). The multinucleate protoplast of the sporangium becomes differentiated, into uninucleated daughter protoplasts by cleavage (C). Each protoplast rounds off to form a nearly spherical or pyriform zoospore. The zoospores move about in an amoeboid manner within the sporangium. The tip of the mature sporangium dissolves to form a circular apical aperature (D) through which the zoospores finally escape one by one. A second sporangium may, sometimes, be formed immediately below the first. The liberated zoospore is a spherical or pyriform uniflagellate cell (E). The single flagellum is much longer than the diameter of the cell. It is of a whiplash type with a painted tip and is inserted at the posterior end. The zoospore with a flagellum attached at the posterior end is **opisthocont.** The zoospore of *Monoblepharis* is thus uniflagellate and opisthocont. On liberation, the zoospores swim about for a while and then come to rest on a suitable substratum. In the quiescent state, the zoospore retracts the flagellum, becomes rounded and secretes a wall around it. The encysted zoospore (F) germinates by two germ tubes (G). One of these grows into a rhizoidal system which fixes the germling to the substratum. The second germ tube forms the mycelium which grows out into the water.

Sexual Reproduction (Fig. 4.17). It is oogamous. The same mycelium that produces the sporangia bears the sex organs (antheridia and oogonia) at a higher temperature (20°C) at the tips of the hyphae (A). Prior to the formation of sex organs, the hypha gets swollen at the tip. The inflated terminal portion is cut off by a basal wall. The cell thus formed matures into an antheridium (B) or oogonium depending upon the species. The position of sex organs thus varies in different species. For example, in *M. polymorphia, M. fasciculata* and *M. insignis,* the antheridium is terminal and the oogonium lies below it on the same hypha (B). During further development, the oogonium increases in size and becomes distended in such a way as to project laterally at right angles to the antheridium. The antheridium becomes pushed to one side and appears to be arising from the surface of the oogonium (C). *M. sphaerica* and *M. hypogyna* have a terminal oogonium (Fig. 4.18 A). The antheridium develops below it (Fig. 4.18 B). *M. macrandra* has the two kinds of sex organs borne on the tips of separate hyphae.

Antheridium (Fig. 4.17A). The antheridium is an elongate, narrow, somewhat club-shaped structure situated at the tip of a hypha. It contains a multinucleate protoplast and is separated from the supporting hypha by a septum (Fig, 4.17 B). Towards maturity its protoplast becomes differentiated into 4-8 posteriorly uniflagellate male gametes or **antherozoids** (B). The single flagellum is of whiplash type and several times longer than the cell itself. When mature the antherozoids are released through a small apical aperture (Fig. 4.17 C). The motile male gametes are smaller in size than the zoospores and show a greater tendency to amoeboid movements, otherwise they resemble the zoospores.

Oogonium (Fig. 4.17). The portion of the hypha immediately below the antheridial septum puts out a small hyphal outgrowth (B). The latter swells into a club-shaped structure which projects to one side ar right

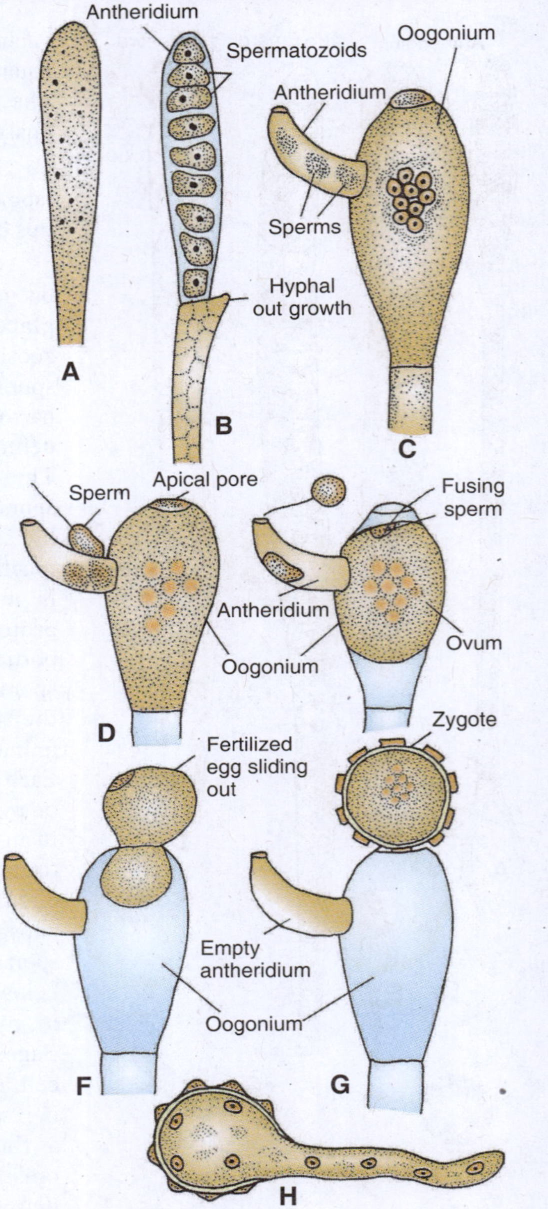

Fig. 4.17 (A-H). *Monoblepharis polymorpha.* A, Hypha inflated at the tip prior to the development of sex organs; B, Terminal antheridium with a hyphal outgrowth below it indicating initiation of oogonium; C, hypha bearing a young oogonium and a dehisced antheridium with three sperms; D-E successive stages in the liberation of sperms, and fertilization of ovum; F, fertilized egg sliding out of the oogonium through the apical pore; G, zygote still attached to the pore in the oogonial wall; H, germination of zygote (A-B after Hawker, C-G based on Sparrow and H after Laibach).

angles to the antheridium (C). The clavate outgrowth is separated by a septum. It is the young **oogonium** which is uninucleate. The oogonium increases in size and becomes globose at maturity. The antheridium appears to be borne on it. The mature oogonium is much larger than the antheridium. Later the oogonial protoplast recedes from the oogonial wall to form a single, central, spherical uninucleate **oosphere** which contains large oil drops (E). The mature oogonium develops a pore at its apex (D).

Fertilisation (Fig. 4.17 D-E). The liberated sperms swim to the oogonium and creep over the wall to reach **apex** (E).One of them crawls through the apical pore (E) and fuses with the oosphere. In some cases the fusion of the male and female nuclei is delayed until oospore is partially clothed with a thick wall.

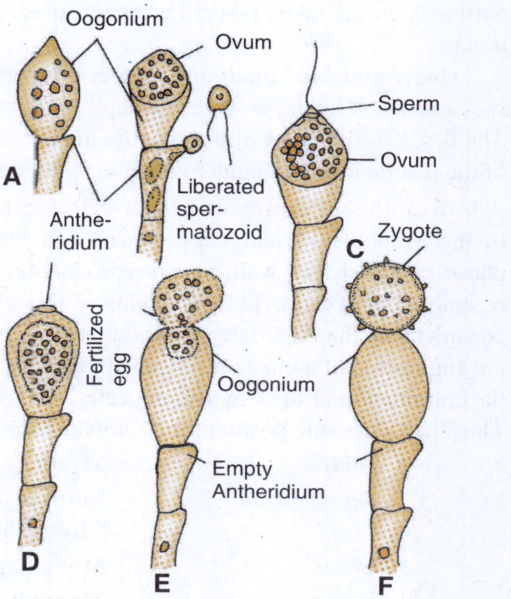

Fig. 4.18 (A-F). *Monoblepharis hypogyna*. A. hypha with the antheridium below the terminal oogonium; B, ripe antheridium with sperms escaping; C, Terminal oogonium with the sperm having entered through a pore at its apex and fusing with the ovum; D, fusion of the sperm and ovum completed; E, fertilized egg sliding out of the apical pore in the oogonial wall; F, zygote still attached to the pore; (A after Hawker and B-E after Woronin).

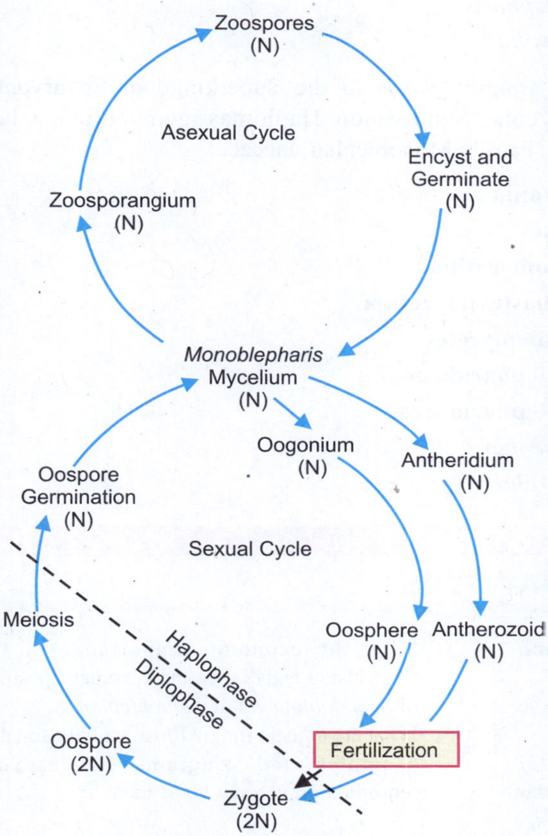

Fig. 4.19. Word diagram of the life cycle of *Monoblepharis* sp.

Oospore (Fig. 4.17 F-G). The fertilized egg, in some species, remains inside the oogonium. In others (*M. polymorpha*), a few minutes after plasmogamy it moves to the tip of the oogonium and comes out to undergo maturation (F). However, it remains attached to the pore in the oogonial wall by a hyaline collar. The male and the female nuclei fuse. The fusion of the male and female nuclei is termed *karyogamy*. After karyogamy, it enters the oogonium, secretes a thick wall around it and becomes the *oospore*. Characteristic blister-like projections develop on the oospore wall (G). Soon it is shed (*M. polymorpha*) and enters upon a period of rest which coincides with summer drought or winter cold. In *M. fasciculata*, the oospore remains attached to the oogonium for a considerable time. Occasionally parthenogenesis takes place. Parthenospores are smooth-walled. They remain within the oogonium.

Under suitable conditions (cooler temperature) the oospore germinates. The thick oospore wall cracks. A hypha emerges through the rupture and grows into a new haploid mycelium (H). The first two divisions of the oospore nucleus are considered to be **meiotic** (Laibach, 1927). Each of the resultant four daughter nuclei may divide once or twice.

Taxonomic Position and Inter-relationship of Monoblepharidales. The inter-relationships of the Monoblepharidales are obscure. In the mycelial or hyphal organisation of the somatic phase, cellulosic cell wall, sexual reproduction advanced as far as oogamy, the Monoblepharidales resemble Oomycetes. The opponents of this view hold that the resemblance is superficial. They consider that the flagellate stages (zoospores and sperms) with one posteriorly inserted flagellum are suggestive of a closer affinity with the Chytridiales and Blastocladiales. In fact they represent the culmination of the Chytridiomycetes. The consensus of opinion favours the second hypothesis. Thus the taxonomic position of *Monoblepharis* according to the majority view is as follows :—

Division	:	**Mycota**
Sub-Division	:	**Eumycotina**
Class	:	**Chytridiomycetes**
Order	:	**Monoblepharidales**
Family	:	**Monoblepharidaceae**
Genus	:	*Monoblepharis*
Species	:	*polymorpha*

Alexopoulos and Mims (1979) place *Monoblepharis* in the Superkingdom Eukaryonta, Kingdom Myceteae, Division Mastigomycota, Subdivision Haplomastigomycotina, Class Chytridiomycetes, Order Monoblepharidales, Family Monoblepharidaceae.

Superkingdom	:	**Eukaryonta**
Kingdom	:	**Mycetae**
Division	:	**Mastigomycotina**
Subdivision	:	**Haplomastigomycotina**
Class	:	**Chytridiomycetes**
Order	:	**Monoblepharidales**
Family	:	**Monoblepharidaceae**
Genus	:	*Monoblepharis*
Species	:	*polymorpha*

QUESTIONS

ESSAY TYPE

1. Describe the general characteristics and classification of class Chytridiomycetes.
2. Give an illustrated account of the life cycle of *Synchytrium endobioticum*.
(*Kumaon, 2004; Rohilkhand, 2002*)
3. With the help of sketches describe asexual phase of *S. endobioticum* or *Monoblepharis*.
4. State the economic importance of the Chytridiales and discuss the systematic position of *S. endobioticum* or *Monoblepharis*.
5. What are aquatic fungi? Illustrate with suitable examples and diagrams the asexual reproduction in any two of them.
(*Kanpur M.Sc. I, 1993*)

6. What do you understand by Holocarpic parasite? Describe the structure and reproduction of any holocarpic fungus studied by you.
7. Describe the life cycle of *Synchytrium* with the help of labelled diagrams. (*Awadh, 1993,1999*)
8. Describe the structure and life cycle of a holocarpic endoparasite studied by you.
9. Describe the thallus structure and reproduction in *Monoblepharis*.
10. Give an illustrated account of the life cycle of *Synchytrium*. Justify that it is a holocarpic fungus. (*Kumaon, 1999; Rohilkhand, 1993*)
11. Give the illustrated life cycle of the fungus causing black wart of potato. (*Kumaon, 1997*)
12. Name the causal organism of black wart disease of potato. Describe the life cycle and methods of control. (*Rohilkhand, 1995*)
13. Describe structure and reproduction of the fungus that causes wart disease of potato. (*Bhagalpur, 1990; Ranchi, 1991*)
14. Discuss the range of thallus in lower fungi. (*Rohilkhand M.Sc., 2006*)

SHORT ANSWER TYPE

15. Write short notes on :
    (*a*) Aquatic fungi (*Kanpur M.Sc. I, 1998*)
    (*b*) *Synchytrium* as a pathogen (*Kanpur, 1996; Awadh 1991, 1996; Kumaon, 2003*)
    (*c*) Wart disease of potato (*Awadh, 1998; Kumaon, 1996;. Magadh, 1993*)
    (*d*) *Monoblepharis*
    (*e*) *Allomyces*
    (*f*) Holocarpic and Eucarpic fungi.
16. Why are chytrids considered most primitive of the phycomycetes? (*Kanpur, M.Sc., 1990*)
17. Descibe the systematic position and diagnostic features of *Synchytrium*. (*Kanpur, 1998, 1996, 2005*)
18. Give the structure and reproduction of any holocarpic fungus. (*Kanpur, 1997*)
19. Explain the structure of thallus in *Synchytrium*? (*Awadh, 1992*)
20. With the help of labelled diagrams, describe the life cycle of *Allomyces*.
21. Compare the sexual reproduction in *Synchytrium, Allomyces* and *Monoblepharis*.
22. With the help of labelled diagrams describe the following :
    (*a*) sorus and prosorus of *Synchytrium* (*Rohilkhand, 1991, 1992*)
    (*a*) life cycle of *Synchytrium*
23. Describe sexual reproduction in a holocarpic fungus with the help of diagrams only. (*Kanpur, 2006*)

OBJECTIVE TYPE

24. **Fill in the blanks :**
    (*i*) The causal organism of the wart disease of potato is ....
    (*ii*) The genus *Synchytrium* reproduces sexually through ....
    (*iii*) The somatic phase of *Synchytrium* is a .... amoeboid mass of protoplasm.
    (*iv*) The zoospores in *Synchytrium* are .... and ....
    (*v*) The sexual reproduction in *Allomyces* takes place through ....
    (*vi*) The sexual reproduction in *Monoblepharis* is through ....
    (*vii*) The members of the class chytridiomycetes are ....
    (*viii*) The thallus in *Allomyces* .... and called ....
    (*ix*) The life cycle in *Allomyces* shows the presence of .... types of sporangia.
    (*x*) *Monoblepharis* is a .... fungus.
25. Select the correct answer :
    (*i*) Which of the following has two types of sporangia in its life cycle?
    (*a*) *Synchytrium* (*b*) *Allomyces*
    (*c*) *Monoblepharis* (*d*) *Albugo*
    (*ii*) Which of the following is a holocarpic endoparasite?
    (*a*) *Synchytrium* (*b*) *Albugo*
    (*c*) *Puccinia* (*d*) *Ustilago*
    (*iii*) The zoospores in *Synchytrium* are
    (*a*) posteriorly uniflagellate
    (*b*) anteriorly uniflagellate
    (*c*) anteriorly biflagellate
    (*d*) laterally biflagellate

(*iv*) The sexual reproduction in lower fungi is usually through
(*a*) planogametic fusion
(*b*) gametangial contact
(*c*) somatogamy
(*d*) gametangial copulation

(*v*) Wart disease of potato is caused by
(*a*) *Synchytrium endobioticum*
(*b*) *Alternaria solani*
(*c*) *Phytophthora infestans*
(d) *Puccinia graminis*

(*vi*) Two types of thalli—the gametothalli and the sporothallic and present in the genus
(*a*) *Synchytrium*
(*b*) *Allomyces*
(*c*) *Monoblepharis*
(*d*) *Albugo*

(*vii*) Sexual reproduction in *Allomyces* takes place through
(*a*) Isogamy (*b*) Anisogamy
(*c*) Oogamy (*d*) Somatogamy

(*viii*) Mitospores in *Allomyces* produce as a result of germination
(*a*) Gametothallus
(*b*) Sporothallus
(*c*) Either gametothallus or sporothallus
(*d*) Neither gametothallus nor sporothallus

(*ix*) The reduction division in the life cycle of *Allomyces* takes place during germination of
(*a*) Spores (*b*) Mitospores
(*c*) Meiospores (*d*) Zygote

(*x*) The reduction division in the life cycle of *Monoblepharis* occurs during the germination of
(*a*) spores (*b*) oospores
(*c*) zygote (*d*) zoospores

(*xi*) Which of the following is intracellular parasite ?
(*a*) *Albugo candida*
(*b*) *Synchytrium*
(*c*) *Saprolegnia*
(*d*) *Aspergillus*

(*xii*) Mycelium is absent in
(*a*) *Cystopus*
(*b*) *Synchytrium*
(*c*) *Aspergillus*
(*d*) *Saprolegnia*

(*xiii*) Holocarpic fungus is
(*a*) *Albugo* (*b*) *Agaricus*
(*c*) *Synchytrium* (*d*) *Puccinia*

(*xiv*) Which of the following is an intracellular parasite ?
(*a*) *Albugo candida* (*b*) *Synchytrium*
(c) *Saprolegna* (*d*) *Aspergillus*

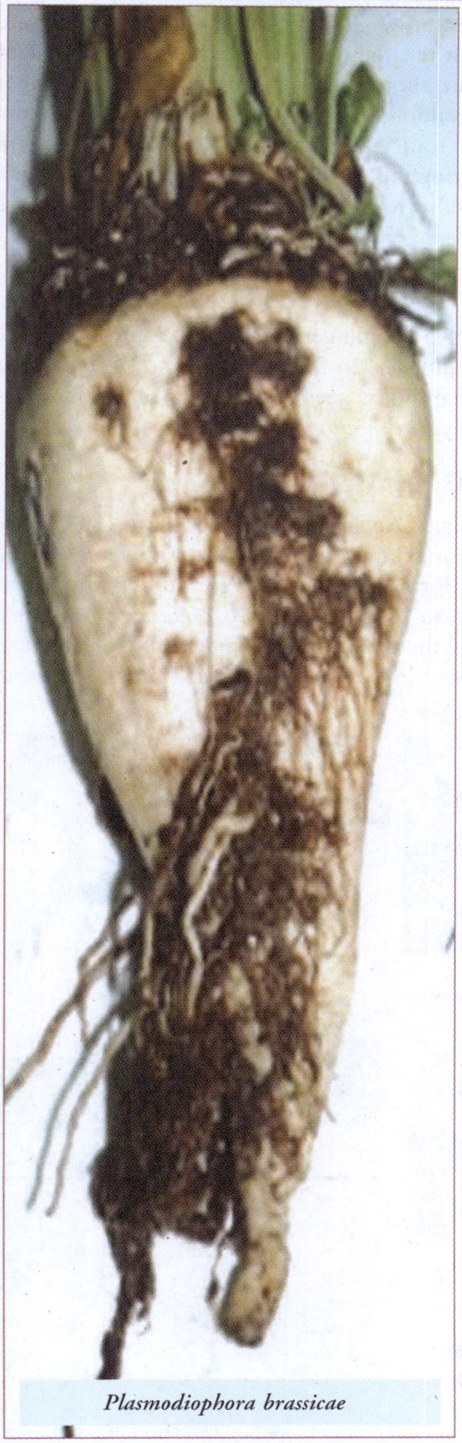

Plasmodiophora brassicae

CHAPTER 5

SUBDIVISION – EUMYCOTINA

Class : Plasmodiophoromycetes

- General Characteristics
- Order : Plasmodiophorales
- Family : Plasmodiophoraceae
- *Plasmodiophora brassicae*
- Life Cycle

General Characteristics. The representatives of this class are obligate endoparasites of the seed plants (Cabbage, *Nasturtium*), ferns (*Azolla*), algae (*Vaucheria*) and Fungi (*Achlya, Saprolegnia*). The somatic phase is a moving, naked, multinucleate mass of protoplasm feeding in a amoeboid fashion. It is known as the *plasmodium*. It grows within the host cells and is *holocarpic*. Two types of plasmodia namely primary or gametangial plasmodia and secondary or cystogenous plasmodia occur in the life cycle. The former are haploid and the latter diploid. The entire protoplasm of the primary plasmodium divides to give rise to thin-walled gametangia, each of which contains a single motile biflagellate, uninucleate structure which functions as a gamete. In older terminology, it is called a zoospore and the plasmodium bearing it as zoosporogenous plasmodium. The cystogenous plasmodium divides by meiosis to form thick-walled resting meiospores. Some scientists call them resting sporangia or cyst. On germination each cyst gives rise to a zoospore or cyst zoospore. Both the gametes and the cyst zoospore are uninucleate, biflagellate and similar in structure. Both the flagella are of whiplash type but of unequal length *(heterokont)* and inserted anteriorly, the shorter near the longer one. The longer flagellum terminates in an end piece. The shorter flagellum lacks the end piece. It has blunt tip. When the swarm cell is in motion, the longer flagellum trails behind and the short blunt flagellum is directed forward. The nuclear division occurring during the somatic phase of both gametangial and cystogenous phases is *cruciform.* Sexual reproduction when present is isogamous and takes place by planogametic copulation. The class includes a single **order Plasmodiophorales.**

ORDER : PLASMODIOPHORALES

The resting spores or cysts also termed the resting sporangia germinate in the soil. The spore or sporangial wall is made of chitin. It contains no cellulose. No fructifications are formed. The occurrence of **akaryote phase** in which the nuclear body seems to disappear, is another characteristic feature. The order includes a single family *Plasmodiophoraceae.*

FAMILY : PLASMODIOPHORACEAE

The family has the same characteristics as the class. According to Waterhouse (1973), the family includes nine genera, identified on the basis of maximum number of resting spores in the cystosorus, shape of the cystosorus and the sporangial forms, size, host range and hypertrophy of the host. The genera are : *Plasmodiophora*, *Tetramyxa*, *Octomyxa*, *Sorosphaera*, *Sorodiscus*, *Spongospora*, *Ligniera*, *Woronina* and *Polymyxa*. All of them constitute a small group of obligate, intracellular parasites (endoparasites) with a complicated life cycle. They attack fungi, algae but

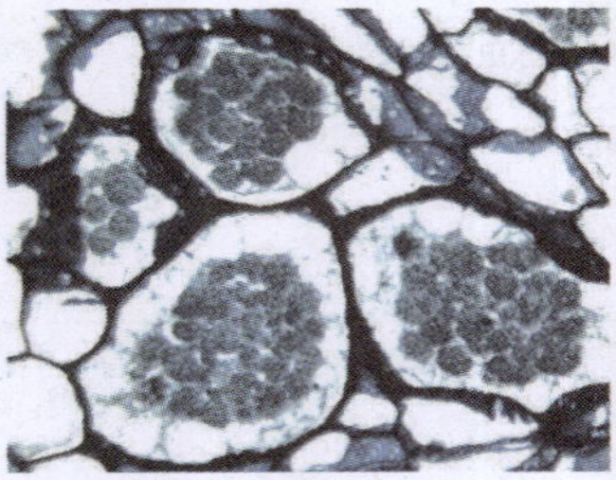
Plasmodiophora

Tetramyxa

Sorosphaera

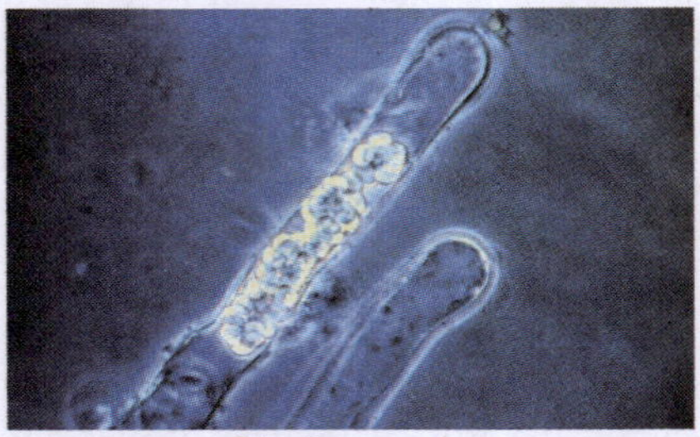
Ligniera

Spongospora,

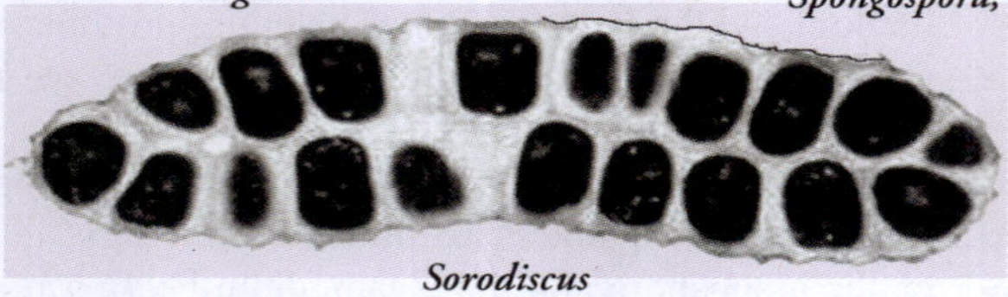
Sorodiscus

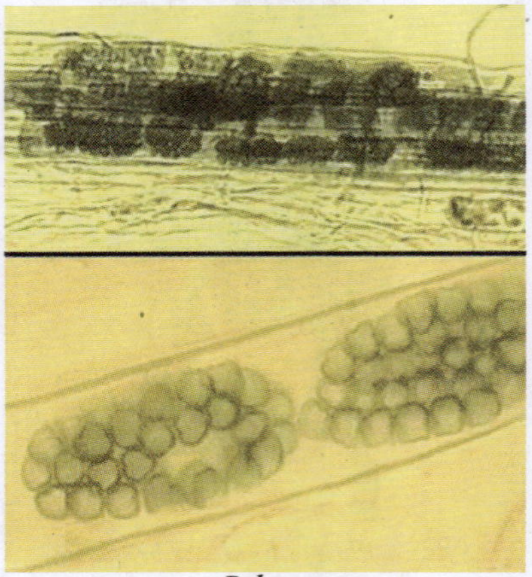
Polymyxa

mostly vascular plants. The presence of the intracellular parasite causes abnormal proliferation of host cell through hypertrophy and hyperplasia. The two important genera of this family are *Plasmodiophora* and *Spongospora. Plasmodiophora brassicae* causes the economically important *club root* disease of cabbage and other members of family Cruciferae both wild and cultivated. *Spongospora subterranea* is responsible for *powdery scab* disease of potato tubers. Both the diseases cause considerable damage to cabbage and potato crops respectively. *P. brassicae* which is perhaps the best known member of family Plasmodiophoraceae is discussed here in detail.

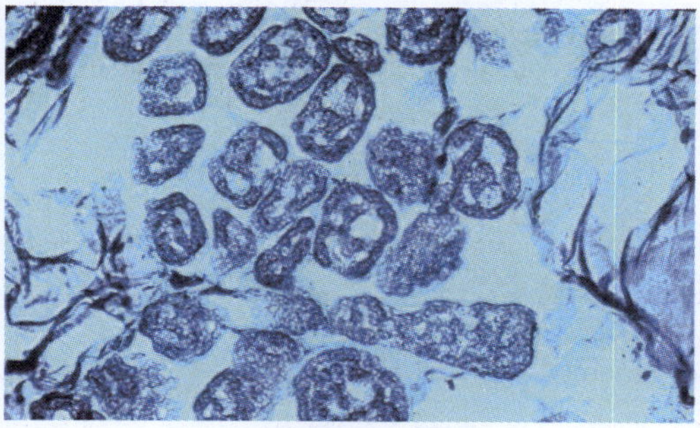

Spongospora subterranea.

Powdery Scab Disease of Potato Tuber.

PLASMODIOPHORA BRASSICAE

Occurrence. The fungus is an obligate endoparasite. It was first discovered, as causal organism of *club root* or *finger* and *toe disease* of all crucifers including many wild and economically important species of *Brassica* by Woronin in 1878. The disease is one of the most widespread and the causal agent (*P. brassicae*) is one of the most intensely studied pathogen.

Symptoms of the Disease. The infected roots bear malformations which take on a variety of shapes. In cabbage and its close relatives, the malformations are spindle-shaped in the case of single point infection. Multiple infections in close proximity induce modifications of spindle type. If an infection takes place at an early stage, the clubs or malformations are formed on the main root and the disease is named *club root disease.* The development of lateral roots is suppressed. The late infections produce galls on the lateral branches suggesting an alternative name of *'Finger and Toe'* disease. The above ground symptoms of the disease are noticeable when it has progressed to a considerable extent. The drooping of leaves at noon or bright days is the first sign. Subsequently it leads to permanent wilting or retarded growth.

Fig. 5.1. *Plasmodiophora brassicae.* Infected roots that have become swollen due to hypertrophy.

LIFE CYCLE

Cook and Swartz (1930) showed that the life cycle of *P. brassicae* comprises two distinct phases, the **haplophase** (primary phase) and the **diplophase** (secondary phase). The former occurs in the *root hairs* and the latter in the cells of the main root cortex of the susceptible host plant. Each phase includes a number of successive stages which are described below:-

A. Haplophase. The resting spores or sporangia (cysts) which are differentiated by meiosis from the diploid protoplast of the sporothallus (secondary plasmodium) are pioneer structures of the haplophase which occurs in the root hairs.

1. *Fine Structure of Resting spore or sporangium* (Fig. 5.2). It is a tiny hyaline, spherical, uninucleate, haploid structure up to 4.0 μ m in dia. According to Williams and Menabola (1967) the sporangial wall is differentiated into 3 layers (A) but in many cases it showed only 2 layers. The outer layer is electron-dense and granular. It included the surface spinules. The middle layer is less electron-dense. Yukawa and Tanaka (1979) described a circular **germ pore** in the sporangial wall with the middle layer (inner in the 2 layered sporangial wall) thickened to form a circular electrolucent plug-like structure in the region of the pore orifice (B). It is surrounded by a ring of electron-dense material contiguous with the inner wall layer. Within the sporangial wall is the sporangial membrane which completely invests the cytoplasm replete with ribosomes. The other organelles present in the cytoplasm are mitochondria, endoplasmic reticulum and dictyosomes. It has a single haploid nucleus. Numerous lipid globules of various sizes is another conspicuous feature of the spore (sporangial) cytoplasm. The mature spores are set free, when the clubs undergo decay. In the soil they germinate to initiate the life cycle.

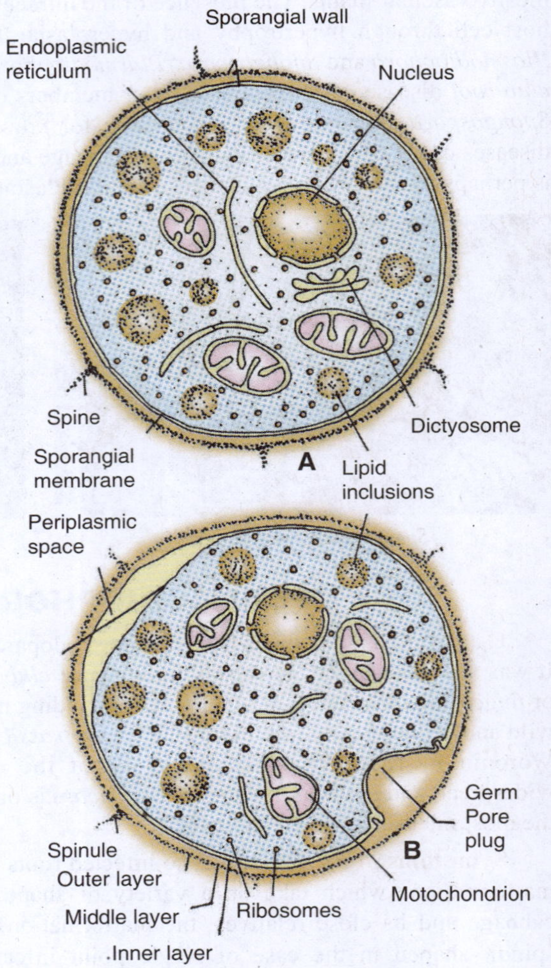

Fig. 5.2 (A-B) *Plasmodiophora brassicae*. Diagrammatic representation of the fine structure of resting spore or sporangium (A, Based on Williams and Menabola and B, on Yakawa and Tanaka.).

2. *Germination of Resting Sporangium or Spore* (*cyst*) (Fig. 5.3). The spores of *P. brassicae* may germinate immediately or remain dormant in the soil for several years.

(*a*) *Conditions affecting germination.* According to Hofing (1930), the most important factor for germination is maturity. Thus the spores or sporangia from older decaying galls germinate faster than those from the young galls. Ellison (1946) noted mutual stimulation of germination by resting sporangia (spores) in masses. Bremer (1924) reported that germination occurred more readily in wet acid than in limed soils. The early suggestion of Chupp (1917) that the presence of host seedling stimulates germination, under certain conditions, has been substantiated by Macfarlane (1959, 67) and Bochow (1965) for *P. brassicae*. According to Macfarlane (1970) presence of host or metabolites secreted by them promote germination. Wellman found 25°C to be the optimum temperature for germination.

(*b*) *Process of germination.* Chupp (1917) and other early workers reported that the spore (A, sporangium) germinated in a favourable medium immediately. It absorbs water and swells considerably (B), particularly on one side along which a slit appears in the spore or sporangial wall. Through the slit, the uninucleate protoplasm emerges (C) as a single swarm cell or zoospore which is spherical or pear-shaped (D) in form. Tommercup and Ingram (1971) observed a single, uninucleate, naked protoplast emerging from an isolated point on the spore wall. Macfarlane (1970) and Aist and Williams (1971) using interference optics observed a *germination pore* in the resting sporangium (or spore) wall through which the protoplast emerges slowly. Recently Yukawa and Tanaka (1979) observed a circular germ pore (ca1 μ m in dia) in the sporangial or spore wall with a thickened plug-like structure (Fig. 5.2 B) in the pore orifice. In their view, this region undoubtedly constitutes the region of the germination pore of the resting spcrangium (or spore) seen by Macfarlane and Aist and Williams. On germination, the plug dissolves to enable the zoospore to emerge through the germ pore. The refractile (lipid) globules in the sporangial cytoplasm gradually disappear as the germination begins. The spore contents become less viscous. A small papilla emerges through the germ pore in the spore wall. It enlarges to form a thin vesicle as the spore contents flow into it. The flagella develop on the spherical or pearshaped zoospore as it completely emerges. It may also be termed as the primary or cyst zoospore.

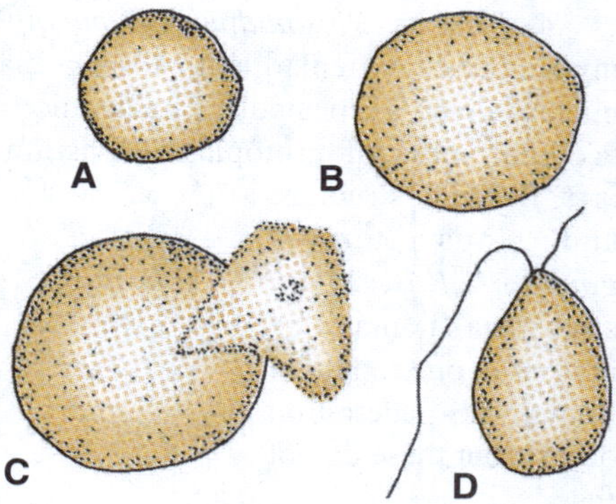

Fig. 5.3 (A-D). *Plasmodiophora brassicae.* Germination of resting spore in the soil. Explanation in the text.

Primary or Cyst Zoospore (D). It is naked and extremely active after emergence for several minutes. Uptill 1934, it was assumed that the zoospore is furnished with a single flagellum inserted at the anterior end (Woronin, 1878, Chupp, 1917, and Wellman, 1930). It is of whiplash type and several times longer than the length of the zoospore. Ledinghan (1934) observed zoospores furnished with two unequal flagella. Ellison (1945) confirmed that the zoospores of *P. brassicae* are biflagellate. Kole and Geilink (1962) observed with electron microscope that each zoospore is furnished with unequal, whiplash type flagella at its anterior end—a long one with an end piece and a short one arising by its side with a blunt end. The latter lacks the end piece.

3. *Primary Infection of Host* (Fig. 5.4.). Infection takes place at the seedling stage in the host. The biflagellate, heterokont (with unequal flagella) primary zoospore (cyst zoospore) swims in a film of water in the soil or creeps over soil particles like an amoeba. According to Ayers (1944), on contact with the root hair of suitable host seedling (Cabbage or other crucifer seedling), the zoospore becomes attached and amoeboid, in this state it is called *myxamoeba* (A_1). The myxamoeba retracts its flagella and enters the root hair directly by a partial dissolution of the root hair wall (A_2). According to Aist and Williams (1971), there is rapid injection of the parasite (encysted cyst zoospored) into the root hairs of the host by means of the special injection apparatus developed in the encysted zoospore. The cyclosis of host cytoplasm sweeps the myxamoebae away from the penetration site soon after injection. The young myxamoebae in the root hairs are separated from the host cytoplasm by two kinds of simul-taneously occurring intrafacial membrane arrangements. One of these is a 7-layered envelope of two closely appressed unit membranes and the second host parasite interface is also a 7 layered envelope.

4. *Primary Plasmodium (Gametothallus,* Fig. 5.4 A) Within the root hair the uninucleate myxamoeba (A_2) enlarges to form a thallus. The protoplasm increases in size and the haploid nucleus divides repeatedly. The divisions are cruciform. The resultant small naked, multinucleate acellular mass of protoplasm constituting the thallus is called the *primary plasmodium*. It is a haploid structure and thus termed *gametothallus*. Many such thalli may be present in one root hair, if multiple penetration occurs, but these do not fuse.

5. *Sexual Reproduction*. (Fig. 5.4 B-C). The haploid primary plasmodium (gametothallus) is concerned with sexual reproduction. After a certain period of growth, the cytoplasm of the gametothallus, while still lodged in the root hair, cleaves into as many uninucleate daughter protoplasts as there are nuclei (usually 4-10) in the plasmodium (B_4). Each daughter protoplast secretes a membrane around it and functions as a **gametangium.** The gametangia are small spherical, thin-walled uninucleate structures. The haploid nucleus of the gametangium undergoes mitosis to form four to eight daughter nuclei (B_5). Each daughter nucleus gathers a small bit of cytoplasm around it and develops into a spindle-shaped motile, biflagellate, isogamete which is smaller in size than the zoospores (C_6). The two flagella are of unequal size and of whiplash type inserted at the anterior end. Thus in one gametangium up to eight isogametes are formed (B_5). They have no cell walls. The whole process takes 2—8 days. The thin gametangium wall bursts (C_6) and liberates the gametes in the root hair. All the stages (A—D) described above occur in the root hairs of the host. These constitute the *haplophase*.

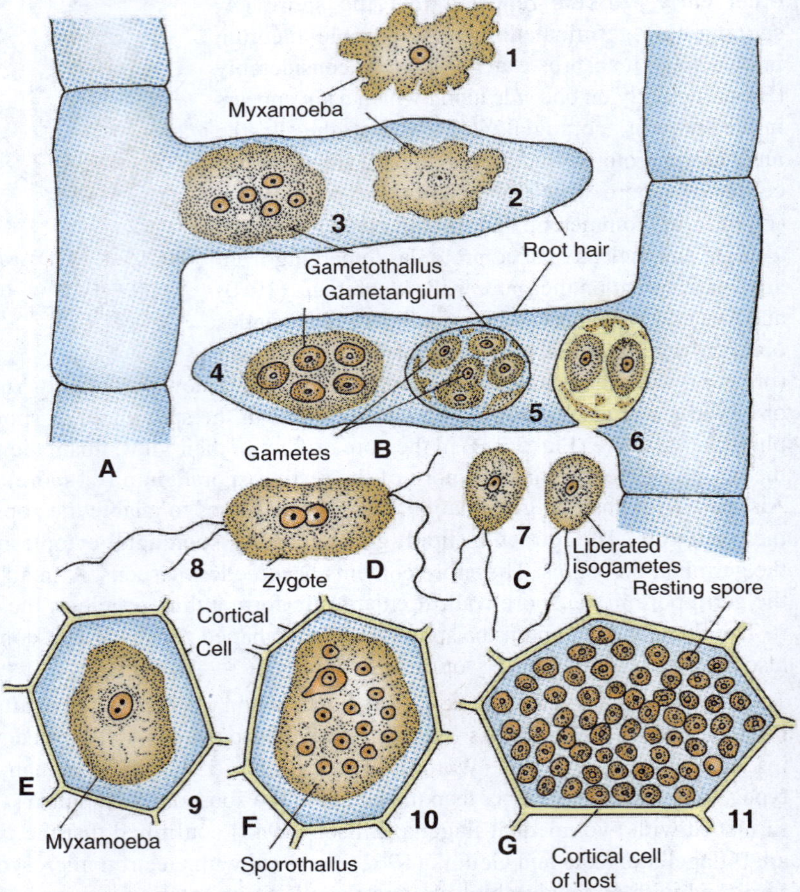

Fig. 5.4. (A-G). *Plasmodiophora brassicae*. A_1, Myxamoeba outside the root hair, A_2, Myxamoeba within the root hair, A_3, Gametothallus; B_4 Differentiation of gametes in the gametangium; B_5, Differentiation of gametes in the gametangium; B_6, liberation of gametes in the soil; C_7, liberated isogametes; D_8, Fusing gametes; E_9 Zygote in the host cortical cell; F_{10}, sporothallus; G_{11}, Resting sporangia or spores in the host cell.

Plasmogamy (Fig. 5.4 B_6, C_7). After liberation, the isogametes copulate in pairs. It has not been established whether gametes from the same gametangium or from different gametangia copulate. It is also not known definitely whether pairing of isogametes takes place in the root hairs or cortical cells after migration. The most widely held view is that the isogametes are released in the soil outside the host through a pore formed at the point of contact of gametangium and the root hair cell wall. Pairing of gametes takes place outside the host in the soil. Pairing of gametes is soon followed by the fusion of their membranes and cytoplasms. This is termed *plasmogamy.* The latter thus results in the formation of binucleate fusion cell (D_8).

Karyogamy. The fusion of two nuclei in the binucleate cell which is termed karyogamy is delayed. It takes place immediately before cyst formation.

B. Diplophase. This phase in the life cycle begins with the formation of quadriflagellate zygotes (D_8). The single nucleus in zygote is diploid.

6. *Zygote* (Fig. 5.4D). The quadriflagellate zygotes are generally found in the cortical cells of the root. As mentioned above it is not definitely known whether the gametes fuse in the root hairs or in the cortical cells after migration. In the former case the flagellated zygotes perforate the epiblema cell walls and reach the meristematic zone of the root. When the quadriflagellate zygote is formed in the soil, it reinfects the root. Once in the root cortex, the zygotes retract their flagella and become amoeboid. The amoeboid zygotes are called **myxamoebae** (E_9). The myxamoebae spread from cell to cell by direct penetration of cell walls. At first they are found in the cortical cells and later in the other tissues of host root and stem.

7. *Secondary Plasmodium (Sporothallus,* Fig. 5.4 F_{10}). Within the invaded cortical cell, the myxamoeba by subsequent growth accompanied by repeated karyokinesis (mitosis) of its diploid nucleus but no cytokinesis, grows to form a relatively large, naked, multinucleate diploid plasmodium. It is called the secondary plasmodium or sporothallus or cystogenous plasmodium. In the mature state, it may occupy the entire lumen of the infected host cell and surround the host cell nucleus (Fig. 5.5 A). It lies immersed in the cytoplasm. The host cell may contain more than one plasmodia.

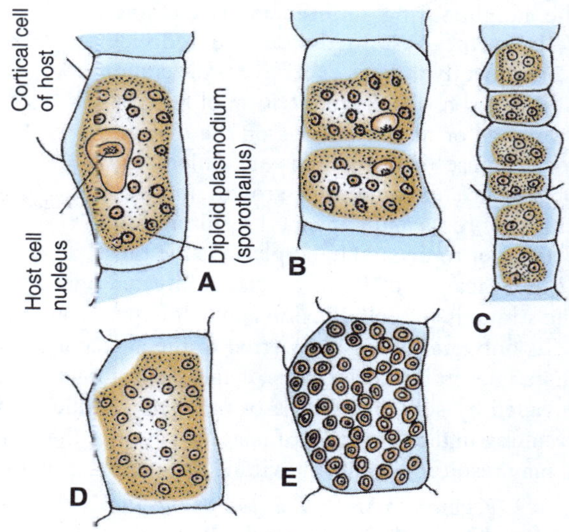

Fig. 5.5 (A-E) *Plasmodiophora brassicae.* A, Multinucleate sporothallus in the host cell, surrounding its nucleus; B, sporothallus divides along with the host cell; C, series of invaded cells forming Krankheitsherd; D-E, Differentiation of resting sporangia or spores.

Ultrastructure of secondary plasmodium (Fig. 5.6). It was investigated by Willams and McNabola (1967). The plasmodium is bounded and thus separated from the host cytoplasm by a smooth, deeply staining osmophilic *membranous envelope* about 250Å thick. It consists of two closely appressed osmophilic membranes. There is no cell wall within the membranous envelope, the growing plasmodial cytoplasm is replete with free **ribosomes.** Besides, it contains abundant *endoplasmic reticulum* and several *mitochondria,* each with a few tubular cristae. *Dictyosomes* occur near the diploid plasmodial nuclei. The presence of numerous prominent *lipid droplets* of variable sizes is another conspicuous feature. Each nucleus of the multinucleate plasmodial cytoplast is bounded by a perforated *nuclear membrane* and contains a single prominent *nucleolus.* A number of microtubules have been reported to occur near each diploid nucleus. The intracellular plasmodium of *P. brassicae* lacks special feeding structures.

8. *Formation of Clubs.* The presence of the secondary plasmodia in the cortical host cells elicits many metabolic responses in the infected cells. One of these is the excessive or abnormal increase in the size of the infected host cells. This is termed *hypertrophy.* Besides hypertrophy, the presence of secondary plasmodia in the cortical cells stimulates them to have active and repeated division. This excessive multiplication of infected or diseased host cells is termed *hyperplasia.* Hypertrophy and hyperplasia of the infected cells stimulates the neighbouring uninfected parenchyma cells of the host to divide and redivide. Consequently there is excessive proliferation of infected host cells in the form of tuberous swellings or malformations on the diseased roots. These malformations are called *clubs* or *galls.* This tumorous growth (galls or clubs) of roots and hypocotyl resulting from cell hypertrophy and hyperplasia is the most striking feature of *Brassica* plants infected with *P. brassicae.* Williams, however, suggested that the stimulus for cell division and cell hypertrophy in the club root galls lies within the parasitized cells only and is not transferred to the adjacent cells. The infected roots act as sinks for growth materials produced in parts of the plant above the ground. The club-tissue is thus likely to be invaded by soft rot bacteria or other rot producing organisms which bring about decay of clubs resulting in the formation of materials toxic to the plant and cause wilting of the tops. Subsequently it may result in permanent wilting or retarded growth.

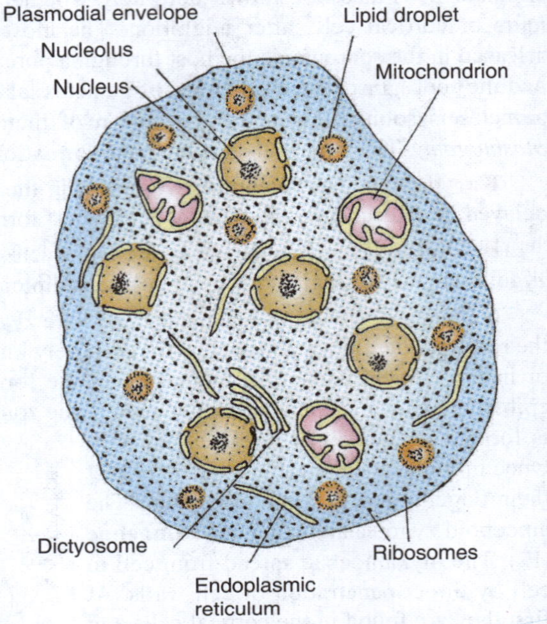

Fig. 5.6. *Plasmodiophora brassicae.* Diagrammatic representation of ultrastructure of sporothallus

9. *Vegetative Multiplication of the Parasite* (Fig. 5.5). At the time of division of infected or diseased host cells (hyperplasia), the secondary plasmodia may also divide (B), a part of it going to each infected daughter cell. The formation of new plasmodia by a vegetative method of fragmentation results in the spread and propagation of the disease. Groups of infected hypertrophied cells become recognisable in the host tissues (C). Woronin (1878) called each such group of hypertrophied cells *Krankhaitsherd* (C). The secondary plasmodium of each enlarged cell of Krankhaitsherd eventually becomes converted into a mass of resting sporangia or cysts (E).

10. *Differentiation of Resting sporangia (Cysts) or spores also called cyst (Sporulation stage* Fig. 5.5 D-E). The mature multinucleate diploid secondary plasmodium (sporothallus) practically fills the entire lumen of the hypertrophied cortical cell (D). It ceases to grow in size. With the cessation of vegetative growth, a new phase in the life cycle ushers in. During this phase, the diploid nuclei of the plasmodium seem to lose their contents and disappear. The chromatin fails to take the usual stain. It is the *akaryote stage.* Soon after the nuclei reappear and the protoplasmic mass of the plasmodium cleaves around each nucleus. The entire plasmodium is thus transformed into a large number of tiny uninucleate, spherical bodies each of which is surrounded by a wall of its own. These are termed the *resting sporangia* (Cysts) or *spores* (E). Till recently the nature of the akaryote stage was an enigma. Lutman (1913) suggested that meiosis takes place during sporangial (Cysts) or spore differentiation. Milovidon (1931) claimed that akaryote stage represents

prophase I of meiosis. Garber and Aist (1979) conclusively showed that the akaryote stage represents the *meiotic prophase I* and the subsequent divisions are remainder of meiosis I and meiosis II. They thus confirmed previous claims both of Lutman (1913) and of Milovidon (1931). According to Garber and Aist's study, the akaryote stage marks the transition between the actually parasitic vegetative stage represented by the diploid plasmodium and the dormant phase represented by the resting sporangia (Cysts) or spores which are haploid structures.

The vacuoles and long slender furrows which are absent in vegetative phase (secondary plasmodium) appear in the plasmodial cytoplasm in the meiotic process. The vacuoles appear in the prophase I and furrows during metaphase I. The latter demarcate the boundary along which cleavage will occur. Both the vacuoles and furrows become prominently conspicuous at the beginning of meiosis II. The plasmodium is still a single unit at this stage. By the end of meiosis II, the plasmodium is divided into a number of tiny daughter protoplasts, each with 2 haploid nuclei. The final step in cytoplasmic cleavage is the division and separation of the tiny binucleate blocks of plasmodial cytoplasm into uninucleate, irregularly-shaped haploid daughter protoplasts which later round up to take on the shape of mature haploid resting sporangia (Cysts) or spores. Each young sporangium or spore is invested by a membranous envelope known as the *sporangial membrane.* The entire secondary plasmodium is thus converted into a sporangial or spore mass. The spores or sporangia (Cysts) in the mass are free from one another but the entire mass of sporangia or spores is surrounded by the common plasma membrane of the host cell. A fine fibrillar matrix termed the *inter-sporangial matrix* remains between the sporangia or spores in the mass. According to Williams and McNabola (1967), granular aggregates of the intersporangial matrix are deposited along the sporangial or spore membrane as small spines termed the *spinules.* Towards maturity, sporangial or spore wall is deposited between the sporangial or spore membrane and the spines. Ultrastructurally, the cyst spore wall is differentiated into 5 layers (Buczacki, *et al.*) (1983). These are: (*i*) an outer layer of loose proteinaceous fibres (l_1), (*ii*) a layer of similar fibres but with lipid granules enmeshed (l_2), (*iii*) a prominantly chitinous wall (l_3), (*iv*) an inner phospholipid containing region (l_4) and (*v*) an inner spore membrane. The chemical analysis of cyst wall shows that the principal components are chitin, protein and lipid. The mature sporangia of spores (Cysts) are set free in the soil when the club roots undergo decay and decomposition by soft rot microoganisms. The chitinous nature of the sporangial or spore (Cysts) wall enable them to

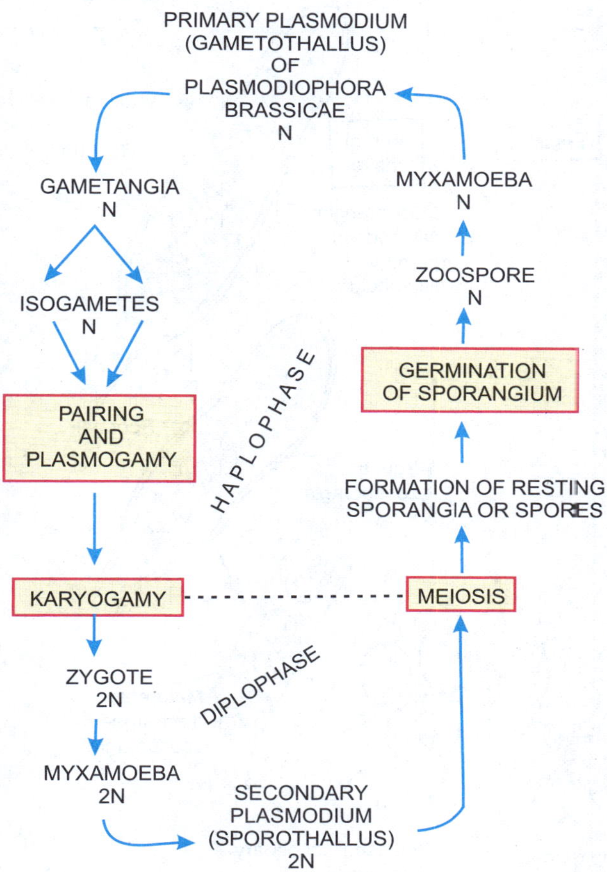

Fig. 5.7. *Plasmodiophora* sp. Graphic representation of the life cycle.

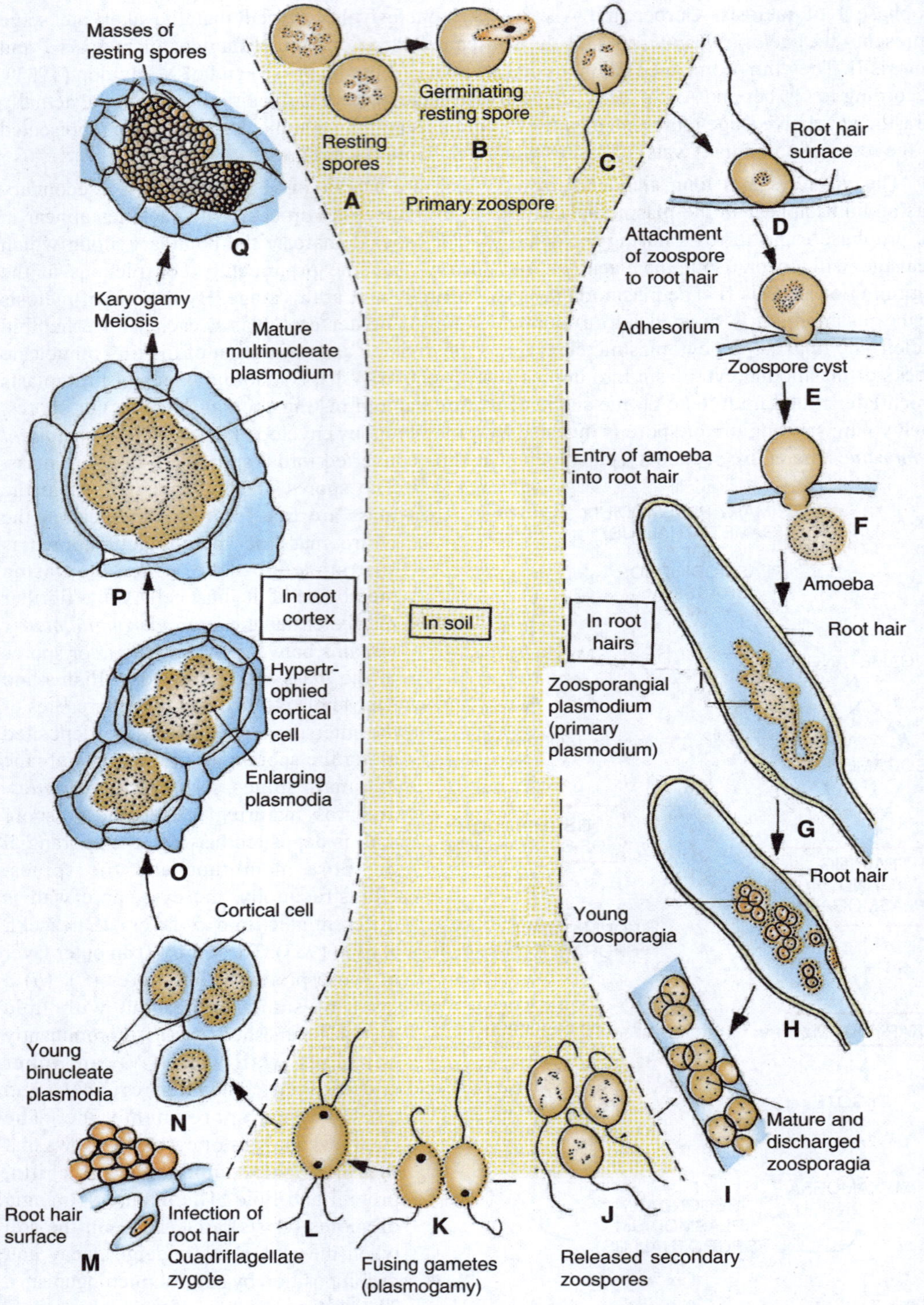

Fig. 5.8. Pictorial diagram of the life cycle of *Plasmodiophora brassicae*.

withstand adverse conditions. The resting spores or cysts (sporangia) thus perennate in the soil. They germinate to produce zoospores on the offset of suitable conditions to continue the fresh life cycle. Macfarlane (1970) reported that the spore or cyst germination is stimulated by the presence of susceptible host seedlings or by certain of their metabolities.

Club Root Disease of Cabbage.

Economic Importance. Two species are of great economic importance. They are *Plasmodiophora brassicae* and *Spongospora subterranea.* The former is the causal agent of *club root disease* of cabbage and other crucifers. *Spongospora subterranea* causes a *powdery-scab disease* of potatoes. Both are destructive diseases. A variety of *Spongospora subterranea (S. subterranea f. nasturtii)* is a pathogen of water cress. These diseases cause a twofold loss. Firstly, there is damage to the crop. Secondly, the contaminated soil becomes unproductive for many years for replanting the same crop plants.

Control Measures. Due to effective control measures, the disease is no longer the scourge that it originally was. The use of resistant varieties and long rotation are effective control measures. Fungicidal transplanting liquids are also employed with varying success. The use of pentachloronitrobenzene and mercuric or mercurous chloride in the transplantic liquid is effective to check infection. The other chemicals which are equally effective are aldrinvapam, sodium dithiocarbamate and tetrachloronitrobenzene. Use of soil free from the pathogen in the transplantation of seedlings is useful preventive measure.

Summary of the Life Cycle. (Fig. 5.7 and 5.8) It comprises two distinct generations or phases namely the *gametophyte generation* or *haplophase* and *sporophyte generation* or *diplophase.* The former occurs in the root hairs and the latter in the cortical cells of the host root (*Brassica sp.*).

The gametophyte generation (haplophase) is initiated by the differentiation of resting spores (sporangia) or cysts by meiosis from the diploid protoplast of the secondary plasmodium in the club roots of the host cortex. The mature cysts or resting sporangia or spores which are haploid are released into the soil when the club roots undergo decay and decomposition by soft rot microorganisms. After liberation, they perennate or germinate immediately if the conditions for growth are favourable. On germination, each resting spore or cyst produces a biflagellate, heterokont, haploid zoospore. The released zoospore swims in a thin film of water in the soil or creeps over the soil particles like an amoeba. Coming in contact with the root hairs of a suitable host, it becomes attached and amoeboid. In this state, it is called a *myxamoeba.* It then retracts its flagella and enters the root hair. Within the root hair it is swept away from the penetration site by host cytoplasmic cyclosis. By growth and repeated cruciform nuclear division, the uninucleate myxamoeba becomes converted into a naked multinucleate thallus like mass of protoplasm. It is acellular haploid and is called the *primary plasmodium* (or gametothallus). Reaching a certain stage of maturity the haploid protoplast of gamete thallus cleaves into 4-10 small, spherical, thin-walled, uninucleate *gametangia.* The haploid nucleus of each gametangium undergoes repeated mitosis to form 4-8 nuclei. Each daughter nucleus gathers a bit of cytoplasm around it to become a spindle-shaped mobile biflagellate isogamete. All the above mentioned haploid stages namely, the resting meiospores (sporangia or cysts), the zoospores, the primary plasmodium (gametothallus), the gametangia and the isogametes (which occur in the root hairs of the host) constitute the *gametophyte generation* or *haplophase.*

The second phase of the parasite life cycle commonly called as the *diplophase* or sporophyte generation starts with the formation of zygotes. The zygotes are found in the cortical cells of the root. It is not definitely known whether the isogametes fuse in the root hairs or in the cortical cells after migration. The consensus favours the view that the isogametes are released in the soil outside the host where pairing followed by plasmogamy takes place. Karyogamy is delayed. A binucleated quadriflagellate zygote thus formed reinfects host root. Within the root cortex, the zygote retracts its flagella and becomes amoeboid in form. It is called *myxamoeba.* Within the invaded cell the myxamoeba by subsequent growth accompanied by repeated cruciform division grows to form a naked, multinucleate diploid mass of protoplasm called the *secondary plasmodium or sporothallus.* It fills the lumen of the cortical cell. It then ceases to grow in size and its diploid protoplast undergoes meiosis. As a result, the multinucleate diploid protoplast of the secondary plasmodium is converted into a large number of uninucleate spherical haploid bodies each of which is surrounded by a wall of its own. These are the resting spores (sporangia) or cysts. The resting spores when mature are set free in the soil by the decay of club by rot microorganisms. In the soil the resting spores or cysts germinate under favourable conditions to continue the fresh life cycle. The zygote and the secondary plasmodium constitute the *diplophase* or *sporophytic generation.*

Alternation of Generations (Fig. 5.7). There is distinct alternation of generations in the life cycle of *Plasmodiophora.* The *zygote* and the secondary plasmodium (sporothallus before spore formation) represent the **sporophyte generation** or **diplophase.** The *resting spores,* the *zoospores,* the *haploid* primary *plasmodium* (gametothallus), the *gametangia,* and the *isogametes* represent **gametophyte generation** or **haplophase.** The two crucial points in the life cycle are meiosis at the time of differentiation of spores (meiospores) and **karyogamy** (fusion of gametic nuclei). At these critical points, the life cycle switches on from one generation to the other. With meiosis the diploid or sporophytic phase ends. The gametophyte phase starts. At the time of karyogamy, the gametophyte phase terminates and the sporophytic phase is initiated with the formation of diploid zygote. These two generations occur regularly one after the other in the life cycle. This is called *alternation of generations.* The life cycle characterised by alternation of generations with sporogenic meiosis is called *haploid-diploid* or *diplohaplontic* (diplobiontic life cycle).

Taxonomic Position and Inter-relationships. The affinities of Plasmodiophoraceae are uncertain. For this reason, they have been assigned different systematic positions in different schemes of classification. The Plasmodiophorales approach the slime molds in two features namely, unequally biflagellate swarm cells and plasmodium as the vegetative phase. These common features point to a relationship between the Myxomycetes and Plasmodiophoraceae and led Schroeter and Saccardo to place the endoparasitic slime molds (plasmodiophorales) as a sub-class under the class Myxomycetes coordinate with the Phycomycetes and other great fungal groups. Their position in the Myxomycetes is questioned by others because of the many differences such as lack of fruiting bodies, absence of cellulose in the spore wall, endoparasitic habit, manner of gametic fusion, cruciform nuclear division, akaryote stage and presence of haploid plasmodia. Many authors (Gaumann and Dodge) placed the non-mycelial fungi in the class Archimycetes. It is a class which includes the lower fungi. Fitzpatrik assigned the Plasmodiophorales the rank of a family and placed them in the order Chytridiales of class Phycomycetes. It is argued that the presence of zoosporangium in some species is a good reason for including Plasmodiophorales in the Phycomycetes. Smith raises the status of the class Myxomycetes to that of a division Myxothallophyta and includes three classes under it These are Acrasieae, Myxomycetae, (slime molds) and Phylomyxineae which includes the Plasmodiophorales. The name Myxothallophyta has been changed to Myxomycophyta. It includes three classes, namely, Myxomycetae, Acrasieae and Plasmodiophoreae. The latter includes the endoparasitic slime molds. The dominance of the plasmodial stage and production of biflagellate motile cells with anterior unequal whiplash flagella are considered sufficient to warrant the inclusion of endoparasitic slime molds as a class Plasmodiophoreae of the division Myxomycophyta. According to this view, the systematic position of endoparasitic slime molds would be as follows :

Division	:	**Myxomycophyta**
Class	:	**Plasmodiophoreae**
Order	:	**Plasmodiophorales**
Family	:	**Plasmodiophoraceae**

Sparrow (1950) placed the endoparasitic slime molds in the group biflagellae (Oomycetes) on account of the discovery of the presence of a second short flagellum on the motile cell. Hawker (1967) opined that the Plasmodiophorales (endoparasitic slime molds) resemble neither the slime molds nor the flagellate Oomycetes in their life cycle and general organisation. He considered them nearer to the Chytridiomycetes in which the zoospore contains a structure which is suggestive of a second abortive flagellum and this included the endoparasitic slime molds under the group Uniflagellatae as a separate class as under

Division	:	**Mycota**
Sub-division	:	**Eumycotina**
Group	:	**Uniflagellatae**
Class	:	**Plasmodiophoromycetes**
Order	:	**Plasmodiophorales**
Family	:	**Plasmodiophoraceae**
Genus	:	*Plasmodiophora*
Species	:	*brassicae*

Karling (1968) suggested that these should be included under a separate division or class of Fungi while Ainsworth (1973) considered Plasmodiophoromycetes as a class of Myxomycotina.

Alexopoulos and Mims (1979) place *Plasmodiophora* in the Superkingdom Eukaryonta, Kingdom Myceteae, Division Mastigomycota, Subdivision Haplomastigomycotina, Class Plasmodiophoromycetes, Order Plasmodiophorales and Family Plasmodiophoraceae.

The controversy will continue till more is known about the sites of Karyogamy and meiosis.

Comparison between the Plasmodiophoromycetes and the Myxomycetes

(*a*) *Resemblances :*

1. The somatic phase is an acellular holocarpic.
2. Differentiation of non-motile uninucleate, haploid resting spores by meiosis.
3. Formation of motile swarm cells by the germination of resting spores or cysts.
4. The swarm cells are uninucleate and biflagellate.
5. The two flagella are of unequal length but both are of whiphash type and inserted at the anterior end.

(*b*) *Differences* :

Plasmodiophora (Plasmodiopohromycetes)	*Physarum (Myxomycetes)*
1. *Plasmodiophora* is an internal obligate parasite (endoparasite).	1. It is a free living amoeboid myxomycete.
2. It is parasitic in its mode of nutrition.	2. It is phagotrophic in its mode of nutrition.
3. The life cycle comprises two distinct phases, the *haplophase* and the *diplophase.*	3. The life cycle consists of a single fairly organized phase,, the haplophase as such.
4. The haplophase starts with the germination of resting spore or cyst in the soil and comprises in addition the zoospores, myxamoebae, haploid, primary plasmodium (gametothallus), gametangia and the isogametes.	4. The resting spore or cyst germinates in the soil to give rise to swarm cells and myxamoebae which function as gametes. The haplophase is thus replaced by a few haploid cells only namely the resting spores (cyst) and the swarm cells or myxamoebae.
5. The zygotes are found in the cortical cells of the host root where each develops into a diploid secondary plasmodium (sporothallus). At maturity it fills the lumen of the host cortical cell.	5. The amoeboid zygote formed in the soil creeps about, feeds on bacteria and other microorganisms and develops into a diploid plasmodium.
6. The diploid secondary plasmodium reaching maturity, forms no fructifications.	6. Reaching maturity, the diploid plasmodium becomes converted into one or more sessile or stalked fruit bodies called the sporangia or sporophores.
7. The diploid protoplast of the secondary plasmodium itself, at maturity, undergoes meiosis to form haploid resting spores or cysts.	7. The diploid protoplast of the sporagium under goes meiosis to form haploid resting spores or cysts.
8. Meiosis takes place at the precleavage stage.	8. Meiosis takes place at the post-cleavage stage.
9. The mature resting spores are released in the soil by decomposition and decay of the root clubs of the host by soft rot bacteria.	9. The peridium disintegrates exposing the spore mass supported by the capillitium. Wind, mites and rain play an important role in dispersing the spores from the exposed spore mass.
10. The thick spore wall is chitinous in nature.	10. The spore wall contains cellulose.
11. Nuclear division is peculiar. The vegetative nuclear division is of cruciform type and meiosis has an akaryote stage.	11. Both of them are absent in the slime molds.

QUESTIONS

ESSAY TYPE

1. Describe the habitat and somatic phase of *Plasmodiophora.*
2. Describe different modes of reproduction met within *Plasmodiophora.*
3. What do you know of sexual reproduction in *Plasmodiophora ?*
4. Describe the life history of any member of the order Plasmodiophorales.
5. Describe cytological peculiarities found in *Plasmodiophora.*
6. Describe the phenomenon of alternation of generations in *Plasmodiophora.* Indicate the length of each generation in a graphic life-sketch.
7. Give an account of the Club Root disease of the Crucifers. What is the causal organism? How does it complete its life cycle within the host tissues ?
8. Discuss phylogeny and systematic position of *Plasmodiophoromycetes* in the light of recent research. *(Kanpur M.Sc., 2005)*

SHORT ANSWER TYPE

9. How do plasmodiophoromycetes differ from Myxomycetes? Explain with the salient features of the life cycles of both of them.
10. List the points on which *Plasmodiophora* resembles and differs from the Slime molds.
11. Compare Myxomycetes with Plasmodiophoromycetes.
12. Write a brief account of *Plasmodiophora.* *(Kanpur M.Sc. I, 1998)*
13. Describe the distinguishing characters of *Plasmodiophora.* *(Allahabad M.Sc., 1999)*
14. Give an illustrated account of life cycle of *Plasmodiophora brassica,* the causal agent of clubroot disease of crucifers. *(Allahabad M.Sc., 2000)*

OBJECTIVE TYPE

15. Fill in the blanks :

(*i*) The causal organism of the clubroot of Crucifers disease is ....

(*ii*) *Plasmodiophora* causes the disease known as .... of crucifers.

(*iii*) *Plasmodiophora* is an .... parasite.

(*iv*) Resting spore is the .... phase of the fungus *Plasmodiophora.*

(*v*) The somatic phase in *Plasmodiophora* is ....

(*vi*) Meiosis takes place in .... plasmodium in the life cycle of *Plasmodiophora*

(*vii*) Zygote in *Plasmodiophora* is initially .... flagellate.

(*viii*) The life-cycle in *Plasmodiophora* consists of .... distinct phases.

(*ix*) Clubroot of crucifers is also known as .... and .... disease of crucifers.

16. Select the correct answer :

(*i*) The life cycle in *Plasmodiophora* consists of
(*a*) one phase (*b*) two phases
(c) three phases (*d*) four phases.

(*ii*) Plasmogamy takes place in ....
(*a*) air (*b*) water (*c*) Soil (*d*) Host

(*iii*) *Plasmodiophora brassicae* causes a diseases in crucifers which is known as
(*a*) Club-root (*b*) white rust
(*c*) early blight (*d*) late blight

(*iv*) In the life cycle of *Plasmodiophora,* meiosis takes place in
(*a*) primary plasmodium
(*b*) Secondary plasmodium
(*c*) sporangium before the formation of resting spores
(*d*) gametangium before the formation of gametes

(*v*) The zygote in the life cycle of *Plasmodiophora* is
(*a*) uniflagellate (*b*) biflagellate
(*c*) triflagellate (*d*) quadriflagellate

Phytophora infestans on host

CHAPTER 6

SUBDIVISION – EUMYCOTINA

Class : Oomycetes

- Order : Saprolegniales
- Family : Saprolegniaceae
- *Saprolegnia*
- *Achlya*
- Order : Peronosporales
- *Pythium*
- *Pythium debaryanum*
- *Phytophthora*
- Family : Peronosporaceae (downy mildews)
- *Plasmopara*
- *Peronospora*
- *Sclerospora*
- Family : Albuginaceae
- *Albugo candida*

General Characteristics. The members of this class are characterised, as a rule, by oogamous sexual reproduction and by gametangial contact. Meiosis is gametangial. The gametes are the only haploid structures in the life cycle. They are non-flagellated. The zoospores are biflagellate. They may be pyriform or reniform in shape. The flagella are inserted apically in the former and laterally in the latter. One of the flagellum is of

tinsel type and the other of **whiplash type.** The former is directed forward and the latter backward when the zoospore is in motion. The zoospores lack cell walls. The mycelium is extensive and coenocytic. The septa remain suppressed in the actively growing hyphae but the reproductive organs are separated by a basal septum. The hyphal wall chiefly consists of **β glucans** and **cellulose** and not chitin.

On the whole the representatives of this group exhibit progressive evolution from the aquatic to the land habitat. The most advanced land forms are obligate parasites which produce wind-disseminated asexual reproductive bodies called the **sporangia.** The sporangia are deciduous. The detached sporangia usually do not produce zoospores. On the other hand, they function like spores (Conidia) and germinate directly by putting out a germ tube. Occasionally the detached sporangium behaves like a zoosporangium and produces zoospores indicating their aquatic ancestry.

Mycelium. The vegetative phase is variable. It ranges from a simple, unicellular thallus to an extensively developed, flamentous structure (mycelium) consisting of branched, coenocytic hyphae. The hyphae grow in and around the medium. The oomycete hyphal wall constains cellulose but no chitin (Frey, 1950, Parker *et al* 1963, Cochrane, 1958). Glucans or mannans also occur in addition. In fact glucans may predominate. A large number of amino acids appear to be typical of Oomycete hyphal walls. Crook and Johnson (1962) reported as many as seventeen in the hyphal walls of *Saprolegnia* and *Phytophtora*. Hunsley and Burnett (1970) reported that all the Oomycetes so far studied have hyphal walls composed of *B*-1, 4 linked cellulose microfibrils and highly branched

Phytophthora on Host

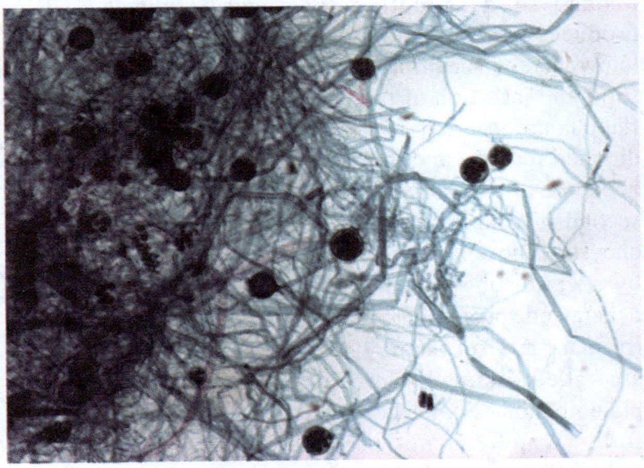

Saprolegnia growing in Water

insoluble *B*-1, 3-glucans which has *B*-1, 6 linkages at the branching residues. Most of the species are **eucarpic** and reproduce both asexually and sexually. During the growing season, the specialized plant pathogens of this class multiply by several successive, asexual generations. Sexual reproduction takes place only once and probably towards the end of the growing season.

Asexual reproduction. In the aquatic species such as *Saprolegnia*, it takes place by the formation of zoospores which are produced inside **zoosporangia.** The latter in the lower members are in no way different from the vegetative hyphae. In the advanced forms the sporangia are definite organs. They have a characteristic shape and are borne on special, reproductive hyphae called the **sporangiophores.** The sporangiophores are not only well developed but are characteristically branched. The zoospores may be formed in the sporangium itself by the cleavage of the protoplast and released through the pore at the apex of sporangium. In some other species, however, the zoospores are differentiated in a vesicle formed at the apex of the sporangium into which the sporangium protoplast migrates. The vesicle then vanishes and the zoospores are set free.

Behaviour of zoospores. Depending upon the species the released zoospores behave in two different ways :

(*a*) In the majority of the species (*Patheoposis*), the liberated zoospores swim about actively. After a while they become quiescent, retract their flagella and round off. Each then secretes a wall around it and thus becomes encysted. The clothed zoospore or cyst then germinates by putting out a germ tube which develops into a new plant. Such species which produce only one kind of zoospores each of which on germination directly develops into a parent plant are called **monomorphic** or **monoplanetic**. This kind of zoospore or behavior is called **monoplanetism**.

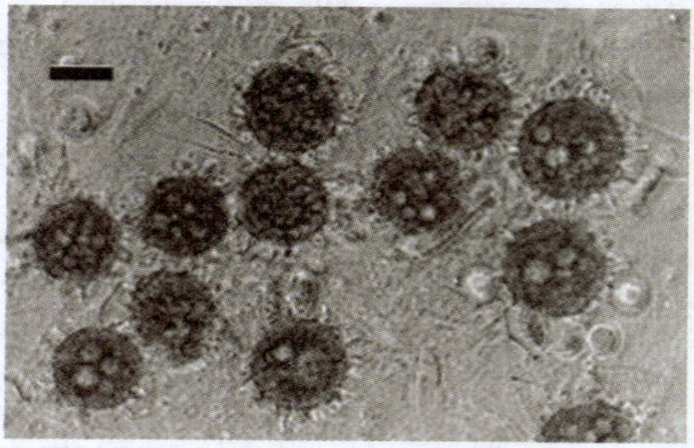

Fungus Leptolegnia

(*b*) There are, however, some simple and primitive genera in this class such as *Sparolegnia* and *Leptolegnia* in which the zoospores released from the zoosporangium behave in a different and quite interesting way. They are pear-shaped and biflagellate. The flagella are inserted at the anterior end. We call these as the **primary zoospores.** Each primary zoospore after a brief period of swarming comes to rest and encysts (secretes a wall around it). The encysted zoospore, instead of germinating by a germ tube, produces a second kind of zoospore. It is known as the **secondary zoospore.** The secondary zoospore differs from the primary zoospore in form and insertion of flagella. It is reniform (kidney-shaped) and is laterally biflagellate. The two flagella are inserted on the concave side, one of these is directed forward and the other backward. The secondary zoospore after emergence from the cyst embarks upon a second swarming period. Thereafter it comes to rest and encysts. The encysted secondary zoospore germinates by a germ tube which produces the mycelium. These genera and species which produce two kinds of zoospores (primary and secondary) in succession in the life cycle are called **dimorphic** or **diplanetic**. This condition or phenomenon is called **diplanetism**. The advanced land forms which do not produce zoospores are called **aplanetic**.

Pythium Seedling Blight.

Evolution of the conidiospore. Another interesting feature of this class is the light the different species throw on the transition from a sporangium to a spore-like conidium.

You have studied above the aquatic genera such as *Saprolegnia* reproduce entirely by the formation of zoospores which are produced within zoosporangia. Many species of *Pythium* are a little more soil-minded. They grow in damp soil and reproduce by zoopores indicating thereby that they are basically aquatic.

Phytophthora and *Pythium debaryanum* are amphibious in their habit. They are adapted to life both in dry and wet habitats. The fungal hyphae in both bear sporangia which under wet conditions poduce zoospores. However, under dry conditions or high temperature the zoosporangia get detached. The detached sporangia are similar in size and shape to the zoosporangia, which produce zoospores, but function differently. Each detached sporangium, on germination, behaves like a **conidium.** It produces a germ tube which develops into a fresh fungus mycelium.

Similarly *Plasmopara viticola* can adjust itself to both dry and wet conditions. Usually it reporduces by the formation of zoozpores. Under dry conditions, however, the detached sporangia behave as conidia, each producing a germ tube.

In *Pythium vexans* sporangia normally function as conidia. The germ tube is formed in the place of a papilla of the sporangium. Rarely, however, the sporangia form zoospores.

Pythium ultimum and a terrestrial species of *Plasmopara* do not produce sporangia. The asexual reproductive organs always behave as conidia. Each germinates by producing a germ tube.

Pythium ultimum growing on host

Plasmopara viticola on grapes

Finally in genera like *Bremia* and *Perononspora* the sporangia bearing hyphae (sporangiophores) are branched. Each branch bears either a single sterigma (*Peronospora*) or a number of them (*Bremia*). Each sterigma bears at its top a single spore-like conidium.

It is evident from the foregoing account that the conidium in this class has evolved from a sporangium, the protoplast of which forms a germ tube instead of producing zoospores. The

Spots of Bremia on leaves

Peronospora spots on leaves

change is an adaptation in response to a terrestrial mode of life which does not permit dispersal by zoospores but favours distribution by air.

Sexual reproduction. Excepting a few most primitive forms which are holocarpic, sexual reproduction is **oogamous** by gametangial contact. The sex organs, which are called **antheridia** and **oogonia,** are terminal or intercalary in position. The fusing gametes are non-flagellated. The protoplast of the oogonium forms one or more oospheres which are uninucleate when mature. Either the entire protoplast (Saprolegniales) is utilised in the formation of oospheres or a part is left behind. The residual protoplast forms a peripheral layer termed the **periplasm,** around the oosphere. It is used up during maturation of oospore.

According to their position and occurrence, the antheridia are grouped into two categories, namely, **androgynous** and **diclinous.** The former develop on the stalk of the oogonium or a branch of it. When developed on a separate hypha the antheridium is described as **diclinous.**

Plasmogamy takes place by gametangial contact in the following ways :—

1. Generally the antheridium gets applied to the side of the oogonium towards maturity.
2. Rarely, as in *Phytophthora,* the oogonium punctures the developing antheridium and grows through it to the top where it forms a rounded structure.

The antheridium, in the first case, is termed **paragynous** and in the second case **amphigynous.** The migration of the male nucleus into the oogonium usually takes place through a fertilisation tube or through a pore in the oogonial wall.

Place of Meiosis in the Oomycetes. It is still under dispute. Stevens as early as 1899 suggested **gametangial meiosis** in *Albugo.* His interpretation was disputed by later workers. Since then it has been widely held that the Oomycetes, as a whole, are haploid and meiosis occurs in the oospore. Sansome (1963) working with *Pythium debaryanum* struck a discordant note. She reported that the two successive divisions which occur in the gametangia constitute **meiosis.** From this one would conclude that the Oomycetes, as whole, would be diploid. During the last two decades various investigators have reported meiosis in the gametangia of *Phytophthora* (Sansome, 1965), *Albugo* and *Peronospora* (Sansome and Sansome, 1974), *Sclerospora* (Sansome, 1966), Bremia (Tommerup, *et al*, 1974), *Achlya* (Barkesdale, 1968), *Saprolegnia* (Bryant and Howard, 1969), and *Apodachlya* (Howard and Bryant, 1971). From the investigations of these workers it would appear that both the Saprolegniales and Peronosporales are diploid.

Albugo on Host

Sclerospora on Bajra

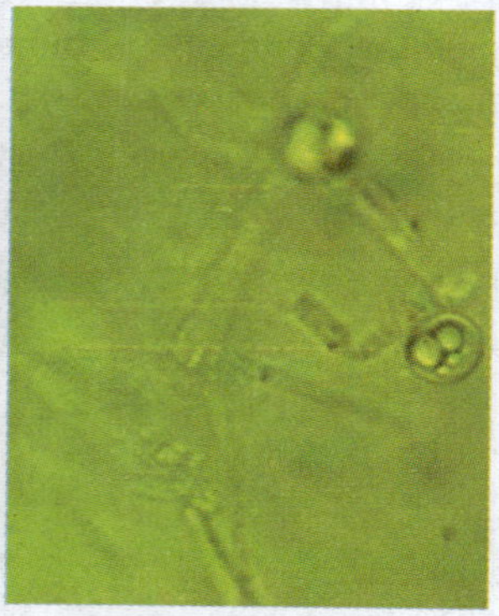
Apodachlya in Water

On the other hand, Timmer, Castro, Irvin, Belsar and Zentmyar (1970) advanced a genetic evidence for **zygotic meiosis** in *Phytophthora capsici.* Stephenson *et al* (1974 *a*, 1974 *B*) working with *Phytophthora capsici* vehemently opposed gametangia meiosis in *Phytophthora.* They reported that meiosis takes place in the first divisions of the oospore nucleus.

Phytophthora capsici on Chillies

More recently Sansome (1976) studied meiosis in *Phytophthora capsici*. She observed pachytene, diplotene and diakinesis stages in the oogonium. She further reported that metaphase and anaphase were followed by a second division without any increase in nuclear and chromosome size between the two divisions. The haploid chromosome number was reported to be nine. She observed an association of 4 chromosomes in one or both parents. The association of 4 chromosomes at diakinesis and metaphase provides an evidence that pairing of homologous parts occurs during the first division and is a proof that the gametangial divisions are meiotic. There is thus overwhelming and conclusive evidence in favour of gametangial meiosis in the Oomycetes according to which the somatic phase constituted by the mycelium as a whole and the sex organs (antheridia and oogonia) would be diploid. The gametes alone represent the haploid structures in the Oomycete life cycle. In spite of this the mycologists, in general, stick to the view that there is zygotic meiosis and the Oomycetes are haploid.

Inter-relationship and origin of Oomycetes. No close relationship seems to exist between the biflagellatae (Oomycetes) and uniflagellatae. In their mycelial and hyphal organisation of somatic phase, cellulose cell wall and oogamous sexual reproduction, the Oomycetes indicate some resemblance with the Monoblepharidales. The resemblance, however, appears superficial because two groups have totally different types of swarmers. Among the Oomycetes themselves, the biflagellate condition of the swarmer with two unequal flagella, the shorter of tinsel and longer of whiplash type, oogamy, transfer of male gamete into female by the fertilisation tube and germination of zygote by a germ tube are some of the common features which all point to the close relationship between the members of the Saprolegniales-Peronosporales series. The Saprolegniales are considered primitive and Peronosporales advanced.

In the siphoneous, coenocytic acellular organisation of the thallus, chemical composition of cell wall, oogamous sexual reproduction, and synthesis of amino acid lysine from diaminopimelic acid, the Oomycetes exhibit a striking resemblance with the Xanthophyceae (Yellow green algae). In their bilateral symmetry, the swarmers of Oomycetes with two anteriorly inserted unequal flagella, the shorter with flimmer hairs and the longer of whiplash type, are suggestive of those of the Heterosiphonales. The recent discovery of tinsel and whiplash types of flagella in the motile stages of *Vaucheria* has given fillip to the belief that the Oomycetes could have evolved by loss of chlorophyll from the *Vaucheria*-like ancestor and probably before it had developed its compound zoospores from many biflagellate zoospores. Whether these resemblances are suggestive of real affinity or the result of converging development is difficult to decide.

Distinguishing features of the Oomycetes from other fungi

1. The mycelium in majority of the members is extensive and coenocytic. Septa appear only in connection with the reproductive organs.
2. The hyphal wall mainly consists of glucans. Cellulose is also present but there is no chitin.
3. Asexual reproduction chiefly takes place by means of biflagellate zoospores. One of the flagellum is of tinsel type and other of whiplash. The former is directed forward and the latter backward when the zoospore is in motion. The sexual reproduction is Oogamous and takes place by gametangial contact.

4. The gametes are non-flagellated and the oospore is thick-walled.
5. Meiosis is gametangial. The somatic phase is thus diploid .
6. The gametes are the only haploid structures in the life cycle.
7. The life cycle is diplointic (haplobiontic-diploid).
8. There is no alternation of generations in the life cycle.

Classification. Martin (1961) and Sparrow (1973) divided the class **Oomycetes** into four orders on the basis of (*i*) the nature and mode of zoospore formation in the zoosporangium proper or in the vesicle and (*ii*) the holocarpic or eucarpic nature of the thallus. The names and characters of the orders are given below :

Order Saprolegniales. Zoospores invariably formed in the sporangium, Diplanectic or monoplanetic or aplanetic. Thalus holocarpic or eucarpic, if eucarpic, hyphae without constriction, not arising from a basal cell, oogonium contains one to several oospheres.

Order Leptomitales. Zoospores invariably formed in the sporangium, diplanetic, monoplanetic or aplanetic; thallus eucarpic, hyphae with constrictions, oogonium contains a single oosphere.

Order Lagenidiales. Zoospores` formed in the sporangium or vesicle, monoplanetic and reniform; thallus primarily holocarpic, mostly aquatic, parasitic or algae or water moulds.

Order Peronosporales. Zoospores formed within the sporangium or vesicle, monoplanetic and reniform; thallus eucarpic, primarily terrestrial living in soil or parasitic on vascular plants.

However, Sparrow (1976) recognised 6 orders in this class, namely Eurychasmales, Thraustochytriales, Labyrinthulales, Saprolegniales, Leptomitales, Lagenidiales and Peronosporales. Of these Saprolegniales and Peronosporales are of importance to man because certain members of these orders parasitize plants of economic importance. They are, therefore, described in some detail in this text.

In 1961 Mrs. Eva Sansome demonstrated that in ***Pythium debaryanum*** meiosis occurs in gametangia and the gametes are haploid. It was found to be true for members of the order Saprolegniales and Peronosporales implying that the vegetative phase of the genera of these orders are diploid. Thererfore, the order Peronosporales is regarded as the most advanced and the order Lagenidiales as the most primitive order of the class Oomycetes.

ORDER : SAPROLEGNIALES

The representatives of this order are primarily aquatic and are popularly called the 'water molds.' They are abundant in well aerated water on certain types of substrata rich in proteins, such as ant's eggs, solid white of egg, bodies of dead insects, dead fish and other animals. The Saprolegniales usually grow best at relatively low temperatures. Certain species, however, are semi-aquatic.They occur on the upper layers of wet or damp soil. Most of the Saprolegniales are saprophytes. Some are parasitic on fish, amphibians and roots of vascular plants causing serious diseases. The chief features of this order are:

(*i*) The coarse usually sparsely branched mycelium
(*ii*) The sporangium not much different from the somatic parent hypha.
(*iii*) Occurrence of diplanetism to polyplanetism.
(*iv*) Slender, hypha-like antheridium in most of the species.
(*v*) Formation of more than one oospheres in the oogonium.
(*vi*) Utilisation of entire protoplast of the oogonium in the formation of oospheres.
(*vii*) Sexual reproduction by gametangial contact and fertilisation generally by means of a fertilisation tube and of paragynous type.

The order includes 25 genera and 250 species. They are placed in five families (Dick, 1973) viz. *Thraustochytriaceae*, *Saprolegniaceae*, *Leptolegniellaceae*, *Ectrogellaceae* and

Haliphthoraceae. Of these *Saprolegniaceae* is regarded as typical of the order and hence, discussed in this text.

FAMILY : SAPROLEGNIACEAE

The mycelium in the members of the family commonly called **water moulds** is well-developed. It consists of branched coenocytic hyphae. The hyphal walls are composed chiefly of cellulose and *B* glucan. Asexual reproduction generally takes place by the formation of diplanetic zoospores which are produced within long, cylindrical zoosporangia, which are terminal in position. The zoosporangia are not much different from the vegetative hyphae. Sporangial proliferation is an important feature of the family. The zoosporangia remain attached to the vegetative hyphae even after the zoospores have been shed. Diplanetism and repeated zoospore emergence are the diagnostic characteristic of this family.

A water mould-Banksia coccinea.

The oogonium produces one or more oospheres and lacks the periplasm. The male gamete migrates into the female through the fertilisation tube. Majority of the species are homothallic. A few, however, are heterothallic. The important genera included in the family are *Saprolegnia, Achlya, Leptolegnia, Dictyuchus* and *Geolegnia*. Of these *Saprolegnia* and *Achlya* are typical genera of the family. They are described here in some detail.

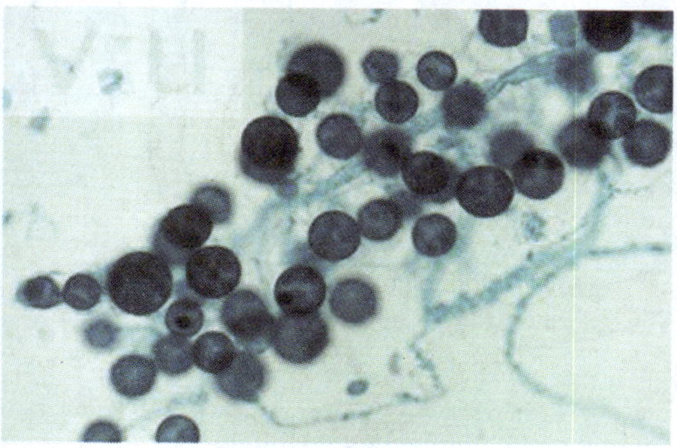

Achlya growing in water

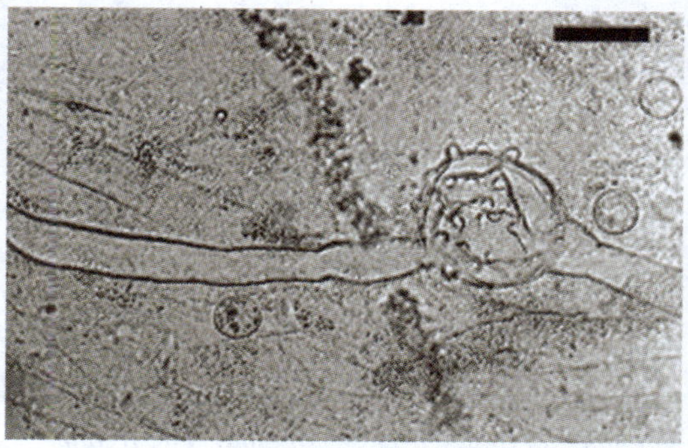

Leptolegnia hyphae

SAPROLEGNIA

Occurrence. *Saprolegnia* is mostly an aquatic genus which includes about 30 species. Except a few all of the them are saprophytic and occur in fresh, clear water. The only exceptions are *Saprolegnia parasitica* and one or two others which are parasitic. The saprophytic species require sulphur and nitrogen in organic form and thus grow on substrata rich in proteins. They grow on the decaying bits of animal tissues in water and also on the bodies of dead insects and other plants or slow flowing streams. A few species grow on moist humus soil. Tiesenhausen in the year 1912 reported a species of *Saprolegnia* from the snow-covered peaks of the Alps. *Saprolegnia parasitica* has been reported to live as a parasite on the gills of fishes. It causes a serious disease known as "salmon disease".

Fungal Culture. The fungus can be grown in the laboratory by placing dead flies (Fig. 6.1), ant's eggs, small pieces of meat, solid white of egg, or three dry split hemp seeds in a dish containing water from a suitable pond. Sesame seeds (4 or 5 in number) can also be used as a bait.

After a couple of days the fungus appears on these substrates in the form of tiny tufts of cotton wool visible to the naked eye. The hyphae grow out from the bait and extend into water where they branch. The transfer of the colony to tap water usually induces the development of zoosporangia. Remaining in the original culture the mycelium may develop sex organs. A pure culture of the fungus can be obtained on agar-agar. The factors which favour its growth are :—

(*i*) A very dilute medium.

(*ii*) Supply of sulphur and nitrogen in the organic form in the medium.

(*iii*) Relatively low temperature. To induce sporulation the medium may be flooded with sterile water.

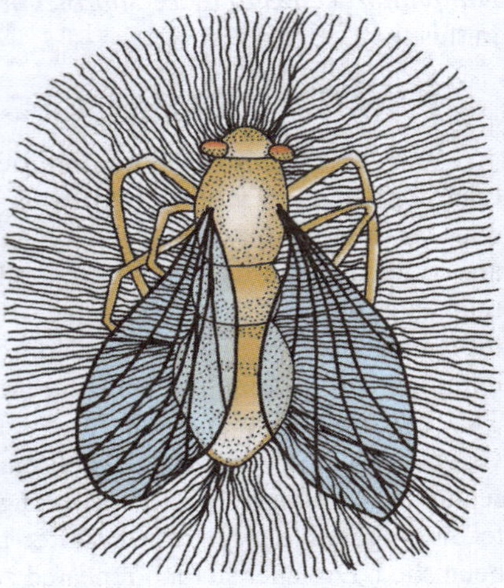

Fig. 6.1. *Saprolegina sp.* growing on a dead insect in water.

Mycelium (Fig. 6.2 A). The mycelium is coenocytic and branched. During the vegetative phase it is composed of two kinds of hyphae :

(*i*) *Rhizoidal* or *intramatrical hyphae*. These are short hyphae which penetrate the substratum. They anchor the mycelium and absorb nutrition.

(*ii*) *Extramatrical hyphae* (Fig. 6.1). These are long hyphae which grow out from the surface of the substratum into the water and extend in all directions. They form the visible portion of the mycelium and produce the reproductive organs (Fig. 6.2A). The extramatrical hyphae are long, slender and extensively branched. They are aseptate and coenocytic. Septa, however, appear in connection with the formation of reproductive organs and rarely in older hyphae. The hyphal wall contains cellulose and not chitin. In addition glucans or mannans are also involved. Next to the hyphal wall is the vacuolate cytoplasm in the form of a thin lining layer containing several nuclei (A). Food is stored in the form of oil globules and glycogen. The mycelium is **eucarpic.**

Reproduction. After a certain period of growth and under suitable conditions of food, temperature and other environmental factors *Saprolegnia* enters the reproductive phase. It bears reproductive organs at the tips of hyphae. *Saprolegnia* reproduces asexually as well as sexually.

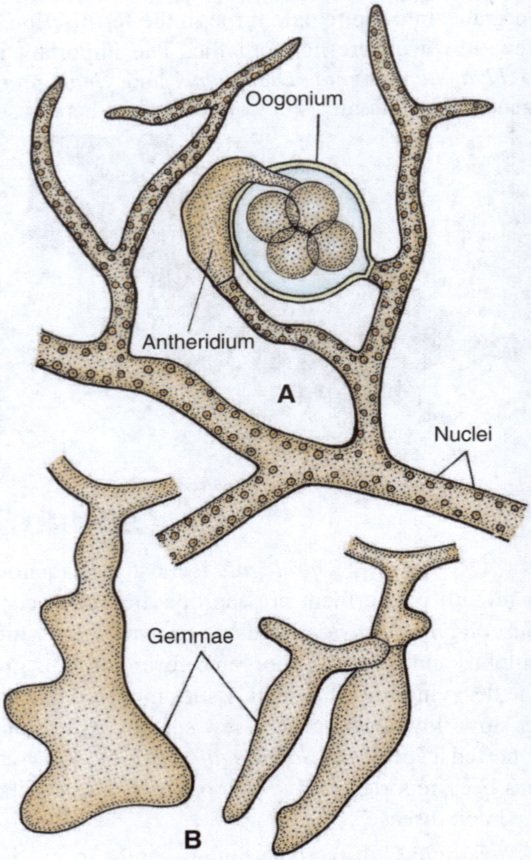

Fig. 6.2 (A-B). *Saprolegnia sp.* A, Hyphae showing structure as seen under the light microscope and bearing sex organs;

1. Asexual reproduction. It takes place by vegetative methods and sporulation.

(*a*) Vegetative reproduction. It takes place by the following two methods :—

(*i*) *Fragmentation.* Under favourable circumstances the hyphae break up into pieces of variable lengths. Each such piece or frangment by further elongation and nuclear division develops into a mycelium.

(*ii*) *Formation of Gemmae* (Fig. 6.2B). They are formed at the tips of hyphae and are unicellular. The foramtion of gemmae is marked by a conspicuous swelling of the hyphal tips. The swollen portion contains abundant food material and possesses several nuclei. It is separated from the hypha by means of a cross wall. The gemmae vary in shape and size. They may be rounded, ovoid or irregular and developed singly or in chains of variable lengths. At maturity the gemmae get detached. The detached gemma germinates by giving out a tubular hypha on a new substrate and develops into a mycelium or short-stalked sporangium typical of the species. Some mycologists call these gemmae as **chlamydospores.**

(*b*) Sporulation (Fig. 6.3). It takes place by means of pip or pear-shaped, biflagellate **zoospores** which are produced in long, cylindrical tapering **zoosporangia.** The zoosporangia are formed at the tips of somatic hyphae which are not differentiated into sporangiophores. The zoosporangia are only slightly greater in diameter than the hyphae bearing them. They are densely filled with protoplasm whereas the somatic hyphae are only lined with a thin layer of protoplasm.

Development of Zoosporangia (Fig. 6.3, A-J). The apical portion of external or extramatrical hyphae shows a slight amount of swelling (A). The swollen portion is conspicuous by its denser granular cytoplasmic contents. A large number of nuclei migrate into this club-shaped apical

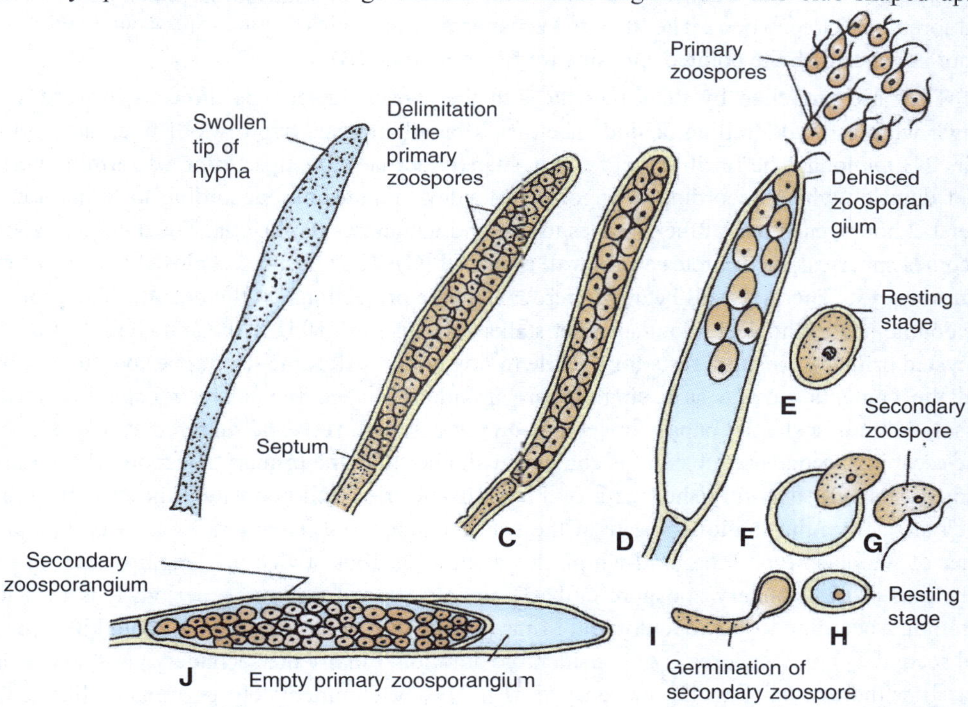

Fig. 6.3. (A-J) *Saprolegnia sp.* A-D, Formation and liberation of primary zoospores; E, Resting primary zoospores; F, Liberation of secondary zoospore; G, Active secondary zoospore; H, Resting stage of the same; I, Germinating secondary zoospore; J, Empty primary zoosporangium showing the development of a secondary zoosporangium inside the old primary one (proliferation of zoosporangia).

portion. Finally it becomes cut off from the remaining hypha by a cross wall and is known as the **zoosporangium** (B) which is usually club-shaped or it may be tubular with ends slightly tapering (cigar-shaped). The mature zoosporangium has a well-defined thin-walled apex.

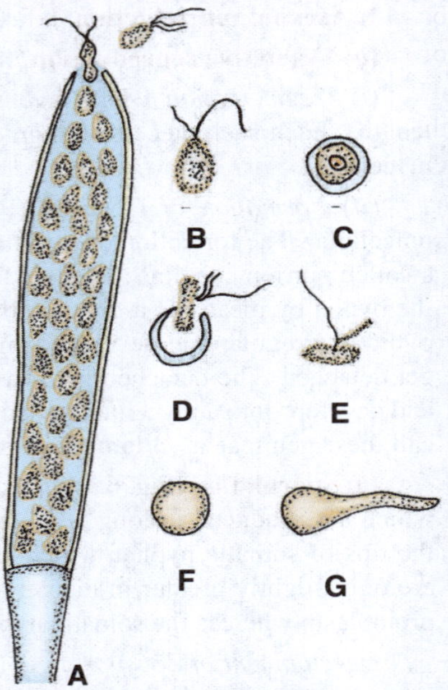

Fig. 6.4 (A-G). *Saprolegnia sp.* Liberation of zoospores and stages showing diplanetism. Explanation in the text.

Differentiation of Zoospores (Fig. 6.4). The colourless multinucleate protoplast of the zoosporangium divides into several uninucleated portions by progressive cleavage of the entire protoplast (A). Each unicleate daughter protoplast rounds off and later assumes a pear-shaped form (A). It is furnished with two apically inserted flagella. One of these is of **tinsel** or **pantonematic** type and the other of **whiplash** or **acronematic** type (B). These pyriform, biflagellate, uninucleate structures are known as the **primary zoospores**. They are released into the surrounding water, through an apical pore that suddenly develops at maturity in the zoosporangium (A). Most of them emerge with a rush as if under pressue within the zoosporangium. The remaining ones come out one by one slowly. The zoospores emerge backwards with the blunt end foremost (Greenwood). The liberated zoospores swim about in water with the pointed end forward for some time (B).

They are propelled by their flagella with the shorter tinsel type directed forward and the longer whiplash type trailing behind. The tinsel flagellum bears two rows of fine, lateral fibrillar hairs 2-3 *um* long. The fibrillar hairs end in shapely defined fine tips. After a swarming period of brief duration which according to Ingold is of a few minutes and according to Manton extends over 1-2 hours, each one of them comes to rest and withdraws its flagella. The quiescent zoospore becomes spherical and secretes a thin wall around it (C). The zoospore enclosed in a sac or cyst is said to encyst. The cyst wall bears numerous delicate projections on the outside. The projections are double headed hooks on long, slender stalks (Martin *et al,* 1951, Kole and Gielink, 1962). The encysted primary zoospore rests for a while (a few hours). After the resting period the cyst breaks and the contents emerge as a single bean or kidney-shaped secondary zoospore through the dissolved tip of a slender papilla developed on the cyst wall (D). The ruptured cyst is left behind. The escaping secondary zoospore is somewhat smaller than the primary zoospore and is reniform (kidney-shaped). It is furnished with two laterally inserted and oppositely directed flagella (E). They arise from the shallow groove of the concave side. One flagellum is of tinsel type and the other of whiplash type. The hind whiplash flagellum is about twice the lenght of the fore-tinsel flagellum. The secondary zoospore embarks upon a second swarming period. It is of a longer duration. According to Alexopoulos and Mims (1979), the swarming period of both kinds (primary and secondary) of zoospores is of considerable duration. Finally the secondary zoospore comes to rest. It withdraws its flagella and encysts (F). The cyst immediately germinates like a fungal spore. There is no emergence of a zoospore in this case. On a new substrate it produces a short germ tube (G). The germ tube grows into a hypha and finally into the mycelium.

Ultrastructure of zoospore. The zoospore, whether primary or secondary has no cell wall. It has a pear-shaped nucleus with a distinct nucleolus. The nuclear membrane has pores. A pair of basal bodies, from which arise the two flagella, lie near the tapered end of the nucleus. A number of microtubules radiate from the basal bodies into the cytoplasm. The zoospore cytoplasm contains numerous ribosomes, mitochondria, lipid bodies and electron dense inclusion bodies scattered in the cytoplasm. Besides there are one or two golgi bodies located near the surface of the nucleus. Contractile vacuoles have also been reported.

Zoospore Behaviour. *Saprolegnia* is peculiar in producing two types of zoospores (**primary** and **secondary**) in succession. Such species are called **dimorphic** in contradistinction to the **monomorphic** which produce only one kind of zoospores. The successive production of two different types of zoospores by a single fungal organism is termed **diplanetism.** The species with two swarming stages separated by a resting stage is called **diplanetic.** The process of repeated encystment followed each time by the production of a further motile stage is called **repeated zoospore emergence.** The species with one swarming stage involving the formation of only one kind of zoospores are called **monoplanetic.** This phenomenon involving the production of one kind of zoospores with one motile stage is called **monoplanetism.** The species with no swarm stage and no production of zoospores are called **aplanetic.** *Geolegnia* is an example of this kind. *Saprolegnia* is an example of a **diplanetic species.** The functional significance of diplanetism (two-phase behaviour of zoospores) and polyplanetism is not definitely known. However, in the absence of suitable substratum the fungus may resort to successive production of zoospores of secondary type separated by the resting stage till it makes contact with a suitable substratum (Salvin, 1940). Hickman (1967) found that repeated emergence of zoospores was suppressed in the presence of nutrients and favoured by cooler temperatures.

Proliferation of Zoosporangia. It is an interesting and characteristic feature in *Saprolegnia.* The basal septum of the empty zoosporangium initiates the development of a secondary zoosporangium (Fig. 6.3 J). It grows and bulges into the empty sporangium to form a new sporangium. The new sporangium matures within the primary one. Sometimes it matures some way beyond the primary zoosporangium. This process may be repeated and several zoosporangia may thus be formed within the walls of the next older one. Each of these matures and releases its contents before the next one is formed. Intercalary and catenulate zoosporangia have also been reported in some species such as *S. torulosa.*

2. Sexual Reproduction (Fig. 6.5). It is **oogamous** and by gametangial contact. The sex organs are called **antheridia** and **oogonia.** They are formed at the tips of somatic hyphae (A) when conditions favourable for sexual reproduction appear or conditions unsuitable for somatic growth set in. Many species are **monecious** or **homothallic.** A few are **dioecious** or **heterothallic.** Following is an account of sexual reproduction in a homothallic species.

1. Oogonium (A). (*a*) *Structure.* The oogonium is generally a more or less a spherical structure with a thick wall. It is delimited from the supporting hypha (stalk) by a septum which is a solid plate. The oogonial wall is usually smooth. In some species it is spiny or papillate. Rarely it has pits (*S. ferax*). Usually the oogonia are produced singly and terminally on short lateral branches of the hyphae projecting from the substratum (A). They may also arise from the main hyphae themselves. Sometimes oogonia are intercalary in position (*S. litorales*). The multinucleate protoplasm of the oogonium which is uniformly granular, at first, becomes cleaved into several **eggs** or **oospheres.** The mature oogonium contains usually four to ten naked, uninucleate, spherical oospheres or eggs (E). Rarely the number may go up to 32. Each oosphere contains a number of oil globules. These oil globules may unite to form a single, larger oil globule. In case this oil globule lies in the centre, the egg is known as **centric.** If it lies toward one side the oospohere is known as **eccentric.** The mature oosphere is dark in colour and is uninucleate.

(*b*) *Development.* The tip of a lateral branch or the end of the main hypha swells. The swelling inflates to form a spherical structure. Thy cytoplasm and a large number of nuclei stream into the swollen portion. Finally a septum appears separating the terminal swelling from the supporting hypha (B). The former is the young **oogonium** and the latter is termed the **stalk.**

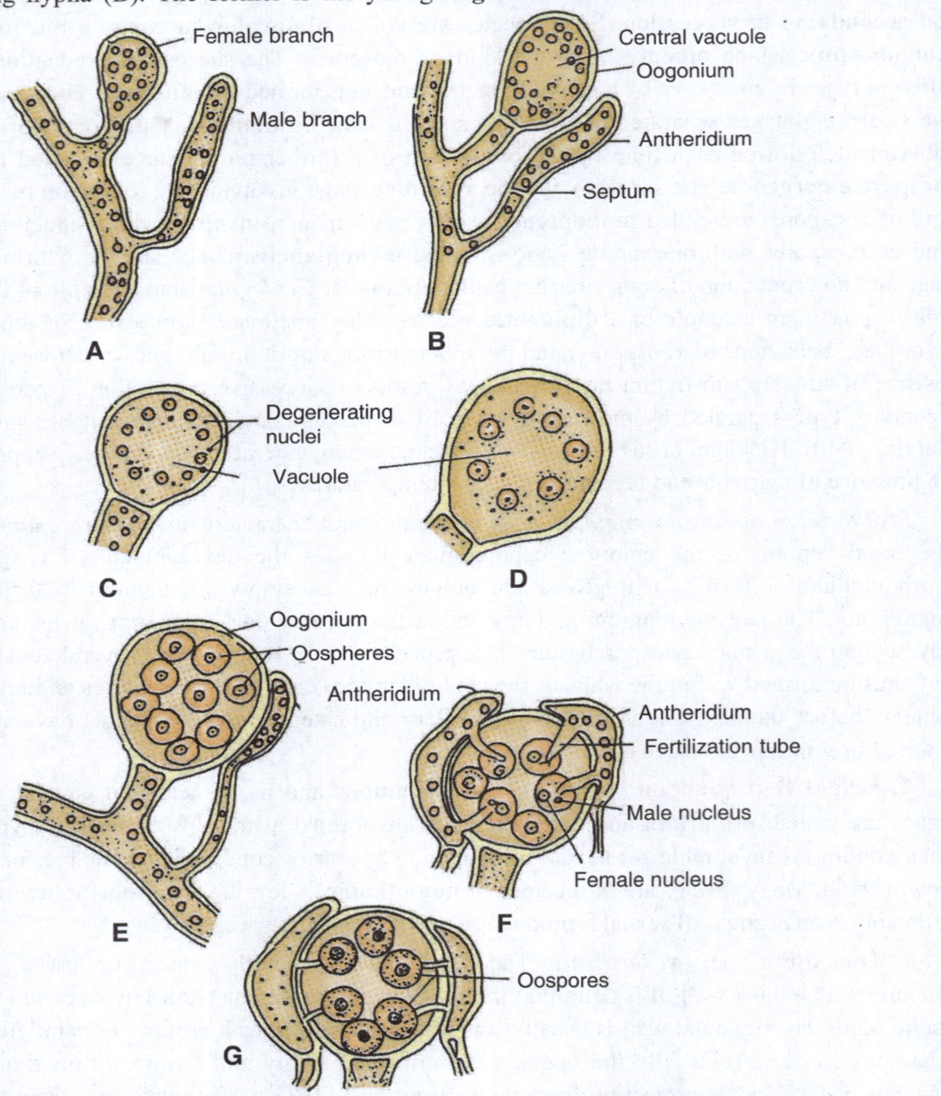

Fig. 6.5 (A-G). *Saprolegnia sp.* Stages in sexual reprodcution and fertilisation. Explanation in the text.

Following oogonium delimitation some of its nuclei and cytoplasm start degenerating. The process starts from the centre towards the periphery of the oogonium and continues till a conspicuous central vacuole forms. The nuclei left over along with the cytoplasm which is packed with lipid globules are restricted, at this stage, to a thin layer along the periphery of the oogonium (C). The nuclei in the peripheral layer undergo another mitotic division which is again followed by the degeneration of some of them. The surviving nuclei gather cytoplasm around them and develop into uninucleate oospheres (D & E). The oospoheres enlarge and become spherical. The entire protplasm of oogonium is used up in the formation of oospheres. The oogonium in *Saprolegnia* thus has no periplasm.

Antheridium (*a*) *Structure.* The antheridia are borne on long or short, slender hyphae called the **antheridiophores** (A). The antheridiophores (one or more) may arise from the same hyphal branch which bears the oogonium and immediately below it or from a different hypha in the neighbourhood or on an entirely different thallus. In the first case the antheridia are said to be **monoclinous** or **androgynous** and in the latter two cases **diclinous.** The mature antheridium is an elongated, tubular, multinucleate structure delimited by a basal cross wall (Fig. 6.5 B). It contains abundant protoplasm and is smaller than the oogonium. No sperms are organised.

(*b*) *Development.* An antheridiophore arises as a lateral outgrowth. As it elongaes its terminal portion becomes inflated to form a tubular or even an irregular, swollen structure. Several nuclei move up into the swelling along with sufficient amount of cytoplasm. Finally a septum separates the terminal portion from the rest of the antheridiophore. This structure is known as **antheridium.** It is stated that the nuclei in the antheridium undergo one or more mitotic divisions.

Fertilisation. At maturity, the antheridiophore grows towards the oogonium till the antheridium becomes closely attached to it (Fig. 6.5 E). One or more antheridia may become closely appressed against the oogonial wall in this way. A fine tubular outgrowth arises from the antheridium at the point of contact (F). It is the **fertilisation tube**. It pierces the oogonial wall and may give out slender branches. Each branch makes its way to one of the oospheres within the oogonium. Sometimes each oosphere in the oogonium receives a separate fertilisation tube from another antheridium (F & G). On coming in contact with the oosphere the fertilisation tube or its branch pierces it and discharges one male nucleus into it. The male nucleus fuses with the female nucleus to accomplish fertilisation. The fertilised egg or oosphere secretes a thick smooth wall around it to become an oospore which is gorged with fatty reserve stored in the form of oil globules or droplets outside the large membrane bound cell inclusion termed the **ooplast**. The reserve food may be stored outside the ooplast in a central (*S. hypogyna*), eccentric (*S. aunisosphorio*), subcentric (*S. tenespora*) or subeccentric (*S. seccentrica*) arrangement. The resultant oospore is a resting structure.

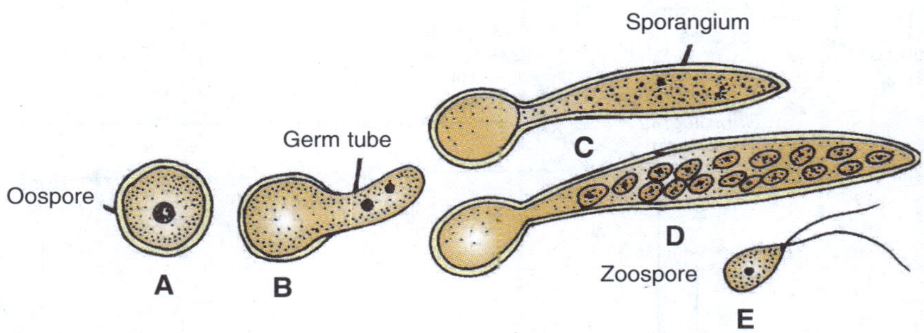

Fig. 6.6 (A-E). Stages in the germination of oospore of *Saprolegnia* sp.

Germination of Oospore (Fig. 6.6 A-E). Mature oospores are liberated by the degeneration of the oogonial wall. The oogonia degenerate after the mycelium bearing them has also undergone degeneration. The liberated oospores may remain inactive for a period of two to five months. In cultures oospores may not be liberated. They remain within the oogonium and germinate in situ (Fig. 6.7). The germ tube grows out through the wall of the old oogonium. At the approach of favourable conditions oospore germinates on a new substrate. It absorbs water and swells. The oospore wall becomes thinner. The protoplasmic contents grow out in the form of a tubular outgrowth, the **germ tube** (B). According to the older view (Fig. 6.8) diploid nucleus undergoes meiosis at the time of zygote germination. The germ tube contains haploid nuclei. During its further growth the germ tube may behave in either of the following ways:—

1. It may directly grow into a new mycelium.
2. It may swell up to form a club-shaped **germ sporangium** typical of the species (C). The terminal germ sporangium is delimited by a basal cross wall. The protoplasmic contents of the germ sporangium produce biflagellate zoospores (D). Each zoospore (E) germinates to form a new mycelium.

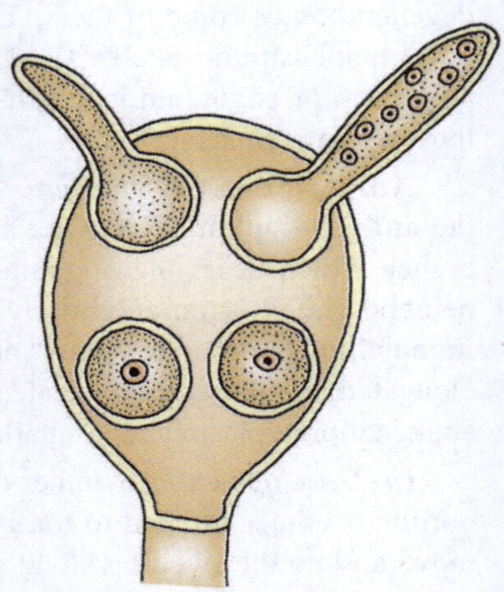

Fig. 6.7. *Saprolegnia sp.* showing *in situ* germination of oospores (within the oogonium).

Oospheres which fail to be fertilized develop **parthenogenetically.** In some species such as *S. ferax* no antheridia are developed next to many oogonia. The eggs of such oogonia function as **parthenospores.** Each parthenospore aparently is similar to an oospore. It develops parthenogenetically into a new mycelium.

Recent investigations have shown that in *Saprolegnia* meiosis occurs at the time of gamete formation (Howard and Moore, 1970) and (Beakes and Gray, 1977) and not at the time of germination of oospore as is generally believed. In case it is confirmed there will be a radical change in our concept of the life cycle of *Saprolegnia.* The somatic mycelium (vegetative phase) according to this view, will then be diploid and the only haploid structures in the life cycle will be the sexual nuclei or the gametes (Fig. 6.8 B).

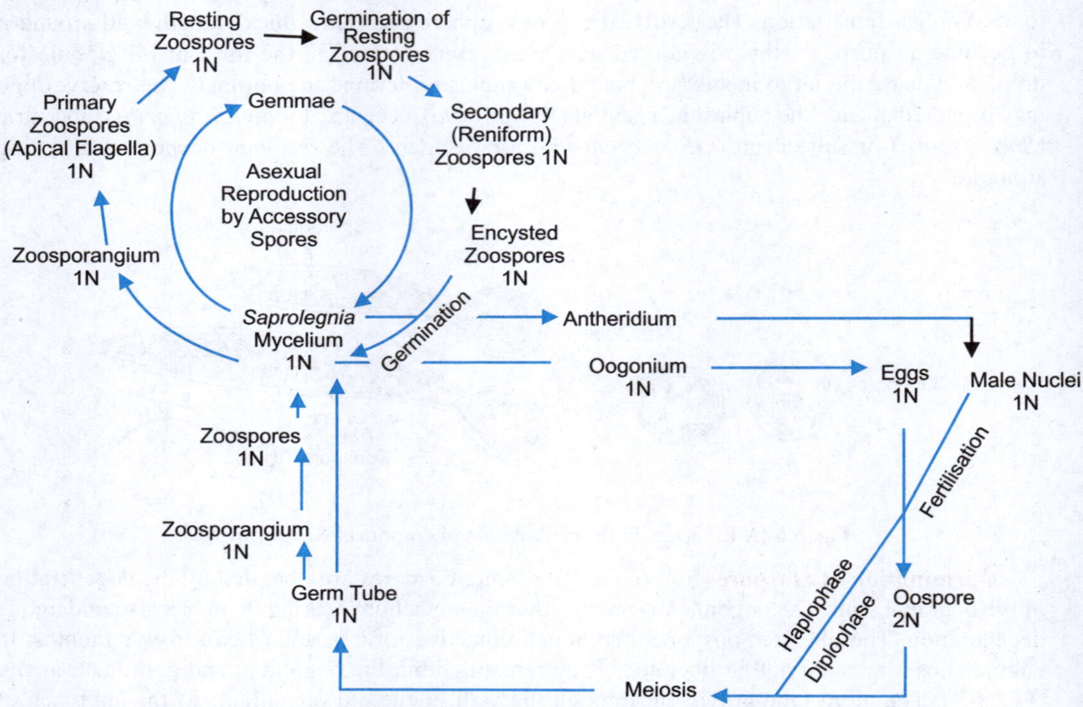

Fig. 6.8 A. Graphic representation of the life cycle of *Saprolegnia* with zygotic meiosis.

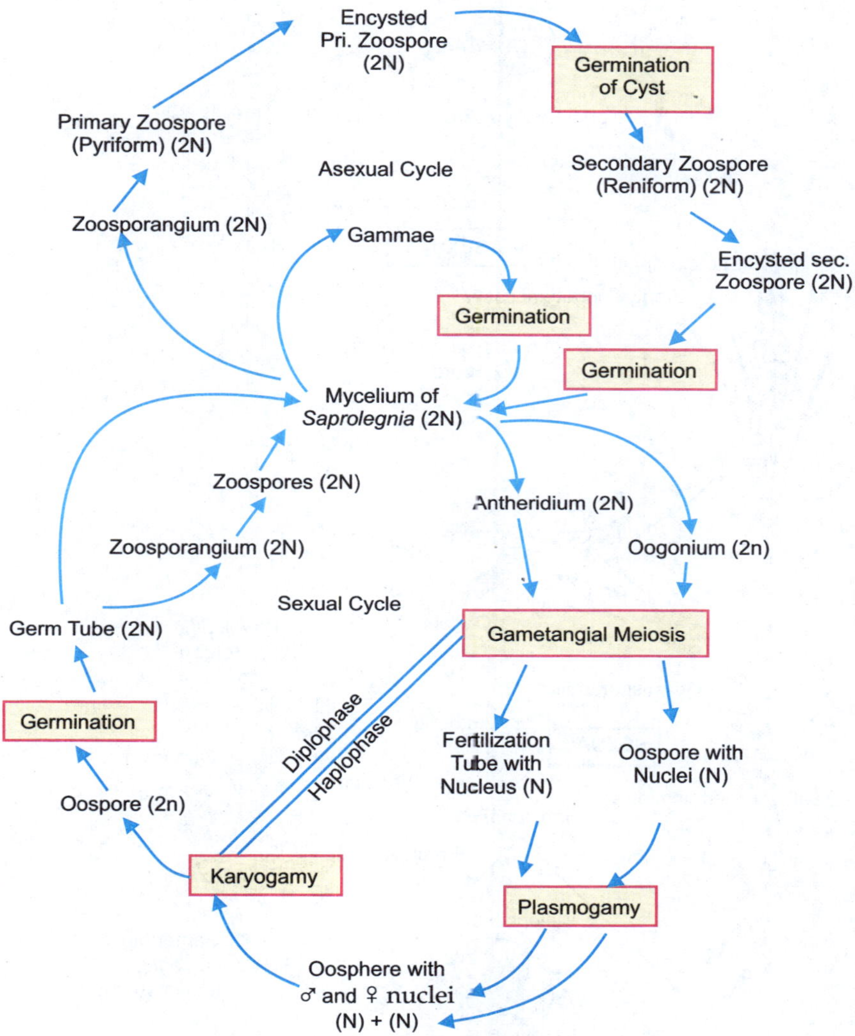

Fig. 6.8 B. Graphic representation of the life cycle of *Saprolegnia* with gemetangial meiosis.

Salient Features

1. The majority of the species of *Saprolegnia* are aquatic saprobes.
2. The mycelium is well developed, profusely branched, and coenocytic. It is often seen as tiny tufts of cotton wool around some bit of decaying plant or animal tissue in water, particularly the dead flies, a piece of meat, cut hemp seeds, etc.
3. The hyphae constituting the mycelium are of two kinds, *rhizoidal* and *extramatrical* hyphae. The former penetrate and ramify within the substratum. They serve to anchor the mycelium and absorb nutrition. The extramatrical hyphae grow on the surface of the substratum and project into the surrounding water. These are concerned with reproduction.
4. The hyphal walls contain cellulose and glucans but no chitin.
5. The mycelium reproduces vegetatively by fragmentation and formation of gemmae or chlamydospores.

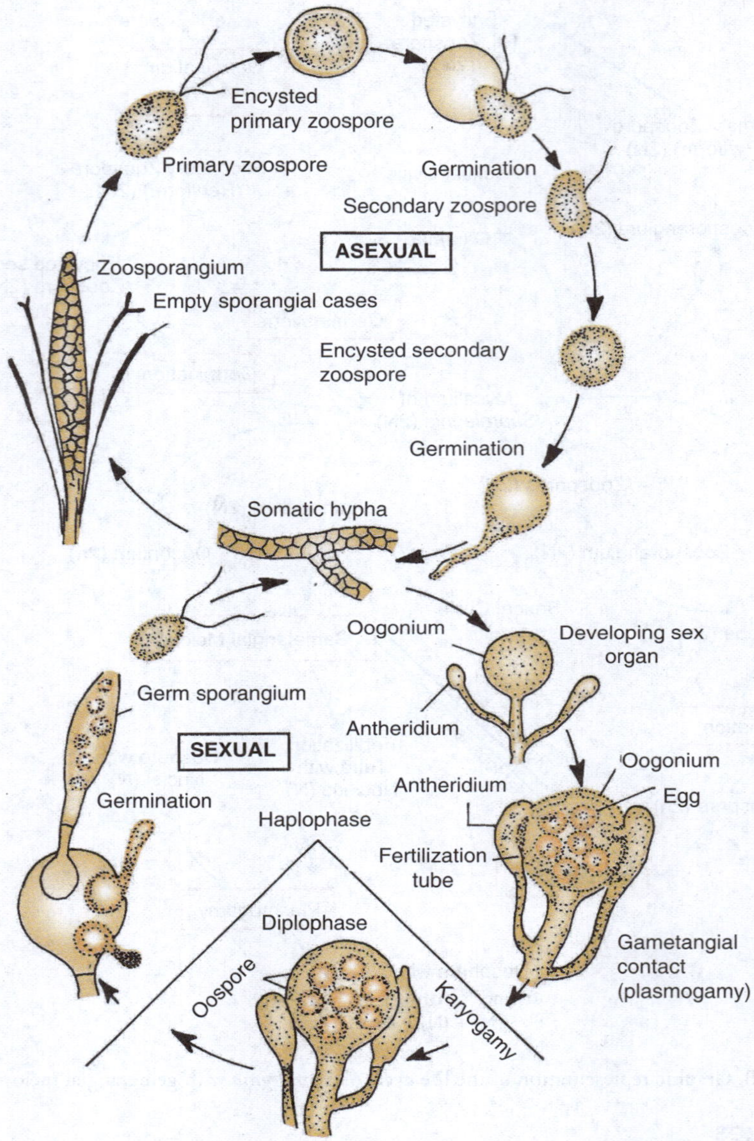

Fig. 6.9 Pictorial life cycle of *Saprolegnia*.

6. Asexual reproduction takes place by means of zoospores which are of two types, **primary** and **secondary.** For this reason *Saprolegnia* is said to be **dimorphic.**
7. The primary zoospores are produced in elongated, cylindrical zoosporangia formed singly at the tips of somatic hyphae. They are pear-shaped, biflagellate, uninucleate structures. The flagella are inserted apically. One of these is of whiplash type and the other tinsel. The zoospores are set free in the surrounding water through an apical pore.
8. After some period of activity the primary zoospore comes to rest, retracts its flagella and secretes a thin wall around it. Instead of the cyst germinaing by a germ tube its contents develop into a single kidney-shaped secondary zoospore with the two oppositely directed flagella inserted laterally in the depression.

9. The liberated secondary zoospores embark upon a second swarming period after which each encysts and germinates by a germ tube to form a new mycelium. *Saprolegnia* is thus **diplanetic.**
10. Proliferation of zoosporangia is a characterisitic feature of *Saprolegnia.*
11. Sexual reproduction is oogamous. The antheridia and oogonia are either formed on the neighbouring hyphae or on the same hypha, the antheridium arising below the oogonium from its stalk. The gametes are aplanogamic
12. The mature oogonium is usually a rounded structure containing several uninucleate oospheres.
13. The mature antheridium is a multinucleate elongated structure much smaller than the oogonium. One or more antheridia become attached to the oogonium by the curvature of its antheridiophore.
14. Fertilisation takes place by means of the fertilisation tube. Entering the oogonium it may branch sending one branch to each oosphere or each oosphere receives a fertilisation tube from another antheridium. Only one male nucleus is discharged into the oosphere. No sperms are organised in the antheridium.
15. After fusion of the male and female nuclei each zygote secretes a thick smooth wall around it to become an oospore.
16. After a prolonged period of rest each oospore germinates by producing a germ tube which in some species directly develops into a new mycelium and in others ends in a terminal germ sporangium in which are produced the biflagellate zoospores.
    Each liberated zoospore germinates in the usual way by a germ tube to form a new mycelium.
17. Recent investigations have shown that meiosis in *Saprolegnia* is gametangial. It occurs at the time of differentiation of sexual nuclei or gametes.
18. The somatic phase in the life cycle is thus diploid.
19. The gamete nuclei are the only haploid structures. There is no haploid phase or gametophyte generation in *Saprolegnia.*
20. Such a life cycle in which there is no alternation of generation and meiosis is gametangial is called diploid or diplohaplontic (haplobiontic-diploid).

Taxonomic Position

Division	:	**Mycota**
Subdivision	:	**Eumycotina**
Class	:	**Oomycetes**
Order	:	**Saprolegniales**
Family	:	**Saprolegniaceae**
Genus	:	*Saprolegnia*
Species	:	*litoralis*

Alexopoulos and Mims (1979) place *Saprolegnia* in the Superkingdom Eukaryonta, Kingdom Myceteae, Division Mastigomycota, Subdivision Diplomastigomycotina, Class Oomycetes, Order Saprolegniales and Family Saprolegniaceae.

Superkingdom	:	Eukaryonta
Kingdom	:	Mycetae
Division	:	Mastigomycotina
Subdivision	:	Diplomastigomycotina
Class	:	Oomycetes
Order	:	Saprolegniales
Family	:	Sapolegnia
Genus	:	*Saprolegnia*

ACHLYA

General. The genus *Achlya* includes about 47 species which are found all over the world in suitable habitats. *A. bisexualis* and *A. racemosa* are common and the best known among them. Most of the species are aquatic. A few grow on damp humus soil. The aquatic species are found in the water of streams, ponds, pools and aquaria growing on floating seeds, dead insects or any bit of organic material of plant or animal origin. A culture of this fungus can easily be raised in the laboratory by placing a cut hemp-seed or dead fly in a petri-dish containing pond water. The fungus makes its appearance on the substrate in a couple of days.

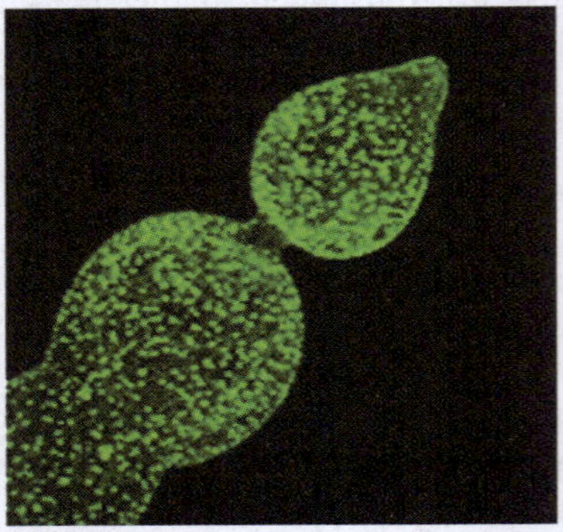

Achlya bisexualis growing on floating seeds

Achlya racemosa growing on dead insects in water

Mycelium. It is in the form of a white, tangled mass consisting of branched hyphae growing in all directions. Some of the hyphae, which are short, penetrate the substratum. They anchor the mycelium and absorb nutrient materials. The hyphae are aseptate, sparingly branched and coenocytic (Fig. 6.9 A). The septa normally remain suppressed during the somatic phase. The hyphal wall contains cellulose. However, it may not form the predominant material According to some investigators, glucans may well be the predominant material. The numerous nuclei lie in the peripheral layer of cytoplasm surrounding a central vacuole.

Reproduction. *Achlya* reproduces by asexual and sexual methods of reproduction.

1. Asexual reproduction. It takes place by vegetative methods and sporulation.

(*a*) Vegetative reproduction. Like *Saprolegnia, Achyla* reproduces vegetatively by fragmentation and gemmae (chlamy-dospore) formation. The gemmae are formed under unfavourable conditions. These are dense aggregations of cytoplasm in certain portions of the coenocytic hyphae. They become cut off by septa. The gemmae may be terminal or intercalary in position and may be produced singly or in chains. They serve as a means of **perennation** during periods of unfavourable conditions.

(*b*) Sporulation (Fig. 6.10 A-E). It takes place by zoospores. They are produced in long, cylindrical zoosporangia borne singly and terminally on somatic hyphae. Sporangium formation starts with a slight inflation of the hyphal tip (Fig. 6.10 A). The swelling grows into a tubular or club-shaped structure by the inflow of cytoplasm and nuclei from the lower part of the hypha. Finally the swelling is cut off as a **zoosporangium** by a cross wall appearing at its base (Fig. 6.10 B). The zoosporangium is poorly differentiated from the somatic hyphae. It can only be distinguished by its dense contents and basal septum. Its diameter may also slightly more than the hypha which bears it. The zoosporangium tapers slightly more than the hypha which bears it. The zoosporangium tapers slightly both towards the base as well as apex. Later a central vacuole appears in the dense greyish contents of the zoosporangium which markedly contrast with the hyaline watery contents of the parent hypha (Fig. 6.10 B). Cleavage of protoplast starts from the centre. Fine, intersecting

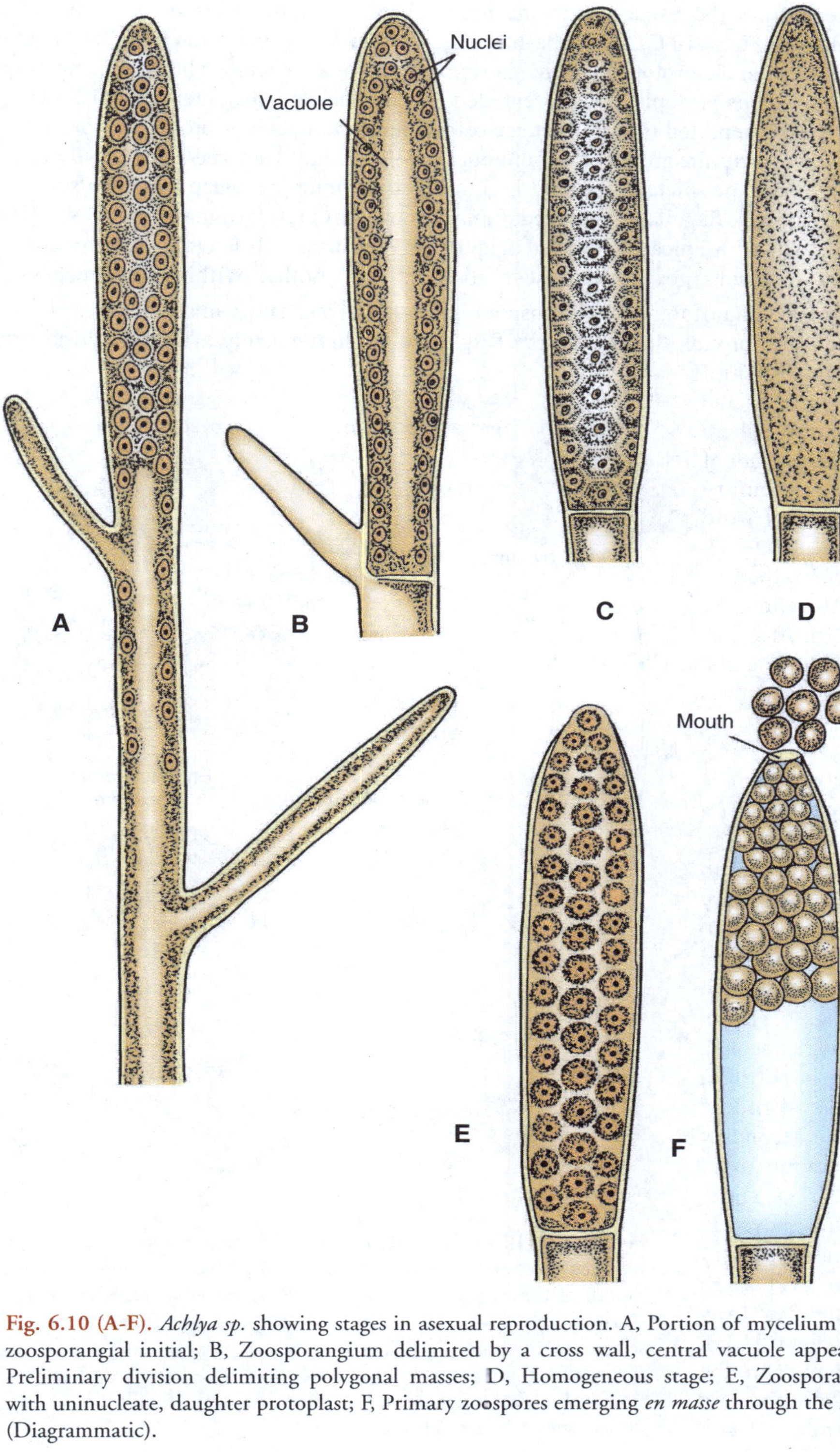

Fig. 6.10 (A-F). *Achlya sp.* showing stages in asexual reproduction. A, Portion of mycelium with a zoosporangial initial; B, Zoosporangium delimited by a cross wall, central vacuole appears; C, Preliminary division delimiting polygonal masses; D, Homogeneous stage; E, Zoosporangium with uninucleate, daughter protoplast; F, Primary zoospores emerging *en masse* through the mouth (Diagrammatic).

lacunae extend from the vacuole into the multinucleate peripheral protoplast dividing it into polygonal masses (Fig. 6.10 C). A papilla-like outgrowth may develop at the tip of zoosporangium. By a peculiar change the protoplast of the zoosporangium again becomes homogeneous (Fig. 6.10 D). The homogeneous protoplast is diferentiated into uninucleate, daughter protoplasts (Fig. 6.10 E). Hawker (1967) reported that in *A. racemosa* uninucleate daughter protoplasts are formed by aggregation of cytoplasm around the numerous nuclei found scattered in the cytoplasm. The daughter protoplasts get metamorphosed into biflagellate **primary zoospores.** The soil dwelling species of *Achlya* lack flagella. Flagellated primary zoospores in *Achlya* may be pyriform (Hawker, 1967) or rounded with apically attached flagella or reniform with laterally inserted heterokont flagella. Before discharge they are seen jostling one another within the zoosporangium.

The apical papilla of the mature zoosporangium dissolves. The primary zoospores escape *en masse* in a cluster through the apical pore (Fig. 6.10 F). Immediately after their discharge, they stop swimming and become arranged in the form of a hollow sphere just outside the mouth of the empty zoosporangium and encyst (Fig. 6.11 A). Thus, in *Achlya* there is a tendency to eliminate the first motile or swarming stage. The first period of diplanetism is extremely short as compared with *Saprolegnia*. After a short resting period of a few hours each cyst (Fig. 6.10 B) germinates to give rise to a secondary zoospore (Fig. 6.11 C-D) which is reniform and has two laterally inserted flagella (Fig. 6.11 D). The flagella are unequal. The shorter is of tinsel type and is directed forward. The secondary zoospores swim about actively for a time, disperse and encyst in turn. From secondary cysts (Fig. 6.11 E-F) emerge the **tertiary zoospores** (Fig. 6.11 G). They are identical to the secondary zoospores in every respect. Each

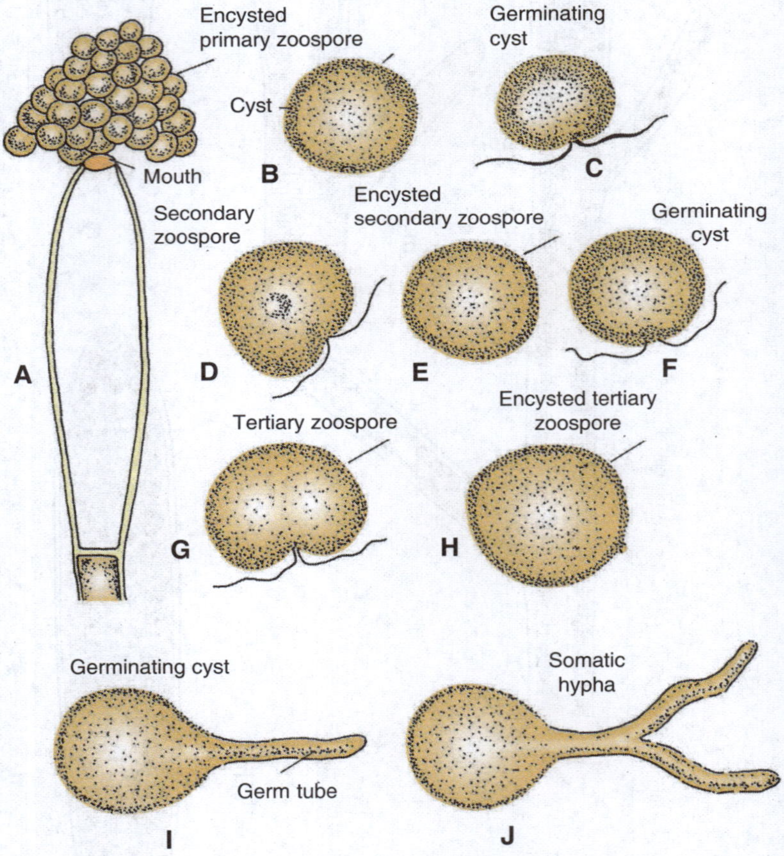

Fig. 6.11 (A-J). *Achlya* **sp.** Liberation of primary zoospores and stages in repeated zoospore emergence. A, Encysted primary zoospores outside the mouth of the empty zoosporangium; B, Encysted primary zoospore; C, Germinating cyst with secondary zoospore emerging; D, Released secondary zoospore; E, Encysted secondary zoospores; F, Germinating cyst with tertiary zoospore emerging; G, Released tertiary zoospore; H, Encysted tertiary zoospore; I, Germinating cyst; J, Germ tube developing into a somatic hypha (Diagrammatic).

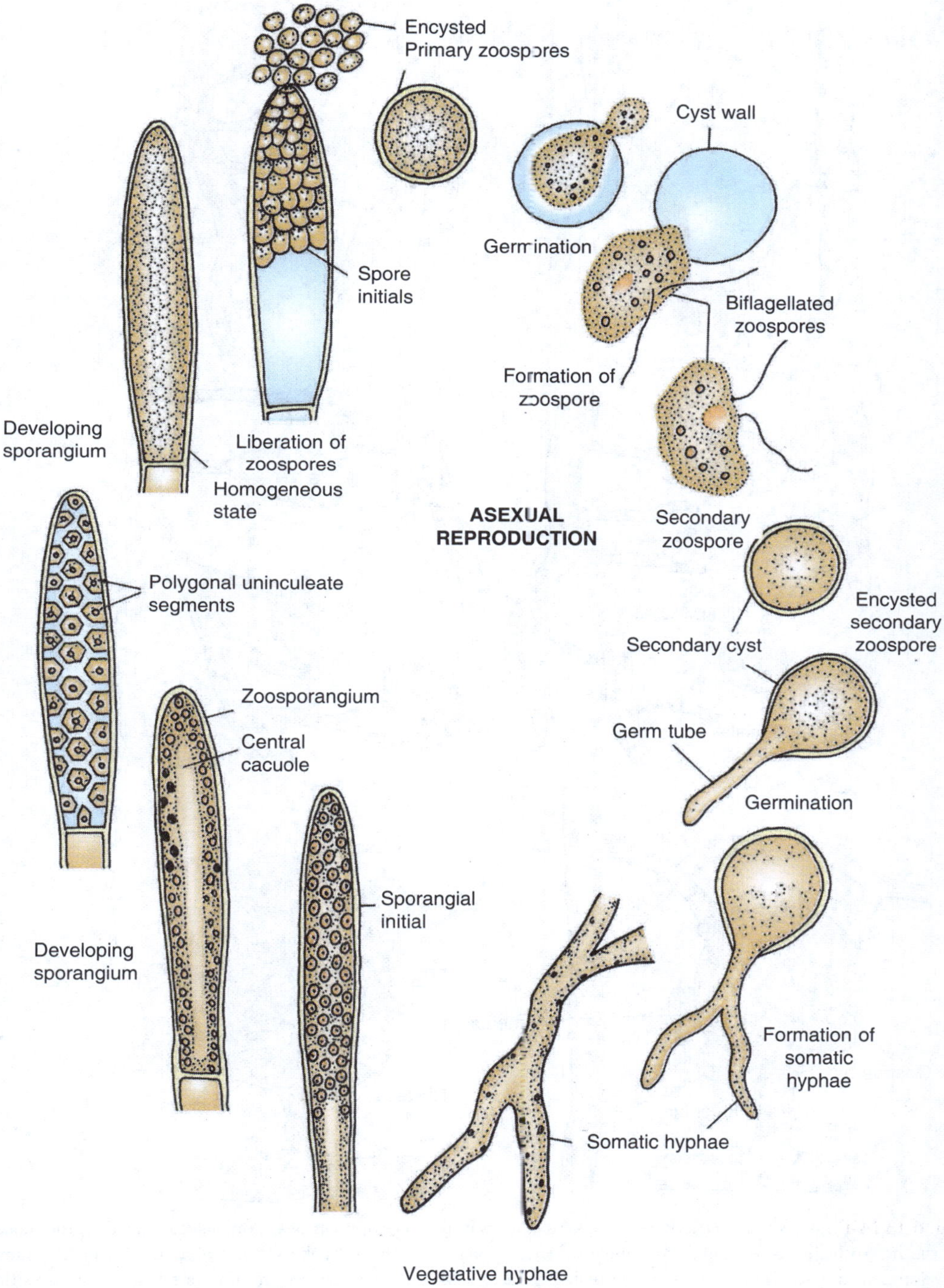

Fig. 6.12. Asexual reproduction in *Achlya*.

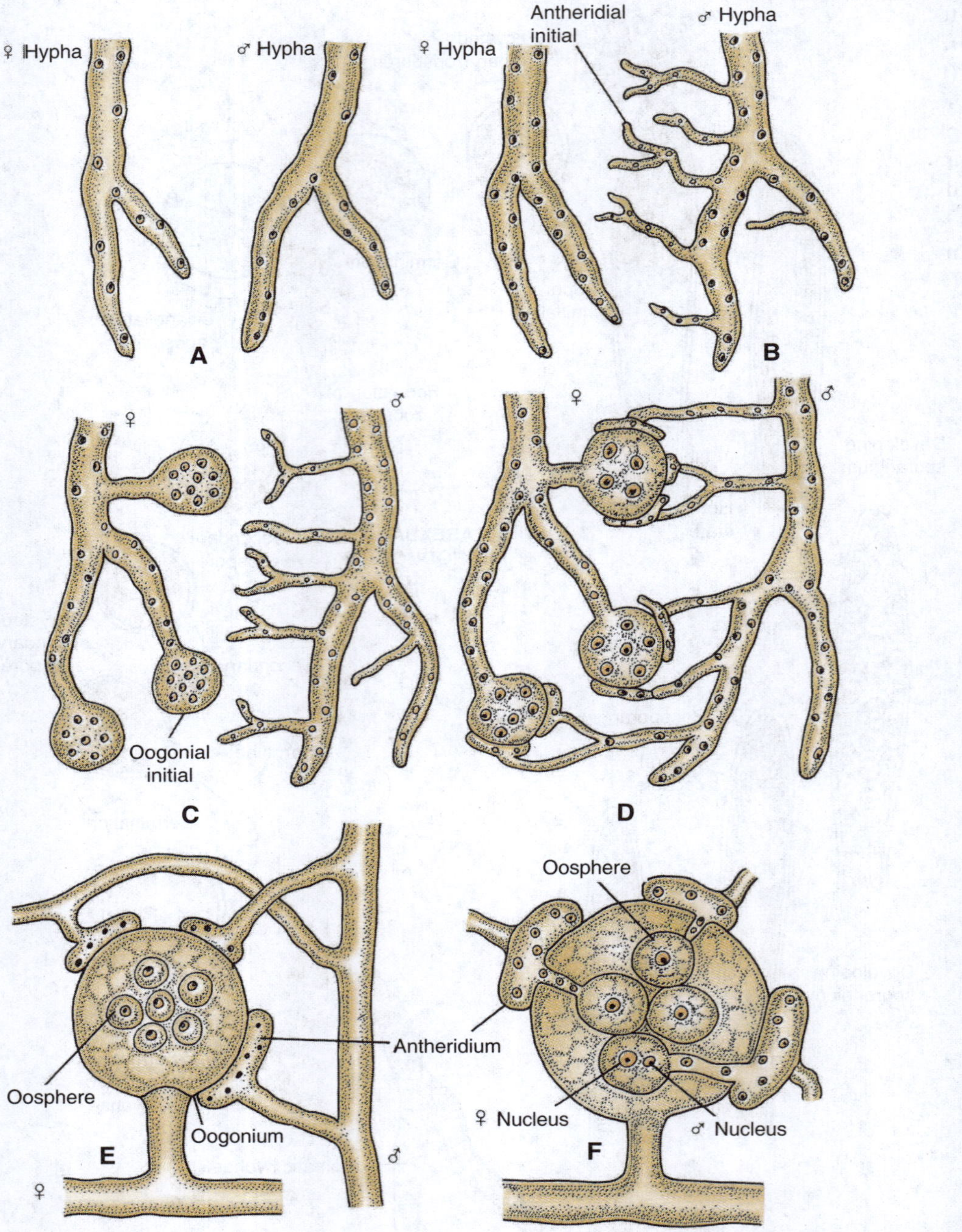

Fig. 6.13 (A-F). *Achlya bisexualis,* Raper. Successive stages in the development of sex organs and sexual reproduction in the heterothallic sp. A, female and male somatic hyphae; B, Female hypha secretes homone A which induces development of antheridial initials on the male hyphae; C, Male hypha now secretes homone B which induces the female filament to form oogonial initials; D, Oogonial initials secrete hormone C which directs the development of antheridial initials towards the oogonial initials and the delimitation of antheridia at their tips; E, Antheridia secrete a final hormone D which causes the delimitation of oogonia and differentiation of eggs within the oogonium; F, Antheridia put out the fertilisation tubes which pierce the oogonium and make contact with the eggs.

tertiary zoospore germinates by a germ tube to develop into a new mycelium (Fig. 6.11 I-J). In *A. racemosa* the cysts of secondary zoospores germinate by germ tubes to develop into new mycelia (Hawker, 1967). Salvin (1940) reported the occurrence of 5 successive swarming stages in a non-sexual *Achlya. Achlya* thus is an example of **polyplanetism** or repeated zoospore emergence. This phenomenon is more marked in the genus *Dictyuchus.* The primary, secondary and tertiary zoospores are all similar in structure and shape and are, therefore, **monomorphic** as compared to the **dimorphic** zoospores of *Saprolegnia.*

Usually there is no internal proliferation of zoosporangia in *Achlya.* A new zoosporangium may, however, grow up from the same hypha at the side of emptied zoosporangium. Hawker (1967) reports that zoosporangia may arise from the tip of the same hypha. Some of these may grow through the old empty ones. The internal proliferation of sporangia is considered a primitive character. It denotes poor differentiation of sporangia from the somatic hyphae.

2. Sexual reproduction. It is **oogamous.** The sex organs (antheridia and oogonia) are developed on some of the somatic hyphae. A few species of *Achlya* are **homothallic** (*A. racemosa*) but the majority such as *A. bisexualis* are **heterothallic** (Fig. 6.13). In the latter a single *Achlya* plant produces either male or female sex organs not both.

(*a*) *Development of sex organs in the Heterothallic species* (Fig. 6.13). Investigations of Raper (1939-1951) of sexual reproduction in the heterothallic species, *A. bisexualis* and *A. ambisexualis* have revealed that the development of sex organs is governed by a set of hormones. When grown separately neither male nor female thalli form sex organs. Instead the hyphae bear sporangia. A male mycelium is or has been growing there and *vice versa.* It means the development of male and female structure is controlled by the hormones that pass between the plants of opposite sexes. The hormones produced by the male plant stimulate the development of oogonia in the female plant. The hormones produced by the female stimulate production of antheridia on the male plant. Raper has termed this kind of mixed sexuality as **gynandromixis.** The production of antheridia and oogonia are governed by four hormones as follows :—

1. The female plant of *A. bisexualis* secretes hormone *A* (*Antheridial*). It induces the male plant to produce antheridial hyphae called **antheridial initials** (B).
2. The male secretes a hormone *B.* It induces the formation of sac-like **oogonial initials** (C).
3. The oogonials now secrete a third hormone *C.* It directs the growth of antheridial initials or hyphae towards the sources (oogonial initials) and plaster themselves around the latter. The antheridia are finally delimited at the tips of antheridial hyphae (D).
4. Upon contact with the oogonial initials the antheridia secrete a fourth and final hormone *D.* It induces the oogonial initials to form oogonia and cause the differentiation of eggs within them (E).

(*b*) *Development of Sex Organs in the Homothallic species* (Fig. 6.14 A-G). The two kinds of sex organs in the homothallic species (*A. racemosa*) are developed on the same mycelium.

Oogonium. The oogonia are developed at the tips either of short lateral hyphae or on a main hypha. The tip of the female branch swells to form a sac-like oogonial initial (A). It is filled with dense homogeneous contents (cytoplasm and nuclei). A central vacuole then appears in the oogonial initial (Fig. 6.14 B). Subsequently the swollen structure is cut off by a cross wall at its base (C). It is the **oogonium.** The oogonia are globose or pyriform in shape and have a thin, smooth hyaline wall. Some of the nuclei in the oogonium degenerate. The surviving ones divide mitotically. The entire multinucleate protoplast now becomes differentiated into one to ten, generally two to six eggs (D). Each mature egg or oosphere is uninucleate and is usually spherical, occasionally ellipsoidal in form. It lacks a surrounding wall. The nucleus is centrally located (D).

Antheridium. The antheridia are developed on thin hyphal branches which arise from the female hypha. The tip of the antheridial hypha enlarges (B). A number of nuclei and some cytoplasm migrate into the inflated tip. The swollen portion is then cut off by a cross wall as an **antheridium** which is more or less club-shaped. Some nuclei in the antheridium degenerate. The survivors divide mitotically. Subsequently the antheridium comes in contact with the oogonial wall which gets thinner at the point of contact.

Fertilisation. The mature antheridium applies itself chaemotropically to the oogoniumm and develops a slender penetration tube at the point of contact. It is known as the **fertilisation tube** (D). The latter pierces through the oogonial wall and reaches an egg. The fertilisation tube may give off branches to the other eggs in the oogonium or more than one antheridia become attached to the oogonium each sending a fertilisation tube. The latter delivers one male nucleus into each oosphere. It fuses with the nucleus of the oosphere. The cytoplasm of fertilised egg secretes a wall (E) which is generally not very thick.

Germination of oospore (Fig. 6.14 F-G). The oospores are released by the degeneration of the oogonial wall. They enter upon a resting period. With the return of favourable conditions the oospores germinate. The outer layer of the oospore wall ruptures. The protoplast still surrounded by the inner layer emerges through the split in the form of a simple

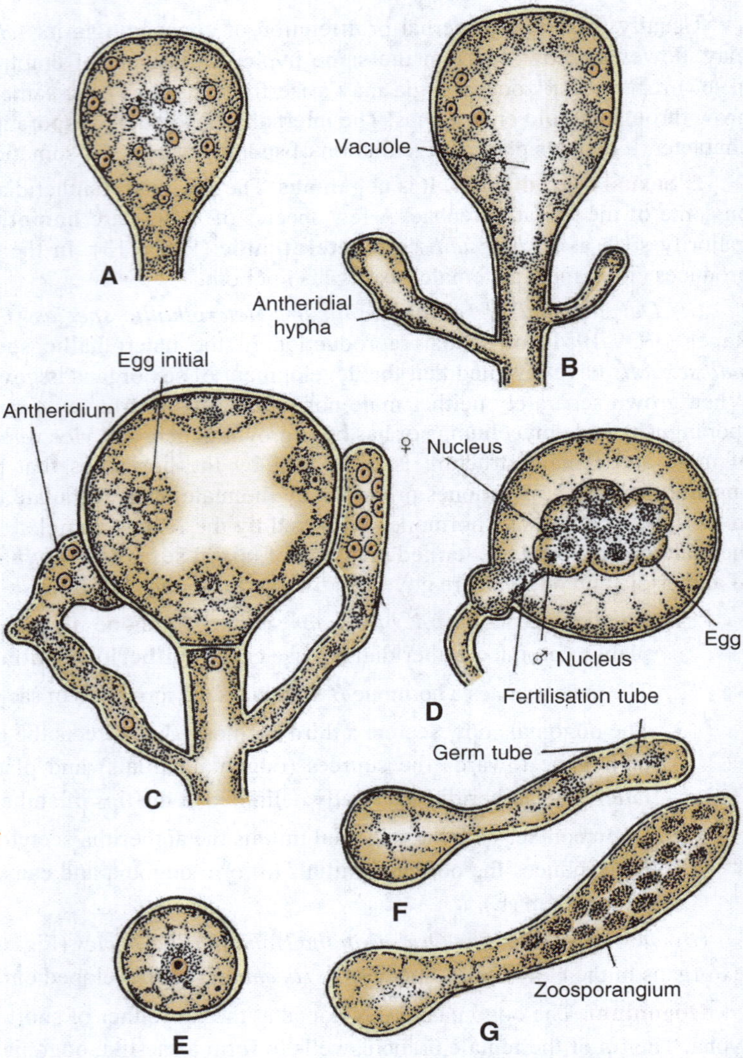

Fig. 6.14 (A-G). *Achlya sp.* Successive stages in the development of sex organs in the homothallic species and germination of oospore. A, Terminal swelling to form oogonial initial; B, Young oogonium with a central vacuole and two antheridial hyphae arising from the oogonial stalk; C, Oogonium with egg initials differentiated, antheridia delimited at the apices of male hyphae; D, Mature oogonium showing fertilisation; E, Oospore; F, Oospore germinates to form the germ tube; G, Germ tube ending in a zoosporangium. (Diagrammatic).

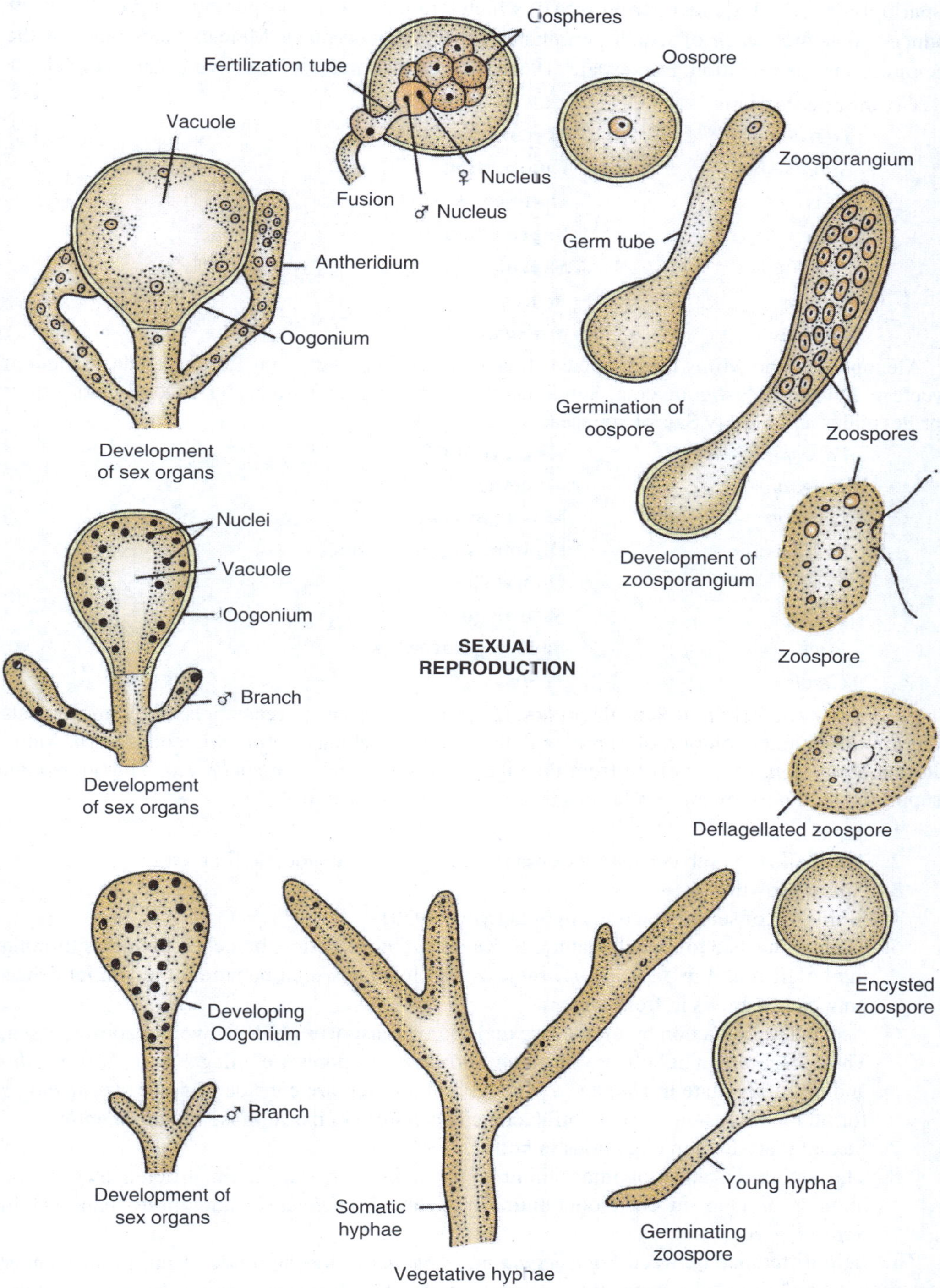

Fig. 6.15. Sexual Reproduction in *Achlya.*

or sparingly branched, slender germ tube (F) which terminates in a zoosporangium (G). The latter produces zoospores, each of which germinates to form a new plant. Meiosis takes place at the time of oospore germination. Barkesdale (1968) showed that meiosis in *Achlya* is gametangial.

Taxonomic Position :

Division	:	**Mycota**
Subdivision	:	**Eumycotina**
Class	:	**Oomycetes**
Order	:	**Saprolegniales**
Family	:	**Saprolegniaceae**
Genus	:	*Achlya*
Species	:	*bisexualis*

Alexopoulos and Mims (1979) place *Achlya* in the Superkingdom Eurkaryaonta, Kingdom Myceteae, Division Mastigomycota, Subdivision Diplomastigomycotina, Class Oomycetes, Order Saprolegniales and Family Saprolegniaceae.

Superkingdom	:	**Eukaryonta**
Kingdom	:	**Mycetae**
Division	:	**Mastigomycotina**
Subdivision	:	**Diplomastigomycotina**
Class	:	**Oomycetes**
Order	:	**Saprolegniales**
Family	:	**Saprolegniaceae**
Genus	:	*Achlya*

Phylogeny and origin of Saprolegniales. Divergent views have been expressed by mycologists about the origin and evolution of Saprolegniales to which belong *Saprolegnia* and *Achlya.* Many hold that *Saprolegnia* has evolved from the algal ancestors much like *Vaucheria.* This viewpoint is supported by the following similarities that exist between the two :—

1. Aquatic habit.
2. The thallus in both consisting of branched, aseptate, coenocytic filaments.
3. Apical growth.
4. Similarity of serum reactions in both (Mez, 1929).
5. Thallus attached to the substratum; In *Vaucheria* by colourless branched filaments forming the holdfast and in *Saprolegnia* by rhizoidal hyphae which penetrate the animal debris on which it grows in fresh water.
6. Asexual reproduction by motile, asexual spores **zoospores** produced within zoosporangia. The zoosporangia of both are similar in form but zoospores are biflagellate in *Saprolegnia* and multiflagellate in *Vaucheria.* The latter, however, are considered to be *synzoospores* formd by the fusion of many biflagellate zoospores similar to those of *Saprolegnia.*
7. Sexual reproduction oogamous in both.
8. The antheridia and oogonia similar in form in both. The main difference is in the number of eggs in each oogonium. It is one in *Vaucheria* and more than one in *Saprolegnia.*

The main difference between *Saprolegnia* and *Vaucheria* are the mode of nutrition, manner of fertilisation and diplanetic zoospores in *Saprolegnia.* The adherents of this view envisaged the evolution of *Saprolegnia* from a Vaucherial ancestor by the loss of chlorophyll and change in the mode of nutrition. This happened before the Vaucheria-like ancestor had developed compound zoospores from many biflagellate zoospores. It is not hard to believe that certain flagellates under certain conditions develop chlorophyll and synthesize their own food or lose it and live as saprobes. Therefore, by the loss of chlorophyll and holdfast would evolve a thallus very similar to an Oomycete like *Saprolegnia.* Diplanetic zoospores, characteristic of *Saprolegnia,* probably existed in the Vaucherial ancestor. The similarity between the zoospores of *Saprolegnia* and the sperms of *Vaucheria* which are colourless in spite of their algal origin tempted some mycologists to consider that the zoospores of present-day *Vaucheria* correspond to the first motile stage and the sperms to the second motile stage of *Saprolegnia* (Gaumann).

The above-mentioned view of the origin of Saprolegniales now receives but little support from the modern mycologists as it emphasises only on morphological resemblance between *Saprolegnia* and *Vaucheria.* It ignores completely the physiological differences and type of flagellation. Many mycologists consider these resemblances to be superficial indicating merely parallel development along different lines rather than close relationship.

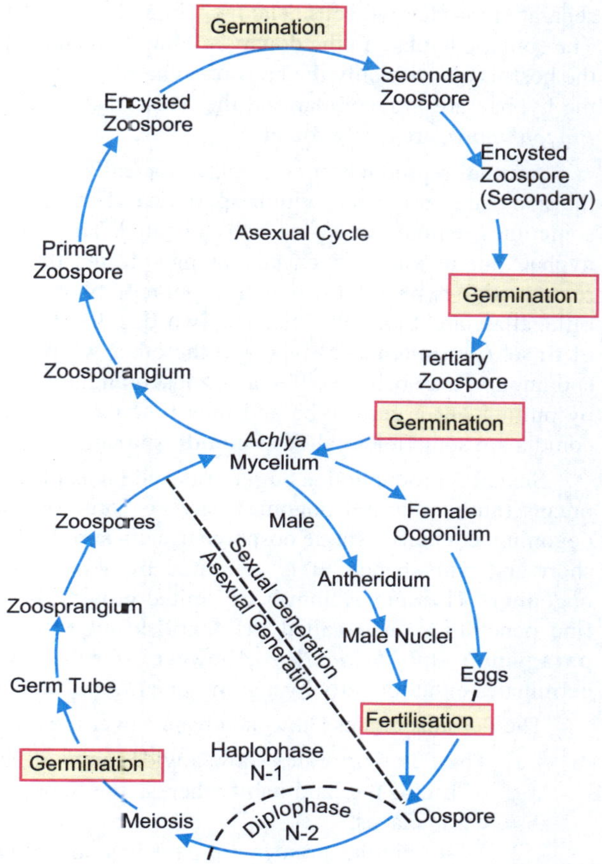

Fig. 6.16. *Achlya* sp. Graphic representation of the life cycle (Zygotic meiosis).

The opponents of the algal origin of primitive Oomycetes like *Saprolegnia* hold that the origin of *Saprolegnia* like ancestors must be sought among the colourless flagellates. This viewpoint is supported by the fact that among the lower fungi there are simple forms which show structural similarities with the Myxomycetes. The latter do not appear to be far removed from the flagellate ancestors. In the Archimycetes there are forms which may be interpreted as leading from the simplest fungi to the simple Oomycetes such as *Saprolegnia.*

ORDER : PERONOSPORALES

General Characteristics. The order is represented by 12 genera and about 325 species. In habit they range from aquatic to the terrestrial ones through the species with amphibious way of life. Many of the terrestrial species are highly specialized obligate plant parasites which cause serious diseases to important economic plants. A few, however, occur as parasites of thallophytes. The rest are saprophytic. Some parasitic species can live as saprophytes after the death of the host (*pythium*). The mycelium is well developed, branched and coenocytic. The septa normally remain suppressed in the vigorously growing vegetative hyphae but appear to wall off injured parts of separate reproductive cells. The mycelium shows no distinction into rhizoidal and aerial hyphae. The somatic hyphae in the downy mildews which are parasitic, inhabit the intercellular spaces of the host tissue with only the haustoria entering the host cells. In most of the Pythiaceae, however, the hyphae are intracellular and the haustoria, with the exception of a few advanced species of *Phytophthora*, are rarely found.

Asexual reproduction takes place typically by the formation of motile asexual spores called the zoospores produced within sporangia. The sporangia are usually globose or oval in form sometimes lemon shaped but not elongated. They are developed at the tips of ordinary somatic hyphae but in more advanced complex forms the sporangia are borne at the tips of special reproductive hyphae known as the sporangiophores. The zoospores are typically kidney-shaped, biflagallate and monoplanetic. The two flagella are inserted laterally on the concave side, one is of **tinsel (pantonematic)** type and the other **whiplash (acronematic)** type. In some species the sporangium may behave either as a zoosporangium and produces zoospores or directly germinates by putting out a germ tube and thus functions as a conidium. The sporangia which behave as conidia are sometimes called as **conidiosporangia**.

Sexual reproduction is **oogamous** and takes place by gametangial contact. Both kinds of sex organs (antheridia and oogonia) may be borne on the same or different hyphae. The globose oogonium contains a single oosphere usually surrounded by **periplasm**. The antheridium is usually short and club-shaped. It is delimited by a basal septum and usually arises from below the oogonium. The antheridium is of limited growth. Fertilisation is usually effected by means of a fine penetration tube called the **fertilisation tube**. This kind of fertilisation is described as **paragynous**. In *Phytophthora*, however, fertilisation is of **amphigynous type**. The oospore germinates either directly by a germ tube (*Peronospora*) or produces zoospores (*Albugo*).

The Peronosporales show an advance over the Saprolegniales in the following respects:-

1. The Peronosporales show a wide range of habit from aquatic, through amphibious way of life to the land habit whereas the Saprolegniales are mostly aquatic. Some occur in damp places.
2. Many Peronosporales are parasites and some are saprophytes. A few parasitic species can live as saprobes after the death of the host. The Saprolegniales are mostly saprophytes rarely parasites.
3. The sporangia in the Peronosporales usually get detached and generally function as conidia. They are globose or oval in form whereas in the Saprolegniales they are tubular or club-shaped in form and do not get detached and never function as conidia.
4. The sporangia in many species are borne on special reproductive hyphae called the **sporangiophores** whereas in the Saprolegniales they are always borne on hyphae usually similar to the somatic hypae which are hardly distinguishable from the sporangia.
5. The zoospores in the Peronosporales are typically reniform in shape and biflagellate with the flagella arising from the concave side. They are thus **monoplanetic** whereas in the Saprolegniales the zoospores are generally **diplanetic** to **polyplanetic**. The primary zoospores are pyriform with the two flagella inserted apically and the secondary ones are reniform with the two flagella arising from the concave side.
6. The oogonium in the Peronosporales contains a single egg which is surrounded by periplasm whereas in the Saprolegniales the oogonium lacks periplasm and contains more than one egg.

Classification. The order is generally divided into three familes on the basis of characteristics of sporangiophore and sporangia. A summary of the salient features and key to these families viz. **Pythiaceae, Peronosporaceae** and **Albuginaceae** are as follows :

Pythiaceae. Sporangiophores undifferentiated from the mycelium, indeterminate in growth, branched, haustoria absent or branched, facultative parasites or saprophytes.

Peronosporaceae. Sporangiophores easily distinguishable from the mycelium, branched, determinate in growth, sporangia singly or in clusters, borne at the characteristically branched sporangiophores; haustoria varied or branched; obligate parasites.

Albuginaceae. Sporangiophores unbranched clavate, each producing a chain of conidia, in basipetal succession; haustoria knob shaped; obligate parasites.

The representatives of these familes are discussed in detail.

FAMILY : PYTHIACEAE

The family includes both aquatic and terrestrial species. Some are amphibious. When arranged in a series they show apparent advance on the previous one in adaptation to a land habit. A few truly aquatic species parasitize algae and small protozoa. Many of the land forms parasitize crop plants and cause serious diseases. Others are saprophytic. Some of the saprophytic species can live as facultative parasites and the parasitic species exhibit saprophytic tendencies. The mycelium is extensively developed and consists of branched, coenocytic hyphae which do not show any distinction into rhizoidal and aerial hyphae. The septa appear only to wall off the injured parts or to separate the reproductive cells. However, theses septa are solid plates and have no pore. The hyphae in most of the species are **intracellular** (penetrate the host-cells) and usually of regular diameter. The haustoria are produced in some species

Asexual reproduction is by means of sporangiospores. The sporangia are borne at the tips of hyphae which may or may not be distinct from the somatic hyphae. The sporangia may remain attached to the parent hyphae and produce zoospores. In most advanced cysts the sporangia are deciduous and get detached. The detached sporangia are dispersed by wind or water. On reaching a suitable substrate they either function as zoosporangia producing zoospores or germinate directly forming a germ tube. Some authors call these detached sporangia as **conidiosporangia**, others liken them to **conidia**.

Sexual reproduction is **oogamous**. It takes place by gametangial contact. The sex organs are formed from the terminal portions of the hyphal branches. The family Pytheaceae includes 8 genera (Waterhouse, 1973). The two best known common genera of the family are *Pythium* and *Phytophothora*. The latter is considered more advanced biologically and morphologically than the former. Both of these are discussed here in some detail.

PYTHIUM

It is the largest genus of the family comprising about 92 species (Waterhouse, 1968) which are cosmopolitan in their distribution. Some are aquatic. They parasitize fresh-water algae. Many are amphibious in their habitat but the majority are soil inhabitants. The amphibious species grow saprophytically in moist humus soil and infect the hypocotyl of seedlings of the various land

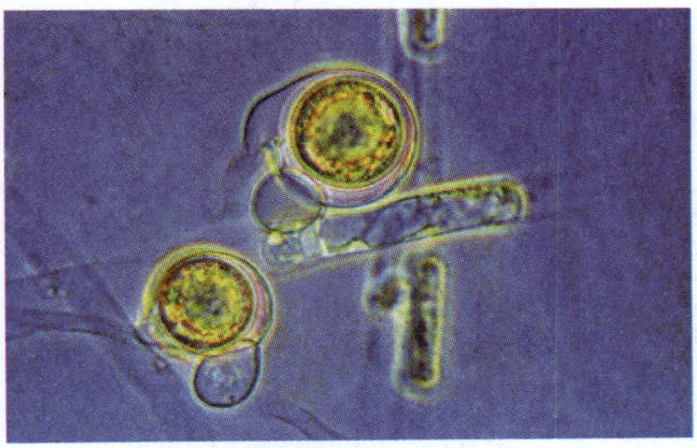

Pythium debaryanum sex organs

Pythium aphanidermatum on inflected seedling

plants such as papaya, mustard, wheat, ginger, tobacco and the like. Thereafter they live as parasites. They often cause great havoc in green houses, seed beds and nurseries. The common and widespread diseases caused by them are the *"damping off"* of tobacco and chillies by *P. debaryanum*, soft rot of papaya and *"damping off"* of potato by *P. aphaonidermatum* or "foot rot" of seedlings of ginger by *P. myriolium*. Most of the species of *Pythium* are root-rot pathogens. Some attack vegetative organs such as cabbage heads and turnip roots. In fact species of *Pythium* constitute one of the most destructive groups of plant pathogens. The saprophytic species play an equally important role in the decay and decomposition of plant remains in the soil. One of the commonest species of *Pythium* is *P. debaryanum.*

PYTHIUM DEBARYANUM

Occurrence. *P. debaryanum* is the classic example of the most common species of *Pythium* which causes the widespread *"damping off"* disease of tobacco, tomato, mustard, chillies and cress seedlings. The fungus lives saprophytically in the moist, humus soil and attacks the seedlings at the soil level. Thereafter it lives as a parasite. Mostly it occurs in tobacco seed beds but it also grows readily on decaying animal and vegetable material floating in water. A luxuriant growth of this fungus can be easily obtained by germinating tobacco, cress or mustard seeds in moist soil or on moist blotting paper under a bell jar.

Predisposing Factors. The three predisposing causes for this disease to appear are :—

(*i*) Excessive moisture or dampness and presence of too much decaying vegetable matter in the soil. These factors promote growth of the fungus mycelium and production of zoospores.

(*ii*) Over-crowding of the seedlings. It increases susceptibility by tending to delay thickening and lignification of the cell walls. The tissue with thin walls of the cells is particularly vulnerable.

(*iii*) Insufficient exposure to air and light. Poor illumination promotes fungal growth.

Mode of Infection. The mycelium appears in the moist substratum. The fungal hyphae attack the young seedlings below the soil line on the hypocotyl (Fig. 6.17 B) or the region of the upper

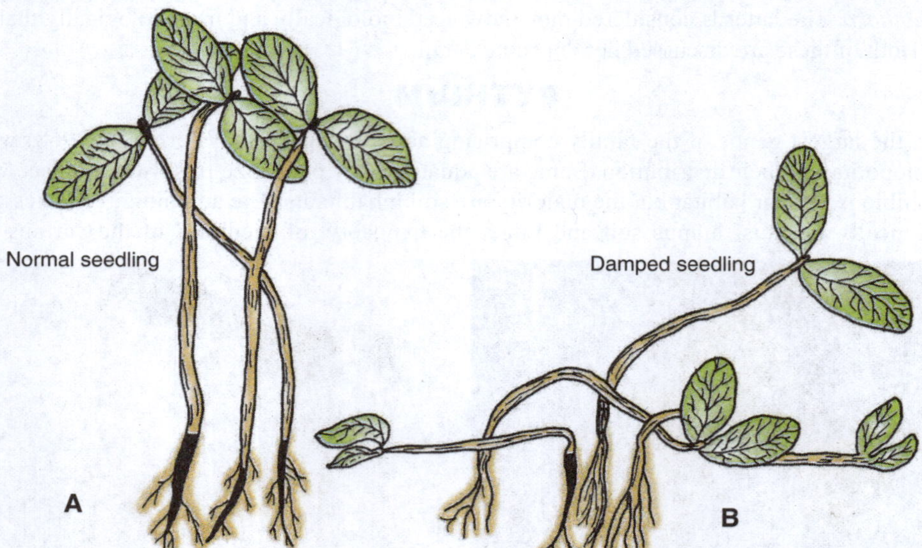

Fig. 6.17 (A-B). *Pythium* sp. A, Normal Seedlings; B, Infected Seedlings.

tap root. The infecting hypha may gain entrance into the host tissue in either of the following two ways :—

(*i*) Through a stoma.

(*ii*) By boring through the walls of the epidermal cells probably by enzyme action. Once within the host the hypha grows vigorously and branches to form the mycelium. The hyphae mostly remain confined to the intercellular spaces but have the power of penetrating into the cells.

Symptoms. The "*damping-off*" disease affects the young seedlins in patches in the seed bed. Once the fungus has gained firm hold there is rapid killing of cells and collapse of the tissue of the hypocotyl. The infected tissue appears soft and water-soaked. The damage may be complete before emergence of the seedling above ground. This is *pre-emergence phase* of the disease. The young seedlings are killed before they come out of the soil. The plumule and radicle undergo complete rotting after coming out of the seed. In many cases the disease may continue even after the emergence of the seedling. This is known as the *post-emergence phase.*The infected seedlings in the patch soon tumble over on to the ground as if pinched off (Fig. 6.17 B). This is due to the fact that at the point of infection the tissues of the host are softened and killed. The infected stem becomes killed at the base. The damage by the parasite is so much that the seedlings can no longer stand upright. They collapse at soil level and fall over. From the point of attack the disease spreads in everwidening circles. The fallen seedlings, at first, lose their colour and soon rot away. Now the fungus continues to live saprophytically upon the dead remains. It spreads by external hyphae and zoospores which attach other seedlings nearby. In this way the disease spreads rapidly.

Somatic phase. The mycelium (Fig. 6.18 A) represents the **vegetative** or **somatic phase** of the fungus. It is well developed, and coenocytic. It presents a white, fluffy appearance and consists of long rather slender, hyaline much branched hyphae which are very much like the filaments of *Vaucheria.* Usually there is a slight constriction at the base of side branch. The hyphae are generally regular in diameter but taper evenly towards their tips. The young vigorously growing hyphae are not divided into cells. The septa, however, appear in conncection with the no distinction into rhizoidal and aerial hyphae. It may live saprophytically or parasitically. Saprophytically it lives on dead organic matter in the soil. Parasitically it grows in the young seedling. The entire mycelium may be within the host tissues or some hyphae may lie within the host and others outside the host. The hyphae which lie within the host tissues are both **intracellular** and **intercellular.** They do not produce haustoria. The external hyphae may infect fresh seedlings growing in the neighbourhood or bear sporangia (Fig. 6.18 B). At first the fungal hyphae are restricted to the cortical region of the hypocotyl. Later they make their way into the vascular bundles and absorb food substances. The whole plant thus comes to be infected by the mycelium. The appropriation of food and production of waste materials by the fungus parasite result in the death of the host. Thereafter the parasite lives on the dead remains as a saprophyte.

Hyphal structure (Fig. 6.18 A). Generally it is held that the hyphal wall contains cellulose. According to Mitchell and Sabar (1966) and Bartnicki-Garcia (1966), cellulose is a minority component or even lacking altogether. Monocha and Colvin (1968) also could not detect the presence of cellulose in the *Phythium* hyphal wall. B glucans is the predominant material. Mature hyphal wall is differentiated into two distinct layers which differ in the arrangement of microfibrils. The outer layer contains randomly dispersed microfibrils and the inner has longitudinal arrangemant. The microfibrils give reflections different from cellulose I or chitin. Within the hyphal wall is the plasma membrane which forms lomasomes. The plasma membrane encloses the hyaline cytoplasm which becomes vacuolate in the older plants. The hyphal cytoplasm contains numerous small randomly dispersed nuclei, mitochondria and dictyosomes. The endoplasmic reticulum and cytoplasmic ribosomes are abundant in the young hyphae. The reserve food is in the form the glycogen and there are oil globules in addition. In the presence of lomasomes and dictyosomes, *Pythium* resembles green algae more than any other member of Mycota (Manocha and Calvin, 1968). The hyphal grwoth is by apical extension.

Reproduction. The reproductive phase sets in much before the death of the host seedling. Reproduction is both **asexual** and **sexual.**

Asexual Reproduction (Fig. 6.18 B-H). It takes place by means of zoospores which are produced in small, globular or oval, sac-like sporangia. The sporangia are formed singly and terminally at the ends of somatic hyphae which project into the damp atmosphere from the mycelium within the host tissue (B). There are no specialized sporangiophores. The sporangia measure 15-26 microns in diameter. Sometimes the sporangia are intercalary (I) The high humidity in the air promotes the growth of the fungal mycelium and the production of zoospores. Each sporangium is a multinucleate structure. It is separated from the rest of the hypha, which still retains the power of elongation, by means of a transverse septum. The sporangiam is pushed aside and the hyphal tip grows to form the second sporangium. The sporangium is filled with hyaline cytoplasm containing numerous nuclei. It is slightly denser than that of the young somatic hyphae.

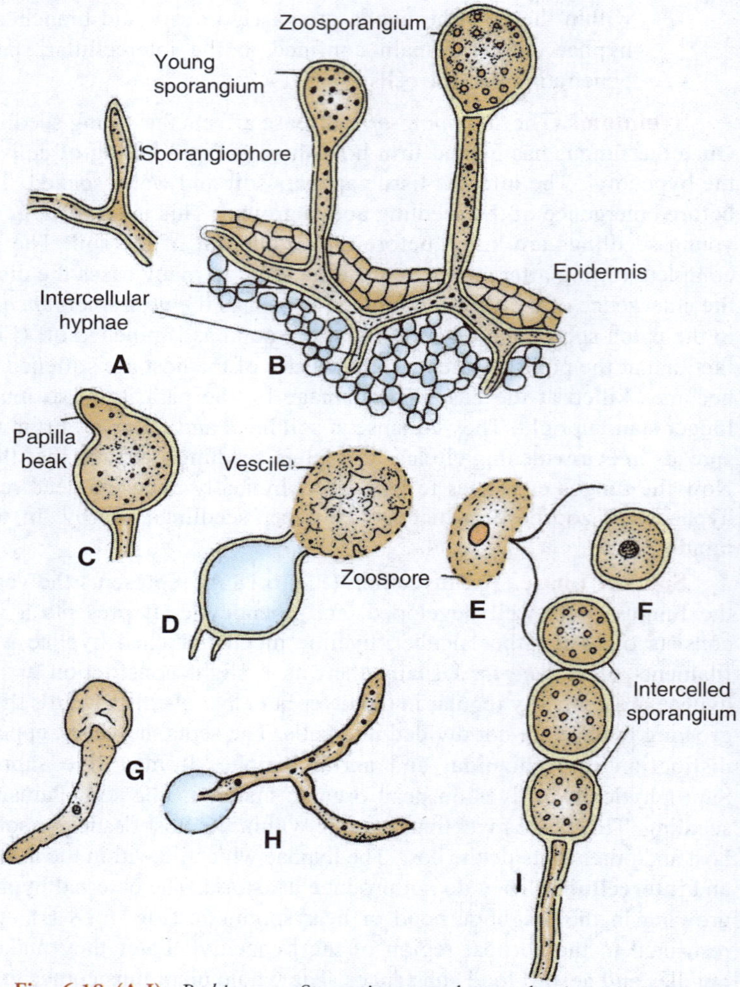

Fig. 6.18 (A-I). *Pythium sp.* Stages in asexual reproduction. A, Young sporangiophore arising from a hypha; B, Mycelium forming sporangiophores and sporangia; C, Sporangium with a beak; D, Empty sporangium with the contents in the vesicle; E, Zoospore; F, Zoospore prior to germination; G-H, Direct germination of sporangium; I, Sporangium in a chain.

(*a*) *Indirect Germination of Sporangia* (Fig. 6.18). Under wet conditions sporangia remain attached to the hyphae bearing them. They function as zoosporangia. The mature zoosporangium puts out from its side or at its apex a narrow, papilla-like outgrowth (C). It is tubular at first and called the *exit tube.* The apex of the exit tube swells up into a thin-walled, tiny, balloon-like vesicle (D). The contents of the zoosporangium by now have divided into a small number of uninucleate, daughter protoplasts. These migrate into the vesicle through the exit tube and become metamorphosed into zoospores. According to some, the undifferentiated protoplast of the zoosporangium migrates through the exit tube into the vesicle where differentiation of zoospores takes place. The zoospores when mature, exhibit rocking motion and bounce on the vesicle wall. Consequently the vesicle suddenly bursts. The exit tube may often persist after the rupture of the vesicle. The released zoospores swim and scatter in all directions (E). The tinsel flagellum is directed forward and the longer whiplash trails behind when the zoospore is in motion in the thin film of water on the host surface or in the soil.

Fine Structure of Zoospore (Fig. 6.19). According to Lunney and Brand (1976), mature swimming biflagellate zoospore of *Pythium proliferum* is typically ovoid in form with a deep longitudinal groove. Viewed on end it presents reniform appearance. The two flagella are inserted laterally. They are attached to a distinct protuberance within the groove region. The anterior flagellum is of *tinsel type* and the posterior one of *whiplash type*. The nucleus which is beaked in the freshly cleaved zoospore is now spherical and has a distinct eccentrically-located *nuclelus*. The dense cytoplasm is rich in ribosomes except the groove region which is ribosome free. There are several perinuclear dictyosomes which proliferate vesicles which fuse resulting in a reticulate, internal membrane system in the groove region. Lunney and Brand (1976) termed it the "water expulsion vesicle". The nucleus is bounded by concentric bands of rough endoplasmic reticulum except towards the groove region where there is a closely apressed packet of ribosomes. Embedded in the cytoplasm are numerous mitochndria with tubular *cristae* and various types of vesicle inclusions such as the lipid vesicles, microbodies, cell vesicles and peripheral vesicles. The rough endoplasmic reticulum cisternae ramify throughout the cytoplasm to the periphery.

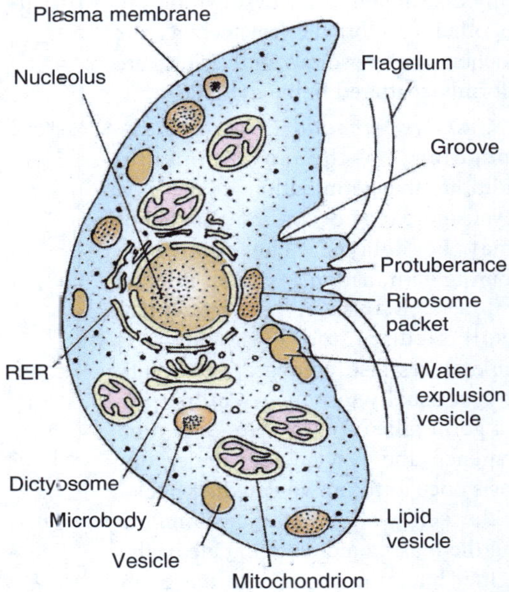

Fig. 6.19 *Pythium sp.* Diagrammatic representation of fine structure of zoospore seen in end view (Based on Lunney and Brand)

The two kinetosomes of the flagella are located at the protuberance (in the groove region) adjacent to the nucleus. An electron opaque rootlet, which is striated, stretches between them. Two more fiber bundles originate from the kinetosome. One of these runs to the anterior and the other to the posterior region of the zoospore along the plasma membrane. Microtubular roots radiate perpendicularly from the rootlets and are distributed into the different parts of the cytoplasm of the zoospore.

Encystment of Zoospores. After swimming actively for a short or long period extending over several hours, depending on the environmental conditions, the zoospores show signs of fatigue. Their movements become slow and jerky. Finally they become quiescent, retract their flagella and become spherical. Soon after the spherical resting zoospore secretes a delicate wall around it to become encysted. The encysted zoospore is usually called a cyst or cystospore (Fig. 6.18F).

Fine Structure of Cyst (Fig. 6.20 A). The mature cyst is spherical in form (A). It has a smooth spore surface. Within the electron-opaque cyst wall is the plasma membrane. The cytoplasm contains a single eccentrically located nucleus with its perinuclear dictyosomes. The concentric layers of rough endoplasmic reticulum no longer surround the nucleus. They appear to have been broken up into smaller rough endoplasmic reticulum packets which are evenly distributed in the cyst cytoplasm. The other cell organelles such as the mitochondria, ribosomes and vesicles are uniformly scattered in the cytoplasm.

Cyst Germination (Fig. 6.20 B). Depending on the environmental conditions the germination of the cyst or cystospore may occur immediately or it may be delayed. The conditions favouring germination are; (*i*) ill-drained and ill-aerated soils, (*ii*) moisture content of soil medium too high and (*iii*) termperatures also comparatively high. The cyst usually germinates by putting out a germ tube (G). Prior to germ tube emergence, the cyst nucleus undergoes mitosis once to form two daughter nuclei and the cell wall vesicles accumulate along the plasma membrane. One of the daughter nuclei migrates into the germ tube. The latter finds its ways into fresh seedlings. The disease spreads rapidly by external hyphae and zoospores under conditions mentioned above.

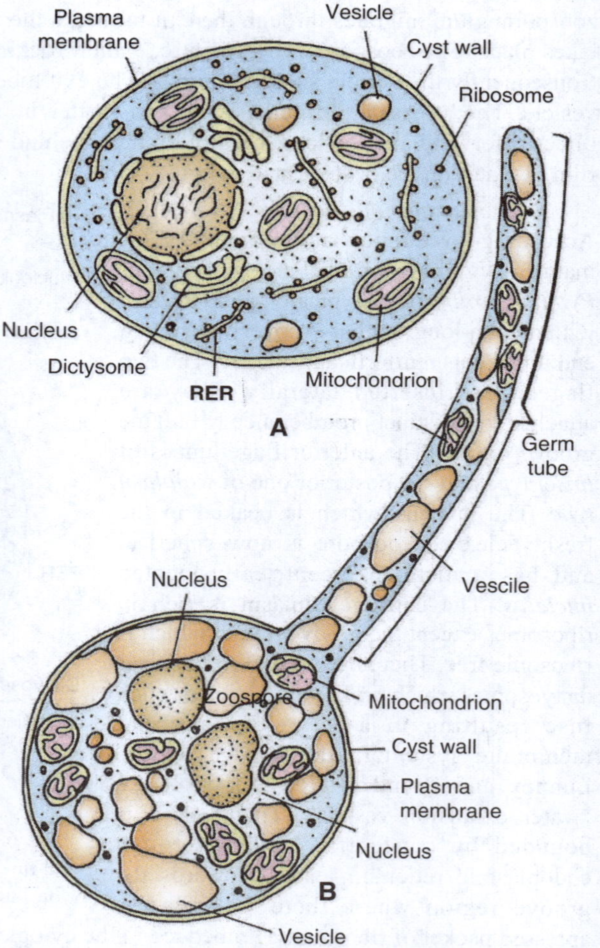

Fig. 6.20 (A-B). *Pythium.* A, Encysted Zoospore showing fine structure; B, germinating cyst (After Lunney and Brand).

Hohnk (1932) reported that the contents of the encysted zoospore may sometimes form a single secondary zoospore as in *Saprolegnia* and *Achlya* but it is of the original type. The zoospore emerges from the cyst and germinates to give rise to a fresh mycelium. This behaviour of zoospores of *Pythium* must not be confused with the diplanetic condition of zoospores of *Saprolegnia.* It is not diplanetism because the primary and secondary zoospores in *Pythium* are morphologically similar.

(*b*) *Direct Germination of Sporangium* (Fig. 6.18 G-H). The sporangia get detached from the hyphae bearing them and are blown by the wind or dispersed by water. On falling on or reaching a suitable host the sporangium germinates like a conidium if the temperature is above the optimum. It directly puts out a germ tube (G) which finds its way into the host either through a stoma or directly through an epidermal cell by secreting an enzyme at its tip. Inside the host it grows and branches to form the mycelium (H). In this case the mycelium develops without the intervention of zoospores. Some mycologists call such sporangia which germinate directly and thus behave like conidia as **conidiosporangia.** Others hold that these sporangia-like structures are in fact conidia. The production of conidiosporangia or conidia permits spread of the disease under relatively dry conditions.

Asexual reproduction in *Pythium* may also take place by means of **gemmae** (Drechsler 1939, 1941). They are intercalary in position and globose in form. Occasionally the gemmae develop thick walls and are called the **chlamydospores.** On germination the gemma or chlamydospore produces a long, tubular hypha which develops a sporangium at its its free, distal end. The protoplast of the terminal sporangium produces zoospores.

Kinds of Sporangia. In *Pythium* the sporangia are of two principal types, **spheroidal** or **globose** and **elongate** or **filamentous.**

(*a*) Globose sporangia. They are more or less rounded in shape and are much broader in diameter than the hyphae and are usually terminal in position. The typical example of this category is *P. debaryanum.* Sporangial proliferation has been reported in the globular sporangia of *P. proliferum.* When the sporangium has shed the zoospores its base bulges up to form a new sporangium which matures within the primary empty sporangium or outside it. *P. intermedium* also produces globular sporangia but in chains (I). They may function as zoosporangia or as conidia. The spherical sporangia of *P. ultimum* always function as conidia and germinate directly by producing a germ tube.

(*b*) Filamentous or elongate sporangia. They are indistinguishable from the vegetative hyphae and have the same diameter as the hyphae bearing them. However, they have denser contents. The filamentous sporangia may be simple as in *P. gracile* and *P. monospermum* or branched as in *P. aphanidermatum* and *P. myriotylum.* In the latter case the sporangia are branched in a digitate manner and are slightly inflated.

Evolution of Conidium. It is evident from the foregoing account that the different species of *Pythium* can be arranged in a series which shows a transition from a zoosporangium to a true conidium. The series starts with *P. gracile* in which the elongate sproangium exclusively produces zoospores. Normally *P. debaryanum* also reproduces asexually by the formation of sporangia which in the damp atmosphere always function as zoosporangia. However, in the dry weather or when the temperature is above the optimum, the sporangium behaves like a conidium and germinates by a germ tube.

In *P. intermedium* the sporangia normally functions as zoosporangia as well as conidia.

P. vexans is an example in which the sporangia usually function as conidia and rarely as zoosporangia. The germ tube is initiated in place of the papilla of a zoosporangium.

In *P. ultimum* the sporangium invariably functions as a conidium and germinates by a germ tube. The zoosporangia are not produced at all.

The gradual transition of the sporangium to a conidium is an adaptation to a subaerial habitat.

Sexual Reproduction (Fig. 6.21 A-E). It is **oogamous** and takes place by gametangial contact at the end of the growing season. At this time the food is nearly exhausted. There is not enough moisture for active growth. At this stage the host has been killed and the fungus is living saprophytically. The sex organs are formed within the dead tissues of the host. The male sex organ is called the **antheridium** and the female **oogonium.** They are developed in close proximity on separate short, lateral hyphae arising from the same mycelium. *Pythium debaryanum* is thus **homothallic.**

Development of oogonium. The tip of the female hypha inflates to form a globular swelling (A). Into the terminal swelling migrate the hyphal cytoplasmic contents. Finally the terminal swelling is cut off from the parent hypha by a basal hyphal plug [Haskins *et al* (1976)] or septum usualy after, but sometimes before antheridial contact. The separated terminal swelling functions as the young oogonium.

Structure of young oogonium. Ultrastructurally the young oogonium has a thin wall. Within the oogonial wall is the plasma membrane enclosing an electron-dense cytoplasm rich in

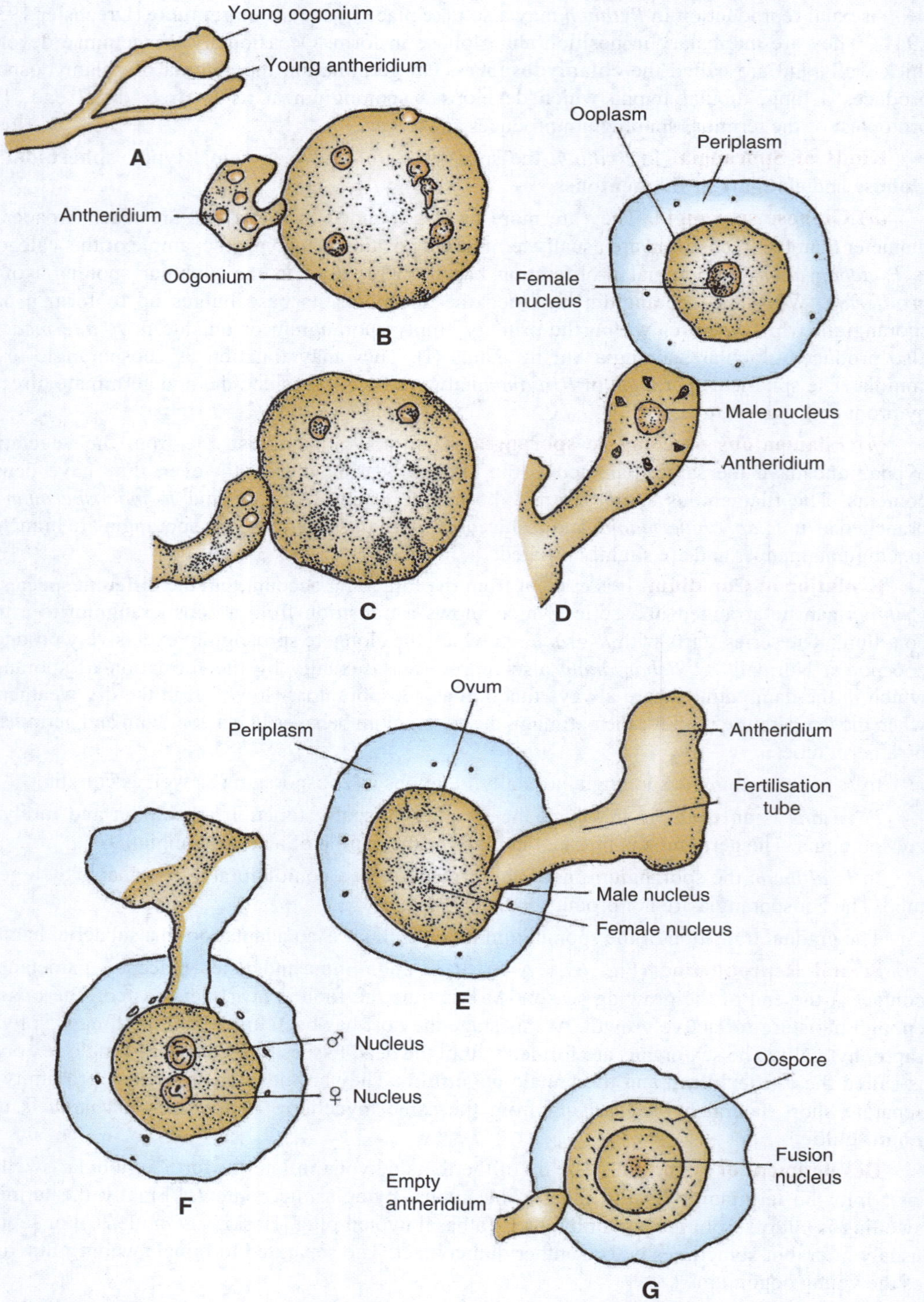

Fig. 6.21. (A-G). *Pythium sp.* Various stages in the development of sex organs and fertilization. (Explanation in the text).

endoplasmic reticulum and ribosomes. Randomly packed in the cytoplasm are the numerous nuclei, mitochondria and dictyosomes. Other inclusions in the cytoplasm are vacuoles and vesicles of various types. Some of these contain a dense spherical storage body and are called the reserve vesicles. Lipid vesicles are small. In many species the oogonial wall is smooth and colourless (*P. debaryanum* and *P. ultimum*) but in some it may be spiny (*P. acanthicum*).

Structure of mature oogonium. At the time of fertilisation the oogonial contents undergo considerable reorganization. The organelles, vacuoles and vesicles move to their destination. The mitochondria, all the nuclei except one and vacuoles migrate to the periphery whereas reserve vesicles and lipid vesicles move inwards and accumulate in the central area of the oogonium. In fact when the fertilization tube penetrates the mature oogonium its protoplast is differentiated into two zones, outer **periplasm** and inner **ooplasm** (D).

(*i*) *Periplasm.* It is the narrow outer zone lining the oogonial wall. It contains spongy or vacuolate protoplast. Apparently it seems to have been formed by shrinking away of oogonial protoplast from the wall. In reality it is formed by the coalescence and deterioration of vacuoles which assembled there after migration (Haskins *et al,* 1976). Much of the lamellar endoplasmic reticulum accumulates around the periphery of the central dense area forming a layer parallel to the oogonial wall. The periplasm which is outside this layer contains surplus nuclei, mitochondria and other organelles.

(*ii*) *Ooplasm.* The large, central portion of the oogonium filled with granular electron dense cytoplasm constitutes the ooplasm. It has a single nucleus and food reserve in the lipid and reserve vesicles. The uninucleate ooplasm functions as an oosphere or egg. Some mycologists hold that the uninucleate condition of ooplasm is attained by disintegration of all but one functional nucleus.

Development of antheridium. It develops terminally on a short male or antheridial branch arising laterally either from the oogonial stalk or a neighbouring hypha (A). In its early development the antheridial branch curves towards the oogonium. It comes in contact with the oogonial wall and flattens against it with the tip getting slightly inflated (C, D). The hyphal contents migrate into the enlarged tip. It is eventually separated from parent hypha by a hyphal plug to function as an **antheridium.** The antheridium in *Pythium* is thus applied to the side of the oogonium and is described as **paragynous.**

(*i*) *Structure of antheridium.* It is an elongated, club-shaped structure much smaller in size than the oogonium. It has a thin wall. Within the antheridial wall is the plasma membrane which in the young antheridium encloses an electron-dense cytoplasm. The cell organelles (ribosomes, endoplasmic reticulum, nuclei, mitochondria and dictyosomes) are randomly packed in the cytoplasm. Besides, there are small vaculoes. At maturity the antheridial protoplast becomes differentiated into a central uninucleate portion which functions as the **male gamete** and the outer periplasm. The surplus nuclei and other organelles migrate into the periplasm where they later degenerate.

Fertilization. The gametangial contact, as metioned above, takes place at an early stage of development usually by the curvature of the male hypha rarely by the female hypha. The tip of the antheridium applies itself closely to the oogonial wall and becomes flattended at the point of contact. The intervening walls of the mature sex organs dissolve at the point of contact (E). The flattened tip of the antheridium puts out a fine, tubular process known as the **fertilisation** or the **conjugation tube.** The latter penetrates the oogonial wall through the pore, pierces the periplasm and dips into the ooplasm. Here it opens and emits (E) a single male nucleus together with a certain amount of cytoplasm. The cytoplasm mingles with cytoplasm of the egg. The male nucleus fuses with the female nucleus.

Post-fertilization Changes. Immediately after fertilization, the egg secretes a thin membrane (McKeen, 1975) or a continuous band of small vesicles (Haskins *et al,* 1976) around it. A thick wall is subsequently deposited at this layer. The organelles involved in the synthesis and supply of material for this wall are reported to be the organelles and cytoplasm in the periplasmic space and the lamellar endoplasm reticulum layer which delimits the periplasm from the ooplasm. With the formation of thick wall, all these structures disappear in the periplasm. Meanwhile the numerous small reserve vesicles in the developing fertilized egg enlarge and coalesce to form a single, large, dense globule in the centre. It is surrounded by the rapidly enlarging and closely packed lipid globules of various sizes. The resultant thick-walled structure gorged with reserve food material and scanty cytoplasm is the mature oospore.

Fine Structure of oospore (Fig. 6.22 A). The thin oogonial wall loosely surrounds the mature oospore. The periplasm having disappeared by now there is empty space between the oogonial wall and the oospore. The oospore thus partially fills the oogonial cavity and is described as **aplerotic.** The mature oospore has a thick smooth wall. The oospore wall is differentiated into two layers, the outer **exine** and the inner **intine.** The exine is electron opaque. The intine is comparatively thicker and electron-transparent. Ruben and Stanghelline (1978) make 3 layers namely (*i*) outer thin electron opaque, (*ii*) the middle thick transparent layer and (*iii*) the thin inner layer. Within the thick oospore wall is a centrally located large, dense spherical area surrounded by closely packed lipid globules of various sizes filling the space between the oospore wall and the central area. The scanty cytoplasm forms a thin layer around the central structure and lies between the lipid vesicles. This central area of the oospore has been differently interpreted by various investigators. Merchant (1968) described it as the large, centrally located *diploid nucleus* in *P. ultimum.* Hemmes and Bartnicki-Garcia termed it the ooplast. Mckeen (1975) termed this central area of the oospore as the *reserve globule* formed by the fusion of the reserve vesicles. According to him, the oospore nucleus is not prominent. It lies sandwiched between the lipid vesicles and reserve globule. Haskins *et al,* (1976) reported that in *P. acanthicum* the reserve bodies in the vesicles expand and display a 'finger print' pattern and the central area in the mature oospore is not as large as depicted by McKeen (1975). It contains one, sometimes two nuclei and much endoplasmic reticulum. The oospore is a resting structure. The nature of the thick oospore wall provides protection and the large amount of lipid globules furnishes energy needed for its long dormancy.

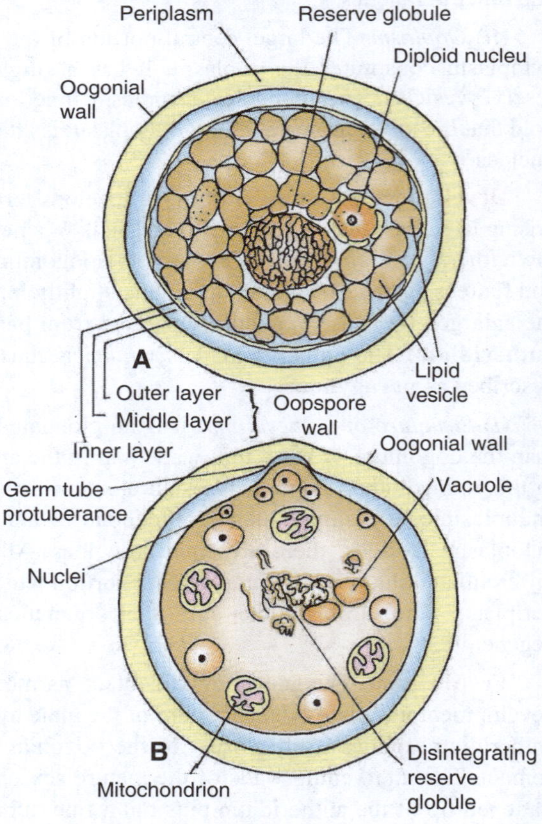

Fig. 6.22 (A-B). *Pythium.* A, Diagrammatic representation of fine structure of the oospore in section; B, germinating oospore (Ruben and Stanghelline)

After fertilization, the antheridium adhering to the oogonial wall retains its characteristic shape as an empty or nearly empty shell. Drechsler (1946) reported *parthenogenesis* in *Pythium*. The unfertilized egg behaves like an oospore and is called *parthenospore*.

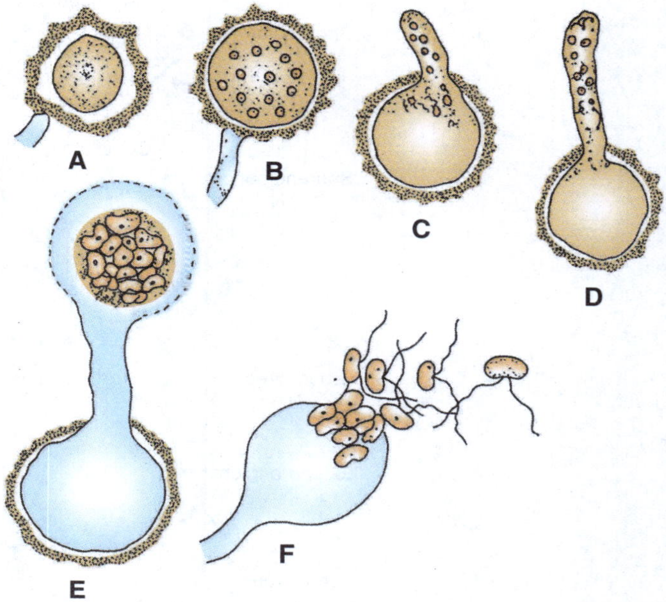

Fig. 6.23. (A-F). *Pythium* sp. Stages in the germination of oospore and differentiation of zoospores (Explanation in the text).

Oospore germination (Fig. 6.23). On the onset of conditions favourable for growth the oospore germinates. Prior to germination the oospore imbibes water and swells completely filling the oogonial cavity. The diploid oospore nucleus undergoes repeated division. The increase in the number of nuclei is accompanied by increase in the number of mitochondria. The central globule begins to disintegrate and becomes surrounded by vacuoles indicating its digestion and absorption. Simultaneously there is erosion and pitting of the middle layer of the oospore wall which becomes thin in the region of germ tube emergence. The emerging germ tube with its wall continuous with the inner layer of the oospore wall pushes through the middle and outer layers, at this site. After emergence from the oospore wall the germ tube presses against the oogonial wall and finally breaks through it behaving in either of the following ways (Fig. 6.23) :—

(*i*) It grows into a small hypha. The latter infects the host and forms the starting point of the fresh mycelium.

(*ii*) The tip of the hypha swells (E) to form a vesicle-like sporangium (*P. debaryanum*). The entire oospore protoplast migrates into the latter. The terminal sporangium may either behave like a zoosporangium and produce the zoospores (F) or gets detached. The detached sporangium behaves like a conidium and germinates directly by putting out a germ tube.

(*iii*) In the third mode of germination (*P. anandrum* and *P. mamillatum*) the contents of the oospore divide to from biflagellate zoospores which are emptied into a vesicle. The vesicle soon vanishes and the zoospores are liberated. Each liberated zoospore germinates and infects the host in the usual manner and forms a new fungus mycelium.

According to Drechsler (1952), at high temperature (28°C) the oospore germinates directly by a germ tube. The latter develops into a new mycelium. At lower temperatures within a range of 10-17°C the germ tube swells at the tip to form a vesicle. The oospore protoplast migrates into the vesicle where differentiation of zoospores takes place.

Pythium life cycle. Prior to germination the diploid nucleus of the oospore divides a number of times. Edison (1915) suggested that the first two successive divisions are meiotic. On the basis of zygotic meiosis scheme (Fig. 6.24) the vegetative thallus (mycelium) of *Pythium* is considered haploid. It bears haploid gametangia (sex organs) containing gametes towards the end of the

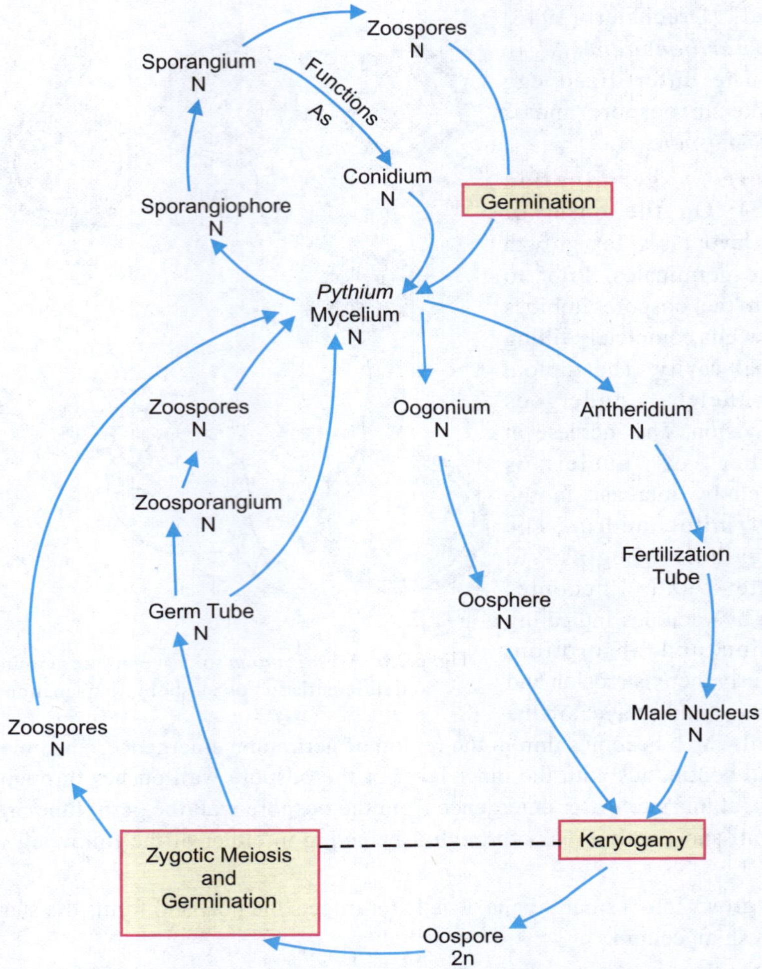

Fig. 6.24. *Pythium.* Word diagram of life cycle with zygotic meiosis.

growing season. The male and female gametes fuse to form the oospore which is the only diploid structure representing the **diplophase** (sporophyte) in the life cycle of *Pythium.* All other structures, the zoospores, the mycelium, the gametangia and gametes represent the **haplophase** (gametophyte). Such a life cycle with a prolonged haploid vegetative phase (haplophase) and a single-celled diploid oospore representing the diplophase is called **haplontic.** It is characterized by zygotic meiosis and haploid adult (mycelium). Till recently this scheme of zygotic meiosis and haplontic *Pythium* life cycle had been held widely by mycologists. Sansome (1961 and 1963) working on *P. debaryanum* struck a discordant note. She advanced evidence indicating that *Pythium* has a diploid vegetative thallus (mycelium) and gametangial meiosis (Fig. 6.25). According to her, the two successive divisions which occur in gametangia (oogonium and antheridium) are meiotic. Dennett and Stanghellini (1977) have also advanced genetic and cytological evidence for diploid life cycle and gametangial meiosis in *P. aphanidermatum.* They have photographed the diplotene-diakinesis figures in both the antheridia and oogonia. On the basis of gametangial scheme of meiosis *Pythium* has a prolonged diploid vegetative phase represented by the zoospores, mycelium and gametangia (oogonia and antheridia). The haploid phase is extremely reduced. It is represented

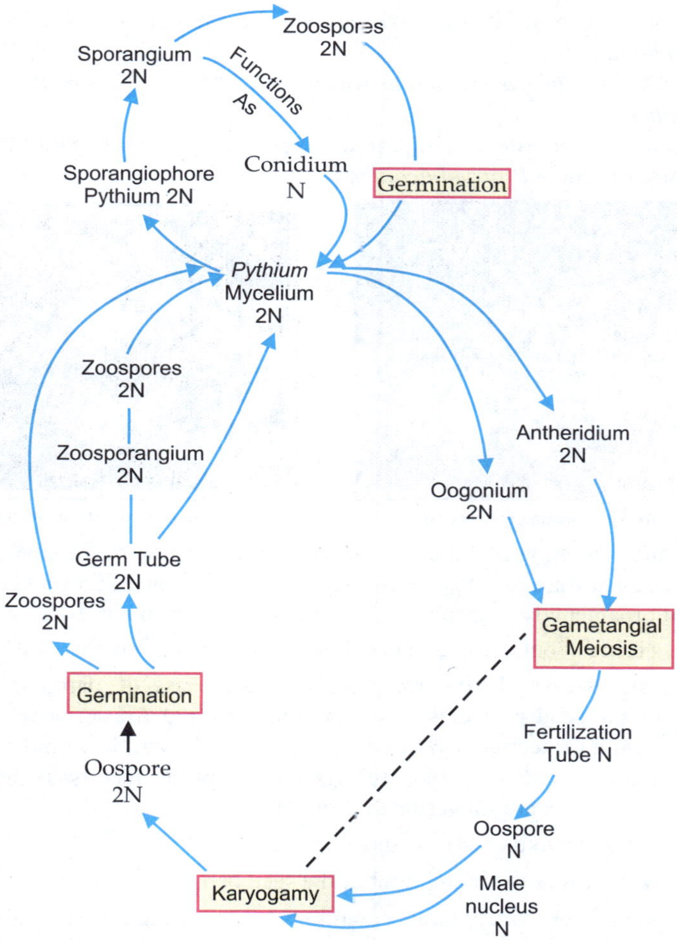

Fig. 6.25. *Pythium.* Word digram of life cycle with gametangial meiosis.

only by the gametes (male nucleus and oosphere). This kind of life cycle is called **diplontic.** It is characterised by gametangial meiosis and diploid adult (mycelium). The vegetative phase whether diploid or haploid is prolonged by the formation of sporangia which germinate directly or indirectly by the intervention of zoospores in the growing season.

Economic Importance. Most of the species are destructive parasites of our economic plants. They cause the following serious diseases, sometimes resulting in a huge loss :

1. "Damping off" of tobacco and chillies. The causal agent is *P. debaryanum.*
2. "Rhizome rot" of ginger. It is caused by *P. myriotylum* and *P. aphanidermatum.*

Damping off of Tobacco

Rhizome Rot of Giner

3. "Soft rot" of Papaya. It was reported by Subramanium in 1920. The causal agent is *P. aphanidermatum.*
4. "Fruit rot" of *Hibiscus esculentus* was reported by Bala Krishnan (1948). The pathogen is *P. indicum.*
5. "Rhizome and root rot" of turmeric was reported by Rama Krishna and Sowmini (1954). The causal agent is *P. graminicolum.*

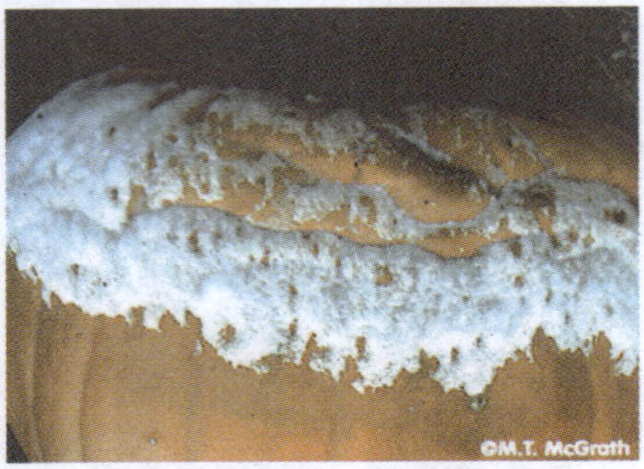

Fruit Rot of Hibiscus esculentus

Rhizome Root rot of Turmeric

'Root rots' and 'damping off' are the serious and extremely destructive diseases caused by species of *Pythium.* They are responsible for poor germination and thin stand of seedlings in field crops and also of ornamentals, vegetable plants and forest trees in nurseries.

Some species bring about decay and decomposition of plant remains in the soil.

Control of the Disease. To control pre-emergence phase of 'damping off', the seeds are treated with a suitable seed protectant such as Blitox-50 and Arasan. The chemical should be applied so as to form a protective layer external to the seed coat. Since moisture content of soil, humidity of air and over-crowding of seedlings are the prime factors in the spread of disease following sanitary measures are suggested to check it :—

(*i*) Thin sowing to avoid over-crowding.

(*ii*) Adequate exposure to air and light of the seed beds.

(*iii*) Prevention of excessively moist conditions by providing nursery beds at a raised level to maintain good soil drainage.

(*iv*) Use of light, more porous and sterile soils.

(*v*) Use of well decomposed manure.

(*vi*) Lowering the humidity of the atmosphere.

(*vii*) Sterilisation of soil by dry heat or steam.

(*viii*) The soil treatment of nursery beds by the use of chemicals such as formaldehyde diluted in proportion of one part to 50 parts of water and Bordeaux mixture (½ gallon per square foot).

(*ix*) Drenching soil with 0.2% Fytolan, 0.5% Perenox and 0.2% Flit 406 have been reported to provide effective check to control the 'damping off' disease.

Salient features

1. *P. debaryanum* parasitizes seedlings of many vascular plants and causes a destructive disease known as "*damping off*".
2. The mycelium is well developed, freely branched and coenocytic. It is located within the tissue of the host. Some hyphae may lie outside the host.
3. The hyphae are both intracellular and intercellular. No haustoria are formed.

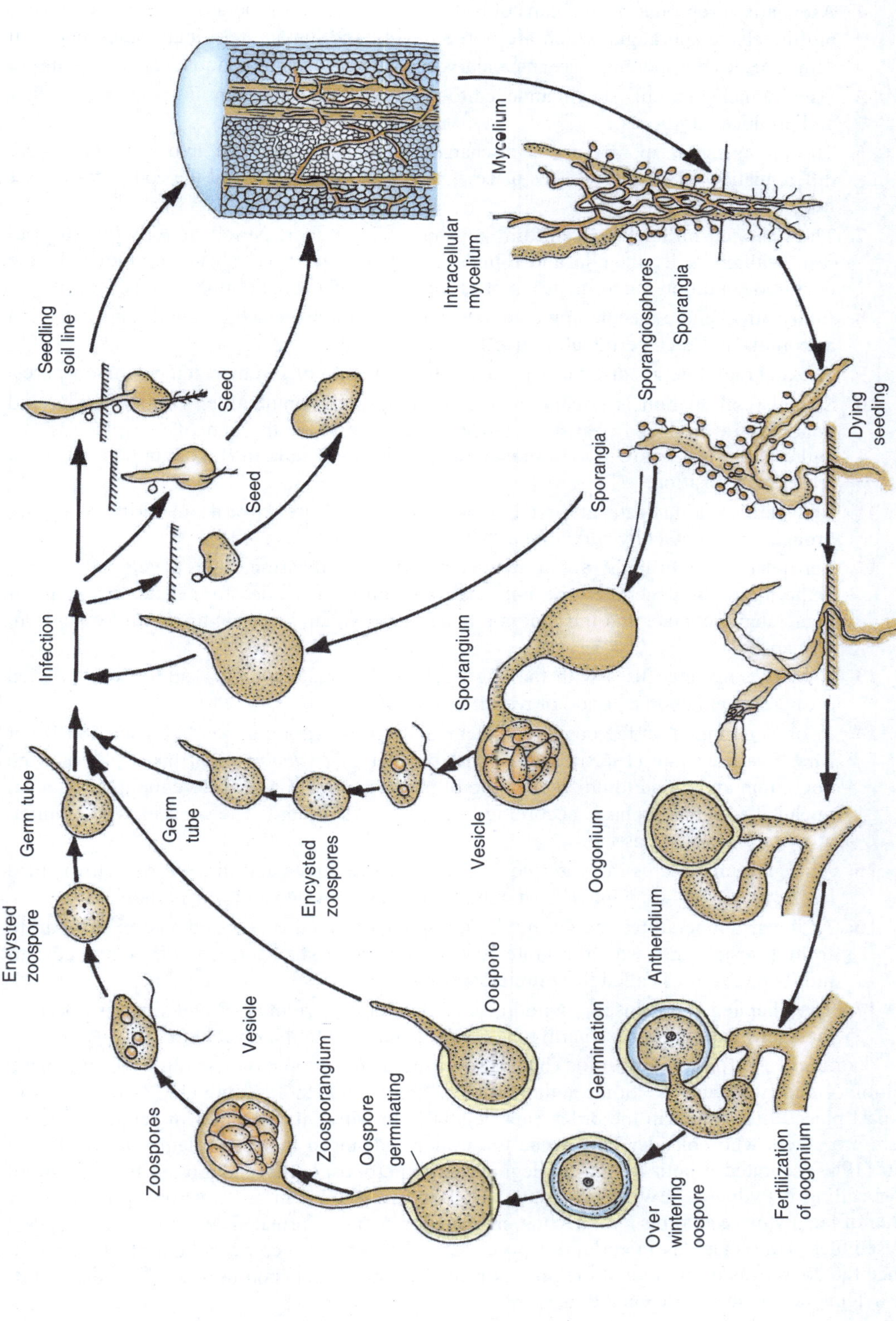

Fig. 6.26. Pictorial diagram of life-cycle of *P. debaryanum.*

4. Asexually it reproduces by means of biflagellate zoospores produced in globose to oval, multinucleate sporangia which are borne on somatic hyphae terminally and singly. In some species the sporangia are intercalary in position. No sporangiophores are organised.
5. Usually the sporangia remain attached to the hyphae bearing them in the damp weather and produce zoospores.
6. The protoplast of sporangium is discharged through an exit tube into a vesicle where differentiation of zoospores takes place. The vesicle wall bursts and the zoospores are set free.
7. The liberated biflagellate, reniform zoospore after a short period of activity comes to rest, withdraws its flagella and rounds off. It then secretes a wall around it. In the encysted condition it germinates to produce a germ tube which infects a fresh seedling.
8. In dry atmosphere the sporangia get detached and are dispersed by wind. They function as conidia and each germinates directly by a germ tube.
9. Asexual reproduction also takes place by the formation of gemmae and chlamydospores.
10. Sexual reproduction is oogamous and the fungus is homothallic. The antheridia and oogonia are produced near each other, the oogonium at the end of a hypha and the antheridium just below it on a branch arising from the stalk of the oogonium. Both are terminal in position.
11. They are multinucleate at first but become uninucleate towards maturity. They are brought into contact by curvature of the hyphae bearing them.
12. Fertilisation is brought about by means of a fertilisation tube developed on the antheridium. It penetrates the oogonial wall through a pore formed at the point of gametangial contact and introduces a male nucleus with a certain amount of cytoplasm into an egg.
13. The male nucleus fuses with the egg nucleus. The resultant thick and smooth walled oospore enters upon a period of rest. It is aplerotic.
14. Prior to germination the oospore nucleus undergoes zygotic meiosis (Edison 1915) but Mrs. Eva Sansome (1963) reported that meiosis in *P. debaryanum* takes place in the oogonium and antheridium at the time of gamete formation or differentiation of sexual nuclei. The somatic phase, according to this view, is diploid. The consensus of opinion, at present, favours this view.
15. At high temperatures the outer coat of the oospore bursts and the contents surrounded by the inner wall grow into a germ tube which develops into a new mycelium.
16. At lower temperatures the germ tube grows to a limited extent and swells at its tip to form a vesicle into which migrates the oospore protoplast and gets differentiated into motile meiospores called the **zoomeiospores.**
17. The liberated zoomeiospore withdraws its two flagella, rounds off and secretes a wall. It then germinates to form a germ tube which develops into the mycelium.

Taxonomic Position of Pythium. There is some difference of opinion regarding the systematic position of family Pythiaceae. Some authors such as Schroeter (1893), Smith (1938) and others in the past placed this family in the order Saprolegniales to which it resembles in certain respects. There are others who consider Pythiaceae to be more advanced than Saprolegniales. Wolf and Wolf (1947) seprated it both from Saprolegniales and Peronosporales and gave it the rank of an order calling it Pythiales. Gwynne Vaughan and Barnes placed it in order Peronosporales. As a matter of fact Pythiaceae forms a connecting link between orders Saprolegniales and Peronosporales. It resembles Saprolegniales in certain respects but shows an advance over them in many ways. Hence the consensus of opinion at the present time favours its inclusion in order Peronosporales. The sytematic position of *Pythium* thus is :

Division	:	**Mycota**
Sub-division	:	**Eumycotina**
Class	:	**Oomycetes**
Order	:	**Peronosporales**
Family	:	**Pythiaceae**
Genus	:	*Pythium*
Species	:	*debaryanum*

Alexopoulos and Mims (1979) place Pythium in the Superkingdom Eukaryonta, Kingdom Myceteae, Division Mastigomycota, Subdivision Diplomastigomycotina, Class Oomycetes, Order Peronosporales, and Family Pythiaceae.

Alexopoulos and Mims (1979)

Super kingdom	:	**Eukaryonta**
Kingdom	:	**Mycetae**
Division	:	**Mastigomycotina**
Sub-division	:	**Diplomastigomycotina**
Class	:	**Oomycetes**
Order	:	**Peronosporales**
Family	:	**Pythiaceae**
Genus	:	*Pythium*

PHYTOPHTHORA

Occurrence. The genus is closely allied to *Pythium*. Waterhouse (1956) reported about 75 species of *Phytophthora*. They are more terrestrial than *Pythium* and thus less dependent on free water for reproduction. Some of the species live as saprophytes but the majority live on flowering plants attacking the aerial parts and causing *leaf blight*, *canker* and *fruit rot diseases*. One of the

Phytophthora Leaf Blight

Phytophthora Stem Canker

Phytophthora Fruit Rot

species *Phytophthora infestans* (Mont.) Debary is of great economic importance and widespread interest. It is called the potato blight organism. *Phytophthora* literally means a plant destroyer. P. *infestans* causes a serious potato disease known as *Potato blight* or *Late blight* of potato resulting in foliage blight and rot of tubers.

The potato blight disease is either absent or occurs in a mild form in the plains because high temperature and drought which immediately follow the potato growing season kill the fungus in

Phytophthora infestans spots on leaves

Phytophthora Leaf Blight of Potato.

the soil and seed tubers stored at ordinary temperature. On the hills above 6000 ft. it occurs on an epidemic scale because the potato crop is sown in the rainy season. The temperature during the day is never above 22-23°C . Since 1943 the disease has been making periodic appearance in the plains of Northern India and is considered a serious disease of potatoes. The disease was first introduced in India in the Nilgiris between 1870-1880. Excess of water coupled with a range of temperature between 22-23°C promotes optimum fungal growth. Above 26°C the mycelium perishes.

Symptoms. The disease manifests itself after the flowering period, usually in the month of January. The first signs are small, black patches which appear at the margins and tips of the leaflets. The patches gradually enlarge and spread over the entire surface of the leaflets. Soon the disease invades the petioles and stems finally killing the entire tops of plants which fall over in a rotten pulp. Subsequently the tubers are affected (Fig. 6.27).

Predisposing causes. The factors which favour the spread of the disease are:—

1. Warm, muggy and moist weather.

2. Continuous rain, heavy dews and ground fog.

Day temperature over above 22-23°C.

Primary Infection. It is suggested that in the temperature and cold regions of the world and at higher altitudes in the Tropics the fungal parasite overwinters in the form of mycelium in the infected potato tubers. With the return of favourable

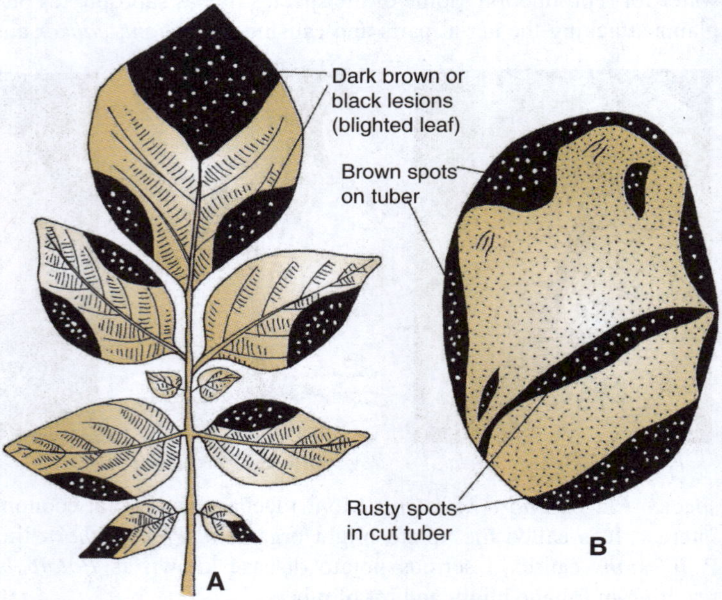

Fig. 6.27. Symptoms of Late Blight disease of potato caused by *Phytophthora infestans.* A, On leaves; B, On potato tuber.

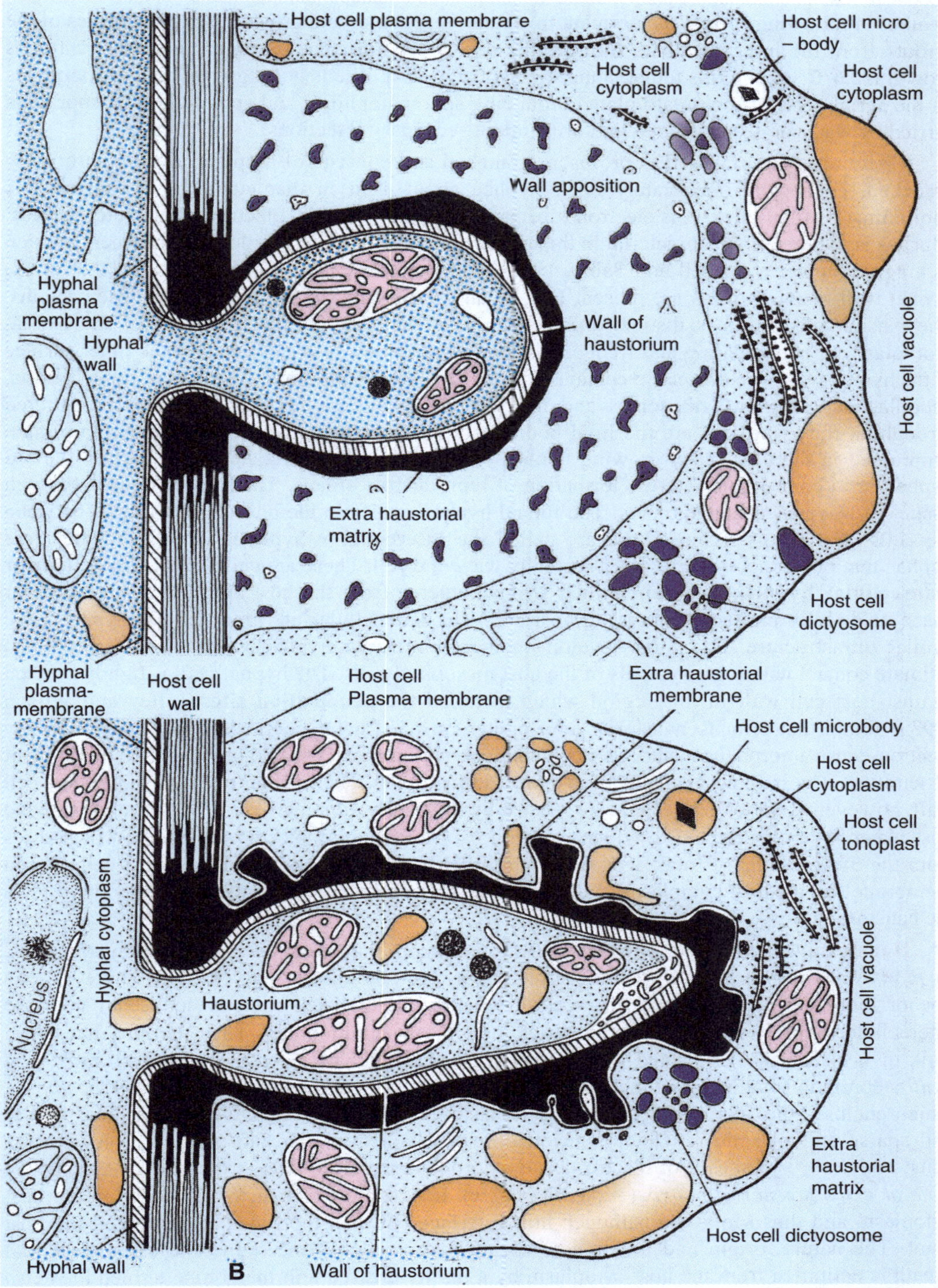

Fig. 6.28 (A-B). *Phytophthora infestans.* Diagrammatic representation of the ultrastructure of the haustorial apparatus in the tuber tissue of resistant variety (A), Eba (which is not completely resistant) and a susceptible cultivar (B) of *Solanum tuberosum* (Based on Hohl and Stossel).

weather in following spring, the dormant mycelium reactivates and grows in the new tissues of the sprouts from the infected tubers forming the new mycelium. The optimum temperature for its growth is 26°C. Above this temperature the hyphae perish. The new fungus mycelium sporulates on the aerial parts of the potato plants producing sporangiophores and sporangia. The sporangia carried by water or dispersed by wind bring about secondary infections.

Mycelium (Fig. 6.29). It is profusely branched and consists of aseptate, hyaline, profusely branched, coenocytic, moderately thick hyphae about 4-8μ in diameter. The hyphal wall is approximately 0.1μ thick. It is electron transparent except for a thin, electron dense outer surface. Glucans is the predominant material in the hyphal wall. Cellulose is a minority component or even lacking altogether (Mitchell and Sabar, 1966). Hunsley and Burnett (1970) reported that in the hyphal wall of *P. infestans* are present B-1, 3- and B-1, 6-glucans as well as cellulose proteins. The plasma membrane is distinct and is seen as a dark line at the periphery of the hyphal protoplasm. It is undulating and frequently forms lomasomes. It is appressed to the inner surface of the hyphal wall. The cytoplasm contains scattered nuclei, dictyosomes, lomasomes, mitochondria, endoplasmic reticulum, ribosomes and many large vacuoles with lipid inclusions. Scattered throughout the cytoplasm are the lipid bodies more numerous in the old hyphae. Septa remain suppressed in the vigorously growing hyphae. They, however, appear here and there in the old hyphae or in connection with the formation of reproductive organs. The fungus spreads through tissues of leaves, stems and tubers. The fungal hyphae ramify in the intercellular spaces between the cells of the host tissue. These are called the **intercellular hyphae**. In addition there are hyphae that penetrate, traverse and eventually leave the cell. These are called the **intracellular** or **transcellular hyphae**. The transcellular hyphae emerge from the host cell into an intercellular space or directly into a neighbouring host cell. The intercellular and transcellular hyphae have similar ultrastructure. Haustoria develop on the intercellular hyphae when the latter come in intimate contact with primary walls of the host mesophyll cells. The hyphal wall is tightly pressed against host cell wall in this region which is called the **penetration site.** Coffey and Wilson (1983) found that prior to haustorial development a lobed hemispherical deposit of moderately electron-dense amorphous material termed the **penetration matrix** is laid in this region of incipient penetration. The bulging hyphal wall is intact in this region. The fibrillar network of the host cell wall, however, is no longer seen. It is replaced by the penetration matrix which is bounded by the host plasma membrane next to which is a thin band of host cytoplasm. An infection peg arises from the hyphal bulge. It makes its way into the host cell pushing in front the host cell plasma membrane and the host cytoplasm. Within the host cell the infection peg expands at its tip to form the haustorium.

Haustoria (Fig. 6.28 B). The haustoria are variously shaped intracellular feeding structures. In *P. infestans* they are small globes, occasionally short, straight or curved pegs. There may be one or more haustoria in each host cell. The intracellular haustorium is connected with the intercellular hypha by a neck-like constriction at the penetration site. Thus the long haustorial stalk or neck usually associated with this organ is lacking. The electron-transparent *haustorial wall* is continuous with that of the intercellular hypha and so is the *haustorial plasma membrane* which encloses the anucleate haustorial cytoplasm containing a few mitochondria, lipid bodies and sparse endoplasmic reticulum. Surrounding the haustorium, in close association with the haustorial wall, is a sheath of electro-dense amorphous material. It is termed the *encapsulation zone or extra haustorial matrix* (= *sheath matrix*). It separates the haustorial wall from the host cytoplasm and thus forms the pathogen host interface between the fungal parasite and the host plant. The nature, origin and function of sheath matrix is still uncertain. The extra haustorial sheath is delimited from the host cytoplasm by a tightly pressed unit membrane termed the *extra haustorial membrane.* The latter is considered the invaginated host plasma membrane. Ehrlich and Ehrlick (1966) did not find any infoldings or deposition of host wall material surrounding the basal portion of the haustorium.

Hohl and Stossel (1976), however, found some haustoria covered by an additional layer of electron-transparent wall material between the extra-haustorial matrix and the extra-haustorial membrane. It is termed the *wall apposition* (A). The wall apposition is continuous with the host cell wall at its base. It may partially or completely surround the haustorium. In the former case it forms a collar-like structure at the base of the haustorium. Hohl and Sutter (1976) supporting Hohl and Stossel (1976) observed that wall apposition may or may not be formed. Coffey and Wilson (1983) reported that at an early stage the haustoria do not possess collars or encasements.

The haustoria are more commonly found in the tubers. In severe cases of infection the entire plant above the ground is killed. The fungus pases winter in the form of mycelium in infected potato tubers.

Asexual Reproduction (Fig. 6.29). With the onset of favourable conditions of warm, damp weather in spring a tuft of slender, branching hyphae arise from the internal mycelium. They push their way out either through a stoma (A) or by piercing through the epidermal cell on the lower surface of the leaf. In the case of the tubers they push their way through the lenticels or injured portions of the skin. These aerial hyphae are hyaline, and sympodially branched and are of indeterminate growth. These special branched aerial hyphae are called sporangiophores (conidiophores).

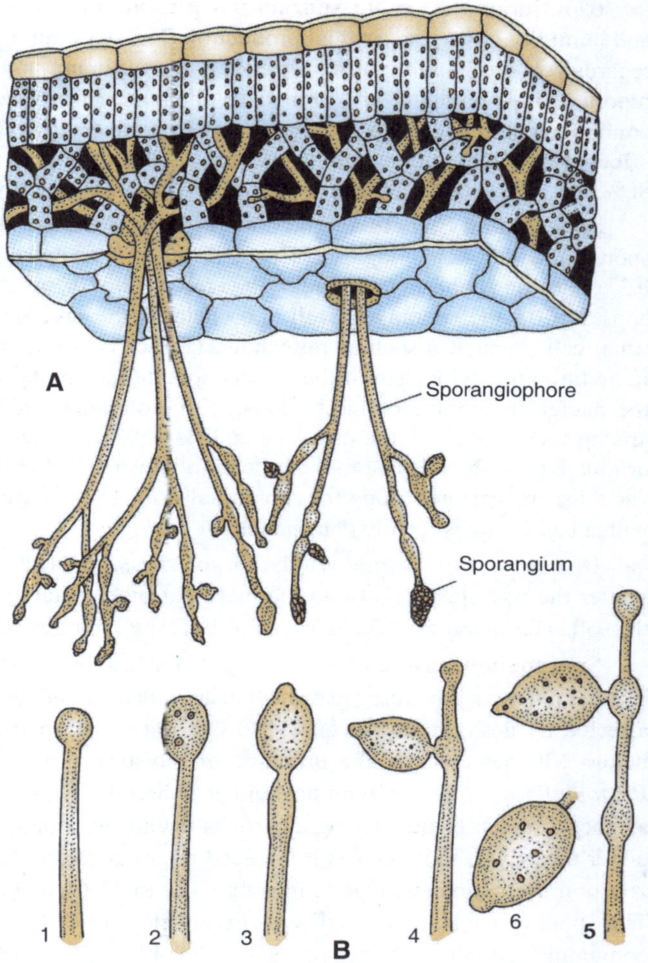

Fig. 6.29 (A-B) *Phytophthora infestans.* A, V.S. infected Potato leaf showing the intercellular mycelium and the emerging sporangiophores; B, the stages in the development of sporangia. (After Ward)

Phytophythora, thus, differs from other members of the family in having sporangiophores distinct from the somatic hyphae.

The sporangium is formed by the inflation of the tip of the side branch of the sporangiophore (B_1). The multinucleate inflation is then cut off by a transverse septum basal plug as a **sporangium**. The hyphal branch bearing the young terminal sporangium continues to grow (B2-3). As the sporangium reaches maturity, the branch tip swells slightly just below the sporangium and proliferates pushing the sporangium to the side as the elongation proceeds (B4). The process may be repeated. The sporangia are, thus, borne terminally but are subsequently shifted to a lateral position. The mature sporangia are lightly attached, the sporangiophore of *Phytophthora* is, therefore, sympodially branched. It bears nodular swellings (B5) which denote the points of detachment of sporangia. These swellings give the branched sporangiophore a jointed appearance characteristic of the genus.

Conditions governing sporan-gial production. According to Crosier (1934) temperature and humidity govern sporangial production. The optimum temperature range is 18°C to 22°C. As regards relative humidity 100 per cent is the optimum and 91% minimum. Cohen *et al* (1975) reported that sporangia are formed in abundance on infected leaves kept in darkness. Under continuous light conditions no sporangia are produced by the fungus. He found that the inhibiting effect of light was temperature dependent. High temperature increased it. At 10°C and 15°C about 80% and 50% sporangia were produced in light whereas only 2-7% were produced at 20-25°C.

Sporangium Structure. The mature sporangium is a hyaline, oval to elliptical. thin-walled spore sac with a basal plug. It has a small stalk and an apical papilla. The sporangial wall is nearly 0.3 μ thick and appears to be faintly layered. Within the wall is the plasma membrane. It is closely appressed to the sporangial well and encloses the multinucleate cytoplasm which contains the usual cell organelles such as mitochondria, dictyosomes, endoplasmic reticulum and ribosomes. In addition the cytoplasm of the mature sporangium contains paired basal bodies associated with the nuclei, membrane bound packets of microtubules and several types of vacuoles. The chief among these latter are the cleavage and flagellar vacuoles. The former which are considered to originate from the dictyosomes are randomly distributed in the cytoplasm. The flagella containing vacuoles are arranged along the sporangial wall. The mature lemon-shaped, pappilate sporangium with a basal plug is readily detachable.

Dispersal of sporangia. Wind, rain splashes or contact with other leaves detach and spread or scatter the ripe sporangia on to the leaves of other potato plants. They may also be washed into the soil. The sporangia lose their viability if they fail to germinate within a few hours.

Sporangium germination. Landing on the healthy leaf of a host plant, the sporangia germinate. The main factors governing germination are moisture and temperature. The sporangia are adversely affected by desiccation. According to Cochrane, they perish if the relative humidity falls much below 100 per cent. In the presence of moisture provided by rain or dew, the sporaniga of *Phytophthora infestants* germinate either indirectly by producing zoospores and thus functioning as zoosporangia or directly by a germ tube and functioning as conidia. The ability to germinate either directly or indirectly is influenced by temperature. Low temperature (12 to 15°C) induces zoospore formation and high temperature (20 to 23°C) favours direct germination by a germ tube. The sporangia lose their vitality to germinate indirectly with increasing age. Young sporangia commonly reproduce by zoospores more readily than old sporangia. According to Elsner, Vander Molen, Horton and Bowen (1970) the loss occurs slowly at low temperature but more rapidly at higher temperature or following a heat shock of 40°C for 7-10 minutes.

(a) Indirect germination (Fig. 6.30). Wet, cool soil favours indirect germination of sporangia. Thus in the presence of moisture at 16°C or below 12-14°C (Sato, 1979) the sporangium behaves as a zoosporangium (A). According to Hohl and Hama moto (1967) the cleavage vesicles which are randomly distributed in the cytoplasm become arranged radially equidistant from the neighbouring nuclei approximating the future planes of cytokinesis. The flagellar vacuoles are arranged along the sporangial wall axially enjoined to the paired basal bodies. At the time of differentiation of zoospores, the margins of flagellar vacuoles fuse with the coalescing cleavage vacuoles. The membranes of this system eventually fuse with the plasma membrane delimiting the multinucleate sporangial protoplast into 5-8 uninucleate daughter protoplasts (B) and freeing the flagella which lie between the plasma membrane and sporangial wall. Each daughter protoplast subsequently metamorphoses into a biflagellate zoospore which is reniform. Of the two flagella one is of whiplash type and the other tinsel. Ferus (1954) reported the development of paddle-like structures at their tips. The flagella arise from the depression on the concave side. The zoospores are set free in a group by the bursting of the apical papilla into a vesicle in *P. cactorum*. No vesicle is formed in *P. infestans* (C).

Indirect germination of sporangium through the intervention of zoospores is an example of the retention of an ancestral character. *Phytophthora* is a terrestrial fungus. Reproduction by zoospores is a primitive feature in its life cycle. Direct germination is considered an evolutionary adaptation. The sporangia which are washed into the soil germinate and infect the tubers. As a result the tubers may rot by harvest time or rot during storage. Under favourable conditions a number of asexual generations may be produced in one growing season. This results in rapid propagation of the pathogen and spread of the disease at an amazing speed causing sudden and disastrous epidemic.

Germination of zoospore and infection of the host (Fig. 6.30 D-G). The sporangia which land on the leaves of potato plants germinate in cool wet environment liberating zoospores. The liberated zoospores (D) swim about actively in a surface film of water for some time and then come to rest. Each quiescent zoospore retracts its flagella and may encyst (E). The encysted zoospore then germinates by putting out a special, short hypha called the germ tube (F). Pristou and Gallegly (1954) reported the formation of an appressorium at the time of infection. The germ tube, as it grows over the epidermis of the leaf and closely adheres to it produces a flattened pressing organ, the **appressorium** at its tip (G_2). From the undersurface of the latter arises a fine tubular, peg-like outgrowth, the **infection hypha**, which pierces the cuticle and forces its way through the epidermis into the host tissue to bring about infection. The infection peg then swells to normal hyphal diameter.

The concept of penetration of foliage leaf direct through the epidermal cell, as suggested by Pristou and Gallegly (1954) was later confirmed by Shimony and Friend, (1975) and Wilson and Coffey (1970). Hohl and Suter (1976), however, reported that the principal method of entry into the host leaf is via stomata. Direct penetration through the epidermis is not common. It is rather rare. After penetration via stomata a plug is often recognisable in the hypha at the level where it enters the leaf. No appressorium is formed in the case of entry through the stoma. Infection is high in cool, wet environment. Within the host the invading hypha grows vigorously and branches to form the

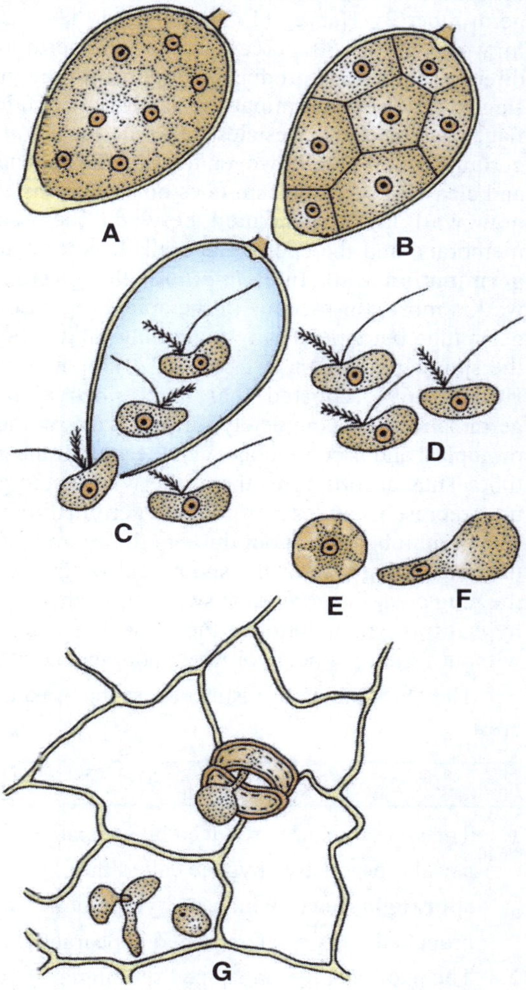

Fig. 6.30 (A-G). *Phytophthora infestans.* Stages in indirect germination of sporangium and germination of a zoospore (A, B E-G after—Ward).

mycelium which is coenocytic and consists of intracellular and intercellula hyphae. In 1983 Coffey and Wilson found direct penetration of the host epidermal cells and the formation of apparessoria as the common method of infection. The infection peg splits the outer epidermal walls. After penetration it swells to form the infection vesicle in the epidermal cell. The infection vesicle subsequently narrows down to form in intracellular infection hypha which grows into the underlying palisade cell. It grows downwards and finally leaves in to enter, intercellular space between the mesophyll (spongy) tissue.

Galactinase is the enzyme produced by *P. infestans.* It might aid the penetration of the tuber tissue (Knee and Friend, 1970).

(b) Direct Germination of Sporangium (Fig. 6.31). In wet warm environment (at higher temperature from 20 to 23°C), the sporangium (1) behaves like a conidium and germinates directly by a germ tube (2). Harvey (1954) reported that at 25°C in water about 50 percent sporangia germinated directly while none produced zoopores. In the initial stages of direct germination vacuoles containing flagella and cleavage vesicles are formed as in indirect germinatiom. Later, however, the flagella degenerate and cleavage of cytoplasm does not occur. Instead a new wall layer is formed between the plasma membrane and the sporangial wall. It is termed the **germination wall.** In *P. infestans,* the germination wall is most conspicuous in the apical region of the germ tube but tapers away to nothing in the body of the sporangium (Elsner, *et al.,* 1970). Hemmes and Hohl (1969) reported that in *P. parasitica* the germination wall completely surrounds the sporangial protoplast and is continuous with the wall of the germ tube. Thus according to them, the wall of the germ tube represents an extension of the germination wall. The germ tube pushes out through the central part of the apical plug area or the sporangial walls. Beyond the plug area, the germ tube swells to form a vesicle or bulge. Form this vesicle arise one or more hyphal tips which continue the growth. Sometimes the emerging germ tube may branch directly without forming a bulge or may not branch at all. Most of the growth takes place at the hyphal tip.

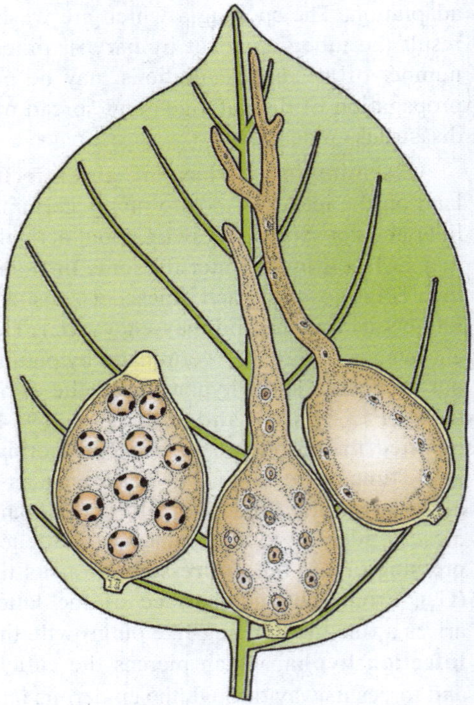

Fig. 6.31. *Phytophthora sp.* showing stages in direct germination of sporangium on a host leaf.

The chief points of distinction in the asexual reproduction between *Phytophthora* and *Pythium* are:-

Phytophthora	*Pythium*
1. The sporangia are produced on special aerial reproductive hyphae called the **sporangiophores** which are sympodially branched and present a joined appearance.	1. The sporangia are produced on the somatic hyphae indistinguishable from the other hyphae of the mycelium.
2. The papillate, lemon-shaped sporangia are always terminal in origin but are subsequently shifted to the side.	2. The globose to oval sporangia are either terminal or intercalary in position.

Phytophthora	Pythium
3. The zoospores are differentiated in the sporangium and are liberated by the bursting of the apical papilla. No vesicle is formed in *P. infestans*.	3. The differentiation of zoospores takes place in the vesicle. They are liberated by the sudden bursting of the vesicular wall.
4. The germ tube enters the host tissue through the wall of the epidermal cells by means of the infection peg that grows from the **appressorium** produced by the germ tube.	4. The germ tube enters the host tissues either through a stoma or by boring through the walls of the epidermal cells probably by enzyme action. No appressoria are formed.

Chlamydospore formation. Various species of *Phytophthora* produce chlamydospores. These develop as oblong intercalary or terminal spores. Most have relatively thick walls but this varies with the species. In a mature chlamydospore the cytoplasm is dominated by lipid and reserve vacuoles (Hemmes, 1983).

Sexual Reproduction in Phytophthora (Fig. 6.32). In *Phytophthora* sexual reproduction is oogamous. Till recently there has been some confusion with regard to the nature of sexuality in the different species of the genus. It has, therefore, been the subject of numerous investigations. Some species were considered *homothallic* while others proved to be *heterothallic*. The former produce sex organs in a single culture but the latter require the presence of two compatible isolates (strains) for oospore formation. The isolates of heterothallic species may be unisexual (Narasimhan, 1930) or potentially bisexual but produce sex organs under normal growing conditions only when two strains of opposite mating types interact (Ashly, 1929). There is no significant morphological difference between the two. They are potentially bisexual but self-sterile. Gallindo and Gallegly (1960), Savage *et al.* (1968) and Brasieur (1972) corroborated the view that the heterothallic species generally are bisexual. From the results of various investigators Shepherd (1978) concluded that all species of *Phytophthora* are monoecious. It means their mycelium is potentially capable of producing both antheridia and oogonia. These monoecious species, however, differ in their mating capabilities. Thus, on the basis of their mating competence. Shepherd (1978) divides them into following three categories :

(*i*) *Self compatible monoecists* (homothallic species). In this category are included the self-fertile species.

(*ii*) *Self-incompatible monoecists* (Heterothallic species). These species produce sex organs only when two strains of opposite mating types (termed A_1, and A_2) interact. They are self-sterile.

(*iii*) *Neuter or Sterile species*. The strains which do not produce sex organs under normal conditions are included in this category.

According to the above classification, *Phytophthora infestans* is considered a self-incompatible monoecist (heterothallic). The isolates are potentially bisexual and morphologically alike. Each strain produces both types of sex organs when the two strains of opposite mating types occur in the same host near each other. Gallegly and Gallindo (1958) used the symbols A_1 and A_2 to designate them. It is worthwhile to note that the mycelium of each strain in *P. infestans* produces both antheridia and oogonia which are self-sterile. Mating, however, takes place between the sex organs of opposite mating types. Oospores in this species are thus formed only if antheridium of A_1 strain copulates with oogonium of A_2 mating type and *vice versa.*

Antheridial incept (Fig. 6.32). It arises as a short lateral hypha from the mycelium with its tip inflated to form a more or less clavate structure known as the young antheridium (A). It is thin-walled and contains non-vacuolated cytoplasm with a single nucleus or more often two nuclei before the entry of oogonial incept (A). Usually the young antheridium is not separated by a septum from the supporting hypha or stalk.

Oogonial incept (Fig. 6.32). It also arises as a short, lateral hypha without any inflation (B). It contains dense, multinucleate cytoplasm.

The antheridial and oogonial incepts of the opposite mating types (A_1 and A_2) grow and curve towards each other. Eventually the tip of the oogonial incept of one strain comes in contact with the young antheridium of the opposite mating strain, punctures and grows through it (B) to emerge on the other side (C) where it swells into a globose structure, the **oogonium** (C). The antheridium, now forms a collar-like structure surrounding the base of the oogonium (C). This type of antheridium is termed **amphigynous.**

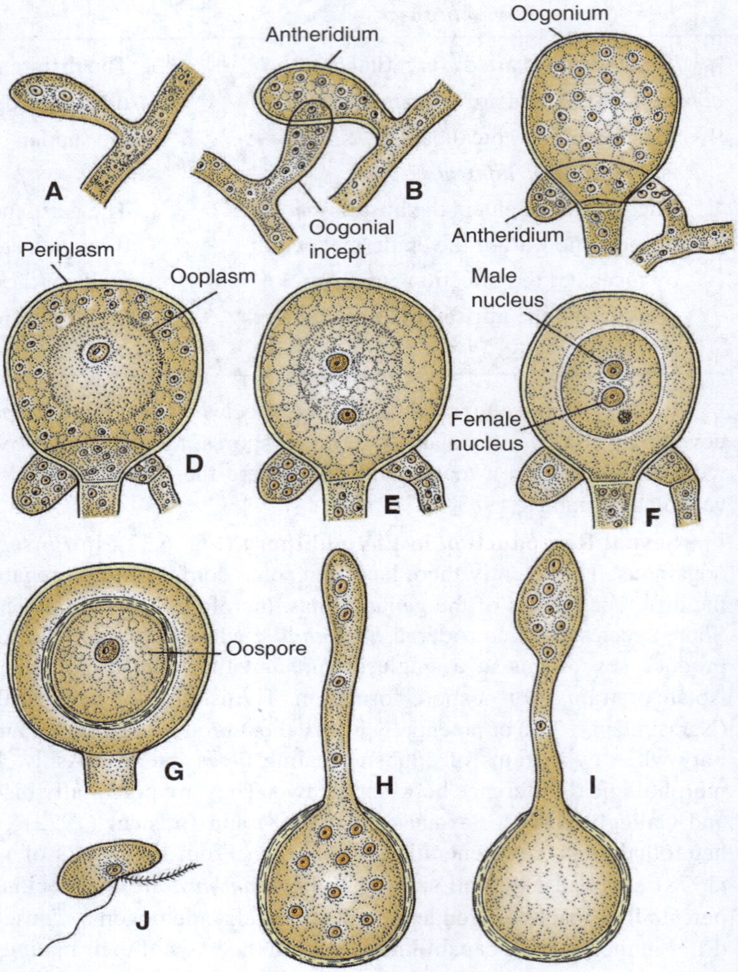

Fig. 6.32 (A-J). *Phytophthora infestans* showing stages in the development of sex organs and fertilisation(A-G) germination of oospore (H-I); liberated zoospore (J).

Nature of amphigyny. Two hypotheses have been put forth to explain amphigynous condition of antheridium in *Phytophthora.* Pethybridge (1914) reported that antheridial and oogonial hyphae (of opposite mating types, A_1 and A_2) grow towards each other. The tip of the **oogonial incept** of one strain comes in contact with the developing antheridium of the opposite mating strain, penetrates it and grows through it without loss of contents (plasmogamy) and breaks out at the top to swell into a spherical or pear-shaped oogonium. The antheridium thus appears sitting like a funnel-shaped collar at the base of oogonium. This view which has come to be known as the **classical concept** was confirmed by Mundkar (1949) using light microscopy, Vujicick (1971) using electron microscopy and many others.

Rosa and Lidegren (1925) and Cooper and Porter (1928), however, pointed out, that this interpretation of amphigyny is an illusion created by the wrapping of the antheridial initial around the oogonial stalk. Stephenson and Irwin (1972) supporting Rosa and Lidergen reported that in *P. capsici*, the antheridial initial encircles the developing oogonial stalk and is not penetrated by the oogonial incept as classically described.

Hemmes and Bartinicki-Garcia (1975) using electron microscope reconfirmed the classical concept of amphigyny of antheridium in *P. capsici*. They observed:

(*i*) contact between oogonial and antheridial hyphae, (*ii*) penetration of antheridial hypha by the oogonial initial, (*iii*) re-emergence of the oogonial initial, and (iv) oogonium expansion at the tip after emergence.

Ho and Zentmyer (1977) advocating the classical concept stated that in *P. cinnamoni* the amphigynous antheridia result from the penetration of antheridial initial by the oogonial initial. The vertical juncture line demonstrated in *P. capsici* by Stephensen and Irwin (1972) to indicate the encirclement of oogonial stalk by the amphigynous antheridium was never observed in *P. cinnamoni* by either light or scanning electron microscopy. It is evident from the above that there is overwhelming evidence in favour of the classical concept of amphigynous condition of antheridium in *Phytophythora*.

Phytophthora capsici blight

Mature Oogonium (Fig. 6.32). It lies above the antheridium and is spherical or pear-shaped in form (B). It has a dense, multinucleate cytoplasm when young (C). Towards maturity it increases in size. The protoplast becomes vacuolated. The nuclei which are spherical and about 40 in number undergo division which according to some workers is haploid mitotic and according to others meiotic. At this stage the oogonial protoplast becomes differentiated into an outer or peripheral, hyaline zone with vacuolate multinucleate cytoplasm surrounding a central, uninucleate region with denser cytoplasm. The former is called **periplasm** and the latter **ooplasm** (D). All the nuclei in the periplasm later disintegrate (E). Prior to fertilization the single nucleus in the ooplasm divides into two daugher nuclei (E). One of these denegerates. The surviving one functions as the **egg** or **oosphere nucleus.**

Mature Antheridium. It is funnel-shaped, sits on the oogonial stalk forming a collar-like structure around the base of the oogonium. This type of antheridium surrounding the base of the oogonium is termed **amphigynous.** It is delimited from its urpporting hypha by a septum. The antheridium contains about 12 nuclei. These are formed from the original two nuclei of the young **antheridium,** after penetration by the oogonial branch by division. The division according to some investigators is haploid mitotic and according to others meiotic. Prior to fertilization all the nuclei except one degenerate. The surviving one functions as the **male nucleus.**

Fertilization. The oogonial wall bulges at one point into the antheridium. This point is called the **receptive spot.** It then dissolves at the receptive spot. Through the opening the antheridium pushes the fertilization tube. Gallegly (1968) reported that in *P. infestans* fertilization tube is formed from the antheridium just above the cross wall of the oogonial stalk. It extends upward into the oogonium. This was confirmed by Laviola (1974) who found that after its origin from the antheridium it penetrated the intervening walls between the antheridium and the oogonial stalk at the point of contact. Within the oogonium it travelled up the oogonial stalk, penetrated the periplasm and entered the oosphere at the nearest point. Here it opens at the tip to deliver the male nucleus and some of the protoplasm. Sometimes the fertilization tube travelled further between the oogonial wall and the oosphere and entered the latter at nearly the mid-equatorial region. In case the antheridium is bicellular the fertilization tube arises from the upper or distal cell above the septum or plug of the oogonial stalk (Ho, Zentmyer and Irvin, 1977). Oospores have also been reported to be formed parthenogenetically.

Oospore. The fertilized egg secretes a heavy wall around it and bcomes an oospore. There is empty space between the oogonial wall and the thick oospore wall. The oospore thus partially fills the oogonial cavity and is described as *aplerotic*. It is a resting spore and thus considered to be an important overwintering structure which plays a significant role as an important propagule in the disease cycle. In cultures sterols are necessary for the formation of oospores. Zentmyer and Irwin (1970) reported that the fully dormant oospore has a thick wall differentiated into two or three layers. There is a large, dense, spherical body in the centre with sharply defined edges known as the **reserve globule.** The cytoplasm surrounding the central body contains several spherical globules perhaps lipid or oil in nature. The male and female nuclei have been reported to fuse very later during maturation of oospore. Backwell (1939) associated oospore dormancy with the nature of the thick oospore wall. It provides protection and is practically impermeable. Germination cannot take place until it is rendered permeable to oxygen and water. The oil or lipid globules provide energy for its long dormancy. Sussman (1966) and Schmitheshner (1968) reported that the dormancy in *Phytophthora* oospore is probably constitutive due to the innate property of the dormant stage. Furthemore, the reported absence of ribosomes (Leary, Roheum and Zentymer, 1979) and of cytochromes in the oospores of *P. capsici* (Rubero, Zentmyer and Erewin, 1976) shows that the metabolic activities of the mature oospore are at very low level.

Germination of oospore. It increased with age. Romero and Irwin (1969) obtained 64% oospore germination when oospores of *P. infestans* were kept at 4°C for 45 days. Germination takes place after the decay of the host tissues and on the onset of conditions suitable for germination. Gough *et al.* (1957) reported that germination of oospores of *P. infestans* occurred at temperatures ranging from 12°C to 25°C. During the pregermination stage *Phytophthora* oospore absorbs water and swells. The diploid oospore nucleus divides repeatedly. The increase in the number of nuclei is accompanied by increase in the number of mitochondria. The central globule becomes surrounded by vacuoles indicating its digestion and absorption. The pitting and erosion of the inner thicker layer which becomes thin at the site of germ tube emergence is indicative of impending germination. At this stage the oospore protoplast covered by the thinned inner layer bulges at the site of emergence and makes its way out through the outer layer of the oospore wall and presses against the oogonial wall. Finally it breaks through the proximal end of the oogonial envelope or emerges by rupturing the oogonial wall at other locations. Growth of the germ tube after emergence was variable. In *P. cactorium* it directly grows into a mycelium. The germ tube in *P. infestans* usually ends in a terminal papillate germ sporangium typical of the species in morphology and methods of germination. The contents of the terminal sporangium may divide to form zoospores. Each zoospore (J) after liberation germinates to give rise to the mycelium. The terminal sporangium may also germinate directly by forming a germ tube [Smoot *et al.*, 1958].

Life cycle and location of meiosis (Fig. 6.33). Till recently the widely held view was that *Phytophthora* has a haploid life cycle (6.33A). The only diploid structure in the life cycle is the oospore. Meiosis takes place in the zygote. The scheme of zygotic meiosis and haploid cycle in *Phytophthora* has been reconfirmed by Timmer *et al.* (1970), Zentmyer and Irwin (1970) and Stephenson *et al.* (1974*a* and 1974*b*) in *P. capsici.* Sansome (1965) sounded a discordant note.

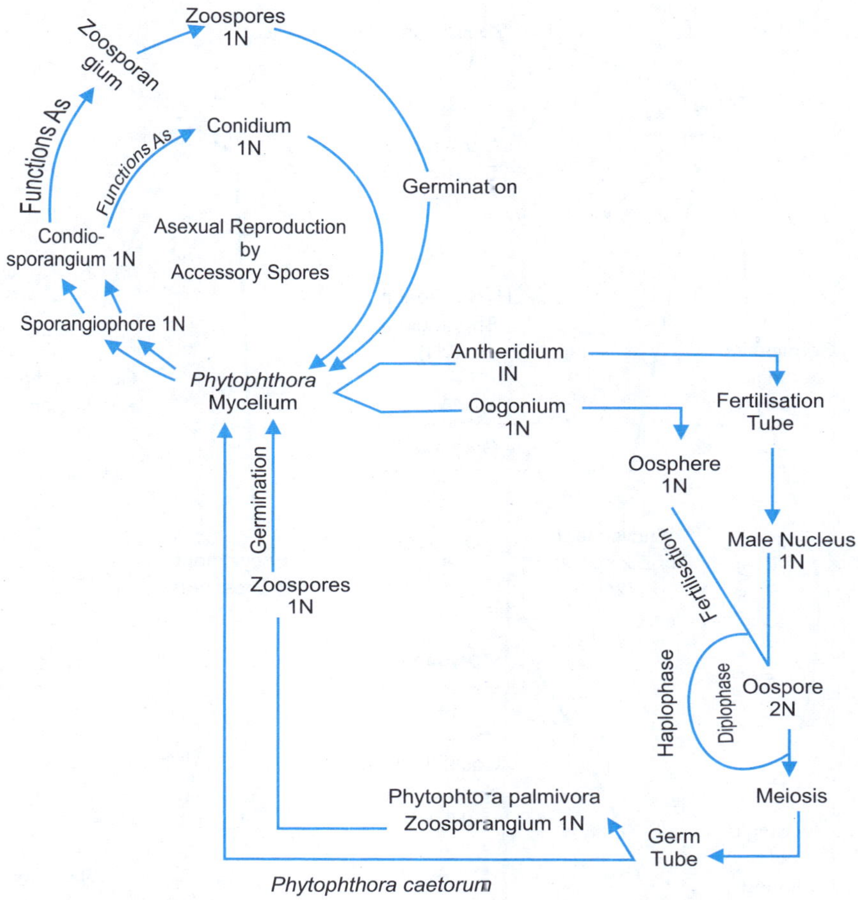

Fig. 6.33 A. Graphic representation of the life cycle of *Phytophthora* with Zygotic meiosis.

She suggested gametangial meiosis in *Phytophthora*. On this basis the haploid phase in this fungus is extremely reduced and restricted to the gametes only. Further evidence in favour of diploid life cycle and gametangial meiosis (Fig. 6.33 B) was furnished by Sansome (1976) in P. *capsici*. She observed pachytene, diplotene and diakinesis stages characteristic of meiosis I in the oogonium thus confirming gametangial meiosis in the life cycle of *Phytophthora*.

Control. The late blight disease of potato can be effectively controlled provided suitable control measures are used at the right time and in a large scale. The methods suggested to control the disease can be divided under the following two categories :

A. Chemical methods. The following methods using chemicals can effectively control the disease :

(*a*) **Spraying :** Spraying the affected plants with fungicide mixtures have been used for quite some time. Bordeaux mixture is the most popular fungicide used to effectively control the disease in the field. The early sprayings with 4 : 4 : 50 strength and late sprayings with 6 : 6 : 50 strength are commonly used. Various formulations of Fytolon, Blue copper and Blitox 50 are gradually replacing Bordeaux mixture. In recent years, Dithane Z-78 or Dithane M-45 and Maneb have been used for spraying with effective results.

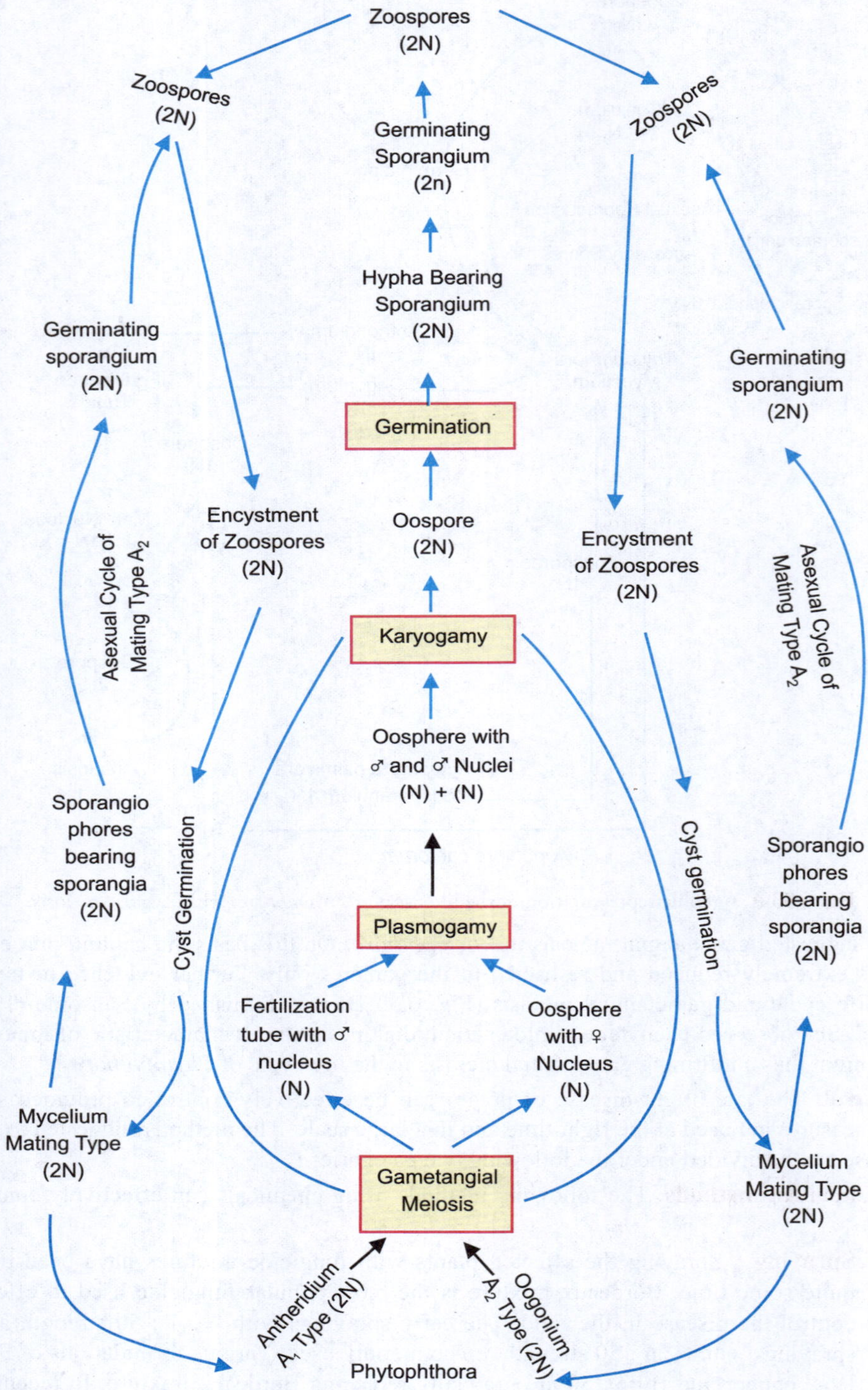

Fig. 6.33 B. Graphic representation of the life cycle of *Phytophthora* with gametangial meiosis.

(*b*) **Treatment of tubers before storage.** This saves the tubers from secondary infection by rot-causing organisms. A-90 minute dip of tubers in Mercuric chloride solution (1 : 1000) prevents the secondary infection.

B. Agricultural methods. The following agricultural methods are commonly used to control the disease :

(*a*) **High ridging.** A 4-6 inches high ridging at the time of earthing helps to reduce the infection of tubers.

(*b*) **Selection of seeds.** Infected tubers should be discarded and only disease-free stock be used for sowing. Although apparently disease-free tubers are no guarantee to loss but it certainly reduces the infection and makes spraying more effective.

(*c*) **Sanitation.** The crop should be clear and all sanitary precautions must be taken to reduce the infection. Weeds should not be allowed to grow. At the time of harvesting, contact between tubers and infected leaves should be avoided.

(*d*) **Sorting of tubers from a blighted field.** Harvesting should be done in clear dry weather and only healthy infection-free tubers should be collected. Before putting these in bins, the potato-tubers should be inspected again and only healthy potatoes should be stored.

(*e*) **Storage of Potatoes.** The tubers should be stored in a cool, dry and well aerated chambers with temperatures ranging between 38°-40°F.

(*f*) **Resistant varieties.** Use of resistant varieties developed throughout the world have proved to be the best method of controlling late blight disease of potato.

Salient Features

1. *Phytophthora infestans* parasitizes potato and causes the *potato blight* or *late blight* of potato.
2. The mycelium is well developed and profusely branched. It consists of aseptate, hyaline, coenocytic hyphae.
3. The hyphae are both intercellular and intracellular. The intercellular hyphae produce usually globular occasionally short, straight or curved peg-like haustoria which penetrate the cell walls of the host tissues to obtain nutrition. The haustoria are more commonly found in the tubers.
4. Asexual reproduction is by means of **sporangia** produced on special aerial hyphae, the **sporangiophores** which arise in tufts from the internal mucelium and emerge to the outside by way of the stomata of the host plant.
5. The sporangiophores are sympodially branched and are characterised by the bodular swelling at each node below the sporangium.
6. The papillate sporangia are lemon-shaped or oval and multinucleate. They are borne terminally and single but are pushed to the side by proliferation and continued growth of the hyphal tips bearing them.
7. The main factors governing germination of sporangia are moisture and temperature. The moisture is supplied by rain or dew.
8. At higher temperature in the presence of water the sporangium germinates directly by a germ tube which produces a mycelium.
9. At low temperature (cooler conditions) in the water the sporangium produces about 5-8 biflagellate **zoospores.**
10. The zoospores are differentiated in the sporangium itself and are liberated by the bursting of the **apical papilla.** No vesicle is formed.
11. The liberated zoospore germinates in the usual manner to produce a germ tube which forms an appressorium and causes infection by means of a fine, infection peg.

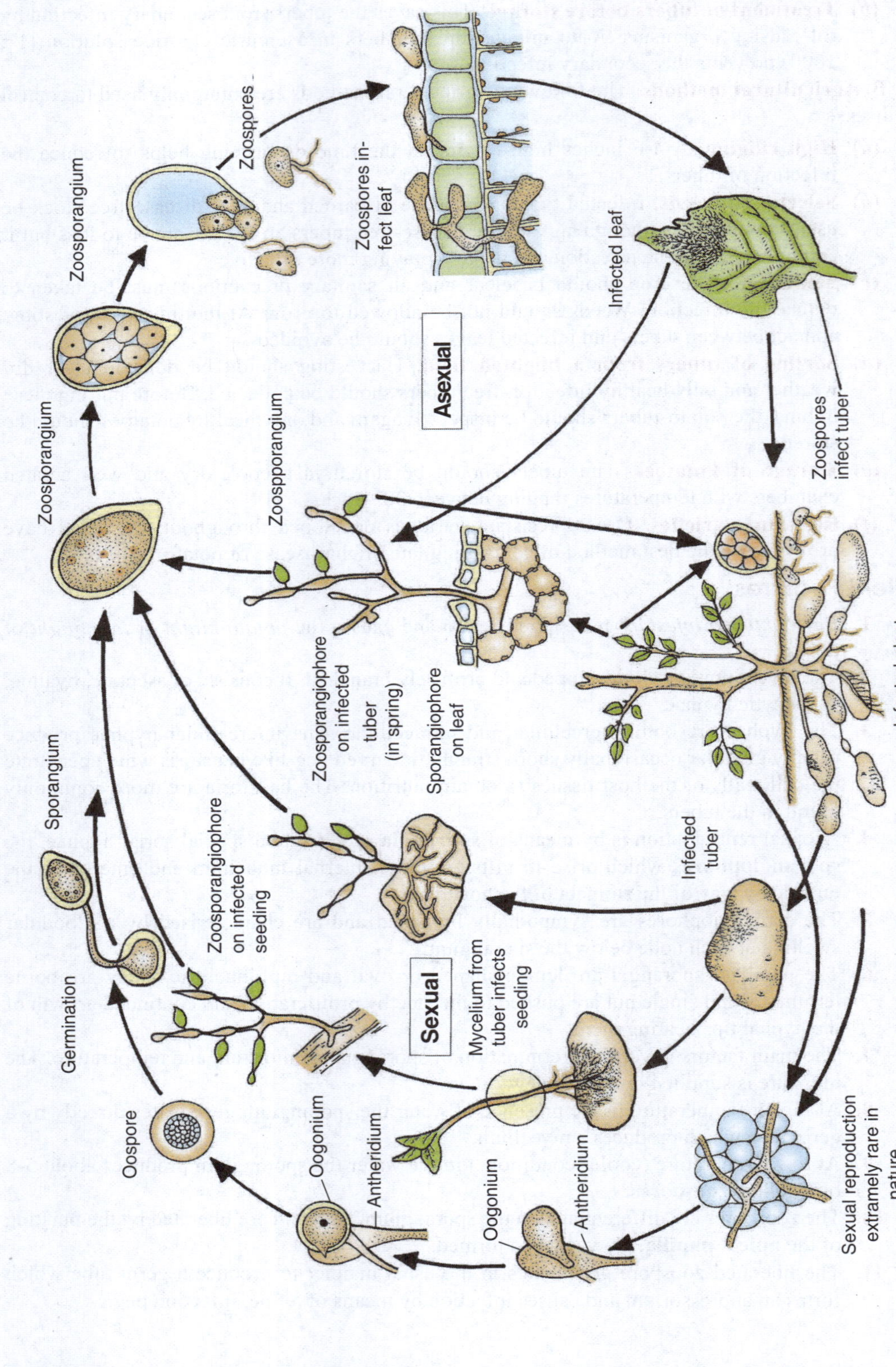

Fig. 6.34. Pictorial Disease cycle of late blight disease of potato caused by *Phytophthora infestans*.

12. Sexual reproduction is oogamous and the fungus is heterothallic. The antheridia and oogonia are produced on two distinct, opposite mating types of hyphae.
13. The antheridial and oogonial hyphae grow towards each other. The tip of the oogonial branch punctures and grows through the antheridium and swells on the other side to form the globose oogonium above the antheridium. The latter forms a funnel-shaped collar around the oogonial stalk at the base of the oogonium.
14. The sex organs at maturity are uninucleate.
15. At the time of germination the oospore produces a germ tube which terminates in a sporangium. The protoplast of the sporangium produces the zoospores but direct germination of the sporangium by a germ tube has been reported by Smoot *et al* (1958).
16. At times the oospore directly germinates by a germ tube.

Taxonomic Position. The genus *Phytophthora* has a chequered taxonomic history. It resembles *Pythium* in its life cycle so much that Fitzpatrick (1930) and others suggested its merger with it (*Pythium*). The opponents of this view held that the differences between the two genera are more marked than the resemblances. *Phytophthora* is more terrestrial than *Pythium* and thus less dependent on water for reproduction than *Pythium.* It completes its life cycle as a parasite on the host tissues and shows no signs of its saprophytic existence in the soil. It overcounters as a dormant mycelium in the infected potato tubers. Haustoria develop on the intracellular hyphae of *Phytophthora* but are absent in *Pythium*. Sporangia in *Phytophthora* are borne on specialised reproductive hyphae called the sporangiophore which are quite distinct from the somatic hyphae whereas in *Pythium* they are formed on hyphae indistinguishable from the somatic hyphae. The antheridum is amphigynous in *Phytophthora* whereas in *Pythium* it is peragynous. Because of these differences mycologists favour the following position of the genus in the plant world:

Alexopoulos (1962)

Division	:	**Mycota**
Subdivision	:	**Eumycotina**
Class	:	**Oomycetes**
Order	:	**Peronosporales**
Family	:	**Pythiaceae**
Genus	:	*Phytophthora*
Species	:	*infestans*

Alexopoulos and Mims (1979) place the genus *Phytophthora* in the superkingdom Eukaryonta, Kingdom Mycetae, Division Mastigomycotina, Subdivision Diplomastigo-mycotina., Class Oomycetes, Order Peronosporates and family Peronosporaceae.

Alexopoulos and Mims (1979)

Superkingdom	:	**Eukaryonta**
Kingdom	:	**Mycetae**
Division	:	**Mastigomycotina**
Subdivision	:	**Diplomastigomycotina**
Class	:	**Oomycetes**
Order	:	**Peronosporales**
Family	:	**Pythiaceae**
Genus	:	*Phytophthora*

Economic Importance. *Phytophthora infestans* or potato blight fungus is of considerable economic and historical interest. It causes a serious disease of potatoes known as *potato blight* or *late blight* of potatoes. In an epidemic form the disease may cause complete crop failure or serious losses in potato crops. This disease caused famines in the past in countries where potato was the mainstay in the diet of the peasantry. An example is the catastrophic potato famine that swept over Ireland in 1845-46. It resulted in the death of over a hundred thousand people from widespread hunger starvation and was the cause of emigration of many to America in the Eastern United States in 1846. The disease caused loss of about 50 million dollars. The potato blight fungus also causes the *late blight* of tomatoes. The tomato fruits are infected and undergo rot at all stages.

FAMILY : PERONOSPORACEAE (DOWNY MILDEWS)

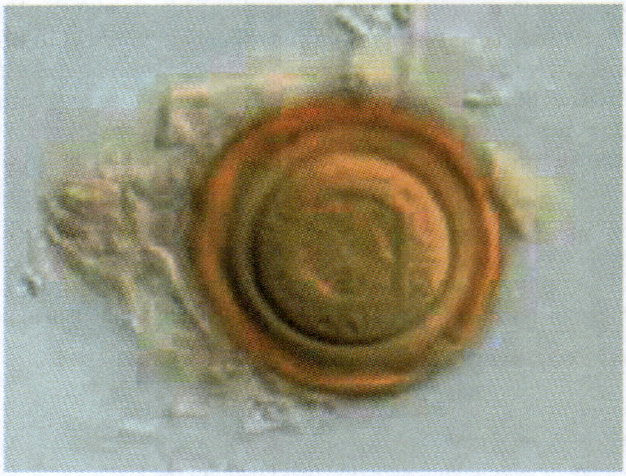

Downy Mildews - Peronosporales.

General. This is the most advanced family of order Peronosporales. It comprises 6 genera with numerous species. All of them are important plant pathogens which are **obligate parasites** of flowering plants such as onion, grape, bajra, jowar, maize, pea, turnip, beet, tobacco and many crucifers. The diseases caused by the Peronosporaceae are called *downy mildews*. The mycelium is branched, tubular and coenocytic. The hyphae are strictly intercellular and occur in the stem but more commonly in the leaf. They absorb nutrition by sending vesicular (Fig. 6.36 A) or branched finger-like haustoria (Fig. 6.40A) into the host cells. The houstoria do not perforate the plasma membrane of the host cell but simply push it in and invaginate. The sporangiophores (or conidiophores) are distinct and of **determinate** growth. They arise from an endophytic mycelium, emerge through the stomata to come into the air when humidity is sufficiently high and branch characteristically. The individual genera of the family are in fact distinguished on the basis of the form of the sporangiophore. Each branch bears a few to many short, delicate branchlets. A single ovoid or pyriform sporangium is formed at the tip of an ultimate branchlet called a **sterigma**.

The sporangia are deciduous and thus get detached when mature. The detached sporangia are disseminated by wind. On landing on a suitable host the sporangium in the genera except *Peronospora* behaves in two different ways. It germinates either discharging a number of small biflagellate, reniform zoospores or directly forming a short germ tube under warmer or dry conditions thus assuming the role of a conidium. The germ tube formed in whatever manner enters the host through a stoma and forms the mycelium.

The sex organs are formed within the tissues of the host on the same or adjacent hyphae. The antheridia are **paragynous.** The protoplast of the oogonium is differentiated into **periplasm** and **ooplasm.** The oospore has a thick wall. The outer layer or epispore is sculputured. The oospore, as a rule, germinates by germ tubes. The only exceptions are *Peronospora tabacina* and *Plasmopara* in which the oospore liberates zoospores in a sporangial vesicle.

The genera placed in the family are known by the common name *Downy mildews hairy fungi.* because the sporangiophores of the members look like hairs emerging from the stomata. Since the oospores and sex organs are not much different in various genera asexual reporductive structures—the sporangiophores have been used as the criterion of delimitation, of genera. Important genera placed in the family include

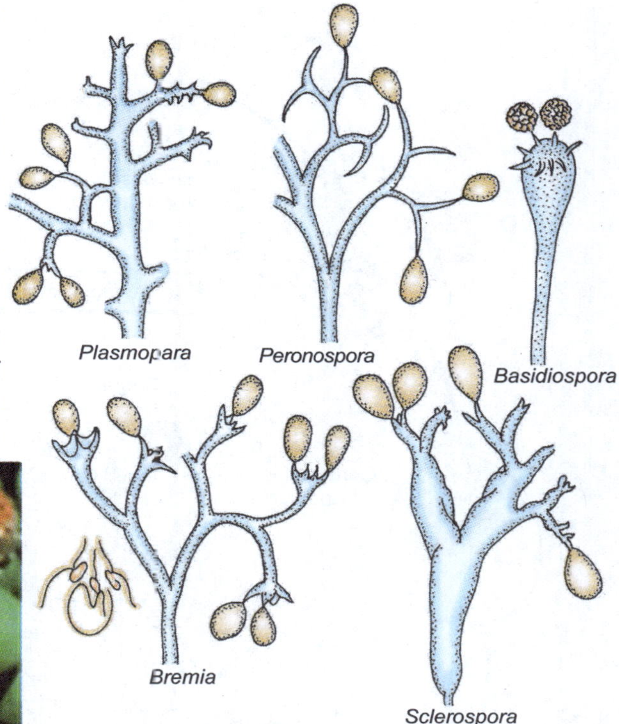

Fig. 6.35. Sporangiophores of Peronosporaceae.

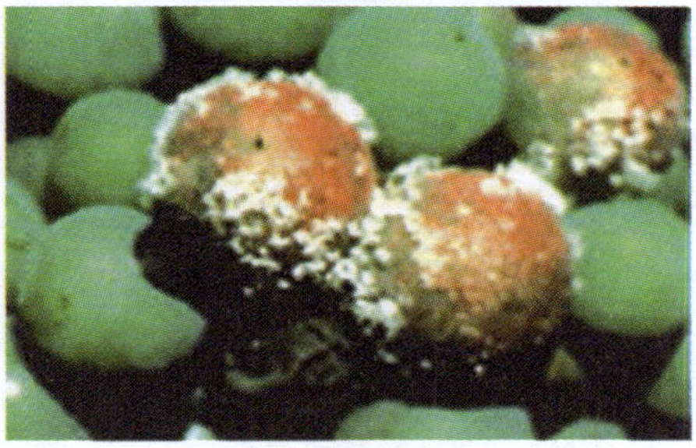

Plasmopara

Plasmopara, *Peronospora*, *Basidiospora*, *Bremia* and *Sclerospora* whose characterisitic sporangiophores are shown in fig. 6.35. Of these *Plasmopara*, *Peronospora* and *Sclerospora* are discussed in detail.

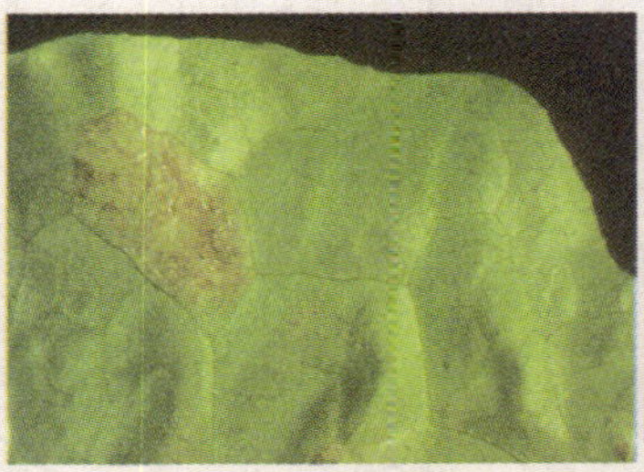

Bremia Spots

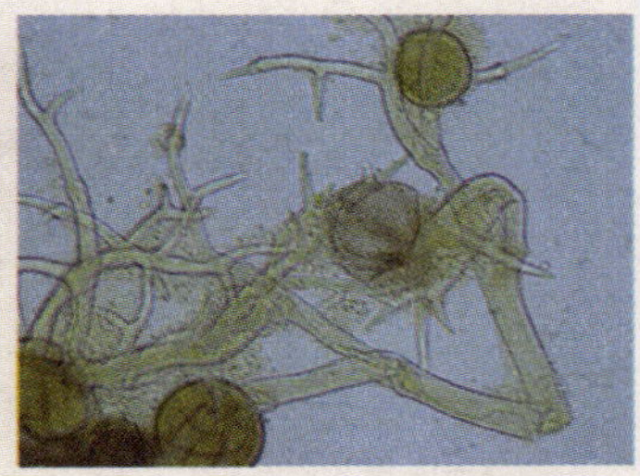

Peronospora hyphae with oospores

PLASMOPARA (FIG. 6.36)

Occurrence. It is a *downy mildew* which comprises about 20 species. All are widely distributed obligate parasites of vascular plants. The best known and the most important from the economic standpoint is *Plasmopata viticola.* It parasitizes the grape vines (*Vitis vinifera*) and causes the *downy mildew* disease of grapes. The disease is reported to occur in several parts of India but not in an epidemic form. However, it used to cause serious damage to vine yards in the West when the fungicides like **Bordeaux mixture** were unknown.

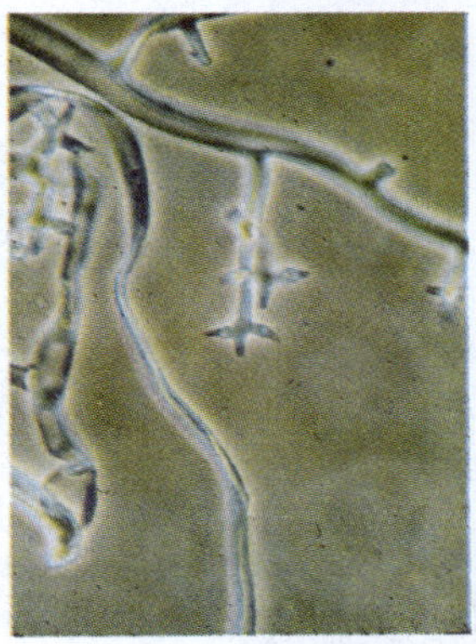

Plasmopara viticola hyphae

Mycelium. It is freely branched and coenocytic. The hyphae are numerous and vary in size. The fungus is strictly intercellular (Fig. 6.36 H). Some of the hyphae give off lateral outgrowths which

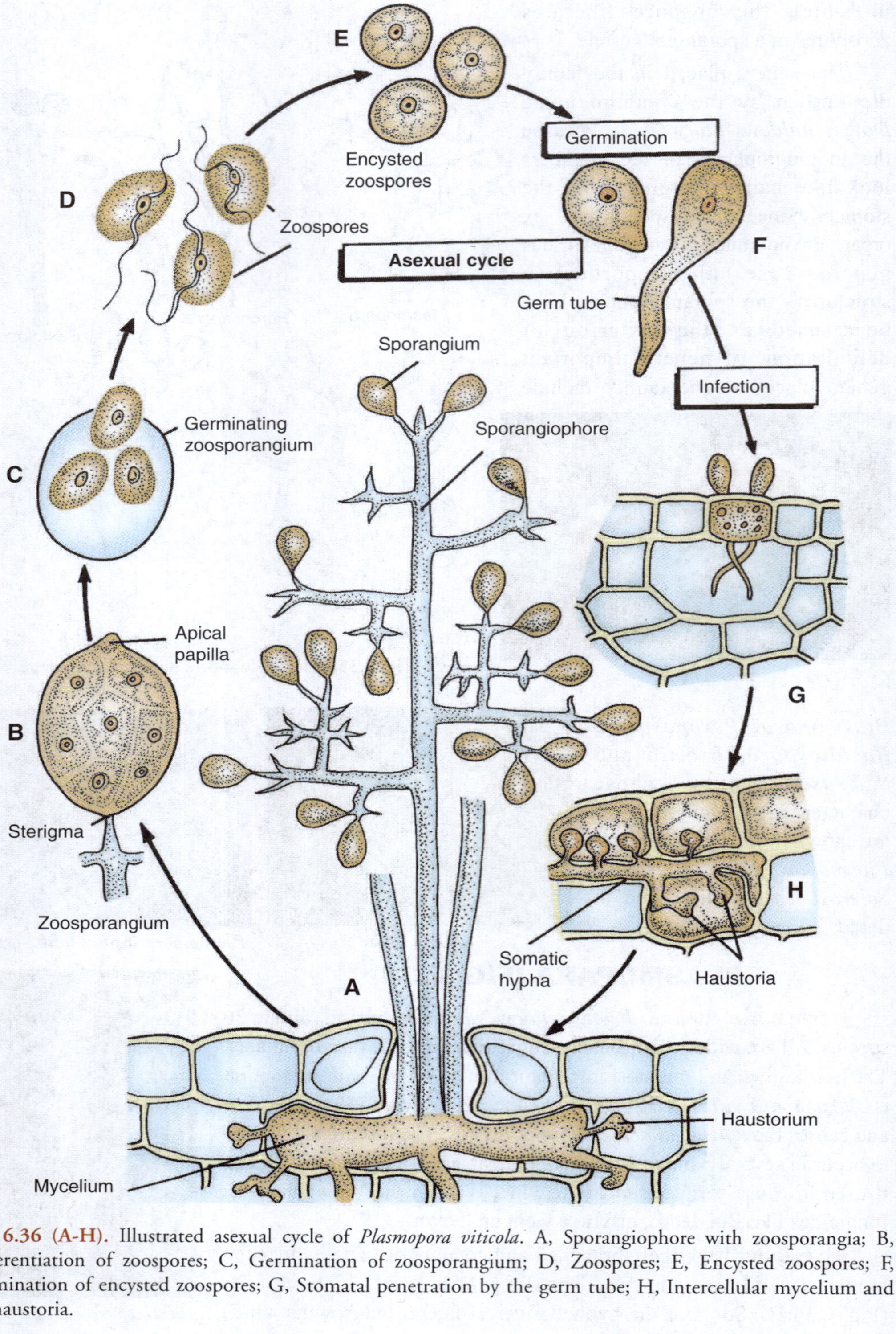

Fig. 6.36 (A-H). Illustrated asexual cycle of *Plasmopora viticola*. A, Sporangiophore with zoosporangia; B, Differentiation of zoospores; C, Germination of zoosporangium; D, Zoospores; E, Encysted zoospores; F, Germination of encysted zoospores; G, Stomatal penetration by the germ tube; H, Intercellular mycelium and the haustoria.

penetrate the host cells and swell inside to form vesicular **haustoria.** The latter secrete enzymes by the help of which they absorb food from the protoplasts of the host cells.

Symptoms. The first sign of the disease is the appearance of small, greenish-yellow spots on the upper surface of vine leaves. Due to increase in size the spots merge into one another and turn brown. In the damp weather frost-like downy patches appear on the lower surface of the leaves. Finally the leaves dry up and fall off. The bunches shrivel and the young shoots become dwarfed and deformed. The petioles, tendrils and other parts may be affected. They present a water soaked spongy appearance. The young berries become covered with downy growth. The older ones may rot and crack.

Asexual Reproduction (Fig. 6.36 A-D). Once established within the host, the mycelium forms pads of hyphae in the substomatal cavities (A). A group of hyphae grows upright from the hyphal pad and emerges through a stoma to function as **sporangiophores** (A). Each sporangiophore is a long, monopodially branched structure. The lateral branches vary in number from three to six. Each branch bears branchlets. The branches and the branchlets lie more or less at right angles to the axis bearing them. The growth of the sporangiophore is determinate and thus bears sporangia when fully developed. A single sporangium is formed at the tip of an ultimate delicate branchlet called the **sterigma** (plural sterigmata). All the sporangia in the crop are of about the same age. They are hyaline and ovate. Each sporangium is attached to its respective branchlet (sterigma) by the small end. At the opposite end, where sporangial wall is the thinnest, is a papilla (B).

The young sporangium is a small ovoid, uninucleate structure. With the increase in size its nucleus undergoes repeated divisions. The multinucleate granular protoplast is then divided into several uninucleate daughter protoplasts (B). The sporangia are shed at this stage and are dispersed by air. If the detached sporangia which float in the air, happen to land on a suitable, moist substrate (leaf of the host), they germinate. A circular pore appears at the papilla by the dissolution of its thin wall. The uninucleate daughter protoplasts emerge through the opening (C). After emergence each protoplast furnishes itself with a pair of flagella (D). The liberated zoospores are plano-convex in form. There is a ridge along the flat side, near which lies the nucleus. The two flagella arise one on either side of the nucleus and are of unequal length. The short flagellum which is of **tinsel type** is directed forward and the longer one which is of **whiplash type** trails behind when the zoospore is in motion. The liberated zoospores swim about in a thin film of water for a while (about half an hour) and then come to rest in the vicinity of stomata. Each quiescent (resting) zoospore rounds up and secretes a thin wall around it (E). Soon the globose cyst produces **a germ tube** (F). The latter gains entry into the host tissue (G) through a stoma (Gregory, 1912). Rarely the sporangium germinates directly by a germ tube like a conidium.

Sexual Reproduction (Fig. 6.37 B-E). It is **oogamous.** Like *Albugo* the sex organs are developed within the host tissue at the tips of adjacent hyphae.

The oogonium arises as a globular swelling at the tip of a female hypha (B). It is then cut off by a cross wall at its base. The young oogonium is uniformly multinucleate. As it advances towards maturity a central, rounded dense **ooplasm** with many nuclei is separated from the peripheral, vacuolated, multinucleate **periplasm.** The mature ooplasm is uninucleate and functions as an **oosphere** or **egg.**

The antheridium is developed at the end of male hypha lying near the oogonium (B). The tip of the male hypha enlarges into a clavate (club-shaped) swelling which is later cut off by a cross wall from the supporting hypha. The clavate cell is the **antheridium.** It contains about a dozen male nuclei and is applied to the side of the oogonium.

Fertilisation. At the point of contact with the oogonium the antheridium sends forth a delicate tube known as the **fertilisation tube** (C). The latter pierces the oogonial wall, pases through the periplasm and enters the oosphere (C). At this stage the tip of the fertilisation tube dissolves. A single male nucleus migrates through it to fuse with the egg nucleus. The fusion between the two gamete nuclei may be delayed until spring. The egg with the two nuclei is called the **zygote.** The zygote soon secretes a thick, rough wall to become an **oospore** (E). The latter enters upon a period of rest. The oospore wall is differentiated into an inner thick smooth **endospore** surrounded by a thin rough **exospore.**

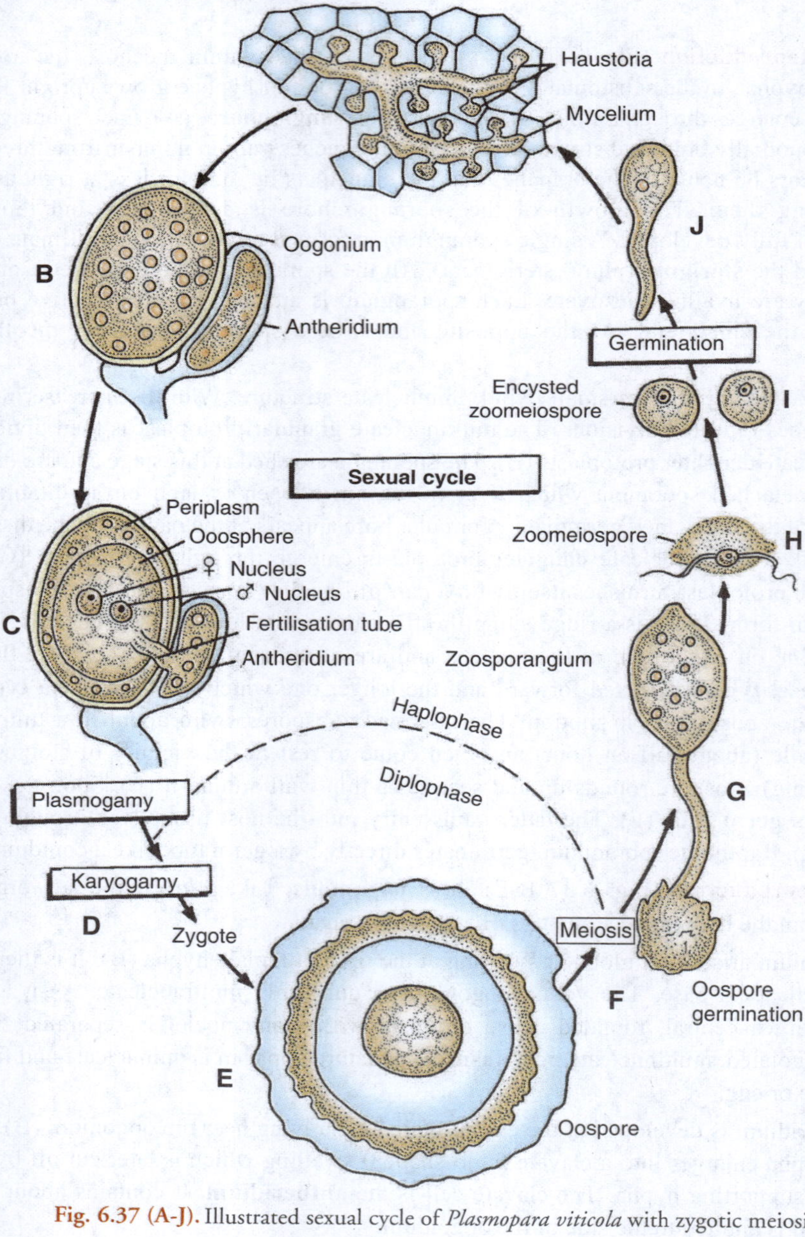

Fig. 6.37 (A-J). Illustrated sexual cycle of *Plasmopara viticola* with zygotic meiosis.

Germination of oospore (Fig. 6.37 G). The following spring when the conditions favourable for growth set in, the oospore germinates. Prior to germination the male and female nuclei fuse. At the time of germination the fused nucleus undergoes repeated division to form 50 or more nuclei. Where meiosis takes place is not definitely known. Some hold that first two divisions are **meiotic.** Bose (1946), however, reported that the reduction of the nucleus to initiate the haplophase takes place in the third or fourth division of the oospore nucleus. The prevalent view favours gametangial meiosis and diploid life cycle. At this stage the oospore wall bursts and germ tube protrudes through the small crack. The tip of the germ tube swells up to form a terminal ovate **sporangium** (G). The multinucleate protoplast of the oospore flows into the sporangium. The latter is then cut off from the germ tube by a septum. Within the sporangium are differentiated 50 or more uninucleate, biflagellate, reniform zoospores. They escape through an apical pore in the sporangial wall. The liberated zoomeiospore germinates in the usual manner on the same or a different host to which it may be carried.

Controls. Spraying the vines with fungicides has proved very effective. Bordeaux mixture, Ferbam and Captan are the best fungicides. Removal of infected parts and their subsequent destruction is another effective control measure. Some success has been achieved in evolving resistant varieties in Europe.

Taxonomic position. Berkeley and Curtis named *downy mildew* of grapes as *Botrytis viticola* and thus was referred to the class Fungi Imperfecti. Later de Bary determined its true affinities. He placed it in the family Peronosporaceae but named it *Peronospora viticola.* Schroeter in 1886 established the genus *Plasmopara.* The downy mildew of grapes was shifted to the genus as *Plasmopara viticola* by Berlese and de Toni. The systematic position of *downy mildew* of grapes in the plant world is as under :—

Division	:	**Mycota**
Sub-division	:	**Eumycotina**
Class	:	**Oomycetes**
Order	:	**Peronosporales**
Family	:	**Peronosporaceae**
Genus	:	*Plasmopara*
Species	:	*viticola*

Alexopoulos and Mimsa (1979) place *Plasmopara* in the Superkindgom Eukaryonta, Kingdom Myceteae, Division Mastigomycota, Sub-division Diplomastigomycotina, Class Oomycetes, Order Peronosporales, Family Peronosporaceae.

Superkingdom	:	**Eukaryonta**
Kingdom	:	**Mycetae**
Division	:	**Mastigomycotina**
Sub-division	:	**Diplomastigomycotina**
Class	:	**Oomycetes**
Order	:	**Peronosporales**
Family	:	**Peronosporaceae**
Genus	:	*Plasmopara*

PERONOSPORA

General. Ainsworth and Bisby (1954) reported about 75 species in this genus. All are aggressive, obligate parasites on higher plants and cause diseases known as downy mildews. Most of the species attack a specific host and thus are said to exhibit **selective parasitism.** Many species parasitize crops and other plants of economic value such as pea, onion, turnip, beet, maize, tobacco, spinach and the like. The well-known species are :—

1. *Peronospora parasitica.* It attacks turnip, radish, cabbage, cauliflower and all *Brassica* species.

2. *P. destructor.* It causes the *downy mildew disease* of onion.

3. *P. pisi.* It occurs on peas (*Pisum sativum*) and causes *downy mildew* of peas. Of late there has been a tendency to split up some of the older species. For example, *Peronospora parasitica* has been split up by Gaumann into a number of distinct species which are specialized to attack single host genera. The fungus attacking wall flower (*Cheiranthus cherii*) is named *Peronospora cheiranthi.* The one which occurs on sarson (*Brassica compestris*) is called *Peronospora brassicae* and so on. All have a similar life cycle and differ only in minor detail.

Peronospora parasitica spots on Mustard

Peronospora destructor on leaves

Peronospora pisi on pea pods

Downy Mildew of Peas

Occurrence. It is an obligately parasitic fungus which commonly attacks *Capsella bursapastoris* and other crucifers causing *downy mildew disease* of the plants it attacks.

Symptoms. Externally the parasite causes grayish-white or purplish brown, downy spots on the lower surface of the lower, older leaves of the host plant. This is due to the covering of spore-producing hyphae. These spots may remain small or enlarge considerably. On the upper side appear tan to yellow lesions. Systematically infected plants may show stunted growth. The greatest deformity appears on the stem in the form of small or several inches long swellings.

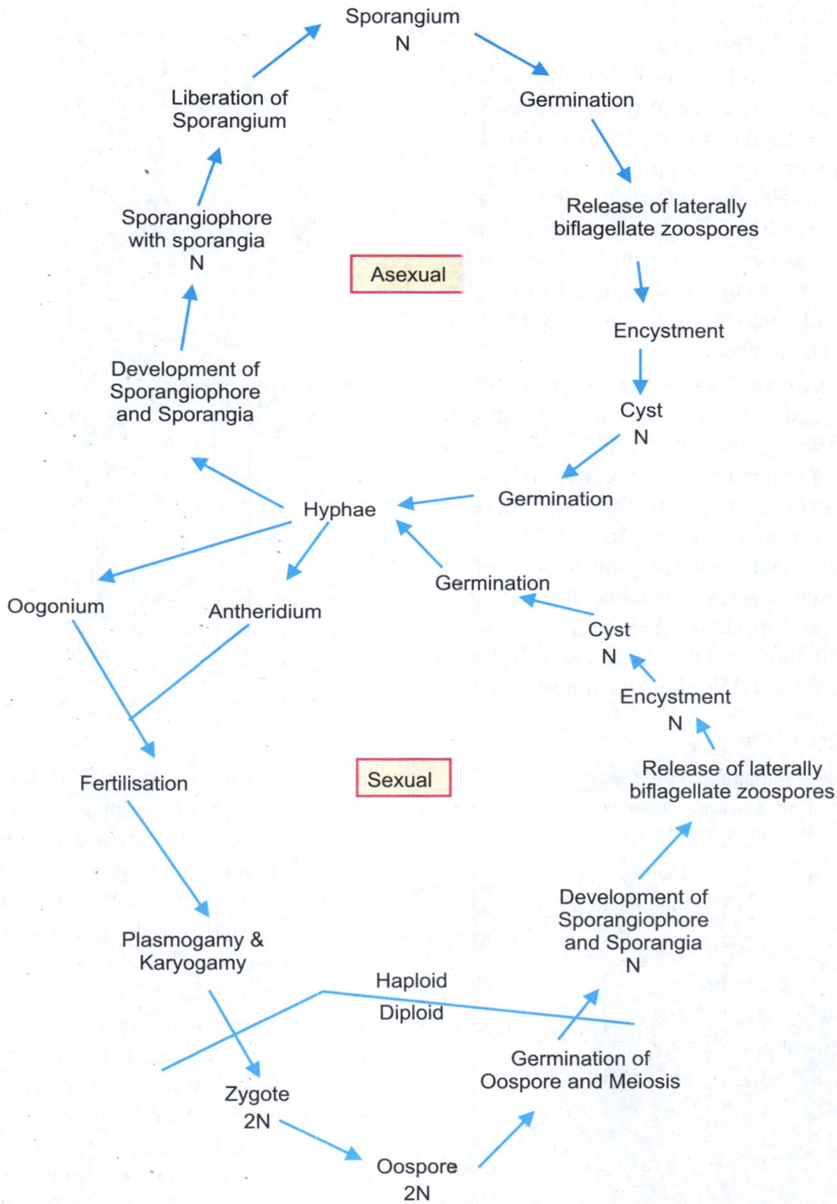

Fig. 6.38. Word diagram of the life cycle in *Plasmopara viticola.*

Thallus (Fig. 6.39). It is a well-developed filamentous, coenocytis endoparasitic mycelium consisting of branched aseptate hyphae. The coenocytic hyphae are strictly intercellular and are abundant in the intercellular spaces of leaves and subapical stem tissues. They draw nourishment from the host cells by means of intracellular **haustoria.** The latter absorb nutrients from the host cells and pass them on to the growing fungus mycelium. The parasitized cells are not killed.

(*i*) *Hyphal structure*. The finely granular hyphal wall is covered by a thin electron dense outer layer. New to the hyphal wall is the undulating plasma membrane. It invests the hyphal cytoplasm which contains the usual cell organelles. The ribosomes are abundant in the granular cytoplasm which also contains the polyribosomes. Cisternae of rough endoplasmic reticulum are sparse. Interspersed randomly in the cytoplasm are the spherical nuclei and mitochondria with tubular cristae. Dictyosomes with associated vesicles containing granular material also occur in the cytoplasm. The other inclusions are the spherical or oval lipid bodies and large and small vacuoles. Storage products like glycogen are absent.

(*ii*) *Haustoria* (Fig. 6.40). Terminal 500 μ of hyphal systems were devoid of haustoria (Davison, 1968). Lower down and adjacent host cells were penetreted and often completely filled with variously shaped haustoria originationg from the coenocytic hyphae filling the intercellular spaces between the host cells. In mature leaves several haustoria may occur in each host cell. In shape they may be clavate, simple digit-like or branched and tubular in nature. The intracellular haustorium may have a

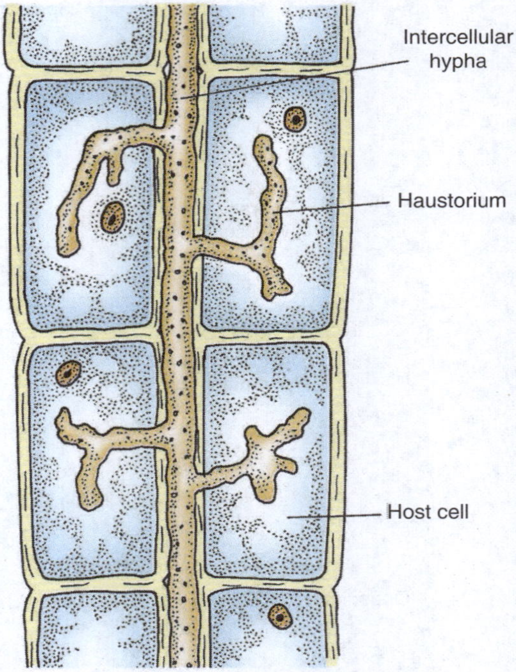

Fig. 6.39. *Peronospora parasitica.* Intercellular hypha with haustoria.

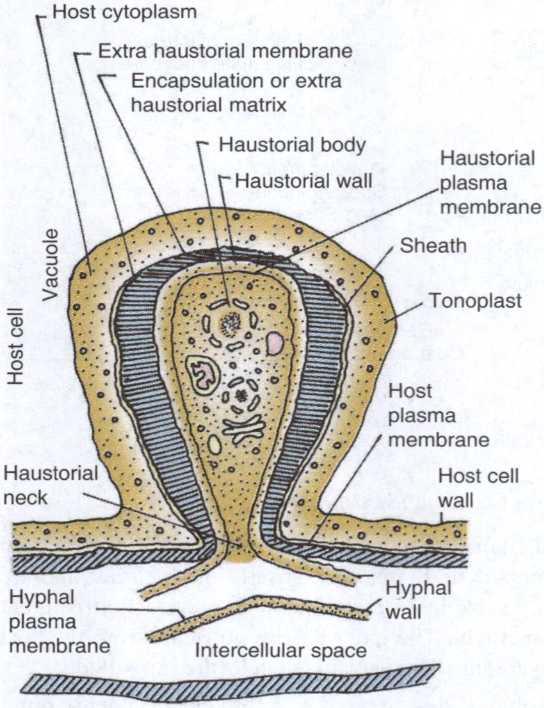

Fig. 6.40. *Peronospora parasitica.* Diagrammatic representation of fine structure of haustorial apparatus.

bulbous portion (haustorial body) connected with the intercellular hypha by a slender stalk or neck constricted at the electron transparent layer constituting the *haustorial wall.* It is continuous with the hyphal wall. Within the haustorial wall enclosing the haustorial cytoplasm is the *haustorial plasma membrane* similar in nature and continuous with the plasma membrane of the intercellular hyphae. The haustorial cytoplasm contains ribosomes and other cell organelles. One to many nuclei, mitochondria with tubular cristae, Golgi bodies and vacuoles with amorphous contents are commonly found as in the hyphae. Cisterne of endoplasmic reticulum are an occasional feature. Contiguous to the haustorial wall and surrounding the haustorium completely is an electron-dense layer of amorphous material. Berlin and Bowen (1964) called it encapsulation whereas other investigators termed it *extrahaustorial matrix* or *sheath matrix.* It is thick at the proximal end of the haustorium but thins towards the distal end.

It is continuous with the similar electron dense material between the host and parasite (fungal) walls at the site of penetration. The extrahaustorial matrix or encapsulation zone is separated from the host cytoplasm by a tightly pressed unit membrane called the *extrahaustorial membrane* which is considered to be the invaginated host plasma membrane. Davision (1968) observed some haustoria to be partially or completely surrounded by a layer or wall of appositional material continuous with and consisting of the similar material as the host cell wall. Berlin and Bowen (1964) termed it the *sheath.* The sheath, however, is much thicker than the host cell wall.

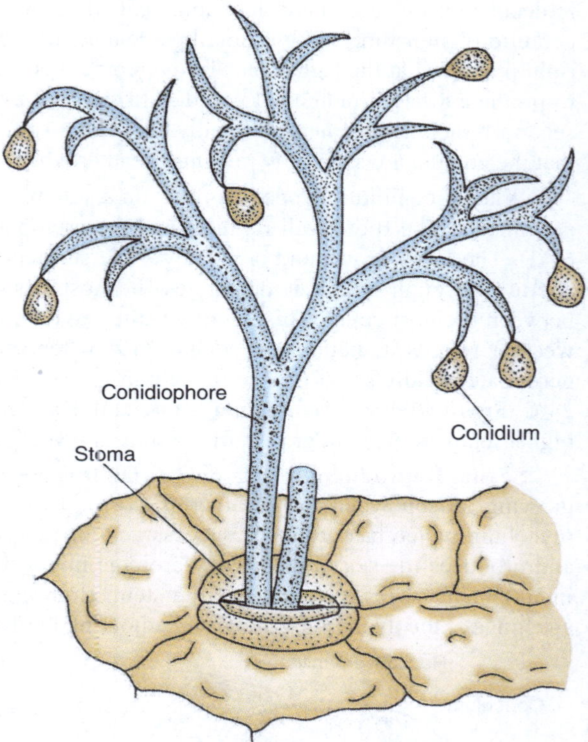

Fig. 6.41. *Peronospora parasitica*, Conidiophore emerging from a stoma. It bears conidia on the branches.

Asexual Reproduction. It shows an advance on the condition found in other members of Peronosporaceae. The deciduous sporangia invariably function as **conidia** and germinate by means of germ tubes. *Peronospora* thus represents the final stage in the evolution of the conidium from a zoosporangium. In fact, plant pathologists use the term **conidia** for such deciduous sporangia and **conidiophore** for the special reproductive hyphae which bear them. The mycologists prefer to call them sporangia and sporangiophore, respectively.

When the fungus mycelium has established itself within the host the hyphae begin to collect in the substomatal cavities. Some of these hyphae emerge through the pores of the stomata on the lower surface of leaves in clusters or singly (Fig. 6.41). The emerging hypha grows vertically. It elongates considerably from 100-300 micron. In its lower portion which forms its major length it is unbranched but branches characteristically in its upper part to function as a **conidiophore.** The conidiophore has a highly distinctive shape. Its base is flattened and somewhat twised where it passes through the stomatal pore. The branching is repeated (6-8 times) and dichotomous. The branches and their subdivisions are more or less reflexed and occur at acute angles. Each ultimate branch or branchlet is long, slender and tapers to a slightly curved, pointed tip. The latter swells to form a colourless, uninucleate, spherical or broadly oval **conidium** which is attached to the branchlet which bears it by a fine, short stalk known as the **sterigma.** The sterigmata are water soluble. The mature conidia thus lose connection with sterigmata in damp weather and are distributed by wind.

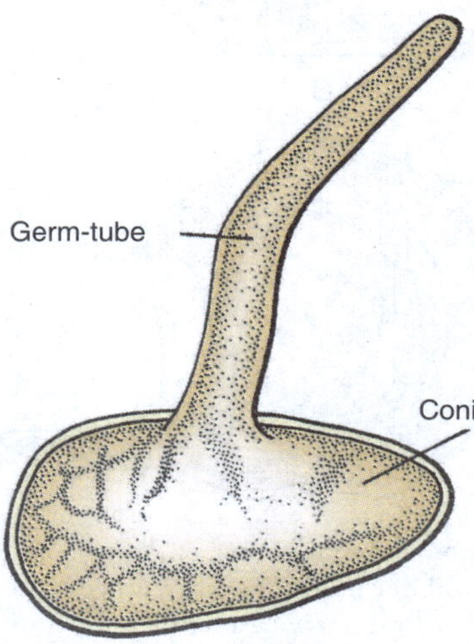

Fig.6.42. *Peronospora parasitica.* Germinating conidium.

Germination of conidia (Fig. 6.42). The conidia remain viable for a brief period. Besides, their germination is dependent on favrouable location (leaf of same species of host) and suitable environmental conditions (moisture). Only a small percentage of conidia produced will ever come to land in the right place and at the right time. To overcome this difficulty the fungus has adopted every device to produce a large number of conidia. In the first place the conidiophores arise in fascicles and secondly each is branched repeatedly. The water soluble sterigma is another device which ensures that the conidia are shed only in damp weather when germination is possible.

When a conidium happens to land on a suitable host under proper environment conditions it germinates. The spore wall ruptures and the protoplast emerges in the form of a germ tube (Fig. 6.42). The latter grows and crawls over the surface of the host leaf until it reaches the stomatal aperture through which it penetrates the host tissue. Inside it ramifies rapidly in the spaces between the host cells. Moist, cool weather favours the spread of the disease whereas dry, warm weather retards it. Felton and Walker (1946) reported that conidia of *P. parasitica* on cabbage germinate rapidly at 16°C. The mycelium grows rapidly at 24°C. The lesions develop rapidly at 24°C. Sporulation and reinfection is rare at this and higher temperature. It is rapid at 16°C than at higher temperatures but growth of the fungus mycelium is slower at 16°C.

Sexual Reproduction (Fig. 6.43 A-D). It is oogamous and takes place towards the end of the growing season when the weather is dry and the environmental conditions are adverse. The mycelium which has produced successive crops of conidia penetrates deeper into the host tissue and produces the sex organs. They are commonly found in the pith on the hyphae in the larger intercellular spaces especially of the stem. Both are multinucleate at first and arise as terminal swellings at the hyphal tips of closely adjoining hyphae (A).

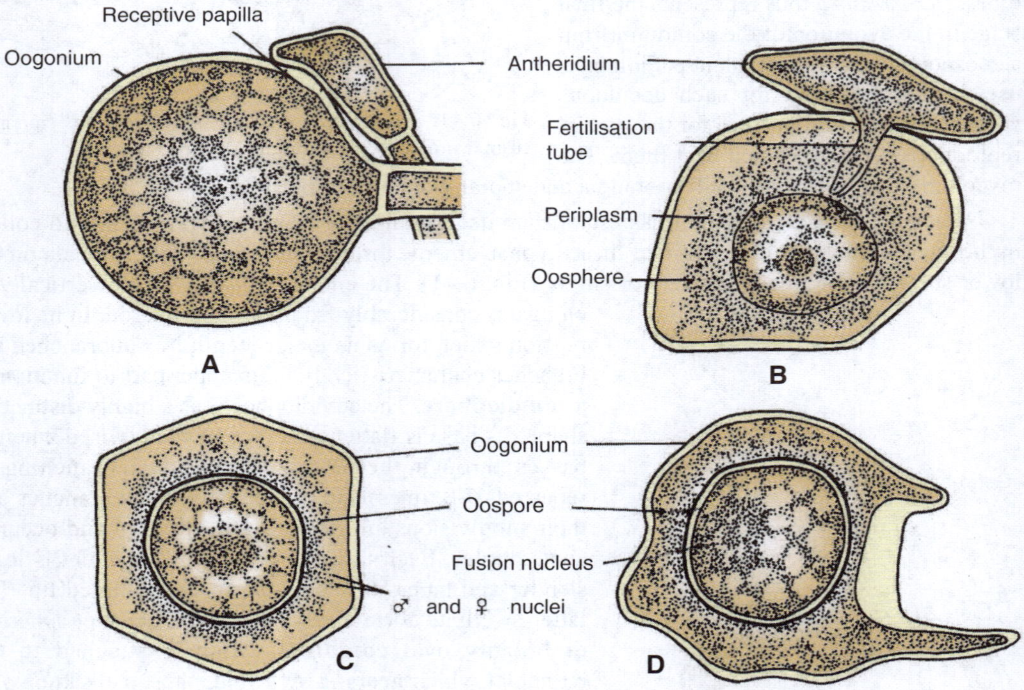

Fig. 6.43 (A-D). *Peronospora parasitica.* Stages in sexual reproduction and fertilisation. A, Young multinucleate sex organs. Note the receptive papilla; B, Oogonium with the protoplast differentited into a central uninucleate oosphere and multinucleate periplasm; Antheridium with a fertilisation tube; C, Oospore with male and female nuclei lying side by side; periplasm degenerating and oospore wall forming; D, Oospore with a diploid nucleus and thick wall (After Wager).

Oogonium. The oogonium is more or less a globose structure. It is separated from the supporting hypha by a cross wall (A). When young it has a dense multinucleate protoplast. The nuclei are evenly distributed. With further increase in size the oogonial protoplast becomes differentiated into :—

(*a*) A central, denser, more or less spherical **ooplasm.**

(*b*) An outer, vacuolate layer, the **periplasm,** lining the oogonial wall.

Associated with the above-mentioned changes rearrangement of oogonial nuclei also takes place. The numerous neclei of the ooplasm excepting one, which is centrally located, pass into the periplasm where they finally abort. The uninuclete ooplasm functions as an oosphere or egg (B). Only a single egg is formed in the oogonium.

Antheridium. The antheridium is paragynous. It is tubular in form and is multinucleate at first. It is separated from the supporting hypha by a cross wall. At maturity all the nuclei excepting one, the **male nucleus**, degenerate.

Fertilisation (Fig. 6.43 B). The antheridium comes in contact with the oogonium and develops a fertilisation tube at the point of contact. The latter pierces through the oogonial wall and the periplasm. Finally its tip dissolves discharging the male nucleus and a small amount of cytoplasm into the oosphere. The male and the female nuclei fuse (C). The zygote thus formed secretes a thick wall. It is laid in the periplasm around the oospore which is globose in form and yellowish brown in colour. The heavy-walled oospore (D) enters upon a period of rest extending over several months. It probably coincides with the period unfavourable for growth. McMeckin (1960) suggested that the fungus parasite survives from one growing season to the next in the sexually produced resting spore (oospore). Meanwhile host tissues undergos decay and oospores are released. They remain in the contaminated soil for some period before germination takes place.

Germination of oospore. After a period of rest and with the return of conditions favourable for growth the oospore germinates. The thick oospore wall bursts and the protoplast emerges in the form of a germ tube which brings about fresh infection. Perhaps the first two nuclear divisions at the time of oospore germination constitute **meiosis.** The place of meiosis in the life cycle however is controversial. According to some (Sansome and Sansome, 1974) meiosis occurs in the gametangia. Evidence in favour of gametangial meiosis in the class Oomycetes as a whole is accumulating.

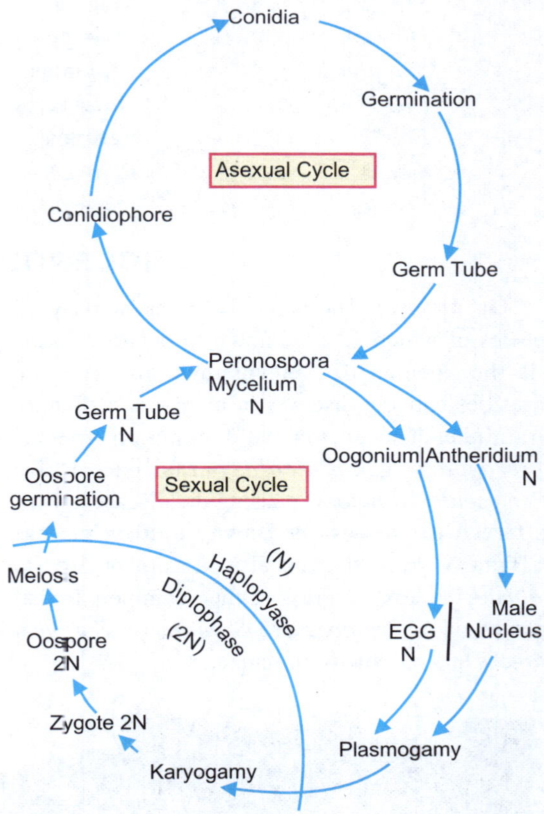

Fig. 6.44. Graphic representation of the asexual and sexual cycles of *Peronospora* with Zygotic meiosis.

Control. Spraying and dusting the plants with fungicides is a very effective method of controlling the *Downy Mildew disease.* Another equally useful practice is the destruction of debris containing the oospores. Rotation of crops with an interval of two to three years in between the successive crops is also considered useful. During this period the oospores in the soil lose their power of germination.

Taxonomic position :

Division	:	**Mycota**
Sub-division	:	**Eumycotina**
Class	:	**Oomycetes**
Order	:	**Peronosporales**
Family	:	**Peronosporaceae**
Genus	:	*Peronospora*
Species	:	*parasitica*

Alexopoulos and Mims (1979) place *Peronospora* in the Superkingdom Eukaryonta, Kingdom Myceteae, Division Mastigomycota, Subdivision Diplomastigomycotina, Class Oomycetes, Order Peronosporales, and Family Peronosporaceae.

Super kingdom	:	**Eukaryonta**
Kingdom	:	**Mycetae**
Division	:	**Mastigomycotina**
Sub-division	:	**Diplomastigomycotina**
Class	:	**Oomycetes**
Family	:	**Peronosporaceae**
Genus	:	*Peronospora*
Species	:	*parasitica*

SCLEROSPORA

Occurrence. The genus is represented by 15 species of which 13 are known to occur in India. All the species of **Sclerospora** are obligate parasities causing diseases in members of Family Graminae. The serious and common species *Sclerospora graminicola* parasitises Bajra (*Pennisetum typhoides*) causing the disease known as **Green ear disease** or **Downy mildew disease** of Bajra. *S. graminicola* is also parasitic on *Setaria italica*—the foxtail grass. Other common Indian species of *Sclerospora* are *S.sorghi* on *Sorghum vulgare* and *S. maydis* on maize.

Downy Mildew of Bajra

Sclerospora graminicola Caused malformed ears

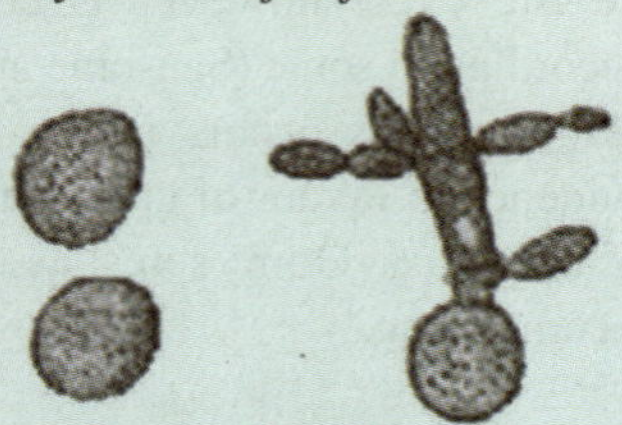

Germination of oospores

Symptoms : Since Butler (1907) studied and reported the Green ear disease caused by *S. graminicola,* the disease has been reported from almost all the Bajra growing areas of the country. Although Butler (1907) considered the disease to be of sporadic in nature, the loss to Bajra crop may be upto 27 or 30 per cent (Nene and Singh, 1976).

The symptoms of the disease appears in two stages viz.

(*a*) Downy mildew stage in which the lower surface of the leaf gets covered with downy growth of the fungus.

(*b*) Green ear stage in which ears or cobs get affected.

According to Singh (1990), Indian strains of *Sclerospora* produce more of oospores and less of sporangiophores. As a result of this, the downy mildew stage on the leaves is not so conspicuous as the green ear stage.

Fig. 6.45. Symptoms of Green ear disease of Bajra. A, On leaf, B, On ears (inflorescences).

The affected plants show stunted and dwarfed growth and become pale yellow in colour giving a sick appearance. Sooner or later the leaves start showing chlorosis in streaks on the upper surface. Just below the streaks on the lower surface fine downy growth of the fungus may be observed. The chlorotic areas turn brown and in advanced stages shredding of the leaves along the streaks/veins may take place. Usually the lateral nodal buds are stimulated to grow and develop into lateral shoots or branching. This abnormality is attributed to the disease because otherwise the plant is unbranched. (Fig. 6.45).

The main symptoms, however are produced in the inflorescence or ears. Abrin ormalities set in at the heading stage and the floral organs of the inflorescence get deformed and transformed into twisted green leafy structures. This gives the ear an appearance of green leafy mass and it is because of this symptom that the name of the disease 'Green ear' has been adopted. Transformed ears may be of three types :

(*a*) The entire ear is transformed into a leafy mass and the cob length is normal;

(*b*) The lower part of the ear is transformed into a leafy mass and the upper part bears grains, the cob length is normal

(*c*) The development of cob is totally suppressed and in its place a bunch of leafy structures develop. (Fig. 6.45)

All the floral parts become hypertrophied and malformed. The bristles of the spikelets become hypertrophied and contorted. The stamens and pistil become sterile and léaf like or may be suppressed altogether. In severaly affected plants stamens and pistil rarely develop and are replaced by small, leafy shoots or horn like outgrowths.

In advanced stages, the affected plants bear green leafy ears alongwith brown twisting and shredding leaves. Side branches on the affected plants may bear leafy masses instead of ears.

Anatomically also the symptoms are apparent. In green ear diseased leaves, the vascular bundles lack the sclerified of epidermal cells and the stomata is missing.

Thallus. The thallus in *Sclerospora* is eucarpic and is made up of well developed filamentous coenocytic endoparasitic mycelium consisting of branched aseptate hyphae. The hyphae are intercellular and freely branched. The fungus draws its nourishment from the host cells by means of haustoria, which are digitate button shaped in the stem cells but are simple branched finger shaped occupying a major portion of the cell cavity of the leaf. The fungus is an obligate endoparasite and therefore never kills the host tissue. (Fig. 6.46)

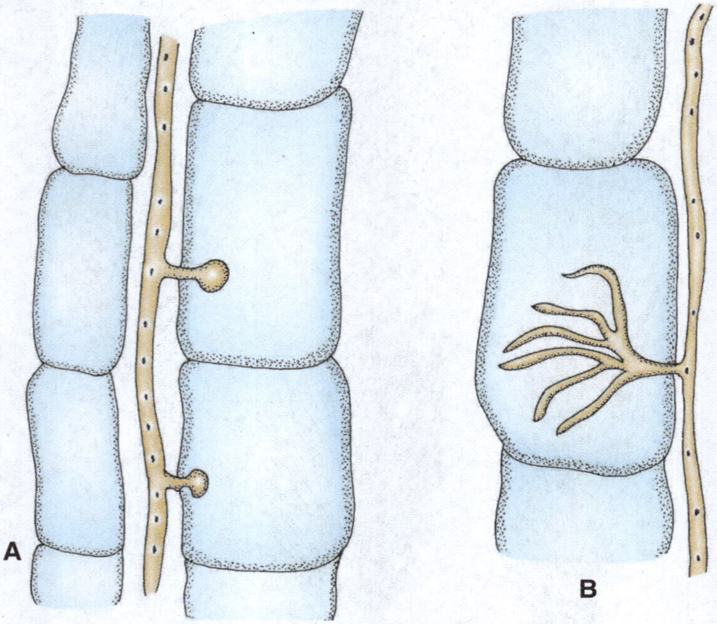

Fig. 6.46. Hyphae showing haustoria (diagrammatic). A, Button shaped; B, Finger shaped.

Asexual Reproduction. Asexual reproduction occurs by means of sporangia produced on sporangiophores. These are generally formed on leaves when temperature and relative humidity are favourable. Prior to the formation of sporangicphores the hyphae accumulate between the cells of mesophyll of the infected leaves. When the sporangiophores are to be formed, tufts of vertical hyphae reach the air space beneath the stomata. The development of sporangiophores is associated with accumulation of nutrition by the fungus drawn from the host. Consequently the chloroplasts of the host cells are wholly or partially destroyed giving the pale-yellow colour to the leaves.

The sporangiophores emerge out of the stomata of the infected leaves. Each sporangiophore is a stout broad hypha unbranched in the lower part but giving out a few (2-6) thick short branches, di—or trichotomously, at the upper part. It is 100 μ in length and 10-15 μ in width. The sporangiophores are swollen to about 30 μ at the tip where branching occurs. The ultimate branches or sterigmata bear sporangia which are formed due to the swelling of the tip of sterigmata and then cut off by a septum after a few nuclei and cytoplasm are transferred to the swelling. When mature, sporangia get detached from the sporangiophores and accumulate on the leaf surface as **Downy mildew**. Each sporangium is hyaline, round or elliptical slightly papillate at the apex and measure 13-34 × 12-23 μ in size. The sporangium has a thin smooth wall.

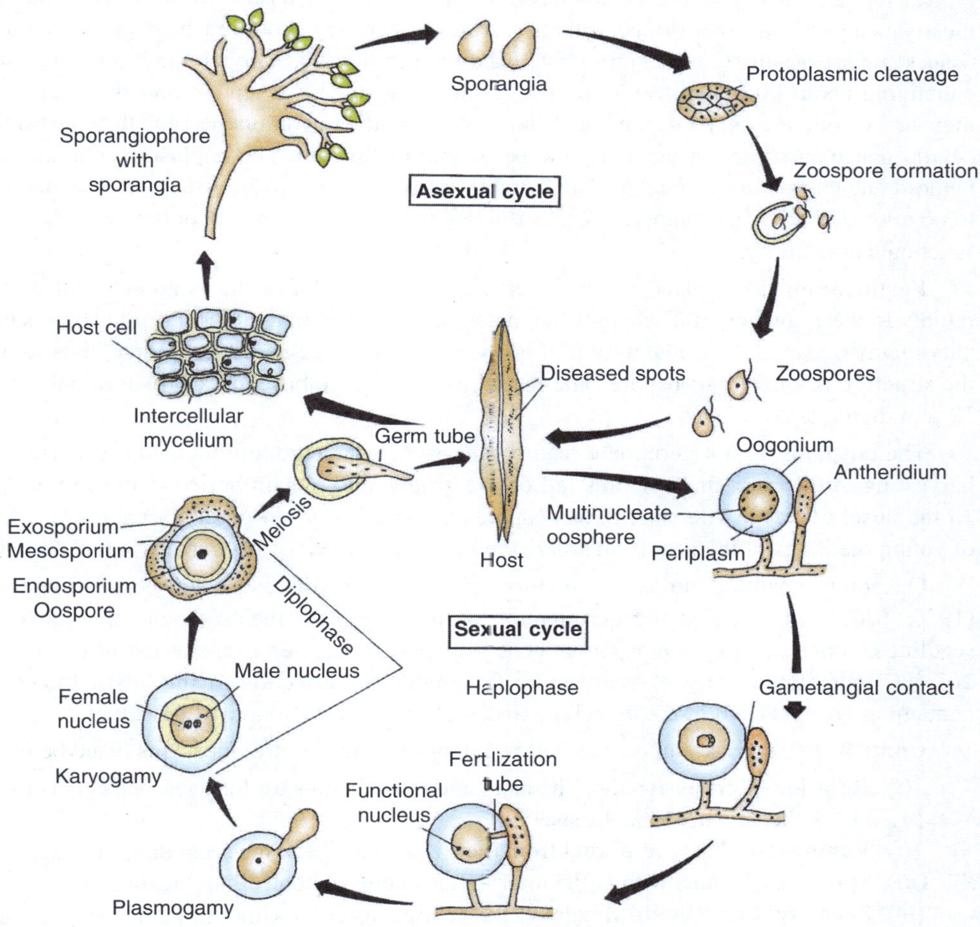

Fig. 6.47. Life cycle of *Sclerospora graminicola*

Mature sporangia fall off and germinate in surrounding dew drops by cleavage of their contents into 4-8-13 biflagellate bean shaped zoospores. These swim for sometime, come to rest and germinate by germ tubes to form new mycelium. The germ tube shows chemotactic response to roots of Bajra.

Asexual reproduction i.e. sporangial formation and its subsequent behaviour in *Sclerospora graminicola* has been studied by a large number of plant pathologists. Accordig to Safeeulla and Thirumalachar (1956), Sporangial formation is favoured by temperatures between 15-25°C, Suryanarayana (1956) concluded that sporangia do not develop at all if the temperature is more than 28°C. Generally optimum conditions for sporangial development in North India occur during August, September and during this period the disease is more common. Although Weston (1924) claimed that sporulation in *S. graminicola* is nocturnal, Safeeulla and Thirumalachar (1956) suggested that if proper and suitable conditions of temperature and relative humidity exist during the day, sporulation can take place. Generally the germination of sporangia occurs within 30 minutes at 20-25°C (Safeulla *et al.* (1963), these remain viable for six hours at 20-23°C (Suryanarayana, 1965).

Sexual Reproduction. Sexual reproduction occurs during unfavourable conditions and is typically oogamous. The sex organs develop within the host tissues in the intercellular spaces, mostly in leaves and malformed spikelets. The oogonia are produced in large numbers in leaf tissues and are usually terminal rarely intercalary in position. Each oogonium is a globose structure containing about 50 nuclei. Soon the oogonium is cut off by a septum and the nuclei undergo meiosis. All but one nuclei degenerate. The contents of the oogonium become differentiated into a central dense part, **the ooplasm** and the peripheral **periplasm.** The ooplasm containing a single haploid nucleus serves as the egg. The Antheridia are usually clavate structures attached laterally to oogonia and contain 4-6 nuclei. Nuclear division also takes place but only one nucleus remains functional at maturity.

Fertilization takes place when fertilization tube penetrates the oogonial wall and passes rapidly to the egg where it discharges the male nucleus and disintegrates. Plasmogamy occurs, but karyogamy occurs only at maturation of oospore. Fertilized egg secretes a three layered wall and the structure is known as oospore. The mature oospore is globose, dark brown in colour and 34-52 μ in diameter.

The oospores do not germinate readily and usually undergo a prolonged rest period. During harvesting of the crop, the oospores fall on the ground and remain buried in the soil of the field. At the onset of favourable conditions oospores germinate (by 1-4 germtube) and infect the roots of young seedlings of the host from where the mycelium spreads systemically in the entire plant.

Oospores constitute the main source of infection as the disease is soil borne. Suryanarayana (1956, 1962) who studied the germination of oospores and their role in infection of young seedlings, concluded that the oospores generally germinate after a rest period of a year between 20-25°C temperature. Dry soils are more favourable for infection of the hosts. Infected seeds containing oospores and/or the mycelium also serve as primary source of inoculum.

Control. The disease can be checked by adoption of the control measures listed below :

(*i*) **Rotation of crops** avoiding the bajra and related crops for long seasons may considerably check the spread of the disease.

(*ii*) **Removal** of **diseased plants** from the fields may check the secondary infection.

(*iii*) Spraying of plants with Dithane M-45 also helps in controlling the disease.

(*iv*) **Seed treatment** with Metalaxyl 25 or Metalaxyl-35 kills the pathogen present in the seeds and helps in controlling the disease. Seed treatment followed with one spray of metalaxyl effectively controls the disease.

(v) Studies on seed treatment with certain fungicides carried out under All India Coordinated Millet Improvement programme suggested that 50% control of the disease could be achieved through treatment of seeds. The fungicides tested were a mixture of 0.1 percent Agrosan GN and 0.4 percent Thiram.

(*vi*) Use of disease resistant varieties is the only effective measure to control and check the spread of the disease. The resistant varieties currently used are : HB-5, PHB-10 and PHB-14.

Taxonomic Position :

Alexopoulos (1962)

Division	:	**Mycota**
Sub-division	:	**Eumycotina**
Class	:	**Oomycetes**
Order	:	**Peronosporales**
Family	:	**Peronosporaceae**
Genus	:	*Sclerospora*
Species	:	*graminicola*

Alexopoulos and Mims (1979) place the genus Sclerospora in Superkingdom Eukaryonta, Kingdom Mycetae, Division Mastigomycotina, Subdivision Diplomastigomycotina, Class oomycetes, order Peronosporales and family Peronosporaceae.

Alexopoulos and Mims (1979) :

Super kingdom	:	**Eukaryonta**
Kingdom	:	**Mycetae**
Division	:	**Mastigomycotina**
Sub-division	:	**Diplomastigomycotina**
Class	:	**Oomycetes**
Order	:	**Peronosporales**
Family	:	**Peronosporaceae**
Genus	:	*Sclerospora*
Species	:	*graminicola*

Features in which Peronosporaceae are more advanced than the Pythiaceae —

1. All the members of the family are obligate parasites.
2. The infected plants do not perish outright, so that the fungus can produce sporangia and in some cases even the resting oospore before the exhaustion of food supply.
3. The hyphae are intercellular and only the haustoria enter the host cell. Even these do not perforate the host plasma membrane but simply invaginate it.
4. The sporangia which are produced simultaneously and in large numbers are invariably deciduous. These can be dispersed in favourable conditions, howsoever short the period may be.
5. There is progressive elimination and subsequent replacement of the readily vulnerable motile stage (zoospores) by the less vulnerable germ tube method of sporangium germination as in *Peronospora.*
6. The sex organs are formed in the host tissue and thus are protected.
7. The oospore secretes a thick, resistant wall around it and is thus able to withstand adverse conditions.

FAMILY : ALBUGINACEAE

Albugo caused white spots

White Rust on mustard leaves

General Characteristics. The family includes a single genus *Albugo* (Pers.) Gray comprising 25 species. They are common obligate parasites of aerial parts of herbaceous terrestrial angiosperms. They cause *white rust* diseases. The unseptate coenocytic mycelium absorbs nutrients by means of knobbed or globular haustoria. Asexual reproduction takes place by means of sporangia produced in basipetal chains on the apices of special, short, stout, unbranched and upright reproductive hyphae called the **sporangiophores.** The sporangiophores are borne in sori formed between the epidermis and mesophyll tissues of hosts. The sporangia of Albuginaceae receive different names from different authors. Some authors call them *deciduous sporangia* as they get detached at maturity. Some prefer to call them **conidiosporangia** because at the time of germination, depending upon the environments, they sometimes function as zoosporangia and produce zoospores. At other times they function as conidia and germinate directly by a germ tube. There are still others who name them **conidia** because they are budded in succession in chains and shed like them. The special hyphae bearing them are termed **conidiophores.** It is true that sporangia in *Albugo* are abstricted in much the same way as conidia. The difference, however, lies in the method of germination. The sporangia usually produce zoospores. The sporangiophores (conidiophores) lie side by side closely packed in a palisade-like layer beneath the epidermis of the host (Fig. 6.52). Sex organs are well developed. Fertilisation is affected by means of a fertilisation tube. The thick-walled oospore after the resting period, germinates in spring.

ALBUGO CANDIDA (SYN. CYSTOPUS CANDIDUS)

Occurrence. The genus *Albugo* is represented by 25 species. All are parasite on land plants. Of these *A. candida* has long been known as an obligate parasite commonly found on a wide range of crucifers. Biga (1955) reported about 241 species in 63 genera are attacked by this fungal parasite. It attacks several annual crucifers such as *Capsella bursa pastoris* (Fig. 6.48) *Lepidium sativum, Brassica campestris, Raphanus sativus* (Fig. 6.49) and several other crucifer weeds. It causes a disease called "white rust" or "blister rust". The fungus attacks all the above ground parts.

Albugo candida infected inflorescence

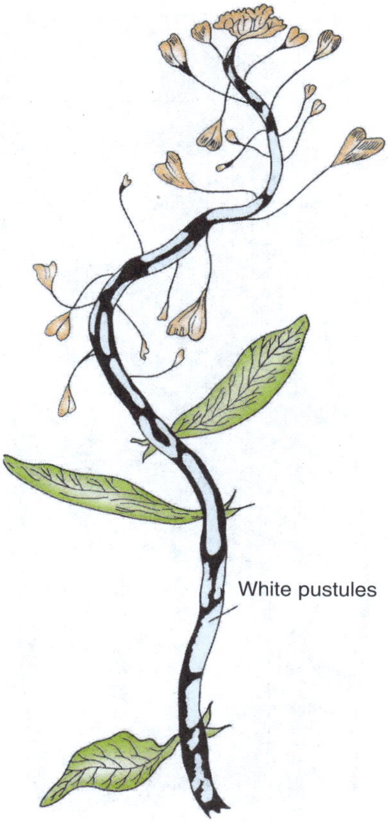

Fig. 6.48. Diseased plant of *Capsella* bearing white pustules of *Albugo candida.* The stem is distorted as a result of infection.

Fig. 6.49. Infected Radish leaf showing pustules of white rust (*Albugo candida*).

Signs and Symptoms. The disease appears in the form of shiny, white, smooth, raised irregular blisters or patches on the leaves (Fig. 6.49) or stems (Fig. 6.48). The root is not attacked. The patches gradually turn powdery. The disease also causes deformation of flowers and fruits of the host. The distortion and hypertrophy of infected inflorescences is the most conspicuous symptom of the disease. These distorted and hypertrophied inflorescences are called the "stag heads". The latter contain thick-walled oospores of the parasite. Sometimes the fungus attack may cause swelling (spindle-shaped galls) and distortion of stem particularly in young plants. Infection takes place through the stomata. Extensive hypertophy, hyperplasia and chlorophyll formationn accompany fungal invasion [Williams and Paund (1963)].

Mycelium (Fig. 6.50). It is well developed and consists of branched, aseptate, coenocytic hyphae. The hyphae live and ramify in the intercellular spaces of the susceptible host tissue. The hyphal wall contains cellulose and not chitin. The hyphal protoplasm is granular and vacuolate in the older parts. It contains numerous nuclei, oil globules and glycogen. Electron micrographs reveal the presence of mitochondria, endoplasmic reticulum and ribosomes. The cytoplasmic membrane which is closely appressed to the hyphal wall forms lomasomes. Septa remain suppressed in the actively growing hyphae but appear to separate reproductive structures and to seal off injured parts. The fungus mycelium grows vigorously. The hyphae branch and ramify within the host attacking the tissues adjoining the point of infection. Sometimes both *Albugo* and *Peronospora* occur on the same host paritcularly *Capsella bursapastoris*. *Albugo* can, however, be distinguished from *Peronospora* by the smaller diameter of its hyphae and more numerous, vesicular haustoria (Fig. 6.50).

The intercellular hyphae of this obligate parasite produce intracellular haustoria in the mesophyll cells of the host (Fig. 6.50). The haustorium arises as a lateral outgrowth at the site where the hyphal wall is tightly pressed against the mesophyll cell wall. An electron-dense amorphous material known as the *penetration matrix* is usually deposited at the site of contact between the host and the hypha cell walls. It is described as the *penetration site* (Fig. 6.51). The slightly crescent-shaped bulge of the haustorial mother cell known as *haustorial initial* perforates the host cell wall at the penetration site and protrudes into the lumen of the mesophyll cell to develop into a haustorium. It is bordered by the invaginated host plasma membrane. With light microscope the haustorium is seen as a small, spherical structure consisting of two parts namely (*i*) the *haustorial stalk* or *neck* and (*ii*) the terminal *haustorial head* or *body*. The haustorial stalk passes through the penetration site to connect the haustorial body to the hyphal wall in the intercellular space between the mesophyll cells. Usually one or two, sometimes more, haustoria are seen in the thin peripheral layer of the host cell cytoplasm adjacent to the chloroplasts.

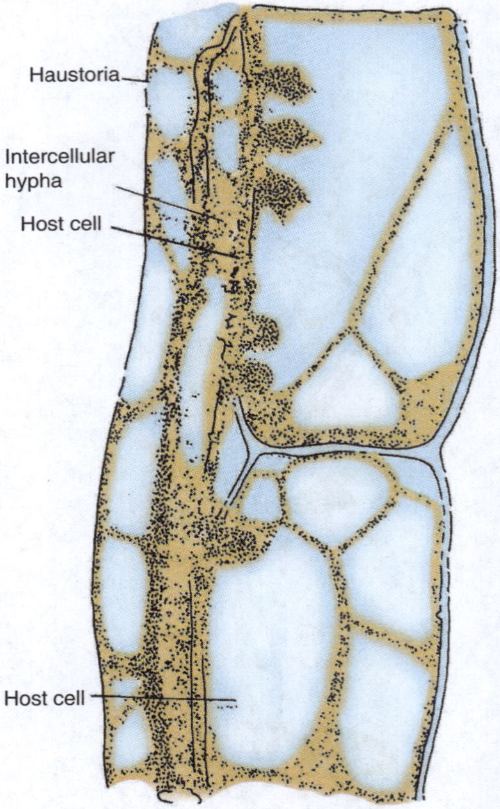

Fig. 6.50. *Albugo candida.* Section through the host stem showing a portion of the mycelium bearing haustoria. (Diagrammatic).

Fine structure of haustorial apparatus (Fig. 6.51 B). It was studied by Berlin and Powen (1964) and Coffey (1975). A dark layer of penetrating matrix known as the *penetration jacket* continuous with the host cell wall surrounds the haustorial stalk or neck at its proximal end. It lies close and external to the haustorial (hyphal) wall of the neck region. External to the penetration jacket clasping the neck is a collar-like structure attached to the host cell wall and made of the same material. The hyphal wall and plasma membrane of the intercellular haustorial mother cell are continuous along the entire length of the neck and body regions of the haustorium. These represent the *haustorial wall* and *haustorial membrane* respectively. Berlin and Powen (1964), however, reported the hyphal wall to be absent in the distal region of the haustorial neck. It was contradicted by Coffey (1975) who found it to be intact in this region. The protoplast of the haustorial neck has ribosomes but lacks other cell organelles. The cytoplasm of the haustorial body contains mitochondria, ribosomes and sometimes cisternae of rough endoplasmic reticulum but lacks the nuclei and perinuclear dictyosomes. An electron dense, granular amorphous material forms the *sheath matrix* on the outer face of the hyphal wall of the haustorial body. The sheath matrix, also known as the extra *haustorial sheath* is absent in the neck region. The invaginated host plasma membrane which completely surrounds the haustorium and remains appressed to it is

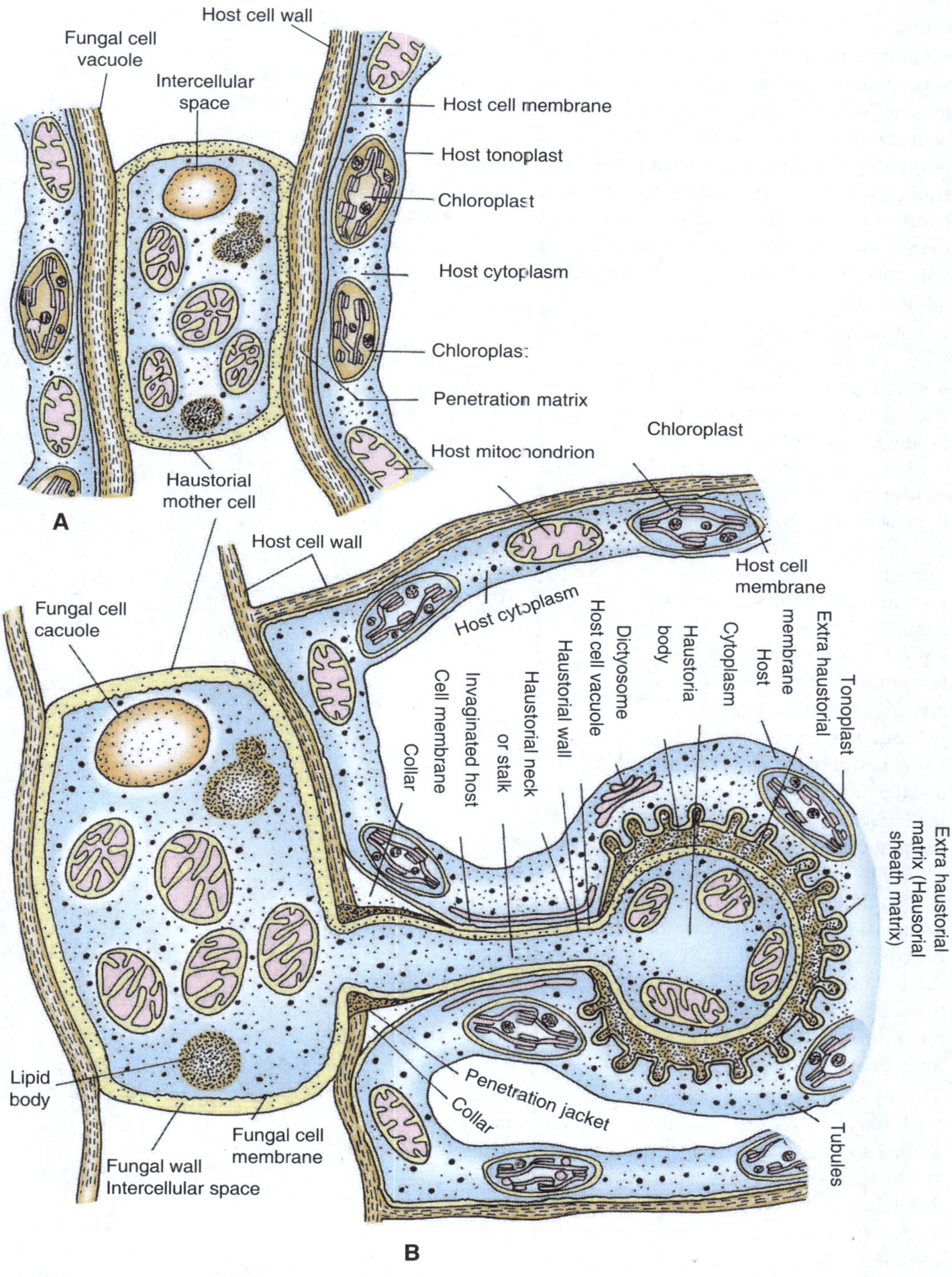

Fig. 6.51. (A-B). *Albugo candida.* Diagrammatic representation of ultrastructure of the haustorial apparatus in section passing through the host mesophyll cell (Based on Coffey)

called the *extrahaustorial membrane.* It is flexible and thrown into a series of tubular projections extending into the host cytoplasm. Each tubule contains a dense core of material similar in appearance to and continuous with the sheath matrix. The extra haustorial or sheath membrane thus delimits the sheath or extrahaustorial matrix from the host cytoplasm and forms the interface between the two. It is obvious from the account given above that the haustorium lies outside the host cytoplasm. It is surrounded by a thin layer of host cytoplasm covered by the tonoplast on the side facing the central vacuole of the host cell and by a sheath membrane on the side facing the haustorium. The sheath matrix forms the physical boundary and the sheath membrane—the key to the physical relationship between the host and the obligate fungal parasite. The transport of all nutrients from the host cytoplasm occurs across the sheath membrane to the surface of the haustorium.

Asexual Reproduction (Fig. 6.52). When the mycelium has reached a certain stage of maturity it produces pads of hyphae at certain areas just below the epidermis. The tips of hyphae constituting the mat grow verticlly into short, upright, thick-walled, unbranched club-shaped hyphae. These are the **sporangiophores** (=conidiophores). They are arranged in a closely packed palisade like layer forming a *sorus* between the epidermis and the mesophyll of the host leaf. Each sporangiophore appears constricted at its junction with the subtending hypha (Khan, 1977). The lower two-third portion of sporangiophore is narrow, thick-walled with a undulating surface whereas the upper one-third is broader, thin-walled with a smoother surface. According to Khan (1977), the sporangiophore wall towads its proximal end consists of two layers, the outer more electron-dense and thicker than the inner layer.

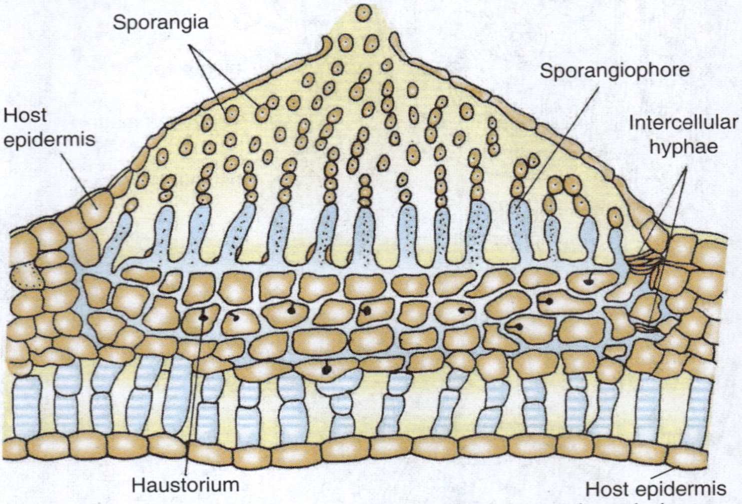

Fig. 6.52. *Albugo candida.* A section of host leaf passing through the sporangial sorus.

(*a*) *Abstriction of sporangia.* In the lower fungi (Phycomycetes) *Albugo* is unique in that its lemon-shaped sporangia are produced in basipetal chains at the tips of clavate sporangiophores. Two different views have been put forth to explain their mode of development. According to one view, the sporangial chains in *Albugo* are abstricted by *percurrent proliferation* (Hughes, 1971). The second view advocates the *blastic mode* of development (Khan, 1977).

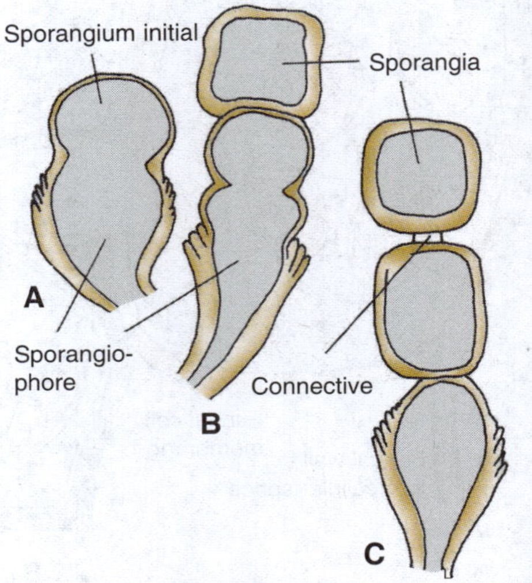

Fig. 6.53 (A-C). *Albugo ipomeae-pandurantae* showing stages in percurrent proliferation of sporangia (After Hughes)

(*i*) *Percurrent Proliferation* (Fig. 6.53). The sporangia in *Albugo* which are cut off in succession are arranged in a basipetal chain on the sporangiophore. According to Hughes (1971), they are produced by successive proliferations of the sporangiophore subtending a sporangium. This mode of development of sporangia is termed *percurrent proliferation.* Generally the sporangiophore increases in length as each successive sporangium is cut off from each successive proliferation at a higher level than the previous one. The first formed sporangium is a *aleuriosporangium.* Reaching a certain size it is delimited from the sporangiophore by a basal septum.The latter eventually splits into two halves so that the subsequent proliferation of sporangiophore involves the exposed half septum. In *Albugo* each successive sporangium is capable of seceding from the sporangiophore or from the young sporangium. The second sporangium is thus formed by proliferation of the sporangiophore with total involovement of the half of the fractured transverse septum exposed by the seceding first sporangium above it. Apart from this, septum is seen at the apex of the young sporangium. The second sporangium is delimited in the same manner as the first. As the second sporangium increases in size it pushes the first upward without disjunction. The process is repeated resulting in a chain of sporangia. Probably the septum seen at the apex of each younger sporangium thickens on both sides to form a connective between the successive sporangia in the chain. According to Hughes, besides increase in length of sporangiophores, this method of sporangium development is accompanied by marked lamination and thickening of the walls of the sporangiophores. The mature sporangiophores are thus longer, more thick-walled and show annellations.

Thakur (1977) corroborated findings of Hughes (1971) on formation of sporangia by percurrent proliferation in *Albugo.*

(*ii*) *Blastic mode of Sporangium development* (Fig. 6.54). According to Khan (1977) the sporangiophore has a fixed sporogenous locus at its apex. The sporangial initial arises as a bud from it (A). It contains about 4-6 nuclei and dense cytoplasm. The two wall layers of the sporangium initial are continuous with those of the sporangiophore wall. Reaching a certain size, the initial is delimited by a basal septum near the sporogenous locus. It becomes the first sporangium and the oldest in the chain (B). The septum is formed by the centripetal growth of the inner layer of the sporangiophore wall (C). A nearly complete septum has a narrow central canal and consists of three layers, upper and lower electron dense and the thick middle one of less electron density (D). After the completion of the basal septum and conversion of the initial into a full-fledged sporangium, a new sporangium initial grows as a bud from the sporogenous locus (B). It pushes the newly-formed sporangium upward. Thus only one sporangium is formed at a time. As the second sporangium initial grows to the normal size, it is also delimited by the formation of a basal septum as the first. The repetition of the process results in the formation of a basipetal chain of sporangia. Soon after the formation of the first sporangium, the

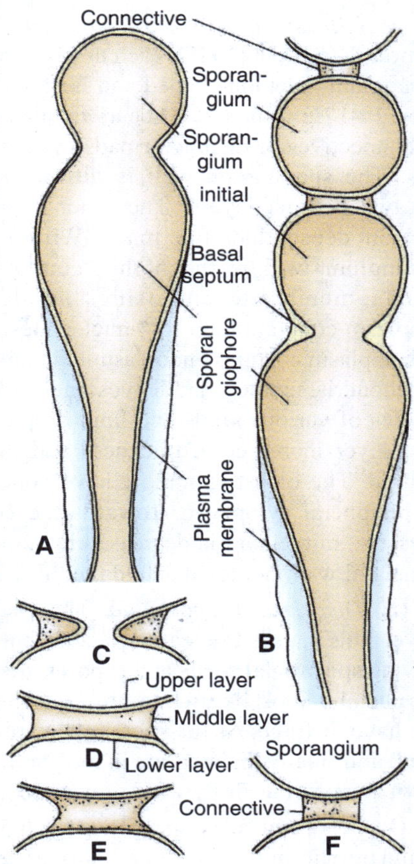

Fig. 6.54 (A-F). *Albugo candida* showing blastic mode of development of sporangia (After Khan)

breakdown of its basal septum begains. It is the middle layer which starts disintegrating from its periphery inwards whereas its upper layer forms the wall of the upper sporangium and the lower layer completes that of the lower sporangium (E-F). The fibrous product of dissolution of the middle layer is held in position by the pellicle which covers both the sporangia and the sporangiophore. It is seen as a connective or disjunctor between the successive sporangia in the chain. Khan (1977) did not notice any increase in length of the sporangiophore during sporangia formation nor did he observe any annellations on the sporangiophore surface.

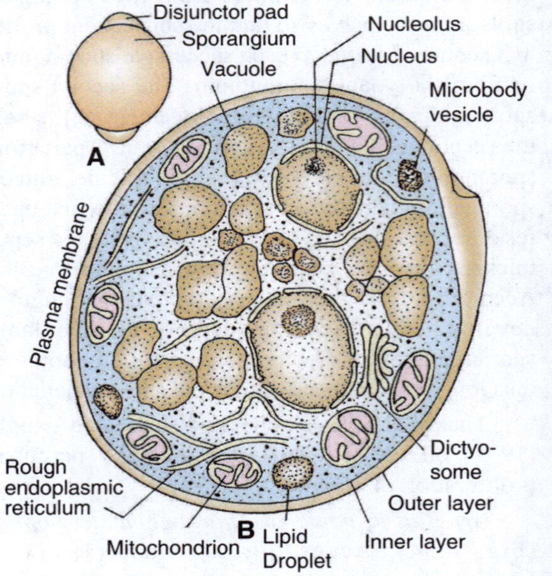

Fig. 6.55 (A, B). *Albugo candida.* A, Sporangium with remains of the disjunctor pads at both ends; B, ultrastructure of sporangium (Based on Khan).

Sporangia (Fig. 6.55 A, B). They are small, hyaline, nearly spherical or lemon-shaped structures with a smooth or somewhat punctate surface. The basipetal arrangement of sporangia in the chain (with the oldest at the top and the youngest at the base of the chain) serves two useful purposes : (i) it permits ready dispersal of the oldest sporangia by air currents or rain water, and (ii) it helps in the proper nourishment of the younger ones.

(*i*) *Ultrastructure of sporangium* (B). According to Khan (1977), each newly formed sporangium is lemon-shaped and is about 19-22 by 14-17 μ m in size. It bears remnants of the connectives or disjunctor pads at both the ends. The sporangium wall is differentiated into two distinct layers. The outer is more electron dense than the inner. Within the sporangium wall is the highly convoluted plasma membrane enclosing the dense cytoplasm containing up to 4 nuclei. Besides, the cytoplasm contains endoplasmic reticulum, mitochondria, perinuclear, dictyosomes, ribosomes both free and attached to endoplasmic reticulum, vesicles of various kinds and lipid droplets. Towards maturity the sporangial wall especially, its inner layer increases in thickness and the number of lipid droplets decrease as the sporangia matured. The oldest sporangia have none. The endoplasmic reticulum becomes accumulated in the peripheral cytoplasm. Towards the end of sporangial maturation, the dictyosomes become quiescent, mitochondria decreased in number and also the amount of endoplasmic reticulum. The sporangial wall increased 3-fold the thickness.

(*ii*) *Dispersal of sporangia.* The chains of sporangia lengthen and press on the epidermis above. This causes the leaf surface to bulge. The overlying epidermis eventually bursts over the growing **sporangial sorus** and exposes the white shining pustules consisting of masses of sporangia. The pustules look like white blisters. The exposed sporangia are white. The distal ones by this time have matured. As the sporangia mature the connectives or gelatinous pads between them dry, shrink and finally disintegrate in moist air. The sporangia in the chain thus separate. They are then blown away in the air by wind or washed away by rain water.

Germination of sporangia (Fig. 6.56). Landing on a suitable host the sporangia begin to germinate within two or three hours under suitable conditions. At the time of germination they behave in either of the following two ways depending on temperature conditions :—

Indirect Germination (Fig. 6.56 B-E). In the presence of moisture and low temperature, the sporangium functions as a zoosporangium (B). The optimum temperature for germination of sporangia is 10ºC. It absorbs water and swells. A few vacuoles appear in its granular cytoplasm. Later the vacuoles disappear and the multinucleate protoplast undergo division. It divides to form five or eight polyhedral uninucleate daugher protoplasts. Meanwhile an obtuse papilla forms on one side of the sporangium. Each daughter protoplast shapes into a slightly concave-convex zoospore (E). It has a disc-like contractile vacuole on one side and is furnished with two flagella, one short and one long. The former is of tinsel type and the later whiplash. The flagella are attached laterally near the vacuole. As the zoospores are differentiated, the papilla swells and opens. The zoospores still immobile, emerge usually one by one (Fig. 6.56 C). According to Vanterpool, the zoospores are, at first, released in a sessile vesicle formed by the swelling of the papilla. The vesicle soon vanishes.

Germinationn of zoospore and infection of host (Fig. 6.56 F-H). Moisture on the surface of the host is essential for germination and infection. The released zoopores swim about in water for a while (E). Finally they settle down on the host, retract the flagella and round off. Each secretes a wall around it (F). The encysted zoospore (cyst) then germinates. It puts out a germ tube (G) which gains entrance into the host throught a stoma (H). Once within the host tissue the germ tube grows and forms the mycelium

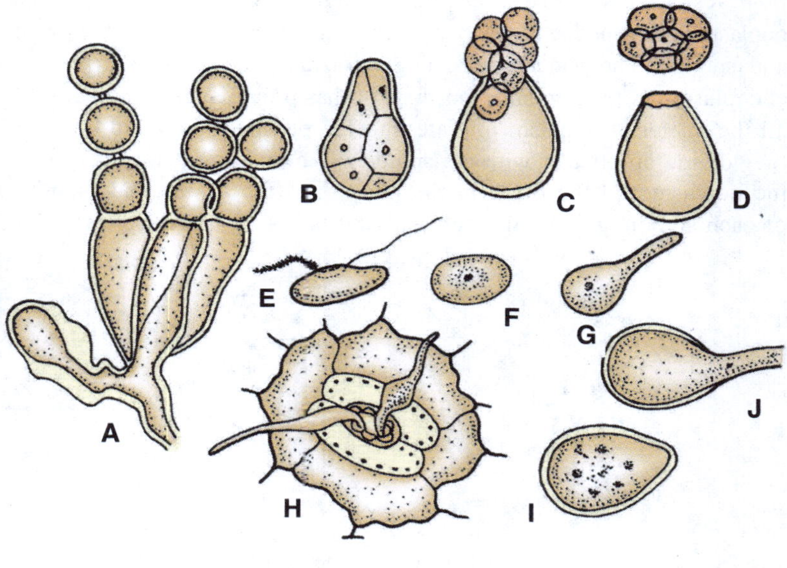

Fig. 6.56 (A-J). *Albugo candida.* A, cluster of sporangiophores bearing sporangia in chains; B-D, differentiation and liberation of zoospores; E, liberated zoospore; F, encysted zoospore; G, germination of cyst to form a germ tube; H, infection through a stoma; I-J, direct germination of sporangium. (After De Bary)

(*ii*) *Direct Germination* (Fig. 6.56 I-J). At high temperature and under comparatively dry conditions the sporangium behaves like a conidium (I). It germinates directly to form a germ tube (J). The conidial method of germination of sporangia in *Albugo* is, however, not common. The germ tube penetrates the host through a stoma or, through an injury in the epidermis. Within the host it develops into a mycelium. Re-infection of the host and infection of other healthy plants in the vicinity goes on by the production of sporangia throughout the growing season.

Sexual Reproduction (Fig. 6.57 A-I). It is **oogamous.** The male sex organ is called an **antheridium** and the female **oogonium** (A). They are developed near each other in the intercellular spaces of host tissues towards the end of the growing season. When the mycelium ages, some hyphae grow deep and lie buried in the intercellular spaces of the tissues of the stem, or petioles. The sex organs arise on separate hyphae called the **male** and the **female** hyphae (A). The two soon establish contact. The antheridium comes in contact with the oogonium at the side. The development of sex organs within the host tissue is externally indicated by **hypertrophy** and distoration in shape in the particular organ.

(*a*) **Oogonium.** It arises as a globular enlargement of the tip of the female hypha. Sometimes the oogonium is intercalary in position. The swelling is multinucleate (A). A cross wall appears below this inflation (B). It separates the terminal oogonium from the rest of the female hypha. The young oogonium has highly vacuolated contents. The numerous nuclei and vacuoles are evenly distributed and the usual cell organelles are dispersed throughout the oogonium. The nuclei divide mitotically and increase in number as the oogonium advanced towards maturity. After the first mitotic division the ooginial cystoplasm shows marked zonation (C). It becomes differentiated into two distinct regions with the rearrangement of the numerous nuclei and other cellular organelles. Most of the original cytoplasm of the oogonium forms the central, rounded dense **ooplasm.** It is multinucleate and contains only a few mitochondria, ribosomes and cisternae of ER. It is rich in lipid vesicles and reserve vesicles containing electron dense inclusions (reserve globules). The ooplasm is surrounded by the peripheral cytoplasm constituting the **periplasm.** It is more vacuolate and spongy. The vacuoles are large. Besides, the periplasm is rich in nuclei, mitochondria, endoplasmic reticulum and ribosomes. It has protoplasm of thinner consistency. Sometimes after, all the nuclei of ooplasm migrate into the periplasm (D) and become arranged in a ring. Here they divide mitotically with the spindles lying in such a way that one pole of each spindle is in the ooplasm and the other in the periplasm (E). At the end of mitosis one daughter nucleus of each spindle goes to the ooplasm and the other to the periplasm (F). However, the ooplasm

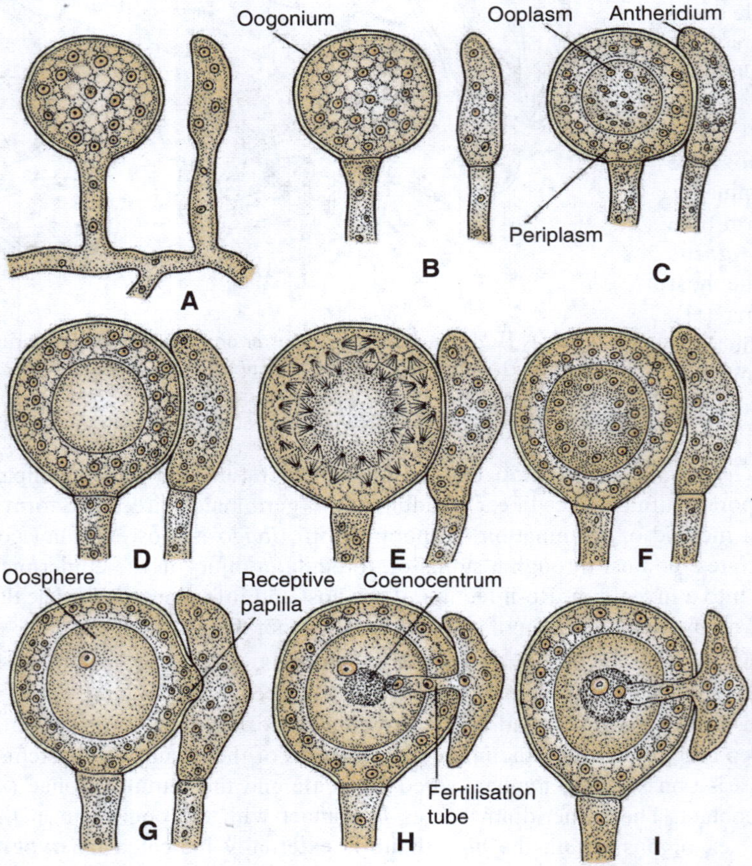

Fig. 6.57 (A-I). *Albugo candida.* Stages in the development of sex organs and fertilisation. Explanation in the text.

at maturity has a single centrally located nucleus (G). There are two views with regard to this uninucleate condition of the mature ooplasm. According to one view, all the nuclei excepting one are extrude from the ooplasm and are deposited in the periplasm. The second view is that in the later stages of development all the nuclei in the ooplasm excepting one degenerate and disappear. The uninucleate ooplasm functions as the female gamete or egg or oosphere (G).

(*b*) Antheridium. It is an elongated club-shaped cell (A). It is multinucleate. The antheridium is developed at the end of a male hypha lying close to the oogonium. The end of the male hypha enlarges into club-shaped swelling. The latter is then cut off by a cross wall from the rest of the male hypha (B). This terminal club-shaped cell is called an **antheridium.** It contains several nuclei (usually 6-12), but only one is functional. The paragynous antheridium comes in direct contact with the oogonium at the side (C).

(c) Fertilisation. At the point of contact of antheridium with the oogonium, the walls become very thin. A portion of the contents of the oogonium surrounded by a thin membrane grows into a papilla-like outgrowth (G). This papilla-like oogonial bulging is called the **receptive papilla.** It is functionless. The receptive papilla bulges (G) into the antheridium but soon disappears. This is followed by the formation of a slender tubular outgrowth from the antheridium. It is the **fertilisation tube** (H). The fertilisation tube passes through the thin spot in the oogonial wall and enters the multinucleate periplasm. It then dips deep into the ooplasm. Prior to this a spherical and granular cytoplasmic body appears in the centre of the oosphere (H). It is known as the **coenocentrum.** The single functional female nucleus is attracted towards it and becomes attached to a point near it. The fertilisation tube finally reaches the coenocentrum and ruptures (I) at the tip to introduce a single male nucleus which fuses with the female nucleus. Thereafter the fertilisation tube collapes but persists and the coenocentrum vanishes. On removing or displacing the attached antheridium, Tewari and Skoropad (1977) observed a clear hole surrounded by some fibrous material.

Oospore (Fig. 6.58 A-C). Tewari and Skoropad (1977) investigated the fine structure and development of *A. candida* oospores. According to them, the young oospore is delimited from the vacuolate periplasm by an electron-dense cell wall. The single layered cell wall of the young oospore encloses dense cytoplasm containing a group of reserve vesicles, lipid vesicles and a few membranous organelles. It is surrounded by periplasm rich in vacuolate cytoplasm containing membranous organelles. Further development of oospore if marked by the deposition of 4 layers, two on the outer and two on the inner side of the first (original) layer of the young oospore. The mature oospore thus has a thick highly differentiated 5-layered wall. External to the oospore wall are the two additional protective investments formed by the persistent periplasm and the oogonial wall. The thick highly differentiated oospore wall together with the two surrounding additional layers contributes to the longevity of *Albugo* oospore. The authors suggest that periplasm plays an active part in deposition of oospore wall layers. Within the fully developed

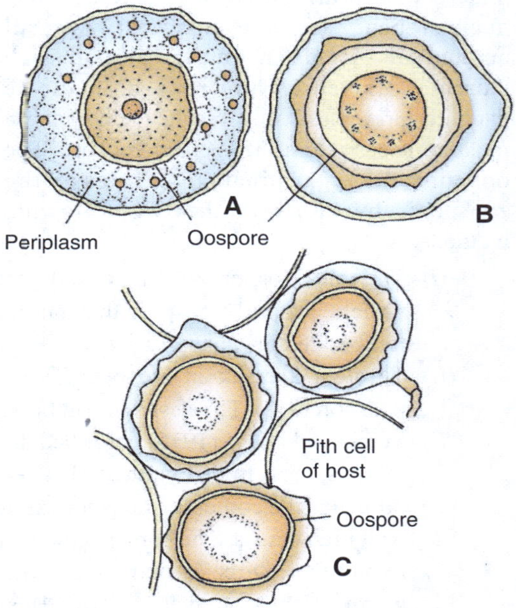

Fig. 6.58 (A-C). *Albugo candida.* Showing structure of oospore as seen under light microscope.

oospore wall is the scanty cytoplasm surrounding a large central reserve globule. Some small bodies resembling the reserve globules in appearance and numerous lipid vesicles occupy most of the space between the oospore wall and the central reserve globule. The highly differentiated thick, oospore wall together with the two additional layers constituted by the persistent perisperm and the oogonial wall provides protection and the numerous lipid vesicles in the oospore cytoplasm furnish energy for the long dormancy or overwintering by oospores in *Albugo*. The outer layer of the oospore wall is comparatively thicker. It is warty or tuberculate. In other species it may have a network of ridges or other patterns.

Location of meiosis. The place of meiosis in *Albugo* is still under dispute. Stevens (1899) suggested that *Albugo* possesses diploid somatic nuclei which undergo meiosis in the gametangia (antheridia and oogonia). His interpretation was disputed by later workers. They held that *Albugo* and other Oomycetes, as a whole, are haploid. Oospore is the only diploid structure in the life cycle. It undergoes zygotic meiosis. Walker (1969) observed that after fertilization when a thick wall is being developed around the oospore, its diploid nucleus divides repeatedly to form 32 nuclei. Probably the two earlier of these divisions constitute meiosis. In this 32 nucleate stage, the oospore enter the resting stage and tides over the period unfavourable for growth.

However, Sansome and Sansome (1974) have advanced evidence in support of gametangial meiosis and diploid life cycle in *Albugo candida*. They hold that the first two divisions of nuclei which occur in the gametangia constitute *meiosis*. This view is gaining ground.

Germination of Oospore (Fig. 6.59). On the onset of conditions favourable for growth, the oospore germinates. The central globule and the lipid droplets gradually disappear. The contents of the oospore assume uniform granular appearance. The diploid nucleus undergoes repeated divisions to form many nuclei (about 100 or even more). A small amount of cytoplasm gathers around each daughter nucleus. Numerous uninucleate daughter protoplasts thus result. Each of these metamorhoses into a biflagellate zoosore. The oospore then germinates to release the zoospores by either of the two following methods:—

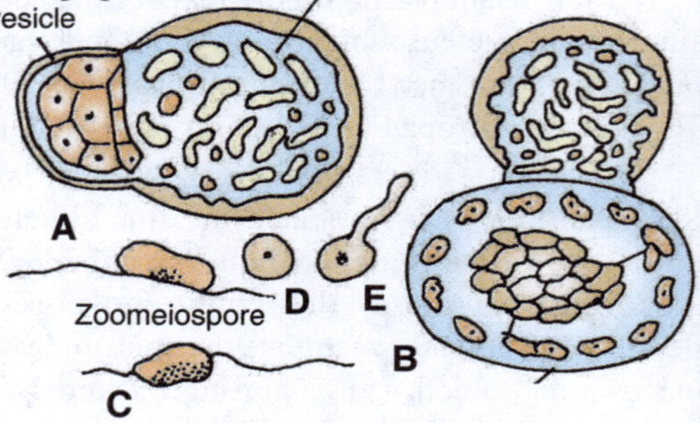

Fig. 6.59 (A-E). *Albugo candida.* A-B, Germination of Oospore; C, liberated zoospore; D-E, germination of zoospore. (After De Bary)

(*i*) The thick oospore wall cracks. A germ tube emerges through the split. It ends in a thin vesicle. From the oospore the zoospores pass into the vesicle. Soon the vesicle perishes to liberate the zoospores.

(*ii*) Through a crack in the oospore wall emerges a thin sessile vesicle containing the zoospores (A). The vesicle soon bursts to liberate the zoospores.

(*iii*) Verma and Petrie (1975) described direct germination of oospore without the intervention of zoospores. In this method after the disappearance of the central globule and the lipid dropletes the contents of oospore assume uniform granular apearance. The thick oospore wall then cracks. The contents then emerge in the form of one or two simple or branched germ tubes. Direct penetration of the host by a germ tube has not been reported. Good germination occured at 10-20°C. They observed that 71 percent of the oospores collected from the field in August germinated within two weeks.

Structure of zoospore. The liberated zoospore is a biflagellate, uninucleate structure which is reniform in shape. The two unequal flagella arise from a depression on the concave side. The shorter flagellum is of tinsel type and the longer one of whiplash type. The former is directed forward and the latter trails behind when the zoospore is in motion.

Germination of Zoospore (Fig. 6.59 D-E). On coming in contact with a suitable host. The flagella are withdrawn. It then rounds off and secretes a wall around it (D). Soon the encysted zoospore (cyst) puts out a germ tube (E) which enters the host tissue through a stoma. Once within the host tissue the germ tube grows vigorously and forms a new mycelium.

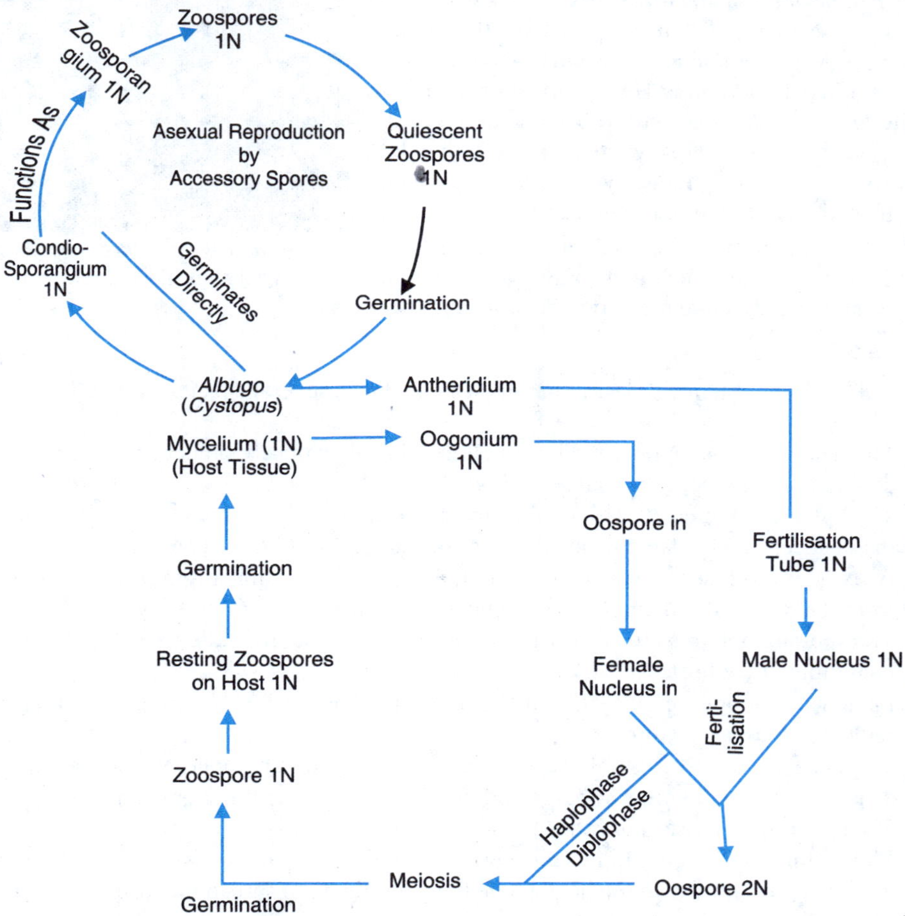

Fig. 6.60A. Graphic representation of life cycle of *Albugo (Cystopus)* with zygotic meiosis.

Primary infection of host. According to Verma *et al.* (1975), the zoospores produced in the germinating oospores serve as a primary inoculum for infection of the susceptible host. The emerging cotyledons are the infection sites. After penetration the first haustorium originates near the tip of the young hypha. The latter then continues to grow leaving the haustorium as a side branch. The formation of the first functional haustorium is the critical step in primary infection. It indicates the establishment of a compatible functional host-parasite relationship. From then onwards hyphal growth increases rapidly. The hyphae grow around the palisade mesophyll cells with haustoria penetrating the adjacent cells. The number of haustoria per cell varies from one to serveral. In the course of time a mycelial base is established inside the host tissue (cotyledons). It produces masses of zoosporangia on the cotyledons which serve as secondary inoculum in initiating systemic infection. The parasite ultimately reaches the inflorescence region where it produces the oospores.

Economic Importance. The fungus is the causal agent of "*white rust*" disease of crucifers resulting in economically significant losses in the yield of turnip, rape and mustard (sarson). The distorted and hypertrophied infected inflorescences, the so-called "stage heads" consist largely of thick-walled oospores. Another species *A. ipomoeae penduraneae* parasitizes sweet potato and *A. bliti grows* as a parasite on some members of the family Amarantaceae. As an economic disease of crucifers, white rust has a limited importance. Heavy infection of parts above the ground may considerably reduce yield. Vanterpool (1959) estimated that white rust infecting older parts of *Brassica napus* may reduce yield up to 20 percent. Heavily infected flowers fail to produce seeds.

Albugo ipomeae penduraneae on sweet potato

Control. Rotation of crops, removal and destruction of infected plants from the field and spraying with fungicides such as with 0.8 percent Bordeaux mixture are the most effective methods of controlling the disease.

Salient Features

1. *Albugo* is an obligate parasitic fungus which attacks several species of crucifers causing the *white rust.*
2. The haplophase is represented by a well developed profusely branched mycelium which consists of aseptate, conenocytic hyphae.
3. The hyphae are intercellular and feed by means of haustoria which penetrate the walls of the host cells and enlarge inside at their tips into button-like or spherical structures.
4. Asexual reproduction takes place usually by wind borne **sporangia** produced in chains from the tips of short, club-shaped hyphae called the **sporangiophores.**
5. The sporangiophores are closely packed forming a solid, palisade-like layer beneath the epidermis of the host.
6. Each sporangiophore cuts off sporangia at its tip one below the other in a long chain with the oldest at the top.
7. The pressure from below of the upwardly growing sporangiophores and the sporangia causes the host epidermis to bulge and finally to burst over the sporangial sorus.
8. The sporangial masses are exposed as a crust of white, blister-like patches.
9. The mature globose, multinucleate sporangia thus exposed are disseminated by wind or washed by rain water to the host where they germinate to spread the disease. The factors governing germination are moisture and temperature conditions.
10. In the presence of moisture provided by rain or dew drops and at low temperature the sporangium germinates by zoospores. In dry air and at high temperature it germinates directly by a germ tube.
11. The zoospores are biflagellate, reniform, uninucleate structures. Of the two flagella one is **tinsel** type and other of **whiplash** type. They are inserted in the depression.
12. The liberated zoospore settles down on the host, withdraws its flagella, rounds off and secretes a wall around it. It then puts out a germ tube.
13. The germ tube, whether produced by the germination of zoospore or the sporangium infects the host tissue gaining entrance through a stoma. It grows and forms the mycelium which is intercellular.
14. Sexual reproduction is **oogamous** and *Albugo* is **homothallic.** The antheridia and oogonia are produced near each other towards the end of the growing season.

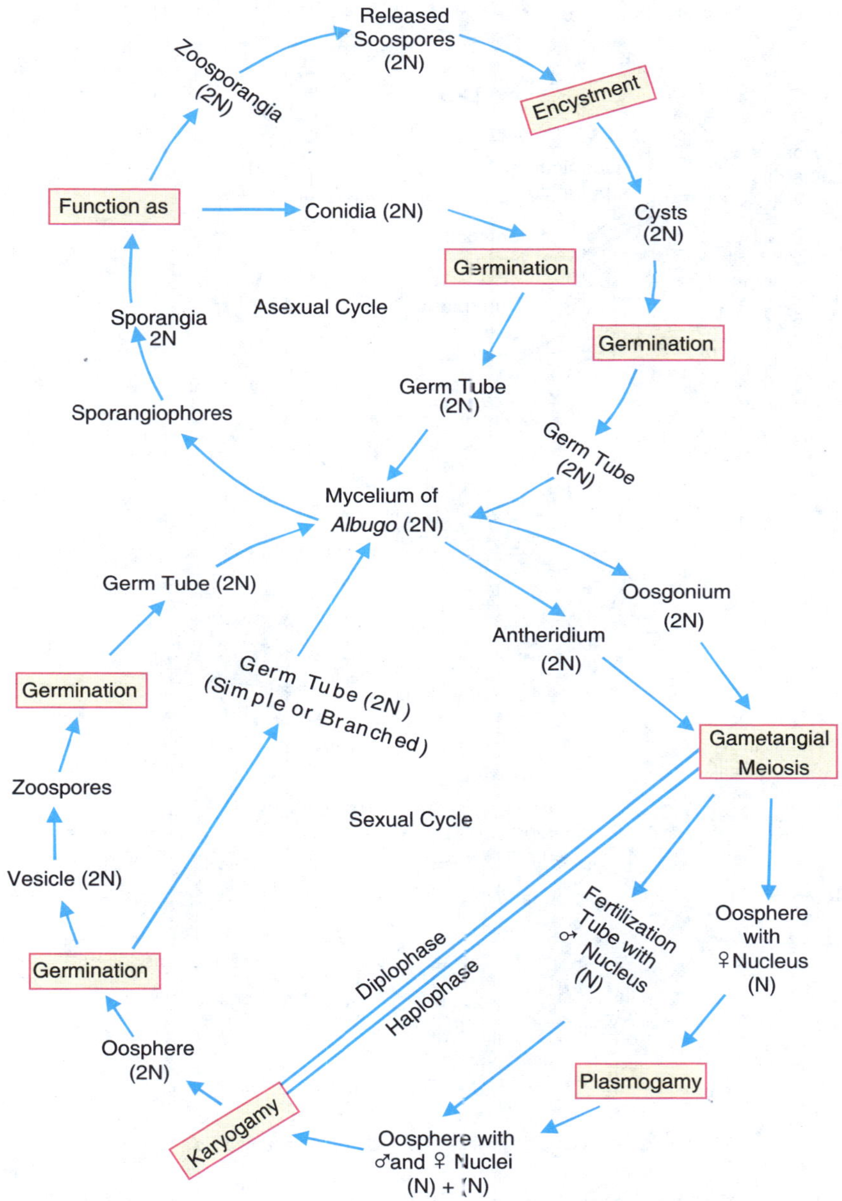

Fig. 6.60B. Graphic representation of the life cycle of *Albugo (Cystopus)* with gametangial meiosis.

15. Both the antheridia and oogonia are multinucleate but the functional nucleus in each is one.

16. The antheridium comes in contact with the oogonium at the side. The double wall at the point of contact dissolves. A fertilisation tube from the antheridium enters the oogonium and introduces a single male nucleus with some cytoplasm. The male and the female nuclei fuse.

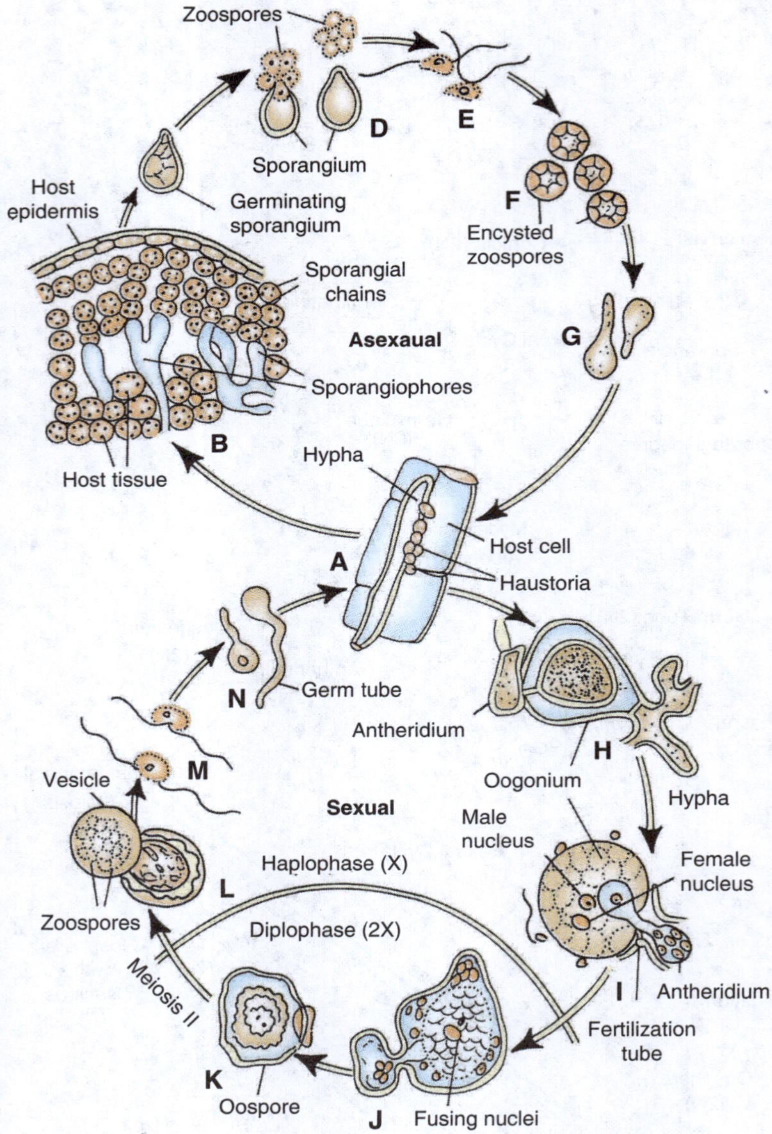

Fig. 6.61. Diagrammatic life cycle of *Albugo candida.*

17. The fertilised egg becomes an oospore by secreting a thick, warty wall around it. The oospore nucleus divides several times to produce about 32 nuclei. In this condition the oospore tides over the unfavourable period.
18. After the resting period the oospore germinates. The nuclei resume mitotic activity. The protoplast divides to form uninucleate daughter protoplast each of which becomes a biflagellate, reniform **zoospore.**
19. The oospore wall cracks and the zoospores pass into a thin vesicle which soon bursts open.
20. The liberated zoospores swim about in water. On coming in contact with a suitable host

each settles down. The quiescent zoospore withdraws the flagella, rounds off and secretes a wall around it. Soon it puts out a germ tube which enters the host through a stoma. Within the host tissue it grows vigorously and forms the mycelium.

Taxonomic Position :

Division	:	**Mycota**
Sub-division	:	**Eumycotina**
Class	:	**Oomycetes**
Order	:	**Peronosporales**
Family	:	**Albuginaceae**
Genus	:	*Albugo*
Species	:	*candida*

Alexopoulos and Mims (1979) place *Albugo* in the Superkingdom Eukaryonta, Kingdom Myceteae, Division Mastigomycota, Subdivision Diplomastigomycotina, Class Oomycetes, Order Peronosporales and Family Albuginaceae.

Superkingdom	:	**Eukaryonta**
Kingdom	:	**Mycetae**
Division	:	**Mastigomycotina**
Sub-division	:	**Diplomastigomycotina**
Class	:	**Oomycetes**
Order	:	**Peronosporales**
Family	:	**Albuginaceae**
Genus	:	*Albugo*
Species	:	*candida*

Evolutionary trends on Oomycetes. The following evolutionary trends are noticeable in Oomycetes :

1. There is progressive evolution from the aquatic to land habitat. Members of orders Lagenidiales, Saprolegniales and Leptomitales are chiefly aquatic. The advanced order Peronosporales includes aquatic, amphibious and terestrial forms.
2. The representatives of this class show a whole range from saprophytes (*Saprolegnia*) to facultative parasites (*Pythium*) and finally to obligate parasites (*Peronospora* and *Albugo*). The advanced land forms are all obligate parasites.
3. In the primative parasitic members, the mycelium is chiefly intracellular but the more advanced forms have intercellular mycelium which draws nourishment by means of haustoria. Of the haustoria the branched ones are considered more advancded than the vesicular or elongated type.
4. *Evolution of conidia.* An interesting evolutionary trend exhibited by the different species of this class is the transition of sporangium from a zoosporangium to a spore-like conidium. The primitive aquatic genera like *Saprolegnia* and *Achlya* reproduce asexually entirely by means of zoospores produced within zoosporangia. In *Phytophthora, Pythium* and *Plasmopara,* the sporangia behave in two ways. In the presence of moisture and at low temperature they behave like zoosporangia and thus produce zoospores but at high temperature the zoosporangia behave like spore-like conidia and germinate directly by a germ tube. *Peronospora* produces sporangia which invariably function as conidia.

5. The Oomycetes are, as a rule, oogamous and the gametes are non-motile. The Saprolegniales (*Saprolegnia* and *Achlya*) exhibit primitive type of oogamy. The number of oospores in the oogonium varies from 2 to several and there is no differentiation of oogonial protoplast into periplasm and ooplasm. *Pythium, Phytophthora* and other Peronosporales produce a single oosphere in the oogonium and zonation of the oogonial protoplast gets well established.

QUESTIONS

ESSAY TYPE

1. (*a*) Describe the salient features of class oomycetes. Give a recent classification alongwith characters of the orders.
   (*b*) Discuss the mode of sexual reproduction in Peronosporales. (*Rohilkhand M.Sc., 2006*)
2. Give an account of the habit, habitat somatic structure and reproduction in *Saprolegnia*. (*Rohilkhand, 1991*)
3. Give an account of sexual reproduction in *Albugo*. (*Punjab, 1991; H.P., 1992*)
4. Describe the structure of mycelium and sexual reproduction in *Phytophthora infestans*. (*Punjab, 1992, 1993*)
5. Name the causal organism of white Rust. Give graphic representation of its life cycle. (*Himachal Pradesh, 1993*)
6. (*a*) With the help of suitable diagrams, describe the sexual reproduction in *Albugo*.
   (*b*) What are the symptoms of white rust disease. (*Himachal Pradesh, 1997*)
7. Describe the life cycle of *Albugo* alongwith control measures of the disease it causes. (*G.N. Deo, 1991; Rohilkhand, 1997; Bihar, 1990; Awadh, 1995*)
8. Describe the structure and mode of reproduction in *Albugo candida*. (*Lucknow, 1994, 1996, 1999*)
9. Some dried flies and insects are placed in a dish, containing pond water. After few days, fungus growths are seen on these dead insects. Name this water mould (fungus). Give asexual and sexual reproduction of this fungus with suitable diagrams. (*Rohilkhand, 1992*)
10. Give a comparative account of sexual reproduction in *Albugo* and *Mucor*. (*Lucknow, 1991*)
11. Describe the life history of *Pythium*. Give names of five diseases caused by it. (*Kumaon , 2003; Lucknow, 1993, 1995*)
12. Give an illustrated account of life-history of *Phytophthora*. (*Allahabad, 2003; Meerut, 1994; Bihar 1991*)
13. What is late blight? Describe the life history of the fungus causing late blight disease of potato. Illustrate your answers with suitable diagrams. (*Allahabad, 1997, 1995, 2000*)
14. What is white rust? Describe the life history of the fungus causing white rust disease of mustard. (*Allahabad, 1996, 1994, 1992; Kanpur, 1999*)
15. Describe a comparative study of structure and reproduction in members of *Peronosporales*. (*Kanpur M.Sc., 1999*)
16. Give a brief note on the life cycle of *Albugo* with neat and well labelled diagrams. (*Kanpur, 1998*)
17. Describe the sexual reproduction in the fungus which causes white rust. What is the difference between *Albugo* and *Cystopus*? (*Kanpur, 1997; Awadh, 1994*)
18. With the help of labelled diagrams only, describe the life history of
    (*a*) *Saprolegnia*
    (*b*) *Pythium* (*Kumaon, 2001*)
    (*c*) *Peronospora*
    (*d*) *Phytophthora*
    (*e*) *Albugo* (*Awadh, 1993*)
19. With the help of diagrams, describe the methods of reproduction in a water mould. (*Gorakhpur, 1992, 1999*)
20. With the help of suitable diagrams, describe the structure and reproduction in *Saprolegnia*. (*Gorakhpur, 1994*)
21. Describe the systematic position, structure and mode of asexual peproduction in *Saprolegnia*. (*Gorakhpur, 1995*)
22. Describe the sexual reproduction in *Saprolegnia*. (*Gorakhpur, 1998*)

23. Give an illustrated account of life cycle of *Albugo*. (Lucknow, 2001; Kanpur, 2005)

24. Describe the life-cycle of *Saprolegnia*. (Kumaon, 2004; Kanpur, 2000)

25. Describe the life history of the fungus causing white rust disease of crucifers. Illustrate your answer with suitable diagrams. Write the methods of its control. (Allahabad, 2001, 2003; Rohilkhand, 2001)

26. Give an illustrated account of sexual reproductive structures in the members of the order Peronosporales. (Allahabad M.Sc., 2000)

27. Write an account of structure, mode of nutrition and reproduction in order Peronosporales. Give the names of any three important plant diseases caused by members of this group. (Kanpur M.Sc. 2004, 2001)

SHORT ANSWER TYPE

28. Name the pathogen and its systematic position of the following :
    (*a*) White rust of crucifers (Himachal Pradesh, 1994, 1996; Kumaon, 1996)
    (*b*) Late blight of potato (Himachal Pradesh, 1994; Kumaon, 1996)
    (*c*) Damping off of seedlings.

29. Name the diseases caused by *Phytophthora infestans* and *Albugo candida.*

30. Give an account of the symptoms, spread and control of white rust of crucifers. (Rohilkhand, 1991)

31. Classify *Albugo* according to Alexopoulos and Mims. Write important characters. (Rohilkhand, 1997)

32. Give classification and diagnostic features of the following :
    (*a*) *Pythium* (Lucknow, 1991, 1994, 1997)
    (*b*) *Saprolegnia* (Kanpur, 1998, 1997)
    (*c*) *Achlya*
    (*d*) *Phytophthora* (Kanpur M.Sc., 1999)
    (*e*) *Albugo or Cystopus* (Kanpur, 1999; 2006; Awadh, 1997, 1996; Lucknow, 1998; Gorakhpur, 1996, Kumaon, 2003)
    (*f*) *Peronospora* (Allahabad, M.Sc., 1999)
    (*g*) *Pythium and Phytophthora* (Allahabad M.Sc., 1999)

33. Distinguish between :
    (*a*) Asexual organs of *Pythium* and *Phytophthora* (Lucknow, 1994)
    (*b*) Sporangiophores of *Peronospora* and *Sclerospora*
    (*c*) Zygospore and zoospore (Lucknow, 1998)
    (*d*) Oospore and Zoospore (Allahabad, 2002)
    (*e*) Zygospore and Oospore (Lucknow, 2005)

34. Describe the symptoms, causal organisms and control measures of the following diseases :
    (*a*) Late Blight of potato (Bihar, 1991; Poorvanchal, 2006; Ranchi, 1991)
    (*b*) Downy mildew of Crucifers
    (*c*) Green ear disease of Bajra
    (*d*) White rust of crucifers
    (*e*) Damping off of seedlings

35. Write short notes on :
    (*a*) Sex organs and process of plasmogamy in *Saprolegnia* (Rohilkhand, 1993, 1994)
    (*b*) Asexual reproduction in *Albugo* (Lucknow; 1995; Allahabad, 1993; Gorakhpur, 1995; Poorvanchal 2001, Rohilkhand, 2003)
    (*c*) Development of conidia in *Albugo* (Bhagalpur, 1994)
    (*d*) Development of sporangia in *Albugo* (Lucknow, 1998, 2005)
    (*e*) Sexual reproduction in *Albugo candida* (Gorakhpur, 1990, 1998)
    (*f*) Diplanetism (Kumaon, 2000; Allahabad, 2001)
    (*g*) Fertilisation tube (Kumaon, 2001)
    (*h*) Oospore (Allahabad, 2001)
    (*i*) Zoosporangium and Zoospores of *Saprolegnia* (Meerut, 1997)
    (*j*) White rust of crucifers (Poorvanchal, 2006)

36. Draw a neat and well labelled diagram of the following :
    (*a*) V.S. of the leaf infected with *Albugo*. (Kanpur, 1996, 1998; Awadh, 1996, 1997, 1998; Lucknow, 1997, 2000; Gorakhpur, 1992, 1998; Poorvanchal, 2003)

(*b*) Asexual reproductive structure of *Albugo* (*Awadh. 1992*)

(*c*) Oogonia of *Albugo* (*Awadh, 1992*)

(*d*) V.S. of host leaf showing conida and conidiophore of *Cystopus (Albugo)* (*Kanpur, 2006*)

(*e*) Sporangiophores of *Sclerospora, Phytophthora, Pythium* and *Peronospora*

(*f*) Sexual life cycle of *Pythium* (*Lucknow, 1998*)

(*g*) Asexual reproduction in *Albugo* (*Lucknow, 2004*)

37. Draw well developed diagrams to show

(a) sporangial development in *Albugo* (*Lucknow*, 2002)

(*b*) sex organs of *Saprolegnia* (*Kanpur, 2001*)

(*c*) antheridia, oogonia and oospores of *Albugo* (*Poorvanchal, 2003*)

(*d*) conidiophores of *Peronospora, Phytophthora* and *Sclerospora* (*Kanpur M.Sc., 2004, 2001*)

(*e*) sexual reproduction in *Albugo* (*Lucknow, 2005*)

(*f*) V.S. host leaf showing conidea and conidophore of *cystopus* (*Kanpur*, 2006)

38. Give a comparative account of asexual reproduction in *Albugo* and *Eurotium.*

39. Name the fungus which causes white rust disease. Give systematic position, symptoms and mode of asexual reproduction in it. (*Gorakhpur, 1997*)

40. Write the asexual reproduction in *Cystopus.* (*Kanpur, 2002*)

41. Describe the asexual reproduction in *Achlya.* (*Poorvanchal*)

OBJECTIVE TYPE

42. Fill in the blanks :

(*i*) Plasmogamy in *Albugo candida* takes place by ....

(*ii*) The name of an aquatic fungus growing on dead insects is ....

(*iii*) Sporangiophores are also known as ....

(*iv*) Monoecious fungus bears male and female sex organs on .... thallus.

(*v*) Sporangial proliferation is found in the fungus ....

(*vi*) Diplanetism is exhibited by ....

(*vii*) *Albugo candida* absorbs its nourishment from the host cells through ....

(*viii*) In *Phytophthora infestans,* the asexual reproductive bodies behave as ....

(*ix*) White rust of crucifers is a pseudorust because the disease is caused by a member belonging to class ....

(*x*) Downy mildews are caused by the members of the order ....

(*xi*) *Pythium* is a .... parasite

(*xii*) *Scleorospora graminicola* causes the .... of Bajra.

(xiii) *Peronospora parasitica* is parasitic on ....

(*xiv*) Zoospores in *Saprolegnia* are produced in ....

(*xv*) Sexual reproduction in *Saprolegnia* is ....

(*xvi*) Oospores are formed as a result of fertilisation between an antheridium and an ....

(*xvii*) Zoospores of *Phytophthora* are produced inside a .... arising from ....

(*xiii*) The mycelium of *Pythium* is ....

(*xix*) In .... the oogonial wall is actually fused with the oospore wall.

(*xx*) Amphigynous antheridium is produced by the fungus ....

(*xxi*) Motile spores produced asexually in sporangia are called ....

(*xxii*) In *Albugo candida* the conidiosporangia are produced in .... succession.

(*xxiii*) The classification of Peronosporales is based on the characters of .... and ....

(*xxiv*) Depending upon the mode of nutrition ,*Pythium* is____________ .

(*xxv*) Damping off of Papaya is caused by ____________.

43. Select the correct answer :

(*i*) Which of the following is diploid?
(*a*) Zoospores (*b*) Spores
(*c*) Conidia (*d*) Oospores

(*ii*) Motile haploid spores are also known as
(*a*) Zoospores (*b*) Aplanospores
(*c*) Conidia (*d*) Zygospores

(*iii*) Haustoria are produced by
(*a*) fungal parasites of animals
(*b*) intercellular hyphae of obligate parasites
(*c*) fungi grown on artificial media
(*d*) intracellular hyphae of parasitic fungi.

(*iv*) Sporangiophores of the determinate growth are produced by fungi belonging to
(*a*) Pythiaceae (*b*) Albuginaceae
(*c*) Peronosporaceae (*d*) None.

(*v*) Which of the following diseases is caused by *Plasmopara*?
(*a*) Green ear disease of Bajra
(*b*) Downy mildew of grapes
(*c*) Late Blight of potato
(*d*) Damping off of Seedlings

(*vi*) Damping off of seedlings is caused by
(*a*) *Albugo candida*
(*b*) *Sclerospora graminicola*
(*c*) *Pythium debaryanum*
(*d*) *Peronospora parasitica*

(*vii*) *Albugo candida* is an example of
(*a*) obligate parasite
(*b*) obligate saprophyte
(*c*) facultative saprophyte
(*d*) facultative parasite

(*viii*) The genera of Peronosporaceae are identified on the basis of the characters of
(*a*) mycelium (*b*) oospore
(*c*) oogonium (*d*) sporangiophores

(*ix*) The genus *Phytophthora* differs from the genus *Pythium* in
(*a*) direct germination of sporangia
(*b*) producing zoospores in the vesicle
(*c*) producing zoospores in the sporangium
(*d*) producing zoospores both within sporangium and vesicle.

(*x*) Sporangiophores of *Plasmopara* are branched at
(*a*) acute angles
(*b*) obtuse angles
(*c*) sparsely branched
(*d*) right angles

(*xi*) In …………, sporangiophores branch dichotomously and end in curiously curved pointed tips.
(*a*) *Sclerospora* (*b*) *Peronospora*
(*c*) *Plasmopara* (*d*) *Albugo*

(*xii*) Thick walled resting spores produced by the members of Peronosporales are called
(*a*) Oospores (*b*) Ascospores
(*c*) Zygospores (*d*) Zoospores

(*xiii*) *Sclerospora graminicola* generally survives in the form of
(*a*) dormant oospores in soil
(*b*) actively growing mycelium on the host
(*c*) dormant structures other than spores in soil and host tissues
(*d*) conidia.

(*xiv*) In *Albugo candida* conidiosporangia are produced in
(*a*) acropetal succession
(*b*) basipetal succession
(*c*) longitudinal manner
(*d*) irregular manner

(*xv*) The successive production of two different types of zoospores by a single fungus is the phenomenon termed as
(*a*) Conjugation (*b*) Autoecism
(*c*) Heterothallism (*d*) Diplanetism

(*xvi*) Which of the following causes white spots on leaves and hypertrophied floral parts?
(*a*) *Phytophthora infestans*
(*b*) *Scleropora graminicala*
(*c*) *Albugo candida*
(*d*) *Pythium debaryanum*

(*xvii*) Symptoms produced by *Sclerospora graminicola* in *Bajra* mainly comprise of
(*a*) Shredding of leaves
(*b*) Downy mildew growth on the leaves
(*c*) both (*a*) and (*b*)
(*d*) Downy mildew, leaf shredding and green ears.

(*xviii*) *Peronospora* is parasitic on both pea and mustard but the conidia of the fungus from mustard will not affect pea because
(*a*) pea and mustard belong to two different families
(*b*) biological specialisation of the conidia prevents the infection

(c) the leaves are different
(d) conidia are not the infective bodies.

(xix) Diplanetism is exhibited by
(a) *Saprolegnia* (b) *Achlya*
(c) *Mucor* (d) *Albugo*

(xx) In *Albugo,* the antheridium is called paragynous because the antheridium
(a) grows beside the oogonium
(b) encircles the oogonium
(c) remains at the base of oogonium
(d) grows on the oogonial wall

(xxi) In *Phytophthora* the asexual reproductive bodies behave
(a) Conidia
(b) Conidiosporangia
(c) Sporangia
(d) both conidia and conidiosporangia

(xxii) White rust of crucifers is a pseudorust because
(a) the causal organism does not belong to rust family
(b) the colour of the pustule is not brown
(c) the disease occurs on members of family cruciferae
(d) the disease has not been observed on wheat

(xxiii) Sporangial proliferation in Saprolegnia is characterised by
(a) germination of spore into a mycelium
(b) the primary sporangium cutting off spores from its apex
(c) production of new sporangia from the vegetative hyphae
(d) development of Secondary Sporangium into the primary sporangium.

(xxiv) The conidiophores in *Cystopus* are produced in
(a) paralleled clusters
(b) transverse clusters
(c) Paralleled clusters in pustules on the surface
(d) irregular fashion.

(xxv) Multinucleate oospores are found in
(a) *Cystopus* candida
(b) *C. blitti*
(c) *C. tragopogonsis*
(d) None

(xxvi) *Saprolegnia* is usually found in
(a) Fresh water bodies
(b) Marine water
(c) Xerophytic conditions
(d) All of the above.

(xxvii) Which of the following fungi is aquatic ?
(a) *Aspergillus* (b) *Saprolegnia*
(c) *Alternaria* (d) *Ustilago*

(xxviii) Diplanetism is present in
(a) *Saprolegnia* (b) *Synchytrium*
(c)r *Peziza* (d) *Puccinia*

CHAPTER 7

SUBDIVISION – EUMYCOTINA Class : Zygomycetes

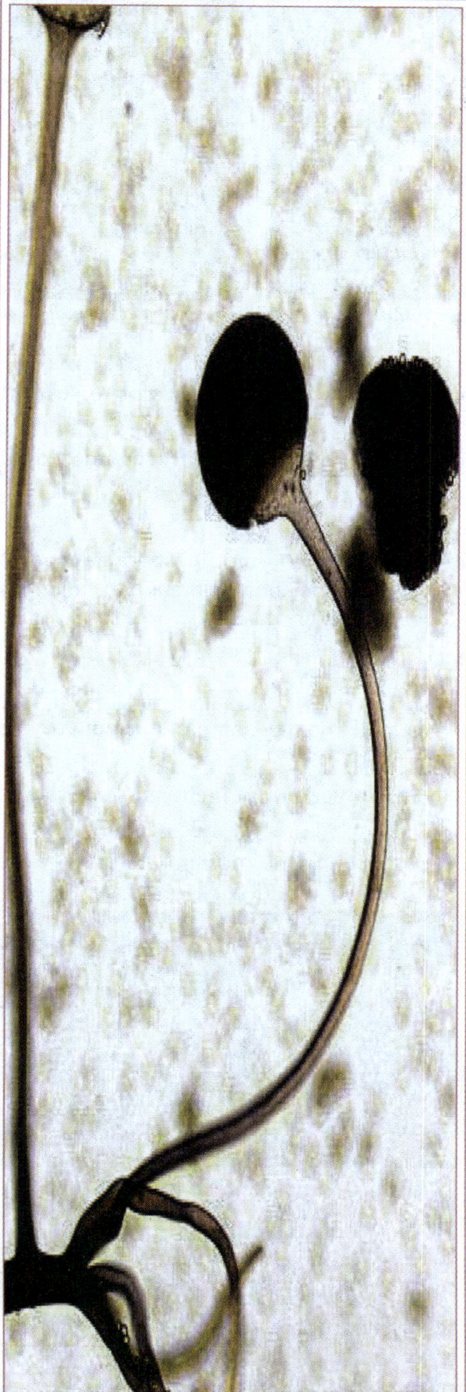

Rhizopus anhizus - Sporangophore with sporangia

- Order : Mucorales
- Family : Mucoraceae
- *Rhizopus stolonifer*
- *Mucor mucedo*
- Family Pilobolaceae
- *Pilobolus kleinii*
- Phycomycetes – Discussion
- Heterothallism

Introduction. The class zygomycetes derives its name from the thick-walled resting spores, the *zygospores* formed as a result of the complete fusion of the protoplasts of two equal or unequal gametangia. It comprises 450 species which are grouped under 70 genera. They all are terrestrial molds which show a wide range in their habit. Most of them are saprobes. Among these some are soil saprophytes and others coprophilous (growing on dung). The parasitic species include pin molds which attack fungi, insects or protozoa. Some are facultative weak parasites (*Rhizopus sexualis*) of higher plants. A few are specialized parasites of animals. A very few occur as obligate parasites of other Zygomycetes.

Terrestrial Molds

Coprophilous Fungi

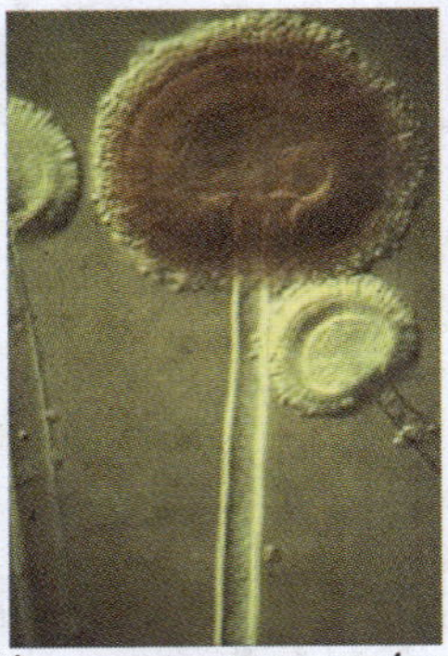

Rhizopus oryzae sporangium

The mycelium is well developed. When young it lacks cross walls but the old hyphae may be septate. Septa also appear in connection with development of reproductive bodies or to seal off injuries.

Distinctive Features of Zygomycetes : 1. The hyphal walls are chiefly composed of chitinchitosan.

2. The motile cells are completely absent in the life cycle.

3. Asexual reproduction typically takes place by means of non-motile sporangiospores commonly produced in large numbers within sporangia. Sometimes the entire sporangium functions as a single spore in the same manner as the conidium.

4. Chlamydospore formation is of frequent occurrence.

5. Sexual fusion involves gametangial copulation.

6. The thick-walled sexually produced zygospore formed by the complete fusion of the protoplasts of two gametangia is a resting structure.

7. The zygospore germinates to produce a hypha, the **promycelium** which bears a terminal sporangium.

Economically the zygomycetes are of significant importance. Some of them are used in the fermentation of food items while a few others are employed to produce enzymes, acids, etc. Saprophytic species spoil our food stuffs. Some zygomycetes are important mycorrhizal fungi and a few others are human pathogens. Some of the zygomycetes are of great interest for having developed peculiar methods of spore dispersed. ``Fungus shotgun" of spore dispersal of *Pilobolus* and forceful propulsion of the asexual spores of the Entomophthorales are the common examples. Animal trapp-mechanism of the zoopagales is of common interest.

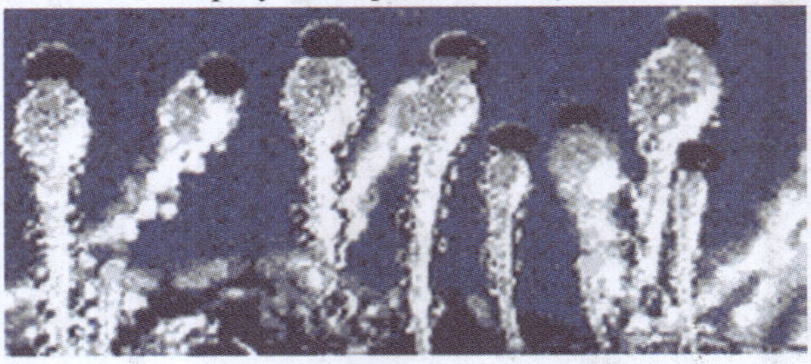

Shotgun spore Dispersal of Pilobolus

Classification of Zygomycetes. Traditionally the class comprises 3 orders: Mucorales, Entomophthorales and Zoopagales. Benjamin (1979) divided the class into 7 orders namely, Mucorales, Demorgaritales, Kickxellales, Endogonales, Entomophthorales, Zoopagales and, Harpellales.

Dube (1993), modifying the original classification of Ainsworth (1973) recognises following three orders under the class Zygomycetes :

Order **Mucorales.** Chiefly Saprophytes, some weak parasites on plants, a few endoparasites of vertebrates; mycelium extensive, asexual reproduction by sporangiospores or rarely by conidia; Zygospore wall may be formed by modification of gametangial walls.

Order **Entomophthorales.** Typically parasitic on animals; rarely saprophytes, mycelium-limited. Asexual reproduction by sporangia turned conidia or true conidia; conidia discharged forcibly, gametangial wall not transformed to Zygospore wall.

Order **Zoopagales.** Typically parasite on animals, rarely saprophytes, mycelium present with complicated haustoria, Asexual reproduction by conidia which are passively discharged; gametangial wall not transformed to Zygospore wall.

Of these, Mucorales have been discussed in some detail.

ORDER : MUCORALES

Occurrence. This order is by far the largest and includes about 250 species which are placed under 45 genera. They are chiefly terrestrial saprophytes. A few species are parasitic on plants and animals. The saprophytic species are commonly found on moist, decaying, organic matter such as dung or decaying plant or animal matter rich in starch and sugars. Mucorales are familiarly known as the 'black molds'.

Black Molds

Somatic Structures. The thallus is a well-developed, branched mycelium which is freely exposed to air. In some species it produces rhizoidal hyphae at the points of contact with a hard surface. Ultrastructurally the hyphal protoplast, besides the nuclei, contains mitochondria, ribosomes, endoplasmic reticulum, and lipid droplets. Dictyosomes consisting of stacks of cisternae and centrioles are absent. Vesicle-like structures are common in the growing hyphal tips. The hyphae are coenocytic and aseptate. Septa, however, are formed at the bases of reproductive structures (sporangia and gametangia) or occasionally in old hyphae. Such septa are solid plates forming complete partitions. The septa, as usual, arise as annular ingrowths beginning at the hyphal wall. The ingrowth advances towards the centre forming a complete portion. In some of the advanced Mucorales, the hyphae are regularly septate. The septa are perforate, each has a central pore. In some forms there are tubular extensions projecting forward in the direction of the protoplasmic flow and others have special type plugs (Benjamin, 1959). Pracker *et al* (1970) reported the occurrence of chitosomes in *Mucor cousil* which they speculated to function in the formation of chitin-chitosan in the walls of mucorraceous fungi.

Some mucorales are *dimorphic*. Under typical conditions they produce a well-developed mycelium but when placed in liquid media under normal environmental conditions they exist as yeast-like forms.

Asexual reproduction. It takes place by means of non-motile spores **(sporangiospores)** which are produced in large numbers within globose usually columellate sporangia borne singly at the tips of simple or branched sporangiophores. Some species produce small, few spored or even monosporous sporangia which are called sporangioles or **sporangiola** (sing. sporangiolum). The sporangiola may be with or without columella. Some species produce both sporangia and sporangiola on the same sporangiophore which is branched. The deciduous monosporous sporangia

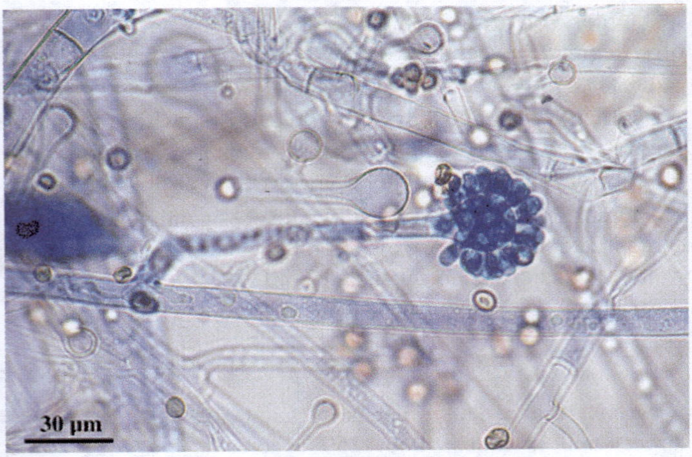

Cunninghamella

or sporangiola are regarded as conidia. In certain species the sporangial and spore walls are completely fused. The term conidia applies appropriately to such one-spored sporangiola. It is obvious from this account that within the Mucorales, we can trace the evolution of **conidium**—a typical asexual spore of the Ascomycetes from a multispored sporangium by reduction in the number of spores from several to one and subsequent fusion of the sporangial and spore walls. The development of true conidia in the genus *Cunninghamella* of Mucorales has led some mycologists to advance the view that the Ascomycetes and Zygomycetes have evolved from a common ancestor. Khan and Talbot (1975), however, showed by TEM that the spore and sporangial walls in the aforesaid organs are closely applied but not fused with one another. The use of term conidium is thus inappropriate.

Spore Structure. The mature spores of Mucorales are typically multinucleate. In form they vary from globose to ovoid but cylindrical in some species. Ultrastructurally the spore wall is bilayered. The outer layer is electron-dense and the inner electron-transparent. The cytoplasm contains one or more nuclei, ribosomes, mitochondria, vacuoles and reserve food in the form of lipid droplets. Endoplasmic reticulum is sparse. The production of asexual spores is determined by environmental factors such as temperature, humidity and light conditions.

Choanephora cucurbitarum on Cucurbita leaf

(*a*) *Effect of temperature.* Investigations of Barnett and Lilly (1950, 1955) on *Choanephora cucurbitarum*, which produces both multispored sporangia and monosporous sporangiola have shown that sporangial production is favoured by high temperature (30–31°C) and 100 per cent relative humidity. Conidial production is inhibited under such conditions. It is favoured by low temperature (25°C) and low relative humidity.

(*b*) *Effect of light.* Under conditions of continuous dark, continuous bright light (60 foot candles) or darkness followed by bright light no conidia are formed. Abundant conidia are produced when a period of darkness is preceded by bright light. The conclusion is that substance which induces production of conidia is synthesized in two steps. The first step requires light and the second step is inhibited by light.

Sexual reproduction. Some species of mucoraceous fungi are homothallic but the majority are heterothallic. Sexual reproduction is isogamous or nearly so and takes place by gametangial copulation. Two short, side branches the **progametangia** arise from the adjacent hyphae. The hyphae bearing the progametangia are called **zygophores**. The progametangia grow towards each other and come into contact end to end. A small, terminal multinucleate cell, the **gametangium** is cut off by a septum at the tip of each progametangium, the remainder of which is called a **suspensor**. In many species the gametangia are similar but in a few others one is larger than the other. The walls between the two gametangia dissolve and their contents mingle in the fusion cell. Karyogamy eventually takes place. A single zygospore is formed within the fusion cell termed the **prozygosporangium** (O'Donell *et at.*, 1978). It enlarges and secretes a thick multilayered pigmented sculptured wall around it to become the *Zygosporangium.* Single zygospore develops within the zygosporangium. After the resting period, the zygospore germinates. The thick zygosporangium wall ruptures and a hypha emerges. It grows to a certain height and forms the terminal *germ sporangium.* The protoplast of the germ sporangium undergoes cleavage to produce non-motile spores called the *germ spores* which are genetically haploid. It means meiosis takes place during zygospore germination. In meiosis segregation of strains takes place. Thus, the germ sporangia in the heterothallic species may contain (i) all the spores of the same mating type (+ or –), or (ii) of three mating types : +, – and ± or (iii) of two types : + and –. How it happens we shall see shortly.

The zygosporangium in *Phycomyces blakesleanus* is protected by sterile, forked bristles arising from the suspensors. In the genus *Mortierella* it is surrounded and protected by a cottony sheath of sterile hyphae. *Endogone* goes a step further. A number of neighbouring zygosporangia are enclosed in a common hyphal sheath or peridium to form a tiny fruiting body resembling the

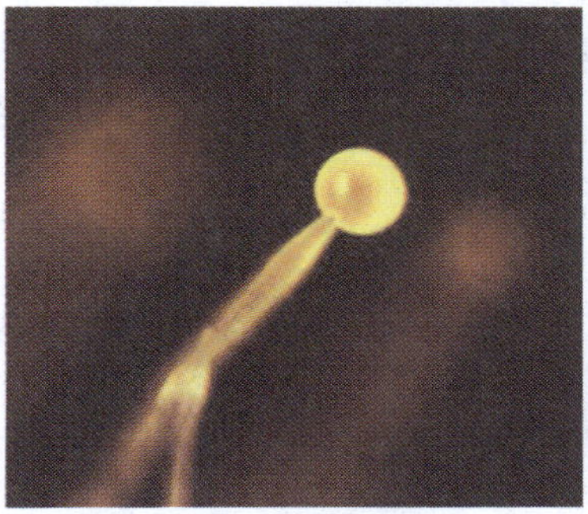

Phycomyces blakesleanus – Sporangium with sporangsphore

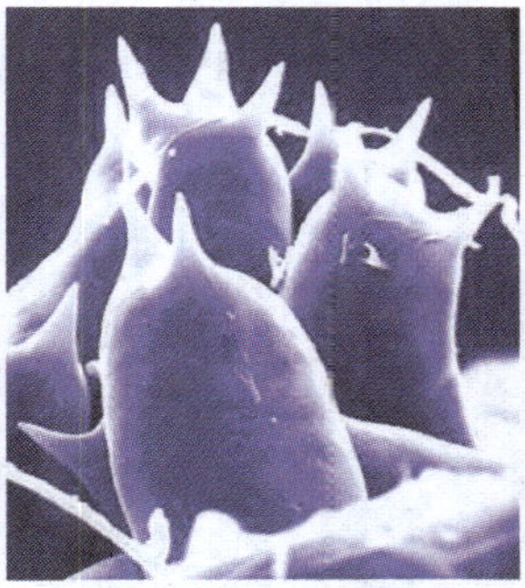

Mortierella – Sterile hyphae

Endogone – Zygosporangia enclosed in a sheath

ascocarp of the Ascomycetes. This tendency to increased protection of fruiting body is a distinct approach to the Ascomycetes and lends support to the view that such species of Mucorales represent the ancestral line from which the Ascomycetes may have developed.

Differentiation of Sex in Mucorales. In many species of Mucorales, the fusing gametangia are borne on the zygophores arising from the some mycelium. Such species are called **homothallic**. Several species, however, have two distinct types of mycelia. These are similar outwardly but function differently in reproduction. They belong to two different mating types. One of these is labelled as plus strain and the other minus strain. The zygospores are produced only when a plus strain zygophore and a minus strain zygophore come into contact. The species which require two distinct types of mycelia (plus and minus strain) to form zygospores are called **heterothallic**. It was Blakeslee who first discovered the phenomenon of heterothallism in Mucorales in 1904.

A detailed account of Heterothallism has been described chapter 17 ahead.

Classification

The order Mucorales has been classified in different ways by different mycologists. Hasseltine and Ellis (1973) divided the order into fourteen families on the basis of nature of sporangium and sporangiophores, habit and sexual reproduction:

1. Choanephoraceae 2. Cunninghamellaceae 3. Dimargaritaceae 4. Endogonaceae

5. Helicocephalidaceae 6. Kickxellaceae 7. Mortierellaceae 8. Mucoraceae 9. Pilobolaceae 10. Piptocephalidaceae 11. Radiomycetaceae 12. Sakseneaceae 13. Syncephalastraceae 14. Thamnidiaceae.

However, Martin (1961) has earlier recognised only nine families listed below :

1. Mucoraceae 2. Pilobolaceae 3. Thamnidiaceae 4. Choanephoraceae 5. Piptocephalidaceae 6. Kickxellaceae 7. Cunninghamellaceae 8. Mortierellaceae 9. Endogonaceae

Of these families, **Mucoraceae** is discussed here due to the fact that mycologists regard the family as primitive as well as advanced. The **primitive characters** of the family are :

(*i*) mycelium aseptate although septa develop to delimit reproductive structures; (*ii*) sporangia large, multisporous and columellate; (*iii*) progametangia highly differentiated and formed opposite; (*iv*) Zygospores formed on aerial hyphae; (*v*) thick walled dark coloured zygospores; (*vi*) simple nutrition and (*vii*) freeliving habit.

The characters of the family regarded as **advanced characters** are : (*i*) mycelium becoming septate; (*ii*) presence of multisporous sporangia to one-spored sporangiola; (*iii*) progametangia least differentiated with parallel suspensors; (*iv*) Zygospores usually produced on or in the substratum; (*v*) Zygospores light coloured; (*vi*) complex nutrition and (*vii*) members being facultative and/or obligate parasites.

In addition, genus **Pilobolus** of the family **Pilobolaceae** is also described because of its unique sporangia.

FAMILY : MUCORACEAE

It is the largest family of the order mucorales. The family includes about 20 genera which grow in different habitats. *Rhizopus* and *Mucor* are the two best known and important genera of this family. *Rhizopus* includes 35 species. All are saprobes. Of these *R. stolonifer* is taken as a type.

Rhizopus Soft Rot

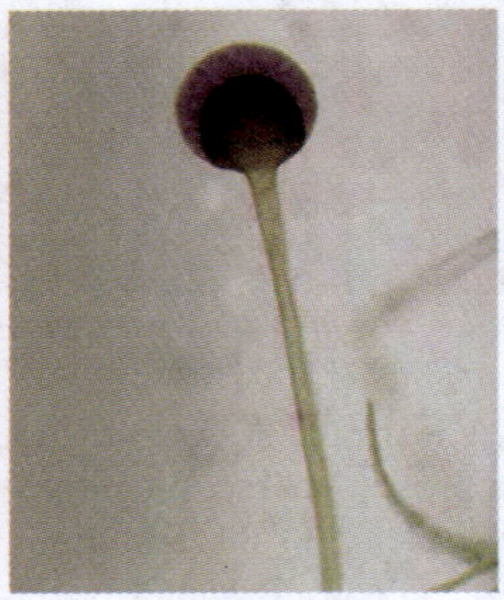

Rhizopus stolonifer

RHIZOPUS STOLONIFER (*syn. R. nigricans*)

Occurrence. It is the commonest and the best known fungus found growing frequently on stale bread and thus is commonly called **bread mold**. It exclusively lives as a saprophyte. Besides it occurs on damp decaying fruits, vegetables and other food material such as pickles, jams and jellies. *Warmth*, *plenty of air* and *moisture* are the three factors essential for its growth. It is frequently found in the rainy season. Under certain conditions *Rhizopus stolonifer* grows as a weak facultative parasite on strawberry fruits causing a transit disease called *'leak'* and also *'soft rot'* of sweet potatoes.

A luxuriant growth of this mold can be easily obtained by placing a piece of bread, which has been moistened and exposed to air, under a bell jar at a moderate temperature for a couple of days. It, at first, appears as small white patches. The patches grow rapidly and run into one another. Finally they cover the entire surface of the bread with white, fluffy mycelia.

A. Somatic phase

(*a*) **Thallus** (Fig. 7.1). It consists of numerous, slender, freely branched filaments, the **hyphae.** The hyphae are tangled and form a fluffy, white mass which makes up the thallus of the fungus. It is called the **mycelium** (Fig. 7.1). It is cottony white during the vegetative phase. However, it soon enters the reproductive phase. Numerous black, pin head-like structures develop on the mycelium giving the entire cowebby mass a blackish appearance, hence the popular name **black molds.** Are all hyphae constituting the mycelium alike? During the early period of growth the mycelium consists of hyphae which are all alike. They remain within the substratum. After a few days of growth this internal mycelium produces some hyphae which come to the surface of the substrate to form an aerial mycelium. The latter shows differentiation of hyphae accompained by division of labour. It consists of three kinds of hyphae :—

(*i*) *Stolons.* These are aerial hyphae that grow horizontally over the surface of substratum. They are comparatively stouter and of larger diameter, slightly arched and less branched aerial hyphae (Fig. 7.1). The stolons grow rapidly in all directions and cause further spread of the

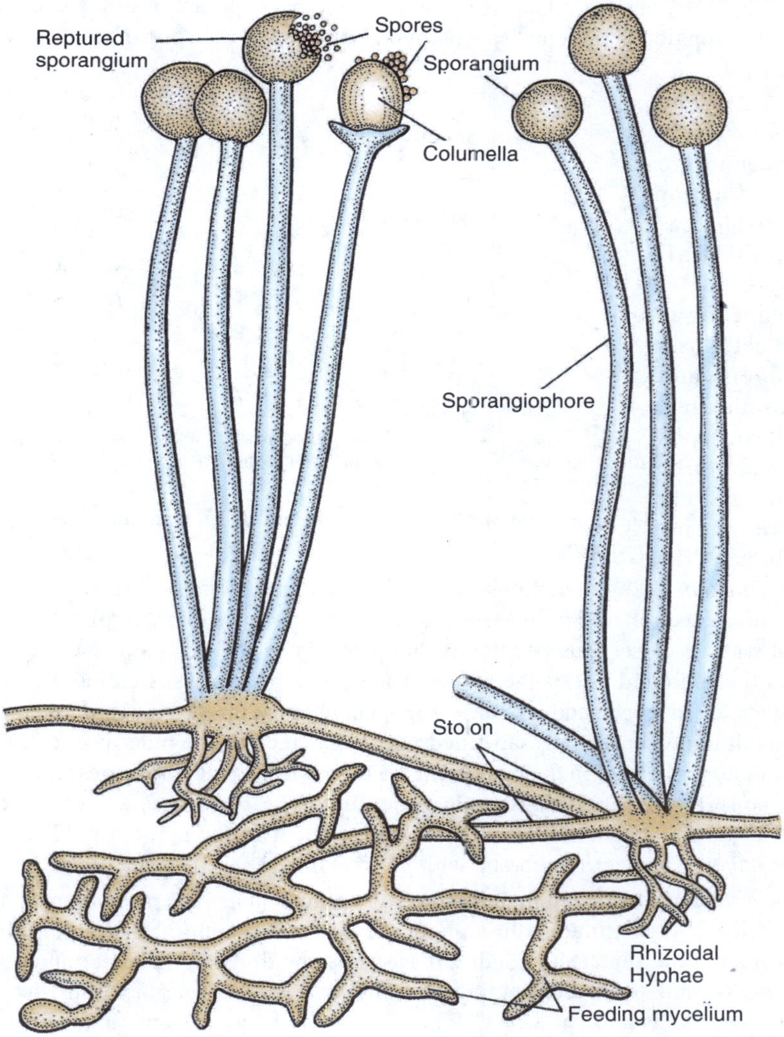

Fig. 7.1. *Rhizopus stolonifer* showing connection of internal mycelium with the aerial mycelium which is differentiated into stolons, rhizoidal hyphae and sporangiophores.

mycelium over the substratum. Each stolon arises from the point of contact of mycelium and substratum. It runs horizontally for some distance and then arches over so that its tip touches the substratum further along to form an apparent node-like structure. From the latter arise fresh stolons which run over the surface of the substratum and repeat the process.

(*ii*) *Holdfast.* It is a cluster of brown, slender, branched rooting hyphae (Fig. 7.1) which arise from the lower surface of the apparent node of each stolon and penetrate the starchy substance of the substratum. They secrete, as they grow down, an enzyme which digests starch. The rooting or rhizoidal hyphae thus serve a double purpose. They anchor the fungus to the substratum and absorb water and nourishment from it for the entire plant.

During the vegetative phase the mycelium consists only of these two kinds of hyphae and is cottony white. It becomes moldy as it enters the reproductive phase and develops the third kind of hyphae.

(*iii*) *Sporangiophores.* Just opposite each holdfast one or more vertically growing hyphae rise into the air (Fig. 7.1). They are reproductive in function and bear sporangia singly and terminally. These negatively geotrophic, unbranched special hyphae are called the **sporangiophores.**

(*b*) Structure of Thallus (Fig. 7.2). The hyphae which collectively constitute the mycelium are transparent, tubular structures. Each hypha has a limiting **hyphal wall** which consists of a substance called **fungus chitin** and not cellulose. Besides hydrogen and carbon it contains nitrogen. In electron micrographs (Fig. 7.12) it is seen to be microfibrillar in structure with the microfibrils running parallel to the surface. The chitinous wall of *Rhizopus* like the cellulose wall of green algae is permeable to water and other substances in solution. Electron microscope studies reveal the presence of a thin **plasma membrane** lining the hyphal wall. It closely invests the hyphal protoplasm and regulates the movements of substances in solution into and out of the hypha. The hyphal protoplast is granular and contains numerous minute nuclei, glycogen and globules. The young hyphal tips are full of dense protoplasm. Further back, in addition, occur small, sap-filled vacuoles which in the older portion coalesce to form a central vacuole. Embedded in the cytoplasm are numerous, small **ribosomes** and a branching system of tubes constituting the **endoplasmic reticulum.** The ribosomes are of uniform size. **Mitochondria** with double membranes and cristale also occur in the cytoplasm. The nuclei are bounded by a two-layered nuclear membrane with pores (Fig. 7.12).

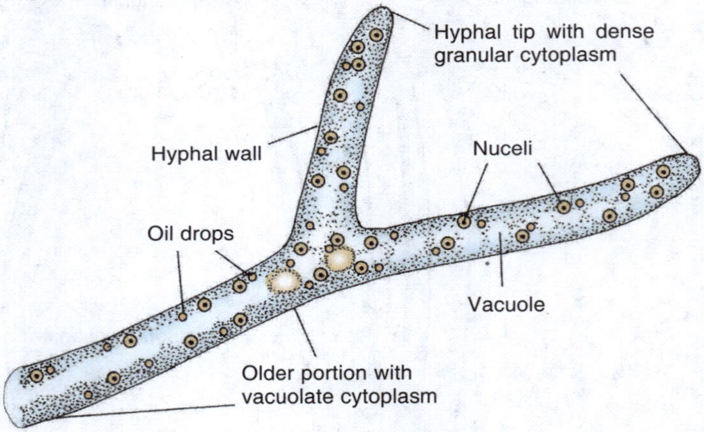

Fig. 7.2. *R. stolonifer.* A hypha showing detailed structure as seen under light microscope.

The cross walls remain suppressed in the actively growing hyphae. The protoplasm is, therefore, not divided into cells. It is continuous throughout the entire mycelium. Such a plant body or a portion of it, in which a multinucleate condition arises by the division of nuclei unaccompained by wall formation, is said to be **coenocytic.** The hyphae in *Rhizopus* are thus aseptate and coenocytic. The cross walls are, however, found in the old hyphae and in connection with reproductive structures. These septa are solid plates. The growth of hyphae is entirely apical.

(*c*) **Nutrition.** *Rhizopus* lacks chlorophyll. Consequently it is unable to synthesize its organic food from carbon dioxide and water. How does then bread mold gets its nourishment? It gets its food readymade from the dead organic matter present in the substratum by means of the root-like, rhizoidal hyphae. The latter secrete **enzymes** as they penetrate the substratum. The enzymes convert insoluble starch and other substances into a soluble form. The solubles enter the rhizoidal hyphae by diffusion through the hyphal walls. In its mode of nutrition bread mold thus is **a heteromorph** and a **saprophyte.** Although the fungus is unable to manufacture sugar from carbon dioxide and water, yet it is able to convert the soluble sugars it absorbs from an external source into complex carbohydrates which are the chief components of the hyphal walls. It can also synthesize proteins from the carbohydrates utilizing inorganic nitrogen compounds such as ammonium salts.

B. Reproductive phase. Reaching a certain stage of maturity the aerial mycelium of *Rhizopus* starts reproducing. It turns moldy. It reproduces by both methods, asexual and sexual.

1. Asexual reproduction. It takes place by vegetative methods and sporulation.

(*a*) *Vegetative Reproduction.* It takes place by the following two methods :—

(*i*) *Fragmentation.* It is the commonest method of propagation of the mycelium stage during the growing season. Fragmentation consists in the separation of hyphae from the aerial mycelium. These are known as the *fragments.* Each fragment by apical growth and branching grows into a new mycelium.

(*ii*) *Chlamydospore formation* (Fig. 7.3). Some species of *Rhizopus* under unfavourable conditions (shortage of waters) produce specially modified thick-walled resting cells called the *Chlamydospores.* The protoplasmic contents of the hyphae accumulate at certain points and lose water (B). The intervening parts remain empty. Each localized protoplasmic accumulation secretes a thick wall around it to become a chlamydospore (C). The chlamydospores in *Rhizopus* are thus *intercalary.* They contain sufficient reserve food and are very resistant to desiccation. They are in fact resting cells. As the substratum, on which the fungus grows, dries out the mycelium perishes. The chlamydospores survive. On the return of conditions suitable for vegetative growth, the surviving (perennating) chlamydospores germinate, each producing a new mycelium. The chlamydospores thus enable the fungus to tide over conditions unfavourable for growth. Chlamydospores are rarely formed in *R. stolonifer* but they are commonly found in other species.

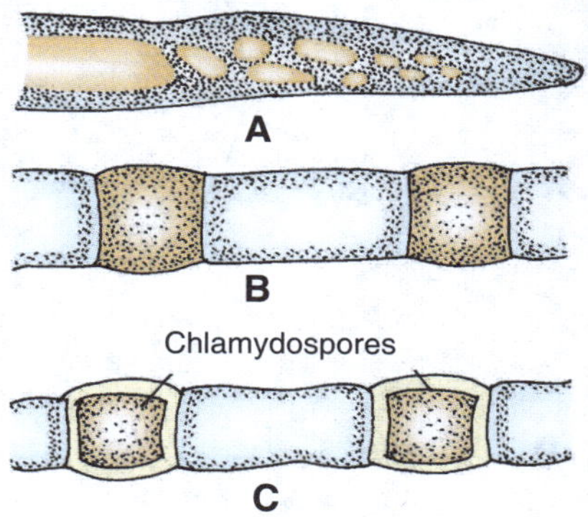

Fig. 7.3 (A-C). *Rhizopus sp.* Showing stages in the formation of Chlamydospores (B-C);

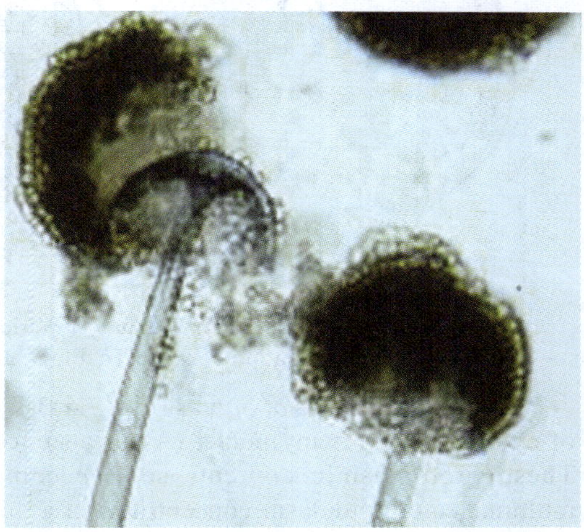

Rhizopus oryzae

(*iii*) *Oidia Formation.* Multiplication by oidia formation is absent in *R. stolonifer* but is common in other species of *Rhizopus* (*R. oryzae*). When the mycelium grows submerged in a nutritive medium such as sugar solution, the young coenocytic hyphae develop septa and divide into short multinucleate segments. These segments separate from each other and become rounded to form **Oidiospores** or **oidia.** The latter increase in number by budding. This is known as the **torula stage** of the fungus. The oidiospores germinate on solid organic substrata and grow into new mycelia.

(*b*) *Sporulation* (Fig. 7.4). During the growing season the fungus reproduces repeatedly by the formation of small, non-motile wind-disseminated spores. They are produced in large numbers within round black structures, the *sporangia.* They are differentiated by mitosis from the protoplast of sporangium endogenously and thus are termed the *sporangiospores.* The sporangiospores serve to multiply the mycelium stage in the life cycle. They belong to the category of *mitospores* and thus play no role in the phenomenon of alternation of generations.

(*i*) *Development of Sporangia* (Fig. 7.4A). As the sporangiophore elongates by linear growth of the region behind the tip, the latter swells into a knob-like vesicle (A). It is more or less spherical in form. The vesicle grows in size due to flow into it of cytoplasm carrying nuclei and food from the sporangiophore. It is the young sporangium.

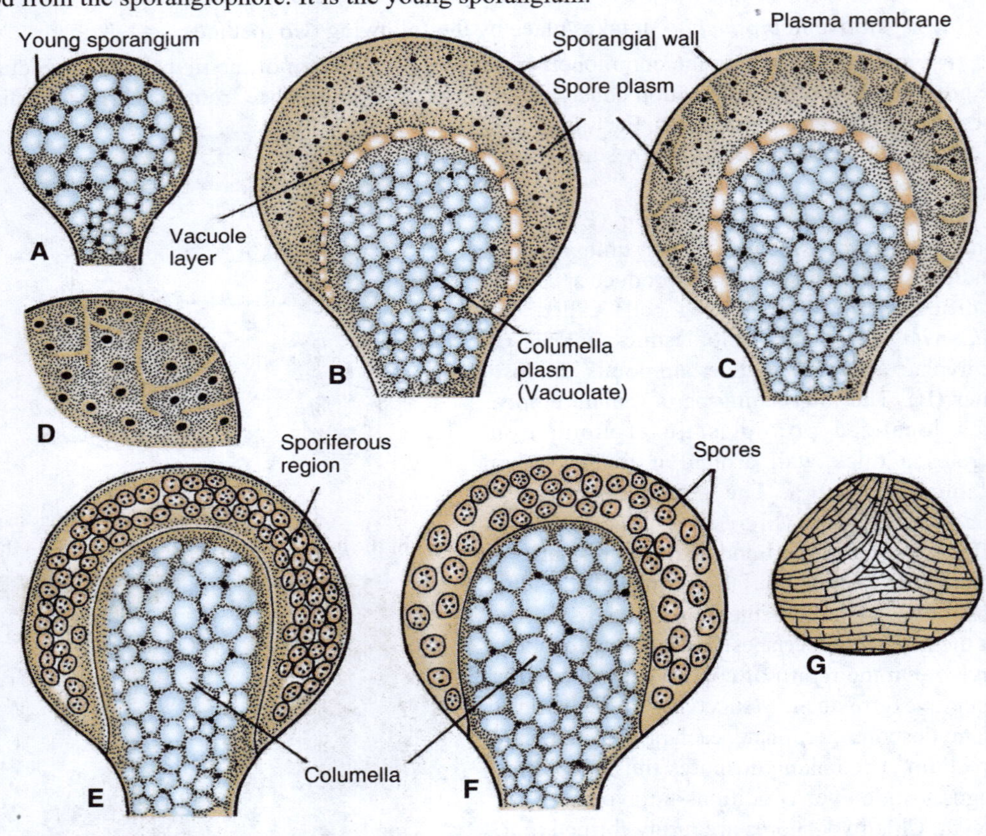

Fig. 7.4 (A-G). *Rhizopus stolonifer.* Various stages in the development of the sporangium and differentiation of spores (A-F); G, a single spore. Explanation in the text.

(*ii*) *Differentiation of Spores* (Fig. 7.4 B-E). The young sporangium contains a certain amount of cytoplasm and many nuclei (A). It also accumulates a considerable amount of reserve food. These protoplasmic contents soon become differentiated into two zones. Most of the multinucleate protoplasm concentrates in a thick dense layer lining the wall (B). This results in the appearance of a highly vacuolated dome-like central zone in the sporangium. The central zone contains dense, vacuolar protoplasm with fewer nuclei. A layer of vacuoles then appears between the two portions (B). The vacuoles flatten (C) and finally fuse laterally to form a cleft between the two zones. A wall finally appears in this cleft (E). The sporangium is now divided into the denser, peripheral, **sporiferous zone** and the central dome-shaped zone, the **columella.** The protoplast of the columella is continuous with that of the sporangiophore. The sporiferous zone undergoes cleavage to form a number of multinucleate segments (D). These segments round off (E) and secrete walls around them to become the **sporangiospores.**

(*iii*) *Structure of Sporangium* (F). The young sporangium is a white spherical structure situated at the free end of the sporangiophore (A). At maturity it turns black. The black colour is due to the presence of sooty spores within. In the centre of the sporangium is a large, dome-shaped structure called the *columella* (E). Surrounding the columella are the numerous tiny sporangiospores which are covered and thus protected by a non-cellular layer constituting the *sporangial wall* (F). The sporangia of *Rhizopus* are haploid. They are produced on the haploid mycelium and contain haploid spores differentiated by mitosis.

(*iv*) *Dehiscence of the Sporangium* (Fig. 7.5A). As the spores mature, there is an increase in the quantity of fluid in the sporangiophore and the columella. It exerts a considerable pressure and makes the columella bulge more strongly into the sporangium. Consequently the fragile sporangium wall cracks into small fragments in dry air or with slightest disturbance (A). The spores which form a dry powdery mass are exposed. A part of the wall remains as a collar-like torn fringe at the base of the distended columella.

(*d*) *Dispersal of Spores* (Fig. 7.5B). The spores in the exposed mass are minute and light. Accoding to Ingold, immediately after the dehiscence of sporangium the mature columella collapses and takes on the form of an upside down bowl with the spores spread over its bulging surface. From there the spores are blown away by the lightest drought to a great distance through the air. In this way, they attain such a wide distribution that they come to be present almost everywhere. The jerking and twisting of the thick-walled sporangiophore also assists in the liberation of spores.

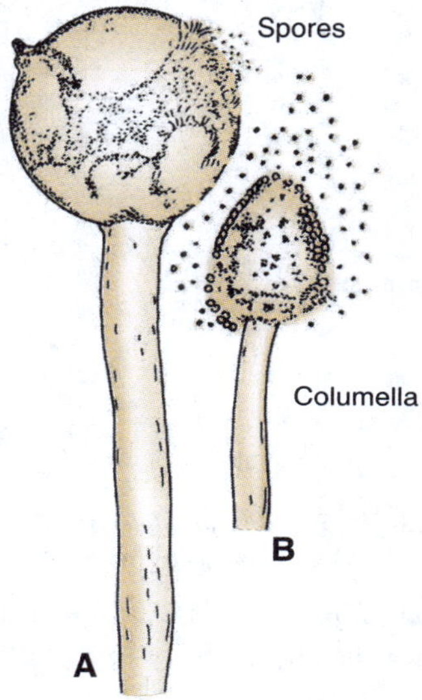

Fig. 7.5 (A-B). *Rhizopus sp.* A, Dehiscence of the sporangium; B, Dehisced sporangium with columella exposed.

(*e*) *Spore Structure* (Fig. 7.6A). The liberated spore is non-motile globose to oval or somewhat angular in form. It is microscopic and consists of a tiny mass of multinucleate protoplast surrounded by a resistant, single layered spore wall which is thick, rigid and electron opaque. It bears numerous prominences representing the ridges. Within the spore wall is the plasma membrane closely investing the cytoplasm which contains several nuclei, mitochondria lipid droplets and ribosome. Endoplasmic reticulum strands are sparse. Dense membrane bound inclusions termed **cytosomes** are abundant. Small vesicles are also present. The spores are coloured bluish or brown but in a mass they appear black.

Spore Germination (Fig. 7.6 B). Under dry or other unsuitable conditions, the spores (A) **remain** viable for long periods. The spores of *R. stolonifer,* when kept dry, are known to germinate after twelve years. On falling on a suitable substratum and under suitable conditions they germinate immediately. Adequate moisture, favourable temperature (between 15° to 20°C) and starchy matter in the substratum are the three conditions essential for their germination. Finding these, the *Rhizopus* spore germinates immediately. It absorbs moisture. The rigid spore wall softens and the protoplast within swells rupturing the spore wall. The spore protoplast emerges through the split in the form of one or more short protuberances called the **germ tubes** (C). There is a suggestion for enzymatic breakdown of the single-layered spore wall at the point of emergence of germ tube rather than a mechanical rupture. The spore wall is apparently weakened by enzyme action. Prior to emergence of the germ tube, a second wall layer known as the **vegetative** or **germination** layer appears only in the germinating spores inside the spore wall. It surrounds the emerging germ tube (Hawker and Abbot, 1963; Buckley, Soneuer and Matsumato, 1968). There is a mass flow of protoplasm towards the germ tube apex in early stages of extension. During spore germination, the nuclei divide, vacuoles increase in size and mitochondria increase in size and number.

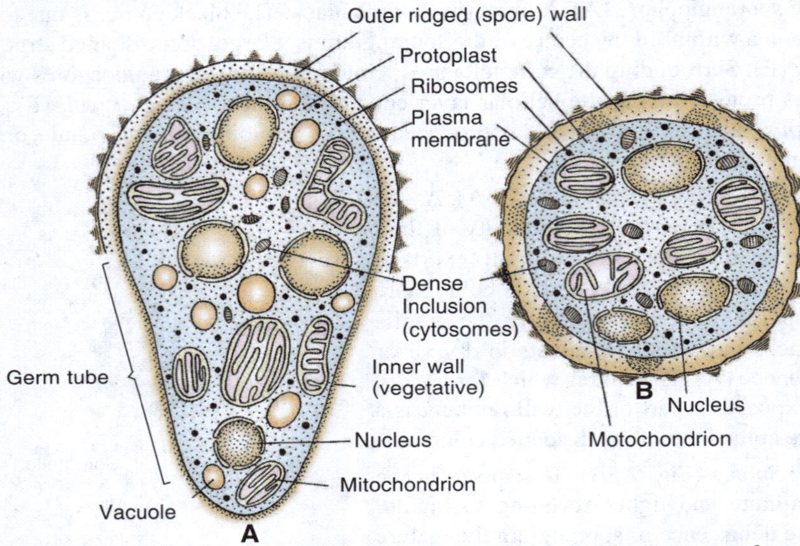

Fig. 7.6 (A-B). *Rhizopus stolonifer.* Diagrammatic representation of fine structure of spore prior to germination (A) and after germination (B). (After Hawker and Abbot)

Development of Mycelium. The germ tube elongates by apical growth with amazing speed and branches extensively to form the internal feeding mycelium. In the young mycelium all the hyphae are alike, multinucleate and asepate. After a few days' growth within the substrate the internal mycelium produces some hyphae which come to the surface of the substrate to form the aerial mycelium which is still connected with the internal feeding mycelium (Fig. 7.1).

Propagation of the vegetative (somatic) phase by vegetative methods of fragmentation and sporulation is quite common. It may continue for several generations under favourable conditions.

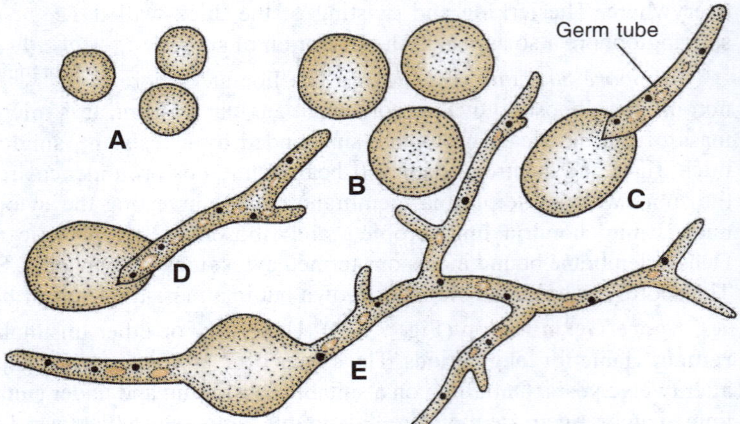

Fig. 7.7 (A-E) *Rhizopus stolonifer.* Various stages in germination of spore and development of mycelium. See explanation in the text.

2. Sexual Reproduction (Fig. 7.8). *Rhizopus* displays a sexual process which is of conjugation type. It consists in the fusion of entire cell contents of the two gametangia. In many respects, it is similar to that of *Spirogyra.* As in *Spirogyra* the gametes are non-motile, isogamous and are produced singly but differ in being multinucleate. Sexual reproduction generally takes place towards the end of the growing season. Some species of *Rhizopus* are **homothallic** (*R. sexualis*) and others are **heterothallic**. To the latter category belong *R. stolonifer.*

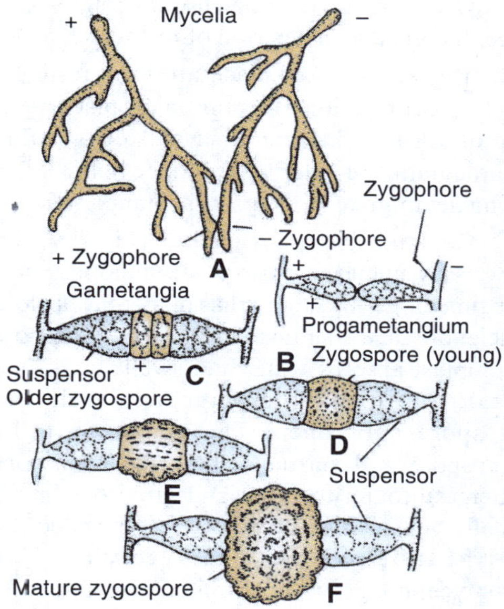

Fig. 7.8 (A-F) *Rhizopus stolonifer.* Various stages in sexual reproduction (conjugation). Explanation in the text. A-D stages in the development of gametangia and plasmogamy; E-F, Zygospore.

In the heterothallic species, the mycelia are of two distinct types. They are similar outwardly but function differently in reproduction. They belong to two different mating types. One of these is labelled as plus strain and the other minus strain. Sexual reproduction takes place only when a plus strain hypha and a minus strain hypha come in contact. Sexual reproduction in *R. stolonifer* is thus possible only when mycelia of plus strain and minus strain grow in the same culture. This, of course, is not of common occur-rence and thus zygospore formation in the heterothallic species is rare.

Sexual reproduction takes place by **gametangial copulation.** The fusing gametangia are multinucleate and similar in every respect but are distinct genetically. The sexual process starts with the formation of two special hyphae termed **zygophores.** The zygophores arise from near the apices of the actively growing somatic hyphae of two neighbouring mycelia of + and – strains. They are chemically induced. According to Burgeff (1924), the process is initiated by the secretion of a diffusible sexual hormone. He showed that zygophore development in each strain begins only when the opposite strain is close enough for the stimulus to reach the receptive hyphae. On the secretion of another diffusible metabolite, the zygophores of each strain bend toward those of the opposite strain preparatory to contact. The zygophores meet (A) and adhere in pairs at their tips to form the fusion septum (B). Thereafter the tips of paired zygophores begin to swell due to the flow of cytoplasm and nuclei into them. The swollen tips are now called the **progametangia.** A transverse septum known as the *gametangial septum* then appears in each progametangium near its tip (C). It cuts off a small terminal cell, the **gametangium** from the proximal part which is now termed the **suspensor** (D). The densely granular, multinucleate protoplast of the gametangium constitutes the **aplanogamete.** The suspensor contains vacuolate cytoplasm with fewer nuclei. In some other species the two gametangia in the pair are markedly unequal in size and suggest **anisogamy.** When the gametangia are mature, the intervening fusion septum between them dissolves. The adjoining aplanogametes intermingle in the fusion cell (D) which is technically known as the **prozygosporangium.** The majority of the nuclei in the combined protoplast (E) fuse in pairs. The two nuclei in the fusing pair belong to opposite strains. The nuclei that do not fuse and are left over, later degenerate. The combined protoplast in the prozygosporangium containing a number of diploid nuclei is called the **zygospore** (E). The prozygosporangium containing the young zygospore increases in size and secretes a thick multilayered wall around it to become the **zygosporangium.** The outermost layer is thick and warty. The single zygospore develops within the zygosporangium.

When mature the suspensors wither and the zygosporangium containing the zygospore is set free. It enters upon a period of rest which may extend over several (5-9) months.

Meiosis. It takes place after the resting period prior to germination of zygospore. During meiosis segregation of strains takes place. According to Cutter (1942), half of the resultant nuclei are of + strain and the other half of – strain. If both the strains of nuclei migrate to the germ sporangium, the latter will produce both kinds of germ spores + as well as –. In case only one kind of nuclei migrate to the germ sporangium it will produce either + germ spores or – germ spores.

Germination of Zygospore (Fig. 7.9). The zygospore germinates under suitable conditions. The early nuclear divisions are meiotic. So at the time of germination it has numerous haploid nuclei of + and – strains. The haploid zygospore protoplast absorbs water and swells (A). As a result the thick zygosporangium wall and exospore split open. The contents of the zygospore still surrounded by the endospore emerge through the split and grow out into a small, stout vertical hypha (B). It is called the **germ tube** or the **promycelium.** The promycelium grows to a limited extent. It soon bears a typical, terminal sporangium (B). It is called the **germ sporangium.** In many cases the germ tube rows horizontally for a short distance over the substratum to form a sparsely branched or unbranched vegetative hypha. From the latter arises a vertically growing branch bearing the **germ sporangium** (C). The latter has the structure of a typical sporangium. The protoplast of the germ sporangium undergoes cleavage to produce non-motile meiospores which are often called the **germ spores.** According to Gauger (1961), the germ spores in *Rhizopus stolonifer* are either all + or – type or of both types + and –. In the latter case half of them contain + nuclei and half – nuclei. How it happens has been explained by Cutter (1942). After meiosis half of the nuclei are of + strain and the other half of – strain. The germ sporangium containing both types of spores receives both + and – nuclei. The promycelium dies after the liberation of the germ spores. In some other Mucorales (*Mucor mucedo*), all the spores in the germ sporangium are of one kind + or –. In still others (*Phycomyces nitens*) each zygosporangium contains three kinds of spores +, –, ± (diploid).

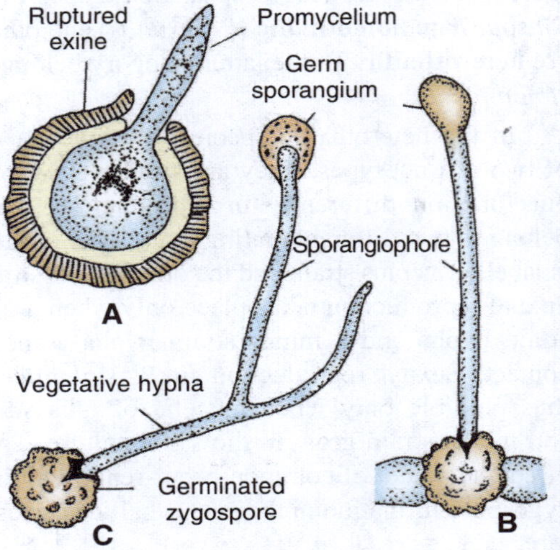

Fig. 7.9 (A-C). *Rhizopus stolonifer.* Germination of Zygospore.

Sometimes the gametangia fail to fuse. Each such gametangium functions as a **parthenospore.** It develops into a thick-walled zygospore-like structure called the **azygospore.**

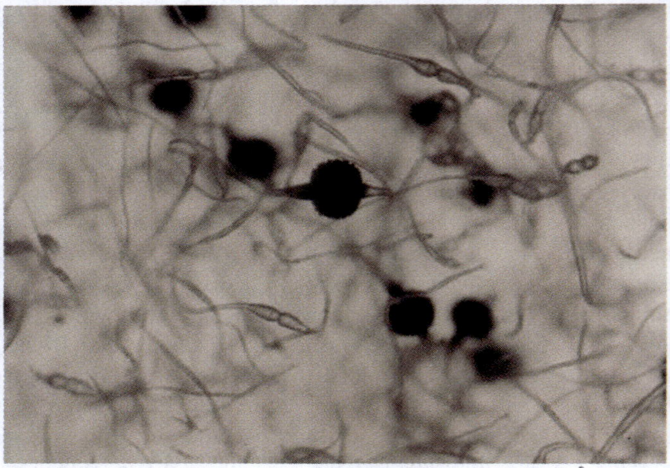

Mucor mucedo

Cytology of the Germ Sporangium. There are two views with regard to the occurrence of meiosis in the life cycle of *Mucorales*. According to some authorities, **meiosis** occurs prior to the germination of the zygospore. Each diploid zygospore nucleus undergoes meiosis. As a result four haploid daughter nuclei are formed from each diploid nucleus. This is a condition similar to that of *Spirogyra.* Two of these haploid nuclei are of plus and two of minus strain. During this division the plus and minus strains are segregated. According to this view, the zygospore is the only diploid structure in the life cycle. The *promycelium, germ or zygosporangium, germ spores, mycelium, gametangia, sporangiophores, sporangia* and the *sporangiospores* all belong to the **haplophase** or **gametophyte generation** (Fig. 7.10). The consensus of opinion favours this view.

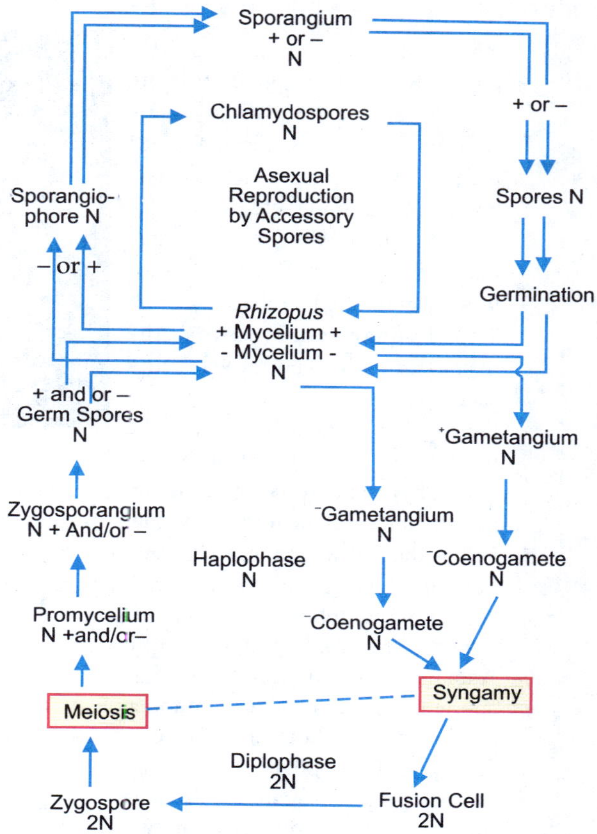

Fig. 7.10. Graphic representation of life cycle of *Rhizopus*.

The other hypothesis is that reduction takes place at the time of the differentiation of spores in the germ sporangium. Taking this hypothesis to be correct the *zygospore, promycelium, germ sporangium* up to the *spore mother cell stage* constitute the **diplophase** or the **sporophyte phase.** The spores of the germ sporangium are the first haploid structures. They together with the mycelium, gametangia, sporangiophores, sporangia and sporangiospores constitute the **haplophase** or **gametophyte generation.** These two phases alternate in the life cycle. This view has been practically discarded by the modern mycologists.

Life cycle (Fig. 7.10). The life history of *Rhizopus* consists of two phases, **asexual** and **sexual.** The structures comprising the asexual phase are the mycelium, sporangiophores, sporangia and the sporangiospores. This phase in the life cycle of *Rhizopus* serves to propagate the haplophase of the fungus during the favourable season.

The **sexual phase** comprises the plus and minus mycelia, plus and minus gametangia and their protoplasts (aplanogametes), zygospore, promycelium, germ sporangium and the germ spores. This cycle is so designed as to enable the fungus to pass uninjured the period which is adverse for vegetative growth. In the sexual phase zygospore is the only diploid structure. All others are haploid. Such a sexual cycle is termed **haploid** or **haplontic.** It is characterised by **zygotic meiosis** and haploid mycelium as the only adult plant.

Salient Features of Rhizopus

Rhizopus – Bread Mold

1. *Rhizopus* is a saprophytic phycomycete. It is found in damp, warm weather specially in the rainy season on dead organic matter of plant or animal origin such as stale bread, animal dung, etc.
2. The mycelium during the vegetative phase is a white fluffy mass of loosely entangled hyphae.
3. The hyphae constituting the mycelium are coencytic and aseptate.
4. During the vegetative phase the mycelium consists of only the stolon hyphae and the rhizoidal hyphae. The former help in the spread of the mycelium and the latter in anchorage and the absorption of nutrients from the substratum.
5. When the reproductive phase sets in, the mycelium turns moldy and develops a third kind of hyphae called the sporangiophores.
6. The sporangiophores arise in tufts in the air from the stolons opposite the rhizoidal hyphae. They are negatively geotropic and unbranched.
7. Asexual reproduction takes place by the formation of non-motile, multinucleate sporangiospores produced endogenously inside the sporangia. The sporangia arise singly and terminally at the tips of sporangiophores.

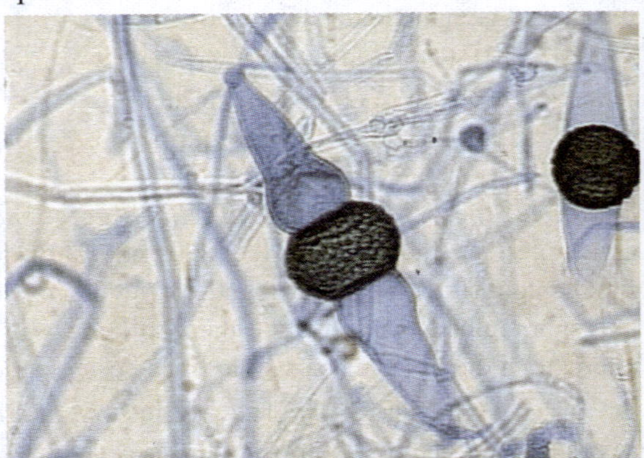
A Mature Zygospore of Rhizopus

8. *Rhizopus* is heterothallic and sexual reproduction is isogamous. It takes place by copulation between two multinucleate gametangia of opposite strains (+ and –) developed by inflation of the tips of special hyphae called **zygosphores.** The gametangia come in contact and fuse.
9. The double wall or fusion septum between the two contacting gametangia dissolves. Their protoplasts mingle through the pore. The + and – nuclei fuse in pairs in the fusion cell called prozygosporangium.
10. The prozygosporangium secretes a heavy wall around it to become a zygosporangium in which a single zygospore develops. The zygospore has many diploid nuclei. The zygospore wall is differentiated into two layers. The outer is called **exospore** and the inner **endospore.**
11. After the resting period but prior to germination meiosis takes place. It is accompained by segregation of strains. Half the resultant nuclei are of + strain and the other half of – strain.
12. On germination the thick zygosporangium wall and exospore crack open. The contents of the zygospore still surrounded by the endospore emerge in the form of a small upright hypha, the **promycelium.**

13. The promycelium bears the germ sporangium at its top.
14. The germ sporangium produces either + or – strain germ spores or both kinds + as well as –
15. Each germ spore on falling on a suitable substratum germinates to form the haplomycelium.

Taxonomic position. The older mycologists like Gwynne Vaughan and Barnes divided the Phycomycetes which included the lower fungi into two sub-classes, (*a*) Oomycetes with **heterogamous** sexual reproduction and (*b*) Zygomycetes with **isogamous** type of sexual reproduction. Later the mycologists who accepted Saprrow's (1943) suggestion of **flagellation** as the important criterion of classification placed the lower fungi with motile cells in their life cycle in the series Uniflagellatae and Biflagellatae. For the sake of uniformity, the Zygomycetes, which lacked motile cells were placed in the series Aflagellatae or Aplanatae. Alexopoulos (1962) took a more logical step and split the Phycomycetes directly into six classes. He elevated the sub-class Zygomycetes to the rank of a class which included three orders, namely, Mucorales, Entomophthorales and Zoopagales. According to Alexopoulos (1962) the taxonomic position of *Rhizopus* would be as under :

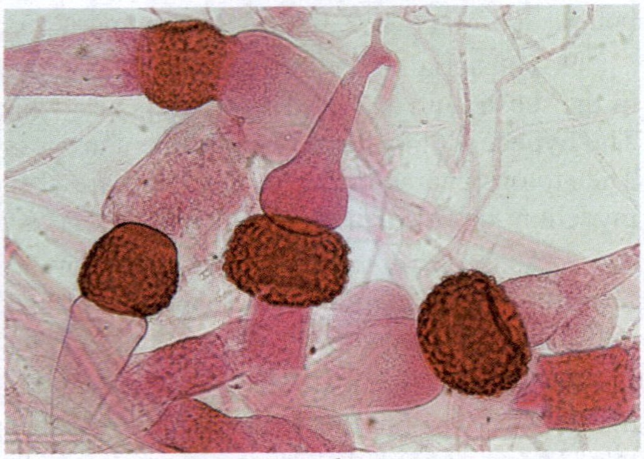

Zygophores

Division	: **Mycota**
Sub-division	: **Eumycotina**
Class	: **Zygomycetes**
Order	: **Mucorales**
Family	: **Mucoraceae**
Genus	: *Rhizopus*
Species	: *stolonifer*

Alexopoulos and Mims (1979) place *Rhizopus* in the Superkindgom Eurkaryonta, Kingdom Myceteae, Division Amastigomycota, Subdivision Zygomycotina, Class Zygmycetes, Order Mucorales and Family Mucoraceae.

Super kingdom	: **Eukaryonta**
Kingdom	: **Mycetae**
Division	: **Amastigomycota**
Sub-division	: **Zygomycotina**
Class	: **Zygomycetes**
Order	: **Mucorales**
Family	: **Mucoraceae**
Genus	: *Rhizopus*
Species	: *stolonifer*

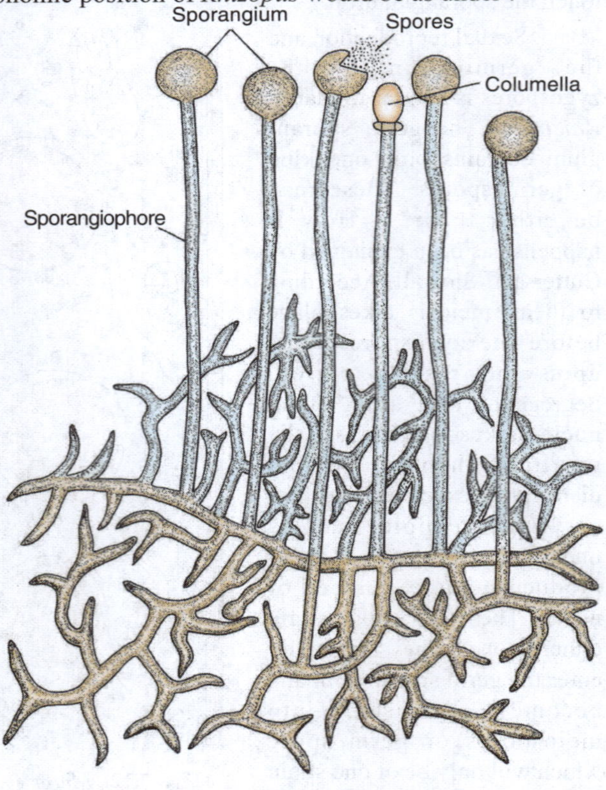

Fig. 7.11. *Mucor sp.* Eucarpic mycelium bearing sporangiophores and sporangia.

MUCOR MUCEDO

In structure, life history and mode of life it closely resembles *Rhizopus*. The more important differences are :—

1. *Mucor mucedo* (Fig. 7.11) appears more readily on damp horse dung when kept under a bell jar for a few days at moderate temperature.

2. The aerial mycelium consists of copiously branched hyphae which ramify over the surface of the substratum and send down branches, which penetrate the substratum and absorb nourishment. The hyphae which absorb nourishment are less specialised than the rhizoidal hyphae of *Rhizopus*. The stolons characteristic of *Rhizopus* are absent in *Mucor*. There is thus no differentiation of the mycelium in *Mucor* into stolons and holdfasts.

3. As in *Rhizopus*, the somatic phase multiplies vegetatively by fragmentation, chlamydospore formtion and formation of oidiospores. Asexually it reproduces by sporulation as in *Rhizopus* but the sporangiophores arise singly and not in clusters of 2 or more.

4. The development of sporangia, differentiation of spores and their germination is similar to that of *Rhizopus* but the dehiscence of mature sporangia and the dispersal of spores tell a different tale. The sporangial wall does not crack into fragments. It turns liquidy and dissolves in the liquid absorbed through the columella. Only a small collar-like narrow rim of the wall persists round the columella at the base as a **collarete.** The spore mass embedded in the liquid constitutes the bright glistening sporangial drop. It is supported by the short collarete. The sporangial drop remains fluid in the humid conditions. The spores are dispersed when any insects such as mites happen to touch the sporangial drop.

5. Sexual reproduction and the germination of the zygospores is similar to that of *Rhizopus*. The germ sporangium contains only one kind of germ spores. These may be either + or –. How it happens has been explained by Cutter and Sjowall. According to them, meiosis takes place before the zygospore enters upon the resting period. Segregation of + and – strain nuclei takes places. All the resultant haploid nuclei disintegrate excepting one. The surviving haploid nucleus undergoes repeated mitosis to produce a large number of nuclei. These are of one strain either + or –. They eventually enter the germ sporangium and become organised into meiospores or germ spore which will only be of one strain + or –.

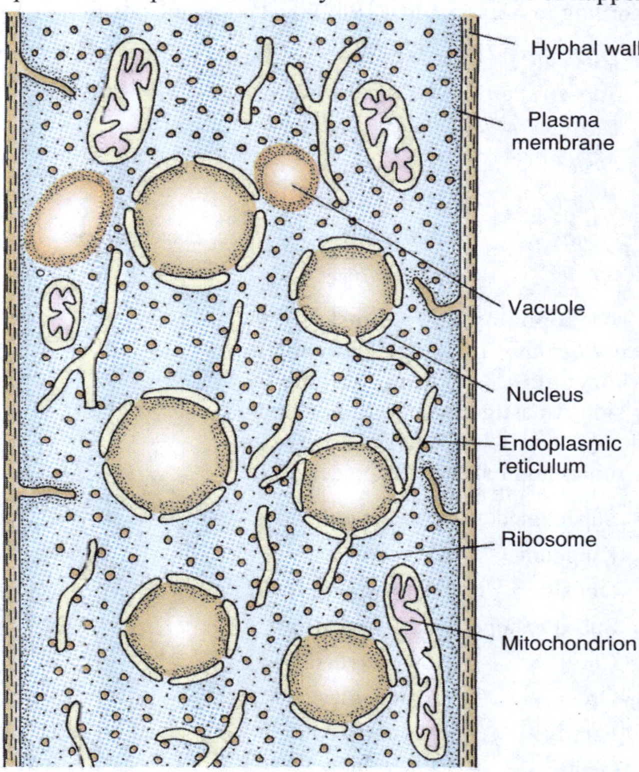

Fig. 7.12. *Mucor sp.* Fine structure of a hypha near the tip based on an electron-micrograph (After Dodge).

Diagrammatic life cycle of *Mucor* has been shown in Fig. 7.13.

Taxonomic position

According to Alexopoulos (1962) :

Division	:	**Mycota**
Sub-division	:	**Eumycotina**
Class	:	**Zygomycetes**
Order	:	**Mucorales**
Family	:	**Mucoraceae**
Genus	:	*Mucor*

According to Alexopoulos and Mims (1979) :

Super kingdom	:	**Eukaryonta**
Kingdom	:	**Mycetae**
Division	:	**Amastigomycota**
Sub-division	:	**Zygomycotina**
Class	:	**Zygomycetes**
Order	:	**Mucorales**
Family	:	**Mucoraceae**
Genus	:	*Mucor*

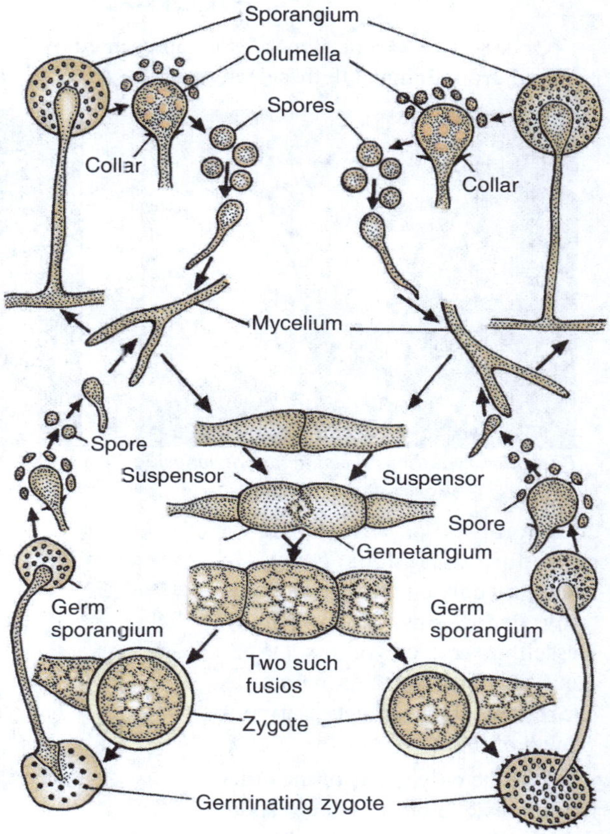

Fig. 7.13. *Mucor* sp. Diagrammatic life-cycle

Differences between *Rhizopus* and *Mucor*

1. In *Rhizopus*, the mycelium is differentiated into stolons, rhizoids and sporangiophores while in *Mucor* the mycelium consists of only one kind of hyphae.

2. Sporangiophores in *Rhizopus* arise from the nodes i.e. junction of rhizoids and stolons while in *Mucor* sporangiophores may arise from any point in the mycelium.

3. Spores in *Rhizopus* are easily blown away while spores in *Mucor* are not easily disseminated and remain adhered to columella from where mites disseminate the spores.

4. Columella in *Rhizopus* becomes dome or umbrella shaped after sporangial dehiscence but in *Mucor* the columella does not undergo change in shape after dehiscence of sporangium.

5. In *Rhizopus,* meiosis occurs during the germination of Zygospores while in *Mucor* meiosis probably occurs before the zygospore undergoes a period of rest.

Pilobolus mycelium with sporangia

Pilularia hyphae

FAMILY : PILOBOLACEAE

The family is represented by three genera : *Pilobolus, Pilaira* and *Ulharomyces.* All the three genera are comprophilous and heterothallic. The phototropic sporangiophores bear dark sporangia with flat columella which is not very conspicuous.

Genus *Pilobolus* is described in detail as a representative of the family.

PILOBOLUS KLEINII

Occurrence. Genus *Pilobolus* is represented by 25 species out of which about 10 species are reported from India. Of these ten species, *P.kleinii, P. longipes* and *P. crystallinus* are most common. *Pilobolus*—a coprophilous fungus grows in laboratory only on media containing dung decoctions. According to Hesseltine and coworkers (1952) dung contains a factor named as **coprogen** which is necessary for growth of the fungus.

Pilobolus crystallinus showing sporangiophores, vesicles and sporangia

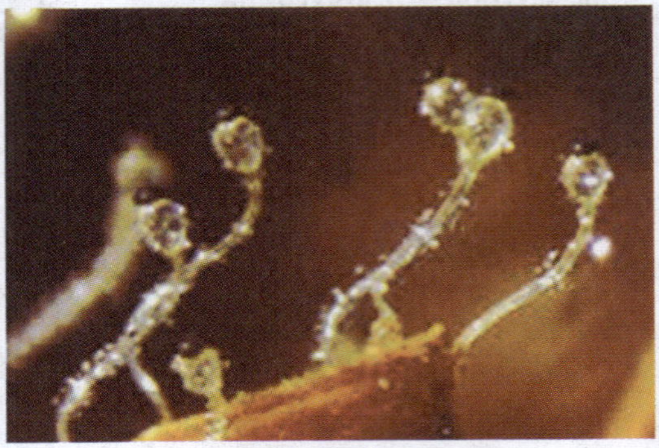

Pilobolus kleinii

It is the only genus of the order Mucorales exhibiting viloent spore discharge.

Somatic phase. The somatic phase of the fungus is a eucarpic thallus which consists of mycelium. The hyphae are coenocytic, aseptate, hyaline branched and remain submerged in the substratum, feeding on dung.

Reproductive phase. The fungus reproduces both by asexual and sexual methods.

1. Asexual reproduction. It occurs mostly by aplanospores produced within the sporangium. Some hyphae which are to produce sporangia become upright and come out of the substratum. Erect sporangiophores develop from a **trophocyst** a caroftene rich

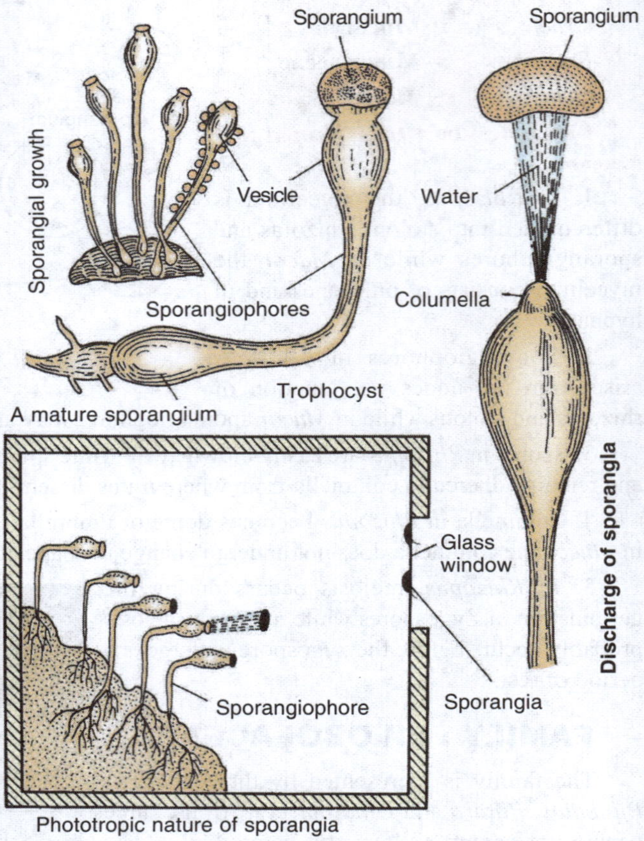

Fig. 7.14. *Pilobolus.* Asexual reproduction

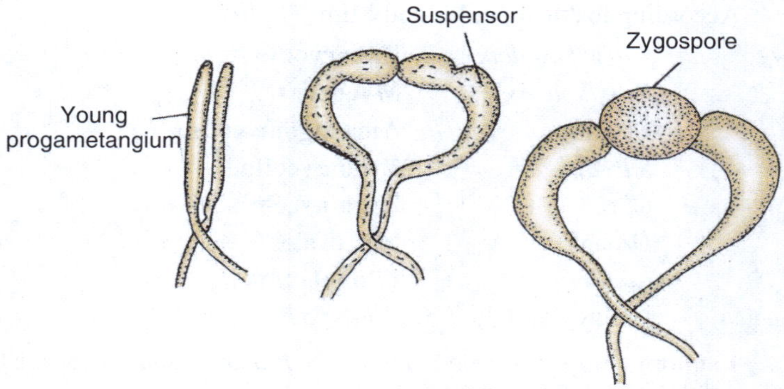

Fig. 7.15. *Pilobolous*. Sexual reproduction

swelling embedded in the substratum. The trophocyst is ovate, globose napiform or cylindrical in shape and is delimited from the mycelium by one or two septa. The trophocyst may be terminal or intercalary in position. The sporangiophore is 0.5 to 1.0 cm long and bears apically a black button shaped columellate sporangium subtended by a much larger cystal clear sub-sporangial vesicle. The sporangiophores are phototropic. Each sporangiophore at maturity consists of a trophocyst, embedded in the substratum, a long straight stalk which is enlarged at its tip to form a sub-sporangial vesicle capped by a small black columellate sporangium containing 15-30 thousand spores.

Dispersal of spores in **Pilobolus** is an interesting phenomenon in which the vesicle plays an active role. The large vacuole in the vesicle has cell sap which has an atmospheric pressure as high as 7 atmospheres. Drops of excreted liquid commonly adhere to the sporangiophore. The vesicle below the sporangium explodes transversely at a line of weakness just beneath the columella. This is accompanied by sudden contraction of the walls of the sporangiophore, the vesicle and trophocyst resulting in the squirting out of the cell sap from the open mouth of the vesicle. The squirting out of the cell sap is in the form of a jet that propels the sporangium alongwith columella to a distance of 2.5 metres. When the sporangium strikes an object it gets attached with the help of a mucilaginous ring and is surrounded by the cell sap. After some time the cell sap dries and the sporangium gets glued to the object say grass leaves. The sporangia reach the alimentary canal of the herbivore animal alongwith the grass leaves and only there the spores are released. The spores get proper nutrition and stimulation to germinate in the alimentary canal which later takes place in the freshly deposited horse dung.

Pilobolus is aptly known as Hat-Thrower or "shot gun fungus" because of the peculiar mechanism of discharge of sporangium.

2. Sexual Reproduction. Sexual reproduction rarely occurs in *Pilobolus* and takes place in adverse conditions. The genus has both homothallic and heterothallic species. Fusion of two multinucleate similar gametangia of opposite strains leads to the formation of thick walled Zygospores. The suspensors are twisted around each other and are then separated like two jaws of a pair of tongs. The zygospore is held by their tips. The zygospore undergoes a period of dormancy and rest after which it germinates and forms a sporangiophore and sporangium. Meiosis takes place during germination and the spores are haploid. Howevere, the zygospore can also germinate by a germ tube directly.

Taxonomic position

According to Alexopoulos (1962)

Division	:	**Mycota**
Sub-division	:	**Eumycotina**
Class	:	**Zygomycetes**
Order	:	**Mucorales**
Family	:	**Pilobolaceae**
Genus	:	*Pilobolus*

According to Alexopoulos and Mims (1979)

Super kingdom	:	**Eukaryonta**
Kingdom	:	**Mycetae**
Division	:	**Amastigomycotina**
Sub-division	:	**Zygomycotina**
Class	:	**Zygomycetes**
Order	:	**Mucorales**
Family	:	**Pilobolaceae**
Genus	:	*Pilobolus*

Economic importance of Mucorales. *Rhizopus* and *Mucor* are the most frequent contaminants of our food stuffs. Some Mucorales form the basis of our industry. *Rhizopus stolonifer* is used in the preparation of fumaric acid and *R. oryzae* in the production of alcohol. Some species of *Rhizopus* yield lactic acid. Some important chemicals such as oxalic acid and citric acids are manufactured by utilising some other Mucorales. Some of the parasitic Mucorales though few in numbers are weak parasites. They cause diseases of fruits and other plant parts in storage and transit. *R. stolonifer* causes leak diseases of strawberries and soft rot of sweet potatoes. Harter and Waimer (1923) reported that *R. stolonifer* produces pectinolytic enzymes in rotting of sweet potatoes. Spalding (1963) showed that *Rhizopus,* while in synthetic media and during decay of sweet potatoes, produces cellulytic enzymes. *Choanephora trispora* and *C. cucurbitarum* are

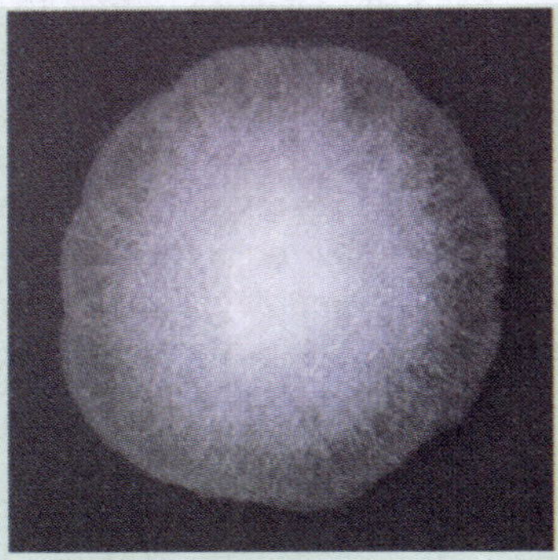

Culture of Absidia corymbifera

Choanephora cucurbitarum on host

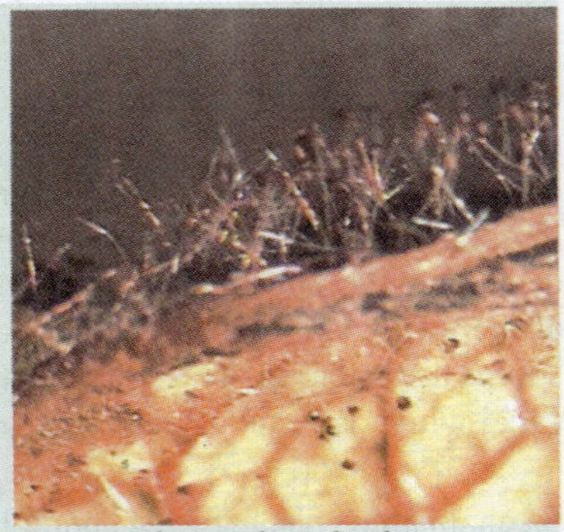

Choanephora hyphae

parasitic on decaying fruits of cucurbits. Some members of the order are parasitic on animals and human beings. Species of *Mucor, Rhizopus* and *Absidia corymbifera* attack the nervous system of man and may prove fatal. A few members of family Piptocephalidaceae are obligate parasites and serve as ideal tools for experimental studies on the physiology of obligate parasitism.

Origin and affinities of Zygomycetes. The origin and affinities of Zygomycetes are obscure. The fact that aplanospores play an important role in the life cycle adds to the mystery. In sexual reproduction by conjugation the Zygomycetes approach the Conjugales among the green algae. The resemblance ends here. The differences between the two groups are very wide.

The well developed, branched coenocytic mycelium in both the Zygomycetes and Oomycetes suggests a relationship between the two groups. In fact a simple soil inhabiting members of Saprolegniales is suggested as an ancestral form of Mucorales. The Entomophthorales are considered descendants of the Mucorales.

The nature of the vegetative mycelium, the gradual increase in its septate condition, presence of perforate septa in the septate members, frequent presence of chlamydospores, gradual transition of the sporangium into a true conidium—a typical spore of the Ascomycetes, tendency of isogamous sexual reproduction to become heterogamous, greater protection of the zygospore and tendency to form a minute fructification—all point to a distinct relationship between the Zygomycetes and Ascomycetes. There is a suggestion that the two have evolved from a common ancestor. This viewpoint has been further strengthened by striking similarities in the fine structure of these two groups of electron microscopy.

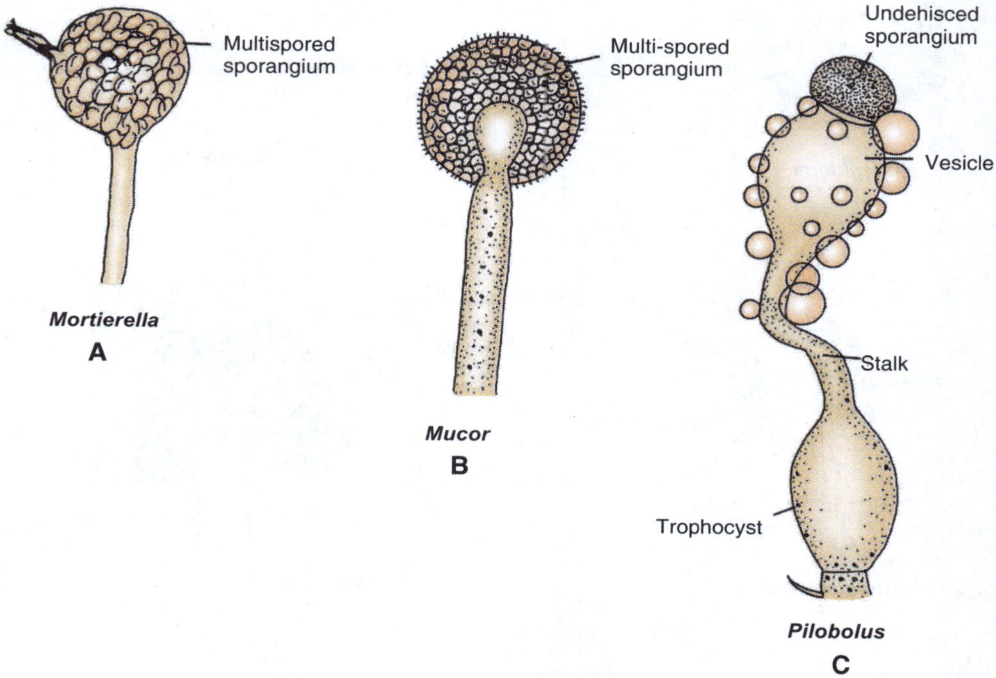

Fig. 7.16 (A-C). *Evolution of conidium from sporangium*

The other viewpoint is that Mucorales are related to the chytrids and both may have evolved from a common ancestor at the flagellate level. This view is based on certain resemblances between the two. The Mucorales resemble the chytrids in their use of chitin and not cellulose as a major cell wall molecule and their synthesis of lysine via the same chemical pathway.

Evolutionary tendencies of the Zygomycetes. Within the Zygomycetes following evolutionary trends are noticeable :—

1. There is a tendency of the somatic phase (mycelium) towards gradual and increased septation in more advanced forms. In the primitive Zygomycetes like *Mucor* and *Rhizopus,* the mycelium is noramally coenocytic and aseptate. The septa appear only at the base of reproductive organs to separate them or seal off damaged parts of the mycelium and also when it ages. These septa are solid plates. In the more advanced Zygomycetes, the mycelium is septate from the very beginning (Entomophthorales). The septa, as in Ascomycetes, are perforate.

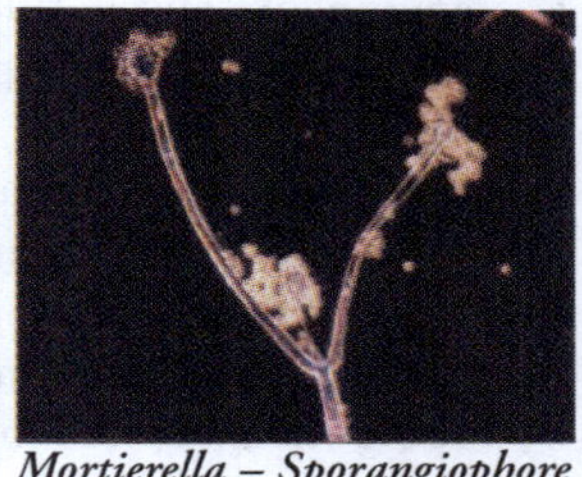

Mortierella – Sporangiophore with sporangia sporangia

2. The gradual transition of sporangium into a conidium is another evolutionary trend traceable in several cases. We shall discuss it briefly here (Fig. 7.14).

The first step in this direction is the progressive reduction in the size of sporangium and reduction in the number of spores produced in it. In *Mortierella,* the sporangium is the simplest and represents the

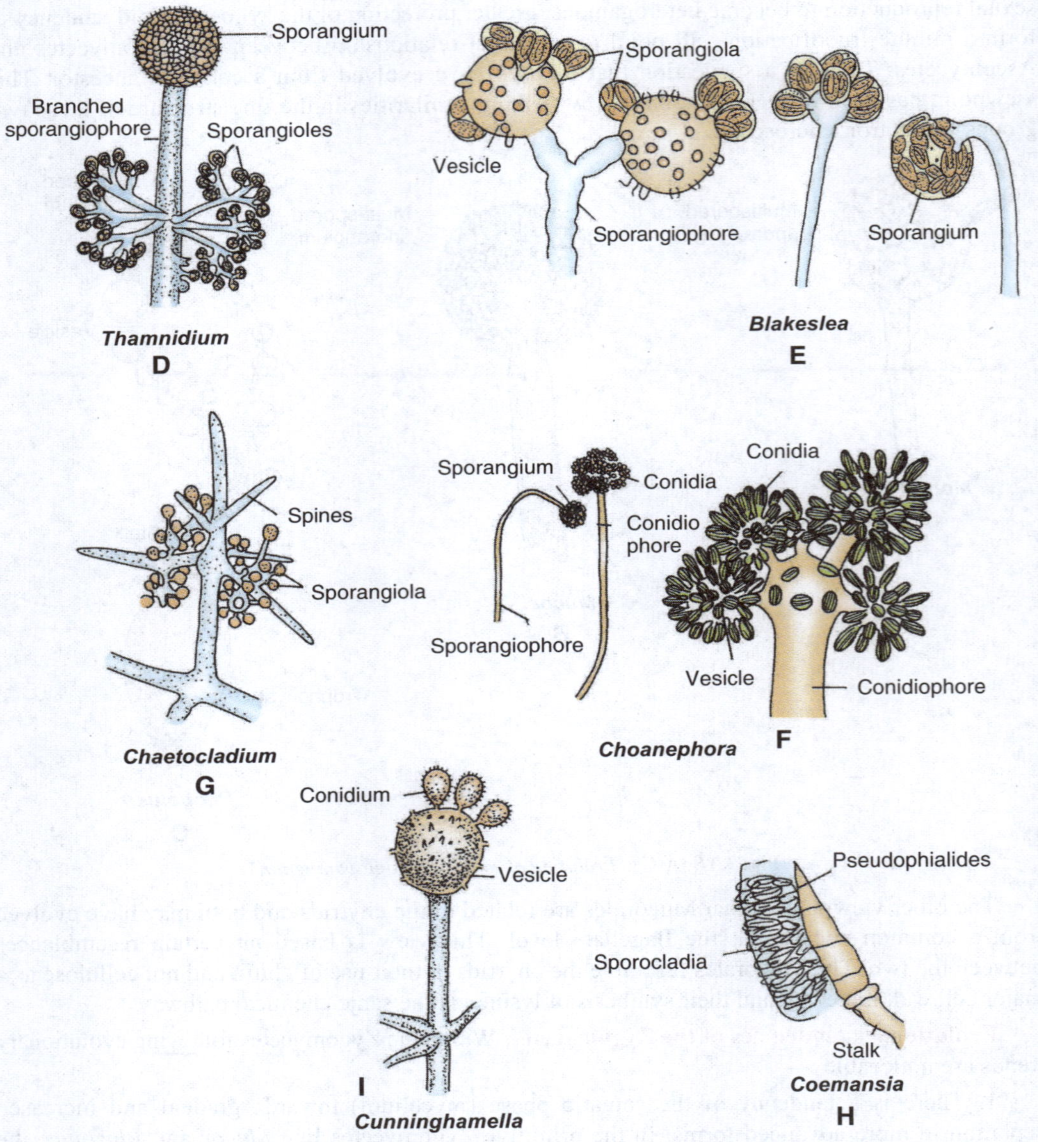

Fig. 7.16 (D-I). Evolution of conidium from sporangium.

most primitive type of sporangium. The sporangium is a spherical enlargement of the apex of the sporangiophore and is filled with spores, all of which are fertile. *Mucor* presents an example of sporangium which develops a columella because of which the number of spores in the sporangium is reduced. *Pilobolus* has a special type of sporangium which is a modification of *Mucor type*. The columella in the sporangium though present is very small and the sporangial wall is thickened. Just below the sporangium, the sporangiophore is enlarged into a vesicle that may be 2-3 times the diameter of flattened sporangium. The tip of the sporangiophore is sensitive to light as a result of which the sporangium is directed towards the source of light. The vesicle enlarges after absorbing moisture. As the vesicle enlarges, the turgor becomes so great that the apex ruptures and the sporangium is blown off alongwith the watery contents of the vesicle, sometimes to a distance of over a meter.

Some species of *Thamnidium* produce tall, erect branched sporangiophores, each branch ending in a large terminal many spored sporangium of *Mucor* type. Some of the basal whorls of lateral branches, however, bear small sporangioles, each with small number of spores. Spores of both the sporangia and sporangioles are similar in form and function.

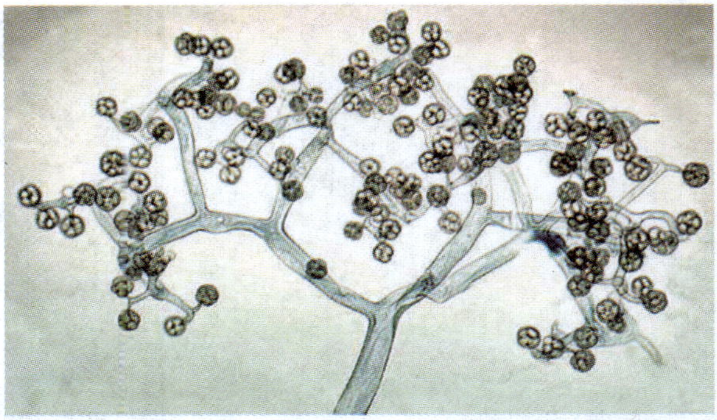

Thamnidium – conidiophores with bunch of conidia

In *Blakeslea,* the sporangial content is dependent upon the amount of nourishment the fungus receives. When the fungus is well nourished, the sporangium is a large columellate structure but when the fungus is poorly nourished, the sporangium is smaller and a few spored without columella. In addition, many two to four spored sporangioles are produced on sterigmata from the surface of large rounded heads clustered at the apex of sporangiophores. These sporangiola are different from the sporangial spores.

Choanephora produces two types of asexual reproductive bodies. While the normal multispored, columellate sporangia are produced at the apex of sporangiophores, the sporangiolis are borne on round heads at the apex of sporangioles. *Chaetocladium* produces deciduous monosporous sporangioles. The sporangial and spore walls are easily distinguishable.

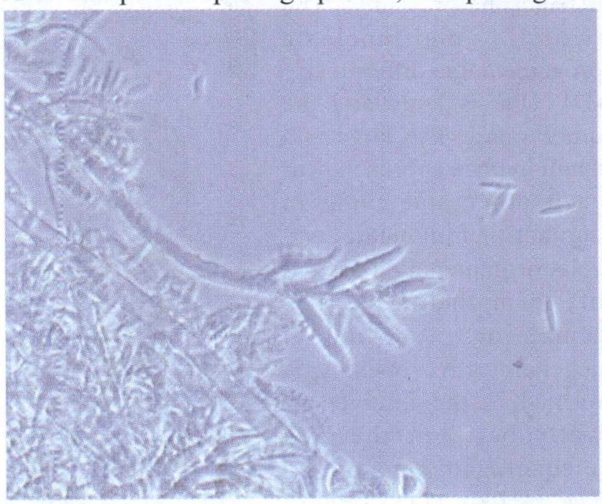

Coemansia – mycelium with one celled structures

In *Coemansia*, the sporangia are reduced to indehiscent one celled structures which are borne on sterigma like structures arranged on one side of lateral several celled sporocladia.

Cunninghamella bears true conidia in which the sporangial and spore walls are inseparably fused. The conidia are produced on globose heads.

It would be clear from the above account that evolution of conidium from sporangium is achieved by the progressive reduction in size of the sporangium to the extent that it becomes equivalent to a conidium. There is no inhibition in the formation of spores. This transition from sporangia to conidia in the Mucorales parallels the *Pythium—Peronospora* series of the Peronosporales (Oomycetes). The difference, however, is that in the latter there is neither progressive reduction in the size of the sporangium nor reduction in the number of spores formed within it. The conidium in this series is evolved abruptly by inhibition in the formation of zoospores in response to unfavourable conditions.

Certainly it is difficult to distinguish monosporous sporangiola from the true conidia. However, recently Khan and Talbot (1975), demonstrated with the help of transmission electron microscopy, that in *Cunninghamella* the spore and sporangial walls although closely applied to one another but can be differentiated from each other. According to them, true conidia do not occur in the Mucorales.

3. Another trend in the Zygomycetes is a tendency of isogamous reproduction to become heterogamous (Fig. 7.17 A-G). In *Mucor* and *Rhizopus* the gametangia are identical (A).

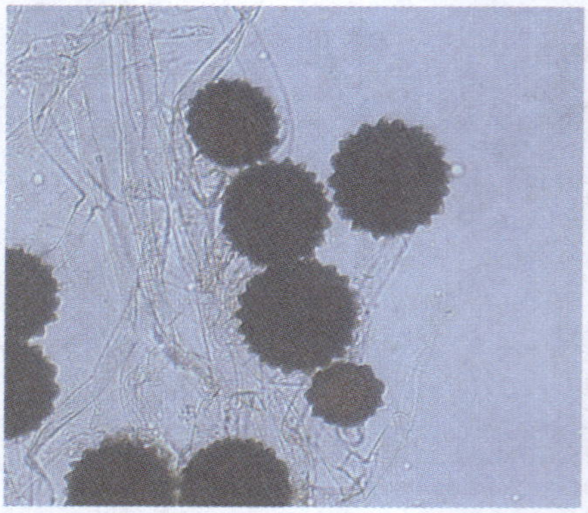

Zygorhynchus – zygospores

Endogone lactiflua

Zygorhynchus which is another genus of Mucorales produces gametangia which are unequal in size (B-C). *Endogone lactiflua* exhibits distinct heterogamy. The two multinucleate progametangia differ in size (D). One is large and the other small. The large and small hyphae approach near each other and come in contact. Uninucleate end cells or gametangia are cut off from their tips. The remaining nuclei degenerate. An opening is formed in the intervening walls between the two unequal gametangia (E). The nucleus of the smaller gametangium (male) migrates through the opening into the larger female gametangium. Both the male and female nuclei then migrate into an outgrowth of the female gametangium where they fuse.

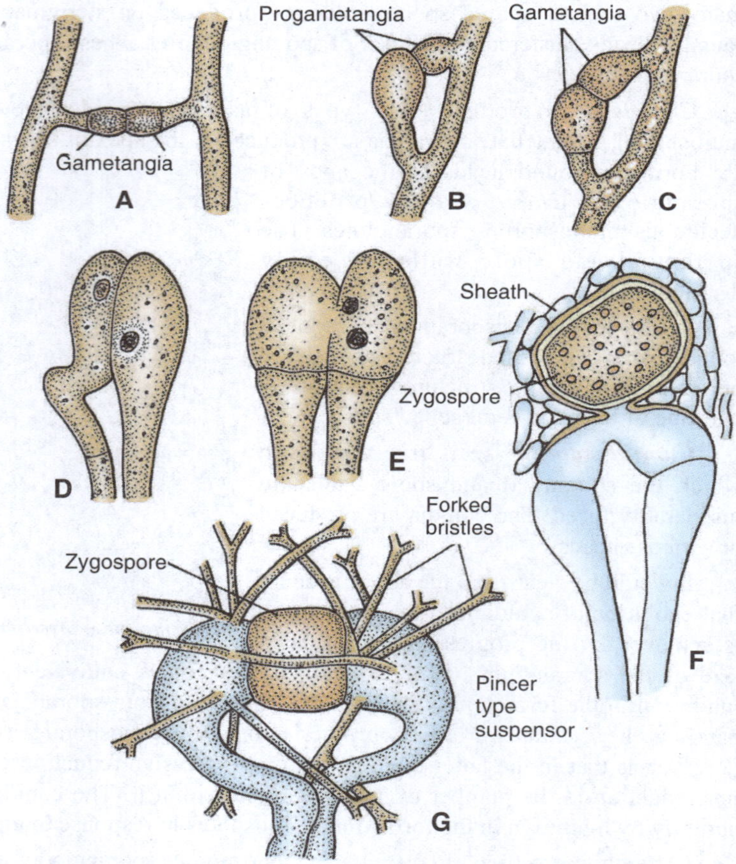

Fig. 7.17 (A-G). *Mucorales* showing stages in transition of isogamous to heterogamous sexual reproduction and increased protection of zygospore. Explanation in the text.

4. Another striking evolutionary tendency in the Zygomycetes is towards greater protection of zygospore and tendency in the advanced forms to form a minute fruit body (Fig. 7.17, F-G). The first step in this direction is the formation of a weft of sterile hyphae forming a distinct envelope or sheath around the zygospore (F) in *Mortierella.*

The whole structure looks like a tiny fruit body. This tendency is carried further in a species of *Endogone* in which a definite fructification or sporocarp is formed. It consists of a number of zygospores enclosed within a common sheath (comparable to peridium) formed by somatic hyphae.

PHYCOMYCETES—DISCUSSION

I. Origin of Lower Fungi (Phycomycetes). It is a debatable question. There are two conflicting views concerning their origin. One of these advocates the algal origin of lower fungi and the other derives them from the colourless flagellates.

1. *Theory of Algal Origin of Phycomycetes.* The algal origin of lower fungi is based on similarities of structure and reproduction between the oogamous lower fungi such as *Saprolegnia* of class Oomycetes and oogamous Xanthophyceae such as *Vaucheria.* In both, the somatic phase consists of aseptate, coenocytic filaments. They have a similar life cycle and oogamy in the sexual phase. There is similarity in the chemical composition of the cell wall of the Oomycetes and *Vaucheria* and the manner of spore production within the sporangia.

This viewpoint was propounded by De Bary (1881). The name Phycomycetes (algal fungi) for the lower fungi emphasizes this viewpoint. De Bary envisaged the evolution of *Saprolegnia* from a vaucherial ancestor by loss of chlorophyll and change in the mode of the nutrition. Probably this took place before the ancestral stock had developed a synzoospore from many biflagellate zoospores. He further held that *Saprolegnia*-like Oomycetes, which evolved from *Vaucheria*-like ancestors, gave rise to the chytrids and to the Zygomycetes by regression.

The opponents of De Bary's theory point out that this hypothesis emphasizes only morphological resemblance between the algae and lower fungi which may represent parallel development along different lines rather than the evolution of lower fungi from forms like *Vaucheria.* The chief weak point of this hypothesis is that it completely ignores the physiological differences (mode of nutrition) and the type of flagellation both of which are features of great phylogenetic importance. This viewpoint of De Bary now receives but little support from the present-day mycologists.

The second school of mycologists led by Charles E. Bessey (1942) advocated the algal origin of lower fungi from some unicellular coccoid Xanthophycean ancestor. It is based on the following similarities between the two :

(*i*) Similarity in structure and anterior position of flagella in the swarmers of both.
(*ii*) Presence of cellulose in the walls of both.
(*iii*) Accumulation of reserve food in the form of glycogen in both.

On the basis of this hypothesis the lower fungi evolved from the unicellular coccoid Xanthophycean ancestor along two divergent lines, one with anteriorly uniflagellate swarmers and the other biflagellate swarmers.

The occurrence of uniflagellate Phycomycetes with a posterior flagellum and absence of cellulose in their walls go against this hypothesis. This shows that the Uniflagellate and Biflagellate Phycomycetes have not evolved from a common ancestor. The lower fungi are thus polyphyletic in their origin.

2. *Theory of Flagellate Origin of Phycomcetes.* Fisher and Dangeard derive the lower fungi from the colourless flagellate ancestors. Similar metabolism and type of flagellation among certain Phycomycetes and protozoans lend support to this viewpoint. This hypothesis emphasizes a polyphyletic protozoan ancestry of the Phycomycetes. The uniflagellate Phycomycetes evolved from the uniflagellate protozoa and the biflagellate Phycomycetes arose from the biflagellate protozoa. The non-mycelial forms were the first to evolve. These, in turn, gave rise to the more advanced mycelial forms by progressive evolution.

To the modern mycologists this theory of flagellate origin of lower fungi seems more probable. They hold that the lower fungi have evolved from the flagellates (flagellate protozoans) via the chytrids. This viewpoint is supported by the fact that the zoospores and the gametes of the chytrids are more like protozoa than the reproductive cells of Algae. Hawker (1967) also advocates a polyphyletic origin of lower fungi along parallel lines. According to him, Chytridiomycetes, Hyphochytridiomycetes and Plasmodiophoromycetes have originated from the flagellates having appropriate types of flagella. The Oomycetes which resemble certain filamentous algae in the composition (cellulose nature) of cell wall, form of sex organs, life cycle, and form of endoplasm and mitochondria may have evolved from an algal ancestor by loss of chlorophyll. He further suggests common ancestry for Zygomycetes and chytrids at the flagellate level.

II. Evolutionary Relationships of Lower Fungi (Phycomcetes). The lower fungi which were formally placed in class Phycomycetes is not a natural group of closely related forms. It is an assemblage of groups of fungi, the sole common feature among which is the relatively simple thallus. The simplest and the most primitive lower fungi have uniflagellate motile cells in the life cycle. The uniflagellate lower fungi in which the motile cell has a posterior flagellum of whiplash type are included in the class Chytridiomycetes. The anteriorly uniflagellate lower fungi with a tinsel type of flagellum are placed in the class Hyphochytridiomycetes. The uniflagellate lower fungi in which the motile cell has a second very short flagellum, in addition, to the usual long flagellum are placed in class Plasmodiophoromycetes. These three classes of lower fungi probably have originated along parallel lines from the colourless aquatic flagellates having appropriate types of flagella.

The Class Chytridiomycetes includes three orders. Of these Chytridiales include unicellular forms which are mostly holocarpic. The Blastocladiales have a slightly more complex thallus which bears specialised reproductive structures. The Monoblepharidales possess a filamentous mycelium and oogamous sexual reproduction which is an unusual feature in a so primitive group of fungi. The Chytridiales which probably had their origin in a flagellate ancestor are considered to have given rise to the Blastocladiales and Monoblepharidales. This evolutionary line of development seems to have ended blindly with the Monoblepharidales.

Some mycologists, however, consider that in the hyphal nature of thallus, elongated hypha-like sporangium and oogamous sexual reproduction Monoblepharidales suggest a close resemblance with the primitive Oomycetes. The opponents of this view argue that the presence of posteriorly uniflagellate motile cells (zoospores and sperms) in the life cycle and occurrence of chitin instead of cellulose in the cell walls of Monoblepharidales indicate that the resemblance with Oomycetes is superficial. The conclusion, therefore, is that Monoblepharidales are more closely related to Chytridiales and Blastocladiales than to Oomycetes.

Hyphochytriales of class Hyphochytridiomycetes resemble Chytridiales in their life cycle, aquatic habit, range of form, body structure and the composition of cell wall. In fact, the resemblance between these two groups is so great that till recently the members of order Hyphochytriales were included in the order Chytridiales. However, the fundamental differnce in the zoospore form, structure and position of flagellum on the zoospore has eventually led the modern mycologists to treat Hyphochytriales as a group distinct from Chytridiales. In the modern schemes of classification it is placed in a separate class Hyphochytridiomycetes with a single order Hyphochytriales.

The affinities of endoparasitic slime molds which are placed in order Plasmodiophorales of class Plasmodiophoromycetes are uncertain. They approach the slime molds (Myxomycetes) with which they have often been classified by older mycologists, in unequally biflagellate swarm cells and plasmodium as the somatic phase. The modern mycologists question the position of endoparasitic slime molds in the Myxomycetes because of the occurrence of zoosporangial stage in some, endoparasitic habit, absence of cellulose in the spore wall, manner of gametic union, the presence of haploid plasmodium and lack of fruiting bodies in the life cycle. They consider the endoparasitic slime molds to be true fungi. Gauman and Dodge thus placed these non-mycelial fungi in the Archimycetes—a class which included the primitive lower fungi. Sparrow (1960) provisionally placed these in the Biflagellatae (Oomycetes) owing to the presence of a second short flagellum on the zoospore.

Hawker (1967) opines that the endoparasitic slime molds neither resemble the slime molds (Myxomycetes) nor the Oomycetes. They are more closely allied to the Chytridiomycetes because of the lack of cell wall in some chytrids (*Synchytrium*) in the early stages of their development and presence of a structure in the zoospore of some chytrids which is likened to a second abortive flagellum. He thus places the endoparasitc slime molds under a separate class Plasmodiophoromycetes under group Uniflagellatae.

Alexopoulos (1962) concurring with Sparrow considered that endoparasitic slime molds have evolved independently of the groups with which they are usually clubbed. He thus places them in a separate class Plasmodiophoromycetes having neither affinities with Chytridiomycetes nor with oomycetes.

The biflagellate zoospores, oogamous sexual reproduction, fertilisation by means of a fertilisation tube and germination of zygote by means of germ tube are the features shared by and thus suggest relationship between the orders Saprolegniales, Leptomitales, Lagenidiales and Peronosporales. They are thus all placed in a single class Oomycetes. Of these, Saprolegniales is considered primitive as it has a relatively simple organisation with sporangia not in any way different from the somatic hyphae, absence of zonation of cytoplasm in the young oogonium and formation of more than one egg in the oogonium. The other three orders, in which the sporangia are clearly differentiated, have a single egg in the oogonium, and show a zonation of cytoplasm into periplasm and ooplasm in the young oogonium, are considered more advanced than the Saprolegniales and probably evolved from the latter. The Saprolegniales, in turn, resemble a Xanthophycean alga *Vaucheria* in certain respects. In both the somatic phase consists of coenocytic filaments and sexual reproduction is by oogamy. These resemblances, in fact. tempted older mycologists to advocate the hypothesis of algal origin of Saprolegniales from a Vaucherial ancestor by the loss of chlorophyll. The opponents of this hypothesis suggest that origin of Saprolegniales is to be sought among the colourless flagellates with an appropriate type of flagelltion via the chytrids.

The relationships of Mucorales and Entomophthorales of class Zygomycetes are still shrouded in mystery. The well developed coenocytic mycelium both in the Oomycetes and Zygomycetes suggests a relationship between the two groups. In fact there is a suggestion that the Mucorales arose from a primitive soil inhabiting ancestor of Saprolegniales by retrogression. The Entomophthorales are considered descendants of the Mucorales.

III. Evolutionary Tendencies in the Lower Fungi (Phycomycetes). The following evolutionary trends are noticeable in the Lower Fungi :—

1. Habitat. There is a gradual and progressive transition from aquatic to purely terrestrial mode of life through the amphibian species. The simplest and the primitive lower fungi such as the Chytridiomycetes are aquatic. They lack mycelium in the somatic phase. The Oomycetes, which are filamentous, exhibit a wide range of variations in habit. *Saprolegnia*, for instance, is a typical saprophytic water mould. *Pythium* is a little more soil-minded. Usually it prefers to grow in damp soil than in water. Asexually it reproduces by zoospores which indicates that basically *Pythium* is an aquatic type. *Phytophthora infestans* is distinctly an amphibious type. Normally the sporangia produce zoospores. Under adverse circumstances they get detached and function as conidia which germinate directly by producing a germ tube. Similarly *Plasmopara viticola* is amphibious. It can adjust to both aquatic and terrestrial habitats like *Phytophthora. Peronospora,* on the other hand, is fully terrestrial. It produces sporangia which invariably function as conidia. The Zygomycetes (Aplanatae) which are the most advanced members of Lower Fungi are typically terrestrial.

2. Habit. In habit, Lower Fungi range from obligate saprophytes to facultative parasites and finally end in highly specialised obligate parasites. The members of orders Hyphochytriales, Blastocladiales and Monoblepharidales are chiefly saprophytes. However, the Chytridiales, which are the simplest and the most primitive among the Lower Fungi are both saprohytic and parasitic. The Saprolegniales and Leptomitales of class Oomycetes are saprophytes whereas Pythiaceae are facultative parasites. The Plasmodiophorales and Lagenidiales are chiefly obligate endoparasites. In the families Peronosporaceae and Albuginaceae obligate parasitism reaches its climax. Among the Zygomycetes the Mucorales are largely saprophytes but most of the Entomophthorales are entomogenous (parasitic on insects).

3. Thallus. The thallus in the primitive Lower Fungi such as Chytridiales is microscopic. Generally it is a uniflagellate motile., holocarpic unicell which in the parasitic species is **endobiotic.** In slightly more complex members of this order the unicell is drawn out at one point into fine branching hairs forming the **rhizomycelium.** The thallus in these species is **epibiotic** and **eucarpic.** The reproductive structures in the epibiotic species are developed outside the host but the rhizomycelium is embedded in the host tissue. The epibiotic eucarpic species are considered advanced. They may have a **monocentric** or **polycentric** thallus. The latter is considered more advanced than the former. In still more advanced forms a scanty mycelium consisting of a few short hyphae is in evidence.

The Monoblepharidales have a mycelium consisting of typical hyphae which are profusely branched and woven into a eucarpic mycelial network. In the Oomycetes the mycelium is well established. It is aseptate and coenocytic. The primitive filamentous forms in the parasitic species are **intracellular** but the advanced species have a intercellular mycelium which absorbs nutrition by putting out haustoria which may be globose, lobed or finger-like simple or branched. The branched haustoria are considered more advanced than the globose or elongated haustoria.

It may not be out of place to mention here that the simple, unicellular thallus of most of the primitive Chytridiomycetes and Hyphochytridiomycetes is motile by a single flagellum whereas in the rest of the lower fungi the thallus is non-motile. The class Zygomycetes has no motile stage in the life cycle. The extensively developed aseptate coenocytic mycelium of the Zygomycetes exhibits a growing tendency towards septation in the advanded forms. Thus in the Entomophthorales the hyphae are normally regularly septate.

4. Holocarpy to Eucarpy. The primitive members of Lower Fungi such as those of families Olpidiaceae and Synchytriaceae of order Chytridiales include truly holocarpic species in which the whole thallus is transformed at maturity into a sporangium. Sexual reproduction in these species is isogamous. The other families of this order (Phlyctidiaceae and Megachytriaceae) include the more advanced chytrids in which the reproductive organs, at maturity, are formed by only a part of the thallus. Such species are called **eucarpic.** The eucarpic species which are considered more advanced exhibit a tendency towards heterogamy. The Oomycetes are mostly eucarpic. The members of class Zygomycetes are chiefly eucarpic.

5. Asexual Reproduction. Most of the Lower Fungi reproduce asexually by means of sporangiospores, which in the primitive orders such as Chytridiales, Blastocladiales, Monoblepharidales and Hyphochytriales are uniflagellate. They are produced in relatively simpler and undifferentiated sporangia. In slightly more advanced members of Lower Fungi such as *Saprolegnia,* the sporangiospores are biflagellate. The advanced Lower Fungi, which are terrestrial, reproduce by means of non-motile sporangiospores or aplanospores (Mucorales).

The sporangia in the Monoblepharidales and Saprolegniales are elongated and little more than enlarged hyphae. They are borne terminally on sporangiophores which are in no way different from the somatic hyphae. The sporangia in these order and in Leptomitales and some species of *Pythium* are undetachable and thus germinate *in situ* by zoospores.

The Leptomitales, Lagenidiales and some species of Pythiaceae have spherical, ovoid or pyriform sporangia which in the more advanced *Peronosporaceae* are oval to lemon-shaped and deciduous. They are borne on distinct sporangiophores.

The sporangial development reaches its culmination point in the production of deciduous sporangia which germinate directly by a germ tube or give rise to zoospores. Finally in the genus *Peronospora* the sporangia behave as conidia.

6. The evolution of a conidium from a sporangium. The primitive members of Lower Fungi belonging to the orders Chytridiales, Blastocladiales, Monoblepharidales, Hyphochytriales, Plasmodiophorales, Saprolegniales, Leptomitales and Lagenidiales invariably reproduce by means of zoospores produced within sporangia. It is considered a primitive method of asexual reproduction. The more advanced members of Lower Fungi exhibit an increasing tendency for the vulnerable motile zoospore stage to be replaced by a less vulnerable non-motile conidia which germinate by a germ tube. The transition from sporangia to conidia in *Pythium-Peronospora* series of class Oomycetes parallels *Blakeslea, Choanephora* and *Cunninghamella* series of Mucorales. However, the change from the sporangial to conidial type of reproduction in the former is abrupt. There is no progressive reduction in the form and size of the sporangium nor any gradual reduction in the number of spores produced. The conidium in this series is evolved by inhibition in the formation of zoospores. The entire protoplast of sporangium forms a germ tube instead of dividing to form zoospores.

(*a*) Pythium-Peronospora series. The aquatic genus *Saprolegnia* reproduces entirely by means of zoospores produced within zoosporangia. *Pythium* which is little more soil-minded and grows on damp soil, reproduces by zoospore indicating that basically it is also aquatic.

In *Pythium debaryanum, Phytophthora infestans* and *Plasmopara viticola* the sporangium at high temperature behaves like a conidium and germinates, by a germ tube but at low temperature produces zoospores.

Pythium vexans and *Bremia* bear sporangia which normally function as conidia but occasionally produce zoospores.

Pythium ultimum and *Peronospora* produce sporangia which are deciduous and dispersed by wind. They always germinate by a germ tube like a conidium.

The *Pythium-Peronospora* series thus illustrates how conidium, a typical asexual spore of Ascomycetes could have evolved from a sporangium. The change is an adaptation in response to a terrestrial mode of life which does not permit dispersal by zoospores but favours distribution by air.

(*b*) Blakeslea, Choanephora and Cunninghamella series. *Rhizopus* and *Mucor*, which are the two primitive genera of Mucorales, bear large, multisposed, columellate sporangia borne singly on unbranched sporangiophores. *Blakeslea* noramally produces Mucor-like sporangia but is made to bear sporangioles by changing conditions of nutrition, temperature and lighting. The branched sporangiophore in *Thamnidium* bears a terminal, multi-spored *Rhizopus* type of sporangium. In addition, it bears sporangioles with a reduced number of spores on the basal whorl of lateral branches of the sporangiophore. The spores of both the sporangium and sporangioles are similar in every respect. *Thamnidium elegans* produces only sporangioles in continuous darkness. *Cokeromyces* normally produces sporangioles. *Chateocladium* produces monosporous sporangia (=sporangioles) which are

Cokeromyces

deciduous. The spore wall is distinct from the sporangial wall. The series culminates in *Cunnighamella* which produces only conidia. The spore and sporangial walls in this genus are indistinguishable.

It is evident from the account given above that the evolution of conidium type of spore from the multispores sporangium in Mucorales is achieved by progressive reduction in size of the sporangium to an extent that it becomes equivalent to a conidium. There is no inhibition in the formation of spores.

7. Sexual reproduction ranges from isogamy to oogamy throuth anisogomy. Evolutionary trends in sexual reproduction are noticeable in Lower Fungi as a group as well as within a particular family or an order.

The Plasmodiophoromycetes and holocarpic members of Chytridiomycetes reproduce by **isoplanogametic conjugation**—a primitive type of sexual reproduction. The more advanced chytrids which are eucarpic and most of the Bastocladiales exhibit anisogamous **planogametic conjugation.** The Monoblepharidales among the Chytridiomycetes are oogamous. A motile male gamete fertilizes a non-motile oosphere produced in an oogonium.

Sexual reproduction in the Oomycetes, on the whole, is oogamous. Motility of gametes vanishes for ever in this class. The well differentiated male and female sex organs are called **antheridia** and **oogonia** respectively. Most of the Saprolegniales which include the water moulds reproduce by a primitive type of oogamy. They have more than one oosphere in each oogonium. The usual number is 4 to 10. There is no zonation of the young oogonial protoplast into periplasm and ooplasm. *Pythiella vernalis,* however, is an exception. It shows differentiation of cytoplasm of young oogonium into periplasm and ooplasm (Couch, 1935). Zonation of the oogonial protoplast gets well established in orders Leptomitales, Lagenidiales, and Peronosporales. They have a single oosphere in the oogonium. The cytoplasm of young oogonium gets differentiated into an ooplasm (oosphere) in the centre and peripheral periplasm. The latter is used up during maturation of oospore.

Sexual reproduction in Mucorales (Fig. 7.15 A) is generally isogamous or nearly so. Two multinucleate gametangia fuse to form a thick-walled resting spore called the **zygospore.** In *Zygorhynchus,* however the fusing gametangia are unequal in size (Fig. 7.17 B-C). *Endogone lactiflua* (Fig. 7.17 D-E) is distinctly heterogamous. Entomophthorales and Zoopagales exhibit degeneration of sexuality—a tendency which continues in the Higher Fungi.

It is obvious that sexual reproduction arose several times in evolutionary development of Lower Fungi. There is no linear series to suggest that all the lower forms are isogamous and higher forms oogamous. In both the Oomycetes and Zygomycetes sexual reproduction becomes less frequent than asexual reproduction and is resorted to towards the end of the growing season when conditions unfavourable for growth are to set in. The product of sexual fusion, which is the zygospore, is a resting structure which helps the species to tide over unfavourable period.

8. There is an increasing tendency towards greater protection of zygospore in the Mucorales culminating in the formation of tiny fruit body (Fig. 7.17 F-G). The series starts with *Absidia* and *Phycomyces.* In the former the thick-walled zygospore is provided with an additional protection by curved and in the latter by dichotomously branched bristles (Fig. 7.17 G) which arise from the suspensors and tend to cover or arch over the zygospore. *Mortierella* goes a step further. In it each zygospore is enveloped in a sheath formed of a weft of sterile hyphae. The zygospore with its enclosing sheath looks like a tiny fruit body (Fig. 7.17 F). This tendency reaces its culminating point in a species of *Endogone* in which definite fructifications or sporocarps are formed. Each sporocarp contains a number of zygospores enveloped in a common matrix of sterile hyphae.

HETEROTHALLISM

Heterothallism. This is a condition where the thalli of a single species are morphologically alike but physiologically different. Physiologically the mycelia are unisexual. But this difference of sex is only rudimentary. There is no apparent distinction between the male and the female mycelia. They are different only in their sexual behaviour. Such a distinction is designated by the signs + and –. The condition of the thallus is called **heterothallism.** It was first discovered by **A.F. Blakeslee in 1904.** In heterothallic species sexual reproduction will not occur between hyphae coming from two plus or two minus mycelia. It will occur only if two hyphae coming from different sexual strains (plus and minus) meet. Heterothallism thus implies the existence in a given species of more than one sexual strain. It is rather the first indication of origin of the dioecious condition of sexual phase of an organism. Blakeslee proved the phenomenon of heterothallism on the basis of experiments.

He took a petri dish containing sterilised agar-agar medium. He then inoculated the agar medium at five different places with spores of known opposite strains (plus and minus) of *Rhizopus.* In due course the spores germinated and produced mycelia. It was observed by him that zygospores were formed only at the places where the hyphae of opposite strains could meet. This indicated that conjugation occurred between the hyphae of two opposite strains. No zygospores were produced at places where hyphae of similar strains came into contact.

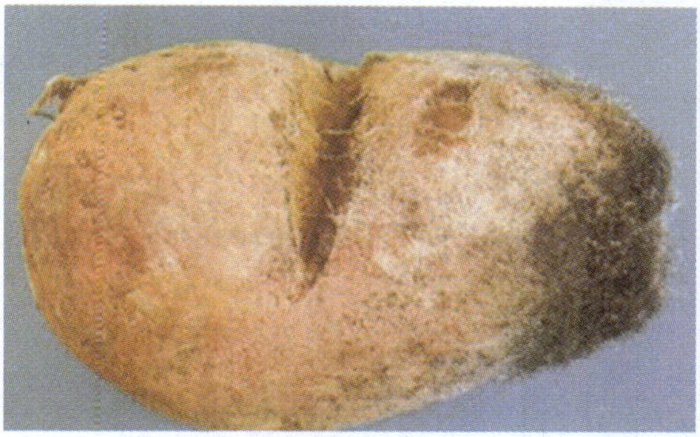

Rhizopus stolonifer causing rot

Economic Importance of Lower Fungi (Phycomycetes). The saprophytic Lower Fungi are frequent contaminants of our foodstuffs (*Rhizopus, Mucor* and others). They spoil our foods and consumer goods, some of them cause rotting of fruits and vegetables in storage. Many of them are of great industrial use. *R. stolonifer* is employed in the preparation of fumaric acid and *R. oryzae* in the preparation of alcohol. Some species of *Rhizopus* such as *R. sinensis, R. stolonifer* and *R. nodosus* yield lactic acid. Some of the Lower Fungi particularly Mucorales are ultilised in the production of important chemicals such as oxalic acid, citric acid and other useful chemicals.

Of the parasitic species *Synchytrium endobioticum* is a serious parasite of potato tubers. It causes the black wart disease. The parasitic species of order Peronosporales are destructive parasites of our crop plants. The common examples of these diseases are potato blight, downy mildew of grapes, brown rot of lemon, root rot or damping off of seedlings. *Plasmodiophora brassicae* of class Plasmodiophoromycetes is the cause of *club root or finger and toe* disease of cabbage and related plants. Some of the Lower Fungi are parasitic on animals including man. Among these certain species of *Mucor, Rhizopus* and *Absidia corymbifera* attack nervous system of man and prove fatal.

Club Root Disease of Cabbage

QUESTIONS

ESSAY TYPE

1. Give an illustrated account of the thallic structure and reproduction in *Rhizopus*.
2. Describe the development of sporangium in *Pilobolus* and mention the mode of dehiscence.
3. With the help of suitable sketches, describe the evolutionary trends in asexual reproduction in Mucorales.
4. Give a comparative account of the mode of asexual reproduction in *Albugo* and *Mucor*. *(Lucknow, 1992)*
5. Give an illustrated account of the life history of *Mucor*. *(Allahabad, 1998, 2002)*
6. With the help of suitable diagrams, explain the asexual reproductive bodies in *Pilobolus* and *Mucor*. *(Kanpur M.Sc., 1996)*
7. What do you understand by Heterothallism? Explain the mechanism of Heterothallism with suitable examples. *(Kanpur M.Sc., 1990, 1987)*
8. Describe the structure of mycelium in *Rhizopus* and itis mode of nutrition.
9. Give an account of asexual reproduction in *Rhizopus*.
10. Discuss evolutionary tendencies in Mucorales.
11. With the help of sketches, describe the transition from sporangium to conidium.
12. Give an illustrated account of life-cycle of *Mucor*. *(Kumaon, 2000)*

SHORT ANSWER TYPE

13. Define Heterothallism. *(Gorakhpur, 1998)*
14. Write short notes on :
    (*a*) Heterothallism and *Mucor.* *(Lucknow, 1994; Kanpur M.Sc., 1989)*
    (*b*) Zygospore formation in *Mucor*. *(Lucknow, 2002)*
    (*c*) Heterothallism in fungi. *(Lucknow, 1999; Kanpur, 1997, 1995; Allahabad, 1996, 1993; Gorakhpur, 1994, 1999; Awadh, 1991, 1992, 1994, 1995, 1996, 1997, 1998)*
    (*d*) Economic importance of Mucorales.
    (*e*) Economic importance of Phycomycetes.
    (*f*) Heterothallism *(Lucknow, 2000)*
    (*g*) Zygospore *(Allahabad, 2000)*
    (*h*) Sporangium *(Allahabad, 2000)*
15. Differentiate between Oospores and Zoospores. *(Rohilkhand, 2002)*
16. With the help of sketches, show range of sporangial structures in the Members of *Mucorales*. *(Kanpur M.Sc., 2002)*
17. Describe the sexual reproduction in Zygomycetes.
18. Compare the sexual reproduction in *Rhizopus* with that of *Spirogyra.*
19. In what respects do *Rhizopus* and *Mucor* differ from each other.
20. What is heterothallism? Describe an experiment to prove the phenomenon in *Rhizopus*.
21. Describe the salient features of Zygomycetes.
22. Give the systematic position and main characteristic of the following :
    (*a*) *Mucor* *(Lucknow, 1996; Kanpur, 1999; Allahabad, 1991; Awadh, 1996, 1997)*
    (*b*) *Pilobolus* *(Allahabad M.Sc,. 2000)*
    (*c*) *Rhizopus* *(Allahabad M.Sc., 1999; Kanpur, 1996)*
23. Illustrate life cycle of *Mucor* with the help of diagrams only. *(Allahabad, 1999)*

OBJECTIVE TYPE

24. Fill in the blanks :

(*i*) Hyphal walls in the members of Zygomycetes are made up of ....

(*ii*) Sexual reproduction in the Zygomycetes results in the formation of ....

(*iii*) The most important salient feature of the zygomycetes is the absence of ....

(*iv*) Heterothallism was discovered by ....

(*v*) The sporangium in *Rhizopus* and *Mucor* is characterised by the presence of ....

(*vi*) The spores in the members of *Zygomycetes* are ....

(*vii*) Sexual reproduction in *Zygomycetes* takes place by gametangial ....

(*viii*) *Pilobolus* is also known as .... fungus

(*ix*) Sporangiophores in *Pilobolus* develop from a basal ....

(*x*) Mycelium in the members of Zygomycetes is ....

(*xi*) *Mucor* is an obligate ....

(*xii*) Heterothallism may be defined as phenomenon of sexual ....

(*xiii*) *Rhizopus* multiplies by the formation of ....

(*xiv*) The gametes taking part in the sexual reproduction of *Rhizopus* are ....

(*xv*) Azygospores in Mucorales are produced from ....

25. Select the correct answer :

(*i*) Which of the following is present both in *Rhizopus* and *Funaria*?
(*a*) Mycelium (*b*) Archegonium
(*c*) Hyphae (*d*) Spore

(*ii*) In Mucorales, azygospores are usually produced from :
(*a*) Zygotes
(*b*) Unfertilised gametangia
(*c*) Vegetative mycelium
(*d*) The female sex orgam

(*iii*) Heterothallism was discovered by
(*a*) E.J.Butler (*b*) J.H.Craigie
(*c*) A.F.Blakeslee (*d*) B.B. Mundkur

(*iv*) Columellate sporangia are characteristic feature of
(*a*) *Synchytrium* (*b*) *Mucor*
(*c*) *Achlya* (*d*) *Albugo*

(*v*) The zygospores are not generally formed in the culture of *Rhizopus* developed from the germination of single spore because
(*a*) of lack of light
(*b*) of shortage of oxygen
(*c*) of the absence of the opposite strain of mycelium (*d*) of nothing

(*vi*) The mode of nutrition in *Pilobolus* is
(*a*) saprophytic (*b*) symbiotic
(*c*) parasitic (*d*) autotrophic

(*vii*) The mycelium is coenocytic in the genus
(*a*) *Mucor* (*b*) *Synchytrium*
(*c*) *Plasmodiophora* (*d*) *Physarum*

(*viii*) Which of the following is the result of sexual reproduction?
(*a*) Zygospore (*b*) Zoospore
(*c*) Conidia (*d*) Chlamydospores

(*ix*) The haploid structures in *Rhizopus* are
(*a*) hyphae, zygote and suspensor
(*b*) zygospore, spores and sporangia
(*c*) mycelium, spores and sporangia
(*d*) mycelium, zygospores and spores.

(*x*) Thick walled resting spores formed asexually are known as
(*a*) Oospores (*b*) Zygospores
(*c*) Oidiospores (*d*) Chlamydospores

(*xi*) Generally the Zygospores are thick walled and its colour is
(*a*) White (*b*) Black
(*c*) Green (*d*) Blue

(*xii*) Zygospores are characteristic of
(*a*) *Penicillium* and *Aspergillus*
(*b*) *Albugo* and *Synchytrium*
(*c*) *Physarum* and *Plasmodiophora*
(d) *Mucor* and *Rhizopus*

(*xiii*) The fungi which have many common features with algae especially in the thallus structure and reproduction have been placed in the class
(*a*) Phycomycetes (*b*) Ascomycetes
(*c*) Basidiomycetes (*d*) Deuteromycetes

(*xiv*) The sporangiophores of *Rhizopus stolonifer* are produced
(*a*) directly above the rhizoids
(*b*) from any point in the aerial stolon
(*c*) from the stolon in the centre of two groups of rhizoids
(*d*) from a root cell.

(*xv*) In *Mucor*, *Rhizopus* and *Pilobolus* the type of sexual reproduction is known as
(*a*) Planogametic fusion
(*b*) Gametangial copulation
(*c*) Gametangial contact
(*d*) Somatogamy

Erysiphe – on grass leaf

CHAPTER 8

SUBDIVISION – EUMYCOTINA (THE HIGHER FUNGI)

Class : Ascomycetes (Sac Fungi)

- The Higher Fungi – Introduction
- Class : Ascomycetes
- Post-Plasmogamy Changes

THE HIGHER FUNGI – INTRODUCTION

It is a large group comprising fungi more complex in structure than the ones we have studied so far. This subdivision includes mainly land forms which vary in form and grow in varied habitats. The thallus is a well-developed, filamentous mycelium which is regularly septate. The septation, however, is incomplete because the septa dividing the hyphae into cells have each a central pore. The cells or segments may each contain one or more than one nuclei. The motile phase in the life cycle of higher fungi is absent. Even the reproductive cells are non-motile. Another feature of importance in the higher fungi is the origin of the **dikaryophase** due to delayed karyogamy (fusion of male and female nuclei).

On the basis of the origin and nature of spores of the sexual stage (meiospores), the higher fungi are divided into two classes as follows :—

Class 1. Ascomycetes. The spores of the sexual stage (meiospores) which are known as the **ascospores** are **endogenous** in origin. They are produced within sac-like structures called **asci** which vary in shape. Globose, clavate or cylindrical are the usual forms. Occasionally the asci are filiform. The number of ascospores produced within each ascus is typically eight. The dikaryophase is of short duration. It is parasitic in its nutrition and occurs within the sexual fruit body.

Class 2. Basidiomycetes. The characteristic spores of the sexual stage (meiospores) are known as **basidiospores.** They are **exogenous** in origin and are borne externally on a structure called the **basidium**. The basidium varies in form through a wide range. The dikaryophase is independent and the main food absorbing (vegetative) phase in the life cycle.

CLASS : ASCOMYCETES

Historical Account. August Corda (1842), a Czech scientist was the first to recognise and delimit the Ascomycetes as a coherent group as known to us at present. Five years later in 1847, a French mycologist Joseph Henri Leveille discovered ascus. The investigations of two French mycologists Louis Rene Tulsane and Charles Tulsane published in a series of papers from 1861-1865 made an outstanding contribution to our knowledge of the morphology and taxonomy of this class.

Occurrence. According to Ainsworth and Bisby (1950), the class includes about 12,000 species. They are grouped under 1,650 genera. Some mycologists report about 15,000 species and 1,700 genera. Ingold (1967) makes above 30,000 species including the Ascolichens. Majority are terrestrial. Some, however, are marine. The former include both saprophytic and parasitic forms. They grow in a wide variety of habitats. The saprophytic species grow on soil rich in humus, on leaf mold, and decaying vegetable or animal matter, on dung, foodstuffs, and

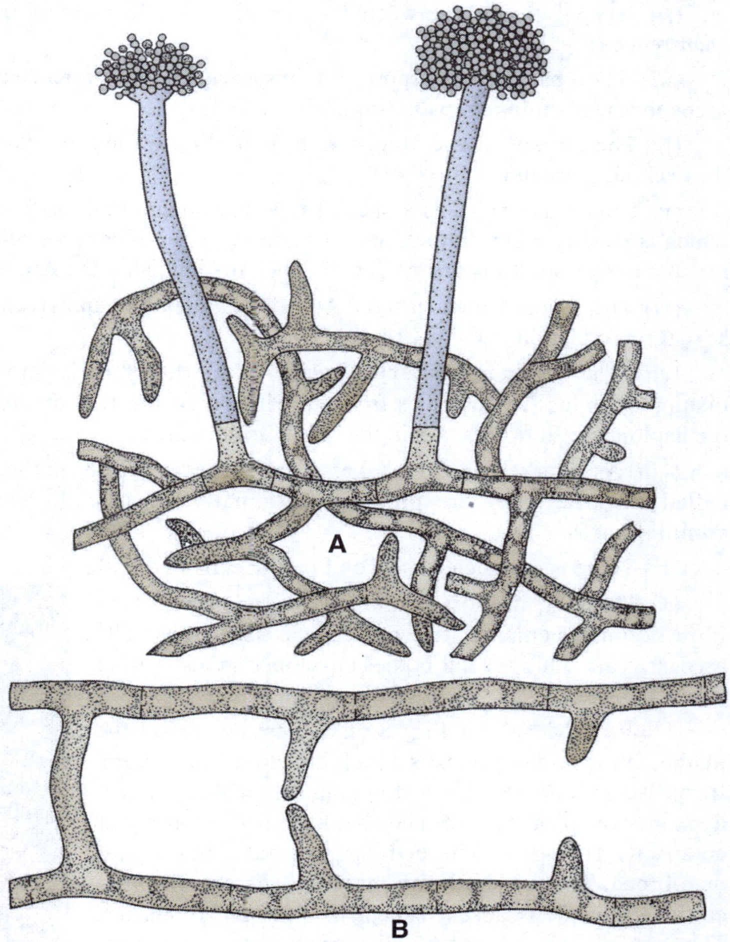

Fig. 8.1 (A-B). *Ascomycetes.* A septate eucarpic mycelium in the asexual (imperfect) stage; B, Sketch illustrating hyphal fusions.

rotting logs. Some species grow only on dung. The dung-loving species are called **coprophilous.** Many, such as blue and green molds and yeasts, grow on the surface of decaying vegetables, fruits, meat, debris of plant and animal origin and on damp leather. A few remain entirely underground and are called **hypogean.** Some are chief contaminants of our foodstuffs. The parasitic species attack both plants and animals including man. They grow within the tissue of their host (**endoparasites**) but a few like the powdery mildews are **ectoparasites.** The marine saprobic Ascomycetes are found on dead organic matter floating or submerged in sea-water whereas the parasitic forms parasitize the algae and higher plants growing there. The class includes widely differing plants. They vary from the inconspicuous, unicellular ascus forming yeasts through filamentous fluffy molds to fleshy mushroom-like morels and some elegantly coloured cup-shaped fungi which grow on rotting logs. All these, however, agree in the formation of sac-like **asci** containing ascospores and have a thallus which is a septate mycelium.

Diagnostic Features of Ascomycetes.

(*i*) The somatic phase consists of a well-developed, profusely branched mycelium with regular septation of the hyphae. The cross walls have each a minute, central, simple pore.

(*ii*) Hyphal fusions between the neighbouring hyphae of the mycelium are of frequent occurrence (Fig. 8.1 B).

(*iii*) The characteristic spores (meiospores) of the perfect stage which are known as the **ascospores** are **endogenous** in origin.

(*iv*) The ascospores are produced in small, thin-walled sac-like, spore-producing units called the **asci** (sing. **ascus**).

(*v*) The ascus, which is a special type of sporangium, has a definite number of ascospores which is usually eight. In fact, the formation of a sac-like ascus cell containing usually a definite number of ascospores is one character which distinguishes the Ascomycetes from all other fungi.

(*vi*) The ascus, which in most Ascomycetes, is an explosive sporangium, is the product of sexual reproduction.

(*vii*) The origin of **dikaryophase** in the life cycle is another important feature which distinguishes the Ascomycetes from the Phycomycetes. It is of short duration and is parasitic on the haplophase. It occurs within the sexual fruit bodies.

(*viii*) Asexual reproduction takes place by means of non-motile, exogenously produced spores called **conidia.** They are produced terminally on special, reproductive hyphae called the **conidiophores.**

(*ix*) There is complete absence of motile cells in the life cycle.

(*x*) The asci, in most of the species, are grouped to form definite, complex fruit bodies, the **ascocarps.** The ascocarps are multicellular bodies enveloped in a sheath of sterile hyphae.

Pyronema

Thallus (Mycelium Fig. 8.1). Excepting yeasts the thallus in the Ascomycetes is a well developed mycelium. It consists of long, slender or stout, much branched, septate hyphae with apical growth. They hyphae may be loosely or compactly entangled. The cells are several times longer than broad. The cell walls are mainly made up of fungus chitin. Cellulose is absent except in one or two species. The segments or cells are usually uninucleate but in some cases as *Pyronema, Dipodascus* and *Morchella* they may be multinucleate. In the vacuolate cytoplasm are found

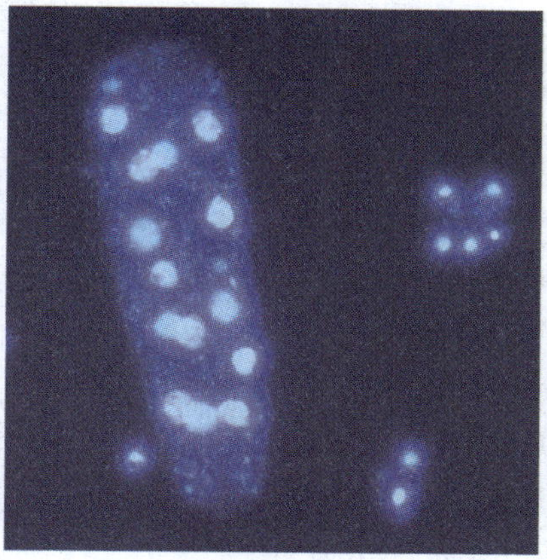

Dipodascus

Morchella growing in nature base

embedded the ribosomes, endoplasmic reticulum and mitochondria (Fig. 8.2). Lomasomes and lipid bodies are also present. The cytoplasmic membrane is distinct. The septum between the cells originates on the inner side of the tubular wall as an annular in-growth of wall material. It grows inwards invaginating the plasma membrane and reducing the diameter of the pore which is not completely closed. Each septum thus has a central pore or perforation which in the Ascomycetes is simple. It enables the protoplasm to flow from cell to cell. Even nuclei may migrate from one compartment to another. The plasma membrane also extends from one cell to the other through the central pores. Elements of crystalline nature known as the **Woronin bodies** are often found near the septa.

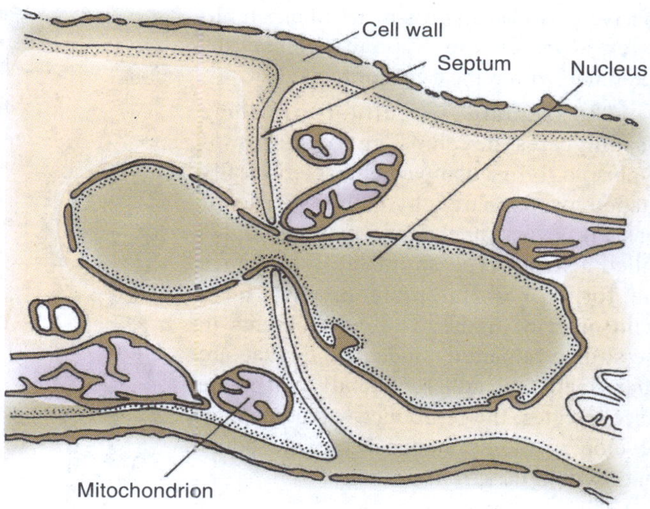

Fig. 8.2. Longitudinal section of the hypha of an ascomycete. Note the perforated transverse septum through which a nucleus is moving.

Kinds of Mycelia.

The mycelia in the Ascomycetes are of the two kinds, namely, homokaryotic and heterokaryotic.

(*i*) *Homokaryotic Mycelium.* It is produced by the germination of uninucleate spores. The cells of the homokaryotic mycelium may be uninucleate or multinucleate. In the multinucleate condition, all the nuclei in the mycelium are genetically similar.

(*ii*) *Heterokaryotic Mycelium.* When genetically dissimilar nuclei exist or are associated in the cells of the same mycelium, it is termed heterokaryotic. The heterokaryotic condition may result from either of the three methods : (i) hyphal fusion, (ii) mutation, or (iii) inclusion of dissimilar nuclei in the same spore after meiosis in heterothallic species.

Fungal Tissues. In some forms all or some of the hyphae of the mycelium become tightly intermeshed so as to lose their individuality. The cells of these compactly aggregated hyphae become isodiametric forming a pseudoparenchymatous structure of macroscopic size (Fig. 8.3 A). Such a structure is called a **sclerotium.**

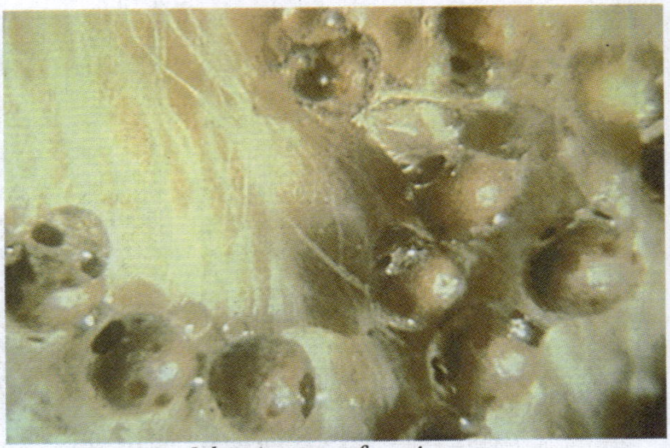

Sclerotium – surface view

The sclerotia are compact resting bodies ranging from the size of a pin head to 1 cm in diameter. In form they may be more or less rounded, cushion-shaped, cylindrical or irregularly shaped objects. The cells in the interior of the sclerotium are filled with stored food. The walls of the outer cells (Fig. 8.3 B) are thickened, almost sclerenchymatous and usually dark in colour. Sclerotia enable the species to survive periods of stress or to hibernate. In a few species the hyphae themselves perennate within the twigs or buds.

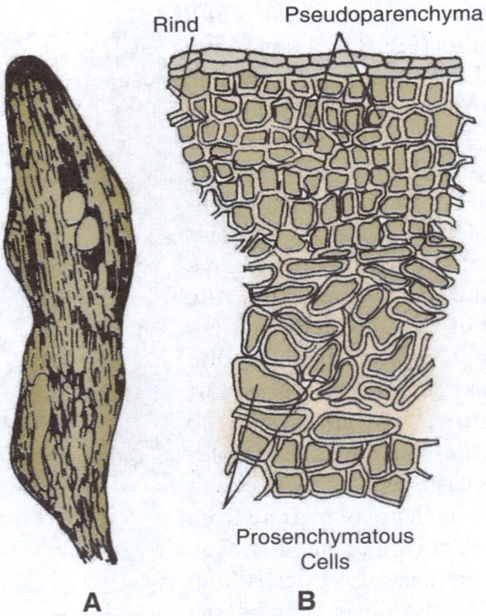

Fig. 8.3 (A-B). Ascomycetes. A, Sclerotium; B, section of A showing structures.

The septate mycelium of some Ascomycetes develops aggregations of hyphae to form a compact characteristically shaped pseudoparenchymatous tissue or mass. It is very much like a mattress and is called a **stroma** (pl. stromata). It may take the form of a flat plate or may be a solid mass of tissue. The cell contents in the stroma remain distinct. The hyphae are interwined and adhere laterally. In some Ascomycetes, the reproductive bodies may develop in the stromata. These are borne on or within the stroma.

Reproduction. The life cycle of an ascomycete fungus, in general, consists of two reproductive stages. These are the **asexual stage** and **sexual stage.** The asexual stage is often called the **imperfect stage** and the sexual is designated as the **perfect** (**ascigerous**) stage. The asexual stage, however, predominates with asexual spores forming the chief means of dispersal during the growing season of the fungus. The sexual or perfect stage occurs only under particular conditions or at a particular season.

Asexual (Imperfect) Stage. In the Ascomycetes this stage in the life cycle is propagated by sporulation (formation of spores) and vegetative methods.

A. Sporulation. Asexual spores called Conidia constitute the chief means of propagation and dissemination in most Ascomycetes in the growing season. It is the most common type of asexual spore produced in the Ascomycetes. Examples are the blue molds and powdery mildews.

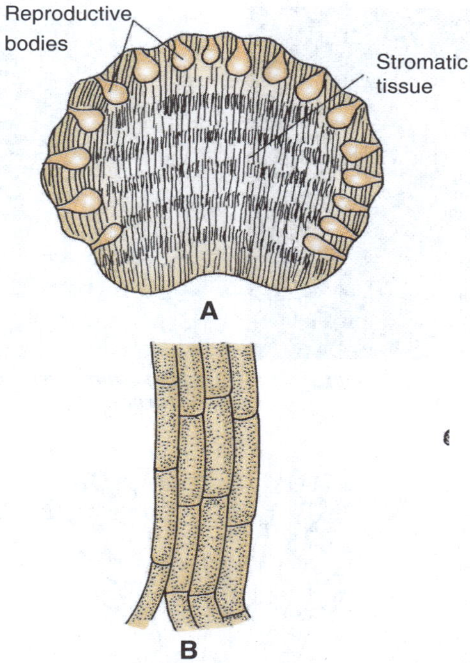

Fig. 8.4 (A-B). Ascomycetes. A, stroma in section; B, a portion of the stromatic tissue showing structural detail.

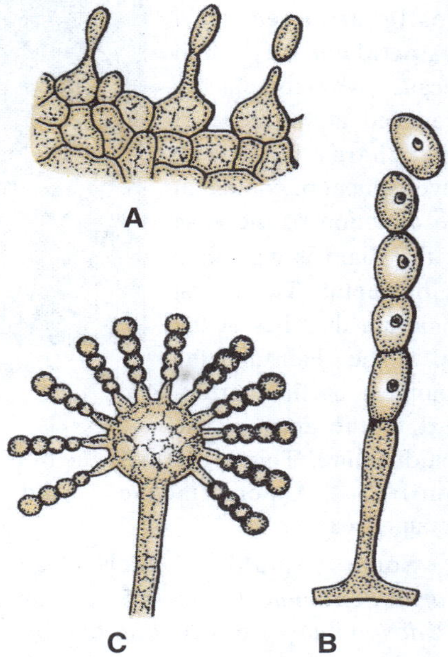

Fig. 8.5 (A-C). Ascomycetes, chief types of conidiophores. A, *Phyllosticta*; B, *Erysiphe*; C, *Aspergillus*.

1. Conidia (Fig. 8.5 A-E). The conidia are exogenously produced, non-motile, deciduous asexual spores. They are thus exposed or unprotected and are abstricted singly or successively from the free, distal end of a special, short, erect, fertile hypha called **conidiophore.** The condiophores may be branched (D) or unbranched (B), unicellular (A) or multicellular (D). In *Penicillium* (D), the conidiophores are long and usually branched. Each ultimate branch bears a cluster of bottle-shaped structures called **sterigmata** each of which produces a long chain of conidia. In some Ascomycetes conidia separate from the tip of the conidiophore as soon as formed (A). In others they cling together in a mucilage drop or remain attached in a chain (B). Some genera such as *Erysiphe* form a single chain of conidia on a conidiophore (B) while others form many chains of conidia on a single conidiophore (C). The

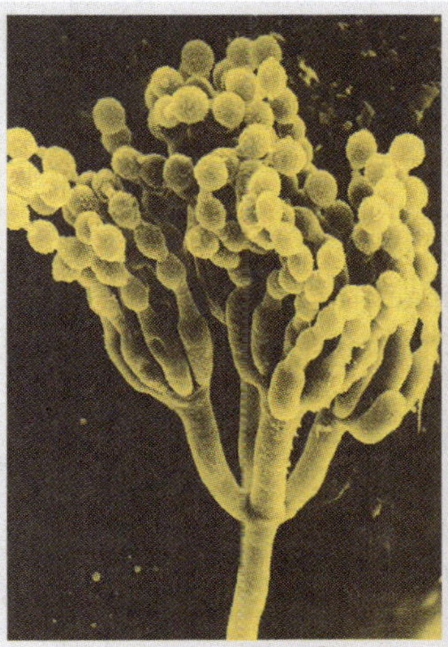

Penicillium – Conidia on penicillius

Erysiphe – condiophores with conidia

conidia in a chain are mostly arranged in a **basipetalous** succession because new conidia are produced at the base of the chain (C). The development of conidia in the direction of the base of the chain is described as **basipetal.** The oldest conidium thus lies at the top of the chain and the youngest at the bottom next to the apex of the conidiophore. This is the case in the families Aspergillaceae and Erysiphaceae.

Hormodendrum causing Lesions

Cladosporium – conidia borne at the top

Some molds such as *Hormodendrum* (Fig. 8.5 E) or *Cladosporium* produce conidia by growth at the tip of the chain (apically) rather than at the base of the chain. In this case the conidiophore abstricts a terminal conidium. The latter pinches off a swelling or bud. The bud in turn becomes the terminal conidium and produces a bud. The process may be repeated. The youngest conidia in these molds are at the tips of the chains and not at the bases of the chains. The development of conidia towards the apex of the chain so that the apical member of a chain is the youngest is described as **acropetal** (E).

Asexual reproduction by conidia frequently takes place in the majority of the Ascomycetes. In fact, conidial development reaches its climax in this class of fungi. Generally conidia are multinucleate. Upon germination, each conidium gives rise to a germ tube (F_3, F_4) which grows into a mycelium (F_5).

Arrangement of Conidiophores. Conidiophores arise in a variety of ways. In some genera they arise singly at intervals from the ordinary hyphae of the mycelium. They stand free from

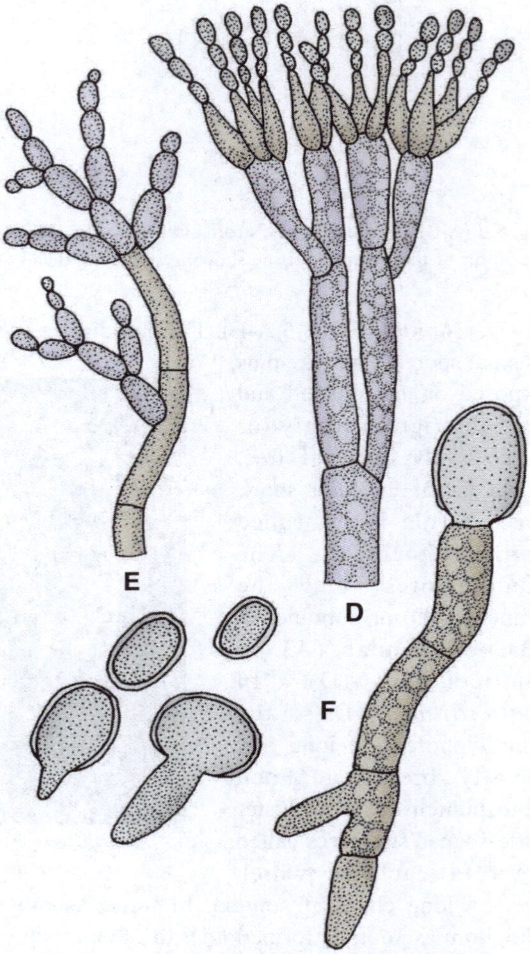

Fig. 8.5 (D-F). Ascomycetes. Chief types of conidiophores (cotd.) and germination of a conidium (F_1-F_5); D, *Penicillium*; E, *Hormodendrum*; F_1-F_5, germination of conidium.

one another and thus appear scattered over the surface of the mycelium (A). Other genera, in which the hypae are densely packed or compacted, have the conidiophores closely packed or laterally abutting one another to form a continuous spore forming layer. In this case the conidiophores are organised into definite asexual, fruiting bodies such as *synnema*, *Acervulus*, *sporodochium* and *pycnidium* which are described in some detail in Chapter 16.

Synnema

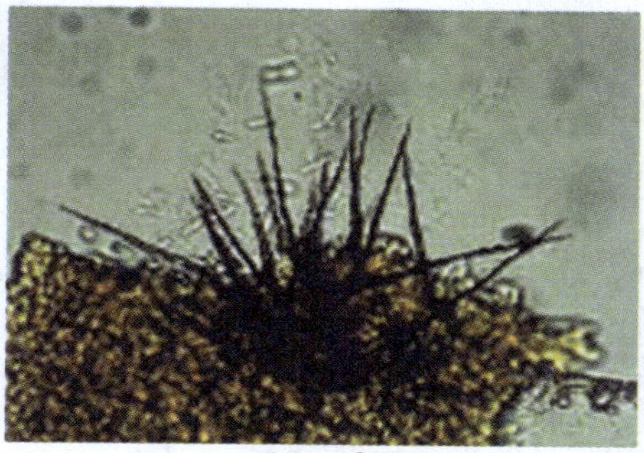

Acervulus

Pycnidium

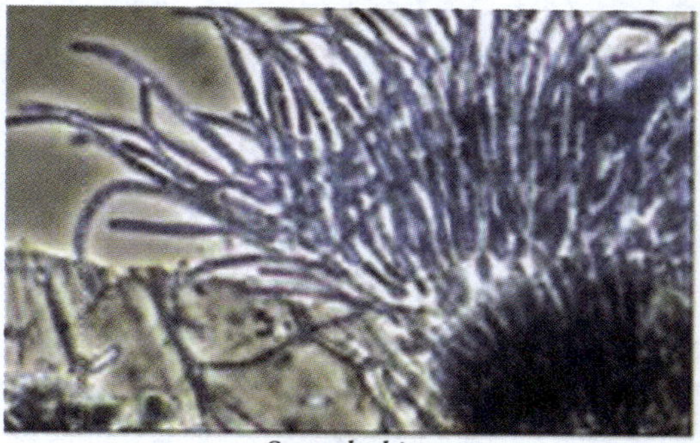

Sporodochium

Significance of Asexual Stage. The significance of asexual or imperfect stage in the life cycle lies in multiplication and dispersal of the species. The somewhat increased resistance of the spores of this stage coupled with their enormous number favours survival of species during unfavourable conditions. However, rarely do conidia overwinter. Besides conidia, two additional types of asexual spores are found in the Ascomycetes. These are **oidia** and **chlamydospores.**

2. Oidia (Fig. 8.6 A). In some Ascomycetes the hyphae do not form spores in succession at their free ends. Instead the septate hyphae may undergo additional septation and break up or fragment under proper conditions into their component cells. These single-celled fragments behave as asexual spores called **oidia.** Oidium is from a Greek root meaning "egg" or an ovoid body. The oidia are formed throughout the length of a hypha. They do not seem to be reproductive structures in the same sense as conidia. There is very little or no difference between the shape and structure of this kind of spores (oidia) and vegetative cells of the hypha producing them. Oidia possess little, if any, greater resistance of drying than any other mycelial fragment. They continue vegetative growth, each starting a new plant.

3. Chlamydospores (Fig. 8.6 B). They are large, thick-walled, spore-like resting cells which contain a lot of reserve food, and resistant to drying and can remain dominant for long periods. Any cell in the mycelium may lose water, accumulate granular reserve materials and develop around it a thick wall to become a **chlamydospore.** The chlamydospores may be terminal or intercalary in position and may be formed singly or in short catenate series. With the approach of adverse conditions (drying and sunlight) the mycelium dies. The chlamydospores remain alive but dormant. With the onset of conditions favourable for vegetative growth, the chlamydospores germinate, each producing a new mycelium.

B. **Asexual reproduction by vegetative methods.** Some Ascomycetes under normal conditions reproduce by vegetative methods of fragmentation, fission and budding.

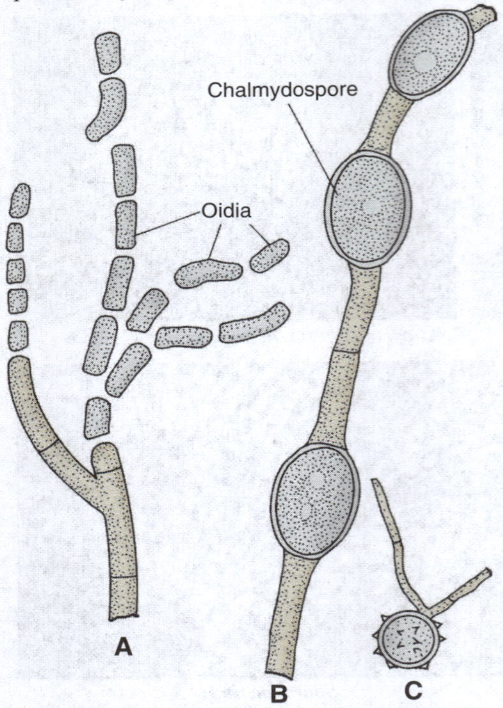

Fig. 8.6 (A-C). Ascomycetes – Asexual reproduction by vegetative methods. A, Septate hyphae forming oidia; B, Intercalary chlamydospores; C, Tuberculate chlamydospore germinating in cell culture.

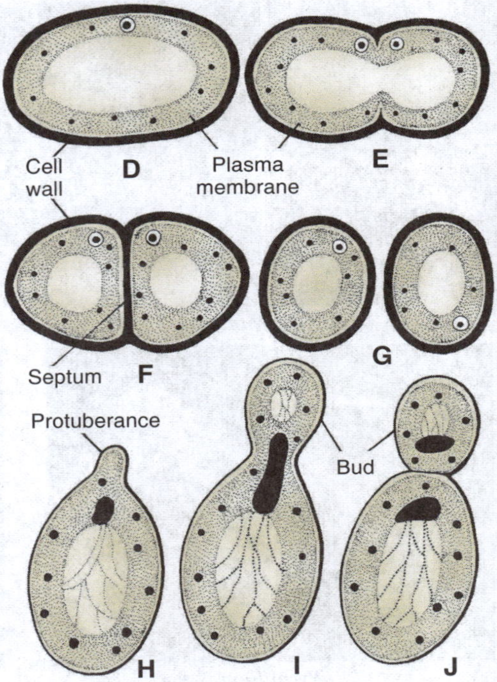

Fig. 8.6 (D-J). Ascomycetes. Asexual reproduction by vegetative methods. D-G, by fission; H-J, by budding.

1. Fragmentation. In this method there is separtion from the mycelium of hyphae or segments of hyphae called the **fragments.** The fragments may consist of one or more living cells. By repeated cell division and apical growth, each fragment produces a new mycelium. Fragmentation thus results in as many new individuals as there are fragments.

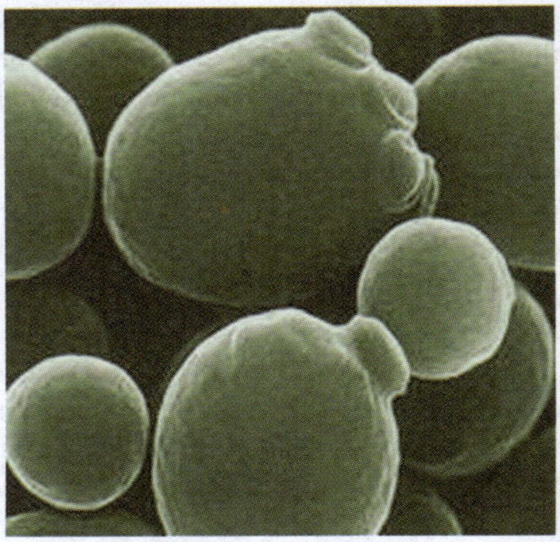

Saccharomyces – cultures

2. Fission (Fig. 8.6 D-G). The Ascomycetes with a unicellular thallus such as fission yeasts usually multiply by this method. It is a kind of cell division. The division is transverse. The mother cell elongates (D). The nucleus divides into two (E). The daughter nuclei move apart. Meanwhile a ring-like ingrowth of the wall material appears at the wall in the middle of the cell (E). It grows inwards covered in front by the plasma membrane. Finally it meets in the centre stretching across the cell forming a complete partition called the **septum** (F). The septum thickens and then splits into two layers, one for each daughter cell before they separate (G).

3. Budding (Fig. 8.6, H-J). *Saccharomyces* (baker's yeast) under normal conditions reproduces by this method only. At the time of budding, the cell protoplast covered by the cell wall bulges out in the form of a small protuberance at or near one pole of the yeast cell (H). The protuberance gradually increases in size and is called the **bud.** Meanwhile the nucleus of the parent cell along with the vacuole divides. One of the daughter nuclei migrates into the enlarging bud (I). The bud grows and becomes constricted at the base. Eventually the opening between the bud and yeast cell closes, double wall forms and the two cells become physiologically distinct (J). They subsequently separate from each other leading independent life.

C. Sexual (Perfect Stage) Reproduction (Fig. 8.7). The eucarpic mycelium in Ascomycetes which produced conidia enters the sexualy or perfect phase under particular conditions or at a particular season usually late fall or early spring. The sexually reproducing mycelia are either **homothallic** or **heterothallic** and sexual reproduction is isogamous or heterogamous. The copulation branches in some of the lower Ascomycetes do not show any distinction and are similar. But in the higher Ascomycetes, they undergo functional and morphological differentiatiation to form male and female sex organs. The male sex organ is called an **antheridium** and the female **ascogonium.** The initial step in sexual reproduction always aims at the fusion of two protoplasts

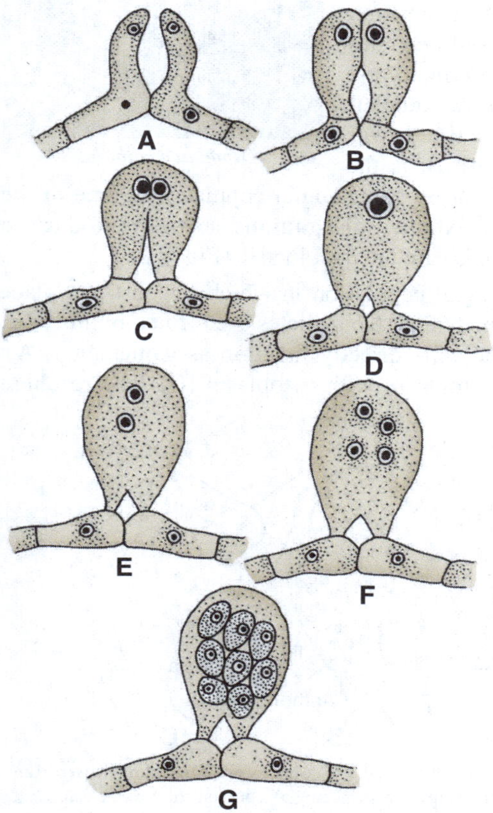

Fig. 8.7 (A-G). Ascomycetes. Stages in isogamous sexual reproduction illustrating gametangial copulation (A-C) and direct development of ascus (D-G) in *Eremascus*. (Based on Guilliermond)

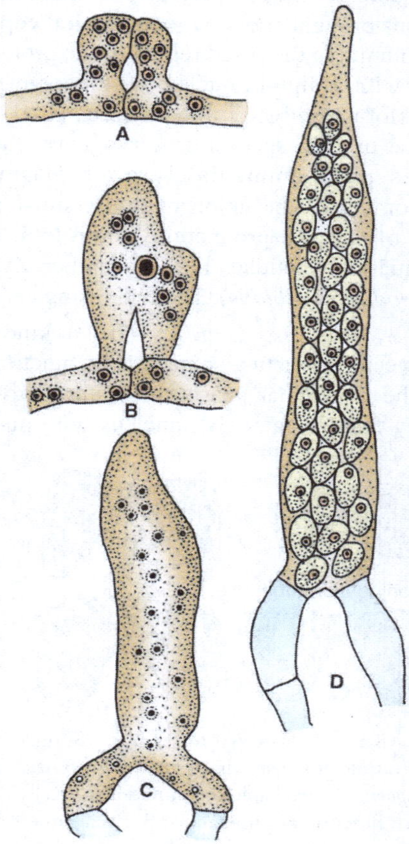

Fig. 8.8 (A-D). Ascomycetes. Stages in anisogamous sexual reproduction illustrating plasmogamy by gametangial copulation and direct development of ascus in *Dipodascus albidus*. (Based on Lagerheim)

which brings the two compatible nuclei close together in the same cell. This process is called **plasmogamy.** We shall, therefore, first study the common methods which the Ascomycetes employ to achieve plasmogamy.

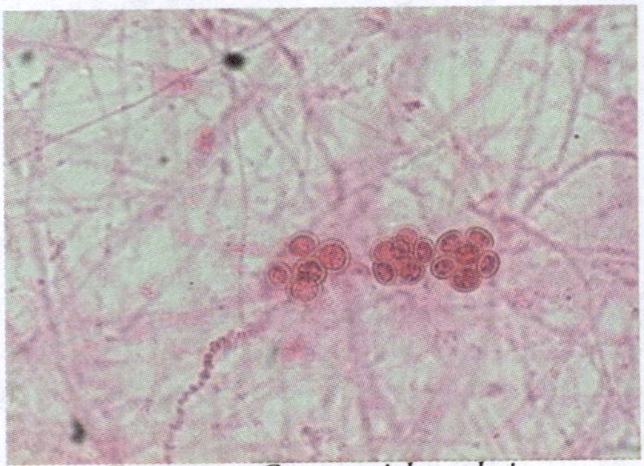

Eremascus – Gametangial copulation

1. Gametangial copulation (Fig. 8.7). In some of the lower Ascomycetes (Hemiascomycetidae) such as *Eremascus*, the copulation branches are indistinguishable from each other as is the case in the Zygomycetes and sexual reproduction is described as **isogamous.** They arise in pairs as small, vertical protuberances, one from each of the two adjoining cells of the same hypha (homothallic) near the separating septum (A). Each sexual branch receives a nucleus from the parent cell and is separated from the latter by a wall appearing at its base. The two sexual branches grow side by side and eventually their tips which may be called **gametangia** become apposed to each other (B). After a time the two gametangia fuse. The intervening walls disappear in the region of contact (C). The union of the two protoplasts, which is known as **plasmogamy**, is thus brought about by **gemetangial copulation.** The two nuclei soon unite in the fused region of the protuberance (D). The fusion cell with a diploid nucleus or the zygote enlarges and becomes transformed directly into an ascus (E-G). There is no dikaryotic phase in such species as karyogamy (fusion of the two nuclei) takes place immediately after plasmogamy. In the lower Ascomycetes the union of two compatible nuclei, which is the aim of sexual reproduction, as a rule, is brought about by gametangial copulation. Some of the Hemiascomycetidae, like the higher Zygomycetes, exhibit **anisogamous** sexual reproduction (*Diopdascus albidus*). The two fusing cells (gametangia) are unequal in size (Fig. 8.8).

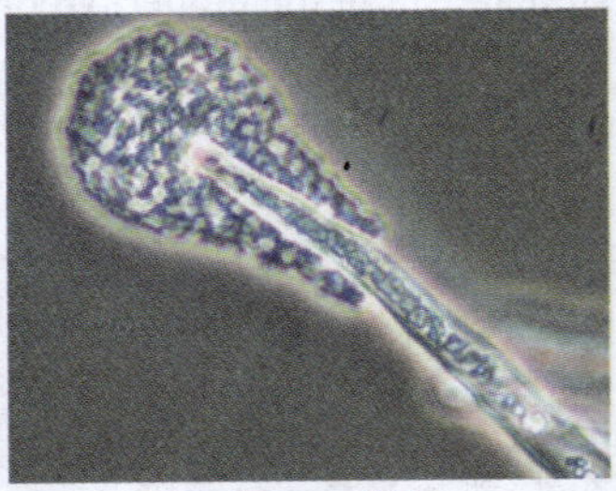

Dipodascus albidus

2. Hologamy (Fig. 8.9). It is a kind of gametangial copulation in which fusion takes place between two mature somatic cells functioning as gametangia. No gametes are formed in this case. In the unicellular yeasts the mature haploid somatic cells directly function as gametangia (A). They fuse in pairs (B), nucleus with nucleus and cytoplasm with cytoplasm (C). The resultant

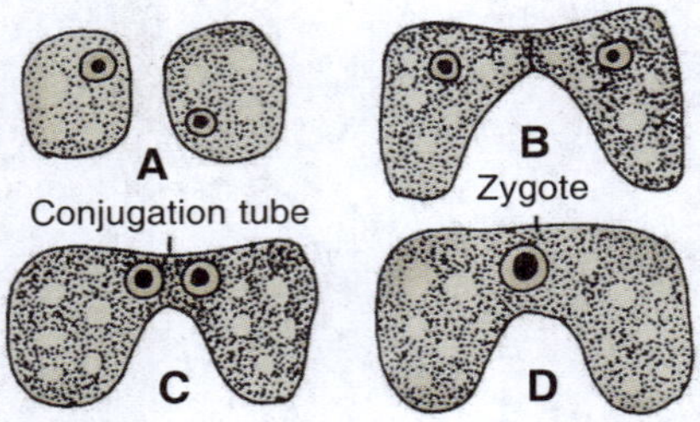

Fig. 8.9 (A-D). Ascomycetes. Stages illustrating hologamy in *Schizosaccharomyces octosporus*. A, two haploid mature somatic cells which function as gametangia; B, Each gametangium putting out a protuberance; C, The two protuberances unite with each other to form a conjugation tube; D, The two sexual nuclei fuse in the conjugation tube.

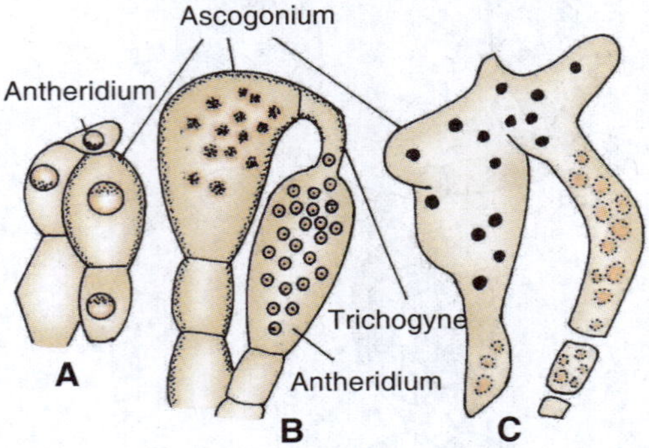

Fig. 8.10 (A-C). Ascomycetes. Diagram illustrating plasmogamy by gametangial contact. A, uninucleate sex organs of *Sphaerotheca*, ascogonium lacks trichogyne; B, multinucleate sex organs of *Pyronema,* ascogonium furnished with a trichogyne; C, multinucleate sex organs of *Claviceps,* ascogonium without trichogyne.

fusion cell, which has a diploid nucleus, is called the **zygote** (D). It directly functions as an **ascus mother cell.**

3. Gametangial contact (Fig. 8.10). The copulation branches in the higher Ascomycetes, as a rule, undergo functional and morphological differentiation to bear distinct sex organs which are unlike each other. Of these the male is called an **antheridium** and female **ascogonium** (Gr. *askos* meaning sac and *gennao* meaning I give birth). Ascognium thus is a female sex organ which eventually gives rise to sac-like asci. Heterogamy thus is the rule in the higher Ascomycetes. Both kinds of sex organs are usually produced adjacent to each other.

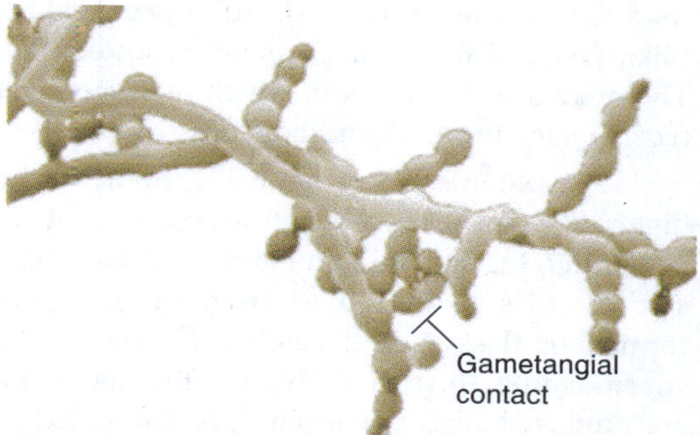

Neurospora – mycelicem showing gametangial contact

The male branch or hypha consists of a stalk bearing a terminal antheridium (A). The latter is smaller than the ascogonium. It is more or less cylindrical in shape and contains one or a few nuclei and in a few cases as many as 400. The female branch in some Ascomycetes becomes the ascogonium. In others it consists of a stalk of one or more cells bearing the ascogonium terminally (B). Sometimes the distal end of the ascogonium is broadly rounded (A). At other times it is prolonged into a tubular receptive process called the **trichogyne** (B). The trichogyne may be septate or unseptate and usually unbranched, rarely branched as in *Neurospora*. The ascogonium is also very variable in form. In some it may be a single uninucleate cell (A). In others a multinucleate segment of narrow hypha. In still others it is a pyriform or globose structure with as many as 1,400 nuclei. The gametes are undifferentiated protoplasts.

Pyronema confluens

The antheridium and the ascogonium come in contact. The double wall at the point of contact dissolves and the male nucleus with some cytoplasm migrates through a lateral pore into the ascogonium. No fertilisation tube is formed. When the ascogonium is furnished with a trichogyne, the latter receives the male nucleus from the antheridium. In some forms at maturity, the tip of the trichogyne grows towards the free end of the antheridium. It then curves down to meet it (B). The walls disappear in the region of mutual juncture. The protoplast of the antheridium containing male nuclei now migrates into the ascogonium (*Pyronema confluens*). These are examples of plasmogamy by **gametangial contact** (C).

4. Autogamy (Fig. 8.11). In some Ascomycetes, the

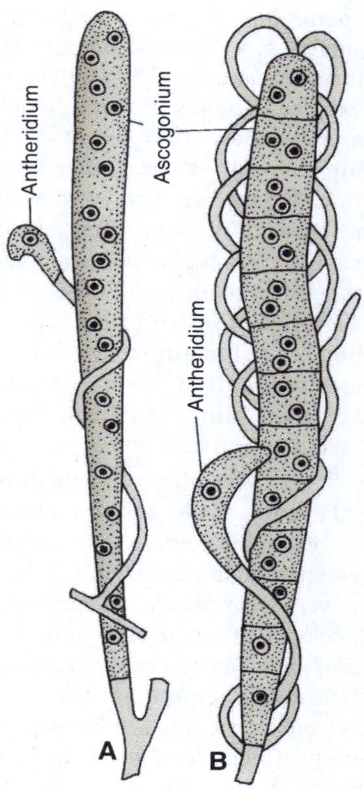

Fig. 8.11 (A-B). Ascomycetes. Stages illustrating autogamy in *Talaromyces vermiculatus* (= *Penicillium vermiculatum*) (After Dangeard)

antheridia are either absent or when formed they are non-functional (A). For instance in *Talaromyces vermiculatus* syn. *Penicillium vermiculatum* the tip of the antheridium touches the tubular multinucleate ascogonium (B). The intervening walls dissolve at the point of contact. Through the common pore the protoplast of two sex organs come in contact but there is no migration of male protoplast or nuclei into the ascogonium. No plasmogamy thus takes place (Dangeard, 1907). Mere contact stimulates the ascogonium nuclei to arrange themselves in functional pairs (dikaryons). This is an example of **autogamy.** There are also instances in which the ascogonia receive nuclei from the trichogynes.

5. Spermatisation (Fig. 8.12). Some of the higher Ascomycetes produce ascogonia but no antheridia. Instead in them the conidia-like male sex cells called **spermatia** are formed in flask-shaped cavities known as the **spermogonia**. In a few species, the spermatia are produced singly or in clusters at the tips of special, reproductive hypae called the **spermatiophores** (*Neurospora sitophila*). The spermatia are specialised uninuclete, small, sporelike, non-motile male cells. When mature the spermatia are detached and carried by some external agency (insect, wind, rain) to the ascogonia where they get attached to the receptive organs such as the trichogynes (A) or other receptive hyphae. A pore is developed at the point of contact. The spermatial nucleus is eventually emptied into the female sex organ through the pore. It migrates through the septa pores down to the ascogonium if the trichogyne is multicellular. This kind of sexual fusion between the spermatium and the female receptive organ is called **spermatisation**. In a species of *Neurospora*, spermatisation has been reported between a conidium and trichogyne.

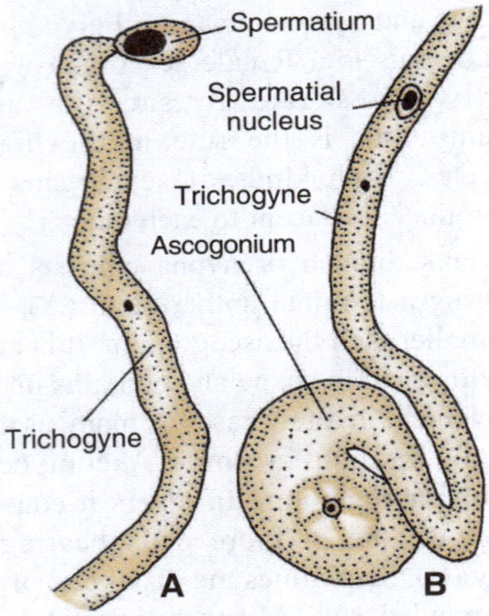

Fig. 8.12 (A-B). Ascomycetes. Plasmogamy by spermatisation in *Mycosphaerella tulipiferae*. A, upper portion of trichogyne with the spermatium adhering to its tip; B, uninucleate ascogonium with the spermatial nucleus passing down the trichogyne. (After Higgins)

By these different methods of plasmogamy the compatible male and the female nuclei become associated in pairs in the ascogonium. The pairs of compatible nuclei are called the **dikaryons**. In the lower Ascomycetes, as mentioned before, the two nuclei in the pair fuse immediately (karyogamy) like the Phycomycetes. No **dikaryotic phase** in the life cycle exists in them. There are other Ascomycetes in which **karyogamy** is considerably delayed after plasmogamy. In them the nuclei remain in close association as a functional pair called a **dikaryon**. There is a well developed **dikaryotic phase** in the life cycle of these species.

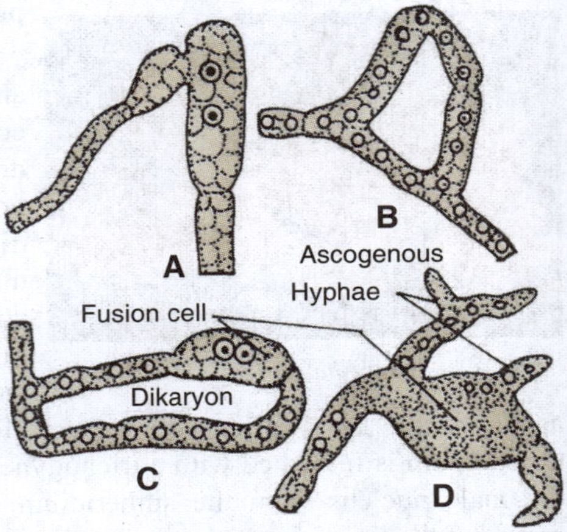

Fig. 8.13 (A-D). Ascomycetes. Plasmogamy by somatogamous copulation. A, conjugation between two vegetative cells in *Tuber aestivum* (After Greis); B-D, stages in somatogamous copulation in *Morchella* sp.; B, conjugation between vegetative cells; C, Fusion cell with a dikaryon; D, Fusion cell giving rise to ascogenous hyphae.

6. Somatogamy (Fig. 8.13). The bringing together of sexually unlike nuclei by vegetative fusions is called **somatogamy** or **somatogamous copulation**. Some of the higher Ascomycetes neither produce antheridia nor spermatia. In these, two vegetative hyphae from mycelia of opposite strains (+ and –) come in contact and fuse. The nuclei of one migrate to the other and travel from cell to cell through septal performations to the ascogonia. This results in the pairing of genetically unlike nuclei to form dikaryons in the ascogonia.

Helvella elastica – growing in nature

There are species in which no gametangia are produced (*Helvella elastica*). Fusion between somatic hyphae of two compatible mycelia takes place (A). The nucleus or nuclei of one migrate to the other throughout. The mycelium into which the nucleus or nuclei have migrated is called **heterokaryotic**. Ascogenous hyphae develop from certain cells of the heterokaryotic mycelium. In *Morchella* somatogamous copulation is a normal feature. Two vegetative hyphae of the mycelium come in contact (B). The intervening walls between the copulating cells dissolve at the point of contact and their multinucleate protoplasts intermingle in the fusion cell (C). Two functional nuclei, one from each copulating cell, form a **dikaryon** (C). All other nuclei in the fusion cell disappear. Ascogenous hyphae later arise from the fusion cell (D).

Compatibility. This term refers to the sexual nature of the mycelium. Literally it means the fusing capability of mycelia. The two fusing elements are said to be **compatible**. With respect to their sexual nature or compatibility (fusing capabilities) the mycelia of Ascomycetes are of two types, namely, **homothallic** and **heterothallic**. The homothallic mycelium is considered **bisexual**. It means sexually it is **self-fertile**. The asci are produced from a single mycelium without the aid of another. Technically the homothallic mycelia are described as **self-compatible**. The heterothallic mycelium, on the other hand, is unisexual. Sexually it is **self-sterile**. It requires the aid of amother compatible or complementary mycelium of a different mating type for sexual reproduction. The two compatible or complementary mycelia may be designated as plus (+) and minus (–) strain mycelia. In the heterothallic species asci are produced only when two compatible mycelia are mated. They are **self-sterile** but **cross fertile**. To elaborate this it means a plus antheridium or spermatium can unite only with a minus ascogonium or *vice versa*. This phenomenon has been termed **heterothallism** by Blakeselee (1904). It was investigaed by him in connection with the sexuality in Mucorales.

In the heterothallic species, compatibility is determined by the genetic make up of the nuclei of the mycelium. The nuclei of one mating type have gene A and the nuclei of the other mating type have gene *a*. The diploid nucleus of the ascus mother cell will have the Mendelian pair A*a*. These genes segregate at the time of meiosis prior to ascospore formation. Consequently of the eight ascospores in the ascus four will carry gene *A* and the other four gene *a*. On germination each ascospore will produce a mycelium in which all the nuclei will have either *A* or *a* gene. The mycelia with nuclei having opposite genes either *A* or *a* are said to be **compatible**. Each compatible mycelium may bear both kinds of sex organs, male as well as female. However, they cannot mate among theselves because they carry the same gene and are self-sterile. Mating takes place only when mycelia of different genetic make up come in contact. The *A* antheridium or spermatium can unite only with an ascogonium and *vice versa*. The genetic make up of the dikaryons produced by the mating of such compatible mycelia will be represented by the formula *A*+*a* and of the zygote nucleus by the formula *Aa*. This kind of compatibility is called **bipolar heterothallism.** Fungi of this category have only two mating types of mycelia. There are **tetrapolar heterothallic** species as well in the Ascomycetes. In them there are four mating types of mycelia whose nuclei carry *A, a, B, b* genes respectively. Compatibility in this case is governed by two Mendelian pairs of genes *Aa* and *Bb*. Mycelia in which nuclei carry opposite genes of both Mendelin pairs are compatible.

POST-PLASMOGAMY CHANGES

1. Development of Asci. The process of plasmogamy in the Ascomycetes initiates a new phase. It is the development of special, sac-like sporangia called **asci** (sing. *Ascus*). The asci are formed in two ways :—

(*a*) *Direct Method of Development of Asci* (Fig. 8.7). In the lower Ascomycetes the young fusion cell has two compatible, haploid nuclei (C). These come together and fuse immediately. The diploid fusion or zygote nucleus enlarges and the cell grows out to form a special, sac-like swollen structure which is spherical in *Eremascus* (D) and elongated in *Dipodascus* (Fig. 8.8 C). It is called the young **ascus.** The binucleate fusion cell which is destined to form the ascus may be called the **ascus mother cell.** The diploid nucleus of the young ascus is larger than the sum of the volumes of the two haploid compatible nuclei which have fused to form it. It immediately undergoes three nuclear divisions. The first two divisions which produce 4 haploid nuclei constitute **meiosis** (F). Each of these haploid nuclei further undergoes a third division which is **mitotic**. The ascus in *Eremascus* finally contains 8 haploid nuclei. A small amount of ascus cytoplasm gathers around each of these 8 nuclei to form 8 uninucleate daughter protoplasts (G). Each daughter protoplast secretes a wall around it to become an **ascospore**. The result is an ascus with eight haploid ascospores which have been fashioned not by cleavage of ascus cytoplasm but by a process known as the **free cell formation.** Not all the ascus cytoplasm is used up in the formation of ascospores. Some remains behind to serve as a nutritive substance for the developing spores. The number of ascospores produced in an ascus varies in different species. It may range from one to hundreds (*Dipodascus*). The usual number, however, is eight. The ascus in lower Ascomycetes is without any enveloping sheath of sterile tissue (G). Thus no definite fructification is formed in the lower Ascomycetes. The asci

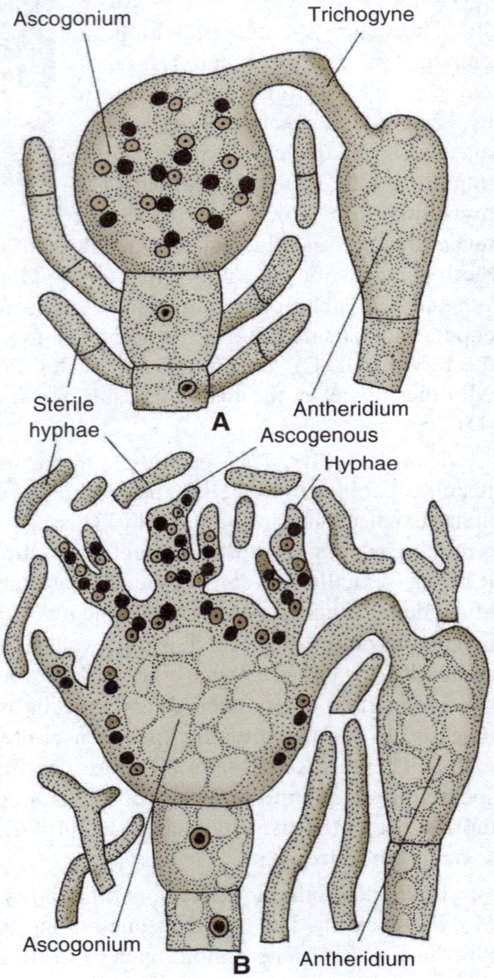

Fig. 8.14 (A-B). Ascomycetes. Plasmogamy by gametangial contact and post plasmogamy changes. A, V.S. sex organs after migration of male nuclei into the ascogonium. B, A later stage showing the arrangement of male and female nuclei in pairs (dikaryons) in the peripheral cytoplasm of the ascogonium, origin of ascogenous hyphae and the passage of paired nuclei into the unseptate ascogenous hyphae. (Diagrammatic).

are produced in a scattered and unprotected manner directly on the vegetative mycelium.

(*b*) *Indirect method of development of asci* (Fig. 8.14-15). The higher Ascomycetes tell a different tale. With the migration of male protoplast each nucleus in the ascogonium becomes paired with one from the antheridium (A). These pairs of male and female nuclei in the ascogonium are called **dikaryons.** The ascogonium containing the dikaryons is said to be dikaryotised. The male and female nuclei in the dikaryon do not fuse at this stage. The dikaryons migrate to the periphery of the ascogonium.

(*i*) *Development of ascogenous hyphae*. From the dikaryotised ascogonium arise papilla-like outgrowths just opposite the peripheral dikaryons (B). These hyphal protrusions are early products of plasmogamy or sexual fusion and are called the **ascogenous hyphae.** The asci are eventually developed from the terminal or sub-terminal cells of the ascogenous hyphae. Thus the development of asci in the higher Ascomycetes is **indirect.** The ascogonium after plasmogamy may remain unicellular or it may become transversely septate (Fig. 8.11 B). In the latter case the ascogenous hyphae arise from one or more of the median segments. Meanwhile the dikaryons undergo several conjugate (simultaneous) divisions. The resultant dikaryons now migrate from the ascogonium or ascogonial segments into the growing ascogenous hyphae which when young are unseptate. The older ones have cross walls. The ascogenous hyphae may be unbranched but more often they are irregularly branched. The branches intertwine with one another. All the cells or only the cells at the distal ends of the ascogenous hyphae are binucleate. The ascogenous hyphae thus represent the binucleate or dikaryotic sporophyte which develops from the dikaryotised ascogonium. The two nuclei in the cell constitute the **dikaryon**. One of these nuclei is of antheridial origin and the other of ascogonial. Meanwhile sterile hyphae grow up from the stalk of the ascogonium and the haploid mycelium around the ascogenous hyphae and intertwine to form a pseudoparenchymatous envelop like a hollow ball. The ascogenous hyphae consitute the **dikaryotic phase** which starts with the establishment of dikaryons in the life cycle. They are dependent for their nutrition on the haploid mycelium and thus are incapable of forming an independent mycelium.

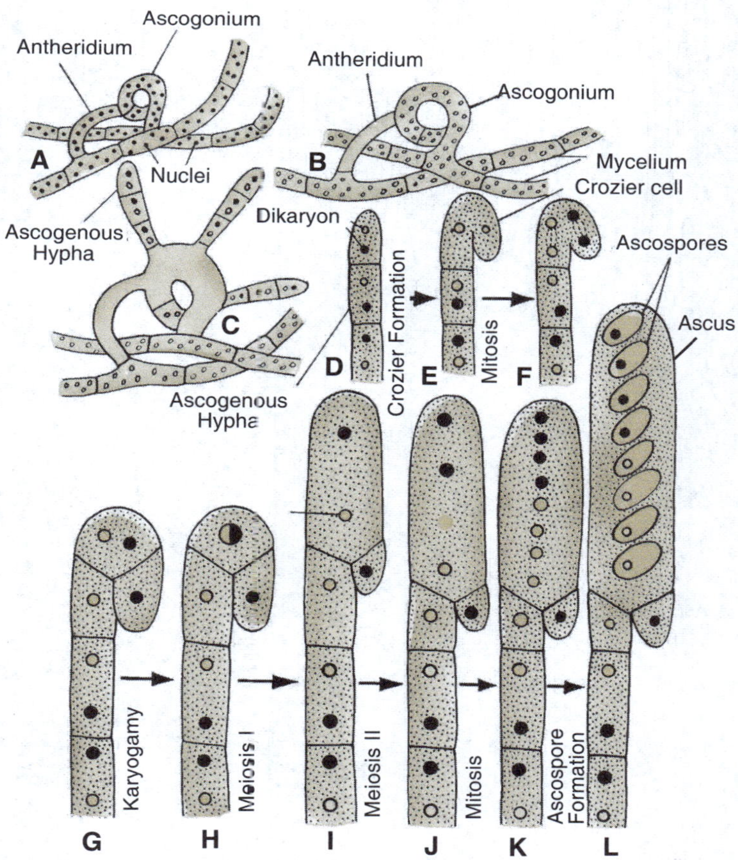

Fig. 8.15 (A-L). Stages in the development of Ascus and Ascospores.

(ii) Development of Asci (Fig. 8.15). The asci are developed from the tips of the ascogenous hyphae or their ultimate branchlets. The tip or end cell which is to form the ascus has a dikaryon (A). It curls over to form a **hook** or **crozier** (B). Both the nuclei in the hook divide simultaneously to form four nuclei. Of these, two lie in the bend or arch of the crozier (hook), one is located at the extreme tip and one near the basal septum of the crozier (C). Walls are now laid between the nuclei in the crozier to cut out a **terminal** or **ultimate** uninucleate hook cell (D_1), a binucleate **arch cell** (D_2) which is also called the **penultimate cell** (as it is last but one in position) and a uninucleate stalk or **antepenultimate cell** (D_3). These three cells compose the characteristic hook or crozier of higher Ascomycetes.

Of the two nuclei in the arch cell, one is male and the other female. The binucleate arch cell enlarges to form the **ascus**. It may also be called the **ascus mother cell** (E). Early in the developments of ascus the male and female nuclei fuse (F). The fusion of a male and a female nucleus is called **karyogamy**. It is equivalent to fertilisation. The fusion nucleus is called a **synkaryon**. The young ascus with a synkaryon represents the transitory **diplophase** in the life cycle.

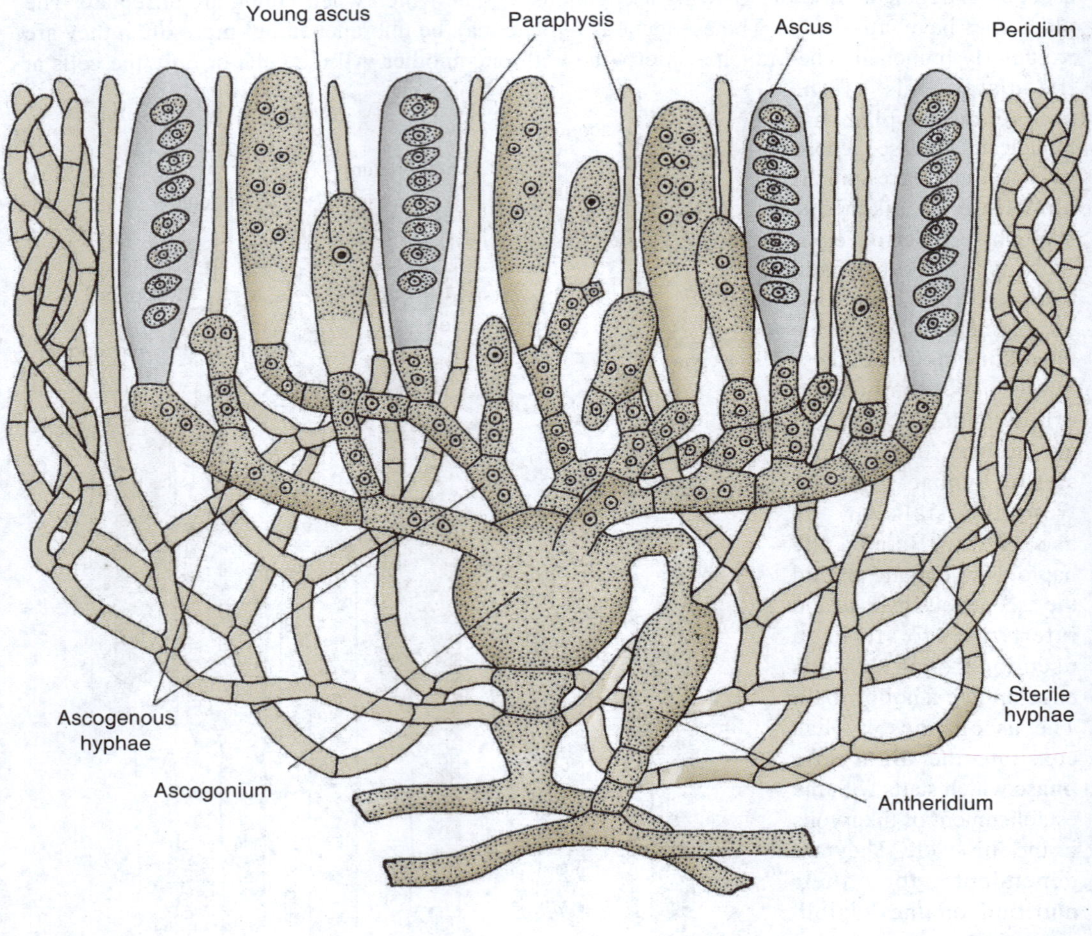

Fig. 8.16. Ascomycetes. Section through a young ascocarp showing the sexual apparatus and arrangement of ascogenous and sterile hyphae (Diagrammatic).

(*iii*) *Ascosporogenesis* (*differentiation of Ascospores*, Fig. 8.15 G-J). The process whereby ascospores are fashioned from the diploid protoplast of the elongating young ascus is termed **ascosporogenesis.** Ascosporogenesis is preceded by **meiosis**. During ascus elongation the synkaryon or zygote nucleus in the ascus undergoes meiotic division typically resulting in the formation of 8 haploid nuclei (G-I), sometimes more. Segregation of strains takes place during the meiotic division. The next step is the formation of the ascus vesicle. It is the double membraned sac formed within the young ascus. It encloses the 8 haploid nuclei embedded in the ascus cytoplasm. The mebrane of the ascus vesicle folds in between the haploid ascus nuclei delimiting 8 ascospores. Each resultant young ascospore is bounded by a bilayered membrane. Towards maturity secondary wall forms around each maturing ascospore. The residual cytoplasm which is not used up in the formation of ascospores is called **epiplasm.** It becomes rich in glycogen and other food substances and supplies nutrition to the developing ascospores. This mode of development of asci from the ascogenous hyphae and not directly from the fusion cell or zygote is described **indirect**. A cluster of asci may be produced at the end of an ascogenous hypha or a branch by the formation of new croziers either from the original arch cell before it matures into an ascus or by fusion of the hook cell and stalk cell. The nucleus of one passes into the other. The so formed binucleate fusion cell can produce another ascus by hook formation. The process may be repeated leading to a whole cluster of asci at the end of an ascogenous branch. The asci are frequently arranged parallel to one another forming a regular layer with the paraphyses intermixed in between. This layer is called hymenium. Meanwhile sterile hyphae grow up in-between the asci to become paraphyses (Fig. 8.16).

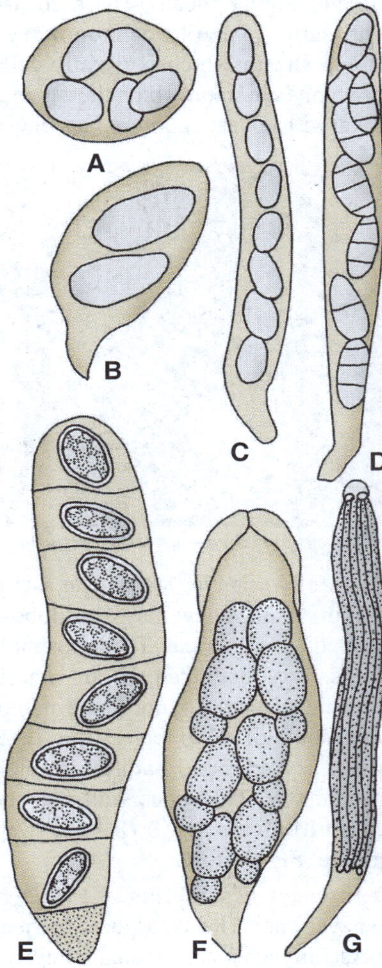

Fig. 8.17 (A-G). Ascomycetes. Different forms of asci. A, more or less a globose ascus; B, broadly ovate ascus with a stalk; C, cylindrical ascus with ascospores in a single row; D, cylindrical ascus with multicellular ascospores; E, septate ascus; F, clavate ascus with ascospores in two rows; G, tubular ascus containing thread-like ascospores in a bundle.

Brachymeiosis. Another view on the development of ascus and ascospores was put forth by Harper (1900) which was supported by Gwynne Vaughan and Williamson (1931) and Foster (1941). According to this view, there are two nuclear fusions and one complete and another incomplete meiosis or broken meiosis. The first fusion of male and female nuclei occurs in the ascogonium itself. This nuclear fusion takes place immediately after plasmogamy. The development of ascogenous hyphae and the migration of nuclei takes place but each cell of the ascogenous hypha contains a pair of diploid nuclei instead of haploid nuclei. The second nuclear fusion as usual takes place in the ascus mother cell resulting in the formation of a tetraploid nucleus. Since the ascospores are haploid, there has to be two meiotic divisions of nuclei, of these the first meiotic division is complete giving rise to four diploid nuclei. But the second meiotic division is incomplete or broken in the sense that during this division mitosis does not take place. Eight haploid nuclei are thus formed. Since the second meiotic division is incomplete or broken it is called **Brachymeiosis**. However, the incomplete meiosis is an impossible biological condition. On this basis the view was rejected by most mycologists.

Formation of Ascocarp (Fig. 8.16). Sexual apparatus including the asci and the ascogenous hyphae are already surrounded by a common ball-like pseudoparenchymatous sheath of sterile hyphae. The latter arise from the stalk of the ascogonium and the haploid mycelium and grow up simultaneously with the ascogenous hyphae. This entire fructification is called **ascocarp.** It consists of the ascogonium with the dikaryons, ascogenous hyphae, a mass of asci containing ascospores and the enveloping sheath. The ascocarp is characteristic of the higher Ascomycetes only. It is made up partly of **haplo** or **monokaryotic** mycelium and partly of **dikaryotic** mycelium. The enveloping sheath which is usually called the **peridium** is haplont in nature. It along with the vegetative mycelium on which the sex organs are borne forms the **haplophase** or **the gametophyte.** The dikaryotised ascogonium, ascogenous hyphae and asci before the differentiation of the ascospores are **diplont**. These constitute the **dikaryotic phase.**

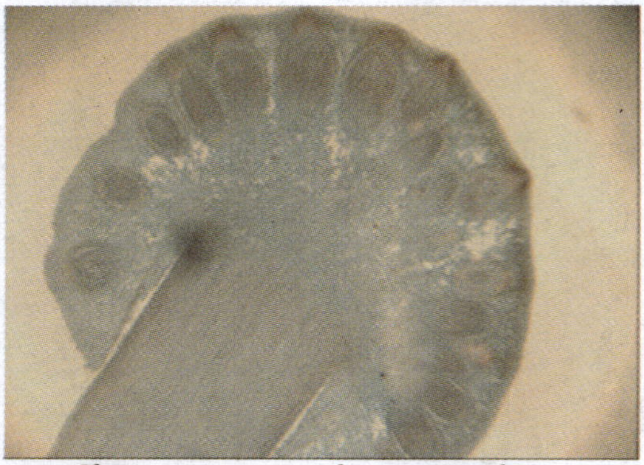

Claviceps purpurea – showing pycnidia

Form of Asci (Fig. 8.17). The asci are very variable in form. They may be globose, oval, club-shaped or cylindrical. The ascospores within the ascus may be arranged in a uniseriate, biseriate, fasciculate or inordinate manner. In a tubular ascus of *Claviceps,* the thread-like ascospores are arranged parallel to one another in a bundle (G). The ascus wall may be single layered (**unitunicate** Fig. 8.18 A) or two-layered (**bitunicate**, Fig. 8.18 B).

Dehiscence of Asci (Fig. 8.19). Dehiscence of asci is variable. The ascospores in the majority of the Ascomycetes are ejected forcibly usually with a puffing action. The glycogen present in the epiplasm becomes rapidly transformed into sugars of high osmotic value. Such a transformation results in the absorption of water. The pressure thus set up within the ascus is responsible for tearing of the ascus wall with a sudden explosion. The ascospores are ejected forcibly into the air where they are caught up by

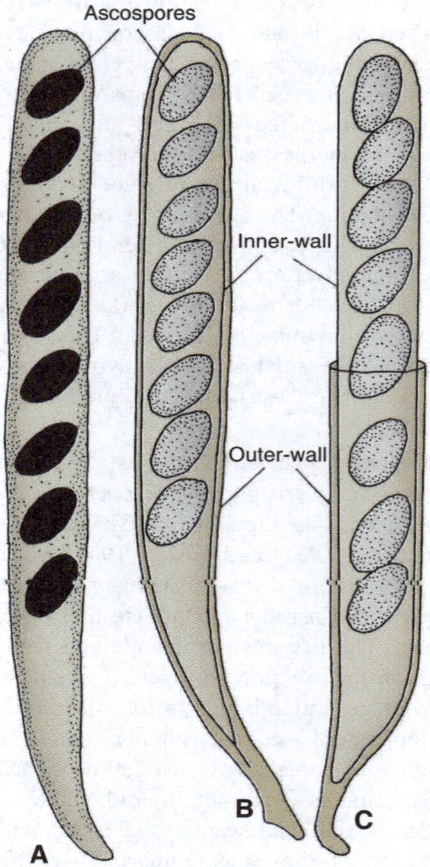

Fig. 8.18 (A-C). Ascomycetes. Two kinds of asci. A, unitunicate; B, bitunicate; C, thick, outer wall of the bitunicate ascus ruptured to allow the ascus with the inner wall to emerge.

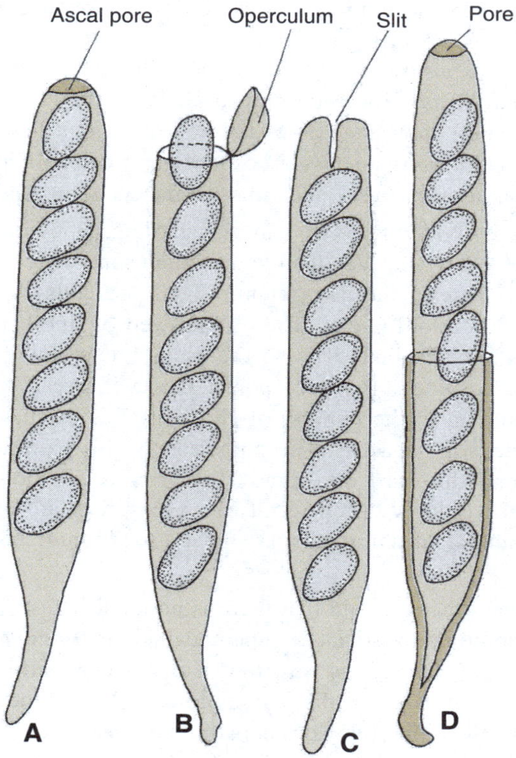

Fig.8.19 (A-D). Ascomycetes. Various types of ascal dehiscence.

the air currents and dispersed. The ascus may dehisce along a definite line called a slit (C) or at a previously differentiated thin apical patch (*Sphaerotheca*). In a few cases there is a separation of lid-like structure at the tip (B). This lid is hinged at one end. It is followed by a violent ejection of the ascospores in a jet of liquid. These explosions may by simultaneous or successive. In some genera the ascus merely

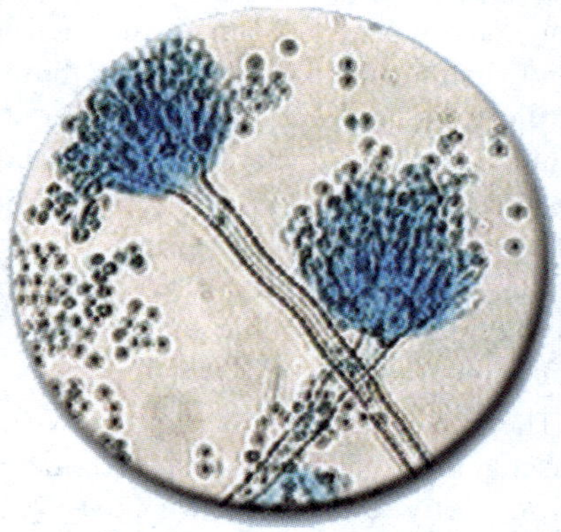

Aspergillus showing conida and conidiophores

ruptures at the tip to form a pore (A). The ascospores escape through the *ascal* pore. In some primitive Ascomycetes (*Aspergillus*) the ascus wall is evanescent and undergoes digestion at maturity. The ascospores are left in the mucilaginous liquid in the cavity of the fruit body. They are dispersed by the insects or washed by rain water at the decay of the surrounding sheath.

Ascospores. They are usually elliptical. In some species, they are spherical. In still a few others they are long and narrow. The ascospores contain densely granular cytoplasm and frequently oil drops. The spore wall is usually thin but in many Ascomycetes it is differentiated into the usual two layers. The epispore may be smooth or thick and sculptured. It may develop a gelatinous sheath or appendage. It is hyaline or opaque, coulurless or variously coloured. The young ascospore is ordinarily unicellular and uninucleate. Towards maturity it may in rare cases become multicellular.

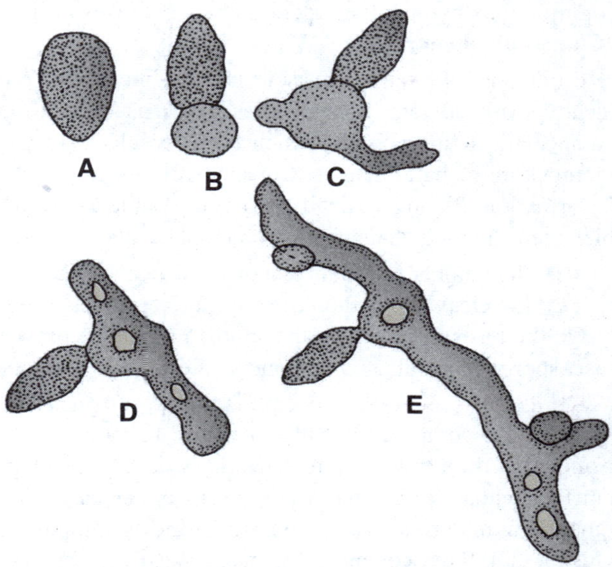

Fig. 8.20 (A-E). Ascomycetes. Germination of ascospore. Explanation in the text.

Germination of Ascospore (Fig. 8.20). Encountering suitable conditions (moisture and temperature round about 20°C) ascospores of most species germinate. The ascospores of some species, however, are induced to germinate after exposure to special treatment (freezing, heating or immersion in acid). After four hours in malt extract the ascospore becomes wrinkled. The contents extrude in the form of a bubble (B). The bubble then gives rise to one or more germ-tubes (C). The latter develop into a septate mycelium. Frequently ascospore germinates directly by the formation of germ-tubes. In a multicellular ascospore each cell may produce a germ tube.

Cytology of the Ascus. The mycologists are divided into two camps regarding the cytology of the ascus. According to some, Harper (1900), Gwynne Vaughan and Williamson (1931) and Foster (1941), the male and the female nuclei fuse in the ascogonium. Thus the cells of the ascogenous hyphae contain a pair each of truely diploid nuclei. This is followed by another or second fusion of diploid nuclei in the ascus mother cell. If it is so the ascus nucleus will be **tetraploid** having a four-fold number of chromosomes. The ascospores, however, are always haploid. There must, therefore, be double reduction in the number of chromosomes during the three successive divisions of the tetraploid nucleus in the ascus. The supporters of this viewpoint hold that reduction takes place during first and the third divisions. This bring about double reduction and the chromosome number is twice reduced. As a result, the ascospores are once more haploid. This double reduction in the number of chromosomes in the ascus is termed as **brachymeiosis.**

The advocates of the second view (Claussen's theory) hold that there is no nuclear fusion in the ascogonium. The male and the female nuclei lie in it associated in pairs called the **dikaryons.** Fusion between them normally occurs in the ascus mother cell. According to this view, the fusion nucleus (synkaryon) in the young ascus is the only diploid nucleus in the life cycle of Ascomycetes. The first two divisions of the diploid nucleus constitute **meiosis**. The third is mitoic. The first theory (brachy-meiosis) has been adversely criticised by the present-day mycologists. Recent cytological investigations definitely show that there is no karyogamy in the ascogonium of the higher Ascomyctes. Chromosome counts by Hiroch (1950) and Olive (1950) have shown that the division of diploid nucleus in an ascus is not brachymeiotic. The consensus of opinion thus favours the second view which was propounded by Claussen in 1912. It is known as **Claussen's theory.**

Phylogeny of Ascus. The ascus is a sac-like, sexual reproductive structure of the Ascomycetes which typically contains 8 meiospores known as the **ascospores.** It is regarded by some mycologists as a specialised sporangium which is the result of karyogamy. The fusion nucleus thus formed undergoes immediate meiosis. Consequently usually eight ascospores are fashioned in it by free cell formation. Phylogenetically it is comparable to the germ sporangium of the Phycomycetes in which the following changes have taken place :—

(*i*) the number of the meiospores has been reduced to eight.

(*ii*) the cleavage method of spore formation has been replaced by free cell formation.

On the basis of this view the spores in the phycomycetous germ sporangium are comparable to ascospores. They also correspond to the spores of vascular plants.

Kinds of Ascocarps. Ascocarp is the spore-fruit or fructification of the higher Ascomycetes. It is usually a complex, ball-like structure formed from the ascogonium usually after the act of plasmogamy. In a few Ascomycetes the ascocarp develops first and the sex organs develop later from the hyphae within the ascocarp. In either case, the peridium is formed from the somatic hyphae. The mature ascocarp consists of the ascogonium with the paired male and female nuclei or fusion cell, the ascogenous hyphae, a mass of asci containing ascospores, paraphyses, and the peridium (enveloping sheath). The two generations (gametophyte and sporophyte) are thus united in a common structure (ascocarp) with the gametophyte almost always strongly developed. The sporophyte is parasitic on it. It never becomes independent. Ascocarp formation is characteristic of all the higher Ascomycetes. Generally the ascocarp contains the asci formed from more than one neighbouring ascogonia. Such a fructification is called a **compound ascocarp.**

The young ascocarps are globose and usually completely closed. In a few Ascomycetes the ascocarp retains this form even at maturity. It opens by the decay or irregular splitting of its wall (**peridium**). In others towards maturity the ascocarp assumes a flask-shaped outline. It opens by a terminal pore called an **ostiole.** In majority of the Ascomycetes, the elongation of paraphysis and growth of sterile hyphae rupture the upper half of the globose ascocarp which thus assumes the form of a cup or saucer. The cavity of the cup is lined by asci which are thus freely exposed. According to their shape, the ascocarps of the Ascomycetes. in general, are classified into four kinds, namely, **apothecia** (sing. apothecium), (*ii*) **perithecia** (sing. perithecium), (*iii*) **cleistothecia** (sing.

Apothecium

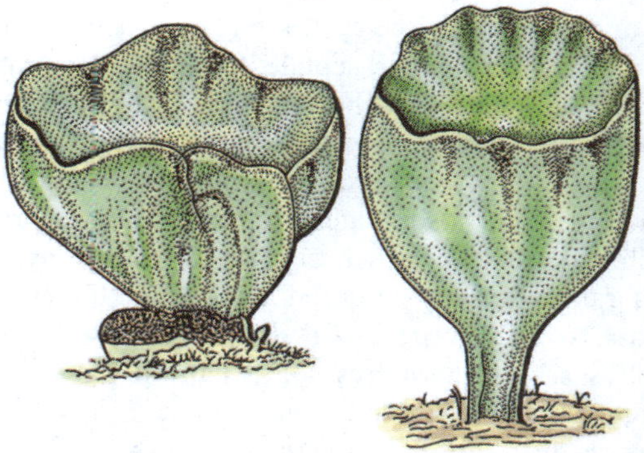

Fig. 8.21 (A-B). Ascomycetes. Two apothecia; A, Sessile apothecium of *Peziza*; B, stalked apothecium of *Urnula*.

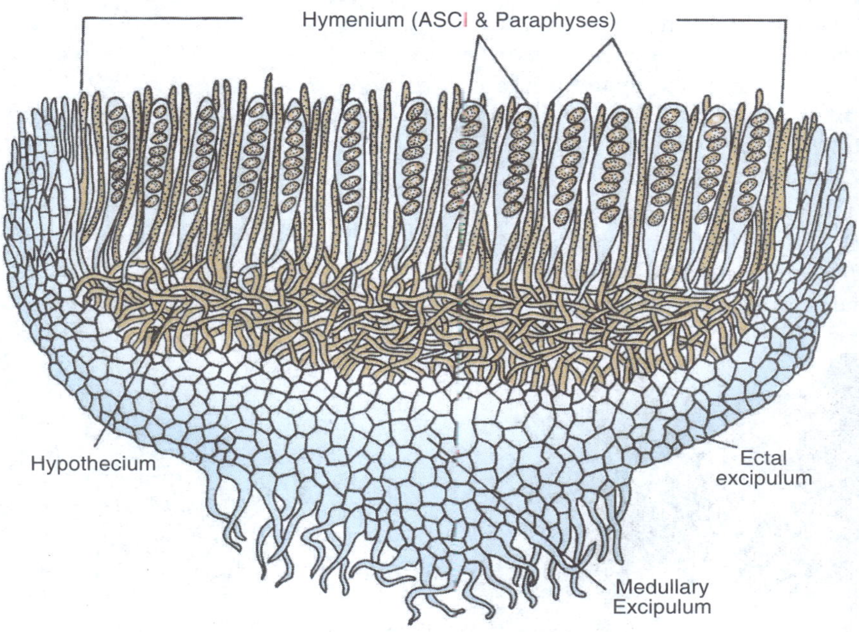

Fig. 8.22. Ascomycetes. V.S. Apothecium showing structural details.

cleistothecium), and (*iv*) **ascostromata** (sing. ascostroma).

1. Apothecium (Fig. 8.21). The typical apothecium is characteristic of the advanced members of higher Ascomycetes. It is a disc or saucer-shaped or even a cup-like structure. In size it usually varies from a few millimetres up to several centimetres in diameter. It varies in colour from pale brown to black sometimes red, orange, yellow or even coulurless. The apothecium in some species is elevated on a distinct stalk (B) and is sessile in others (A).

Structure of apothecium (Fig. 8.22). Chiefly it consists of three parts, namely *hymenium, hypothecium* and *excipulum*.

(*i*) Hymenium. It is the fertile layer that lines the cavity of the cup or surface of the disc and is also called the **thecium**. It consists of closely packed elongated sac-like cells, the **asci** (Fig. 8.22). The asci are typically uninucleate and stand perpendicular to the surface like a palisade. The mature asci are generally tubular cells. Each is bounded by a stretched, thin elastic wall which appears to be single but in reality it consists of two layers. Within the cell wall is a lining layer of cytoplasm. In the centre is a large vacuole. In it usually lie the eight ascospores. Intermingled with the asci are slender, septate, hyphae, the **paraphyses**. Generally they are of the same length as the asci or somewhat shorter. In the fructification they arise earlier than the asci. As they grow they force open the sheath of the closed, young ascocarp. The paraphyses have a protective, supporting and perhaps a distributing role. They bring up food material which is absorbed by the developing asci. In some genera the paraphyses project beyond the asci. The projecting tips may branch. The tips of branches unite to form a layer which spreads over the tips of the asci and forms a thin, brightly coloured layer called the **epithecium**. The bright colour of many apothecia is due to pigmented oil drops in the paraphyses. The paraphyses are haploid in character. In some forms they are pointed, in others club-shaped, and in still others cylindrical. They have no connection with the ascogenous hyphae.

(*ii*) Hypothecium. Immediately beneath the hymenium is a thin layer of interwoven hyphae. It is known as the **hypothecium**. It consists of a network of hyphae running horizontal to the hymenial layer. The paraphyses arise directly from this layer.

(*iii*) Excipulum (meaning receptacle). It is the haploid fleshy part of the ascocarp or in fact the apothecium proper which supports the hypothecium and hymenium. It makes the basal

Perithecium

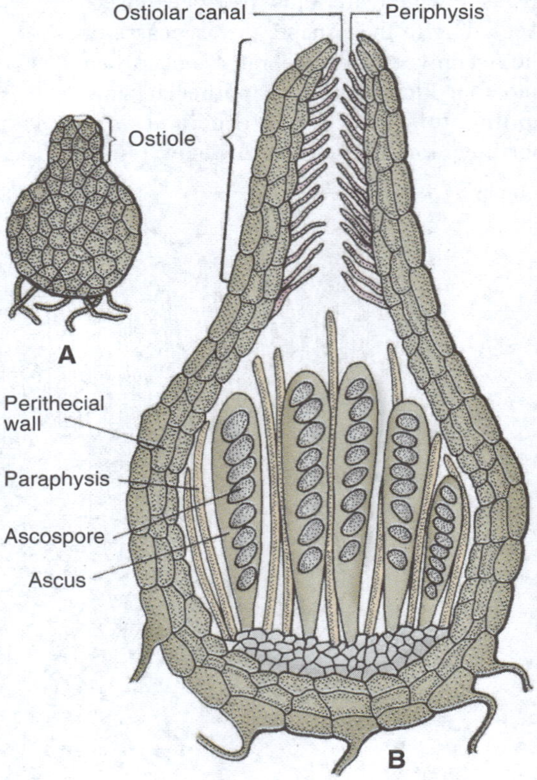

Fig. 8.23 (A-B). Ascomycetes. A. Entire perithecium; B, V.S. perithecium showing structure.

part and sides of the apothecium and consists of pseudoparenchyma or of closely compacted interwoven hyphae. The outer layers of the excipulum consistute the **ectal excipulum** and inner portion **ental** or **medullary excipulum.**

2. Perithecium (Fig. 8.23). It is a small rounded or flask-shaped ascocarp (A) usually less than a millimetre in diameter. At maturity it has a definite apical pore or opening, the **ostiole** (B) through which the ascospores escape. The ostiole which is a characteristic feature of the perithecium may simply be a short papilla opening by a circular pore. In many species it is in the form of a long neck with an apical pore. The ostiolar canal is lined by slender, short, delicate, hair-like sterile growths called the **periphysis**. They point outwards. In many species the perithecial neck is phototropic. The perithecial wall is called the peridium. It encloses a cavity lined by the hymenium either throughtout or only at the base. The hymenium consists of club-shaped or cylindrical asci intermingled with paraphyses which in some species may be absent.

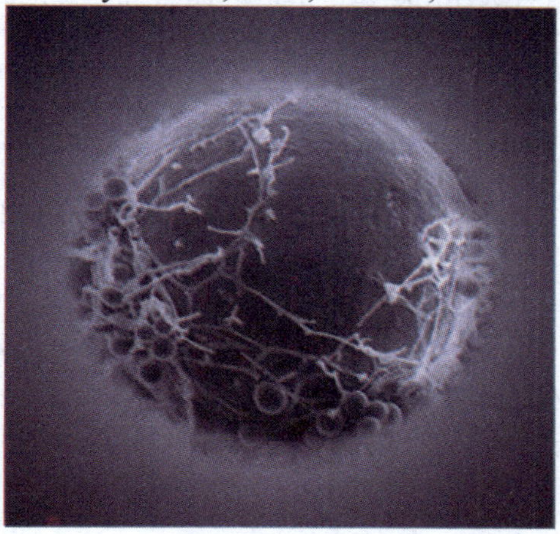

Cleistothecium

3. Cleistothecium (Fig. 8.24). It is a small hollow, more or less globose, indehiscent ascocarp without an opening. The fruit wall which is called the **peridium** may simply consist of loosely interwoven hyphae or pseudoparenchyma. It may be smooth but usually it is covered with hyphal outgrowths called the **appendages** (A) which may be of various types. The globose or ellipsoidal asci lie scattered at different level within the peridium (B). In some they are arranged regularly and form distinct hymenium (C). The ascospores escape by the decay or irregular splitting of the wall of the asci and the cleistothecium. The cleistothecium may be sessile or stalked and may contain a single ascus or several asci.

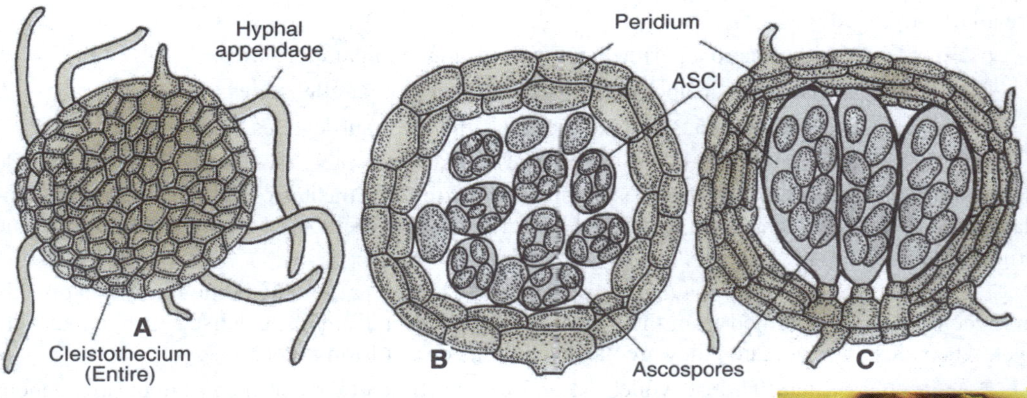

Fig. 8.24 (A-C). *Ascomycetes.* A, Entire cleistothecium bearing hyphal appendages; B, section of a cleistothecium of *Aspergillus* with scattered asci; C, L. S. cleistothecium of *Erysiphe* with asci forming a regular layer.

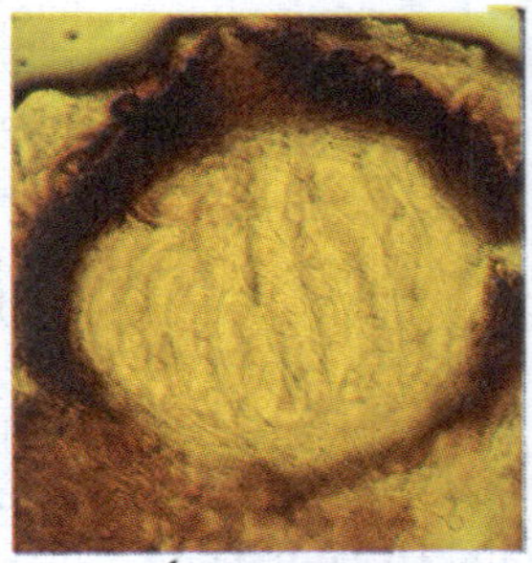

Ascostroma

4. Ascostroma (Fig. 8.25 A). It is a characteristic fructification of sub-class Loculoascomycetidae. In this case bitunicate asci are formed directly in cavities called **locules** within a stromatic ascocarp. The stroma is a mattress or cushion of compactly woven somatic hyphae.

The asci are not organised into a hymenium or distinct fruit body. The stroma itself forms the wall of the fructification. There may be a single ascus in each locule (**monascal**) or more than **one polyascal** (A). The monascal locules are arranged in an irregular manner within an ascostroma and the polyascal locules are arranged in a difinite basal layer. In some species the stroma is small and the partitions between the two individual locules break down towards maturity. It becomes difficult to distinguish this type of ascostroma from a perithecium. The locules may open at the surface of the stroma by pin-point openings. In many species the upper part of the stroma disintegrates and the asci are exposed. The species with unitunicate asci developed in a loculate stroma are retained in the sub-class Euascomycetidae.

5. Hysterothecium (Fig. 8.25 B-C). It is a small, black, elongated, boat-shaped fructification with leathery or hard texture. It opens at maturity by a single long, narrow slit which runs parallel to the long axis of the fruit body nearly from top to the bottom (B). It is characteristic of the order Hysteriales.

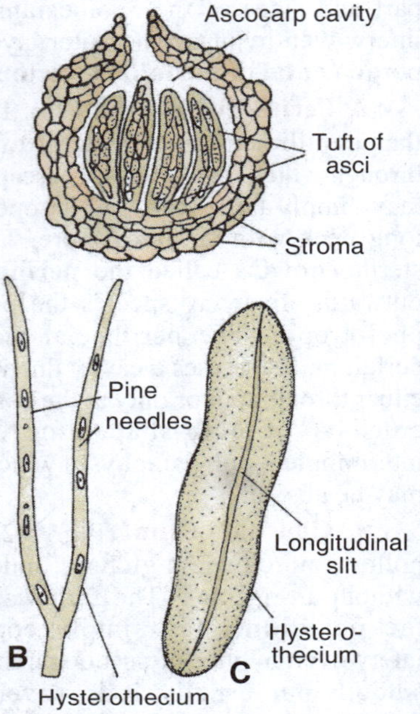

Fig. 8.25 (A-C). Ascomycetes. A, L. S. mature ascostroma of *Mycosphaerella tulipiferae* showing apical pore and basal group of asci in the locule; B, pine needles with hysterothecia; C, single hysterothecium (A after Higgins).

Classification. The classification of Ascomycetes is difficult and complex.

According to Gwynne Vaughan and Barnes (1937), the class is divided into the sub-classes :

Plectomycetes. Ascocarp a cleistothecium with asci irregularly arranged.

Pyrenomycetes. Ascocarp a perithecium, asci arranged in parallel series.

Discomycetes. Ascocarp an apothecium, asci arranged in parallel series.

However, Alexopoulos (1962) divided the class into three sub-classes namely.

Hemiascomycetidae, Euascomycetidae and **Loculoascomycetidae.** Some mycologists add **Laboulbeniomycetidae** which includes epiphytes or external parasites on insects, to the list. This classification is based on three features, kind of ascocarp, method of ascus development and structure of asci.

1. Hemiascomycetidae (=Protoascomyce-tidae). The asci are usually borne singly and directly on mycelium and are formed from the zygote, but sometimes from a specialised ascogenous cell. Even when closely aggregated they are naked and thus do not form ascocarps.

2. Euascomycetidae. The asci which are typically **unitunicate**, are formed from the ascogenous hyphae. They are closely aggregated and surrounded by a common sheath of haploid, vegetative hyphae to form a characteristic fructification called an **ascocarp** which may be a perithecium, apothecium or cleistothecium.

3. Loculoascomycetidae. The asci are bitunicate. They are produced singly or in small groups in separate locules in a stroma. This kind of fructification is called ascostroma (pl. ascostromata).

In 1973, Ainsworth divided the subdivisions Ascomycotina into six classes namely **Hemiascomycetes, Loculoascomycetes, Plectomycetes, Laboulbeniomycetes, Pyrenomycetes** and **Discomycetes.** This classification is based on the presence or absence of ascocarp, the type of ascocarp and uni-or bitunicate nature of the ascus.

1. Ascocarp and Ascogenous hyphae lacking, thallus mycelial or yeast like : class **Hemiascomycetes.**
2. Ascocarp and ascogenous hyphae present, thallus mycelial
 2 (*i*) Asci bitunicate, ascocarp an ascostroma : class **Loculoascomycetes.**
 2 (*ii*) Asci unitunicate typically
  (*a*) Asci evanescent, scattered within the ascocarp which is a cleistothecium, ascospores aseptate : class **Plectomycetes.**
  (*b*) Asci arranged within the ascocarp in regular or parallel layers.
   (*i*) Exoparasites of arthropods, thallus reduced, ascocarp a perithecium, asci inoperculate : class **Laboulbeniomycetes.**
   (*ii*) Not exoparasites of arthropods.
    (A) Ascocarp typically a perithecium which is ostiolate, asci inoperculate with an apical pore or slit : class **Pyrenomycetes.**
    (B) Ascocarp an apothecium or a modified apothecium, frequently macroscopic, epigean or hypogean, asci inoperculate or operculate : class **Discomycetes.**

QUESTIONS

ESSAY TYPE

1. Give an account of the economic importance and list the diagnostic features of class Ascomycetes.
2. Describe the somatic structures and the various modes of asexual reproduction met with in the Ascomycetes.
3. Give brief account of sexual reproduction and formation of ascocarp in the higher Ascomycetes.
4. Describe the various stages in the development of ascus in the Ascomycetes and its different modes of dehiscence.
5. What is an ascus? Discuss the Cytology and Phylogeny of an ascus.
6. What is an ascocarp? Describe the various kinds of ascocarps met within the Ascomycetes.
7. With the help of a sketch give a detailed account of the structure of an apothecium.
8. Discuss the various views put forth to account for the origin of Ascomycetes.
9. Define the term 'ascus'. Describe the process of ascospore formation in any fungus you have studied. *(Agra 1962)*
10. Discuss the formation of ascocarp in Ascomycetes. *(Kanpur M.Sc., 1998)*
11. Discuss the degeneration of sexual structure in ascomycetes studied by you. *(Lucknow, 1995)*
12. Explain the phenomenon of degeneration of sexuality in Ascomycetes. *(Lucknow, 2000)*
13. Give an account of different types of fruiting bodies in Ascomycetes. *(Poorvanchal , 2003)*
14. Discuss various criteria on which different classification of Ascomycetes have been proposed. Outline a recent classification of this group. *(Kanpur M.Sc., 2002)*
15. Describe the formation of Ascocarp in Ascomycetes. *(Kanpur M.Sc., 2003)*

SHORT ANSWER TYPE

16. **Write short notes on :**
 (*a*) Fructifications in Ascomycetes *(Kanpur, M.Sc., 1989, Rohilkhand, 2001)*
 (*b*) Ascocarps in Ascomycetes *(Kanpur, M.Sc., 1990)*
 (*c*) Fruiting bodies of Ascomycetes *(Gorakhpur, 1997)*
 (*d*) Brachymeiosis *(Kanpur, 1996, 2003, 2006, Lucknow, 2004, Meerut, 1994)*
 (*e*) Apotheceium *(Bihar, 1990)*
 (*f*) Types of Ascocarp *(Rohilkhand, 1992)*
 (*g*) Ascus and Ascospores *(Kumaon, 1997, 1999, 2003)*
 (*h*) Ascogenous Hyphae *(Kumaon, 2000)*

(i) Cleistothecium (Allahabad, 2002)
(j) Ascus (Allahabad, 2004)
(k) Crozier formation (Kanpur M.Sc., 2004)

17. In what respects, the apothecium, perithecium and cleistothecium differ from each other. (Gorakhpur, 1993)

18. What is an ascus? Describe its structure in detail. Name some fungi of this group. (Kanpur, 1995)

19. What is cleistothecium and how is it formed? (Punjab, 1993)

20. Differentiate between
(a) Cleistothecium and apothecium (Rohilkhand, 1991, 1994, Kanpur, 2001)
(b) Cleistothecium and perithecium
(c) Apothecium and perithecium

21. Describe the ascospore formation in ascomycetes. (Poorvanchal, 2001)

22. Describe different kinds of fruiting bodies in Ascomycetes. (Kanpur M.Sc., 2004; Kanpur, 2005)

OBJECTIVE TYPE

23. Fill in the blanks :

(i) Ascomycetes are also known as ....
(ii) The characteristic spores of Ascomycetes are ....
(iii) Asexual reproduction generally takes place by means of externally produced structures called ....
(iv) The mycetium in Ascomycetes is ....
(v) An .... is a sac like structure produced after sexual reproduction.
(vi) An apothecium is .... shaped ascocarp.
(vii) The closed ascocarp without any opening is called ....
(viii) The ascocarp which is flask-shaped with an ostiole is called ....
(ix) Karyogamy takes place in ....
(x) During the development of ascospores, the first meiotic division is complete while the second meiotic division is incomplete. This is known as ....
(xi) Generally the ascus has .... ascospores.
(xii) Cleistothecium is a fruiting body of the class ....
(xiii) Ascogenous hyphae develop only after .... reproduction.
(xiv) Ascocoarps are produced in the members of the class ....
(xv) Fungi producing sac like structures which contain 8 Spores have been placed in ....

24. Select the correct answer :

(i) A fungus producing 8 spores in a sac like structure should rightly be placed in:—
(a) Phycomycetes
(b) Ascomycetes
(c) Basidiomycetes
(d) Deuteromycetes

(ii) Generally during the formation of ascospores and asci there may be
(a) single nuclear fusion and single meiosis,
(b) single nuclear fusions and double reduction division,
(c) two nuclear fusions and two complete reduction divisions,
(d) two nuclear fusions and one complete reduction and one incomplete reduction division.

(iii) A closed ascocarp is named as
(a) Perithecium (b) Apothecium
(c) Cleistothecium (d) Hypothecium

(iv) Ascogenous Hyphae are present in the life cycle of
(a) *Aspergillus* (b) *Mucor*
(c) *Rhizopus* (d) *Pilobolus*

(v) Ascocarps are the characteristic features of
(a) Phycomycetes (b) Ascomycetes

(*c*) Basidiomycetes (*d*) Deuteromycetes

(*vi*) Stroma is
(*a*) compact somatic hyphae with fruit bodies
(*b*) loosely inter-woven hyphae
(*c*) small hyphal branch
(*d*) a group of spores

(*vii*) The protective covering of sterile hyphae around an ascocarp is known as—
(*a*) periderm (*b*) peridium
(*c*) appendages (*d*) epiderm

(*viii*) The asexual reproduction in Ascomycetes is generally through
(*a*) Conidia (*b*) Zoospores
(*c*) Zygospores (*d*) Aplanospores

(*ix*) In Ascomycetes, Karyogamy occurs within the
(*a*) ascogonium (*b*) antheridium
(*c*) ascus mother cell (*d*) ascogenous hyphae

(*ix*) An ascocarp which is flask-shaped with an ostiole in the centre is
(*a*) Apothecium (*b*) Perithecium
(*c*) Cleistothecium (*d*) Hypothecium

(*x*) Perfect stage of fungus implies that
(*a*) the fungus ia perfectly healthy
(*b*) the fungus reproduces asexually
(*c*) the fungus reproduces sexually
(*d*) the fungus produces resting spores.

(*xi*) The term sclerotium refers to a modified mycelium which is
(*a*) an underground structure
(*b*) a hard resting body
(*c*) a food storing organ
(*d*) easily carried off by wind

(*xii*) Ascospores are formed in the structures known as
(*a*) basidium (*b*) ascus
(*c*) sporangium (*d*) gametangium

(*xiii*) Asexual reproductive structures formed directly at the tip of hyphae are
(*a*) Spores (*b*) Aplanospores
(*c*) Conidia (*d*) Zygospores

(*xiv*) Who was the first to recognise and delimit the Ascomycetes as a distinct group of fungi?
(*a*) August Corda
(*b*) E.J. Butler
(*c*) Antonie von Leeuwenhoek
(*d*) Carolus Linnaeus

(*xv*) In ascocarp the structure not found is
(*a*) Ascus (*b*) Paraphyses
(*c*) Hyphae (*d*) Chlamydospore

Taphrina deformans causing leaf curl disease

CHAPTER 9

SUB CLASS – HEMIAS-COMYCETIDAE
Yeast

- General characteristics
- Order : Taphrinales
- Order : Endomycetales
- Family : Saccharomycetaceae
- *Saccharomyces cerevisiae*
- Economic Importance of Yeasts.

General. The sub-class Hemiascomycetidae includes the simplest Ascomycetes which some mycologists consider primitive and others degenerate. They have very simple structure. Sexual reproduction takes place by the conjugation of two cells (gametangia) which in some species are alike (*Eremascus*) and in others unlike (*Dipodascus*). The fusion cell may be directly transformed into an ascus or grows out to produce an ascus. The chief characteristics of this class are :–

(*i*) The mycelium either poorly developed or lacking.

(*ii*) Absence of ascogenous hyphae.

(*iii*) Complete absence of ascocarps.

(*iv*) Direct development of asci without the intervention of ascogenous hyphae.

(*v*) Asci borne directly on the mycelium or formed from a specialised ascogenous cell.

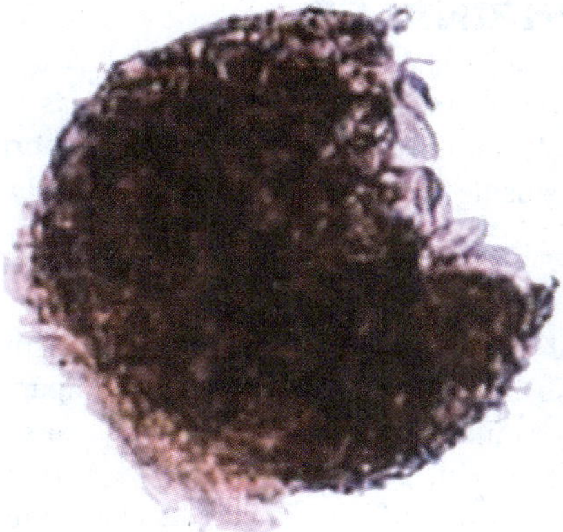

Eremascus

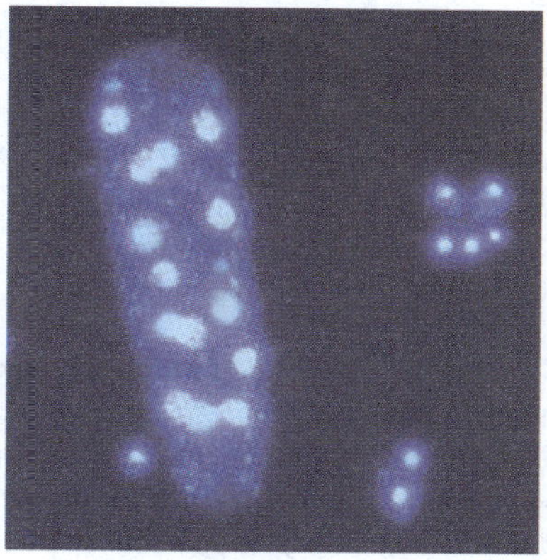

Dipodascus

Classification

The sub-class includes about 250 species assigned to nearly 50 genera which are placed under the following two orders :–

1. Order Taphrinales. It is characterised by the presence of a true mycelium. The asci are developed from special, binucleate ascogenous cells derived from the binucleate mycelium and not from the zygote. They develop in the same manner as the chlamydospores. The asci lie in a palisade-like layer parallel to one another but without any enclosing sheath or peridium. No fruit body is thus organised. Some mycologists recognise only a single family Taphrinaceae with a single genus *Ascomyces* (=*Taphrina*) in this order. There are others like Kramer (1958) who include family Protomycetaceae in addition. They make about 6 genera with 125 species in this order. The best known species of this order is *Ascomyces* (=*Taphrina*) *deformans*. It causes a disease of the peach plant known as *peach leaf curl*. The parasitised leaves become curled, thickened and distorted.

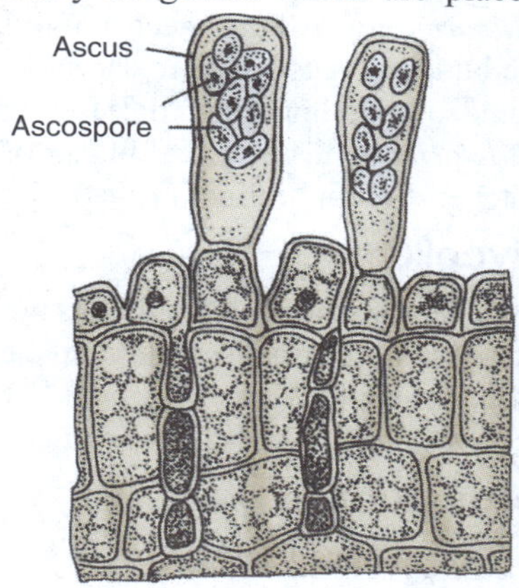

Fig. 9.1. *Taphrina deformans.* Naked Asci forming a hymenial layer on the surface of the host but there is no protective sheath. (After Miss Martin)

2. Order Endomycetales. (Saccharomycetales). The asci develop directly from the fusion cell (zygote) or sometimes parthenogenetically. The zygote is formed by the fusion of the protoplasts of two gametangia or two ascospores or two haploid vegetative cells. The plasmogamy is immediately followed by karyogamy so that there is no dikaryophase in the life cycle. This order is discussed here in detail.

Taphrina deformans

Peach Leaf Curl

ORDER : TAPHRINALES

General account. The order Taphrinales was established by Gaumann (1926) includes two families Protomycetaceae and Taphrinaceae. Later the family Protomycetaceae was elevated to the rank of order Protomycetales (Martin, 1961; Karamer, 1973). The family Taphrinaceae is monogeneric includes only single genus *Taphrina* causing thickening, blisters, lesions and witches brooms on many angiospermic plants.

Family : Taphrinaceae

With a saprobic haploid uninucleate yeast phase that probably undergoes plasmogamy to produce the infective mycelial phase in the presence of the host. Mycelial compartments are delimited by septa with simple pores, that appear similar to those of filamentous aascomycetes. Mycelium is reported as dikaryotic or multinucleate with paired nuclei (*Taphrina*). Within the plant terminal or intercalary cells of the mycelium become differentiated as ascogenous cells and undergo development, eventually to produce ascospores.

Taphrina

Occurance: There are about 100 species of Taphrina are known, the most important one is *T.deformanas*. All the species of this genus are dimorphic : the parasitic phase that is mycelia and the budding yeast stage in saprobic phase. Asci are formed during parasitic phase of the fungus. Some of the important diseases caused by Taphrina Peach Leaf Curl or leaf blister disease (*T.deformans*), yellow leaf blister of popular (*T.populina*), Witches' brooms of Birch (*T.betulina*), witches' broom of Plum (T.insititiae) and many diseases of fruits eg. T.pruni.

Mycelium

The mycleium of Taphrina is multicellular and branched. Cells are typically binucleate however, evidence from serial sections and electron microscopy showed multinucleate hyphal compartments with the nuclei arranged in pairs. The hyphae may be intercellular or subcuticular and sometimes grow within the walls of the epidermal cells of the host.

Reproduction

Asexual reproduction:

Asexual reproduction tales place by small oval or spherical uninucleate yeast like conidia called Blastospores, which develops by budding of ascospores while they are still within the ascus. The blastospores germinate either by producing germ tube which develops into mycelium, or also by budding of blastospores to produce secondary blastospores.

Sexual Reproduction:

Taphrina reproduce sexually as well but there is no specific sex organ develops in the species. In most of the species of Taphrina, one celled thick compact mycelia layer develops in the subcuticular region of the host layer. The cells in this region at this level are more or less rounded and function as ascogenous cells which are binucleate. Karyogamy occurs, apparently between the paired nuclei present within each ascogenous cell, and the cell begins to elongate as the ascogenous cells enlarge and elongate, they exert pressure on the host cuticle from below, and eventually break through to form a compact surface layer of asci "the hymenium" – on the epidermis of the host. While elongation is proceeding, the diploid nucleus divides mitotically, and the daughter nuclei move toward opposite ends of the cell. A septum develops between these two nuclei, separating the ascogenous cell into a short basal stalk cell and a large upper ascus cell. Now a meiosis followed by subsequent mitotic division result in the formation of eight nuclei. Each nuclei surrounds some cytoplasm around it and develop into eight somewhat rounded/ oval ascospores. Ascus of taphrina is unitunicate. Soon after they are formed, the ascospores produce small, round or avoid blastospores by budding. This process may begin before the ascospores are released from the asci. Release of ascospore takes place by simple bursting of ascus tip. Blastospores which are produced by budding germinate by producing germ tube which infects host.

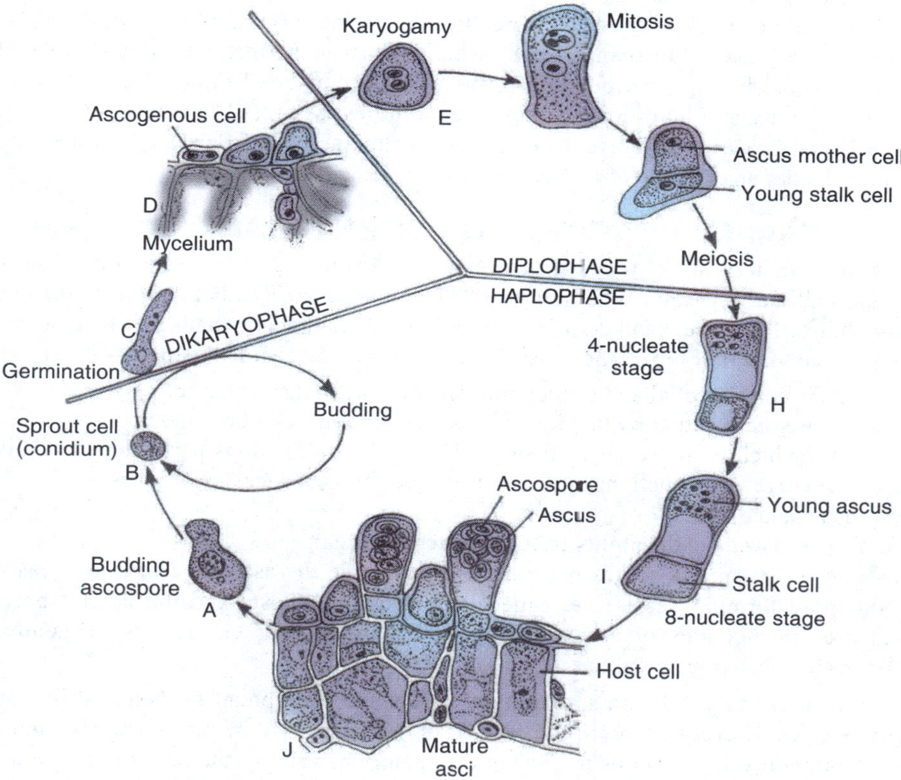

Life cycle of Taphrina deformans (Berk) Tul. Redrawn & revised from Alexopoulos (1952)

Control measures:

There are various ways to control peach leaf curl. Like, timely use of Bordeaux mixture is quite effective in controlling the disease. There are other fungicides like fytolan (.3%), blitox, captan, dithane are also used for controlling the disease. Field sanitation practices, burning of the infected leaves, use of resistant varieties are the other control measures.

Classification

Alexopoulos (1962)	**Alexopoulos and Mims (1979)**
Division : Mycota	Superkingdom: Eukaryonta
Sub-division : Eumycota	Kingdom: Mycetae
Class: Hemiascomycetidae	Division : Amastigomycotina
Order: Taphrinales	Sub-division : Ascomycotina
Family: Taphrinaceae	Class : Ascomycetes
Genus : *Taphrina*	Sub-class: Hemiascomycetes
Species : *deformans*	Order : Taprinales
	Family: Taphrinaceae
	Genus: *Taphrina*
	Species: deformans

ORDER : ENDOMYCETALES

General account. According to Smith, the order includes about 140 species which are placed under 45 genera. Some of these are saprophytic and others parasitic. The thallus is greatly reduced often to a single cell. Copulation of two cells leads to the formation of a zygote which directly develops into an **ascus.** In many forms, ascus is formed parthenogenetically. In still others, the diploid zygote by repeated budding forms a large number of diploid cells, each of which develops into an ascus. The order includes three families. Martin (1961) recognises four (Ascoideaceae or Dipodascaceae, Endomycetaceae, Spermophthoraceae and Saccharomycetaceae). Of these we shall consider an example of family Saccharomycetaceae.

FAMILY : SACCHAROMYCETACEAE

This family includes a group of reduced or degenerate Ascomycetes (Guilliermond, 1940). Popularly they are called the **yeasts.** What are yeasts? They are non-mycelial Ascomycetes that form no fruiting bodies. Their usual and dominant growth form is a unicell which may be globose, ellipsoid or shortly cylindrical in form. The yeasts have the following characteristics :–

(*i*) The thallus is non-mycelial and of predominantly a unicellular structure.

(*ii*) Reproduce asexually either by budding or fission, sometimes by both methods.

(*iii*) The zygote, which is the result of fusion of two cells or develops parthenogenetically from a vegetative cell, functions directly as an ascus and produces ascospores.

(*iv*) The ascus is naked.

(*v*) Most of the yeasts have the ability to ferment certain sugars.

Some yeasts do not form ascospores but otherwise resemble the ascospore forming yeasts. The former (non-spore forming yeasts) are called **asporogenous yeasts** and the latter (spore-forming yeasts) **ascosporogenous** or **ascosporic yeasts.** Some authors exclude **asporogenous** yeasts from the family Saccharomycetaçeae.

Occurrence. Mostly the yeasts are **saprophytes.** They are widespread in their distribution and like bacteria are found everywhere on the face of the earth usually in substrates containing sugar. Many are commonly found on or in the liquid organic material in the soil and in animal excreta. From there, they are scattered and blown on the surface of fruits and grains. In nature, the yeasts occur in the sticky sugary exudate on the surface of sweet fruits such as apples, grapes and dates and in the nectar of flowers. They are thus particularly abundant in vine yards and orchards. They also occur in milk and other foods subject to fermentation. Some saprophytic forms have been domesticated. A few species of yeasts are parasitic. They parasitize plants and animals including man. As parasites they cause very little damage.

According to Lodder and Kreger van Rij (1953), the ascosporic yeasts or "true yeasts" comprise about 68 species which are placed under 14 genera. Of these three genera are animal parasites. The important genera of ascosporic yeasts are *Saccharomyces, Schizosaccharomyces, Debaromyces, Saccharomycodes, Nodosonia, Pichia* and *Hansenula*.

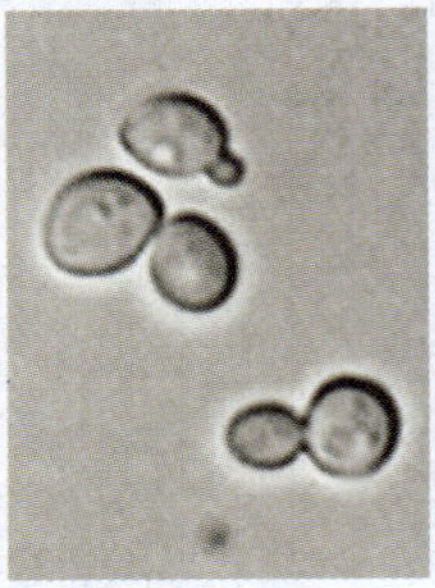

Debaromyces

Hansenula polymorpha

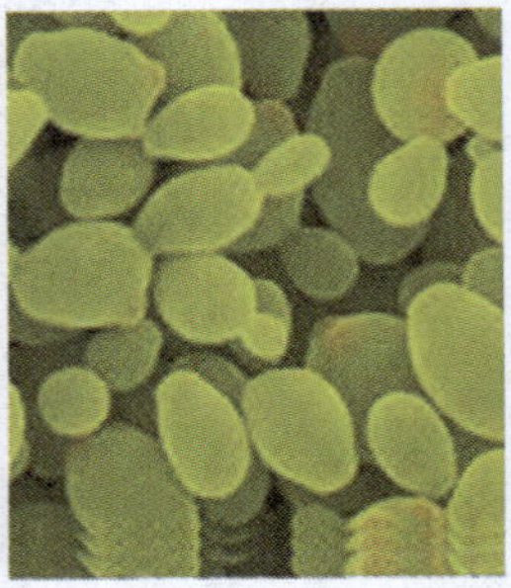

Pichia

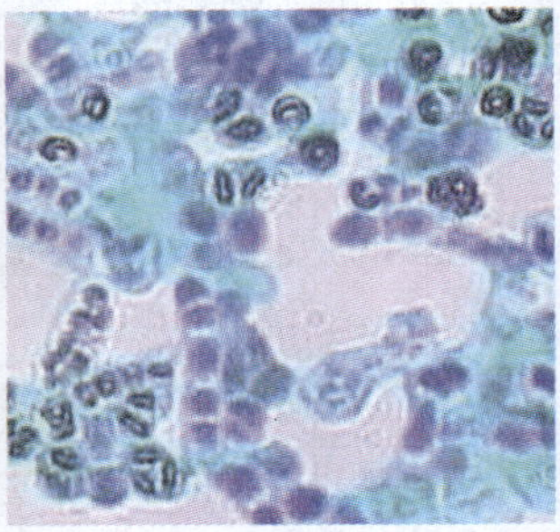

Schizosaccharomyces

Kinds of yeasts

On the basis of asexual methods of reproduction yeasts are roughly divided into three categories as follows :–

1. Budding yeasts (Fig. 9.3 A–C). They reproduce asexually by the method of budding in which the parent cell buds off a small daughter cell called the **bud**. The common example of *budding yeasts* are members of the genus *Saccharomyces*. It has 30 species. Of these, two are of great commercial importance. *S. cerevisiae* is the principal agent of alcoholic fermentation and bread making. *S. ellipsoideus* is employed in wine making.

Budding yeast – Saccharomyces cerevisiae

Budding yeast – Saccharomyces ellipsoideus

2. Fission yeasts (Fig. 9.5 A–D). They reproduce asexually in which the parent cell divides by transverse division into two daughter cells of equal size. The process is called **fission**. The best known genus of fission yeasts is *Schizosaccharomyces*. It includes three species. Of these, *S.octosporus* is well known.

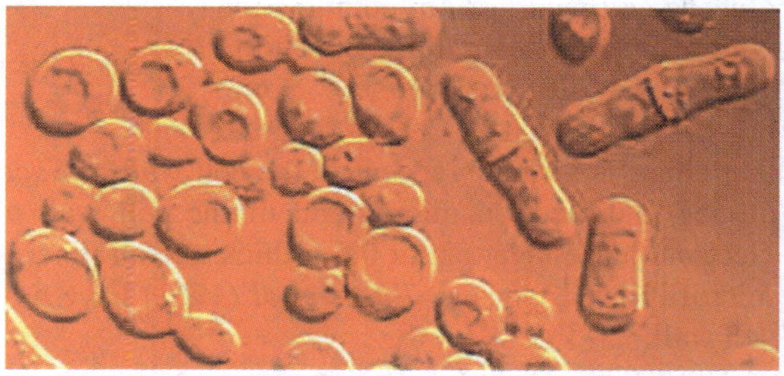

Fission yeast – Schizosaccharomyces

3. Helobial yeasts. Yeasts with asexual reproduction intermediate between budding and fission are called **helobial** as they combine these two methods. The parent cell pushes out a small daughter or the bud. The bud separates from the former not by a constriction but by the formation of a septum. The genus *Saccharomycodes* is an example of this category. *S. ludwigii* is the commonly known species of this genus.

The most thoroughly investigated of all the yeasts is the yeast of commerce called *Saccharomyces cerevisiae*. We shall study its life cycle in detail.

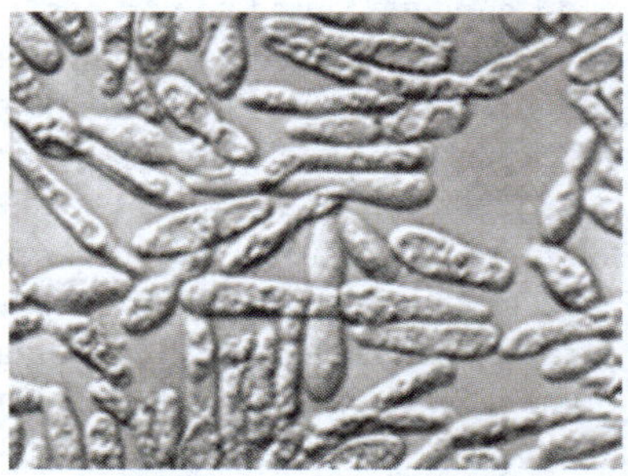

Helobial yeast S. ludwigii

SACCHAROMYCES CEREVISIAE

It is commonly employed in bread making and beer brewing. Bacause of this it is popularly called the brewer's or baker's yeast.

Yeast culture. A culture of yeast can be prepared in the laboratory by placing a piece of yeast cake or a few grains of dried yeast in a solution of molasses and water or sugar solution. The yeast cells will multiply rapidly in any well-aerated, nutrient solution containing adequate amount of sugar. Millions of yeast cells will be available for study in a day or so.

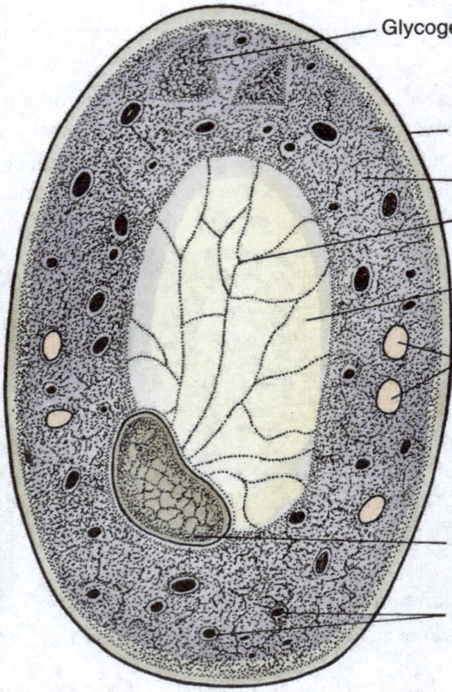

Fig. 9.2 A. Saccharomyces cerevisiae. A diploid yeast cell showing structure.

Thallus (Fig. 9.2). Unlike other sac fungi the thallus is non-mycelial. It is not made up of hyphae. Instead it consists of a single minute oval or spherical cell (Fig. 9.2). In some other species, yeast cell is elongated, cylindrical or rectangular in shape. In size it ranges from 2 to 8 μ in diameter by 3 to 15 μ in length. Individually the yeast cell appears hyaline in colour. Like other plant cells, it consists of a tiny mass of protoplast surrounded by a cell wall.

The cell wall. It is thin, delicate, firm and chitinous in nature. It is made up of two polysaccharides, namely, *glucans* and *mannan B* (polymers of glucose and mannose) in combination with traces of protein, lipid and chitin. There is no cellulose. Some mycologists (Walker, 1967) reported that chitin is virtually absent in the cell wall of *Saccharomyces*. According to Agar and Douglas (1955), the electron micrograph of the cell wall of *S. cerevisiae* shows two layers, an outer electron dense layer about 0.5 μ thick and an inner less dense layer about 0.2 μ thick containing microfibrils of glucans. Vitols *et al* (1961) have not been able to demonstrate two layers in the cell wall.

The Protoplast. It mainly consists of **cytoplasm** surrounding a minute **nucleus**. The cytoplasm is differentiated into two portions, the *outer* and the *inner*. The outer portion is a thin, cytoplasmic membrane (**ectoplasm**). The inner portion is dense or granular. It is the **endoplasm**. The endoplasm contains a single, small dense **nucleus** less than 1 μ in diameter. It can be made out by the use of special staining methods. Associated with the nucleus and occupying much of the space in the yeast cell is a single large **vacuole**. The vacuole contains a solution or suspension of **volutin** which is a complex material consisting of RNA, lipoprotein and poly-phosphates. The reserve food products in the cytoplasm are in form of granules of glycogen, oil globules, and protein particles. Embedded in the cytoplasm are the living inclusions, ribosomes and **mitochondria**. Some people regard the vacuole as the nucleus and the nucleus as nucleolus. Electron microscope studies carried on by Agar and Douglas (1957) and Hashimo *et al* (1950) revealed that the nucleus and vacuoles are separate entities. The nucleus has a nuclear membrane and is distinct from the vacuole. It is situated at one side of the vacuole. The nuclear membrane has pores. Dark staining strands extend from the nucleus around the vacuole.

Nutrition. Like other-fungi the yeasts lack chlorophyll. They are thus **heterotrophic** in their mode of nutrition. They obtain readymade food from an external source by living usually as **saprophytes**. The brewer's yeast obtains nutrition in the form of simple sugars in solution. It gets

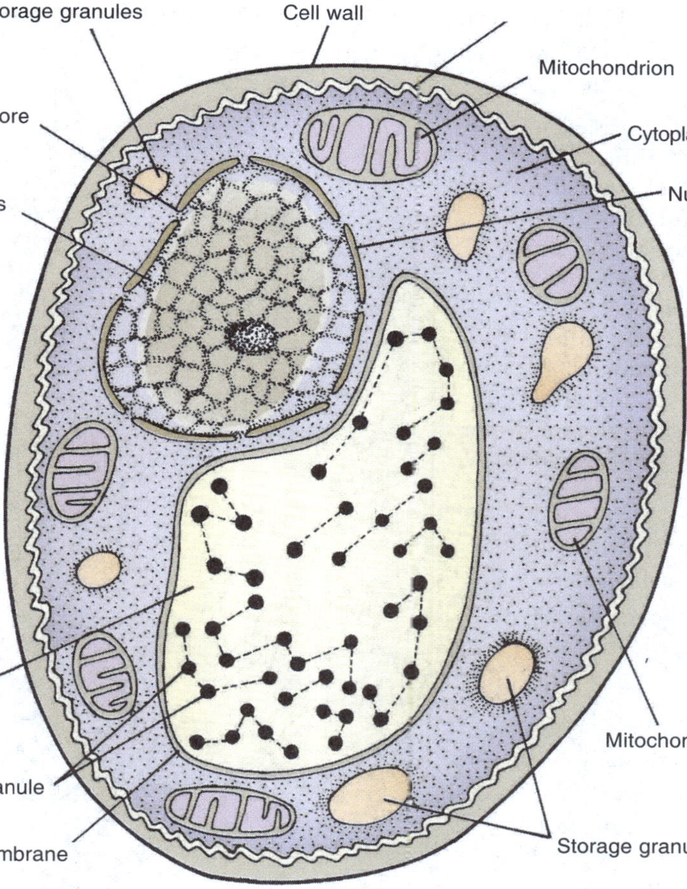

Fig. 9.2 B. *Saccharomyces cerevisiae.* Cell structure as seen under electron microscope.

it from the substratum on which it grows such as the bruised ripe fruits, fruit juices, sugar solutions or crushed moistened grains. The nutrient elements required are carbon, hydrogen, oxygen, nitrogen, phosphorus, potassium, sulphur, magnesium, iron, zinc, manganese, copper and molybdenum. The last five metallic elements function as enzyme activators or as components of enzymes in minute traces. Carbon is secured chiefly from sugars or starch. In addition a certain vitamin-like growth factor is also essential for growth. It is called "bios". It is active in extremely low concentrations.

The yeast cells secrete exzymes on the substratum. These enzymes are collectively called the **zymase**. The zymase changes the starch or complex sugars present in the substratum into simpler sugars. The latter diffuse into the endoplasm through the thin, delicate cell membrane. A small percentage of the absorbed sugar is used as food and assimilated. When yeast is growing in a well aerated nutrient medium the rest of the sugar is completely oxidised in normal respiration. The reaction is shown by the following overall equation :–

$$C_6H_{12}O_6 + 6O_2 \rightarrow 6CO_2 + 6H_2O + \text{energy}$$

In such a solution, the yeast plants multiply rapidly. In the absence, or poor supply (scarcity) of oxygen the yeast plants multiply slowly, the major protion of sugar is converted into carbon dioxide and ethyl alcohol within the cytoplasm. The reaction is shown by the following equation :–

$$\underset{\text{glucose}}{C_6H_{12}O_6} + \text{yeasts} \rightarrow \underset{\text{ethyl alcohol}}{2C_2H_5OH} + CO_2 + \text{energy}$$

The products of this reaction (CO_2 and C_2H_5OH) diffuse out, through the cell wall into the surrounding liquid. The breakdown of carbohydrates (starch and sugars) by yeast into carbon dioxide and alcohol in the absence of oxygen is called alcoholic **fermentation**. This process is utilised in the production of industrial alcohol, in wine making, in brewing and in bread making. The energy released during respiratory processes is utilized by the cell in carrying out the vital functions.

Growth. It occurs by an overall increase in cell size.

Reproduction

Yeasts reproduce both asexually as well as sexually.

1. Asexual Reproduction. Most of the yeasts multiply asexually by a vegetative method called **budding**. A few known as the fission yeasts undergo **fission**. *Saccharomyces* normally reproduces by budding.

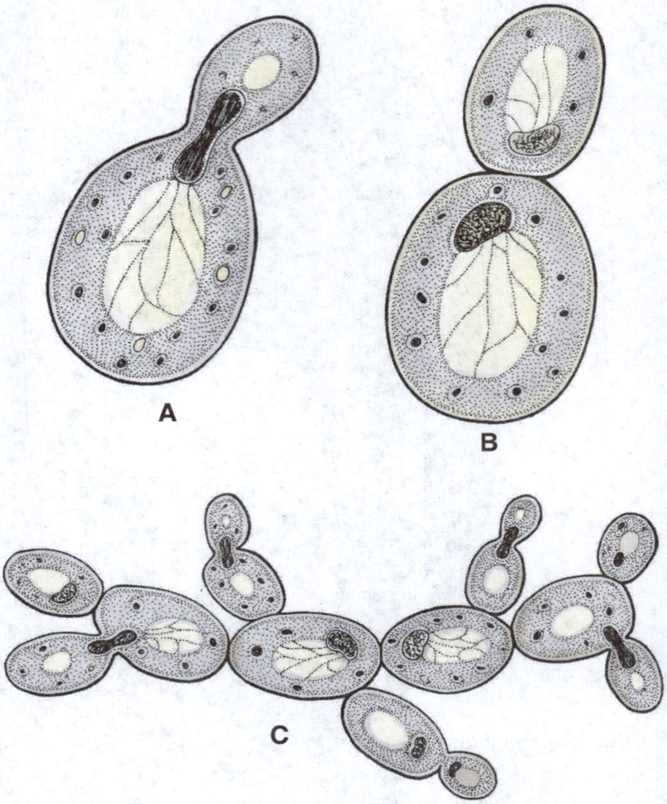

Fig. 9.3 (A–C). *Saccharomyces cerevisiae* showing vegetative reproduction by budding.

(*a*) Budding (Fig. 9.3 A–C). Under favourable conditions *Saccharomyces* exclusively reproduces by this method. The favourable environment for a yeast plant is the presence of nutritive solution such as sugar solution or starch mixture. At the time of budding, a small portion of the cell wall, at or near one pole of the yeast cell softens and thins. The protoplast in this region, covered by the thin softened membrane, bulges out in the form of an outgrowth or protuberance (A). The protuberance gradually increases in size. It is known as the **bud**. As the bud is forming, the nucleus of the parent yeast cell divides and with it the vacuole. Some believe that the nuclear division is amitotic. Others consider it to be mitotic. One of these daughter nuclei alongwith its associated daughter vacuole migrates into the enlarging bud. The bud grows and becomes constricted at the base. The cytoplasm of the bud and mother cell remains continuous for some time. Eventually the opening between the two cells closes. A double cross wall forms and the two cells become physiologically distinct and may separate leaving a scar on both cells. After the separation of the daughter cell a scar with a convex surface remains on the mother cell. The daughter cell as well retains a corresponding concave birth scar (Fig. 9.4). The detached bud or yeast cell grows and again starts budding. A new bud arises on the surface of the parent cell at the opposite end to the first and not from the same area. This is known as bipolar budding. In the presence of abundant food supply, the process of budding is quickened. It becomes so rapid that

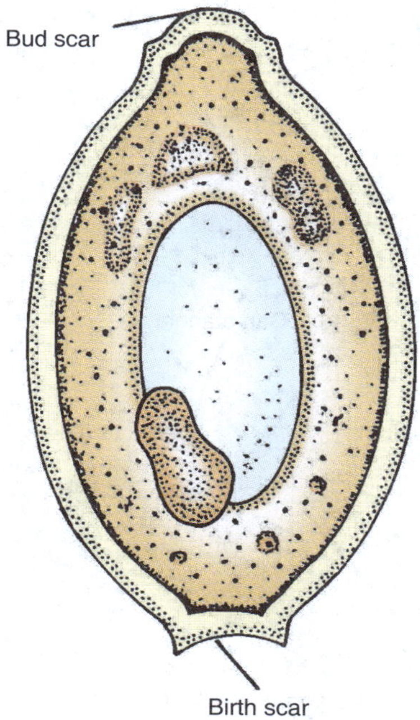

Fig. 9.4 Yeast cell *(Saccharomyces)* showing birth and bud scars.

the buds often produce new buds before separation from the mother cell. The process is repeated. In this way a large number of buds are formed without being detached from one another. This results in the formation of branched or unbranched chains of cells constituting the **pseudomycelium** (C). There may be as many as 64 connected cells in the pseudomycelial mass. The cells in the chain or pseudomycelium are loosely joined together. Soon the chains break. The constituent cells are separated from each other.

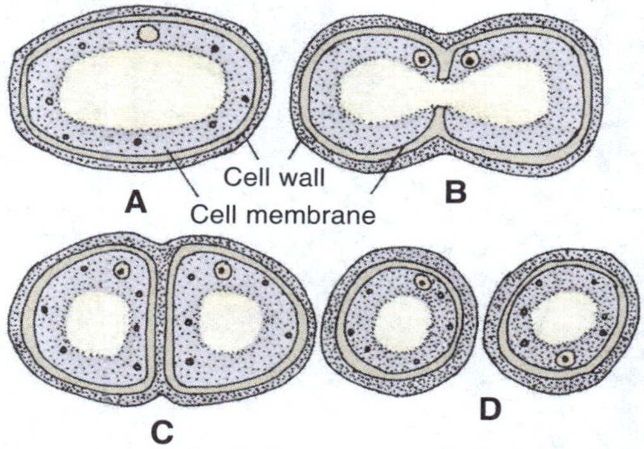

Fig. 9.5 (A–D). *Schizosaccharomyces* (Fission yeast) showing vegetative propagation by fission (Diagrammatic).

(*b*) **Fission** (Fig. 9.5). Vegetative reproduction by **fission** has also been reported but it occurs in some other yeasts and not in *Saccharomyces* (budding yeasts). They are called the **fission yeasts** for this reason. The division is transverse. The mother cell elongates (B). The nucleus divides into two. The daughter nuclei move apart. Meanwhile a ring-like ingrowth appears at the wall of the yeast cell in the middle. Covered in front by the cytoplasm, it grows inward toward the centre of the cell. Finally, it stretches across the cell forming a complete partition called the septum (C). The septum thickens and then splits into two layers, one for each daughter cell before they separate. Fission is, however, not as common as budding.

2. Sexual Reproduction

A. *Saccharomyces cerevisiae* (Fig. 9.6 C–E). Recent investigations have shown that like other Ascomycetes the formation of ascospores in yeasts is contingent upon a sexual process. Of course, no sex organs like antheridia and oogonia are produced. Instead, the sexual process is extremely simplified. It consists of three phenomena characteristic of the sexual process, namely **plasmogamy** (sexual fusion), **karyogamy** (fusion of two nuclei) and **meiosis**.

(*a*) In a culture of **brewer's or baker's** yeast (*Saccharomyces*) there occur intermixed two kinds of somatic cells namely, small dwarf strain and large strain cells.

(*i*) **'Dwarf strain' yeast cells** (Fig. 9.6 A). They are small haploid yeast cells which are spherical in shape. Reid and Hartwell (1977) reported that the small haploid dwarf yeast cells which are of two opposite mating types (+ and – strain) behave in any of the 3 ways depending upon the environmental conditions :–

(1) Under normal conditions (plenty of nutrient supply) there is absence of cells of the opposite mating type, these multiply by budding and increase in number.

(2) When food supply is scarce, they enter a dormant phase and become resistant to killing by enzymatic degradation and high temperature.

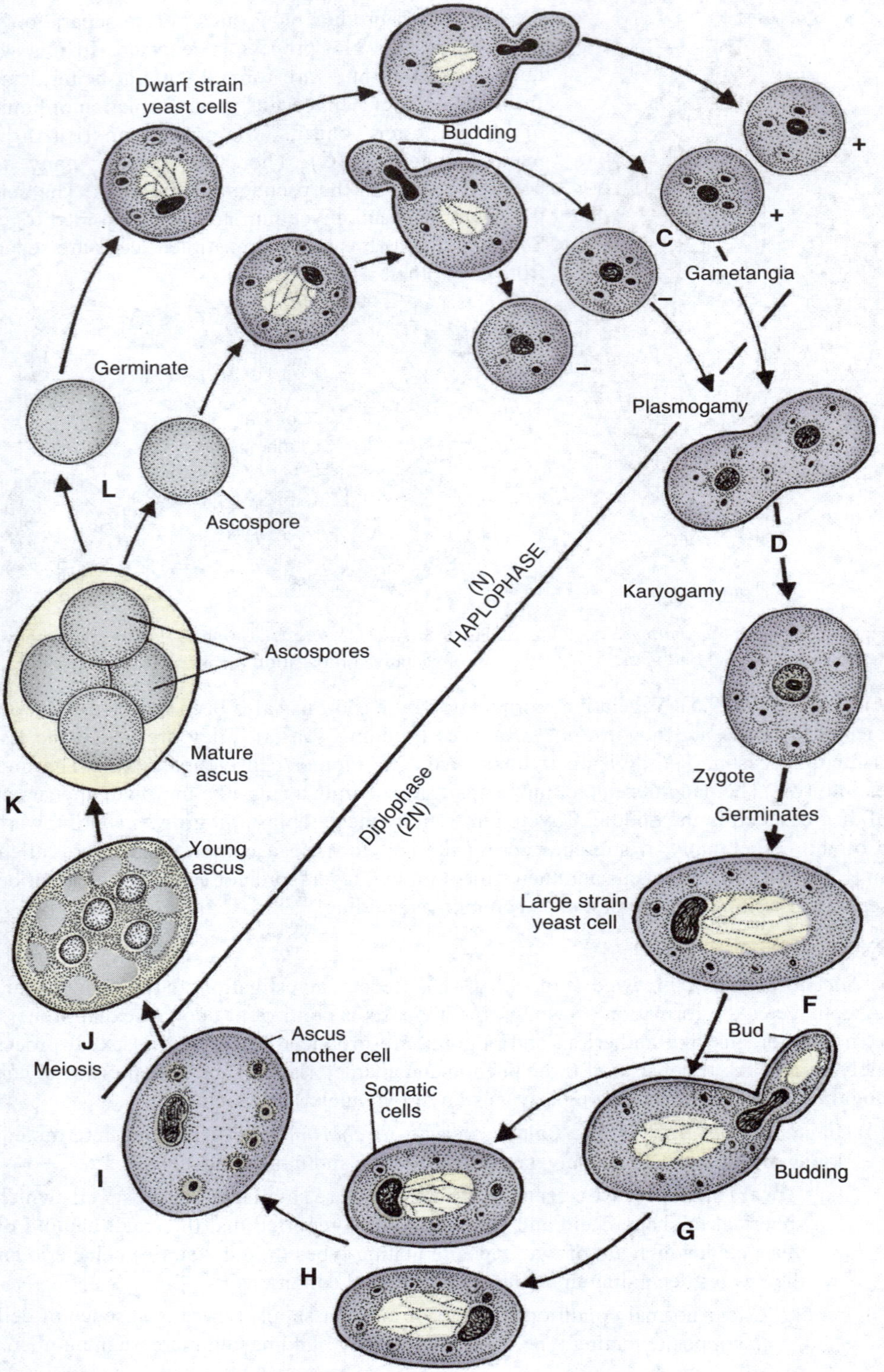

Fig. 9.6 (A–L). *Saccharomyces cerevisiae.* Sexual Cycle.

(3) Under normal conditions and presence of the small haploid cells of the opposite mating strain they function as gametangia and resort to sexual or gametangial conjugation.

Gametangial Conjugation (Fig. 9.7). During the conjugation process the + and – strain haploid dwarf yeast cells agglutinize to form clusters. In this condition each mating type secretes a sex hormone (Tkacz and Mackay 1977) which induces the haploid spherical cell (A) and (B) of the opposite strains to elongate to a pear-shaped form (A^2 and B^2) and also causes alterations in the cell walls of the newly formed regions. These surface alterations facilitate fusion.

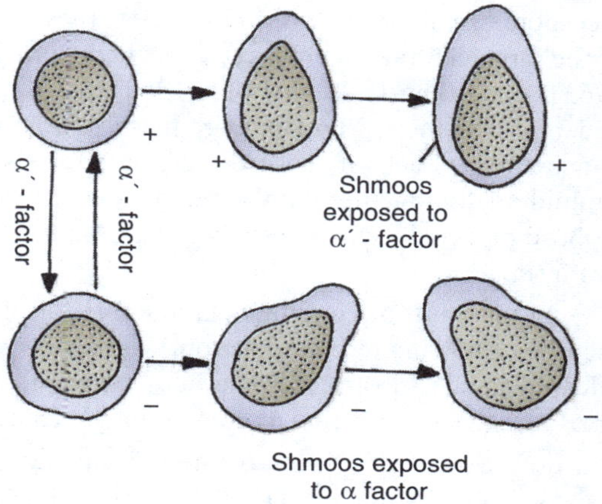

Fig. 9.7 (A–B). *Saccharomyces cerevisiae.* Early stage in gametangial conjugation showing secretion of sex hormones by the dwarf yeast cells of opposite mating types (Based on Tkacz and Macky)

Plasmogamy (Fig. 9.7 C-D). At the sites with modified cell walls, the two haploid cells of the opposite mating types bend towards each other and fuse to form a **conjugation bridge** (C). Fusion between the protoplasts of the + and – strain takes place through the conjugation bridge (D). It is called **plasmogamy**. A large fusion cell (D) with two nuclei (+ and –) lying side by side to form a *dikaryon* results.

Karyogamy. The two nuclei of the dikaryon finally fuse. It constitutes **Karyogamy and** is equivalent to fertilisation. The diploid fusion nucleus is called the synkaryon and the cell containing it is the zygote. It is evident from the account given above that the small or dwarf strain yeast cells represent the **haplophase** in the life cycle of *Saccharomyces cerevisiae*.

(*ii*) Zygotes ('Large strain' yeast cells, Fig. 9.6 E-F**)**. The zygotes (E) formed as a result of sexual fusions between two haploid yeast cells of opposite mating types are, in fact the large strain yeast cells. They are diploid, larger in size than the dwarf strain haploid yeast cells and are ellipsoidal in form (F). These are commonly employed in industrial processes. Under favourable conditions they multiply by budding (G) and thus increase in number. At the onset of conditions unfavourable for vegetative growth such as shortage of food, water and low temperature, the diploid large strain yeast cells (zygotes) resort to ascospore formation. They become spherical in shape (I) and directly function as **asci**.

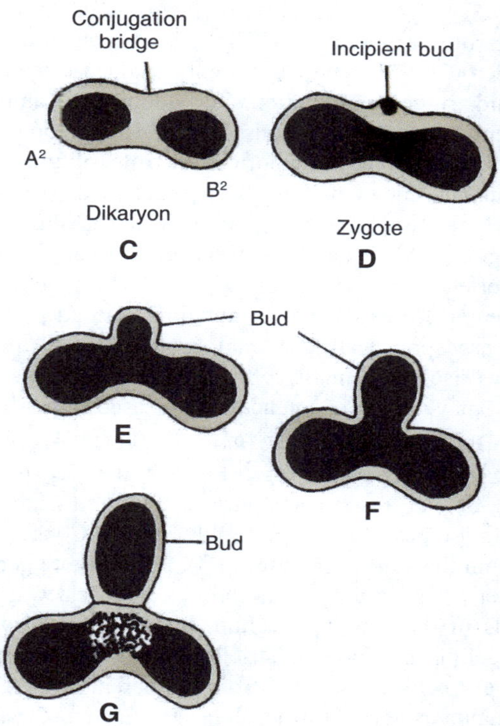

Fig. 9.7 (C–G). *Saccharomyces cerevisiae.* Later stages in gametangial conjugation (C-D) and budding of zygotes (E-G). (Based on Tkacz and Mackay)

Ascospore Formation

The ascus mother cell (I) is spherical in shape. Glycogen and aeroboic metabolism favour ascospore formation (Pinon, 1970). The earliest change in the ascus mother cell is the disappearance of the large central vacuole which is characteristic of the vegetative cell (Moor and Muhlethaler, 1963). According to Hashimoto (1960), the diploid nucleus of the spherical ascus mother cell undergoes **meiosis**. Meiosis I produces two haploid nuclei, one of plus strain and the other of minus strain. Each of these divides at Meiosis II (J). An envelope encloses each of the four haploid nuclei together with some cytoplasm, a few mitochondria and other organelles to form 4 globose or ovoid ascospores (Lynn and Magae, 1970). Two of these are of + strain and two of - strain (K).

On the onset of conditions favourable for growth, the ascospores swell and press against the ascus wall. Owing to pressure from within the ascus wall ruptures and the haploid ascospores are released in the substrate. Sometimes sexual fusion between two ascospores of opposite strains takes place. They function as copulating gametangia, unite and form a zygote.

Fine Structure of Ascospore. The mature ascospore is a thick-walled, globose structure. The ascospore wall is differentiated into three layers. Of these, the outer and inner are electron dense and the middle electron-transparent and thick. The cytoplasm contains a single nucleus with a poriferous nuclear membrane, a few mitochondria and some other organelles (Lynn and Magae, 1970). The ascospores of *Saccharomyces* are acid-fast.

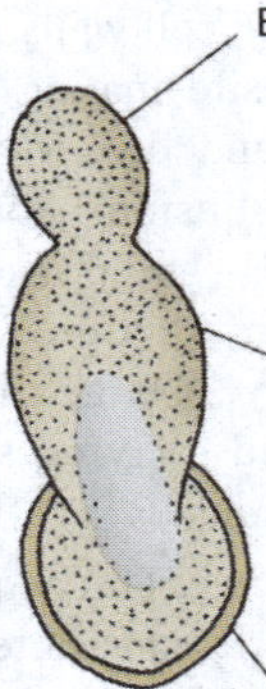

Fig. 9.8. *Saccharomyces cerevisiae.* Germination of ascospore.

Ascospore Germination (Fig. 9.7). Ascospore germination is anaerobic (Sevarese, 1974). As regards nutritional requirements, Palleroni (1961) reported that only D-glucose and other monosaccharides could trigger germination. The addition of ammonium sulphate and vitamins enhances it. Rousseau and Halvorson (1973) concurring with Palleroni strongly supported the role of glucose as a germination stimulant. They found that the optimal conditions for germination require a temperature of 30ºC, glucose in the medium and a *p*H within the range 5-4-8.2. No additional supplements are required. Sevarese (1974) reported that in addition to glucose and ammonium sulphate addition of vitamins and minerals is essential for extensive germination and bud formation. Bud formation involves the extension of the ascospore wall as a small protuberance. The latter increases in size to become a bud which gives rise to a somatic cell typical of the species. During early stages of bud formation, the ascus wall may remain in contact with the ascospore wall (Fig. 9.8).

B. *Saccharomycodes ludwigii* (Fig. 9.9). In another species of yeasts of a different genus *Saccharomycodes ludwigii* (Fig. 9.9) the ascospores directly function as gametangia and copulate (A). Sexual fusion takes place within the ascus between the two adjacent ascospores of opposite mating types, *A* and *a* or + and - strains. Consequently two diploid cells or zygotes are produced within the wall of the ascus (B). Each zygote germinates *in situ* to form a small **germ tube**. The latter grows and emerges through the ascal wall and functions as a **sprout mycelium** (C). The cells of the sprout mycelium bud off diploid yeast cells. The latter separate from the sprout mycelium not by a constriction but by the formation of a septum at the base. Soon the diploid **sprout cells** get severed from the parent cells. The detached diploid somatic or sprout cells (D) function as **ascus mother cells** (E). Each of these becomes converted into an **ascus** as its diploid nucleus undergoes **meiosis** (F) to produce four ascospores. Two of these are of A mating type and two of *a* type (G). They can also be called as plus and minus strains. It is obvious that the haplophase (gametophyte) in *Saccharomycodes ludwigii* is extremely reduced. It is represented only by the ascospores.

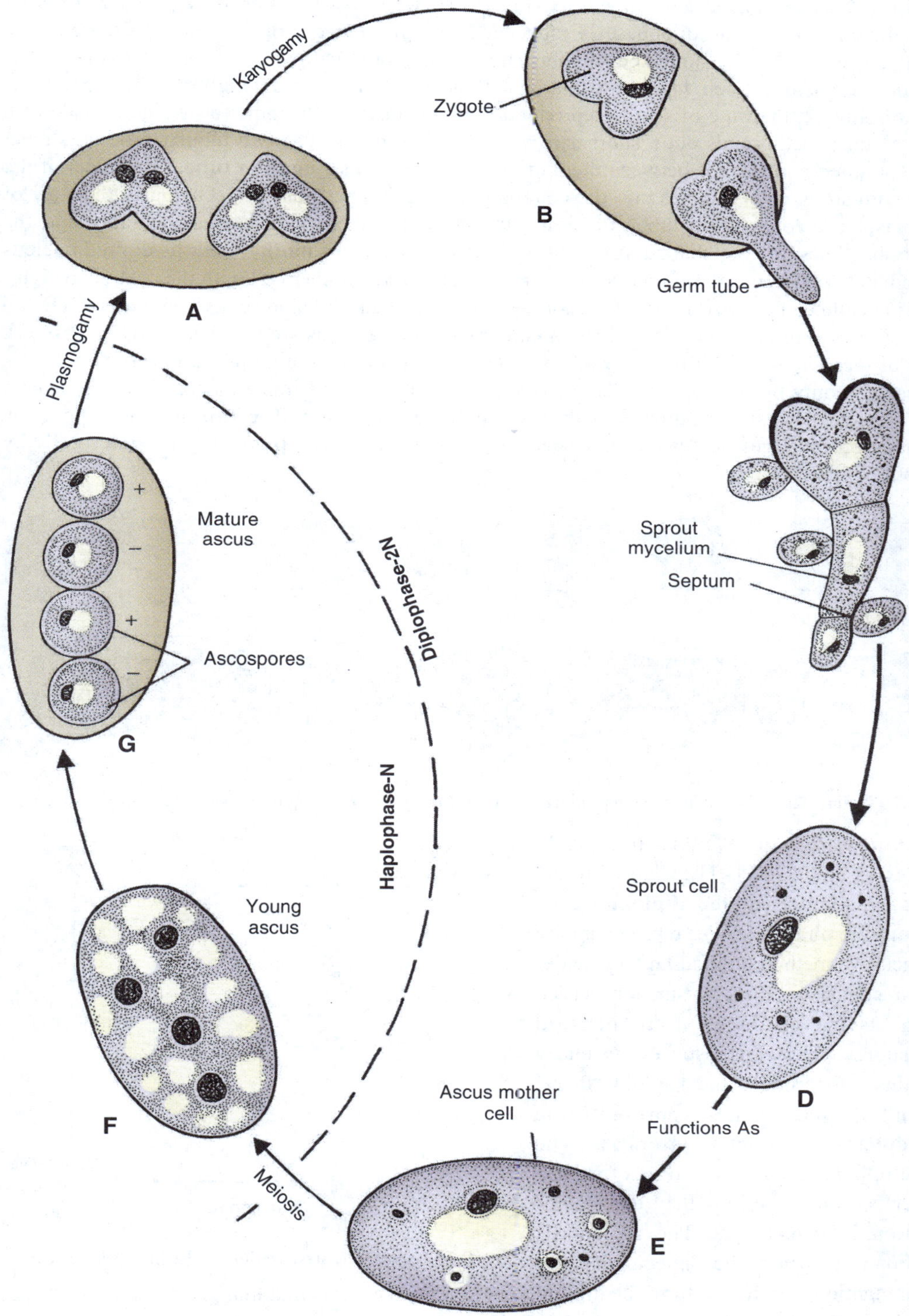

Fig. 9.9 (A–G). *Saccharomycodes ludwigii.* Sexual cycle.

C. *Schizosaccharomyces octosporus* (Fig. 9.10) which is a fission yeast, tells a different story. In this species the dilophase is reduced. It is represented, only by the **zygote cell.** The vegetative cells (A) are elongated, uninucleate and haploid. They multiply vegetatively by the method of **fission**. Each haploid somatic cell is a potential gametangium. This species is **homothallic**. At the time of sexual reproduction two somatic cells come to lie side by side (A). Each of these cells sends out a short narrow beak-like process. The two tubular processes meet (B). The intervening walls between them dissolve to form a **conjugation tube.** The nuclei of the two conjugating cells migrate into the conjugation tube (C) and fuse (D). Owing to the flow of cytoplasm the conjugation tube containing the diploid nucleus enlarges considerably and the **zygote** becomes dumble-shaped. It functions directly as an **ascus mother cell**. Its diploid nucleus immediately undergoes three successive divisions. The ascus mother cell with the contained eight-nuclei is called an **ascus** (G). The first and the second nuclear divisions constitute **meiosis** (E and F). The third is **mitotic** (G). The resultant eight nuclei in the ascus are thus haploid (Fig. 9.10 G). Each of these is organised into an **ascospore** (H). In this case the diplophase is very **short.** It is represented only by the zygote cell (D) which undergoes meiosis immediately after karyogamy. The ascal wall finally ruptures and the ascospores are released. The liberated ascospore, on germination, enlarges somewhat and behaves like a somatic cell producing daughter cells by fission.

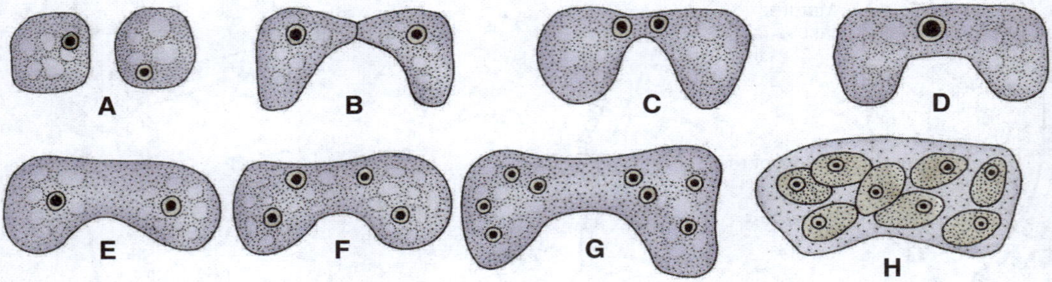

Fig. 9.10 (A–H). Yeasts. Stages in sexual reproduction and formation of ascospores in *Schizosaccharomyces octosporus.*

Alternation of Generations in *S. cerevisiae* (Fig. 9.11). The "large strain" yeast cells represent the **diplophase or sporophyte phase**. Besides reproducing by a vegetative method of budding they, under certain conditions, directly function as asci. Each ascus produces four haploid ascospores. These ascospores normally germinate to produce the haploid "dwarf strain" of yeast which represents the **haplophase** or **gametophyte** phase. The gametophyte cells bear no sex organs. Under certain conditions they themselves function as gametangia. The protoplasts function as gametes. The gametangia resort to conjugation and fuse to form the diploid **zygote**. The zygote by budding gives rise to the "large strain" diploid yeast cells representing the **diplophase**.

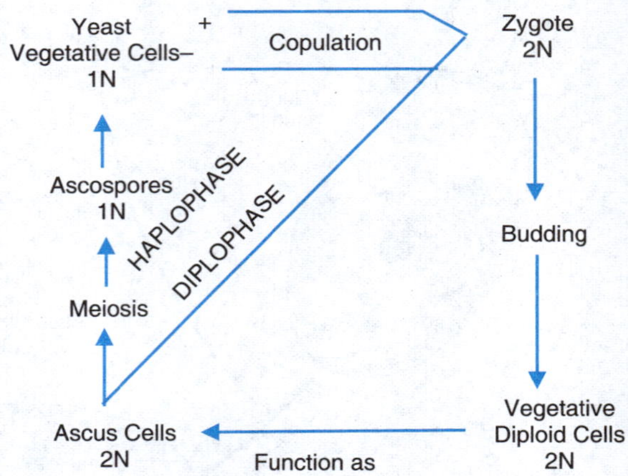

Fig. 9.11. Graphic representation of the life cycle of yeast.

It is evident from this account that the life cycle of yeast consists of two distinct phases, the **sporophyte phase** or **diplophase** and the **gametophyte phase** or **haplophase** (Fig. 9.10). The former consists of the "large strain" diploid cells (zygotes) and the asci. The latter (gametophyte phase) consists of the ascospores and the "dwarf strain" yeast cells. These two phases alternate in a single life cycle. It is interesting to note that in the life cycle of *Saccharomyces cerevisiae* both the haplophase and the diplophase are of equal importance. Each phase is perpetuated by budding. Sometimes the gametophyte stage is reduced. The haploid cells conjugate to form the 'large strain' yeast cells. In such cases the haploid ascospores are the only gametophyte structures in the life cycle. *Saccharomycodes ludwigii* is an example of this type.

In *Schizosaccharomyces octosporus*, the condition is reversed. In this case, the diplophase is short. It is represented only by the zygote cell. It undergoes meiosis immediately after karyogamy to produce 8 ascospores.

ECONOMIC IMPORTANCE OF YEASTS.

The importance of yeasts to man lies in their ability to ferment carbohydrates hence the name *Saccharomycees* (Gr. Saccharon meaning sugar and mykes meaning fungi) is applied to them. During respiration the yeast cell oxidises sugar to form a simple organic acid and releases energy. When the supply of free oxygen is restricted the organic acid is split into carbon dioxide and alcohol. These products are of no use to the yeast plant but are of immense value to the **baking** and **brewing industries**. In baking carbon dioxide is an industrial product and the alcohol is waste. The former is responsible for raising of dough and giving spongy texture to the bread. The alcohol produced during the baking process is driven off as a by-product. In the brewery, on the other hand, alcohol is a valuable product. Commercial manufacture of ethyl alcohol is a large industry. Carbon dioxide, which is a by-product in the process, can be compressed into solid form.

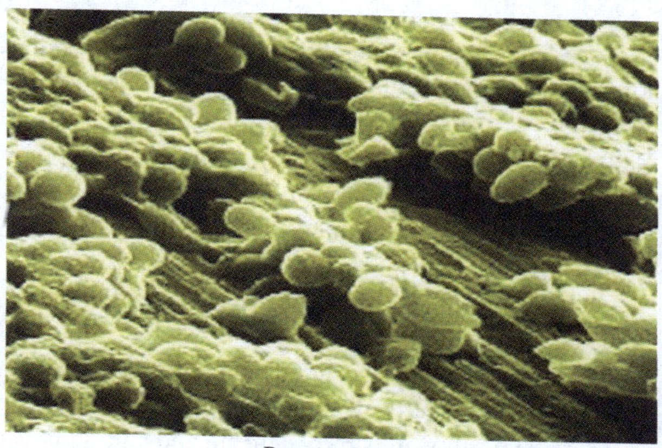

Brewing Yeast

In view of their industrial importance, the yeasts have been hybridized and certain domesticated strains, varieties and species have been developed by selection and breeding. These are superior to the wild yeasts industrially as they differ in their ability to synthesize vitamins and to ferment sugars. Some strains are better from the standpoint of nutrition or keeping qualities. The common among them are the brewers' yeast, bakers' yeast, wine yeast, distillers' yeast and the like.

The extensive use of yeast in brewing and baking industries has given great impetus to the establishment of another important industry. It is the preparation of **yeast cake** on commercial scale. Commercially the yeast cakes are prepared by pressing into cubes a large number of yeast cells together with starch which is an inert substance. The yeast cakes are doubly useful, in the home and in the baking industry. They may be eaten directly as health giving food. It was a fad sometimes ago. In the laboratory the yeast cakes are placed in a warm sugary solution to demonstrate the fermentation process The yeast cells are activated. The tiny yeast cells are rich in protein. Some strains afford considerable fat. They are, therefore, mingled with other foodstuffs to increase their food value. Fresh yeast cells are excellent source of vitamins *b* and *g* and thus form an important source of these highly valued substances. Compressed yeast is also used as a source of vitamins and of enzymes useful in the manufacture of syrups and confectionery products. *Ashby gosypii*, a filamentous yeast, is used in the production of vitamin **riboflavin**. Yeasts are also employed to impart flavour to the cocao beans. Besides the above-mentioned uses of yeasts, they are utilized in medicine, as a source of enzymes, in the production of alcohol and in many other ways.

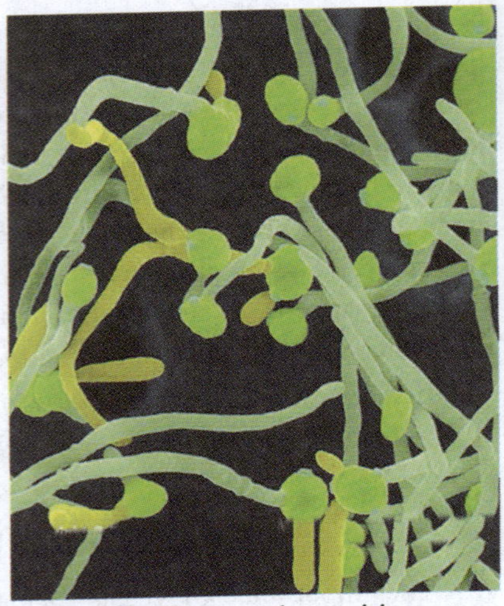

Candida albicans – A parasitic yeast

The yeasts have a negative value as well. Some species are troublesome contaminants in brewing and wine making. They release unpleasant flavours or odours in their metabolic activities. Yeasts may spoil foods containing sugar especially fruits. They impart a yeasty flavour. A few species are parasitic on man. One of them causes a serious disease known as cryptococcosis. *Candida albicans* is the cause of thrush. Some produce superficial infections of man.

Salient Features

1. The yeasts are unicellular, degenerate, non-mycelial saprobic fungi that form no spore fruit. They are found in nature in the liquid organic material in the soil, animal excreta, surface of ripe fruits and nectar of flowers.
2. They are capable of fermenting carbohydrate solutions producing ethyl alcohol and carbon dioxide.
3. The thallus is a minute hyaline oval or spherical cell which is differentiated into a distinct cell wall enclosing the protoplast.
4. The cell wall is thin, delicate and chitinous in nature.
5. The cell protoplast consists of granular dense cytoplasm, a minute nucleus, a large vacuole and mitochondria.
6. Vegetative reproduction takes place chiefly by *budding*. In some yeasts it takes place by the method of *fission*.
7. Ascospore formation is the usual method of reproduction. Their formation involves a sexual process. The enlarged, diploid, old yeast cells under adverse circumstances function as *asci*.
8. The diploid nucleus of the ascus cell undergoes meiosis to form four haploid nuclei which are organised into non-motile meiospores known as ascospores. The mature ascus thus contains 4-8 ascospores.
9. The liberated ascospores under favourable conditions germinate, each producing a small, haploid spherical yeast cell. It is the **dwarf strain** of yeast.
10. The *dwarf-strain* yeast cells multiply by budding and constitute the haplophase in the life of *Saccharomyces cerevisiae*.
11. Sexual reproduction by the formation of sex organs is absent. The simplified sexual process is represented by **plasmogamy** immediately followed by **karyogamy** and **meiosis**.
12. Plasmogamy in *S. cerevisiae* consists in the fusion of the two haploid **dwarf strain** yeast cells to form the **large strain** diploid yeast cell which normally reproduces by budding.

13. In *Saccharomycodes ludwigii* the two haploid ascospores fuse to form a diploid zygote. The zygote germinates to produce the sprout mycelium.
14. The cells of the sprout mycelium bud off diploid yeast cells.
15. The diploid vegetative cells function as asci and undergo meiosis to form the ascospores.
16. The yeasts thus exhibit alternation of generations in their life cycles which is of three types :–
    (*a*) In *Saccharomyces cerevisiae* both the alternating generations are propagated by budding and thus are of equal importance. (Haplo-diplobiontic)
    (*b*) In *Saccharomycodes ludwigii* the haploid phase is short and represented by the haploid ascospores only. (Diplobiontic)
    (*c*) In *Schizosaccharomyces octosporus* the diploid phase is represented by the zygote only which immediately after its formation undergoes meiosis. (Haplobiontic)

Inter-relationships and Taxonomy of yeasts. The yeasts in their life cycle normally do not form a mycelium. The usual and dominant growth form is a unicell. The unicellular thallus is very simple in its structure. To some mycologists the unicellular nature of the yeast thallus and its simple structure imply that the yeasts are primitive. The other equally simple and primitive organisms are the bacteria. In the thallus being a unicell the yeasts resemble bacteria. Furthermore, like most bacteria, they are non-motile, lack chlorophyll and absorb food materials in a soluble form through the cell wall only by diffusion. The cell physiology of yeasts is basically like that of bacteria. The resemblance, however, ends here.

In their structure and methods of reproduction, the yeasts resemble fungi closely. Like the fungi the yeast cell has a definite nucleus and possesses mitochondria and endoplasmic reticulum which are lacking in the bacterial cells. The occurrence of chains of yeast cells (pseudomycelium) in a nutritive medium and the development of hyphal filaments in many species of yeasts under certain cultural conditions are indicative of the fact that the yeasts may be related to mycelial Ascomycetes. The production of ascospores in sac-like asci by a process similar to that of the mycelial Ascomycetes is a further proof in support of this contention. It has been experimentally proved that yeast cells grown in the presence of chemicals such as penicillin and other suitable media may produce a true mycelium. Some produce ascospores by a process similar to the ascomycetous molds. These observations indicate close relationships between the yeasts and ascomycelous molds. On the basis of above-mentioned experimental evidence, most of the mycologists consider yeasts as ascomycetous molds whose usual and dominant growth form is a budding unicell. It is connected with their habit of growing in sugary solutions. Many normally filamentous fungi can be induced to develop budding stages in media with high concentration of nutrients and by reducing the oxygen content of the culture medium. In *Saccharomyces*, this habit has become permanent. On the basis of these experimental data the yeasts are now considered reduced or degenerate forms. The taxonomic position of yeasts, thus is :–

Alexopoulos (1962)

Division	:	**Mycota**
Sub-division	:	**Eumycotina**
Class	:	**Ascomycetes**
Sub-class	:	**Hemiascomycetidae**
Order	:	**Endomycetales**
Family	:	**Saccharomycetaceae**
Genus	:	*Saccharomyces*
Species	:	*cerevisiae*

Alexopoulos and Mims (1979)

Superkingdom	:	**Eukaryonta**
Kingdom	:	**Mycetae**
Division	:	**Amastigomycotina**
Subdivision	:	**Ascomycotina**
Class	:	**Ascomycetes**
Sub class	:	**Hemiascomycetidae**
Order	:	**Endomycetales**
Family	:	**Saccharomycetaceae**
Genus	:	*Saccharomyces*
Species	:	*cerevisiae*

QUESTIONS

ESSAY TYPE

1. With the help of a graphic life history sketch describe the life cycle of *Saccharomyces cerevisiae*.
2. Where are the yeasts found in nature? Discuss their economic importance.
3. Describe the somatic phase, mode of nutrition and various modes of asexual reproduction met within the yeasts.
4. Describe sexual reproduction in three different types of yeasts.
5. Give an account of alternation of generations in yeasts.
6. Describe the structure of *Saccharomyces* and discuss its economic importance. *(Allahabad, 1996, 2002; Lucknow, 1993, 1997)*
7. Describe the haplobiontic, diplobiontic and haplo-diplobiontic types of life cycles in yeasts with examples.
8. With the help of diagrams, describe the life cycle of *Saccharomyces cerevisiae*. *(Guru Nanak Deo, 1992)*
9. Give an account of sexual reproduction in different yeasts. *(Lucknow, 2001)*
10. Describe structure, life-cycle and usefulness of *Saccharomyces*. *(Kumaon, 2002)*

SHORT ANSWER TYPE

11. **Give well labelled diagrams of the following:**
    (*a*) Yeast cell *(Allahabad, 1992, Lucknow, 1995)*
    (*b*) Structure of yeast cell *(Lucknow, 1992)*
    (*c*) Budding
    (*d*) Reproduction in yeast
    (*e*) Electron microscope structure of yeast cell. *(Lucknow, 2000)*
12. **Write short notes on the following :**
    (*a*) Sexual reproduction in *Saccharomyces* *(Lucknow, 1994)*
    (*b*) Cell structure of *Saccharomyces* *(Lucknow, 1998)*
    (*c*) Budding
    (*d*) Fission
    (*e*) Life cycles in yeasts.
    (*f*) Economic importance of yeast. *(Lucknow, 2001)*
    (*g*) *Taphrina* *(Allahabad M.Sc., 2000, Poorvanchal, M.Sc., 2003)*
13. Give four differences in the structure of phytoplasma and yeast cell. *(Lucknow, 2002)*
14. Describe sexual reproduction in a holocaspic fungus with the help of diagrams only. *(Kanpur, 2006)*

OBJECTIVE TYPE

15. **Fill in the blanks :**
    (*i*) An example of unicellular fungus producing ascospores is ....
    (*ii*) In *Saccharomyces*, ...is the most common method of vegetative reproduction.
    (*iii*) The name of the fungus used in alcoholic fermentation is ....
    (*iv*) Sexual reproduction in yeast takes place by ....
    (*v*) .... is popularly known as brewers yeast.

(*vi*) Yeast is an important source of ....

(*vii*) Fermentation of sugars takes place by ....

(*viii*) yeasts are obligate ....

(*ix*) The fungus used for flavouring cheese is ....

(*x*) In *Taphrina*, the asci are ....

16. Select the correct answer :

(*i*) One of the important characters of Hemiascomycetidae is

(*a*) asci produced in cleistothecium

(*b*) asci produced in apothecium

(*c*) asci produced in perithecium

(*d*) asci arise naked.

(*ii*) Naked asci are produced by

(*a*) *Taphrina*

(*b*) *Mucor*

(*c*) *Aspergillus*

(*d*) *Penicillium*

(*iii*) Which of the following reproduces through hologamy?

(*a*) *Saccharomyces* (*b*) *Mucor*

(*c*) *Aspergillus* (*d*) *Penicillium*

(*iv*) Latin binomial of brewers yeast is

(*a*) *Rhizopus stolonifer*

(*b*) *Aspergillus flavus*

(*c*) *Saccharomyces cerevisiae*

(*d*) *Penicillium notatum.*

(*v*) Haplo-diplobiontic type of alternation of generation is characteristic of

(*a*) *Saccharomyces cerevisiae*

(*b*) *Schizosaccharomyces octosporous*

(*c*) *Aspergillus flavus*

(*d*) *Penicillium notatum*

(*vi*) Unicellular holocarpic fungus is

(*a*) *Mucor* (*b*) *Saccharomyces*

(*c*) *Aspergillus* (*d*) *Neurospora*

(*vii*) The vegetative cells of *Saccharomyces* are identified by the presence of

(*a*) chloroplasts

(*b*) a distinct cell wall

(*c*) a large vacuolated nucleus

(*d*) a small nucleus without a nuclear membrane.

(*viii*) Fermentation of sugars occurs by

(*a*) *Mucor* (*b*) *Saccharomyces*

(*c*) *Aspergillus* (*d*) *Penicillium*

(*ix*) Yeasts is an important source of

(*a*) Sugars (*b*) Proteins

(*c*) Vitamin C (*d*) Riboflavin

(*x*) *Taphrina deformans* causes

(*a*) Rust of wheat

(*b*) White rust of crucifers

(*c*) Stem gall of coriander

(*d*) Peach leaf curl.

(*xi*) Vacuolated nucleus is found in

(*a*) *Aspergillus* (*b*) *Saccharomyces*

(*c*) *Mucor* (*d*) *Morchella*

Aspergillus Ear Rot

CHAPTER 10

SUB CLASS – EUASCOMYCETIDAE
Series : Plectomycetes

- General characters
- Series : A. Plectomycetes
- Order : Plectascales or Aspergillales (Eurotiales)
- Family : Aspergillaceae (Eurotiaceae)
- *Aspergillus* (*Eurotium*)
- *Penicillium* (*Talaromyces*)

General Characters. The sub-class Euascomycetidae includes the higher Ascomycetes the chief characteristics of which are :–

1. The mycelium is septate and extensively developed.
2. The septa between the cells have each a central pore.
3. The ascogenous hyphae develop from the ascogonium after plasmogamy.
4. The asci typically develop from the ascogenous hyphae and in a few species from a group of ascogenous cells.
5. The asci are **unitunicate** and borne on ascogenous hyphae on or in the spore fruit called the **ascocarp**.
6. There is a trend towards gradual degeneration of the sexual apparatus particularly the antheridium.

Species in which sex organs are absent, the ascogenous hyphae develop from the binucleate cells and become surrounded by sterile hyphae. The asci in some species develop from the terminal cell of a branch of an ascogenous hypha but usually they are formed from the penultimate cell by a process called hook or crozier formation (Fig. 8.15).

Classification

According to Ainsworth and Bisby (1950), the sub-class includes about 12,000 species which are placed under 1600 genera. The classification of Euascomycetidae is a hard nut to crack. The lack of knowledge of the morphology of most of the members has stood in the way of arriving at a natural classification of this group. Till 1962, it was considered convenient to divide the sub-class directly into four series, **Plectomycetes, Pyrenomycetes, Discomycetes** and **Laboulbeniomycetes.** This classification was chiefly based on the type of the ascocarp. The main drawback of this classification which is based on superficial similarity in the appearance of the fruit body is that it separates related species and groups, puts them together with distantly related ones. Walker (1967) suggested a scheme which he considers more natural. It is chiefly based on the form of asci, their distribution within or on the ascocarp and gross morphology of the ascocarp. On this basis he divides the class Euascomycetidae into two series :–

Series A : **Plectomycetes.** The asci are typically globose, and evanescent. They are arranged irregularly (at various levels) within a spherical, cleistothecium or flask-shaped perithecium. It includes two orders *Plectascales* (*Aspergillales* or *Eurotiales*) and *Microascales*.

Series B : **Hymenoascomycetes**. The asci are clavate or cylindrical. They are arranged in a palisade-like layer forming a hymenium within or on an ascocarp. This series includes two sub-series, namely, **Pyrenomycetes and Discomycetes**.

SERIES : A. PLECTOMYCETES

The chief characteristics are :–

(*a*) The ascocarp is either of cleistothecium type (*Plectascales*) or a long necked perithecium opening by an ostiole (**Microascales**).

(*b*) The asci within the ascocarp are formed at different levels and thus are irregularly arranged within the cleistothecium.

(*c*) The asci are globose or broadly oval.

(*d*) The ascus wall dissolves to release the ascospores in the ascocarp.

(*e*) The peridium either decays or splits irregularly to liberate the ascospores. In some species the ascospores are set free through an ostiole.

The series **Plectomycetes**, on the basis of gross morphology of ascocarp, is subdivided into two orders.

(1) *Plectascales* (*Aspergillales*). Ascocarp is a spherical cleistothecium.

(2) *Microascales*. The ascocarp is a long-necked perithecium with an ostiole. Of these order Plectascales (Aspergillales) is discussed in detail here.

ORDER : PLECTASCALES OR ASPERGILLALES (Eurotiales)

General characteristics. The order *Aspergillales* (*Eurotiales*) is also called **Plectascales**. The sex organs in this order develop before the envelope surrounding the ascogenous hyphae and the gametangia. Sterile hyphae develop only with the formation of the ascogenous hyphae. The order includes several hundred species which are placed under 50 genera. Majority of them are saprophytes. A few species, parasitic on plants and animals, also occur. The saprophytic species such as the blue and green molds are of great economic importance. Some of these are sources of antibiotic drugs, organic acids and commercial enzymes. A few are essential in the ripening of cheese. The saprophytic forms are found almost on any decaying organic matter. The chief means of multiplication are the conidia. They are produced in enormous numbers. It is the conidial mass which gives hyphal mats their characteristic tint. In fact they owe their success to two features :

(*i*) Production in enormous numbers of light, dry, air dispersed conidia.

(*ii*) High osmotic pressure of the cell sap which enables them to germinate in comparatively drier habitats and in concentrated solutions of salts and sugars in which most other fungi and bacteria are unable to grow. The conidial stage in these fungi is predominant. The ascal stage is rarely seen.

The asci are globose or broadly club-shaped. They arise at different levels and thus are scattered irregularly within a round closed ascocarp called the **cleistothecium**. There is thus no hymenial layer. The ascospores are released in the cleistothecium by the dissolution of ascus wall. Subsequently they are liberated by the decay or breaking of the peridium. Bessey (1950) recognises seven families in the order, Walker makes four (*Gymnoascaceae, Aspergillaceae, Elaphomycetaceae* and *Onygenaceae*) but Alexopoulos (1962) reduced the number to three. He raised *Onygenaceae* to the rank of an order *Onygenales*. Martin (1961), however, classified the order into two families. Of these Eurotiaceae or Aspergillaceae is considered here in detail.

FAMILY : ASPERGILLACEAE (Eurotiaceae)

The family is characterised by its sub-aerial and sessile **cleistocarps**. The peridium is made up of pseudoparenchyma formed by the closely interwoven sterile hyphae. It may consist only of a single layer. Conidia when mature are echinulate. The immature ones are smooth walled. The family comprises a number of genera but is best known through *Eurotium* (=*Aspergillus*) and *Talaromyces* (=*Penicillium*) which are popularly known as bluish, greenish and yellowish molds.

Aspergillus flavus

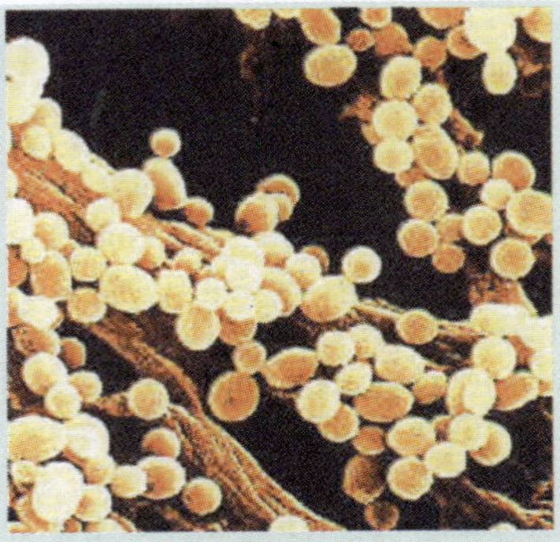

Penicillium notatum

ASPERGILLUS (EUROTIUM)

Occurrence.

This fungus is also called *Eurotium*. It is chiefly a saprophytic fungus which is widely distributed. It grows on decaying vegetables; on fatty media such as butter and ghee; on starchy media as bread and rice; on preserved fruits as jams and jellies. It is very commonly found on rotting oranges and phyllanthus fruits. It thus grows on a wide variety of substrata by virtue of the large number of enzymes it produces. All that the fungus requires are some organic matter and little of moisture.

Aspergillus appears in the form of greenish, smoky patches along with *Mucor, Rhizopus* and *Penicillium* on moist bread when kept under a bell jar for a couple of days. The other common shades are yellow, black and blue. In all *Aspergillus* includes about 200 species (Raper and Fenell, 1965). Majority of these are known only in the conidial stage. A few produce cleistothecia (perfect stage). At first when the connection between the two stages was not fully established the sexual or perfect stage was called *Eurotium* and the conidial or imperfect stage as *Aspergillus*. Now both the stages are usually denoted by the common generic name *Aspergillus* because conidial stage is common and predominant and moreover generic name *Aspergillus* was introduced first.

Economic Importance

About 33 species of this genus have been reported from India mostly from the soil. Their role in the economy of nature has not been fully investigated. *Aspergillus* causes *rot* of figs and dates, decay of tobacco and cigars. It spoils nuts, bread and other food-stuffs. In humid weather it grows even on leather and cloth fabrics. To cut short this cosmopolitan genus is one of the most frequent contaminants of food. *Aspergillus niger*, popularly known as the black mold is considered a 'weed of laboratory' as it often contaminates the bacteriological and mycological cultures. *A. niger* and several other species cause decay of foodstuffs. Some of these during the process produce poisonous substance known as **mycotoxins**. The most important among these is aflatoxin. Of the saprophytic species *A. oryzae* is employed in making alcohol by the saccharification of rice starch. *A. niger* is utilized in the production of citric and other organic acids. Some species yield antibiotics. The cultures of *A. niger* and *A. oryzae* yield a wide

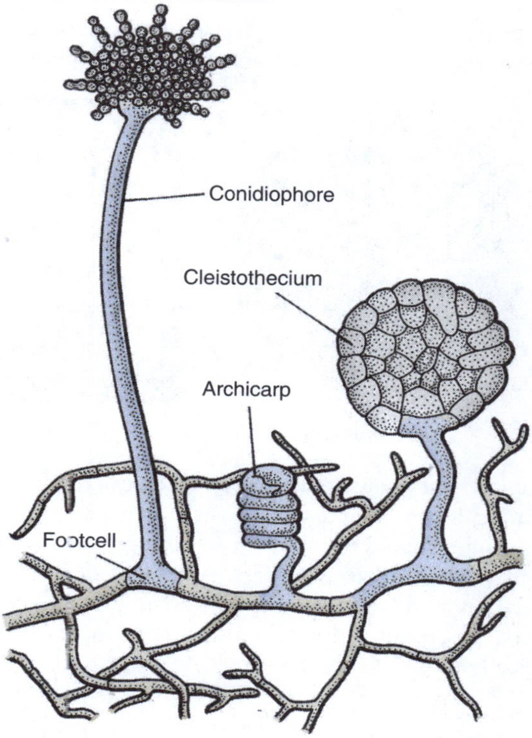

Fig. 10.1 (A). *Aspergillus sp.* A. plant consisting of a portion of mycelium bearing a conidiophore, an archicarp and cleistothecium.

Aspergillus Ear Rot

Aspergillus Rot

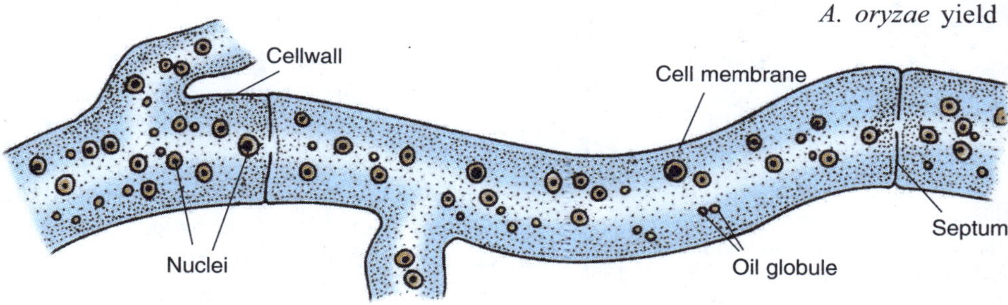

Fig. 10.1 (B). *Aspergillus sp.* A portion of a hypha enlarged to show details of structure as seen under light microscope. (After Miss Dale).

range of enzymes used in industrial fermentation. A few species of *Aspergillus* are parasitic. They cause diseases in animals and plants. *A. fumifatus, A. flavus,* and *A. niger* parasitize men and

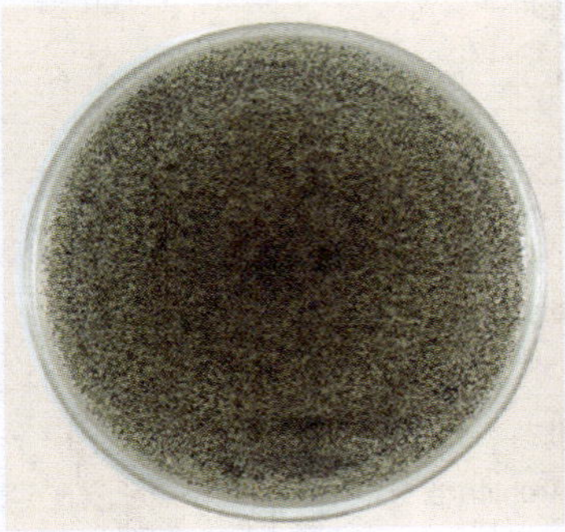

Aspergillus niger

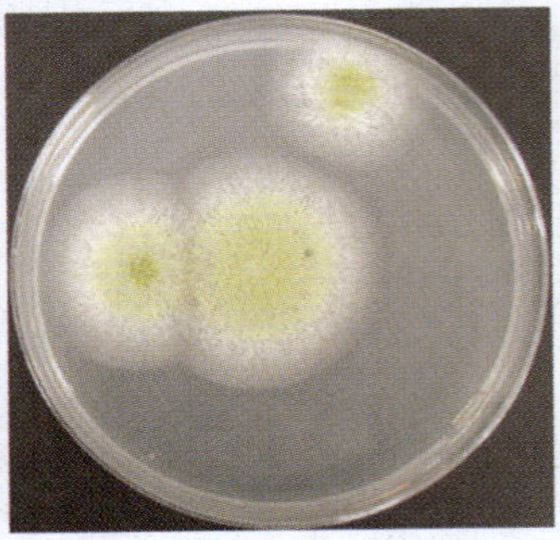

Aspergillus oryzae

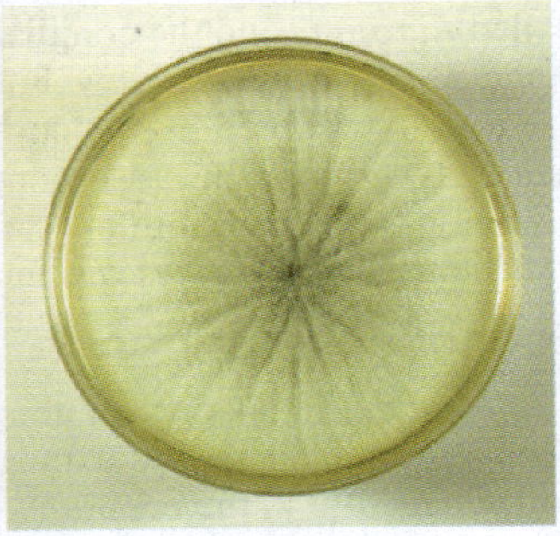

Aspergillus fumigatus

animals. They cause a number of diseases grouped under the name **Aspergilloses** (sing. aspergillosis). They also occur in the human ear and cuase **Otomycosis**. *A. niger* causes rot disease of pomegranates, dates and figs. Azuma *et al.* (1968) reported the isolation, purification and characterization of an immunologically active glycoprotein for skin test and serological reaction from the mycelium of *A. fumigatus*.

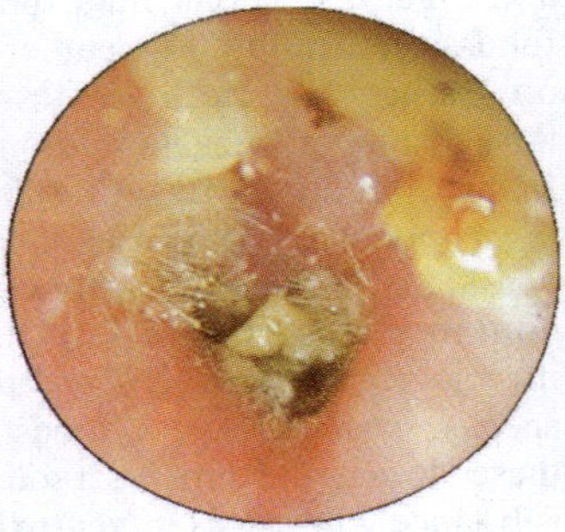

Otomycosis

Mycelium (Fig. 10.1 A). It is well developed and made up of a loosely interwoven mass of hyaline, bright or pale coloured, extensively branched, septate hyphae. Some of the hyphae ramify superficially upon the substratum. Others penetrate deeply into the substratum. The latter absorb food for the entire mycelium. The hyphae are freely branched and form dense mats on the substratum. The slender, delicate hyphae have thin smooth walls. Campbell (1970) found no evidence for multilayered nature of the hyphal wall but observed the presence of only diffuse flocculent material on the outer wall surface of *A. fumigatus*. The hyphae are divided into segments by cross walls. The cross walls have each a simple, central pore which permits the streaming of cytoplasm and migration of nuclei from one cell into the next. Each segment or cell contains one to many nuclei. The nuclear membrane has typical nuclear pores. The nuclei are embedded in a mass of granular cytoplasm which also contains the mitochondria, ribosomes lying free in the cytoplasm, endoplasmic reticulum and vesicles. Lipid globules constitute the principal reserve food.

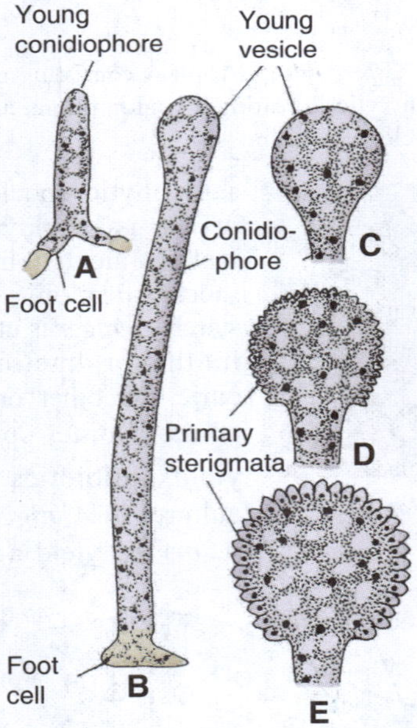

Fig. 10.2 (A–E). *Aspergillus sp.* Asexual reproduction. A, young conidiophore arising from the foot cell; B, older conidiophore with the tip slightly swollen to form a young vesicle; C, older vesicle; D-E, vesicle in optical section showing the development of primary sterigmata (Based on Thom and Raper).

Under certain conditions, the mycelium develops into a **sclerotium**. Agnihotri (1968) reported that the sclerotia of *A. niger* are fairly large ranging from 0.8 to 1.2 mm in dia. Normally they are globose to subglobose bodies. Each sclerotium is surrounded by a cream to buff-coloured, thick-walled interwoven hyphae forming a pseudoparenchyma layer.

Asexual Reproduction (Fig. 10.2). In addition to vegetative propagation by **fragmentation,** the fungus reproduces asexually by means of asexual spores known as **conidia** which are produced exogenously in chians at the tips of certain vertically growing aerial hyphae called the **conidiophores** (Fig. 10.3).

(*a*) *Conidiophores* (Fig. 10.2). Certain cells in the older parts of the young, vigorously growing prostrate mycelium become thick-walled. These thick-walled, T-shaped cells are characteristic of the genus *Aspergillus* and are known as the **foot cells**. Each foot cell produces a special erect branch as an outgrowth (A). These upright hyphal branches are the young **conidiophores**. Each conidiophore grows to a length of about 2.5 mm and swells at its tip (B) to form usually a globose or elliptical bulbous head. This swollen end of the conidiophore is called the **vesicle** (C). The vesicle may, in some species, be hemispherical or clavate. The lumen of the vesicle is continuous with the upper part of the conidiophore and is multinucleate. A. Hanlin (1976) reported that the vesicle in *A. clavatus* is surrounded by a thick wall. Besides the numerous nuclei, it contains mitochondria and other cell organelles. The conidiophores, at maturity, are long, stout, unbranched and commonly unseptate, very rarely septate structures. They arise singly from the foot cells of the mycelium and are not organised into any kind of asexual fruit bodies.

From the surface of the multinucleate vesicle arise numerous, radially arranged, tubular outgrowths (E). These specialized conidiogenous outgrowths or cells are the **sterigmata** (Sing. sterigma) or **phialides**. The sterigmata are arranged compactly side by side and thus completely cover the entire surface of the vesicle. The species with conidia bearing phialides arising directly from the vesicle are termed *uniseriate species* (*A. flavus,* Fig. 10.3). In some other species, the phialides are borne on intermediate cells, the **metullae** (primary sterigmata) which are attached to the vesicle. These are termed biseriate species (Fig. 10.4 A). In biseriate species such as *A. fonsecaceus* the conidia bearing sterigmata are called **secondary sterigmata**. Generally species tend to be either uniseriate or biseriate. Raper and Fenell (1965), however, reported cases in which both types of arrangements occur in the same species or even in the same conidial head. The whole structure consisting of the foot cell, the upright hypha, the vesicle, the metullae and the phialides constitutes the **conidiophore** (Fig. 10.3).

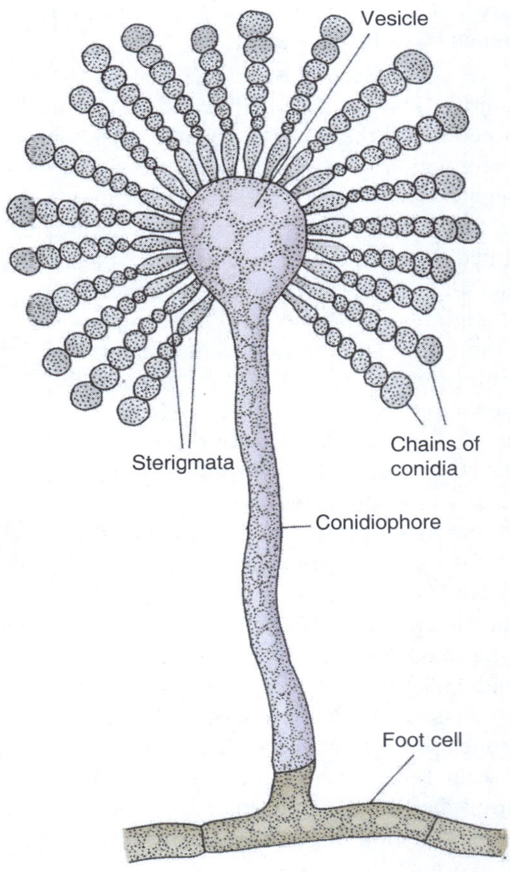

Fig. 10.3. *Aspergillus.* A mature conidiophore bearing sterigmata and chains of conidia.

(*b*) *Development of phialides.* It was studied by Hanlin (1976) in *A. clavatus*. Phialide formation starts with the appearance of numerous thin areas in the otherwise thick vesical wall. The thin areas are formed by the dissolution of wall material. The cytoplasm adjacent to these areas pushes out synchronously to form broadly oval, peg-like outgrowths, each with an attenuated base. These are the **phialides**. As the immature phialides increase in size they come in contact with one another. Into each phialide migrates a single nucleus, mitochondria and other cell organelles from the vesicle. At maturity each phialide is cut off from the vesicle by a basal septum. It is usually flask-shaped with a tapering apex. The opening between the vesicle and the young phialide is relatively narrow. The nucleus migrating from the vesicle into the phialide becomes elongate as it passes through this narrow opening.

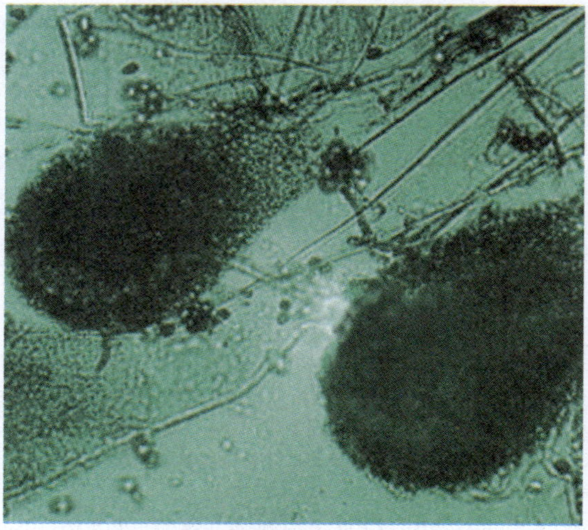

Aspergillus clavatus

(*c*) *Abstriction of conidia* (Fig. 10.4 B-C). The entire broadly oval phialide tip pushes out at maturity. The mature phialide thus has a narrow, tapering, tubular apex (B_1). The conidia are formed inside the tip of the phialide. The phialides or sterigmata are uninucleate. At the time of the formation of first conidium, the single nucleus in the phialide divides by mitosis into two daughter nuclei. One of these migrates into the narrow, tubular phialide tip which enlarges to form the first conidium (B_2). Towards maturity the first conidium is delimited by a basal septum formed just inside the mouth of the phialide apex (B_3). The basal septum has a central pore (Hanlin, 1976). It is too narrow to permit the migration of organelles. The protoplast of the first conidium thus delimited by a septum rounds off and secretes a wall of its own to become a conidium which soon after assumes its characteristic shape and size (C_4). The conidial wall may fuse partially or completely with the parent cell wall or remain free from it. Meanwhile the tip of the sterigma immediately below the first conidium elongates to form a tube (C_5) which is again cut off as a uninucleate cell (C_6). The latter develops into a second conidium in the same manner as the first and pushes the latter outward without disjunction (C_6). With the delimitation of the second conidium, the septum separating it from the first conidium thickens greatly on both sides forming a connective between the two conidia. This series of events is repeated. The phialide thus continues to grow and cuts off conidia one below the

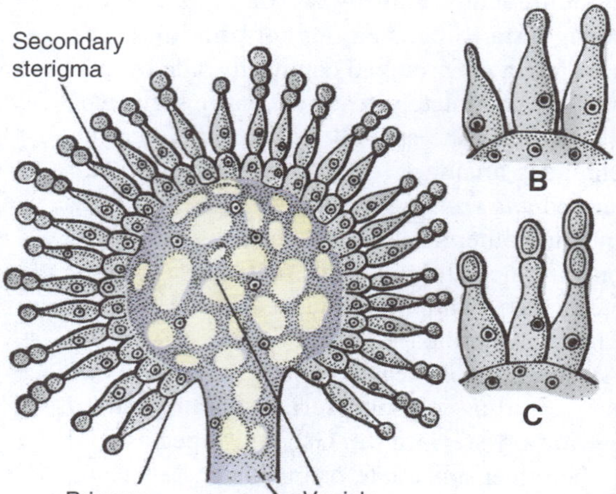

Fig. 10.4 (A–C). *Aspergillus sp.* Asexual reproduction (cotd.). A vesicle with secondary sterigmata bearing conidia; B–C, stages in the abstriction of conidia. (Based on Thom and Raper)

other. Consequently a chain of conidia is formed at the tip of each sterigma. All the conidia in the chain are held by similar connectives. A fine channel remains open through the thickened septal pore (Hanlin, 1976). The conidia are formed in basipetal chains with no increase in the length of the phialide. The youngest conidium is at the base of the chain next to the tip of the conidiophore. The oldest is at the top. This basipetalous arrangement of conidia in the chain has the following advantages :–

1. Permits ready dispersal of mature conidia by air currents.
2. Proper nourishment of the younger conidia.

As the conidial chain increases in length the connectives begin to break down separating the conidia. The conidia are thus continually shed and new ones formed below.

The conidia are uninucleate at first and remain so in the majority of species but become multinucleate in some others by successive nuclear divisions. In *A. repens* the conidia have 12 nuclei and in *A. herbariorum* 4 each. The conidia of *A. niger* are binucleate and of *A. nidulans*

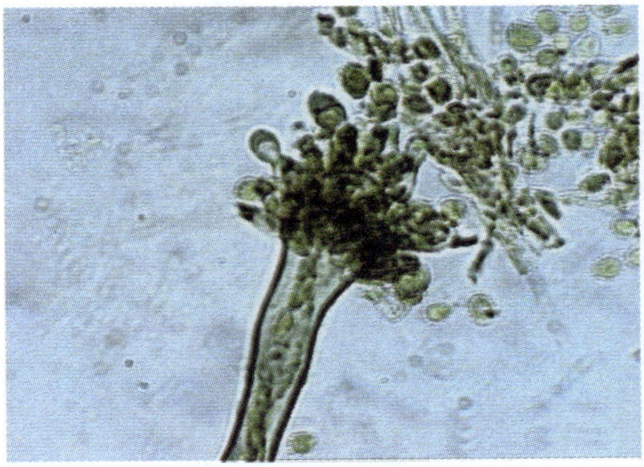

Aspergillus repens

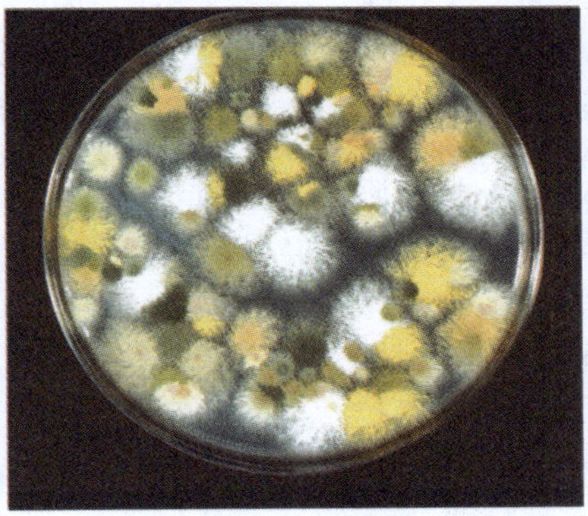

Aspergillus nidulans

uninucleate. They are black, green, brown, blue or yellow in colour, according to the species and the medium on which the fungus is growing. The mature conidia are globose and unicellular. They have a surface which is wetted only with difficulty. The conidial wall is thick and usually differentiated into two layers, the outer **epispore** and the inner **endospore**. The epispore in a mature conidium is finely spiny. The young conidia have a smooth epispore layer. The study of fine structure of the mature conidium revealed the spore wall to be composed of three layers. There is an electron dense inner layer, a darker middle layer of equal thickness and a slightly thinner somewhat wavy, dark outer layer. The conidium often contains electro-transparent areas (Hanlin, 1976). The conidia are very small, light and dry and thus are dispersed by wind. They are always present in the air or in dust.

On falling on a suitable substratum, each conidium germinates. It, at first, produces a germ tube which grows into a new mycelium. During germination a third wall layer forms within the two original spore wall layers in *A. niger*. The germ tube wall is continuous with this third spore wall layer only. In *A. oryzae* the germ tube wall is an extension of the inner layer of the two-layered spore wall (Hawker, 1966 and Bracker, 1967).

Under the microscope the conidiophore in a fresh, undisturbed preparation is seen as a long hypha tipped by an indistinct black body. The vesicle is not seen because the conidia are tightly packed and are opaque *en masse*. They hide from view the vesicle which can be seen only by removing the conidia mechanically. The conidial stage is very prominent and commonly seen in nature and cultures.

Sexual reproduction (Fig. 10.5 A–C). The sexual or perfect stage is rare. Possibly the species lacking it have lost it during the course of evolution. This view is supported by the fact that even in species that form asci there is evidence of sexual degeneration. Different species of this genus show variation in their sexual behaviour. The antheridium is absent in some species or if present it is nonfunctional. The asci, however, develop from the female sex organ (ascogonium). In others it is well developed and functional. *Aspergillus* is **homothallic**. *A. heterothallicus* is the only species that has been reported to be truly heterothallic (Kwon, Fennel and Raper, 1964). A second heterothallic species, *Aspergillus fishri* has been reported by Chung and Sang (1974). The male and female sex organs, in the homothallic species, are developed close together on the same hypha or on separate nearby hyphae of the same mycelium (A). Both are elongate, multinucleate and generally coil around each other.

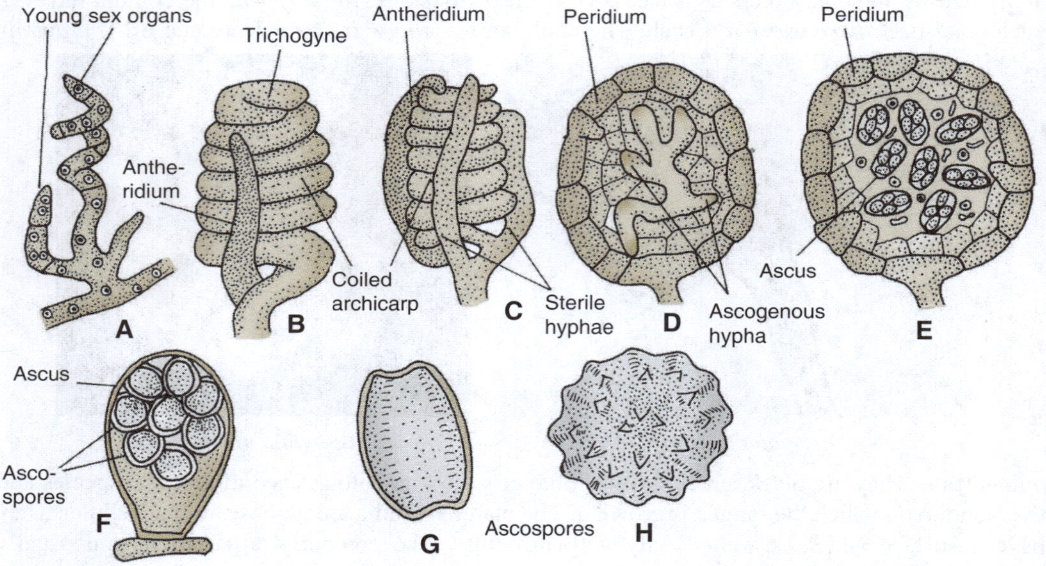

Fig. 10.5 (A–H). *Aspergillus sp.* Sexual Reproduction. A, Young sex organs; B. Coiled archicarp and an antheridium; C, archicarp enclosed by sterile hyphae, arising from its base; D, section of young cleistothecium showing ascogonium with ascogenous hyphae surrounded by peridium; E, section of mature cleistothecium showing asci surrounded by peridium; F, a single ascus containing 8 ascospores; G, ascospore from side view; H, ascospore (surface view).

(*a*) *Ascogonium*. A small, loosely coiled septate hyphal branch, the **archicarp** arises from a vegetative hypha (A). It is differentiated into three parts. The terminal segment is generally the longest and single celled. It contains up to 20 nuclei and is called the **trichogyne**. It functions as the receptive part of the female sex organ. The segment below the trichogyne functions as the female gametangium and is called the **ascogonium**. The protoplast of the ascogonium is also multinucleate. It does not round off to form the egg. Below the ascogonium is the stalk consisting of a few cells. All the parts of the archicarp thus are multinucleate. The archicarp, at first, is a loosely coiled hyphal branch (A) but later the coiling becomes tight and close (B).

(*b*) *Male Sex Organ*. Shortly before or after septation of the archicarp, a male branch, the **pollinodium** grows up beside the archicarp from the same hypha or from another adjacent one

(A). The pollinodium winds spirally round the tightly coiled archicarp once or several times. Finally it arches over its apex and then cuts off a unicellular antheridium at its tip (C). The antheridium is a multinucleate slightly swollen structure. The lower septate portion of the pollinodium below the antheridium constitutes the **stalk**. All the segments or cells are multinucleate.

(*c*) *Plasmogamy*. Fusion may take place between the antheridium and the trichogyne. The tip of the antheridium arches over the apex of the trichogyne and fuses with it. The intervening walls dissolve. The contents of antheridium then pass through the opening into the trichogyne. An open communication between the antheridium and the trichogyne has, however, been denied by many. The antheridium is considered by them to be rudimentary or non-functional. In some species the male nuclei have been reported to degenerate before the antheridium has reached maturity. In still other species the antheridium may entirely be absent.

LIGHT THAT FORMS LIKE ASPERGILLUS THROW ON THE DEGENERATION OF SEX ORGANS IN THE ASCOMYCETES

The different species of *Aspergillus* illustrate the way in which elimination of male sex organ (antheridium) in the Ascomycetes has taken place. These species can be arranged in a series. The series indicates the stages or steps in which the progressive degeneration of the male sex organ (antheridium) has been accomplished. These steps are :

1. In *A. herbariorum* the pollinodium cuts off a terminal antheridium. The tip of the antheridium, however, do not migrate and fuse with the contents of the ascogonium.
2. In some species the antheridium may develop late. By that time the ascogenous hyphae have already been developed from the ascogonium.
3. In a few species the male nuclei in the antheridium are reported to degenerate before the antheridium has reached maturity.
4. The antheridium in some species such as *A. flavus* and *A. fisheri* and *A. fumigatus* is not developed at all.

(*d*) *Development of Ascocarp* (Fig. 10.5, D–E). Whether the antheridium is functional or not the ascogonium in all cases develops into a fruit body or the fructification which is called the **ascocarp**. The haploid male and female nuclei in the ascogonium come to lie in pairs. In species in which plasmogamy does not take place, the female nuclei themselves approach each other and come to lie in pairs called the **dikaryons**. The two nuclei in the pair may lie in the same membrane or in separate membranes. After pairing of the nuclei, the ascogonium may become septate. The segments are binucleate. From the dikaryotic segments arise the **ascogenous hyphae** (D). At the same time sterile hyphae grow up from the base of the archicarp and invest the sexual apparatus and the ascogenous hyphae. They intertwine to form a pseudoparenchymatous **sheath** shaped like a hollow ball of nearly the size of a pin head around the sexual apparatus. The ascogenous hyphae branch frequently. The branches are of different lengths. The small, thin-walled asci which develop at their tips, lie at different levels presenting a scattered arrangement. The whole structure enclosed by the sheath is the fruit body or fructification. It is called an **ascocarp** (D). The ascocarp in *Aspergillus* is a small, rounded, yellow hollow ball about 150–200 μ in diameter with smooth walls. Even at maturity it remains closed. An ascocarp with a hollow, closed form is called the **cleistothecium** or **cleistocarp**. The sheath is differentiated into a single layered yellow **peridium** and one or more ill defined inner layers constituting the **tapetum** (E). The tapetum is nutritive in function. The outer layer provides protection to the developing asci which are formed at different levels and thus lie in a scattered arrangement within the cleistothecium. It secretes a yellow substance which makes the cleistocarp readily recognisable. It is yellow and soft.

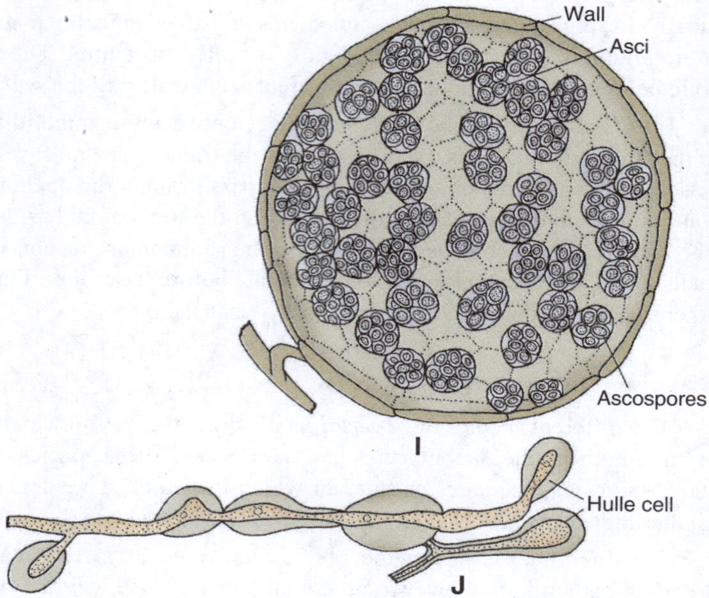

Fig. 10.5 (I-J). *Aspergillus*. I. A mature cleistothecium containing asci and ascospores; J. Hülle cells.

Development of asci and differentiation of ascospores. The swollen terminal cells of the ascogenous hyphae or their branches give rise to asci. They are binucleate and directly function as **ascus mother cells.** No croziers are formed. The two nuclei in the ascus mother cell fuse. The young ascus with the fusion nucleus enlarges. It represents the transitory **diplo-phase** in the life cycle. Its diploid nucleus or the synkaryon under-goes three successive divisions. The first and the second divisions constitute **meiosis**. The third is **mitotic**. This results in the formation of eight haploid daughter nuclei. Each haploid nucleus becomes surrounded by a certain amount of cytoplasm. A wall is then secreted around each uninucleate proto-plast to organise it into a meiospore which is botanically called an **ascospore**. In this way eight haploid ascospores are formed from the contents of each ascus by a process known as **free cell formation**. The cytoplasm left over in the ascus is known as **epiplasm**. The mature asci may be globose, ovoid or pear-shaped (F). They do not survive long and are not explosive. When the ascospores are formed the ascus wall, the ascogenous hypha and the tapetum all undergo degeneration to form a nutritive fluid. The ascospores lie free in the hollow cleistothecial cavity. The nutritive fluid provides nutrition to the developing ascospores. When the ascospores are mature the peridium also undergoes decay. The ascospores are thus liberated.

Structure of Ascospores. The liberated ascospore is a haploid, uninucleate structure (G). It is somewhat, lens-shaped with a small groove round the edge and thus shaped like a pulley wheel. The spore wall is differentiated into two layers, the outer **epispore** and the inner **endospore**. The epispore is usually thick and sculptured. In side view the ascospore is seen like a pulley wheel but in surface view it may appear round or star-shaped (H). On falling on a suitable substratum each ascospore germinates to form a germ tube which develops into the characteristic haploid mycelium.

Alternation of Generations (Fig. 10.6). From the above account, it. is evident that in a single life cycle of *Aspergillus* there occur two phases or generations. Of these, one is more conspicuous. It is independent so far as its nutrition is concerned. This independent phase is the **gametophyte**.

It is characterised by the haploid number of chromosomes in the nuclei. It bears the sex organs and is concerned with sexual reproduction. In the life cycle, it starts with the formation of ascospores and ends with the formation of dikaryons in the ascogonium. It consists of the ascospores, parent mycelium, antheridia when present and young ascogonia. There are accessory means of multiplication of this phase in the life cycle. It is done by the formation of conidia. Conidia play no role in the alternation of generations.

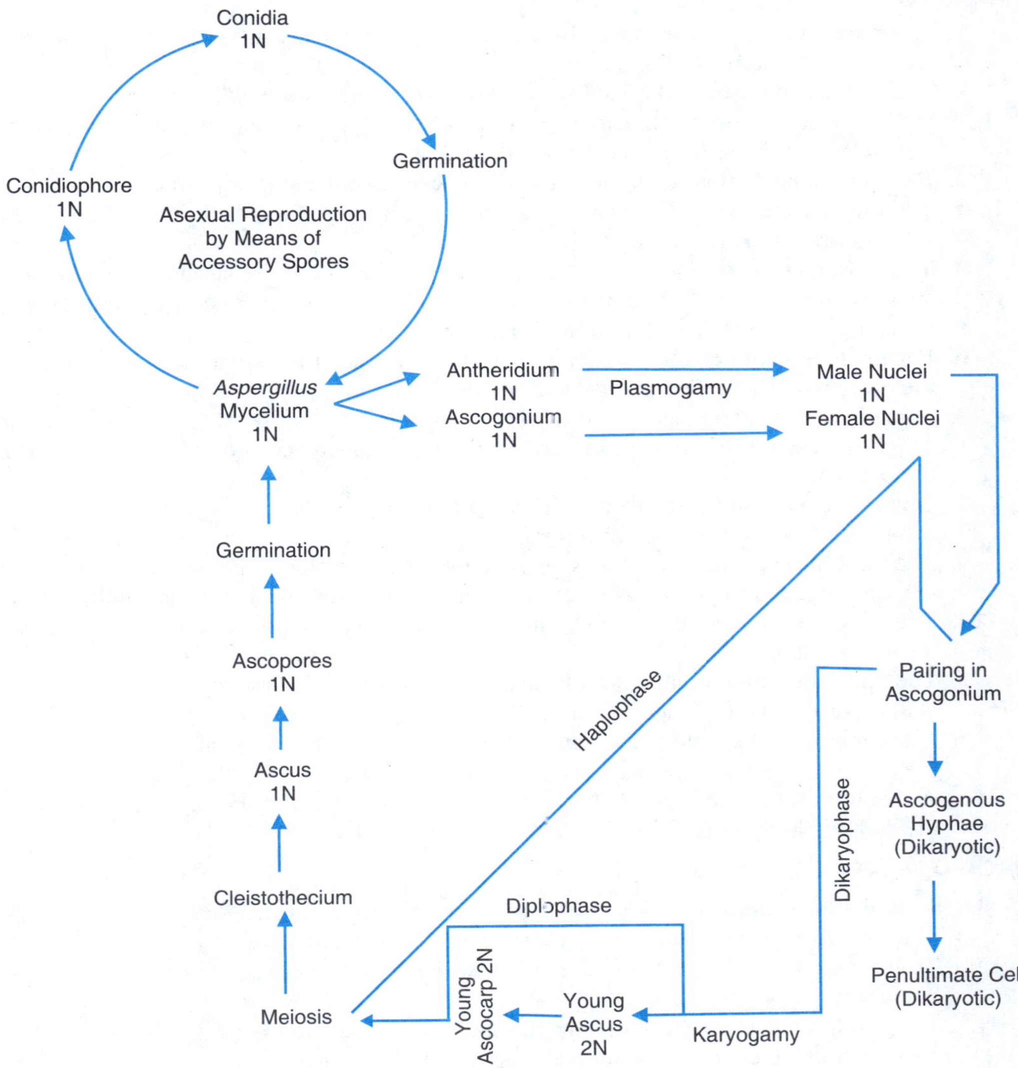

Fig. 10.6. Graphic representation of the life cycle of *Aspergillus* (*Eurotium*).

The second phase in the life history of *Aspergillus* is the *sporophyte*. It is not conspicuous and is dependent for its nutrition on the parent gametophyte (*haplomycelium*). The sporophytic phase starts with the formation of dikaryons in the ascogonium. It ends with meiosis of the synkaryons in the asci. The sporophytic phase thus consists of (*i*) **dikaryophase** consisting of the ascogonium with dikaryons, the ascogenous hyphae, and (*ii*) the transitory **diplophase** represented by the young ascus before meiosis.

These two phases sporophyte and gametophyte normally alternate in a single life cycle. Hence *Aspergillus* is said to exhibit **alternation of generation** in the life cycle.

Salient Features. The life history of *Aspergillus* consists of three phases. They are the **haplophase,** the **dikaryophase** and the transitory **diplophase**.

Haplophase

1. It is represented by the extensively branched mycelium consisting of septate, colourless hyphae. The cells are multinucleate.
2. Besides fragmentation, the haplomycelium is propagated asexually by means of non-motile, wind disseminated, exogenously produced spores called the **conidia** which are uninucleate at first but soon become multinucleate.
3. Each conidium is globose. It has thick, echinucleate wall and is unicellular.
4. The conidia are formed one below the other in a long chain at the top of somewhat bottle-shaped **sterigma**.
5. The sterigmata arise side by side completely covering the entire surface of the globose, swollen head, the **vesicle** developed at the top of a special long, erect unbranched, unseptate, stout hypha called the **conidiophore**.
6. Each conidiophore arises as an outgrowth from a special thick-walled T-shaped cell of the mycelium. It is called the **foot cell**.
7. Sometimes the sterigmata are produced in two layers. Those of the upper layer are called **the secondary sterigmata** and of the lower layer as **primary**. The former bear the conidia.
8. On falling on a suitable substratum each conidium germinates by a germ tube which soon develops into the haplomycelium.
9. The sex organs **antheridia** and **ascogonia** are produced close together on the hyphae of the same mycelium. The female branch is called the **archicarp** and the male **pollinodium**.
10. The sex organs are elongated, multinucleate structures. The pollinodium is coiled round the ascogonium.
11. The tip of the antheridium comes in contact with the trichogyne. The intervening double wall dissolves at the point of contact.
12. If the migration of male nuclei takes place they pair with the female nuclei, otherwise the ascogonial nuclei approach near each other to lie in pairs in the ascogonium. Each such pair is called a dikaryon. With the formation of dikaryons the haplophase ends and the dikaryophase starts.

Dikaryophase

13. After the establishment of dikaryons, the ascogenous hyphae arise from the ascogonium. They are septate and branched. The branches lie at different levels.
14. The asci are formed from the terminal binucleate cells of the ascogenous hyphae or their branches.
15. The young binucleate ascus mother cell is the last structure of the dikaryophase which consists of the ascogonium with dikaryons and the ascogenous hyphae.
16. The dikaryophase is parasitic on the haplomycelium.
17. The whole sexual apparatus in the meantime is enclosed in a compact layer of sterile hyphae forming the **sheath** or **peridium**.
18. The fructification thus formed is a small, globose ascocarp. The wall is smooth and yellow. The ascocarp in *Aspergillus* is soft. It has no opening and thus is closed and called the **cleistothecium**.

Diplophase

19. With the fusion of the two nuclei in the ascus mother cell the dikaryophase ends and the transitory *diplophase* starts.

Future Haplophase

20. During further development, the young ascus enlarges and the synkaryon undergoes meiosis. The resultant four haploid nuclei divide mitotically to form eight nuclei which are organised into ascospores. With the establishment of ascospores starts the new haplophase.

21. The walls of the evanescent asci dissolve, the ascospores are released into the cavity of the cleistothecium. Eventually the wall of the cleistothecium decays to liberate the ascospores.
22. Each ascospore germinates under suitable conditions to form a new characteristic haplomycelium.

Taxonomic Position

Alexopoulos (1962)

Division	:	**Mycota**
Sub-division	:	**Eumycotina**
Class	:	**Ascomycetes**
Sub-class	:	**Euascomycetidae**
Series	:	**Plectomycetes**
Order	:	**Plectascales (Aspergillales)**
Family	:	**Aspergillaceae**
Genus	:	*Aspergillus*
Species	:	*repens*

Alexopoulos and Mims (1979)

Superkingdom	:	**Eukaryonta**
Kingdom	:	**Mycetae**
Division	:	**Amastegomycota**
Subdivision	:	**Ascomycotina**
Class	:	**Ascomycetes**
Sub-class	:	**Plectomycetidae**
Order	:	**Eurotiales**
Family	:	**Eurotiaceae**
Genus	:	*Aspergillus*
Species	:	*repens*

PENICILLIUM (=TALAROMYCES)

Occurrence. The penicillia are commonly known as the green or blue molds. However, their colour varies through a wide range. They are cosmopolitan in their distribution. Usually *Penicillium* grows as a saprophyte on decaying fruits and vegetables. Decaying oranges, lemons and other fruits of the family Rutaceae usually exhibit a bluish growth on them. This is due to the blue coloured spores of *Penicillium*. It also grows on the preserved fruits and jellies. Some species of *Penicillium* have been reported to grow on the forest floors and a few of them grow on cultivated and manured ground. *Penicillium italicum* and *P. digitatum* are well known plant pathogens which attack citrus fruits causing the familiar soft rot disease. *P. funiculosum* induces the pine apple

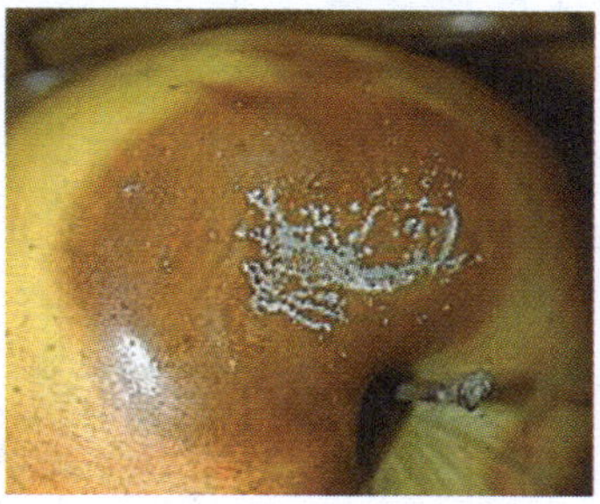

Penicillium digitatum

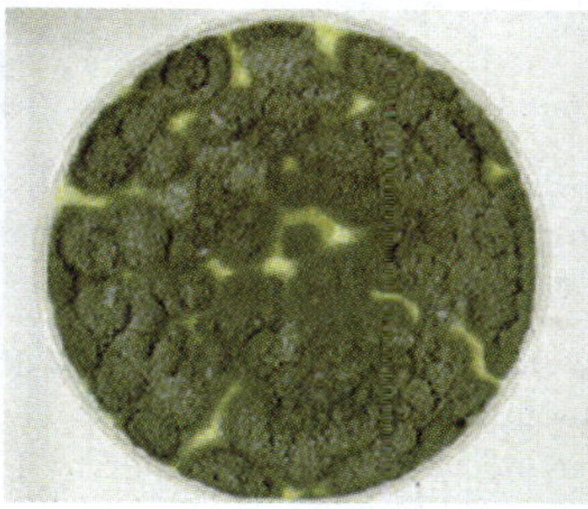

Penicillium funiculosum

Penicillium italicum

(*Ananas comosus*) fruit disease termed *fruitlet core rot* (Rohrbach and Pfeffer, 1975). Wild species of *Penicillium* that grow upon damp wood have been known to digest hemicelluloses. Some species (*P. roqueforti* and *P. camemberti*) cause fermentation in cheese and lend it characteristic aroma and consistency. Raper and Thom (1949) in their monograph of the genus reported 136 species. Some mycologists make 140 species.

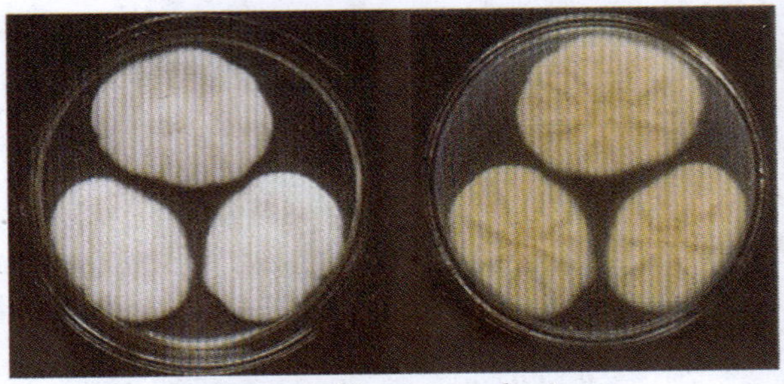

Penicillium camemberti

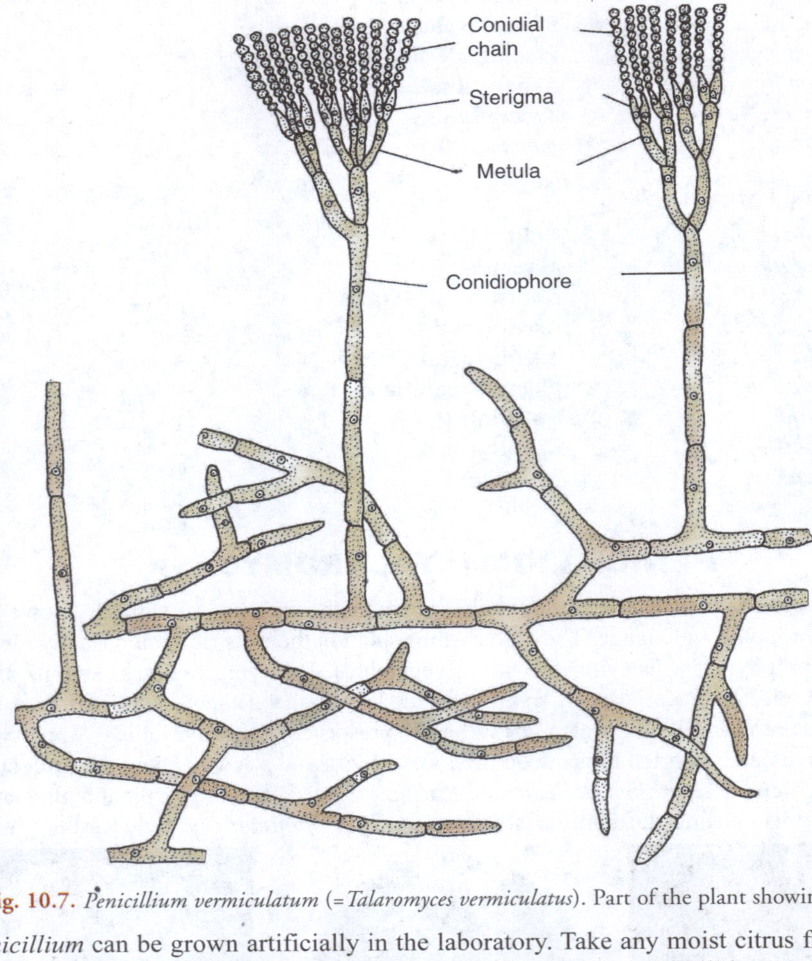

Fig. 10.7. *Penicillium vermiculatum* (=*Talaromyces vermiculatus*). Part of the plant showing habit.

Penicillium can be grown artificially in the laboratory. Take any moist citrus fruit or a piece of cheese. Cover it with a bowl and place it in a damp, warm place. It will become moldy. Small velvety patches of blue or sky blue or bluish green colour will appear on the substrate.

Mycelium (Fig. 10.7). The mycelium is well developed and copiously branched. It is composed of colourless, slender, tubular, branched and septate hyphae. The hyphae run in all directions on the substratum and become intertwined with one another to form a loose network of hyphae constituting the mycelium. Some of the hyphae may even grow into the interior of the substratum

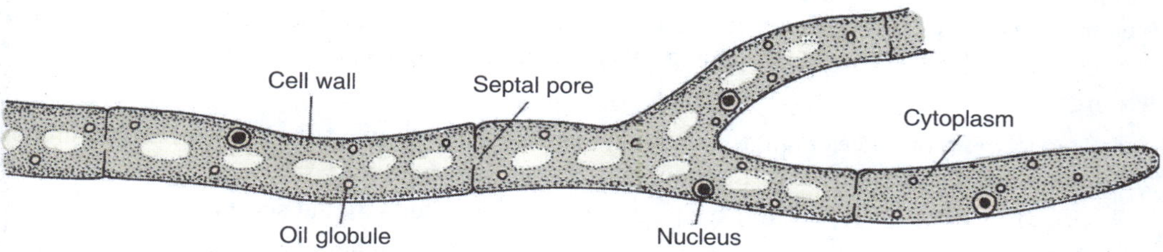

Fig. 10.8. *Penicillium sp.* A portion of a hypha enlarged to show structure.

and the rest spread on the surface. The former secrete enzymes and absorb food materials from the substratum. These are the **haustorial hyphae**. The aerial hyphae receive nourishment through the **haustorial** hyphae and produce reproductive structures. Baker (1944) observed anastomosing between hyphae of two mycelia resulting in a **heterokaryotic mycelium**. The mycelium in a few species may develop into a **sclerotium**. The hyphae constituting the mycelium are septate and the cells are short. The septa between the cells have each a central pore. Through the pores the protoplasm flows from cell to cell.

Cell Structure (Fig. 10.9). The thin hyphal or cell wall is microfibrillar in structure with the microfibrils running parallel to the surface. In *P. notatum* it is reported to consist of 3 or 4 layers. The outermost layer (W_1) is made up of glucans. It is irregular and electrondense. The second electron layer (W_2) is composed of proteins and has a regular contour. The third layer (W_3) is electron-transparent and formed of fibril material arranged parallel to the plasma membrane. The fibrils consist of chitin and are embedded in a granular matrix. The fourth layer (W_4) is probably composed of pectic or hemicellulose material. When present it directly abuts on the plasma membrane which closely invests the granular cytoplasm. Embedded in the cytoplasm are the living inclusions namely mitochondria, ribosomes and endoplasmic reticulum. The cells in *P.*

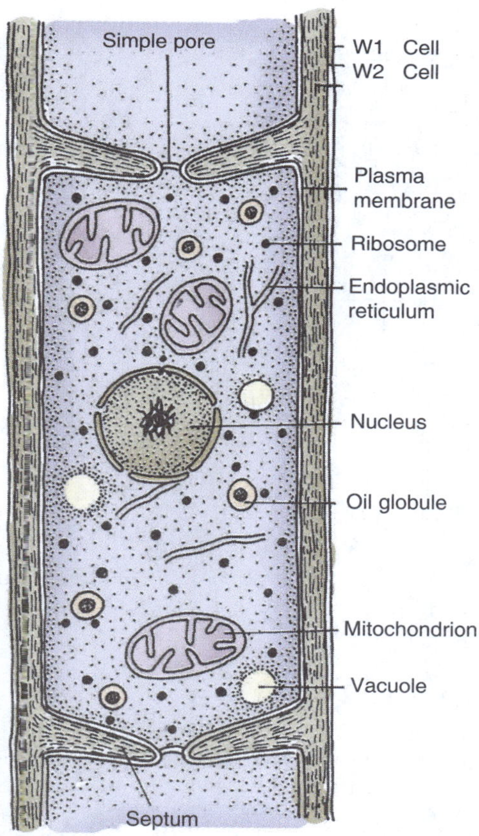

Fig. 10.9. *Penicillium sp.* Diagrammatic representation of fine structure of a cell.

Penicillium notatum

vermiculatum are uninucleate but in some other species they are bi- to trinucleate. The nuclear membrane has pores. Oil globules may also be present as reserve food. During the vegetative phase, the mycelium is hyaline or pale. The colour appears when the fungus enters the reproductive phase.

Reproduction

Penicillium reproduces both asexually and sexually. The asexual stage however, is dominant and constitutes the usual mode of reproduction. Sexual stage is rare.

1. Asexual Reproduction. It takes place by vegetative methods and sporulation.

(*i*) *Vegetative Reproduction*. It is accomplished by the most common method of **fragmentation**. The hyphae break up into short segments. Each segment or fragment grows by repeated division into a full-fledged mycelium.

In some species, the mycelium forms compact resting bodies, the **sclerotia**. Sclerotia enable the species to survive periods of stress or to hibernate. On the onset of conditions favourable for growth each sclerotium germinates to form a new mycelium. The sclerotia thus serve primarily as a means of perennation rather than multiplication.

(*ii*) *Sporulation*. Normally it takes place by the formation of non-motile, asexual spores, the **conidia** which are produced exogenously at the tips of long, erect special septate hyphae called the **conidiophores**. *Penicillium* multiplies repeatedly by this method during the growing season.

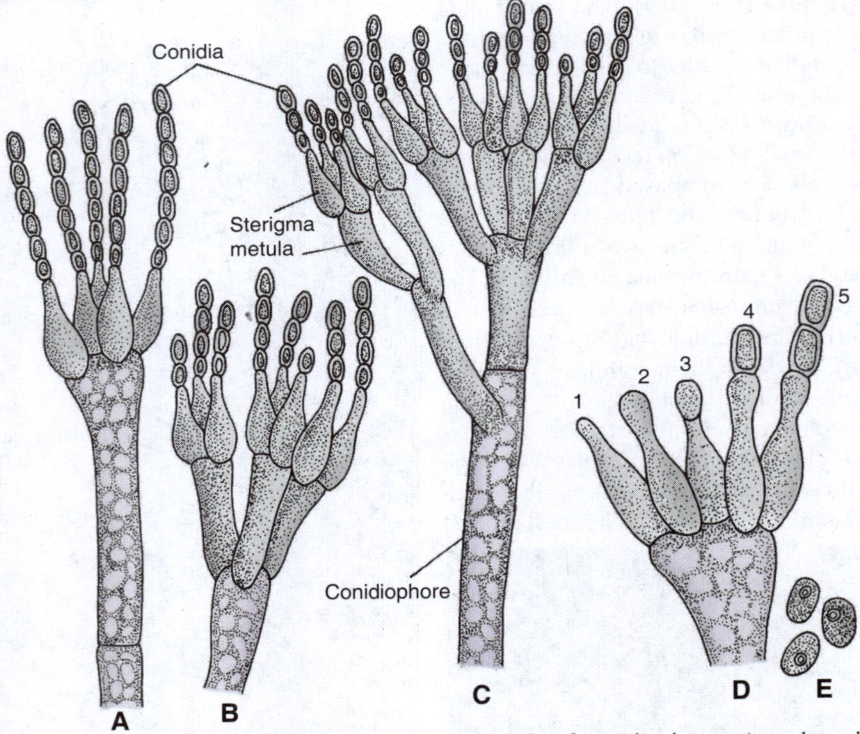

Fig. 10.10 (A–E). *Penicillium* sp. Different kinds of conidiophores. A, unbranched conidiophore; B–C, branched conidiophores; D_{1-5}, stages showing abstriction of conidia; E, mature conidia.

Conidiophores (Fig. 10.10 A–C). A conidiophore arises as an erect, tubular hyphal outgrowth from any cell of the mycelium and not from a specialised cell (foot cell) as in *Aspergillus*. After some period of vegetative growth upright hyphae arise from the older portions of the mycelium.

Penicillium thomii

They are negatively geotropic and arise singly from any cell of the mycelium. Each grows up in length vertically. Reaching a certain height the septate conidiophore branches once or twice or even more times. These are termed as **pirmary, secondary** or **tertiary branches,** respectively. Only rarely are the conidiophores unbranched (*Penicillium thomii*). The unbranched axis of the latter bears a tuft of flask-shaped **sterigmata** (A). In the species with branched conidiophores, the ultimate branches which bear tufts of flask-shaped **sterigmata** or the **phialides** are called the **metulae**. The lower branches which support the metulae when short and form a part of the penicillus are called the **rami**. Conidia are abstricted from the tips of the phialides or sterigmata and are borne in long, unbranched chains. Baker (1944) reported that the phialides and the upper cells of the conidiophore are uninucleate. The phialide wall is electron-transparent with a thin electron opaque surface layer. The apical portion of the conidiophore with its branches (metulae), sterigmata and chains of conidia looks like a small artist's brush known as the **penicillus**. The generic name *Penicillium* is derived from 'penicillus,' which reflects the form of conidial chains arising from the sterigmata supported on the metulae.

Development of conidia (Fig. 10.11 A-E). The conidia are formed within the narrow tips of the flask-shaped phialides (A). The conidium initial is formed by the distension of the tubular tip of the phialide (B). The phialide nucleus undergoes mitosis. One daughter nucleus remains in the phialide and the other migrates into the swollen tip (conidium initial). The conidium initial

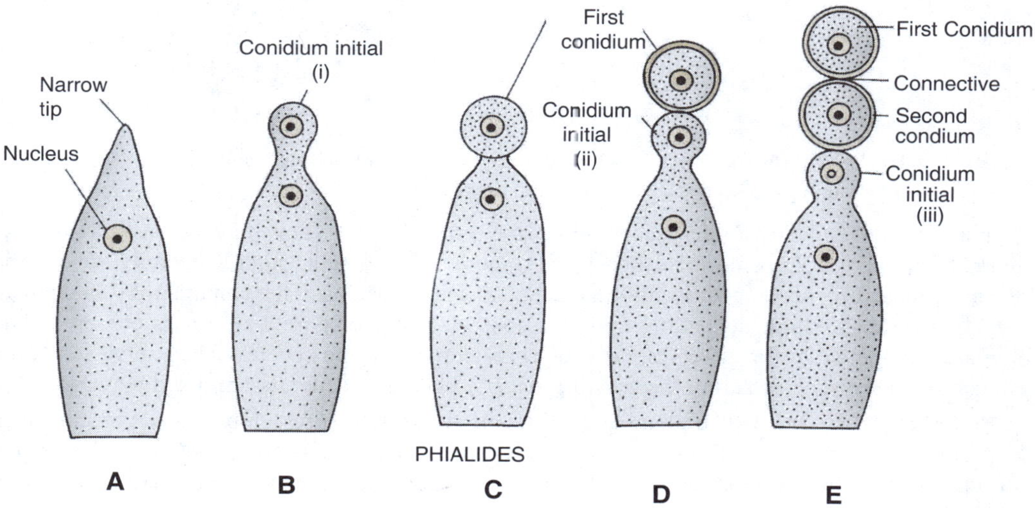

Fig. 10.11 (A–E). *Penicillium sp.* Stages showing development of conidia in chains.

protoplast is then cut off from the phialide protoplast by a thin perforate septum (C). The perforation remains as a channel between the successive conidia in the chain. It is filled with electron-opaque material. The newly delimited conidial protoplast secretes a wall around it distinct from the phialide wall and functions as the first conidium (D). The spore or conidial wall during further development may remain distinct or fuse partially with phialide wall. The tip of phialide below the first conidium again elongates and swells (D). A second conidium is formed by repeating the process (E). Like this, one below the other, a long chain of conidia is formed. In older conidium chains, the septum region remains only as a narrow connecting strand between conidia. The conidium chains are enclosed within an electron-opaque surface layer which appears to be continuous with the surface layer of the phialide wall (Fletcher, 1971). The conidia in the chain are arranged in a basigenous manner. The youngest conidium lies next to the tip of the sterigma and the oldest away from it. The basigenous arrangement of conidia serves two useful purposes. It permits ready dispersal of mature conidia from the tips. Secondly, it aids the proper nourishment of younger conidia which are nearest to the tips of the sterigmata. As the conidial chain increases in length, the connectives between the older conidia break down resulting in the separation of mature conidia. The conidia are thus shed continuously. Being small, light and dry they are dispered by air currents.

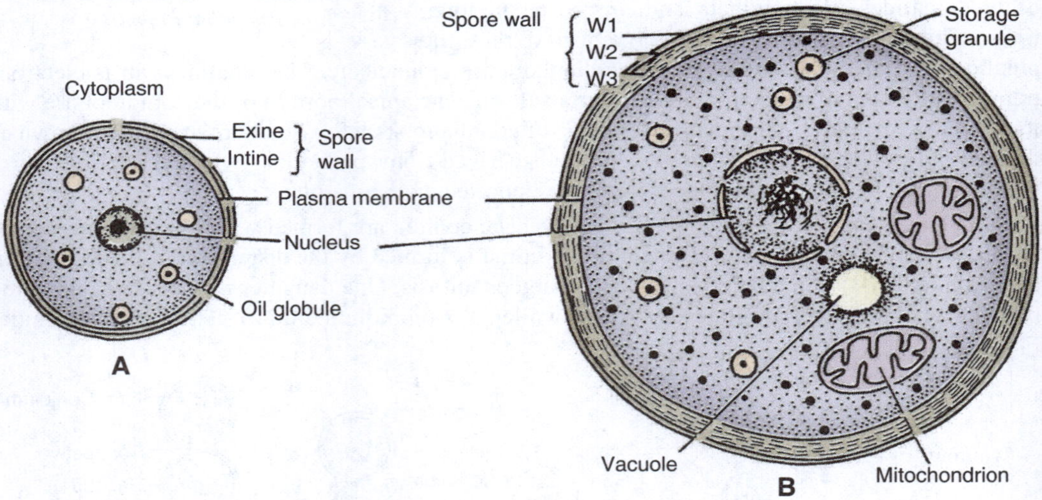

Fig. 10.12 (A–B). *Penicillium sp.* Showing structure of resting conidium. A, under light microscope, B, under electron microscope.

The conidia under the microscope look like small beads. They may be ovoid, globose, elliptical or pyriform. In some species they are smooth and in others rough. Generally the conidia are uninucleate but may become multinucleate in some species. The conidia may be bluish, greenish or pale in colour depending upon the species. It is the spore wall that is coloured. The conidia are responsible for the colony colour characteristic of the species. They serve for quick propagation of the species during the growing season. They are easily disseminated by wind.

Structure of Conidia. The conidia are tiny spore-like structures globose to ovoid in form. The pigmented spore wall is differentiated into two layers, outer **exine** and inner **intine**. The exine is comparatively thick, smooth or spiny. The intine is thin. Under electron microscope (Martin *et al.,* 1973) it appears to consist of 3 or 4 layers. There is the outermost layer (W_1) with an irregular, undulating contour. It is electron-dense. Empty speaces are often seen between it and the next wall layer. Next to it is a slightly thin inner electron-opaque layer W_2 with a more regular contour. This layer (W_2) gradually merges into the next wall layer W_3 which is thick and electron-transparent. It usually abuts on the plasma membrane. However, in some cases, there is another slightly electron-dense layer (W_4) present directly above the plasma membrane. Within the spore wall is the plasma membrane. Embedded in the cytoplasm of the conidium are the mitochondria and ribosomes. The endoplasmic reticulum strands are not discernible. The vacuoles are absent. However, Martin *et al.* (1973) reported the presence of a single large vacuole in the resting conidium of *P. notatum*. The conidium cytoplasm contains oil globules. Usually the conidium contains a single nucleus. The nuclear membrane is two-layered and is poriferous.

The conidial stage in *Penicillium* is more dominant than in *Aspergillus*. Majority of the species of *Penicillium* are known only in the conidial stage. Some (about 20 species) are now known to produce cleistothecia. At first when the connection between the two stages was not fully established the conidial or imperfect stage was given the name form or genus *Penicillium* and the sexual or perfect stage as *Talaromyces*. The discovery of sexual or perfect stage in these form-species places them in the true Ascomycete genus *Talaromyces*. Many mycologists, however, still adhere to the old name *Penicillium* because the conidial stage is predominant. Moreover, the generic name *Penicillium* was introduced first.

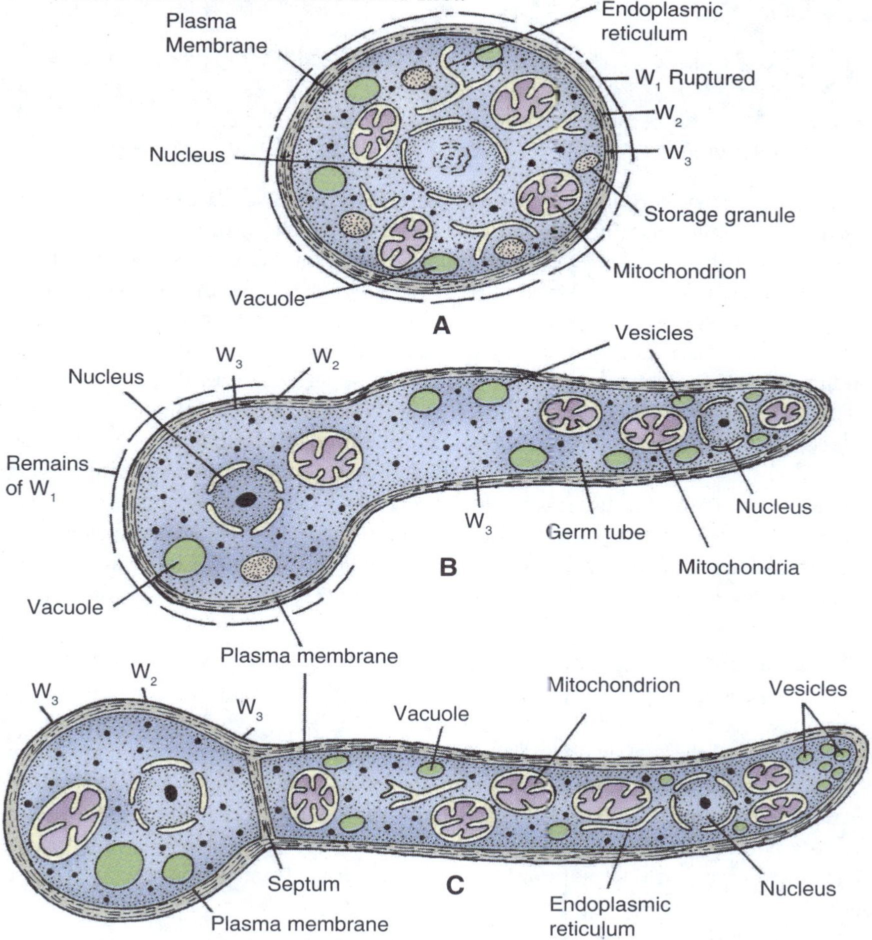

Fig. 10.13 (A–C). *Penicillium sp.* Diagrammatic representation of the stages in the germination of conidium. A, swollen conidium; B, emergence of germ tube; C, formation of primary septum (Based on Martin *et al.*)

Germination of Conidia (Fig. 10.13). On falling on a suitable substratum and under suitable conditions (moisture, suitable temperature and food), the conidium absorbs moisture and swells. The swollen conidium germinates by putting out a germ tube. Fletcher (1971) studied the fine structural changes during germination of conidia in *Penicillium griseofulvum.* He reported that the ungerminated conidium has a two-layered spore wall. The protoplast contains a single nucleus and mitochondria. During germination a third layer appears inside the two-layered spore wall. It is continuous with the germ tube wall. In *P. frequentans*, the germ tube wall is considered to be an

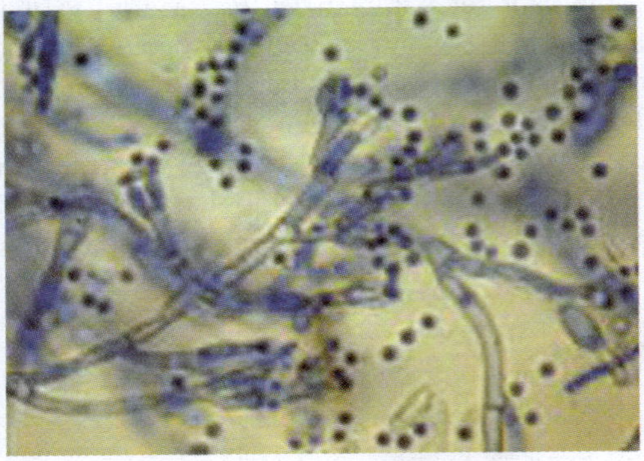

Penicillium griseofulvum

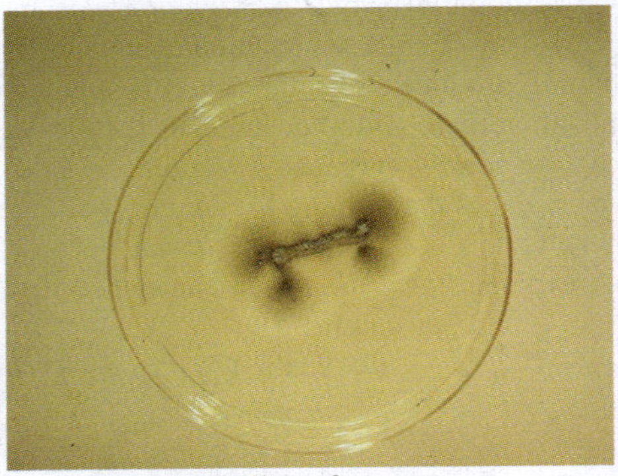

Penicillium frequentans

extension of the inner layer of the original spore wall (Hawker, 1966; Bracker, 1967). During swelling, the mitochondria increase in size and become lobed, endoplasmic reticulum becomes visible and vacuoles are formed. The septa formed in the germ tube are perforate and have Woronin bodies associated with them. Marchant (1968) suggested enzymic breakdown of the two-layered spore wall at the point of germ tube emergence rather than mechanical rupture. Martin *et al.* (1973) reported that during swelling of the conidium (A), the outer layer (W_1) of the wall of the conidium becomes broken at several places and separated from the underlying spore wall layer W_2. The smooth surface layer W_2 with a regular contour thus becomes exposed. At this stage the swollen conidium puts out a germ tube (B). During spore swelling the large mitochondria of the resting spore divide to produce smaller mitochondria. The size of the single vacuole decreases. Endoplasmic reticulum reappears in the swollen and germinating spores. The wall of the emerging germ tube has been reported to be continuous with the electron transparent wall layer W_3. The electron-dense wall layer W_2 remains at the base of the germ tube and around the spore. The single nucleus of the resting spore divides early in germination. One of these migrates into the germ tube.

With the elongation of the germ tube most of the mitochondria migrate into it and become concentrated near the tip. Small vesicles reappear along the plasma membrane. Some were seen at the tip of the germ tube. Later, a septum is formed at the point of emergence of the germ tube. It is continuous with the transparent layer W_3 of the spore wall (C). During further development, layers W_2 and W_1 of the hyphal wall are formed by deposition of electron-dense material on the outer surface of the germ tube. The latter elongates and divides by septa. Septum formation is preceded by nuclear division. Each septum has a central pore. There are Woronin bodies associated with the septa. By further growth, septation and branching mycelium is formed. It consists of uninucleated cells.

Oidia Formation. The mycelial hyphae of **Penicillium**, when made to grow immersed in a sugary solution, they divide by additional septa into short uninucleate segments. The latter become rounded and separates as thin-walled spore-like structures called the **oidia** or **oidiospores**. Often the oidia increase in numbers by budding. This is known as the **torula stage**. The fungus in the torula condition brings about fermentation of sugar into alcohol. This process is called alcoholic fermentation. On a solid medium each oidium germinates to produce a normal mycelium.

Similarly if conidia of *Penicillium* happen to fall in a sugary solution they fail to produce normal mycelia. Instead each conidium germinates to produce a thin-walled yeast-like cell which starts budding like the oidiospore to produce the torula stage.

Penicillium griseofulvum

2. Sexual Reproduction (Fig. 10.14). Sexual process in *Penicillium* has been studied in a few species such as *P.vermiculatum* (=*Talaromyces vermiculatus*), *P. glaucum*, *P. brefeldianum* and a few others. All of them are reported to be **homothallic**. Derx (1925) reported heterothallism in one species (*P. luteum*) but his observation has not been confirmed by subsequent investigators. The structure of sex organs varies from species to species. In *Talaromyces vermiculatus* (*Penicillium vermiculatum*) sexual reproduction is oogamous. It was studied in detail by Dangeard in 1907. The male and female organs are known as **antheridia** and **ascogonia** respectively.

(*a*) *Ascogonium*. The mature ascogonium is a long, erect multinucleate, unseptate, tubular structure. At its upper end it may be curved like the handle of an umbrella (E). It arises as a lateral outgrowth from any cell of the vegetative mycelium. When young the ascogonium is uninucleate (A). As it elongates the single nucleus within it divides and redivides to give rise to a definite number of daughter nuclei which is either 32 or 64 (B).

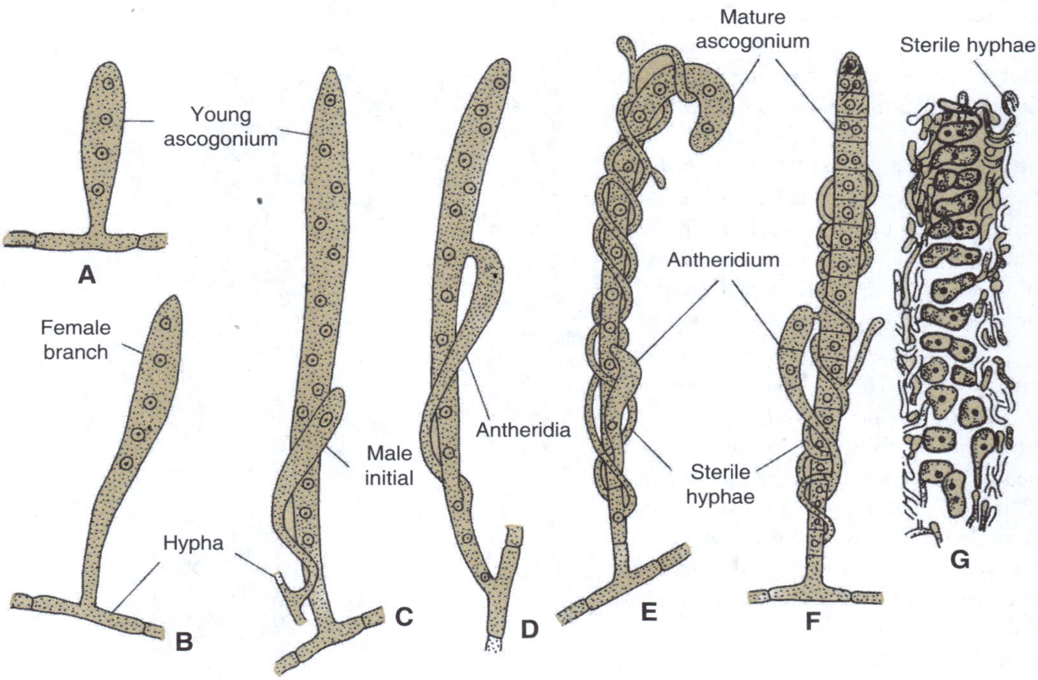

Fig. 10.14 (A–G). *Penicillium vermiculatum.* Showing stages in sexual reproduction and post-autogamy changes. A–D, stages in the development of sex organs; E, sexual apparatus surrounded by sterile hyphae arising from its base; F, establishment of dikaryons and septation of the ascogonium; G, later stage. (After Dangeard)

(*b*) *Antheridium*. Meanwhile a slender uninucleate hyphal branch originates either from an adjacent cell of the same hypha which gives rise to the ascogonium or from a separate neighbouring hypha. It is the **antheridial branch** (C). It grows up coiling loosely around the ascogonium making several turns about it (D). The distal end of the male branch becomes slightly inflated and is eventually cut off as an **antheridium** by a septum (E). The antheridium is a short, terminal, club-shaped, uninucleate structure.

(*c*) *Plasmogamy*. It is the union of two protoplasts which brings the compatible nuclei close together in the same cell. The tip of the antheridium comes in contact with the ascogonium. At the point of contact, the double wall dissolves. Through the common pore the two protoplasts come in contact. According to Dangeard, the migration of the male nucleus into the ascogonium does not take place. This fact has been confirmed by a number of other workers. Mere contact of the antheridium protoplast stimulates the ascogonium. Beyond that the antheridium plays no role. The female nuclei in the ascogonium, however, arrange themselves in pairs. The pairing of the female nuclei in the ascogonium is called **autogamy**. Each pair is called a **dikaryon**. With the establishment of dikaryons the haplophase ends and the **dikaryophase** starts in the life cycle.

In some other species of *Penicillium*, both antheridium and ascogonia are claimed to be functional. The intervening walls between the two sex organs dissolve and their protoplasts come in contact. The male nucleus migrates into the ascogonium. Further details of the male nucleus are not known fully. The suggestion, however, is that it divides a number of times. The male nuclei then come to lie by the side of the female nuclei, each to each. Each pair of nuclei consisting of one male and the other female is called a **dikaryon**.

Sexual reproduction has been worked out in *P. glaucum*. Two short lateral hyphae arise from the mycelium close together and become coiled about each other. One of them is the **antheridium** and the other **ascogonium**. Plasmogamy thus takes place by gametangial contact. Dikaryons are established in the ascogonium. This is followed as usual by the septation of the ascogonium into binucleate segments.

Post Plasmogamy or Autogamy changes (Fig. 10.14 E-G), (*i*) *Septation of Ascogonium. Plasmogamy* or *Autogamy* is followed by septation of the ascogonium. Each segment has a pair of nuclei (dikaryon, F). Meanwhile entangled sterile hyphae grow up around the sexual apparatus and form an investment of loosely interwoven hyphae and sexual apparatus and afford protection to the structures developing within (E-F).

(*ii*) *Development of Ascogenous Hyphae* (Fig. 10.14 F). As a result of the stimulus of plasmogamy or autogamy

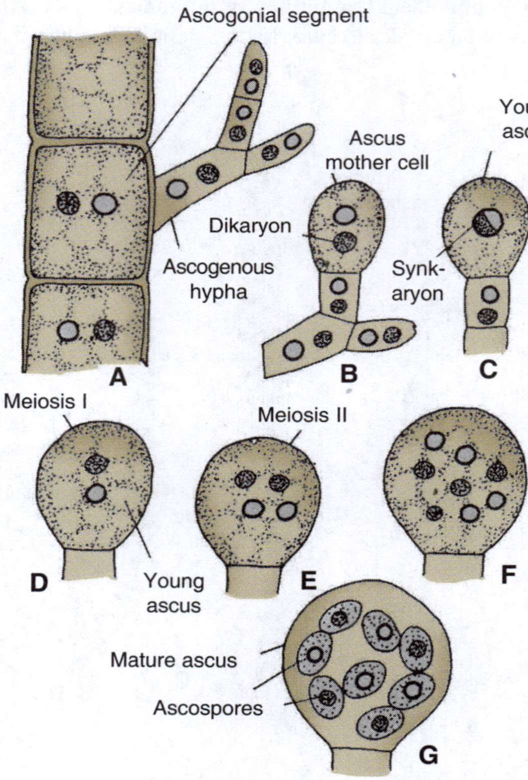

Fig. 10.15 (A-G). *Penicillium vermiculatum.* Direct method of ascus development (A-C); D-G, of ascospores by meiosis (Diagrammatic).

one or more lateral outgrowths arise from some of the binucleate segments situated in the middle of the septate ascogonium. Each outgrowth is called an **ascogenous initial**. It is binucleate. Each ascogenous initial develops into a branched **ascogenous hypha** composed of binucleate cells. The branches are of different lengths. In some other species of *Penicillium* the lateral branches of ascogenous hyphae are one-celled.

(*iii*) *Formation of Asci*. According to Emmons (1935), asci in many species are developed directly from the binucleate cells toward the tips of the branches of the ascogenous hyphae. Consequently they are arranged in short chains. In *P. vermiculatum* almost all the cells of the ascogenous hyphae develop into asci directly without crozier formation. Rarely the asci are formed from the croziers.

1. *Direct Method of Ascus Development*. This is illustrated by *P. vermiculatum* (Fig. 10.15). All the binucleate cells of the ascogenous hyphae are capable of directly developing into ascus mother cells. No croziers are formed. The binucleate cell enlarges into a sac-like structure. It is the **ascus mother cell**. It is globose or pear-shaped in form. The two nuclei of the ascus mother cell eventually fuse. The cell containing the fusion nucleus (synkaryon) is called the young ascus.

2. *Indirect* (*crozier*) *method of ascus formation* (Fig. 10.16). In this case, the terminal binucleate cells of the ascogenous hyphae or their branches curl over to form a hook-like structure called the **crozier**. For further details of the process refer to prior pages.

The binucleate cell in both cases enlarges to function as the **ascus mother cell**. It is the last structure of the **dikaryophase**. The two nuclei in it eventually fuse. This is **karyogamy**. Karyogamy

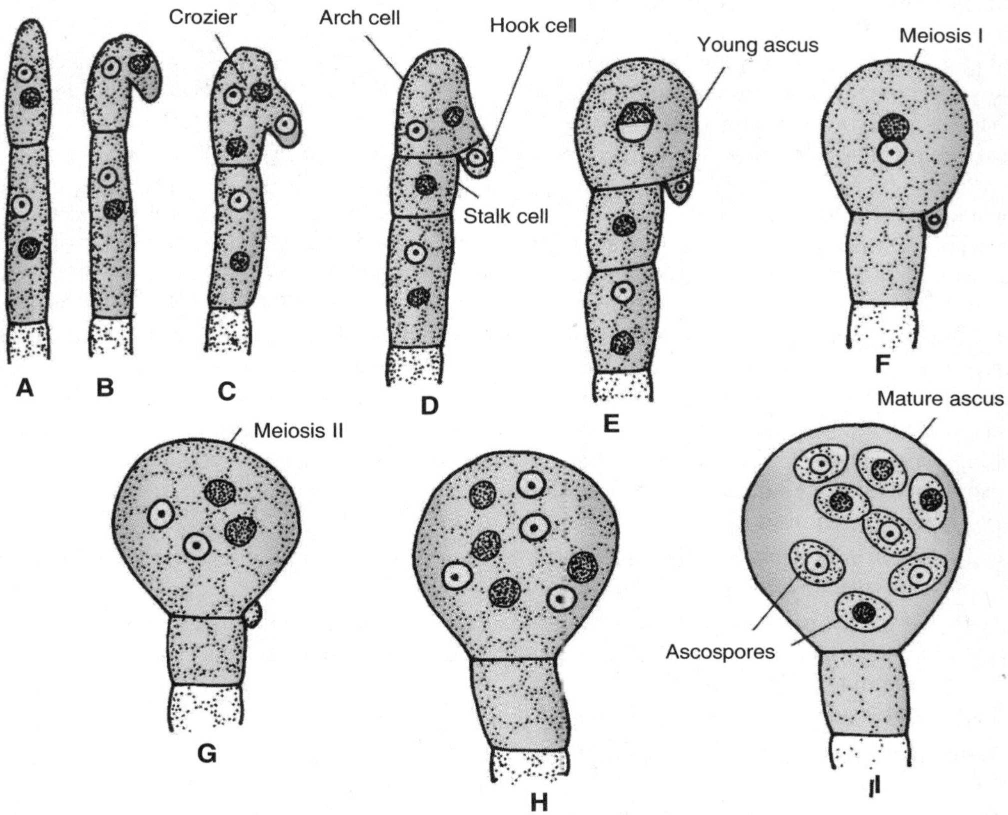

Fig. 10.16 (A–I). *Penicillium sp.* Indirect (crozier) method of ascus development. A-E, development of ascus; F-I, differentiation of ascospores by meiosis (Diagrammatic).

is thus equivalent to fertilisation. With karyogamy, the dikaryophase in the life cycle ends and the **diplophase** begins. The septate ascogonium, ascogenous hyphae and the ascus mother cells constitute the **dikaryophase**. It is parasitic. The cell containing the fusion or diploid nucleus (synkaryon) is the **young ascus**. It is the only diploid structure in the life cycle and thus represents the short-lived, transitory **diplophase**.

(*iv*) *Differentiation of Ascospores* Fig. 10.16 (F-I). The young ascus enlarges. Its diploid nucleus undergoes three successive divisions. The first and the second divisions constitute **meiosis**. The third is **mitotic**. The resultant 8 nuclei are thus **haploid**. A small amount of cytoplasm gathers around each daughter nucleus. The uninucleate protoplasts secrete their own walls and are fashioned into uninucleate **ascospores**. Eight ascospores are thus formed by the method of free cell formation in each mature ascus which may be spherical or pear-shaped in appearance.

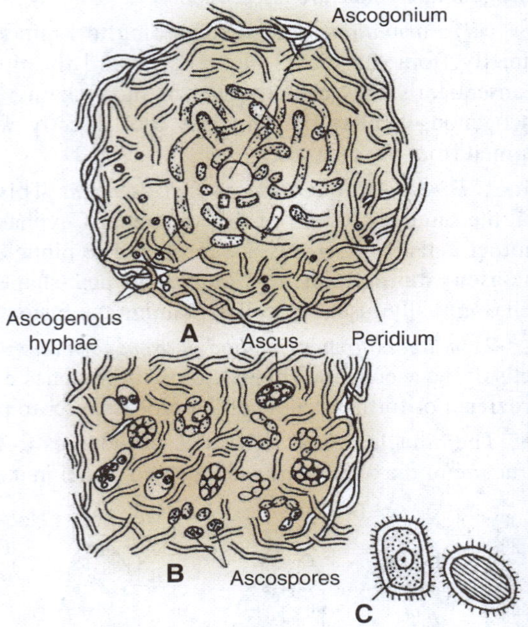

Fig. 10.17 (A-C). *Penicillium vermiculatum*. A, T.S. young cleistothecium. Note the ascogonium surrounded by ascogenous hyphae and peridium made up of loosely interwoven hyphae; B, Portion of a mature cleistothecium. Note the scattered asci intermingled with sterile hyphae; C, ascospores. (After Dangeard)

(*v*) *Formation of Ascocarp* (Fig. 10.17). With the septation of ascogonium and development of ascogenous hyphae, a large number of sterile hyphae grow up around the sexual apparatus. The ensheathing sterile hyphae get interwoven to form a hollow **ball-like** structure, the **peridium** which surrounds and protects the ascogenous hyphae as they grow and branch within. This is the ascocarp. The asci within the ascocarp are scattered. The ascocarp of *Talaromyces* is of indefinite growth. It continues to increase in size even after the ascospores begin to mature. The peridium or sheath is thicker than that of the ascocarp of *Aspergillus* and consists of loosely interwoven hyphae. This entire structure is spherical and has no opening (A). Such a closed fruit or ascocarp is called the **cleistothecium**. The asci are borne in chains within the ascocarp. In some other sp. of *Penicillium*, the peridium is compact and pseudoparenchymatous. The asci are borne singly and terminally on one-celled lateral branches of the ascogenous hyphae within it. The cleistothecium of *Penicillium* or *Aspergillus* represents the following three generations :–

(*i*) *Parent haplophase*. It is represented by the sheath or peridium made up of loosely interwoven hyphae.

(*ii*) *Dikaryophase*. It consists of the binucleate cells of the ascogonium, ascogenous hyphae and the ascus mother cells. The transitory diplophase is represented by the young ascus containing a diploid nucleus.

(*iii*) *Future haplophase*. It is represented by the ascospores in the asci.

Discharge of Ascospores. At maturity the walls of the asci dissolve. The liberated ascospores, each shaped like a pulley wheel (side view), float in the nourishing fluids, formed by the degeneration of the inner layers of the peridium, walls of asci and the ascogenous hyphae. The ascospores absorb nutrition and mature. The mature ascospores are finally released by the decay of the other wall of the peridium.

Ascospore Structure (Fig. 10.18). The liberated ascospore is a haploid uninucleate structure. It is lens-shaped with a small groove around the edge and thus appears like a pulley wheel in side view (B). The spore wall may be smooth or sculptured. It is differentiated into two layers, the outer **epispore** and inner **endospore**. The epispore is usually thick and sculptured. In face view the ascospore is round to star-shaped (A).

Germination of Ascospores. The liberated, uninucleate ascospores remain dormant when temperature and moisture conditions are low and air supply is restricted. On falling on a suitable substratum and under suitable conditions (suitable temperature, moisture and plenty of air) each ascospore germinates. It absorbs moisture and produces a germ tube. The latter by growth, septation and branching forms the mycelium.

Distinction between *Penicillium* and *Aspergillus*. The two genera can easily be distinguished from each other by the following facts :–

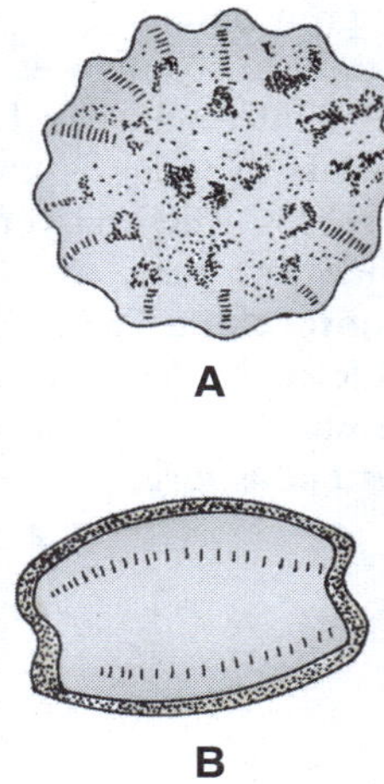

Fig. 10.18 (A–B). *Penicillium sp.* Ascospores. A, surface view; B, side view.

1. The conidiophore in *Aspergillus* is unseptate and unbranched. It arises from a specialized, T-shaped thick-walled **foot cell**. In *Penicillium* it is long, slender, septate and branched. It arises from any vegetative cell of the mycelium. There are no foot cells.

2. In *Aspergillus* the conidiophores enlarge into a vesicle at its tip which bears the finger-like sterigmata. The whole structure has the form of a spherical head. The conidiophore in *Penicillium* ends in a whorl of branches which give it a broom-like appearance.

3. The peridium of cleistothecium of *Penicillium* is thicker and gererally consists of loosely interwoven hyphae than that of *Aspergillus* which is yellowish in colour and quite soft.

Penicillium brefeldianum (Fig. 10.19). In this species the sexual apparatus (antheridia and ascogonia) is absent. The vegetative hyphae give rise to two similar protuberances. They are known as the **copulation branches**. The two copulation branches coil around each other. Their tips come in contact. At the point of contact the walls between the tips dissolve. The contents of one migrate into the other to establish a dikaryon in the fusion cell. This species of *Penicillium* provides an example of isogamous sexual reproduction. After the migration of the contents the copulation branches are surrounded by sterile hyphae. The latter arise from the adjacent cells of

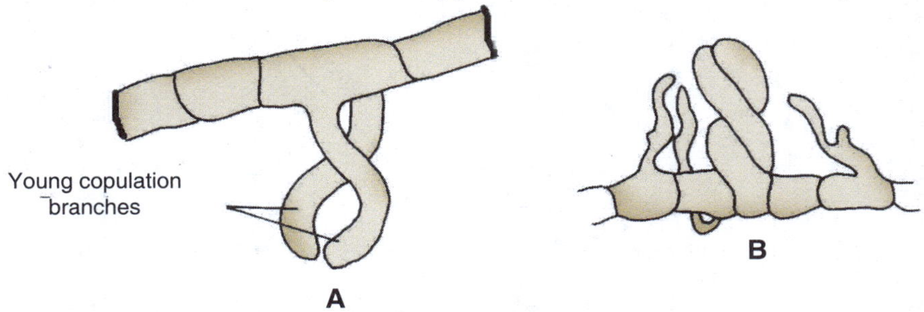

Fig. 10.19 (A-B). *Penicillium brefeldianum.* Sexual reproduction. (After Brefeld)

vegetative hyphae. These sterile hyphae form a hyphal knot. The latter subsequently assumes a plectenchymatous character. After about a week the outer membrane of the hyphal knot thickens to form a hard outer covering. The fusion or dikaryotic cell, in the meantime, sends out certain outgrowths. They are unseptate and are known as the **ascogenous hyphae**. The dikaryon in the dikaryotic cell undergoes conjugate division. A pair of nuclei passes into each ascogenous hypha. The ascogenous hyphae become divided into short and cylindrical binucleate cells. Some of the cells grow out into buds with spirally coiled tips. From the convex surface of these spirally coiled tips there arise branches that curve upwards and coil. From these branches there again arise short multicellular branches. Each cell of these last formed branches becomes spherical. It is known **as an ascus**. Nothing is known about the nuclear fusions in the ascus. Each ascus contains eight ascospores. The asci containing ascospores, the ascogenous hyphae and the surrounding sterile hyphae with their outer hardened rind constitute a fructification. It is closed and is called the **cleistothecium**. The walls of the asci degenerate and the ascospores are liberated in the centre of the cleistothecium. Later the outer hard rind of the cleistothecium also breaks. As a result the ascospores are released. Each ascospore possesses a double membrane the outer **exospore** and the **inner endospore**.

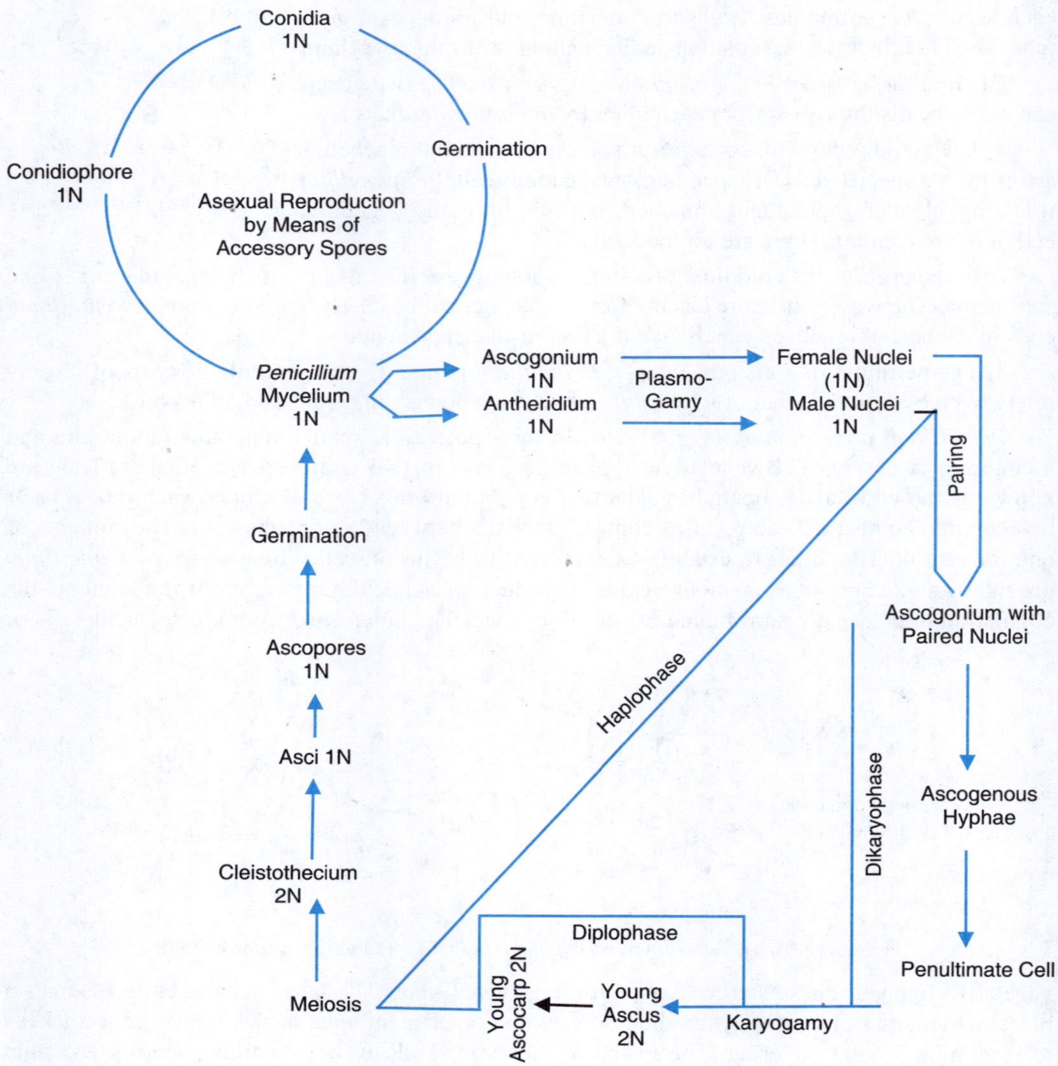

Fig. 10.20. Graphic representation of the life cycle of *Penicillium sp.*

Economic Importance

Penicillia are of tremendous economic importance. They cause considerable damage to leather goods, fabrics and furniture. Under damp conditions the wood of doors, windows and household furniture gets covered with a profuse growth of this fungus which spoils the wood. During damp and warm weather this fungus appears on bread, cheese, butter, jam, jelly, and causes spoilage of these foodstuffs. However, the fungus is not poisonous. Some species such as *Penicillium digitatum* and *P.expansum* attack citrus or other fruits in storage but most others are saprophytes. They cause rot and decay of lemon, orange and other fruits and vegetables in transit and storage causing loss worth of thousands of ruppes each year. *P. expansum* is the common agent of decay of apples in storage. The fungus gains entry through bruises and wounds. Much of the damage

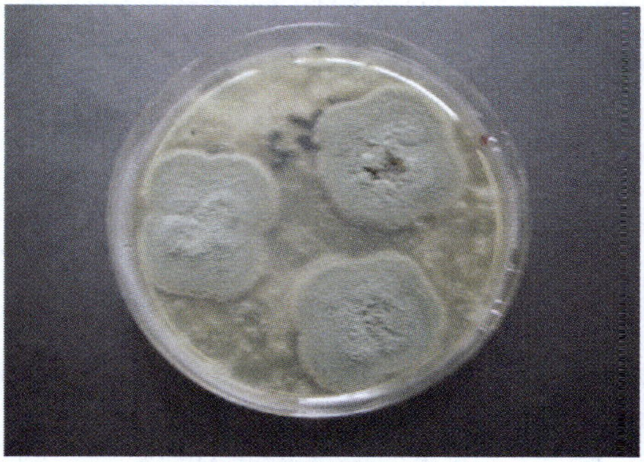

Penicillium expansum

Penicillium expansum-decay of Apples

can, however, be prevented. The spores cannot germinate under conditions of restricted air supply, low temperture and dry conditions. Thus by proper cooling, restricted ventilation or maintaining low humidity these substances can be prevented from becoming moldy. Careful handling of fruits and vegetables so as to avoid bruises and wounds is another useful preventive measure.

Many species of *Penicillium* are important industrially in the production of organic acids such as oxalic, fumaric, gluconic, and citric acids. Some species such as *P. roqueforti* and *P. camemberti* are of value in food manufacture. They are used in cheese making to which they give not only colour and soft texture but also impart distinctive flavour and odours. *P. notatum* and P. chrysogenum are a source of the wonder drug **Penicillin** which for many has been life saver. It destroys or inactivates many pathogenic micro-organisms in persons. Some persons, however, are allergic to it. To them penicillin can be as toxic as a poison. **Griseofulvin** is another antibiotic produced by *P. griseofulvum*. It inhibits the growth of some other fungi and is thus used to control ringworm disease of animals and some fungal diseases of plants.

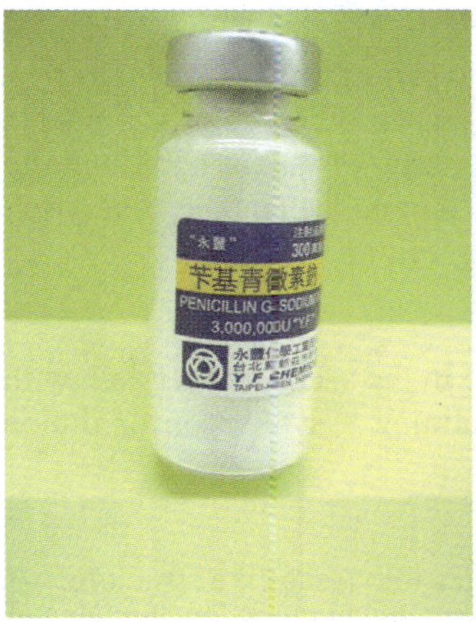

Penicillin

Penicillium roquefortii in Cheese Making

Salient Features. The life cycle comprises three phases, the *haplophase*, the *dikaryophase* and the transitory *diplophase*.

Haplophase

1. It is represented by a well developed, saprophytic mycelium consisting of colourless, septate, branched hyphae. The cells are uninucleate to bi-or multinucleate and the septa between the cells have each a central pore.
2. The haplomycelium during the growing season multiplies by means of light, dry, non-motile, wind borne, spores called the **conidia** which are uninucleate, sometimes multinucleate.
3. The conidia arise in chains from the tips of bottle-shaped sterigmata and are thus **exogenous** in origin. They are arranged in a basigenous manner.
4. The sterigmata are borne in tufts on the ultimate branches (metulae) of the conidiophores.
5. The conidiophores are long, erect, septate special hyphae of the mycelium. They branch in a broom-like manner in the upper portion. This brush-like portion of the conidiophore is called the **penicillus**.
6. The haplomycelium is **homothallic**. Both the antheridia and the ascogonia are borne on the same mycelium.
7. The ascogonium in *P. vermiculatum* at maturity is an erect, tubular, aseptate, multinucleate structure.
8. The antheridium is a short, club-shaped uninucleate cell, surmounting the antheridial branch which coils spirally round the ascogonium.
9. The tip of the antheridium comes in contact with the ascogonium. The double wall at the point of contact dissolves.
10. The male nucleus, however, does not migrate into the ascogonium in *Talaromyces vermiculatus* (*P. vermiculatum*).

Dikaryophase

11. Pairing of female nuclei takes place in the ascogonium. Each pair is called a **dikaryon**. With the establishment of dikaryons the haplophase ends and the dikaryophase starts.
12. Establishment of dikaryons is followed by the septation of the ascogonium, each segment possessing a dikaryon.
13. From the middle segments arise the ascogenous hyphae which branch. The branches are of different lengths.
14. The terminal, binucleate cell of each ascogenous hypha or its ultimate branches develops into an ascus. The ascus mother cell is binucleate. It is the last structure of the dikaryophase which consists of the ascogonium after pairing of nuclei, the ascogenous hyphae and the ascus mother cells. The dikaryophase is, however, 'dependent for its nutrition on the haplomycelium.

Diplophase

15. Karyogamy takes place in the ascus mother cell, with this the dikaryophase ends.
16. The young ascus containing a synkaryon or diploid nucleus represents the transitory **diplophase**.

Future Haplophase

17. The young ascus enlarges into a sac-like structure. The synkaryon immediately undergoes meiosis followed by mitosis. The 8 haploid nuclei are organised into ascospores which are the pioneer structures of the future haplophase.

18. The mature, sac-like asci, each containing eight ascospores are spherical or pear-shaped in form.
19. The sexual apparatus in the meantime is surrounded by loosely interwoven envelope of sterile hyphae in some species and compact in others. It is called the **peridium**. The entire ascus fruit consisting of the septate ascogonium, ascogenous hyphae, asci containing ascospores and the surrounding peridium is called the **ascocarp**.
20. The mature ascocarp of *Penicillium* is a round, closed structure. It is known as the **cleistothecium**.
21. At maturity the ascus wall dissolves away releasing the ascospores into the cleistothecium. The wall of the cleistothecium disintegrates to liberate the ascospores.
22. The liberated uninucleate ascospore has sculptured spore wall. It is pulley-like in appearance in side view and round to star-shaped in face view.
23. Each ascospore, on falling, on a suitable substratum germinates to produce the haplomycelium.

Taxonomic Position

Alexopoulos (1962)

Division	:	**Mycota**
Subdivision	:	**Eumycotina**
Class	:	**Ascomycetes**
Subclass	:	**Euascomycetidae**
Series	:	**Plectomycetes**
Order	:	**Plectascales (Aspergillales)**
Family	:	**Aspergillaceae**
Genus	:	*Penicillium (Tadaromyces)*
Species	:	*vermiculatum (vermiculatus)*

Alexopoulos and Mims (1979)

Superkingdom	:	**Eukaryonta**
Kingdom	:	**Mycetae**
Division	:	**Amastigomycotina**
Subdivision	:	**Ascomycotina**
Class	:	**Ascomycetes**
Subclass	:	**Plectomycetes**
Order	:	**Eurotiales**
Family	:	**Eurotiaceae**
Genus	:	*Penicillium*
Species	:	*vermiculatum*

QUESTIONS

ESSAY TYPE

1. Describe the habit, habitat and economic importance of *Aspergillus* and *Penicillium*.
2. (*a*) Give an account of the structure of mycelium and asexual reproduction in any one of the following :–

   (*i*) *Aspergillus* (*Eurotium*).

   (*ii*) *Penicillium*.

   (*b*) Give an account of asexual reproduction in *Aspergillus*. (*Poorvanchal, 2004, 2006*)

3. Describe the methods of reproduction and propagation of *Eurotium* (Aspergillus). (*Poorvanchal, 2005*)

4. Describe the life history of a typical ascomycete laying emphasis on the nuclear changes.

5. Describe sexual reproduction and development of fructification of *Eurotium*.

6. What light do forms like Aspergillus throw on the degeneration of sex organs in the *Ascomycetes*? Discuss the phenomenon of alternation of generations in *Eurotium*.

7. Give an account of sexual reproduction and fruit formation in *Penicillium vermiculatum* or *P. brefeldianum*.

8. Give an illustrated account of the life history of *Eurotium* or *Penicillium*. (*Kumaon, 2004*)

9. Describe with diagrams the development of the fructification and structure of the ripe fructification of either *Aspergillus* or *Penicillium*.

10. Define the term ascus. Describe the process of ascospore formation in any fungus you have studied.

11. With the help of suitable sketches describe the life history of *Eurotium* (*Aspergillus*). (*Lucknow, 2003*)

12. With the help of suitable diagrams, describe the asexual and sexual cycle of 'Black mold' studied by you. (*Rohilkhand, 1992*)

13. Describe life-cycle of *Aspergillus*. (*Kanpur, 2002*)

14. Describe sexual reproduction in *Aspergillus* and comment upon the economic importance of the fungus. (*Rohilkhand, 2002*)

SHORT ANSWER TYPE

15. How does the mycelium of *Rhizopus* differ from that of *Penicillium*? (*Lucknow 1994; 1997, Guru Nanak Deo, 1991*)

16. Give diagnostic features and assign the following to their systematic position :
   (*a*) Eurotium (*Gorakhpur, 1991, 1996; Kumaon, 1998*)
   (*b*) *Aspergillus* (*Kumaon, 2003 ; Kanpur, 1997; Lucknow 1996, 1998, 2000; Awadh, 1991; Rohilkhand, 1996*)
   (*c*) *Penicillium* (*Awadh, 1995, 1996; Kumaon, 1998*)

17. Write short notes on
   (*a*) Structure of Cleistothecium of *Eurotium* (*Gorakhpur, 1994, 1996*)
   (*b*) Fruiting bodies of Ascomycetes (*Gorakhpur, 1997*)
   (*c*) Sexual reproduction in *Eurotium* (*Gorakhpur, 1998*)
   (*d*) Cleistothecium of *Eurotium* (*Lucknow, 1993*)
   (*e*) Asexual reproduction in *Aspergillus* (*Lucknow, 1995*)
   (*f*) Asexual reproduction in *Penicillium* (*Awadh, 1992*)
   (*g*) Economic importance of *Penicillium* (*Himachal Pradesh, 1992*)
   (*h*) Economic importance of *Aspergillus* (*Guru Nanak Deo, 1992*)
   (*i*) Imperfect stage of Aspergillus (*Kumaon, 2001*)
   (*j*) *Leaf Curl of Peach* (*Kanpur M.Sc., 2001*)

18. Give a comparative account of asexual reproduction in *Albugo* and *Eurotium*. (*Gorakhpur, 1999*)

19. Describe the method of asexual reproduction in *Aspergillus*. (*Kanpur, 1997, 2001*)

20. What is ascus? Describe its structure in detail. Name some fungi having asci.(*Kanpur, 1995*)

21. Give an account of diagrammatic life cycle of a saprophytic ascomycete.

22. Describe the methods of asexual reproduction in *Aspergillus* including details of development of conidia. (*Punjab, 1991*)

23. Discuss post-fertilisation changes that take place during the sexual reproduction in *Penicillium*. (*Himachal Pradesh, 1991, 1997*)

24. Discuss briefly the formation of cleistothecium in *Penicillium*. (*Himachal Pradesh, 1992*)

25. Explain the cell structure and asexual reproduction in *Penicillium* with the help of suitable diagrams. (*Himachal Pradesh, 1992, 1996*)

26. Describe conidia formation in *Aspergillus*. (*Lucknow, 2002*)

27. How are conidia formed in *Penincillium* ? Describe with diagrams.

28. Draw labelled diagrams of :

(*a*) Cleistothecium of *Aspergillus* (*Kanpur, 2001*)

(*b*) Asexual reproduction in *Aspergillus* (*Poorvanchal, 2002; Kanpur, 2006*)

(*c*) *Penicillius*

(*d*) Ascospores of *Aspergillus*

(*e*) Cleistothecium of *Penicillium*

(*f*) Life cycle of *Aspergillus*

(*g*) Life cycle of *Penicillium*

(*h*) Conidial apparatus of *Aspergillus niger* (*Lucknow, 2005*)

29. Differentiate between :

(*a*) *Aspergillus* and *Penicillium* (*Allahabad M.Sc.; 1999*)

(*b*) Penicillus and Conididophore.

30. Describe the life cycle of *Penicillium*.

OBJECTIVE TYPE

31. Fill in the blanks :

(*i*) *Aspergillus* is also known as ....

(*ii*) The cleistothecia are the characteristic nature of the series ....

(*iii*) The species of *Aspergillus* have been named after the .... of conidia.

(*iv*) Aflatoxin is produced from ....

(*v*) The latin name of Blue mold is ....

(*vi*) The wonder drug *Penicillium* is produced from ....

(*vii*) Haploid spores produced in ascus of *Aspergillus* are called ....

(*viii*) .... is a sac like structure containing light haploid ascospores formedas a result of Plasmogamy, Karyogamy and meiosis.

(*ix*) In *Penicillium*, the conidia are borne on ....

(*x*) The fruit body of *Penicillium* is called ....

32. Select the correct answer :

(*i*) In *Penicillium*, conidia are produced

(*a*) in sori consisting of several conidiophores

(*b*) on unbranched conidiophores

(*c*) in branched conidiophores

(*d*) on both branched and unbranched conidiophores.

(*ii*) Penicillin was discovered by

(*a*) Ian Fleming

(*b*) Alexander Fleming

(*c*) Edward Jenner

(*d*) Louis Pasteur

(*iii*) Which of the following diseases is caused by *Aspergillus*?

(*a*) Cholera (*b*) T. B.

(*c*) Tetanus (*d*) Otomycosis

(*iv*) The latin name of the perfect stage of *Aspergillus* is

(*a*) *Eurotium* (*b*) *Aspergillus*

(*c*) *Penicillium* (*d*) *Talaromyces*

(*v*) The basal cell from where the conidiophore of *Aspergillus* arises is known as

(*a*) Trophocyst (*b*) Tip cell

(*c*) Foot cell (*d*) Terminal cell

(*vi*) The conidium usually germinates

(*a*) directly (*b*) indirectly

(*c*) by germ tube (*d*) unusually

(*vii*) The female sex organ is normally called

(*a*) Oogonium (*b*) Ascogonium

(*c*) Both a and b (*d*) Neither a nor b.

(*viii*) The ascocarp produced in *Penicillium* is known as

(*a*) Perithecium

(*b*) Cleistothecium

(*c*) Apothecium

(*d*) Hypothecium

(*ix*) The reproductive organs of *Aspergillus* are

(*a*) antheridium and oogonium

(*b*) spermatium and oogonium

(*c*) antheridium and Ascogonium

(*d*) spermatium and Ascogonium

(*x*) In *Aspergillus*, hyphal cell is

(*a*) always uninucleate

(*b*) always multinucleate

(*c*) uninucleate in many and multinucleate in some

(*d*) uninucleate when young but multinucleate when mature.

CHAPTER 11

SUB CLASS – EUASCOMYCETIDAE Series : Hymenoascomycetes A. Sub-series : Pyrenomycetes

Claviceps purpurea on a common grass

- Introduction
- Sub-Series : Pyrenomycetes
- Order : Erysiphales
- Family : Erysiphacae
- *Erysiphe*
- *Phyllactinia*
- *Sphaerotheca*
- Order : Clavicipitales
- Family Clavicipitaceae
- *Claviceps*

INTRODUCTION

The series Hymenoascomycetes includes higher Ascomycetes in which the unitunicate asci are usually clavate or cylindrical in form. They are usually arranged regularly in a palisade-like layer to form a distinct fertile layer, the **hymenium** within or on the mature ascocarp. The series comprises numerous species which are grouped under two sub-series, namely, **Pyrenomycetes** and **Discomycetes** on the basis of the gross morphology of the ascocarp.

SUB-SERIES : PYRENOMYCETES

The club-shaped or cylindrical asci are usually borne in a parallel series forming a well defined fertile layer called the **hymenium**. It lines the base and the sides of the mature ascocarp. Occasionally the asci are borne in a basal tuft. The ascocarp may either by (i) a globose cleistothecium or (ii) a flask-shaped perithecium.

(a) **Pyrenomycetes with globose cleistothecia**. This section of the Pyrenomycetes includes two orders, *Erysiphales* (Perisporiales) and *Meliolales*. Of these, the former is important and thus discussed here.

ORDER : ERYSIPHALES

This order includes obligate ectoparasites which usually grow on the surface of the host plant. Popularly these fungi are called the *powdery mildews* because of the powdery appearance of their white conidiospores which form a coating on the host surface. The ascocarps are minute, closed usually spherical cleistothecia with a pseudoparenchymatous peridium. They are generally black when mature, often dark in colour and bear hyphal appendages of various types which serve to anchor the cleistothecia to the host and also assist in their dispersal. The order *Erysiphales* resembles order *Plectascales* in two features, namely, cleistothecium type of ascocarp and scattered arrangement of asci in the young cleistothecium. In the mature cleistothecium of Erysiphales, the large globose or ovoid, often stalked asci are arranged parallel to one another forming a well-defined **hymenium** at the base of the cleistothecial cavity. Occasionally there is only a single ascus. The order includes a single family **Erysiphaceae.**

FAMILY : ERYSIPHACEAE

The family comprises 90 species assigned to 13 genera. Bessey (1950) reported 14 genera. The chief among them are *Erysiphe, Uncinula, Podosphaera, Phyllactinia, Sphaerotheca,* and

Erysiphe

Phyllactinia

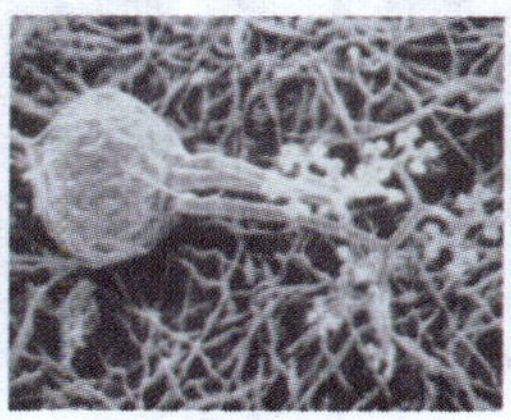

Podosphaera

Uncinula

Microsphaera

Microsphaera. All of them are highly specialised obligate ectophytic parasites of Angiosperms causing *powdery mildew* disease. Some of them are serious pathogens of our economic plants such as cereals (barley, oat, rye, wheat), grasses, grape vines, pea, apples and cucurbits. The mycelium is superficial and grows over the surface of the host absorbing nutrition by putting out uninucleate haustoria which in most species, penetrate only the epidermal cells but in few species, for instance, *Leveillula taurica*, the hyphae penetrate into the sub-epidermal tissues (mesophyll).

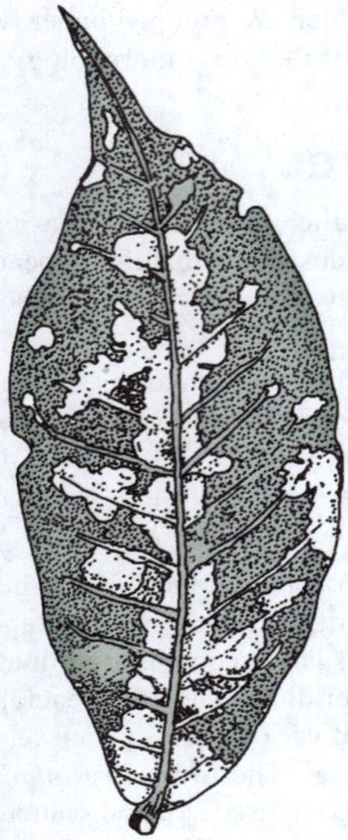

Fig. 11.1. *Powdery mildew* on Phlox leaf. Note the white mildew areas on the infected leaf surface.

Asexual reproduction takes place by hyaline, barrel-shaped, one-celled, air-dispersed conidia which are borne in short upright chains on simple short, erect conidiophores. The conidiophores are usually one-celled. The conidia germinate by one to three germ tubes in moist air. The genera included in the family are distinguished on the basis of the form of the hyphal appendages on the peridium and the number of asci (solitary or numerous) in the cleistothecium. The globose to ovoid asci have a short stalk in some species. Typically the ascospores are unicellular and oval in form. They vary in number from 2 to 8 in each ascus. *Erysiphe* is taken as a type to illustrate the pattern of life cycle while a brief account of *Phyllactinia* and *Sphaerotheca* have also been described.

ERYSIPHE

The type genus *Erysiphe* includes several species of which about 22 species are reported from India. All of them are obligate ectophytic parasites which infect a large number of species of angiosperms particularly the grains (cereals) and grass throughout the world. They occur on the aerial lparts causing *powdery mildew disease*. These powdery mildew fungi are generally epidermal cell penetrating parasites. The commonest species among these are :—

1. *E. cichoracearum*. It is the causal agent of *powdery mildew disease* of cucurbits.
2. *E. polygoni*. It causes *powdery mildew* of peas.
3. *E. graminis*. It causes *powdery mildew* of cereal crops such

Erysiphe cichoracearum

Erysiphe polygoni

Erysiphe graminis

as wheat, barley, rye, oats and also many of the grasses. It is the only species which causes *powdery mildew* on a monocotyledonous host. The powdery mildew disease of barley (*Hordeum vulgare*) and wheat are caused by *E. graminis. f.* sp. *hordei* and *E. graminis f.* sp. *tritici* respectively. According to Yarwood (1957), many species of powdery mildews do not form cleistothecia in warm climates. They perpetuate themselves by means of conidia. The latter germinate immediately on falling on a suitable host. They are not long lived.

Primary Infection (Fig. 11.2). The conidium on falling on a suitable host (leaf) germinates immediately by putting out one to three germ tubes. The germ tube, soon after emergence bends down, comes in contact with the cuticle and widens to form an *appressorium* with a small lobe in the centre. A fine tube-like outgrowth, the **infection peg** arises from the appressorial lobe, penetrates cuticle as well as the epidermal cell wall and grows into the lumen of the epidermal cell where a haustorium develops. A collar of amorphous material deposits around the penetration or infection peg. After the formation of the functional haustorium in the host epidermal cell, an elongating **secondary hypha** develops from the appressorium. The development of the secondary hypha is dependent on the formation of a functional haustorium in the host epidermal cell. It is a critical step in primary infection. Haustorium is the only structure of the fungal parasite which develops inside the host. It has been reported that for every functional secondary hypha there is a corresponding haustorium in the leaf epidermis. The secondary hypha, at the expense of the food material derived from the host plant, through the intracellular haustorium grows rapidly over the surface of the affected part (leaf or stem) of the host and branches rapidly to form the mycelium.

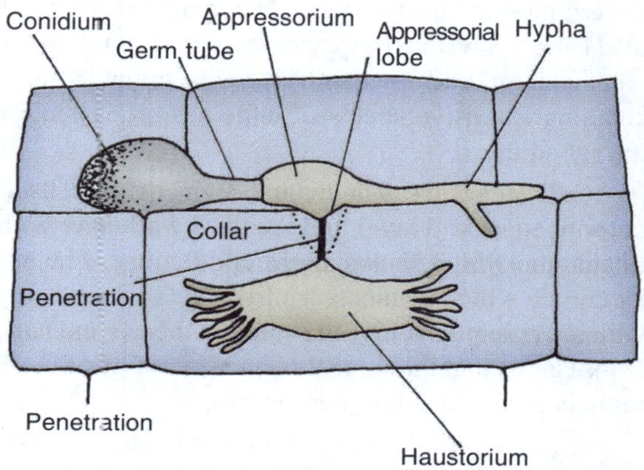

Fig. 11.2. *Powdery mildew.* Germination of conidium and infection of the host (After Johnson, Bashnell and Zeyen).

Mycelium. It grows over the leaf surface of the host and thus is superficial. It is known as the **ectomycelium**. The ectomycelium consists of hyaline, branched septate hyphae. The septa between the cells have each a central pore. The hyphae interlace to form a web-like white

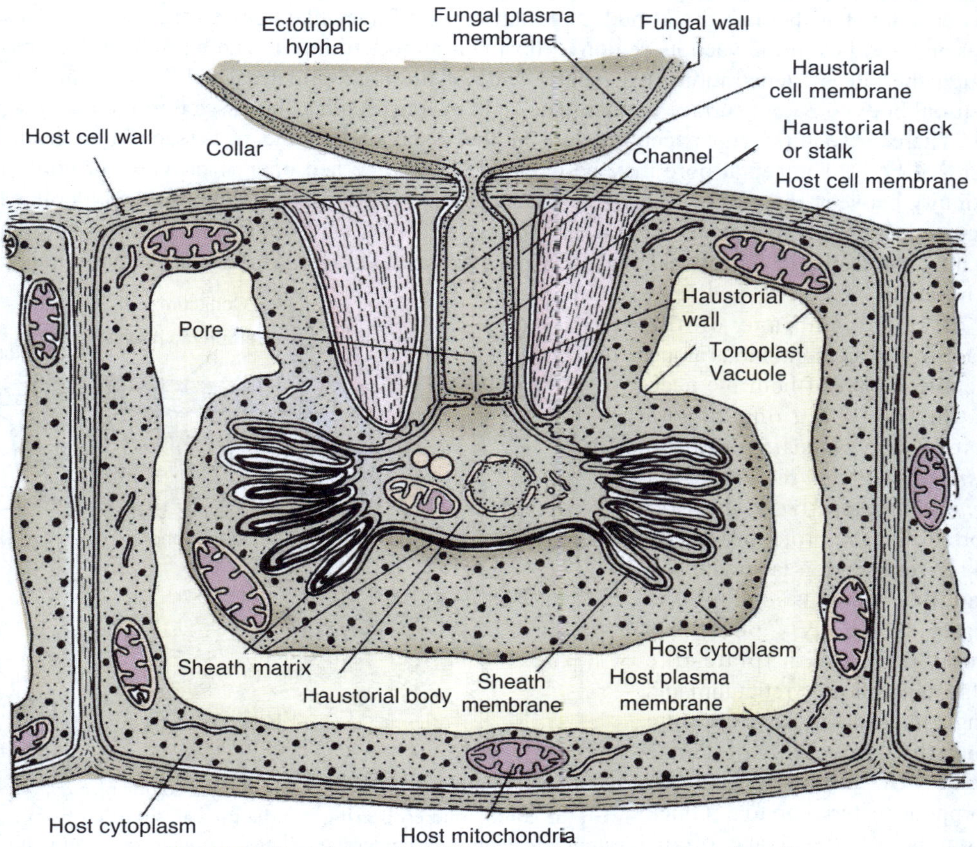

Fig. 11.3. *Erysiphe graminis.* Diagrammatic representation of the fine structure of the haustorial apparatus in section (Based on Bracker).

covering on the host surface. The hyphal walls are thin. The young hyphae, according to Martin *et al.* (1983), have denser cytoplasm and smaller vesicles than the more mature ones. The latter contain large vacuoles and glycogen granules in rosettes. Just behind the hyphal tip, cytoplasm contains extensive sheets of endoplasmic reticulum (ER) together with many mitochondria and microbodies.

Cell Structure. The hyphal segments or cells are short. The cell wall is tripartite. There is the outer opaque wall layer and the third wall layer with a fine granular texture adjacent to the plasma membrane which shows slight undulations. The cell walls of the superficial hyphae are bounded by cuticle which connects the hyphae. Each cell contains a single nucleus usually centrally located with a single nucleolus. The nuclear membrane has pores. The nucleus is embedded in a vacuolate cytoplasm containing glycogen as food reserve. Besides, embedded in the cytoplasm are the mitochondria, ER and microbodies.

Haustorium. Each hyphal colony which is the product of a single elongated secondary hypha has a single haustorium.. All interactions between the parasite and the host occur through this region (haustorium). It lies in the lumen of the epidermal cell surrounded by the host cell protoplast. The haustorium may be simple and globular or pyriform in shape as in *E. cichoracearum*. In *E. graminis* it is more complex.

Fine Structure of the haustorium of E. graminis (Fig. 11.3). It has been studied by Bracker (1968) and later by Sullivan *et al.* (1974). The haustorium consists of a small, slender, tube-like hypha called the **haustorial neck** or **stalk** ending in an elongated, swollen plate like central structure known as the **haustorial body**. The latter bears finger-like bodies (digitate processes) at either end and lies in the vacuole region of host cell almost filling it. The haustorial neck passes through the site of penetration in the outer wall of the host epidermal cell and conncects the haustorial body to parent surface hyphae. The stalk is constricted as it passes through the host cell wall. There is septum with a central pore at the distal end of the neck separating from the haustorial body. The septal pore may be open or plugged. When open it provides protoplasmic continuity between the two regions of the haustorium. According to Bracker (1968) a collar like structure made up of the same material as the host cell wall surrounds the haustorial neck at its proximal end leaving a narrow channel-like space between the two. This collar is continuous with the host cell wall. The wall and the plasma membrane of the surface hypha are continuous along the entire length of both the neck and the body regions of the haustorium constituting the *haustorial wall* and the *haustorial membrane* respectively. Bracker reported that the protoplast of the neck or stalk region lacks the cell organelles whereas that of the haustorial body possesses mitochondria with plate-like cristae, endoplasmic reticulum and a single nucleus but lacks the golgi apparatus. An electron dense matrix of finely granular amorphous material up to 5 μ thick surrounds the outer face of the body region of the haustorium and

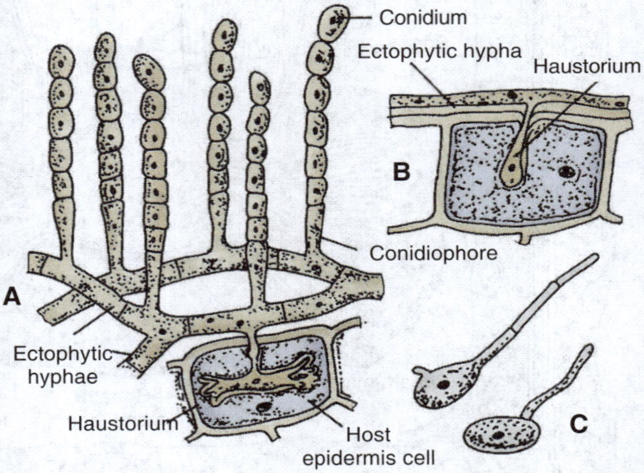

Fig. 11.4 (A-C). *Erysiphe.* A, part of the hyphal colony of *E. graminis* with the conidiophores, bearing conidia in chains, arising from the ectophytic hyphae; B, epidermal cell of *E. cichoracearum* containing a pyriform haustorium; C, germinating conidia.

extends around the finger-like haustorial lobes. This sheath-like matrix is called the *sheath matrix* or *extra haustorial matrix*. It is absent in the neck region. A thin membranous covering known as the *sheath membrane* or *extra haustorial membrane* (sometimes referred to as encapsulation membrane) delimits the sheath matrix from the host cytoplasm and thus forms the interface between the two. The sheath membrane is considered a derivation of the host plasma membrane. This view is based on the hypothesis that the host plasma membrane is invaginated by the penetrating haustorium. The haustorium thus lies outside the host cytoplasm. It is surrounded by a thin layer of host cytoplasm covered by the tonoplast on the side facing the central vacuole and by a sheath membrane (invaginated host plasma membrane) on the side facing the haustorium. The sheath matrix forms the physical boundary and the sheath membrane the key to the physical relationship between the host and the obligate fungal parasite. All interactions between the host and the parasite occur through this region.

Asexual Reproduction (Fig. 11.4). With the establishment of the ectomycelium on the surface of the host, the fungal parasite starts reproducing asexually during the growing season by producing asexual spores known as **conidia**. The conidia are formed exogenously in chains at the tip of the conidial apparatus.

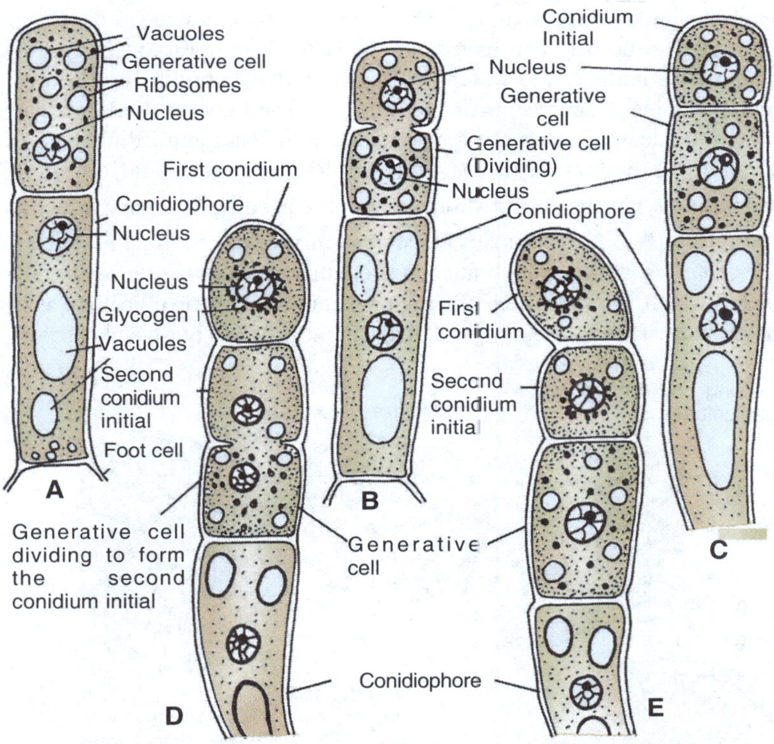

Fig. 11.5 (A-E) *Erysiphe*. Development of conidial apparatus and abstriction of conidia (Based on Martin *et al.*).

Development of Conidial Apparatus (Fig. 11.5). According to Martin *et al.* (1983), it arises as an erect outgrowth from a cell in the older part of the superficial hypha. The outgrowth grows vertically approximately at right angle, to the surface of the host. It is known as the conidiophore initial. The cell subtending the conidiophore initial is called the **foot cell**. A cross wall appears in the apical region of the elongated conidiophore initial dividing it into an upper **conidiogenous** or **generative cell** and lower **stalk** or **conidiophore** cell. The latter in some species (*E. graminis* var. *hordei*) may be swollen at the base in a characteristic manner. In a fully developed conidial apparatus, the conidiophore cell cytoplasm contains a few large vacuoles. The single nucleus is located in its upper part (A). It may as well contain glycogen granules. The conidiogenous or generative cell, on the other hand is filled with dense cytoplasm containing numerous ribosomes, small scattered vacuoles and no glycogen. It is uninucleate. The whole system borne on the foot cell and comprising the conidiophore, conidiogenous or generative cell and the conidium initials constitute the conidial apparatus.

Development of Conidia (Fig. 11.5 A-E). The terminal generative cell undergoes division (B) and cuts off the first conidium initial at its tip (C). It is followed by the extension of the generative cell (D) and another division of the latter to produce the second conidium initial below the first which is pushed above (E). This periodic extension and division of the generative cell forms a succession of conidia which cling together in an upright chain (Fig. 11.4A). The conidia in the chain represent successive stages in sequence of differentiation. They ripen from the apex downwards and then get abscissed. The abscissed conidia are readily dispersed by air currents. The disease spreads widely in summers by means of conidia which are produced in abundance in relatively cool, moist environment throughout most of the host's growing season.

Conditions for germination of conidia. The germination of *powdery mildew* conidia is dependent on temperature and relative humidity. They germinate over a wide range of temperature. The minimum is reported to be 0°C and maximum 30-36°C. The optimum is the range of 12-25°C. Regarding humidity, neither free water nor high relative humidity is essential for their germination. In fact moistening of spores impairs germination. Their germination in dry atmospheric conditions is attributed to adequate internal water in the conidia. Nair and Ellingboe (1968) found that conidia of wheat powdery mildew showed highest germination at 17°C and 65 percent relative humidity. For the germination, refer to the heading "Primary Infection".

Sexual Reproduction (Fig. 11.6). It takes place in summers. Excepting one species all others are **homothallic.** *E. graminis* lacks sexual branches altogether. In *E. polygoni* sex organs lie at the ends of sexual branches which arise near each other in pair from the somatic mycelium as lateral short erect uninucleate outgrowths. They lie closely parallel to and more or less twisted about each other.

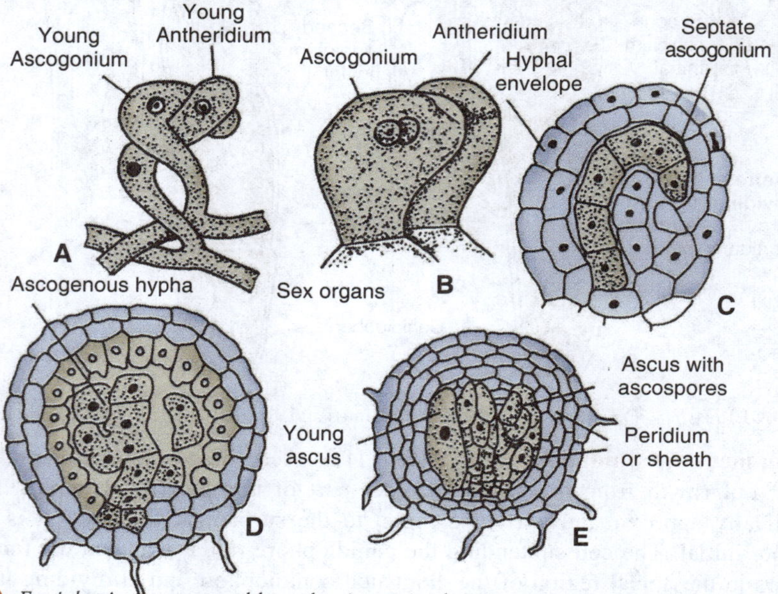

Fig. 11.6 (A-E). *Erysiphe.* A, young sexual branches interwined; B, Ascogonium with male and female nuclei forming a dikaryon; C, dikaryotised ascogonium becoming septate; D, Young, cleistothecium with the ascogenous hyphae arising from the binucleate penultimate cell; E, Section of mature cleistothecium.

(*a*) Antheridium. The nucleus of the male branch which is usually slender divides into two. The two nuclei are separated by a septum. The terminal uninucleate cell thus delimited functions as the **anheridium** and the lower supporting cell as the **stalk cell.**

(*b*) Ascogonium. The female branch also consists of a terminal, uninucleate, slightly swollen, club-shaped **ascogonium** and the basal stalk cell.

(*c*) Plasmogamy. The antheridium makes contact with the ascogonium at its apex and closely presses against it. In the region of contact, the separating walls between the two sex organs break down and a **pore** is formed. The male nucleus passes into the ascogonium through the pore. A dikaryon is thus established in the ascogonium (B). According to some investigators (Harper, 1896 and Hein, 1929), fusion of male and female unclei takes place immdieately in the ascogonium but this is denied by several others (Allen, 1936; Beatus, 1948, 1950; and Bergman, 1941), who hold that the fusion of male and female nuclei is delayed until the ascus mother cell is formed. The consensus of opinion favours the latter view. At this stage sterile hyphae grow up from the stalk cell of the ascogonium and form a sheath around the sexual apparatus (C). The sheath or peridium is thick. It consists of three to many layers of compacted hyphae.

Post-plasmogamy changes. The dikaryotised ascogonium elongates and divides by cross walls into a row of 4 or 5 cells (C). The penultimate cell or segment of the row contains two or more nuclei. Ascogenous hyphae bud out from this cell (D) and branch to form a dense interwoven mass of hyphae which develop slowly within the peridium through the winter. Binucleate, intercalary cells of the ascogenous branches about 5 to 20 in number function as **ascus mother cells.** The two nuclei in the ascus mother cell fuse to form a **synkaryon.** The young ascus with a synkaryon increases greatly in size. The fusion nucleus (synkaryon) in it undergoes meiosis followed by a single mitosis to form eight haploid nuclei. This is the **octonucleate stage** in the

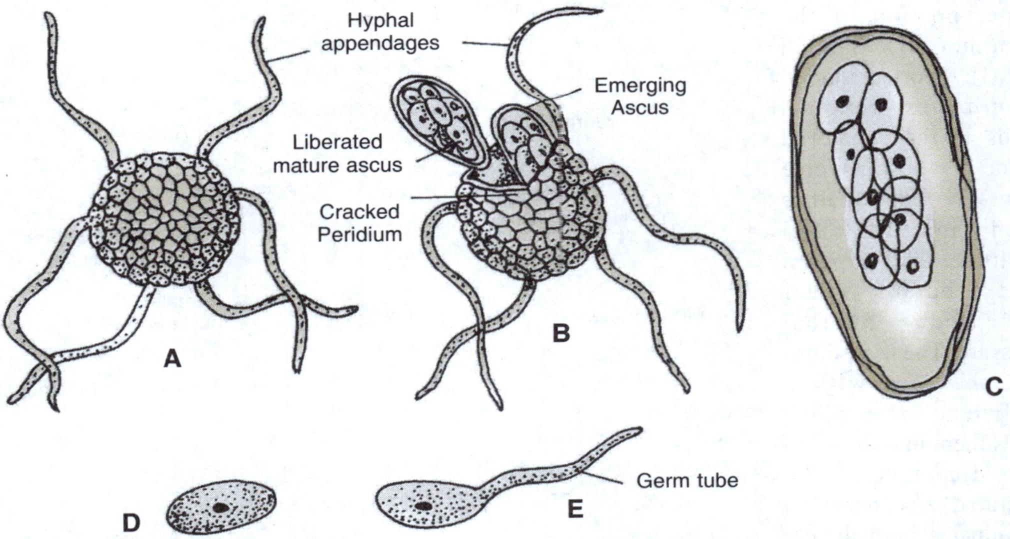

Fig. 11.7 (A-D). *Erysiphe.* A, Cleistothecium with unbranched hyphal appendages arising from its surface; B, Dehisced clesitothecium with asci emerging; C, An ascus with ascospores; D, Ascospore; E, Germinating ascospore.

development of an ascus. In spring the asci in the cleistothecium mature to produce ascospores by **free cell formation.** Depending upon the species the number of ascospores produced in each ascus after the octonucleate stage varies from two to eight. The remaining nuclei lie in the epiplasm and degenerate. The ascospores are not fully ripened until the following spring.

Cleistothecium (Fig. 11.7). The cleistothecia are globose structures without an opening. They appear as black specks on the surface of white mycelium. In some species, they are immersed in the mycelium. The peridium may consist of 6 to 10 cells in thckness. The component cells are polygonal in shape. The cells in the outer layers are without protoplast and are thick-walled. Certain superficial cells of the peridium develop into long hypha-like, unbranched (A) or slightly branched appendages. The number of these appendages varies from 2 to 8 in *E. polygoni* but it goes up to 25 in *E. graminis*. All the asci originate from the base of the cleistothecial cavity, and spread out in a fan-shaped manner (Fig. 11.6 E). They are ovate and almost sessile in *E. polygoni* but cylindrical or ovoid in some other species. The number of ascospores in each ascus varies from two to eight. In *E. cichocearum* there are only two rarely three ascospores in an ascus. The ascospores are hyaline, one-celled, unicellular structures elliptical in shape.

Dehiscence of Cleistothecium (Fig. 11.7 B). The mature cleistothecia remain dormant during winter. The thick peridium remains intact and enables them to resist adverse (winter) conditions. With the onset of spring, the epiplasm in the asci undergoes a change which results in increase in the osmotic pressure within them. Consequently, the asci absorb water and swell. The swelling of the contents causes an irregular rupturing of the upper portion of the peridium (B). Salmon (1903) reported that in *E. graminis*, peridium splits in the equatorial plane by a transverse split. The upper half is shed and the asci protrude. The exposed asci burst under increased internal pressure. The ascospores are shot out with a sufficient force which hurls them into the air to be dispersed. The liberated ascospore (C) germinates immediately on falling on a suitable host (D) to start new foci of infection.

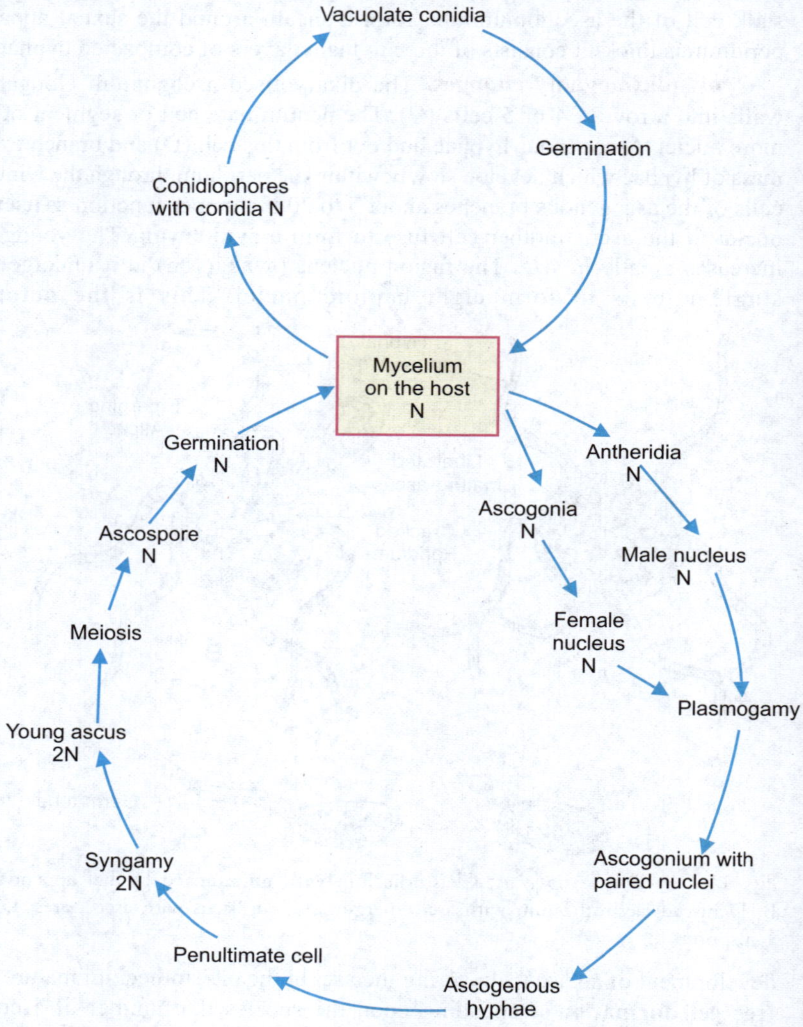

Fig. 11.8. Diagrammatic life cycle of *Erysiphe*

Disease aspect. *Erysiphe* is the causal agent of *powdery mildew* disease. The disease does not kill the host plant but it is reduced in vigour owing to constant loss of food. There is reduced yield and reduction in normal plumpiness of the Kernel.

Symptoms. The disease appears on the aerial parts of the plant as white mildewy patches (Fig. 11.1). At a later stage cleistothecia are visible as black specks over the white mycelial felt.

Host-Parasite Relationship. Baenziger and Kilpatrick (1979) reported that yield losses caused by *powdery mildew* of wheat (*Triticium aestivum*) can be as high as 45 percent. This is attributed to the decrease in the photosynthetic activity of the leaf tissue as a result of the death of the mesophyll cells adjacent to the epidermal cells containing functional haustoria. Last (1962) reported that the infected barley plants showed reduced yield, grain size and number of tillers. The root growth was reduced more than shoot growth. These harmful effects are attributed to increased respiration and decreased photosynthetic activity which result in deprivation to host of necessary metabolites for normal growth.

Overwintering. According to many workers, overwintering takes place in the cleistothecial stage. Mundkar (1949) reported that the disease perpetuates through the dormant mycelium. According to Walker (1969), the parasite overwinters as a mycelial mat. The cleistothecia appear to him of secondary importance in overwintering. There is evidence to show that the mycelium in some species of powdery mildews may overwinter in the dormant buds of perennial host plants. Yarwood (1957) is of the opinion that in warm climates the powdery mildews do not form cleistothecia. They perpetuate themselves by means of conidia. This view appears to be more plausible.

Control Measures. Sowing resistant varieties of crop plants provides the best control of the disease. Rogueing of the diseased plants and burning them in the field is important. Dusting the crop with sulphur dust or spraying with karathane is effective in cotrolling the disease. Powers (1958) reported that the antibiotic Anisomycin is effective to eradicate powdery mildew disease on wheat leaves. It is effective primarily by direct contact in the form of sprays.

Taxonomic position. The order Erysiphales resembles the Plectascales in the following features:

(*i*) Ascocarp is a minute, closed cleistothecium.
(*ii*) The asci in the young cleistothecium are scattered.
(*iii*) Sterile hyphae investing the sexual apparatus develop with the formation of ascogenous hyphae.

On the basis of these resemblances, the earlier mycologists placed Erysiphases along with Plectascales in the series **Plectomycetes.** The cleistothecium of Erysiphales, however, is more advanced in its internal differentiation. The mycologists thus now lay more emphasis on the structure of asci and their arrangement in the mature fruit rather than on the form of the fructification. On the basis of these features the order Erysiphales falls in the series **Hymenoascomycetes** (**Ascohymeniales**), sub-series Pyrenomycetes. The taxonomic position of *Erysiphe* thus is :—

Alexopoulos (1962)

Division	:	**Mycota**
Sub division	:	**Eumycotina**
Class	:	**Ascomycetes**
Sub class	:	**Euascomycetidae**
Series	:	**Hymenoascomycetidae**
Sub series	:	**Pyrenomycetes**
Order	:	**Erysiphales**
Family	:	**Erysiphaceae**
Genus	:	*Erysiphe*

Alexopoulos and Mims (1979)

Superkingdom	:	**Eukaryonta**
Kingdom	:	**Mycetae**
Division	:	**Amastigomycotina**
Sub division	:	**Ascomycotina**
Class	:	**Ascomycetes**
Sub class	:	**Hymenoascomycetidae**
Series	:	**Pyrenomycetes**
Order	:	**Erysiphales**
Family	:	**Erysiphaceae**
Genus	:	*Erysiphe*

PHYLLACTINIA

The genus **Phyllactinia,** represented by about 100 species, is worldwide in distribution. The fungus generally parasitises trees. Two commonly occuring species of *Phyllactinia* are *P. dalbergiae* parasitic on *Dalbergia sissoo* (the shishamtree) and *P. guttata* parasitic on hazel trees.

Phyllactinia guttata

P. dalbergiae causes the powdery mildew disease of shisham which is the most common disease. It is a hemiendophytic pathogen producing profuse mycelium on the surface of the host.

Mycelium. The mycelium is partly internal and partly found on the surface of the host. The mycelium spreads on both the surfaces of the host. The hyphae are branched, septate and uninucleate. Although haustoria are not present in epidermal cells, are a small 5-7 celled branch produced by the superficial hypha penetrates the leaf and enters through stromata. This may develop in substomatal chamber and the intercellular spaces of the adjacent cells. This hypha may give rise to haustoria which are saccate. Since the mycelium is partly found on the surface of the host and is also partly internal, it is known as **hemi-endophytic mycelium**.

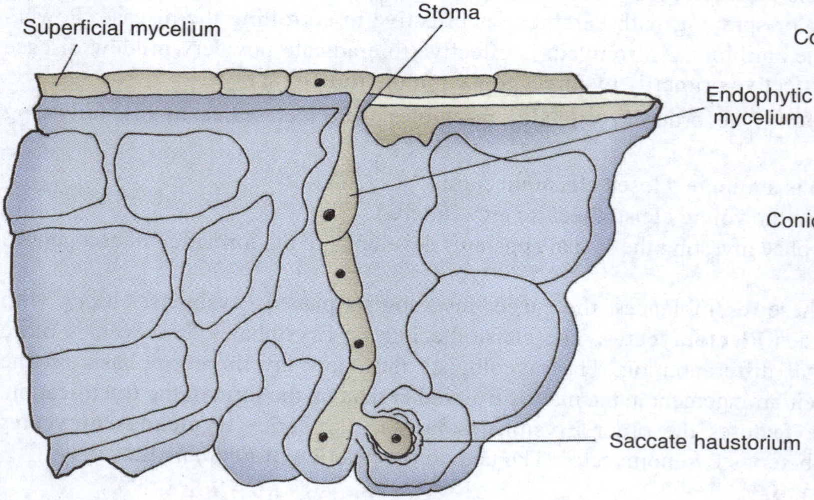

Fig. 11.9. *Phyllactinia.* Hemiendophytic mycelium showing haustorium.

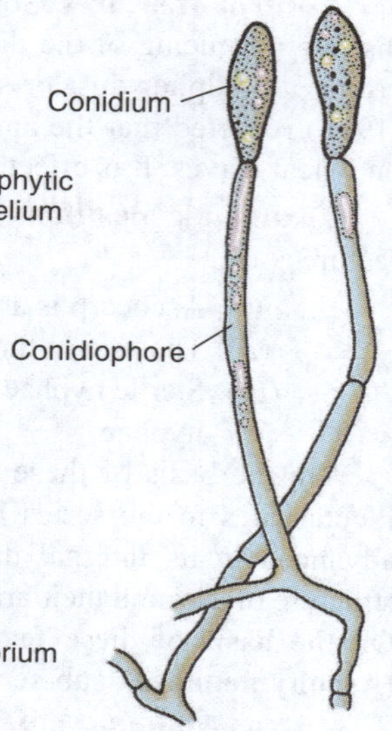

Fig. 11.10. *Phyllactinia sp.* Conidiophores and Conidia.

Asexual Reproduction. After the fungus has established itself in the host, it starts reproducing asexually during favourable conditions of growth. The asexual reproduction takes place by the formation of **conidia** produced exogenously in chains at the tip of coridial apparatus.

Development of Conidia. The conidia are formed on conidial apparatus known as conidiophores. These develop vertically from the superficial mycelium. In the beginning these are formed on both the surfaces of leaves but later they are confined to the lower surface only. The young conidiophore divides transversely to produce two cells—the basal *stalk* or *stipe cell* and the upper *conidium mother cell*. The latter again divides to form two cells—the upper cell acts as a *conidium* and the lower cell as *subterminal cell*. When the first formed conidium is shed, the subterminal cell divides again to form another conidium. Generally the conidia are formed singly but in large numbers forming a white powdery mass on the leaf surface.

Structure and germination of conidium. The conidium is single celled, clavate, uninucleate, thin walled and hyaline. The conidia are disseminated by wind and upon reaching a suitable substratum germinate to form a new mycelium.

Sexual Reproduction. It takes place under unfavourable conditions. The fungus is **heterothallic** and the gametangia are formed on two closely lying hyphae. The male gametanium is **antheridium** and the female gametanium is **ascogonium.**

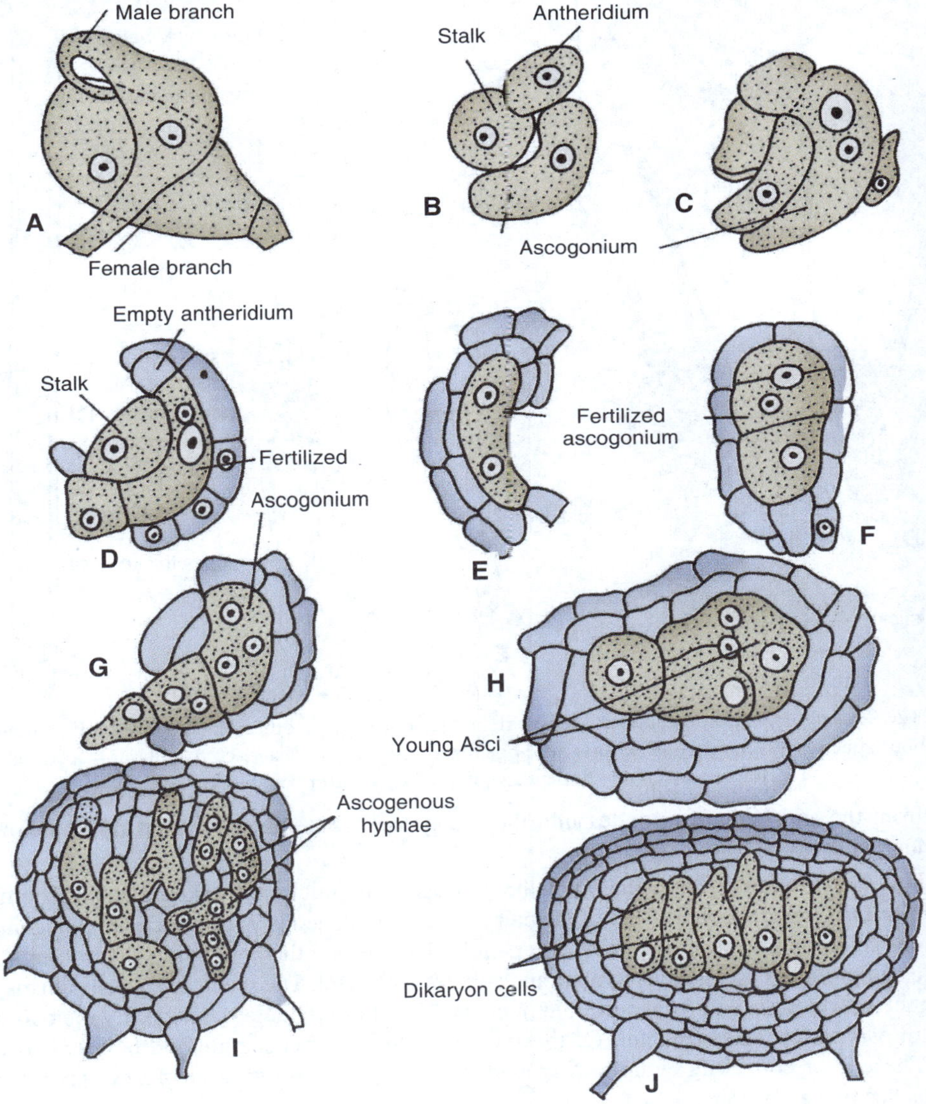

Fig.11.11. *Phyllactinia*. Sexual Reproduction; A-B. Development of Antheridium and ascogonium; C, migration of male nucleus to ascogonium; D-Pairing of nuclei, E-G. Stages in the development of ascogenous hyphae, H, young cleistothecium with ascogenous hyphae; I, young asci within the cleistothecium.

(a) Antheridium is small clavate, one celled and uninucleate structure.

(b) Ascogonium is thick, ovoid, one celled, uninucleate and closely applied to or coiled around the antheridium.

(c) Plasmogamy. When mature, the wall between the two sex organs dissolves and the

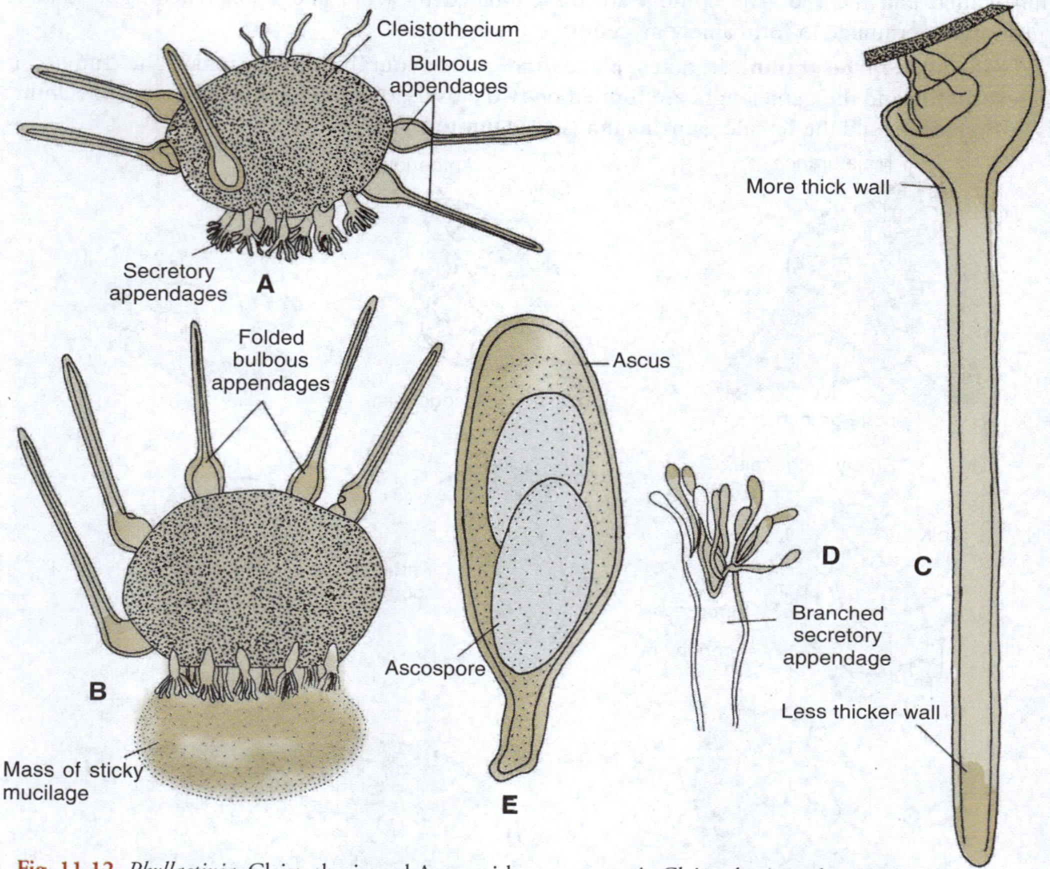

Fig. 11.12. *Phyllactinia.* Cleistothecia and Ascus with ascospores; A, Cleistothecium showing both secretory and bulbous appendages; B, Cleistothecium with appendages showing mucilage; C, A bulbous—appendage; D, Branched secretory appendage; E, An ascus with two ascospores.

nucleus from the antheridium migrates into the ascogonium. Plasmogamy takes place followed with nuclear pairing.

(d) Post Plasmogamy Changes. After the pairing of the nuclei has taken place, a common binucelate cell is formed. Each nucleus of the pair divides simultaneously. The fertilised ascogonium also divides to form a row of 3-5 cells, the penultimate cell of the row being dikaryotic. This dikaryotic cell produces ascogenous hyphae and subsequently asci. The young ascus has a dikaryon i.e. two nucelei which fuse to form a diploid nucleus. This diploid nucleus divides by meiosis and then by mitosis to form eight nuclei. Of these eight nuclei, six degenerate and only two nuclei survive. These after collecting cytoplasm around them give rise to two **ascospores.** At maturity each ascus has two ascospores.

Simultaneously, the vegetative hyphae grow upwards from the basal cell of the fertilised ascogonium to form a 2-3 layered sheath around the developing ascocarp (*Cleistothecium*).

Cleistothecium. The mature cleistothecium is brown or black in colour, globose with several layered wall and bears appendages. The appendages are of two types : (a) an equatorial group of radiating appendages with bulbous swollen bases and (b) a crown of repeatedly branched mucilage secreting appendages which help in adhesion of cleistothecium to the substratum.

After completion of dormancy, the cleistothecium ruptures irregularly by elongation of asci. The asci come out and the ascospores are liberated from the asci.

The **ascospores** are ovate to elliptical, haploid, uninucleate and smooth walled. Upon reaching a suitable host, the ascospore germinates to form new mycelium which soon becomes ready to produce comidiophores and conidia.

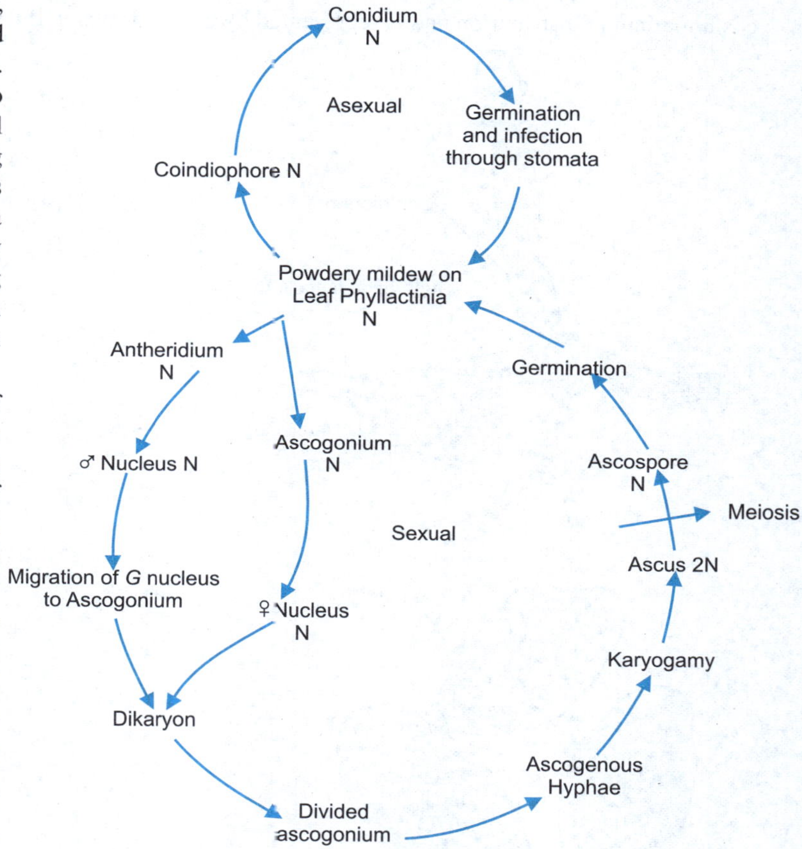

Fig. 11.13. Diagrammatic Life Cycle of *Phyllactinia.*

Taxonomic position

Alexopoulos (1962)			Alexopoulos and Mims (1979)		
Division	:	**Mycota**	*Superkingdom*	:	**Eukaryonta**
Sub division	:	**Eumycotina**	*Kingdom*	:	**Mycetae**
Class	:	**Ascomycetes**	*Divisions*	:	**Amastigomycotina**
Sub class	:	**Euascomycetidae**	Sub division	:	**Ascomycotina**
Series	:	**Hymenoascomycetidae**	Class	:	**Ascomycetes**
Sub series	:	**Pyrenomycetes**	Sub class	:	**Hymenoascomycetidae**
Order	:	**Erysiphales**	Series	:	**Pyrenomycetes**
Family	:	**Erysiphaceae**	Order	:	**Erysiphales**
Genus	:	*Phyllactinia*	Family	:	**Erysiphaceae**
			Genus	:	*Phyllactinia*

SPHAEROTHECA

Genus **Sphaerotheca** is another important member of the family Erysiphaceae. The genus is cosmopolitan in distribution and is represented by eight species in this country.

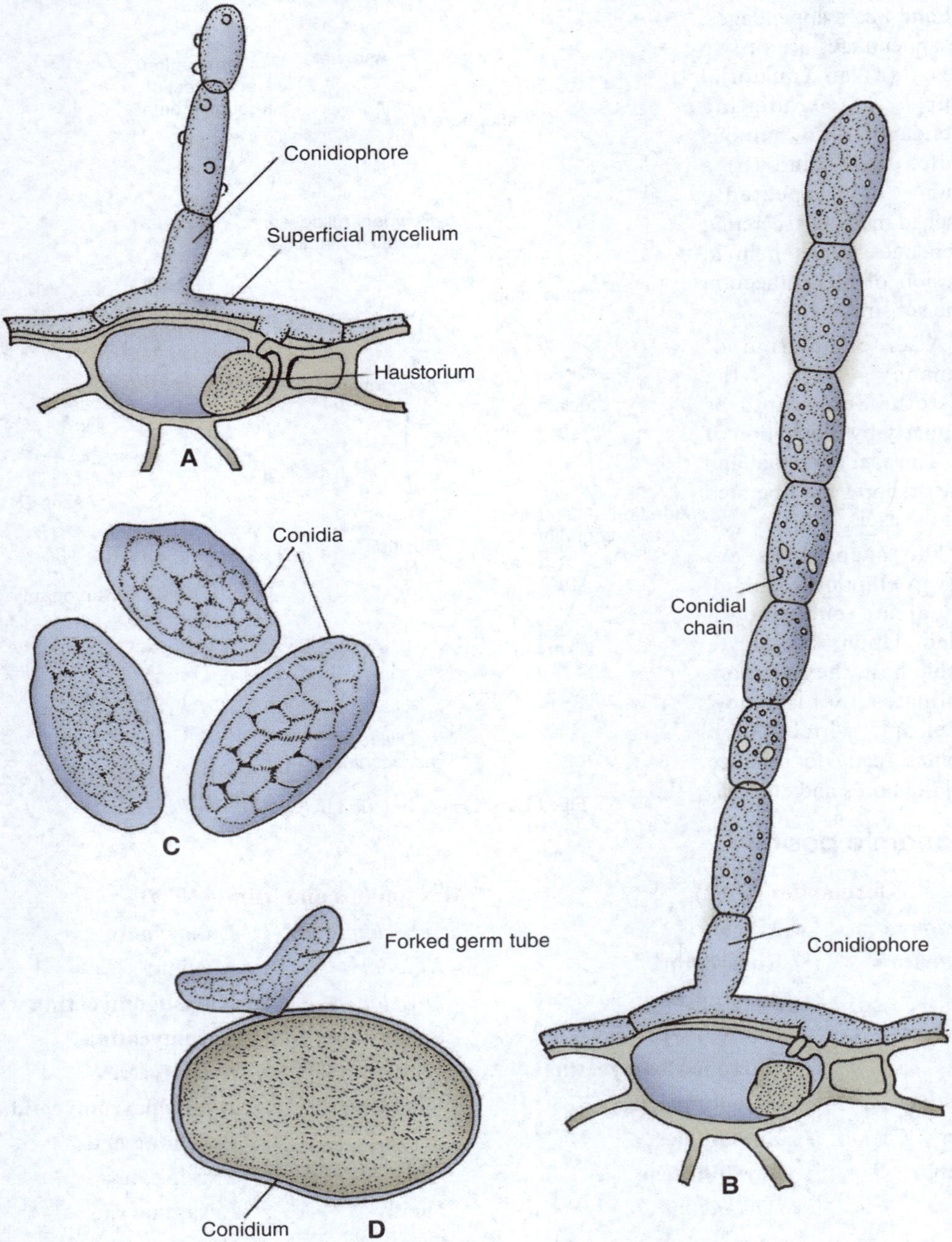

Fig. 11.14. *Sphaerotheca.* A, An ectophytic (superficial) mycelium; B, Conidiophore and chain of conidia on superficial mycelium; C, Conidia; D, Conidial germination.

Sphaerotheca is a biotrophic parasite of flowering plants causing **powdery mildew** disease. The most common species is *S. fuliginea* causing powdery mildew disease of the members of the family Cucurbitaceae like *Luffa, Lagenaria* and *Cucurbita*. Another species *S. pannosa* sauses powdery mildew disease of roses. In Europe and USA the common species are *S. mascularis (S.humuli)* and *S. mors-uvae*.

Sphaerotheca fuliginea

S. pannosa

Sphaerotheca mors-uvae

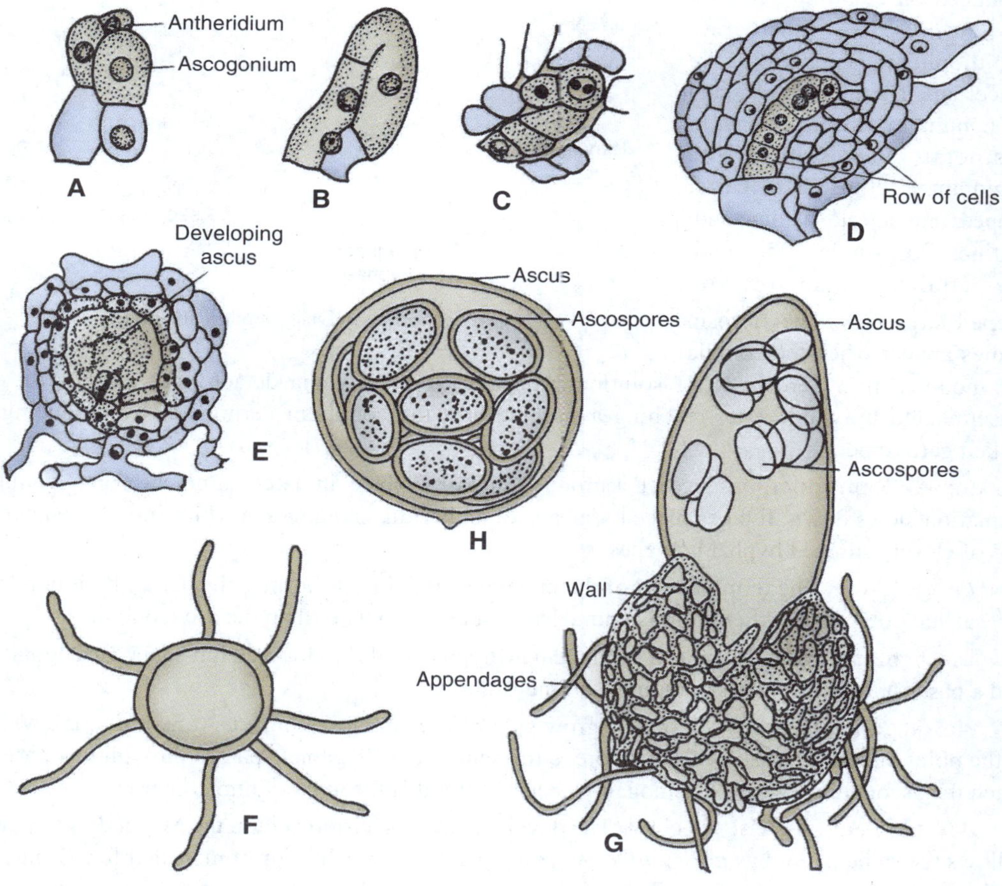

Fig. 11.15. *Sphaerotheca.* Sexual reproduction. A, Gametangia; B, Plasmogamy; C, Ascogonium showing a pair of nuclei; D, A row of cells with a binucleate cell; E, Developing ascus and ascocarp; F, A cleistothecium under surface view; G, A ruptured cleistothecium; H, An ascus with eight ascospores.

Mycelium. The mycelium is superficial *i.e* ectophytic and is made up of branched septate hyphae, the cells being uninucleate. The mycelium forms a white cottony mass on the surface of the host. The parasite draws its nourishment from the host by means of globular or lobed haustoria which are formed by the hypha at the point of contact with the host.

Asexual Reproduction. Asexual Reproduction takes place by means of conidia produced on conidiophores. As in *Erysiphe*, the short erect conidiophores arise from the mycelium. The mother cell cuts off conidia in a vertical chain in basipetal succession. The conidium is ellipsoid to barrel shaped, uninucleate, hyaline with distinct fibrosin bodies which are conical, disc or rod shaped structures. The fibrosin bodies appear when the conidia are mounted in 3 percent KOH solution. At maturity, the conidia detach very easily and are disseminated by wind. After reaching a suitable host the conidium germinates by producing a forked germ tube.

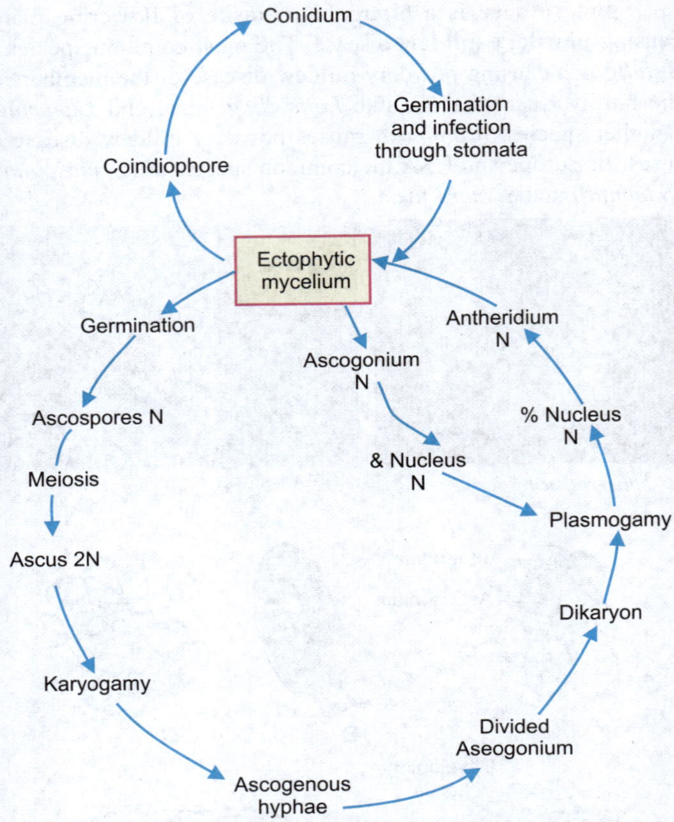

Fig. 11.16. *Sphaerotheca.* Diagrammatic life cycle.

Sexual Reproduction. Sexual reproduction takes place in later summers when conidia formation slows down. It takes place by means of antheridia and oogonia which are formed at the tips of closely situated hyphal branches.

Antheridium. The terminal cell of the male hypha divides to form a slender antheridium and a basal stalk cell. The antheridium is uninucleate and more slender than the ascogonium.

Ascogonium. The terminal cell of the female hypha divides to form a club-shaped ascogonium and a basal stalk cell. The ascogonium is uninucleate.

Plasmogamy. Both the sex organs grow side by side in close contact. At maturity, the walls, at the point of contact, dissolve and a pore is formed. The male gamete passes through this pore to reach the ascogonium where plasmogamy occurs. After this the antheridium collapses.

Development of Cleistothecium. The development of cleistothecium, the asci and ascospores follows the same pattern as in *Erysiphe*. After the plasmogamy, the pair of male and female nuclei settle down in the ascogonium. The ascogonium now produces a row of cells by successive divisions. The penultimate cell of this row of cells must be dikaryotic, whereas the remaining cells may be uninucleate. This dikaryotic cell acts as ascus mother cell in which karyogamy occurs resulting in a diploid nucleus. This diploid nucleus undergoes meiosis and subsequent mitosis forming eight nuclei. The ascus mother cell also elongates and forms the ascus. Each of the eight nuclei is enveloped by cytoplasm and eight ascospores are formed. Each ascus is ovoid, broadly elliptic to subglobose and contains eight ascospores. A hyphal sheath develops and protects the developing asci which are enclosed in an ascocarp—the cleistothecium.

Cleistothecium. The mature cleistothecium is globose and dark brown in colour. The wall is two layered thick, each several cells thick. The outer layer is made up of darker, melanised, polygonal and uninucleate cells whereas the inner one is made up of thin walled binucleate cells. The outer layer is protective while the inner one provides nourishment to developing ascus. The cleistothecium also bears a variable number of simple myceloid appendages. These are as long as the diameter of cleistothecium and they are usually interwoven with the mucelium.

The cleistothecium usually has a single ascus with eight ascospores. Each ascospore is ellipsoid to nearly spherical, uninucleate and haploid.

Dehiscence. The ascus and ascospores are released only after the disintegration of cleistothecium. The ascospore germinates on a suitable host and forms new mycelium.

Gordon (1966) is of the view that although *Sphaerotheca* follows a more or less similar pattern in sexual cycle as in *Erysiphe*, generally, fertilised ascogonium might not be playing any functional role in the development of asci and ascospores. Thus, the ascogonium may be designated as pseudoascogonium and antheridium as pseudoantheridium.

Taxonomic position.

Alexopoulos (1962)			**Alexopoulos and Mims (1979)**		
Division	:	**Mycota**	*Superkingdom*	:	**Eukaryonta**
Sub division	:	**Eumycotina**	*Kingdom*	:	**Mycetae**
Class	:	**Ascomycetes**	*Divisions*	:	**Amastigomycotina**
Sub class	:	**Euascomycetidae**	*Sub division*	:	**Ascomycotina**
Series	:	**Hymenoascomycetidae**	*Class*	:	**Ascomycetes**
Sub series	:	**Pyrenomycetes**	*Sub class*	:	**Hymenoascomycetes**
Order	:	**Erysiphales**	*Series*	:	**Pyrenomycetes**
Family	:	**Erysiphaceae**	*Order*	:	**Erysiphales**
Genus	:	***Sphaerotheca***	*Family*	:	**Erysiphaceae**
			Genus	:	*Sphaerotheca*

(*b*) Pyrenomycetes with flask-shaped perithecium. This group includes Ascomycetes in which the ascocarp is a flask-shaped perithecium. The members of this group are often called the flask-fungi. The club-shaped or cylindrical asci are usually borne in a parallel series forming a fertile layer called the **hymenium.** The latter lines the base and generally the sides of the inner wall of the perithecium. Occasionally the asci are borne in basal tufts. The ascospores in the majority of the species are liberated throught a circular pore at the top of the fruit. It is termed the **ostiole.** The perithecia may be free or embedded in a stroma formed from the hyphae of the fungus. In either case the perithecium has a definite wall of its own. Lining the ostiole are short hair-like hyphal growths arising from the inner wall of perithecial neck. These are the **periphyses.**

There is considerable difference of opinion among the mycologists as to the number of orders included in this series. Most of the mycologists enlist *Chaetomiales, Sphaeriales, Diaporthales, Hypocreales, Clavicipitales, Coronophorales and Coryneliales.* The genus *Claviceps* is usually selected as an example of the Pyrenomycetes with perithecium type of fructification as it received considerable attention of the mycologists. It is placed in the order **Clavicipitales.**

ORDER : CLAVICIPITALES

General characters. It is a small group of stromatic genera. The mycelial hyphae are closely packed to form a pseudoparenchymatous, cushion-like structure called the **stroma.** It is of soft, fleshy texture. Embedded in the stromatal head are numerous perithecia, each with its own perithecial wall enclosing a cavity. Each perithecium opens by a pore at the stromatic surface. It is called an **ostiole.** The asci are long, slender, cylindrical structures each of which is furnished with a thick cap pat its apex. They arise in a tuft from the base of the perithecium. Each ascus contains eight filiform **ascospores** which escape through a cylindrical pore in the **ascal cap.** After liberation the ascospores, in some species, break into short segments. Each segment functions as a spore. From the sides of the inner wall of the perithecium arise slender, sterile hyphae called the **paraphyses.** Chaudefaud (1960) reported that the paraphyses are closely appressed to the perithecial wall to form a part of it. They do not occur on the tuft of asci. The order includes a single family Clavicipitaceae. It has the same characteristics as those of the order.

FAMILY : CLAVICIPITACEAE

The two important genera included in this family are *Claviceps* and *Cordyceps.* Of these the former is taken as a type.

CLAVICEPS

It parasitizes grasses (Gramineae) and infects the gynocecium causing a disease known as **ergot.** The genus includes about a dozen species of cosmopolitan occurrence. The best known is *C. purpurea* whose full life history was established in 1863 by Kuhn.

CLAVICEPS PURPUREA, *(The ergot of rye)*

Occurrence. This fungus ergotizes certain susceptible wild grasses and cereals such as rye (*Secale cereales*) and barley and occasionally wheat. The infection takes place when the plants are blooming. On the rye plant *C. purpurea* is responsible for causing a disease known as **ergot of rye.**

Distribution. It is a disease of the temperate zone and occurs in abundance in Canada, U.S.A., Russia and many other countries of European continent. In India, it is of rare occurrence and has been reported from a few places. Pushkar Nath and Padwick (1941) and Kamat (1953) recorded the

Claviceps purpurea

Ergot of Rye

fungus in the hilly regions of North. Kulkarni (1942) reported the ergot disease from Bombay, and Karnatak on Jowar crop.

Primary Infection. It takes place by the wind dissemintated ascospores (Fig. 11.19 A). Stager (1912), however, reported that the conidia are resistant to cold and thus remain viable during winter. The surviving conidia possibly bring about infection in the following growing season. The consensus of opinion still favours the view that the primary source of inocculum are the ascospores liberated as a result of natural germination of sclerotia. Infection takes place when the plants are blooming. Watkins and Littlefield (1976) reported that the incidence of ergot infection is greater at or before anthesis than after anthesis (time when anthers are first extruded). Both the stigma and ovary wall are the sites of infection (Shaw and Mantle, 1980). Penetration occurs via both. Colonization is intercellular.

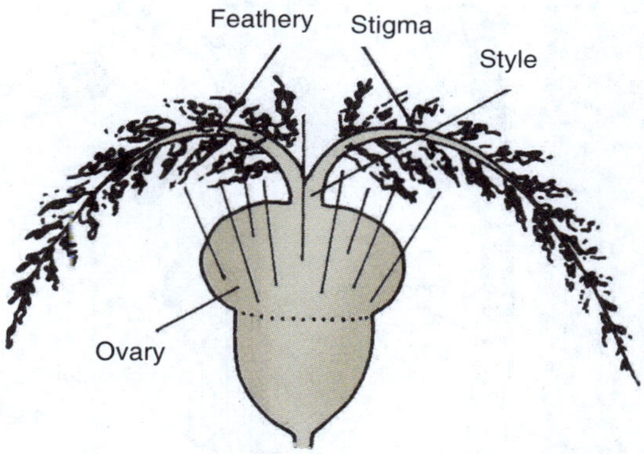

Fig. 11.17. Gynoecium of *Secale cereale*.

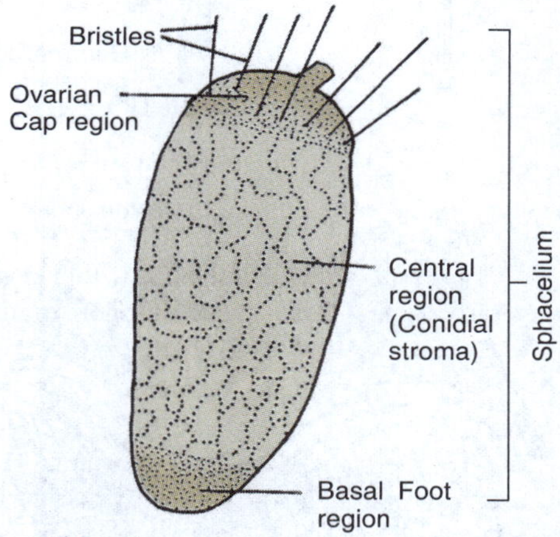

Fig. 11.18. *Claviceps purpurea*. Sphacelial stage. Note the three regions of the sphacelium (Based on Luttrell)

Landing on the feathery stigmas or ovary wall of *Secale cereale* (Fig. 11.17), the ascospore germinates by several germ tubes (Fig. 11.19B). The germ tubes penetrate between the cells of the stigma. Their growth is directed towards the ovule. The intercellular hyphae grow downwards between the cells of the style and extend into the apical tissue of the ovary. From there the hyphae grow around the ovule in the inner layers of the ovary wall and reach the tip of the rachilla (vascular tissue in the base of the ovary). The fungal hyphae do not extend into the rachilla but form a narrow interface with the vascular tissue. To the naked eye the interface often appears as a brown line.

Hyphal Structure (Fig. 11.19 C). The intercellular hyphae are hyaline, septate and about 3 μ m in dia. Within the hyphal wall is the plasma membrane. The latter encloses granular cytoplasm containing mitochondria, vesicles, small lipid globules and several nuclei.

Sphacelium (Fig,. 11.18). Within about a week's time after infection, the hyphal mass increases to such an extent that only few cells of the ovary could be seen intact. The ovule is completely surrounded and penetrated by the proliferating hyphae but there is no penetration into the rachilla. Finally the ovule is destroyed and is replaced by a thick soft cottony hypha mass or mat (D) of nearly the same form as the floral cavity itself by the 10th day. It is known as the **sphacelium** (Fig. 11.18). The latter replaces the ovarian tissue and consists of loose plectenchyma. It is differentiated externally into 3 regions, namely, (i) the basal foot region with a smooth, white surface located at the base of the host floret, (ii) a central larger portion constituting the conidial stroma region of the sphacelium; and (iii) the sparsely invaded ovarian apical or cap region covered with bristles.

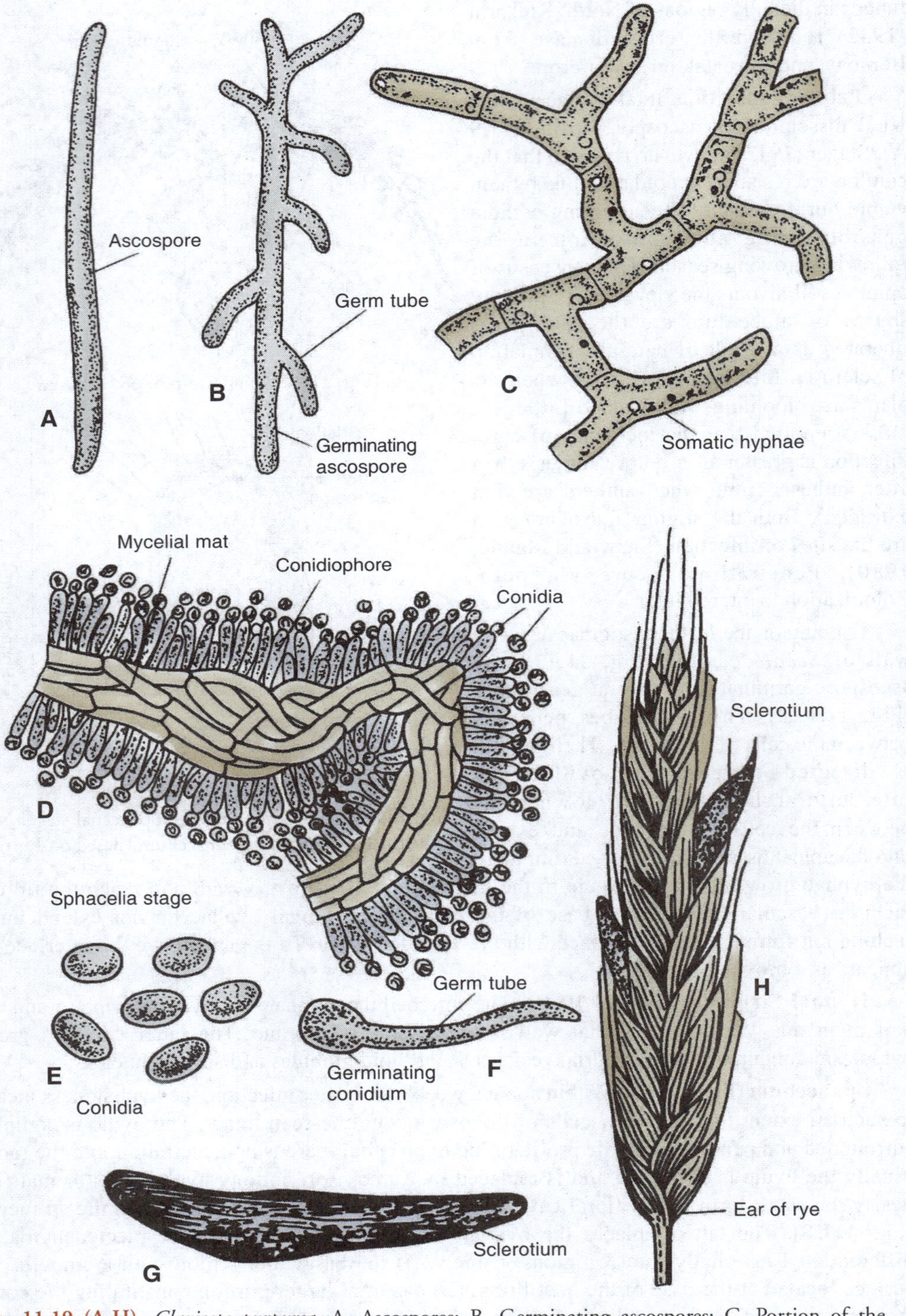

Fig. 11.19 (A-H). *Claviceps purpurea*. A, Ascospores; B, Germinating ascospores; C, Portion of the mycelium; D, Sphacelial stage; E, Conidia; F. Germinating conidium; G, Resting sclerotium; H, Ear of Rye with sclerotia.

Sporulation (Fig. 11.19D). The first sign of sporulation which provides external evidence of infection appears between 4-6 days after infection in the form of white ridges of conidiophores on the surface of the ovary. Certain hyphae from the mycelial mat (stroma of sphacelium) grow outwards, emerge between the ovarian epidermal cells, rupture the cuticle and produce a compact, palisade-like superficial layer of *phialidic conidiophores*. At its tip each conidiophore bears a succession of small, spore-like structures, the **conidia**. They are abstricted in a basipetal succession. The conidia are small, hyaline, uninucleate structures spherical to oblong in form. The earlier mycologists mistook these conidia as an imperfect fungus *Sphacelia segetum*. So the conidial stage of the fungus came to be known as the **sphacelial stage** in the life cycle. It is characterised by the freely sporulating conidiophores all over the external surface of the ovary and internal pockets (D). At about this stage a sticky, dense yellow, sugary fluid termed the *honey dew* begains to exude between the glumes thereby making the infected florets conspicuous. For this reason some mycologists prefer to call the sphacelial stage as the honey dew stage. Conidia accumulate in the sugary exudate and ooze from the florets in drops of honey dew by the 6th or 7th day (Luttrell, 1980). Insects are attracted to the honey drops. While feeding upon honey they pick up the conidia and carry them to other healthy florets on the same or another rye plant to cause secondary infections. In this way the disease spreads during the growing season. Abundance of sugar in the honey dew makes it an excellent substrate for microbial growth. According to Cunfer (1976), the honey dew contains a factor(s) which inhibits the germination of conidia of certain fungi that come in contact with the sphacelial stage. It is, however, ineffective against the conidia of those fungi (*Fusarium heterosporum*) which are capable of colonizing sphacelial stage and prevent maturation of ergot sclerotia. According to some mycologists, honey is secreted by the hyphae and according to others by the flower nectary. The latest view is that profuse exudation of honey dew is the result of the fungal parasite tapping the host's vascular system which remains uninvaded. The production of conidia is profuse from 7 day to 11 day and ceases by 13 day (Luttrell, 1980). By the 10th day after infection of the ovary, all the host tissues, to which the parasite has access, are penetrated by the hyphae. The path of infection is complete. There are no further invasion of the host tissues. Further development of the fungus takes place by the development of the sclerotial stage.

Differentiation of Sclerotium (Fig. 11.20). Milovidov (1954) observed that sclerotial cells are formed by transofrmation of sphacelial cells. The process starts at the base of the sphacelium. Shaw and Mantle (1980) concurring with this view reported that the sphacelial tissue has the capacity to differentiate into sclerotial form. As the transition starts, the fungal growth at the base of the sphacelium forms hyphae mainly of the sclerotial type and not of sphacelial type. As the conidial production ceases and growing season is about to be over, hyphal differentiation into sclerotial form becomes evident in the lower part of the sphacelium.

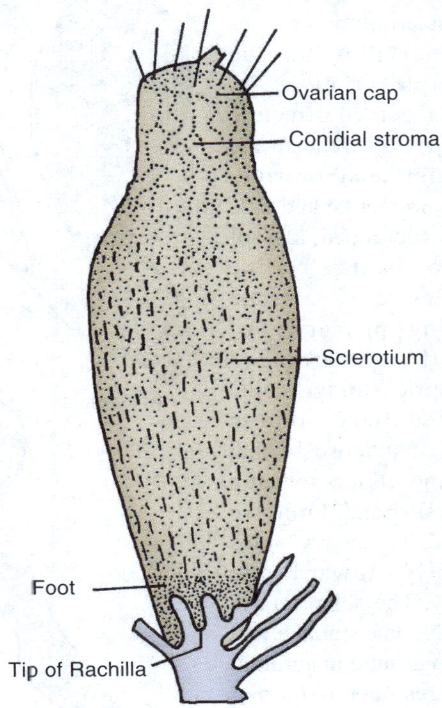

Fig. 11.20. *Claviceps purpurea*. Sclerotial stage.

According to Luttrell (1980), the hyphae across the top of foot region of the sphacelium assume a vertical orientation and form a palisade-like layer. These hyphae differ from the adjacent sphacelial tissue in being short, bulbous, wide and frequently septate. They are composed of cylindrical multinucleate, lipid rich cells containing small vacuoles and tend to form prosenchyma. This palisade-like layer of short hyphae is termed the **generative layer.** According to Luttrell, the generative layer delimits the foot and gives rise to the tissues of the ergot sclerotium. Hyphal tip growth of the generative layer results in increase in length of the developing sclerotium. Branching and septation of the resulting sclerotial hyphae and lipid deposition in the sclerotial cells determine its compact plectenchymatous nature. As the differentiation of sclerotial tissue proceeds from below upwards it pushes up beneath and within the conidial stroma. Consequently the remnants of the conidial stroma are sloughed from its surface. The increase in sclerotium length pushes the conidial stroma (loosely packed sphacelial hyphae) upward with the sparsely infected bristly ovarian cap at its top. By this time the sclerotium is conspicuously developed. It protrudes through the tips of the florets with the **sphacelial cap** consisting mainly of the remains of the ovary cell walls, stigmas, remnants of the ovule and degenerting sphacelial hyphae, still persisting at its apex. The host tissues are no longer alive. The selerotium is considerably firmer than the sphacelium which consists of loosely packed hyphae. The hyphae around the periphery of the sclerotium darken and become coalesced to form the **rind.** The sclerotium which has destroyed the ovarian tissue and replaced the grain lies between the glumes of the spikelet. It protrudes out of the glumes as a dark spur on the spike (Fig. 11.19 H). It is much longer and broader in size than the grain which it has replaced. The mature ears of rve bear both the sclerotia and the healthv rye grains.

The sclerotium matures in about 4 weeks time. The mature sclerotium is a hard pseudoparenchymatous dark-purple or blackish, usually a curved structure (Fig. 11.19 G). Beneath the dark outer coat or rind, which is protective and thus prevents too much loss of moisture, is the hyaline (colourless), tissue consisting primarily of compactly arranged isodiametric storage cells. The medullary region consists of a loose hyphal conducting tissue forming the central strand. From the latter radiate narrow veins irregularly toward the periphery. The sclerotia are produced in late summer and fall to the ground in autumn. They are resistant to freezing and thus possess a long viability enabling the fungus

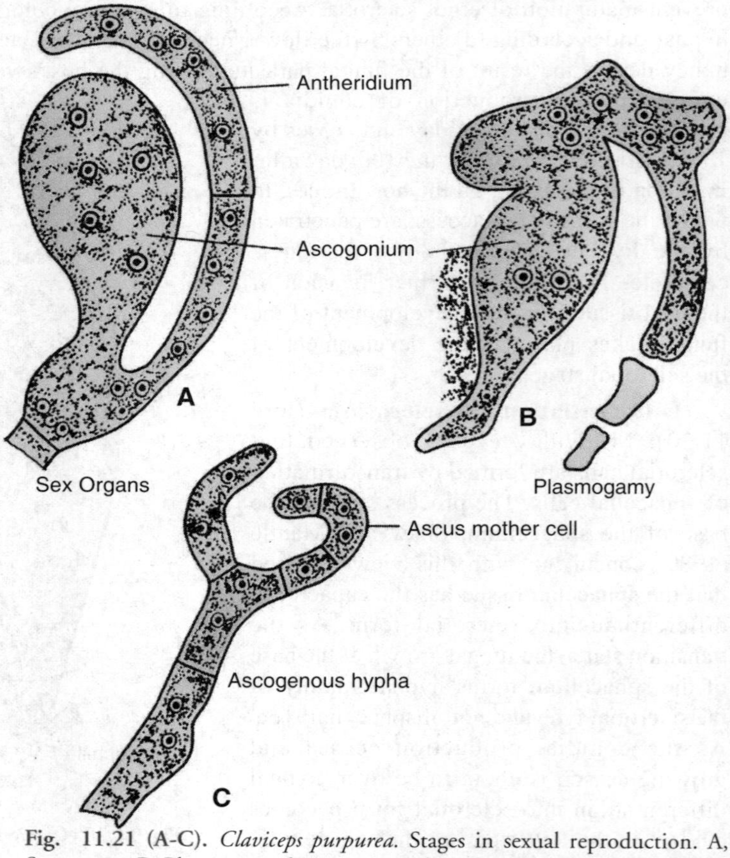

Fig. 11.21 (A-C). *Claviceps purpurea.* Stages in sexual reproduction. A, Sex organs; B, Plasmogamy by gametangial contact; C, Ascogenous hypha.

to over winter in this condition. Sclerotia are, thus, the structures of survival of ergot fungi and the source of primary inoculum for infection. However, Cunifer and Seckenger (1977) reported that sclerotia which mature on plants and fall to the ground at crop maturity cannot survive until the following spring. Six months is the longest surviving period for any sclerotia.

Dispersal of Sclerotia. The sclerotia reach maturity at about the same time as the healthy grains in other ovaries on the spike. Many of the mature sclerotia fall to the ground (G). Some become intermingled with the grains during harvesting operations. They are returned to the field at the time of fresh sowing and may be sown together with the grain.

Germination of Sclerotium (Fig. 11.22A). The sclerotia which have fallen to the ground hibernate during winter. The following spring they germinate under suitable moisture conditions.

About half a dozen or even more small pink or dark purple, drum stick-like outgrowths grow from its surface and turn towards light (A). Each of these is about 10-12 mm long and is called a

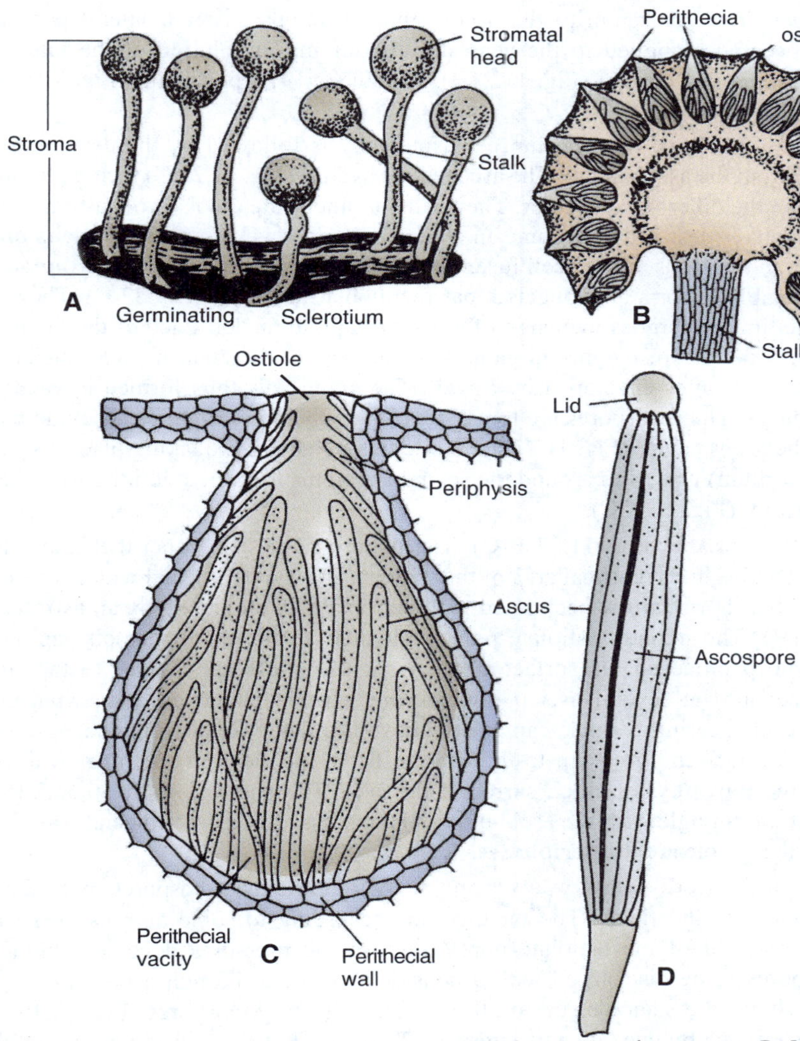

Fig. 11.22 (A-D). *Claviceps purpurea.* A, Germinating sclerotium with stromata; B, V.S. Stroma showing perithecia in a peripheral layer; C, V.S. Locule containing the perithecium; D, Ascus with ascospores enlarged considerably.

stroma (pl. stromata). The stroma consists of a slender stalk ending in a flattered spherical or capitate head. Both the stalk and the stromatal head are made up of compactly interwoven hyphae.

Sexual Reproduction

(a) *Sex organs* (Fig. 11.21). Embedded in the pseudoparenchymatous tissue of each stromatal head are a number of small, flask-shaped cavities arranged in a peripheral layer (Fig. 11.22 B). Some authors prefer to call them the **perithecial cavities.** Each cavity contains both kinds of sex organs (Fig. 11.21A). They are developed from the terminal cells of lateral branches arising from the same hypha at the base of the perithecial cavity. The fertile hyphae have richer protoplasmic contents. The male gametangium is called the **antheridium** and female **ascogonium.** Both are multinucleate at maturity. The ascogonium is relatively stouter and broader than the antheridium which is slender and elongated. There may be one or more antheridia arising from the base of the ascogonium.

(*b*) *Plasmogamy.* Towards maturity the ascogonium develops a small, lateral papilla-like outgrowth which bceomces applied to the nearest antheridium, Dissolution of the intervening walls occurs in the region of contact (Fig. 11.21B). Killian (1919) reported the migration of the male nuclei into the ascogonium.

Development of Ascocarp (Perithecium). Plasmogamy is followed by the development of ascogenous hyphae from the ascogonium. The ascogenous hyphae (Fig. 11.21 C) recurve terminally to form croziers as in the other Ascomycetes. The resultant binculeate crook or penultimate cell of each ascogenous hypha is destined to become an elongate ascus. It is known as the **ascus mother cell.** The two nuclei in the ascus mother cell fuse. The cell with fusion nucleus is the **young ascus.** It elongates considerably to form an elongated, narrow tubular **ascus** (Fig. 11.22 C). The diploid nucleus at the same time undergoes meiosis to form four haploid nuclei. Each of the latter again divides mitotically. A portion of cytoplasm gathers round each of the eight haploid nuclei. Each unincleate protoplast becomes enveloped by a wall. The eight cells thus formed in each ascus mature into **ascospores.** The ascospores are long, thread-like, hyaline structures. They lie parallel to one another in the ascus cavity (Fig. 11.22D). While these changes are taking place a soft thin, perithecial wall (peridium) grows up around the sexual apparatus to form a definite fructification called the **perithecium** (Fig. 11.22C).

Structure of the Perithecium (11.22 B-C). The mature flask-shaped perithecia are deeply sunk in cavities or locules in the cortical area of the stroma (B). The perithecial wall is one or two cell layers thick with a short, narrow neck lined with periphyses and opening by an **ostiole** at the stromatic surface (B). The minute, ostiolate pore is situated at the tip of a small papilla. The presence of ostiolate papillae on the surface gives a slightly warty appearance to the stroma. Within each perithecium lie several asci (C) which are narrow, elongated, somewhat curved tubular structues, each bearing a thick cap at its apex (D). They arise from the base of the perithecial cavity in a tuff and grow up to the ostiole. From the sides of the inner wall of the perithecium arise the **paraphyses** which surround the asci. The paraphyses are thus lateral in origin. They do not intermingle with the asci but remain appressed to the perithecial wall. Lining the short neck and the ostiole are the **periphyses.**

Dehiscence of asci and dispersal of ascospores. By the time the ascospores reach maturity the rye plants blossom in the fields. The mature asci, according to some authors, grow up in succession and extend out of the ostiolate pore. Hygroscopic pressures develop within them causing the ascospores to be discharged with a considerable force through a pore in the thick ascal cap into the air for a distance of up to 50 mm. The filiform ascospores (Fig. 11.19A) are thus forcefully ejected, one by one, at short intervals. They are shot out only when the perithecia are veritcally upright. The direction is adjusted by the slow twist and turn movements of the stalk just below the stroma. These movements bring each perithecium in an upright position for a short while. The discharged ascospores are borne by the air currents to the flowering ears of rye plants to start the life cylce afresh. Stager, however, claims that the ascospores are disseminated by insects. The ascospores are thus disseminated when the rye flowers are open with the feathery stigmas spread out.

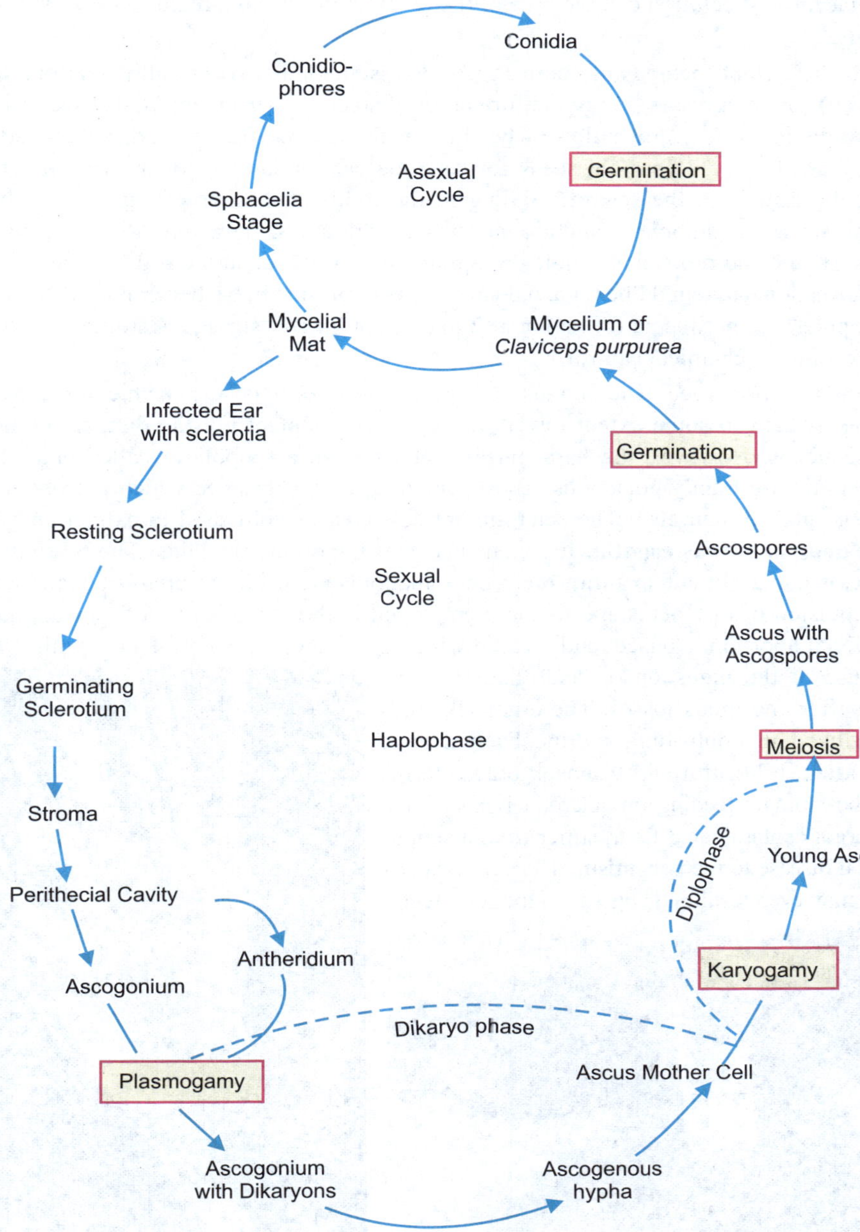

Fig. 11.23. *Claviceps purpurea.* Word diagram of the life cycle.

Host-Parasite Relationship. Tulsane (1950) pointed that *ergot of rye* caused by *C.purpurea* is a replacement disease. The fungal parasite disintegrates the host tissues in the floral cavity to make room for the development of it sphacelial and sclerotial stages which constitute the replacement structures. It colonizes the floral tissue. The physical presence of the intercellular hyphe of the parasite appear to cause only slight distortion of the host cells (Shaw and Mantle, 1980). While selectively colonizing the ovary, the parasite maintains a compatible relationship with the adjacent host cell tissue at the base of the ovary in the rachilla region as a source of water and nutrients.

Control. The chief methods of control ergot disease include : (*i*) burial of sclerotia by deep ploughing, (*ii*) sowing grasses before the formation of sclerotia, (*iii*) planting disease resistant or non-susceptible crops, (*iv*) prevention of heading in fence rows or other adjacent waste areas, (*v*) sowing ergot free or 2-year old seeds for sclerotia cannot survive for more than 6 months, (*vi*) killing the fungus in the sclerotia falling in the fields at the harvesting time by burning, (*vii*) burning straw and stubble in and around the adjacent areas. Hardison (1972) recommended the use of systemic and protectant fungicides applied over sclerotia at the soil surface to suppress perithecia formation. He found Phenyl 5, 6 dichloro-2-trifluoro-methyl-1-benzimidazole carboxylate useful in suppressing perithecia formation at 2 mg/92 cm^2 of soil surface. Cardium chloride may provide an economic chemical cantrol.

Economic Importance. The fungus (*Claviceps purpurea)* is responsible for causing the **ergot disease** which to some extent lowers the yield of rye plant but the disease seldom is of much consequence. However, the hard purple-black **sclerotia** (popularly called ergots) which replace the grain are highly poisonous. They contain a toxic substance which poisons and kills animals. It is fatal to man also. The sclerotia are collected or cultivated to extract an officially recognized **drug** known as **ergotin.** It contains several important alkaloids. The best known are **ergonovine, ergometrin** and **ergotamine.** These alkaloids are useful in pregnancy and labour to promote expulsion of the foetus and thus are employed in the manufacture of a medicine which causes artificial abortion. Consequently, cattle which graze on grassland infected with *Claviceps* abort because of the ingestion of sclerotia. Thus the cattlemen suffer enormous losses. The ergot alkaloids are also utilized in controlling uterine haemorrhage (bleeding) after child birth and to check other uterine troubles. The animals feeding on sclerotia lying on the ground or ergot contaminated flour, suffer from a serious physiological disease termed **ergotism.** The severe attack of ergotism may even result in gangrene of toes and feet.

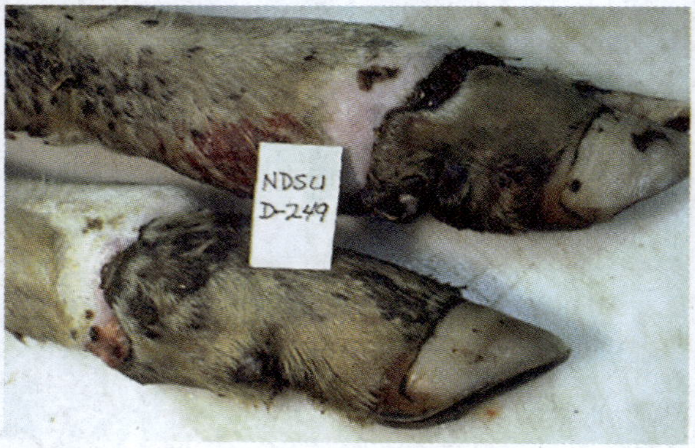

Ergotism

Ergots

Taxonomic Position. There is considerable difference of opinion as regards the taxonomic position of *Claviceps*. Some mycologists including Miller (1949) and Luttrell (1951) placed it in the order Sphaeriales. Bessey (1950) and Martin (1961) included it in the order Hypocreales. Nannfeldt (1932), Gaumann (1952), Dennis (1960) and Alexopoulos (1962) classified it in a separate order Clavicipitales. The consensus of opinion at the present time favours the latter viewpoint. Accordingly, the systematic position of *Claviceps* is as under :

Alexopoulos (1962)			Alexopoulos and Mims (1979)		
Division	:	**Mycota**	*Superkingdom*	:	**Eukaryonta**
Subdivision	:	**Eumycotina**	*Kingdom*	:	**Mycetae**
Class	:	**Ascomycetes**	*Division*	:	**Amastigomycota**
Sub-class	:	**Euascomycetidae**	*Subdivision*	:	**Ascomycotina**
Series	:	**Hymenomycetes**	*Class*	:	**Ascomycetes**
Sub-series	:	**Pyrenomycetes**	*Sub-class*	:	**Hymenoascomycetidae**
Class	:	**Clavicipitales**	*Series*	:	**Pyrenomycetes**
Family	:	**Clavicipitaceae**	*Order*	:	**Clavicipitales**
Genus	:	*Claviceps*	*Family*	:	**Clavicipitaceae**
Species	:	*purpurea*	*Genus*	:	*Claviceps*
			Species	:	*purpurea*

QUESTIONS

ESSAY TYPE

1. Give the name and symptoms of the disease caused by *Sphaerotheca.* With the help of labelled diagrams only, explain its life cycle. *(Allahabad, 1991)*
2. Give an account of structure, reproduction and economic importance of *Sphaerotheca*. *(Allahabad, 1995, 2000, 2001, 2004)*
3. What is powdery mildew? Give an account of any fungus causing a powdery mildew. *(Allahabad, 1997)*
4. Describe structure and methods of reproduction in Erysiphales and name parasitic members of the group. *(Kanpur, M.Sc., 1990, 1994, 1998)*
5. Describe structure, reproduction and parasitic importance of any member of Clavicipitales. *(Kanpur M.Sc., 1999)*
6. Give an account of the symptoms, life cycle and economic importance of *Claviceps purpurea*. *(Rohilkhand, 1994)*
7. With the help of labelled diagrams, describe the structure and life cycle of *Phyllactinia.*
8. Give an illustrated account of the structure and life cycle of *Erysiphe* or *Sphaerotheca.*
9. Describe the life-cycle of *Erysiphe*. *(Kumaon, 2004)*
10. (*a*) Describe the life cycle of any ascomycetous fungus studied by you. Also write the control measures and economic importance. *(Rohilkhand, 2000)*

    (*b*) Describe the causal organism, symptoms, disease cycle and control measures of powdery mildew of cucurbits. *(Rohilkhand M.Sc., 2006)*
11. Give a general account of Erysiphales and the characteristics on the basis of which its different genera can be identified. *(Allahabad M.Sc., 1999)*

SHORT ANSWER TYPE

12. Describe the most important characteristics of the following their taxonomic position and the diseases caused by them :
(*a*) *Erysiphe*
(*b*) *Phyllactinia*
(*c*) *Sphaerotheca* (*Allahabad, 2003*)
(*d*) *Claviceps* (*Rohilkhand, 1996*)

13. Give an account of Erysiphales. How would you distinguish the various genera on the basis of their ascocarps.

14. How would you identify the following in your laboratory?
(*a*) *Claviceps* (*Allahabad M.Sc., 1999*)
(*b*) *Erysiphe*
(*c*) *Phyllactinia* (*d*) *Sphaerotheca*

15. Write short notes on :
(*a*) *Claviceps*
(*Kanpur M.Sc., 1989; 1990; 1996, 2002*)
(*b*) *Phyllactinia* (*Kanpur M.Sc., 1997*)
(*c*) *Sphaerotheca* (*d*) *Erysiphe*
(*Kanpur, M.Sc., 1991; Bihar, 1991*)
(*e*) *Ergot of rye*
(*Kanpur, M.Sc., 1991; Rohilkhand, 1992, 1996*)

16. Differentiate between Downy mildews and Powdery mildews.
(*Kanpur M.Sc., 1989; 1993, 1996; 1998*)

17. Describe the symptoms and control measures of powdery mildew of pea or *ergot* disease of rye.

18. Describe the economic importance of *Claviceps* and *Erysiphe.*

19. Compare the ascocarps of *Erysiphe* and *Claviceps*. Name the causal organism of powdery mildew of pea.

20. Describe the life cycle of the fungus causing the disease.

21. Give labelled diagrams of the following :
(*a*) Cleistothecia of *Sphaerotheca*
(*Allahabad, 1992*)
(*b*) Sclerotia of *Claviceps*
(*c*) Cleistothecia of *Erysiphe*

22. Draw diagrams to show various types of appendages present on ascocarp of members of powdery mildew pathogens.
(*Kanpur M.Sc., 2002*)

OBJECTIVE TYPE

23. Fill in the blanks :

(*i*) The most important characteristics of the fungi belonging to Pyrenomycetes is ....

(*ii*) *Erysiphe* causes the disease known as ....

(*iii*) Ergot disease of rye is caused by ....

(*iv*) *Claviceps* produces an alkaloid which is used as a ....

(*v*) The ascus in *Phyllactinia* has .... ascospores.

(*vi*) The fungus used in genetical studies is ....

(*vii*) *Sphaerotheca* has cleistothecia with external ....

(*viii*) The asci in Pyrenomycetes are ....

(*ix*) *Erysiphe* grows mycelium which is .... and sends .... into the host cell.

(*x*) The characteristic feature of the genus *Claviceps* is the presence of ....

24. Select the correct answer :

(*i*) Powdery mildew disease is caused by
(*a*) *Erysiphe polygoni*
(*b*) *Sclerospora graminicola*
(*c*) *Neurospora sp.*
(*d*) *Claviceps purpurea*

(*ii*) The fruit body of *Erysiphe polygoni* is called
(*a*) Apothecium (*b*) Hypothecium
(*b*) Perithecium (*d*) Cleistothecium

(*iii*) Ergot of medicine is extracted from :
(*a*) *Erysiphe polygoni*
(*b*) *Phyllactinia sp.*
(*c*) *Claviceps purpurea*
(*d*) *Synchytrium endobioticum*

(*iv*) Which of the following has asci enclosing only two ascospores?

(*a*) *Erysiphe polygoni*

(*b*) *Phyllactinia sp.*

(*c*) *Claviceps purpurea*

(*d*) *Synchytrium endobioticum*

(*v*) The fungus which is important for its use in genetic studies is

(*a*) *Aspergillus* (*b*) *Rhizopus*

(*c*) *Penicillium* (*d*) *Neurospora*

(*vi*) The character of cleistothecium of *Phyllactinia* is that the appendages are

(*a*) mycelium like, dicholomously branched at the tip

(*b*) curled at apex

(*c*) provided with bulbous base

(*d*) absent

(*vii*) The ascocarps of which of the following families are stromatic?

(*a*) Erysiphaceae

(*b*) Claviceptaceae

(*c*) Melampsoraceae

(*d*) Sphaeropsidaceae

(*viii*) Which of the following is the fruit body of *Erysiphe* and *Penicillium*?

(*a*) Perithecium

(*b*) Cleistothecium

(*c*) Apothecium

(*d*) Stroma

(*ix*) Sclerotia are produced in

(*a*) *Erysiphe*

(*b*) *Sphaerotheca*

(*c*) *Claviceps*

(*d*) *Phyllactinia*

(*x*) In *Claviceps,* the ascocarp is

(*a*) Perithecium

(*b*) Apothecium

(*c*) Cleistothecium

(*d*) Hypothecium

Morchella deliciosa (Edible fungus)

CHAPTER 12

SUB CLASS – EUASCOMYCETIDAE Series : Hymenoascomycetes B. Sub-series : Discomycetes

- General Characteristics
- Order : Pezizales
- Family : Morchellaceae
- *Morchella*
- Family : Pezizaceae
- *Pyronema*
- *Coprobia* (Syn. Humaria)
- *Ascobolus*
- Discussion and Conclusion
- Phycomycetes and Ascomycetes Compared

GENERAL CHARACTERISTICS

Popularly the members of this group are called the **cup fungi**. The most distinctive feature is the cup or saucer-shaped usually fleshy ascocarp which is botanically known as the **apothecium**. The apothecia vary in colour through a wide range and may be red, yellow, orange, brown, black or even colourless. They vary in shape as well. Often they are typical cups or discs, but other forms such as sponges, saddles, tongues, and bells are not in anyway uncommon. The ascocarps invariably are open. The asci are club-shaped or cylindrical in form and are arranged in columnar disposition (a palisade-like layer) on the surface or in open cavities. The only exceptions are the truffles in which the underground ascocarps remain closed. Paraphyses usually occur intermingled with the asci. In habit the Discomycetes are mostly saprobes found growing in high humidity on soil, humus, dead logs of wood and dung.

Cup Fungi

Apothecium

The conidial stage appears to be non-existent in most cases. Gradual and progressive reduction in sexuality is another feature of this sub-series. It is accomplished by stages. In some genera such as *Pyronema* (Fig. 12.6) both the sex organs are functional. Fusion between them takes place. The intervening walls at the point of contact dissolve. The protoplasts of the two sex organs unite. The male and the female nuclei arrange themselves in pairs (dicaryons). Same is the case with *Lachnea stercorea*. Both the sex organs are fully developed. The antheridia are thus functionless. In *Lachnea cretea* the antheridium is totally absent. *Humaria granulata* goes a step further. The antheridia are absent. This is accompanied by the disappearance of the trichogyne from the ascogonium (Fig. 12.15 A). Finally in *Morchella, Humaria rutilans* and *Peziza vesicolusa* both the sex organs are totally absent. The whole sexual apparatus is lacking. The sexual process simply consists in the union of two vegetative nuclei in a pair.

Classification. In most of the Discomycetes the ascocarps are above ground. They are called **epigean**. The epigean discomycetes are usually divided into the following two groups :–

1. Inoperculate Discomycetes. The ascospores escape through an apical circular pore at th tip of the ascus or by splitting of ascus at the apex. This group includes two orders, namel *Ostropales* and *Helotiales*.

2. Operculate Discomycetes. The ascospores are released through a preformed **apical p** which opens by a lid-like structure at the apex. The lid which is also called the **operculun** hinged at the top of the ascus. The operculate Discomycetes comprise a single order *Peziz* which is discussed here in some detail.

ORDER : PEZIZALES

General. The order includes about 500 species grouped under 57 to 60 genera. Almost all are saprobes and prefer to grow on dead, rotting wood such as dead trunks, branches and stumps of trees, leaf mold, humus soil and dung. A few of them are interesting as they grow on the waste matter of a particular animal. For instance, *Ascodesmis porcina* grows only on hog dung and *Ascobolus magnificus* on horse manure. The mycelium ramifies usually beneath the surface of the

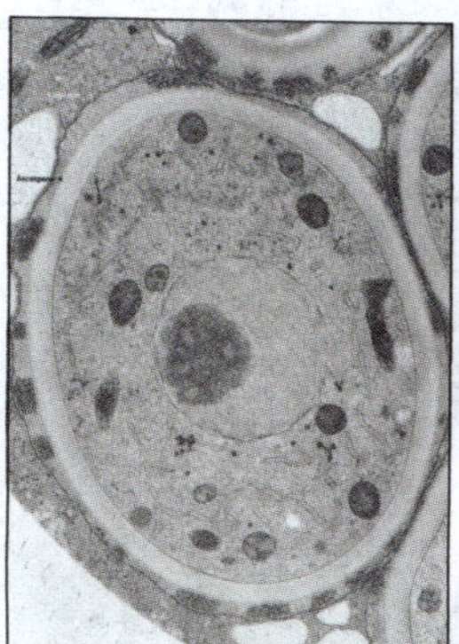

Ascodesmis porcina

Ascobolus

substratum forming extensively branched hyphal masses from which develop fleshy, often large apothecia which are always above ground (**epigean**) often brilliantly coloured and may be sessile or stalked. The bright colour may be due to pigmented oil drops in the paraphyses. The apothecia vary in size through a wide range. The large cups may be 100-125 mm in diameter and the smallest one may be less than 1 mm. across. The asci are club-shaped or cylindrical in form with a lid-like operculum hinged at the apex. They are phototrophic and dehisce in a characteristic manner puffing spores out in clouds. The apices of mature asci project above the level of hymenium and turn towards light. The apical pore opens by a lid and the ascospores are shot out explosively. The spore discharge is usually accompanied by a hissing sound. The single-celled, uninucleate ascospores are oval to spherical in shape. Asexual reproduction by conidia is uncommon, in fact, not known in the *Pezizales*.

It is a large order. Kimbrough (1970) subdivided it into 9 families but Korf (1973) listed only six. However, Alexopoulos and Mims (1979) recognised only three families : Sarcoscyphaceae, Morchellaceae and Pezizaceae. The last two are discussed here in some detail.

FAMILY : MORCHELLACEAE

Morchella

Verpa

This soil dwelling family includes the morels. It is characterised by its arge, stalked pileate uctificatons (apothecia). fertile region is distinct the stalk. It is cup- d only when young. further growth the m becomes convex. re turned outwards from honey- (sponge-like bell-shaped caps (*verpa*) terminating a stalk. The hymenium may be smooth or y folded by unequal growth of the hymenial surface which is thrown into folds. by the separation of a lid from the top. The ascospores are multinucleate. The mpletely lacking. Copulation occurs somatogamously or even autogamously. genera of this family are *Disciotis* and *Morchella* commonly called true hich is a highly prized edible fungus is taken here as a type.

MORCHELLA

a number of species. The important ones are *Morchella conica, M. elata,* a, *M. hybrida* and *M. crassipes*. All the species grow as **saprophytes**.

a hybrida

Morchella crassipes

Morchella elata

Morchella esculenta

e y, ore n is ales

nagnificus

black or even colourless. They vary in shape as well. Often they are typical cups or discs, but other forms such as sponges, saddles, tongues, and bells are not in anyway uncommon. The ascocarps invariably are open. The asci are club-shaped or cylindrical in form and are arranged in columnar disposition (a palisade-like layer) on the surface or in open cavities. The only exceptions are the truffles in which the underground ascocarps remain closed. Paraphyses usually occur intermingled with the asci. In habit the Discomycetes are mostly saprobes found growing in high humidity on soil, humus, dead logs of wood and dung.

Cup Fungi

Apothecium

The conidial stage appears to be non-existent in most cases. Gradual and progressive reduction in sexuality is another feature of this sub-series. It is accomplished by stages. In some genera such as *Pyronema* (Fig. 12.6) both the sex organs are functional. Fusion between them takes place. The intervening walls at the point of contact dissolve. The protoplasts of the two sex organs unite. The male and the female nuclei arrange themselves in pairs (dicaryons). Same is the case with *Lachnea stercorea*. Both the sex organs are fully developed. The antheridia are thus functionless. In *Lachnea cretea* the antheridium is totally absent. *Humaria granulata* goes a step further. The antheridia are absent. This is accompanied by the disappearance of the trichogyne from the ascogonium (Fig. 12.15 A). Finally in *Morchella, Humaria rutilans* and *Peziza vesicolusa* both the sex organs are totally absent. The whole sexual apparatus is lacking. The sexual process simply consists in the union of two vegetative nuclei in a pair.

Classification. In most of the Discomycetes the ascocarps are above ground. They are called **epigean**. The epigean discomycetes are usually divided into the following two groups :–

1. Inoperculate Discomycetes. The ascospores escape through an apical circular pore at the tip of the ascus or by splitting of ascus at the apex. This group includes two orders, namely, *Ostropales* and *Helotiales*.

2. Operculate Discomycetes. The ascospores are released through a preformed **apical pore** which opens by a lid-like structure at the apex. The lid which is also called the **operculum** is hinged at the top of the ascus. The operculate Discomycetes comprise a single order *Pezizales* which is discussed here in some detail.

ORDER : PEZIZALES

General. The order includes about 500 species grouped under 57 to 60 genera. Almost all are saprobes and prefer to grow on dead, rotting wood such as dead trunks, branches and stumps of trees, leaf mold, humus soil and dung. A few of them are interesting as they grow on the waste matter of a particular animal. For instance, *Ascodesmis porcina* grows only on hog dung and *Ascobolus magnificus* on horse manure. The mycelium ramifies usually beneath the surface of the

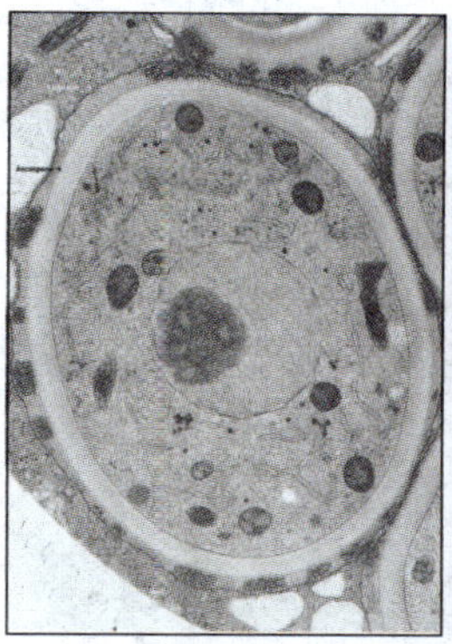

Ascodesmis porcina

Ascobolus magnificus

substratum forming extensively branched hyphal masses from which develop fleshy, often large apothecia which are always above ground (**epigean**) often brilliantly coloured and may be sessile or stalked. The bright colour may be due to pigmented oil drops in the paraphyses. The apothecia vary in size through a wide range. The large cups may be 100-125 mm in diameter and the smallest one may be less than 1 mm. across. The asci are club-shaped or cylindrical in form with a lid-like operculum hinged at the apex. They are phototrophic and dehisce in a characteristic manner puffing spores out in clouds. The apices of mature asci project above the level of hymenium and turn towards light. The apical pore opens by a lid and the ascospores are shot out explosively. The spore discharge is usually accompanied by a hissing sound. The single-celled, uninucleate ascospores are oval to spherical in shape. Asexual reproduction by conidia is uncommon, in fact, not known in the *Pezizales*.

It is a large order. Kimbrough (1970) subdivided it into 9 families but Korf (1973) listed only six. However, Alexopoulos and Mims (1979) recognised only three families : Sarcoscyphaceae, Morchellaceae and Pezizaceae. The last two are discussed here in some detail.

FAMILY : MORCHELLACEAE

Morchella

Verpa

This soil dwelling family includes the morels. It is characterised by its large, stalked pileate fructificatons (apothecia). The fertile region is distinct from the stalk. It is cup-shaped only when young. During further growth the hymenium becomes convex. The asci are turned outwards and arise from honey-combed (sponge-like *Morchella*) or bell-shaped caps (*verpa*) terminating a stalk. The hymenium may be smooth or becomes grossly folded by unequal growth of the hymenial surface which is thrown into folds. The asci dehisce by the separation of a lid from the top. The ascospores are multinucleate. The sex organs are completely lacking. Copulation occurs somatogamously or even autogamously. The two important genera of this family are *Disciotis* and *Morchella* commonly called true morels. *Morchella* which is a highly prized edible fungus is taken here as a type.

MORCHELLA

The genus includes a number of species. The important ones are *Morchella conica, M. elata, M. esculenta, M. deliciosa, M. hybrida* and *M. crassipes*. All the species grow as **saprophytes**.

Morchella conica

Morchella hybrida

Morchella crassipes

Morchella elata

Morchella esculenta

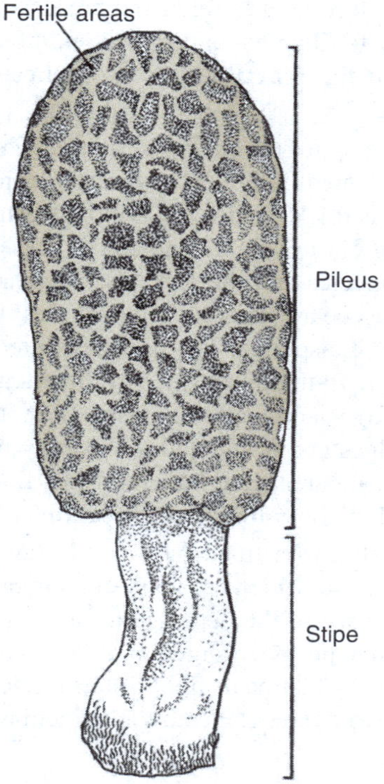

Fig. 12.1. *Morchella esculenta* (morel). A network of white areas on the pileus represents the sterile ridges and the honey comb-like shaded ones are the depressions or pits which are lined by hymenium.

There is a suggestion that some species of *Morchella* are mycorrhizas but there is no proof to substantiate this. The commonest and the best known is *M. esculenta* (Fig. 12.1). It is also known as the *common morel* or *sponge mushroom* as its ridged, reticulate fruiting body looks like a mushroom with a sponge-like cap. The morels in reality are the ascocarps of the fungus whose mycelium is underground and thus not seen. In India the edible morel which is considered a table delicacy, is known as *Guchhi*.

Occurrence. *Morchella* mostly grows in soil rich in organic matter usually in the dense, deciduous forest litter and fruits in the spring. Occasionally it is seen growing on the stumps and logs of decaying wood. On the ground it grows in humus soil and among decaying leaves taking nourishment from the dead organic materials. The epigeal fructifications (ascocarps) often appear in clusters in nature during spring in the cool areas of the temperate zone when soil temperature is near about 10ºC. In India, the edible morel is found in Kashmir.

Economic Importance. It is one of the delicious, edible fungi, which are tasteful articles in human diet. The edible part is the fructification. The fructifications are cooked with rice and vegetables and eaten all over India particularly in the Punjab. It is a more common dish with the vegetarians and is considered as good as meat or fish. Unfortunately no method has yet been found to grow morels commercially.

Mycelium (Fig. 12.3C). It is inconspicuous but extensive and subterranean. It grows a few inches deep in the soil and thus hidden from view. The mycelium consists of a mass of loosely interwoven branched hyphae. The hyphae ramify through the interstices of the soil nourishing themselves on the organic materials. They are septate. Each segment or cell contains several nuclei. Under adverse circumstances the mycelia produce sclerotia.

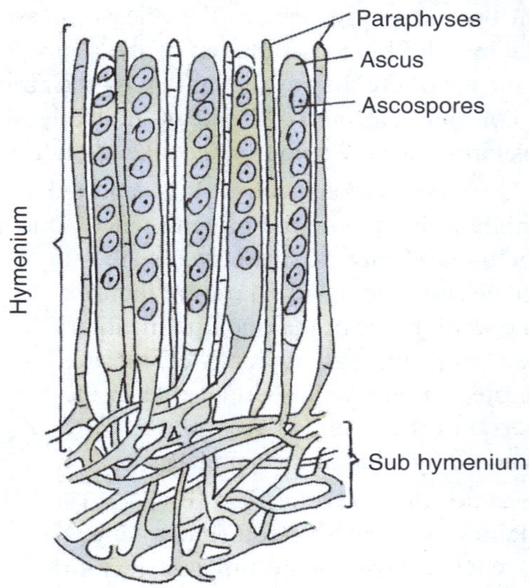

Fig. 12.2. *Morchella sp.* Section through depression of the pileus showing paraphyses and asci.

Development of Ascocarp. It begins as a dense knot of hyphae. Under favourable conditions of food and moisture, the mycelial hyphae grow rapidly and branch repeatedly. They get interwoven to form dense, compact masses. These masses of hyphae are called the **hyphal knots**. They are formed just beneath the surface of the soil. Each hyphal knot in the presence of abundant food and moisture (after rain) develops into an aerial stalked fructification called an **ascocarp**. It is cup-shaped only when young. During further growth the hymenium becomes convex. The asci thus face outwards. The smooth hymenium later becomes complicated by unequal growth of the hymenial surface which is thrown into folds and presents a honey-combed or sponge-like surface. This device increases the fertile area and thus the number of asci. Since the stalk and the terminal honey-combed cap of the **fructification** are delicate structures the morels grow best under moist conditions and shade.

Ascocarp (Fig. 12.1). The fructification or the ascocarp is a spore producing structure. The mature ascocarp consists of a stalk-like portion, the **stipe** surmounted by a hollow, conical, cap-like object, the pileus (Fig. 12.1). The stipe is cream coloured. It is thick, fleshy and hollow with a waxy surface. The fructification ranges from 2.5 cm to 12 or 14 cm in height. The colour varies from a dirty greyish white to a dark brown depending on the species and the age. The terminal conical cap or the **pileus** constitutes the fertile portion of the ascocarp. It is hollow and fleshy. The honey combed surface of the pileus gives it the appearance of a sponge. The pileus is pitted, sometimes ridged and is about the same length as the stipe. When young the surface of the pileus is quite smooth. Towards maturity there appears a network of ridges and depressions (pits) on its surface due to unequal growth of the hymenial surface. The ridges are sterile and are usually of the same colour as the stalk. The depressions or pits are the fertile areas and have a brown tint. The ridges unite the fertile areas which form an irregularly distributed **hymenium** or **thecium**.

A thin section (Fig. 12.2) through the depression reveals that it is lined by a fertile layer called the **hymenium** or **thecium**. The latter consists of numerous **asci** which are long, cylindrical, operculate cells. They stand side by side perpendicular to the surface of the depression forming a palisade-like layer. Each ascus usually contains eight hyaline, **ascospores** arranged in a uniseriate manner. The asci are positively phototrophic. They bend towards the source of light. Intermingled with the asci are elongated, sterile slender hyphae called the **paraphyses**. Beneath the hymenium is the **sub-hymenium** or **hypothecium**.

A cross-section through the stipe reveals that it is made up of pseudoparenchymatous tissue. This tissue in the region of the depression of the pileus forms the hypothecium.

Dehiscence of Asci. The tips of mature asci project above the level of the hymenium and turn towards light being positively phototrophic. The residual cytoplasm within each ascus changes into sugars of high osmotic value. The latter absorb water. As a result a considerable amount of pressure is set up in the asci. The pressure set up within the asci aided by the turgor pressure set up in the paraphyses causes the release of ascospores through a preformed apical pore which opens by a lid hinged at the top of the ascus. The internal force projects the ascospores high up into the air where they are caught up by the air currents. The explosions of the asci are successive and not simultaneous. The pileus gradually shrinks as it dries up. Eventually it withers. The ascospores are large colourless, oval and multinucleate when mature (Korf, 1973).

Dispersal of Ascospores. It is reported that the catabolic activities in the ripe fruiting body are high during the period of dehiscence. This is indicated by the marked rise in the temperature of the tissue. Conventional currents are set up in the surrounding air. These are induced by the warm tissue of the ripe fructification. These conventional currents of air are suggested to play an important role in the dispersal of the liberated ascospores. They are carried by wind to distant places.

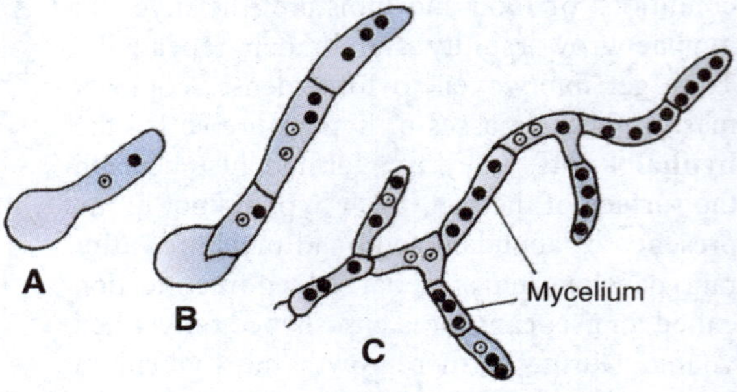

Fig. 12.3. *Morchella sp.* (A-C). Different stages in germination of ascospore.

Ascospore germination (Fig. 12.3). On falling on a suitable soil (moist humus soil) each ascospore germinates (A) to produce a new mycelium soon after release (Harvey *et al*, 1978). The ascospores that happen to fall on unsuitable soil perish. Ascospores do not remain viable after one year near the soil surface (Schmidt, 1983). According to the latter, the ascospores germinate and develop extensive mycelium at 15°C under highly nutritious non-competitive conditions. However, the ascospore germinates at a temperature as low as 2°C but rate of germination and hyphal growth is slower at lower temperature. Above 15°C germination is inhibited.

Sexual Reproduction. The sex organs which constitute the accessory parts of sexual process are completely suppressed in *Morchella*. Consequently the sexual process is extremely simplified. It involves two distinct processes, namely, *Plasmogamy* and *Karyogamy*. The latter is immediately followed by **meiosis**.

(*a*) Plasmogamy. It consists in the union of cytoplasmic contents of two cells without nuclear fusion which is delayed. This results in a sequence of binucleate cell generations constituting the **dikaryophase**. Geis (1940) reported that plasmogamy occurs at a very late stage during the development of the ascocarp in the subhymenial layer. It takes place just before the formation of asci. Plasmogamy in *Morchella* takes place by the following two methods :–

(*i*) *Somatogamous Copulation* (Fig. 12.4). Two vegetative hyphae of the subhymenium region of the pileus come in contact. The intervening walls between the copulating cells dissolve at the point of contact (A). The two multinucleate protoplasts intermingle in the fusion cell (B). Two functional nuclei, one from each copulating cell, form a **dikaryon**. All other nuclei in the fusion cell disappear. Later ascogenous hyphae arise from the fusion cell containing the dikaryon (C). Each young ascogenous hypha receives a pair of nuclei. These are the derivatives of the parent dikaryon. The ascogenous hyphae afterwards become septate. The terminal cells of these hyphae

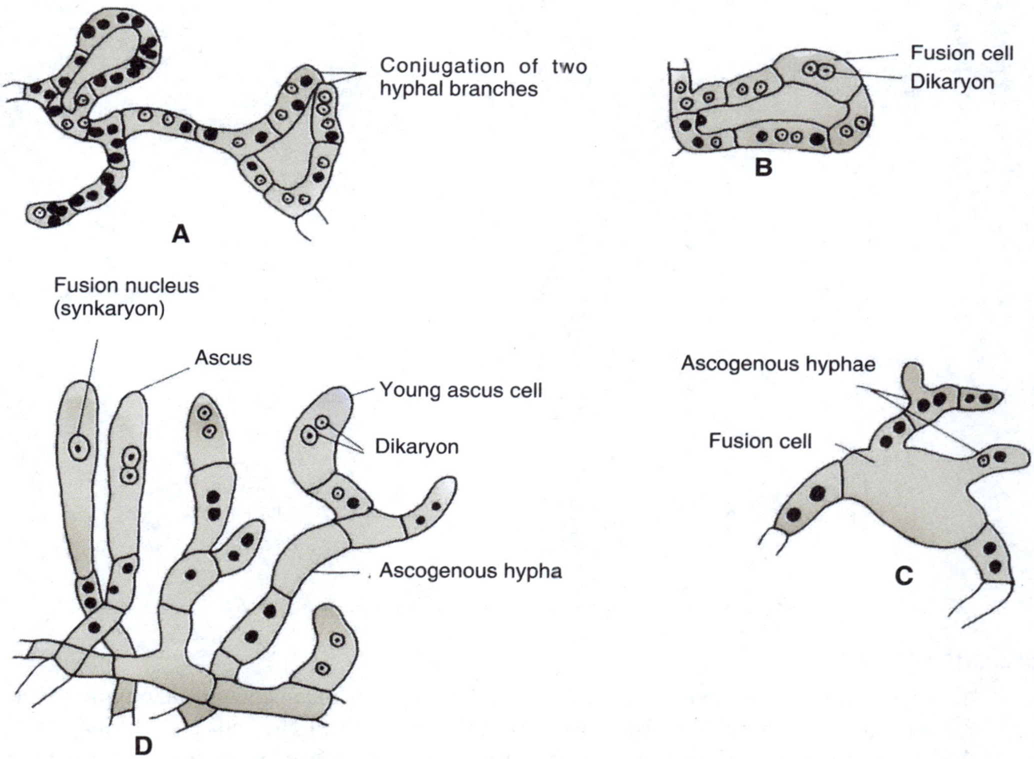

Fig. 12.4 (A-D). Sexual Reproduction in *Morchella*. A, Somatogamous copulation; B, Fusion cell with a dikaryon; C, Fusion cell giving rise to ascogenous hyphae; D, Ascogenous hyphae with developing asci.

function as **ascus mother cells** (D). They are binucleate. The nuclei are the derivatives of the parent dikaryon.

(*ii*) *Autogamous Pairing*. It has been reported in *M. elata*. In this case any vegetative cell of the sub-hymenium (hypothecium) may establish a dikaryon. It is accomplished by pairing of two of its own vegetative nuclei. It is called **autogamous pairing**. The remaining nuclei in the cell degenerate. The cells with dikaryons (dikaryotic cells) develop ascogenous hyphae as in somatogamous copulation.

(*b*) Karyogamy. The two nuclei in the ascus mother cell fuse. The fusion cell with a diploid nucleus is called the **young ascus**. It represents the transitory **diplophase** in the life cycle. The asci in *Morchella* thus develop from the tip cells of ascogenous hyphae without the formation of croziers.

(*c*) Meiosis. The young ascus cell elongates. The synkaryon in the ascus undergoes two successive divisions. These constitute **meiosis**. During meiosis the number of chromosomes in the resultant four daughter nuclei is reduced to half the diploid number. The four haploid nuclei undergo the third division. It is **mitotic**. In this way eight haploid daughter nuclei are formed in each ascus cell. Each of these is fashioned into an ascospore. With the formation of ascospores the haploid or gametophyte phase is again initiated in the life cycle of *Morchella*.

It is speculated that the overwintering stage in the life cycle of *Morchella* is the *sclerotium*.

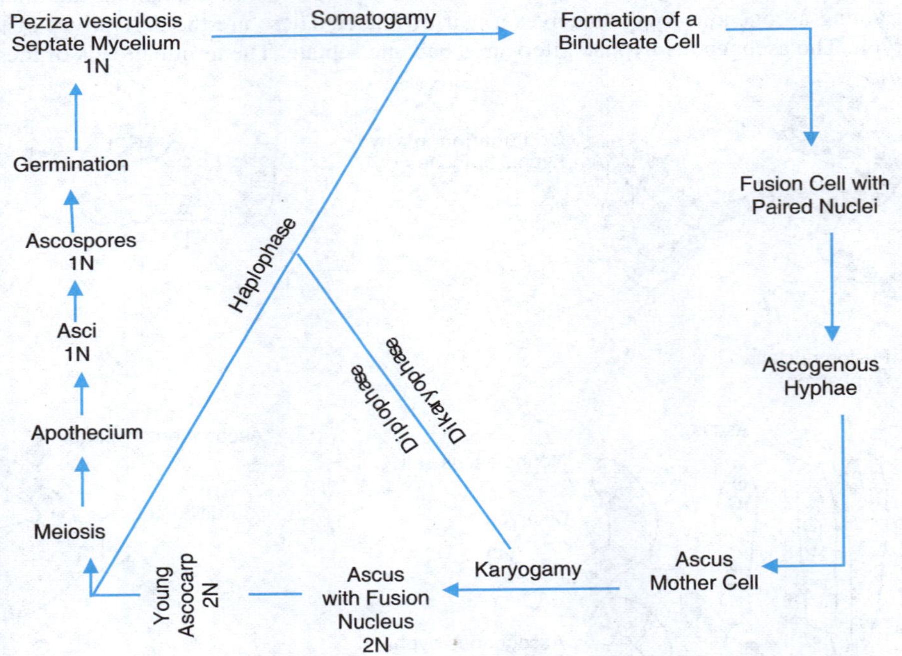

Fig. 12.5. Graphic representation of the life cycle of *Morchella*.

Salient Features

1. *Morchella* is a saprobic discomycete. The mycelium is inconspicuous and underground. It consists of a mass of loosely interwoven septate hyphae. The cells are multinucleate.
2. The so-called morels (guchhi) are the aerial fructifications or ascocarps of the fungus.
3. The mature ascocarp varies in height from 25 to 100 or 125 mm and is differentiated into a thick fleshy, hollow stipe surmounted by a conical cap called the **pileus**.

4. The pileus has a pitted surface. The pits or the depressions are lined by a fertile layer consisting of the asci and paraphyses.
5. Asexual reproduction by spores (conidia) is lacking.
6. No sex organs are developed. The sexual process is extremely simplified. It consists of plasmogamy and karyogamy. The latter is immediately followed by meiosis.
7. Plasmogamy in some species is accomplished by **somatogamous copulation** between the cells of two somatic hyphae or in some by **autogamous pairing**.
8. Ascogenous hyphae arise from the fusion cell containing the dikaryon and become septate.
9. The asci are developed from the terminal binucleate cells of the ascogenous hyphae which directly function as ascus mother cells. The two nuclei in the ascus mother cell are the derivatives of the parent dikaryon.
10. No hook formation takes place in *Morchella*.
11. Karyogamy takes place in the ascus mother cell. The young ascus contains the synkaryon.
12. The young ascus grows in size and the synkaryon undergoes meiosis. The haploid nuclei divide mitotically to form 8 nuclei which are subsequently organised into meiospores known as the **ascospores**. They are multinucleate. They are multinucleate when mature.
13. Each ascus dehisces by the separation of its apex as a lid. The ascospores are thus liberated.
14. Each ascospore under suitable conditions germinates by a germ tube to form a new mycelium.

Taxonomic position

Alexopoulos (1962)

Division	:	**Mycota**
Subdivision	:	**Eumycotina**
Class	:	**Ascomycetes**
Sub-class	:	**Euascomycetidae**
Series	:	**Hymenoascomycetes**
Sub-series	:	**Discomycetes**
Order	:	**Pezizales**
Family	:	**Morchellaceae**
Genus	:	*Morchella*
Species	:	*esculenta*

Alexopoulos and Mims (1979)

Superkingdom	:	**Eukaryonta**
Kingdom	:	**Mycetae**
Division	:	**Amastigomycota**
Subdivision	:	**Ascomycotina**
Class	:	**Ascomycetes**
Sub-Class	:	**Hymenoascomycetidae**
Series	:	**Discomycetes**
Order	:	**Pezizales**
Family	:	**Morchellaceae**
Genus	:	*Morchella*
Species	:	*esculenta*

FAMILY : PEZIZACEAE

General. It includes saprophytic species which commonly grow on rotten wood or bark, among mosses and manure heaps. Some exclusively grow on dung and are called **coprophilous**. Many grow on the ground rich in organic matter. The conidial stage is absent in most of them. The cup or saucer-shaped apothecia are coloured bright to dark brown, vary in size from the minute cups ranging from 0.4 mm. across to 125 mm. in diameter. They are invariably **epigean** and may be stalked or sessile. This family includes a number of genera. The best known among them are *Pyronema, Peziza, Coprobia* (*syn. Humaria*) and *Ascobolus* and these are discussed here.

Pyronema confluens

Peziza

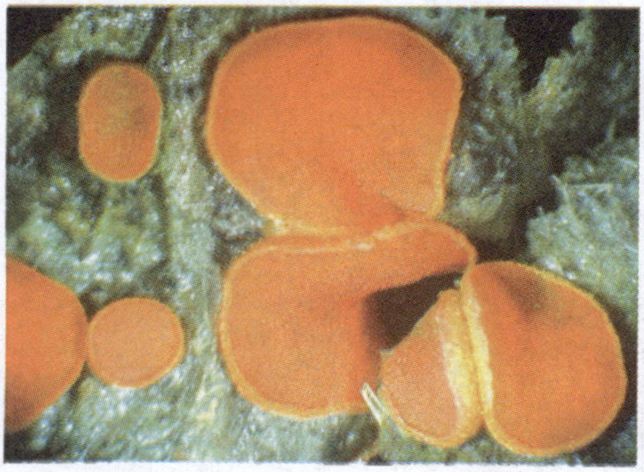

Coprobia

Ascobolus

PYRONEMA

It is a saprophyte which is usually found growing on decaying leaves or on burnt ground. Gilbert (1913) reported it growing in green houses and seed beds on soil sterilized by steam. The genus includes three species. Of these *Pyronema confluens* (*P. omphalodes*) is the commonest and widely distributed species. It grows commonly on half charred wood and soil that has been burnt over. The other commonly known species is *P. domesticum*.

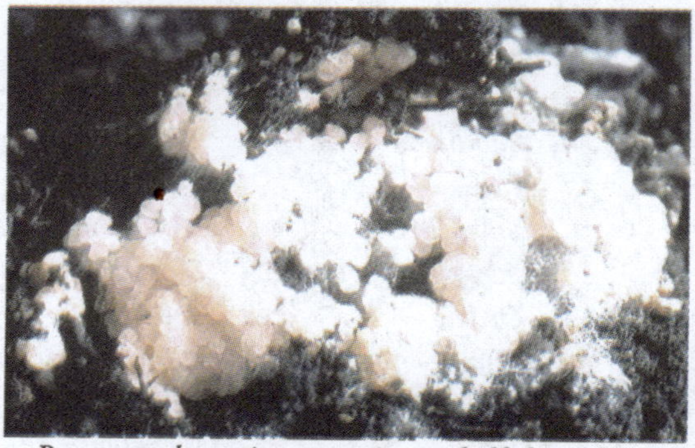

Pyronema domesticum growing on half charred wood

Mycelium. It is well developed, filamentous and superficial. The hyphae are hyaline and profusely branched. They ramify over the surface of the substratum to form a cottony layer. The glistening white mycelium forms frost-like masses on the substratum. The septate hyphae consist of short, multinucleate cells. Claussen (1912) recorded 6 to 12 nuclei in each cell.

Asexual Reproduction. Asexual reproduction by means of asexual spores is usually lacking. L. R. Tulasne and C. Tulasne (1865) however reported that asexual reproduction may sometimes take place by the formation of **oidia** which are produced in chains on certain erect hyphae of the mycelium.

Sexual Reproduction (Fig. 12.6). It is **oogamous** and the species is **homothallic**. The male sex organ is called an **antheridium** and the female **ascogonium**. Both are multinucleate. The hyphae bearing sex organs occur closer together on the mycelium and are short and erect (A). They fork usually once or twice or even more times. Each branch is 2 to 4 cells in height. The end cell of one of the paired branches develops into an antheridium and the other into an ascogonium. *P. domesticum* requires exposure to light to initiate gametangia formation and for the continued maturation of apothecia under normal cultural conditions.

(*a*) *Antheridium*. The male branch consists of two to four cells. The terminal cell of this branch becomes enlarged into a clavate (club-shaped) structure called the **antheridium** (A). It is multinucleate. The male nuclei divide and redivide to form a hundred or more nuclei. The lower two or three cells of the branch constitute the **stalk**. All the cells are multinucleate. The antheridium is closely appressed to the ascogonium (A).

(*b*) *Ascogonium*. It is developed on a two to four-celled sister branch of the dichotomy (A_3). The terminal cell of the female branch swells up to form a rounded or subglobose **ascogonium**. The lower cells of the branch constitute the **stalk**. The ascogonium is multinucleate. It contains about 100 or more nuclei. From the upper end of the sp`herical ascogonium arises a short, slender outgrowth. It elongates to form a slender, tubular, curved structure called the **trichogyne**. The trichogyne is a unicellular but multinucleate structure. It is separated from the ascogonium by a cross wall formed at its base.

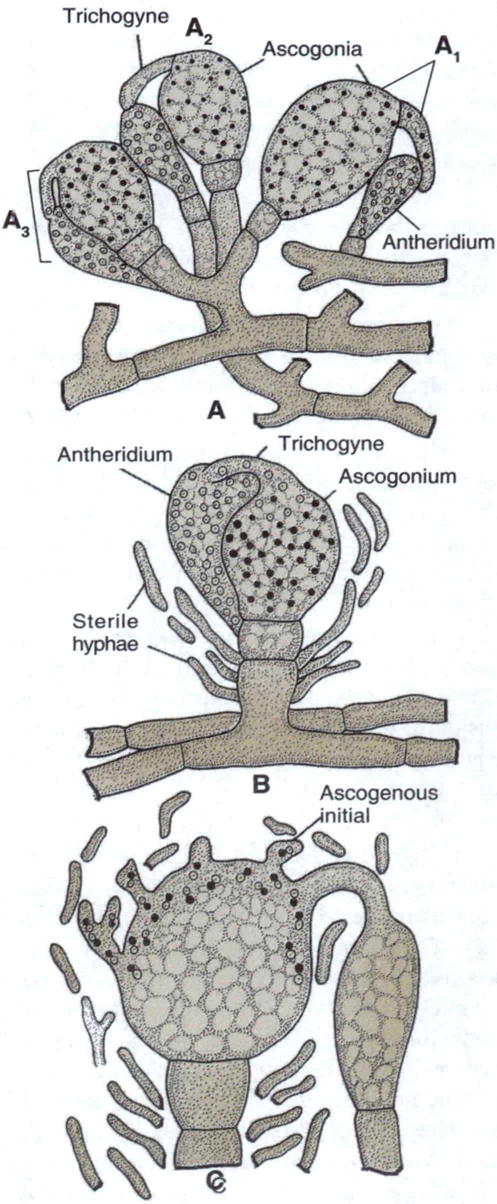

Fig. 12.6 (A-C). *Pyronema confluens*. Stages in sexual reproduction. A_{1-3}, a cluster of sex organs; A_1, ascogonium with the trichogyne in contact with the antheridium. A_2, tip of trichogyne just fused with the antheridium and the nuclei in it have degenerated; A_3, Septum between the trichogyne and antheridium has disappeared and the male nuclei have just started to migrate into the trichogyne; B, a later stage with the male nuclei migrating into the trichogyne; C, a still later stage. (Based on Claussen)

(*b*) *Plasmogamy*. The trichogyne grows towards the antheridium and makes contact with its apex (A_2) or the flank (A_1). The intervening walls at the point of contact dissolve. A pore is thus formed (A_3). The nuclei of the trichogyne disappear. A majority of male nuclei with some cytoplasm pass into the trichogyne through the pore. They pass down the trichogyne and finally enter the ascogonium through the central pore in the septum at its base. Thereafter, the pore becomes blocked by some granule.

According to some, the male and the female nuclei in the ascogonium become arranged in pairs to form the **dikaryons** (C). They do not fuse. Fusion between them is delayed. It takes place only in the ascus mother cell. The young ascus comes to possess the fusion nucleus or the synkaryon. It is **diploid**.

Gwynne Vaughan, Williamson (1931) and others (Harper, 1900; Tandy, 1927) hold that the male and the female nuclei in the ascogonium fuse to form diploid nuclei. These diploid nuclei or their derivatives pass into the ascogenous hyphae. Another fusion between the two diploid nuclei takes place in the ascus mother cell. The young ascus thus comes to possess a **tetraploid nucleus**.

The third view is advanced by Dangeard (1907) and his followers (Claussen, 1912; Moreau and Moreau, 1927; Hirsch, 1950). They hold that no communication between the trichogyne and the antheridium is established. The male nuclei do not pass into the trichogyne. The antheridium is functionless. The male nuclei degenerate and disappear in the male organ. The female nuclei in the ascogonium arrange themselves in pairs to form the dikaryons.

The sexual process in *Pyronema confluens* thus needs a thorough and a critical study to solve the riddle. Anyway the consensus of opinion is against fusion of male and female nuclei in the ascogonium. They simply arrange themselves in pairs to form dikaryons.

Post Plasmogamy Changes (*a*) Development of ascogenous hyphae (Fig. 12.6 C). Following plasmogamy a number of usually 15-20 tubular outgrowths arise from the sides of the ascogonium (C). These outgrowths develop into **ascogenous hyphae**. Into each ascogenous hypha migrates a pair of nuclei (dikaryon) from the ascogonium. During their further development the ascogenous hyphae become multicellular by the appearance of cross walls but the nuclei in the cells remain in pairs (Fig. 12.7 B). The cells of the ascogenous hyphae particularly those towards the tips are binucleate (Fig. 12.7 B). The ascogenous hyphae may fork at their tips.

(*b*) Formation of Asci (Fig. 12.8). The binucleate terminal cells of the ascogenous hyphae or their branches become crozier-like (A). The two nuclei in the crozier divide to form four daughter nuclei (B). Later cell walls appear in the crozier so as to form a terminal uninucleate hook cell, a sub-terminal binucleate **arch** cell, and an anti-penultimate uninucleate stalk cell (C). The binucleate arch cell functions as the **ascus mother cell**. The two nuclei in

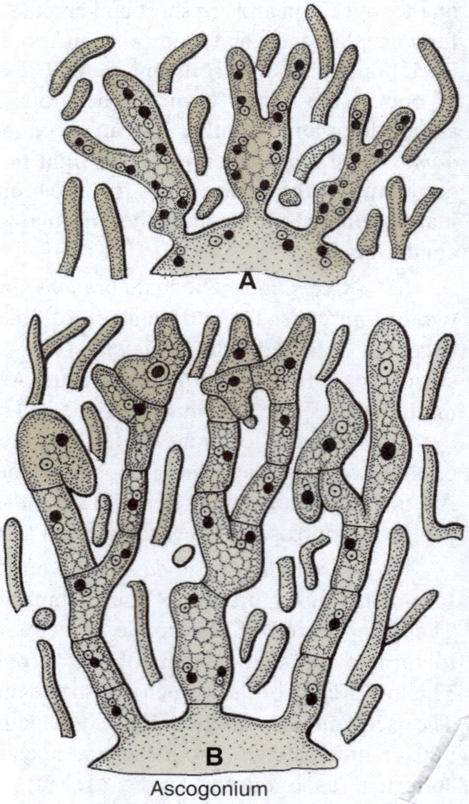

Fig. 12.7 (A-B). *Pyronema confluens* showing development of ascogenous hyphae (A) and initiation of ascus development (B) (After Claussen).

namely (*i*) inner *endospore*, (*ii*) middle *epispore* and (*iii*) the outer *perispore*. Concurrently with the development of the ascospore wall, an apical ring differentiates around the ascus just below the tip. The small portion of the ascus above the apical ring constitutes the **lid** or **operculum**. The apical ring around the ascus serves as a weak area that eventually ruptures to allow the escape of ascospores.

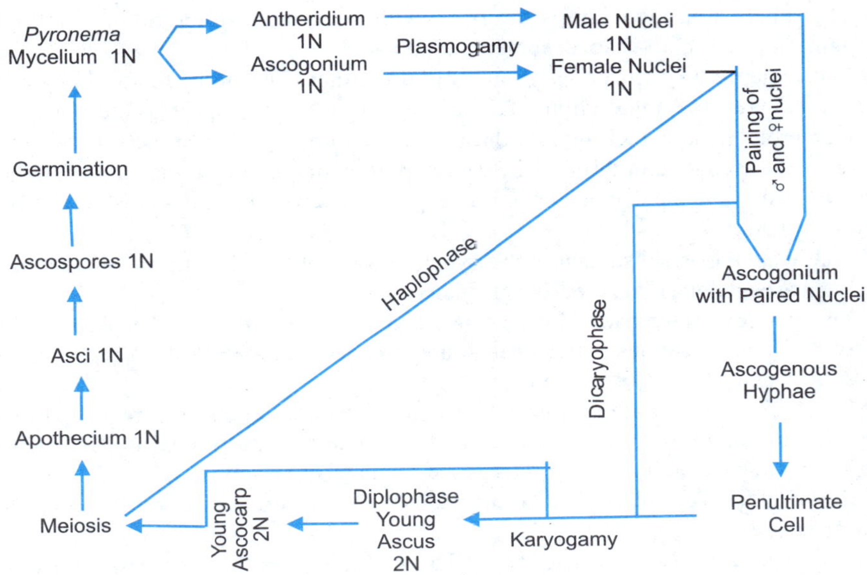

Fig. 12.11 Graphic representation of the life cycle of *Pyronema*.

Dehiscence of Asci. Hung (1977) reported that just prior to the liberation of mature ascospores the perispore membrane starts blebbing. The vesiculation process continues until the perispore completely disappears. Thus at the time of liberation the ascospore wall consists only of two layers—epispore and endospore. At this stage the ascus epiplasm degenerates. In the vacuole thus formed ascospores are arranged either in a uniseriate manner (*P. confluens*) or suspended in the upper portion of the vacuole (*P. domesticum*). The glycogen released from the disintegrated epiplasm changes into sugars of high osmotic value. This transformation results in the absorption of water and setting up of high pressure in the ascus. According to Ingold, the walls of the mature asci are highly distended under the hydrostatic pressure thus set up. The apical ring breaks, with the lid either dispersed or hanging backwards at the top providing an open pore for ascospore liberation. With the release of pressure the ascus wall contracts both longitudinally and transversely ejecting ascospores violently through the apical pore into the air. Hung (1977) reported that in *P. domesticum* all the ascospores in the ascus are ejected at the same time by a single spurt of force. In *P. confluens* the ascospores escape one by one. There is a minute fraction of a second between the escape of successive ascospores and thus they appear to be discharged simultaneously. Under suitable conditions each liberated ascospore germinates to form a new haploid mycelium of *Pyronema*.

Salient Features

1. *Pyronema* is a saprobic discomycete usually found inhabiting half charred wood or burnt places.
2. The mycelium is well developed and ramifies over the surface of the substratum forming

a cottony white, glistening mass of a loose network of septate, branched hyphae. The cells are multinucleate.

3. Asexual reproduction by means of spores is generally lacking.
4. Sexual reproduction is **oogamous** and *Pyronema confluens* is **homothallic**.
5. The two kinds of sex organs, **antheridia** and **ascogonia** are developed terminally on separate, sister branches of the dichotomy and are multinucleate.
6. The spherical ascogonium is furnished with a receptive structure, the **trichogyne**. The antheridium is club-shaped, sometime globose.
7. Plasmogamy takes place by **gametangial contact**. The tip of the trichogyne comes in contact with the antheridium. The intervening walls disappear at the point of contact. Numerous male nuclei migrate through the common pore into the trichogyne and later enter the ascogonium through the central pore in the septum at the base of trichogyne.
8. The male and the female nuclei arrange themselves in pairs—the **dikaryons** in the ascogonium.
9. Following the establishment of dikaryons, ascogenous hyphae arise from the ascogonium. Each ascogenous hypha receives a dikaryon.
10. The ascogenous hyphae become septate and may fork at the tips. The terminal binucleate cell of the ascogenous hypha or its branches become **crozier like**. The two nuclei in the crozier undergo conjugate division.
11. Septa are formed in the 4-nucleate **crozier cell** so as to cut of a **terminal** uninucleate cell, a **subterminal** binucleate **arch cell** and an **anti-penultimate** uninucleate cell.
12. The binucleate arch cell functions as the **ascus mother cell**.
13. Karyogamy takes place. The ascus mother cell with a synkaryon is called the **young ascus cell**.
14. The ascus cell elongates and the synkaryon undergoes three successive divisions. The first two constitute **meiosis**.
15. The resultant 8 haploid nuclei are eventually organised into 8 **ascospores** in the mature ascus.
16. Meanwhile sterile hyphae arise from the mycelium to form the sub-hymenium and also from the stalk of the ascogonium to form the sheath of the open ascoscarp called the **apothecium**. The latter lacks a rim or peridium.
17. The ascospores are shot off violently with a hissing sound by the separation of an apical lid.
18. Each liberated ascospore germinates under suitable conditions to form a new haploid mycelium.

Taxonomic Position. In spite of the fact that the study of *Pyronema* has provided valuable information about ascus development and ascosporogenesis its taxonomic position is still controversial. The older mycologists place it in the family Pezizaceae as under :–

Alexopoulos (1962)

Division	:	**Mycota**
Subdivision	:	**Eumycotina**
Class	:	**Ascomycetes**
Sub-class	:	**Euascomycetidae**
Series	:	**Hymenoascomycetes**
Sub-series	:	**Discomycetes**
order	:	**Pezizales**
Family	:	**Pezizaceae**
Genus	:	*Pyronema*
Species	:	*confluens*

Alexopoulos and Mims (1979)

Superkingdom	:	**Eukaryonta**
Kingdom	:	**Mycetae**
Division	:	**Amastigomycota**
Subdivision	:	**Ascomycotina**
Class	:	**Ascomycetes**
Subclass	:	**Hymenoascomycetidae**
Series	:	**Discomycetes**
Order	:	**Pezizales**
Family	:	**Pezizaceae**
Genus	:	*Pyronema*
Species	:	*confluens*

Dennis (1977) places *Pyronema* in the family Ascobolaceae. Kimbrough (1970) and Korf (1973) place *Pyronema* in the family Pyronemataceae.

PEZIZA

Peziza vesiculosa cups in nature

It is saprophytic, very often coprophilous in habit. It frequently grows on dung, rotting wood or rich humus of forest soil. Some species of this genus grow on burnt or charred wood. The species commonly found is *Peziza vesiculosa*. It is one of the larger cup fungi and is cosmopolitan in its distribution.

Mycelium. It is well developed, frequently perennial and consists of a dense network of hyphae. The hyphae are branched and septate. The cells are uninucleate. The hyphae are hidden from view as they ramify within the substratum. They from a complex system which extracts nourishment from the substratum. The fruiting bodies are above ground.

Asexual Reproduction. It takes place by the formation of conidia and chlamydospores. The conidia are exogenously formed spores. They are abstricted from the tips of conidiophores. Each conidium germinates to form a new mycelium.

The chlamydospores are thick-walled resting cells. They are intercalary in position. They may be formed singly or in series within the cells of the hyphae. Under suitable conditions each chlamy-dospore germinates and gives rise to a new mycelium.

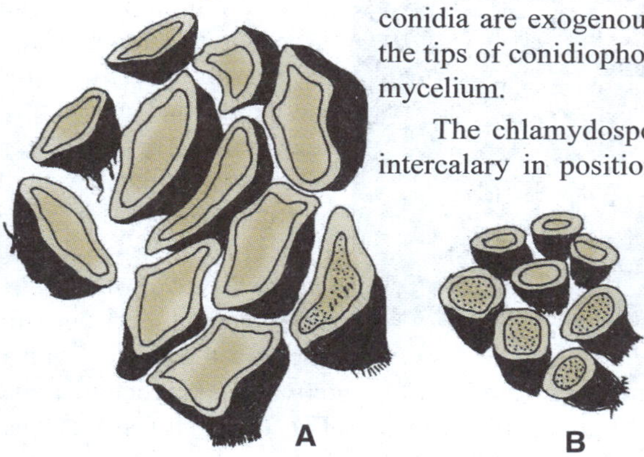

Fig. 12.12 (A-B). Apothecia of Pezizaceae. A, Apothecia of *Peziza*. B, Apothecia of *Humaria*.

Sexual Reproduction. The sexual apparatus is wholly lacking in *Peziza vesiculosa*. This does not prevent the development of a fructification or the **apothecium** (Fig. 12.12A) which is aerial and relatively a short-lived structure. The sexual process does take place. It is extremely simplified and consists in the association of two purely vegetative nuclei in a pair. The adult mycelium consists of a tangled mass of hyphae. Certain vegetative cells in the centre of the tangled hyphal mass have been seen to possess nuclei which become associated in pairs. These pairs of nuclei are called the **dikaryons**. The dikaryotic condition is brought about either by **autogamous pairing** or by **somatogamous copulation** between the vegetative cells of the adjacent hyphae of the tangled hyphal mass. The cells with the dikaryons give rise to the ascogenous hyphae which become multicellular by cross walls. Their cells are binucleate. The terminal binucleate cell of each ascogenous-hypha functions as an **ascus mother cell**. Formation of croziers in the development of asci has not been reported in *P. vesiculosa*. The *ascogenous hyphae* and dikaryotic cells from which they are developed together with the ascus mother cells represent the dikaryophase in the life cycle of Peziza. The two nuclei of the ascus mother cell fuse to form the synkaryon. The young ascus with the synkayon represents the transitory **diplophase** (Fig. 12.14). The synkaryon undergoes three successive divisions. Of these the first and the second constitute meiosis. This results in the formation of eight haploid nuclei which become organised into ascospores. The mature ascus is an elongated, cylindrical cell (Fig. 12.13B). The ascus wall is lined by a thin layer of cytoplasm (epiplasm) which encloses a central vacuole filled with sap. In the vacuole lie the oval ascospores. The erect asci lie side by side lining the cavity of the cup-shaped apothecium (Fig. 12.13A). The asci near the margin of the cup bend towards the source of light being pisitively phototropic. Interspersed between the asci are the sterile hyphae called **paraphyses**. The rest of the apothecium consists of densely interwoven, branched hyphae forming a pseudoparenchymatous tissue which supports the hymenium (Fig. 12.13A). The apothecia (Fig. 12.12A) are sessile or shortly stalked cup-shaped structures regular in form and large in size varying from 2 cm. to several inches in diameter. In *P. vesiculosa* the apothecium is of pale fawn colour but *P. aur antia* has brilliant orange apothecium.

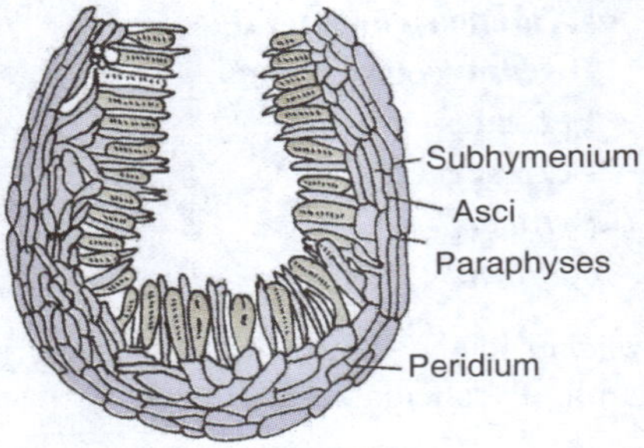

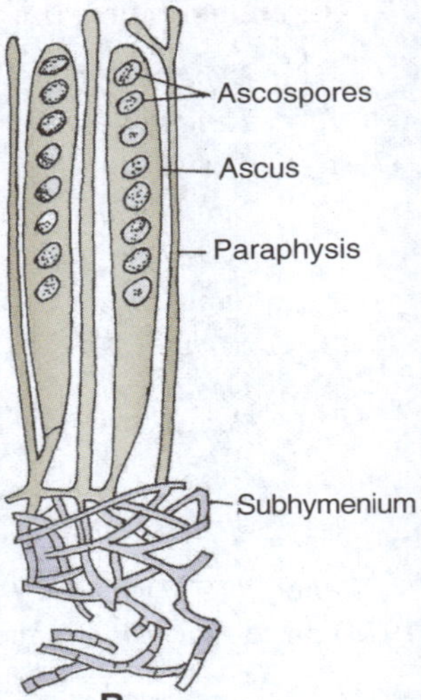

Fig. 12.13 (A–B). *Peziza.* A, V. S. entire fructification (apothecium); B, Portion of the same to show details. (Diagrammatic).

Peziza aurantia apothecia in nature

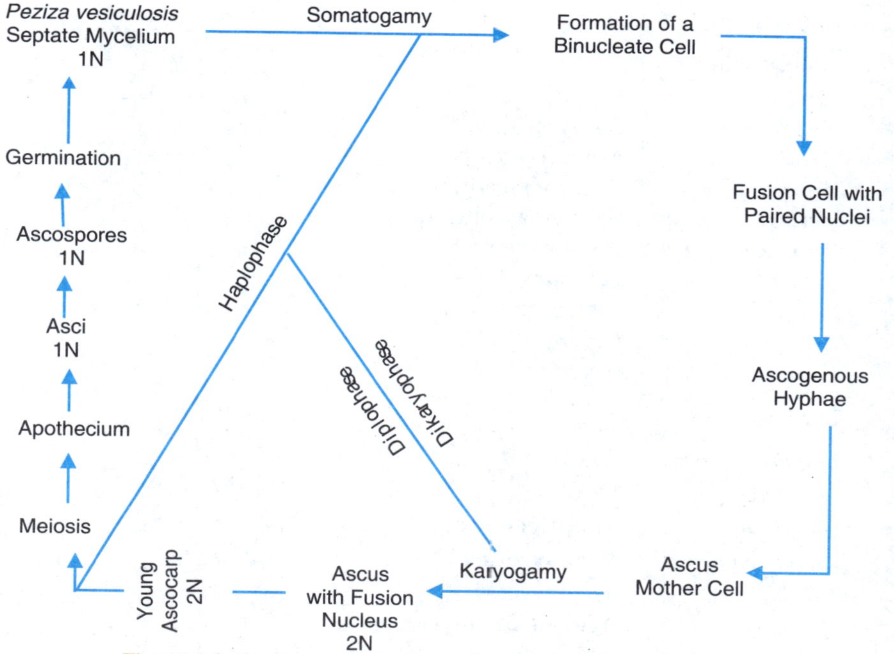

Fig. 12.14. Graphic representation of the life cycle of *Peziza*.

Salient Features

1. *Peziza* is a saprophytic discomycete which is often coprophilous in habit. Species growing on rotting wood and rich humus are also plentiful.
2. The mycelium is inconspicuous as the branched septate hyphae constituting it ramify within the substratum. The cells are uninucleate.
3. Asexual reproduction takes place by means of non-motile spores of two kinds, namely, **conidia** and **chlamydospores**.
4. The sexual apparatus is lacking in *Peziza vesiculosa*. The sexual process is extremely simplified. It consists in the association of two vegetative nuclei to form a dikaryon by somatogamous copulation of adjacent hyphae or by autogamous pairing.
5. Ascogenous hyphae arise from the dikaryotic cells. They become septate and branched. The cells are binucleate.
6. The terminal, binucleate cells of the ascogenous hyphae or their branches function directly as ascus mother cells.
7. Formation of croziers has not been reported in *P. vesiculosa*.
8. The ascogenous hyphae, the dikaryotic cells from which they are developed and the binucleate ascus mother cells constitute the **dikaryophase** in the life cycle. It is parasitic.
9. The two nuclei in the ascus mother cell fuse to form the synkaryon.
10. The young ascus with a synkaryon represents the transitory **diplophase**.
11. As the ascus cell advances towards maturity the synkaryon undergoes three successive divisions, the first two constitute **meiosis**.
12. The resultant haploid nuclei are organised into 8 ascospores in the mature ascus.

The ascocarps in *Peziza* are large, fleshy, red to orange, cup-shaped, epigean, sessile or subsessile apothecia.

Taxonomic Position

Alexopoulos (1962)

Division	:	**Mycota**
Subdivision	:	**Eumycotina**
Class	:	**Ascomycetes**
Sub-class	:	**Euascomycetidae**
Series	:	**Hymenoascomycetes**
Sub-series	:	**Discomycetes**
Order	:	**Pezizales**
Family	:	**Pezizaceae**
Genus	:	*Peziza*
Species	:	*vesiculosa*

Alexopoulos and Mims (1979)

Superkingdom	:	**Eukaryonta**
Kingdom	:	**Mycetae**
Division	:	**Amastigomycota**
Subdivision	:	**Ascomycotina**
Class	:	**Ascomycetes**
Subclass	:	**Hymenoascomycetidae**
Series	:	**Discomycetes**
Order	:	**Pezizales**
Family	:	**Pezizaceae**
Genus	:	*Peziza*
Species	:	*vesiculosa*

COPROBIA (SYN. HUMARIA)

General. All the species are saprophytic. They grow on cowdung or soil rich in humus. The apothecia are small just the size of a pin head. They are dull red or orange in colour. A rich culture of *Coprobia* (=*Humaria*) can be obtained in the laboratory by placing fresh cowdung cake under a bell jar. After about two weeks when the dung has not yet become dry and the first crop of black molds has died down, the tiny apothecia of *Coprobia* (*Humaria*) make their appearance. The mycelium is extensively branched. The hyphae are septate and colourless. The cells have comparatively large nuclei and granular contents. At a certain stage of maturity the reproductive phase ushers in the life cycle. Reproduction by the formation of asexual spores is totally absent.

Coprobia granulatus apothecia in nature

Sexual Reproduction (Fig. 12.15). It varies in different species of *Humaria*. In *Coprobia granulatus* only the female sex organ is developed. It is called the **ascogonium**. The antheridium is lacking. The mature ascogonium is a multinucleate, oval structure. It has no **trichogyne**. It is borne at the end of a side branch, the **archicarp** (female branch) arising from a mycelial hypha. The archicarp consists of a long multicellular stalk bearing the ascogonium at its top. The mature ascogonium is reported to contain about 600 to 1400 female nuclei. Gwynne Vaughan and Barnes have reported that young ascogonia in a single spore culture perish. They undergo no

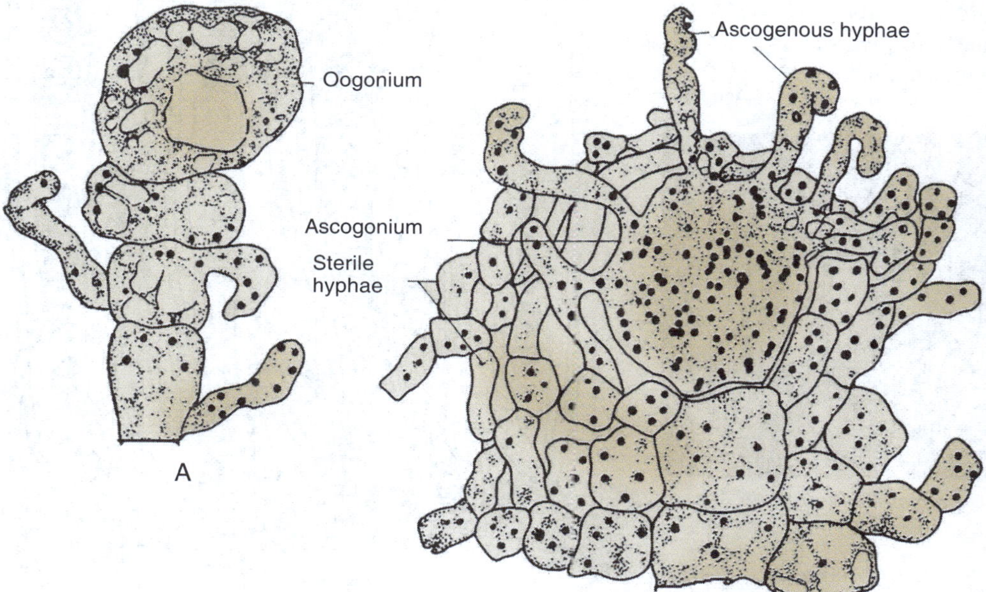

Fig. 12.15 (A-B). *Coprobia (Humaria) granulatus.* Sex organs. A, Young ascogonium; B, ascogonium bearing ascogenous hyphae.

further development. It is only when fusion between two vegetative hyphae of opposite strains of mycelia takes place that the ascogonium develops further. The nuclei resulting from the hyphal fusions migrate to the ascogonium through the septal perforations. They come to lie by the side of female nuclei in the ascogonium. These pairs of nuclei are called the **dikaryons**. They do not fuse. Some mycologists are in favour of their complete fusion in the ascogonium to form diploid nuclei. This assumption is fiercely criticised by the modern mycologists. There seems to be no significance of double fusion and double reduction. After the formation of the dikaryons the ascogonium buds out **ascogenous hyphae** (B). The whole structure now becomes surrounded by a protective sheath of vegetative hyphae.

A pair of nuclei migrates into each ascogenous hypha. These nuclei fuse in the ascus mother cell to form a synkaryon (Fig. 12.16B). The synkaryon is diploid. After three nuclear divisions of the synkaryon eight haploid nuclei are formed

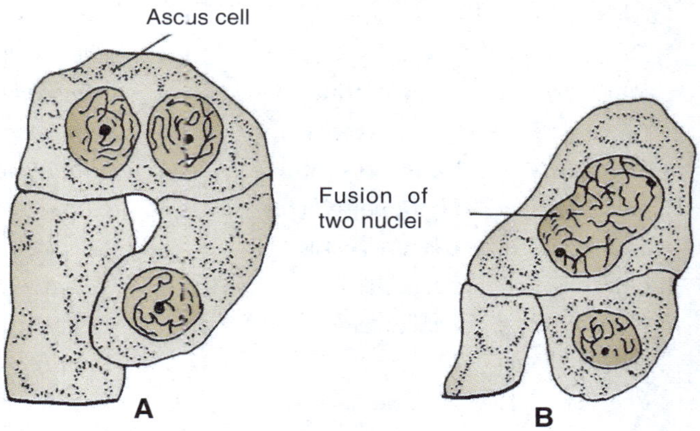

Fig. 12.16 (A-B) *Coprobia (Humaria) granulatus.* A, Formation of ascus mother cell; B, Fusion of nuclei in the young ascus.

(Fig. 12.17). These nuclei by further changes become **ascospores**. Each ascospore on falling on a suitable substratum germinates to give rise to a fresh mycelium.

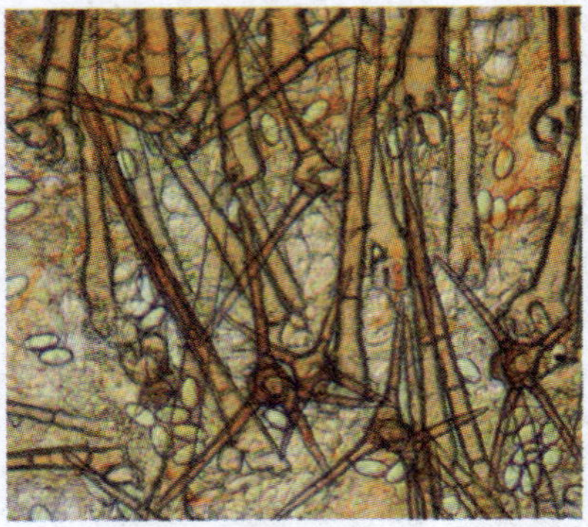

Coprobia (Humaria) rutilans

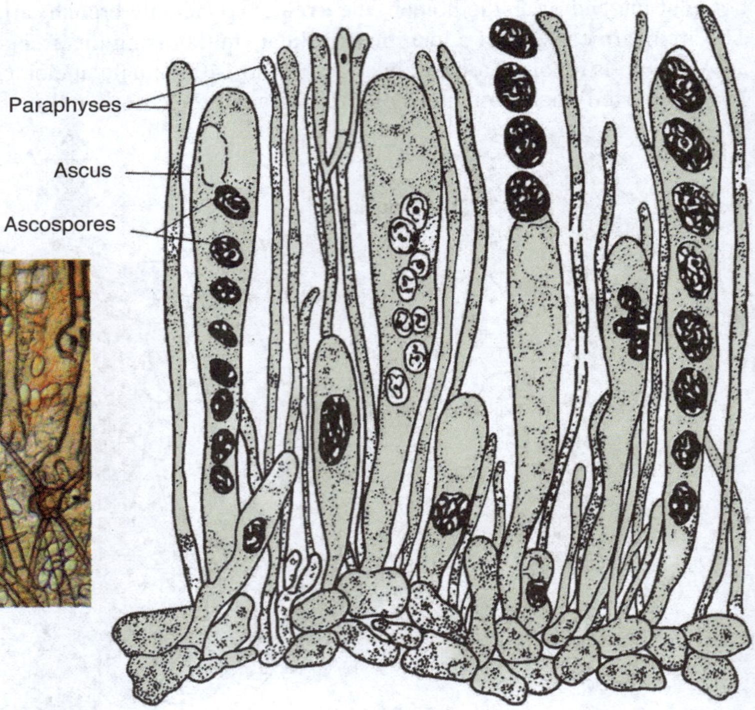

Fig. 12.17. *Coprobia (Humaria) granulatus.* A portion of the hymenium showing paraphyses and asci containing ascospores.

Coprobia (Humaria) rutilans develops no sex organs. The entire sexual apparatus is absent. The sexual process is extremely simplified. Sexual fusion takes place between any two vegetative cells of the hyphae of opposite strains. The nuclei of one of the cells migrate into other. There they come to lie in pairs with the nuclei of that cell and form **dikaryons**. The cells containing a dikaryon or **dikaryons** bud out ascogenous hyphae. The subsequent stages of development resulting in the formation of apothecia are the same as in *Humaria granulatus*.

Taxonomic Position

Alexopoulos (1962)

Division	:	**Mycota**
Subdivision	:	**Eumycotina**
Class	:	**Ascomycetes**
Subclass	:	**Euascomycetidae**
Series	:	**Hymenomycetes**
Subseries	:	**Discomycetes**
Order	:	**Pezizales**
Family	:	**Pezizaceae**
Genus	:	*Coprobia*
Species	:	*granulatus*

Alexopoulos and Mims (1979)

Superkingdom	:	**Eukaryonta**
Kingdom	:	**Mycetae**
Division	:	**Amastigomycota**
Subdivision	:	**Ascomycotina**
Class	:	**Ascomycetes**
Subclass	:	**Hymenoascomycetidae**
Series	:	**Discomycetes**
Order	:	**Pezizales**
Family	:	**Pezizaceae**
Genus	:	*Coprobia*
Species	:	*granulatus*

ASCOBOLUS

Ascobolus apothecea

Genus *Ascobolus* with 48 species is cosmopolitan. Most species of the genus are coprophilous, producing apothecia on dung of herbivores. From India about a dozen species of *Ascobolus* are reported growing on dung of herbivores. The common Indian species are *A. magnificus, A. lignata, A. denudatus* and *A. viridis.*

Mycelium. The mycelium of the fungus is well developed and ramifies in the substratum forming complex hyphal masses. The mycelium is septate with multinucleate cells and branched. The complex hyphal mass absorbs the food material from the substratum and supplies to fruiting bodies. Thus it acts as an absorptive organ for the aerial apothecia developing on the substratum.

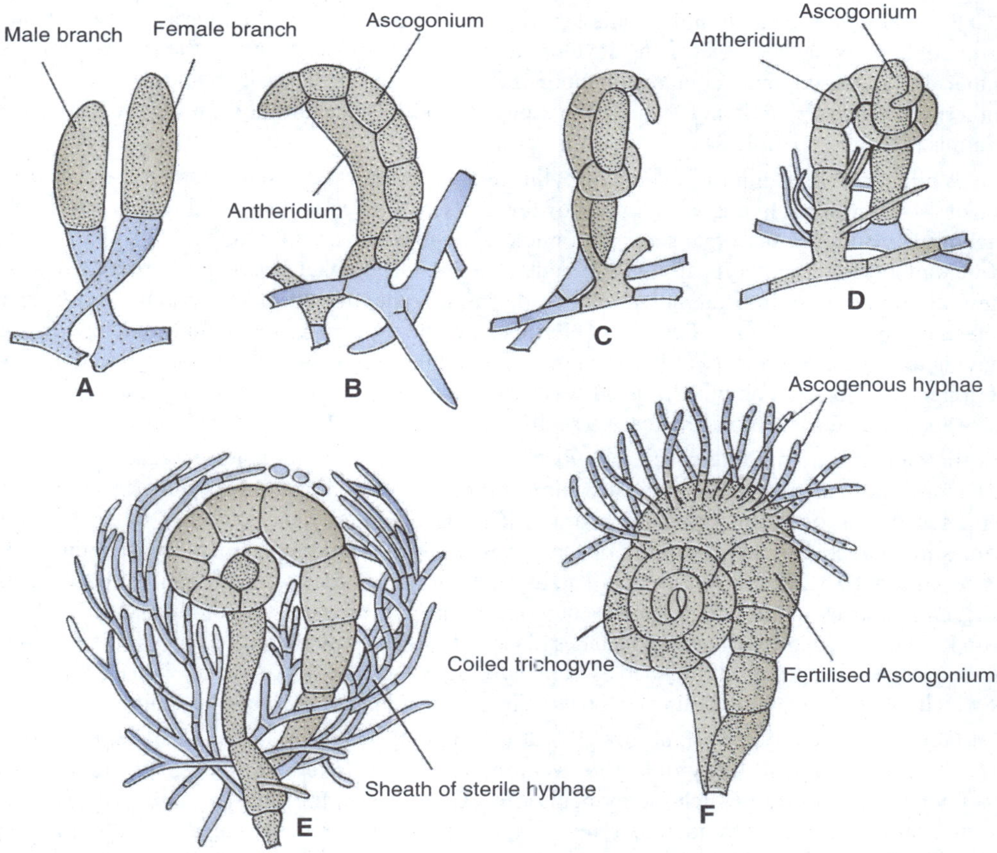

Fig. 12.18. *Ascobolus*. Stages in sexual reproduction. A, young sex organs, B-D, stages of maturation of sex organs and plasmogamy. E, fertilised ascogonium, F, development of ascogenous hyphae from fertilised ascogonium.

Asexual Reproduction. Asexual reproduction in *Ascobolus* is not of common occurrence. In some species like *A. furfuraceus,* asexual reproduction takes place by oidia or arthrospores produced in chains on small hyphal branches. The oidia are thin walled structures and germinate to form new mycelium. The oidia are also reported to play an active role in sexual reproduction.

Sexual Reproduction. Species of *Ascobolus* may be heterothallic or homothallic. Common heterothallic species are *A. magnificus, A. furfuraceus,* and *A. immersas* while *A. crenulatus* is homothallic.

Ascobolus furfuraceus – apothecia

Ascobolus immersas

In *A. magnificus* male and female sex organs i.e. ascogonia and antheridia develop on two compatible mycelia. The sex organs develop on different branches. Usually the two sex organs i.e. antheridia and ascogonia develop on both the strains i.e. each strain being bisexual. But the antheridia of one strain do not fertilise the ascogonia of the same strain meaning thereby that each strain is self-incompatible.

While the antheridium is clavate or cylindrical and erect, the female element i.e. ascogonium consists of a coiled—hypha with a stalk of few cells, with swollen oogonial region and an apical trichogyne. Both the sex organs are multinucleate. The trichogyne elongates till it reaches the tip of the antheridium and coils around the body. The wall of contact between the two dissolves and the male nuclei pass into the apical cell of the trichogyne and ultimately reach the ascogonium where plasmogamy occurs. Female nuclei also devide and male and female nuclei form pairs moving towards the periphery of the ascogonium. Soon the ascogenous hyphae are given out from the periphery of the ascogonium. In usual way, asci develop, each containing eight ascospores. The ascospores are ellipsoidal or spherical, one celled, uninucleate, and purple to dark brown in colour.

In another heterothallic species, *A. furfuraceus* the antheridium is absent and the function of the male component is performed by oidia produced in chains. The oidium from one strain fertilises the ascogonium of another strain. The ascogonium develops only when the oidium comes in contact with the mycelium of the opposite strain. The development of ascogonium takes place within 10-12 hours of contact. The ascogonium is made up of a broad coiled base and a long narrow trichogyne which shows chemotropic growth towards the oidium and finally fuses with it. The development of apothecia takes place in the usual way within 10 days of the fertilisation. In the apothecium asci are surrounded by paraphyses. Each ascus is long, clavate with eight ascospores of which four belong to one strain and the remaining four belong to the opposite strain.

Generally two types of apothecial development in *Ascobolus* has been recognised : (*i*) **Cleistohymenial** in which the hymenium is enclosed in its early development; (*ii*) **Gymnohymenial** in which the hymenium is exposed from the very beginning. *A. furfuraceus* is an example of cleistohymenial type of apothecial development while *A. magnificus* is an example of gymnohymenial type.

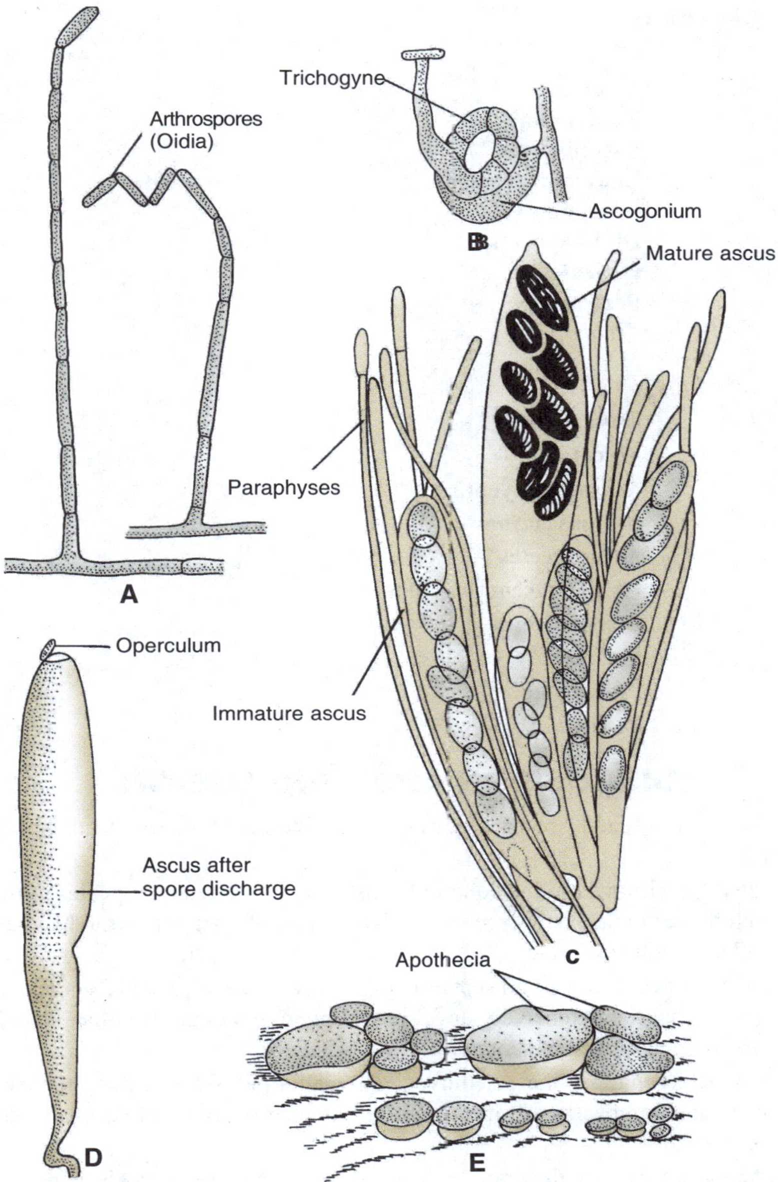

Fig. 12.19. *Ascobolus furfuraceus*. A, arthrospores (oidia); B, ascogonium with trichogyne; C, asci with ascospores, and paraphyses; D, dehisced ascus; E, apothecia on substratum.

In *A. citrinus* both the antheridia and oidia are absent. The development of apothecium from the female branch is parthenogenetic with copulation.

The apothecia are generally yellowish in colour, saucer shaped and about 5 mm in diameter. At maturity the surface of the apothecium is studded with purple dots which signify the ripe asci. As the asci mature, they elongate above the general level of hymenium.

Asci are long cylindrical structures intermixed with paraphyses. Each ascus has an operculum at its apical end. The ascospores are ellipsoid with mucilaginous epispore. In *A. immersus*, the ascospores are very large measuring 70μ × 30μ. On germination, each ascospore gives rise to new mycelium.

Taxonomic Position

Alexopoulos (1962)

Division	:	**Mycota**
Subdivision	:	**Eumycotina**
Class	:	**Ascomycetes**
Subclass	:	**Euascomycetidae**
Series	:	**Hymenomycetes**
Sub-series	:	**Discomycetes**
Order	:	**Pezizales**
Family	:	**Pezizaceae**
Genus	:	*Ascobolus*

Alexopoulos and Mims (1979)

Superkingdom	:	**Eukaryonta**
Kingdom	:	**Mycetae**
Division	:	**Amastigomycota**
Subdivision	:	**Ascomycotina**
Class	:	**Ascomycetes**
Sub-class	:	**Hymenoascomycetidae**
Series	:	**Discomycetes**
Order	:	**Pezizales**
Family	:	**Pezizaceae**
Genus	:	*Ascobolus*

DISCUSSION AND CONCLUSION

A. Evolutionary tendencies in the Ascomycetes. The evolutionary trends noticeable in the Ascomycetes are :–

(*i*) The gradual transition from the aquatic forms to purely terrestrial species noticeable in the Phycomycetes culminates in the Ascomycetes which are fungi of drier, terrestrial habitats completely lacking motile cells in their life cycle.

(*ii*) The tendency towards increased septation of hyphae in the advanced Zygomycetes becomes firmly established so that the extensively developed septate mycelium becomes the characteristic feature of the somatic phase of the Ascomycetes.

(*iii*) The septate hyphae of the Ascomycete mycelium show a further tendency to come together and intertwine compactly to form fungal tissue characteristic of compact bodies such as the stroma and sclerotia.

(*iv*) From the primitive to the more advanced Zygomycetes there is a tendency for the sporangia to be transformed into sporangioles and these into conidia which are the typical asexual spores of Ascomycetes. The conidia play an important role in the spread of Ascomycetes and thus form a striking feature in the life cycle. In the lower Ascomycetes the conidia are formed in exceptional abundance and throughout the period of mycelial growth. In fact in some species of Ascomycetes conidia alone appear to be sufficient for survival. These fungi have thus lost the capacity to produce sexual spores. Such Ascomycete species are allotted to the class Imperfect Fungi.

(*v*) The conidia bearing hyphae (conidiophores) in many Ascomycetes become organised into definite, asexual fruiting bodies such as acervuli, sporodochia, pycnidia and synnemata.

(*vi*) **Reduction of Sexuality.** Among the lower Ascomycetes the sexual apparatus is well developed. Both the sex organs, **antheridia** and **oogonia** or **ascogonia** are well marked and functional. They are usually developed on the same mycelium. In few cases as in the family Laboulbeniaceae, they occur on distinct individuals. There is, however, a gradual and progressive degeneration of the sex organs as we proceed from the lower to the higher Ascomycetes. In the most advanced Ascomycetes there is complete suppression of the sex organs. In them the sexual apparatus is wholly lacking. This progressive reduction in sexuality can be well illustrated by a comparison of species. They can be arranged in a series illustrating the gradual reduction and gradual elimination of the sexual apparatus.

Firstly, the antheridium ceases to function. In some species of *Aspergillus* both the sex organs are well developed. Sexual fusion takes place. Same is the case in *Lachnea stercorea.* There is, however, no migration of the contents of the antheridium into the oogonium in this species. It shows that the antheridium has become non-functional. There is another species of the same genus, which goes a step further. It is *Lachnea cretea.* In this species the antheridium is absent. The ascogonium is well developed and functional. *Coprobia* (*Humaria*) *granulatus* illustrates another step towards degeneration of sexuality. It is characterised by the absence of trichogyne from the ascogonium. The antheridium is also lacking. *Coprobia* (*Humaria*) *rutilans* illustrates the final stage in the elimination of sex organs. The antheridia are absent and so are the ascogonia or the oogonia. The sexual process in these advanced forms becomes extremely simplified. It takes place by the copulation between the cells of two vegetative hyphae. Finally even this impulse towards sexuality disappears. The two vegetative nuclei of adjacent cells of the same homothallic mycelium come together to establish a dikaryon. The two nuclei of the dikaryon fuse in the ascus mother cell to form a synkaryon. The latter by meiosis followed by mitosis gives rise to 8 haploid ascospores within the ascus. Hence in the final stage the development of the sexual apparatus in the life cycle is totally abandoned.

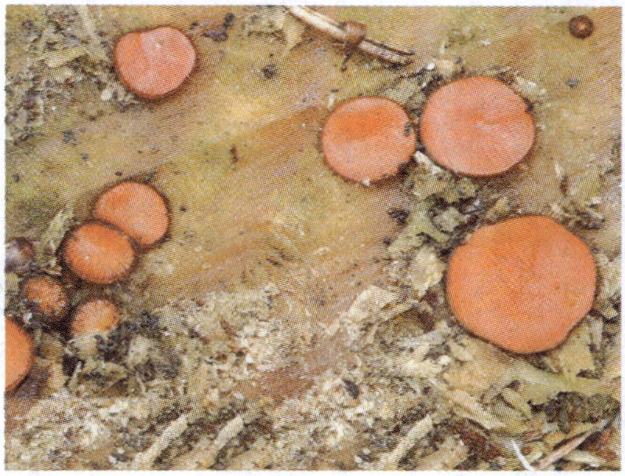

Lachnea – apothecia

(*vii*) **Gradual and Progressive Differentiation (complexity) of Ascocarp.** The degeneration of the sexual apparatus in the Ascomycetes is accompanied by a corresponding morphological differentiation of the ascocarp. In other words the ascocarp becomes more and more complex as we proceed from the lower to the higher Ascomycetes. This evolutionary series starts with the primitive Ascomycetes such as the yeasts in which the ascospores are produced in typical vegetative cells which are diploid. In *Eremascus* the zygote itself enlarges to form a globose ascus. *Taphrina* goes a step further. Special asci are formed from the binucleate ascogenous cells. But both in *Eremascus* and *Taphrina* the asci are not grouped in ascocarps. The next step is illustrated by *Aspergillus* and *Talaromyces* (*Penicillium*). In both these genera the asci are grouped in a small primitive type of ascocarp called the **cleistothecium**. The asci in the cleistothecium are distributed at random. They do not form a definite hymenial layer. The series then passes through the sub-series *Pyrenomycetes* and *Discomycetes*. The former is characterised by the presence of **perithecium** type of ascocarp and the latter with a shield type called the **apothecium**. In both the asci are organised into a definite fertile layer called the **hymenium**. The series finally ends in the complex fructification of *Morchella*. The ascocarp of *Morchella* is very complex. It is differentiated into a stalk-like portion, the **stipe** surmounted by a fertile cap-like, conical portion called the **pileus**. The *Morchella* ascocarp with its pileus and stipe can be morphologically compared with the stipe and pileus of the **basidiocarp** of the Basidiomycetes.

(*viii*) Another evolutionary trend is the origin of dikaryophase in the life cycle which in the advanced Ascomycetes increases in duration and importance but never becomes independent. It remains dependent on the haploid maycelium for its nutrition and occurs only inside the fruit body. The diplophase correspondingly becomes transitory and insignificant represented only by the fusion nucleus of the young ascus.

B. PHYCOMYCETES AND ASCOMYCETES COMPARED

1. Important Differences

TABLE : V

		Phycomycetes	*Ascomycetes*
Mycelium	1.	Either lacking or aseptate and coenocytic when present.	With the exception of the yeasts, the mycelium is always well developed and septate. The hyphae have a tendency to aggregate to form fungal tissues.
	2.	There is gradual transition from the primitive aquatic forms to purely terrestrial species.	These are fungi of drier terrestrial habitats which lack motile cells.
Asexual reproduction	3.	It takes place by means of sporangiospores, motile or non-motile, conidiosporangia and chlamydospores.	Asexual reproduction takes place by means of fission, budding, fragmentation, chlamydospores and oidia but the usual method is by means of exogenously produced asexual spores called conidia. No sporangiospores (motile or non-motile) are produced.
Sexual reproduction	4.	It is either isogamous or heterogamous. There is a gradual and progressive increase in complexity of the sex organs from the primitive to the advanced forms.	There is a gradual and progressive simplification and ultimate elimination of the sexual apparatus from the lower to the higher forms.
Dikaryophase	5.	Plasmogamy without any exception is immediately followed by karyogamy so that there is no dikaryophase in the life cycle of the Phycomycetes.	Plasmogamy is generally not immediately followed by karyogamy so that the two nuclei do not fuse for a considerable time in the life cycle. They lie in a pair in close association to form a dikaryon. The resultant dikaryotic cell by successive conjugate divisions, elongation and septation forms a distinct dikaryo-phase in the life cycle.
Diplophase	6.	It is comparatively long-lived and represented by the zygospore which usually enters upon the resting period.	The increased importance of dikaryophase has reduced diplophase to insignificance. It is transitory and represented only by the young ascus cell with a diploid nucleus.
Ascocarp	7.	The resultant single celled structure (zygospore) is small and unicellular. Usually it is without any sheath and thus no fruit body or ascocarp is formed.	The resultant fructification is generally large and multicellular. In the vast majority of Ascomycetes it is covered by a distinct sheath, and is called the **ascocarp.**

2. Important Similarities. (*i*) In their modes of life and methods of nutrition the Phycomycetes and Ascomycetes are similar.

(*ii*) The conidia and conidiophores which play an important role and are a striking feature in the life cycle of Ascomycetes are similar to those of the advanced Zygomycetes.

(*iii*) The germ sporangium of Phycomycetes is considered homologous to the ascus of the Ascomycetes. The supporters of this view look upon the ascus as a specialised sporangium which has been derived from the germ sporangium by reduction in the number of spores till it has become fixed at eight and the cleavage method of spore formation having been replaced by free cell formation.

C. Origin and Evolution of Ascomycetes. The phylogeny of Ascomycetes is still uncertain. The supporters of polyphyletic origin of fungi from the various classes of algae derive the Ascomycetes from the red algae (presumably the Florideae). The floridean hypothesis was first proposed by Sachs in 1875. The adherents of this hypothesis such as Thaxter (1896); Bessey, (1914, 1925, 1942, 1950); Dodge, (1914); and Jackson, (1900) base phylogenetic speculation on the striking similarities (between the sex organs and fructifications of these two groups) which include :–

(1) Female gametangium with a trichogyne.
(2) Occurrence of non-motile male cells called **spermatia**.
(3) Origin of sporogenous filaments (**ascogenous hypha**) in Ascomycetes and **gonimoblast filament** in Florideae as outgrowths from the female gametangium.
(4) Similarity between the ascocarp and cystocarp.

In addition to the above, lack of motile cells in the life cycle, filamentous thallus with apical growth and a central pore in the transverse septa are other features which the Ascomycetes share with the red algae.

To some phylogeneticists, the above-mentioned common features are an indication of direct relationship between the two groups. They hold that the Ascomycetes have arisen from the red algae by the loss of photosynthetic pigments coupled by a change in the mode of nutrition to parasitism and saprophytism. The adoption of land habit by the Ascomycetes resulted in the development of **ascus** as the spore producing structure and **ascospores** as the type of spores replacing tetrasporangia and naked tetraspores, respectively.

On the basis of floridean hypothesis, the lower Ascomycetes with the simplest sexual apparatus are considered degenerated forms. The immediate descendants produced spermatia and ascogenous hyphae. Their original progenitors in the red algae were forms in which gonimoblast filaments developed directly from the carpogonium. The next in order, probably, were the Ascomycetes in which the male sex organ (Antheridium) developed near the female sex organ (ascogonium) bearing ascogenous hyphae. From these evolved the lower Ascomycetes with the loss of ascogenous hyphae and zygote functioning directly as an ascus.

In the eyes of opponents, the floridean hypothesis of origin of Ascomycetes does not stand critical examination. The chemical nature of cell wall is different in the two groups. The method of ascospore formation in the Ascomycetes is unique and has no parallel in the red algae. They consider these morphological resemblances illusory. At the most they consider it a case of parallel evolution of similar structures between two fundamentally different groups occurring in similar habitats.

To many mycologists Phycomycetes particularly the advanced Zygomycetes appear to be the ancestral stock from which the Ascomycetes have evolved. De Bary (1887) proposed the theory of monophyletic origin of Fungi. He advocated that the Ascomycetes (along with the Basidiomycetes) arose from the Phycomycetes. The ardent supporters of this hypothesis are Dangeard, (1907); Claussen, (1912); Atkinson, (1915); Guilliermond, (1928), Fitzpatrick, (1930),

Gaumann, (1952) and Hawker, (1967). This hypothesis is based on the homology between the germ sporangium of Zygomycetes (Mucorales) and the ascus of the Ascomycetes. The supporters of this view regard the Ascus as a specialised sporangium derived from the germ sporangium of Mucorales by the following changes among others :–

(*i*) The number of meiospores has been reduced till it has become fairly constant at eight.

(*ii*) The cleavage method of spore formation has been replaced by **free cell formation**.

The affinities with the Zygomycetes seem reasonable in some other respects as well. The tendency among the Zygomycetes towards the development of the conidium–a typical asexual spore of the Ascomycetes and a tendency towards greater protection of the zygospore suggests a relationship with the Ascomycetes. The conidia of many Ascomycetes such as Erysiphales, Aspergillales and others are similar to those of the Mucorales. Further the mycelium in both the Ascomycetes and Zygomycetes is usually filamentous and highly developed. The only difference is that it is septate in the Ascomycetes and aseptate in the Zygomycetes. But in the order Entomophthorales of class Zygomycetes the hyphae are normally regularly septate. Sexual reproduction by conjugating gametangia in some of the Hemiascomycetidae (lower Ascomycetes) is very similar to the higher Zygomycetes. There is fusion of unequal cells (gametangia) in each of which there is a single functional sexual nucleus. The female gametangium in *Dipodascus* (Hemiascomycetidae) elongates to form an ascus wheras in some of the higher Zygomycetes it grows out to form a thick-walled zygospore. In fact the transition from the gametangiogamous Zygomycetes to the Hemiascomycetidae is so gradual that some mycologists interpret these facts as showing further development of evolutionary tendencies found in the Zygomycetes. Further support for De Bary's hypothesis is furnished by the following similarities between the members of these two groups :–

(*i*) Similar glycogen-like type of stored food

(*ii*) Cell walls constructed mainly of chitin

(*iii*) Sexual fusion of the nuclei tends to be delayed until some time after mating, both in some of the advanced Zygomycetes and lower Ascomycetes.

From the above-mentioned similarities between the members of these two groups one may infer that the ancestor of the Ascomycetes may have been an ancient member of the Zygomycetes. On the basis of this hypothesis the lower Ascomycetes which lack ascogenous hyphae and have the zygote directly functioning as an ascus, are considered the immediate descendants of the ancestral stock. These gave rise to forms in which plasmogamy by gametangial contact was not immediately followed by karyogamy and asci developed indirectly from the ascogenous hyphae arising as outgrowths from the ascogonium. Next in order evolved the higher Ascomycetes with spermatia instead of antheridia.

Hawker (1967) emphasized that similarities in fine structure (sparse smooth endoplasmic reticulum) and mitochondria with groups of parallel plate-like cristae and other similarities between the Zygomycetes and Ascomycetes make it highly probable that the Ascomycetes may have evolved from some ancient member of the Zygomycetes not closely similar to any extant Zygomycete.

In an attempt to explain the origin of hooks which are characteristic structures of the Asomycetes some phylogeneticists have recently proposed a new theory about the origin of Ascomycetes (Euascomycetidae). They consider hooks as degenerate sex organs. According to them, sexual reproduction in the ancestors of Ascomycetes proceeded by the formation of lateral conjugation bridges between the adjacent cells of the same hypha. During the course of evolution this process was replaced by conjugation of gametangia (gametangiogamy). The original process, however, persisted in the living Ascomycetes in the form of hooks which no longer function as they once did. On the basis of this theory the Hemiascomycetidae which lack hooks have ended blindly leading from the Phycomycetes. They have nothing to do with the higher Ascomycetes. The latter originated directly from some very primitive extinct Phycomycete stock and developed later parallel to the living Phycomycetes.

D. Economic Importance of Ascomycetes. The saprophytic Ascomycetes such as yeasts, blue molds and green molds which are cosmopolitan in their distribution are of great value to man. They play an important role in the production of **food** and **medicine**. The yeasts ferment sugar to produce alcohol and carbondioxide. The entire brewing industry (wine and liquor making) and baking industry (bread making) are built around this simple plant. Food yeast is important as diet supplement. Yeasts are also a source of many vitamins. Species of *Penicillium* are employed in the production of cheese and high grade citric acid. Some species of blue and green molds are employed in the production of many other important organic acids. Some Ascomycetes yield antibiotic drugs such as penicillin and flavicin. Production of **penicillin** from *Penicillium notatum* and *P. chrysogenum* has become a large industry in India and other countries. Ergot, a valuable drug, is obtained from an ascomycete (*Claviceps purpurea*) which infects cereals. The drug is used at child birth because it stops bleeding and checks other uterine disturbances. Included in the Ascomycetes are the popular edible fungi, the truffles and morels, which are considered table delicacies. *Neurospora* has become an important experimental organism in genetic research.

The negative role the Ascomycetes play in the economy of nature is equally significant. Some spoil foods and other consumer goods. The celluloytic Ascomycetes destory fabrics containing cellulose. Some of the parasitic species cause serious diseases of our economic plants. Powdery mildews such as *Erysiphe graminis* infect various grains. *Erysiphe cichoracearum* affects fruits and vegetables causing loss by low production. *Peach leaf curl, brown rot* of stone fruits, *black rot* of plums and *bitter rot* of apples are the common diseases of fruit trees caused by parasitic Ascomycetes. The parasitic Ascomycetes also include several severe though uncommon human pathogens. Several species are the causative agents of diseases of ear and lung. Some of them cause a respiratory disease called **aspergillosis**. Body ring worm is another fungal disease caused by an ascomycete.

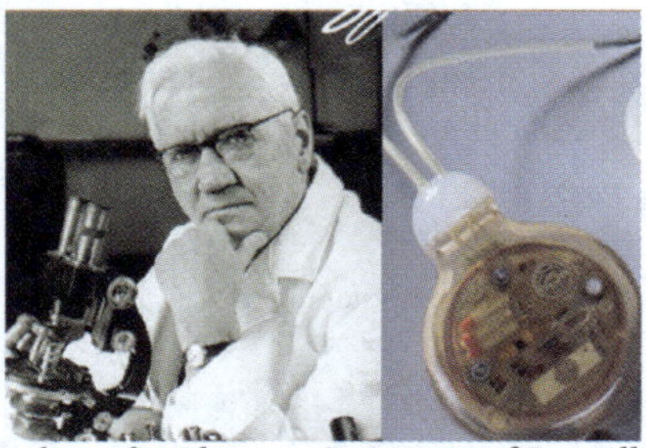

Alexander Fleming, Discoverer of Penicillin

Liquor Making by Yeasts

Neurospora as Experimental Organism

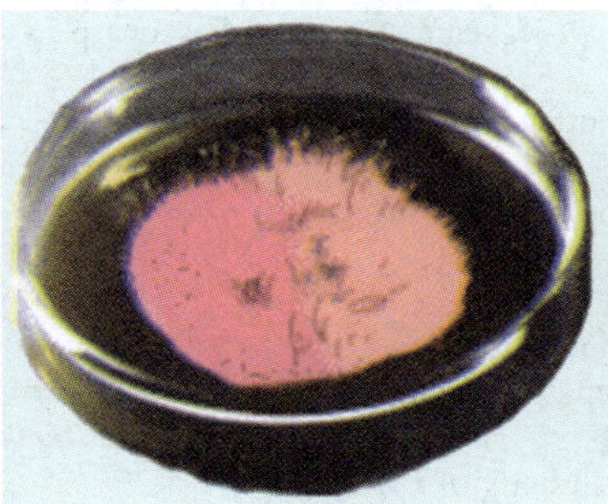

Neurospora

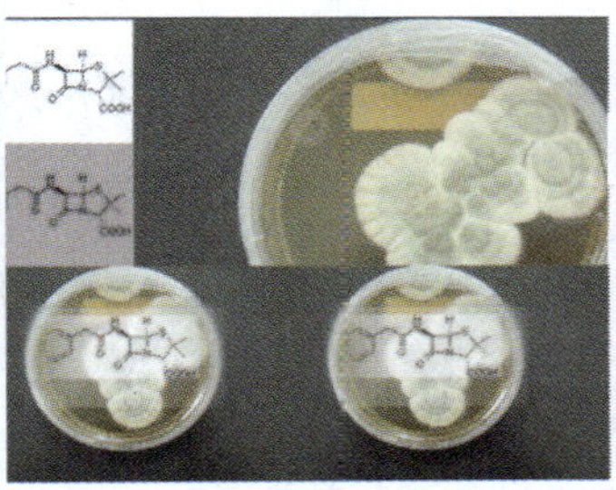

Penicillin

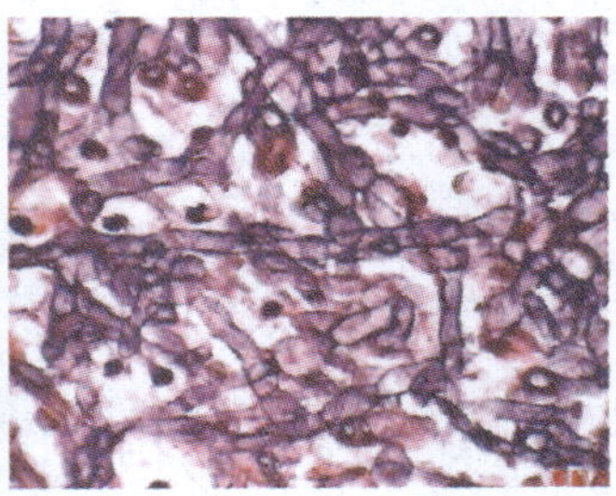

Aspergillosis

QUESTIONS

ESSAY TYPE

1. Give an illustrated account of the life history of any member of the order Pezizales studied by you.
2. Discuss gradual and progressive reduction in sexuality as illustrated by the members of order Pezizales.
3. Describe sexual reproduction and development of ascocarp in *Pyronema*.
4. Discuss briefly important evolutionary tendencies recognisable in the class Ascomycetes.
5. Describe the development and structure of the ripe fruit of any edible Ascomycetous fungus.
6. Give an illustrated account of the life history of any typical Ascomycetes studied by you.
7. Give an account of sexual reproduction in *Morchella*.
8. What is an ascocarp? Describe the development of ascocarp in *Ascobolus*. *(Lucknow, 1999)*
9. Describe the sexual reproduction in *Peziza*. *(Awadh, 1994, 1998)*
10. Give an account of the habitat, morphology, and reproduction in *Peziza*. *(Gorakhpur, 1991; Allahabad, 1995)*
11. Give an account of structure and development of apothecium in *Peziza*. *(Allahabad, 1998; Poorvanchal, 2006)*
12. With special reference to *Morchella*, describe the structure and development of ascus and ascospores. Compare Ascomycetes with Phycomycetes. *(Kanpur M.Sc., 1992)*
13. With the help of labelled diagrams describe the life cycle of *Pyronema*.
14. Give an illustrated account of the structure and reproduction in *Coprobia*.
15. Give the systematic position and habit of *Peziza*. Describe its apothecium in detail with the help of diagrams.
16. Give in detail the life history of *Peziza*.
17. Give an account of sexual reproduction in *Pyronema*.
18. Describe the structure and reproduction in *Pyronema* or *Coprobia*.
19. Where will you find *Morchella/Peziza*? With the help of suitable diagrams describe the process of sexual reproduction in it.
20. Explain the phenomenon of "Degeneration of Sexuality in Ascomycetes." *(Lucknow, 2000)*
21. Describe the structure and reproduction in *Peziza*. *(Poorvanchal, 2002)*
22. Describe in detail the structure and mode of reproduction in *Morchella*. *(Meerut, 1997)*

SHORT ANSWER TYPE

23. **Give the diagnostic characters and systematic position of the following genera:**
    - (*i*) *Morchella* *(Kanpur, 1999)*
    - (*ii*) *Peziza* *(Awadh, 1991; 1996; 1997; Gorakhpur 1992; 1996; Kanpur M.Sc., 1995; Rohilkhand, 1995; 1997)*
    - (*iii*) *Pyronema* *(Kumaon, 1998)*
    - (*iv*) *Ascobolus* *(Lucknow, 1993; 1997, 2000)*
    - (*v*) *Coprobia*
24. **Give only labelled diagrams of the following:**
    - (*a*) V. S. of ascocarp of *Morchella* *(Kanpur, 1997;2005)*
    - (*b*) Internal structure of *Morchella* fruit body *(Kanpur, 1994)*
    - (*c*) V. S. Apothecium of *Ascobolus* *(Lucknow, 1995; 1996, 2002)*
    - (*d*) V. S. *Peziza* apothecium *(Awadh, 1993; 1995; 1996; 1997; 1998; 1999; Gorakhpur, 1995; 1997, Poorvanchal, 2001, 2003)*
25. **Write notes on the following :**
    - (*a*) *Ascobolus* and structure of ascocarp *(Lucknow, 1992)*
    - (*b*) Apothecium of *Ascobolus* *(Lucknow, 1993)*
    - (*c*) Apothecium of *Peziza* *(Allahabad, 1992; Bihar 1991; Magadh, 1993; Andhra, 1992)*
    - (*d*) Apothecium *(Allahabad, 1996)*
    - (*e*) Somatogamy in *Peziza* *(Rohilkhand, 1993)*
    - (*f*) Fruit body of *Pyronema*
    - (*g*) Fruit body of *Coprobia*

(h) Sexual reproduction in *Peziza* (*Poorvanchal, 2004*)

26. Draw well labelled diagrams of the following :

(a) V.S. Apothecium in *Ascobolus* (*Lucknow, 2004*)

(b) Sex organs in *Pyronema*

(c) Apothecium in *Peziza*

(d) Ascocarp of *Morchella*

27. Discuss the degeneration of sexual structures in Ascomycetes studied by you. (*Lucknow, 1995*)

28. Differentiate between :

(a) *Peziza* and *Morchella* (*Kanpur, M.Sc., 1985*)

(b) *Claviceps* and *Peziza* (*Kanpur, M.Sc., 1988*)

(c) *Cleistothecium* and *Apothecium* (*Kanpur, 2006*)

29. Compare and contrast Phycomycetes and Ascomycetes.

30. Discuss briefly the origin and evolution of Ascomycetes.

31. Give a brief account of the economic importance of Ascomycetes.

32. Give a brief account of fructification in *Morchella*.

33. Where will you find *Morchella*? With the help of suitable diagrams, describe its spore fruit.

34. Give a graphic representation of life cycle of *Peziza*. (*Guru Nanak Deo, 1992*)

35. What is an ascus? Describe the formation of ascus in *Peziza*.

36. Compare *Peziza* with *Morchella*.

OBJECTIVE TYPE

37. Fill in the blanks :

(i) *Ascobolus* is a .... fungus because it grows on horse dung.

(ii) *Morchella deliciosa* is commonly known as ....

(iii) Apothecium is the characteristic feature of fungi belonging to ....

(iv) *Pyronema* is .... growing on decaying leaves.

(v) The sexual reproduction in *Pyronema* takes place by ....

(vi) Somatogamous copulation takes place in ....

(vii) *Peziza* is a member of ....

(viii) Discomycetes is so named because the ascocarp is .... or ....

(ix) The members of the series Discomycetes are ....

(x) Apothecium is developed in ....

38. Select the correct answer :

(i) Cup-shaped ascocarps are characteristic features of
(a) Plectomycetes (b) Pyrenomycetes
(c) Discomycetes (d) Ascomycetes

(ii) Which of the following is a coprophilous fungus?
(a) *Phytophthora* (b) *Morchella*
(c) *Ascobolus* (d) *Aspergillus*

(iii) In *Pyronema confluens* the sexual reproduction takes place by
(a) Hologamy
(b) Somatogamy
(c) Gametangial Copulation
(d) Gametangial Contact

(iv) The sexual reproduction is somatogamy in
(a) *Pyronema confluens*
(b) *Lachnea cretea*
(c) *Humaria granulata*
(d) *Morchella deliciosa*

(v) Which of the following is an edible fungus?
(a) *Morchella deliciosa*
(b) *Pyronema confluens*
(c) *Aspergillus nidulans*
(d) *Phytophthora infestans*

(vi) Asci are produced alternating with
(a) Elaters (b) Paraphyses
(c) Mycelium (d) Nothing

(vii) Fertile layer in Pileus of *Morchella* is called
(a) Hymenium (b) Subhymenium
(c) Trama (d) Cortex

(viii) Reduction division in the life cycle of *Pyronema* takes place in
(a) ascogonium (b) ascus mother cell
(c) young ascus (d) mature ascus

(ix) Apothecium in *Peziza* is
(a) aerial and short lived
(b) underground and short-lived
(c) aerial and persistent
(d) underground and persistent

(x) Which of the following spores are produced by *Morchella*?
(a) Oospores (b) Ascospores
(c) Basidiospores (d) Chlamydospores

(xi) The fruit body of *Peziza* is
(a) Cleistothecium (b) Apothecium
(c) Perithecium (d) Basidiocarp

Grain Rusts

CHAPTER 13

THE HIGHER FUNGI
Class: Basidiomycetes (Club Fungi)

- General Features
- Mycelium
- Asexual reproduction
- Sexual reproduction
- Salient features of Basidiomycetes
- Classification

General Features. It is the most advanced of all the fungal classes. It includes about 15,000 species with more than 500 genera. This big group of fungi includes both saprophytic and parasitic species. The mycelium is usually, perennial. It ramifies in the substratum which in the case of parasitic species may be wood, roots and stems of living trees. The saprophytic species live in moist, decaying wood, under the bark of rotten logs, dung, wet dead leaves or humus of the soil. As in the Ascomycetes motile structures are lacking in this class and the mycelium is septate. The basidiomycetes produce highly complicated fruiting bodies called the **basidiocarps** (Fig. 13.7). The basidiocarps in the higher forms are massive and constitute the conspicuous and the only visible portion of the fungus. They are concerned with reproduction and bear reproductive structures called **basidia**. The development of basidium is a unifying character common to all members of the class Basidiomycetes which otherwise have a great variety of form, habitat, structure and physiology. The form of the basidium is more varied than that of the ascus. The development and cytology up to the initiation of basidiospores is, however, similar to that of the ascus. This evidently supports the view that the two classes (Ascomycetes and Basidiomycetes) have a common origin even if remote. The basidium typically bears four spores **exogenously**, each usually at the tip of a minute stalk known as **sterigma**. These characteristic spores of the Basidiomycetes are called the **basidiospores**. Like the ascospores they are the result of plasmogamy, karyogamy and meiosis. The last two occur in the basidium. The basidiospores are dispersed by wind, insects or other agencies. The common examples of this class are the *smuts*, the *rusts*, the *mushrooms*, the *toad stools*, the *puff balls* and the *pore fungi*.

Corn Smuts – Symptom

Puffballs – Basidiocarp

Rusts on leaf

Pore Fungi – Basidiocarp

Mushrooms Fruit bodies

Toadstools Fruit bodies

Historical. Leveille (1837) discovered the basidium. The groups now known as Homobasidiomycetidae was established as a natural taxon in 1842 by Corda. It was Fries who in 1821 recognised rusts and smuts as a natural group. It now forms a major part of the present sub-class—Heterobasidiomycetidae.

Mycelium (Fig. 13.1). The well developed, filamentous mycelium consists of a mass of branched, septate hyphae generally spreading in a fan-shaped manner. The cell wall is chitinous in nature. Within the cell wall is the plasma membrane. The cytoplast contains a complement of usual cell organelles except the chloroplasts. The septum, as in the Ascomycetes, originates as an annular outgrowth on the inside of the tubular wall. It grows inwards like a narrow shelf reducing diameter of the pore. The hyphae ramify in the substratum and absorb food. The mycelium generally is a weft of interlacing and anastomosing hyphae. In a few genera, however, the mycelial hyphae run parallel to one another and get bundled together to form definite and conspicuous thick cords of macroscopic size. These are the **rhizomorphs**. The rhizomorph is covered by a sheath (cortex) and behaves like a unit. The mycelium in different species varies in colour and may be white, yellow or orange. It is generally perennial.

The mycelium of Basidiomycetes passes through three distinct stages namely, the primary, the secondary and the tertiary before the fungus completes its life cycle. The first stage is represented by the primary **mycelium or homokaryon** (B) which is formed by the germination of a basidiospore (A). The latter, finding conditions of temperature, food supply and moisture congenial for growth, germinates to form a hypha consisting of uninucleate cells. It represents the **primary mycelium** and constitutes the **haplophase**. It does not bear any sex organs–a feature in sharp contrast to the *Phycomycetes* and *Ascomycetes*, nor does it bear any basidia or basidiospores. The primary mycelium may multiply by **conidia** or sometimes by **oidia**.

The second stage is the **secondary mycelium** (Fig. 13.2). The cells of the secondary mycelium are binucleate. It represents the **dikaryophase** in the life cycle. The Basidiomycetes, in fact, differ from the Ascomycetes in the increased prominence of the dikaryophase which is independent, long-lived and thus plays a prominent role in the life cycle. The secondary mycelium originates from the primary mycelium as follows :

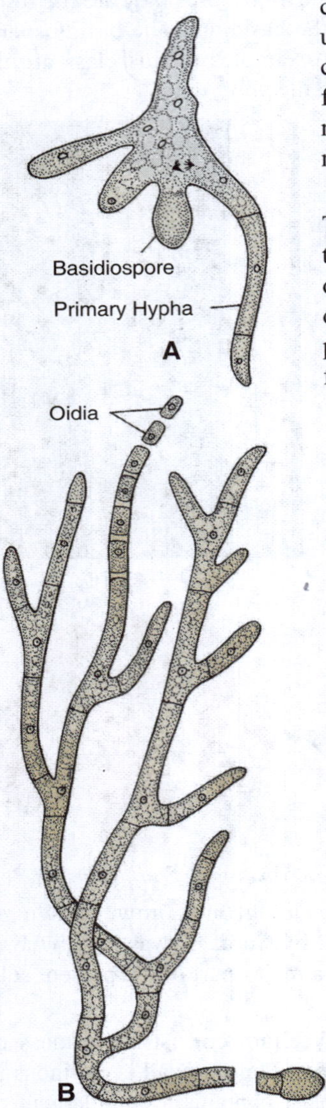

Fig. 13.1 (A-B). *Basidiomycetes.* A, Germinating basidiospore of *Coprinus lagopus* (after Bensaude); B, primary mycelium with one of the hyphae breaking into oidia. (Diagrammatic).

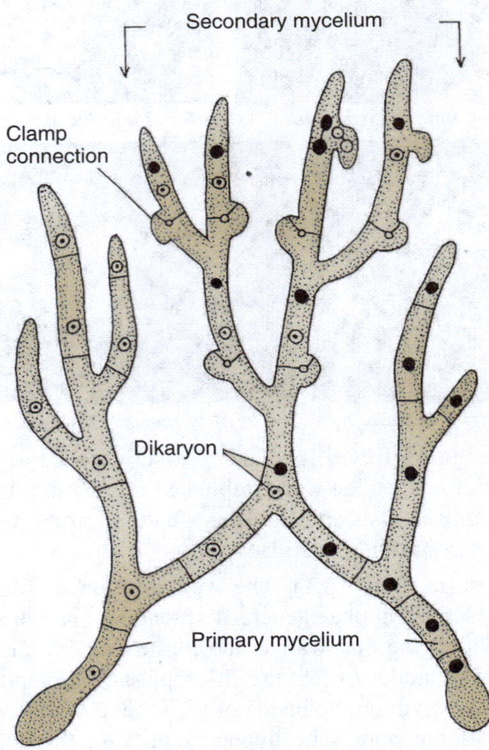

Fig. 13.2. *Basidiomycetes.* Sketch showing the formation of a secondary mycelium from a dikaryotised cell produced by somatogamous copulation between two uninucleate cells of primary mycelia of opposite strains.

Most of the Basidiomycetes are heterothallic. It means the primary or homokaryotic mycelium in them is of two distinct strains which are called + and – strain. The first step in the formation of the secondary mycelium involves the interaction between two compatible primary mycelia (Fig. 13.2). Two compatible hyphae (+ and – strain) from the neighbouring mycelia meet. The intervening walls between the two adjacent cells at the point of contact dissolve. The protoplasts of the uninucleate cells intermingle in the fusion cell (plasmogamy). The two nuclei in the fusion cell do not fuse. They lie side by side constituting a **dikaryon**. One of these is of + strain and the other of – strain. The binucleate cell thus established is known as the dikaryotised cell. Both the uninucleate and binucleate cells may be found in the same mycelium. The first formed portions of a mycelium may be uninucleate and the later formed portions binucleate. The dikaryotic cell divides repeatedly by conjugate divisions to give rise to a **secondary** or **dikaryotic mycelium** (Fig. 13.2). It consists of binucleate cells. During nuclear divisions of the dikaryotic cell special structures called the **clamp connections** are formed (Fig. 13.3). These clamp connections ensure that the sister nuclei of the dikaryon, at each division, separate into daughter cells. The secondary mycelium differs from the primary mycelium in being long-lived, presence of clamp connections, vigorous growth, mode of branching, binucleate condition of cells, and development of fructifications in the Homobasidiomycetidae and teleutospores or brand spores (probasidia) in the Heterobasidiomycetidae.

Clamp Connections (Fig. 13.3). When the dikaryotic cell (A) is ready to divide a pouch-like outgrowth arises from its wall (B), it arises midway between the two nuclei of the dikaryon. The two closely associated nuclei of the cell now divide simultaneously. This is called **conjugate division**. One of the four daughter nuclei, generally the lower one of the upper pair, passes into the pouch (C). A septum appears at the base of the pouch (D). As a result the pouch is cut off from the main cell. It may now be called a **clamp cell**. The clamp cell grows into hooklike structure. Its tip bends over and finally fuses with the lateral wall (E) of the parent cell. The clamp cell now forms a bridge. It is called the **clamp connection**. Another septum is laid down vertically under the bridge usually at about the level of the upper end of the clamp connection (E). It divides the parent cell into two daughter cells. The terminal daughter cell has two nuclei. Each of these is a sister nucleus of the parent dikaryon. The lower or basal daughter cell possesses one nucleus. The fourth nucleus lies in the clamp connection. The nucleus of the clamp connection now migrates into the basal daughter cell. The latter also becomes binucleate (F). The clamp connection thus simply functions as a bypass. It ensures that the sister nuclei formed by the conjugate division of the dikaryon separate into two newly formed daughter cells. The clamp connections are usually formed on the terminal cells of the hyphae of the secondary mycelium. They generally persist on the old dikaryotic hyphae. The presence of hooklike clamp connections is a safe criterion for distinguishing a **secondary** or **dikaryotic** mycelium from the **primary** or **monokaryotic** mycelium. The clamp connections by some mycologists are considered homologous to the hooks of the ascogenous hyphae of the Ascomycetes. This view is supported by the general occurrence of a clamp connection at the base of a basidium. Moore and Mclear (1962) studied the fine structure of septa of the dikaryotic hyphae of Basidiomycetes.

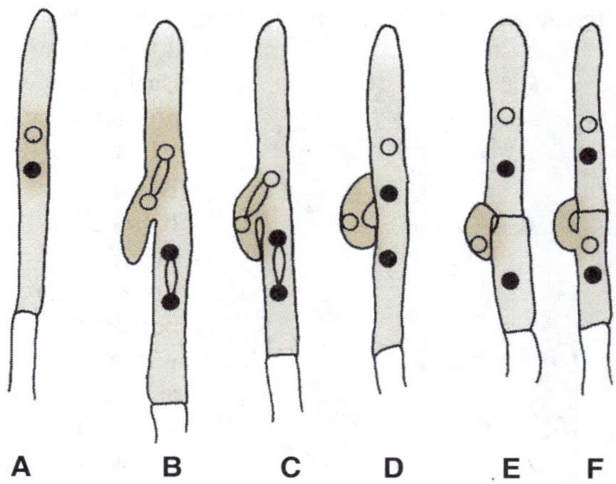

Fig. 13.3 (A-F). *Basidiomycetes.* Diagram illustrating the formation of clamp connections.

They found that the septum which is a cross wall flares sharply and broadly at the centre of the hypha to form a barrel-shaped structure with open ends. This is the actual pore surrounded by a swollen rim which is a part of the annular septum. This type of septum is termed **a dolipore septum**. The septum and the swelling are covered by the cell membrane. The opening on either side of the dolipore septum is guarded by a curved or crescent-shaped double membrane pore cap which in section looks like a parenthesis (round bracket). The septal pore-cap is thus given the name **parenthesome**. The parenthesome is similar to the endoplasmic reticulum. The lower parenthesome is seen to be interrupted by gaps in the form of pores. The upper parenthesome may be continuous. The dolipore parenthesome septum complex is unique to the Basidiomycetes. It maintains cytoplasmic continuity but prohibits nuclear migration.

Fig. 13.4. *Basidiomycetes.* Diagram showing dolipore septum in section. Based on an electron micrograph by Backer and Butler.

The secondary mycelium in the fruit bodies of the higher Basidiomycetes becomes organised into specialised tissues. It is called the **tertiary mycelium**. The fructifications are thus formed of the tertiary mycelium. The cells of the tertiary mycelium are also binucleate. The distribution of the dolipore parenthe-some septum complex in the Basidiomycetes is widespread. It is characteristic of the primary, secon-dary, and generative tertiary mycelium of the Homo-basidiomycetes as well as of the basidiocarpic myce-lium of the Heterobasidiomycetes. The major ex-ceptions are the rusts and smuts (Moore and Marchant, 1972).

The formation of **basidia**, the **dikaryotic mycelium,** the **dolipore septum** with the **parenthesomes** guarding the pore on both sides and **clamp connections** are the four diagnostic features of this class.

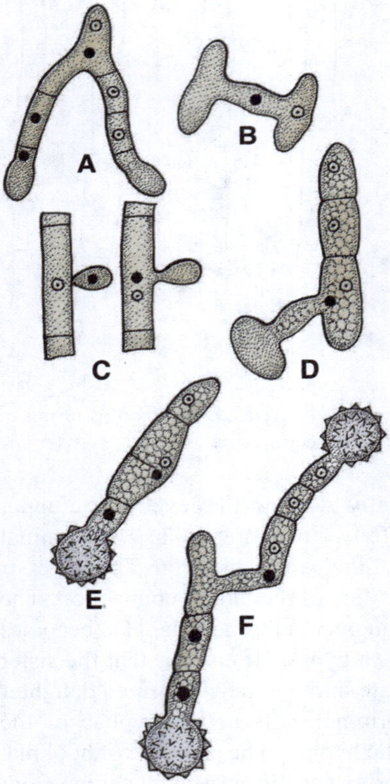

Fig. 13.5 (A–F). *Basidiomycetes.* Diagram illustrating chief methods of diploidisation or dikaryotisation. (A and C diagrammatic; B and D after Harper, and E after Rawitscher).

Diploidisation or Dikaryotisation (Fig. 13.5). The process by which the primary mycelium is converted into secondary or dikaryotic mycelium is called **diploidisation** or **dikaryotisation**. The first step in diploidisation is the establishment of a dikaryon in the fusion cell (Fig. 13.2). The dikaryotised cell through repeated divisions by clamp connections gives rise to a secondary mycelium in which each cell possesses a dikaryon (two neclei). Diploidisation takes place by the following methods :

1. *By Hyphal Fusions.* In this case fusion occurs between the vegetative cells of two neighbouring hyphae of the primary mycelia of opposite sexual strains (A).

2. *By conjugation of basidiospores* (B). Two basidiospores of opposite strains (*Ustilago anthearum*) meet and conjugate. The binucleate basidiospore formed in this way germinates to give rise to a secondary mycelium.

3. By the fusion of a germinating oidium of one strain with a cell of the primary mycelium of the opposite strain (C). The binucleate cell formed in this way by elongation and division by clamp connections develops into a secondary mycelium.

4. By fusion between a germinating basidiospore and a haploid cell of the basidium (D) as in *U. violacea*.

5. By fusion between the two haploid cells of opposite strains of the basidium (E) as in *U. carbo*.

6. By fusion between basidia formed by the germination of two smut spores (F).

Basidia are always produced from the binucleate cells of the secondary mycelium (Fig. 13.8).

Asexual Reproduction (Fig. 13.6). It takes place by the following methods :–

(*i*) By Conidia (Fig. 13.6 A-B). The production of conidia is not of so common occurrence in the Basidiomycetes. They are produced in the rusts, smuts and some other Basidiomycetes. In the smuts, they are budded off from the basidiospores and the mycelium. The uredospores of rusts are also of conidial nature and function. The conidia in the Basidiomycetes are produced by the dikaryotic mycelium (A). They serve to propagate the dikaryophase in the life cycle.

(*ii*) By Oidia (Fig. 13.6 C). These are small, hyaline thin-walled unicellular sections or fragments of the mycelium. They may be uninucleate or binucleate accordingly as they are produced by the breaking up of the primary or secondary mycelium. They usually do not round up or secrete thick walls to become spore-like. They germinate by means of germ tubes. The latter grow into new mycelia. In some species the oidia are segmented from special, short lateral hyphal branches called the **oidiophores** (C). They are segmented from the tip of the oidiophore in succession towards the base (basigenous succession). The oidia serve a double function. They may either germinate to form primary mycelia or bring about diploidisation. In the latter case the germinating oidium acts as a spermatium and fuses with the hyphal cell of an opposite strain (Fig. 13.5 C).

(*iii*) Budding and fragmentation constitute the unimportant vegetative means of asexual reproduction in Basi-diomycetes.

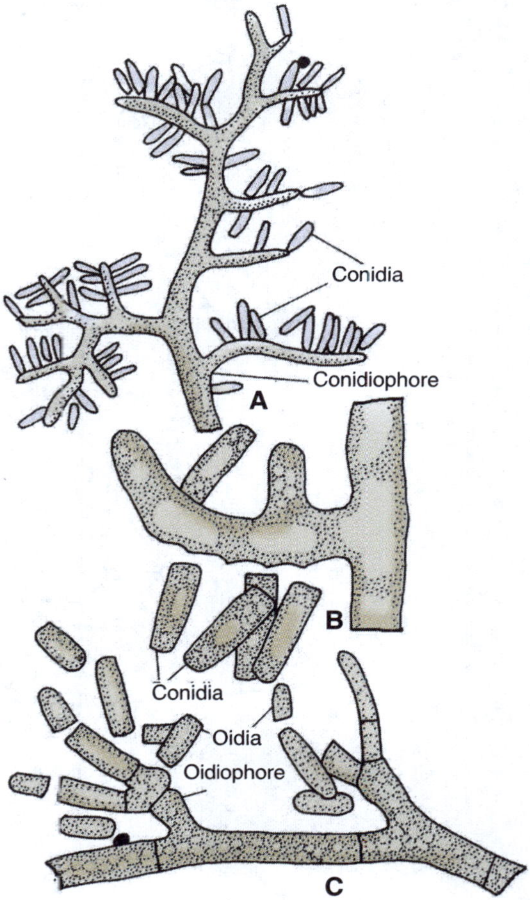

Fig. 13.6 (A-C). *Basidiomycetes*. Asexual reproduction by conidia (A and B) and oidia (C). A-B, *Corticum alutaceum* (after Lyman and Brefeld); C, *Coprinus filamentarius* primary mycelium with oidia and oidiophores (after Bensaude).

The Basidiocarps (Fig. 13.7). In the higher Basidiomycetes (class Homobasidiomycetidae) the secondary mycelium develops fruiting bodies called **basidiocarps**. The basidiocarps are usually

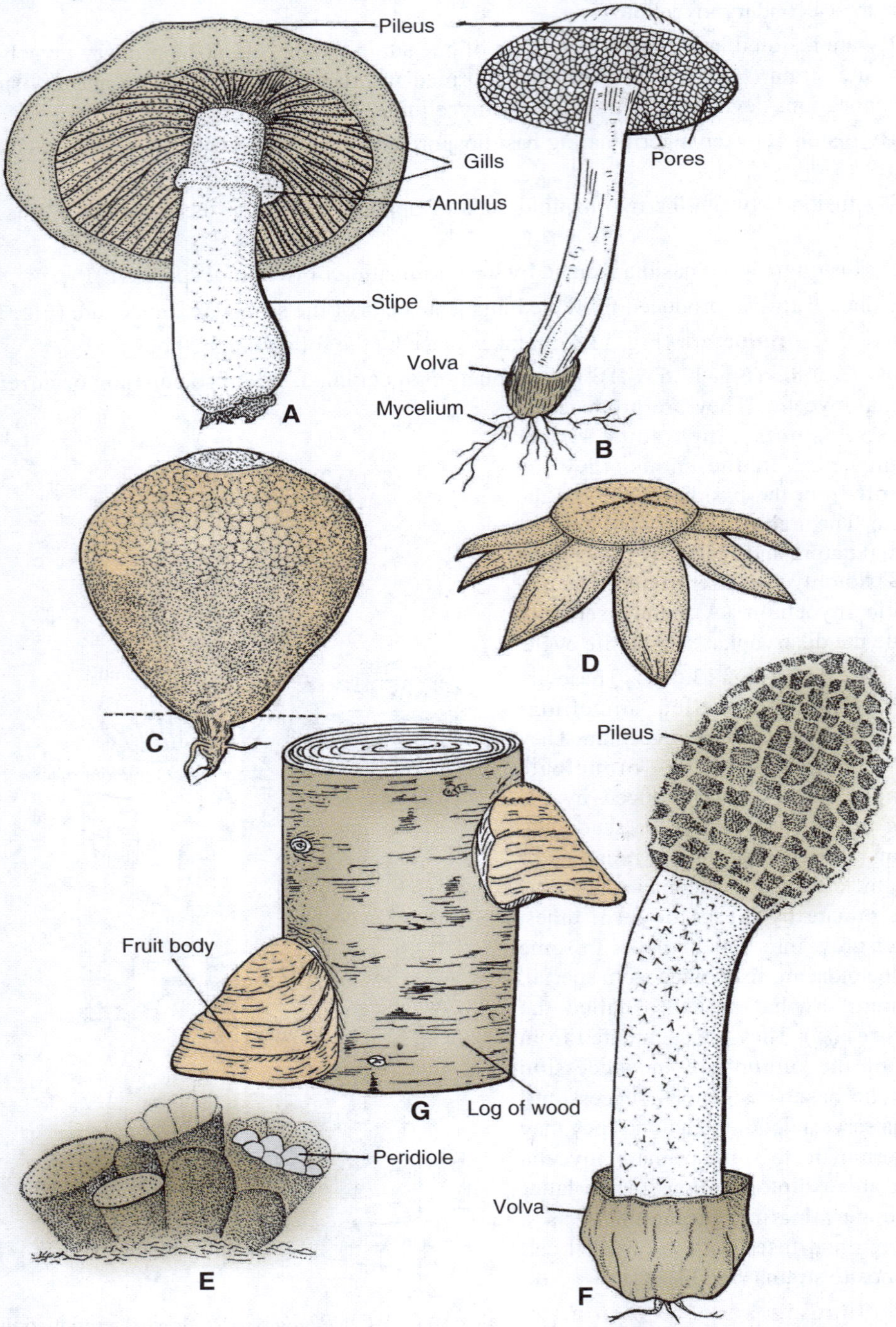

Fig. 13.7 (A-G). *Basidiomycetes.* Common types of basidiocarps. A, basidiocarp of *Agaricus* with gills on the underside of pileus; B, basidiocarp of *Boletus* with pores on the underside of pileus; C, basidiocarp of *Lycoperdon*; D, basidiocarp of *Geaster*; E, fruit bodies of *Cyathus*; F, mature fruit body of *Phallus*; G, fruit bodies of *Fomes*.

massive aerial sporophores which bear basidia. They are of various sizes, types, textures and forms. In texture they may be thin. crust-like, gelatinous, papery, thick and fleshy, leathery, corky, woody and spongy. In size they range from small microscopic objects to macroscopic bodies 3 feet or more in diameter. In form they may be umbrella-shaped, fan-shaped, round and the like.

The portion of the secondary mycelium which forms the fructifications (basidiocarps) is, sometimes, called the **tertiary** or **generative** mycelium. It is dikaryotic.

Basidia. The basidia which are characteristic reproductive structures of this class are of two types in general, the **holobasidia** and the **phragmobasidia**. The former are aseptate and thus unicellular and the latter are septate structures. The holobasidia are characteristic of most of the Basidiomycetes particularly the gilled or fleshy fungi. They are developed in a palisade-like layer on the basidiocarp. This fertile layer is called the **hymenium**. Interspersed among the basidia are the sterile hyphae known as the paraphysis. The hymenial layer may be exposed from the beginning or exposed towards maturity or remains closed throughout.

(*a*) Development of holobasidium (Fig. 13.8). The simple, club-shaped or more or less cylindrical **holo** or **homobasidium** lacks septa, and has a rounded apex. It originates as a terminal cell of a binucleate hypha of the secondary or tertiary mycelium in the basidiocarp (Fig. 13.8). The narrow elongated, binucleate young basidium is separated from the supporting hypha by a septum (a). A clamp connection is generally found at the basidium over the separating septum. During further development,

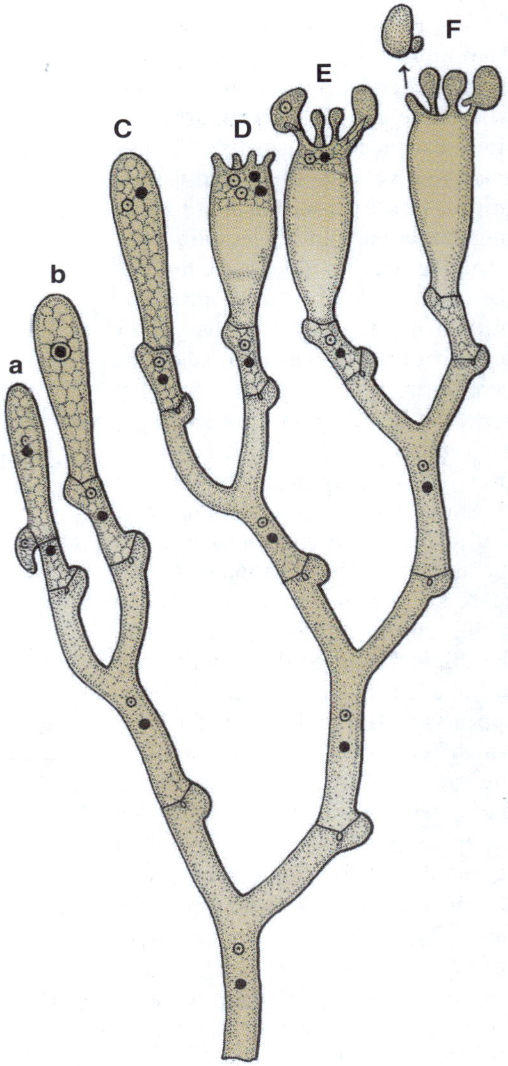

Fig. 13.8 (A-F). *Basidiomycetes.* Diagram illustrating the development of a holobasidium from the terminal cell of the dikaryotic mycelium.

the young basidium increases in size and becomes broader. Its two nuclei fuse (**karyogamy**) to form a fusion nucleus or **synkaryon** (*b*). Soon after, the fusion nucleus (synkaryon) undergoes meiosis to form four haploid nuclei (*d*). In the higher Basidiomycetes the basidium remains unseptate or single celled (*e*). It is thus called the **holobasidium** (homobasidium). The holobasidium is not morphologically differentiated into probasidium (hypobasidium) and metabasidium (epibasidium). In this case the two terms denote the two different stages of development of the same structure (basidium). The early stage of development of the holobasidium when **karyogamy** takes place represents the probasidium and the later stage when **meiosis** takes place represents the metabasidium. The holobasidium closely resembles the ascus in its development and cytology up to the initiation of basidiospores. It differs from the ascus in two respects, namely, the production of exogenous spores and their number which is four instead of eight.

(*b*) **Formation of Basidiospores.** The club-shaped holobasidium pushes out four slender projections at its top (*d*). These are the **sterigmata**. The tip of each sterigma swells to form a sac-like swelling, the **basidiospore initial**. Into each basidiospore initial migrates a haploid nucleus from the basidium. During its passage through the narrow neck of the basidiospore initial the haploid nucleus in the mushrooms is said to assume vermiform shape (*e*). It again becomes spherical in the spore. Subsequently each basidiospore initial is separated from its respective sterigma by a wall. Hawker (1967) reported that no cross wall is formed after the migration of the nucleus. The protoplast of the spore secretes a new wall around it within and in initimate contact with the original wall of the basidiospore initial. The basidiospore wall thus appears two-layered. The outer layer which represents the parent wall of the spore initial is known as **perispore**. The inner layer is called the **epispore**. It is the true spore wall secreted by the spore protoplast. Usually the perispore and epispore are fused. Typically the basidia are four spored structures (F). The basidiospores are borne externally. Each basidiospore has a small lateral outgrowth near the juncture with the sterigma. It is called the **hilum**. Of the four basidiospores two are of plus strain and two of minus strain.

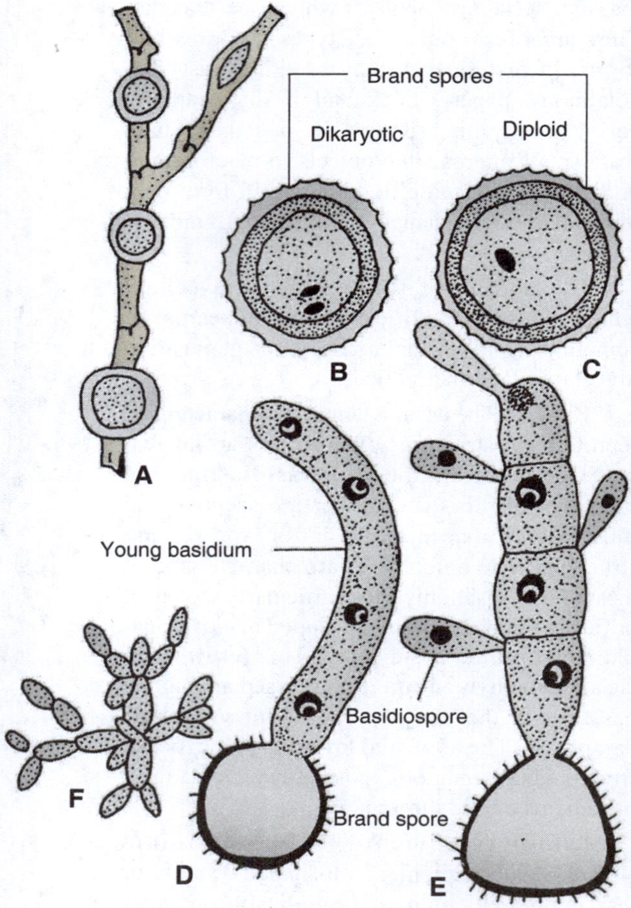

Fig. 13.9 (A–E). *Basidiomycetes. Ustilago* sp. A, Dikaryotic hypha forming brand spores; B-C, brand spores (dikaryotic and diploid); D, Germinated diploid brand spore to form an epibasidium with four haploid nuclei arranged in a row; E, mature four-celled epibasidium bearing basidiospores; F, Brand spore of *Ustilago* budding in a nutrient solution (after Brefeld).

Development of Phragmobasidium (Heterobasidium, Fig. 13.9). The septate basidium or phragmobasidium is typical of the rusts and smuts which usually do not form any fructification or basidiocarps. The phragmobasidia are formed by the germination of spores produced by the rounding up of the binucleate cells of the dikaryotic mycelium in smuts (A) to form spores. These spores are usually thick-walled and are known as the **smut** or **brand spores** or **teleutospores**. Initially they are binucleate (B). These binucleate teliospores are sometimes called the **probasidia** (B). Later the two nuclei fuse to form a zygote nucleus or **synkaryon**. The thick-walled smut spore with a synkaryon is called encysted probasidium or **hypobasidium** (C). The latter germinates to produce a short germ tube, the **promycelium** or **epibasidium**. The zygotic nucleus undergoes meiosis and the resultant four haploid nuclei become uniformly distributed in the epibasidium (D). Septa are laid between the nuclei dividing the epibasidium into four uninucleate cells. The basidium at this stage is divisible into two parts, the first formed basal hypobasidium (or probasidium) and the latter formed distal epibasidium (or metabasidium). The phragmobasidium being septate is less like a typical ascus but the cytological events are again similar. Each epibasidial cell gives out a small, slender lateral projection, the **sterigma**. The tip of the sterigma enlarges to form a sac-like swelling, the **basidiospore initial**. Meanwhile the haploid nucleus in each epibasidial cell divides mitotically into two. One of these migrates into the developing basidiospore through its respective sterigma and the other remains in the basidial cell. The uninucleate basidiospore initials mature into basidiospores. A few species lack sterigma. In them the basidiospores are sessile. Each basidium typically bears four basidiospores (E). The number, however, varies from one to many. The basidiospores generally are unicellular, uninucleate, haploid, tiny structures. To begin with they are hyaline, single cells and may remain colourless or become pigmented. In form they may be oval, round or elongated. *Dacromyces* has septate basidiospores.

(*c*) *Discharge of Basidiospores* (Fig. 13.10 A-D). The basidiospores which are exposed on hymenium are usually perched in an oblique manner (asymmetrically) on the tips of sterigmata. As a rule they are discharged forcibly and in quick succession by the "water drop mechanism". As the basidiospore matures, the turgid basidium forces out of the sterigma tip a liquid which begins to collect in the form of a droplet at the base of the **basidiospore** (A). The droplet gradually grows bigger till it attains a certain size (B) and suddenly **pushes** off the basidiospore forcibly into the air to a short distance (C). The surface tension is said to provide the necessary force. The long distance dispersal is, however, dependent on air currents. The basidiospore carries the water drop with it.

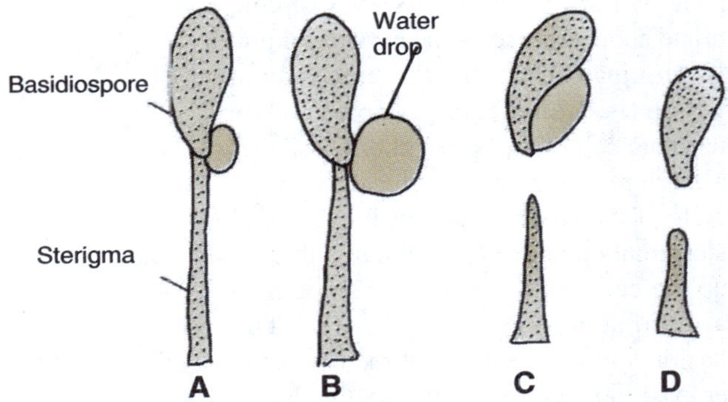

Fig. 13.10 (A-D). *Basidiomycetes*. Dispersal of basidiospores. (After Buller, 1922).

(*d*) *Germination of Basidiospores* (Fig. 13.1). On falling on a suitable substratum the basidiospore germinates. It puts out a germ tube (Fig. 13.1 A). The latter develops into a primary mycelium (Fig. 13.1 B). In a few species the basidiospores bud off into secondary basidiospores or conidia.

Kinds of Basidia (Fig. 13.11). The basidia vary in form in different groups of Basidiomycetes. In general they are of two types, namely, unseptate or **holobasidia** (D) and septate or **phragmobasidia** (A).

1. Holobasidium (D). It is a single-celled, unseptate, club-shaped structure with a rounded apex. The holobasidia are formed from the terminal binucleate cells of the secondary mycelium (Fig. 13.8) which enlarge to form club-shaped basidia borne in a palisade-like layer on the basidiocarp. Probasidium is the name given to the young basidium in which nuclear fusion occurs. At its top the basidium usually bears four, sometimes only two, basidiospores, each at the end of a short, slender process called the **sterigma**. Two of these are of one strain and two of the other. Holobasidium is characteristic of the order Agaricales (mushrooms and toad stools).

2. Phragmobasidium (Fig. 13.11 A). It is septate and is further divided into three kinds accordingly as the division is by transverse or vertical septa or have a deeply incised apex.

(*a*) Stichobasidial type. The phragmobasidium in this type is cylindric and transversely septate. It is differentiated into two parts, the first formed portion and the latter formed portion.

The former is called **probasidium** or **hypobasidium** and the latter **metabasidium** or **epibasidium**. The septa in some genera are formed in the hypobasidium (A). Each cell of the hypobasidium produces an unseptate epibasidium laterally. It bears terminally a basidiospore on a short sterigma. This kind of basidium is typical of the Auriculariales. The epibasidia are formed only on one side of the hypobasidium (A). In *Ustilago* (Fig. 13.9 E) diploid brand spore represents the hypobasidium. It germinates to form the epibasidium which is transversely septate. The terminal cell of the epibasidium produces a sterigma at its apex. The three other cells of the epibasidium push out lateral sterigmata one each more than half way up the cell. In *U. maydis,* the sterigmata are absent (Fig. 13.9 E). The basidiospores are borne directly on the epibasidial cells. In *Tilletia* the brand spore (hypobasidium) germinates to form a transversely septate four-celled epibasidium which bears an apical cluster of sickle-shaped, septate sporidia. In rusts also the transverse septa are formed in the epibasidium. Each cell of the epibasidium bears a basidiospore on a lateral sterigma.

(*b*) **Chiastobasidial type** (Fig. 13.11 B). In this case, the phragmobasidium is vertically septate. The septa are formed in the hypobasidium which is more or less rounded. Each cell of the hypobasidium is prolonged into a long slender branch at its apex. It is the **epibasidium**. The latter bears a basidiospore at its tip on a slender sterigma. This type of basidium is found in *Exidia nucleata* (Tremallales).

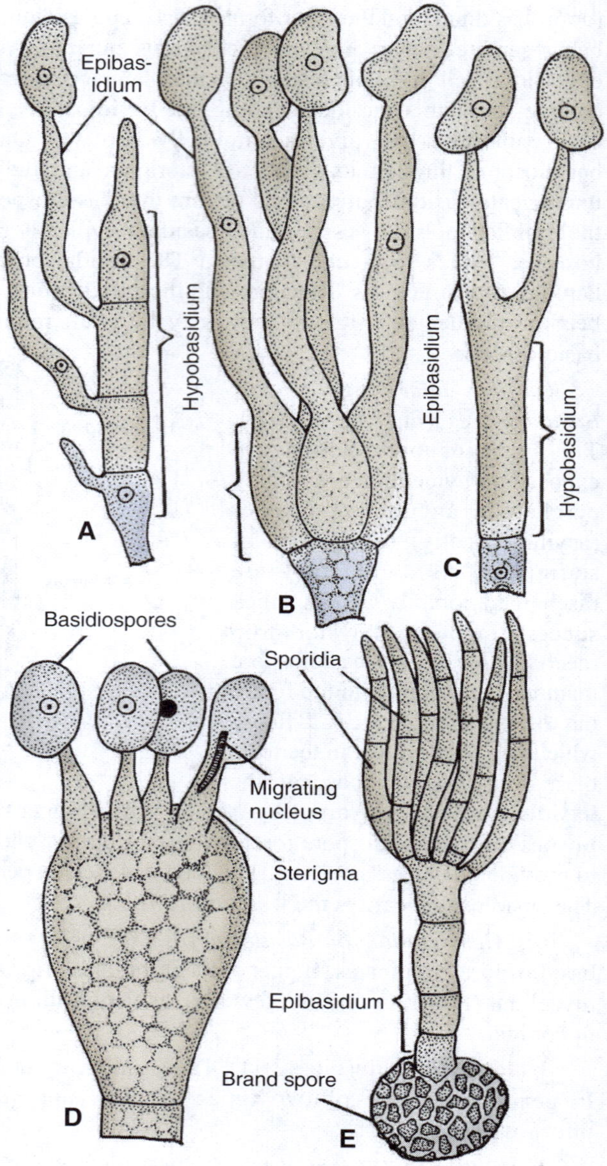

Fig. 13.11 (A-E). *Basidiomycetes.* Different types of basidia. A, Stichobasidial type; B, Chiastobasidial type; C, Tuning-fork type; D, Holobasidium; E, Stichobasidial type with a terminal cluster of septate, sickle-shaped sporidia.

(*c*) **Tuning Fork Type** (Fig. 13.11 C). The probasidium or hypobasidium is borne at the end of a binucleate hypha. It is narrow and elongated with a wall thicker than the parent hypha. Its tip is prolonged into two long arms known as the **epibasidia**. The whole structure thus resembles a **tuning fork type**. Each epibasidium at its tip bears a small pointed sterigma supporting an obliquely perched basidiospore. This type of basidium is of universal occurrence in the family

Dacrymycetaceae (*Dacrymyces deliquescens*). The phragmobasidium in this type lacks septation. The epibasidia grow from the tip of the hypobasidium.

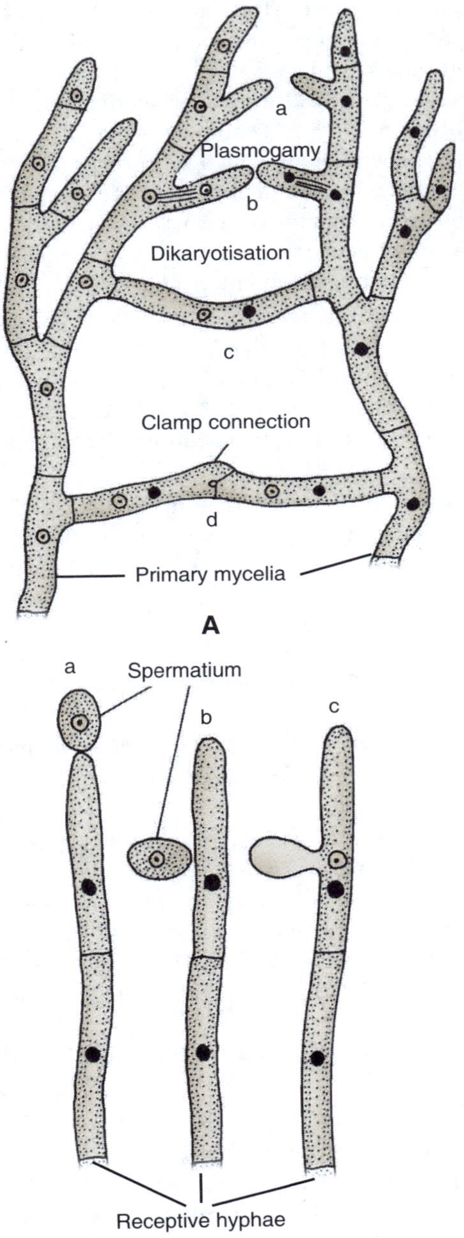

Fig. 13.12 (A-B). *Basidiomycetes*. Plasmogamy by somatogamy (A) and spermatisation (B).

Dacromyces deliquescens growing on wood

Sexual Reproduction (Fig. 13.12). The development of sex organs, antheridia and ascogonia, is universally absent throughout the class. Majority of the species are **heterothallic**. Morphologically the mycelia are alike but they are different in their sexual behaviour. This rudimentary difference in sex, shown at the time of sexual fusion, is designated by *plus* and *minus* signs. These signs are called the **sexual strains**. Either of these mycelia, if cultured artificially, remains sterile. They form no fructifications. Fructifications are formed only if two mycelia of opposite strains come into contact. The sexual process is thus extremely simplified. It consists of three fundamental processes characteristic of sexual reproduction, namely, *sexual fusion* or *plasmogamy, karyogamy* and *meiosis*.

1. Plasmogamy (Fig. 13.12 A-B). It means the union of two protoplasts whereby the sexual nuclei of opposite strains come close together in a pair within the same cell. In Basidiomycetes plasmogamy is accomplished either by *somatogamy* (A) or by *spermatisation* (B).

(*a*) *Somatogamy* (Fig. 13.12 A). Two somatic hyphae of the primary mycelia of opposite strains come in contact (b). The walls between the adjacent cells at the point of contact dissolve (c). The two nuclei come to lie side by side in the fusion cell which thus becomes binucleate (c). This sexual union or plasmogamy by fusion of somatic cells is called **somatogamy**. The binucleate or dikaryotic cell thus formed, by successive divisions and clamp formation (d) at each division, develops into a secondary mycelium. Plasmogamy thus is basically the means to initiate the dikaryophase in the life cycle. In the homothallic species plasmogamy takes place by the formation of tubular connections between the somatic cells of the same mycelium.

(*b*) *Spermatisation* (Fig. 13.12 B). It is another method whereby plasmogamy occurs. Plasmogamy by spermatisation exclusively takes place in the rusts. The rusts produce numerous, tiny, uninucleate, non-motile, spore-like bodies called the **spermatia**. They are formed in flask-shaped organs, the **spermagonia** (Fig. 14.15) developed on the upper surface of the leaf of the second host. The spermatia are carried by the various agencies. Generally, they are carried by the insects to the special **receptive hyphae** of opposite strain produced in another spermagonium. The spermatia adhere to these hyphae at the tips (*a*) or laterally (*b*). At the point of contact the wall dissolve and a pore is formed (*c*). The contents of the spermatium, which function as a male gamete, migrate through the pore into the receptive hypha and make it binucleate (*c*). The receptive hypha thus functions as a female organ. Plasmogamy by the union of a spermatium with a receptive hypha is known as **spermatisation**. Some mycologists use this term in a wider sense to refer to the contact of detached non-motile cells such as spermatia, microconidia, conidia and oidia with trichogyne or a receptive hypha.

2. Karyogamy. The terminal binucleate or dikaryotic cells of the hyphae of the secondary mycelium develop into basidia (Fig. 13.8). The two nuclei in the dikaryotic cell fuse. This fusion of the two nuclei is called **karyogamy** (Fig. 13.8 b). The resultant diploid fusion nucleus is called a **synkaryon**. The young basidium containing the synkaryon is called the **probasidium**. It represents the transitory diplophase.

3. Meiosis. The synkaryon in the probasidium soon undergoes two nuclear divisions (Fig. 13.8 c-d). These divisions constitute **meiosis**. Meiosis restores the haploid condition in the life cycle. **Karyogamy** and **meiosis** take place in the **basidium** at different stages of development. The basidium is thus homologous to the **ascus** of Ascomycetes.

Salient Features of Basidiomycetes

1. The somatic phase consists of a well developed, septate, filamentous mycelium which passes chiefly through two stages :
   (*a*) **Primary mycelium.** It is formed by the germination of a basidiospore and contains a single haploid (n) nucleus in each cell. It bears neither sex organs nor any basidia and basidiospores. It is short-lived.
   (*b*) **Secondary** or **dikaryotic mycelium.** It constitutes the main food absorbing phase and consists of cells each containing two haploid nuclei (n+n). It is long-lived and plays prominent role in the life cycle. In the Homobasidiomycetidae it may continue to grow for years producing fructifications every year by the interweaving of hyphae. The fructifications bear basidia and basidiospores. In the Heterobasidiomycetidae it forms teleutospores or brand spores which germinate to produce basidia bearing basidiospores.
2. Except in rusts and smuts the septal pore in the Basidiomycetes is complex. It is dolipore parenthesome type.
3. The motile cells are absent in the life cycle.
4. The clamp connections on the dikaryotic hyphae are of universal occurrence.
5. Asexual reproduction by spores plays an insignificant role in the life cycle. The Homobasidiomycetidae do not form any asexual spores. The Heterobasidiomycetidae form them in the dikaryotic mycelium. The latter produces uredospores and aeciospores in the rusts.
6. The sex organs are lacking in the Basidiomycetes. The sexual process is represented by plasmogamy and karyogamy. Karyogamy is immediately followed by meiosis.

7. Basidium is the characteristic reproductive organ of Basidiomycetes in which both karyogamy and meiosis take place.
8. Typically the basidium bears four basidiospores exogenously. The number, however, varies from one to many depending on the species.
9. The basidiospore germinates to produce the primary mycelium.

The Basidiomycetes and Ascomycetes resemble each other in the following respects :–

1. The Basidiomycetes and Ascomycetes are similar in their habit as both include parastitic as well as saprophytic species.
2. The purely terrestrial mycelium consists of septate, filamentous hyphae both in the Basidiomycetes and Ascomycetes.
3. The septa have each a central pore in both.
4. The motile cells are completely lacking in the life cycle of both the classes.
5. The sex organs which are completely absent in the Basidiomycetes have gradually been eliminated from the life cycle in the advanced Ascomycetes.
6. The sexual process in both the classes comprises plasmogamy and karyogamy. The latter is immediately followed by meiosis.
7. The delayed fusion of nuclei of opposite strains after plasmogamy has resulted in the origin and establishment of a binucleate or dikaryophase in the life cycle of both Basidiomycetes and Ascomycetes.
8. The characteristic reproductive organ, **basidium** of Basidiomycetes and **ascus** of Ascomycetes resemble each other in development and cytology till the initiation of spores. Both arise from the binucleate cells, the basidium from the dikaryotic hyphae and **ascus** from the ascogenous hyphae. Karyogamy and meiosis both occur in the basidium as in the ascus.
9. The basidiospores are usually uninucleate and haploid as are the ascospores in most of the Ascomycetes.
10. The dikaryotic mycelium in the Basidiomycetes is comparable to the ascogenous hyphae of the Ascomycetes.
11. The **basidiocarps** of Basidiomycetes are comparable to the **ascocarps** of Ascomycetes but the two are not homologous structures.
12. A clamp connection of the Basidiomycetes is considered homologous in structure and analogous in function to the hook of the Ascomycetes.

The Basidiomycetes differ from the Ascomycetes in the following respects :

1. The septal pore in most of the Basidiomycetes is not a simple hole as in Ascomycetes but is a complex structure known as a **dolipore**. The actual pore is barrel-shaped. It is surrounded by a swollen rim which is a part of the annular septum. The opening of the pore on either side is guarded by a curved **pore cup** called **parenthesome**.
2. The primary mycelium consisting of cells, each with a haploid (n) nucleus is short-lived. On the other hand the primary mycelium in the Ascomycetes is dominant and long-lived.
3. The primary mycelium in the Basidiomycetes bears neither the sex organs nor basidia or basidiospores.
4. The conidia (uredospores and aeciospores) are borne on the secondary mycelium whereas in the Ascomycetes they are borne on haplomycelium.
5. Excepting the Uredinales all traces of sexual apparatus have been lost throughout the class.
6. Presence of clamp connections is a characteristic feature of the secondary mycelium.
7. The dikaryotic (secondary) mycelium is a long-lived, independent structure whereas the ascogenous hyphae of the Ascomycetes which are homologous to the former are short-lived, dependent and occur only inside the fruit body.

8. The dikaryotic mycelium which is the result of a single mating of compatible hyphae may produce a large number of fructifications instead of only one as in the Ascomycetes.
9. The fruit bodies in the Basidiomycetes consist entirely of dikaryotic hyphae whereas in the Ascomycetes the basal hyphae, peridia and paraphyses are haploid.
10. The basidium bears a definite number of basidiospores, which is usually four whereas the ascus of the Ascomycetes usually produces eight ascospores.
11. The basidiospores are produced externally or exogenously on the basidium whereas the ascospores are produced inside or endogenously in the ascus of the Ascomycetes.

Economic Importance

Many of the Basidiomycetes are of great economic importance because of their beneficial as well as harmful nature. Some of them are the causative agents of most destructive diseases of our cereal crops. To this category belong the **smut diseases** of corn, wheat, oats, and barley as well as the **wheat rusts**. They destroy several million rupees worth of crops every year. Some of the higher Basidiomycetes such as the **pore fungi** are the common wood rotters. They destroy lumbar and timber. *Mushrooms* which also belong to this group are of great economic value as food. They are regularly cultivated for being delicious. The young fleshy sporophores of many species of puff balls (*Lycoperdon* and *Clavatia*) are also edible. *Clavatia* contains an anticancer substance

Lycoperdon – Fruit bodies

Clavatia – Fruit bodiy

calvacin. The **toad stools**, however, are poisonous. Some of these such as *Amanita* are fatally poisonous whereas others cause only discomfort. Many members of this class form ectotrophic mycorrhizal associations with the roots of forest trees. The association is mutually beneficial. The fungus obtains sugars and other organic substances from the roots of the tree partner whereas the mantle of the fungal partner serves to pass on nitrogen, phosphorus, and other elements absorbed by the mycelium to the root. The saprophytic Basidiomycetes play a significant role in decomposing the dead fallen leaves and other forest litter converting waste material and returning it to the soil.

Amanita – A fruit body

Origin and Phylogeny. Two hypotheses have been put forth to explain the origin of the Basidiomycetes. They are the **Ascomycetean hypothesis** and **Phycomycetean hypothesis**. The consensus of opinion now favours the first view.

Ascomycetean Hypothesis. According to this hypothesis, Basidiomycetes have evolved from Ascomycetes. This view is based on the fact that the Basidiomycetes resemble Ascomycetes more closely than any other class of fungi. These resemblances are :–

1.. Their hyphal structure is similar.

2. In both the classes plasmogamy is not immediately followed by karyogamy. It is delayed for some time so that the resultant fusion cell contains two nuclei.
3. Secondary mycelium is homologous to the ascogenous hyphae of the Ascomycetes.
4. Clamp connections of the Basidiomycetes are homologous to the croziers of the Ascomycetes.
5. Asexual reproduction takes place by means of conidia which are produced in the same way in both the classes.
6. Basidia are homologous to the asci. Both have a similar origin and early development. In fact basidium is considered an early evolutionary development from an ascus of the Ascomycetes.
7. Basidiospores are homologous to the ascospores.
8. Basidiocarps are comparable to the ascocarps.

From the above-mentioned similarities mycologists conclude that the Basidiomycetes have arisen from the Ascomycetes. In fact they believe that a basidium is an evolutionary development from an ascus. The fact that in some Ascomycetes the ascus produces only four ascospores gives further support to the close affinity between the two classes.

The views regarding evolution within the Basidiomycetes are rather speculative. Some mycologists think that the Basidiomycetes with septate basidia (phragmobasidia) are nearer to the ancestral Ascomycetes. On this basis the Heterobasidiomycetidae such as rusts (Uredinales) are considered primitive. The rust spermagonium suggests a relationship to the Ascomycetes. According to this view, the unseptate basidium (holobasidium) of the Homobasidiomycetidae is more advanced and derived from the septate basidium by simplification in which the septa disappeared. The massive fructifications of the Homobasidiomycetidae support the view that they are more advanced than the Heterobasidiomycetidae.

Some mycologists, however, maintain that the Heterobasidiomycetidae are more advanced. They have arisen from the simpler members of the Homobasidiomycetidae much like the ancestors of *Agaricus*. This viewpoint is supported by the fact that the unseptate basidium (holobasidium) is very much similar to the unseptate ascus. According to this interpretation, Homobasidiomycetidae are more primitive and closer to the ancestral Ascomycetes.

Classification. Alexopoulos, Smith, Hawker and other modern mycologists classify Basidiomycetes on the basis of origin and structure of basidium as follows :–

Basidiomycetes

- Basidium partitioned or deeply divided
  - Sub-class 1. **Heterobasidiomycetidae**
    - Order 1. *Tremellales*
    - Order 2. *Uredinales*
    - Order 3. *Ustilaginales*
- Basidium not septate or deeply divided
  - Sub-class 2. **Homobasidiomycetidae**
    - Series 1. Hymenomycetes (Hymenium exposed)
      - Order 1. *Exobasidiales*
      - Order 2. *Polyporales*
      - Order 3. *Agaricales*
    - Series 2. Gasteromycetes (Hymenium enclosed)
      - Order 4. *Hymenogastrales*
      - Order 5. *Lycoperdales*
      - Order 6. *Sclerodermetales*
      - Order 7. *Phallales*
      - Order 8. *Nidulariales*

Ainsworth (1973) proposed a system of classification which is as follows :

Basidiomycotina

- Basidiocarp absent, teliospores grouped in sori or scattered within the host tissue organised in a hymenium; saprophytic or rarely parasitic
  Class: **Teliomycetes**
- Basidiocarps present and well-developed; basidia typically organised in a hymenium; saprophytic usually rarely parasitic
  - Basidiocarps gymnocarpous or hemiangiocarpous, basidia phragmobasidia or holobasidia, basidiospores ballistospores
    Class : **Hymenomycetes**
  - Basidiocarps angiocarpous; basidia holobasidia, basidiospores not ballistospores
    Class : **Gasteromycetes**

QUESTIONS

ESSAY TYPE

1. Give the distinguishing characters of the class Basidiomycetes and discuss the chief features in which it differs from the Ascomycetes.
2. List in a tabular form the resemblances and differences between the Basidiomycetes and the Ascomycetes.
3. Describe habit, habitat and economic importance of the Basidiomycetes.
4. Describe the somatic organisation of Basidiomycetes.
5. List the various ways in which diploidisation takes place in the Basidiomycetes.
6. Write notes on diploidisation, clamp connections and spermatisation.
7. Describe the various modes of asexual reproduction met with in the Basidiomycetes.
8. Give an account of sexual reproduction in the Basidiomycetes.
9. From the examples known to you discuss as fully as you can the sexuality found in the Basidiomycetes.
10. Give a comparative account of the development of a germ sporangium of *Rhizopus*, an ascus and a basidium.
11. "There has been a progressive degeneration of sex organs during evolution of Fungi". Justify this statement with the help of suitable examples. *(Kanpur M.Sc., 2001)*

SHORT ANSWER TYPE

12. Describe the development of a basidium and give a list of the various kinds of basidia met with in the class Basidiomycetes.
13. Describe the significance of the role played by plasmogamy, karyogamy and meiosis in the life cycle of Basidiomycetes.
14. Distinguish between conidia, ascospores and basidiospores. Also throw some light on the role played by these spores in the life cycle of Ascomycetes and Basidiomycetes.
15. Describe in detail the structure and developmental stages of Basidium. How will you differentiate between a basidium and an ascus. *(Kanpur M.Sc, 1991, 2003)*
16. Explain the significance of the role played by plasmogamy, karyogamy and meiosis in the life cycle of Basidiomycetes. *(Gorakhpur, 1993)*
17. Describe the various methods of dikaryotisation in Basidiomycetes. *(Lucknow, 1998; Punjab, 1993)*
18. How does the development of the ascus differ from that of a basidium?
19. Write short notes on :
    (*i*) Basidium and basidiospores *(Kumaon, 1999; Lucknow, 1996)*
    (*ii*) Basidiospores *(Kumaon, 1997)*
    (*iii*) Clamp Connection

(Kumaon, 2001; Kanpur M.Sc, 1991; 1994; 2003; Allahabad, 1992; 1995; 1998 Awadh, 1992; 1997; 1998; Lucknow, 1993; 1996; 1999; Allahabad M.Sc. 2000; Kanpur, 2002)

(*iv*) Dikaryotisation
(Kanpur, 1992; Allahabad, 1998; Awadh, 1996; Bihar, 1991; Poorvanchal M.Sc., 2003, Rohilkhand M.Sc., 2006)

(*v*) Dolipore Septum
(Kumaon, 2000, Allahabad, 2000, 2002; Kanpur M.Sc, 1996; 1998; Allahabad, 1993)

(*vi*) Basidium Formation (Kanpur M.Sc., 1999)

(*vii*) Distinguishing characters of Basidiomycotina (Poorvanchal, 2002)

20. How will you distinguish between :

(*a*) Diploidization and dikaryotisation, (Allahabad, 1991)

(*b*) Ascus and Basidium
(Allahabad, 1991; 1995; 1998; 1999; Gorakhpur, 1991; Awadh, 1993; Lucknow, 1998, 2005 ; Magadh, 1993, Rohilkhand, 2002)

(*c*) Clamp Connection and Crozier formation (Allahabad, 1993; Awadh, 1999)

(*d*) Phragmobasidia and Holobasidia (Allahabad, 1999)

(*e*) Dikaryotisation and Diplanetism (Rohilkhand, 1995)

(*f*) Ascomycetes and Basidiomycetes (Kanpur, 2006)

21. Explain the following:

(*a*) Basidiocarp (Kanpur, 1995)

(*b*) Clamp Connection
(Lucknow, 2004; Kumaon, 2004; Kanpur, 1995)

(*c*) Phragmobasidium

(*d*) Holobasidium

(*e*) Holobasidium and Heterobasidium (Kumaon, 2001, 2004)

22. Describe cell division in Basidiomycetes. (Kanpur, 2002)

OBJECTIVE TYPE

23. Fill in the blanks :

(*i*) .... discovered the basidium.

(*ii*) The somatic phase in Basidiomycetes is made up of .... mycelium.

(*iii*) The characteristic spores of Basidiomycetes are known as ....

(*iv*) Basidiospores are produced on ....

(*v*) The fruit body in Basidiomycetes is known as ....

(*vi*) The process by which primary monokaryotic mycelium becomes secondary dikaryotic mycelium is known as ....

(*vii*) .... connection indicates that the mycelium is ....

(*viii*) Septum in the mycelium of Basidiomycetous fungi is .... shaped and is known as .... septum.

(*ix*) Aseptate unicellular basidia are known as ....

(*x*) .... are also characterised by the absence of sex organs.

(*xi*) Genetically .... phase dominates in Basidiomycetes.

(*xii*) Formation of Basidiospore is .... white that of .... Ascospore is ....

(*xiii*) Basidiospore is uninucleate and ....

(*xiv*) Clamp connections are observed in ....

24. Select the correct answer :

(*i*) The number of basidiospores produced on each basidium is —
(*a*) 2
(*b*) 4
(*c*) 6
(*d*) 8

(*ii*) Basidiospores are characteristic spores of class
(*a*) Phycomycetes
(*b*) Ascomycetes
(*c*) Basidiomycetes
(*d*) Deuteromycetes

(*iii*) The process of dikaryotisation involves
(*a*) plasmogamy only
(*b*) plasmogamy associated with transfer of nuclei
(*c*) karyogamy
(*d*) plasmogamy and karyogamy

(*iv*) Somatic mycelium in Basidiomycetes is
- (*a*) Monokaryotic aseptate
- (*b*) Monokaryotic and septate
- (*c*) Dikaryotic and aseptate
- (*d*) Dikaryotic and septate

(*v*) Secondary dikaryotic mycelium is identified on the basis of the presence of
- (*a*) Clamp Connections
- (*b*) Antheridia
- (*c*) Oogonia
- (*d*) Ascogonia

(*vi*) Basidiospores are different from Ascospores because the former are produced
- (*a*) externally
- (*b*) internally
- (*c*) both externally and internally
- (*d*) neither internally nor externally

(*vii*) Septate structures bearing sterigmata on which basidiospores are produced are called
- (*a*) Holobasidia
- (*b*) Phragmobasidia
- (*c*) Basidiocarps
- (*d*) Ascocarps

(*viii*) Dolipore septum is characteristic feature of
- (*a*) Ascomycetes
- (*b*) Basidiomycetes
- (*c*) Discomycetes
- (*d*) Phycomycetes.

(*ix*) A phragmobasidium means
- (*a*) a septate basidium where both cells are fertile
- (*b*) a septate basidium where only one cell is reproductive
- (*c*) a septate basidium which for all purposes is one structure
- (*d*) an entire basidium which is fully reproductive

(*x*) The dikaryotic mycelium is characterised by the presence in each of its cells of
- (*a*) a single diploid nucleus
- (*b*) a single haploid nucleus
- (*c*) two haploid nuclei belonging to opposite strains
- (*d*) two haploid nuclei belonging to similar strains.

Grain Rusts

CHAPTER 14

SUB CLASS – HETEROBASIDIOMYCETIDAE

GENERAL CHARACTERS

This sub-class includes species which are specialised parasites of higher plants. The diagnostic features of this subclass are:

(*i*) Lack dolipore septa and clamp connections. Recently, however, clamp connections have been recorded in many species.

(*ii*) Formation of thick-walled binucleate spores known as the teleutospores.

(*iii*) Basidium partitioned transversely or longitudinally into two, three or four cells (**phragmobasidium**).

(*iv*) The basidia result from the germination of a thick-walled, resting spore (teleutospore).

(*v*) The ripe basidium divisible into two parts, the basal part called the **hypobasidium** and the distal part termed the **epibasidium.**

(*vi*) Basidiospores borne on the epibasidium may be sessile or stalked.

(*vii*) Number of basidiospores per basidium may be four or more than four.

Classificiation. There is a great controversy with regard to the classification of this sub-class into orders and families. Martin (1971) divided the sub-class into the following three orders :

(1) *Tremellales.* Most of the members are saprophytic but some are parasitic on higher plants, mosses or insects. The saprotrophs are common on dead tree branches and decaying logs. The mycelium produces a well-developed fructification called the **basidiocarp.** It has a jelly-like, waxy or cartilaginous texture. The basidia are arranged in a hymenial layer over the surface of the basidiocarp. Popularly the Tremellales are called the jelly-fungi.

(2) *Ustilaginales.* It includes the **smut fungi** which mostly parasitize higher plants. Basidiocarp is absent. Basidia are developed by the germination of thick-walled resting cells (teleutospores) which are usually black and called the **smut** or **brand spores.** The basidiospores are sessile and irregular in number. They are not discharged forcibly.

(3) *Uredinales.* They are obligate parasites of certain seed plants and ferns and are called the **rusts.** The basidiocarp is absent. The thick-walled resting cells are two-celled structures called **teleutospores.** They are brown to black in colour and on germination give rise to basidia. The basidiospores, which are stalked, are discharged forcibly. Each basidium usually produces four basidiospores.

In the Ainsworth's (1973) system of classification, the class **Teliomycetes** has been divided into two orders : **Ustilaginales** and **Uredinales.**

While in the order Ustilaginales, the basidiospores are sessile and passsively released, in the order Uredinales, basidiospores are produced on short pointed sterigmata and are forcibly discharged.

The orders Ustilaginales and Uredinales are discussed in some detail.

(I) ORDER : USTILAGINALES (THE SMUTS)

The Ustilaginales include the smut fungi. The smut fungi do not form fructifications (basidiocarps). They are highly specialised parasites of angiosperms. More than 1,100 species are known. They cause destructive diseases known as the **smut diseases.** They receive this name from the black dusty masses of spores resembling soot or smut formed within the tissues of the host plants. The smuts cause serious diseases of economic plants particularly cereal crops destroying usually reproductive organs such as the flower and seed. Examples of important smut diseases in the cereals are the *bunt disease* of wheat, *loose* and *covered smut* of barley, oat, corn, rye, sorghum, millet and wheat. The smut diseases cause millions of rupees worth damage annually to

Loose Smut of Barley

Loose Smut of Oats

Loose Smut of Rye

Covered Smut of Barley

Covered Smut of Oats

Covered Smut of Wheat

food crops and ornamentals. The smut spores may also produce allergic effect on the labourers working with the diseased grain. The mycelium in the host is intercellular and dikaryotic. The terminal or intercalary binucleate cells of the hyphae round off to form thick-walled resting spores variously named as chlamydospores, brand spores, smut spores or teleutospores. They are usually black in colour. The smuts which form spores in the ovaries of grasses are called *grain smuts*. The grain smuts of cereal grasses comprise two families *Ustilaginaceae* and *Tilletiaceae*. We shall deal only with the former in the text.

FAMILY : USTILAGINACEAE

The promycelium or epibasidium formed by the germination of the thick-walled resting spore (teleutospore) is trasversely septate. It is a four-celled structure. Each segment bears a single, sessile basidiospore which is discharged passively. The important genera of smut fungi included in this family are *Ustilago*, *Sphacelotheca* and *Tolyposporium*. *Ustilago* is taken as a type and discussed in detail while the genus *sphacelotheca* is discussed briefly.

Ustilago

Sphacelotheca

USTILAGO

General. *Ustilago* is the genus of smut fungi on which Ustilaginaceae and Ustilaginales are based. The genus claims about 400 species. All are parasitic. The species included in this genus destroy host tissues of the organs in which sporulation takes place. Virtually the entire mycelium at these sites is converted into spores after necrosis of the host tissues. No columellae or peridia of fungal origin are formed. All species of the genus complete their life cycle on one host and thus are called the **autoecious** parasites. The most common hosts are the members of the grass family particularly the cereals such as Corn, Oat, Barley, Wheat, Millet, Bajra and Rye. The smuts cause destructive diseases reducing the yield of grains by 20 to 50 per cent. They are thus responsible for several thousands of rupees worth loss annually in our country. In Punjab smut disease is called *Kangiari* and in U.P. *Kandua*.

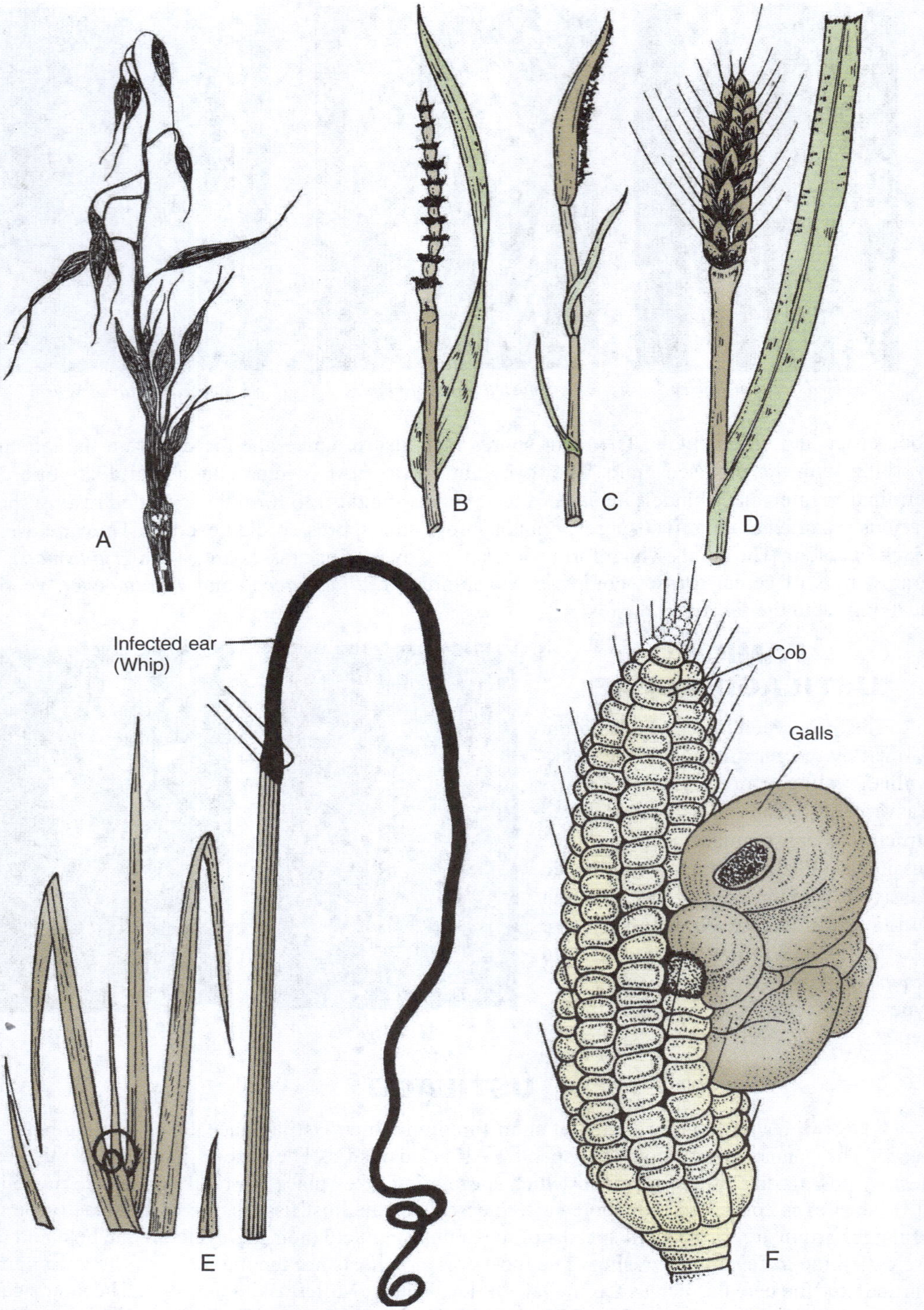

Fig. 14.1. Smuts caused by *Ustilago sp.* A, smut of oats; B, Loose smut of wheat; C, Doob smut; D, Covered smut of Barley; E, Smut of sugarcane; F, Corn smut.

Maclean and Cook described the Smut fungi as obligate parasites. Alexopoulos, however holds that all smuts are not obligate parasites. He claims to have obtained mycelia of a number of species on artificial media. Hawker (1967) stated that smut fungi are facultative saprophytes and not obligate parasites. They are capable of growth as saprophytes on agar media but grow only in the yeast-like form and do not form septate hyphae. In nature, however, the dikaryotic mycelium on the angiosperms appears to be obligately parasitic. The smut diseases are of world-wide occurrence. They are common in all parts of India.

Kinds of Smut diseases. The smut diseases are of two types. They are the **Loose smuts** and **Covered smuts.** In the former the black sooty mass of spores is exposed at the flowering time.

The spores are blown away by wind which carry them to a healthy inflorescence of host species where infection of the young ovary or of the inner surface of the enveloping pales takes place. The ripe grains of these secondarily infected ears contain internal dormant mycelium. In the covered smuts the spore masses remain covered by the wall of the grain and the glumes. They are not exposed. They are freed or liberated only during threshing when the wall of the grain ruptures under pressure. In this case the healthy grain is superficially contaminated with smut spores. Both the loose and covered smuts have little or no effect on the vegetative growth of the host. Oats and Barley suffer from both Losse and Covered smuts. A list of diseases caused by species of *Ustilago* is given in table VI below.

Each species attacks its own host and produces the disease characteristic of it. The species which normally cause the Loose smut will not produce the Covered smut or *vice versa.*

Mycelium. According to the nuclear behaviour, the mycelium of *Ustilago* passes through two distinct stages of development. These are the **primary** and **secondary mycelia.** The primary mycelium consists of hyaline, slender, septate hyphae with a single haploid(n) nucleus in each cell. This kind of mycelium is also called **monokaryotic mycelium** or **haplomycelium.** It is formed by the germination of a basidiospore (fig. 14.2 A). It might be of a plus or a minus strain according as it is developed from a plus or a minus strain basidiospore. It seldom develops very extensively. In most species the primary mycelium soon becomes converted into a secondary mycelium. The primary myeclium is thus of very short duration.

TABLE I

Name of disease with host	Causal organism
1. Loose Smut of oat.	*Ustilago avenae*
2. Loose smut of wheat	*U.nuda—var. tritici (U. tritici)*
3. Loose smut of Barley	*U.nuda*
4. Covered smut of Barley	*U.hordei*
5. Covered smut of oats	*U.kolleri*
6. Smut of maize	*U.maydis*
7. Whip smut of sugarcane	*U.scitaminae*
8. Anther smut of caryophyllaceae	*U.violacea*
9. Smut of doobgrass	*U.cynodontis*

The secondary mycelium consists of hyphae with two haploid (n+n) nuclei in each cell. Such hyphae are called dikaryotic hyphae. These dikaryotic hyphae are septate and extensively branched. The septa between the cells have each a central pore. The dolipore septal complex is, however, absent in the smuts. Through these pores the adjacent cells communicate with each other. The mycelium of most **species** of *Ustilago* found within the host is generally **dikaryotic** or **secondary**

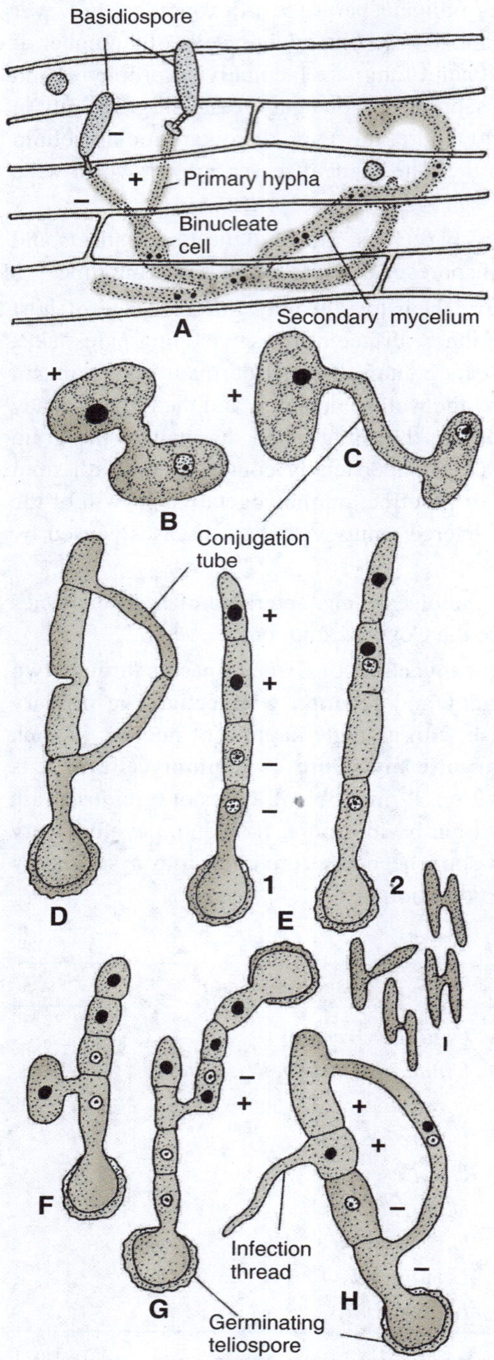

Fig. 14.2 (A-I). *Ustilago.* Various methods of dicaryotisation. Explanation in the text (A after Hanna, B-C after Harper, D after Dickson, E_1-E_2 after Rowitscher, G-H diagrammatic and I after Brefeld)

mycelium. It develops extensiely within the tissues and spreads to the various parts of the host. In fact the secondary mycelium constitutes the most conspicuous and important part of the somatic or vegetative phase of the majority of species of *Ustilago.* In many species, septa develop clamp connections. The hyphae ramify in the spaces between the host cells. They are thus **intercellular.** The intercellular hyphae may develop distinct **haustoria** which penetrate the walls of the host cells and absorb nutrition. The host cells, however, are not destroyed. In some species the haustoria are absent. In *Ustilago maydis*, the hyphae are **intracellular.** They penetrate the cells and obtain nutrition directly from the protoplasm of the host cells. The growth of the parasitic mycelium within the host tissues causes little or no disturbance to the vegetative development of the host plant. In some species, the mycelium is scattered throughtout the various parts of the host. It is said to be **systemic.** In others it spreads near the point of infection and is called **localised.**

Ustilago maydis

Dicaryotisation or diploidisation in Ustilago. (Fig. 14.2). The process whereby the primary mycelium produced by the germination of basidiospores changes into a secondary mycelium is called *discaryotisation* or *diploidisation.* The process is initiated by the pairing together of two haploid cells of opposite strains of a species. They copulate and one of them becomes binucleate. The two nuclei in the fusion cell constitute a **dicaryon**. They do not fuse in the vegetative phase. The resultant dicaryotic or binucleate cell develops into a dicaryotic hypha which by further growth forms the dicaryotic or secondary mycelium. The formation of a dicaryotic cell is a prerequisite to normal infection in *Ustilago.* In *U. maydis* (corn smut) copulation to form the dicaryotic cell occurs inside the host tissue (corn plant) but in all other species, in gereal, it occurs outside the host. The various methods of diploidisation in *Ustilago* are detailed below :—

1. *By hyphal fusions (somatogamy) between primary mycelia* (A). In *U. maydis,* the basidiospores or sporidia fall on the surface of the host (corn plant) and germinate to produce haploid mycelia. The latter penetrate the host epidermis and grow horizontally beneath. Dicaryotisation takes place within the host by means of hyphal fusions [somatogamy between hyphae of suitable mating (opposite) strains]. It may take place immediately after penetration within the host or after there has been some growth of primary mycelia. Subsequent migration of nuclei into the fusion cells initiates the dicaryotic phase. Binucleate cells thus are formed by elongation and repeated cell division by clamp connections from the secondary mycelium.

2. *By Fusion between the Germ Tubes of two germinating basidiospores* (B-C). As the basidiospores germinate the germ tubes of the basidiospores of opposite strains meet and fuse. The intervening walls at the point of contact dissolve. The nucleus of one germ tube migrates into the other. The latter becomes binucleate. It grows into a secondary mycelium. Example of this type is *U.hordei.*

3. *By Conjugation between the basidiospores.* In some species the basidiospores multiply by budding to produce secondary spores (sporidia). The secondary spores or sprout cells of opposite strains (copulate). The common wall between them dissolves at the point of contact or they send copulation tubes towards each other. The nucleus of one migrates into the other throught the connecting link (I). The binucleate sporidium, or sprout cell, on germination, produces the secondary mycelium. *U. receptacularum* and *U. violocea* are common examples.

4. *By union of the basidiospores of one strain with the germ tube of the basidiospores of another strain.*

5. *By the union of infection threads. U. tritici* is an example (H). The promycelium or basidium does not bear the basidiospores. Its haploid cells grow into small, slender hyphae one each. These are called the **infection threads.** Two neighbouring infection threads of the opposite strains fuse. The nucleus of the one passes into the other. As a result one of the infection threads becomes binucleate. It grows to form the secondary mycelium. Similarly in *U. nuda* fusion between the compatible cells of the epibasidium takes place by conjugation tubes (D). The conjugated binucleate cell forms a binucleated hypha which infects the host.

6. *By fusion between the two haploid cells of the same epibasidium* (E1, E2.) In this case fusion takes plce between two haploid cells of opposite strains of the same basidium. *U. hordei* and *U. carbo* are the examples.

Ustilago nuda

7. *By fusion between two basidia* formed by the germination of smut spores of opposite strains (G). *U. nuda* is an example.

8. In *U. violacea* the binucleate cell may arise by the union of a basidiospore with one of the basidial cells of the opposite strain (F).

Reproduction. Sex organs are absent in *Ustilago.* Plasmogamy, karyogamy and meiosis, the three fundamental events of the sexual process do occur. Plasmogamy takes place by the fusion of two haploid compatible cells. The binucleate cell thus formed by repeated divisions produces the secondary or dikaryotic mycelium. The intercellular hyphae of the latter feed on the host plant, accumulate reserve food materials and reaching a certain stage of development enter the *sporulation stage.*

1. Sporulation (Fig. 14.3). In wheat, oat and barley, the invading secondary mycelium becomes active at the flowering time of the host. It grows vigorously and reaches the inflorescence region where it branches profusely and infects embryonic spikelets. The parenchymatous tissue in the embryonic spikelets is destroyed and occupied by the hyphal masses. By the time, the head or ear emerges from the host leaf, it is generally completely destroyed (B).

Sporulation starts in the centre of the hyphal mass and progresses outwards as hyphal proliferation continues. The hyphae divide by additional septa into shorter binucleate segments called the *spore fundaments.* These hyphae are called the **sporogenous hyphae.** They are closely interwined. The binculeate protoplast of each segment functions as the spore initial. The spore formation in *Ustilago* is thus endogenous and the sproes are formed singly inside the hyphal segments. Sporulation is preceded by the thickening of the hyphal walls and their subsequent gelatinisation. The sporogenous hyphae thus lose their identity. The spore initials (binucleate cell protoplasts) lie in a hyaline, gelatinous matrix. They are, at first, variously shaped but become globular as they enlarge. Each secretes a new wall around it to become a **teliospore** or **brand spore.** By the time the spores are morphologically mature, the gelatinous material disappears. The spores are closely appressed into a hard, compact mass called a **smut ball** or **sorus.** The spores in the sorus are readily separable by slight pressure. The sorus or smut ball is covered by a peridium of host cells in *U. hordei.* Virtually all the hyphae in hyphal mass are converted into spores after necrosis of the host tissues. No peridia or columellae of the fungal origin are formed. In *U. hordei* the group of spikelets at each node of the rachis forms a single irregularly shaped spore mass or sorus.

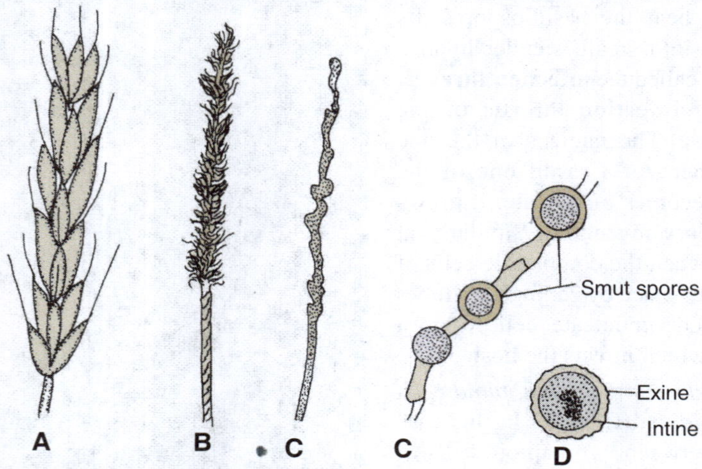

Fig. 14.3 (A-C). *Ustilago tritici.* A, Normal ear of wheat; B, Smutted ear of wheat; C, Naked, rachis left after the spores are blown away by wind; C, A portion of a hypha with smut spores; D, A binucleate smut spore.

U. avenae

U. tritici

U. hordei

U. nuda

The differentiation and development of spores in other species of *Ustilago* such as *U. avenae, U. tritici* and *U. nuda* may closley follow the pattern of events described above in *U. hordei.* The only difference is that in *U. avenae* and *U. tritici,* the fungal hyphae grow profusely between the anticlinal walls of the host epidermis and destroy the latter. The mature sori in these two species are thus naked, *U. nuda* with a growth pattern similar to *U. hordei* has the smut balls or sori enclosed in a fragile peridium derived from the host tissue. The smuts in which the sori are covered by the membranous covering or peridium are called **covered smuts.** In loose smuts, the sori are naked. Each smut ball or sorus contains numerous, thick-walled, rounded spores. The thick spore wall is differentiated into two layers (D). The outer **exine** or **exosporium** is thick. It may be smooth, reticulate, or spiny. The inner **intine** or **endosporium** is always thin. The binucleate smut spores are generally the resting spores. They remain dormant under adverse conditions. Some mycologists call the smut spores as **teleutospore.** The older mycologists termed them **chlamydospores.** The use of the term chlamydospores for the smut spores of *Ustilago* appears to be inappropriate. The smut spores are binucleate structures produced only by the binucleate cells of the secondary mycelium which originates as a result of plasmogamy (sexual fusion). They are thus reproductive in nature and homologous to the teleutospores of rusts rather than to the chlamydospores.

The smut spores are dispersed by wind, insects, or washed by water. When all the spores are blown off small rachis are left behind on the infected ear. In *U. tritici*, the smut spores do not function as resting spores. They serve as means of propagating the disease during the growing season. They fall on the stigmas of the flower and soon germinate to infect the ovules in the ovaries of the healthy plants. The spores of the covered smuts are liberated by the rupture of the walls of the grains at the threshing time.

2. *Germination of smut spore to form the Basidium* (Fig. 14.4). Ths smut spores carried by the wind may fall on the soil, on the grain, and other favourable places. Under suitable conditions such as warmth and moisture they germinate. Duran and Safeeulla (1968) reported that in most smuts optium temperature for smut spore germination ranges from 20 to 30°C. Light also stimulates

smut spore germination. (*ii*) *Karyogamy.* Prior to germination the two nuclei (one of plus and the other of minus strain) in the smut spore fuse to form a **synkaryon.** It is diploid. The thick-walled smut spore with a synkaryon represents the encysted **probasidium** or **hypobasidium** stage (A). It absorbs moisture and swells up. The exosporium or the epispore layer ruptures. The endospore or the endosporium protrudes in the form of a short, cylindrical hypha, the **promycelium.** The promycelium is also called the **epibasidium** or **metabasidium.** (*iii*) *Meiosis.* The diploid nucleus migrates into the epibasidium and divides twice. These two divisions constitute **meiosis** (B) and (C). The resultant four nuclei in the epibasidium are thus haploid. Since segregation of the sexual strains takes place during meiosis two of these nuclei are of **plus** strain and two of **minus** strain.

They are arranged in a row (C). Septa

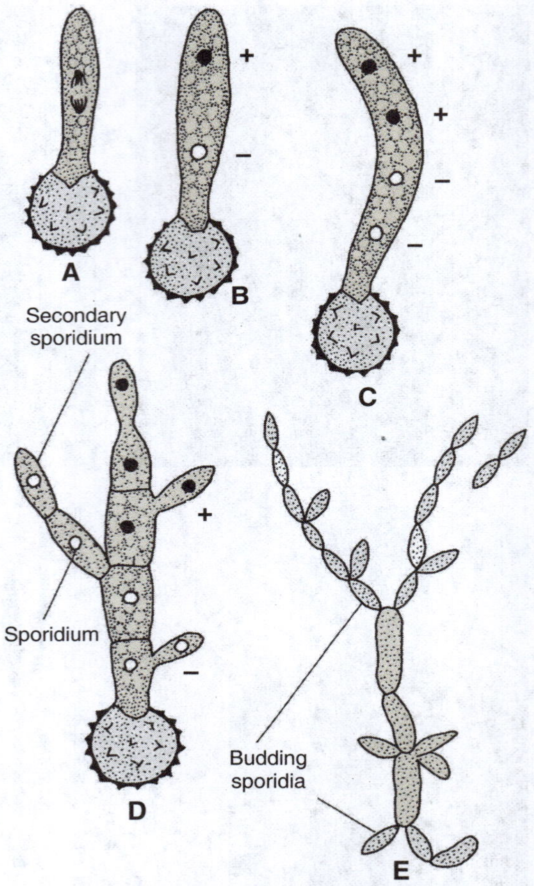

Fig. 14.4. *Ustilago maydis.* A-D, stages in the germination of smut spore and development of a phragmobasidium which bears the basidiospores or sporidia; E, promycelium budding to form numerous secondary sporidia or sprout cells.

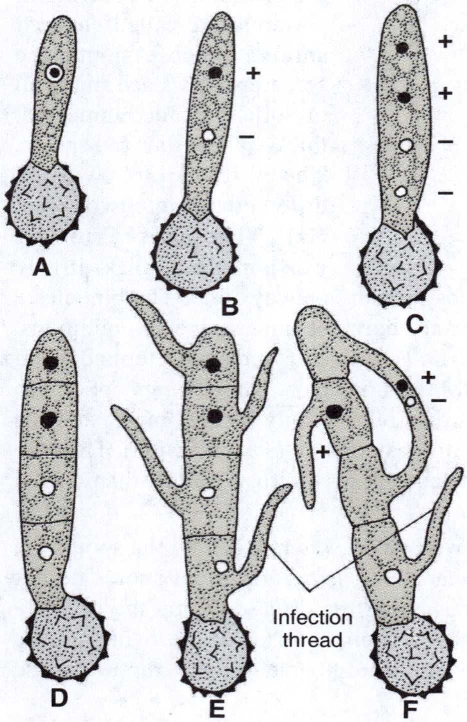

Fig. 14.5 (A-F). *Ustilago tritici.* Stages in the germination of smut spore and development of basidium and formation of infection threads instead of basidiospores.

are laid between the nuclei (D). The epibasidium at this stage is composed of four haploid cells.

The haploid nucleus in each cell of the epibasidium divides mitotically into two daughter nuclei. One of these remains in the cell. The other migrates into the developing bud which arises laterally from each epibasidial or promycelial cell (E). The haploid uninucleate buds later separate as **basidiospores,** also called the **sporidia.** Two of these are of plus strain and two of minus train. The basidiospores of *Ustilago* are not borne on sterigmata. Consequently the sessile basidiospores of smuts are discharged passively. They are dispersed by wind and thus may happen to fall on the soil or on the host plant.

The basidiospores of some species such as *U. maydis* are capable of multiplying by budding like the yeast cell (F). The new spores formed by budding are called **secondary spores** or **conidia.** In *U. tritici* which parasitizes wheat the basidiospores are lacking. The haploid cells of the epibasidium or the promycelium, instead produce slender, short hyphae (Fig. 14.5 E). These are called the **infection threads.**

Germination of basidiospores and infection of the Host. The basidiospores or the secondary sporidia produced from the them by budding germinate either on the soil or on the young host plant (*U. maydis*) itself. Each basidiospore produces a fine germ tube, also called the **infection tube.** The germ tube is haploid (monokaryotic). In most speices it cannot infect the host tissues. Exception is *U. maydis.* Infection is generally brought about by the dikaryotic germ tube. Dikaryotisation or diploidisation of the germ tubes if brought about differently in different species of *Ustilago.* The following examples will illustrate the point :

1. *Ustilago hordei.* It causes *Covered smut* of Barley. The basidiospores germinate in the soil or on *barley* grains as the latter are sown. The germ tubes produced by them are unable to bring about infection. Diploidisation is brought about by fusion between the haploid germ tubes of the two basidiospores of opposite strains (Fig. 14.2 B-C). As a result one of the tubes becomes dikaryotic (binucleate). The dikaryotic germ tube is capable of infecting the young barley seedling at a very early stage as it emerges from the grain. It gains entry into the host seedling through the hypocotyl and reaches the coleoptile. *U. hordei* thus provides an example of infection at the seedling stage. Loose smut of Oats caused by *U. avenae* is also an example of infection at the seedling stage.

2. *Ustilago tritici* (Fig. 14.5). The smut spores germinate on the feathery stigmas of the flower. Each produces a four celled **promycelium** or **epibasidium** (D). The cells of the epibasidium do not bear basidiospores. Instead each basidium cell produces a slender tubular outgrowth, the **infection thread** (E). It is haploid. The infection threads of the same basidium with nuclei of opposite strains fuse to form a binucleate (dikaryotic) hypha (F). The latter grows through the style until it reaches the ovary which it penetrates. In the ovary it ramifies in the intercellular spaces of the ovary tissue. By the tenth day of its origin it gains entry into the ovule. It is an example of infection through the flower. The mycelium lies dormant in the grain (Fig, 14.6 A), and is again activated when the grain germinates (Fig. 14.6 B). It spreads and grows along with the seedling (Fig. 14.6 C) till the latter matures and produces flower. The mycelium finally invades the

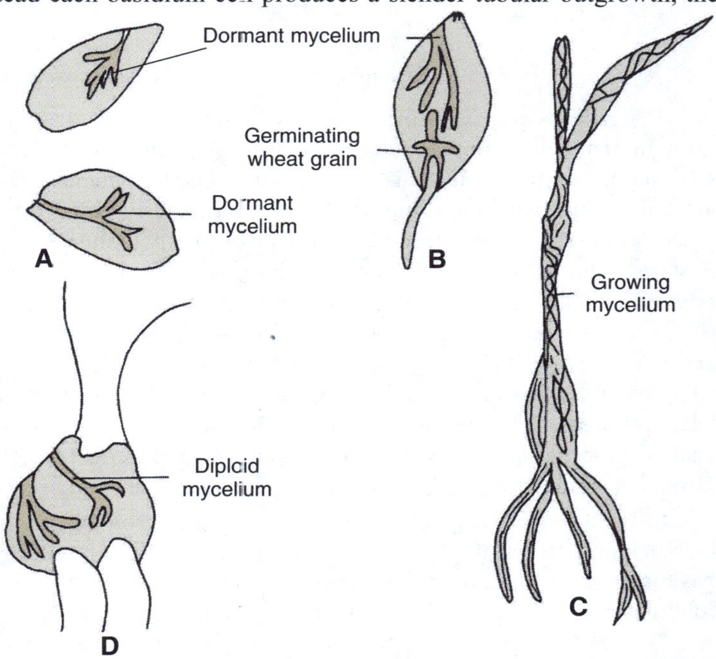

Fig. 14.6 (A-D). *Ustilago tritici.* A, Infected wheat grain with a dormant mycelium; B, Germinating wheat grain with an activated mycelium; C, Wheat seedling with the myceliium growing and spreading; D, Diploid mycelium has finally invaded the ovary.

ovaries (Fig. 14.6 D) and ovules. Inside the ovaries it produces millions of **smut spores** which are exposed by the decay of host tissues. When the wind blows the spores are carried away leaving the naked rachis (Fig. 14.3 C).

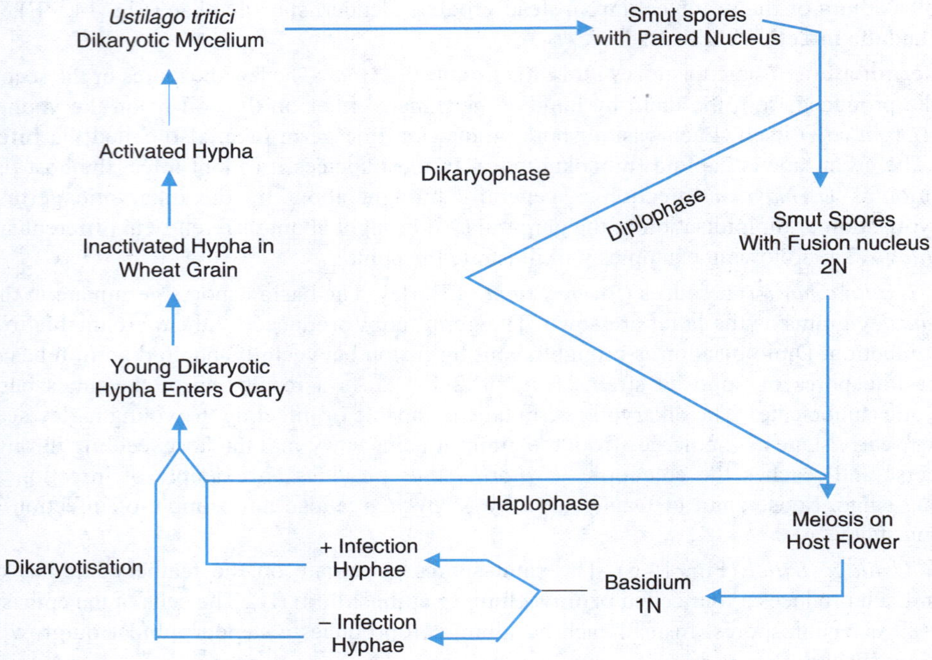

Fig. 14.7. Graphic representation of the life cycle of *Ustilago tritici.*

3. *Ustilago maydis (Corn smut* Fig. 14.4) It is an example of general primary infection through many embryonic tissues of the host. The smut spores (teliospores) which are produced in summer are, at first, binucleate when young. The two nuclei eventually fuse. The mature spores are thus uninucleate and diploid. They lie dormant in winter on corn debris or other favourable places in the soil. They germinate in the following planting season of maize. The thick spore wall ruptures. Through the split emerges a short cylindrical germ tube known as the *promycelium* or *epibasidium.* (A). Immediately the diploid nucleus migrates into the latter and undergoes meiosis (B-C). The resultant four haploid nuclei are distributed uniformly and are arranged in a row. Septa are laid between the nuclei in the epibasidium (D). Each cell of the latter bears a haploid basidiospores. The basidiospores are capable of budding. These basidiospores or the secondary basidiospores are carried by the wind. They happen to fall on a young corn plant. There each basidiospore germinates to produce a uninucleated germ tube (Fig. 14.2 A). It enters the host through a stoma or pierces the wall of the epidermal cell and brings about primary infection.

Infection can take place any time during the growing season and through any young and meristematic part of the host (stem, leaves, ears, tassels, etc.). The haploid germ tubes from two basidiospores of plus and minus strain fuse in the tissue of the host and produce a binucleate cell (dikaryotic cell). This is **diploidisation** by somatogamy or somatogamous copulation (Fig. 14.2 A). The resultant binucleate or dikaryotic cell grows by elongation and cell division by clamp connections to form a full-fledged secondary mycelium. The cells of the secondary mycelium (dikaryotic mycelium) are binucleate. The secondary mycelium plays a dominant role and carries on the life cycle of the fungal parasite. It ramifies intercellularly and even intracellularly throughout the tissues of host. It is reported that some of the hyphae of the secondary mycelium that reach the surface of the host, produce several crops of binucleate conidia during the growing season. The mature binucleate conidia are dispersed by wind. Falling on the host the conidia initiate new or secondary infections. The disease spreads in this way. Eventually the secondary mycelium develops extensively at certain points. At these points the extensive development of the mycelium causes swellings called galls or tumours (Fig. 14.8). These tumours can appear on any portion of the host *e.g.* stem, leaves, ears, tassels. Each swelling contains an indefinite number of smut spores.

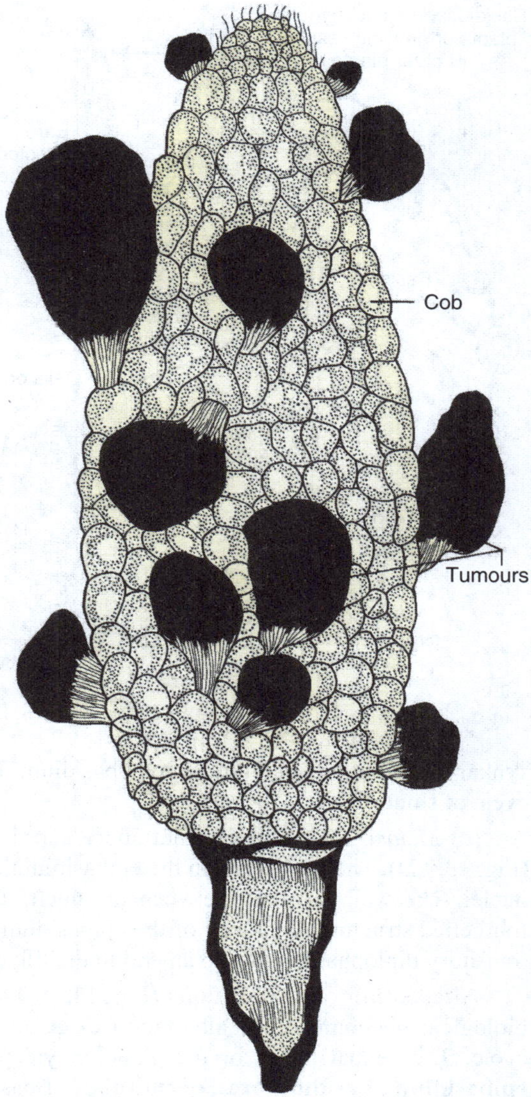

Fig. 14.8. *Ustilago maydis.* Infected cob of Maize showing tumours.

Sexuality in Ustilago. No sex organs are developed in *Ustilago*. The sexual process is represented by three fundamental phenomena characteristic of it, namely, **plasmogamy, karyogamy** and **meiosis.**

(*a*) *Plasmogamy* (Fig. 14.2). Heterothallism is common in the genus *Ustilago*. The mycelia though morphologically alike are different physiologically. Physiologically they are unisexual. There is, however, no apparent distinction into male and female mycelia. They are different only in their sexual behaviour. The difference of sex is thus very rudimentary. It is denoted by the signs *plus* and *minus*. Such mycelia are said to be **heterothallic.** Plasmogamy in heterothallic species is brought about by different methods of diploidisation. It may be accomplished by conjugation between basidiospores of opposite strains (B-C). Union may as well take place between a basidiospore of one strain and a cell of the basidium of opposite strain (F). There may be fusion between basidia of different smut spores (G). Diploidisation is also brought about by somatogamous copulation between vegetative cells of the two hyphae of opposite strains. In either case a binucleate condition is established in one of the conjugating cells. The binucleate cell is also called the **dikaryotic cell.** The dikaryotic condition once established is maintained for a considerable period in the life cycle. Plasmogamy therefore initiates **dikaryophase** in the life cycle. The binucleate cell by elongation and division generally by clamp formation develops into a secondary mycelium.

(*b*) *Karyogamy.* With karyogamy the dikaryophase ends. The two nuclei in the smut spore **fuse**. This fusion between the two nuclei may be regraded as a culmination of the sexual process begun at the time of diploidisation. It is equivalent to the fertilisation process.

The diploid nucleus formed in this way is called a **synkaryon.** The smut spore with a

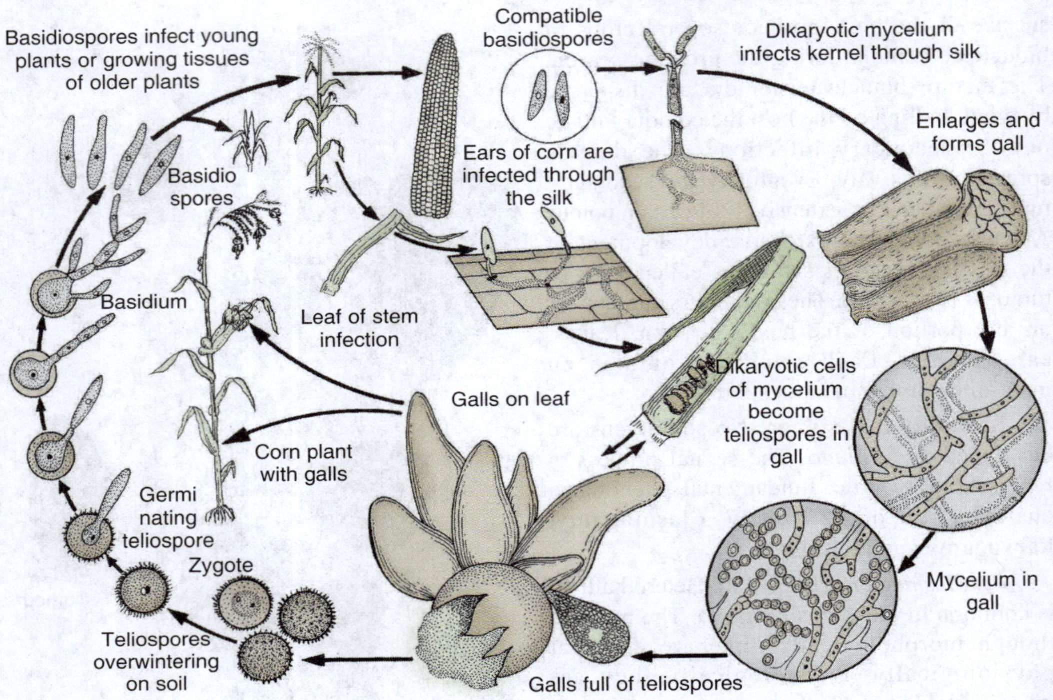

Fig. 14.9. *Ustilago maydis.* Diagrammatic disease cycle.

synkaryon is the probasidium or hypobasidium. It represents the transitory **diplophase** in the life cycle of smuts.

(*c*) *Meiosis.* The diploid smut spore germinates to form the **promycelium** or **epibasidium** (Fig. 13.9 D). The synkaryon in the epibasidium undergoes **meiosis** to form four halpoid daughter nuclei. The walls are laid between the nuclei (Fig. 13.9 F). The epibasidium thus becomes a fourcelled structure. Each cell of the epibasidium bears a haploid **basidiospore.** With meiosis the transitory diplophase comes to an end in the life cylce of *Ustilago.*

Alternation of Generations (Fig. 14.7). The life cycle of *Ustilago* illustrates the important biological phenomenon of alternation of generations. There are two distinct phases in the life cycle. The sexual phase or the gametophyte phase is represented by the haploid four-celled **epibasidium, basidiospores, germ tubes** of basidiospores and the **haplo** or **primary mycelium** in some species (*U. maydis*). It ends with plasmogamy which initiates the **dikaryophase** in the life cycle. The dikaryophase in smuts consists of the **dikaryotic** or **secondary** mycelium and the binucleate **smut spores.** With karyogamy which consists in the fusion of the two nuclei in the smut spore ends the dikaryotic phase. The smut spore with the synkaryon (probasidium) represents the transitory **diplophase.** The dikaryotic phase together with the transitory diplophase consitutes the **sporophyte phase.** The sporophyte phase ends with meiosis. With meiosis starts the future gametophyte. These two phases alternate with one amother in the life cycle of *Ustilago.* One regularly follows the other. Hence *Ustilago* is said to exhibit alternation of generations in its life cycle.

Control. Smut diseases can be controlled by crop rotation, seed disinfection and by growing seeds of disease resistant or immune varieties.

Taxonomic Position

Alexopoulos (1962)			Alexopoulos and Mims (1979)		
Division	:	**Mycota**	*Super kingdom*	:	**Eukaryonta**
Sub division	:	**Eumycotina**	*Kingdom*	:	**Mycetae**
Class	:	**Basidiomycetes**	*Division*	:	**Amastigomycotina**
Sub class	:	**Heterobasidiomycetidae**	*Sub division*	:	**Basidiomycotina**
Order	:	**Ustilaginales**	*Class*	:	**Basidiomycetes**
Family	:	**Ustilaginaceae**	*Sub class*	:	**Teliomycetidae**
Genus	:	*Ustilago*	*Order*	:	**Ustilaginales**
			Family	:	**Ustilaginaceae**
			Genus	:	*Ustilago*

SPHACELOTHECA

General. *Sphacelotheca* causes smut of *Sorghum* (Jowar). In India, three species of the genus attack Jowar plants causing serious diseases. The grain smut or covered smut of Jowar is caused by *S. sorghi*, the loose smut of Jowar is caused by *S. cruenta* and *S. reiliana* causes the head Smut of Jowar. These diseases are very common in the states of Andhra Pradesh, Madhya Pradesh, Karnataka, Tamilnadu and Uttar Pradesh. Of the three smut diseases of Jowar, covered smut or grain smut causes serious losses to the crop and is the most destructive.

Sphacelotheca cruenta

Sphacelotheca reiliana

Symptoms. The fungus attacks the ovaries and all the ovaries in the head are turned into sori of the smut. The normal grain is thus replaced by an oval or cylindrical dirty grey sac with a slightly conical tip. In most cases stamens are absent but the stigma is not transformed. However, other floral parts of the host are not affected.

Mycelium. The mycelium is mostly intercellular without haustoria, sparsely branched, septate and hyaline. The primary mycelium formed by the germination of basidiospores enters the host through seedlings. When the primary mycelium of the opposite strain also happens to enter the same seedling, dikaryotic mycelium is formed through anastomosis and fusion. This dikaryotic mycelium grows with the seedling and produces the smut sori in the ovaries.

Reproduction. During threshing, the teleutospores are released from the infected ovaries and get lodged on the surface of seeds. These remain viable for a maximum of period of five years. Generally in the next season, in presence of water the teleutospores germinate to produce a three

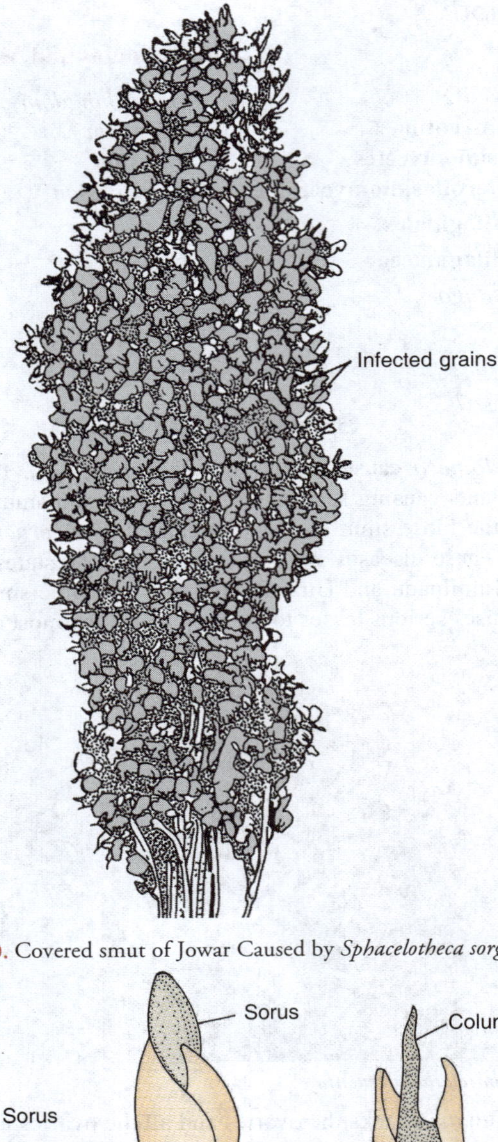

Fig. 14.10. Covered smut of Jowar Caused by *Sphacelotheca sorghi*

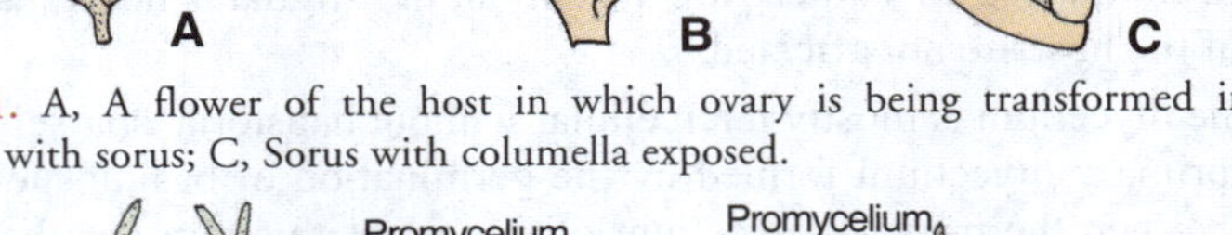

Fig. 14.11. A, A flower of the host in which ovary is being transformed into sorus; B, A grain with sorus; C, Sorus with columella exposed.

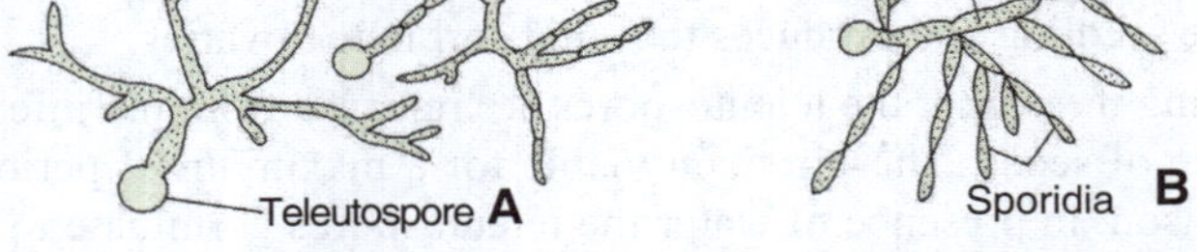

Fig. 14.12. *Sphacelotheca sorghi*. A, Germination of Teleutospore and formation of promycelium; B, Development of sporidia on promycelium.

celled promycelium which through budding produces terminal as well as lateral **sporidia.** The promycelium may also act as infection hyphae which may enter the seedlings of the host.

Generally sporidia fuse in pairs and form dikaryotic mycelium. The dikaryotic mycelium enters the seedlings, following mainly the apical meristem of the seedlings. This mycelium ultimately reaches the ovaries of the flowers which ultimately gets filled with dikaryotic mycelium. Individual cells of the hyphae round off, nuclei of the cell fuse and the cells develop a thick wall around them. These thick walled, uninucleate diploid cells are **teleutospores.** A large number of these spores are formed which fill the grain which is transformed into **a sporesac** or **sorus.**

Each sorus is dark grey in colour, enclosed by a wall made up of fungal tissue, oval or cylindrical and conical at the tip and measure 125 mm in length and 40 mm in width. At the base the sorus is surrounded by unaltered glumes. At the centre of the sorus is present **Columella** which is a slender column of hard tissue made up of fungal mycelium and host tissue. The sorus is filled by teleutospores but hyaline empty cells are also present in the sorus. Individiual teleutospores are smooth, globose to oval and brownish olive in colour which appears dark brown in mass. After release from the sorus, the teleutospores come to lie on the surface of seeds and empty sorus with protruding columella is left.

In the next season, the teleutospores germinate in presence of water. The outer exospore ruptures and the inner thin endospore comes out in the form of a germ tube. The diploid nucleus undergoes meiosis to produce four nuclei. The germ tube becomes a three celled promycelium which buds off haploid sporidia.

Taxonomic position

Alexopoulos (1962)			Alexopoulos and Mims (1979)		
Division	:	**Mycota**	*Super kingdom*	:	**Eukaryonta**
Sub division	:	**Eumycotina**	*Kingdom*	:	**Mycetae**
Class	:	**Basidiomycetes**	*Division*	:	**Amastigomycotina**
Sub class	:	**Heterobasidiomycetidae**	*Sub division*	:	**Basidiomycotina**
Order	:	**Ustilaginales**	*Class*	:	**Basidiomycetes**
Family	:	**Ustilaginaceae**	*Sub class*	:	**Teliomycetidae**
Genus	:	*Sphacelotheca*	*Order*	:	**Ustilaginales**
Species	:	*sorghi*	*Family*	:	**Ustilaginaceae**
			Genus	:	*Sphacelotheca*
			Species	:	*sorghi*

Other Important Smut Genera

Entyloma. *E. oryzae* causes leaf smut of rice.

Entyloma oryzae

Leaf Smut of Rice

Neovossia. *N.indica* causes karnal bunt of wheat and *N.horida* causes bunt of rice.

Neovossia indica

Neovossia horrida

Bunt of Rice

Tilletia. *T.caries* and *T.foetida* cause stinking smut of wheat or coverd bunt of wheat. The sori of the fungus emit smell of rotten fish. The teleutospores of *Tilletia* are H-shaped.

Tilletia caries

Stinking Smut of Wheat

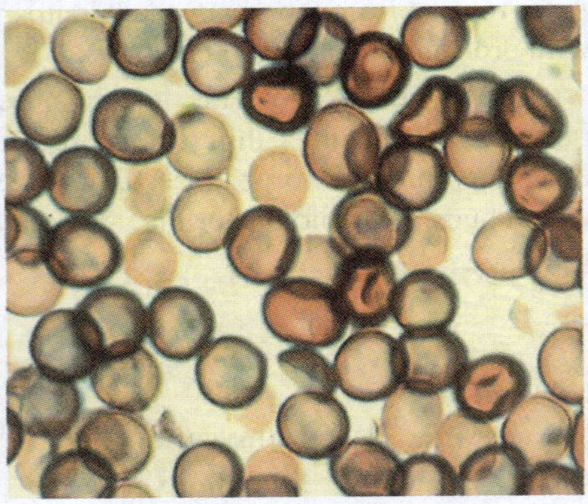
Tilletia foetida

Tolyposporium. *T. penicillarie* causes smut of Bajra—a very common disease. Charcteristic spore-balls are produced by the fungus and the sorus lack the columella.

Tolyposporium ehrenbergii causes long smut of Jowar.

Urocystis. Flag smut of wheat is caused by *U. tritici*; produce characterisitc spore balls in which the central three to four fertile cells are surrounded by a sheath of flatened strile cells. *U. brassicae* causes gall smut of mustard Roots while *U. cepulae* causes onion smut.

Flag Smut of Wheat

Onion Smut

QUESTIONS

ESSAY TYPE

1. What are smuts? Describe the somatic organisation of the smuts.
2. Describe the various methods of diploidisation met with in the smuts whereby the secondary mycelium originates.
3. Describe the mode of development, structure and function of the various kinds of spores met within the smuts.
4. Give an account of the loose smut of Wheat. *(Punjab, 1957, 1958)*
5. Describe in detail the life history of any species of *Ustilago* (smut) you have studied. *(Punjab Uuiv . 1994; Allahabad 199 0)*
6. Write what you know about sexuality in *Ustilago* and alternation of generations.
7. Describe the germination of basidiospores of three different species of *Ustilago* and modes of infection of the host.
8. Give a list of the species of *Ustilago* causing loose smuts and covered smuts of the important cereal crops.
9. With the help of diagrams only, illustrate the life cycle of *Ustilago maydis*. *(Kanpur, 1996)*
10. Describe the causal organisms, symptoms and methods of control of the following ;
    (*a*) Loose smut of wheat *(Kanpur, 1998; Lucknow, 1992, 1996, 1998)*
    (*b*) Loose smut of Barley *(Kanpur, 1999)*
    (*c*) Smut of Jowar (*d*) Smut of maize
11. What are smuts? Describe the structure and reproduction in *Ustilago*. *(Awadh, 1993; Bihar, 1992)*
12. Illustrate the life cycle of causal organism of Rust or Smut disease. *(Awadh, 1996)*
13. Describe the various modes of infection in the smuts and suggest the methods of control. *(Kanpur M.Sc., 1984, 1992)*
14. What are smuts? Describe the life history of one of the smuts laying emphasis on cytological changes. *(Kanpur, M.Sc., 1988, 1996)*
15. Give an illustrated account of the life cycle of any species of *Ustilago* studied by you. *(Allahabad, 1992)*
16. What is loose smut? How does it differ from covered smut? Describe the life history of the fungus causing loose smut of wheat. *(Allahabad, 1993, 1997)*
17. What is a covered smut? Describe the life history of the fungus causing covered loose smut. *(Allahabad, 1995, 2001)*
18. Explain the role played by plasmogamy, karyogamy and meiosis in the life-cycle of Basidiomycetes. *(Gorakhpur 1993)*
19. Describe the method of reproduction in a smut fungus studied by you. *(Gorakhpur, 1996)*
20. Describe the structure and reproduction in *Sphacelotheca.*
21. Name the causal organism of Jowar smut. Describe the life cycle of the fungus causing the disease.
22. Name the pathogen causing 'loose smut of wheats. Give its systematic position and illustrate its life cycle. *(Lucknow, 2000)*
23. (*a*) Give an account of various modes of infection in smuts by giving suitable examples and methods of their control. *(Kanpur M.Sc. 2002)*
    (*b*) Elaborate the methods of sexual reproduction in Ustilaginales. *(Rohilkhand M.Sc., 2006)*

SHORT ANSWER TYPE

24. **Write short notes on :**
    (*a*) Sexuality in *Ustilago* *(Kanpur, 1996; Kanpur M.Sc., 1991)*
    (*b*) Smuts *(Kanpur, 1997)*
    (*c*) Loose smut *(Awadh, 1994)*
    (*d*) Smut disease *(Awadh, 1995, 1998; Rohilkhand, 1996)*
    (*e*) Smut diseases and their control *(Lucknow, 1995)*
    (*f*) Modes of infection in smuts *(Kanpur M.Sc. 1990; Gorakhpur, 1995, 99; Magadh, 1995)*
    (*g*) Symptoms of loose smut of wheat *(Gorakhpur, 1994; 1997)*
    (*h*) *Ustilago* *(Bihar, 1991)*
    (*i*) Diploidisation in *Ustilago* *(Magadh, 1993)*
    (*j*) Method of dikaryotisation in smuts *(Rohilkhand, 1992)*

(*k*) Ear of wheat infected with *Ustilago*. (*Lucknow, 2002*)

(*l*) Systematic position and characteristic features of *Ustilago*. (*Kanpur, 2006*)

(*m*) Smut diseases of important cereal crops and their mode of infection. (*Lucknow, 2004; Kanpur, 2006*)

(*n*) Smut disease of wheat. (*Poorvanchal, 2006*)

25. Describe the various methods of dikaryotisation in Basidiomycetes. (*Lucknow, 1998*)

26. Give the most important characterisitics of the following :

(*a*) *Ustilago* (*Awadh, 1991, 1996, 1997; Gorakhpur, 1992*)

(*b*) *Sphacelotheca*

27. Describe the similarities and differences between *Ustilago* and *Sphacelotheca*.

28. Give systematic position and main characteristics of the following:

(*a*) *Ustilago* (*Lucknow, 2000 ; Allahabad ,2003, Kanpur, 2005, 2006*)

29. Differentiate between Loose and Covered Smut. (*Kanpur, 2005*)

30. (*a*) Describe the life cycle of *Ustilago* with the help of diagrams only. (*Kanpur, 2005*)

(*b*) Describe disease cycle in loose smut of wheat. (*Poorvanchal, 2003*)

OBJECTIVE TYPE

31. Fill in the blanks :

(*i*) The members of the order Ustilaginales are commonly called ....

(*ii*) Loose smut of wheat is caused by ....

(*iii*) Covered smut of Barley is caused by ....

(*iv*) *Ustilago maydis* causes the smut of ....

(*v*) Whip tail disease of sugarcane is caused by ....

(*vi*) The smut spores are actually ....

(*vii*) Monokaryotic mycelium in the smuts is formed by the germination of ....

(*viii*) *Sphacelotheca sorghi* causes the smut of ....

(*ix*) Smut of Bajra is caused by ....

(*x*) Bunt of wheat is caused by ....

32. Select the correct answer :

(*i*) The name smut diseases is given to those caused by Ustilaginales because

(*a*) the mycelium is black in colour

(*b*) it mostly parasites the important crops

(*c*) the host becomes completely black

(*d*) the fungus produced black sooty spore masses.

(*ii*) Whip tail smut disease of sugar cane is caused by

(*a*) *Ustilago maydis*

(*b*) *Ustilago scitamineae*

(*c*) *Ustilago hordei*

(*d*) *Ustilago nuda*

(*iii*) Smuts are regarded as

(*a*) obligate parasites which can be cultured on axenic culture media

(*b*) obligate parasites which can be cultured on synthetic culture media

(*c*) obligate parasites which can not be cultured at all

(*d*) obligate parasites which can infect only living host.

(*iv*) Loose smut of wheat is caused by

(*a*) *Albugo candida*

(*b*) *Ustilago tritici*

(*c*) *Puccinia graminis*

(d) *Synchytrium endobioticum*

(*v*) Smut spores act as

(*a*) Uredospore (*b*) Teleutospore

(*c*) basidiospore (*d*) Aeciospores.

(*vi*) Production of Spore balls is characteristic of

(*a*) *Tilletia* (*b*) *Ustilago*

(*c*) *Tolyposporium* (*d*) *Puccinia.*

(*vii*) In the life cycle of smuts uninucleate basidiospores are produced as a result of germination of

(*a*) Uredospore (*b*) Smut spores

(*c*) Basidiospores (*d*) Zygospores.

(*viii*) *Ustilago* produces in large quantities the spores known as

(*a*) Uredospores (*b*) Zoospores

(*c*) Zygospores (*d*) Teleutospores.

(*ix*) Covered smut of Barley is caused by *Ustilago*

(*a*) *hordei* (*b*) *nuda*

(*c*) *scitamineae* (*d*) *avenae.*

(*x*) Mycelium in *Ustilago* is

(*a*) intercellular without haustoria

(*b*) Intercellular with haustoria

(*c*) Intracellular

(*d*) absent

(*xi*) Motile structures are absent in

(*a*) *Cystopus* (*b*) *Ustilago*

(*c*) *Synchytrium* (*d*) *Saprolegnia*

(II) ORDER : UREDINALES

Economic Importance and General Characteristics

The order included a group of some 4,500 species of highly specialized and economically most important internal parasites which attack higher plants and sometimes the ferns. They are economically important because majority of them cause disease of cereals and other crops resulting in enormous losses. In spite of the damage they cause, they seldom kill the host plant immediately. These organisms receive the popular name **rusts** or **rust fungi** from the rusty colour of the spores (uredospores) which appear on the surface of the host. In nature the rust fungi appear to be **obligate parasites.** Till recently all attempts to grow them in the laboratory apart from the host or hosts failed. However, a considerable success in culturing some of them on artificial media has been attained within recent years. Wolf (1974) stated that so far 9 species belonging to 4 different genera of rust fungi have been grown on pure culture in the laboratory. Hotson and Cutter (1951) and later Cutter (1951 and 1959) were the pioneers to meet success in their efforts to cultivate *cedar apple rust fungus* on an artificial medium. Brown (1967) divided rust fungi into two categories namely, *physiologically obligate parasites* and *ecologically obligate parasites*. The former would not grow apart from the host or hosts. The latter to which the wheat rusts belong fail to grow as saprobes in nature but can be grown in the laboratory on the artificial media. In the rusts the mycelium lacks the dolipore septal complex. It spreads intercellularly forming haustoria. More than one type of spores are formed in the life cycle. The rusts are thus polymorphic. The basidia are transversely septate. The basidiospores are produced on sterigmata and are discharged forcibly. The rusts like Smuts lack basidiocarps. Karyogamy takes place in the teliospore.

Rust Fungi

Rusts infect ferns, gymnosperms and angiosperms and are known since ancient times. According to Pliny (29-79 AD) Rust is the greatest enemy of cereals. This statement holds true even today because the damage caused by rusts is severe but slow and prolonged. Rusts have attracted both the Mycologists and the plant pathologists.

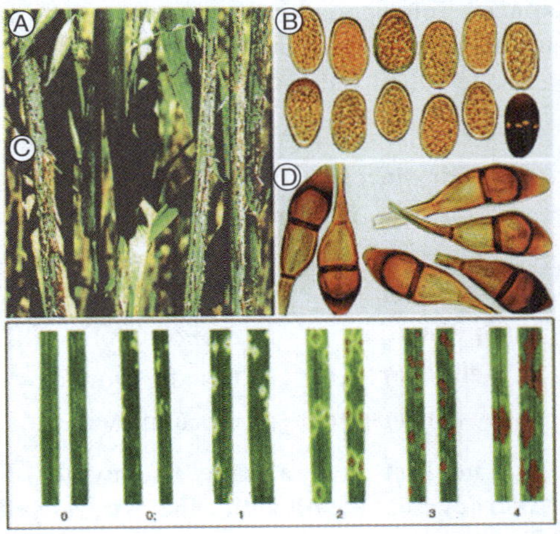

A. Field view of wheat plants infected with *Puccinia graminis;* B. Uredospore (Microscopical view); C. Teleutospore (Microscopical view); D. Events tracing percentage cover area of pustule infection on the infected wheat leaf.

Puccinia graminis

Life Cycle and the Spore Stages

Some of the rusts need two different unrelated hosts to complete their life cycles. These are known as **Heteroecious** (Hetero = different; cion = Home)rusts and the phenomenon exhibited by them is called **Heteroecism.** An example of heteroecious rust is *Puccinia graminis* which requires two hosts to complete its life-cycle. These are wheat (*Triticum vulgare*)—a monocot and Barberry *(Barberis vulgaris)*—a dicot. The host-wheat bearing the teleutospore (teliospore) stage is called the **Primary host** and the host-Barberry—that bears the spermatial and aecial stages is known as **Alternate host.** When some of the spore stages of primary host is formed on a related host, this host is called the **Collateral host.**

In some other rusts, only one host is required to complete the life cycle. Such rusts are called **Autoecious** (Auto = me; cion = Home) **rusts** and the phenomenon exhibitied by them is known as **Autoecism.** *Puccinia butlerii* is an autoecious rust because it complete its life-cycle on one host, *Launaea asplenifolia.*

In the life cycle of a rust, the following spore stages are formed:

Stage 0	:	**Pycnium** bearing spermatia and receptive hyphae
Stage I	:	**Aecium** bearing aeciospores,
Stage II	:	**Uredinium** bearing urediniospores.
Stage III	:	**Telium** bearing teleospores
Stage IV	:	Basidiospores

While Basidiospores and pycnium producing spermatia and receptive hyphae are uninucleate, aeciospores, urediniospores and teliospores are binucleate. When these spore stages were assigned Roman numerals, the function of Pycnia was not known. Hence, these were named as **Stage 0.** It was Craigie (1927) who discovered that Pycnium is the gametic stage bearing sex cells which are self sterile. It is the stage where plasmogamy and dikaryotisation take place when spermatia and receptive hyphae of the compatible mating type come in contact.

Stage I Aecium is produced on the same leaf simultaneously with pycnia. Aecia produce Aeciospores which are called binucleate and cannot infect the host on which they are produced. **Stage** II **Uredinium** (pl. Uredia) is the sorus which produces one celled binucleate repeating spores. These spores are called **Urediniospores** and are generally produced in the early part of growing season. **Stage III Telium** or teleutosorus produces thick walled, two celled binucleate **teliospores** which serve as the site of Karyogamy and meiosis. These also serve as **resting spores.** Teliospores are the characterisitic spores on the basis of which the rust genera have been identified. On germination, the teliospores produce **Basidiospores** which are thin walled, uninucleate and one-celled and constitute the **Stage IV**.

Modern mycologists have given definitions of these spore stages which are as follows:

Aeciospores	:	non repeating vegetative spores produced usually after dikaryotisation, which germinate to initiate dikaryotic mycellium, usually associated with pycnia (spermogonia).
Basidiospores	:	monokaryotic spores produced on basidia usually after meiosis.
Spermatia	:	monokaryotic cells behaving as gametes.
Teliospores	:	Basidia producing spores.
Urediniospores	:	repeating vegetative spores usually produced on dikaryotic mycellium.

The rust genera have life cycles designated as **Macrocyclic**, **Demicyclic** and **Microcyclic** depending upon the spore stages produced. **Macrocyclic** rusts are those which produce all the above mentioned five spore stages. These rusts are also known as **long-cycled rusts**. These spore stages may be produced in the same host as in case of autoecious rusts or on two different unrelated hosts as in heteroecious rusts. *Puccinia graminis* is an example of heteroecous macrocyclic rust while *P. helianthi*

Hemileia – A microcyclic rust

represents the autoecious macrocyclic rust. **Demicyclic** rusts are those which lack the uredinial stage and may be autoecious or heteroecious. *Gynoconia peckiana* is an autoecious demicyclic rust and *Gymnosporangium juniperivirginiane* is a heteroecious demicyclic rust.

Gymnosporangium – a heteroecious demicyclic rust

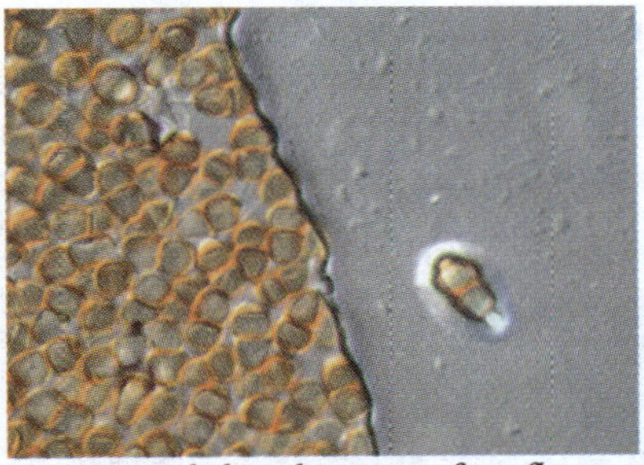

Puccinia helianthi – rust of sunflower

Microcyclic rusts are those which produce only the telial stage. *Hemileia* is an example of a microcyclic rust.

Classification. On the basis of the teleutospore structure, the Uredinales have been classified by Arthur (1934) into two families : Pucciniaceae and Melampsoraccae. Alexopoulos (1962) and Alexopoulos and Mims (1979) recognise three families : Pucciniaceae, Melampsoraceae and Coliosporaceae. Saville (1978), however, recognise five families : Pucciniaceae Melampsoraceae, Phragmidiaceae, Revenaliaceae and Pucciniastraceae. However, modern mycologists recognise only two families : **Pucciniaceae** and **Melampsoraceae.**

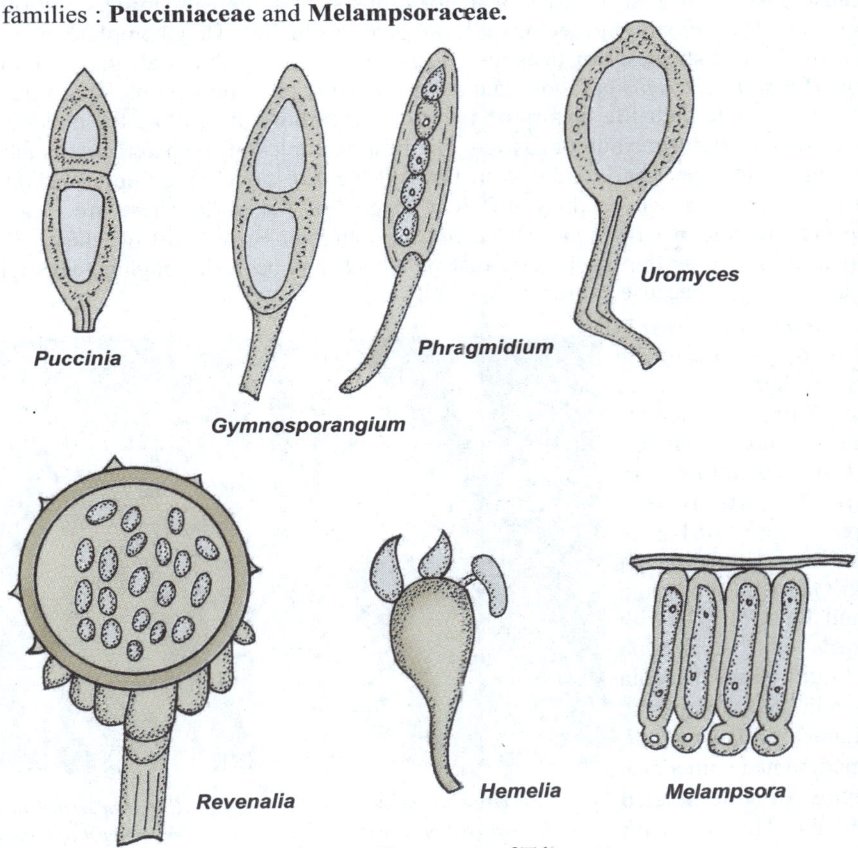

Fig. 14.13. Different types of Teliospores.

FAMILY : PUCCINIACEAE

It is characterised by the presence of stalked teleutospores. The stalk is well developed in many species. In a few it is short and in others it becomes detached at a very early stage. The spore wall is thick and smooth or variously sculptured. The most important and widely distributed genus belonging to this family is *Puccinia.* We study it as a type. The other important genera included in the family are *Uromyces, Hemileia* and *Phragmidium.*

Uromyces – tea rust

Hemileia – coffee rust

Phragmidium – rust

PUCCINIA

The genus includes about 3,000 species. Of these 147 have been reported from India. All are internal parasites attacking angiosperms. Only the spore masses are visible externally. Some of them such as *Puccinia graminis tritici* (*Black stem rust of wheat*) are of great economic importance as they cause destructive rust diseases of major crops such as cereals, millets and other crops. Many species of *Puccinia* are **autoecious** (*P. obtegens* on thistle). They complete their entire life cycle on a single host species. For instance *P. asparagi* (D-C) produces all four types of sori on *Asparagus*, *P. major* on *Crepis paludosa* and *P. pulverulenta* on willow herbs. There are, however, others which complete their life history on two unrealated hosts much like the malarial parasite. These are known as **heteroecious parasites.** The best examples of the heteroecious parasites are the **wheat rusts.** This peculiar condition of the wheat rusts is known as **heteroecism.** In India wheat rust disease is caused by three different species of *Puccinia.* These are *P. graminis, P. striiformis* (yellow rust of wheat) and *P. recondita* (Orange or Brown rust of wheat). *P. graminis tritici* which is known as the black stem rust of wheat, has been thoroughly investigated. It is usually taken as a type because its life cycle is fully known.

Cultivation of wheat rust in the Laboratory. Since the discovery of heteroecism in rusts by de Barry, all attempts to cultivate the *black stem rust of wheat* in the laboratory failed. It was, therefore, considered an **obligate parasite.** Investigations of Fuchs and Gartner (1958) and Gartner and Fuco (1962) with Uredospores of *P. graminis tritici* on liquid or solid media resulted in a limited saprophytic growth. Later Williams *et al,* (1960) experimented with a race of *P. graminis tritici* designated 126—ANZ-67. He met with

Puccinia recondita (Brown rust of wheat)

Puccinia striiformis (Yellow rust of wheat)

great success. There was unlimited growth from the uredospores. The hyphae thus produced bore a fresh crop of uredospores. The ingredients of the medium used by Williams *et al.* were mineral salts, sucrose, 0.1% yeast extract and 1% agar and the pH of the medium was maintained at 6.4. This experiment showed that the *black stem rust of wheat* could no longer be considered strictly obligate in its parasitism. Brown (1967) put it in the category of *ecologically obligate parasites.* Recently Bushnell (1976) was able to grow successfully isolates of races 38 and 17 of *P. graminis* of sp. *tritici* which had grown poorly on earlier trials on 1% peptone media. They were grown successfully on media containing higher concentration of peptone combined with an acid hydrolysate of casein.

PUCCINIA GRAMINIS

P. graminis sp. tritici is a heteroecious parasite. It causes rust disease known as the 'black stem rust' of wheat in all wheat growing areas of the world. Till recently it had been considered strictly obligate in its parasitism. However, Williams *et al.* (1966) and Bushnell (1976) have successfully grown this parasite on culture media as a *facultative saprophyte.* Normally the fungal parasite passes a part of its life cycle constituting the **dikaryophase** on wheat (*Triticum aestivum*) and a part constituting the **haplophase** on barberry (*Barberis vulgaris*) or *Mahonia.* The dikaryotic mycelium is unable to grow on barberry and *vice versa.* The full life cycle is thus completed only when both hosts are present. *Puccinia graminis* is thus an example of a **heteroecious** parasite. The wheat plant on which the fungus parasite passes its dikaryophase is called **primary host** and the *Barberry* is called the **secondary** or **alternate host.** The fungus parasite survives in the absence of the alternate host (barberry). However, it can produce new races by hybridisation only on barberry.

Dikaryophase. This phase in the life cycle of *Puccinia* is confined to the primary host which is wheat. It consists of dikaryotic mycelium and two spore stages, **uredineal** and **telial.**

1. Dikaryotic Mycelium. It is internal and thus invisible until it is ready to reproduce. .The binucleate secondary or dikaryotic mucelium is filamentous, well developed and branched. The hyphal branches, which are septate, ramify in the intercellular spaces of the tissues of the stem and leaves of the host plant (wheat Fig. 14.14 B). The septal pore is simple (Ehrlick and Ehrlik, 1963). Each cell contains a pair of nuclei ($n+n$) constituting a dikaryon. The nuclear membrane is double layered and perforated. Besides the two nuclei, the cytoplasm contains free ribosomes mitochondria, glycogen particles, lipid bodies and other unidentified particles. To obtain nutrition the intercellular hyphae develop intracellular food absorbing organs called **haustoria.**

List of important diseases caused by species of Puccinia

Name of the disease	Causal organism
1. Rust of Onions	*P. allii*
2. Rust of Asparagus	*P. aspargi*
3. Safflower Rust	*P. carthami*
4. Corn Rust	*P. coronata*
5. Black stem of Rust	*P. graminis tritici*
6. Sunflower Rust	*P. helianthii*
7. Maize Rust	*P. sorghi*
8. Mint Rust	*P. menthae*
9. Orange or Brown Rust of wheat	*P. recondita*
10. Yellow or stripe Rust of wheat	*P. striiformis*
11. Sorghum Rust	*P. purpurea*
12. Holyhock Rust	*P. malvacearum*
13. Rust of oats	*P. avenae*
14. Rust of Rye	*P. graminis secalis*

Haustorial apparatus. Allen (1923) reported that the tips of certain hyphae of the intercellular dikaryotic mycelium come in contact with the cell walls of the host cells and become separated from the remainder of the mycelium by septa. The resultant small, binucleate tip cells swell somewhat and function as **haustorial mother cells.** At the point of contact, each haustorial mother cell develops a buldge-like thickening which is closely appressed to the host cell wall. A fine pore appears at the point of contact extending through the cell walls of both the pathogen and the host. A delicate infection peg from the haustorial mother cell passes into the host cell through the pore. After penetration, the infection peg elongates to form the haustorial neck. A wall is formed around it by the fungal pathogen. The plasma membrane of the host invaginates as the developing neck elongates. The latter eventually ceases to elongate. Its apical portion expands to form the body region of the haustorium. Dickinson (1949, 1955) stated that the formation of haustoria by the infection hyphae is induced by contact stimulus. The intercellular mycelium probably by enzymatic degradation enters the host cells. The haustoria may be knob-like or finger-like, rarely convoluted. Rykenberg and Truter (1973) reported that the intracellular saccate or extensively lobed body of the haustorium is connected by a narrow penetration tube or the stalk to the extracellular haustorial mother cell. The hyphal wall and the plasma membrane of the haustorial mother cell are continuous along the entire length of the penetration tube and haustorial body constituting the haustorial wall and haustorial membrane respectively. A collar-like structure attached to the host cell wall and made of the same material is produced by host cell in response to the presence of the penetration tube. It surrounds the penetration tube at its proximal end. The haustorial body is binucleate. On the outer face of the hyphal wall (or haustorial wall) in the body region only is formed an electron dense sheath matrix consisting of a granular amorphous material. The sheath matrix is also designated as the extrahaustorial sheath. Since the fungus obtains nutrition from the host the infected wheat plants exhibit stunted growth. Despite the serious damage that the rust causes the host plant never succumbs completely. The yield of wheat crop is, however, reduced. Each cell of the fungus mycelium has a pair of nuclei (dikaryon). The transverse septa between the cells have each a central pore without a dolipore parenthesome complex. The clamp connections. however, have not been reported on the dikaryotic mycelium. The mycelium is generally localised. It grows and ramifies near the point of infection extending only short distance into the host tissue. The fully developed dikaryotic mycelium enters upon the reproductive phase.

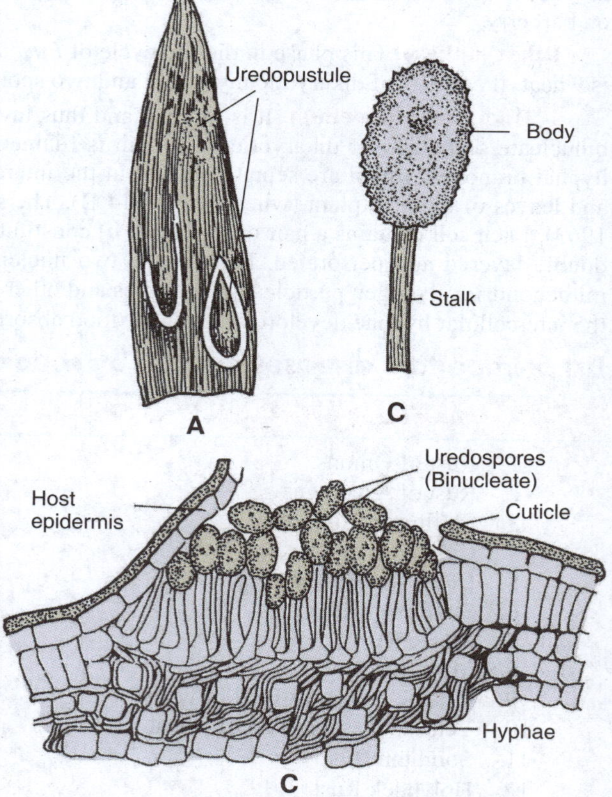

Fig. 14.14 (A-C). Uredineal stage of *Puccinia graminis tritici*. A, Uredosori on wheat leaf; B, A section through the uredosorus; C, A single uredospore.

Reproduction. The dikaryotic mycelium of *Puccinia graminis* reproduces by **sporulation.** The spores produced are of two kinds, the

uredospores and the **teleutospores** or **teliospores.** They are produced near the surface of the host tissues. When mature they break through in slits or pustules called the **sori.** The mature spores are thus seen externally.

(*a*) Uredineal Stage (Fig. 14.14). Prior to uredospore formation, the hyphae of the dikaryotic mycelium begin to aggregate near the surface of the infected organ such as stem, leaf, leaf sheaths or glumes to form a hyphal mass which surrounds isolated host cells (B). The mycelial hyphae then produce masses of cells subepidermally. These are called **uredia.** From the uredia arise vertically growing slender, stalk-like hyphae arranged in a close palisade-like layer. The tip of each hyphal stalk or uredospore mother cell swells to form a single binucleate oval **uredospore** or **uredeniospore.** The uredospores are thus formed in groups. Each such group is called a **uredosorus** or **uredinium** (B). The developing uredosori are seen in the wheat leaf as pale streaks. They exert pressure on the overlying epidermis which is, at first, lifted but finally ruptured in the form of slits or blisters (A). It is through these slits that the rusty coloured masses of uredospores are seen externally. They are stalked, unicellular, oval, binucleate structures (C). In masses the uredospores appear rusty red in colour. The uredospore wall is thick. It is differentiated into three layers (Williams and Ledengam, 1964) : outermost **pellicle** the middle **exosporium** and the inner thin **endosporium**. The outermost layer (pellicle) is faintly echinulate and generally with four equatorially arranged thin areas on it. These thin areas on the otherwise thick wall are called the **germ pores.** Harder *et al* (1986) however, consider the uredospore wall to probably be composed of 5 layers including the pellicle as the outermost layer. The uredospore wall contains silica deposits. The binucleate uredospores function as conidia and are capable of germination as soon as they are produced. They are carried in the air to healthy wheat plants which they infect immediately. The uredospores thus serve as dispersal agents of rust rapidly multiplying the dikaryophase during the growing season. However, unlike the conidia of Ascomycetes, the uredospores are binucleate. They are able to infect wheat plants only and not the alternate host barberry.

Germination of Uredospores and infection of the host (Fig. 14.15 B-C). On falling on another wheat plant and under suitable conditions the uredospore germinates within a few hours. The endosporium comes out in the form of a slender tube through the germ pore (B). More than one germ tubes may be produced by the same uredospore. They emerge through different germ pores.

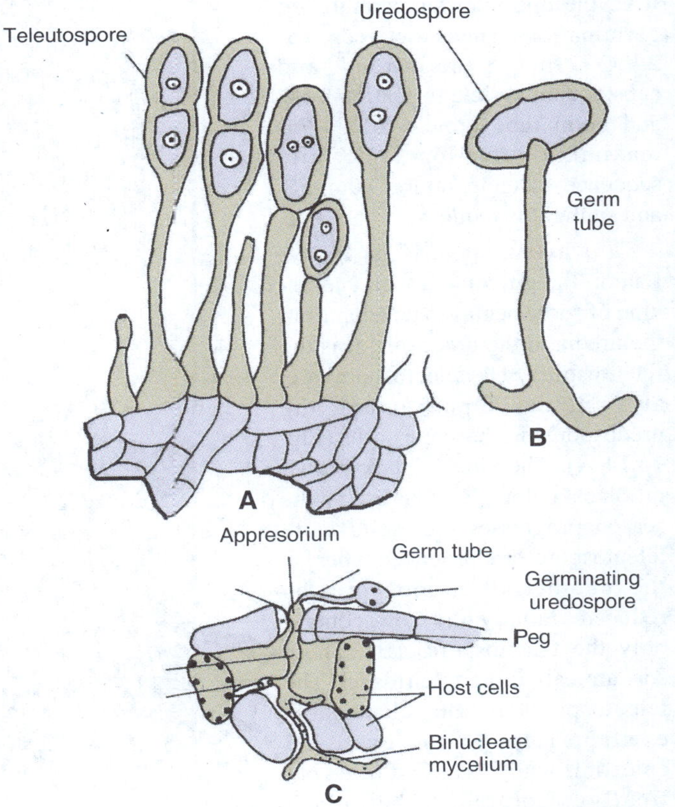

Fig. 14.15 (A-C). *Puccinia graminis tritici.* A, Sorus with both uredo and teleutospores; B, Germination of uredospore; C, Infection of wheat leaf by a germ tube through a stoma.

The germ tube by elongation grows over the surface of the host leaf till it reaches a stoma where its tip swells to form an **appresorium** (C). The two nuclei and the protoplasm of the germ tube migrate into the appresorium. The appresorium is then cut off from the empty germ tube by a septum. Both the germ tubes and the appresoria walls are composed of 2 layers and appear to be coated with mucilaginous-like substance (Harder *et al.* 1986). A peg-like outgrowth, the **infection peg** arises from the centre of the free end of the appresorium. It enters the stomatal aperture. Reaching the substomatal cavity the tip of the infection peg swells up into a **vesicle**. The contents of the appresorium pass through the infection peg into the sub-stomatal vesicle. Infection hyphae, one or more in number, then develop from the vesicles, grow towards the neighbouring host cells and ramify in the intercellular spaces between them. The hyphae grow vigorously and form the new dikaryotic mycelium between the mesophyll cells of the host tissue. In a few days time the new mycelium produces a new crop of uredospores. This repeated cycle recurs several times in a single growing season. As a result the disease spreads in a few days from plant to plant and field to field over a large area provided the environmental conditions are favourable. Temperature and humidity are the two climatic factors which facilitate infection and thereafter the development of the dikaryotic mycelium of black stem rust of wheat. The source of hymidity may be the dew drops, guttation drop, light rainfall or irrigation water. In a thin film of water and light intensity below 300 ft. C, the optimum temperature for germination ranges between 15-24°C for penetration and substomatal vesicle formation 29°C and germ tube growth 20°C. For maximum infection a definite sequence of temperature, humidity and sunlight is required.

(*b*) Telial Stage (Fig. 14.16). Late in the growing season another kind of spore begins to develop from the uredia in the uredosori. It is the **teleutospore.** The teleutospores are, at first, developed among the uredospores in the same sorus (Fig. 14.14 A). They are of dark brown or black colour. Gradually as the season progresses more and more teleutospores are produced whereas the number of uredospores is reduced. Finally the sori contain only the teleutospores (B). These sori are called the **teleutosori.** The teleutospores in the teleutosorus exert pressure on the overlying epidermis which is at first lifted but finally ruptured. The black teleutospores are exposed (A). The cells of the mycelium producing the teleutospores are called the **telia.**

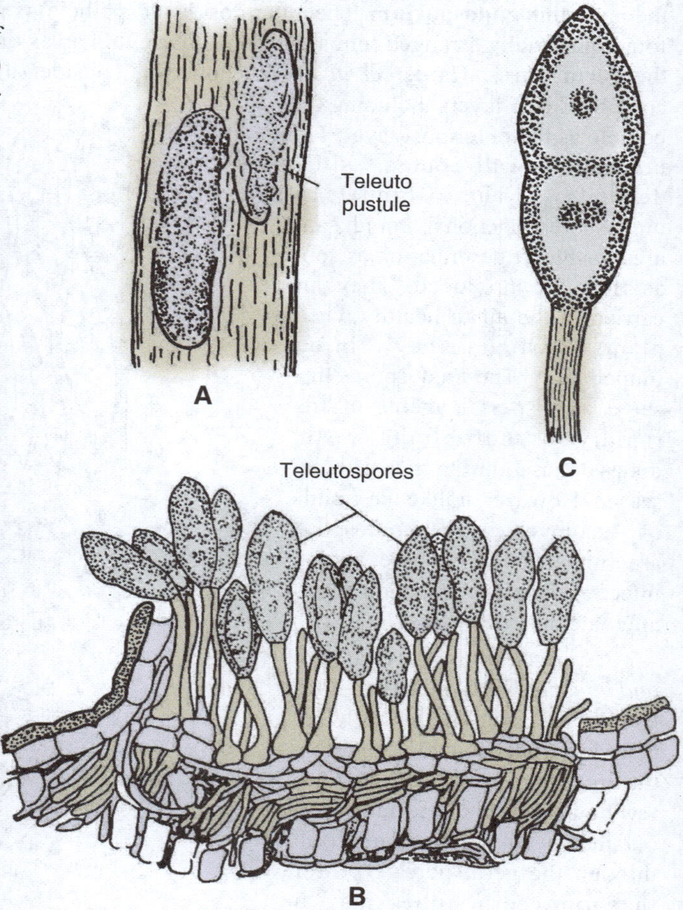

Fig. 14.16 (A-C). Telial stage of *Puccinia graminis tritici*. A, Teleutosori on wheat; B, A section through the teleutosorus; C, Single teleutospore with a fusion nucleus in each cell.

This stage in the life cycle in which the teleutospores are produced is called the **telial stage** or the **black spore stage**. It is considered the **perfect stage** of the Uredinales because it is in the teleutospores that **karyogamy** and **meiosis** take place

The teleutospores (Fig. 14.16 C) are black or dark brown, stalked, two-celled, spindle-shaped structures slightly constricted at the septum. The black spore wall is thick and smooth. The free end of the spore may be rounded to slightly pointed with a thick wall at the apex. There is a single germ pore in the wall of each cell. It is at the apex in the upper cell of the spore and a little below the septum in the lower cell. Each cell of the teleutospore has two nuclei (one of plus strain and the other of minus strain). Unlike the uredospores, the teleutospores are borne on persistent pedicels and do not germinate immediately.

During maturation period of teleutospore, karyogamy takes place. The two nuclei in each cell of teleutospore fuse to form a diploid nucleus. Overwintering takes place in the uninucleate diploid condition of the bicelled teleutospore which represents the reduced **diplophase** in the life cycle. After the harvesting period, mature teleutospores remain dormant on straw, stubble and out of season self-grown plants in the fields and survive even the severest winters. After the resting period the teleutospore germinates if the conditions are favourable. The conditions favourable for their germination are high atmospheric humidity, presence of moisture and freezing condition prior to germination. The teleutospores germinate to give rise to the **basidial stage** in the life cycle of *Puccinia.*

Basidial Stage (Fig. 14.17). After the resting period and under favourable conditions the teleutospores germinates *in situ* to produce the **basidial stage** in the life cycle. The characteristic structures of this stage are the **basidia** and the **basidiospores.** The basidium in *Puccinia* is divisible into three parts namely the (*i*) *probasidium* or *hypobasidium* (*ii*) the *epibasidium* or *metabasidium* and (*iii*) the *sterigmata* which bear the basidiospores. Each cell of the teleutospore containing a diploid nucleus represents the probasidium or hypobasidium. At the time of germination a short, slender hyaline hypha of limited growth, the **promycelium** or **epibasidium** grows out through the germ pore from each cell (probasidium) of the teleutospore (C). This is followed by the migration of the diploid nucleus into the epibasidium. There the diploid nucleus undergoes two divisions constituting **meiosis** (C, a). Segregation of strains takes place. The four haploid nuclei thus produced in each epibasidium come to lie at more or less equal distances. Septa appear between the nuclei dividing the

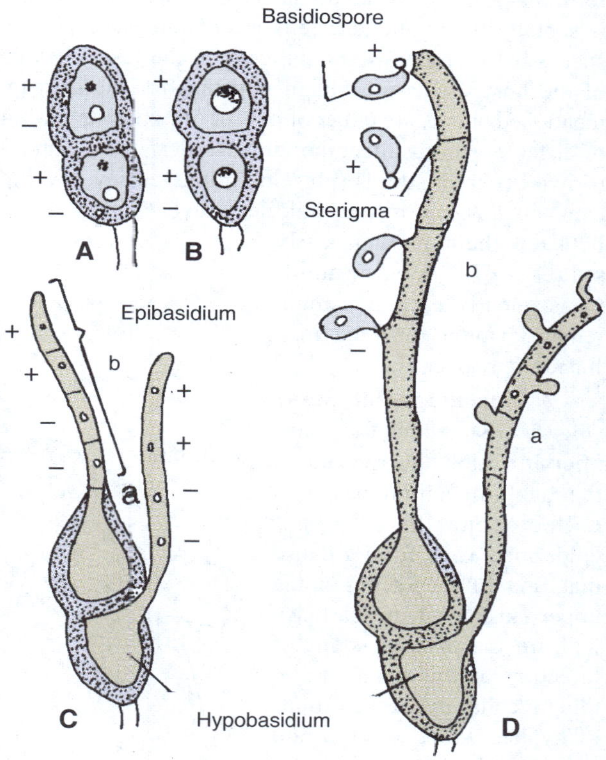

Fig. 14.17 (A-D). *Puccinia graminis tritici.* A, Young teleutospore; B, mature teleutospore; C, germinating teleutospore and meiosis; D, basidial stage.

epibasidium into four uninucleate haploid cells (C, b). From each of the three lower epibasidial cells then grows a short narrow tube, the **sterigma.** The sterigma usually arises from just below the upper septum of the cell. From the terminal cell of the epibasidium, the sterigma arises from the apex. The free tip of each sterigma swells to form a **basidiospore** (D, a). In the meanwhile the haploid nucleus from each epibasidial cell passes up its respective sterigma into the developing basidiospore. Two out of the four basidiospores on each epibasidium are of plus strain and the other two of minus strain. They are small, nuicellular, uninucleate haploid structures (D, b). When mature the basidiospores are discharged by the water droplet method or spore drop mechanism with a force into the air. They are carried by wind to the leaves of alternate host barberry which they infect. The basidiospores remain viable only for a few days. They cannot bring about infection of wheat and thus perish soon if the suitable (alternate) host is not available.

Haplophase. This phase in the life cycle is confined to the alternate host *Barberry (Berberis vulguris).* It starts with the basidiospores. The haplophase consists of **primary** or **haplomycelium,** and the two spore stages. These are the **spermagonial** (also called **pycnidial**) and **aecial (aecidial)** stages (Fig. 14.22 A).

1. Haplomycelium. The basidiospore germinates (Fig. 14.24) on the leaf of the alternate host (*Berberis*) provided the moisture and temperature conditions are suitable. It puts out a germ tube also called the **primary hypha.** The primary hypha infects the young leaf of the alternate host by piercing through the epidermis directly. The cuticle of older leaves is too thick to allow penetration by the delicate germ tube. Once within the host tissue it grows vigorously and branches freely to form the primary mycelium, also called **monokaryotic** or **haplomycelium.** The primary mycelium is septate. Its component cells are uninucleate. The mycelial hyphae ramify in the intercellular spaces between the mesophyll cells of the leaf. They produce haustoria which penetrate the cells of the host tissue and obtain nutrition for the growth of the mycelium. The basidiospores, as mentioned above, are either of plus or of minus strain. It may often happen that several basidiospores of different strains infect the same *Barberis* leaf. Naturally the newly formed haplomycelia will be of two different strains (plus and minus). They develop side by side. The two strains of mycelia can thus co-exist in the same leaf. Even if they are very close to each other no fusion occurs between them in their early stages of development. Consequently each mycelium remains haploid for some time. Fusion may occur later.

2. Spermagonial Stage (Fig. 14.18). About four days after infection the hyphae of the haplomycelium begin to collect beneath the upper epidermis. They form a dense mat. Small light patches on the upper surface of the barberry leaf are the external signs of infection at this stage. From this mycelial mat arise groups of hyphae which fashion into small, flask-shaped structures (Fig. 14.18) called **spermagonia** (also called the **pycnidia**). The pycnidia or

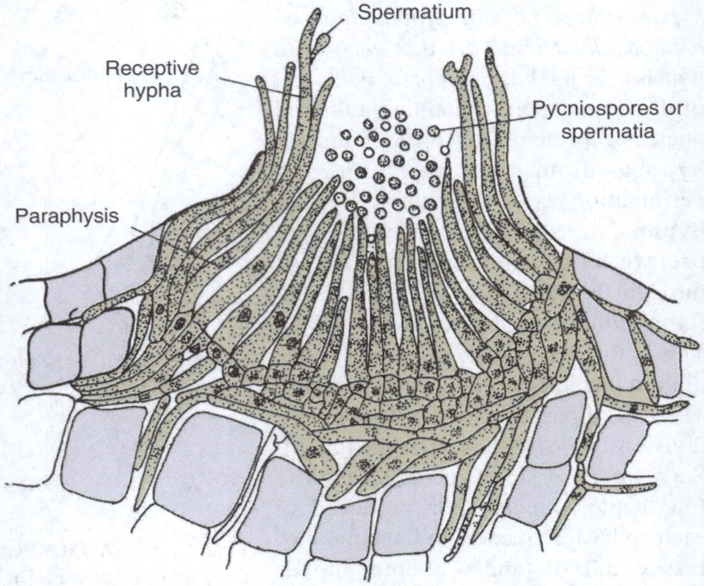

Fig. 14.18. *Puccinia graminis tritici.* A mature spermagonium in section.

spermagonia, like the mycelia from which they arise, are of plus or minus strain. When spermagonia mature, the infected areas becomes swollen and are seen as small, orange yellow bumps on the upper surface of infected leaf. The spermagonia are sub-epidermal in position. They are buried in the mesophyll tissue of the area which gets thickened as a result of the presence of the fungus mycelium.

Each spermagonium (Fig. 14.18) consists of a wall surrounding a cavity. It opens on the upper surface of the host leaf through a small pore called an **ostiole.** From the wall of the spermagonium arise three kinds of hyphae :—

(*i*) *Spermatiophores.* Numerous fine, elongated, uninucleate cells or short hyphae arise from the cells of the wall (Fig. 14.18). They project into the cavity of the spermagonium and are called **the spermatiophores.** They are closely packed and arranged in a palisade-like layer lining the cavity. Each spermatiophore (Fig. 14.19) by successive divisions of its nucleus abstricts at its free tip a number of small cells one after the other. These are the **spermatia.** Each spermatium has a single nucleus and very little cytoplasm. The mature spermatia fall into the spermagonial cavity. They are non-motile and cannot infect either host.

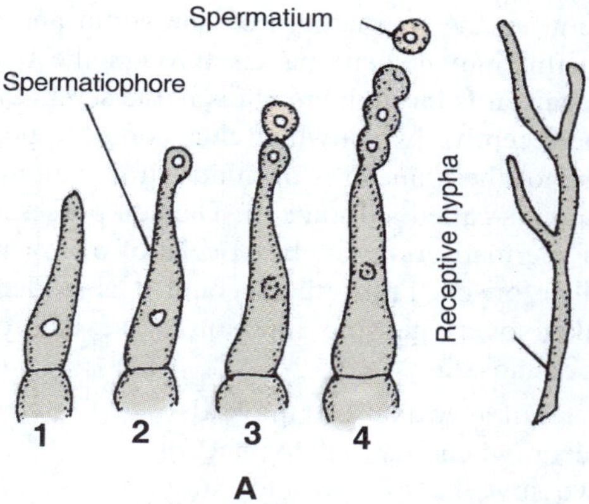

Fig. 14.19 (A-B). *Puccinia graminis tritici.* A, steps in the development of spermatia at the tips of spermatiophores; B, receptive or flexuous hypha (Based on Allen).

(*ii*) *Periphysis* (Fig, 14.18). They are long, delicate sterile hyphae which develop near the ostiole from the spermagonial wall. At first all the periphyses converge towards a central point.

From there they curve upwards in a cluster towards the ostiole. Finally they project through and beyond the ostiole end.

(*iii*) *Flexuous receptive hyphae* (Fig. 14.18). Adjacent to the ostiole and below or among the stiffer tapering periphyses, develop another kind of hyphae. They are slender, delicate, cylindrical with blunt free tips. Being flexuous these are named the **receptive** or **flexuous hyphae.** They are septate and may be simple or branched. Each septum has a central pore.

The nature of spermagonia and spermatia has long been a debatable point. It was in 1927 that Craigie, Buller and their associates demonstrated that spermatia function as male cells. The receptive or flexuous hyphae are supposed to function as trichogyne hyphae. They develop late and are spermatised by the spermatia of opposite strain.

Spermatisation (Fig. 14.20). The mechanism of plasmogamy in *Puccinia* is known as **spermatisation.** The mature **spermatia** exude from the ostiole of spermagonium in a drop of sticky, thick liquid called **nectar.** The nectar with its scent and sugary content attracts the insects

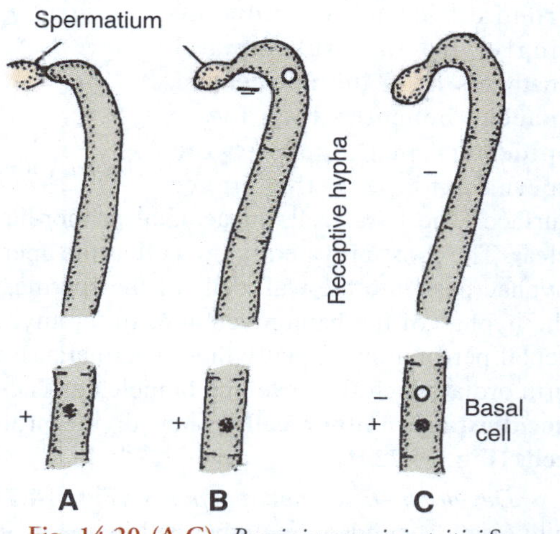

Fig. 14.20 (A-C). *Puccinia graminis tritici.* Stages in spermatisation of flexuous or receptive hyphae (Diagrammatic).

particularly the flies, to the leaf. The spermatia stick to the legs and proboscis of the visiting insects and thus are dispersed from leaf to leaf or from one spermagonium to another on the same leaf. By the time the drop of nectar exudes, the receptive hyphae grow out through the ostiole into the nectar. These are spermatised by spermatia of opposite strain brought there by visiting insects (A). The intervening walls at the point of contact between the spermatium of one strain and the receptive hypha of another strain dissolve. The spermatium nucleus passes through the opening into the receptive hypha in which it moves downwards through the pores in the septa. Finally the spermatial nucleus reaches the basal cell of the receptive hypha which thus comes to possess two nuclei which lie side by side in a pair (C). One of these nuclei is of minus strain and the other of plus strain. This pair of nuclei of opposite strain is called a **dikaryon**. The cell possessing the dikaryon is said to be diploidised. As a result of spermatisation, the basal cells of one or more receptive hyphae in the spermagonium become dikaryotised. From this account it is evident that the spermatia and the receptive hyphae have taken over the sole function of sexual reproduction in which plasmogamy takes place by spermatisation.

Dikaryotisation may also occur when haploid hyphae of two mycelia of opposite strain come into contact.

3. Aecidial or Aecial Stage (Fig. 14.21 A-B). The haplomycelium, which has built up spermagonia beneath the upper epidermis, has in the meanwhile penetrated the entire leaf and reached its lower surface. There the mycelial hyphae accumulate and form spherical masses of uninucleate cells. These are called the **aecidial** or **aecial primordia.** The aecidial primordia or protoaecidia lie singly in the substomatal chambers or in the intercellular spaces beneath the lower epidermis just opposite the spermagonia on the upper surface. The basal cells of aecidial primordia become **binucleate.** How this takes place is not clear. The most prevalent view is that the spermatial nuclei from the basal cells of the receptive hyphae pass into the wall cells of the spermagonium (pycnidium). From there they migrate into the hyphae of the haplomycelium. In the mycelial hyphae they travel passing down through the septal perforations. Finally these spermatial nuclei reach the cells of the basal layer of the aecial primordia which thus become binucleate. These binucleate cells of the basal layer are called the **aecidiospore mother cells.** They divide producing alternately **aecidiospores** and **intercalary cells** (Fig. 14.22 B).

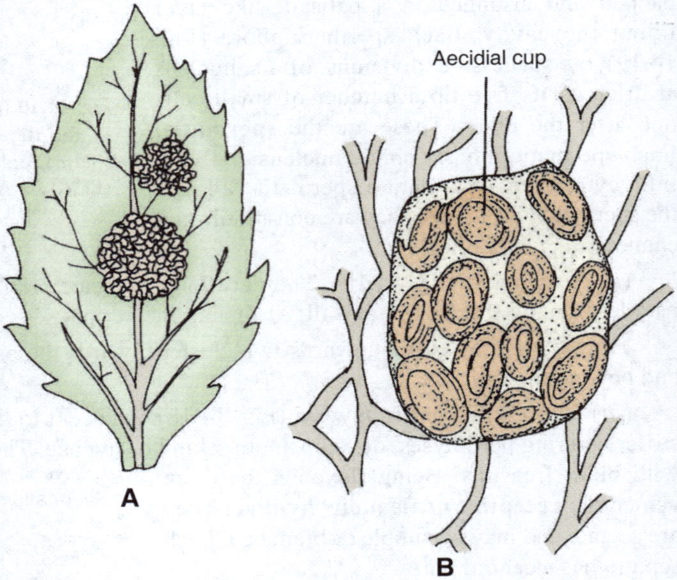

Fig. 14.21 (A-B). *Puccinia graminis tritici.* A, An infected leaf of *Berberis* with aecidial cups on the lower surface; B, The same enlarged.

Development of aecidiospores (Fig. 14.22 B). At the time of aecidiospore formation, the binucelate aecidiospore mother cell increases in length. The two nuclei in it divide conjugately. A small daughter cell is then cut off at its terminal end. Two of the nuclei remain in the aecidiospore mother cell and the other two pass into the daughter cell. The daugher cell increases in size and divides into two, upper bigger binucleate **aecidiospore** or **aeciospore cell** and the lower smaller binucleate, sterile **disjunctor** or **intercalary cell**. Each aecidiospore mother cell undergoes a series of such divisions. In this way closely packed chains of cells are formed at the tips of the aecidiospore mother cells. The oldest cell is at the top of the chain and the youngest adjacent to the tip of the aecidiospore mother cell.

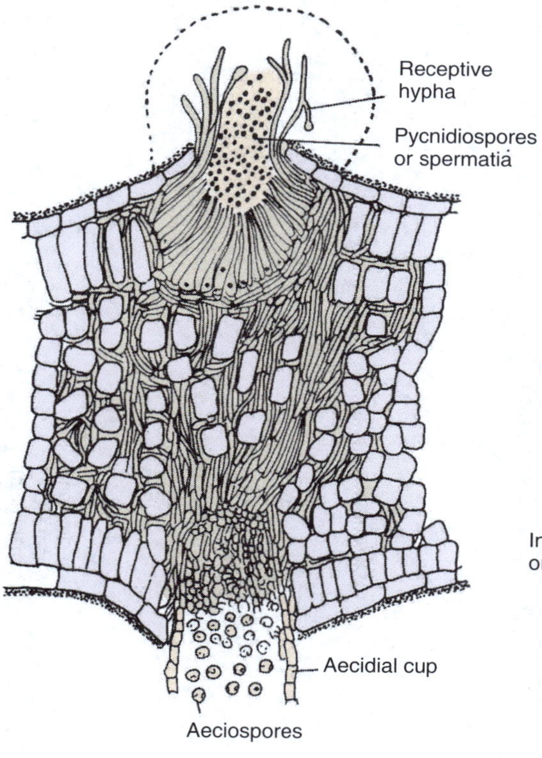

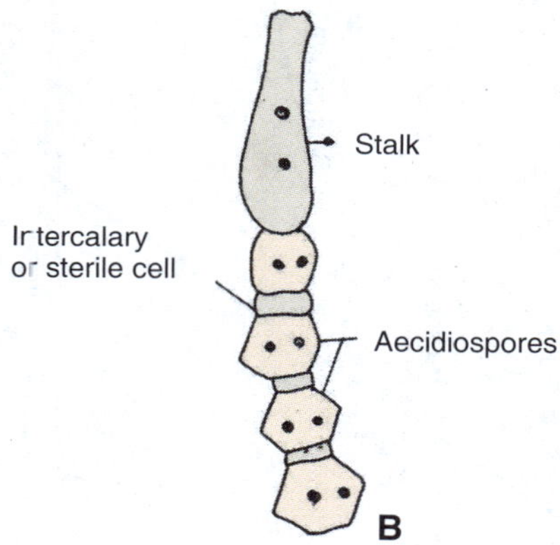

Fig. 14.22 (A-B). *Puccinia graminis tritici*. Spermagonial and aecidial stages. A, V. S. Berberis leaf showing a spermagonium (pycnidium) on the upper surface and an aecial cup on the lower surface; B, showing mode of development of aecidiospores.

In the meantime the marginal cells at the base of the young aecidium divide repeatedly to form a wall which completely surrounds the cells in the aecidium. This wall is called the **peridium**. As the aecidia mature the chains of cells grow up. As a result of this, pressure is exerted from within. At first the overlying lower epidermis ruptures and then the aecidium splits open to form a cup-like structure (Fig. 14.22 A). The aecidial cup is partly within the leaf tissue and partly projects above it. The torn roof of the peridium often forms a lip around the edge of the aecidial cup. By this time the intercalary cells have disorganised. The aecidiospores in the chains become separated from each other and fill up the aecidial cup (Fig. 14.22 A). The aecidiospores filling the cup are now exposed. They are, at first, polyhedral, thin-walled, unicellular binucleate structures which are orange coloured. The polyhedral spores absorb water, round off suddenly and thus are jerked out of the aecidium to be disseminated by the wind. The epispore may be smooth or echinulate and has one or more germ pores. The aecidiospores are unable to germinate on the host (Barberry) on which they are produced.

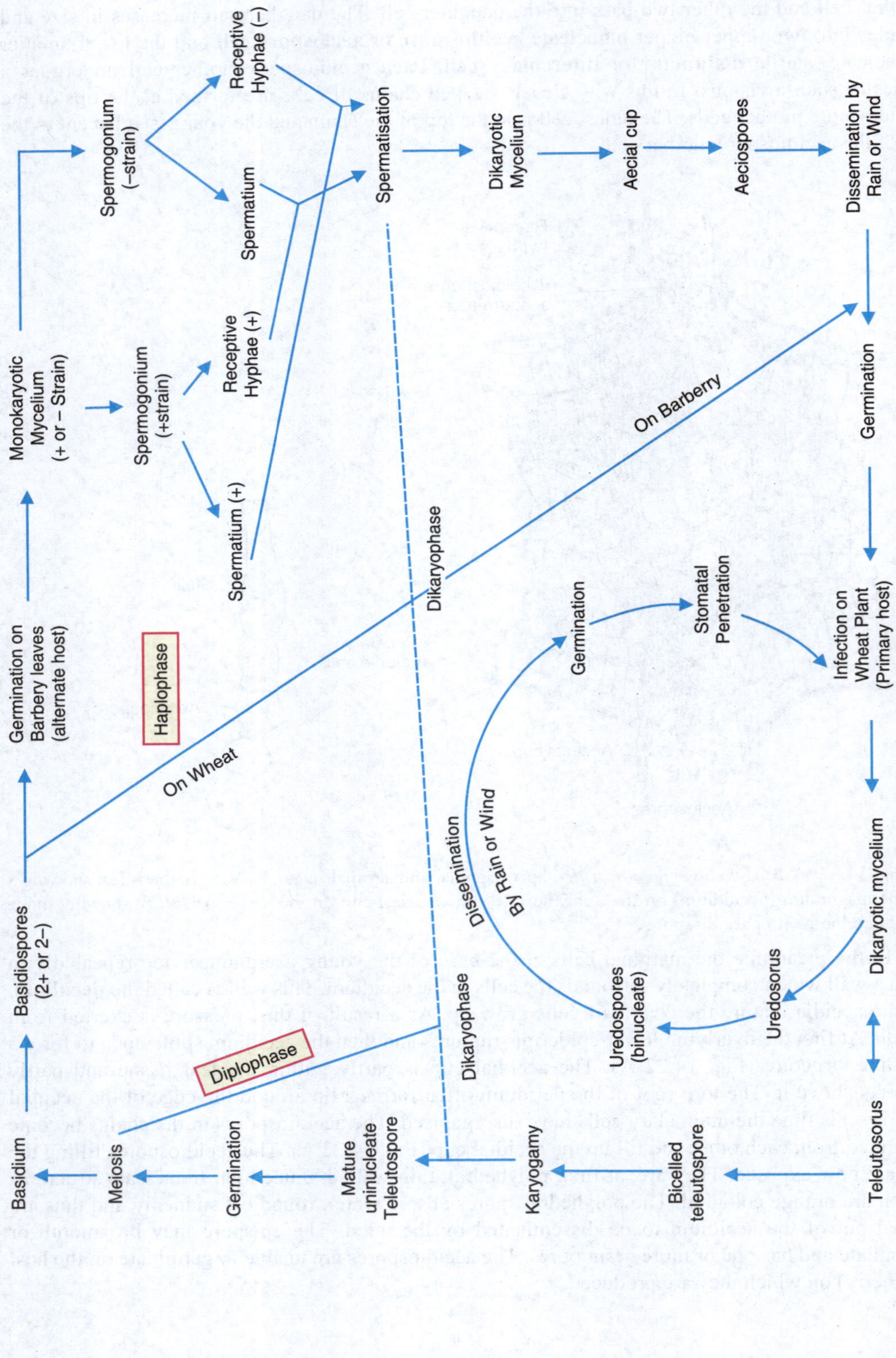

Fig. 14.23. *Puccinia graminis*. Schematic representation of life cycle.

Germination of Aecidiospore. The binucleate aecidiospores are carried by air currents to the primary host (wheat) on which they germinate under suitable conditions. At the time of germination the aecidiospore puts out a **germ tube** or **primary hypha** which emerges through the germ pore. The germ tube crawls over the surface of the wheat leaf and finds its way into the host generally through a stoma. On reaching the stoma the tip of the germ tube swells up to form an appresorium and a vesicle in the substomatal cavity. From the vesicle arise hyphae which grow intercellularly. They branch and form an extensive dikaryotic mycelium. It is chiefly intercellular. Only haustoria enter the host cells. The mycelium soon begins to form masses of uredospores and repeats the life cycle.

Life Cycle Pattern (Fig. 14.23). It is evident from the account given in the preceding pages that five different types of spores are produced in a single cycle of *Puccinia*. These are :—

1. Spermatia (uninucleate) in spermagonia on Barberry.
2. Aecidiospores (binucleate) in aecidia on Barbarry.
3. Uredospores (binucleate) on the dikaryotic mycelium on wheat.
4. Teleutospores (binucleate) on the dikaryotic mycelium on wheat.
5. Basidiospores (uninucleate) on epibasidia.

The rusts thus have a polymorphic life cycle because of the presence of different types of spores in the life cycle. These spore forms are produced in a regular sequence. The sequence is controlled by inherent internal factors. It is never reversed. Because of this long life cycle the rust, in which all the five spore stages are produced, is called **macrocyclic** or **long cycled** rust. The best example of this is *Puccinia graminis tritici.* There are, however, examples of rusts in which the life cycle is generally reduced. This is accomplished by dropping out from the life cycle some of the spore types. Such a life cycle is said to be **microcyclic** and the rusts are called **microcyclic** or **short cycled** *rusts.* All the wheat rusts in India have been reported to be short cycle rusts. The haplophase which is ordinarily passed on the alternate host is altogether absent. There are no Barberry bushes or *Mahonia* plants in the plains. The spermagonial or aecidial stages found on Barberry and *Mahonia* in the Indian hills have been found to have no connection with the wheat rust in the plains. The wheat rust in the plains, therefore, produces only two types of spores, the **uredospores** and the **teleutospores.**

Recurrence of Wheat Rust in the Plains of India. The black stem rust of wheat is a short cycled rust in India. The part of the life cycle involving the alternate host Barberry is cut out. It thus produces only uredospores and teleutospores. The latter have lost power of germination. The uredospores cannot withstand the high temperature of summer months in the plains. There is thus no inoculum left in the fiels to infect the next wheat crop. How does then the disease reoccur every year in the plains with equal vigour? Till recently Mehta's hypothesis (1952, 1954) that the primary inoculum of black rust of wheat is introduced in the plains from the Himalayas in the North and Nilgiri and Pulney hills in the South held the sway. According to Mehta's hypothesis, at the higher altitudes (3,000-7,000 ft), the uredospores can oversummer in the congenial temperature on self-sown plants, out of season crops, tillers and possibly on grasses. The surviving uredospores serve as primary inoculum for the next wheat crop near or at the foothills. From there the infection spreads. The uredospores are blown by wind from the infected plants in the foothills to the healthy wheat plants in the plains. Recent investigations of Joshi *et al* (1971-1974) and other workers do not support Mehta's hypothesis so far as the role of Himalayas is involved. They hold that the spring movement of black stem rust uredospores is from the South to North as is the general pattern of movement of black rust in most cuntries of Northern Hemisphere. They argue

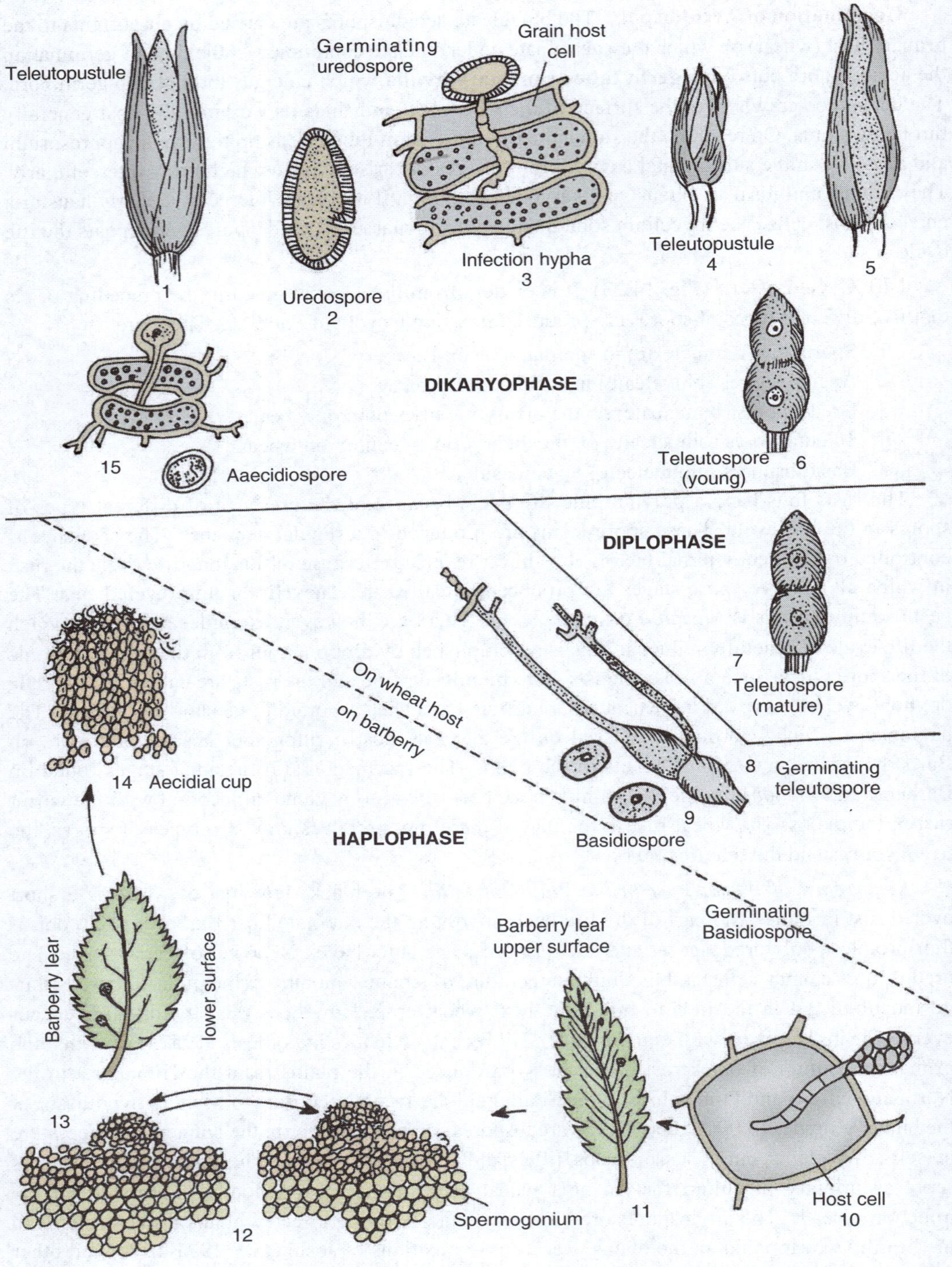

Fig. 14.24. Diagrammatic representation of the life cycle of *Puccinia graminis*.

that in the hills of Northern India, low temperature conditions prevailing during November to February are unfavourable for sporulation, infection, establishment and spread of the disease. Rust pustules become inactive at low temperature (below 14°C). When the conditions become favourable in the month of March, there is very little time left for its spread as the wheat plants reach maturity. In their view the principal source of infection of black stem rust of wheat in the plains of North is the dissemination of inoculum from the South and Central India.

The inoculum in the Himalayas is considered to be reintroduced from the plains in spring first in the foothills, then the lower elevations and finally at higher elevations.

Alternation of Generations. The typical life cycle (Figs. 14.23 and 14.24) of *Puccinia* illustrates the important biological principle of **alternation of generations.** There are two distinct generations or phases in the life cycle. One of these is the **dikaryophase** confined to the wheat plant. The other is the **monokaryotic** or **haplophase** passed on the alternate host (Barberry). The former is characterised by the binucleate condition of the cells. It corresponds to the **sporophytic** phase. The dikaryophase in the life cycle is initiated by the germination of aecidiospores (aeciospores) on wheat and multiplied during the growing season of wheat by the formation of uredospores. Both are binucleate. The dikaryotic mycelium first produces the binucleate uredospores (Fig. 14.24; 1 and 2) and later the binucleate teleutospores (Fig. 14.24; 4-6). Each cell of the young teleutospore has a pair of nuclei. These however fuse (karyogamy) in the ripe teleutospore (14.24; 7). The cells of the mature teleutospore, each of which is equivalent to a probasidium, are the only diploid structures in the entire life history. They represent the reduced **diplophase.** The young teleutospores are the last structures of the dikaryophase. The mature teleutospores represent the diplophase. The dikaryophase and the diplophase together constitute the **sporophyte phase** in the life cycle. It starts with the aecidiospores and consists of the dikaryotic mycelium, uredospores and teleutospores.

Each diploid cell of the teleutospore germinates to produce the epibasidium (Fig. 14.24; 8). The diploid nucleus of each cell of the teleutospore (probasidium) migrates into its respective epibasidium and undergoes meiosis to form the haploid basidiospores. This shows that karyogamy after disploidisation is considerably delayed but meiosis follows karyogamy immediately. Meiosis precedes the haplophase or the monokaryotic phase which is characterised by the uninucleate condition of the cells. It starts with the basidia. The basidiospores (Fig. 14.24; 9) are the first haploid uninucleate spores in the life cycle. They germinate on the leaves of the alternate host barberry to produce the haplomycelium (Fig. 14.24; 10). The latter produces the spermagonia and spermatia (Fig. 14.24; 12) and also the aecidial primordia. The aecidial primordia are the last structures of the haplophase. The latter ends with the diploidisation of the basal cells in the aecidial primordia. With this the binucleate condition is re-established and dikaryophase again starts.

From the account given above it is evident that in the life cycle of *Puccinia* the dikaryophase and the diplophase constitute the **sporophyte** generation. It regularly alternates with the haplophase or **gametophyte** generation and *vice versa.* One follows the other in a regular sequence. This means that the two generations constituting a single life cycle alternate with each other. In other words, *Puccinia* exhibits the phenomenon of alternation of generations.

Comparison between P. graminis, P. recondita and P. striiformis—Causing Wheat Rust

S.No.	*P. graminis*	*P. recondita*	*P. striiformis*
1.	Causes black rust of wheat attacking the stems most severely.	Causes brown rust attacking the leaf lamina exclusively	Causes yellow rust attacking the leaf lamina most severely along-with stem and ears.
2.	Generally appears in March in Northern India and the last week of November in South India	Appears in January in the Northern and Eastern states of India	Generally appears during January-February in Northern and Eastern states.
3.	Uredosori appear on leaves, stems and ears.	Uredosori are restricted to leaves.	Uredosori appear on leaf sheaths and ears.
4.	Uredopustules are red or brown in colour, elongated and large.	Uredopustules are bright orange in colour scattered, small, oval or round.	Uredopustules always arranged in long stripes, lemon yellow in colour, small, oval
5.	Uredospores brown and oval with four equatorial germ pores.	Uredospores bright orange in colour, oval with 3-4 germ pores.	Urediospores yellow coloured, nearly round with 6-10 germ pores.
6.	Teleutosori confined to stem, black in colour, elongated.	Teleutosori rarely formed, if formed, develop on lower surface of leaves, brown in colour.	Teleutosori confined to the lower surface of the leaf, generally yellow coloured.
7.	Teleutosori burst on maturity	Teleutosori do not burst at maturity.	Teleutosori do not burst at maturity.
8.	Teleutospores black coloured, bicelled round or conical at the apex.	Teleutospores brown coloured, bicelled, rounded and thickend at the apex.	Teleutospores yellow coloured, bicelled, curiform, conical at the apex.
9.	Primary Host : Wheat Alternate Host : *Berberis vulgaris*	Primary Host : Wheat Alternate Host : *Muehlenbergia hugeli*	Primary host : Wheat Alternate Host : *Thalictrum fulvum*
10.	Recurrence due to urediniospores blown from Hills.	Same as in *P. graminis*	In addition to urediniospores inoculum comes from collateral Hosts.

Control Measures. Since the alternate host (*Barberry* or *Mahonia*) is not of economic importance, the disease can be controlled by the eradication of the alternate host. Of course in the case of short cycled rusts, in which the part of life cycle involving the alternate host is cut out, this remedy is of no avail. In any case the eradication of barberry bushes is of some advantage. It prevents the origin by hybridisation of new races. In the case of short cycled rusts the following control measures may be adopted :—

1. The best remedy is to grow the rust resistant varieties.
2. Rotation of crops may prove useful in some cases.
3. Destruction of self-sown wheat plants, green stubbles and tillers, which carry over the rust in the uredospore stage, is useful.
4. Early sowing.
5. Cultivation of early maturing varieties of wheat.

6. *Proper irrigation.* High humidity favours the germination of rust spores. Hence light irrigations at proper intervals particularly during the months of January and February will lessen the incidence of infection.
7. Dusting the crop four times during the critical period with finely powdered sulphur dust at an interval of ten days provides a good control. For a detailed account of the control measures refer to Chapter XII.

Comparison between Rusts and Smuts

Rusts and smuts can be compared as follows :—

(*a*) Points of resemblance

(*i*) The mycelium in both passes through two stages, the monokaryotic (primary) stage and the dikaryotic (secondary) stage.

(*ii*) The dikaryotic mycelium is of greater importance. It is usually perennial in both. It is intercellular and obtains nourishment by means of haustoria. It produces teleutospores.

(*iii*) Motile cells in the life cycle of both smuts and rusts are lacking.

(*iv*) The teleutospores in both are resting spores and on germination produce basidiospores. *Ustilago tritici* is the only exception.

(*v*) The tertiary mycelium and basidiocarps are lacking in both.

(*vi*) The dolipore parenthesome septal complex, which is a hallmark of the Basidiomycetes, is absent in both the rusts and smuts.

(*vii*) In nature both appear to be obligate in their parasitism but recently the scientists have been able to grow both on culture media as facultative saprophytes.

(b) Points of difference

TABLE : II

Rusts	Smuts
1. Wheat rusts are **heteroecious** and others are autoecious.	1. All smuts are **autoecious.**
2. The rusts are intercellular and obtain their nutrition by means of haustoria. Clamp connections on the secondary mycelium are rare.	2. The smuts may be intercellular or intracellular (*U. maydis*). Haustoria are present. Clamp connections are common.
3. The dikaryotic mycelium produces three kinds of binucleated spores; **uredospores** and **teleutospores** on the primary host and **aeciospores** on the alternate host.	3. It produces only one kind of binucleate spores called the **smut pores** which are comparable to the teleutospores of rusts.
4. The teleutospores are developed from the terminal cells of the mycelium.	4. Smut spores are formed from the intercalary cells.
5. The teleutospores are stalked, 2-celled and each cell is binucleate.	5. The brand spores (teleutospores) are unicellular and binucleate.
6. Each cell of the 2-celled teleutospores produces an epibasidium which bears four basidiospores. They are borne on sterigmata and are discharged violently by the water drop method.	6. The single-celled, brand spore which is equivalent to teleutospores, germinates to produce a single epibasidium which bears a variable number of basidiospores. They are not borne on the sterigmata nor are they discharged violently.

Taxonomic Position

Alexopoulos (1962)

Division	:	**Mycota**
Sub-division	:	**Eumycotina**
Class	:	**Basidiomycetes**
Sub-class	:	**Heterobasidiomycetidae**
Order	:	**Uredinales**
Family	:	**Pucciniaceae**
Genus	:	*Puccinia*
Species	:	*graminis*

Alexopoulos and Mims (1979)

Super kingdom	:	**Eumycotina**
Kingdom	:	**Mycetae**
Division	:	**Amastigomycota**
Sub-division	:	**Basidiomycotina**
Class	:	**Basidiomycetes**
Sub-class	:	**Teliomycetidae**
Order	:	**Uredinales**
Family	:	**Pucciniaceae**
Genus	:	*Puccinia*
Species	:	*graminis*

Other important Genera of family Pucciniaceae, (Fig. 14.13)

Gymnosporangium. Represented by 57 species, the genus is restricted to temperate regions. Most of the species are heteroecious and demicyclic, the telial stage occurs on Gymnosperms while aecial cups are formed on dicotyledons. *G. juniperi-virginiane* forms pycnial and aecial stages on apples and telial stage on Junipers. The teliospores (teleutospores) are bicelled and pedicellate.

Gymnosporangium

Hemileia. The genus parasitises Angiosperms mostly the members of Family Rubiaceae. Of the 40 species known, *H. vastatrix* is parasitic on coffee plants causing coffee rust. Pycnial and aecial stages are not known, the uredinia are formed in sub-stomatal chambers. The telia are formed along with uredinia. Teliospores (teleutospores) are one celled, reniform, thinwalled and pale.

Hemileia vastatrix

Phragmidium. The genus known by 60 species is an autoecious macro—or demicyclic rust parasitising the members of the family Rosaceae. The teliospores (teleutospores) are stalked, transversely septate and many celled. *P. mucronatum* and *P. tuberculatum* are important parasites of ornamental Roses.

Phyllanthus emblica

Ravenalia. Represented by 150 species, this autoecious rust parasitises legumes in mostly tropical countries. The teliospores (teleutospores) fuse to form a teliospore head in which each teliospore has a separate stalk. Each teliospore head is subtended by a sterile cyst cell. *R. emblicae* parasitises *Phyllanthus emblica*.

Uromyces. Species of the genus cause important rust diseases on members of family Leguminosae. Rust of beans caused by *U. phaseoli*, Rust of gram caused by *U. ciceris-arietini*, Rust of pea caused by *U. pisi* and *U. fabae* and carnation rust caused by *U. dianthi* are some of the exmples of the genus which is macro or microcyclic and heteroecious or autoecious. The teliospores (teleutospores) are one celled, binucleate and thickened at the apex. In the Gram rust caused by *U. ciceris arietini*, the aecial and pycnial stages are not known, only the telial and uredionial stages are known and produced on leaves. *Trigonella polycerata* acts as collateral host.

Ravenalia

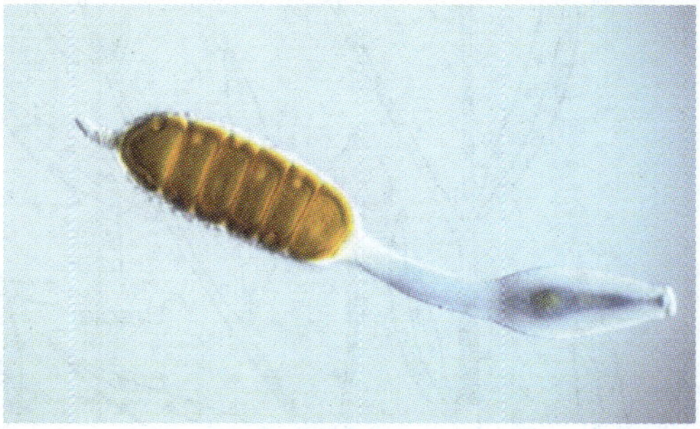

P. tuberculatum

Phyllanthus emblica – Leaves showing pustules

U. fabae – pea rust

FAMILY : MELAMPSORACEAE

The characteristic feature of the family is the teleutospore which are sessile, united into layers, single celled and sub-epidermal. The type genus is **Melampsora** which is discussed here.

MELAMPSORA

General. The genus *Melampsora* is represented by 80 species and is an obligate parasite. It may be autoecious or heteroecious. One species *Melampsora lini* causing rust of Linseed *(Linum utissimum)* is the best known. In infected plants, the leaves and stems get covered with bright orange pustules which are uredia and contain uredospores (urediniospores). Usually these pustules are produced on both the sides of the leaves. After some time the pustules on the

Melampsora – Rust of Linseed

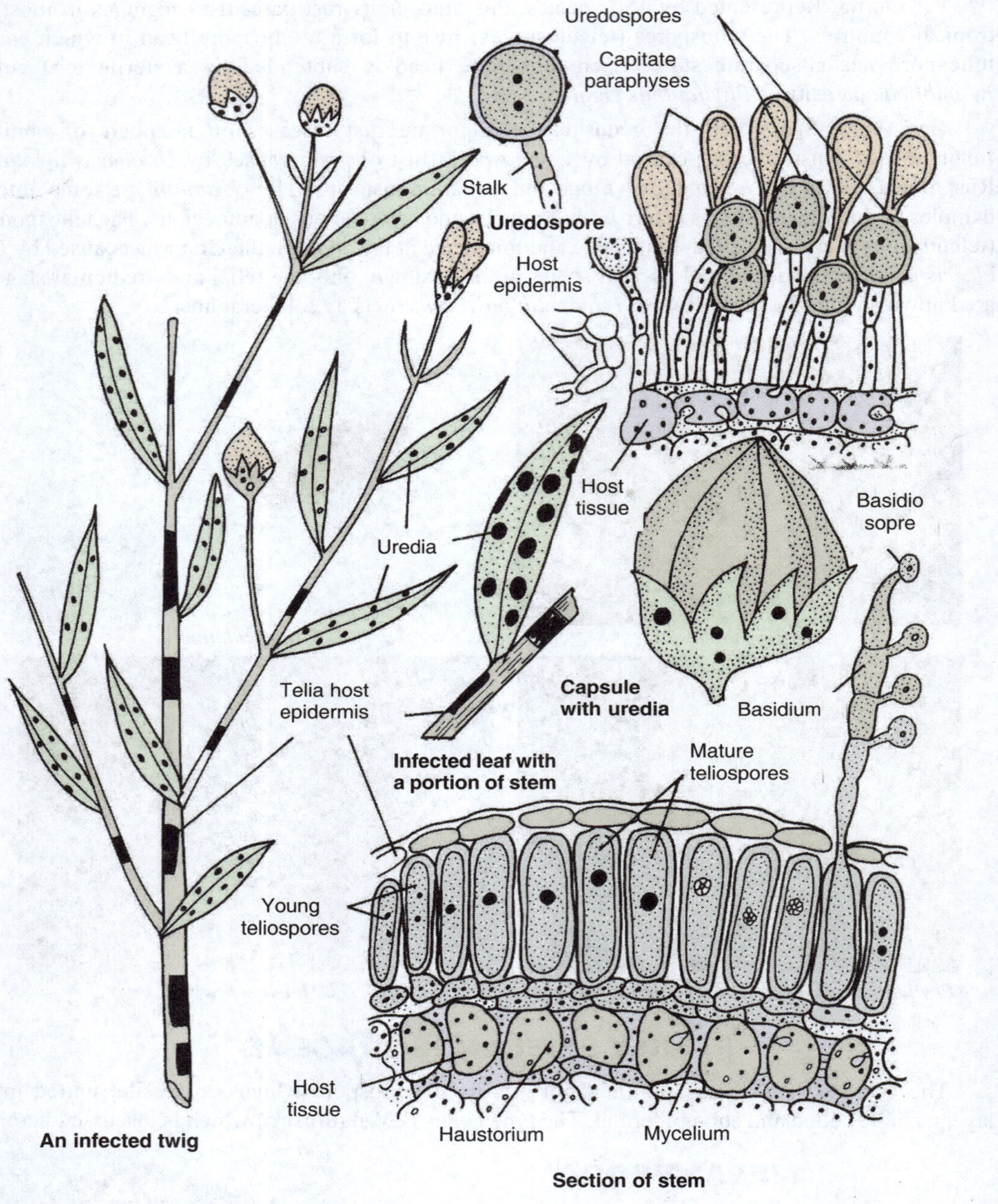

Fig. 14.25. *Melampsora lini* : Rust of Linseed showing symptoms, uredospores and teleutospores.

stem turn black. These black pustules contain teleospores (teleutospores) and are called **telia**. The pods also get infected but the telia are not produced on leaves. Roots are not affected although all parts of the plants get infected.

Mycelium. The mycelium is intercellular, septate, branched, dikaryotic, subepidermal and usually found within the parenchymatous tissues of the host.

Life-cycle. *M. lini* infecting several species of *Linum* is an autoecious rust. Although uredinial and telial stages are very common, the pycnidial and aecidial stages are not commonly recorded in nature. However, these stages have been produced in experimental plants by Prasada (1940).

The initial infection is due to urediniospores blown down from Hills or produced locally on collateral hosts. These spores imitiate infection in freshly grown flax plants and by December-January, bright orange or reddish brown pustules appear on leaves, stems and petioles. While on leaves the pustules are formed in circles, on stems the pustules are elongated. These pustules are uredia which are small scattered or in groups and contain urediniospores (uredospores). The uredospores are globose and echinulate and formed at the tips of long multicellular sporophores.

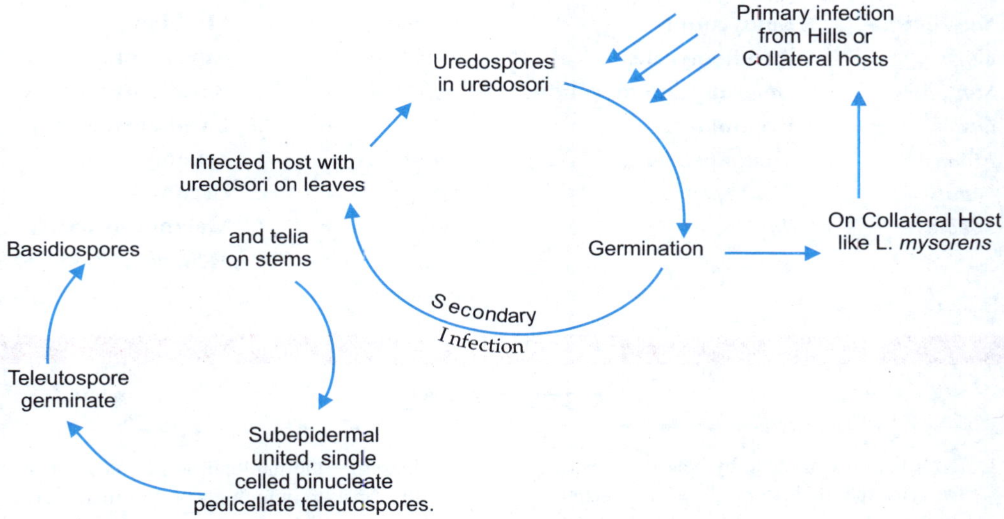

Fig. 14.26. Schematic representation of life cycle in *M.lini* in plains.

These sporophores remain intermingled with capitate paraphyses. The pressure of the uredospores and paraphyses breaks open the epidermis and the uredospores are released, disseminated by wind and cause infection upon reaching a suitable host. Usually these uredospores serve to spread the disease and the fungus until the end of the host season. Since the uredospores can not tolerate the high temperature of the plains, they die or they are blown away to hills where they survive on flax or collateral host.

At the end of the season, black or reddish brown crust like telia are formed on the stems. These telia consists of a palisade layer of black to brown coloured, sessile, cylindrical, unicellular, binucleate teleutospores. These can withstand the adverse environmental conditons for a long period of time and they also have a long dormant period. The teleutospores are released during harvesting of the crop and remain buried in the soil of the field. According to Prasada (1948), the teleutospores can not tolerate the scorching heat of India and die.

When exposed to freezing temperatures, the teleutospores germinate to produce four uninucleate basidiospores on a promycelium. In India the basidiospores do not play any role in the recurrence of the disease and die.

In places, where environmental conditions are favourable and in green house experiments the basidiospores germinate to produce small, rounded or oval yellow pycnia on both the surfaces of the leaf. The pycnia contain paraphyses and pycniosporophores producing pycniospores at their apical ends. After about fifteen days of the formation of pycnia, aecia develop on the lower surface of the leaf. The aecia are small, orange coloured and scattered and contain polygonal, binucleate aeciospores. But the aecia do not contain paraphyses and peridia. On germmation the aeciuspores produce dikaryotic mycelium and ultimately uredia.

Linum mysorens usually acts as a collateral host for the fungus.

Control Measures. The disease can be controlled through a number of measures, for the details Chapter 22 may kindly be referred to.

Taxonomic position

Alexopoulos (1962)

Division	:	**Mycota**
Sub-division	:	**Eumycotina**
Class	:	**Basidiomycetes**
Sub-class	:	**Heterobasidiomycetidae**
Order	:	**Uredinales**
Family	:	**Melampsoraceae**
Genus	:	*Melampsora*
Species	:	*lini*

Alexopoulos & Mims (1979)

Super kingdom	:	**Eumyconta**
kingdom	:	**Mycetae**
Division	:	**Amastigomycota**
Sub-division	:	**Basidiomycotina**
Class	:	**Basidiomycetes**
Sub-class	:	**Teliomycetidae**
Order	:	**Uredinales**
Family	:	**Melampsoraceae**
Genus	:	*Melampsora*
Species	:	*lini*

QUESTIONS

ESSAY TYPE

1. (*a*) What do you mean by heteroecious rust ? Describe the life history of *Puccinia graminis*. (*Poorvanchal, 2001*)

   (*b*) What is heteroecism? Give the life history of any heteroecious fungus prescribed in your course. (*Bhagalpur, 1991*)
2. Give an account of the life history of *Puccinia*. Give a life sketch to show the nuclear changes.
3. Give an account of *Puccinia* with special reference to sexuality in the genus. (*Bihar, 1991*)
4. Describe different kinds of spores produced by *Puccinia* and indicate the part that they play in the life history.
5. Describe the life cycle of a heteroecious rust fungus. How would you explain the annual occurrence of the rusts in the plains of India? Suggest some practical methods for combating the disease. (*Agra., 1995; Rohilkhand, 1996*)
6. Describe the somatic organisation of the dikaryophase in the life cycle of *Puccinia*. Name the host on which it is passed. List the different kinds of spores produced in this phase and describe their functions.
7. Describe the haplophase of *Puccinia*. Name the host to which it is confined. Give an account of the spore stages produced during this phase.
8. Write briefly about the spore types of *Puccinia*. (*Bharatidasan, 1995*)
9. Explain the life cycle of *Puccinia*. (*Bhartidasan, 1995*)
10. Describe the structure and functions of various spore forms occurring in the life cycle of *Puccinia*. (*Punjab, 1999*)
11. Describe in detail the stages in the life cycle of *Puccinia graminis*. What means have been suggested to control the disease? (*Bundelkhand, 1995; Awadh, 1998*)
12. Differentiate between Autoecious and Heteroecious Rust. Write the life-cycle of a Heteroecious Rust. (*Bundelkhand, 1997*)
13. (*a*) With the help of labelled diagrams only, describe the life-cycle of *Puccinia graminis* causing black rust of wheat. (*Bundelkhand, 1998; Gorakhpur, 1995, 1997, 1999; Lucknow,1993; Awadh, 1994, 1996; Kanpur, 1998, 2003; Rohilkhand, 2002*)

(*b*) Illustrate the life cycle of a heteroecious fungus with the help of labelled diagrams only. (*Lucknow, 2000*)

(*c*) Describe disease cycle and symptoms of Black Stem rust of wheat (*Poorvanchal, 2002, 2005*)

14. What is heteroecism? Describe the stages of *Puccinia graminis* on its alternate host. (*Bundelkhand, 1999*)

15. Mention the reasons why the life cycle of *Puccinia graminis tritici* is considered macrocyclic and heteroecious. Give suitable diagrams. (*Gorakhpur, 1990*)

16. Give an illustrated account of life-cycle of *Puccinia*. (*Kumaon, 2000; Allahabad, 2004*)

17. What is heteroecism ? With the help of diagrams, describe only that part of the life history of *Puccinia graminis* found on Barberry plant. (*Allahabad, 2000, 2002*)

18. (*a*) Give an account of the order Uredeniales and describe three important genera of this group. (*Allahabad, M.Sc. 1999*)

(*b*) Which is called as heteroecious fungus? Write the name of any polymorphic fungus and in brief mention the different kinds of spores found in sexual reproduction. (*Kanpur, 2006*)

SHORT ANSWER TYPE

19. Give an account of that part of life history of *Puccinia* that occurs on wheat. Explain its annual recurrence in plains of Northern India. (*Allahabad, 1991, 1998; Awadh, 1999; Kanpur, 1996, 1997; Himachal Pradesh, 1994*)

20. What is heteroecism? With the help of diagrams only describe the life history of fungus causing black rust of wheat. (*Allahabad, 1994, 1996*)

21. What are the characteristics of Uredinales? Illustrate with example the life history of any macrocyclic rust. (*Kanpur, M.Sc. 1993, 1998*)

22. Describe the life history of *Puccinia graminis tritici* and the control of the disease caused by it. (*Lucknow, 1995, 1997*)

23. What are rusts? How are they different from smuts? With the help of only labelled sketches describe the life cycle of the fungus causing black rust of wheat. (*Lucknow, 1999*)

24. What are different types of spores found in the life cycle of *Puccinia* with the help of diagrams, describe their structure. (*Kanpur, 1995*)

25. Explain briefly the life history of *Puccinia graminis tritici*. Support your answers with suitable sketches.

26. Describe briefly the nuclear changes that take place in *Barberis* leaf in the life history of *Puccinia graminis tritici*. (*Himachal Pradesh 1990; Rohilkhand, 1993*)

27. Name the host to which haplophase of *Puccinia* is confined. Give a brief account of spore stages produced during this phase. (*Himachal Pradesh, 1993*)

28. Name the pathogens causing rust of wheat in India. Which of them is most destructive? Describe the annual recurrence of the disease. (*Rohilkhand, 1991*)

29. With the help of labelled diagrams, describe the methods of plasmogamy in *Saprolegnia*, *Peziza* and *Puccinia*. (*Rohilkhand, 1994*)

30. Give an account of life-cycle of *Puccinia* on Wheat plants. Describe control measures. (*Magadh, 1993, 1994; Bhagalpur, 1990*)

31. Represent labelled life-cycle of *Puccinia*. Which of the phase represents the haploid phase of its life-cycle. (*Magadh, 1990*)

32. Describe the symptoms, etiology and control measures of Rust of Linseed. (*Ranchi, 1991*)

33. With the help of labelled sketches describe the life-cycle of *Melampsora lini* causing rust of linseed.

34. Draw labelled diagrams only of the following:

(*a*) T.S. Wheat stem infected by *Puccinia* (*Bundelkhand, 1996*)

(*b*) T.S. Barberry leaf infected by *Puccinia* (*Bundelkhand, 1996; Gorakhpur, 1998; Lucknow, 1996, 2001; Awadh, 1997; Rohilkhand, 1995; Agra, 1995, Kanpur, 2006*)

(*c*) T.S.Wheat leaf through teleutosorus and uredosorus (*Gorakhpur, 1992; Awadh, 1995; Kanpur, 1997, 1999*)

(*d*) Teleutispores found in different kinds of rust of wheat and rust of linseed (*Kanpur, M.Sc., 2003*)

(*e*) Teleutospore of *Puccinia* (*Awadh, 1992*)

(*f*) Life cycle of *Puccinia*/Casual agent of rust of wheat (*Rohilkhand, 2003; Meerut, 1992; Himachal Pradesh, 1992; Lucknow 2005*)

(*g*) Aecial and Pycnial cups of *Puccinia* (*Rohilkhand, 1997*)

(*h*) T.S. wheat leaf infected with *Puccinia*. (*Lucknow, 2001, 2002*)

(*i*) V.S. of barberry leaf passing through Aecial Cup. (*Gorakhpur, 2003*)

(*j*) V.S. of Barbeny leaf infected with *Puccinia* showing pycinidial stage. (*Kanpur, 2003*)

35. (*a*) Name one rust which is macrocyclic.

(*b*) What are different rusts on wheat?

(*c*) How does stem rust of wheat recur in the plains of India. (*Kanpur, M.Sc. 1985, 1987; Himachal Pradesh, 1997*)

(*d*) Describe in detail the formation of accidial cups in *Puccinia*. (*Kanpur, 2003*)

(*e*) Define Heteroecism. (*Poorvanchal, 2003*)

36. Differentiate between

(*a*) Rust and Smuts (*Kanpur M.Sc., 1994; Kanpur, 1995; Rohilkhand, 1996; Magadh, 1993*)

(*b*) *Puccinia* and *Uromyces*

(*c*) Black Rust and White Rust (*Agra 1993, 1994*)

(*d*) *Melampsora* and *Uromyces*

(*e*) Macrocyclic and Microcyclic (*Kanpur M.Sc. 1995*)

(*f*) Autoecious and Heteroecious

(*g*) Basidium and Ascus

(*h*) Uredospores and Teliospores (*Awadh, 1993; Gurunanak Deo, 1992*)

(*i*) Primary host and Secondary host (*Rohilkhand, 1994*)

(*j*) *Puccinia* and *Ustilago* (*Gurunanak Deo, 1991*)

(*k*) Dikaryotisation and Diploidisation (*Agra, 1993*)

(*l*) Teleutospore and basidiospore (*Lucknow, 2003*)

(*m*) Heterothallism and Heteroecism (*Allahabad, 2003*)

37. Describe the symptoms and general methods of control of the following :

(*a*) Black Stem Rust (*Lucknow, 1991, 1996, 1998; Himachal Pradesh, 1995*)

(*b*) Rust of Linseed (*Bihar, 1991*)

38. Compare in a tabulated form the rusts and the smuts.

39. Describe the formation of teleutospores in the life history of *Puccinia graminis*. (*Himachal Pradesh, 1990*)

40. Write short notes on ;

(*a*) Control of wheat rust (*Bundelkhand, 1993*)

(*b*) Autoecious and Heteroecious rust (*Bundelkhand, 1994; Gorakhpur, 1991*)

(*c*) Prof. K.C.Mehta (*Bundelkhand, 1995; Poorvanchal, 2003; Kanpur M.Sc., 1984, 1989; Agra, 1991; Rohilkhand M.Sc., 2006*)

(*d*) Black Rust of wheat (*Bundelkhand, 1996*)

(*e*) Physiologic races (*Bundelkhand, 1999*)

(*f*) *Melampsora* (*Kanpur M.Sc., 1991, 1996*)

(*g*) Uromyces

(*h*) Teleutospores (*Bihar, 1992; Magadh 1990*)

(*i*) Pycniospores

(*j*) Uredospores

(*k*) Basidiospores

(*i*) Aeciospores

(*m*) Uredosori (*Himachal Pradesh, 1990*)

(*n*) Dikaryotic phase of the life cycle of *Puccinia*

(*o*) Work of Prof. K.C.Mehta on annual recurrence of rust in India. (*Rohilkhand, 1992, 1995*)

(*p*) Heteroecism (*Allahabad 2001; Allahabad M.Sc. 1999, 2000*)

(*q*) Pycnial stage (*Kanpur, 2001*)

(*r*) *Ravenalia* (*Poorvanchal M.Sc., 2003*)

(*s*) Recurrence of rust (*Kanpur M.Sc., 2002*)

(*t*) Systematic position and important characters of *Puccinia*. (*Kanpur, 2006*)

41. Give one example each of autoecious and heteroecious rust. (*Lucknow, 2003*)

42. Write a note on annual recurrence of wheat rust in India. (*Kanpur,2002, 2006*)

43. Define Heteroecious rust. (*Poorvanchal, 2002*)

44. Name four diseases caused by rust fungi. (*Kanpur. M.Sc.2001*)

OBJECTIVE TYPE

45. Fill in the blanks :

(*i*) Aecidial stage of *Puccinia* develops on ....

(*ii*) Black stem rust of wheat is caused by ....

(*iii*) Non repeating bicelled binucleate spore stage in *Puccunia* is known as ....

(*iv*) The famous famine of Bengal was caused due to the disease ....

(*v*) The spore stage which is uninucleate and thin walled and produced on phragmobasidia is called ....

(*vi*) *Melampsora lini* causes the rust of ....

(*vii*) The primary host of rust of wheat is ....

(*viii*) The spores that are produced on Barberry plants but can infect only the wheat plants is called ....

(*ix*) An example of heteroecious rust is ....

(*x*) Macrocyclic rusts are those which produce all the .... spore stages.

46. Select the correct answer :

(*i*) Teleutospores are stalked, bicelled each cell binucleate with a papilla. Such Teleutospores are present in
(*a*) *Puccinia* (*b*) *Uromyces*
(*c*) *Phragmidium* (*d*) *Urocystis*

(*ii*) Rust of gram is caused by
(*a*) *Puccinia* (*b*) *Uromyces*
(*c*) *Phragmidium* (*d*) *Urocystis*

(*iii*) Common name of the disease caused by the members of uredinales is
(*a*) Rust (*b*) Smut
(*c*) White rust (*d*) Wart

(*iv*) In a macrocyclic rust, the stage 'o' refers to
(*a*) Aecium (*b*) Uredinium
(*c*) Pycnium (*d*) Basidium

(*v*) Teleutospores are sessile column shaped in
(*a*) *Puccinia* (*b*) *Uromyces*
(*c*) *Melampsora* (*d*) *Hemileia*

(*vi*) Repeatedly produced spores in the life cycle of rust are called
(*a*) Aeciospores (*b*) Uredospores
(*c*) Basidiospores (*d*) Pycnidiospores

(*vii*) When a fungus requires two different hosts to complete its life cycle, the fungus is referred to as
(*a*) Homothallic (*b*) Heterothallic
(*c*) Autoecious (*d*) Heteroecious

(*viii*) The host on which rust produces its telial stage is called
(*a*) Primary Host (*b*) Secondary Host
(*c*) Alternate Host (*d*) Collateral Host

(*ix*) The great Bengal famine of 1942-43 was caused by the infection of
(*a*) *Puccinia graminis tritici*
(*b*) *Phytophthora infestans*
(*c*) *Helminthosporium maydis*
(*d*) *Synchytrium endosporium*

(*x*) Which of the following plant pathologists of India is famous for his researches on Wheat Rust?
(*a*) S.N.Das Gupta (*b*) B.B.Mundkur
(*c*) E.J.Butter (*d*) K.C.Mehta

(*xi*) A macrocyclic rust is one which
(*a*) needs two different hosts to complete its life cycle
(*b*) produces five types of spores to complete its life cycle
(*c*) does not show any asexual reproduction
(*d*) has a prolonged life cycle

(*xii*) Phragmobasidium is produced in
(*a*) *Albugo* (*b*) *Agaricus*
(*c*) *Synchytrium* (*d*) *Puccinia*

(*xiii*) When two different hosts are required by a fungus to complete its life cycle, the phenomenon exhibited by them is known as
(*a*) Autoecism (*b*) Heteroecium
(*c*) Homothallism (*d*) Heterothallism

(*xiv*) The structure which produces basidiospores in *Puccinia* is called
(*a*) Holobasidium
(*b*) Phragmobasidium
(*c*) Ascus
(*d*) Gametangium

(*xv*) In *Puccinia graminis*, sexual reproduction takes place through
(*a*) Somatogamy
(*b*) Spermatisation
(*c*) Gametangial copulation
(*d*) Gametangial contact

(*xvi*) *Macrocyclic* fungus is
(*a*) *Ustilago* (*b*) *Puccinia*
(*c*) *Agaricus* (*d*) *Morchella*

(*xvii*) Recurrence of rust was first studied by
(*a*) K.C. Mehta
(*b*) K.S. Thind
(*c*) C.V. Subramanaiam
(*d*) A.F. Blackslee

(*xviii*) K.C. Mehta is associated with the study of
(*a*) Smut of Sugarcane
(*b*) Rust of wheat
(*c*) Loose smut of wheat
(*d*) All of the above

(*xix*) Diploidisation in *Puccinia graminis* occurs in
(*a*) Uredospores (*b*) Teleatospore
(*c*) *Basidiospores* (*d*) Arcidiospores

(*xx*) In *Puccinia*, monokaryotic mycelium develops after germination of
(*a*) Acciduisoires (*b*) Uredospores
(*c*) Basidiospores (*d*) Pycnidiospores

(*xxi*) Heteroecious fungus is
(*a*) *Ustilago* (*b*) *Puccinia*
(*c*) *Albugo* (*d*) *Alternaria*

Polyporus, the Pore Fungi

CHAPTER 15

SUB CLASS – HOMOBASIDIOMYCETIDAE

Agaricus

- General Features
- Series 1: Hymenomycetes
- Order : Agricales
- Family : Agaricaceae
- Order : Polyporales (Aphyllophorales)
- Family : Polyporaceae
- *Polyporus*
- Order : Lycoperdales
- Family : Lycoperdaceae
- *Lycoperdon*

General Features. The fungi included in this sub-class comprise about 7,500 species which are grouped under 300 genera. They grow in varied habitats such as forest litter, grassland, on dung (coprophilous), on sand dunes and a few even on termite mounds. Most of them are saprophytes. Some are important plant parasites and good many wood destroying fungi. They are cosmopolitan in their distribution and are best known to every one by their large, conspicuous fructifications called the **basidiocarps.** Because of the massiveness and complexity of structure of the basidiocarps the Homobasidiomycetidae are probable considered the most advanced of all the fungi. The basidiocarps are borne on the mycelial strands (rhizomorphs) of the secondary mycelium. The septal pore is more complex than the Ascomycetes and is called a **dolipore.** The Homobasidiomycetidae posses the following distinctive features :

Basidiocarps

1. Form large conspicuous fructifications called **basidiocarps.**
2. The septal pore is a complex dolipore septum with parenthesomes.
3. The basidium develops directly from a terminal cell of a dikaryotic hypha.
4. It is simple, unseptate, typically club-shaped or cylindrical.
5. With a few exceptions the basidium shows no morphological differentiation into probasidium and Epibasidium.
6. Commonly the basidia are arranged in a regular fertile layer, the **hymenium** which extends over the surface of the basidiocarp.
7. The hymenium in some is exposed from the very beginning or becomes exposed at or before maturity. The former are called gymnocarpic and the latter angiocarpic.
8. Both karyogamy and meiosis take place in the basidium itself at different stages of its development.
9. The basidium bears generally four, sometimes one, two or even eight basidiospores.
10. Basidiospores are borne singly at the tips of small, delicate stalks called **sterigmata** which arise directly from the basidium.

Asexual reproduction plays a minor role in the life cycle. It takes place by means of oidia, arthrospores and rarely by conidia.

Classification. The sub-class Homobasidiomycetidae is divided into many orders. Their number varies with different authors. These orders are sometimes grouped under the following two series :

Series 1. *Hymenomycetes.* The hymenium is exposed even when the basidiospores are young and undeveloped. The basidiosores are borne obliquely on the sterigmata and are discharged forcibly. This series includes a single order *Agaricales* represented by the mushrooms, and toad stools. Martin (1961) and Alexopoulos (1962) added *Polyporales* to the list.

Series 2. *Gastromycetes.* The hymenium is enclosed and never exposed. It disintegrates when the spores are mature. The basidiospore discharge is not explosive. This series includes the puff balls and earth stars. The most important orders included in this series are *Hymenogastrales, Lycoperdales, Sclerodermatales, Phallales* and *Nidulariales.*

SERIES 1 : HYMENOMYCETES

According to some mycologists, the series Hymenomycetes includes a single order Agaricales but latest trend is to divide it into two orders, Agaricales and Polyporales. Both are described here in some detail.

ORDER : AGARICALES

It is a large order including more than 4,000 species placed under about 197 genera. All are saprophytes. The common examples are the mushrooms and toad stools. Together the two are called **agarics.** The basidia are arranged to form a definite hymenial layer exposed on plate-like structures called the **gills.** The order includes a number of families. Of these Agaricaceae is discussed here.

FAMILY : AGARICACEAE

General Characters. Hymenium lines the surface of plates of tissue called **gills** which are generally borne on the underside of a cap-like or an umbrella-shaped structure called the **pileus.** From the middle of the undersurface of the pileus springs a stalk-like **stipe.** The cap and the stalk together constitue the fruiting body called the **basidiocarp** or **sporocarp.** Because of the presence of characteristic gills the members of this family are generally called the **gill** or **gilled fungi.** They are generally fleshy and are also known as **fleshy fungi.** Some of these gilled or fleshy fungi are edible and others poisonous. Ordinarily the edible fleshy (or gilled) fungi are called the **mushrooms** and the poisonous ones **toad stools.** Four edible species of mushrooms have been reported from India. These are *Agaricus campestris, Lentinus subnudus, Calocybe indica* and *Volvariella volvacea.* The fifth *Termitomyces eurhizus* has recently been added to the list by Purkayastha and Chandra (1974). The family *Agaricaceae* is cosmopolitan in its distribution and occurs from the temperate region to the tropics. The most important and well known genus of the family is *Agaricus.* It is studied as a type genus.

Edible Mushrooms

Agaricus campestris

Termitomyces eurhizus

Calocybe indica

Poisonous Toadstools

Volvariella volvacea

AGARICUS (PSALLIOTA, FIG. 15.1)

Introductory. *Agaricus campestris* is the field mushroom. It is an edible gilled fungus which is cosmopolitan in its distribution and grows wild. It is known by different names in different parts of our country. In the Punjab it is called *dhingri.* The species is exotic to India. Its edible fructifications are imported from Afghanistan. According to Rolf Singer (1962), It is doubtful that this species has ever been grown commercially. The most widely cultivated species for food purposes is *Agaricus bispora.* It is grown in different parts of India. Solan in Himachal Pradesh is an important centre of mushroom cultivation.

Agaricus Bispora for Food Purpose

Habit and Habitat. *Agaricus* is a saprophyte and grows in the open fields, grass lands and in soil in cellulose and lignin materials. It is also found growing in the decaying litter on forest floors or in the humus deposited at the surface and below the surface of the ground. It grows best in moist shade and is commonly seen during the rainy season in the wild state. It has been suggested that the mycelium of this fungus forms association with the roots of grasses among which it grows. The fact that it lives as a physiologically well balanced parasite on these roots has not been well established. What we ordinarily call the mushroom is not the whole fungus. It is simply the fructification which is technically called basidiocarp. The vegetative part of the organism remains hidden in the soil. It is the inconspicuous mycelium.

The cultivated mush-room has appeared under a number of names in literature. The earliest name seems to be *Agaricus campestris* var. *hortensis* (1885). Since then many varieties and forms of the cultivated mushroom such as *Psalliota arvensis* var. *hortensis* (1908), *Psalliota hortensis* (1926), *Agaricus hortensis* (1821), *Psalliota bispora* (1938), *Agaricus bisporus* (1946), *Agaricus campestris* (1943) and *Agaricus brunnescence* (Peck, 1900) have been proposed. *Mulloch* (1976) suggested that *Agaricus brunnescence* is the most appropriate and correct name for the cultivated mushroom (*A. bisporus*). Since it is also one of the oldest, it must be accepted according to the rules of priority.

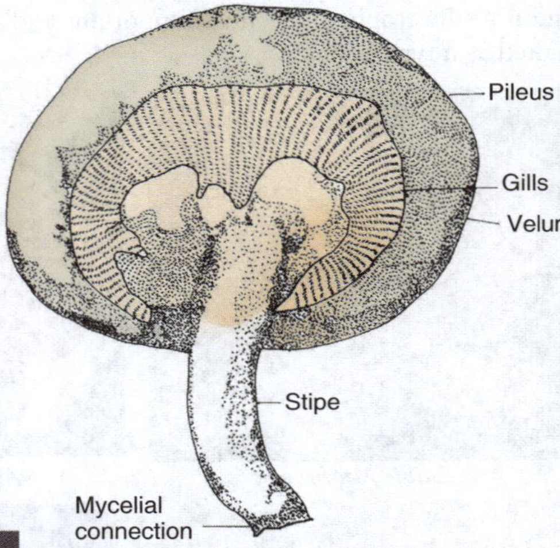

Fig. 15.1. Basidiocarp of *Agaricus sp.*

Agaricus brunescens

Mycelium (Fig. 15.2). The primary mycelium produced by the germination of basidiospore is of short duration. It is haploid and either of plus strain or minus strain. The hyphae are septate. The cells contain oil globules, vacuoles and are short and uninucleate. Soon as a result of hyphal fusions the primary mycelium becomes **binucleate** usually without clamp connections rarely with clamp connections. The mycelium with binucleate cells is called **secondary** or **dikaryotic mycelium.** It is long-lived and abundant. It produces mushrooms year after year. The secondary mycelium forms a branching network of hyaline hyphae. The hyphae are long, branched and short celled. The cells are binucleate and communicate with one another by means of central pore in the septum. It is a typical Basidiomycete **dolipore septum** in which the opening is guarded on both sides with its **parenthesomes.** The hyphae traverse the soil in all directions from a central point and form a circular colony covering an extensive area. It may remain in this condition for a considerable period before it forms the overground fructifications. Commonly the hyphae interlace and twist to form thick, white hyphal cords called the **rhizomorphs** which bear the fruiting bodies (Fig. 15.3A). In many mushrooms the fructifications develop at the tips of the hyphae of the circular colony forming a ring. This process is repeated with further growth of the hyphae so that widening circles of basidiocarps are seen. The appearance of basidiocarps in a ring from the invisible mycelium is called a *"fairy ring"* because of an old superstition that the mushrooms growing in a ring indicate the path of dancing fairies.

Mushroom Cultivation. The edible mushrooms are cultivated on a large scale. They form the basis of a great industry in many foreign countries. The principal growth requirements of this fungus are a suitable supply of moisture, moderate temperature and abundant organic matter. In India the mushrooms are cultivated in Himachal Pradesh, South India and other places. Himchal Pradesh offers ideal Agro-climatic conditions for growing edible mushrooms. A special mention may be made of Himachal University College of Agriculture, Solan in this connection. It has achieved commendable success in mushroom cultivation. The Mushroom Research Laboratory at Solan has not only excelled in standarizing the technique for raising white button mushrooms and Oyster mushrooms but has also helped in establishing mushroom growing centres in different parts of the state such as Chail, Simla and Kasauli, and 116 other mushroom farms all over the country by supplying mushroom spawns to the growers. At Solan the mushrooms are cultivated in chopped paddy straw and compost. Freguson (1978) reported that a mixture of horse dung, chopped wheat straw, gypsum and additives constitute the **compost.** It is the usual substratum for the commercial propagation of the mushrooms. Compost is prepared by a complicated process called **composting.** It takes about 12 days to prepare compost.

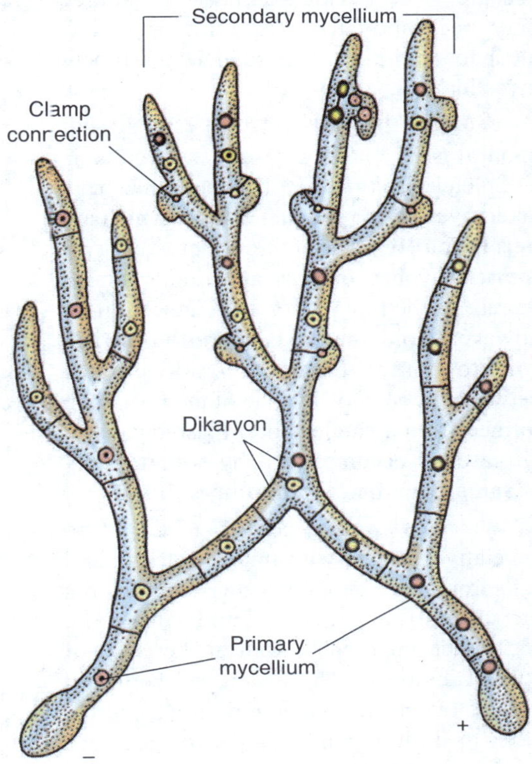

Fig. 15.2. *Agaricus*, diagram showing somatogamous copulation.

The beds for cultivation of mushrooms are specially prepared. They contain substratum of compost rich in nitrogenous compounds. These beds are then inoculated with crumbs of *mushroom spawn.* The mushroom spawns are usually of the following types :—

1. Blocks of richly manured soil containing the fungus mycelium and basidiospores in abundance.

2. Pure culture of the fungus mycelium on some nutritive medium.

3. **Grain spawn.** It is a sterilized cereal grain colonized by the mushroom mycelium. The mycelium will grow from the grains and spread throughout and on to the surface of the compost as white hyphae and white mycelial strands visible to the unaided eye. As the mycelium grows it induces the compost to change from black to light reddish brown colour. A compost temperature of 25-28°C during spawn growth is optimum. If the temperature rises above 33°C, mushroom mycelium will not grow. After mycelial colonization of compost sporophores are formed. Mander (1943) observed the presence of a volatile substance secreted by the mushroom mycelium which inhibits strand and fruit body formation if allowed to accumulate in mushroom beds. A layer of soil or heat chalk mixture known as the "casing layer" is necessary to stimulate fruit body formation. It is suggested that bacteria in this layer are responsible for the induction of fruiting though the exact effect of bacteria remains obscure. Recent evidence indicates that they may be responsible for removing self-inhabitors of fruiting located in the mushroom mycelium.

Sexual Reproduction. The sexual apparatus in the form of sex organs is completely lacking. Their function has been taken over by the somatic hyphae which are **heterothallic.** The fusion between two somatic hyphae of plus and minus strains represents the first phase. Some mycologists, however, hold that it is **homothallic** and somatogamous copulation takes place between the somatic hyphae of the mycelium formed from a single spore. Plasmogamy in all cases is accomplished by **somatogamy** or **somatogamous copulation** as follows :

1. *Plasmogamy* (Fig. 15.2). Two vegetative hyphae with uninucleate haploid cells from mycelia of opposite strains or from the same mycelium come into contact (Fig. 15.2). The intervening wall at the point of contact dissolves. The fusion cell comes to possess two nuclei which move towards each other to lie in a pair. This pair of nuclei in the fusion cell constitutes a **dikaryon**. The dikaryotised cell by successive divisions gives rise to the binucleate or dikaryotic

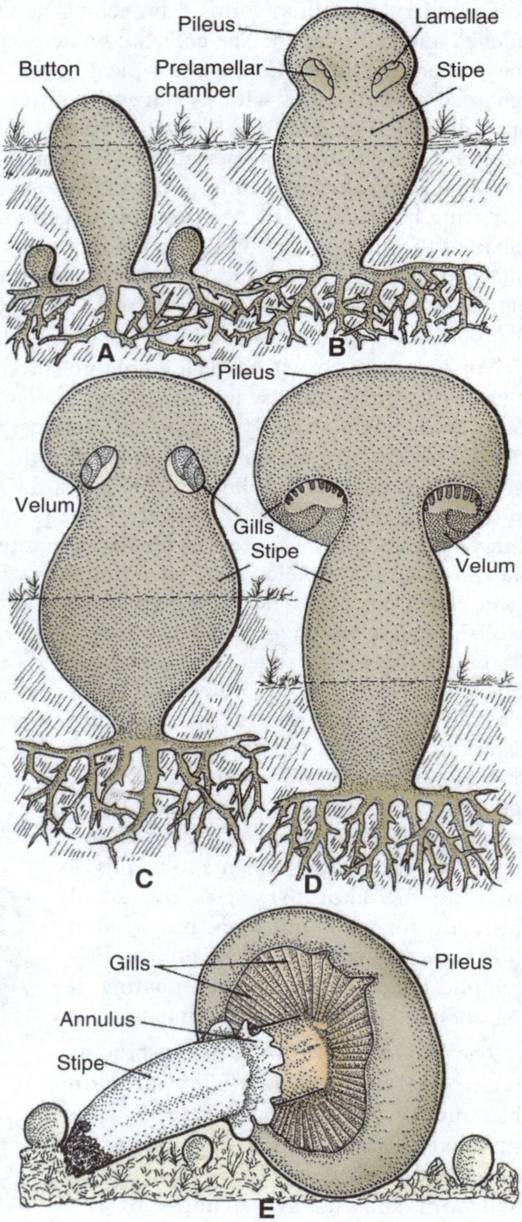

Fig. 15.3 (A-E). *Agaricus campestris.* Stages in the development of basidiocarp. A, button stage; B, showing the fomation of prelamellar chamber in the pileus region; C, differentiation of gills in the prelamellar chamber; D, a later stage; E, shows rupture of velum and formation of annulus. B-D, vertical section of developing basidiocarp. (Based on Smith).

mycelium. At every division the two nuclei of the dikaryotic cell divides conjugately into four daughter nuclei. This cell then develops a clamp connection which ensures that the sister nuclei separate into two daughter cells. The process is repeated several times resulting in the formation of a mycelium with binucleate cells. Some mycologists hold that clamp connections are absent in *Psalliota.* The binucleate mycelium thus originates from the uninucleate mycelium as a result of plasmogamy by somatogamous copulation. Plasmogamy in fact initiates dikaryophase in the life cycle.

The binucleate or the dikaryotic mycelium is perennial. It forms the chief food absorbing phase of the fungus and at the appropriate time bears fructifications called the **basidiocarps** (Fig. 15.3). The basidiocarps are formed when the mycelium has absorbed and accumulated abundant food supply. They are developed in groups from the mycelium under suitable temperature and sufficient moisture.

2. Karyogamy. It means the fusion of the two nuclei of the dikaryon. It is considerably delayed. It takes place in the young basidium (Fig. 13.8 B).

3. Meiosis. Karyogamy is immediately followed by meiosis and takes place in the basidium prior to basidiospore formation (Fig. 13.8 C-D). The basidiospores are thus haploid.

Development of the Basidiocarp (Fig. 15.3). The basidiocarps arise as tiny white apical swellings on the branches of the subterranean mycelial strands which are known as the **rhizomorphs.** Each swelling consists of a dense knot of dikaryotic hyphae. The tiny white knots enlarge into round or ovoid structures which break through the surface. These tiny balls represent the **button stage** of the basidiocarp (A). A longitudinal section of a button shows differentiation into a bulbous basal part representing the **pileus** (B). At the junction of these is a ring-like cavity, the **gill chamber.** It is formed by drawing of tissue radiating outwards from the stipe. These are the **gills.** They are already forming. The margin of the button is connected with the stalk by a membrane called the partial or **inner veil** or **velum** (C). By further elongation of the stalk, the button projects above the soil and enlarges considerably in size. The growth proceeds more rapidly at the upper portion of the button. It is slow at the lower portion. As a result the button opens into an umbrella-like cup or the **pileus** (D). At the same time the inner veil which covers the lower surface of the pileus ruptures exposing the hymenium on the gills. The mushroom thus is **hemiangiocarpic.** Remnants of the inner veil remain attached to the stalk or the stipe in the form of a ring or **annulus** (E). The latter encircles the stipe close to the attachment of the pileus. The exposed young gills are white, at first, but later turn pink.

In some gill fungi such as *Amanita* the button when young is completely covered by a membrane called the **universal veil.** As the button grows and enlarges into the pileus and the stalk the universal veil ruptures leaving a cup-like structure, the **volva** surrounding the base of the stipe. Volva is absent in *Agaricus campestris.*

In dry weather when the substratum is hard and dry the young basidiocarps (buttons) do not increase in size. They remain underground in a partially developed condition until the soil above become thoroughly wet by rain. In a moist soil they enlarge rapidly due to the extension of existing cells rather than to cell division. This accounts for the sudden overnight appearance of numerous mushrooms (basidiocarps) above ground after a heavy shower in the rainy season. The phrase 'mushroom growth' refers to this phase of rapid expansion.

Mature Basidiocarp. (Fig. 15.3 E). It is a massive structure which consists of a stalk-like portion, the **stipe** supporting at the top a broad umbrella-shaped cap, the **pileus.** The stipe is a thick fleshy, cylindrical structure, pinkish white in colour. It is usually broader and swollen at the base and is centrally attached to the pileus. The remains of the strands of the mycelium are sometimes seen still sticking to the base of the stipe. More than half way up, the stipe, bears a membranous ring called the **annulus.** It represents the remains of the ruptured inner veil. The stipe supports and raises the pileus into the correct position for spore dispersal. It also transports nutrients and water from the mycelium to the pileus and gills for their development. The mature pileus is 2.4 inches in diameter. It is umbrella-shaped. The convex upper surface of the pileus is white, cream coloured or brown. From the underside of the pileus hang numerous thin, vertical strips or plates of tissue, the **gills** or **lamellae** (Fig. 15.4 B). They look like partitions which hang down in a vertical position. The negative geotropism of the stipe, diageotropism of pileus and positive geotropice nature of gills help keep the latter in an accurate vertical position. The gills are flesh-coloured or pink known as the **hymenium** or **thecium.** The specialized fertile cells constituting the hymenium are the **basidia** which produce the **basidiospores**. The umbrella-shaped pileus affords protection to the gill tissue. The function of the basidiocarp is to produce and disperse the numerous basidiospore produced by the basidia when young. The mature gills are dark brown or purplish black. The colour changes with the pigmentation of spores. They radiate from the margin of the pileus towards the stipe. However, they are not joined to the stipe. All the gills are not of the same length. Between the longer gills there are shorter gills. The surface of the gill is covered on both sides with a fertile layer of spore producing tissue. It is called the **hvmenium** or **thecium.**

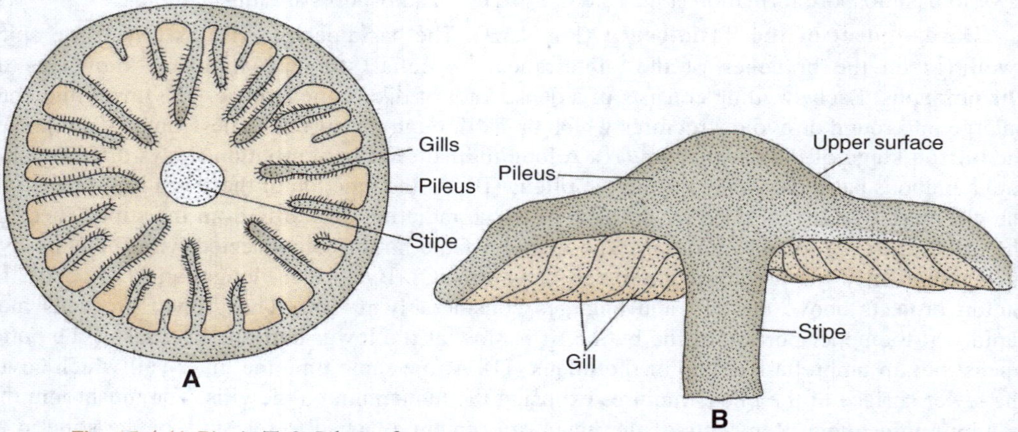

Fig. 15.4 (A-B). A, T. S. Pileus of *Agaricus*; B, Pileus cut lengthwise to show gills hanging from the under-surface and radiating towards the stipe.

Anatomy of Basidiocarp (Fig. 15.5). The entire fructification, as mentioned above, is composed of a fundamental tissue of an interlacing mass of tertiary hyphae. The hyphal cells are multinuucleate. The septa between the cells have one central pore each with a septal pore appartus.

(*a*) Stipe. The hyphae run longitudinally parallel to each other in the stipe region. In the peripheral portion they are compactly arranged to form a pseudoparenchymatous tissue. It is called the **cortex.** Towards the central region of the stipe these hyphae are loosely interwoven so that intercellular spaces are easily recognisable between them. The central region of loosely arranged hyphae is called **medulla.** The cells constituting the hyphae are large and elongated. Within the cell wall is the thin electron dense plasma membrane. According to Manocha (1965), the cytoplasm contains a few mitochondria. The ER is scanty and generally inflated and fragmented into vesicles. There are a large number of dense, granular glycogen particles. Interspersed between these large fundamental cells are the second type of cells known as the **connective cells.** They are long and thread-like. Both the type of cells are multinucleate. According to some workers, the rapid elongation of the upper portion of stipes during the later developmental stages of the basidiocarp is due to cell elongation alone. However, Craig *et al* (1977) showed that both cell elongation and cell division are responsible for elongation of the upper part of the stipe region.

***(b)* Pileus.** The structure of the pileus is like that of the stipe. The two regions, **cortex** and **medulla** of the stipe extend right up to it. The hyphae of the stipe region extend into the pileus and fan outwards. Some of these pass down into the gills.

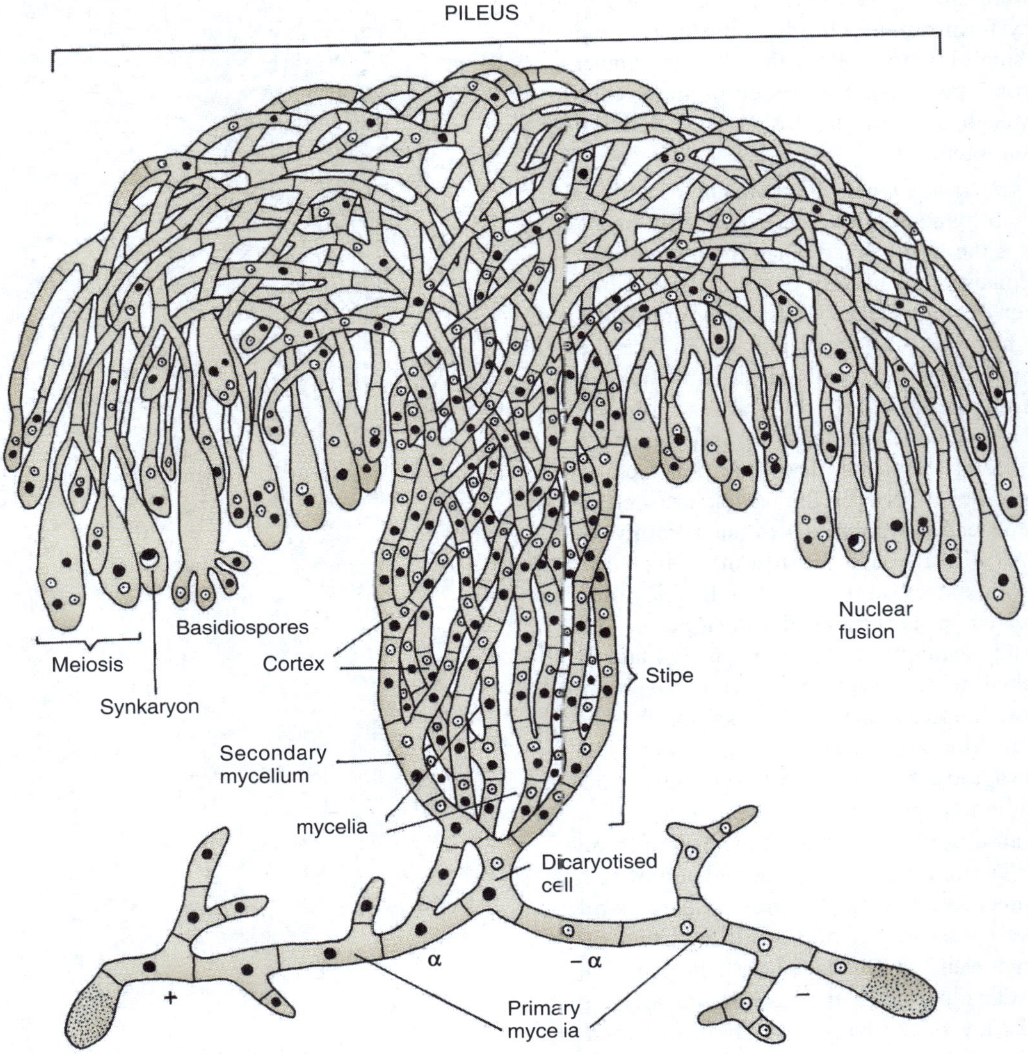

Fig. 15.5. *Agaricus campestris*. A scheme representing diagrammatically the anatomy of the basidiocarp developed from a secondary mycelium originating by the fusion of two haploid hyphae (+ and –) of opposite strains of mycelia.

***(c)* Structure of a Gill** (Fig. 15.6). The gill exhibits a complex structure. It is a sheet of interwoven hyphae which are more closely compacted and denser in this region. They extend from the pileus to form the undersurface of the gill. In a tangential section of the pileus, the gills are cut transversely. In a transverse section of gill we can make out the following regions :—

1. *Trama.* It is the innermost or central part of the gill between the two hymenial surfaces. It is made up of extensions of the hyphae of the pileus. The hyphae in this region anastomose and are irregularly interwoven. They run more or less longitudinally. The cells contain fewer nuclei than the cells of the hyphae constituting the stipe and the pileus.

2. *Subhymenium* or *Hypothecium.* The hyphae constituting the trama give off short lateral branches. The latter curve outwards towards the two surfaces of the gill where they form a more closely compacted tissue of small inflated cells with still fewer nuclei (2 or 3 per cell). This tissue of small cells on each side of the trama is called the **subhymenium.**

3. *Hymenium* or *Thecium.* On the surface of sub-hymenium covering both sides of the gill is the fertile layer called the **hymenium.** It consists of a closely packed palisade-like layer of club-shaped cells. They are called the **basidia** (A). The basidia are, in fact, the terminal elongated cells of the same hyphae which constitute the trama and the subhymenium. The basidia are the spore producing elements. They are non-septate. Interspersed between the basidia are sterile, more slender hyphhae called the **paraphyses** or **cystidia.** Each basidium bears four basidiospores at its free end. They are elevated on short stalks, the **sterigmata.** The young basidium has a **dikaryon** (B_1) and is packed with cytoplasm. As the basidium grows the two nuclei of the dikaryon fuse to form the **synkaryon** (B_2). This is **karyogamy.** It ends the **dikaryophase.** The basidium containing the synkaryon in the mature basidium undergoes meiosis to form four haploid nuclei (B_3). Segregation of sex strains takes place at this time so that two of these nuclei are of plus strain and two of minus strain. At this stage four delicate small peg-like outgrowths, the **sterigmata,** develop at the top of the basidium. The tip of each sterigma swells to become a **basidiospore initial,** which is small, and bead-like in appearance (B_4). Meanwhile a vacuole appears at the base of the developing basidium. It progressively enlarges pushing one of the 4 daughter nuclei along with some cytoplasm into each basidiospore initial (B_5). The basidiospore in *A. brunnescence* is thus uninucleate. During its passage through the narrow neck of the basidiospore, the haploid nucleus assumes a vermiform shape (15.6 B_5). It again becomes spherical in the basidiospore. The young basidiospore is subsequently separated from its respective sterigma by a septum. The mature basidium has a large vacuole with the cytoplasm. When young the basidiospores are unpigmented but towards maturity the spore wall turns pinkish purple. Each mature basidiospore is seated slightly obliquely at the top of the sterigma with the spore point

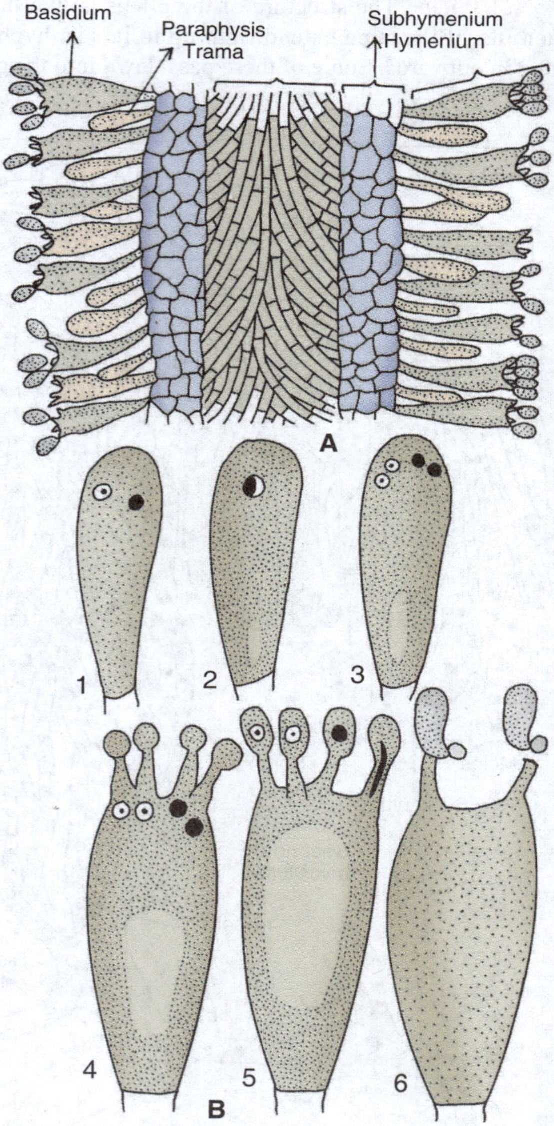

Fig. 15.6 (A-B). *Agaricus campestris.* A, cross section of the gill showing structure; B_{1-6}, stages in the development of the holobasidium.

which is a minute projection (**hilar appendix**), a little to one side near its attachment to the sterigma (B_6).

Morphology of Basidiospore (Fig. 15.7). The haploid basidiospore of *A. campestris* is a violet-brown, thick-walled oval or ellipsoid structure with a minute, eccentrically placed spine-like projection—the **hilar appendix** on the adaxial side of the basidiospore wall at the basal end called the **hilum.** The latter denotes the former point of attachment of the spore to the sterigma. Because of its small size the morphology of hilar appendix has been under dispute. Buller (1922) and Olive (1964) described it as the specialized part of the adaxial wall of the basidiospore which is the site of formation of the liquid or gaseous droplet associated with the violent "spore discharge". There are a few reports dealing with the electron microscope study of this structure. Olak and Resinger (1974) reported that the hilar appendix in the Agaricales contains two orifices. One of these, the *hilar orifice* (orifice hilare) is a pore-like scar. It results from the rupture of the sterigma-spore interface. The second which is associated with droplet formation and its subsequent forcible discharge, is the *hilare droplet orifice.* According to Rast and Hollenstein (1977), the adaxial supra hilar region of the spore surface shows on special differentiation of hilar appendix (B). It is an open pore type hilum. The innermost layer of the spore wall thickens to plug the latter.

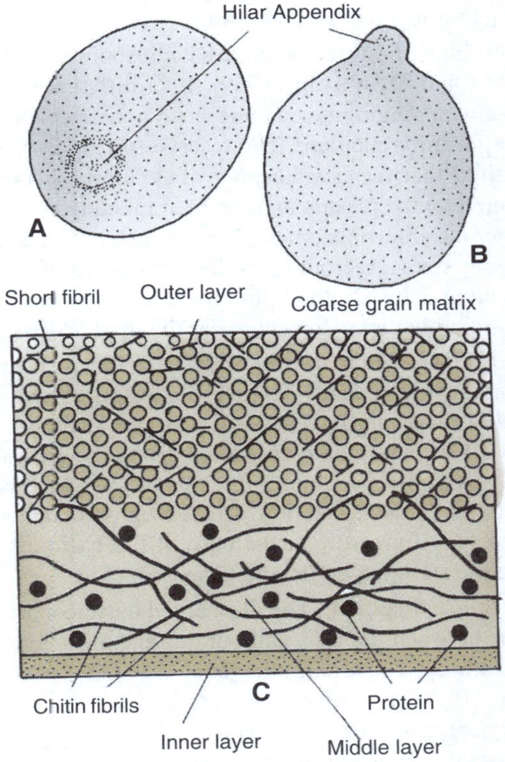

Fig. 15.7 (A-C). *Agaricus bisporus.* A, Basidiospore with smooth surface; B, Basidiospore with adaxial supra hilar region showing no special differentiation; C, ultrastructure of spore wall (A-B based on Greuter and Rast; C, based on Rast and Hollenstein).

Yoon and McLaughlin (1979) consider hilar appendix to be a specially differentiated part of adaxial spore wall at the junction of the sterigma and the basidiospore. In the young basidiospore when it is less than half its normal size and the nucleus has not yet migrated into it, a change in the adaxial wall structure in the site of hilar appendix takes place. It thickens and an eletron light zone develops. When the electron light zone is fully developed and the spore reaches maturity, the hilar appendix swells outwards and takes on a drop-like appearance. No plug is seen at this stage between the sterigma and the spore. Prior to spore discharge, the cytoplasm adjacent to the hilar appendix is replaced with an eletron-light plugging material.

Fine structure of basidiospore (Fig. 15.8). The basidiospore in the dormant state has a thick spore wall which lacks any ornamentation. Thus the spore has a smooth surface. Even the preformed germ pore is not detectable. Ultrastructurally the spore wall is differentiated into three layers, the outer, middle and the inner (Manocha, 1965, Rast and Hollenstein, 1977). The former two are thicker than the innermost which is next to the cytoplasm. The outer smooth layer is electron opaque. It is comparatively thicker than the middle layer and is chiefly granular amorphous. In addition it contains some fibrillar material. The middle layer is fibrillar.

The chitin with numerous evenly dispersed fibres are embedded in a B-glucan-protein matrix. According to Rast and Hollenstein (1977), it is relatively less electron-dense than the outer layer which stains deeply. The thin inner layer is almost electron-transparent. It is amorphous and seems to be composed of a mucilaginous type of substance. It thickens at the junction of the sterigma and the spore to plug the open pore type hilum [Rast and Hollenstein (1977)]. Lipid occurs throughout the spore wall.

Within the three-layered basidiospore wall is the spore protoplast. According to Manocha (1965) and O'Sullivan (1969), it is differentiated into a cell membrane and cytoplasm which contains a single nucleus and a few other cell organelles such as mitochondria, ribosomes and endoplasmic reticulum. The reserve food is in the form of oil globules, Greuter and Rast (1975) reported that the basidiospore of *A. bisporus* is normally binucleate. The cytoplasm is closely covered by the plasma membrane. Embedded in it are the mitochondria with plate-like cristae and numerous free ribosomes. The cytoplasm is packed with one large or several lipid droplets of varying sizes. Accroding to these investigators, the cytoplasm of *A.bisporus* lacks endoplasmic reticulum. Close to the lipid droplets occur clusters of granules probably glycogen. Mitochondria with microbodies also occur near the lipid bodies.

Spore discharge (Fig. 15.6 B_6). The mechanism of spore discharge is still uncertain. The common view is that when the basidiospore reaches maturity, hilum secretes a tiny drop of watery fluid. It gradually increases till it attains a certain size (B_6). At this stage basidiospore is suddenly shot off the sterigma perpendicularly to the gill surface with a force just enough to carry it about half way (1/10 to 1/8 mm.) into the narrow space between the two gills. As regards the mechanism of discharge the latest view is that drop is contained in a limiting membrane. With its rupture there is release of pressure at the base of the basidiospore. Consequently the basidiospore is shot off from the sterigma by jet propulsion.

Spore dispersal. The discharged basidiospores falling vertically downwards from the narrow space between the gills and out into the air below the cap are carried away by the currents of air. The long stipe facilitates the process as it raised the pileus into correct position for spore dispersal above the soil surface providing sufficient space underneath, which allows air currents to carry away the falling spores. The basidia mature in successive crops. The older ones which have discharged the basidiospores wither away. New ones grow up to take their place until all the basidial initials have been used up. The basidiospores are thus discharged successively over short intervals for several days. Each basidium bears a single crop of basidiospores.

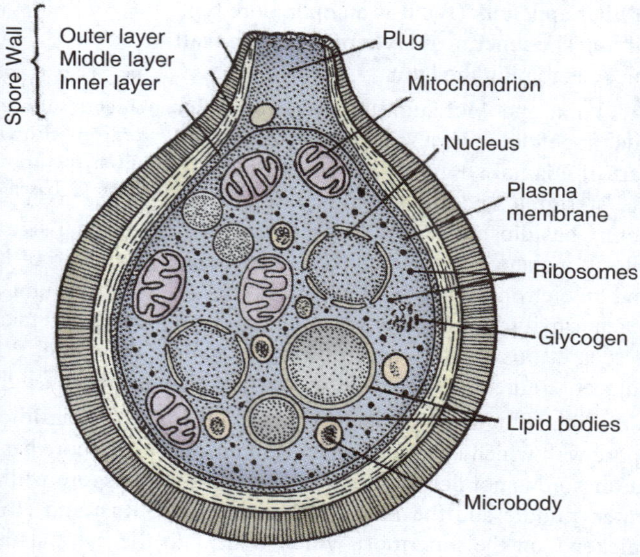

Fig. 15.8. *Agaricus bisporus.* Diagrammatic representation of the basidiospore showing fine structure (Based on Greuter and Rast).

Agaricus bisporus, the mushroom which is cultivated commercially is distinguished from *A. campestris* the field mushroom by its darker and more firmer pileus, inferior flavour and two basidiospores per basidium. The mature basidiospores are purplish black in colour. The number of spores produced by a single basidiocarp is enormous, about 1,800,000,000 spores. They are discharged successively over short intervals for several days. The umbrella-shaped pileus affords protection to the fertile layer from rain.

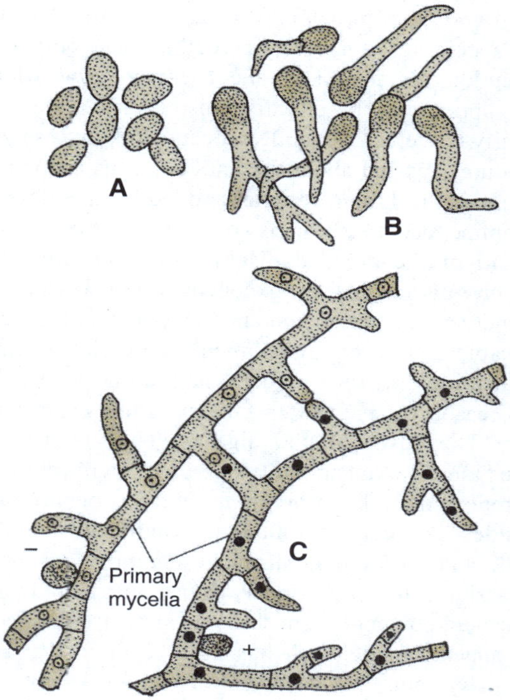

Fig. 15.9 (A-C). A, Basidiospores; B, germinating basidiospores; C, Hyphal fusions between primary mycelia of opposite strains.

Germination of basidiospores (Fig. 15.9). The basidiospores (A) which chance to fall on a suitable substrate germinate. An adequate moisture supply and favourable temperature are essential for germination. Mc Teague, Hutchinson and Read (1959) found that the presence of the vegetatively growing mycelium of *A. campestris* in the medium containing spores of this fungus stimulates spore germination. The stimulating factor is a volatile metabolite secreted by the mycelium. The presence of olefin which is probably 2 : 3 dimethyl - I-pentene was shown to be present by analysis with spectrometer. Each basidiospore puts out through some point on the spore wall a protoplasmic sac which elongates into a narrow germ tube (B). The latter develops into a primary mycelium which is either of plus or minus strain. It consists of uninucleate cells. The binucleate condition is soon achieved by somatogamous copulation between hyphae of two primary mycelia of opposite strains growing in the vicinity (C). The binucleate mycelium starts producing the basidiocarps.

Asexual Reproduction. The field mushroom is reported to multiply by the formation of **chlamydospores** and **conidia.**

Taxonomic Position

Alexopoulos (1962)

Division	:	**Mycota**
Subdivision	:	**Eumycotina**
Class	:	**Basidiomycetes**
Sub-class	:	**Homobasidiomycetidae**
Series	:	**Hymenomycetes**
Order	:	**Agaricales**
Family	:	**Agaricaceae**
Genus	:	*Agaricus*
Species	:	*brunnescence*

Alexopoulos and Mims (1979)

Super kingdom	:	**Eumyconta**
Kingdom	:	**Mycetae**
Division	:	**Amastigomycotina**
Subdivision	:	**Basidiomycotina**
Class	:	**Holobasidiomycetidae**
Series	:	**Hymenomycetes**
Order	:	**Agaricales**
Family	:	**Agaricaceae**
Genus	:	*Agaricus*
Species	:	*brunnescence*

Economic Importance of Agaricus. (*a*) *Positive.* All species of *Agaricus* are edible. The only two exceptions are *A. silircola* and *A. xanthodermis* which are supposed to be slightly poisonous at times but usually not lethal. Several years ago, Tokyo Research Institute reported about the anticancerous properties of certain mushroom varieties and the Japanes doctors recommended mushrooms to patients to check the spread of the dreaded disease. In food value the mushrooms range from excellent to fair. In spite of 88 percent of water content the mushrooms are a valuable source of nourishment being rich in high quality proteins (5-7 percent) and amino acids. For this reason they are often known as the vegetarians meat. They also contain a digestive enzyme that aids digestion. The nutritional value of Oysters (dhingri) is reported to be more than button mushrooms. Besides they contain folic acid, vitamin B-complex and C and several minerals like Calcium, Phosphorus, Potassium, Iron and Copper. Their mineral content is higher than that of mutton. Because of high folic acid content the mushrooms are extremely beneficial for blood formation and thus are good for anaemic patients. It has been reported that glycogen content of *A. bisporus* sporophore is normally about 2 to 4 percent of the dry weight at the immature button stage. It increases to 5-8

A. Xanthodermis-Non Edible Mushroom

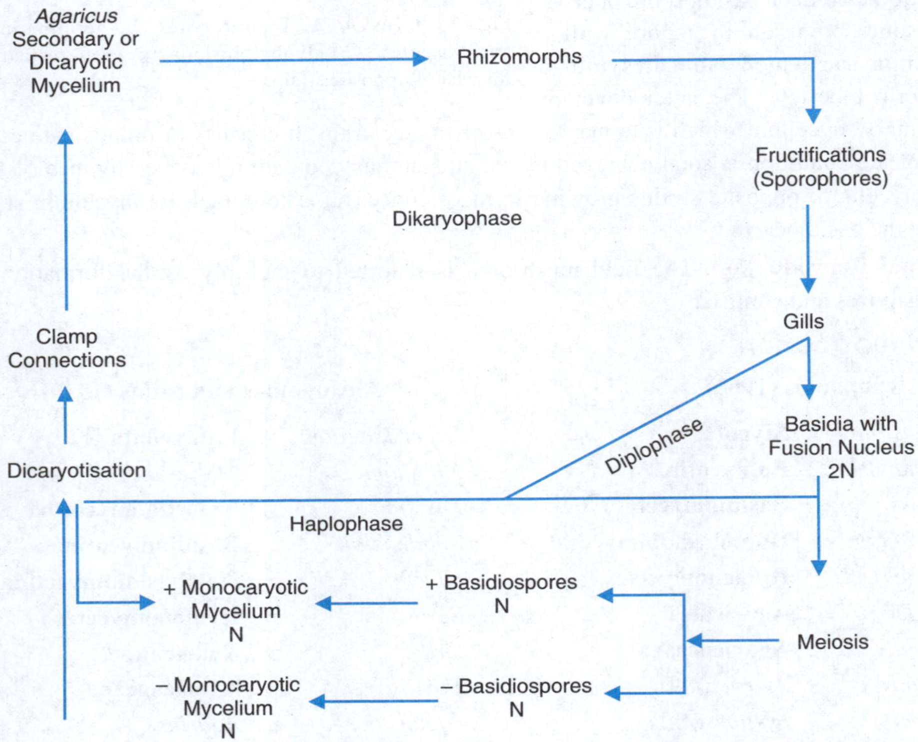

Fig. 15.10. Graphic representation of the life cycle of *Agaricus*.

percent at the mature flat stage. The mushrooms (Cysters) being devoid of fat cholesterol and starch, are prescribed as excellent diet for diabetes patients and for those who suffer from high blood pressure. Dr. Chandramani reported that mushrooms are rich in vitamin C, niacin and pantothenic acid (a constituent of vitamin B_2). All these are considered beneficial for diabetes, anaemic and heart patients. The mushrooms thus being of considerable economic importance are grown commercially and form the basis of food industry. *Agaricus bispora,* the white spored mushroom, is cultivated to a greater or lesser extent all over the world. The other commonly cultivated species are *A. bitorquis* and *A. arvensis* *Cortinellus berkeylyanus* is cultivated in Japan. Another mushroom *Volvariella* is commercially grown in the tropics. The two commonly cultivated species of this genus are *V. volvacea* and *V. diplosia.* They are commonly grown in China, Indonesia, India, Burma, Indo-China and Madagascar. *Pholiota praecox* is another edible mushroom but it has no flavour. Purple-spored agarics such as *Agaricus campestris, A. rodmani* and *A. placomyces* are also edible. *Agaricus xanthodermis* and a few others contain bacteriostatic substances which inhibit development of bacteria. The hallucigenic mushrooms of the genus *Psilocybe,* which are considered sacred by certain Mexican tribes, produce ecstatic effects. These sacred mushrooms are now used experimentally in medicine to assist in the treatment of disease by psychological methods.

A. bitorquis

A. arvensis

Cortinellus berkeylyanus

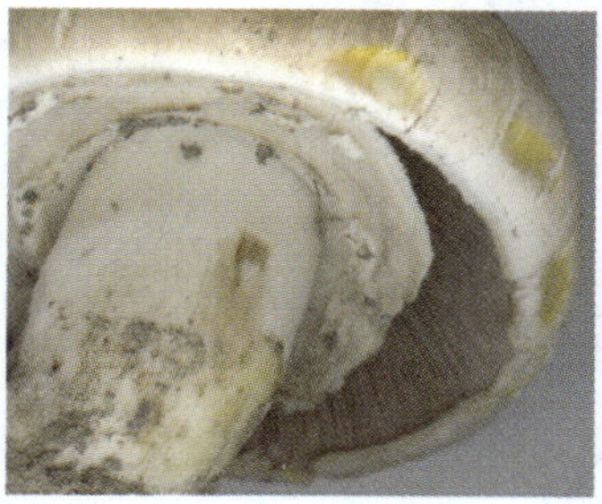
A. placomyces

Pholiota praecox

Volvoriella volvocea

(*b*) *Negative.* A few members of the order are important wood rotters. *Armillaria mellea* is one of them. It is called the **honey mushroom**. It causes foot rot of apple and some forest trees. Some species of this family are supposed to be of possible use in the production of antibiotics. Of course, no new and useful substance has so far been discovered. A member of species form mycorrhizal associations with the trees. Many agarics are highly poisonous. To this category

Armillaria mellea

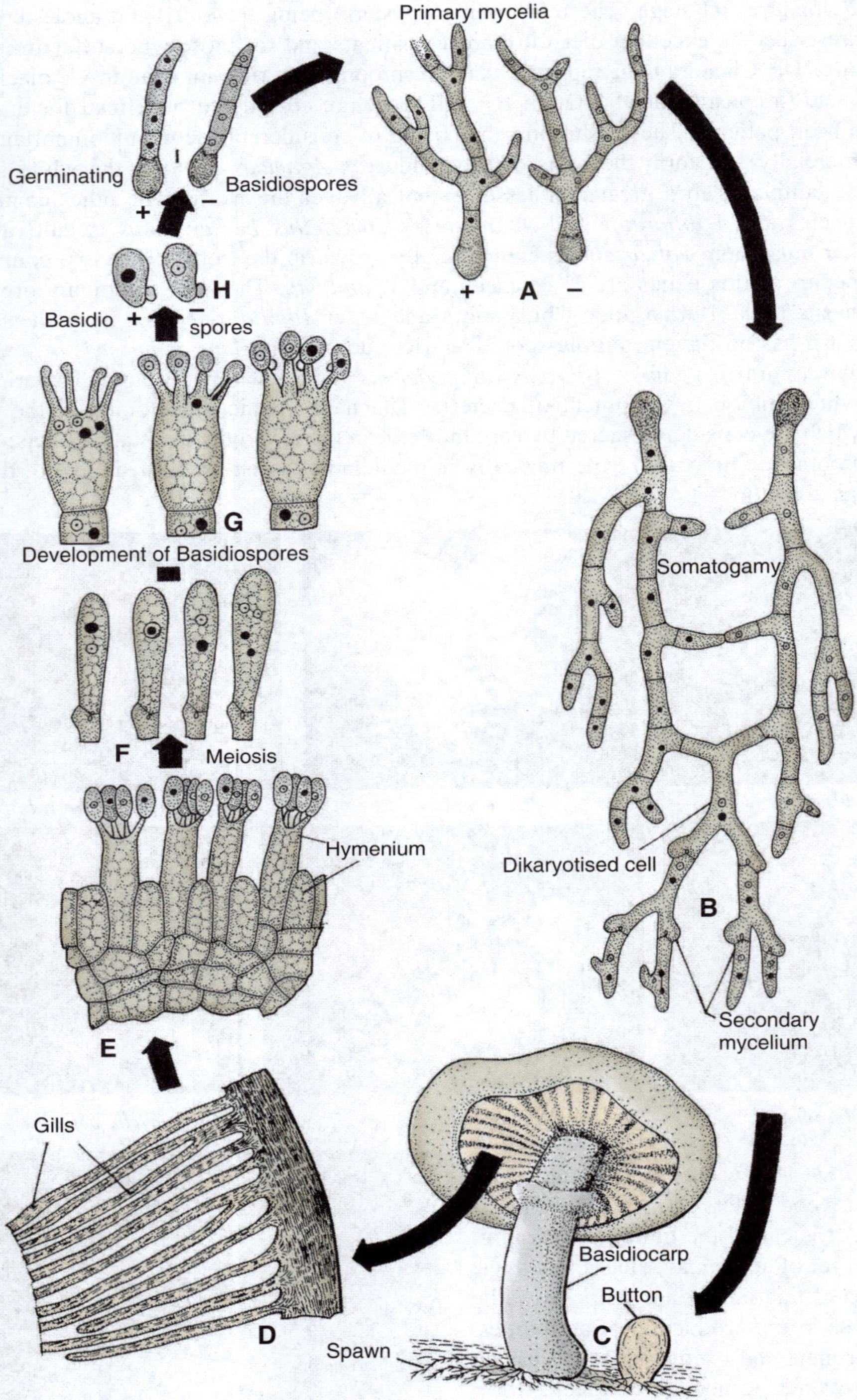

Fig. 15.11 (A-I). Pictorial life cycle of *Agaricus*. Some mycologists hold that *Psalliota* is homothallic. In that case secondary mycelium which bears the basidiocarps will be formed by fusion between hyphae of the same mycelium.

belongs *Amanita muscaria,* the fly mushroom. *Amanita verna* is the deadliest of all known poisonous mushrooms. Some members of this order aid in the decomposition of dead bodies and their wastes. In this way they restore simple substances to the air and soil.

Amanita muscaria

Amanita verna

Salient Features of Agaricus (Fig. 15.11)

1. The primary mycelium produced by the germination of basidiospores is of short duration. It consists of cells each with a haploid (n) nucleus (A).
2. Soon as a result of hyphal fusions of compatible strains, a binucleate condition established in one of the cells of the primary mycelium (B).
3. The binucleate (dikaryotised) cell by repeated divisions usually without and rarely with clamp connections forms a new mycelium with binucelate (n+n) cells. It is the secondary or dikaryotic mycelium.
4. The secondary mycelium is abundant, and long-lived (perennial). It produces mushrooms (basidiocarps) year after year (C).
5. As in the other Basidiomycetes, the mushrooms lack specialised sex organs.
6. Plasmogamy which brings the elements of a sexual union together is accomplished by somatogamous copulation (hyphal fusions) of compatible strains.
7. Karyogamy or the actual fusion of two nuclei of the dikaryon is delayed and does not occur till the formation of the basidium.
8. The plus and minus strain nuclei of dikaryons co-exist in all the cells of the secondary mycelial strands and the fruiting bodies (basidiocarps) which it bears.
9. The basidiocarps arise as tiny, white apical swellings on the branches of the underground mycelial strands called rhizomorphs (C).
10. Each enlarges to form a 'button' which comes above ground and rapidly expands into a mushroom.
11. The mature mushroom is a massive structure consisting of a stalk-like **stipe** supporting at its top a broad umbrella-shaped cap, the **pileus.** More than half way up, the stipe bears a membranous ring, the **annulus.**
12. From the undersurface of the pileus hang down plates of fungal tissue in a vertical position. These are the **gills** (D).
13. The surface of the gill, on both sides, is covered with a fertile layer, the **hymenium** (E).
14. The hymenium consists of a closely packed palisade-like layer of club-shaped cells, the **basidia.** Interspersed between the basidia are sterile hyphae, the paraphyses or **cystidia.**
15. The basidium mother cell or protobasidium (F) has two nuclei which function as gametes and fuse. The fusion of the two nuclei (karyogamy) in the protobasidium ends the dikaryophase in the life cycle.

16. The probasidium containing the diploid nucleus or synkaryon is called the young basidium. It represents the short-lived diplophase in the life cycle.
17. As the young basidium grows in size the synkaryon within undergoes **meiosis** during which segregation of strains takes place. Of the four haploid nuclei two are of plus strain and the other two of minus strain.
18. In the mature basidium the haploid nuclei migrate into the basidiospores through their respective sterigmata (G).
19. With the completion of spore production the basidiocarp collapses and dies.
20. The liberated basidiospores which are true meiospores (H), germinate to give rise to the haploid (primary) mycelia (A).

ORDER : POLYPORALES (APHYLLOPHORALES)

General Characteristics. The order Polyporales, also known as the Aphyllophorales, is characterised by single-celled club-shaped basidia (holobasidia) forming a distinct hymenium layer borne on gymnocarpic basidiocarps. In these type of basidiocarps, the hymenium is exposed. It is not enclosed by the veil of the basidiocarp even when the basidiospores are immature. The hymenium layer is lamellate or porous. In some it may be toothed, ridged or even smooth. The forms having pores are termed polypores. In contrast to the Agaricales, the basidiocarp is not soft and putrescent. The basidiospores are borne on several sterigmata and are discharged explosively. Martin (1961) and Alexopoulos (1962) recognise six families in this order. Of these we shall discuss Polyporaceae.

FAMILY : POLYPORACEAE

General characteristics. The members of this family are the wood inhabiting fungi and are commonly called the **pore fungi.** They receive this name because of the nature of the hymenial layer. It lines the inner surface of the tubes. These tubes open on the under surface of the pileus giving it a porous appearance. The pores may be circular or angular or even elongated. The basidiocarp in some species, is composed only of generative hyphae. This type of basidiocarp is termed **monomitic.** In the **dimitic type** the hyphal system is composed of two types of hyphae, generative and either binding (connective) or skeletal hyphae. The hyphal system of **trimitic** type of basidiocarps is composed of all the three types of hyphae, *generative*, *binding* (connective) and *skeletal*. The mature basidiocarps may be hard and woody, or tough and leathery and are either sessile or stipitate. Whey young the basidiocarp is soft and pliable. One member known as *chicken mushroom* has an edible, orange-yellow fructification. The polypores are the common wood rotters and notorious destroyers of timber. They are a common sight on the stumps and dead or dying, standing or fallen trees, sometimes covering the entire trunk. They grow either as saprophytes or parasites. Some pore fungi grow in soil. There is great disagreement as to the delimitation of genera of Polyperaceae. Bessey (1950) reported 48 genera. Overholts (1953) recognised only 13 genera. The common ones which have been recorded from India are *Polyporus, Fomes* and *Polystictus*. *Polyporus* is considered here as a type of the pore fungi.

Polyporus

Polystictus

POLYPORUS

Brown Wood Rot

Habit and Habitat. It is a cosmopolitan genus including about 50 species (Peglar, 1973). All grow parasitically on the roots, trunks and larger branches of trees. They are common on dead or dying, standing or fallen tree trunks. Rarely do they occur on grass and other herbaceous roots. They cause *brown wood rot* of various forest and shade trees and are known as the wood rotters. Some attack and destroy timber. *Dalbergia, Albizza*, Mulberry, birch, oak and other timber trees are the common hosts.

Mycelium. It is inconspicuous and subterranean. At first, it grows in the soil near the host roots. Eventually it attacks the roots and grows beneath the bark. The mycelium is formed by the germination of basidiospores. It consists of numerous white, slender, branched and septate hyphae. The cells of the primary mycelium are uninucleate. Soon the cells become binucleate by hyphal fusions. The binucleate or dikaryotic cells by elongation and division by clamp connections form the dikaryotic mycelium. The secondary or dikaryotic mycelium is long-lived. As the parasitism becomes more severe the mycelium forms a complete, thick layer of hyphae around the central woody cylinder. The hyphae secrete enzymes which digest the lignified walls of the wood cells.

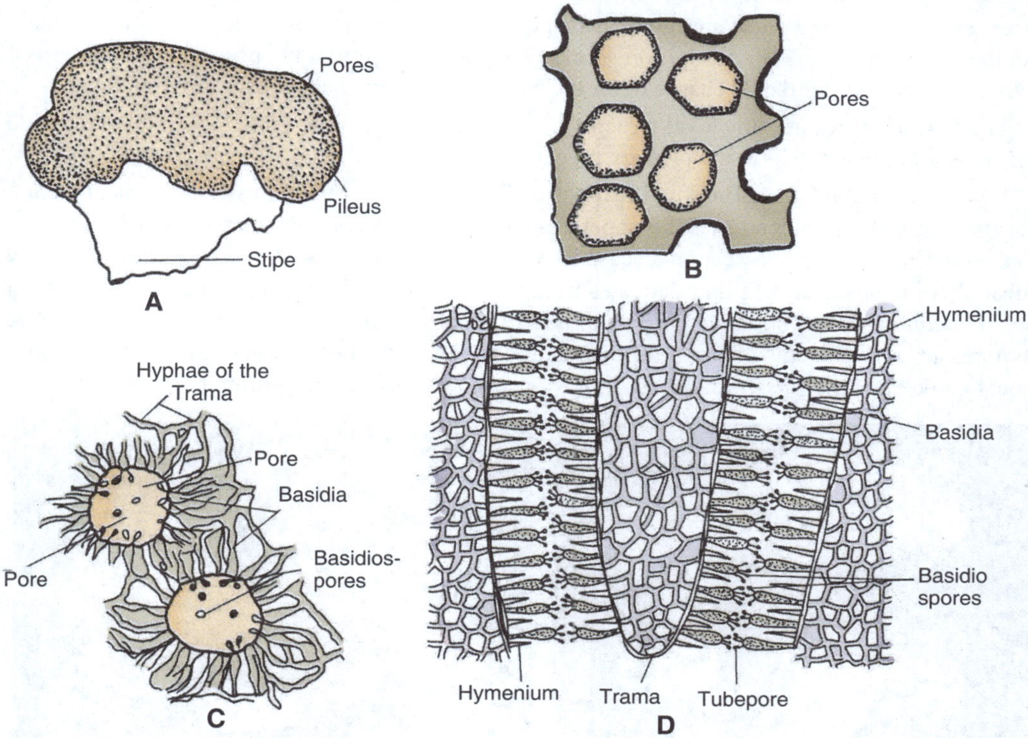

Fig. 15.12. (A-D). *Polyporus sp.* A, Ventral surface of the pileus; B. Surface view of pileus from the ventral side showing pores; C, A portion of the hymenium; D, V. S. through a part of pileus.

Basidiocarp (Fig. 15.12 A). From the dikaryotic mycelium develop the basidiocarps. Each basidiocarp appears as a rounded dense knot of hyphae. It arises from the subterranean mycelium. As the knot grows it bursts through the bark. It gradually increases in size. Soon it becomes differentiated into a short stalk, or the **stipe** terminated by a rounded cap-like structure, the **pileus.** At this stage the basidiocarp is soft and pliable. Towards maturity it becomes woody, corky or leathery. These fructifications in *Polyporus* last for a single season (annual) and constitue the only visible part of the fungus. In the wood inhabiting species they grow horizontally from the surface of the trunk. Many species of shelf fungi *(Fomes)* have perennial basidiocarps which produce spore forming hyphae in annual layers.

The stipe is brown or dark brown in colour, about 2-6 inches in length. It is made up of numerous hyphae compacted together. At its top it bears an expanded structure, the **pileus.** The attachment of the stipes to the pileus varies from a central to an excentric position in different species. The upper surface of the pileus is flat with a slightly undulating or smooth surface. It is brown or lightly coloured. It is marked with a few concentric brown or whitish brown rings near the periphery. *Polyporus versicolor* has a zonately marked pileus. The lower surface of the pileus is smooth and flat. It has no radiating gills. Instead it bears numerous, fine pores in the form of minute circular holes. They are the openings of many small tubes (Fig. 15.12B). They give a porous appearance to the undersurface of the pileus. Each pore leads inwards into a hollow tube lined by a fertile layer called **hymenium.** The hymenium is made up of sterile cells, the **cystidia** and the spore producing club-shaped cells, the **basidia** (Fig. 15.12C). The latter project slightly into the cavity of the tube and are somewhat larger than the cystidia. Each basidium bears four spores, each raised on a short sterigma. The spore output is enormous rather fantastic. A single basidiocarp may produce several billion spores. They escape through the pores and are dispersed by air and insects. After the discharge of all the basidiospores, the basidiocarp degenerates.

The basidiospores are tiny, oval structures, white or brownish in colour. Each under suitable conditions germinates to form the primary mycelium.

Economic Importance. The fructifications of *Polyporus betulinus* are used in the manufacture of charcol crayons. In olden times the upper tissue of the pileus was considered an efficient razor strop. Some parasitc species cause disease of forest and shade trees. A few attack and destroy lumbar. *P. squamosus* attacks and damages living trees of several genera and destroys structural wood. *P. betulinus* is a destructive parasite of beach, oak, birch and other conifers. The saprophytic polypores are active agents of wood decomposition. A few soft, fleshy species are edible and sold in local markets. *P. tuberaster* and *P. tunetanus* are even cultivated in a primitive way.

Polyporus tuberaster

Polyporus squamosus

Taxonomic position

Alexopoulos (1962)

Division	:	**Mycota**
Subdivision	:	**Eumycotina**
Class	:	**Basidiomycetes**
Sub-class	:	**Homobasidiomycetidae**
Series	:	**Hymenomycetes**
Order	:	**Polyporales**
Family	:	**Polyporaceae**
Genus	:	*Polyporus*
Species	:	*betulinus*

Alexopoulos and Mims (1979)

Superkingdom	:	**Eukaryonta**
Kingdom	:	**Mycetae**
Division	:	**Amastigomycota**
Subdivision	:	**Basidiomycotina**
Class	:	**Basidiomycetes**
Sub-class	:	**Holobasidiomycetes**
Series	:	**Hymenomycetes**
Order	:	**Aphyllopharales**
Family	:	**Polyporaceae**
Genus	:	*Polyporus*
Species	:	*betulinus*

SERIES 2 : GASTROMYCETES

General. These are Homobasidiomycetidae in which the fertile portion of the fruiting body, which is known as the **gleba,** is enclosed by a wall known as the **peridium.** The fruit body is thus **angiospermic** in its development. It either remains closed permanently or opens only when the mature basidiospores have fallen from the holobasidia into the cavity. In the former case they are discharged by the disintegration of the peridium and in the latter case the peridium opens in various ways to release the basidiospores. The basidiospores are borne symmetrically on the non-tapering sterigmata and thus are not discharged forcibly. They simply fall off carrying a portion of sterigmata. All Gastromycetes are saprophytes and are found growing in diverse habitats such as in the soil, dead wood, or dung or rarely forming mycorrhizal association. The septate hyphae usually have dolipore septa. The best known members of the series are puff balls, earthstars, stinkhorns and bird's nest fungi. This series is represented by about 525 species allotted to 120 genera.

Classification. Zelter (1949) makes nine orders but Smith (1951) recognised as many as ten orders in this series. Martin (1961) lists only five. Of these we deal briefly with the order Lycoperdales in this text.

ORDER : LYCOPERDALES

General Characters. The basidiocarp (Fig. 15.13 A) is usually globose, pear-shaped or ovoid and fleshy when young. It may be sessile or have a sterile stalk-like base. It has two to four layers. The thick conspicuous surface layer is the **peridium.** At maturity the basidia and the tramal hyphae undergo autodigestion. The gleba is thus turned into a powdery mass consisting of light coloured basidiospores intermixed with the capillitium. It is surrounded by the thick peridium. The latter ruptures or open by one or more pores or splits into lobes at maturity to allow the basidiospores to escape. Puff balls and certain earthstars are the common representatives of this order. Smith (1951) makes five families in this order but Martin (1961) lists only three, namely Arachniaceae, Geastraeae and Lycoperdaceae. Dring (1973) added another Mesophillaceae thus raising the number to four. Of these, Lycoperdaceae is being briefly described.

FAMILY : LYCOPERDACEAE

It includes the puff balls which grow as saprophytes in the grassy open fields, city lawns, pastures, in the woods, on the ground, tree stumps or decaying logs of wood. The peridium that encloses the gleba is differentiated into two distinct layers, outer and inner. The outer peridial layer is known as the **exoperidium** and the inner **endoperidium.** The former is fragile and may be spiny, warty or granular. It weathers away towards maturity leaving the membranous endoperidium intact. The two common and important genera of this family are *Lycoperdon* and *Clavatia.*

Polyporus squamosus

Lycoperdon

LYCOPERDON

Mycelium. The primary mycelium produced by the germination of the basidiospore consists of short, uninucleate cells. It grows on the rotting wood or in soil. It is short-lived and soon changes into the dikaryotic mycelium by hyphal fusion and repeated division of the dikaryotic cells by clamp connections. The secondary mycelium is extensive and consists of binucleate cells. The hyphae are often bundled into hyphal strands or **rhizomorphs** which form a branching system in the substratum. In a section the rhizomorph consists of the following three zones :

1. *Outer cortical zone.* It consists of loosely arranged hyphae.
2. *Inne subcortical zone.* It is made up of closely packed hyphae.
3. *Central core* or *Medulla.* The hyphae run parallel in this region.

Basidiocarp (Fig. 15.13) (a) *Development.* It arises as a tiny, hyphal, knot-like outgrowth from the medulla or central core at the tip of the rhizomorphs. Occasionally it may be lateral in position. The young basidiocarp primordium is a homogenous mass of interlaced hyphae. There is no distinction into **gleba** and **peridium.** With further growth, a palisade-like layer of closely arranged hyphae forms all over the basidiocarp primordium. Later it becomes differentiated into an outer pseudoparenchymatous portion termed the **exoperidium** and the inner palisade-like **endoperidium.** With the differentiation of these two peridial layers around the young basidiocarp the hyphae in the cetre of the embryonic gleba begin to pull away from one another. Consequently numerous small irregular cavities are formed in the central region of the gleba (B). However, their formation spreads towards the periphery. Thus, the gleba of the basidiocarp of *Lycoperdon* is of **lacunose** type. With the increase in the size of the cavities, club-shaped **basidia** are formed at the tips of the newly formed hyphal branches. These arise from the encircling hyphae which form the walls of the cavities. The component cells of the hyphae are binucleate and so are the young basidia. The basidia are arranged in a regular fertile layer constituting the **hymenium** (C) which lines each cavity. The fusion of the two nuclei in the basidium is immediately followed by meiosis. With the formation of four haploid nuclei in the basidium four long sterigmata develop at its apex. They produce uninucleate basidiospores singly at their tips. The mature basidiospore is globose in shape and has a thick variously ornamented wall. It becomes binucleate by the division of a single nucleus. The mature basidiospores fall off the basidia and completely fill the cavities. Subsequently the basidia and most of the hyphae constituting the gleba disintegrate. The remaining hyphae become thick-walled to constitute the **capillitium** which is intermingled with the powdery mass of spores.

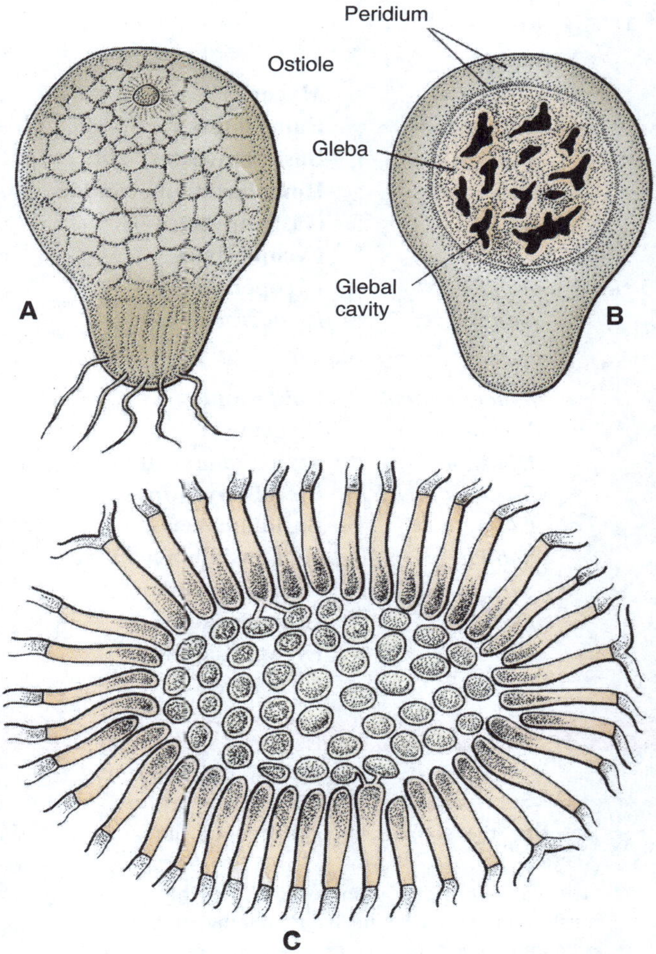

Fig. 15.13 (A-C). *Lycoperdon sp.* A, Basidiocarp; B, Longitudinal section of a young basidiocarp; C, Old glebal cavity (Enlarged).

(*b*) *Structure.* The mature basidiocarp is a globular, pear-shaped or ovoid structure which may be sessile or may have a sterile, stalk-like base (A). It consists of an outer dry sheath or jacket which encloses the spore bearing tissue called **gleba** (B). The sterile, dry, thin jacket is the endoperidium. The exoperidium weathers away towards maturity. There is a small, circular opening, the **ostiole** in the centre at the top of the endoperidium. Through it puffs of brown spores are released when the basidiocarp is hit by rain drops or wind borne debris or when any object touches the basidiocarp. Within the endoperidium is a powdery mass of spores intermingled with the sterile remains of branched hyphal threads of the gleba constituting the **capillitium.** The basidiocarps vary in size from 2 cm across to the giant puff balls which may be 3 or 4 feet in diameters.

Economic importance. The young fleshy puff balls of most spepcies are edible. North American Indians hold that eating the giant puff balls (*Clavatia)* prevents stomach tumours. Recent investigations have shown that *Clavatia* contains an anti-cancer substance known as *calvacia.*

Taxonomic Position :

Alexopoulos (1962)

Division	:	**Mycota**
Subdivision	:	**Eumycotina**
Class	:	**Basidiomycetes**
Sub-class	:	**Homobasidiomycetidae**
Series	:	**Gastromycetes**
Order	:	**Lycoperdales**
Family	:	**Lycoperdaceae**
Genus	:	*Lycoperdon*

Alexopoulos and Mims (1979)

Superkingdom	:	**Eukaryonta**
Kingdom	:	**Mycetae**
Division	:	**Amastigomycotina**
Subdivision	:	**Basidiomycotina**
Class	:	**Basidiomycetes**
Sub-class	:	**Holobasidiomycetidae**
Series	:	**Gastromycetes**
Order	:	**Lycoperdales**
Family	:	**Lycoperdaceae**
Genus	:	*Lycoperdon*

QUESTIONS

ESSAY TYPE

1. Describe the somatic organisation of the Agaricales.
2. Give a botanical name of the edible mushroom. Describe its habit, habitat and economic importance.
3. Give an account of the life history of a saprophytic basidiomycete you have studied *(Agaricus)*.
4. Describe the structure of the fructification (basidiocarp or sporophore) of *Agaricus* laying emphasis on the nuclear changes that take place in its development.
5. Describe the structure of the sporophore of *Agaricus* and give the development of the basidium.
6. Describe the life cycle of Gill Fungi/*Agaricus*. *(Kumaon, 2003; Agra, 1993, 1996; Kumaon, 1999)*
7. Describe sexual reproduction and development of fructification in *Agaricus*. *(Himachal Pradesh, Univ., 1992)*
8. Give a detailed structure of gill of *Agaricus*. *(Himachal Pradesh, 1993, 1994)*
9. With the help of diagrams describe various types of fruiting bodies in fungi studied by you.
10. Give the structure and reproduction of an edible fungus studied by you. *(Bundelkhand, 1999)*
11. Give an account of the structure, reproduction and economic importance of *Lycoperdon*.
12. Give an illustrated account of the structure and reproduction in *Polyporus*.
13. Write in detail about the fruitifications of *Peziza and Agaricus*. *(Allahabad, 2003)*
14. Describe the important features of the life history of *Agaricus*. Draw T.S. of Gill. *(Poorvanchal, 2003)*
15. (*a*) What is dikaryotisation ? What is its significance ? Give an illustrated account of various methods of dikaryotisation in higher fungi. *(Allahabad M.Sc., 2000)*
(*b*) With the help of diagrams, describe the life cycle of *Agaricus*. *(Poorvanchal, 2003)*

SHORT ANSWER TYPE

16. Write short noes on :
    (*a*) Structure of Pileus of *Agaricus* (*Agra, 1991; Awadh, 1995*)
    (*b*) *Polyporus* (*Lucknow, 1999*)
    (*c*) Reproduction in *Agaricus* (*Gorakhpur, 1992*)
    (*d*) Dikaryotisation in *Agaricus* (*Gorakhpur, 1996*)
    (*e*) Fruit body of *Agaricus* (*Poorvanchal, 2002; Awadh, 1994*)
    (*f*) *Agaricus* Gill (*Rohilkhand, 1992*)
    (g) Sexual Reproduction in *Agaricus*. (*Meerut, 1999*)
    (*h*) *Agaricus* (*Kumaon, 2002 ; Allahabad, 2004*)
    (*i*) Fairy rings (*Allahabad, 2000*)
    (*j*) Economic importance of *Agaricus* (*Kanpur, 2001, 2003, 2006*)
    (*k*) Basidiocarp of *Agaricus* (*Kanpur, 2003*)
    (*l*) *Lycoperdon* (*Kanpur M.Sc., 2001*)
    (*m*) Differences between internal structure of *Agaricus* and *Polyporus* (*Lucknow, 2005*)
    (*n*) *Dikaryotisation* (*Rohilkhand 2001*)
17. Draw labelled diagrams of the following :
    (*a*) *Agaricus* Gill (*Agra, 1995; Kanpur, 1998, 2001; Awadh, 1992; Rohilkhand, 1996*)
    (*b*) Sporophore of *Agaricus* (*Lucknow, 1995; Awadh, 1997*)
    (*c*) V.S.Gills of *Agaricus* (*Lucknow, 1997, 2002; Gorakhpur, 1990, 1995, 1997; Kanpur 1999; Awadh, 1998*)
    (*d*) T.S.*Polyporus* (*Lucknow, 1997, 1998*)
    (*e*) T.S. Gills of *Agaricus* (*Kanpur, 2005; Gorakhpur, 1998*)
18. With suitable sketches distinguish between the internal structure of *Agaricus* and *Polyporus*. (*Lucknow, 1994*)
19. Give an illustrated account of V.S. of basidiocarp of *Agaricus*. (*Alahabad, 1999*)
20. Draw labelled diagram of the internal structure of *Agaricus* fruit body. (*Kanpur, 1995*)
21. Where does *Agaricus* grow and how would you identify it? (*Kanpur, 1997*)
22. Classify the following genera and mention their important characters :
    (*i*) *Agaricus* (*Rohilkhand, 1995, 1997*)
    (*ii*) *Polyporus* (*iii*) *Lycoperdon*
23. Draw a well labelled diagram of T.S. Gill of *Agaricus*. Describe its structure. (*Himachal Pradesh, 1999*)
24. Briefly describe the characters of the sub-class Homobasidiomycetidae.
25. Describe the structure of the basidiocarp in *Polyporus*.
26. Compare the structure of Basidiocarps in *Agaricus*, *Polyporus* and *Lycoperdon*. Comment on their economic importance.
27. Differentiate between
    (*i*) Fruit bodies of *Agaricus* and *Morchella* (*Kanpur, 2001*)
    (*ii*) *Agaricus* and *Polyporus*
    (*iii*) *Polyporus* and *Lycoperdon*
28. (*a*) What are fairy rings? How are they formed ? (*Kanpur, 2002*)
    (*b*) Name any edible fungus. (*Poorvanchal, 2004, 2005*)
    (*c*) Describe the structure of Gill in *Agaricus*. (*Poorvanchal, 2004, 2005*)

OBJECTIVE TYPE

29. **Fill in the blanks :**
    (*i*) The characteristic feature of Homobasidiomycetidae is the presence of ....
    (*ii*) Basidiocarps are the fruit bodies of ....
    (*iii*) Basidia are usually arranged in a regular fertile layer called ....
    (*iv*) Basidium generally bears .... basidiospores
    (*v*) Karyogamy and meiosis takes place in ....
    (*vi*) Mushrooms and Toad stools have been placed in the series ....
    (*vii*) Members of the series Gastromycetes are commonly called ....
    (*viii*) The edible mushroom is ....
    (*ix*) The secondary dikaryotic mycelium forms the .... phase of *Agaricus*.

(*x*) *Polyporus* causes .... disease of forest and shade trees.

(*xi*) Agaricus differs from Polyporus in having .... basidiocarp.

30. Select the correct answer :

(*i*) The mushroom is
(*a*) a plant made up of fine green threads
(*b*) an edible fungus
(*c*) a bryophyte without roots
(*d*) a flowering plant.

(*ii*) The fertile portion in the fruit body of *Lycoperdon* is called
(*a*) Gleba (*b*) peridium
(*c*) hymenium (*d*) gills.

(*iii*) In *Agaricus* the reduction division takes place in
(*a*) basidiospore (*b*) basidium
(*c*) chlamydospore(*d*) basidiocarp.

(*iv*) The fruit body in *Agaricus* is made up of
(*a*) primary mycelium
(*b*) secondary mycelium
(*c*) tertiary mycelium
(*d*) diploid mycelium.

(*v*) Basidia in *Agaricus* and produced on
(*a*) gills (*b*) pileus
(*c*) stipe (*d*) Rhizomorph.

(*vi*) The brown gills of mushrooms
(*a*) are meant for respiration
(*b*) have no function to perform
(*c*) help the plant to float after heavy rains
(*d*) bear basidiospores which help in reproduction.

(*vii*) In fairy rings, reproduction takes place by
(*a*) spores (*b*) flowers
(*c*) gills (*d*) seeds.

(*viii*) Toad stools can not manufacture their own food because
(*a*) they have no chlorophyll
(*b*) they have no roots
(*c*) they have no leaves
(*d*) they have no need for food.

(*ix*) In which of the followings gills are present?
(*a*) *Aspergillus* (*b*) *Alternaria*
(*c*) *Agaricus* (*d*) *Mucor.*

(*x*) Which of the following is an edible fungus?
(*a*) *Aspergillus* (*b*) *Alternaria*
(*c*) *Agaricus* (*d*) *Mucor.*

(*xi*) Which of the following are wood rotters?
(*a*) *Agaricus* (*b*) *Polyporus*
(*c*) *Morchella* (*d*) *Mucor*

(*xii*) Wood decaying fungi are :
(*a*) *Polyporus* (*b*) *Fusarium*
(*c*) *Both (a) & (b)* (*d*) *Morchella*

(*xiii*) Which of the following is the fertile portion in the fruiting body of *Agaricus*?
(*a*) *Pileus* (*b*) Hymenium
(*c*) *stipe* (*d*) *Gill*

(*xiv*) Fairy rings are formed by
(*a*) *Rhizopus* (*b*) *Mucor*
(*c*) *Penicillium* (*d*) *Agaricus*

(*xv*) Somatogamy occurs in
(*a*) *Agaricus* (*b*) *Aspergillus*
(*c*) *Cyatopus* (*d*) *Alternaria*

(*xvi*) Fruit body of *Agaricus* is called
(*a*) *Ascocarp* (*b*) *Basidiocarp*
(*c*) *Stipe* (*d*) *Ergot*

Alternaria Blight (Lesions)

CHAPTER 16

CLASS – DEUTEROMYCETES (*Fungi Imperfecti*)

- General Features
- Form-Order : Moniliales (Hyphomycetes)
- Form-Family : Tuberculariaceae
- Form-Genus: *Fusarium*
- Form-Family : Dematiaceae
- Form-Genus : *Alternaria*
- Form Genus : *Cercospora*
- Form-Order : Sphaerospidales (*Photomitales*)
- Form-Genus : *Phyllosticta*
- Form-Order : Melanconiales
- Form-Genus : *Collectotrichum*

General Features. This class comprises a group of fungi in which only the asexual or **imperfect stage** is known. The sexual stage also called the **perfect stage** is unknown. The latter is either non-existent or has not been discovered so far. The life cycle of these forms is thus incompletely known. Such fungi with imperfectly known life cycles are included in the class *Deuteromycetes* or *Fungi Imperfecti.* Their number goes up to thousands. Many of them live as saprophytes and many more as parasites. The latter are the causative agents of diseases in plants and animals including man. Some cause spoilage of stored products.

The somatic phase in the majority of these fungi consists only of the haploid mycelium. It is septate and profusely branched. Reproduction takes place chiefly by the formation of exogeneously developed asexual spores. They are of the nature of **conidia.** Formation of oidia and chlamydospores has also been reported. Sporangiospores, meiospores such as ascospores and basidiospores, and sexual organs are either *non-existent* or have not been observed so far. In many fungi which were formerly described as fungi imperfecti, sexual stages have been found. One such example is furnished by *Aspergillus* or *Eurotium.* The sexual stage of this fungus was called *Eurotium* and the asexual or the conidial stage as *Aspergillus.* These two stages were considered to belong to two differnt fungi.

Whenever the perfect stage of the imperfect fungus is known it is transferred to its systematic position among the fungi in the classes Ascomycetes and Basidiomycetes. The name applied to the perfect stage has priority.

The conidial stage is very variable in the different members of this class. The classification is itself based on the peculiarities of conidial development. Hence, before taking up the classification we describe the various methods of conidial development.

Conidia and Conidial Development. The conidia are exogenously produced spores. They are non-motile and are borne externally or vertically growing hyphae called the **conidiophores.** Generally, the condiophores are simple or branched hyphae with special region of conidia formation. The apical region of conidiophore may produce a single conidium or a chain of conidia usually in basipetal succession.

Ellis (1971) has recognised two types of conidial developments :

1. Blastic. In this type of conidial development marked enlargement of the recognisable conidial initial takes place before it is delimited by a septum. It may be of two types :

(*i*) **Holoblastic.** In the formation of conidia both the inner and outer walls of the conidia forming cell of the conidiophores take part. In *Cladosporium* the conidial chain is branched due to formation of two buds at the apex of a conidium.

(*ii*) **Enteroblastic.** Only the inner wall of the conidiogenous cell takes part in the formation of conidia. It may be of two types :

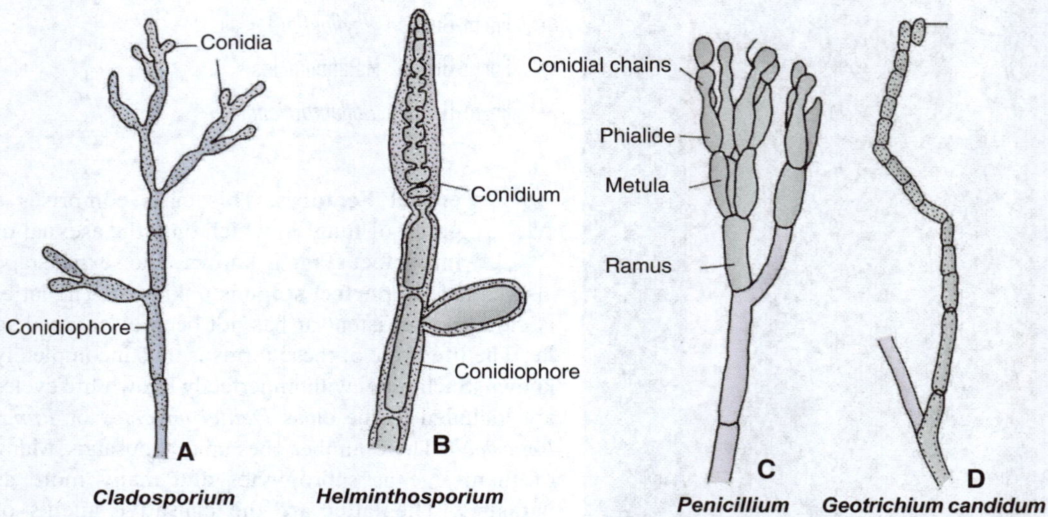

Fig. 16.1. Types of conidial development. A, Holoblastic; B, Enteroblastic Tretic; C, Enteroblastic phialidic; D, Thallic.

(*a*) **Tretic.** The conidium develops when the inner wall protrudes through a channel in the outer wall. *Helminthosporium velutinum* is an example.

(*b*) **Phialidic.** Conidial development takes place from a specialised conidiogenous cell known as **phialide** conidia are formed in basipetal succession. *Aspergillus* and *Penicillium* are examples.

2. Thallic. In this type of conidial development there is no enlargement of conidial initial, or when such development does occur, it takes place after the initial has been delimited by a septum or septa. *Geotrichum candidum* is an example of thallic development of conidia.

1. Types of conidia on free conidiophore

The conidia may be of two types :

A. Thallospores. These are asexual spores formed by the transformaton of existing cells of the thallus. These spores are set free by decay or disarticulation of the parent hyphae but not by a process of abstriction. Thallospores may be terminal or intercalary in position and may be of two types :

(*i*) Arthrospores. Arthrospores arise by close septation of the distal part of hypha, the septa being formed in basipetal succession. Proceeding from the apex to the base of the hypha each cell is rounded off and is set free as an **arthrospore.** *Coremiella ulmariae* produces arthrospores.

(*ii*) Chlamydospores. These are formed from a solitary or neighbouring intercalary cells of a hypha, which round off, enlarge and develop a thickened often pigmented wall and dense contents containing food reserves. Chlamydospores generally function as resting spores and are formed under unfavourable environmental or nutritional conditions. In *Trichoderma viride* the mature chlamydospores are more or less spherical or oblong and wider than long.

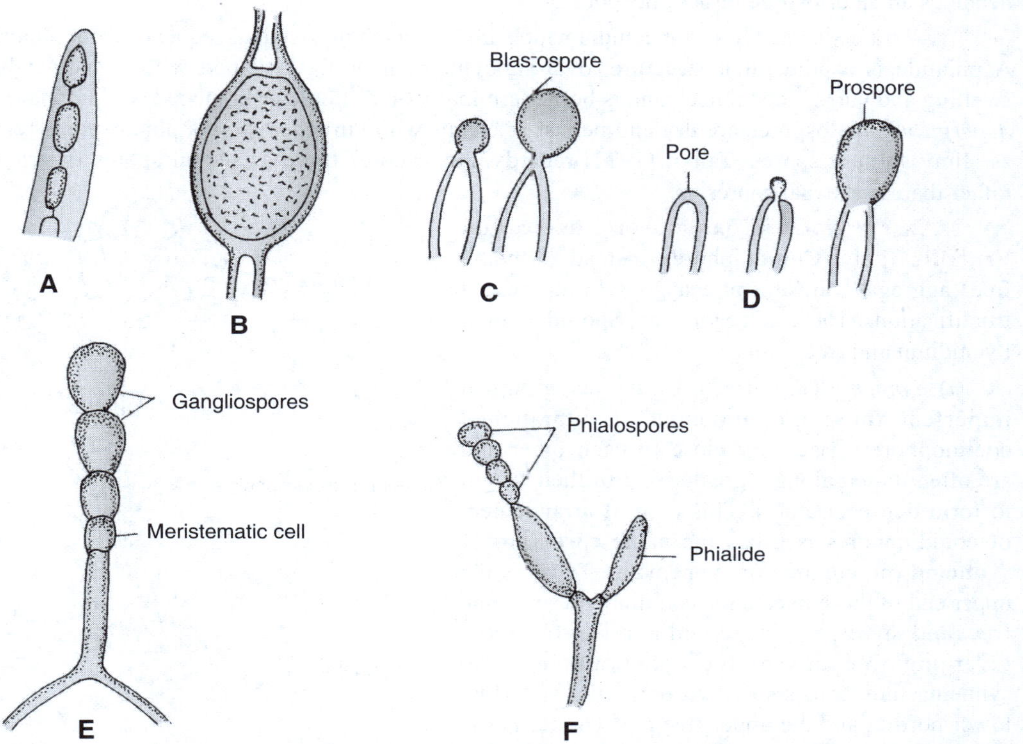

Fig. 16.2. Diifferent types of spores. A, Arthrospores; B, Chlamydospore; C, Blastospore; D, Porospore, E, Aleuriospore or Gangliospore; F, Phialospore.

B. Conidiospores. These are asexual spores formed singly or in chains on a **conidiophore.** The sporogenous part of the conidiophore is usually apical but it may be lateral as well. These spores or conidiospores or conidia as they are commonly called, may be produced in basipetal or ascopetal succession on either free conidiophores or their aggregations in the form of fructifications.

Conidiosphores produced on free condiophores may be of four types. These are (*i*) Blastospores, (*ii*) Porospore, (*iii*) Aleuriospores or Gangliospores and (*iv*) Phialospores.

(*i*) Blastospores. A blastospore or bud spore is formed by budding of somatic cells of a hypha or conidiophore or by budding the cells of other types of spores. Each bud first appears as a small globular outgrowth from a parent cell and then enlarges while still attached with the parent, finally separting by abstriction to form an independent spore. Blastospores may be formed singly or in ascopetal succession and rarely in basipetal succession. Genera belonging to family Torulaceae like *Torula, Bispora, Septomema* etc. form typical blastospores.

(*ii*) Porospores. Porospores are modified **blstospores** and are produced as a bud extruded through a distinct pore in the wall of conidiophore. At the tip of the conidiophore, a pore is enzymatically dissolved in the outer layer of the wall while the inner wall blows out through this pore like a bud and develops into the conidium. Examples of porospores are found in *Alternaria, Helminthosporium* etc.

(*iii*) Aleuriospores or Gangliospores. These spores are formed by the apex of the conidiophore becoming inflated and delimited by a septum at an early stage. In *oidium* aleuriospores or gangliospores are formed in a basipetal chain by successive division of a meristematic cell at the base of the chain. The proximal daughter cell remains meristematic while the distal daughter cell becomes an aleuriospore or gangliospore.

(*iv*) Phialospores. These are conidia which are abstricted in basipetal sequence from phialides. A phialide is a unicellular structure, oval subcylindrical or flask-shaped with a distinct basal swelling and narrow distal neck and is borne terminally on a simple or branched conidiophores. In *Aspergillus* phialospores are dry and persist in chains while in *Stachybotrys* phialospores remain as slimy solitary spores. Talbot (1972) regards phialospores more as sexual spores (spermatia) rather than as asexual spores.

2. Aggregation of Conidiophores or Asexual Fructifications. Conidiophores instead of being free, aggregate in several ways to form asexual fructifications. These are: Synnema, Sporodochium, Pycnidium and Acervulus.

(*i*) *Synnema* (Fig. 16.3). In another group of imperfect fungi the branched or unbranched conidiophores arise very close to each other and are often united along a greater part of their length to form dense fascicles. This type of arrangement of conidiophores is given the name **corenium** or **synnema** (pl. coremia or synnemata). Towards the upper end of the synnema the conidiophores become free and diverge. The conidia arise from the diverging free tips of the conidiophores. The synnema thus consists of two parts, the stalk-like lower portion and the upper free top. The former is often longer than the latter. The whole fructification resembles a long feather cluster or a brush.

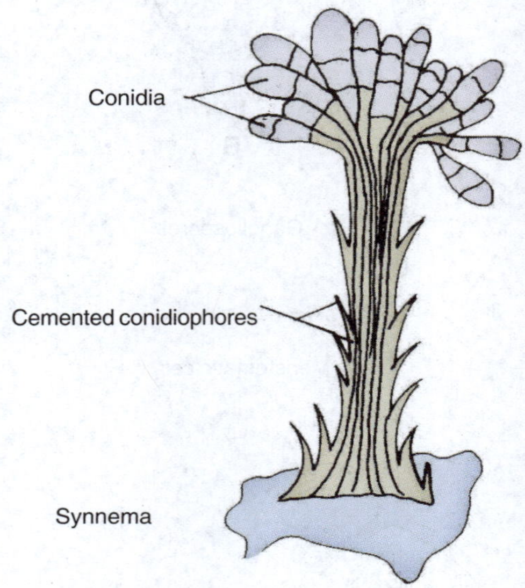

Fig. 16.3. *Fungi Imperfecti.* Synnema.

(*ii*) *Sporodochium* (Fig. 16.8). The sporodochium is a hemispherical or barrel-shaped asexual fruiting body found in some imperfect fungi. It consists of two parts, the upper and the lower. The lower or the basal part is a cushioned stroma-like mass of hyphae. It breaks through the host tissue. From the exposed upper surface arise the conidiophores which bear conidia at their tips. The conidiophore consitute the upper part of the **sporodochium.**

(*iii*) *Pycnidium* (Fig. 16.4) It is a flask-shaped structure.The cavity of the flask is lined by a sterile pseudoparenchymatous tissue comprising its wall. It is called the **pycnidial wall.** The fertile layer lies in the cavity. They whole structure consisting of the cavity, the wall, and the fertile layer is called the **pycnidium.** The fertile layer consists of very short conidiophores. They arise from the pycnidial wall. The conidia abstricted from the tips of the conidiophores are called the **pycniospores** or **pycnidiospores.** The pycnidia may remain closed or open at the surface by a small pore called an **ostiole.** The pycnidia vary greatly in their structure and development, size, shape and colour in the different genera.

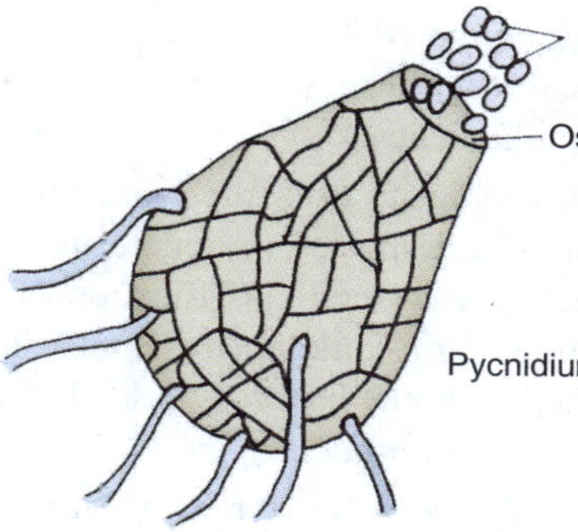

Fig. 16.4. *Fungi Imperfecti.* Pycnidium.

(*iv*) *Acervulus* (Fig. 16.5). It is a saucer-shaped structure consisting of stromatic mass of hyphae and a fertile layer of conidiophores. The conidiophores are short. They arise from a mat of closely crowded hyphae forming a small disc called **stroma** and lie side by side to form a continuous layer. The conidia are abstricted from the tips of the conidiophores. Intermingled with the conidiophores are sometimes found dark, long, stiff bristles. These are called the **setae.**

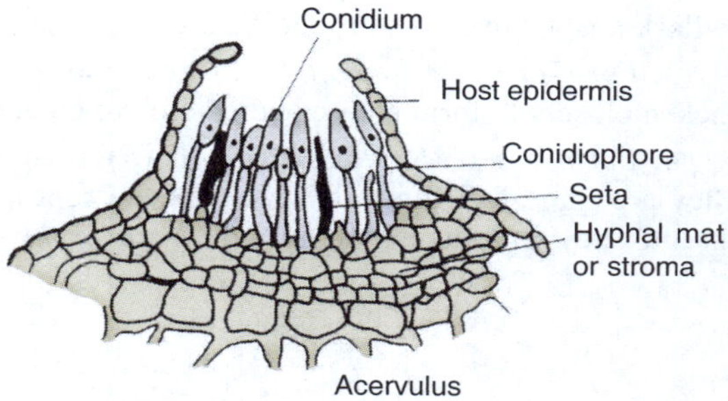

Fig. 16.5. *Fungi Imperfecti.* Acervulus.

Classification. The classification of Fungi Imperfecti is completely arificial. It separates many genera, which had their sexual stages been known, would have been near each other. It is based on their conidial peculiarities and bears no connection to their sexual stage not to their origin or evolution. It cannot, therefore, be called a natural classification. Such a system shows no relationship. It is better to call it a key which helps in cataloguing or identifying them. Deuteromycetes thus is an artificial class created for convenience to include fungi which do not fit in the other groups of fungi.

This system of classification is therefore known as Artificial system of classification and is mainly based on conidial characters. The naming of the conidial stages is so perfectly convenient that International Botanical Congress at its 1950 meeting held at Stockholm decided to legalise it, still recognising of course the name of the perfect stage for the entire organism as and when discovered.

In accordance with conidial characters, P.A. Saccardo (1880-1886) proposed a system of classification which has been used ever since it was proposed and is still popular. He divided the class into four orders viz. **Moniliales** (Conidia produced directly on hyphae or on conidiophores), **Melanconiales** (conidia produced in acervuli), **Sphaeropsidales** (conidia produced in pycnidia) and **Mycelia Sterilia** (conidia lacking). Moniliales were further divided into four families, Melanconiales into one and Sphaeropsidales into four families. Each family was also divided into sections based on septation and shape of conidia. These sections are : **Amerosporae** (conidia one celled), **Allantosporae** (conidia cylindric, curved), **Hyalosporae** (conidia hyaline), **Phaeosporae** (conidia coloured), **Didymosporae** (conidia ovoid to oblong, one septate, hyaline or coloured), **Phragmosporae** (conidia with 2 or more cross septa), **Dictyosporae** (conidia muriform, hyaline or coloured), **Scolecosporae** (conidia filiform), **Staurosporae** (conidia star shaped) and **Helicosporae** (conidia helical).

Subsequent mycologists used the word "form" as a prefix before the subdivisions, class, order, family, genus and species to denote the artificial and temporary nature of this class of Fungi. The characters on which classification of the *Form-class* Fungi Imperfecti is based are, (*i*) presence or absence of conidia, (*ii*) colour, shape and septation of conidia, and (iii) kind of asexual fruiting body. On the basis of these characters the form-class Deuteromycetes is divided into the folloing four *form orders:*

1. *Form-order Moniliales* (or Hyphomycetes). Conidia are usually borne freely on the mycelium or on free conidiophores arising from the mycelium. In some cases conidiophores are grouped into sporodochia or synnemata. Reproduction by budding and oidia formation also takes place. This form-order includes both saprophytes and parasites.

2. *Form-order Sphaeropsidales* (or Photomitales). Conidia are produced within small, globose or flask-shaped pycnidia or modifications of such structures. It includes plant parasites.

3. *Form-order Melanconiales.* Conidia are produced singly or in chains from conidiophores packed closely to form an asexual fructification called an *acervulus.* It includes plant parasites.

4. *Form-order Mycelia sterilia.* This form-order included 20 genera and 200 form species. They are imperfect fungi in which special reproductive stages including the conidia are totally absent. They reproduce by *sclerotia, rhizomorphs* and other vegetative means.

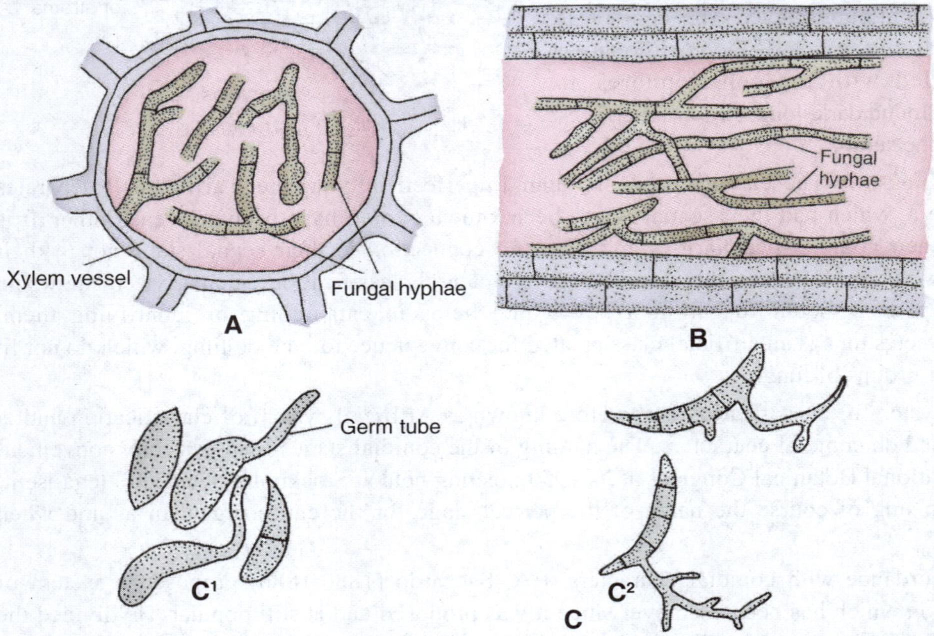

Fig. 16.6 (A-C). *Fusarium oxysporum.* A, xylem vessel in cross section containing mycelial hyphae; B, xylem vessel in L.S. containing mycelium; C, germinated microconidia (C_1) and macroconidia (C_2)

Ainsworth (1973) proposed a system of classification which is now accepted by most mycologists. According to this classifiction, the Subdivision Deuteromycotina has been divided into three classes as follows :

Subdivision Deuteromycotina

- Budding cells with or without pseudomycelium; true mycelium lacking or not well developed
  Form class **Blastomycetes**
- Mycelium well developed assimilative, budding cells absent
  - Mycelium sterile or bearing spores directly or on special branches (sporophores) which may be variously aggregated but not in pycnidia or acervuli
    Form class : **Hyphomycetes**
  - Spores produced in pycnidia or acervuli
    Form class **Coelomycetes**

However, the system of classifying the Form class Deuteromycetes into four Form orders has been followed here.

FORM-ORDER : MONILIALES (HYPHOMYCETES)

General. This form-order includes about 10,000 form-species, both parasitc and saprophytic. The parasitic forms include both plant and animal pathogens. They casue serious fungal diseases of man as well as plants. Many of the saprophytic forms form the basis of industry. The common examples are *Penicillium roqueforti, P. chrysogenum* and *P. notatum.* Many saprobic forms inhabit soil and are said to play an important role in soil economy. The imperfect stage consists of a mycelium which is septate and branched. The conidia are produced at the tips of conidiophores which may be free or assembled into synnemata or sporodochia. A number of form-families are included in this form order. The important ones are, *Moniliaceae, Dematiaceae, Stilbaceae, and Tuberculariaceae.* In addition *form-family Cryptococcaceae* which includes the false yeasts is also added to the list by some mycologists. The three important genera of this **form-order** are *Alternaria, Fusarium* and *Cercospora.* All the three are parasitic and cause important diseases of crop plants. We study their life histories in detail. *Fusarium* belongs to form-family *Tuberculariaceae* whereas *Alternaria* and *Cercospora* are included in *form-family Dematiaceae.*

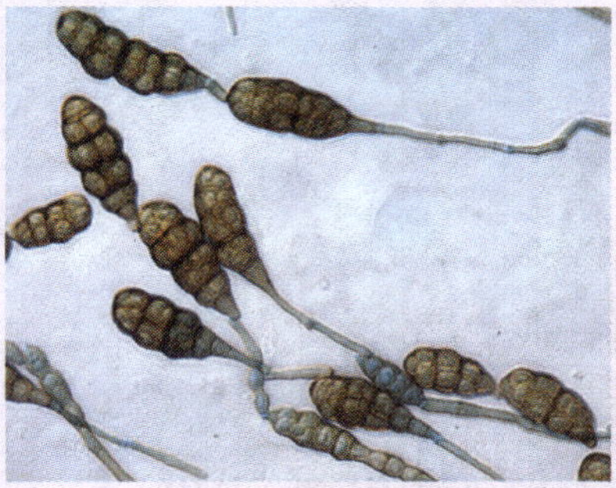

Alternaria – Conidia

Fusarium – Conidia

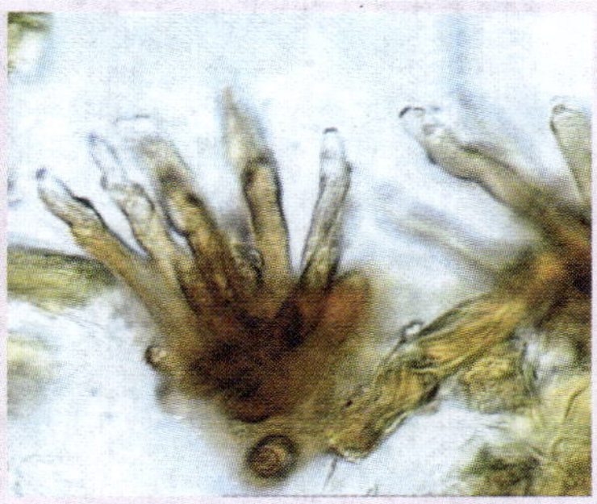

Cercospora – Conidia

FORM-FAMILY : TUBERCULARIACEAE

The characteristic spore fruit is of sporodochium type. The family includes about 152 *form-genera* (Bender, 1931). *Fusarium* and *Tubercularia* are the two commonly known among these. *Fusarium* which is the largest form-genus of this family is taken as an example.

FORM-GENUS : FUSARIUM

General. The form-genus *Fusarium* includes a large number of species and many forms within species. Many of these are saprobic or saprohytic. Some are only mild facultative parasites. Others are parasitic. All, however, have a saprophytic stage. Some of the parasitic species are **soil inhabitants** and can live indefinitely as soil saprohytes but some are **soil invaders.** The latter soon die if they are not able to infect a suitable host plant. The parasitic form-speices cause rot of stores fruits, vegetables and other commodities and are responsible for dry rot of potato tubers of colocasia corms (Tandon *et al,* 1956) or soft rot of rhizomes of ginger (Mehrotra, 1952) during storage. Some are mild root parasites and some primarily cortical invaders which cause stem cankers, foot rots and pre-emergence damping off diseases. The most destructive species, however, are :—

1. *Fusarium udum* which casues *wilt* of Arhar (*Cajanus cajan*).
2. *F. lycopersi* which causes *wilt* of tomatoes (*Lycopersicum esculentum*).
3. *F. lini.* It is a causative agent of *wilt* of Flax.
4. *F. orthaceras.* It causes wilt of gram (*Cicer aurietinum*).
5. *F. vasinfectum.* It produces *wilt* of cotton.
6. *F. cubense.* It is parasitc on Banana and causes the *wilt disease.*

Fusarium udum- wilt affected arhar plant

F. lycopersici – on tomato

F. lini on Linseed

F. lycopersici - wilt of tomato

F. cubense – on Banana

All the above-mentioned spcies are vascular parasites, and are thus often referred to as **vascular fusaria.** They are now placed under a single species *Fusarium oxysporum* as forms (Snyder and Toussoum, 1965).

Mycelium. It is extensive. The hyphae are septate and branched. They are both intercellular and intracellular. When young they may be colourless or with a tinge of pink, purple or yellow and become dark-coloured at maturity. The dark mycelium produces thick bands which plug the vascular tissues (Fig. 16.6 A-B) and produce toxic secretions The toxins are carried up in the xylem vessels. As a result the plant wilts and dies.

Asexual Reproduction. (Fig. 16.7). It takes place by the formation of three kinds of asexual spores, **microconida, macroconidia** and **chlamydospores.** Sclerotia are also formed.

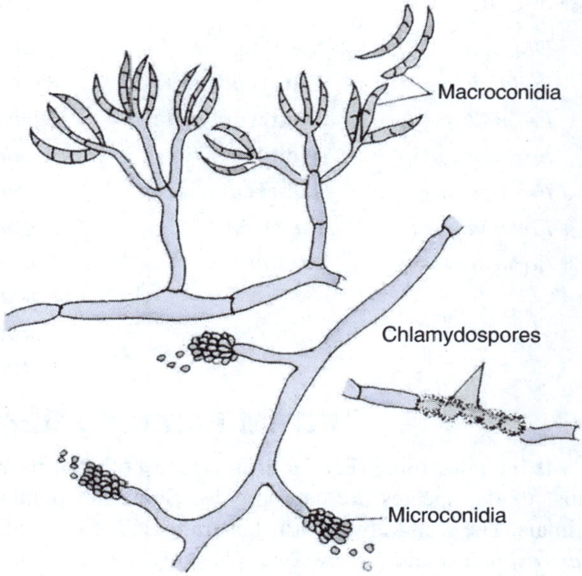

Fig. 16.7. *Fusarium.* Macroconidia and clustered microconida. Portion of a hypha bearing chlamydospores is also shown.

Microconidia (Fig. 16.7). They are very small conidia abstricted from the tips of simple or branched conidiophores. The conidiophores are distinguishable from the vegetative hyphae.

The microconidia vary in form from the rounded to oval. They are often held in small masses. At times they are elongated or crescent-shaped.

Macroconidia (Fig. 16.7). They are large, multicellular usually two to four-celled conidia. In form they are elongated, sickle-shped or crescent-shaped. They are produced at the tips of simple or sparingly branched conidiophores which are assembled to form a sporodochium type of fructification (Fig. 16.8). The microconidia and macroconidia are produced in vast numbers. They are distributed by the wind. On falling on a suitable substratum they germinate and initiate new infections.

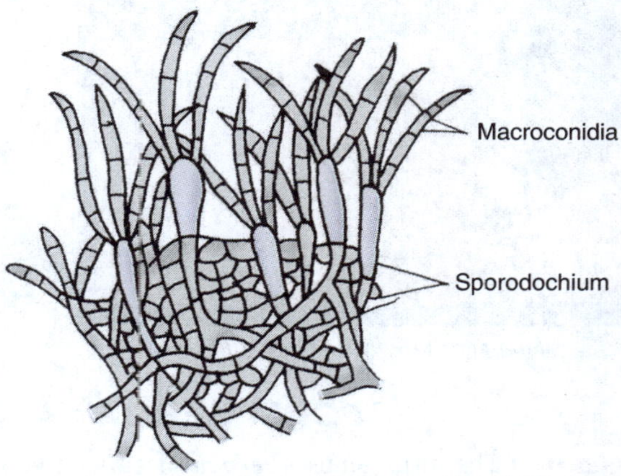

Fig. 16.8. Sporodochium and Macroconidia of *Fusarium.*

Chlamydospores (Fig. 16.7). They are rounded, oval, thick-walled cells formed in the hyphae. They may be formed singly or in chains of two or more. They become separated from the parent hyphae after maturing and function as resting spores. Under suitable conditions, the chlamydospores germinate by means of germ tubes to form a fresh mycelium.

Sclerotia. The mycelium often forms compact, resting bodies of thick-walled hyphae. These are called the **sclerotia.** They function as storage organs and also serve as means of perennation and vegetative propagation.

The conidia and mycelia are short-lived. The fungus persists as chlamydospores and sclerotia which germinate under suitable conditions. Infection usually takes place through the fibrous root system near the root tip region or through wounds. The parasite then invades the root cortex and finally establishes itself in the xylem vessels. It does but little damage to the cortex. The vascular fusaria are favoued by relatively high soil temperature. The disease aspect has been discussed in the last chapter.

Taxonomic Position :

Alexopoulos (1962)

Division	:	**Mycota**
Subdivision	:	**Eumycotina**
Form class	:	**Deuteromycetes**
Form order	:	**Moniliales**
Form family	:	**Tuberculariaceae**
Form genus	:	*Fusarium*
Form species	:	*oxysporum*

Alexopoulos and Mims (1979)

Superkingdom	:	**Eukaryonta**
Kingdom	:	**Mycetae**
Division	:	**Amastigomycota**
Subdivision	:	**Deuteromycotina**
Form class	:	**Deuteromycetes**
Form sub-class	:	**Moniliales**
Form order	:	**Moniliales**
Form family	:	**Tuberculariaceae**
Form genus	:	*Fusarium*
Form species	:	*oxysporum*

FORM-FAMILY : DEMATIACEAE

It includes imperfect fungi in which both the mycelium and the conidia are dark-coloured. Most of the species are saprophytes. Some are plant pathogens and a few parasitize men and animals. The well known form-genera included in this form-family are *Alternaria, Cladosporium, Heterosporium* and *Cercospora.* We study the life-histories of *Alternaria* and *Cercospora.*

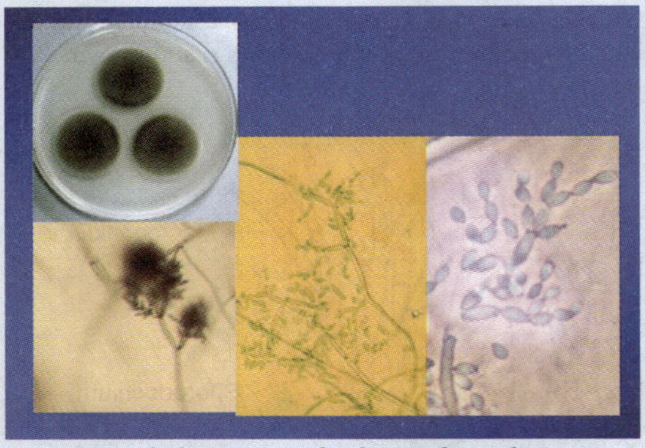

Cladosporium – hyphae and conidia

Heterosporium – on host

FORM GENUS : ALTERNARIA

General. The form-genus *Alternaia* occurs universally. It has several form-species. Many of these are encountered with as contaminants in laboratory media and on petri-dish cultures. In nature the alternarias grow mostly as saprophytes on plant debris and dying plant parts and also in the soil. Several form-species are parasitic on plants. The important plant pathogens are :—

1. *Alternaria solani.* It causes **early blight** (not to be confused with **late blight** caused by *Phytophthora infestans*) of potato and other members of family Solanaceae.

2. *Alternaria brassicae* and *A. brassicicola* cause leaf spot disease in crucifers such as mustard, cabbage, cauliflower, and knol kohl.
3. *Alternaria tenuis*. It is the causative agent of leaf blight of wheat seedlings. Agarwal (1959) reported that *A. zinneae* parasitizes *Ageratum conyzoides*.

Alternaria solani – on potato tubers

A.brassicae – on Mustard leaves

Mycelium. It is not large and extensive but is short, septate, branched, light brown but becoming darker with age. The colonies of *Alterncria* are woolly but more compact with the underside very dark-coloured. In the parasitic species, the hyphae are intercellular at first, but later penetrate cells of the invaded tissues and thus become intracellular. The cells are usually multinucleate.

Reproduction. Alternarias have no sexual or perfect stage. They multiply asexually by the method of sporulation. The characteristic asexual spores which are produced exogenously are the conidia (Fig. 16.9).

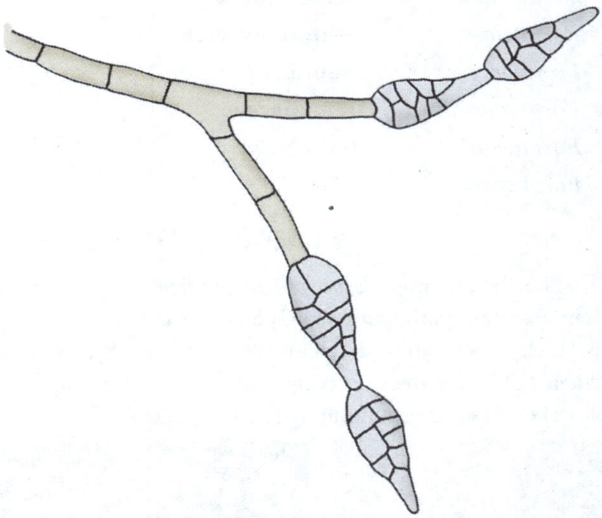

Fig. 16.9. *Alternaria sp.* Hyphae bearing multicellular muriform conidia.

The conidia are produced at the tips of ordinary hyphae which are comparatively short and dark coloured. Special hyphae termed conidiophores are not recognisable. The conidia are large, dark coloured, several celled and beaked. The number of cells varies from 8-14 or even more. The component cells occur packed in muriform, conical masses. The septa dividing the spore into cells are both transverse and vertical and their number is not fixed. According to Tandon and Srivastava, it is affected by environmental conditions. The conidia of *Alternaria* thus belong to the category of **dictyospores.** Usually they are borne end to end in chains of two or three. Occasionally they may occur singly at the tip of a hypha. Each conidium originates as a bud from the terminal cell of a hypha (Fig. 16.10A). The conidia are wind borne and thus invade laboratories where they contaminate cultures. They are also abundant in the house dust. The conidia germinate readily in the presence of moisture and suitable temperature by putting out a number of germ tubes (B). The disease aspect is discussed in the last chapter.

Taxonomic Position :

Alexopoulos (1962)

Division	:	**Mycota**
Subdivision	:	**Eumycotina**
Form class	:	**Deuteromycetes**
Form order	:	**Moniliales**
Form family	:	**Dematiaceae**
Form genus	:	*Alternaria*
Form species	:	*solani*

Alexopoulos and Mims (1979)

Superkingdom	:	**Eukaryonta**
Kingdom	:	**Mycetae**
Division	:	**Amastigomycota**
Subdivision	:	**Deuteromycotina**
Form class	:	**Deuteromycetes**
Form order	:	**Moniliales**
Form family	:	**Dematiaceae**
Form genus	:	*Alternaria*
Form species	:	*solani*

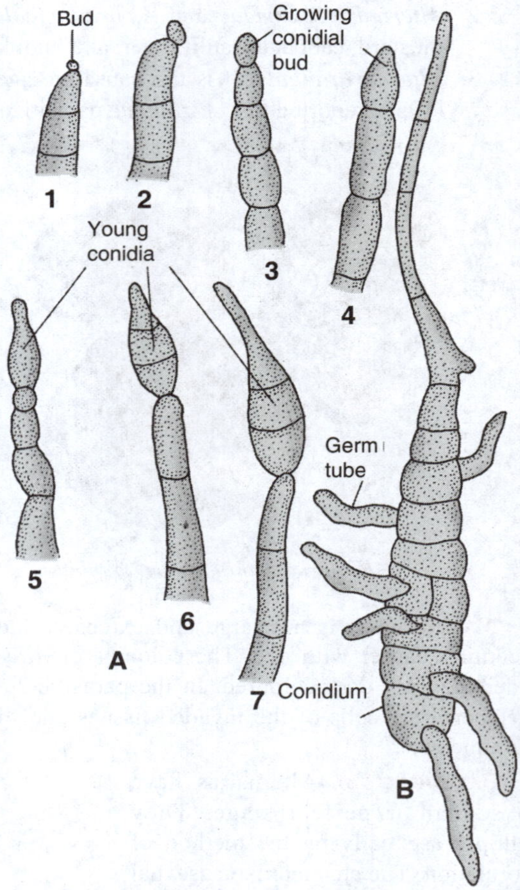

Fig. 16.10 (A-B). *Alternaria solani.* A_{1-7}, stages in the development of conidium; B, germinated conidium. (After Rands).

FORM GENUS : CERCOSPORA

The form-genus *Cercospora* includes about 3,800 form-species (Chupp, 1953). Majority of them are plant pathogens which cause **leaf spot** diseases of higher plants of economic value (Fig. 16.11A). Commonly the leaf spot disease is called the **tikka disease.** *C. personata* and *C. arachidicola* are the two commonly known form-species which are responsible for the leaf spot (tikka) disease of groundnut (*Archis hypogea*).

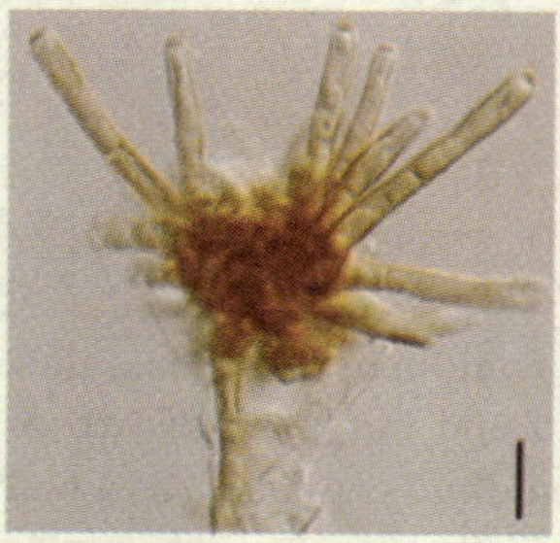

Cercospora – acervulus

Cercospora arachidicola on ground nut leaf

Tikka Disease of Groundnut

Mycelium. The mycelium in many species (*C. personata*) is entirely internal. The hyphae ramify in the intercellular spaces between the mesophyll cells of the host leaf obtaining nutrition by sending branched haustoria into the spongy and palisade cells. In some form species (*C. arachidicola*) the mycelium consists of both external and internal hyphae. The latter in the beginning are intercellular but later on become intracellular. They do not produce haustoria. Before the mycelium enters the reproductive phase, the hyphae accumulate and become compacted to form brown to black globular mass of hyphae, the **stroma** immediately beneath the epidermis of the host leaf in a substomatal cavity (Fig. 16.11 B).

Asexual reproduction (Fig. 16.11 B-C). It takes place by means of long, cylindrical usually hyaline, multiseptate conidia (C) which are abstricted successively at the tips of unbranched, dark conidiophores (B). The latter arise in tufts from a stroma lying in a substomatal cavity and emerge by rupturing the overlying epidermis. The conidiophores are geniculate (Knee-jointed) and 1-2 septate. The conidium, as it falls off, leaves a scar on the conidiophore.

The perfect stage of both *C. personata* and *C. arachidicola* is known. It is *Microsphaerella berkleyi* in the former.

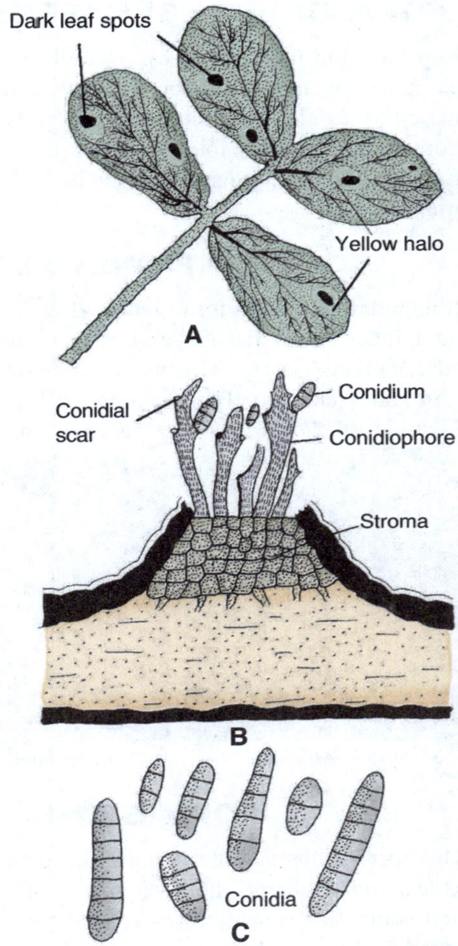

Fig. 16.11 (A-C). *Cercospora sp.* A, Leaf spots on the leaflets of *Arachis hypogea;* B, L.S. acervulus with the geniculate conidiophores emerging; C, conidia.

Taxonomic Position

Alexopoulos (1962)		
Division	:	**Mycota**
Subdivision	:	**Eumycotina**
Form class	:	**Deuteromycetes**
Form order	:	**Moniliales**
Form family	:	**Dematiaceae**
Form genus	:	*Cercospora*
Form species	:	*personata*

Alexopoulos and Mims (1979)		
Superkingdom	:	**Eukaryonta**
Kingdom	:	**Mycetae**
Division	:	**Amastigomycota**
Subdivision	:	**Deuteromycotina**
Form class	:	**Deuteromycetes**
Form sub-class	:	**Hyphomycetidae**
Form order	:	**Moniliales**
Form family	:	**Dematiaceaeae**
Form genus	:	*Cercospora*
Form species	:	*personata*

FORM-ORDER : SPHAEROPSIDALES (*PHOTOMITALES*)

Reproduction takes place by pycnidiospores which are produced in variously shaped fruit bodies called **pycnidia** with or without ostioles. The pycnidia may be separate or jointed by stromatic tissue. The pycnidium may be globose, or beaked. In some spcies it has a long neck. The form-order includes four form-families (Sphaeropsidaceae, Zythiaceae, Leptostromataceae and Excipulaceae) recognisable on the basis of pycnidial charaters. Of these Sphaeropsidaceae is important.

FORM-FAMILY : SPHAEROPSIDACEAE

It includes about 568 form-genera (Bender 1931) but many more have been discovered since then and added to the list. Several of them are saprophytes but some are parasites which cause destructive plant diseases. The important among these are *Phyllosticta, Macrophoma, Phomopsis, Septoria* and *Ascochyta.* The *pycnidia* are brown or black and usually flask-shaped often with a small ostiole. *Phyllosticta* is taken as an example.

Septoria – leaf spots

Macrophoma – leaf spots

Phomopsis – leaf spots

FORM-GENUS : PHYLLOSTICTA

This form-genus includes pathogenic species which parasitize leaves. *Phyllosticta acericola* causes *leaf spot* of maple and *P. solitaria* causes apple blotch. Tandon and Bilgrami (1954, 1957) reported some *leaf spot* diseases caused by *Phyllosticta* on *Atrocarpus* and *Cycas revoluta* in Allahabad. The fungus has a branched, septate mycelium.

Asexual reproduction. (Fig. 16.12). It takes place by means of small, hyaline, unseptate pycnidiospores (D) produced at the tips of very short conidiophores (C) arising from the wall of the pycnidial cavity (B). The pycnidia are ostiolate, dark-coloured and globose in form (B). They are embedded in the tissue of the host.

Phyllosticta acericola Causes leaf spot of Maple

Phyllosticta solitaria Causes Apple Blotch

Taxonomic Position

Alexopoulos (1962)			Alexopoulos and Mims (1979)		
Division	:	**Mycota**	*Super kingdom*	:	**Eukaryonta**
Sub-division	:	**Eumycotina**	*Kingdom*	:	**Mycetae**
Form class	:	**Deuteromycetes**	*Division*	:	**Amastigomycota**
Form order	:	**Sphaeropsidales**	*Sub-division*	:	**Deuteromycotina**
Form family	:	**Sphaeropsidaceae**	*Form class*	:	**Deuteromycetes**
Form genus	:	*Phyllosticta*	*Form sub-class*	:	**Coelomycetidae**
Form species	:	*acericola*	*Form order*	:	**Sphaeropsidales**
			Form family	:	**Sphaeropsidaceae**
			Form genus	:	*Phyllosticta*
			Form species	:	*acericola*

FORM-ORDER : MELANCONIALES

It is a small group of plant pathogens. The mycelium is internal and asexual fruiting bodies are of **acervulus** type. The acervuli are sub-epidermal and develop beneath the host cuticle or epidermis. Usually they burst through the host surface as the pigmented conidia mature. The conidiophores which are usually simple, rarely branched are aggregated in a palisade-like layer on the surface of the stroma. The conidia are abstricted singly or in chains. This form-order includes a single form-family Melanconiaceae.

FORM-FAMILY : MELACONIACEAE

The important genera included in this form-family are *Colletotrichum, Pestalotia,* and *Gloeosporium.* Form genus *Colletotrichum* is briefly described here.

Colletotrichum – Red rot of sugarcane

Pestalotia – showing acervuli on leaf

FORM GENUS : COLLETOTRICHUM

Mycelium. It is internal, septate and branched. The slender hyphae are both intercellular and intracellular. They are usually hyaline when young. With age the hyphal wall becomes darker and thicker and the cytoplasm, which contains droplets of oil, becomes denser. The thick-walled hyphae finally collect, intertwine and become compacted to form small usually black stromata beneath the cuticle or epidermis (Fig. 16.13A).

The acervulus consisting of a dense, palisade-like layer of short, hyaline conidiophores, is formed on the stromatic surface (A). The conidiophores are unseptate and usually unbranched, rarely branched (B). The conidia are abstricted singly, sometimes in chains at the tips of conidiophores acrogenously. As the conidia mature the overlying cuticle or epidermis ruptures.

The conidia are usually hyaline to slightly ochreoous, continuous (non-septate or one-celled),

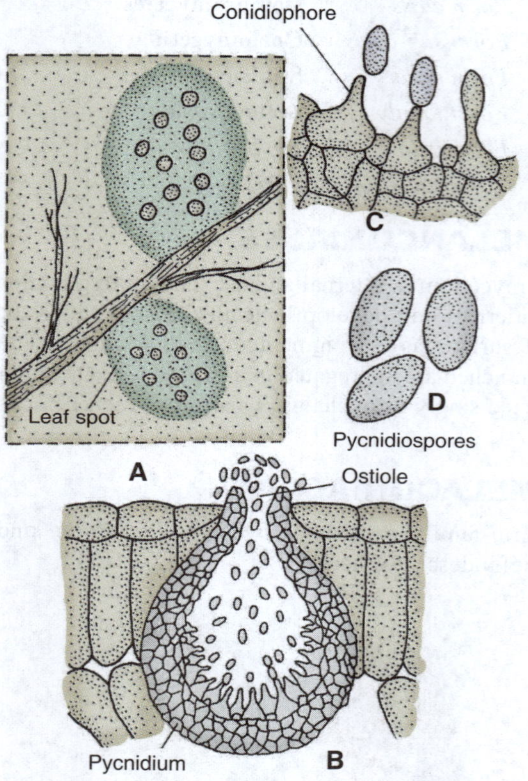

Fig. 16.12. (A-D) *Phyllosticta sp. A*, portion of infected leaf showing leaf spots; B L.S. Pycnidium; C, conidiophores C, conidia; bearing pycnidiospores; D, pycnidiospores.

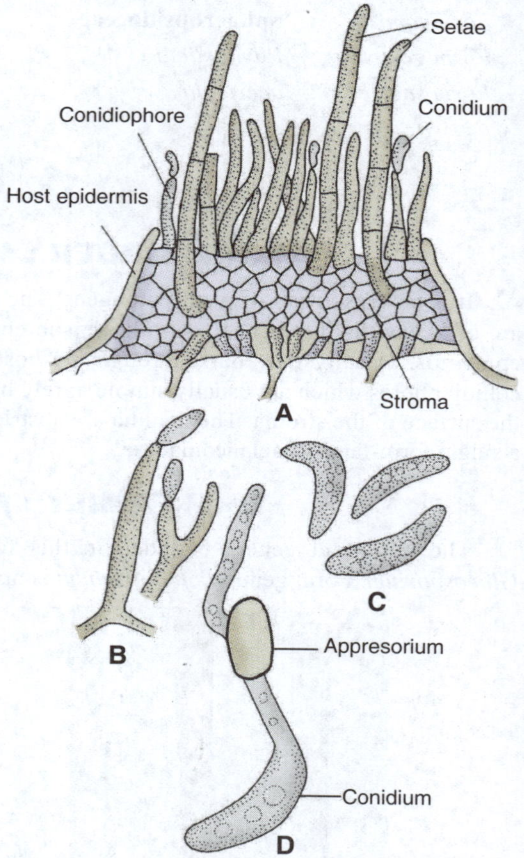

Fig. 16.13 (A-D). *Colletotrichum sp.* A, vertical section of acervulus; B, conidiophores; D, germinated conidium (B-C after Buller).

oblong to fusoid usually with a large oil globule in the centre, somewhat curved with rounded ends or pointed at one end.

The acervuli (A) are characterised by the presence of long, rigid, dark brown bristles or setae which either form a fringe around the acervulus or are intermixed with the conidiophores. The setae are septate and attenuated towards their upper lightly coloured ends. Moist weather is essential for conidia formation. They are dispersed by wind, spattering rain or irrigated water. The conidia are short-lived and thus germinate readily under suitable conditions of moisture and temperature by one or more germ tubes (D) to form the mycelium. More often an appresorium is formed when the germ tube comes in contact with the host surface. Penetration through the host cuticle occurs by means of a slender infection tube or peg which arises from the side of appresorium in contact with the host.

Terminal or intercalary thick-walled chlamydospores are formed by the septation of hyphae. They persist in the soil for a long time.

The form-genus includes a number of species which cause serious diseases to economically important plants. The common examples are :—

1. *Colletotrichum falcatum* incitant of red rot of sugarcane (*Saccharum officinarum*).
2. *C. capsici* incitant of ripe fruit rot and die back of chillies (*Capsicum anuum*).
3. *C. gloeosporioides* incitant of mango anthracnose. The disease aspect is considered in the last chapter.

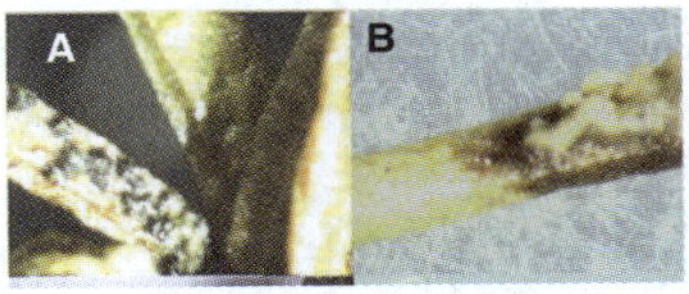

A - B– Leaves exhibiting synptoms

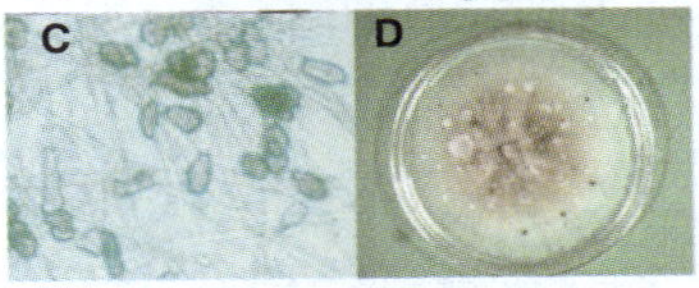

C - Conidia of the Fungus,
D- Fungus in Culture

Colletotrichum gloeosporioides

Colletotrichum – spots on leaf

Taxonomic Position

Alexopoulos (1962)

Division	:	**Mycota**
Sub-division	:	**Eumycotina**
Form class	:	**Deuteromycetes**
Form order	:	**Melanconiales**
Form family	:	**Melanconiaceae**
Form genus	:	*Colletotrichum*
Form species	:	*falcatum*

Alexopoulos and Mims (1979)

Super kingdom	:	**Eukaryonta**
Kingdom	:	**Mycetae**
Division	:	**Amastigomycota**
Sub-division	:	**Deuteromycotina**
Form class	:	**Deuteromycetes**
Form sub-class	:	**Coelomycetidae**
Form order	:	**Melanconiales**
Form family	:	**Melanconiaceae**
Form genus	:	*Colletotrichum*
Form species	:	*falcatum*

QUESTIONS

ESSAY TYPE

1. With the help of suitable sketches, describe asexual reproduction in the Deuteromycetes.
2. Give a broad outline of the classification of Fungi Imperfecti. Why is it not appropriate to call it a natural classification?
3. Describe the life history of *Fusarium.*
4. What are imperfect fungi? Give main characters and systematic position of *Cereospora*? Describe any disease caused by its fungus (*Allahabad, 1992*)
5. (a) What are fungi Imperfecti? Give an illustrated account of the life cycle of *Fusarium.* (*Lucknow, 1999*)
5. (b) Elaborate the method of reproduction in moniliales . (*Rohilkhand M.Sc, 2006*)
6. Describe briefly the structure and modes of reproduction of *Alternaria*. Mention the diseases caused by it and methods of its control. (*Punjab, 1998*)
7. With the help of suitable sketches, describe the various types of spores in imperfect fungi studied by you.
8. Describe general account of fungi imperfect with special reference to parasexuality. (*Poorvanchal M.Sc., 2003*)
9. Give an illustrated account of morphology and reproduction of *Cercospora*. Describe the symptoms and control measures of the disease caused by this fungus. (*Rohilkhand, 2002*)

SHORT ANSWER TYPE

10. Give an illustrated account of the following :
    (*a*) Arthrospores
    (*b*) Chlamydospores
    (*c*) Blastospores (*d*) Porospores
    (*e*) Conidiospores (*f*) Phialospores
    (*g*) Pycindium
    (*h*) Acervulus
    (*i*) Imperfect fungi (*Kumaon, 2002*)
11. Write notes on *Alternaria* and *Phyllosticta.*
12. What are imperfect fungi? Give their general characteristics.
13. Explain the following with the help of diagram only :
    (*a*) Spores of *Alternaria* (*Awadh, 1992; Kanpur, 1998, 1999*)
    (*b*) Conidiophore and conidia of *Alternaria* (*Awadh, 1995, 1997; Kanpur, 1996*)
    (*c*) Dictyospore type of conidia with conidiophore (*Awadh, 1996*)
    (*d*) Conidiophore and Conidia of *Cercospora.* (*Rohilkhand, 1991*)
    (*e*) Conidia of *Alternaria* and *Cercospora* (*Rohilkhand, 1997*)
    (*f*) Asexual reproduction in *Fusarium.* (*Lucknow, 2002*)
    (*g*) Asexual reproduction in Deuteromycetes. (*Lucknow, 2004*)
    (*h*) Conidium (*Allahabad, 2002*)
    (*i*) Acervulus (*Gorakhpur, 2003*)
14. Give the most important charcteristics of the following and their systematic position :
    (*i*) *Alternaria* (*Awadh, 1991, 1996, 1997; Kanpur, 1997, 1998; Allahabad, 1991; Gorakhpur, 1991, 1993, 2003; Rohilkhand, 1996; Kumaon, 1998; Meerut, 1998; Bundelkhand 1993,1998*)
    (*ii*) *Colletotrichum* (*Allahabad M.Sc., 1999, 2000; Gorakhpur, 1992*)
    (*iii*) *Fusarium* (*Kanpur M.Sc., 2002, 2004; Lucknow, 1991, 1993, 1994, 1996, 1997, 1998*)
    (*iv*) *Cercospora* (*Kumaon, 2000; Allahabad, 2003, 2004*)
15. Write short notes on the following :
    (*i*) T.S. leaf infected with *Alternaria* (*Awadh 1993, 1996*)
    (*ii*) Acervulus of *Colletotrichum* and the disease caused by it (*Allahabad, 1990, Gorakhpur, 1990*)
    (*iii*) Reproduction in *Alternaria* (*Gorakhpur, 1994, 1998*)
    (*iv*) *Alternaria* (*Kumaon, 2001; Punjab, 1997, Bihar, 1991*)

(*v*) Imperfect Fungi(*Bundelkhand, 1996*)

(*vi*) *Fusarium* (*Lucknow 2000*)

(*vii*) Structure and reproduction in *Alternaria* (*Poorvanchal, 2005*)

16. How will you identify the following pathogens?

(*a*) *Alternaria* (*Kanpur, 1995*)

(*b*) *Fusarium* (*c*) *Cercospora*

(*d*) *Phyllosticta*

(*e*) *Colletotrichum*

17. Draw labelled diagrams of the following :

(*a*) Acervulus of *Colletotrichum* (*Gorakhpur, 1995*)

(*b*) Pycnidium

(*c*) Conidia of *Fusarium*

18. Give the general characters of Deuteromycotina (*Punjab, 1997*)

19. Give a brief account of the classification of Deuteromycetes

20. (*a*) Describe the structure and reproduction in Alternaria. (*Punjab, 1999*)

(*b*) Describe disease symtoms of *Alternaria*. (*Kanpur, 2006*)

SHORT ANSWER TYPE

21. Mention briefly the differences in haustoria and appressoria. (*Lucknow, 2003*)

22. Differentiate between

(*a*) *Micro and Macroconidia* (*Lucknow, 2003*)

(*b*) *Alteruaria and Fusarium*

(*c*) *Fusarium and Cercospora*

(*d*) Conidia of *Alternaria* (*Kanpur, 2001*)

(*e*) Hypha of *alternaria* with conidiophore and conidia (*Kanpur, 2003*)

23. What are fungi imperfecti ? (*Poorvanchal, 2001*)

24. Describe important features of the life cycle of *Alternaria*. (*Poorvanchal, 2002*)

25. Draw suitable sketches to differentiate various conidial frutifications in Deuteromycetes. (*Kanpur M.Sc. 2002*)

26. Mention systematic position and points of identificiation of *Alternaria*. (*Kanpur M.Sc. 2006*)

OBJECTIVE TYPE

27. Fill in the Blanks :

(*i*) Deuteromycetes are also known as fungi, imperfecti because .... reproduction is absent.

(*ii*) Chlamydospores are .... spores produced on the mycelium

(*iii*) In *Alternaria* the conidia are ....

(*iv*) When the branched or unbranched conidiophores arise very closely and are often united along a greater part of their length to form dense fasicles, such structures are known as ....

(*v*) *Colletotrichum falcatum* produces ....

(*vi*) Macro Conidia and **microconidia** are formed in the genus..........................

(*vii*) The classification of Deuteromycetes is artificial because .... reproduction is absent.

(*viii*) Wilt of Arhar is caused by ....

(*ix*) *Alternaria solani* produces .... directly on ....

(*x*) Multicellular conidia are produced in the genus ....

(*xi*) Tikka disease of groundnuts is caused by ....

(*xii*) Pycnidia are produced in the genus ....

(*xiii*) Red rot of sugarcane is caused by ....

(*xiv*) Early blight disease of potato is caused by ....

28. Select the correct answer :

(*i*) Multicellular muriform in conidia are formed in

(*a*) *Alternaria* (*b*) *Aspergillus*
(*c*) *Mucor* (*d*) *Fusarium.*

(*ii*) Macro and microconidia are produced in

(*a*) *Alternaria* (*b*) *Aspergillus*
(*c*) *Mucor* (*d*) *Fusarium.*

(*iii*) Fungi imperfecti are so named because of the absence of

(*a*) Asexual reproduction

(*b*) vegetative reproduction

(*c*) sexual reproduction
(*d*) reproduction.

(*iv*) Which of the following causes early blight disease of potato?
(*a*) *Alternaria solani*
(*b*) *Phytophthora infestans*
(*c*) *Synchytrium endobioticum*
(*d*) *Albugo candida.*

(*v*) Tikka disease of ground nuts is caused by
(*a*) *Aspergillus* (*b*) *Puccinia*
(*c*) *Cercospora* (*d*) *Fusarium.*

(*vi*) *Fusarium udum* causes the disease
(*a*) Early blight of potato
(*b*) Wilt of Arhar
(*c*) Tikka disease of ground nuts
(*d*) Red rot of sugarcane.

(*vii*) Sexual reproduction is absent in the members of the class
(*a*) Phycomycetes
(*b*) Ascomycetes
(*c*) Basidiomycetes
(*d*) Deuteromycetes.

(*viii*) Fungal spores produced at the tips of hyphae are called
(*a*) Sporangiospores
(*b*) Porospores
(*c*) Conidia
(*d*) Chlamydospores.

(*ix*) Only asexual reproduction is found in
(*a*) *Mucor* (*b*) *Synchytrium*
(*c*) *Fusarium* (*c*) *Puccinia.*

(*x*) Conidiophores and conidia are produced in Acervuli in
(*a*) *Agaricus* (*b*) *Cercospora*
(*c*) *Colletotrichum*(*d*) *Alternaria.*

(*xi*) *Colletorichum falcatum* causes the disease known as
(*a*) Early blight (*b*) Late Blight
(*c*) Wilt (*d*) Red rot.

(*xii*) Conidiophores and conidia aggregated in pycnidia are present in
(*a*) *Alternaria* (*b*) *Colletotrichum*
(*c*) *Phyllosticta* (*d*) *Fusarium.*

(*xiii*) In *Cercospora* the conidia are
(*a*) unicellular
(*b*) bicellular
(*c*) multicellular with only transverse septa
(*d*) multicellular with transverse and longitudinal septa.

(*xiv*) *Fusarium vasinfectum* causes the disease
(*a*) Wilt of Arhar (*b*) Wilt of Cotton
(*c*) Wilt of gram (*d*) Wilt of flax.

(*xv*) Conidia of *Alternaria* are
(*a*) Unicellular
(*b*) Multicellular
(*c*) Both Unicellular and Multicellular
(*d*) Neither unicellular nor multicellular.

(*xvi*) Geniculate conidia are present in
(*a*) *Cercospora* (*b*) *Alternaria*
(*c*) *Aspergillus* (*d*) *Saprolegina*

(*xvii*) Numerous Conidia form a cushion like structure in *Colletotrichum*. These are called
(*a*) *Apothecium* (*b*) *Acenulus*
(*c*) *Cleistothecium*(*d*) *Coremia*

CHAPTER 17

FUNGI – GENERAL TOPICS AND DISCUSSION

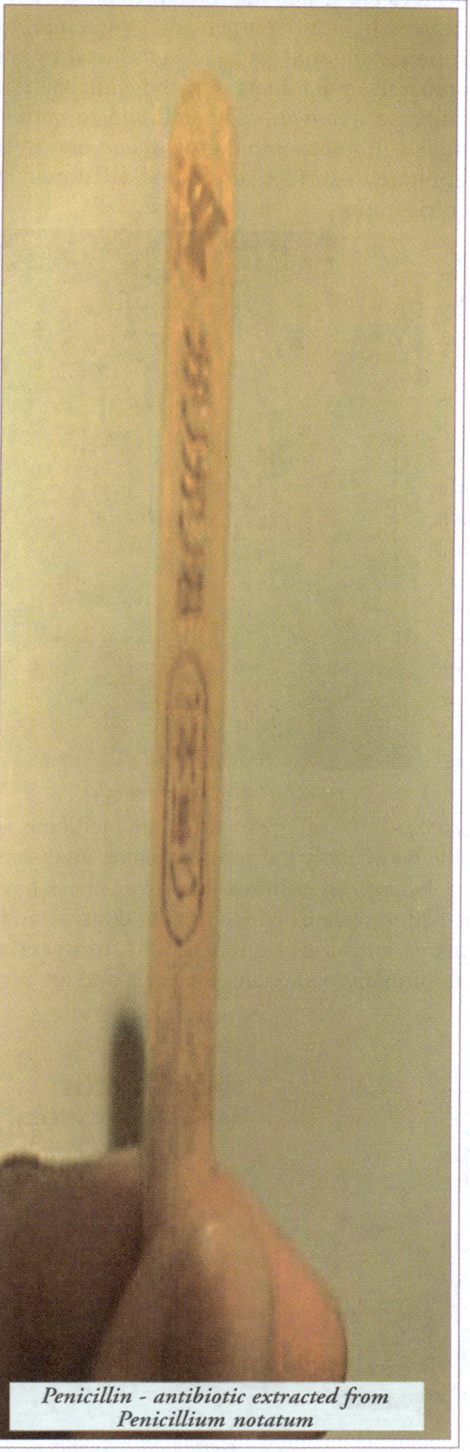

Penicillin - antibiotic extracted from Penicillium notatum

- Economic Importance of Fungi
- Mushroom Cultivation in India
- Heterothallism
- Sex hormones
- Parasexual cycle or Parasexuality
- Culture of Fungi
- Mycorrhiza
- Rhizosphere & Phyllosphere
- Ecological groups of fungi
- Fossil Fungi
- Pigmentation in Fungi

ECONOMIC IMPORTANCE OF FUNGI

Fungi include hundreds of species which are of tremendous economic importance to man. In fact our lives are intimately linked with those of fungi. Hardly a day passes when we are not benefited or harmed directly or indirectly by these organisms. They play an impotant role in medicine yielding antibiotics, in agriculture by maintaining the fertility of the soil and causing crop and fruit diseases, forming basis of many industries and as important means of food. Some of the fungi are important research tools in the study of fundamental biological processes. Some of the fungi particularly molds and yeasts play a negative role by causing spoilage of stored goods such as foodstuffs, textiles, leather, rubber, plastics, timber and even glass.

1. Role of Fungi in Medicine. Some fungi produce substances which help to cure diseases caused by the pathogenic microorganisms. These substances are called the **antibiotics.** The term antibiotic, therefore, denotes an organic substance, produced by a microorganism, which inhibits the growth of certain other microorganisms. The most important antibiotics are produced by the moulds, actinomycetes or bacteria. They are used to combat the evil effects of pathogenic bacteria and viruses. The use of antibiotics is not limited to disease treatment. The addition to certain antibiotics in small amounts to the feed of slaughter animals promotes rapid growth and improves the quality of the meat products. Application of an antibiotic to surface of freshly killed poultry preserves the fresh-killed taste during long periods of refrigeration.

The discovery of antibiotic agents as drugs is comparatively a recent history. The role of fungi in producing antibiotic substances was first established by Sir Alexander Fleming in 1929. He extracted the great antibiotic drug **Penicillin** from *Penicillium notatum.* It was the first antibiotic to be widely used. Penicillin is an organic substance lethal to microbes. It is far more effective than ordinary drugs and germicides. It has no adverse effect on human protoplasm but kills bacteria especially gram-positive type. Penicillin is now produced on a commercial scale all over the world including India from the improved strains of *P. notatum* and *P. chrysogenum.* There is a Penicillin factory at Pimpri in India. The success of penicillin as an antibiotic was later found to be limited. Naturally this led to further research for new antibiotics which would act on pathogenic bacteria and viruses not affected by penicillin. This research resulted in the discovery of a number of other antibiotics. Of these, **streptomycin** is another.

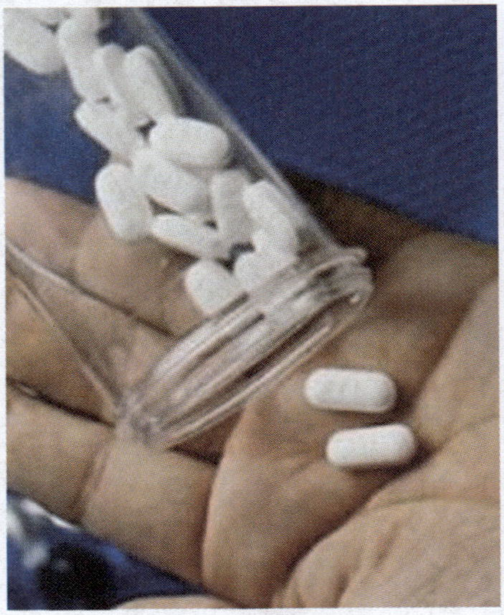

Sreptomycin - Samples

Streptomycin is obtained from *Streptomyces griseus.* It is of great value in medicine. It destroys many organisms which are not killed by penicillin particularly the gram-negative organisms. A number of antibiotics have also been extracted from *Aspergillus* cultures. However, these have not been proved so effective as penicillin. Some of the actinomycetes which are not considered to be true filamentous bacteria are the sources of many antibiotics such as **chloromycetin, aureomycin, terramycin,** etc. They inhibit the growth of many pathogenic bacteria and are also

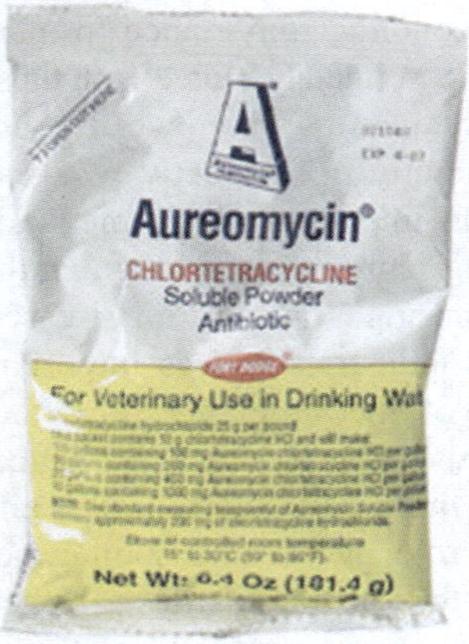

Aureomycin - Samples

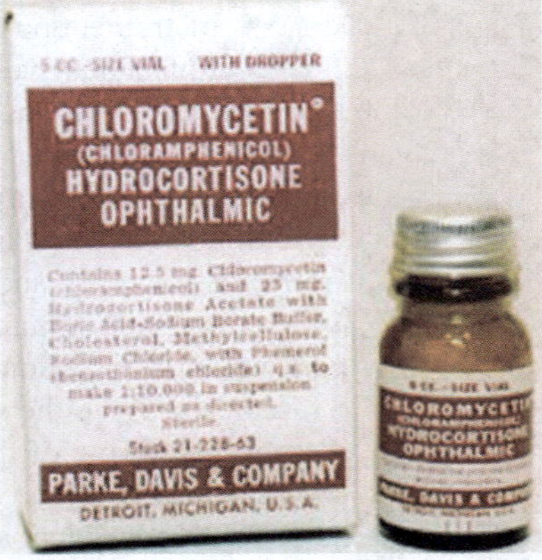

Chloromycetin - Samples

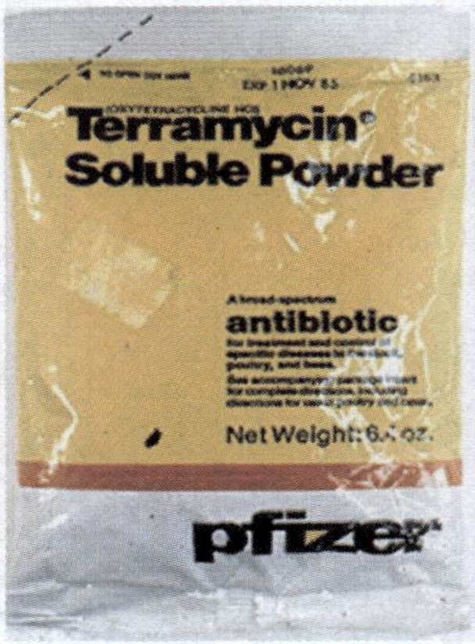

Terramycin - Samples

used successfully in the treatment of various virus diseases. Many animal and human diseases which do not respond readily to other antibiotics are effectively cured by aureomycin. The plasmodia of certain species of *Myxogastres* have been reported to yield soluble antibiotics. These check the growth of certain bacteria and yeasts in culture. The antibiotics play an important role to combat plant diseases as well.

Griseofulvin which is recovered from mycelium of *Penicillium griseofulvum* and many other species has antifungal properties. It acts on the hyphae by interfering with wall formation. Consequently the hyphal tips curl and cease to grow. When administered orally it is absorbed into the body where it accumulates in the keratinized tissues of the epidermis and hair. It is thus effective against fungal skin diseases such as ringworms and athlete's foot disease.

Claviceps purpurea produces sclerotia in the ovaries of the flowers of grasses such as rye. The sclerotium is called the **ergot** of rye. Ergot is used in veterinary and human medicine. It contains a mixture of alkaloids which cause rapid and powerful contractions of the uterus. The medicine is thus used to control bleeding during child birth. Ergot is highly poisonous. A derivative of ergot known by the name of **lysergic acid** (**LSD**) is used in experimental psychiatry.

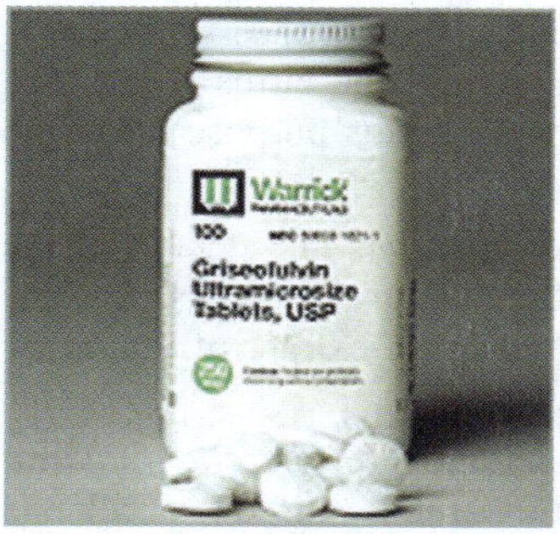

Griseofulvin - antifungal

Sclerotia of Ergot

Claviceps purpurea (source of LSD)

The giant puff ball *Clavatia* contains an anti-cancer substance *calvacin.* The eating of these fungi prevents stomach tumours.

Role of Fungi in Industry. The industrial uses of fungi are many and varied. In fact the fungi form the basis of many important industries. There are a number of industrial processes in which the biochemical activities of certain fungi are harnessed to good account. A brief sketch of some of the most important of these processes is given below :—

(*i*) *Alcoholic fermentation.* It is the basis of two important industries in India or rather all over the world. These are **brewing** and **baking.** Both are dependent on the fact that the fermentation of sugar solutions by yeasts produces ethyl alcohol and carbon dioxide. In brewing or wine making industry alcohol is the important product. The other by-product which is carbon dioxide was formerly allowed to escape as a useless thing. Now carbon dioxide is also considered a valuable by-product. It is collected, solidified and sold as "dry ice". In the baking or bread-making industry CO_2 is the useful product. It serves two purposes :—

(*i*) causes the dough to rise.

(*ii*) makes the bread light.

The other by-product, which is alcohol, is incidental. The yeasts secrete the enzyme complex called **zymase** which brings about conversion of sugar into alcohol. Many excellent yeast strains are now available. The yeasts lack diastase. So they cannot break starch into sugar. There are a number of fungi popularly known as the *moulds.* They secrete a whole range of enzymes and thus bring about fermentation of complex carbohydrates. In producing industrial alcohol moulds are employed as *starters* to bring about saccharification of the starch. At the second stage, *yeast* is employed to act on the sugar. Although mould can complete the conversion to sugar but the yield is better if yeast is employed for the second stage. The moulds commonly used for purpose of saccharification are *Mucor racemosus. M, rouxii* and some species of *Rhizopus. Aspergillus flavus* is used in the production of African native beer.

(*ii*) *Enzyme preparations.* Takamine on the basis of his intensive study of the enzymes produced by *Aspergillus flavus-oryzae* series has introduced in the market a few products of high enzymic activity. These are **Digestin, Polyzime, Taka diastase,** etc. They are used for dextrinization of starch and desizing of textiles.

Cultures of *Aspergillus niger* and *A. oryzae* on trays of moist, sterile bran yield a well known amylase which contains two starch splitting components.

Invertase is extracted from *Saccharoymces cerevisiae.* It has many industrial uses. It hydrolyses sucrose to a mixture of glucose and fructose.

(*iii*) *Preparation of organic acids.* The important organic acids produced commercially as the result of the biochemical activities of moulds are oxalic acid, citric acid, gluconic acid, gallic acid, fumaric acid, etc. Oxalic acid is the fermentation product of *Aspergillus niger.* Citric acid is made by mould fermentation. Many species of *Penicillium* are used for the purpose. The acid is produced on a commercial scale and is cheaper than the acid made from the citrus fruits. The gluconic acid is prepared from sugars. The moulds chiefly employed for this purpose are some species of *Penicillium* and *Aspergillus.*

Gallic acid is prepared on a commercial scale in Europe and America. The details of the method employed, however, are not known. It may be a modification of Calmete's process. Calmette (1902) obtained the gallic acid as the fermentation product of an extract of tannin by *Aspergillus gallomyces.*

(*iv*) *Gibberellins.* These are plant hormones produced by the fungus *Gibberella fujikuroi* which causes a disease of rice accompained by abnormal elogation. Gibberellin is used to accelerate growth of several horticultural crops.

Gibberellin - Samples

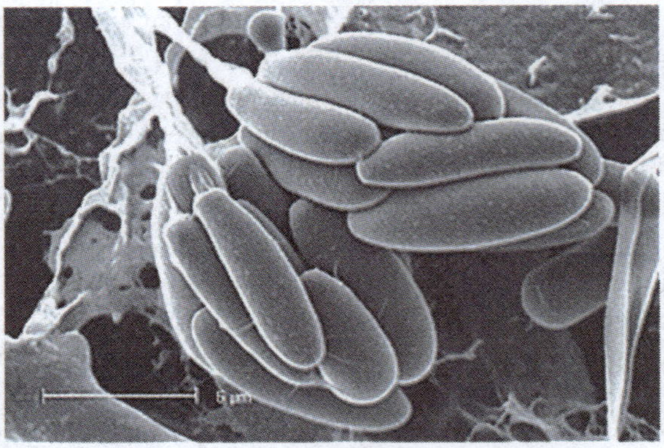

Gibberella fujikuroi (cause of foolish seedling disease of rice)

A PLANT HORMONE

(*v*) *Cheese Industry.* Certain fungi popularly known as the cheese moulds play an important role in the refining of cheese. They give cheese a characeristic texture and flavour. The two chief kinds of mould refined cheese are :—

(*a*) Camembert and Brie types. They are soft.

(*b*) Roquefort Gorgonzola and Stilton types. They are green or blue veined cheese. The moulds concerned are *Penicillium camemberti* and *P. caseicolum* in the first type and *P. roqueforti* in the second type.

Camembert Cheese - Samples

Roquefort Cheese - Samples

(*vi*) *Manufacture of Proteins*. As a supplement to the normal diet, some fungi particularly the yeasts are employed to synthesize proteins. The yeast (*Saccharomyces cerevisiae* and *Candida utilis*) contain high percentage of protein of great nutritive value. They are grown with ammonia as the source of nitrogen and molasses as the source of carbon. The manufactured product is called *Food Yeast*. It contains 15% protein and B group of vitamins.

(*vii*) *Vitamins*. The yeasts, are the best source of vitamin B complex. A number of preparations of high potency have been made from the dried yeast or yeast extracts and sold in the market. A number of moulds and yeasts are utilised in the synthesis of Ergosterol which contains Vitamin D. Riboflavin—another vitamin useful both in human and animal food—is obtained from a filamentous yeast, *Ashby gossypii*.

(*viii*) A good many fungi synthesize fat from carbohydrates. *Endomyces vernalis, Penicillium javanicum* and *Oidium lactis* have a high fat content. The microbiological production of fat is, however, too costly for use.

(ix) *Antibiotics*. For details refer to role of fungi in medicine.

Certain fungi form an important basis of fermentation of Cocaobeans. Mention must also be made here of the use of Lichens in yielding certain dyes and reagents. An important substance is extracted from *Roccella* lichen. It forms the basis of litmus paper which is used as an indicator to determine the acidity or alkalinity of a solution.

Dyes (extracted from Roccella)

Role of Fungi in Agriculture. The fungi play both a negative role and a positive role in agriculture.

(A) *Negative role*. They have a negative value because they are the causative agents of different diseases of our crop, fruit and other economic plants. These *fungal diseases* take a heavy toll and cause tremendous economic losses. The modest estimate is that about 30 thousand different diseases (including bacterial and virus) attack the economic plants grown for food or commercial purposes. The more important of these diseases are :

(*i*) *Damping off disease*. The seedlings of almost every type of plant grown as a commercial crop such as tomatoes, corn, cotton, mustard, peas, beans, tobacco, spinach, etc., are prone to this disease. It is caused by a species of *Pythium*.

(*ii*) *The potato blight* (*Late blight* of potatoes) is another destructive crop disease. It does a great damage to the potato tubers. A heavy attack of this disease in Ireland in 1845 destroyed the entire potato crop and caused so severe a famine that over a million Irish people migrated to U.S.A. Besides potatoes, it infects egg plants, tomatoes, etc.

Late Blight on Potato Tuber

Late Blight on Potato Leaves

(*iii*) *Downy mildews of grapes*. It ruins the vine yards and thus causes heavy losses to the crop. When the disease was first introduced into France from U.S.A, it caused a havoc to the vine yards. Almost all the French vine yards were destroyed before Bordeaux mixture, which proved an effective fungicide against this disease, was discovered.

Downy Mildew of Grapes

(*iv*) *Ergot disease of rye*. It is an important disease of a cereal crop—rye. It results in the formation of poisonous sclerotia in the rye kernel. It is called *ergot* of rye. Ergot is highly poisonous to man. Ergot poisoning causes hallucinations, insanity and finally death.

(*v*) *Apple scab*. It is a serious disease of the apple crop. It lowers the quality as well as quantity of the fruit.

Apple Scab on fruits

Brown Rot of Stone Fruits

Ergot on Rye

(*vi*) *Brown rot of stone fruits*. It causes enormous losses in the fruit crop of apricots, cherries, plums and peaches.

(*vii*) *Smut diseases* of corn, wheat, oat and other cereal crops cause serious reduction in the yield and quality of grain.

(*viii*) *Red rot disease of sugarcane*. It is a serious disease of sugarcane whose incidence has increased during the last few years particularly in the northern parts of the country.

(*ix*) *Rust diseases*. They attack our cereal crops and forest timber. Some of them such as black stem rust, yellow rust and orange rust are a serious threat to our wheat crop.

Orange Rust on leaves

Red rot of Sugarcane

(*x*) *Blackarm, Wilt and root rot of cotton.* These diseases of cotton, which is a very important commercial crop of our country, take a heavy toll of the crop every year.

(*xi*) *Pore fungi.* They are the common wood rotters. They destroy timber and lumbar.

Wilt of Cotton

Root rot of Cotton

Wood Rotting Pore Fungi

The above-mentioned diseases caused by fungi are thus responsible for a huge loss to our crop and other economic plants. The pathogenic fungi are always a nuisance to the agriculturists. They affect the agricultural economy of our country seriously. The farmer and the Agriculture department must wage a constant war against them.

It will not be out of place if a brief mention is made here of some of the human diseases caused by the fungi. Some species of *Aspergillus* such as *A. fumigatus, A. flavus,* and *A. niger* are human pathogens. They cause disease collectively known as *aspergilloses* such as aspergilloses of lungs, external ear, etc. Many parasitic Fungi Imperfecti live in the mucous membranes of throat, bronchi and lungs and cause infection of mouth and lungs. A few fungi cause skin discoloration. Others (Trichophytoneae) are the causative agents of a disease known as *athlete's foot.* The well known skin disease 'ring worm' and barber's itch are also fungal diseases. *Monilia*—a member of the class Fungi Imperfecti—causes a throat or mouth disease known as *thrush.* A few fungi cause serious diseases of domestic animals. Some fungi produce diseases among annoying insects harmful to crop and thus help to destroy them and keep them in check.

Entomopathogenic Fungus-Pandora

Thrush

Aspergilloses

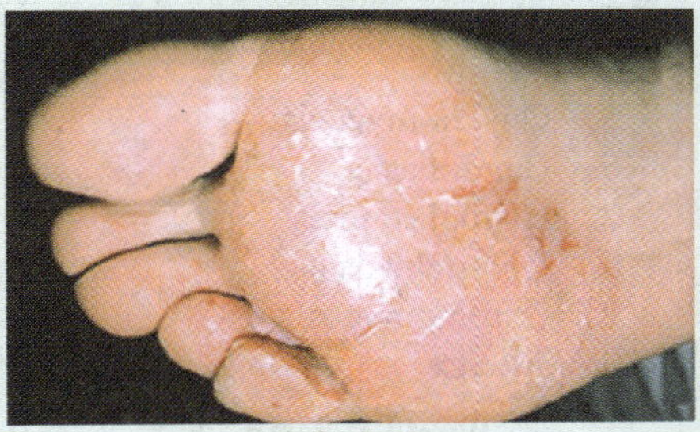

Athelete's Foot

TABLE I

List of some important plant diseases caused by fungi

	Name of the disease	Pathogen
1.	Club root of Crucifers	*Plasmodiophora brassicae*
2.	Wart disease of potato	*Synchytrium endobioticum*
3.	Stem Rot of Papaya	*Pythium aphanidermatum*
4.	Damping off of seedlings	*Pythium sp.*
5.	Late blight of potato	*Phytophthora infestans*
6.	White rust of crucifers	*Albugo candida*
7.	Downy mildew of peas	*Peronospora pisi*
8.	Green ear disease of Bajra	*Sclerospora graminicola*
9.	Powdery mildew of peas	*Erysiphe polygoni*
10.	Powdery mildew of wheat	*Erysiphe graminis*
11.	Leaf curl of peaches	*Taphrina deformans*
12.	Stem gall of coriander	*Protomyces macrosporous*
13.	Ergot disease of rye	*Claviceps purpurea*
14.	Rust of wheat	*Puccinia graminis*
15.	Rust of pea	*Uromyces pisi*
16.	Rust of gram	*Uromyces ciceris-arieteni*
17.	Rust of Linseed	*Melampsora lini*
18.	Covered smut of barley	*Ustilago hordei*
19.	Loose smut of wheat	*Ustilago nuda*
20.	Bajra smut	*Tolyposporium penicillariae*
21.	Grain smut of Jowar	*Sphacelotheca sorghi*
22.	Bunt of wheat	*Tilletia tritici*
23.	Early blight of potato	*Alternaria solani*
24.	Wilt of pigeon pea	*Fusarium oxysporum*
25.	Red rot of sugarcane	*Colletotrichum falcatum*
26.	Tikka disease of groundnuts	*Cercospora personata*
27.	Stripe disease of barley	*Helminthosporium graminieum*

TABLE II

List of some common disease of human beings caused by fungi

	Name of the disease	Pathogen
1.	Aspergillosis	*Apsergillus flavus, A. niger*
2.	Blastomycosis	*Blastomyces dermatidis*
3.	Otomycosis	*Aspergillus fumigatus*
4.	Neuritis	*Mucor pusillus*
5.	Onychomycosis	*Trichophyton purpureum*
6.	Candidiasis	*Candida albicans*
7.	Histoplasmosis	*Histoplasma capsulatum*
8.	Geotrichosis	*Geotrichum candidum*
9.	Chromomycosis	*Cladosporium immitis*
10.	Allergy	Spores of *Aspergillus, Chaeotomium etc.*
11.	Dermatomycosis	*Trichoderma viride*

TABLE III

List of some animal diseases caused by fungi.

Name of the disease	Pathogen
Penicillosis	Spp. of *Penicillium*
Aspergillosis	Spp. of *Aspergillus*
Athelete foot	*Tinea rubrum*
Ringwarm	*Trichophytom, Microsporum*
Mucomycosis	*Mucor, Rhizopus*

In addition to causing diseases in plants, human beings and domestic animals as described above, the fungi also play the following harmful roles :

(*a*) **Destruction of timber.** Several fungi such as *Polyporus, Serpula lacrymans, Fusarium negundi, Coniophora cerebella, Lentinus lapidens* and *Penicillium divaricatum* cause destruction of valuable timbers by reducing the mechanical strength of the wood.

(*b*) **Destruction of textiles.** Several fungi are able to grow on cotton and woolen textiles causing their destruction. These include spp. of *Alternaria, Penicillum, Aspergillus, Mucor* and *Fusarium.* Spp. of **Stachybotrys** causes destruction of cotton in storage. *Chaetomium globosum* is reported to cause greatest damage to textiles.

Timber Destruction

Stachybotrys Damaging Paper

(*c*) **Destruction of Paper.** Paper pulp wood is destroyed by the growth of *Polyporus adustus, Polystictus hirsutus* etc. several fungi such as species of *Chaetomium, Aspergillus, Stachybotrys, Alternaria, Fusarium, Dematium, Mucor, Cladosporium* etc. cause extensive damage to paper of books, newspapers and paper industry.

(*B*) *Positive Role.* Some soil fungi are beneficial to agriculture because they maintain the fertility of the soil. Some saprophytic fungi particularly in acid soils where bacterial activity is at

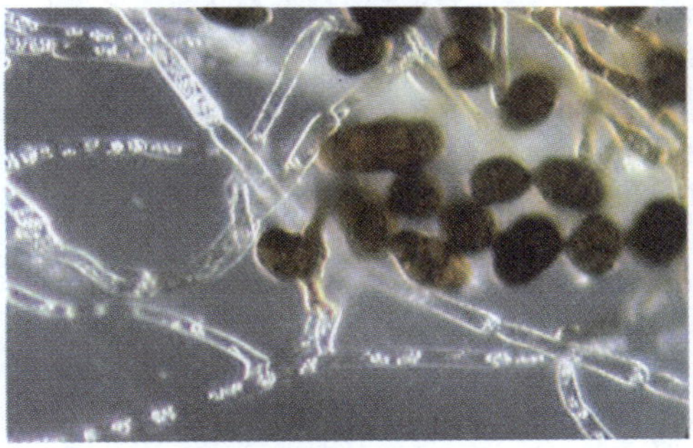

Soil Fungi

Soil mushrooms enhancing soil fertility

its minimum cause decay and decomposition of dead bodies of plants and their wastes taking up the complex organic compounds (cellulose and lignin) by screting enzymes. The enzymes convert the fatty, carbohydrate and nitrogenous constituents into simpler compounds such as carbon dioxide, water, ammonia, hydrogen sulphide, etc. Some of these return to the soil to form humus and the rest of the air from where they can again be used as raw material for food synthesis. There are fungi in the soil which produce more ammonia from proteins than the ammonifying bacteria.

Besides, many saprophytic fungi of decay maintain the never ending cycle of carbon dioxide which is a most important raw material for plant photosynthesis in nature. They also bring about rot, decay and decomposition of animal and plant remains releasing plant nutrients in a form available to green plants as food. The soil fungi utilize many inorganic salts. These are prevented from being lost from the soil by leaching. Some fungi form mycorrhizal association with the roots of certain plants and help them in their nutrition. Such plants will grow satisfactorily only when the mycelium of the appropriate fungal partner is present in the soil. The fertile soil contains twice as much living fungus cell material as the material from bacteria and other soil microorganisms.

Gibberellin spray accelerates plant growth

Giberrrelin—producued by *Gibberella fujikuroi* is used as growth hormone accelerating plant growth.

Many insect pests can be controlled by the growth of fungi such as *Empusa sepulchrasis, Metarrhizium anisopliae, Cordyceps melothac* etc.

Metarhizium - an entomopathogenic fungus

Some common fungal inhabitants of the soil help to combat diseases caused by soil borne fungi. *Trichoderma lignorum* and *Gliocladium fimbriatum* are found in damp soils. They have an inhibitory effect on the growth of the mycelium of *Pythium.* They serve to suppress fungi causing the damping off disease of the seedlings and thereby influence favourably the growth of crops.

Drechsler (1937) reported that there are some **predacious fungi** (Fig. 17.1 A-C) in the soil. They trap and destroy the nematodes (eel worms). Some species of these predatory soil fungi form loops on the mycelium (A). These loops act as nooses. They tighten and strangle the nematodes as they try to pass through (B). The mycelium later sends out special hyphae (C) to absorb nutrition from the captive. Some

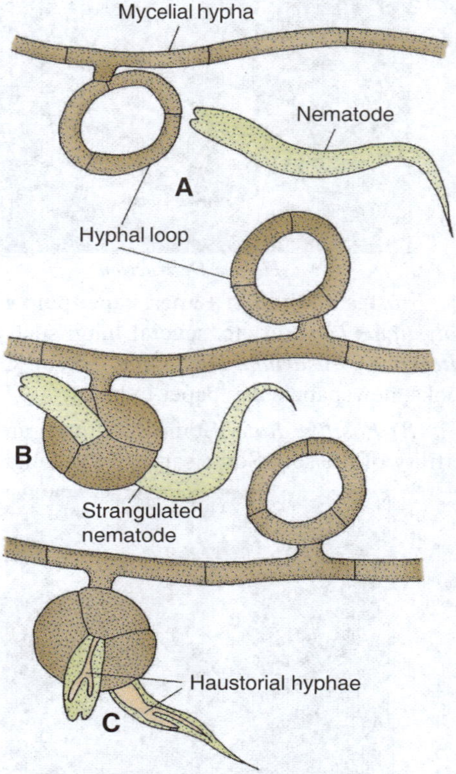

Fig. 17.1. *Predacious Fungi.* Explanation in the text.

predatory soil fungi produce conidia which are sticky. As the nematodes pass through the soil the stickly conidia stick to their bodies. There the conidia germinate to produce hyphae which penetrate into the tissues of the host and absorb nourishment.

At the National Botanical Research Institute, Lucknow and at several other national institutes, fungi are being tested as biopesticides especially as nematicides and as fungicides. An important fungus being used as nematicide is *Beauveria bassiana* against borers, thrips, and aphids. *Trichoderma viride* and *T. harzianum* are other examples which are used against a large number of soil - borne pathogens.

Role of fungi as food and as food producers. Many species of fungi are edible. about 2000 species of them have been reported from all over the world. Of these, about 200 are said to occur in the Western Himalayas. Many edible fungi are of great economic value as food. They are regarded as delicacies of the table. There are said to be over 200 species of edible fungi. The fructifications of some fungi such as the field mushroom *Agaricus campestris* (dhingri), *Podaxon podaxis* (Khumb), the honey coloured mushrooms, the fairy ring mushrooms, the puff balls (*Lycoperdon* and *Clavatia*), morels (*Morchella, guchhi*), and truffles are edible. The content of available food in them is not high but they supply vitamins and are valuable as appetisers. Yeasts and some filamentous fungi are valuable sources of vitamins of the B-complex.

Amanita

A few of the mushrooms are fatally poisonous, some cause only discomfort. To the former category belong *Amanita.*

The fungi are also important as producers of foodstuffs. Certain species of *Penicillium* are active in the refining of certain kind of cheeses. A mention has already been made in the useful fermenting activities of yeasts in bread making and in the preparation of alchoholic and other drinks. They are also a source of vitamins.

Some fungi, such as red bread mold, *Neurospora sitophila* and others, complete their sexual life cycle in a few days and thus make ideal organisms for the study of the laws of heredity. The slime molds (*Physarum polycephalum*) are now widely used in research. *P. polycephalum* has proved an excellent experimental organism for the study of DNA synthesis, meiotic cycle and the mechanism of protoplasmic streaming.

Penicillium digitatum on Bread

Aspergillus on Bread

Many fungi are responsible for spoilage of food stuffs. *Penicillium digitatum* causes rotting of *citrus fruits.* Milk and milk products are spoiled and made unfit for human use due to the growth of several fungi such as *Mucor, Aspergillus, Penicillium, Oidium* and *Fusarium. Mucor mucedo* and spp. of *Aspergillus* grow on bread and spoil it. *Oidium lactis* develops the fishy odour of butter causing damage to the butter.

In tropical conditions, many fungi such as *Mucor* sp., *Penicillium, Neurospora, Fusarium, Aspergillus* etc., grow on meat causing sufficient spoilage.

Aflatoxins—the most potent carcinogenic agent—are produced by *Aspergillus flavus, A. fumigatus, A. parasiticus* and *Penicillium islandicum* on dried foods and groundnut meal. Aflatoxins are reported to bind with DNA and prevent its transcription arresting protein synthesis. These are responsible for liver cancer in animals and human beings.

Mushroom toxins are produced by several poisnous mushrooms. These cause diarrhoea, vomiting, liver damage, complete unconsciousness etc. Mushroom toxins are commonly produced by *Amanita phalloides,* spp. of *Helvella* and some species of *Inocybe.*

Ergot toxins— produced by *Claviceps purpurea* contain poisonous alkaloids like ergotamine, ergometrimine, ergocrystinine, ergocrystinin and ergonovin. These cause diarrhoea, abdominal pain, vomiting and psychiatric disturbances.

Fungi As Test Organism—During the last two decades, the fungi has been used to test various biological processes. Since they grow very fast and require a short period to complete their life cycles, the fungi are best suited for use as test organisms. Fungi form very good research material for genetical studies and other biological processes Genus *Neurospora* has become very good material for genetic studies while *Physarum polycephalum* is used to study steps in DNA synthesis, morphogenesis and *mitotic cycle*.

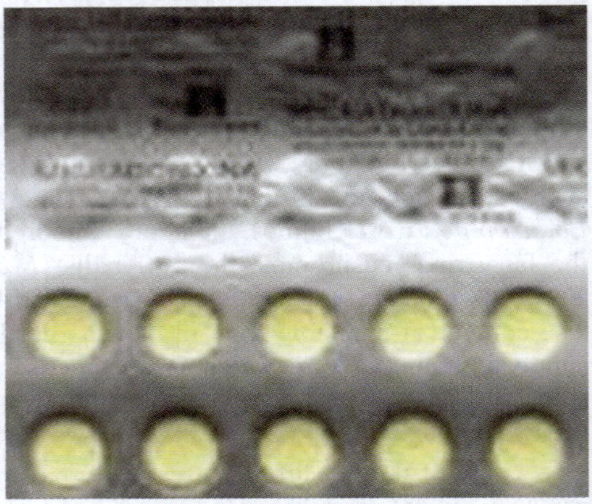

Ergotamine - a poisonous alkaloid produced by Claviceps purpurea

To detect the presence and quantity of vitamin B in given sample, *Neurospora crassa* is commonly used. Similarly *Aspergillus niger* is used for detection of trace elements like zinc, nickel and copper even when they are present in very minute quantities.

ECOLOGICAL GROUPS OF FUNGI

In ecological terms, the main types of fungi are saprotrophs and pathogens and mutualists.

Saprotrophic fungi (also known as saprobes, saprophytes) obtain their energy and nutrients by breaking down dead organic matter such as in soil, litter, dung, and wood (*Rhizopus*, *Mucor Polyporus*, *Morchella* etc) Pathogenic fungi derive their energy and nutrients from living organisms. These fungi invade and commonly kill trees, insects and other

Symbionts or mutualists undertake a specialised mode of life — fungi that form mutually beneficial associations with other organisms. Both partners benefit in some way, and each on its own maybe unable or less likely to survive (pp. Lichens & Mycorrhiza).

Garett (1951) has described ecological group of soil fungi which may be defined as assemblage of species characterized by some peculiar advantage for pioneer clonization of a particular substrate. Substrate of soil fungi may be consists of living or dead organic matter laying on the soil fungi may be consists of living or dead organic matter laying on the soil. Garrett (1951) has described ecological groups of soil fungi as **Saprophytic sugar fungi, Root inhabiting fungi, Lignin decomposing fungi.** The term sugar fungi was first used by Burges, (1939) for those fungi characterized by its capacity to utilizing readily available sugars, and possibly hemicelluloses, but not the cellulose or lignin. Examples of saprophytic sugar fungi are the Phycomycetes. *Fusarium spp. Pythium spp., Phymatotricum omnivorum* etc are some of the root inhibiting fungi. This ecological group is highly spcialized root parasites, root inhibiting fungi has the capacity for indefinite mycelia spread along the living roots of the host plant may be as ectotrophs. Lignin decomposing fungi belong to mostly higher basidiomycetes. Wood decaying fungi may be classified as

White rot fungi–fungi that can decompose lignin

Brown rot fungi–fungi that can decompose only cellulose

Soft rot fungi–fungi that can decompose only cellulose, but they attack damper wood and decay usually occurs near the wood surface.

FOSSIL FUNGI

Fossil Fungi

Fungi being microsopic, are often difficult or impossible to identify in the fossilized state.. The earliest record of fossil fungi is from chytridiomycetes from Vendian strata (Late precambian period 650 to 544 million years ago) from north Russia. Fungal fossils do not become common and uncontroversial until the early Devonian, when they were abundant in the Rhynie chert beds in Scotland which host to a plethora of fossil fungi. Some of well recognized examples are endotrophic mycorrhizae from the stem of *Aglaophyton major* named *Glomites rhyniensis*. Such a association in Fossil fungi study has shown that fungi and land plants were forming symbiotic relationships even at that very early stage in terrestrial evolution. Well preserved perithecium, an enclosed reproductive structure, possibly the species of *Savoryella* have been reported from decayed stems, rhizomes and microphylls of *Asteroxylon mackiei*. The perithecia are of about 400 μm in size and often to have devlop in the stomatal chamber of the plant. Aquatic chytridsare the most common forms of fossil fungi reported from the Rhynie chert beds. *Aglaophyton major* stem also harbor *Palaeoblastocladia milleri*- a blastocladean fossil fungus. The reconstructed form showed gametothalli with gametangia and sporothalli bearing terminal zoosporangia and pitted thick walled resting sporangia. The fungus filament is dichotomously branched (Remy et. al., 1994). *Milleromyces rhyniensis* and *Lyonomyces pyriformis* are the other aquatic chytirds from the same area on a charophyte Palaeontiella cranii (Taylor et. al., 1992). In fact, all four major groups of modern fungi have now been found in Devonian strata, showing that the fungi had successfully invaded the land and begun to diversify before the first vertebrates crawled out of the sea.

PIGMENTATION IN FUNGI

Many fungi produce pigments during their growth which are substantive as indicated by permanent staining that is often associated with mild growth on textile and plastics. Fungal pigments are typically composed of many complex chemical substances that are formed during their metabolic processes. These pigments may be encrusted in spores, present in mycelium or secreted to a substrate such as on paper and leather. Some fungal pigments have been shown to be anthraquinone derivatives, resembling the important group of Vat dyes. Little is known that mushrooms were commonly used textile dye before the invention of synthetic dyes because of,the vast range of vivid and rich colours achievable,(Mussak & Bechtold 2009). The chromophores of mushroom dyes contain a variety' of fascinating organic compounds. Their pigmentation may vary with the age and some undergo distinctive colour changes on bruising; therefore, the colours *capitalensis* (a foliar endophyte) produced black pigment Melanin. Melanin help to protect the protoplast against damaging effects of UV radiations. Perhaps melanin in the fungal hyphae of *P. capitalensis* may be responsible for the successful life cycle as cosmopolitan endophyte. The striking reddish pink colour of phylloplane yeast *Sporobolomyces roseus* is due to three major caroteniods β-carotene, torulene and torularhodin. Some betalain like pigment (red) are known from fungi like Muscaaurin ll, Muscaaurin I and muscaflavin are example of fungal pigments, *Protomyces pachyderma and P.inounjii contain* β-g carotene and lycopene with β-carotene predominating.

QUESTIONS

ESSAY TYPE

1. Write an essay on the economic importance of fungi. *(Poorvanchal, 2001, 2004, 2006; Bundelkhand, 1996, 1997)*
2. Discuss the role played by fungi in agriculture, industry and medicine.
3. Write an essay on the "Fungi as friends and as foes of human beings".
4. In what ways are fungi important economically to man?
5. Describe the economic importance of fungi. *(Agra, 1996, Kanpur, 1999)*
6. (a) Describe in brief the significance of fungi industry. *(Kanpur M.Sc., 2004)*
   (b) Describe the importance of fungi in human life . *(Poorvanchal, 2005)*

SHORT ANSWER TYPE

7. In what ways are fungi economically useful to man? *(Gorakhpur, 1994)*
8. Describe the role of fungi in agriculture. *(Himachal Pradesh, 1992)*
9. Discuss the role of fungi in industry. *(Himachal Pradesh, 1993, 1996)*
10. Describe the role of fungi in medicine and soil fertility.*(Himachal Pradesh, 1995, 1997)*
11. Give the beneficial effects of fungi. *(Gurunanak Dev, 1991)*
12. (*a*) What are the harmful effects of fungi? Give a brief account.
    (*b*) Name an edible mushroom. *(Poorvanchal, 2002)*
13. Write short notes on "Economic importance of fungi". *(Awadh, 1991, 1993, 1994, 1995, 1996, 1997; Kanpur, 1996; Gorakhpur, 1992, 1996)*
14. Name the microorganisms used in the following :
    (*a*) Baking Industry
    (*b*) Citric acid fermentation
    (*c*) Antibiotics *(Kanpur, 1994)*
15. Write notes on the following :
    (*i*) Importance of fungi in medicine industry or role of fungi in medicine. *(Kanpur, M.Sc., 2006; Poorvanchal, 2002)*
    (*ii*) Role of fungi in biological control. *(Allahabad, 1993)*
    (*iii*) Useful activities of fungi *(Kumaon, 1999)*
    (*iv*) Economic importance of fungi. *(Lucknow, 2004; Meerut, 1998)*
    (*v*) *Fungicides* *(Kanpur, 2003)*
    (*vi*) Edible fungi. *(Rohilkhand, 2002)*
    (*vii*) Fermentation *(Lucknow, 2005)*
16. Describe the role of microorganisms in bakery industry/brewery industry. *(Poorvanchal, 2003)*
17. (a) Write giving examples, use of fungi as test organisms. *(Kanpur M.Sc., 2002, 2004, 2006)*
    (*b*) Who discovered penicillin?
    (*c*) Name any biofungicide.

OBJECTIVE TYPE

18. Select the Correct Answer :

(*i*) The wonder drug Penicillin was discovered by Alexander Fleming from
(*a*) an alga (*b*) a fungus
(*c*) a bacterium
(*d*) an actinomycete (actinomycete)

(*ii*) Rust of wheat and smut diseases are caused by
(*a*) bacteria (*b*) fungi
(*c*) viruses (*d*) protozoa

(*iii*) Taka diastase is isolated and prepared from
(*a*) *Penicillium notatum*
(*b*) *Aspergillus oryzae*
(*c*) *Saccharomyces cerevisiae*
(*d*) *Puccinia graminis*

(*iv*) Which of the following is used in bakery and brewery industries?
(*a*) *Penicillium notatum*
(*b*) *Aspergillus oryzae*
(*c*) *Saccharomyces cerevisiae*
(*d*) *Puccinia graminis*

(*v*) *Claviceps purpurea* produces an alkaloid which is used during
(*a*) child birth (*b*) fever
(*c*) menstruation (*d*) old age

(*vi*) Dermatomycosis is a disease of
(*a*) Heart (*b*) Liver
(*c*) Kidney (*d*) Skin

(*vii*) Gibberellin—a growth hormone is isolated from
(*a*) *Gibberella fujikuroi*
(*b*) *Penicillium notatum*
(*c*) *Aspergillus niger*
(*d*) *Puccinia graminis*

(*viii*) Which of the following are used as a food?
(*a*) *Agaricus* and *Morchella*
(*b*) *Aspergillus* and *Penicillium*
(*c*) *Mucor* and *Rhizopus*
(*d*) *Puccinia* and *Ustilago*

(*ix*) Which of the following is used in genetical studies?
(*a*) *Neurospora sitophila*
(*b*) *Puccinia graminis*
(*c*) *Mucor mucedo*
(*d*) *Alternaria solani*

(*x*) Aflatoxins—the most potent Carcinogenic agent was isolated from
(*a*) *Penicillium notatum*
(*b*) *Aspergillus flavus*
(*c*) *Mucor mucedo*
(*d*) *Alternaria solani*

(*xi*) Fungi growing on Dung are called
(*a*) coprophilous (*b*) lichens
(*c*) bacteriophage (*d*) predacious

HETEROTHALLISM

The term **Heterothallism** was first used by an American geneticist A.F. Blakeslee in 1904 when he observed that zygospores could develop in some spp. only when two mycelia of different strains were allowed to come in contact with each other. Blakeslee made these observations as a result of his studies on zygospore formation in Mucorales. He found that in some species of *Rhizopus,* zygospores were formed freely while in others like *R. nigricans,* zygospores were formed rarely. In *Mucor hiemalis,* too zygospores were formed rarely. On the basis of his studies, he divided the various species of Mucorales into two groups: Heterothallic and Homothallic. Thus, heterothallic species are those which require mycelia of two different strains to interact to enable the zygospores to be formed while the homothallic species are those which require mycelia of only one strain to interact for the formation of Zygospores. According to Blakeslee (1904) Heterothallic condition is "essentially similar to that in dioecious plants and animals and although in this case the two complimentary individuals which are needed for sexual reproduction are in general not so conspicously differentiated morphologically as in higher forms, such a morphological difference is often distinctly visible." He concluded that the Zygospore formation is a sexual process. In homothallic species, the mycelium is bisexual while the mycelium in heterothallic species is unisexual, (+) and (–) strains represent the two different sexes.

Heterothallism may therefore be defined as the condition in which Zygospore formation takes place only when mycelia arising from **asexual** spores of two genetically different mating types, (+) and (–), are allowed to interact. On the other hand the condition, in which one individual orginating from a single asexual spore is capable of forming zygospores independently, is known as **Homothallism.**

Blakeslee proved the phenomenon of heterothallism on the basis of the experiments he conducted using several species of Mucorales. He inoculated spores of two different strains of *Mucor hiemalis* on a petridish containing synthetic agar medium. After a few days it was observed that the zygospores were formed along the zone of contact of two mycelia. In another experiment he inoculated spores of only one strain on a petridish containing synthetic agar medium and after some time it was observed that absolutely no zygospores were produced in this experiment. From these experiments, he concluded that zygospores could be formed only when mycelia of two different strains were allowed to come in contact.

Blakeslee and his coworkers (1928) examined different genera of Mucorales to test whether these were homothallic or heterothallic. Most of the genera tested were homothallic while only a few genera were found to be heterothallic. In heterothallic species of *Mucor mucedo,* the zygospores upon germination produced germsporangia which contain spores of only one strain (either + or –). In this case, zygospores could be formed only when mycelia formed from spores of (–) strain formed in germsporangia produced on germination of zygospores are allowed to come in contact with the mycelia produced from the spores of (+) strain. But in the heterothallic species of *Phycomyces nitens*, spores of (+) and (–) strains are produced in the same germsporangium. Blakeslee thus regarded the (+) and (–) strains of heterothallic species as differing in sex and the term heterothallism may therefore be treated as equivalent to dioecism in haploid organisms.

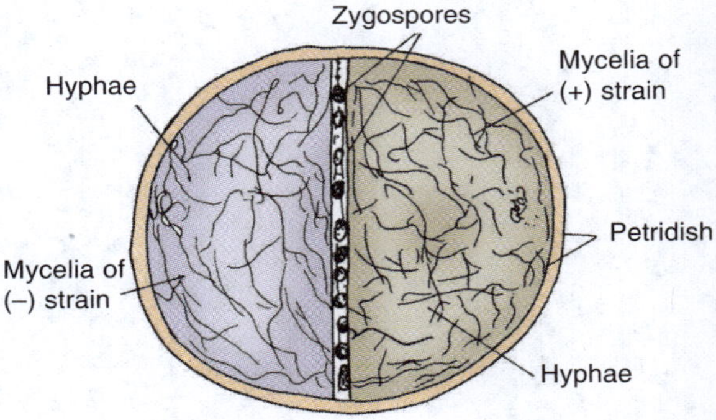

Fig. 17.2 *Mucor hiemalis.* Zygospore formation indicating Heterothallism.

Since the discovery of heterothallism in Mucorales by Blakeslee, the phenomenon has been reported in several groups of fungi. Though variations may occur, but all heterothallic species share one common feature of intermycelial contact. While in *Dictyuchus monosporous* it is dioecism, in *Ascobolus magnificus,* it is expressed as self-sterility or self incompatibility. Both the fungi resemble the heterothallic Mucorales in that sexorgans or gametangia are formed only when opposite strains come in contact. In *Dictyuchus monosporous* the two sex-organs are formed by different strains but in *Ascobolus magnificus,* each strain produces sexorgans but these are self-sterile. On the basis of these observations, the heterothallism may be of two types : **Morphological heterothallism** and **Physiological heterothallism.**

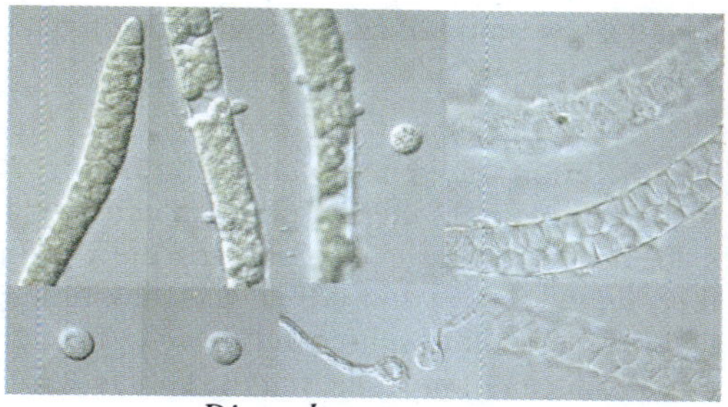

Dictyuchus monosporus

Morphological heterothallism. Morphological heterothallism may be defined as the condition when morphologically different male and female sex crgans are produced in two closely associated mycelia. The two sex organs or gametes are so morphologically different that it is easier to term one of them as male and the other as female-examples of such type of morphological heterothallic fungi are : *Achlya ambisexualis, A. bisexualis, Blastocladiella variabilis, Dictyuchus monosporus, Phytophthora palmivora* and *Peronospora parasitica* (photographs given below). However, in *Blastocladiella variabilis* the male and female gametangia are morphologically distinct, the male being smaller than the female.

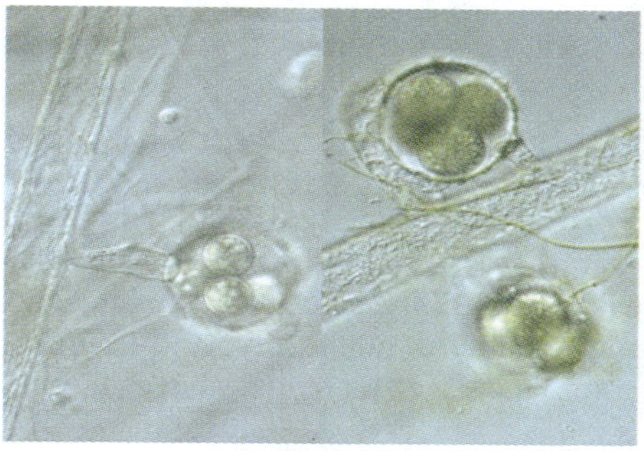

Achlya ambisexualis

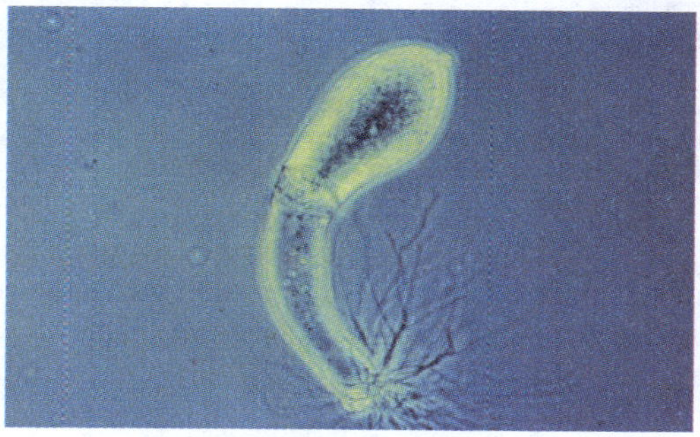

Blastocladiella variabilis

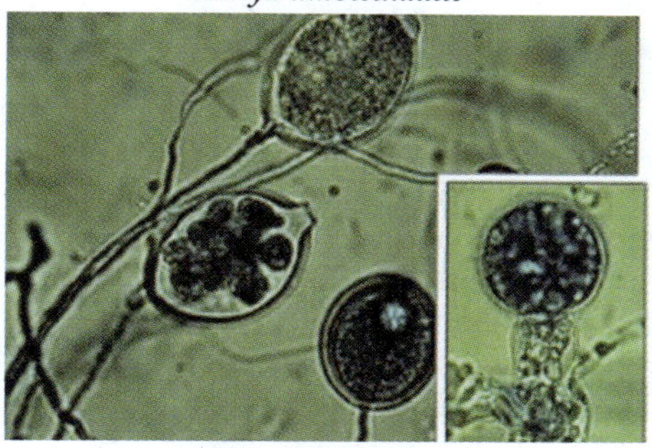

Phytophthora palmivora

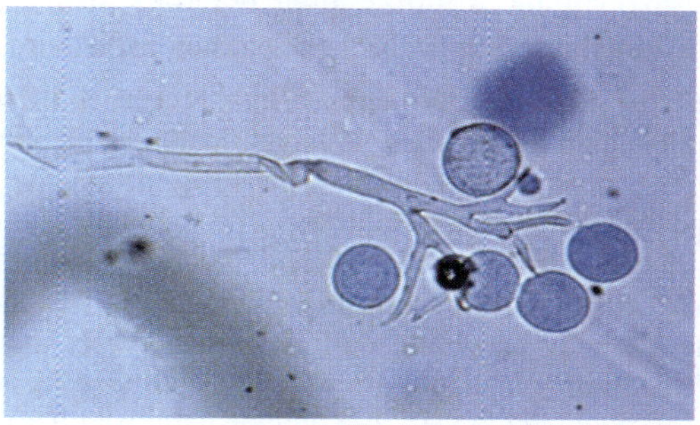

Phytophthora parasitica

Whitehouse (1949) also used the term **haplodioecious** for morphologically heterothallic species of fungi.

Physiological Heterothallism. In physiological heterothallism, the interacting thalli differ in mating type or incompatibility, irrespective of the presence or absence of the sex organs or gametes. This means that sexual reproduction takes place by two morphologically similar but physiologically different hyphae in physiological heterothallism. The gametangia as well as gametes do not show morphological differentiation but physiologically they behave differently. Physiological heterothallism may be of two types :

***(i)* Two Allelomorphs or Two-Allele Heterothallism.** When nuclei of both the mating types are different in genetic characters, this type of Heterothallism is known as **Two-Allele heterothallism.** In these types compatibility is governed by a pair of Alleles represented by A and a located at single same locus of the chromosome. Due to the dominauce of **A** over **a**, **A** is represented by (+) and **a** by (–). At the time of meiosis, separation of the chromatids take place. Half of the haploid spores thus have (+) and the other half (–) allele. The spores bearing (+) allele will produce (+) mycelia and the spores with (–) allele will give rise to (–) mycelia. The mycelia of (+) and (+) and (–) and (–) are self-sterile or self-incompatible. Thus, two complimentary mating types (+) and (–) are essential for sexual reproduction.

Two-Allele heterothallism has been reported in several fungi of like *Ascobolus magnificus, Puccinia graminis, Neurospora sitophila, Mucor mucedo, Ustilago kolleri* etc. (photographs given below)

Puccinia graminis

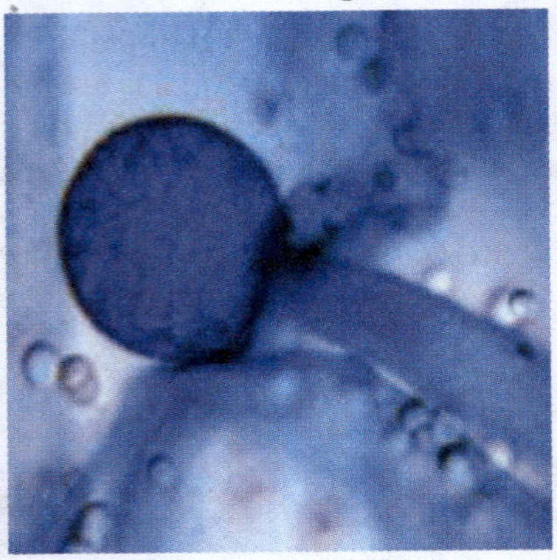

Neurospora sitophila

Mucor mucedo

***(ii)* Multiple Allelomorph or Multiple Allele Heterothallism.** In this type of heterothallism, more than two (multiple) alleles determine the sexual compatibility. These may be located at one (bipolar) or two (tetrapolar) loci. Because of the larger number of alleles involved in this type of heterothallism, chances of mating of compatible strains increase. As stated above, the multiple allele heterothallism may be of two types :

(*a*) Bipolar Multiple-allele heterothallism

(*b*) Tetrapolar multiple-allele heterothallism.

(*a*) Bipolar Multiple-Allele Heterothallism. This type of heterothallism is controlled by multiple alleles at a single locus, instead of a pair of Alleles. For example, if the locus is named as L, the multiples alleles will be designated as L_1, L_2, L_3, L_4—Ln and these are present on the single locus L. The meiotic division will give rise to thalli which may be of several mating types, generally equal to the number of alleles. The thallus containing the allele L_1 can mate with a thallus of any mating type except L_1. Similarly L_2 can mate with any thallus except that containing L_2 allele and so on. In this type of heterothallism, incompatibility factors are more commonly involved. Bipolar multiple allele heterothallism is characteristic of Basidiomycetes except rusts and smuts (Fig. 17.3)

	L_1	L_2	L_3	L_4
L_1	N	C	C	C
L_2	C	N	C	C
L_3	C	C	N	C
L_4	C	C	C	N

Fig. 17.3. Bipolar multiple allele heterothallism. C-compatible, N-non compatible

(*b*) Tetrapolar Multiple Allele heterothallism. This type of heterothallism is characteristic of Basidiomycetes except rusts. In this type of heterothallism, which is very similar to bipolar multiple allele heterothallism, compatibility is determined by **two loci.** Multiple allele—the compatible factor is present on two loci L_1 and L_2 of two chromatids of a chromosome. At the time of meiotic division, both the loci are separated with chromatids. It is estimated that at least 100 alleles are present on each locus. In *Schizophyllum commune,* 122, alleles of factor A and 61 of B have been identified in the laboratory. According to rough estimates, the number of alleles may be even more, about 350-450 of L_1, and 65 of L_2. Any two mating types, which differ in allele present on L_1 and L_2 are compatible. If the allele composition of mating type is A_1 B_1, it would be compatible with any other type of allele composition except A_1 B_1. But the mating type with allele composition is not fully compatible with allele composition A_2 B_2 or A_2 B_2. Figure 17.4 fully explains type of heterothallism which has been reported in *Ustilago maydis* and *Comprinus firmaterius.*

	A_1B_1	A_1B_2	A_2B_1	A_2B_2
A_1B_1	N	NFC	NFC	C
A_1B_2	NFC	N	C	NFC
A_2B_2	NFC	C	N	NFC
A_2B_2	C	NFC	NFC	N

Fig. 17.4. Tetrapolar multiple allele heterothallism. N-non compatible, NFC—Not fully Compatible, C-Compatible.

This type of heterothallism encourages out-breeding. Whereas in bipolar multiple Allele heterothallism, the out-breeding is 25%, in tetrapolar, it is 100%. This may be due to enormous increase in the number of possible mating types of thalli.

According to Garrett (1963), "heterothallism promotes the out-breeding and therefore subserves the same end as the sexual process, which it renders most efficient. Hetrothallism is not the same as sex, it is refinement super imposed upon it."

FLAT AND BARRAGE REACTION

If the heterokaryons formed have common A ralleles and dissimilar B alleles, the mating patterns is termed as **flat reaction.** Such heterokaryons show poor hyphal growth having curly and irregular branching patterns. Presence of common A alleles prevents formation of clamp connections, disrupt nuclear distribution pattern causing morphological and metabolic abnormalities. Similarly, if the heterokaryons have common B alleles and dissimilar A alleles, the mating patterns is termed as **barrage reaction.** The hyphal growth of such heterokaryons ceases at the point of contact between the colonies of two thalli thus forming a zone of aversion (no growth) or barrage. It is reported that common B factors are responsible for such behaviour that prevents karyogamy and meiosis.

GENETIC DETERMINANTS OF HETEROTHALLISM

The two factors responsible for the phenomenon of heterothallism are sex factor and incompatibility factor. In absence of sex factor sex organs or gametes are not formed on the thallus and the sexuality (maleness/femaleness) is determined by the direction of nuclear migration during plasmogamy. The recipient of nucleus is considered as female and the donor as male. The incompatible factor determines the mating capacity in presence or absence of sex factor. In most of the higher ascomycetes and basidiomycetes sex factors are absent and thus incompatibility factor alone determines the sexual fusions. Segregation of these factors during meiosis determine the sexual status and mating behavior. Three types of segregation is possible

1. segregation of sex factor
2. segregation of incompatible factor
3. no segregation of any factor

The spores of first two types containing meiotic nuclei germinate to give rise to self sterile heterothallic thalli whereas the spores of third type gave rise to homothallic thalli.

HOMOTHALLISM

Homothallism is the phenomenon in those fungi which are self-fertile and self-compatible. Such fungi are known as **Homothallic fungi.** The thallus in these fungi is of one and **single strain**, thus two mycelia developing from a single spore are capable of interacting and forming diploid structures after syngamy. These fungi are of two types :

(*i*) Primary Homothallic fungi are those in which spores are uninucleate and the nucleus has only one genotype. Examples of such fungi are *Allomyces javanicus, Pyronema omphalodes* and *Puccinia malvacearum* besides many members of the order **mucorales.**

(*ii*) Secondary Homothallic fungi. In these fungi, nuclei of compatible mating types are present in each spore. In other words, the spores of these fungi are heterothallic and therefore sexual reproduction can take place in mycelia produced from a single spore under laboratory conditions in artificial cultures. *Neurospora tetrasperma* and *Coprinus ephimerus* are examples of secondary Homothallic fungi.

However, there are certain fungi which are neither **Homothallic** nor **Heterothallic.** Several mycologists like Korf (1952), Hartman (1956), Burnett (1956) and Esser (1963) have strongly advocated to abandon the use of these terms. Esser (1963) urged to use the terms **monoecious** and **dioecious** for the fungi as well, According to him, monoecious fungi are those in which an individual can act as **donor** and **recipient** of nucleus. While dioecious fungi are those in which individuals can either act as **donor** or **recipient** only. In his opinion, monoecious fungi may possess both the sex organs on the same thallus or may have none but are capable of receiving or donating nuclei. These may be self-fertile like *Glomerella* or self-sterile like *Neurospora.* The dioecious fungi may be morphologically and physiologically dioecious. The morphologically dioecious fungi may bear male and female sex organs on separate thalli while physiologically dioecious species either bear sex organs which are morphologically dissimilar or the sex organs are absent. In spite of this, the terms **Homothallism** and **Heterothallism** are **widely used even now.**

QUESTIONS

ESSAY TYPE

1. Write a short essay on Heterothallism.
2. Give a brief account of the modern concept of Heterothallism.
3. (a) Who was Blakeslee? Describe his contributions.

   (b) Write an essay on compatibility of fungi with special reference to heterothallism. *(Rohilkhand M.Sc., 2006)*
4. What is Heterothallism ? Write the name of any two fungi in which it occurs. Describe in brief. *(Kanpur, 2006)*

SHORT ANSWER TYPE

5. What is Heterothallism? Explain the phenomenon in fungi. *(Bundelkhand, 1994)*
6. Define Homothallism. *(Poorvanchal, 2001; Gorakhpur, 1998)*
7. Explain Heterothallism. *(Kanpur, 1995)*
8. Write short noes on :

   (*i*) Heterothallism *(Kanpur, 1996, 1997, 2005; Awadh, 1991, 1992, 1994, 1995, 1996, 1997, 1998; Lucknow, 1999; Allahabad, 1996, 2000; Himachal Pradesh, 1990; Rohilkhand, 2003; Allahabad M.Sc. 2000)*

   (*ii*) Heterothallism in fungi *(Kanpur, 1999; Agra, 1994, 1996; Gorakhpur, 1994, 1999)*

   (*iii*) Heterothallism in *Mucor* *(Lucknow, 1994)*

(*iv*) Heterothallism and sex in fungi (*Rohilkhand, M.Sc., 1997*)

9. Differentiate between Homothallism and Heterothallism. (*Awadh, 1999; Gorakhpur, 1993*)

10. Explain the difference between Heteroecism and Heterothallism. (*Gorakhpur, 1991; Bundelkhand, 1995*)

11. Define heterothallism. Give example. (*Kanpur M.Sc., 2001*)

12. Define the phenomenon of heterothallism in fungi with the help of examples studied by you. (*Kanpur M.Sc., 2003*)

13. Write an essay on compatibility in fungi with special reference to heterothallism. (*Rohilkhand M.Sc., 2006*)

OBJECTIVE TYPE

14. Fill in the blanks.

(*i*) The term heterothallism was first used by ....

(*ii*) Heterothallism may be defined as the condition when zygospore formation takes only in the presence of .... strains.

(*iii*) When zygospore formation takes place from one individual arising from a single asexual spore, the condition is known as ....

(*iv*) Heterothallism was discovered in the fungi belonging to the order ....

(*v*) When morphologically different sex organs take part in sexual reproduction the condition is known as .... Heterothallism.

(*vi*) When the sex organs are morphologically identical but physiologically different, this type of heterothallism is known as ....

(*vii*) The example of heterothallic fungus is ....

(*viii*) For morphologically heterothallic species, the term .... is also used.

(*ix*) Homothallic species may also be ....

15. Select the correct answer.

(*i*) Blakeslee is associated with
(*a*) homothallism
(*b*) white rust
(*c*) heterothallism
(*d*) loose smut of wheat

(*ii*) Who is credited with the discovery of process of heterothallism ?
(a) B.B. Mundulus
(*b*) Alexander Fleming
(*c*) E.J. Butler (*d*) A.F. Blakeslee

(*iii*) Heterothallism in fungi was discovered by
(*a*) Alexopolus (*b*) Anton deBary
(*c*) A.F. Blakeslee
(*d*) Alexander Fleming

SEX-HORMONES

It was de Bary (1881) who first reported the presence of sex hormones in dioecious species of *Achlya*. In 1908, Kaufmann supported the contention of de Bary on the basis of his observations on *Saprolegnia hypogyna*. According to Kaufmann, the development of antheridial hyphae is stimulated by certain hormones. Studies conducted by Couch (1924) and Bishop (1957) on *Dictyuchus* and *Sapromyces reinschii* and Raper (1954) on *Achlya* have substantially confirmed the role of hormones in sexual reproduction in fungi. Therefore the sex hormones may be defined as a diffusible substance playing a specific role in the sexual reproduction that produces it. Machlis (1972) has classified the sex hormones into the following three types :

(1) Erotactins

(2) Erotropins

(3) Erogens

Erotactin is a sexual hormone which attracts motile gametes, **erotropin** is the sex hormone that induces chemotropic growth of sexual structures and **erogen** is the sex hormone that controls the induction and differentiation of sexual structures. However, the sex hormone may perform more than one functions, as in *Achlya*, the hormone that controls development of sexual structures also determines the direction of the sex organs.

Description of some important Sex-Hormones

Although a large number of sex hormones have been reported, only a few have been chemically characterised and extensively studied. These are : **Sirenin, antheridiol and oogoniol, trisporic acid** and **yeast D factor.** These are briefly described here.

SIRENIN. It is a sperm attracting hormone produced by water mould *Allomyces, A. macrogynous* and *A. arbuscula*. It is the female gametes which release sirenin to attract the male gametes. The synthesis of sirenin by female gametes and their function to attract male gametes was demonstrated by Machlis (1958). The empirical formula of sirenin is $C_{15}H_{24}O_2$ and the molecular weight is 236. It is an oxygenated sesquiterpene and is species specific. Although both D– and L– forms have been isolated, L– form is biologically effective.

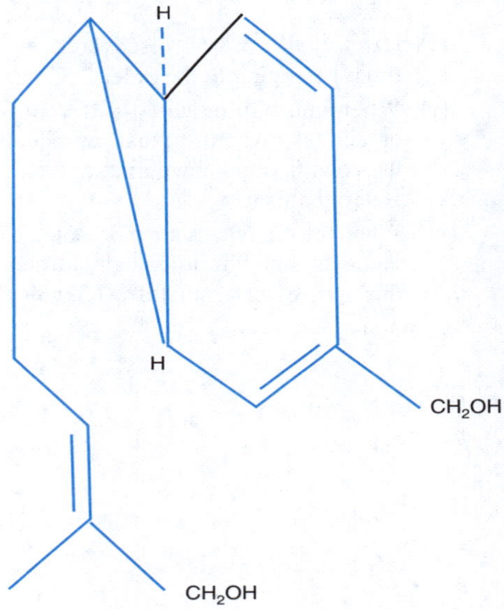

Fig. 17.5. Chemical formula of Sirenin and is species specific. Although both D- and L- forms have been isolated, only L- form is biologically effective.

ANTHERIDIOL AND OOGONIOL

Raper (1939-1959), on the basis of his classical studies on sex hormones in *Achlya*, demonstrated the presence of multi hormonal regulartory system in *A. ambisexualis* and *A. bisexualis*. He envisaged the presence of four hormones A, B, C and D but now it has been conclusively proved that only two hormones named as **Antheridiol** and **oogoniol** are involved. Hormones A and C are **antheridiols** and hormones B and D are **oogoniols.**

Antheridiol is a crystalline compound having emperical formula as $C_{29}H_{42}O_5$ and molecular weight as 470. The hormone was isolated by Mc Morris and Barksdali (1967) and its structure was proposed by Arsenault *et al.* (1969).

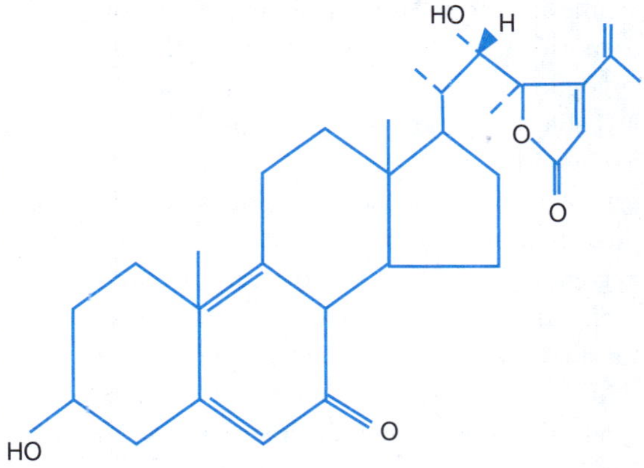

Fig. 17.6 Antheridiol

Antheridiol is reported to stimulate four types of reactions : (*a*) initiation of antheridial hyphae on male plant, (*b*) chemotropic stimulation of antheridial hyphae, (*c*) stimulation of male hyphae for production of oogoniol and (*d*) delimiation of antheridia.

Oogoniol. The hormone is synthesized by male hyphae of *Achlya ambisexualis* only in the presence of antheridiol. However, Barksdali *et al.* (1974) reported that oogoniol is synthesised by some hermaphrodite strains without the stimulus of antheridiol. Two crystalline compounds, possessing hormone B activity have been isolated from culture filtrates of *Achlya heterosexualis* by Mc Morris and his coworkers (1975). These have been named oogoniol-1 and oogoniol-2. Oogoniol is a crystalline steroid with molecular weight as 500. The hormone stimulates the development of oogonium on female hyphae.

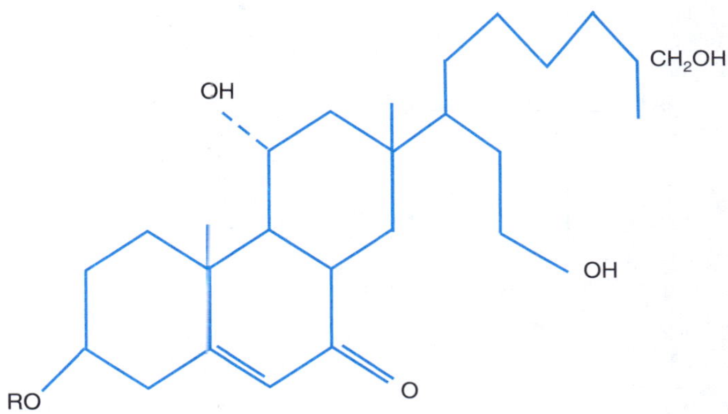

Fig. 17.7. Oogoniol

TRISPORIC ACID. Trisporic Acid is a sex hormone which has been isolated from *Blakeslea trispora* and *Mucor mucedo.* It has been found to play active role in sexual reproduction of several members of the order Mucorales. Though the concept of sex hormone in mucorales was developed in 1924 by Burgeff, it was only in 1956 that Burgeff could isolate a sex hormone in cell free extract and this sex hormone was named as Trisporic Acid. It is an unsaturated and oxygenated form of trimethyl cyclo-hexane. Three kinds of trisporic acid have been identified, trisporic acid A, B and C. Trisporic acid C plays the major role (80%) as a sexhormone, followed with trisporic acid B with 15% activity and trisporic acid A is least active with 1-2% activity. It appears that trisporic acid A does not have the functional group in the acid chain.

Empirical formula of trisporic acid is $C_{18}H_{26}O_4$ and the molecular weight is 306.

The role of sex hormones in *Mucor mucedo* was studied by Plempel *et al.* (1957-61). The sex

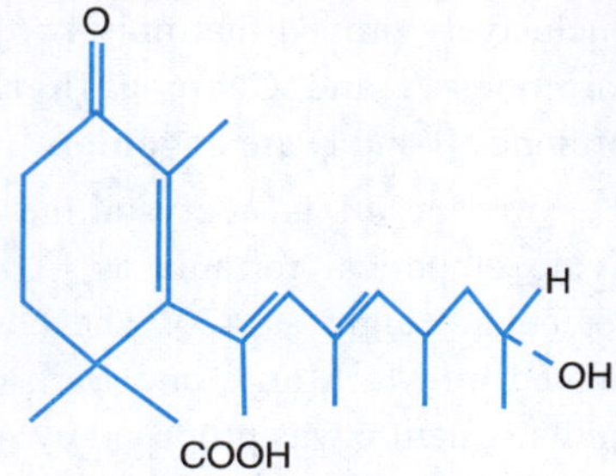

Fig. 17.8. Trisporic Acid

hormone trisporic acid, present in this species is synthesized from B-carotene. In the heterothallic mycelia, trisporic acid B and C stimulate the development of zygophores. This hormone is synthesized only when the mycelia of (+) and (–) strains grow in a normal continuous diffusable medium. The hormone trisporic acid synthesized in (–) strain induces the development of progametangium in (–) strains and vice-versa. The zygophores of (+) and (–) strain are attracted towards each other due to chemotropic response of a volatile substance which also promotes the synthesis of trisporic acid in two opposite mating types. Fig. 17.9 explains the role of trisporic acid and its collaborative synthesis by the compatible mating types of *Mucor mucedo*.

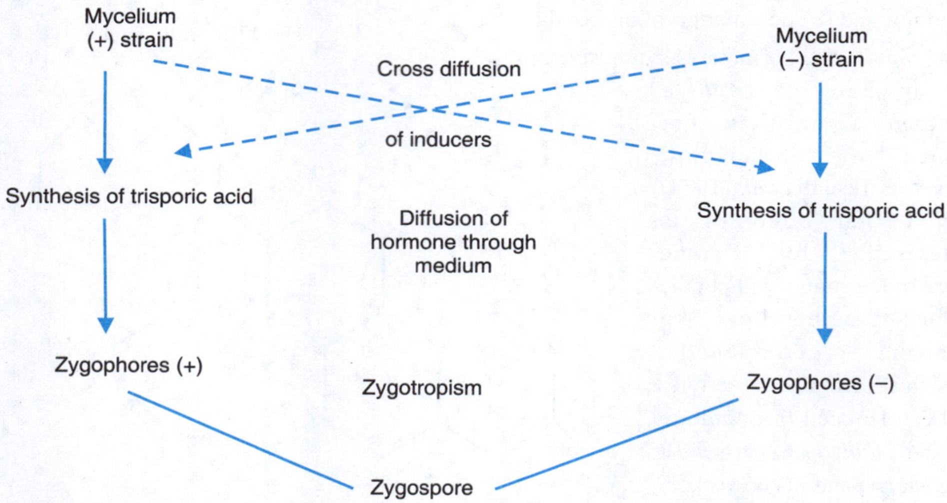

Fig. 17.9. Collaborative synthesis of trisporic acid in *Mucor mucedo.*

It is believed that synthesis of trisporic acid as described for *Mucor mucedo* takes place in other members of Mucorales as well.

Yeast α factor. The presence and the involvement of some hormone in the sexual reproduction of *Saccharomyces cerevisiae* was suggested by Levi in 1950. In this fungus, the haploid cells are of two mating types **a** and α which conjugate to form diploid cells. In 1956, Levi showed that the α haploid cells produce a diffusible chemical which induces the formation of copulatory process by compatible **a** cells. These **a** cells, due to influence of the chemical substance produced by α cells, stop their growth and reproduction by budding. Instead, these **a** cells swell in size and form giant cells of various shapes. These giant cells are 30 or more times heavy in dry weight in comparison to normal haploid cells. The α factor acts only on **a** cells and has no effect on α cells. The α factor is reported to inhibit DNA replication in the **a** cells.

Duntze *et al* (1970-1973) isolated the α factor in pure form. It is a peptide complexed with copperion and has molecular weight of 1400.

QUESTIONS

ESSAY TYPE

1. Write a short essay on Sex hormones in fungi.
2. Describe in detail the various sex hormones found in fungi.
3. Describe the role of trisporic acid in sexual reproduction in ***Mucor mucedo***.

SHORT ANSWER TYPE

4. Write short notes on :
   (*a*) Yeast α factor (*b*) Trisporic acid
   (*c*) Sirenin (*d*) Sex hormones.
   (*e*) Role of hormone in sex control in fungi.
   (*Poorvanchal, 2003*)
5. Write a brief account of sex hormones in *Achlya*.
6. Write in brief about sex hormones in fungi.
   (*Kanpur M.Sc., 2003*)

OBJECTIVE TYPE

7. **Select the correct answer :**
   (*i*) Sex hormones are produced during
   (*a*) vegetative reproduction
   (*b*) asexual reproduction
   (*c*) sexual reproduction
   (*d*) all of the above.
   (*ii*) Yeast α factor hormone is produced by
   (*a*) *Mucor* (*b*) *Penicillium*
   (*c*) *Saccharomyces* (*d*) *Aspergillus*
   (*iii*) The empirical formula of Sirenin is
   (*a*) $C_{15}H_{24}O_2$ (*b*) $C_{29}H_{42}O_5$
   (*c*) CH_3COOH (*d*) $C_{18}H_{26}O_4$
   (*iv*) Trisporic acid is a sex hormone produced by
   (*a*) *Mucor mucedo*
   (*b*) *Allomyces arbusculus*
   (*c*) *Aspergillus nidulans*
   (*d*) *Penicillium notatum*
   (*v*) The hormone oogoniol induces
   (*a*) development of oogonium
   (*b*) development of antheridium
   (*c*) meiosis
   (*d*) production of zygospores

PARASEXUAL CYCLE OR PARASEXUALITY

In some fungi, true sexual cycle comprising of nuclear fusion and meiosis is absent. These fungi derive the benefits of sexuality through a cycle known as **Parasexual Cycle.** The **Parasexual Cycle** is defined as a cycle in which plasmogamy, karyogamy and meiosis (haploidisation) take place but not at a specified time or at specified points in the life-cycle of an organism. Generally parasexual cycle occurs in those fungi in which true sexual cycle does not take place. The members of class **Deuteromycetes** (**Deuteromycotina**) in which sexual cycle does not occur, exhibit parasexual cycle generally.

Parasexual cycle was first discovered by **Pontecarvo and Roper** of University of Glasgow in 1952 in *Aspergillus nidulans,* the imperfect stage of *Emericella nidulans.* Since then parasexual cycle has been discovered not only in several members of Deuteromycetes but also in fungi belonging to Ascomycetes and Basidiomycetes. So far it has not been reported in Phycomycetous fungi. Parasexual cycle also known as somatic recombinations has been reported in *A. nidulans* (Pontecarvo and Roper, 1952), *A. niger* (Pontecarvo *et al*, 1953), *A. fumigatus* (Stromnaes and Garber, 1963), *A. oryzae* (Ishitani, 1956), *A sojae* (Ishitani *et al* 1957), *Penicillium chrysogenum* (Pontecarvo and Sermonti 1969), *P. expansum* (Barron, 1962), *P. italicum* (Stromnaes et al, 1964), *Cephalosporium mycophyllum* (Tuveson and Coy, 1961), *Cochliobolus sativus* (Tintine, 1962), *Fusarium oxysporum* f.sp. *pisi* (Buxton, 1956), *F. oxysporum* f.sp. *cubense* (Buxton, 1962), *F. oxysporum* f. sp. *callistephi* (Haffman, 1967), *Coprinus fumetorius* (Gans and Prud, Homme, 1958), *Ustilago maydis* (Holliday, 1961), *U.violacea* (Day and Jones, 1968), *U. hordei* (Dinoor and Persoon, 1969), *Pyricularia oryzae* (Yamasaki and Niizeki, 1965) and *Verticillium alboatrum* (Hastie, 1962). Probably, *Puccinia graminis tritici* also exhibits parasexual cycle.

According to Pontecarvo (1958), parasexual cycle in *A. nidulans* involves the following steps :

(*i*) Formation of heterokaryotic mycelium
(*ii*) Fusion between two nuclei (Karyogamy)
 (*a*) Fusion between like nuclei
 (*b*) Fusion between unlike nuclei
(*iii*) Multiplication of diploid nuclei
(*iv*) Occasional Mitotic crossing over.
(*v*) Sorting out of diploid nuclei
(*vi*) Occasional haploidisation of diploid nuclei.
(*vii*) Sorting of new haploid strains.

A brief account of these steps are being presented below :

(*i*) Formation of heterokaryotic mycelium. Heterokaryotic mycelium is formed in several ways. The most common is by the anastomosis of somatic hyphae of different genetic combinations. The foreign nucleus or nuclei introduced into a mycelium nultiplies and its progeny spreads through the mycelium rendering it heterokaryotic. Mutation in one or more nuclei of a homokaryotic mycelium also makes it heterokaryotic. It happens in some of the fungi belonging to Ascomycetes. Still a third way is by the fusion of some of the nuclei and their subsequent multiplication and spread among the haploid nuclei. In this type of heterokaryotic mycelium a mixture of haploid and diploid nuclei occur.

***(ii)* Fusion between two nuclei (Karyogamy).** The fusion of nuclei in the mycelium has been demonstrated. The nucelar fusion may be of two types : (a) fusion between like nuclei and (b) fusion between unlike nuclei. The nuclear fusion results in the formation of homozygous or heterozygous diploid nucleus respectively. If the genotype of unlike nuclei present in the heterokaryotic mycelium is A and B, then five types of nuclei can be formed by their fusion : two types of haploid nuclei (A and B), two types of homozygous diploid nuclei (AA and BB) and one type of heterozygous diploid nucleus (AB).

***(iii)* Multiplication of diploid nuclei.** The above mentioned five types of nuclei multiply at about the same rate but the diploid nuclei are present in much smaller number than the haploid nuclei. Portecarvo (1958) estimates a proportion of one diploid heterozygous nucleus to 1000 haploid nuclei.

***(iv)* Occasional mitotic crossing over.** During multiplication of diploid nuclei, mitotic crossing over may take place. This results in the formation of new gene combinations. These recombinations, which are dependent on the existence of heterokaryosis, give the fungus some of the advantages of sexuality within the parasexual cycle. According to Pontecarvo's (1958) estimates, the amount of recombinations which may be expected to occur in an ascomycete through its parasexual cycle is 500 times smaller than through its sexual cycle. However, in *Penicillium chrysogenum* and *Aspergillus niger,* diploidisation and mitotic crossing over occur more frequently indicating the importance of parasexual cycle in evolution of new strains.

***(v)* Sorting out of Diploid nuclei.** In those fungi which produce uninucleate conidia, sorting out of the diploid nucleus occurs by their incorporation into conidia which germinate to produce diploid mycelia. Diploid strains of several important imperfect fungi have been isolated. Roper (1952) first synthesized and isolated diploid strains of *Aspergillus nidulans.* The conidia of diploid strains are some what larger than those of haploid strains.

***(vi)* Occasional haploidisation of the diploid nuclei.** Occasionally, some hyphae of diploid mycelium form haploid conidia which form haploid mycelia on germination. The formation of haploid conidia by diploid mycelium indicates that haploidisation occurs in some diploid nuclei.

***(vii)* Sorting of new haploid strains.** Some diploid nuclei undergo haploidisation in the mycelium and are sorted out by incorporation of haploid nuclei in the uninucleate conidia. Some of these haploid strains are genotypically different from their parents because of their mitotic recombinations.

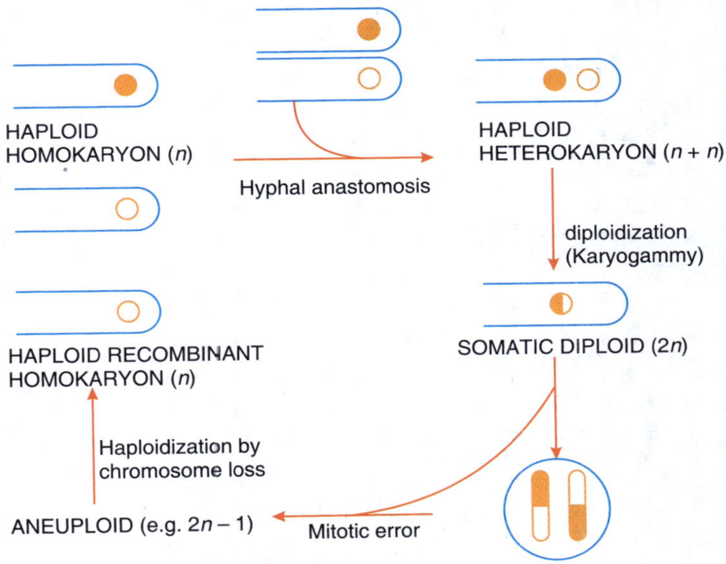

Thus, after the parasexual cycle has operated for sometime, the mycelium may contain the following types of nuclei : (a) haploid nuclei like those of both the parents, (b) haploid nuclei with various new genetic recombinations, (c) several types of diploid homozygous nuclei and (d) several types of diploid heterozygous nuclei.

Significance of Parasexual cycle

Parasexual cycle is of importance in industrial processes. Several fungi which are used in various industrial processes belong to fungi imperfecti or Deuteromycetes and in these fungi only parasexual cycle operates. New and better strains of these fungi are obtained by mutation through parasexual cycle. The strains of desirable characters can be developed through mitotic recombinations. Parasexuality can also be applied in the analysis of genetic and physiological processes of perfect and imperfect fungi. Parasexual cycle has also been successfully employed in genetic control of pathogenicity and host-range in several species of *Fusarium.*

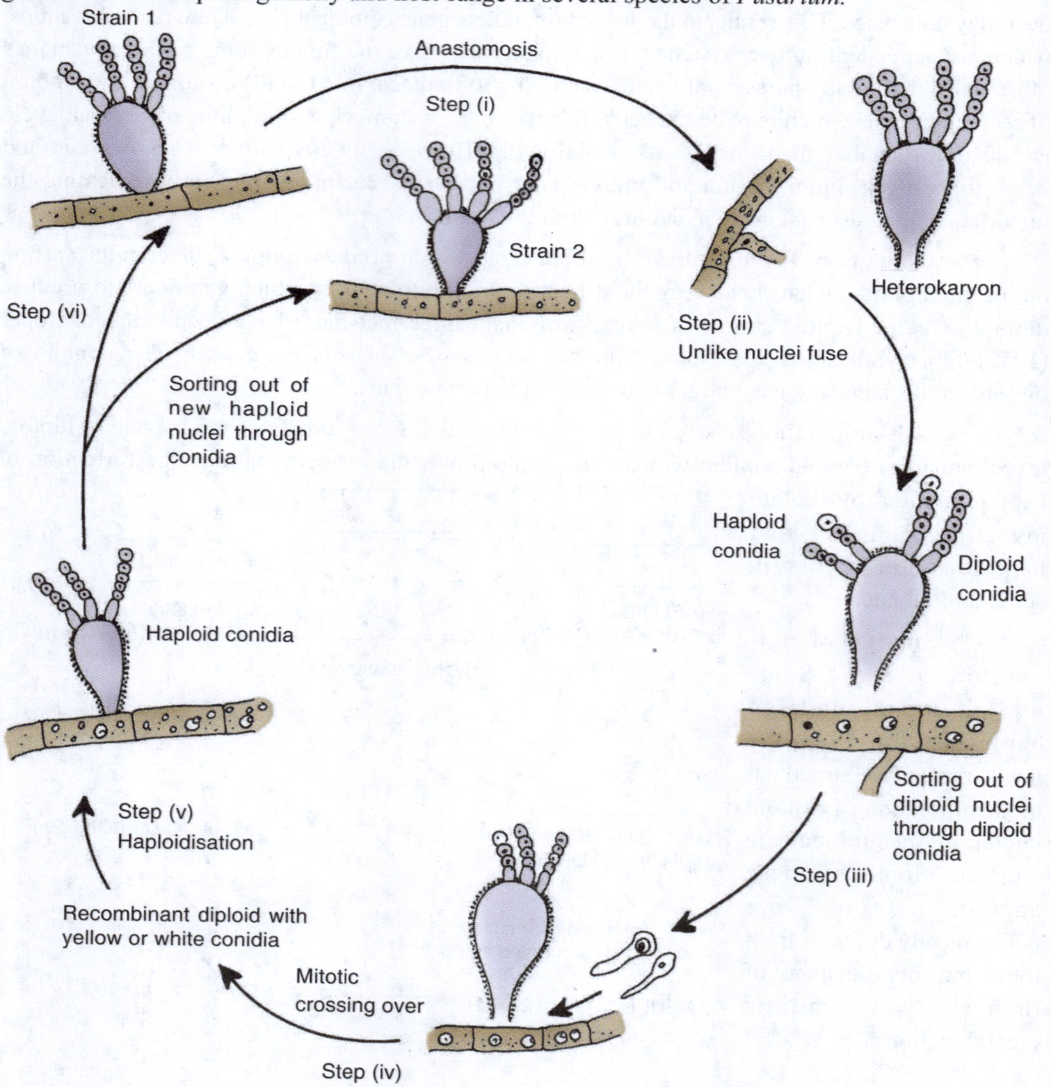

Fig. 17.10. Pontecarvo's (1958) idea of Parasexual cycle

Comparison between sexual and parasexual cycle.

S.No.	Sexual Cycle	Parasexual cycle
1.	Nuclear fusion takes place in specialised structures.	Nuclear fusion takes place rarely in vegetative cells.
2.	Zygotes usually persist one nuclear generation.	Zygote persists through many mitosis.
3.	Recombination by meiosis; crossing over in all chromosome pairs, of reduction of chromosome number, and random assortment of members of each chromosome pair are characteristics of this cycle.	Recombination by rare accidents of mitosis; mitotic crossing over at each event usually confined to one exchange in a single chromosome arm and haploidisation independent of crossing over are characteristics of this cycle.
4.	Products of meiosis can easily be recognised and isolated.	Products can be recognised only by application of suitable genetic markers.

QUESTIONS

ESSAY TYPE

1. What is parasexuality? Comment upon its importance.
2. Write a short essay on "Parasexual cycle".
3. Give a detailed account of parasexual cycle in an imperfect fungus studies by you.
4. Write about various methods of parasexual reproduction in fungi. *(Kanpur M.Sc., 2002)*
5. What is parasexuality ? Describe briefly parasxeuality in fungi giving proper examples. *(Kanpur M.Sc.,2004)*
6. Compare parasexual and sexual cycles in fungi. *(Kanpur M.Sc., 2006)*

SHORT ANSWER TYPE

7. Compare the parasexual cycle with sexual cycle.
8. Write notes on the following :
   (*a*) Parasexuality *(Allahabad M.Sc., 2000; Lucknow, 1992, 2005; Rohilkhand, M.Sc., 2000)*
   (*b*) Parasexual reproduction
   (*c*) Parasexual cycle *(Rohilkhand, M.Sc., 1993; Bihar, 1991)*
   (*d*) Parasexuality in fungi *(Lucknow, 2004)*
   (e) Explain parasexuality *(Rohilkhand M.Sc. 1990, 2005)*

OBJECTIVE TYPE

9. **Fill in the blanks :**
   (*a*) Parasexual sexual cycle was discovered by ....
   (*b*) .... are the group of fungi in which parasexual cycle is not reported.
   (*c*) Parasexuality was first reported in ....
   (*d*) Mitotic crossing over takes place in .... cycle.
   (*e*) Products of parasexual cycle can only be recognised through .... markers.

The mycologist raises pure culture of a fungus to study the fungus in detail regarding its reproduction, physiology and genetics. The pure culture may be on liquid solid/media and in obtaining the pure culture of the organism, the mycologist employs certain procedure, the details of which form the subject matter of this chapter.

PREPARATION OF CULTURE-MEDIA

Cleaning of Glass wares

All glass wares used in microbiological or pathological studies should be rigidly cleaned. Initial cleaning should be done with soap or vim using a brush. They can finally be washed with the following mixture :

Potassium dichromate	—	100 gms.
Conc. H_2So_4	—	500c.c.
Water	—	1000 ml.

The same mixture can be used several times. After washing with this mixture, the glasswares are washed for sometimes in running tap water, then finally with distilled water and allowed to air dry.

Preparation of Media

The choice of medium depends on the type of microorganism to be grown taking into consideration the nutritional requirements of the particular form. Out of a large number of such media, some are described below :

(*i*) Based on the consistency, the culture media are of two types :

(*a*) Liquid media or Broth : These media are liquid in consistency and are used in cases where microbial growth in pure form is to be separated out for further studies.

(*b*) Solid Media : The liquid media can be solidified by the addition of Agar-Agar, a jelly like substance extracted from the sea-weed, *Gelidium.* Addtion of about 1.5% of this substance at or near the neutral pH is sufficient to make the medium solid.

(*ii*) Based on the composition, the culture media may be of two types :

(a) Natural Media : These media incorporate certain naturally available substances of "Somewhat" unknown composition. Some examples of such media are : Malt extract agar, Soil extract agar, Oat meal agar and Potato dextrose agar. Potato dextrose agar is commonly used for phytopathological studies. The composition of Potato Dextrose agar (PDA) is given below :

Potatoes (Peeled and Sliced)	—	200 gms.
Dextrose	—	20 gms.
Agar-Agar Powder	—	15-20 gms.
Distilled water	—	1000 ml.

Potato slices are allowed to simmer in 500 ml. distilled water, the extract filtered by means of a muslin cloth, dextrose and agar-agar powder are added, the volume made up to 1000 ml and made to boil.

(b) Synthetic Media : These media have known constituents. Examples of synthetic media are Standard synthetic and Czapek-Dox media, composition of which are given below :

(*i*) Standard synthetic Agar (Brown's medium)

$MgSo_4.7H_2O$	—	0.5 gms.
$K_2H\ PO_4$	—	0.5 gms.
KNO_3	—	2.0 gms.
Starch soluble	—	20.0 gms.
Agar-Agar	—	15-20 gms.
Distilled water	—	1000 ml.

(*ii*) Czapek-Dox Agar

$NaNO_3$	—	2.0 gms.
$KH_2\ PO_4$	—	1.0 gms.
KC1	—	0.5 gm.
$MgSO_4.7H_2o$	—	0.5 gms.
$FeSO_4.7H_2o$	—	0.01 gm.
Sucrose	—	30 gms.
Agar-Agar	—	15-20 gms.
Distilled water	—	1000 ml.

All the constituents are weighed out distilled water added and the mixture heated to boiling. It will be better to add K_2HPO_4 or KH_2Po_4 in the end to avoid precipitation. The pH is adjusted as desired by adding NaOH or Citric acid solutions in required quantity.

Filteration of Media and dispensing into culture tubes or Flask

To a large funnel, a rubber tubing with a stop-cock is fitted. The medium is filtered through a piece of muslin cloth and the appropriate quantities are dispensed into culture tubes (6"× 3/4") or flasks (250 ml capacity), by adjusting the stop-cock.

All the glasswares in which the medium it is to be dispensed must be plugged with cotton wool. Culture tubes meant to be agar slants should receive approximately 10 ml of the medium while those ultimately to be used for plating (in flasks) approximately 50ml. The remaining medium should be put in flasks, plugged with cotton wool and used as required.

Sterilization of the Medium : Autoclaving

The media thus dispensed with should be steam-sterilized in an autoclave for 15 mts at 16 lbs/inch pressure. For convenience, the culture tubes are arranged in a wire basket and then put in the autoclave. Before opening the autoclave, the pressure is allowed to come down to zero.

Slanting : The culture tubes with 10 ml of medium and meant for use as tube slants should be taken out from the autoclave while hot, slightly rotated, put on a slanting position on a slanting rack and allowed to cool. The medium due to the presence of agar-agar would solidify on cooling. The agar slants are now ready for culture work.

Sterilization of Glass-ware : It may be done in 2-ways

1. Dry heat sterilization : This involves heating the glasswares (especially petridishes meant for planting work) in an electric oven at 150°C for 30 mts. The petridishes in suitable batches should be wrapped in bamboo paper and tied with a thread and then placed in the oven.

2. Steam sterilization : For certain type of culture work, this method should be used. The glassware should be plugged with cottonwool and steam sterilized in an autoclave at 15 lbs/inch pressure for 15 mts.

Plating : It should be done in a culture room sterilised through sulphur or formaline fumigation. The table on which the process is to be carried out should be cleaned with a swab of cotton dipped in alcohol.

All these steps will help to check the contamination of plates by unwanted microorganisms. The medium in the flask should be used for plating. If the medium in the flasks has solidified, it has to be melted in a water bath. The flasks should be taken to a culture room, the cotton plug opened, the mouth gently heated on a flame, the upper lid of the petridish slightly raised, the medium is poured into the pertridish, the lid replaced and the medium allowed to cool. The process of plating the medium should be carried out in as little time as possible in order to avoid any contamination. Now, the petridishes with the culture medium are ready for inoculation with the microorganism.

ISOLATION OF MICRO-ORGANISMS FROM SOIL

Various methods have been devised to isolate pure cultures of microorganisms from soil. Methods which yield a relatively broad spectrum of micro organisms are commonly used by plant-pathologists for isolating soil borne pathogens and are described here :

Direct inoculation

A crumb of soil, about 1cm in diameter is transferred with a sterilized forceps to the centre of a sterile petridish containing solidified agar medium. The petri-dish is incubated at 28-30°c and the mycelia radiating out from soil crumb are cut and transferred to agar slants.

Soil-Plate Method

A small amount of soil is transferred into a sterile petridish with the help of a sterilized spatula. 20 ml of melted and cooled agar medium is added to the petri-dish containing soil and the pertridish is gently shaken so that the soil-particles are dispersed through out the agar medium. The petridish thus obtained is incubated at 28°-30°C for a few days after which the fungi begin the appear. Pure cultures of the desired fungus is obtained by transferring that particular fungus to agar slants.

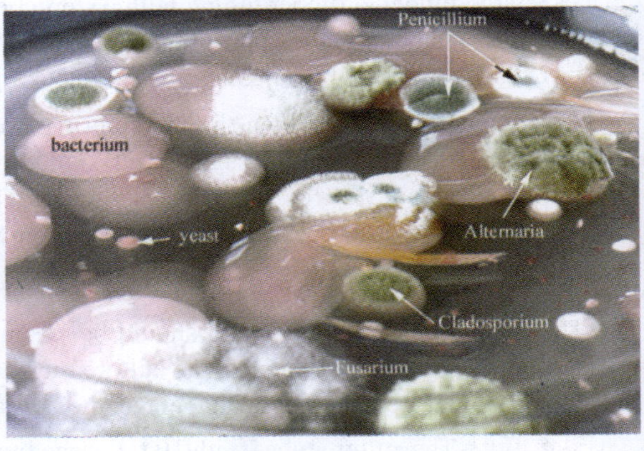

Soil Plate Method of isolation of microbes

Soil-dilution Plate Methods

10 gms. of soil sample is taken in a 250 ml conical flask. To this is added 100 ml of sterile water and the flask is vigorously shaken for a few minutes so that soil solution is obtained. This will represent 1:10 or 1/10. 10ml of the supernatent of this 1/10 solution is taken and 90 ml of sterile water is added and the resultant solution will be 1/100. In the same way, other dilutions like 1/1000, 1/10,000 and 1/100000 are made.

One ml of the desired dilution is poured into a sterile petri-dish and to this is added 20 ml of melted and cooled agar medium. The petri-dish is rotated by hand to disperse the medium and the soil suspension. The petri-dishes are then incubated at 28°-30°C for a few days after which the growth of fungi takes place. Pure cultures of the fungi are then obtained by transferring them to agar slants.

Fungal culture of pure form is obtained by inoculating a single spore or a piece of mycelium on the agar medium in tube slants or in petridishes as shown in Fig. 17.11. The inoculated agar plate is covered and the tube slant is plugged with non-absorbent cotton wool. They are then placed in an incubator which is usually maintained at a temperature of 30-37°C for a couple of days after which the agar plates and slants will show fungal growth in pure form.

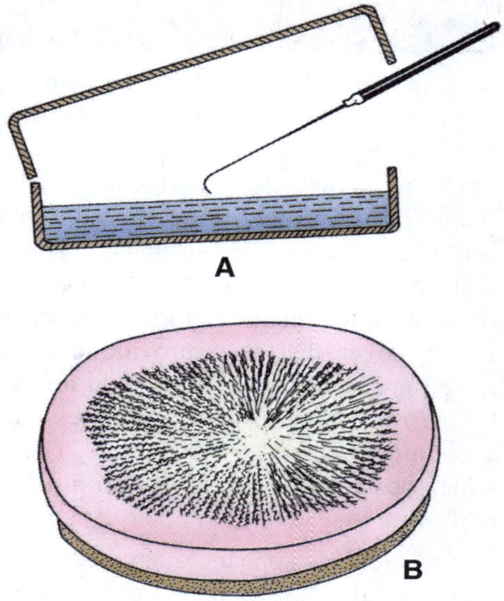

Fig. 17.11 (A-B). Fungal culture. A. Noculation; B. Culture of fungus

QUESTIONS

ESSAY TYPE

1. Give an account of the procedure of obtaining fungal cultures in pure form.
2. Describe the composition of two nutrient media.
3. Give a brief account of various types of culture media used in fungal cultures.
4. Give an account of culture media and describe the methods of sterilisation and inoculation. *(Kanpur, 2002)*

SHORT ANSWER TYPE

5. Write short notes on :
   (*a*) Czepek-Dox medium
   (*b*) Natural Media
   (*c*) Isolation of fungi.
   (*d*) Soil dilution plate method.
   (*e*) Methods of isolation of microorganisms *(Poorvanchal, 2004, 2005)*
6. Write a note on pour plate method of isolation and culture of microorganisms. *(Poorvanchal, 2002)*
7. What is pure culture ? *(Poorvanchal, 2002)*
8. Write a short note on culture of fungi. *(Poorvanchal, 2003)*
9. Write ingredients and methods of preparation of any one cultural medium. *(Poorvanchal, 2003)*

OBJECTIVE TYPE

10. Fill in the blanks.

(*i*) The cultures of fungi are needed for .... studies.

(*ii*) Natural culture media are those in which .... products are used.

(*iii*) In liquid culture media .... is not added.

(*iv*) The nitrogen source in Brown's synthetic medium is ....

(*v*) .... is added as a source of Carbohydrate in Czepeck-Dox medium.

MYCORRHIZA

The term '*Mycorrhiza*' is a composite word made up of two words – *mycos* meaning fungi, *rhizos* meaning roots. Fungi in association of roots of flowering parts form the non pathogenic or feebly pathogenic association which is generally mutually symbiotic. In this association both the fungi and higher plants derive benefit from each other. When the other component is weak, the fungus becomes pathogenic. Both the components are associated in a relationships that is nutrition and often termed as *mycotrophic*. Generally, this word should be applied to association of fungi with thalli of lower plants, rhizomes and roots of higher plants. The association of fungi with roots of higher plants should be termed as *Mycorrhiza* and that of fungi with thalli as *mycothalli*. It is believed that roots provide an excellent habitat for the growth of these fungi. While the fungal presence might be essential for the growth of higher plants assisting in the mineral up take by the roots, the host plant supplies the nutritional requirements of the fungi. In this way, this association helps both the partners. The important feature of mycorrhizal organs is that the plants which possess them do not suffer any serious disease, may be because of the presence of fungal hyphae.

Mycorrhiza appears to have been known to the mankind since fourth century B.C. Fossils of mycorrhizal association in *Lepidodendron* sp. (Clubmosses) appear to have occurred but no clear picture emerged except the presence of intracellular hyphae and a few infection points. It was Frank (1885) who gave definite information about the composite fungus root organs of plants *Cupuliferae*. Frank also introduced the term *mycorrhiza* for this association. He also showed that seeds of certain orchids germinated only in the presence of specific type of fungi. It was also observed that in most of the forest trees, mycorrhizas may be formed with a number of different fungi but in others, the association is specific. However, in 1936, Burges established that mycorrhizal association is feebly parasitic and has no symbiotic value.

The physiological and biological significance of mycorrhizal association has been a matter of debate for a long time. Some botanists regarded fungal components as parasities upon the roots. Others believed the relationship between the fungi and roots in this association is more complex than fungal parasitism and that both the roots and fungi benefit from each other. Experimentally also it has been shown that mycorrhizal roots have greater absorption rate and disease resistance than the non-mycorrhizal ones.

Mycorrhizas could be (*a*) Ectomycorrhiza and (*b*) Endomycorrhiza

(a) *Ectomycorrhiza* (Ectotrophic mycorrhiza). In this type of mycorrhiza, the root system is completely surrounded by a distinct mantle or sheath of fungal tissue from which hyphae penetrate the outermost cell layer (a) of the root of all seeded plants. Sometimes the light pseudoparenchymatous fungal tissues send branches between the cortical, cells of the root and outward in the soil thus isolating the root from the soil and comprising about half of the total weight and volume of the organ. The hyphae penetrate the cells of the cortex to form a network of hyphae *Hartignet*. The hyphae which move into the soil ramify in the soil particles. In some species, the *Hartignet* is formed only internally while in majority of the species, an external sheath is present. New lateral roots formed by the mother root also get the mycorrhiza as they pass through the cortex. The presence of mycorrhiza also prevents growth in length of the roots as a result of which swollen or coralloid roots are formed.

The ectomycorrhizas are also called sheathing *mycorrhizas*. It is estimated that about 3% of all seeded plants are have ectomycorrhizas being common in northern temperate forest trees especially in Fagaceae and Pinaceae but also in Betulaceae and Tilliaceae. In tropics, the ectomycorrhizas are well represented in Myrtacceae and Dipterocarpacae. The trees in which ectomycorrhizas are common include *Pinus, Picea, Cedrus*. *Abies, Fagus* and *Quercus*.

Majority of fungi forming ectomycorrhizas are from Basidomycotina especially Hymenomycetes. Over 100 species are known to form mycorrhizas. Majority comes from the general *Amanita, Boletus* and *Tricholoma*. Many species in the genera *Cortinarius, Lactarius* and *Russla* also form Ectomycorrhizas. Some of the fungi like *Amanita muscaria* and *Bletus edulis* have wide host range while other show host specificity. *Boletus elegans* is one such example which forms mycorrhiza only with larch.

The fungus depends on the host for carbon source. The mycorrhizas are responsible for enhanced mineral uptake and hydrolysis of organic phosphorus in soil. The mycorrhizae sheath serves as a reservoir of mineral ions.

Mycorrhizal trees have greater vigour than the non-mycorrhizal trees. In some cases the fungus produces extra hyphal enzymes which break down lignin and hydrolyze cellulose of humus and plant debris making suitable for the plants to utilise. This has been shown in *Boletus subtomentosus* with *Pinus montana* as also in *Lactarius delicious* with *Poxius sylvesrius*. Mycelium of the fungus in the soil attached to mycorhiza is capable of tapping greater volume of soil than the lost root. The nutrients absorbed by the host pass through the fungal mycelia enabling these to grow.

Basically ectomycorrhizas are either *coralloid* or *ball types*.These have also been differentiated on the basis of external forms into three types (Melin and others) : (*i*) long thin unbranched structures which grow upon roots, (*ii*) short dichotomously branched coloured roots the colour depending upon the fungus and (*iii*) knots or tuber like growths developed by fusion of flaps produced by merger of clusters of forked roots.

(*ii*) *Endomycorrhiza* (*Endotrophic mycorrhiza*). In endomycorrhiza, the fungal hyphae penetrate the root epidermis and enter the cortex establishing themselves in the cortical cells. The hyphae may be present on the root surface externally as well. The fungal hyphae may be septate and aseptate and grow intra-or-inter-cellularly. These do not invade green tissues and they also do not enter the storage tissues. The endomycorrhizas are not harmful to the plants but these are associated with the process of root development. Roots becomes dichotomously branched increasing the surface area of absorption. In saprophytic angiosperm *Monotropa*, the endomycorrhiza are shown to be associated with the process of absorption of nutrients. The germination of seeds and spores and early development of young and adults are also influenced by endomycorrhizal fungi as these provide food materials and enzymes essential for metabolism. In addition, these fungi break down the cellulose and lignin in the soil and supply carbon compound to their hosts.

The endomycorrhizas are ubiquitous present in arctic, temperate and tropical regions. These are found in almost all flowering flowering plants including gymnosperms as also ferns and liverworts. This type of mycorrhiza has also been found to be present in plants of economic importance such as grasses, cereals and legumes. In these plants, the external mycelium is absent but present in *Monotropa* and members of family Ericaceae. In orchids, the external mycelium is scanty. Generally the fungal component is present on the root surface externally forming a loose mass of hyphae without entering it. The branches of these hyphae enter the root cortex and establish themselves in cortical cells.

There are two main types of endomycorrhizas.

(*a*) Endomycorrhiza due to aseptate hyphae belonging to *Phycomycetes*. The fungal component belong to the genera *Pythium* and *Endogone* and *Rhizofagus – Endogone* complex and the non-fungal component comes from the families *Graminae*, *Palmae*, *Leguminosae*, and gymnosperms excluding Pinaceae, pteridophytes and liverworts.

Vascular arbuscular mycorrhizas (VAM) are considered as the most important type of endo-mycorrhiza. These are produced by aspetate myceliae fungi which penetrate the root and forming *vesicles* and arbuscles in plants roots after infection. These are found in bryophytes and pteridophytes and gymnospersms except Pinaceae and in virtually all families of angiosperms. VAM are of general occurence in Gramineae, Palmae, Rosaceae and Leguminosae but are absent in Chenopodiaceae, Cruciferae, Cyperaceae and Reseadaceae. The fungal hyphae produce very thick walled brown to black balloon like Chlamydospores which range between 10-250 μ in diameter. These are produced singly or in clusters as vesicles at the loose end of hyphae. Loose aggregations of spores may also be formed within a poorly developed reproductive structure called sprorcarp. Chalmydospores are most common in soil though Zygospores are sometimes also formed. These are produced by the intercellular hyphae, form *vesicles*, intercalary or apical in position and produced as large thick walled swellings. The vesicles are known to function as storage organs or as resting spores. They remain in soil when the roots decay and infect the new root when produced. The vesicle containing mycorrhizas are known as *Vesciular mycorrhiza.*

Arbuscules (*Haustoria*) are formed by branches of smaller diameter which pass into cortical cells. These arbuscules are often dichotomously branched and they function as complex haustoria. Arbuscules become swollen at their apices and form sporangioles which become free within the cells of the host. The sporangioles disintegrate later. Such structures are known as *arbuscule-sporangiole* complex and are of common occurrence and in Papilionaceae and Graminae families.

The intercellular hyphae may produce a combination of vesicles and arbuscules. Such mycorrhizas are known as *Vesicular–Arbuscular mycorrhiza* (VAM) and are common in plant kingdom. VAM are also known as phycomycetous mycorrhiza and has also been reported from carboniferous and Devonian periods.

In roots of orchids, form species of *Rhizoctonia* form intracellular hyphae which become coiled and digested in the host cell. These coiled intracellular hyphae are known as pelotons. Mostly VAM fungi comes from Endogonaceae family belonging to order Mucorales of class Zygomycetes. Generally the genera belong to *Gloius*, *Glomerulus*, *Gigaspora* and *Sclerocystis*. Spores of these isolated from soil have been used to infect healthy plants. So far these have not been grown on artificial culture media. Although spores or chlamydospores do germinate on nutrient media, the hyphae grow till food supply is available in the nutrient agar and these hyphae cannot be subcultured. Spores from infected maize plants have been successfully used to produce typical VAM in soyabean, onion, redclover and strawberry.

(*b*) *Endomycorrhiza* caused by fungi with septate hyphae belong to mainly Basidiomycetes or Deuteromycetes having affinities with Basidiomycetes. The fungal component comes from *Armillaria mellea* and species of *Rhizoctonia* while the non fungal component comes from Ericaceae, Orchidaceae and Gentiamaceae and a few bryophytes and pteridophytes.

Other types of Mycorrhizas

(*i*) *Ericoid mycorrhiza* is another type of endomycorrhiza in which different kind of fungi is involved. These mycorrhizas are found in plants of family Ericaceae where the rootlets are covered with a very sparse loose mass or weft. Some branches penetrate the cortical cells forming intercellular hyphae coils enclosed by host plasmalemma. After sometime, the host cells disintegrate allowing these parts to collapse, such mycorrhizas are called Ericoid mycorrhizas which help in phosphorus and nitrogen uptake by host plants.

(*ii*) *Ectendomycorrhiza* – This type of mycorrhiza is a little different from the above and has been reported in a number of plants of family Ericaceae and plants of other families of Ericales. The short roots of the host are swollen and are invested with hyphal sheath. Although Harting net is absent in the hyphae sheath but Intercellular hyphal corls develop in outer cortical cells which ultimately get lysed. Since the mycorrhiza fungus is present both externally and internally, this type of mycorrhiza is known as ectendomycorrhiza. *Arbutus* and *Arctostaphylos* are the best examples of this type of mycorrhiza.

(*iii*) *Pseudomycorrhiza*. This term was introduced by Melin and others to include short roots attacked by certain root parasites in the absence of any other fungi. This resulted in slow growth and branching was rare or absent. Mostly the fungus involved was *Rhizoctonia sylvestris* which is a weak pathogen.

Nature and Imporance of Mycorrhiza

There has been conflicting views about the nature of mycorrhiza. These views can be grouped in two categories :

(*a*) Mycorrhizas are pathogenic in nature and cause harm to the host.

(*b*) Mycorrhizas are symbiotic in nature of benefitting both the partners.

Most of the scientists are the view that the mycorrhizas are symbiotic in nature as evidenced by the following :

(*i*) Mycorrhizas induced increasing in branching of roots followed by increasing surface of absorption of roots. Mycorrhizal roots receive better nutrition than the non-mycorrhizal ones.

(*ii*) Mycorrhizas are equally important in plant nutrition in deficient soils. The mycorrhiza may increase the availability of phosphorus and several other nutrients. Reports have been published to show that mycorrhiza enhance the uptake of N_2, P, Ca, Na, Fe, Cu, Bo, Zn, Al and strontium. According to Brown (1968) Sanders and others (1975), mycorrhizal infection may be a precondition for nodule formation in phosphorus deficient soil.

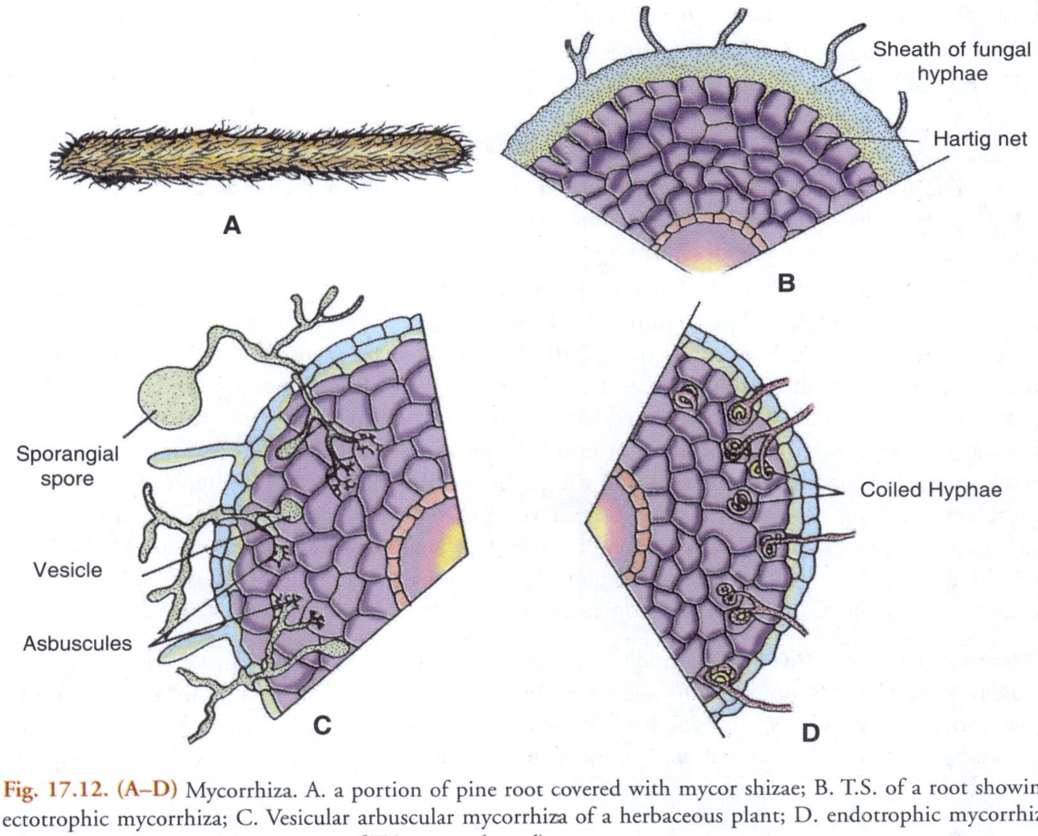

Fig. 17.12. (A–D) Mycorrhiza. A. a portion of pine root covered with mycor shizae; B. T.S. of a root showing ectotrophic mycorrhiza; C. Vesicular arbuscular mycorrhiza of a herbaceous plant; D. endotrophic mycorrhiza of an Orchid root. (B–D. A portion of T.S. root enlarged)

(*iii*) Mycorrhizal plants show greater dryweight than the non-mycorrhizal plants growing on nutrient deficient soils because the former absorb nutrients more and disproportionately than the latter.

(*iv*) Mycorrhizal trees have greater vigour because the infection release several inhibitory compounds to establish barriers against pathogenic infection.

(*v*) The fungal partner gets suitable habitat for the growth on the root surface where nutrients required for the fungal growth are available in plenty. Roots also excrete certain substances which promote spore germination and stimulate root infection.

(*vi*) Mycorrhizas are universally present in orchids because the fungal mycorrhiza provides sugars and vitamins essential for germination of seeds and their early development.

(*vii*) In forest plants like *Pinus montara* and *Pinus sylvestris,* the interdependence of both the partners is well understood. In these plants, the fungal partner produces extra hyphae enzymes which break down lignin and hydrolyse cellylose of plant debris and humus making these available for use by the host.

(*viii*) Mycorrhizal infection has morphogenetic effect on the host leading to prolonged life growth and branching of booklets. It has been shown that mycorrhizal infection enchanced auxin production due to which dichotomous branching increased.

(*ix*) Mycorrhizas play an important role in Afforestation. It has been experimentally shown that spruce and pine seedlings die due to starvation if they are not invaded with mycorrhizal fungi *Boletus viscidus* – Of all the mycorrhizal fungi tested, this fungus produced the best results. Seedlings with mycorrhizal infection are immune to infection.

(*x*) VAM fungi are obligate biotrophs but are not host specific. Therefore they can be used as biological fertilisers.

RHIZOSPHERE

Varieties of organic compounds are released by the plant roots in their vicinity as root exudates and root secretions. These organic compounds diffuse in the soil closer to root surface forming a specific zone with intense microbial activity.

The rhizosphere refers to the *zone of soil which is influenced by the secretions of the adjacent plant roots*. It is the region of place of interaction involving plant roots, belowground soil microbes and plant pathogens. The zone has been further subdivided by some as the outer rhizosphere, inner rhizosphere and the rhizoplane i.e. the root surface itself (Fig.) Inner rhizosphere is the zone of more intensive change, generally considered to be around 15-20 μm. Rhizosphere soil is characterized by greater concentration of organic solutes released by plants as root exudates and root secretions. Thus the characteristics of rhizosphere soil is substantially different from those of non-rhizosphere soil (Rovira et. al. 1983). The term *rhizosphere effect* indicates overall influence of plant roots on soil microbial population. The extent of the rhizosphere effect is quite ill defined and variable depending upon the age of plants, soil conditions affecting diffusion of the solutes or volatiles from the roots and their biological activity; it can vary from 1 mm to 10 mm.

The root 'exudates' (leaking passively from cells) are low molecular weight compounds like sugars, organic acids, amino acids and several other compounds. The secretions are present in the form of high molecular weight polysaccharides covering root surface in the form of mucilages. Cells which are senescing are also major source of organic losses, called as lysates.

The movements of low molecular weight substrates are not used by microbial populations at root surface and they diffuse from the root surface in the soil. This leads to occurrence of concentration gradient in the soil away from the root surface which is depend on soil porosity, moisture, charge on clay particles, pH etc. Presence of greater concentration of organic compounds in the rhizosphere soil attracts comparatively more microbial population of bacteria, fungi and actino-mycetes, nematodes, protozoans and algae than present in non-rhizosphere zone. Microorganisms also influence the root hair and lateral root developments in many plants. Plant decomposers, ammonifers, fermenters, nitrogen fixers, motile bacteria are more active in this zone, therefore oxygen consumption and CO_2 production is also more as compared to non-rhizosphere soil. Microbial activity decreases as we move from rhizosphere to non-rhizosphere soil. Rhizosphere effect is more pronounced during flowering of plants than during seedling stage, more in sandy soil and least in humus soil. This 'R:S Ratio', the ratio of microbial population in rhizosphere and non-rhizosphere zone gives good picture of relative stimulation of microorganisms in the rhizospheres of different plant species.

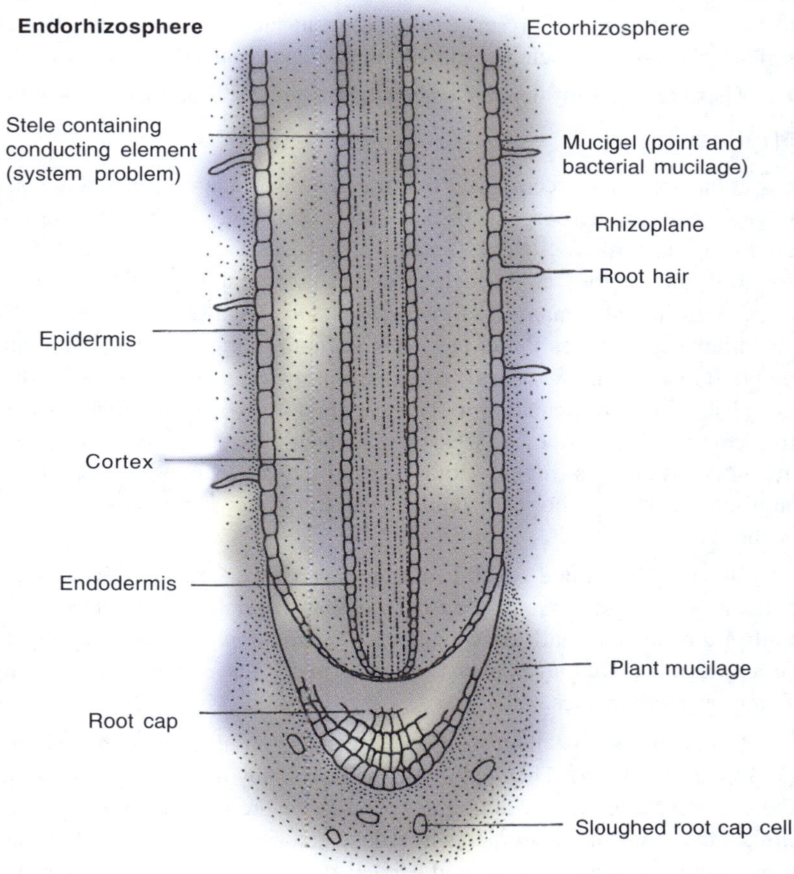

Root exudates greatly influence the life cycle and activity of many microorganisms. Such as spores of many fungi like *Fusarium, sclerotium, Nerticilium* germinate only when stimulus is provided by root exudates. Root exudates also provide food base for growth of several antagonistic organisms which could suppress growth of phytopathogenic organisms in the soil. For example number of *Streptomyces* and *Trichoderma* were more in the rhizosphere of resistant plant varieties than those found rhizosphere of susceptible varieties. Roots of Allium release certain volatile stimulators which help release fungistasis of sclerotia of *Sclerotium cepivorum*.

of wilt on certain susceptible varieties of plants. Glycosides and hydrocyanic acid of root exudates sometimes inhibit the growth of fungal pathogens. For example, Bison (a Flax variety) is resistant to *Fusarium* because of presence of hydrocyanic acids in its root exudates. Root exudates neutralize soil pH, alters soil microclimate of rhizosphere zone through liberation of CO_2 and H_2O.

These changes influence the activity of root infecting plant pathogens in the soil.

Phyllosphere

It has long been recognized that the aerial surface of green plants also provide habitats for a large number of microorganisms. Like roots, the leaves of the plants also release certain chemicals on its surface called 'Leaf diffusates'. These leaf diffusates are rich in amino acids, glucose, fructose and sucrose etc. which provide nutritional requirements of the leaf microflora. Leaf microflora includes many bacteria viruses and fungal flora belonging to both actinmycetes and basidiomycetes. It was Potter (1910) who first reported presence of a number of microorganisms on phylloplane (surface of leaf) when he simply pressed leaves on solidified cultural medium in a petridish and then incubated it. The systematic ecological studies was done later on cereal and tropical tree leaves by Last (1955) and Ruinen (1961), respectively. However most of the studies are concentrated to fungal flora including yeasts found on leaves, some of them has got saprophytic and pathogenic tendencies. *Erysiphe* is perhaps the best example of epiphytic pathogen.

Burri (1903) found several million bacteria per gram of leaf tissue and not that these were actively growing and are distinct from soil and aerial flora. Bacteria recorded on leaf surface are both gram positive and gram negative. *Pseudomonas*, *Xanthomonas*, *Flexibacter* and *Erwinia* (gram negative), *Staphylococcus*, *Bacillus* and *Micrococcus* (Gram positive) are very frequently found on phyhoplanes.

Chemistry of host plant's surface affects the number and composition of microbial ephiphytes (Jeffree et. al., 1976). Then epiphytic microbes they themselves can influence the development of their associates either by acting as a source of their nutrients or by producing inhibitors which limits the growth of other organisms (Blakeman and Brodie, 1976). Besides microclimatic condition around phyllosphere (area around leaf) also influence the phylloplane flora. Foliar pathogens can also influence the quality of the surface environment as a consequence of their effect on the physiology of the host plant.

Mushroom Cultivation in India

Button Mushroom is the most popular mushroom variety grown and consumed all over the world over. In India, its production earlier was limited to the winter season, but with technology development ,these are produced almost throughout the year in small, medium and large farms, adopting different levels of technology. The species *Agaricus bisporus* (Button Mushroom) belong to class Basidomycetes and family Agaricaceae.

Food Value

Edible Musrooms are popularly known as " boneless vegetarian meat" or 'white vegetables'. They are highly nutritious contain ample amounts of proteins, vitamins and fibre apart from having certain medicinal properties. It contains around 20%-35% protein (dry weight) which is higher than those of vegetables and fruits and is of superior quality. Because of its nutrient and medicinal value it is being cultivated viz; button, shitake, oyster, wood ear and paddy straw mushrooms contribute 99% of the world production.

India produces more than 100,000 tonnes of mushrooms annually of which Haryana contributes around 6000 tonnes, Punjab 60,000 tonnes, U.P. 7,000 tonnes and H.P. 6000 tonnes. Rest of the production comes though J&K, Tamil Nadu, Karnataka, Kerala and other States. (RMCU, DMR, Solan, 2010).

Around 60 Species of poisonous, and of these only a few can be lethal. Most of poisonous mushrooms cause anything from a mild stomachache to severe physical distress-including vomiting, diarrhea, cramps and loss of co-ordination. Poisonous mushrooms are referred to as 'toadstools'. Common poisonous mushroom species are *Amanita phllodies, (death cup), Armillariella mellea* (honey mushroom) and *Omphlotus olearius (Jack o' Hantern).*

Method of Cultivation

Most of the production of white button mushroom in our country is seasonal. If cultivation is done by conventional method using unpasteurized compost the yield of mushroom is low. However, in recent years, improved agronomic practices has resulted in greater yield mushrooms. Cultivation of the common white button mushroom requires technical skill. Temperature and humidity management is the foremost requirement for cultivation of button mushrooms. It requires two different temperatures (temperature for spawn or vegetative growth: 22-28° C and for fruit body formation: 15-18° and humidity level 85-95% good ventilation is needed for substrates that are sterilized. It could easily be contaminated unless spawned under very aseptic conditions. Therefore steaming at 100° C (Pasteurization) is more acceptable steamer.

Compost Preparation

Any cellulose waste like agriculture by products (cereal straw, maize stalks, hay, sugarcane bagasse) Old gunny bags can be used for compost formation. These materials provide cellulose, hemicellulose and lignin, which are utilized by the mushrooms mycelium as the carbon source. These materials also provide physical structure to the substrate with proper aeration for composting on which micro flora developed which is essential for the fermentation. Rice and barley straw are very soft, degrade very quickly during composting and also absorb more water as compared to wheat straw. While using these substrates, proper management is needed for periodic watering, quantity of water to be used, schedule of turning and adjustment to the rates. Since these by products do not have adequate nitrogen and other components required for the fermentation process, compounding mixture is supplemented with the nitrogen and carbohydrates is needed to start this process.

Spawning

Spawn is pure culture of the fungus mycelium, which is grown on special media. It is always prepared with great care to avoid contamination. Spawning is mixing of spawn for optimum and yield. Optimum dose for spawn between 0.5 and 0.75% of fresh weight of compost.

The optimum temperature for growth of *A. bisporus* is 23+-2° C. Relative humidity of the cultivation room should range from 85-90% during spawn-run.

Mushroom (Button mushroom) Cultivation

Harvesting

Mushrooms primordia begin to form 3 to 4 days after opening the bags. Another 2 to 3 days is required to mature mushrooms when it become ready for harvesting. For harvesting the mushrooms, they are grasped by the stalk and gently twisted and pulled. Knife is not used for mushroom harvesting. The mushrooms remain fresh for up to 3 to 6 days in a refrigerator/cool place.

Most type of the mushroom cultivation follow the similar pattern except that there is difference in specific requirements for water, temperature and humidity management.

SOME IMPORTANT CONTRIBUTIONS IN FUNGI AND PLANT PATHOLOGY

De Bary (Heinrich Anton De Bary,1831-1888)

The Tulsane brothers (France) and De Bary (Germany) are considered the founders of modern mycology. De Bary was born in Frankfurt am-Main, Germany in 1853 and received his M.D. degree from the University of Berlin in 1953. His contribution to mycology is immencely outstanding. His famous publication of the book '*Untersuchungen Über die Brandpilze*' in 1953 established beyond doubt that fungi are the causes of disease not the result of plant disease. On the basis of his studies and observations he concluded that 'It has been shown that the smut and rust fungi originate not from the cell content or from the secretion of diseased cells and that they are not the result, but the cause of pathological processes.'

De Bary and his school worked in many fields of mycology ranging from morphology, physiology, sexuality and natural relationship of fungi. His contributions are of great value in the field of mycology and plant pathology. He elaborated the broad classification of fungi in his book "Morphologie und Physiologieder Plize, Flechten und Myxomyceten" (1866) that was followed by another book "Vergleichende Morphology und Biologie der Plize, Mycetozoen und Bacterien" in 1884. An English translation of the book "Comparative Morphology and Biology of the Fungi, Mycetozoa and Bacteria" appeared in 1887. His pathological studies in the family Pernosporaceae (1881) and the disease they incite especially potato disease (*Phytopthora infestans*), discovery of heterothallism in rusts and his work on *Sclerotinia sclerotiorum* (1886) opened the new field of investigation into the physiology of parasitism. These contributions of De Bary has led him to be called **Father of Plant Pathology**.

De Bary's school has produced some of all time great mycologist and plant pathologist like Woronin of Russia, Brefeld of Germany, Millardet of France, Ward of England, Farlow of U.S.A and Fischer of Switzerland.

Robert Koch (Koch's Postulates)

Robert Koch (1843–1910) was a German Physician and a well known bacteriologist. He was the first one to experimentally prove and discover the causal agent of anthrax disease is *Bacillus anthracis* in 1876, also reported that tuberculosis and cholera are also caused by bacteria. Through his findings he was able to confirm the germ theory of disease which was earlier proposed by Louis Pasteur. Robert Koch has done a lot of work on microbial techniques particularly developing the pure culture technique and use of agar-agar as solidifying media for microbial culture. Through his experiments with diseases and isolation of their microorganisms from the diseased plant, animal, or human Robert Koch (1887) was able to set four steps rule to prove the causal agent of the disease. Koch considered that these rules must be satisfied before any microorganism can be regarded as a pathogen. This was later known as Koch's Postulates/rules. These four steps of the Koch' postulates are:

1. The suspected causal agent (Pathogen) must be consistently associated with the same symptom in the diseased organism to be examined.
2. The suspected causal organism must be isolated from the diseased plant and then grown in pure culture away from the host plant.

3. From this isolated suspected causal organism it is re-inoculated into a healthy susceptible host plant. The host plant must reproduce the same symptom of the disease.
4. The same suspected causal organism must be re-isolated/recovered again from the experimentally inoculated and infected host *i.e.,* the characteristics of the recovered organism and the isolated organism of step number 2 must be exactly same.

Koch's Postulate are very useful in the field phytopathology to develop host–pathogen relationship be it fungal, bacterial, viral, plasma and nematode diseases. However in some cases like plant diseases caused by obligate pathogens (mildews and rust fungi), phloem inhibiting bacteria and protozoans culture or purification of the organism is not possible, therefore Koch's Postulate rule are not possible to apply. But despite these limitations Koch's Postulates are very useful in identifying the new disease-pathogen relationship. Robert Koch is regarded as **Father of Microbial Technique** because of his contribution in the refining the method of pure culture of microorganisms.

The suspected causal organism must be isolated from the diseased plant organism and grown in pure culture.

Millardet (*Pierre Marie Alexis, 1838*)

Famous for his Bordeaux Mixture, born in Montmerey-la-Ville, France, in 1838 has worked under Hofmeister at Heidelberg and De Bary at Freiburg. Though he has doctorate degree both in Medicine and Science, but was more interested in botany. In 1876 he was called to chair of Botany at university of Bordeaux, where he served till his retirement in 1899. Millardet's work was wide spread in many directions, ranging from his early studies on the morphology, physiology, and systematic relationships of plants, investigations of Phylloxera of vine, including introduction of resistant stocks, hybridization experiment for obtaining disease resistant varieties and on control measures of downy mildews of vine through the application of Bordeaux mixture, etc.

Millardet became pioneer in suggesting the two most important ways to control plant diseases *i.e.* use of disease resistant varieties (1878, 1891a,b) and application of fungicidal sprays to plants leaves (1885a,b,c; lodeman,1896). Downy mildew of vine was first reported in Europe in 1878 and the disease spread so rapidly that it threatened to ruin the vine yards of Europe. During course of his phytopathological studies he found that vines that had been treated with a mixture of copper sulfate and slaked lime deter mildew infection in their leaves where as those untreated ones gets infected and defoliated. He performed many spraying experiments with various combinations of copper, calcium and iron salts to control the downy mildews in vines during 1883–1884. Finally he published his findings in May, 1885 and recommended spraying of **mixture of copper sulfate and slaked lime** for control of mildews, which became famous as **Bordeaux mixture**. Later Bordeaux mixture was emerged as most popular fungicidal spray for a quite long period, until less injurious fungicides came into picture.

Mathieu Tillet (1714)

Millet was born in Bordeaux (France) sometimes about,1714. Very little is known about his early life and qualification, he was 'an amateur experimenter'. After a well designed field experiments of over three year period on bunt of wheat (caused by *Tilletia tritici*) he published a paper on its findings in 1775. He proved that application of black dust (spores of fungus) from bunted wheat plants to seeds from bunt-free plant greatly increased bunt in the crop they produced. While emphasizing that the bunt was contagious disease, he proposed seed treatments, especially with saltpeter solution and lime could partially prevent the disease. Though he did not knew that bunt is caused by organism still he was awarded a prize by Royal Academy of Literature, Science and Arts of Bordeaux for his brilliantly designed experiment and dissertation on it.

Ivanowski (1892)

First successfully experimentally demonstrated that tobacco mosaic disease have been caused by agents which successfully passed the Chamberland filter that retains even the smallest bacteria. Though he was unable to isolate any bacterium, he still believed that it is either the pathogenic bacterium which has somehow passed the chamberland filter or it is some toxic secretions by it that has made infectious filtrate.

Beijerinck (1898)

He tested the Ivanowski's hypothesis and concluded that it is not the pathogenic bacterium or any of its toxins that cause tobacco mosaic disease but it is '*contagium vivum fluidum*' which he referred as Virus (poison), cause the tobacco mosaic disease. He also mentioned that it multiplies *in vivo.* For this contribution Beijerinck regarded as **"Father of Virology."**

Prof. S.N. Dasgupta, F.N.A

He started research in mycology and plant pathology in 1935. He was Professor of Mycology and Plant Pathology in Lucknow University during 1949–1961. His work on **saltation in fungi** earned him to great fame. He demonstrated that disease black tip of mango is due to deficiency of Boron and could be controlled by spraying Borax. Indian phytopathological society constituted an award in the name of Prof. S.N. Dasgupta for contributions in the field of mycology and plant pathology.

Prof. K.S. Bilgrami

He was Prof. and head of botany Dept., Univ. of Bhagalpur. He has published about 100 research papers on physiology and pathological aspects of fungi in reputed journals. He received his D.Phil. and D.Sc. degree from Allahabad University. Prof. Bilgrami along with Prof. H.C. Dubey published a book 'A Textbook of Modern Plant Pathology' (Vikas Publishing House Pvt. Ltd., New Delhi. Bilgrami have given a comprehensive survey of leaf spot diseases of ornamental plants and fruit trees respectively. An important publication of Prof. Bilgrami was "Studies on growth response of some pathogenic species of Phyllosticta in different nitrogen media. (1964) Flora Bd.154."

Prof. C.V. Subramanium

Prof. Subramanium was born in 1924 and did his graduation from Presidency college, Madras in 1944; Ph.D and D.Sc from Madras University in 1948 and 1957, respectively. He started his career as senior lecturer (Botany) in 1951, then as Professor in the Centre for Advance studies at Madras University. He has established an active school of research in taxonomy and biology of fungi mainly fungi imperfecti, especially the Hyphomycetes. His studies led to discovery of many new genera and species, proposed a new system of classification of Hyphomycetes. Prof. Subramanium is also famous for his well known work on soil mycology and soil born plant diseases and plant pathogens like *Fusarium* and *Drechslera.* Prof. Subramanium is recipient of Shanti Swaroop Bhatnagar Memorial Award for biological sciences (1965), Birbal Sahni Medal of Indian Botanical Society in 1972. He is Fellow of Indian Academy of Sciences and of Indian National Science Academy. Internationally Prof. Subramanium was convener of first international symposium on taxonomy of fungi in 1973, visited university of Bristol in 1968, has chaired sessions at Edinburgh and Seattle Botanical Congress and 2nd International Marine Mycology, Bremerhaven, 1972, delivered keynote address on fungi imperfeti at Kananaskis Conference in 1969. He has been invited to organize and chair the symposium on fungal taxonomy at International Botanical Congress, Leningra,1975.

INDIAN PHYTOPATHOLOGICAL SOCIETY

The society was established by eminent plant pathologist Dr. B.B. Mundkur at Indian Agriculture Research Institute, New Delhi on February 28, 1947 under the Chairmanship of Prof. S. R. Bose. Dr. B. N. Uppal, Dr. A. Sattar and Dr. R.S.Vasudeva were the founder members of the society with Dr. B.B. Mundkur as convener of the society. The Indian Phytopathological Society is considered to be the third largest society of plant pathologist in the world. The society also publish a journal *Indian Phytopathology*. The society is an international organization devoted to the stud of plant diseases and their control.The society provides latest information in the field of plant pathology research and latest developments. The society focuses on the field of Mycology, Fungal Pathology, Bacteriology, Virology, Phytoplasmology and Nematology. It provides a forum for scientific interactions among national and international scientists working in the field of plant pathology research education and extention work. Society also organizes national and international seminars, symposia and conferences in the field of plant pathology. Currently society has more than 2000 members located in around fifty countries. Members of the society includes research scientists, students, teachers, extention professionals, private consultants, administrators and pest management personnels. All have a common goal to promote the knowledge about plant diseases and their control. As per societies document the society has following objectives:

- Advance the cause of Mycology and Plant pathology in India.
- Encourage and promote mycological and plant pathological study and research in the country.
- Disseminate the knowledge of mycology and plant pathology.
- Facilitate closer association and relations among members and other scientific workers in India and abroad.

SOME IMPORTANT JOURNALS IN MYCOLOGY AND PLANT PATHOLOGY

International

Nature: An interdisciplinary science journal first published on November 4th, 1869. It is ranked as world's most cited journal. Published by Nature Publishing Group, U.S.A.

Science popular as science magazine, is a academic journal of American Association for Advancement of Science (AAAS) and is one of the top scientific journal. It was first published in 1880.

Fungal Biology (Former title: ***Mycological Research***): An International Journal of British Mycological Society, Published by Elsevier Publishers.

Mycologia: A peer reviewed scientific Journal of The Mycological Society of America. HighWire Press.

Journal of Applied Microbiology published by The Society for Applied Microbiology, Wiley online Library.

Plant Pathology: A journal of British Society for Plant pathology, John Wiley & Sons.

Fungal Ecology: Published by Elsevier on behalf of The British Mycological Society.

Plant Pathology Journal *of Science Alert*

Journal of Yeast and Fungal Research: An Academic Journal.

European Journal of Plant Pathology: Published by Springer Netherland.

Journal of Plant Pathology: International journal of Italian Phytopathological Society.

National

Phytopathology: A journal of Indian Phytopathological Society.

Indian Journal of Mycology and Plant Pathology: Published by Indian Society of Mycology and Plant Pathology, Udaipur

Journal of Indian Botanical Society: Published by Indian Botanical Society, India.

Proceedings of Indian Academy of Sciences*: *Biological Sciences. Published by Indian Academy of Sciences.

Journal of Basic and Applied Mycology: Published by Society of Basic and Applied Mycology, Bhopal

QUESTIONS

ESSAY TYPE

1. Write a brief account of Mycorrhiza.
2. Comment upon the nature and importance of mycorrhiza.

SHORT ANSWER TYPE

3. Draw well levelled diagram of VAM
4. Write short notes on
   (*a*) Endomycorrhiza
   (*b*) Mycorrhiza
   (*Allahabad, 2000 ; Poorvanchal, 2001*)
   (*c*) Importance of Mycorrhiza.
   (*Kanpur M.Sc., 2006*)

OBJECTIVE TYPE

5. **Fill in the blanks.**
   (*a*) Mycorrhiza is an association between the .... and a fungus.
   (*b*) .... types of Mycorrhiza commonly occur.
   (*c*) The full form of VAM is ....
   (*d*) Generally the roots of angiosperms and .... are associated in the formation ....

SECTION II

SCHIZOMYCOTA, VIRUSES AND MYCOPLASMA

Bacterial Canker

CHAPTER 18

SCHIZOMYCOTA – BACTERIA

- Introduction
- Structure of bacterial cell
- History of Development of Bacteriology
- Definition of Bacteria
- Occurrence, Distribution, size shape and structure of Bacteria cell.
- Nutrition
- Reproduction
- Life History of *Bacillus subtilis*
- Economic Importance of Bacteria

Introduction

A Dutch man, Anton van Leeuwenhoek, was the first to discover the bacteria on June 10, 1675, but he thought them to be animalcules. For about two centuries, thereafter, the bacteria did not receive much scientific attention. Later Louis Pasteur in 1864 and Koch in 1876 reported that these mocroorganisms could cause disease. But it was in 1890 that an earnest study of these primitive, mostly unicellular, organisms began. During the last few decades much has been added to our knowledge of bacteria. Now the study of bacteria forms the subject of a special science called *bacteriology.* A brief history of development of Bacteriology has been given in Table 1. It is possible to touch here only the more general aspect of this science.

TABLE I
History of Development of Bacteriology

Year	Scientist	Contribution
1674	Antony van Leeuwenhoek	Discovered microscope, provided the first illustrated description of microorganisms described as animalcules.
1796	Edward Jenner	Discovered the vaccine against smallpox.
1822-1895	Louis Pasteur	Contradicted the theory of spontaneous generation, established that fermentation is a microbial activity, discovered aerobic and anaerobic microbes.
1843-1910	Robert Koch	Established the relation between causal agent and disease by proposing laws known as Koch's postulates, developed pure culture teachniques for the isolation of microorganisms, described etiology of anthrax, isolated bacteria causing tuberculosis and cholera.
1842-1909	E.C. Hansen	Started the use of pure cultures of fungi and bacteria in industrial fermentation
1876-1885	Sergius N.Winogradsky and M.W.Beijerinck	Proposed and established the role of micro-organisms in biogeochemical cycles such as carbon, nitrogen and sulphur cycle.
1882	Elie Metchinkoff	Proposed phagocytic theory of immunity,
1884	Hans Christian Gram	Discovered a technique for differential staining of bacteria for identification purposes, known as gram's stain.
1886	H.Hellriegel and H.Wilfarth.	Established that root nodules in legumes have relation with nitrogen fixing bacteria.
1888	M.W.Beijerinck	Isolated and cultivated the root nodule bacteria.
1890	Sergius N. Winogradsky	Isolated the first nitryfying bacteria *Clostridium*; discovered *Azotobacter* in butyric acid fermentation.
1898	Kiyoshi Shigo	Established diagnosis of *Bacillus dysentrica* causing dysentry.
1919	Sawaksinan and Jacob Joffe	Isolated an autotrophic bacteria *Thiobacillus thiooxidans*.
1924	Abert Calmette	Developed B.C.G. Vaccine.
1928	Alexander Fleming	Discovered Penicillin
1928	F.Griffith	Discovered transformation in Bacteria
1933	Martin Schlesinger	Successfully isolated bacterial virus for the first time.
1943	S.E.Luria and M.Delbuck	Discovered spontaneous mutation in Bacteria.
1944	O.T.Avery, C.M. MacLeod and M.Mc Carty	Successfully purified the transforming principle in bacteria and confirmed its identification as DNA.
1946	J.Lederberg and E.L. Tatum	Suggested the existence of recombination in bacteria.
1952	N.Zinder and J.Lederberg	Discovered the process of transduction in bacteria.
1971	B.Pierson and K.Castenholz	Discovered the green non-sulphur bacteria–*Chloroflexus*.

Bacteria may be defined as procaryotes without chlorophyll

Occurrence and distribution. The bacteria constitute a highly specialised group of one-celled plants. There are about 2,000 known species They have a closer association with man because of their universal distribution. They are among the most numerous of all living things, being present in almost every conceivable environment. They occur in abundance in both fresh and salt water. The air we breathe is full of them. They are present in the food we eat and the objects we touch. They flourish in our mouth and intestines. They live in the bodies of other organisms and their dead remains. Some live in hot springs and others in ice. They are found in all soil and above the earth. In fact they live in all conditions not fatal to living matter. The features which contribute to their universal distribution are :

(*a*) Extremely simple structure.
(*b*) Small size and consequent large surface-to-volume ratio. In order to maintain their small size and this favourable ratio, cell division must occur rapidly.
(*c*) Resistance of vegetative cells to adverse environmental factors.
(*d*) Formation of highly resistant endospores.
(*e*) Diversity of their modes of nutrition.

Plant Characteristics. The bacteria are microorganisms that possess rigid cell walls and when motile have flagella. Mostly they are unicellular organisms lacking true nucleus and typical plastids. The plant characteristics of the bacteria are :

1. Presence of a definite and rigid cell wall which in a few species contains cellulose.
2. The tendency of some to grow in filaments.
3. The ability of autotrophic bacteria to synthesize organic food from inorganic materials such as carbon dioxide and water.
4. The absorption of food in soluble form through the cell wall.
5. Structure of the bacterium cell and reproductive methods essentially similar to some of the thallophytes.
6. Ability to synthesize all amino acids from inorganic mitrogen.
7. Existence of transitional forms between bacteria and fungi.

Size. Bacteria are exeedingly small unicells. In fact they are the smallest of all known cellular organisms and are visible only with the aid of the microscope. Some of them are so small that they are hardly visible under the high power of ordinary light microscope. Their size is usually expressed in *microns,* a smallest unit of the metric system. A micron (μ) is equivalent to 0.0001 mm. or about 1/25,000 of an inch. An average spherical bacterium cell is 1/2 micron (μ) in diameter. In comparison, a grain of salt is about 500 microns across. Of course the size of bacteria varies considerably ranging from about 0.2μ (micron) to 2 micron (μ) in width and less than 2-10 microns in length. The coccus forms vary from 0.5 to 2.5 microns in diameter. The variations in the *Bacillus* forms are still greater from 0.3 to 15 μ in length and from 0.2 to 2.0μ in width. *Escherichia coli* is 0.4 to 0.7 μ in diameter and 1.0 to 3.0 μ long. The average bacillus size is 3 microns long. A single drop of liquid can contain about 50 million bacteria. A few *Bacillus* and *Spirilla* forms are much larger.

Forms or shapes of Bacteria (Fig. 18.1) The shape of these tiny organisms is their only readily visible characteristic. Thus it is used as a basis for describing them. The bacteria vary in their shape which is usually constant for each species. The three fundamental forms among the bacteria are **rods, spheres** and **helices.** True bacteria belong to one of these three types :

1. Rod-shaped Bacteria. The majority of bacterial cells look like hyphens (-) or tiny rods (A) under the microscope. The rods may be cylindrical and straight or slightly curved or wavy with rounded or blunt ends. These are the *bacilli* (sing. *bacillus*). Bacillus is a Latin word meaning "Little stick or rod". The bacillary forms are by far the commonest and range around 1.5μ in diameter and 10μ in length. Some bear flagella and others not. The bacilli occur singly

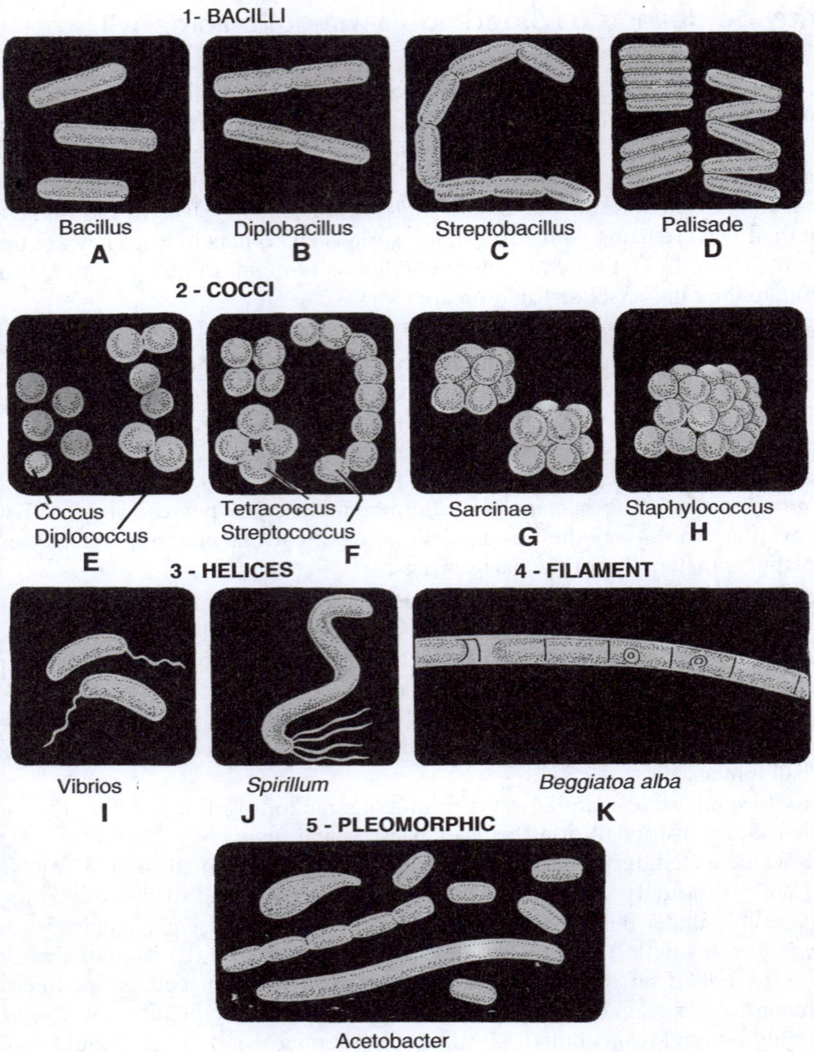

Fig. 18.1 (A-L). Forms or shapes of Bacteria.

(A), in groups of two (*diplobacillus,* B), or in a chain (C) forming a filament (*streptobacillus*) or occasionally in parallel or palisade arrangement (D). *B. subtilis, B. anthracis* and *Lactobacillus* are the common examples of rod-shaped bacteria. In some bacilli the rods are so short as to be indistinguishable from spherical forms.

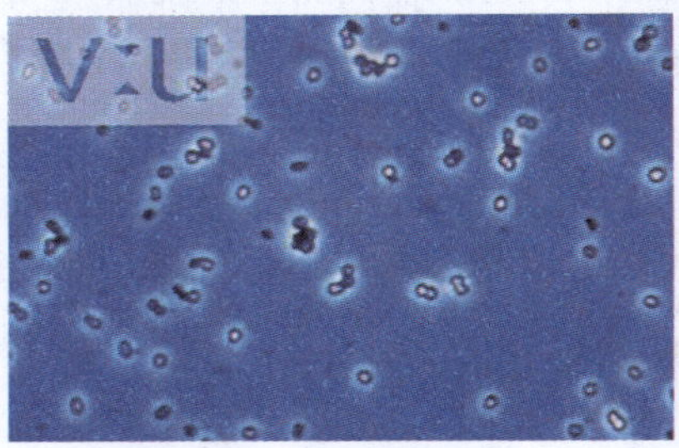

Diplococcus pneumoniae

2. Spherical Bacteria. The second largest group of bacteria have spherical or ellipsoidal cells varying from 0.5 to 1.25μ in diameter. These are called **cocci** (sing. **coccus**). Almost all coccal forms lack flagella and are non-motile. The spherical cells may occur singly but often tend to stick together and various cell arrangements occur. *Diplococcus* sp. tend to occur in groups of two (E) or multiples of two (*D. pneuomoniae*). Tetrads of cocci (F) are formed by *Tetracoccus.* In *Streptococcus* sp. the cocci occur in chains (*S. lactis*). The coccal cells form loose masses of variable size and irregular in shape like bunches of grapes (H) in *Staphylococcus* (*S. albus* and *S. aureus*). In *Sarcina* sp. the spherical cells divide regularly in three planes producing box-like cubical packets of eight or multiples of eight with the cells arranged in rows (G).

3. Helical Bacteria. The helical forms are of two types, **spirilla** and **vibrios.**

Spirilla (J). The bacteria having the body twisted spirally like a cork-screw through 1-5 complete turns are the *Spirilla* (sing. *Spirillum*). The *Spirilla* are larger than the *cocci* and *bacilli* ranging form 10-50μ in length though only 0.5μ to 3μ in diameter. They are smaller in number probably with fewer than 100 species. The spirilla forms usually bear two or more flagella at one or both poles. They occur as single cells or in short chains but do not form colonies. *Spirillum undulum* and *S. volutans* are the common examples of this type.

Vibrios (I). The bacterium cell in the *vibrios* is short and looks like a curved rod (I) or twine of a spiral. In fact it is less than the full turn of a helix and shaped much like a slightly twisted comma. The vibrios have flagella at one pole only and seldom exceed 10μ in length and 1.5 to 1.7μ in diameter. *Vibrio comma* is a typical example of this category.

4. Filament (K). A fourth type of form may occur in a few species. It is a *thread* or *filament.* The number of filamentous forms, as compared with others, is very small. A group of filamentous forms including *Beggiatoa* and *Thiothrix* oxidize hydrogen sulphide. Granules of sulphur accumulate within the cells.

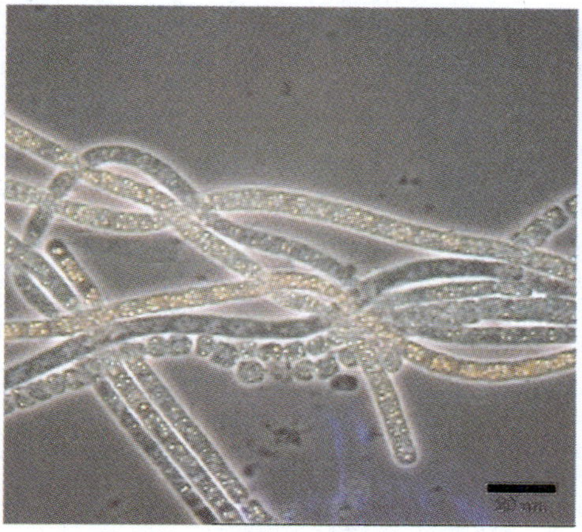

Beggiatoa - Filamentous bacteria

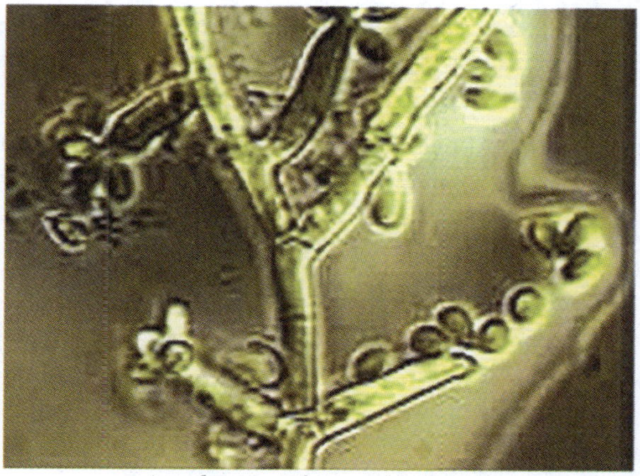

Acetobacter - Pleomorphic bacteria

5. Pleomorphic Bacteria (L). Some bacteria are able to change shape and size with changes in the environmental conditions. These changes are ususally temporary. In a few species, however, several types of cell forms may occur in the life cycle. For example, *Acetobactor,* a group of vinegar bacteria occur as small rods, long rods, ellipsoids or a chain of small rods. Such bateria are called *pleomorphic.*

Staining Bacteria. (Gram Reaction). The most important and commonly used of bacterial stains is the Gram stain. It is a differntial stain which has a simple staining procedure. It was devised by the Danish bacteriologist, Gram, in 1884 and thus named, the **Gram stain** after its discoverer. The procedure involves the staining of a bacterial smear containing a variety of bacteria with crystal violet solution.

The dye is prepared by dissolving 2.0 gm of Crystal violet (80% dye content) in 20 ml of Ethyl alcohal (95%). The dye is applied for 30 seconds to the bacterial smear. It is then gently rinsed off. The smear is next with an aqueous solution of iodine which is prepared by adding 1.0

gm Iodine and 2 to 5 gm potassium iodine to 300 ml of distilled water. Before use the solution is allowed to stand for 24 hours for iodine to dissolve. The iodine solution, in turn, is rinsed off the smear after 30 seconds.

Examination of the smear with the microscope reveals that all the bacteria have taken the purple colour. The next step in the staining procedure is to apply and renew 95 percent ethyl alcohol to the smear until all except the thickest layers of the smear have ceased to give off the dye. Examination of the smear at this stage reveals that many of the bacteria still retain the original purple colour whereas from others it has been washed off. The bacteria that keep the original purple stain are called **gram-positive.** Those which yield it to the solvent (alcohol) and become decolourised are called **gram-negative.** The decolourised bacteria (gram-negative) are nearly invisible. To study their details the final step in the staining procedure is to apply the counter stain such as eosine or safranine to the smear. It gives a good colour to the gram-negative bacteria and at the same time contrast with the purple of the gram-positive bacteria.

The division of bacteria into gram-positive and gram-negative forms, on the basis of the reaction to gram-stain, is correlated with other characterisic differences between them. All spore-producing bacteria are gram-positive. The polarly flagellated bacteria are gram negative.

A comparison between Gram positive and Gram Negative bacteria has been described in the Table 2.

TABLE 2

Comparison between Gram Positive and Gram Negative bacteria.

S.No.		Gram Positive	Gram Negative
1.	Nutritional Requirements	Complex, in many species	Relatively simple
2.	Cell wall		
	(*a*) Structure	Single layered, 150-200°A thick.	Triple layered, 75-120°A thick.
	(*b*) Chemical composition	80% Peptidoglycans and the rest are polysaccharides	Peptidoglycans only 3-12%, mainly composed of lipoproteins and lipid polysaccharides.
		Teichoic acid Present	Teichoic acid Absent
		Lipids Absent	Lipids Absent
		Resistant to alkalies and insoluble in 1% KOH solution	Sensitive to alkalies and soluble in 1% KOH solution
	(*c*) Rigidity	Very rigid due to high percentage of peptidoglycans	Elastic due to plastic nature of lipo protein—polysaccharide mixture

Structure of Bacterial Cell (Fig. 18.2). Bacteria are unicellular organisms. The unicell is exceedingly small in size. Little structural detail can be made out in such a small body with an ordinary light microscope. The use of an electron microscope has revealed details of "fine structure" hitherto unknown (Fig. 18.3). Like other living plant cells, the bacterial cell comprises a **cell wall** and **protoplast.** External to the cell wall may be present a thin layer of **slime.**

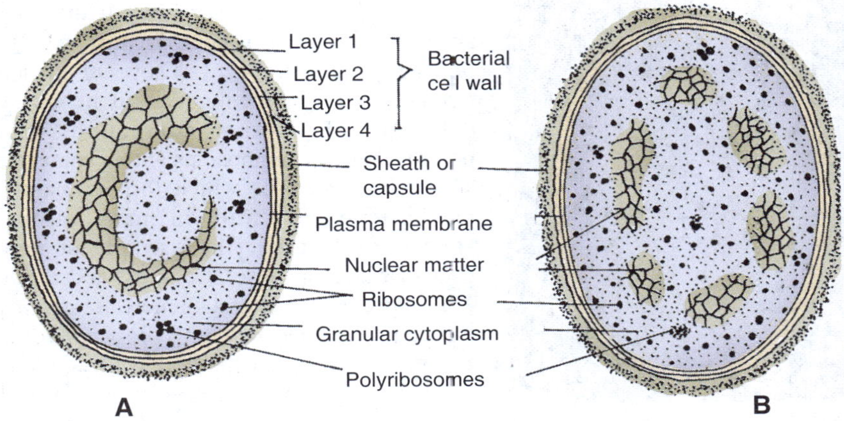

Fig. 18.2 (A-B). *Bacteria.* Diagrammatic representation of the ultrastructure of a spherical bacterial cell. (Based on Mazanek and Martinec).

1. The Slime. It is a viscous or gelatinous substance secereted by the cell protoplast. The slime diffuses through the cell wall and deposits in the form of a thin extracellular, viscous layer external to the cell wall. It is usually composed of polysaccharides or of polypeptides of one or two different aminoacids. Under certain conditions of growth the slime accumulates form a thick conspicuous layer around the cell wall. It is called the **sheath** or **capsule** (A). The sheath protects the enclosed organism against desiccation and antibodies. The encapsulated pathogenic bacteria can withstand phagocytosis. Some bacteria produce diffuse slime which spreads into the environment and encloses bacterial colonies. This is known as loose slime. The bacterial cell with a capsule normally does not bear flagella.

2. The cell wall (Fig. 18.3A). In the electron micrograph the cell wall is seen as a thin, sharply defined envelope around the protoplast. It ranges in thickness around 0.02μ. The cell wall is tough though flexible. The inert and somewhat rigid bacterial cell wall limits the volume occupied by the protoplast and thus gives rigidity and shape to the bacterial cell. It shows granular structure and lacks microfibrils.

(*a*) *Structure of cell wall.* The bacterial cell wall proper is composed of 4 layers (Mazanec and Martinec). Of these two are of lower electron density. The other two layers are of higher electron density and thus darker. The outer darker layer (L_4) is wavy. Within it is the lighter layer of low electron density (L_3). Next comes the inner dense or darker layer (L_2) followed by the innermost layer of low electron density (L_1). The L_2 darker layer of higher electron density is considered to be mucopeptide.

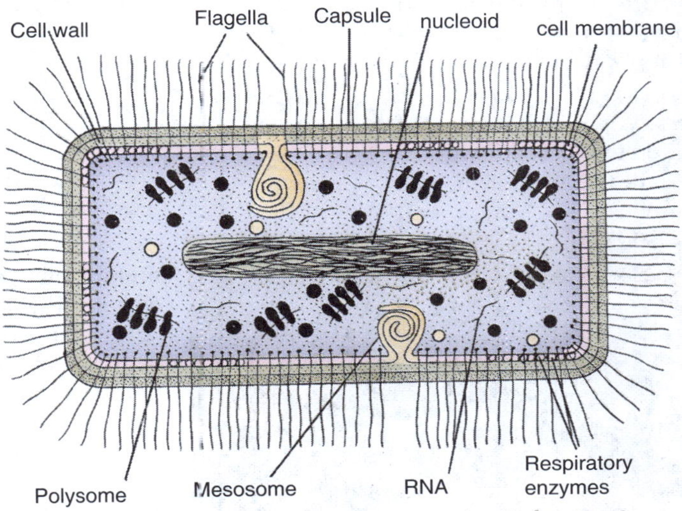

Fig. 18.3 A. *Bacteria.* Rod-shaped bacterial cell showing fine structure (Diagrammatic).

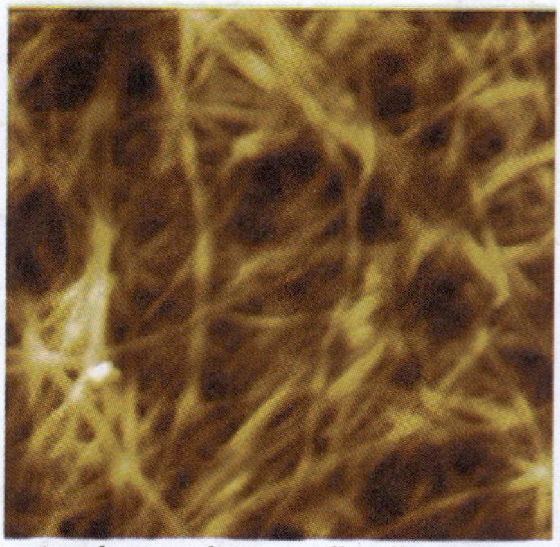

Acetobacter xylinum - cell wall contains cellulose

(b) *Chemical Composition.* It differs in different species. Basically it is composed of acetylglucosamine and acetylmuramic acid. They are present in equal amounts and form the main component of the bacterial cell wall. The other compounds included are polysaccharides, aminoacids and in some species lipids. The polysaccharides which form a major fraction of the bacterial cell wall, in many cases, consist of aminosugars rather than simple sugars (glucose, galactose and mannose). Cellulose and chitin are absent. Three species of bacteria, namely, *Acetobacter xylinum, A. acetigenum* and *Zymosarcina venticuli* are the only exceptions in which the cell wall contains cellulose. Recently microbiologists have reported that the bacterial cell wall contains an aminoacid **diaminopimellic acid.** It commonly occurs in the bacterial cell walls and those of some blue green algae. It is a polypeptide attached to muramic acid, a derivative of glucose.

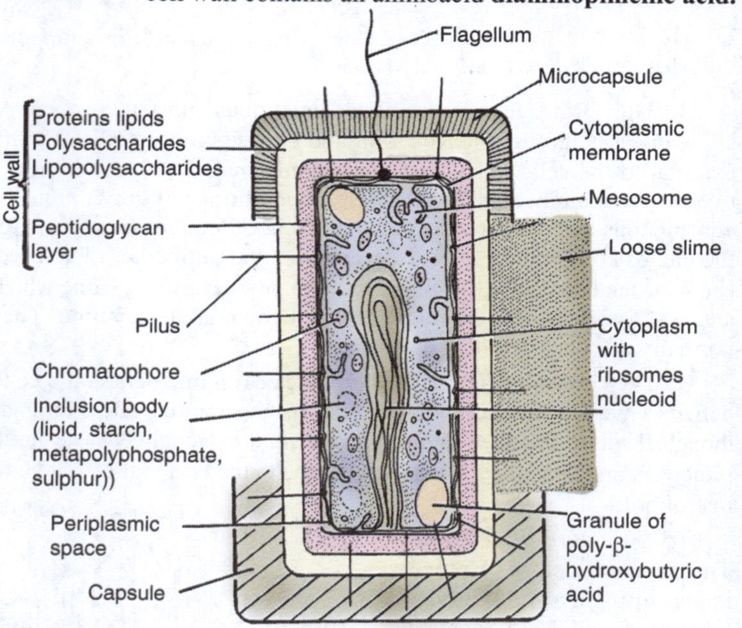

Fig. 18.3 B. A composite diagram of Bacterial cell showing various structures.

3. The protoplast. Within the cell wall is the living stuff of the bacterial cell. It is called the **protoplast.** The protoplast is a clear watery or slightly viscous substance. It is differentiated into the **cell** or **cytoplasmic membrane** (also called **plasmalemma) cytoplasm and chromatin** or **nuclear body.**

(a) *The cytoplasmic membrane.* At the surface of the protoplast is differntiated a delicate, fine membrane known as the **cell membrane** or **cytoplasmic membrane.** It completely encloses the protoplast and lies close against the cell wall pressed by the turgor of the protoplast. Chemically it consists of phospholipids with proteins and some polysaccharides. It is a differentially permeable membrane which controls the passage of dissolved substances in and out of the cell when active metabolism is going on in the protoplasm. In addition, the cytoplasmic membrane serves as a matrix upon which many respiratory and

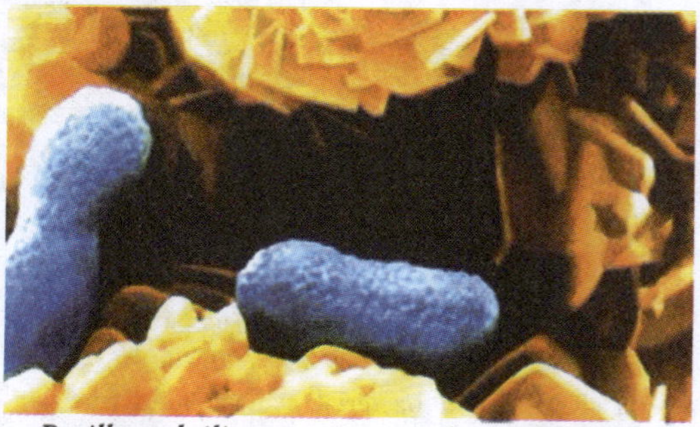

Bacillus subtilis - contains cytoplasmic infoldings called mesosomes

other enzymes are organised. In *Bacillus subtilis* and other gram-positive bacteria the cytoplasmic membrane, at certain places, is seen to form infoldings which in some cases form complex structures within the cytoplasm. These are called the **mesosomes.**

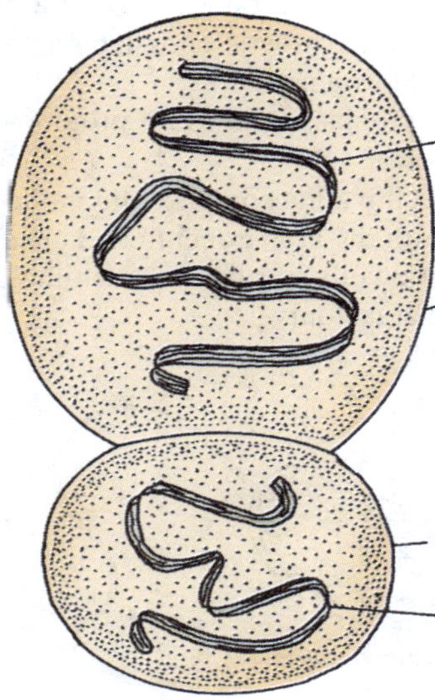

Fig. 18.4. Fine structure of ribosome.

(b) *The Cytoplasm.* It is a complex mixture of proteins, carbohydrates, lipids, minerals, nucleic acids and water. The latter form as much as 90 percent of the cell. Organic matter is in the colloidal state. The non-living inclusions are the storage granules of volutin, glycogen, lipid globules or protein crystals. Sulphur and iron are also found in some bacteria. The cytoplasm shows no streaming movements and contains no obvious vacuoles. It appears rather dense and highly refrigent under light microscope and is fairly homogeneous when free from inclusions. In electron micrographs the bacterial cytoplasm is granular and contains numerous minute, nearly spherical, hollow bodies called the **ribosomes.** Ultrastructurally, the ribosome consists of subunits which are nearly spherical but of unequal sizes (Fig. 18.4). Each subunit is composed of roughly equal amounts of **RNA** and **protein**. The ribosomes lie free in the cytoplasm but sometimes they occur in small groups (**polyribosomes**) held by a strand of messenger **RNA.** Thr ribosomes are the sites of protein synthesis. Other cell organelles such as mitochondria, endoplasmic reticulum and dictyosomes have not been seen in the bacterial cytoplasm. The chloroplasts are absent and so is a definitely organised nucleus. The pigments in the photosynthetic bacteria (green and purple bacteria) have long been considered to be diffused throughout the cytoplasm. Recent investigations with the electron microscope have revealed the presence of numerous ultramicrosopic chromatophores in the cells of certain photosynthetic bacteria. In rare cases lamellar structure has been revealed.

(c) *The Nuclear Apparatus* (Fig. 18.2). It is the area in which all the chromatin or genetic material (DNA) of the cell is concentrated. Usually it is centrally located in resting bacterial cells and is equivalent of the nucleus in other organisms. However, it lacks the nuclear membrane and nucleoli. It does not divide by mitosis. Nucleoid or gonophore is the term applied to the nuclear body of bacteria. Lacking the bounding nuclear membrane, the shape of the nuclear body varies. Often it is irregular, sometimes spherical or oval but elongated (rod-shaped) or dumb-bell-shaped in the dividing cells. The cell in the phase of rapid growth may contain one to four unclear bodies or nuceloids. In *Micrococcus cryophillus* the nuclear material is scattered in the form of islands separated by cytoplasm (B) or it may be arranged in the form of a ring (A) enclosing islands of cytoplasm. In contrast to the granular cytoplasm, the nucleoids appear as light on low density areas containing fine fibrils of DNA (genetic or chromatin material) variously interwoven. The **DNA** in the nucleoid of *Escherichia coli* (Fig. 18.3) occurs in the form of a single, two stranded, thread-like molecule about 1,000 microns (=1mm) long. It is ring-shaped (circular) and tightly packed. No ends are seen. The **DNA** molecule contains all the genetic information of the cell. It is often but erroneously referred to as a chromosome.

In its structure the bacterial cell resembles the cyanophycean (blue-green algae) cell in many respects and differs from the eukaryotic cells of other plants in the following respects :—

(*i*) The nuclear membrane and nucleoli are absent in the nuclear body.
(*ii*) The mitochondria, dictyosomes and endoplasmic reticulum are lacking in the cytoplasm.

(*iii*) The nuclear body does not divide by mitosis.

(*iv*) The photosynthetic lamellae when present (phyotosynthetic bacteria) are not organised into chloroplasts.

(*v*) The cytoplasm shows no protoplasmic streaming movements and contains few, if any, vacuoles.

(*vi*) The cell wall is complex and multi-layered.

The cell with the above-mentioned characteristics is termed **prokaryotic** in contradistinction to the eukaryotic cells of plants and animals. The prokaryotic cells are characteristic of bacteria and blue-green algae only.

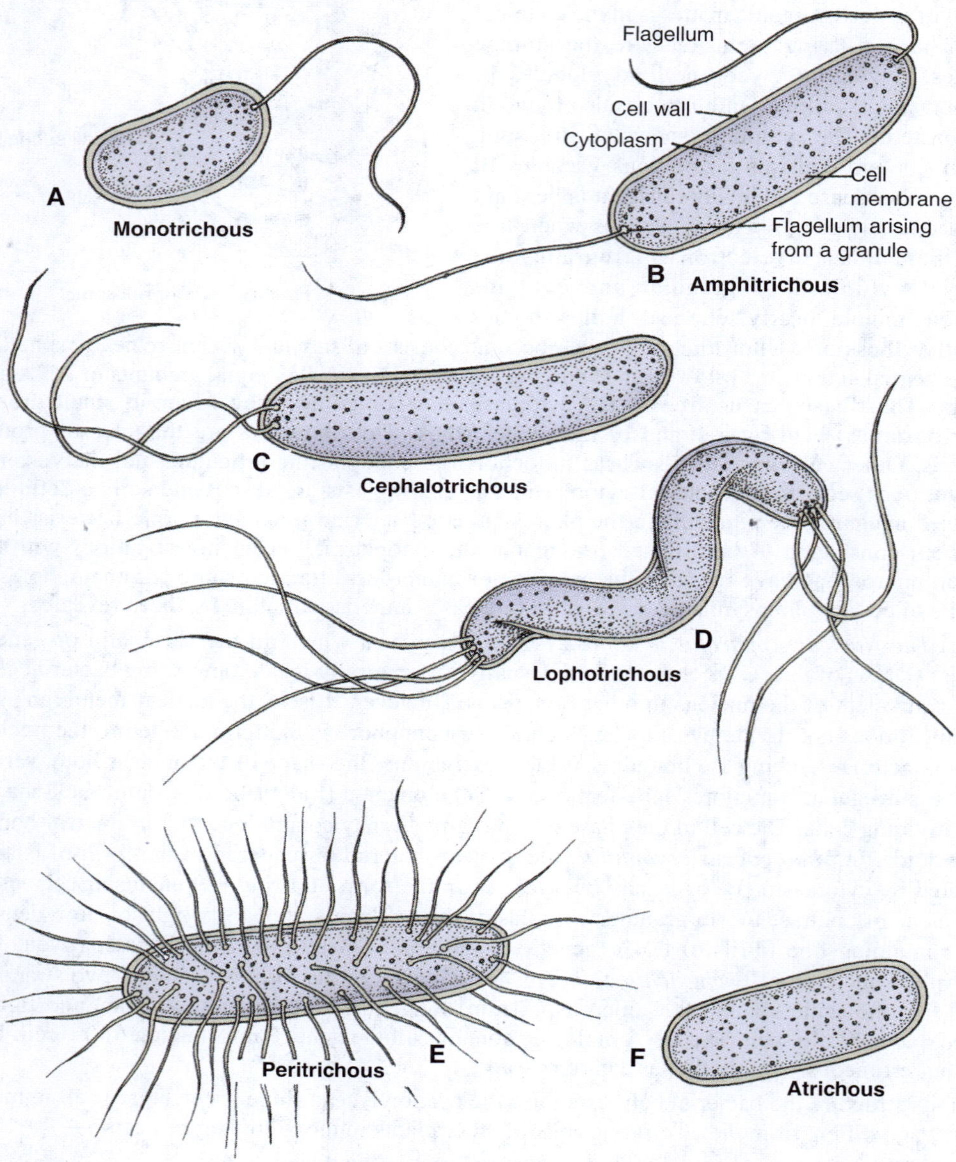

Fig. 18.5 (A-F). *Bacteria*. Different kinds of flagellation. Explanation in the text.

Bacterial Motility. Some bacteria are self-motile. They are able to swim through the liquid in whch they live. They cannot crawl over dry surfaces or fly through the air. Motility is universal among the spirilla, common among the bacilli but lacking or rare in coccal forms. The organs of locomotion are small whips or hair-like appendages called the **flagella.** These fine filaments are of uniform thickness. They are about 120A°-150A° thick and 4 or 5μ long. They are cytoplasmic in nature. Each flagellum arises within the cytoplasmic membrane but very close to it from a granule called a **blepharoplast** and passes out through the cell wall.

1. Flagellation (*Distribution of Flagella*). The flagella are distributed over the surface of the bacterial cell in a characteristic manner. Their number, position and arrangement varies with the species. They *may* be restricted to one or both the ends of the bacterium cell (**polar flagellation**) or may be distributed uniformly all over the body surface (**non-polar flagellation**). Bipolar flagellation is of rare occurrence. On the basis of number and arrangement of flagella, bacteria are classified into the following categories :—

A. Polar Flagellation (Fig. 18.5 A-D). This type of flagellation is restricted to a rather homogenous group of bacilli and spirilla. They are all gram-negative. Polar flagellation is of the following four types :—

(*a*) *Monotrichous*. In this case the bacterium cell bears only single flagellum which is inserted at or near one pole of the cell (A). The flagellum has more than two curves. All *vibrios* (*Vibrio cholerae* and *V. metschnikovii*) are monotrichous.

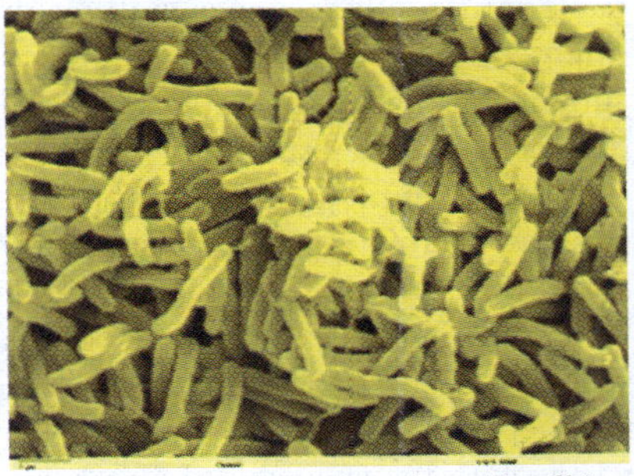

Vibrio cholerae (*Monotrichous flagella*)

(*b*) *Amphitrichous*. There is one flagellum at each pole of the bacterium cell (B). This type of flagellation is of are occurrence.

(*c*). *Cephalotrichous* (Thimman, 1959). There are two or more flagella in a bunch at one pole of the bacterium cell (C). Rod-shaped bacterium, *Pseudomonas fluorescens* is an example of this category. The flagellum shows more than two curves.

(*d*) *Lophotrichous* (Thimman, 1959). There are two or more flagella at both the poles of the bacterium cell (D). The flagella show one or two curves. The spirilla are the common examples (*Spirillum volutans*).

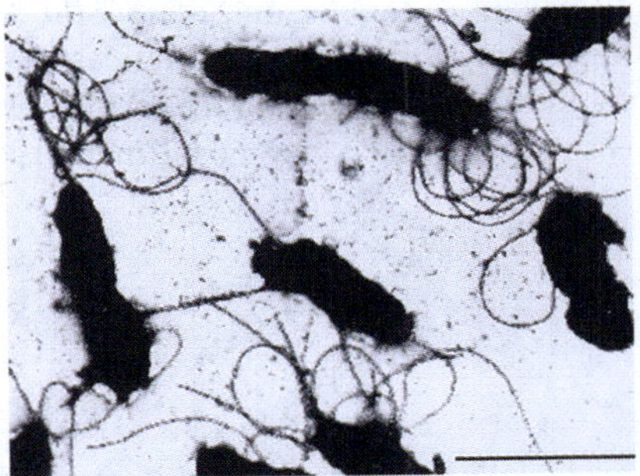

Pseudomonas fluorescens (*Cephalotrichous flagella*)

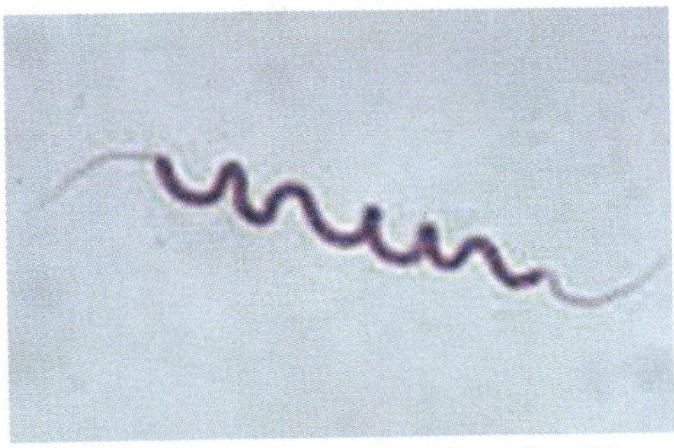

Spirillum volutans (*Lophotrichous flagella*)

B. Non-polar Flagellation (Fig. 18.5 E). In this case the flagella are found all over the cell surface. It is of one type only.

(*e*) *Peritrichous* (E). The bacterial cell bears a large number of flagella which are distributed evenly all over the cell surface. *Proteus vulgaris* and *Bacillus typhosus* are the common examples. In fact, peritrichous flagellation is more commonly found among the bacteria than polar flagellation.

(*f*) *Atrichous* (Fig. 18.5 F). All bacteria which lack flagella are called atrichous. *Diptheria bacilli* and *Lactobacillus*—a milk souring group of bacteria—are entirely without flagella and thus non-motile. The passive type of movement is exhibited by most of the bacteria. It is purely physical and is also known as *Brownian movement.* This type of movement is shown by all small particles suspended in a liquid. The movement of bacterial flagella is screw-type motion in three dimensions.

2. *Structure of Bacterial Flagellum* (Fig. 18.6 A-C). The bacterial flagella are the most primitive of all motile organs. Structurally they are unique. Each is composed of a single thin fibril as against the 9+2 pattern of fibrillar structure characterisitc of eucaryotic motile cell types. The bacterial flagellum is thus equivalent to one of the peripheral half fibrils (microfibrils) of the flagellae of all motile eukaryotic organisms. Electron microcopy reveals that the bacterial flagellum is helical in shape. Because of its spiral patterned surface it apparently seems to consist of a few fine fibrils twised tightly into a rope-like structure coiled in an open wave form (A). The flagellum is composed entirely of protein which is **Flagellin.** It gives a typical helical diffraction pattern in the electron microscope. Its molecular weight is about 40,000. According to Lowy and Hanson (1965), the bacterial flagellum, under high resolution of electron microscope, is seen to be composed of globular subunits arranged in helices of various kinds. The diameter of each subunit is approximately 40-50Å (Angstrom units). The flagellum is composed of several chains of subunits (flagellin molecules) forming a hollow cylindrical thread. There is no axoneme. Whether the hollow core is empty or filled with a similar or different protein is not known. The number and orientation of chains in the flagellum varies with the species. It is not a uniform feature of all bacterial flagella. In *Salmonella,* for example, the flagellum consists of four helices or spiral chains of subunits as shown in figure 18.6 B. One of the chains is shown by black globules. Ringo (1967) investigated

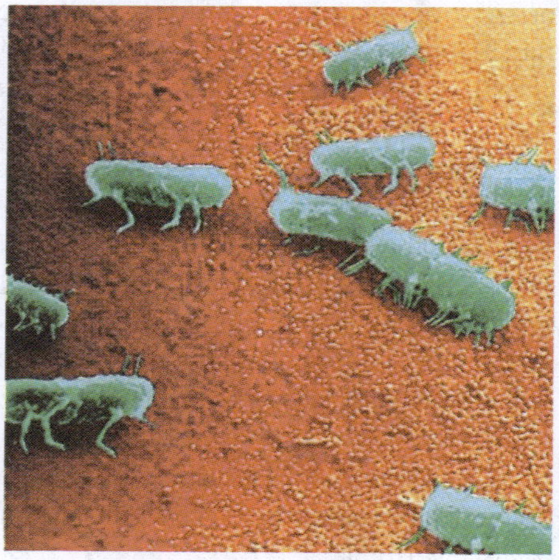

Salmonella (*Flagellum constitutes flagellin*)

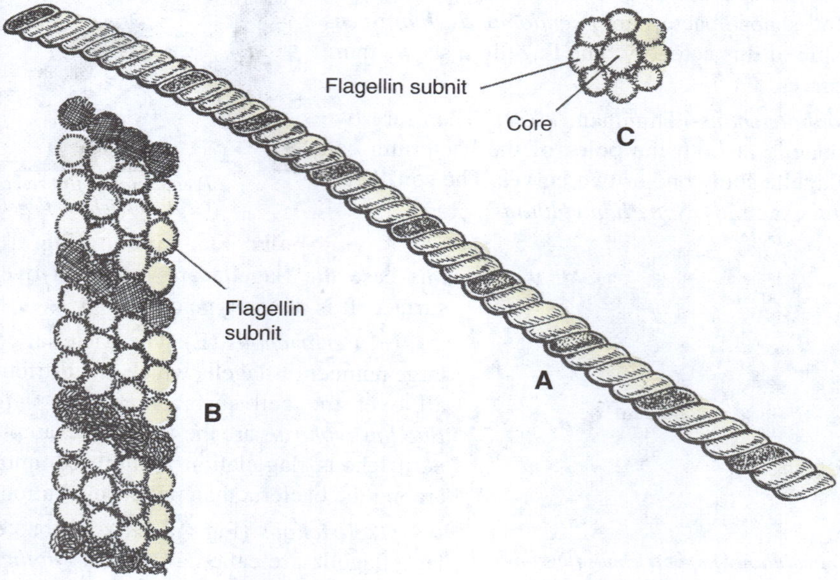

Fig. 18.6 (A-C). *Bacteria.* Molecular structure of flagellum. A, Flagellum as seen in electron microscope; B, Portion of flagellum under high resolution of electron microscope; C, end view of flagellum.

the molecular structure of the flagellum of a species of *Chlamydomonas.* He reported that in a cross section each of the peripheral half fibrils (microfibril) consists of a ring of 13 subunits surrounding a homogeneous core. The diameter of each subunit is 40Å which corresponds to a protein with a molecular weight of approximately 40,000.

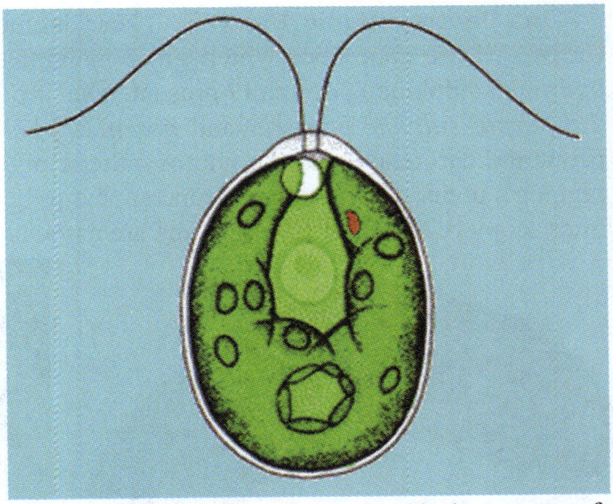

Chlamydomonas (Flagellum microfibrila comprise of 13 subunits

Pili (Fig. 18.7). These are minute, straight, hair-like, non-flagellar appendages found extending out from the cell wall of many gram-negative bacteria. Like flagella they can only be seen in electron micrographs. They are somewhat rigid and composed entirely of a protein known as **pilin.** They may be several microns long but are smaller in diameter which is around 30-50Å. There may be one hundred to four hundred **pili** or **fimbriae** distributed over the surface of the bacterium cell. Pili are considered as the organelles of attachment.

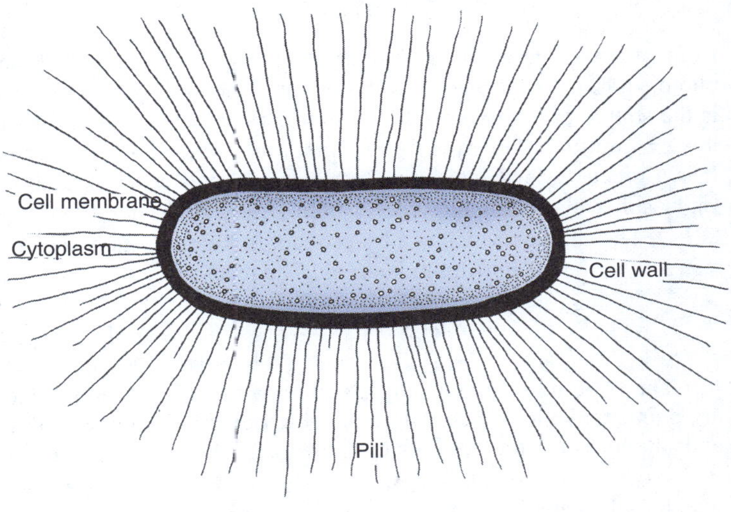

Fig. 18.7. *Bacteria.* Cell showing the distribution of pili.

Nutrition. Most of the bacteria lack chlorophyll. They are thus unable to synthesize organic compounds which they need as food from simple inorganic substances. They obtain these from the external sources readymade. For this reason these organisms are put in the category of **heterotrophic** plants. However. there are some bacteria which are able to synthesize organic compounds necessary for their structure and metabolism from simple inorganic compounds. These are called **Autotrophic** bacteria. The latter live on a strictly inorganic diet. The autotrophic bacteria resemble green plants in their mode of nutrition but with a difference. The green plants utilize water as one of the raw materials in photosynthesis (food manufacturing process). The autotrophic bacteria, in contrast use, other compounds of hydrogen such as hydrogen sulphide (H_2S) and methane (CH_4) and not water with the result that no oxgen is evolved as a by-product in bacterial food manufacturing process. The energy used is this case in either obtained from the sunlight (photosynthetic bacteria) or chemically by the oxidation of some inorganic substances present in the environment such as iron, sulphur, nitrogen compounds (Chemosynthetic bacteria).

1. Autophytic or Autotrophic Bacteria. The autophytic bacteria are thus of two types, **photosynthetic** and **chemosythetic.** The former are also called **photosynthetic autotrophs** and the latter **non-photosythetic autotrophs.**

(a) Photosynthetic Bacteria. These include all the three, coccal, bacillary and spirillum forms. All are anaerobes which grow in light and are usually found in sulphur springs where hydrogen sulphide is normally present. The two common examples of photosynthetic bacteria are the *purple sulphur bacteria* and *green sulphur bacteria.* The former contain **bacteriochlorophyll** and carotenoids ad the photosynthetic pigments as the latter **chlorobium chlorophyll.** The pigments under the light microscope appear to be uniformly distri-buted in the cytoplasm. Electron micriscopy reveals that the pigments are associated with membranes which arise as invagi-nations from the plasma membrane and form a network of small spherical vesicles connected by narrow constrictions. These bacteria are unique because they are the only organisms which are capable of synthesizing carbohydrate food without chlorophyll *a.* Moreover, photosynthesis in photosynthetic bacteria like *Rhodospirillum* is different from photosynthesis in green plants. No free oxygen is released as a byproduct in bacterial photosynthesis. Hydrogen is provided by donor substances other than water. Hydrogen sulphide is the source of hydrogen. The by-product is sulphur and not oxygen. The process takes place at low expenditure of energy. *Chlorobium limicola* is another example of photosynthetic autotrophs. It is a green sulfur bacterium. The hydrogen sulfide is the donor of hydrogen in this case also. Light splits hydrogen sulphide in both cases. Hydrogen combines with carbon dioxide to form CH_2O. The sulphur which is a by-product is stored as globules in the cells of sulphur bacteria but excreted in the green sulphur bacteria. The reaction is :

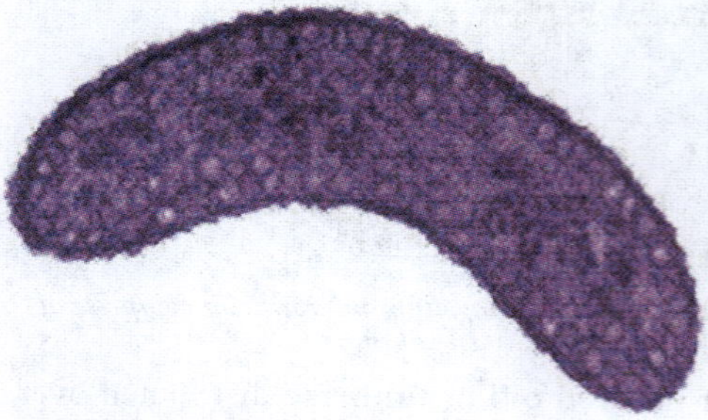

Rhodospirillum

Chlorobium limicola

PHOTOSYNTHETIC BACTERIA

$$2H_2S + CO_2 \xrightarrow[\text{Carboydrates}]{\text{light}} (CH_2O) + 2S + H_2O$$

There are non-sulphur purple and brown bacteria found in the mud and stagnant water. They are photosynthetic and contain **bacteriochlorophyll** pigment. They use organic hydrogen donors and sulphur is not the by-product in their case. Light is still the source of energy.

(b) Chemosynthetic Bacteria. These are non-photosynthetic autotrophs (nitrifying bacteria) which lack pigments. They get energy for food synthesis by the oxidation of certain inorganic substances such as ammonia, nitrites, nitrates, ferrous iron, hydrogen sulphides and a number of other metallic and non-metallic materials available in the environment. The bacteria absorb inorganic molecules of the substances into the body where a chemical reaction takes place. In this reaction the chemical bonds are broken and energy is released. This energy is used up by the bacterial cell to combine CO_2 and water into food molecules. This process of manufacturing food is called **chemosynthesis.** It is a chemical process and is evidently a form of respiration. No light is involved in chemosynthesis. The reactions are all exothermic. These bacteria are thus independent both of light and of organic materials. They are unique because they can live even under conditions unfavourable for plant growth. Examples of chemosynthetic bacteria are *sulphur bacteria, iron bacteria, nitrifying bacteria* and *hydrogen bacteria.*

Sulphur bacteria oxidise sulphur compounds present in water in which they live. The energy released in the reaction is utilized for food synthesis. Sulphur remains behind as a residue in the cytoplasm of the cell. The reaction is represented by the following equation :

$$H_2S + 1½O_2 \longrightarrow S° + H_2O + \text{energy}$$

So is the case with iron bacteria. They inhabit water containing iron compounds. The iron bacteria oxidise the ferrous compounds to ferric form liberating energy. The energy yielding reaction is :—

$$Fe^2 \longrightarrow Fe^3 + e-$$

The ferric iron is deposited as insoluble ferric hydroxide.

Nitrosomonas europaea and *Thiobacillus denitrificans* are the other examples of chemosynthetic autotrophs. The former obtains energy for food synthesis by the oxidation of ammonia. The reaction is as follows :

$$\underset{\text{(Ammonia)}}{NH_3} + 1\tfrac{1}{2}\, O_2 \longrightarrow H_2O + \underset{\text{(nitrous acid)}}{HNO_2} + \textbf{energy}\ \text{(ATP)}$$

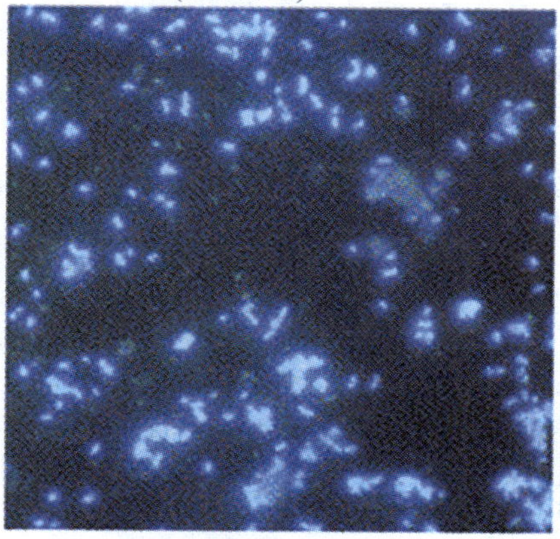

Thiobacillus denitrificans

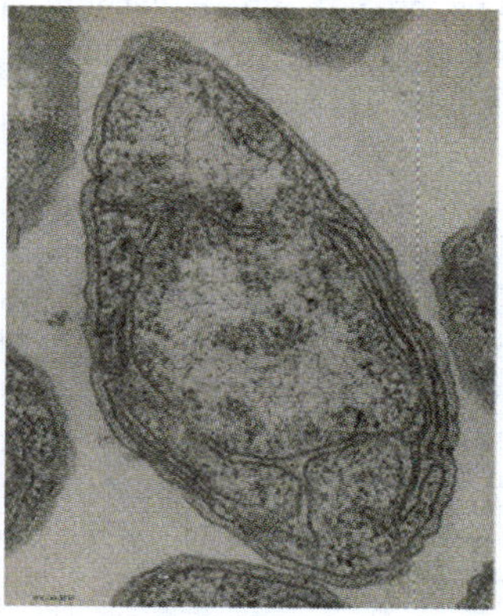

Nitrosomonas europaea

CHEMOSYNTHETIC AUTOTROPHS

In other nitrifying bacterium *Nitrobacter*, the only source of energy is the oxidation of nitrite. It is represented by the following equation :

$$NO_2 + 1\tfrac{1}{2}\, O_2 \longrightarrow NO_3 + \textbf{energy}\ \text{(ATP)}$$

The oxidation of hydrogen gas by hydrogen bacteria is represented by the following equation :—

$$H_2 + 1\tfrac{1}{2}\, O_2 \longrightarrow H_2O + \textbf{energy}\ \text{(ATP)}$$

The chemical energy released in the form of ATP in the above-mentioned examples is used by the bacterial cell to synthesize food molecules from the inorganic substances. In all chemosynthetic bacteria the energy required for the food making process is thus obtained from their oxidative activities. The two processes of food synthesis in bacteria may be contrasted as under.

	Chemosynthesis	Photosynthesis
1.	The amount of energy available in chemosynthesis is much less as compared with photosynthesis.	The amount of energy available in photosynthesis is much more than in chemosynthesis.
2.	There is no gain of energy from outside the planet.	There is distinct gain of energy from outside the planet.
3.	No light is involved in this process.	It takes place in the presence of light.
4.	Reactions are all exothermic.	The reactions are endothermic.
5.	Energy required for the process is obtained by the oxidation of certain inorganic substances available in the environment.	Solar energy trapped by the pigments is used in the process.

2. *Heterotrophic or Heterophytic Bacteria.* The colourless heterotrophic bacteria, which form the majority, cannot synthesize organic compounds from simple inorganic substances. Lacking the pigment, they cannot capture solar energy which is essential to synthesize the substances they need as food. Perforce they secure their food ready made from external sources. Thus, the heterotrophic bacteria live in places where organic food is readily available either from living organism or their dead remaings and waste products. The former are called *parasites* and the latter *saprophytes* or *saprobes*.

(*a*) *Saprophytic bacteria.* They grow in dead, decaying organic material, and live by digesting and absorbing them. In the course of obtaining food for themselves, the saprophytic bacteria gradually break down complex organic compounds into simpler products. This they do by secreting enzymes. The breakdown of carbohydrates is called *fermentation.* The breakdown of protein materials is called *putrefaction.* These simpler products are rendered into a soluble form and are absorbed as food. These two processes constitute the useful chemical activities of saprophytic bacteria.

Fermentation. The word fermentation comes from a Latin word meaning "to boil". It refers to bubbling due to the carbon dioxide given off. The fermentation of sugar by yeast is the classic example. Certain species of bacteria are able to ferment glucose and galactose. Lactic acid fermentation is principally carried out by bacteria. It takes place without the evolution of carbon dioxide. In its place an organic acid (Lactic acid) is formed. This acid causes the souring and curding of milk. It is carried out by lactic acid bacteria. The process takes place in two steps. First the lactic sugar in the milk is changed into glucose by the enzyme lactose. It is secreted by milk souring bacteria. The glucose is later changed to lactic acid by the lactic acid bacteria. The formation of lactic acid causes the souring and curding of milk. Curding is the result of coagulation of the milk protein (casein).

Putrefaction. Various species of bacteria are principally concerned in the decomposition of proteins into simpler compounds. The earlier stages in protein putrefaction take place in the absence of oxygen. They are brought about by an enzyme produced by the anaerobic bacteria concerned in the process. As a result the complex protein compounds are reduced to proteosis, peptones, polypeptides, peptides and amino acids. Some of these substances have foul or unpleasant odour. The later stages of decomposition usually require oxygen. The amino acids are further decomposed by various saprophytic bacteria resulting in the production of much simpler substances such as methane (marsh gas), carbon dioxide, hydrogen sulphide, water, nitrogen, hydrogen and ammonia. The formation of ammonia resulting from the decomposition of amino acids is called **ammonification.** A few species of saprohytic bacteria are able to decompose fats into glycerine and fatty acids. A few putrefying bacteria are phosphorescent. They produce light causing rotten fish, meat or wood to shine in the dark.

(*b*) *Parasitic Bacteria.* They live on or within living organisms both plants and animals. They obtain their organic food from the host on which they grow. Not all parasitic bacteria cause disease. Some may be harmless to their hosts and others cause serious diseases in them. To the former category belong the symbiotic species which give something useful to the host and receive shelter and food in returen. The numerous *coli* bacteria inhabiting the intestines in man and most other organisms are the typical examples of this kind. They do not cause disease or any other injury to the host but help in the digestion of cellulose by breaking down the cell walls through the secretion of various enzymes. The also deposit vitamins in the intestines which can be absorbed and used by the host. In nature they live on the otherwise unusable food. The harmless parasitic bacteria are said to be **non-pathogenic** and the disease causing ones **pathogenic.** Many diseases of plants and animals including man are caused by the pathogenic bacteria. The disease is the result of either the direct attack of the parasite upon the host tissues or liberation of toxic substances which may be either caustic in effect or a protein to which the body reacts. This disturbs the host metabolism. The common examples of human diseases which are bacterial in origin are cholera, pneumonia, diphtheria, tuberculosis, tetanus, typhoid and many others. There are relatively few bacterial pathogens of plants. The unicellular condition of the bacterial cells is the handicap. It cannot be readily transported inside the multicellular plant body as it lacks the circulatory system. Cabbage rot, pear blight, cirtus canker and a few others are common bacterial diseases of plants.

Respiration. Like other living things bacteria respire. They oxidize food materials present in the cytoplasm to obtain energy. Most bacteria make use of the free oxygen of the atmosphere or oxygen dissolved in the liquid environment. They are called the **aerobes** or **aerobic bacteria.** They are so called because they can live only in the presence of free oxygen. Free oxygen is necessary for their respiration. The free oxygen diffuses in through the bacterial cell wall and oxidizes the food materials present in the cytoplasm. The reaction takes place in two steps. The first step includes the oxidation of food materials with the removal of pairs of hydrogen atoms. The second step comprises the oxidation of hydrogen atoms by oxygen with the liberation of energy. The released energy is captured in ATP (Adenosine triphosphate) formed from Adenosine diphosphate (ADP) and phosphoric acid. The oxidation of food thus results in the production of CO_2, H_2O and formation of ATP. Carbon dioxide diffuses to the exterior through the body surface. The important feature of respiration is not so much the gaseous exchange but the formation of ATP. It is a compound in the body of a living organism which captures and stores the energy released in respiration which otherwise would simply have produced heat. The energy entrapped in *ATP* is used gradually as the need arises to run many reactions in the bacterial cell. Aerobic respiration is represented by the following equation :—

$$C_6H_{12}O_6 + 6O_2 \longrightarrow 6CO_2 + 6H_2O + 686 \text{ K. Cal. (energy in the form of ATP)}$$

Anaerobes or Anaerobic bacteria. There are a considerable number of bacteria which are able to live and multiply in the absence of free oxygen. In fact they perish in the presence of free oxygen. These peculiar bacteria obtain oxygen for their respiration from organic compounds such as sugar. They are called the **anaerobes** or **anaerobic bacteria.** A good example of this type are the bacteria which decompose glucose to form alcohol and carbon dioxide.

$$\underset{\text{Glucose}}{C_6H_{12}O_6} \longrightarrow \underset{\text{Alcohol}}{2C_2H_5OH} + 2CO_2 + 29 \text{ k calories of energy}$$

Anaerobic respiration is accomplished by the secretion of certain oxidizing enzymes. The latter bring about breakdown of foods. It is followed by the rearrangement of atoms within the organic molecule. Certain molecular groups take up the contained oxygen from the others. The amount of energy available from this type of respiration is much less than when free oxygen is used. This is because glucose molecules are not completely oxidised. Much of the energy stored in glucose remains in alcohol. Some of the anaerobic bacteria do not require free oxygen. They are actually poisoned or killed by its presence. They can live and multiply only in the absence of free oxygen. Examples are syphilis and tetanus bacteria. Such bacteria are called **obligate anaerobes.** There are other anaerobic bacteria which can live and grow whether oxygen is present or not. They are called the **facultative anaerobes.**

Reproduction. Asexual reproduction is characteristic of all bacteria. Sexual reproduction was long thought to be absent but investigations with the help of electron microscope have clearly demonstrated the exhchange of genetic material in some species of bacteria. This is an essential feature of the sexual process and is denoted by the term **genetic recombination.**

A. Asexual Reproduction. It takes place by two methods (1) vegetative and (2) spore formation.

1. *Vegetative Reproduction in Bacteria.* The most common means of vegetative reproduction in bacteria is cell division or fission.

(*a*) Fission (Fig. 18.8 A-F). Bacteria multiply by this method when there is sufficient amount of water, fair supply of nutrients and favourable temperature. Fission consists in the splitting of the mother cell into two or more daughter cells. In bacteria the division of the mother cell into two daugher cells of approximately the same size is the rule. The process is thus called **binary fission.** The bacterial cells of the *bacillus* and the *spirillum* forms typically divide in a transverse plane which is at right angles to the long axis of the cell. The coccal forms may divide in one or more than one plane. Mazanec and Martinec studied cell division in *Micrococcus cryophillus* using electron microscope. According to them, binary fission in this bacterium involves the following :—

(*i*) *Divsion of nuclear material* (Fig. 18.8 A). Reaching the maximum size which is fairly constant for the species, the bacterial cell elongates. This is followed by the division of the nuclear material by a process analogous to mitosis. However, there is no spindle formation and resolution of chromatin material into chromosomes. The divsion of nucleus precedes the divsion of cytoplasm of the cell.

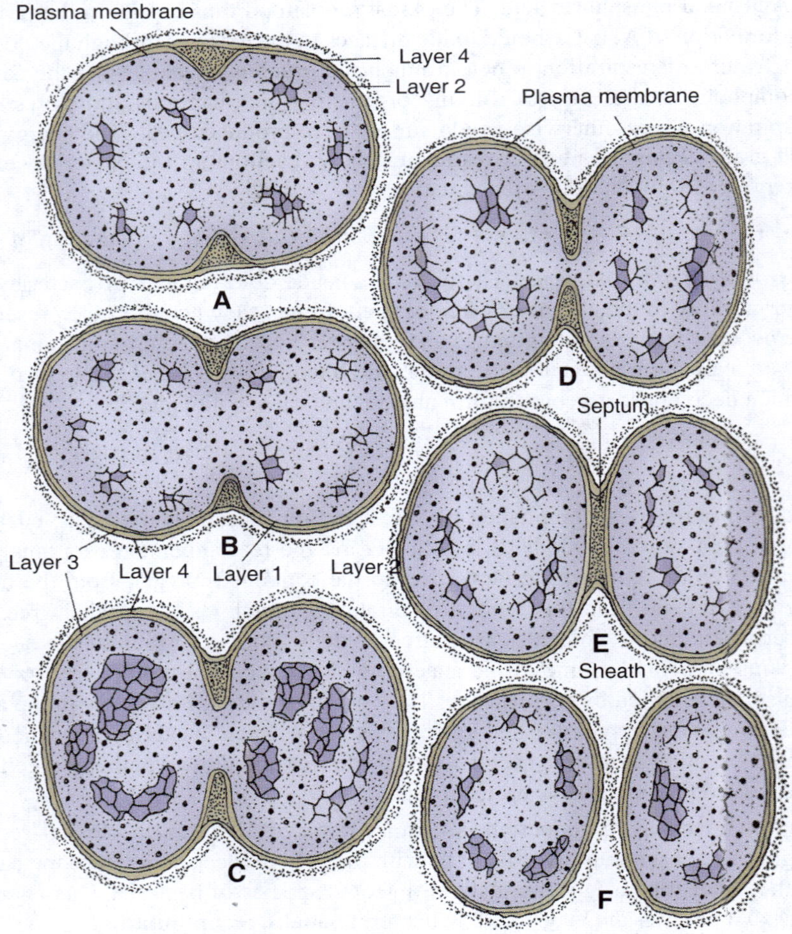

Fig. 18.8 (A-F). *Bacteria*. Stages in the cell division of *Micrococcus cryophillus*. Diagrammatic(Based on Mazanec and Martinec)

(*ii*) *Septum formation.* The division of the nuclear matter is followed by the formation of a septum between the two halves of the nuclear material. The process starts with a ring-like invagination of the plasma membrane. Concomitant with this is the thickening of the cell wall layer L_2, which is electron dense and consists of mucopeptide, to form the *septum initial* (A) in the region of the plasma membrane sulcus. Layer L_1 which is electron transparent, bulges in front of the septum initial (B). The peripheral layers L_3 and L_4 of the cell wall are not involved in septum formation. They simply pass over this region or show a small median constriction at the early stages of cell division. The continuous centripetal growth of the ring-like invagination of the plasma membrane, septum initial (L_2) and layer L_1 eventually form a complete partition (C-D), which divides the parent protoplast into two sister protoplasts. The partition consists of a middle thin electron-dense mucopeptide septum enclosed on either side by the electron-transparent layer L_1 and plasma membrane (E).

(*iii*) *Splitting of the middle septum.* The middle mycopeptide septum thickens and finally splits centripetally into two separate thin electron-dense layers. The two sister cells are now completely separated but still enclosed within the common cell wall layers L_3 and L_4 as a single structural unit (E). At this stage each sister protoplast has a continuous plasma membrane, electron-transparent layer L_1 and electron opaque mucopeptide layer L_2. The ends of the latter (Split septum) are still continuous with the common peripheral parent wall layers L_3 and L_4.

(*iv*) *Separation of sister cells (F).* To achieve complete separation of the two sister cells which is the last step in binary fission, the small median constriction, that developed at the surface of the dividing cell at the early stage, gradually deepens during the process making fission between the daugher cells visible externally. The growth of the sister cells in addition sets up turgor pressure which pulls on their adjacent walls in the region of the contact. Consequently the separation of the two sister cells starts at the periphery (E). It slowly proceeds inwards towards the axis of the mother cell separating the two sister cells (F). The peripheral cell wall layers L_3 and L_4 which do not take part in the division of the cell, are formed *de novo* between the daughter cells before they actually separate.

Fission in bacteria often occurs as often as every 18-20 minutes under favourable conditions of temperature, food and moisture supply. Rarely the doubling period is as little as twelve and half minutes. A cholera bacterium dividing at the rapid rate of 20 minutes will produce in 24 hours a progeny numbering 4,700,000,000,000,000,000,000 individuals. It would weigh about 2,000 tons. Such a fantastic rate of multiplication is only theoretically possible. Actually it does not happen so. The process of division slows down or may soon cease. The causes are space limitations, competition for oxygen in the case of aerobic bacteria, inadequacy or lack of food supply and production of substances unfavourable to growth.

Bacteria do not multiply at a uniform rate by fission. In the beginning the rate of multiplication is very slow. After this short initial period the rate of multiplication increases rapidly and reaches a peak. During the peak period the rate of increase remains constant for a time and then the rate of multiplication falls down rather gradually. Normally all bacteria reproduce by this vegetative method of fission. This feature earned for them the name **fission fungi.** Hence, the older mycologists placed bacteria in the class **Schizomycetes** (Schizo, a Greek word meaning to split or cut, *mycetes* meaning fungi) coordinate with the other four classes of true fungi.

(*b*) Budding (Fig. 18.9). Some bacteria multiply by budding. At the time of budding a small portion of the bacterial cell wall at or near one pole softens and thins. The cell protoplast in this region covered by the thin wall bulges out in the form of a small protuberance. The protuberance gradually increases in size. It is termed the **bud.** Meanwhile the parent genetic material divides into two portions. One of these migrates into the bud. The bud grows and may become separated by a constriction at the base as in Bifidibacterium. The bud cell subsequently grows to the normal size and may in turn produce a bud. In one group of bacteria (Hyphomicrobeales) commonly called the budding bacteria, a branch strand of cell wall material may be initiated prior to the separation of a bud. Consequently many bud cells may be joined end to end by such thread-like secreted strands (Fig. 18.9).

(c) Cysts. In certain bacteria the entire protoplast of the cell recedes from the cell wall and becomes rounded. A thick wall is then secreted around it to form a resistant structure somewhat similar to the endospore. It is called the cyst. The cysts are formed in certain species of *Azobacter.* Under suitable environmental conditions the cyst germinates to produce a new bacterium cell.

(d) Segmentation. Some other species of bacteria reproduce vegetatively by a process called **segmentation.** In this case the protoplast of the bacterium cell, at some stage, divides to form very tiny bodies called **gonidia.** The cell wall ruptures and the liberated tiny gonidia grow into new bacterial cells under suitable conditions. To some bacteriologists vegetative reproduction by means other than binary fission seems doubtful.

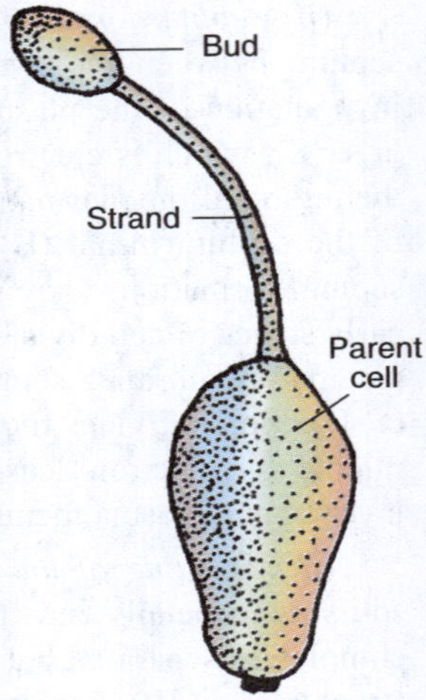

Fig. 18.9. *Bacteria.* Showing budding in the Hyphomicrobiales.

2. *Sporulation in Bacteria.* Some bacteria produce non-motile spores which are of the following types :—

(*a*) Conidia (Fig. 18.10A). *Streptomyces*—an important member of a Actinomycetes which are now considered true gram-positive bacteria, has a filamentous branched thallus. It cuts off tiny, oval or rounded, spore-like bodies, the **conidia** in chains at the tips of special aerial branches (conidiophores). Each conidium under suitable conditions germinates to give rise to new bacterial filament.

(*b*) Oidiospores (Fig. 18.10B). In an another member *Actinomyces* (order Actinomycetes) the hypha instead, of obstructing spores in succession at the free end, undergoes additional septation throughout its length to form numerous small reproductive units known as the *oidiospores*. Each oidiospore, on germination produces, a filamentous bacterium.

(c) Sporangiospores (Fig. 18.10C). In some branching bacteria sporangia-like structures may develop at the end of certain hyphae. The protoplast of the sporangium may divide to form tiny sporangiospores. On liberation the sporangiospores germinate under suitable conditions, each producing a filamentous branching *bacterium.*

(*d*) Endospores. These are highly resistant, physiologically dormant, single-celled structures formed usually inside a bacterium mother cell.

(*i*) *Occurrence of Endospores.* All bacteria do not form endospores. Only a limited number of the larger rod-shaped forms (bacilli), which

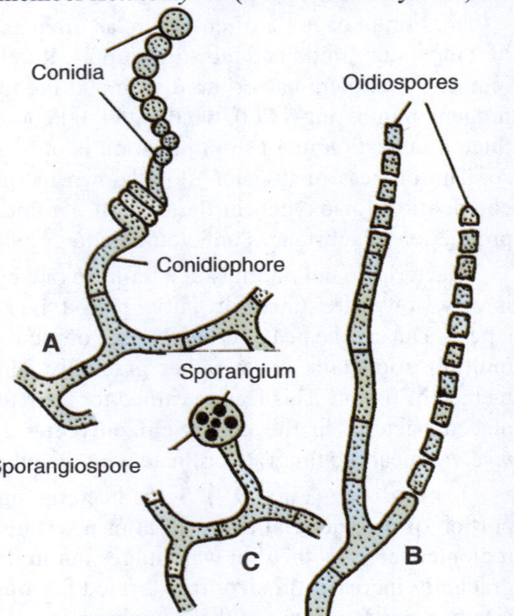

Fig. 18.10 (A-C). *Bacteria.* Actinomycetales (branching bacteria) producing three types of spores.

are all included in the family Bacillaceae, produce endospores. The common examples are species of aerobic genus *Bacillus* and anaerobic genus *Clostridium.* None of the smaller *bacilli, cocci* with one exception (gram-positive *Sporosarcina*) and *Spirilla* produce endospores. The spore-forming species are generally gram positive. Some bacteriologists, however, hold that a few spore producing species occur in the gram-negative genera *Vibrio* (*Sporovibrio*) and *Spirilla.*

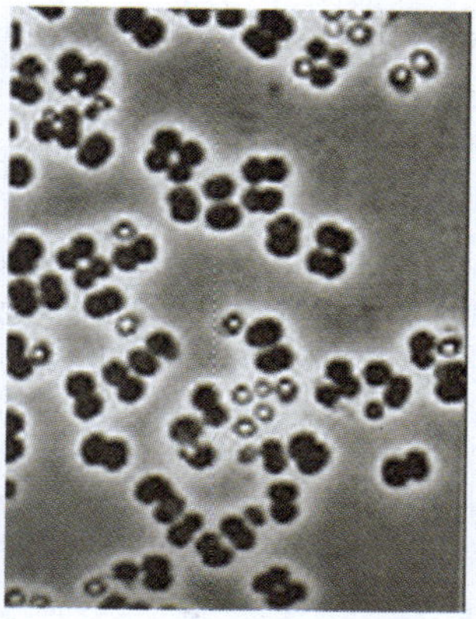

Sporosarcina

(*ii*) *Form and Position of Endospores* (Fig. 18.11). The endospores are usually formed singly. There are, however, instances in which two endospores are regularly formed in each bacterium cell. The bacterium cell with a mature endospore is called a **sporangium.** The size, shape and position of the endospores within the sporangium varies from species to species.

In form the endospores may be spherical or elliptical (oval) rearely cylindrical and may be central, sub-terminal or terminal inposition in the sporangium. The endospore occupies one quarter to one half of the area of the cell. The diameter of the endospore may be slightly less than, equal to or greater than that of the rest of the parent cell. The ovoid or spindle-shaped sporangium with a central enlarged endospore is known as **clostridium** (C). **Plectridium** is the term applied to the racquet-shaped sporangium with an enlarged terminal endospore (E). The mature endospore is a highly refractive body impervious to ordinary stains.

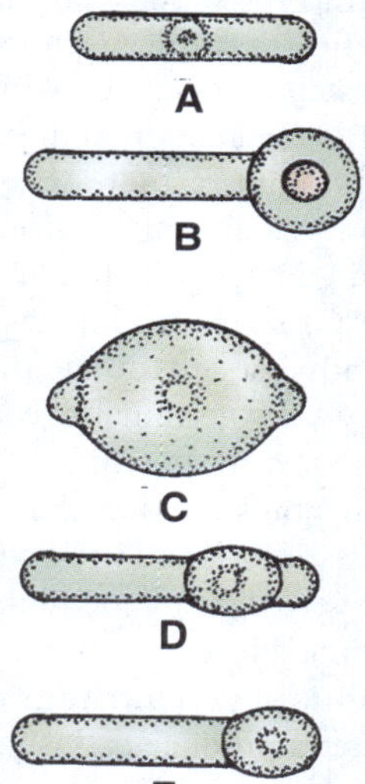

Fig. 18.11 (A-E). *Bacteria.* Different types of endospores and their position in the sporangium.

(*iii*) *Structure of Endospore* (Fig. 18.12). The structure of endospore varies somewhat from species to species. There is a tiny core of prtoplast consisting largely of chromatin or nuclear material (**DNA**) surrounded by dense cytoplasm. The cytoplasm has low water content but the protein content may be as high as 90 percent. There is, in addition, the usual stabilizing agent **dipicolinic acid.** It contains fewer active enzymes and in small amounts. The chromatin material lies near the surface (B) or centre of the spore cytoplasm (Fig. 18.13D). The protoplasmic mass is invested by a delicate **spore wall** which usually becomes the cell wall of the vegetative cell. The contents of the endospore within the spore appear packed in a very orderly manner. Electron micrograph reveals a thick layer of relatively low density outside the spore wall. It is known as the **cortex.** External to the cortex is the **spore coat** which is generally differentiated into layers. It consists largely of a protein which contains a disulphide rich component (Vinter, 1961) with physical structure resembling that of Keratins (Kadotal *et al.* 1965). It is located in the outer layers of the spore coat. Presumably it plays an important role in the survival of spores in adverse environments. The spore surface in electron micrographs shows different surface patterns. The surface may be smooth, wrinkled, ribbed or sculptured in geometric patterns. Sometimes the whole spore body is invested by an exosporium (*Clostridium sporogenes*) The size, form, location and surface patterns are characters important for classification.

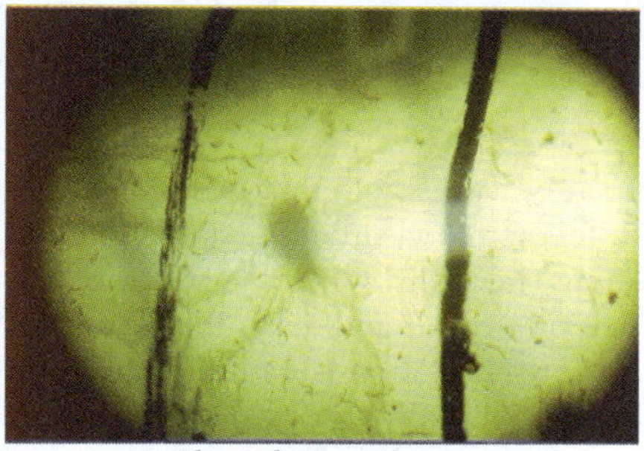

Clostridium sporogenes

(*iv*) *Physiological characteristics of Endospores.* The endospores show astonishing tenacity of life being highly resistant to extremes of heat and cold, light, desiccation, disinfectants, toxic chemicals and other conditions which would readily kill a vegetative cell. So resistant are some bacterial endospores that they can survive for several years, decades and probably for centuries. Endospores of *Anthrax bacillus* can withstand boiling water for an hour whereas those of other species survive prolonged boiling. Endospores of one bacillus resist 100°C for over 20 hours and some can withstand pressure cooking at 120°C or remain uninjured for an hour or more in a hot oven. Even immersion in liquid air such as Helium (at-269°C) for several hours has no adverse effect on them. They can germinate readily after immersion. The endospores are thus the most resistant forms of life known. It is very difficult to kill them. Fortunately very few disease producing **(pathogenic)** bacteria produce endospores. However, not all endospores are so resistant. Some are killed at temperatures slightly higher than those lethal to vegetative cells (60°C-70°C). The remarkable resistance of endospores is attributed to :—

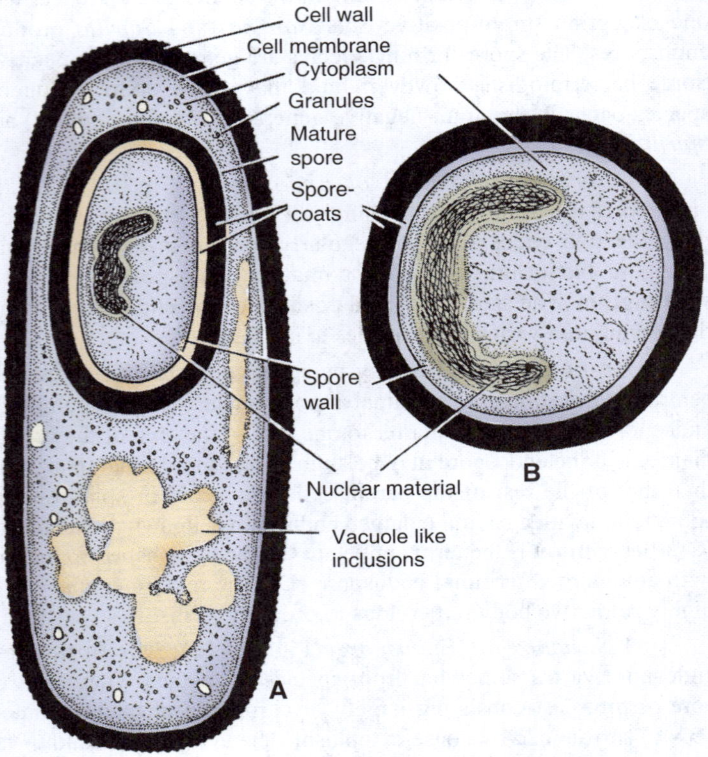

Fig. 18.12 (A-B). *Bacteria.* Structure of endospore. A, Sporangium containing a single sub-terminal endospore: B, cross-section of endospore (Diagrammatic).

(*i*) Low water content.
(*ii*) Low metabolic actitivty.
(*iii*) Protective nature and impermeability of relatively thick spore coat.
(*iv*) Presence of fewer active enzymes.
(*v*) Presence of unusual stabilizing agent dipicolinic acid.
(*vi*) Increased calcium content of endospore.
(*vii*) Physico-chemical nature of the spore protoplast.

(*v*) *Factors Favouring Endospore Formation.* Until recently it was held that endospores are produced under adverse circumstances such as extremes of temperature, drought, starvation and accumulation of toxic substances in the surrounding medium. The well-known fact that aerobic spore-forming bacilli do not form endospores in the absence of oxygen and anaerobic forms do not sporulate in the presence of oxygen makes it less likely that sporulation is entirely a reaction to unfavourable conditions. Endospores, in fact, are produced by healthy bacterial cells under favourable conditions usually when multiplication has ceased. More frequently it takes place at a temperature favourable for vegetative growth. According to Forbisher (1957), endospore formation appears to require some energy food. It may be glucose or manganese ions depending upon the species. Presence of iron and manganese in culture medium increases endospore formation. Carpenter is of the opinion that endospore formation is inhibited by certain metabolic by-products such as straight chain organic (fatty) acids with 10-14 carbon atoms. Knayse (1945) stated that drying does not favour sporulation to any appreciable extent. It is generally held that adequate food supply favours vegetative growth but appears to delay spore formation. Experimentally it has been shown that cultivation in dilute media promotes sporulation in bacteria. This lends support to the suggestion that exhaustion of nutrients is a factor in spore formation. Some bacteriologists hold that substances that interfere with aminoacid metabolism inhibit sporulation.

The conclusion, therefore, is that endospore formation depends upon organism's protein metabolism. Hardwick and Foster (1952) envisaged that sporulation is due either to the conversion of vegetative proteins into spore proteins or formation of fresh spore proteins from the culture medium. Substances in the culture medium which oppose conversion of vegetative proteins into spore proteins inhibit sporulation.

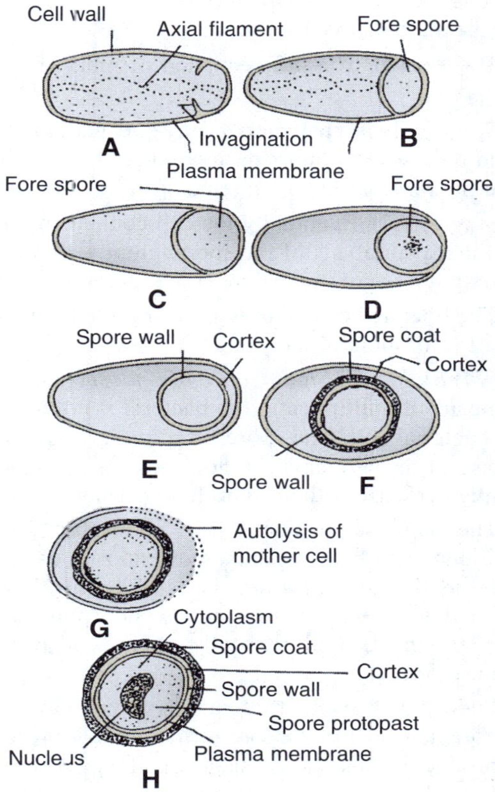

Fig. 18.13 (A-H). *Bacteria.* Stages in the development of endospore; C-D, show laterally advancing digit-like processes of the mother cell, plasma membrane which finally meet and fuse to engulf the fore-spore; F, engulfed fore-spore which has secreted the spore wall.

(*vi*) *Development of Endospores* (Fig. 18.13). The development of spores in the two spore-producing genera of bacteria (bacilli and clostridia) is in general similar. The sequence of changes is distinguishable into the following seven stages.

Stage 1 *Formation of an axial filament of nuclear material* (A). The nuclear matter of the bacterial cell by condensation become arranged in the form of an axial thread (A). It nearly extends from one pole to other.

Stage 2 *Development of fore-spore* (B). A ring-like invagination of the plasma-membrane arises near one pole of the cell (A). It grows centripetally and finally fuses in the centre of the cell to form a septum which completely separates the fore-spore from the mother cell (B). A small amount of nuclear material is entrapped within the forespore.

Stage 3 *Eugulfment of fore-spore* (C-D). This is brought about by the insinuation between the cell wall and fore-spore plasma membrane of mother cell. Cytoplasm is bounded on both outer and inner sides by new membrane (Mackay and Morris 1971). The fore-spore embedded in the material cytoplasm secretes a cell wall around it. It is termed the **spore wall** (E).

Stage 4 (E). It is characteriszed by the deposition of the cortical layer probably differentiated from the material cytoplasm.

Stage 5 (F). The differentiation of multi-layered **spore coat** around the cortex represents this stage.

Stage 6 It is characterized by the maturation of the spore inside the mother cell. It may extend over several hours during which the spore changes in size and form. The mother cell now called the *sporangium* often remains alive for a short time after maturation of **endospore.**

Stage 7 (G). Eventually the cell contents not utilised in spore formation begin to disintegrate (G). Finally autolysis of the sporangium wall and remains of cytoplasm sets the endospore free. The lytic release of the mature spore represents stage 7.

The liberated endospores are dispersed by wind and thus occur everywhere. They may remain quiescent for months and years.

(*vii*) *Biological Significance of Endospores.* The generally held view that endospore formation is a means of multiplication in bacteria is not tenable. Usually a single endospore is produced by each bacterial cell. The spore, in turn, germinates to give rise to one bacterium cell. There is no increase in the progeny. Further, most of the species of bacteria which do not form endospores multiply perfectly without sporulation. Thus endospore formation is not a reproductive mechanism.

The other suggestion is that endospore formation is a means to tide over unfavourable conditions. The high resistance of endospores to heat, cold, drought and disinfectants tempted some scientists to put for this thesis. The objection to this view is that setting in of such conditions in the environment does not induce endospore formation. Contrary to this endospore are formed under favourable conditions when there is adequate food supply and at a temperature suitable for vegetative growth.

Endospores are chiefly found in several days' old, well matured cultures. Some bacteriologists, therefore, consider endospore formation to represent a mature stage of bacterial cell development. The product of this mature stage which is the endospore, represents a dormant and resistant stage in which there is little metabolic activity and low rate of respiration. However, it is alive but dormant. With its properties of low water content, low metabolism, realtively thick impervious wall, increased calcium content, physico-chemical nature of the protoplast, presence of dipicolinic acid and small amounts of active enzymes, the endospore functions as a resting structure and provides resistance to adverse environmental conditions typical of spores. According to this view, therefore, endospore is evidently a resting stage in the life cycle of bacteria. It is a resting cell that remains viable for extremely long periods. Primarily it enables the organism to withstand rigorous conditions of life and secondarily serves as a means of multiplication particularly when more than one endospores are formed per bacterial cell.

(*viii*) **Germination of Endospore** (Fig. 18.14). The highly resistant endospores germinate immediately (within a few hours) when placed in suitable environment conditions. The conditions which favour germination are the presence of moisture, food supply (nutrients), suitable temperature and oxygen tension. The addition of certain germinating agents such as an amino acid, glucose, and adenosine to the culture medium accelerates the process in the case of aerobic bacilli. Certain chemical compounds and heavy metals such as copper, zinc, nickel, and cobalt retard germination.

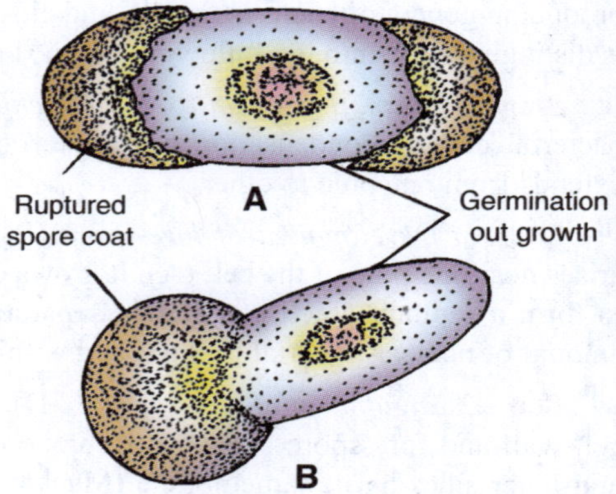

Fig. 18.14. *Bacteria.* Germination of endospore.

The spore coat imbibes water and softens. The intake of salts and nutrients causes swelling of the spore contents which lose their refractility. Subsequently the softened spore coat or coats rupture at the equator (A, *Bacillus maycoides*) or at one end (*Clostridia, B*). In the latter case the new growing cell emerges at the pole of the endospore. Usually the spore wall becomes the cell wall of the young vegetative cell. In the case of equatorial splitting, the remnants of the cracked spore coat appear as caps at each pole of the emerging vegetative cell (Fig. 18.14A). Eventually these fall off. The first step after the initiation of germination but prior to the initiation of the germination outgrowth is the initiation of germination messenger RNA which is completely absent in the dormant state of the endospore. During germination the endospores lose about 25 to 30 percent of their dry weight. Heat shocking within the range of 80°C to 85°C for a few (8-10) minutes promotes germination in the endospores of bacilli.

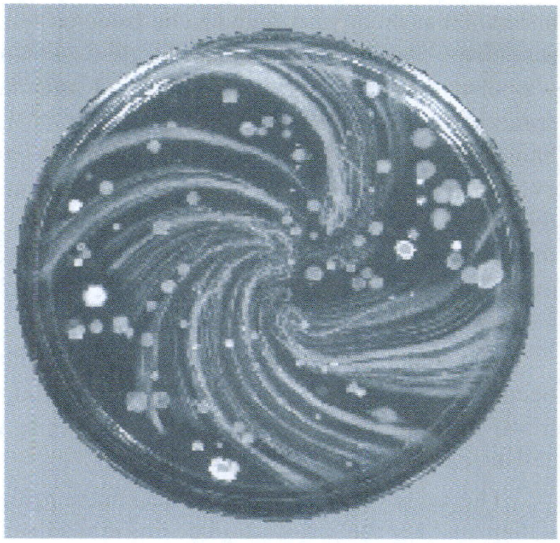

Bacillus mycoides

B. Genetic Recombination (Sexual Reproduction). Sexual reproduction has long been thought to be absent in Bacteria. Experiments with the help of eletron microscope, however, have demonstrated the exchange of genetic material followed by genetic recombination in some species of bacteria. This is an essential feature of a sexual process. Bacteria are able to exchange genetic material by **exotic means** as well as by a straightforward **conjugation process** which closely resembles sexual reproduction in other organisms. There are two principal methods of bringing about genetic changes by exotic means in bacteria. In both these there is transfer of DNA from one organism to another without the two ever coming into contact but the technique employed is dissimilar. These two phenomena are **transformation** and **transduction.**

(*i*) Transformation. The first transformation experiment was carried out with *Pneumococcus* which causes pneumonia. The bacterium is oval in shape and is surrounded by a capsule. On the basis of the nature of capsule polysaccharides, the bacteriologists have grouped pneumococci into several types. Occasionally a *Pneumococcus* cell loses the capacity to secrete a capsule around it. This inability is transmitted to all its progeny. In this way a clone of noncapsulated bacterial cells is produced. These clones of pneumococci have lost their virulence and do not cause disease. On a solid medium these non-capsulated, non-pathogenic pneumococci form colonies which appear **rough.** The capsulated or encapsulated pneumococci, on the other hand, form colonies which appear **smooth.** It has become customary with the bacteriologists to apply terms 'rough' and 'smooth' to pneumococci without or with capsules. The former are denoted as *R* type and the latter *S* type pneumococci.

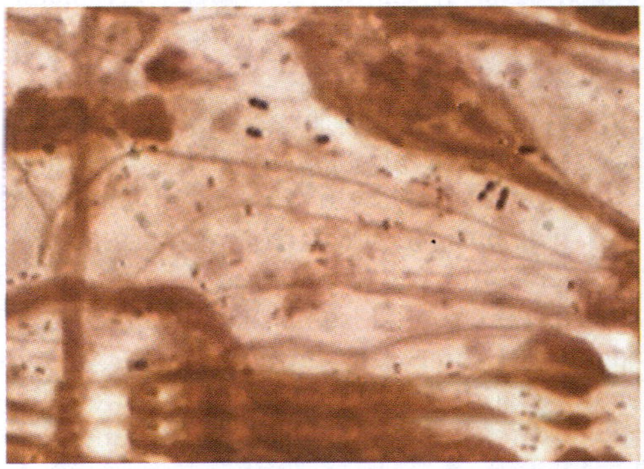

Pneumococcus (*Established bacterial transformation*)

A British bacteriologist Griffith in 1928, injected mice simultaneously with a mixture of *R* type (rough, non-pathogenic) and heat killed *S* type (smooth, pathogenic) bacteria. Neither of these would have been harmful to the mice. On the contrary the mixture killed many animals. The animals that died were found to be infected with capsulated (smooth) pneumococci of *S* type. The result was astonishing because *S* type organisms which were injected had been killed by heat. It was obvious from this experiment that association with the dead smooth *S* type bacteria had in some mysterious way restored virulence of living, rough *R* type bacteria. On examination, the bacteria from the dead mice were found to possess capsules and formed smooth colonies. It is evident that the association trnasformed living *R* type bacteria into *S* type bacteria. The fact that the progeny of the latter were all *S* type showed that it was not due to their having surrounded themselves with capsules of the dead ones. In all similar experiments, the acquired property was that of the dead organism. In other words the dead, smooth bacteria transformed the living rough bacteria into their own capsular type.

The substance in dead, capsulated bacteria responsible for the transformation has subsequently been isolated and is found to be deoxyribonucleic acid (**DNA**). It has also been found that transformation is not dependent on the presence of dead bacteria. It could equally take place with cell-free extracts from smooth bacteria.

The cells of the transformation strain are said to be **competent cells.** The **DNA** portion, in which resides the whole transforming ability, is called the **transforming principle.** The initial steps in transformation are **adsorption** and **penetration** of the transforming principle (**exogenous DNA**) into living bacterial cell which is assumed to possess areas in the wall. After penetration the subsequent steps consist in, (*i*) the incorporation of the transforming principle (DNA) into the genetic material (chromosome) of the recipient cell, (*ii*) replication of the modified chromosome with new information, and (iii) formation of a transformed cell population. From the account given above it is clear that transformation may be defined as the transmission of heritable property from one cell to another by way of free, naked DNA without the recipient and donor cells ever coming into contact. The genetic change in transformation is permanent.

(*ii*) Transduction. It is a widespread phenomenon which differs from transformation in the mode of genetic transfer. In transduction, genetic transfer is carried out by the agency of a bacterial virus known as the **bacteriophage.** Only temperate as opposed to virulent bacteriophages can act as agents of this transaction. Transduction thus is a genetic transfer mediated by a bacteriophage. The phenomenon of transduction has been studied in *Escherichia coli* with the phage vector P_1 and *Salmonella typhimurium* with its vector P_2. The bacterial cell infected with a temperate phage is termed **Lysogenic.**

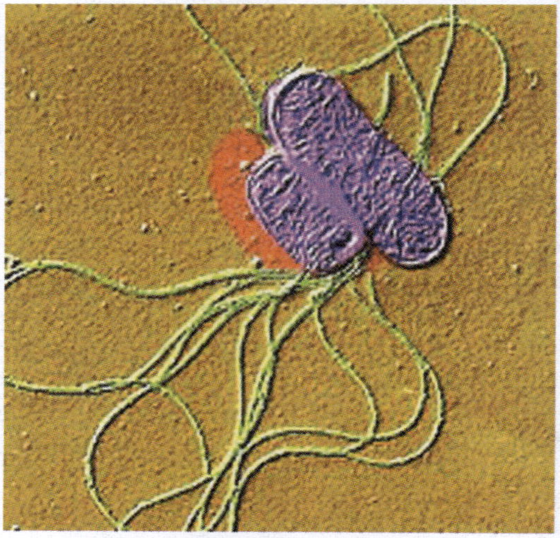

Salmonella typhimurium
(Exhibits transduction)

Zinder and Lederberg (1952) reported the transfer of several characters between strains of *Salmonella.* They cultivated a lysogenic mutant of *S. typhimurium* which required *tryptophan* with another strain which required *histidine* but was susceptible to the same bacteriophage. They found that some cells of the lysogenic organism acquired one of the characteristics of second organism. They required neither tryptophan nor histidine. As the phage multiplies within its host cells, the fragments of the host DNA are incorporated in the newly formed phage particles. The host cell disintegrates and the phage particles are released. The host cell disintegrates and the phage particles are released. The released phage particles, which carry a small amount of the host DNA, may infect the cells of the susceptible strain. In this way the host DNA is transferred to susceptible cell (new host) and incorporated into genome of the recipient. This is transduction.

Lederberg and Tatum (1944) furnished a clear experimental evidence of genetic recombination mechanism which closely resembled sexual reproduction in other organism. It is termed **bacterial conjugation.** On the basis of this startling discovery Lederberg and Tatum were awarded a Nobel prize in 1959.

***(iii)* Bacterial conjugation** (Fig. 18.15). It has been studied in colon bacillus (*Escherichia coli*). There are two mating types of cells in *E. coli.* Both are haploid. One of the these is the **donor** or **fertile cell** (F+) and the other **recipient** or **receptor cell** (F^-). The cells of the opposite types come together in pairs. The donor cell has a special kind of *pili* known as the *sex pili* which help to attach it to the cell wall of the recipient cell. A conjugation bridge or tube develops between the two conjugant at the point of contact. Through it the genetic material (**DNA** strand) of the donor cell enters the recipint cell in a linear series.

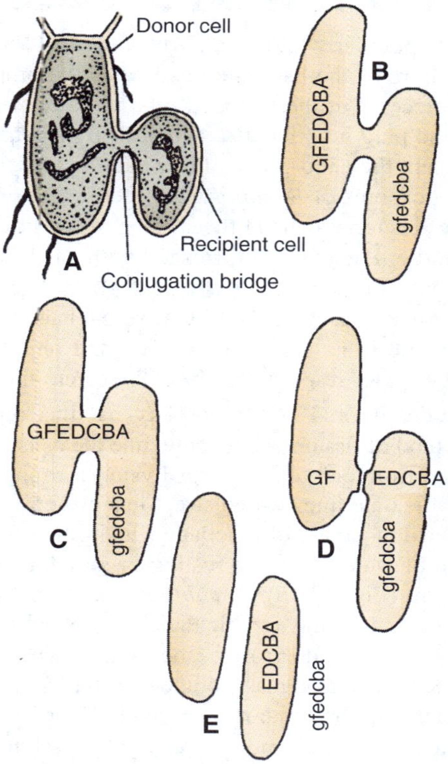

Fig. 18.15 (A-E). *Bacteria.* Bacterial conjugation in colon bacillus (*Escherichia coli*) A, formation of conjugation bridge; B-E, stages in the transfer of genes from the donor to the recipient cell.

It appears that during the conjugation process (Fig. 18.15) the ring-shaped **DNA** strand breaks at a definite point and becomes a filament (B). Without this the transfer of genetic material cannot take place. However, it still remains double stranded. The genes ABCDEFGHI are arranged on it in a linear series. Wolman and Jacob (1956) experimentally demonstrated that the first gene **A** of the donor cell moves into the recipient cell at about seven minutes. This is followed by gene **B** at 9 minutes, gene C at 10 minutes, D at eleven minutes, gene E at 18 minutes and F at 25 minutes (D). The conjugants stay joined for a short time so that only a portion of the DNA strand of the donor cell could get transferred to the recipient cell through the conjugation bridge (D). The recipient ex-conjugant or zygote is thus partly or incompletely diploid. It possesses a complete recipient "chromosome" (DNA) but only a part of a donor "chromosome" (D). When the progeny of zygote cell are grown it is found that the cell is a **recombinant.** It grows under conditions that will support neither of the parents. This partly diploid state is maintained only for a short period. The zygote cell (recombinant) shortly discards the fragment of the donor 'chromosome' and reverts to the haploid state. Thus, bacterial conjugation differs from true sexual reproduction in two respects, absence of meiosis and formation of partly diploid zygote. After transfer of genetic material the donor cell must perish. Mandlelstam and Mcquillen (1968) suggested that only one strand of the donor cell DNA enters the recipient cell. As it does, synthesis of complementary strand takes place. At the same time a new DNA strand is synthesised on the donor DNA to replace the strand which has been removed.

LIFE HISTORY OF BACILLUS SUBTILIS (FIG. 18.17)

It is popularly known as *Hay bacillus*. It is a common example of a rod-shaped bacterium. A rich culture of the bacterium can easily be prepared in the laboratory. Take hay and chop it into small pieces. Put the chopped bits in water to prepare an infusion. Decant the clear fluid into a flask and plug it with cotton wool. Boil the water in the flask for some time to kill other forms of life except the endospores of hay bacillus which are resistant. Allow the boiled decoction to cool. The endospores of *B. subtilis* will germinate on incubation. The decoction will now contain thin films of bacterial cells at the surface of the liquid. They are self-motile. Each cell is furnished with 8-12 flagella. They are evenly distributed all over the body. The cell has the same structure as described for a typical bacterium cell. It is saprophytic in its mode of nutrition and requires plenty of free oxygen for respiration. The cells secrete slime copiously.

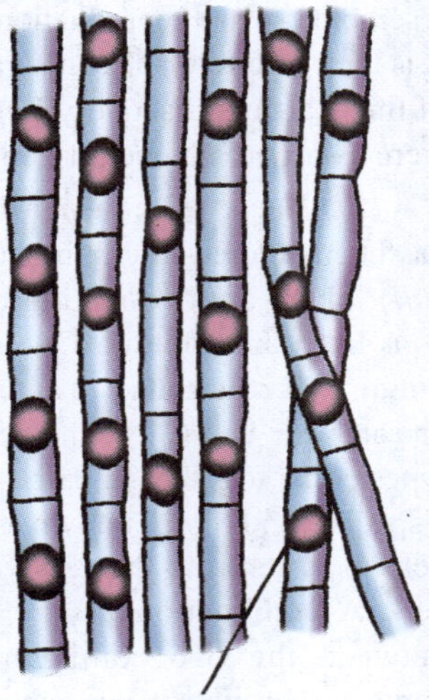

Fig. 18.16. *Bacteria*. Zoogloea stage.

Under normal conditions hay bacillus reproduces by the method of fission. After some time the cells often become attached from end to end after divsion forming long chains of bacilli sometimes called streptobacilli. The filaments are embedded in a mass of mucilage which forms a scum on the surface of water. It is the resting or **zoogloea stage** of the bacterium (Fig. 18.16). *B. subtilis* being motile, the chains of cells swim. Being aerobic the chains of cells have avidity for oxygen and thus grow in the scum at the surface of the fluid medium. During the zoogloea stage hay bacillus forms endospores in the same manner as described for the typical bacterium. The endospores are highly resistant to adverse conditions of life. With the return of favourable conditions the endospores germinate. Each gives rise to a new bacterium cell. *B. subtilis* is important as a source of antibiotic, **subtelin.**

ECONOMIC IMPORTANCE OF BACTERIA

The bacteria are best known to the general public as the causative agents of disease. We must not forget that the bacteria do not always mean disease. Many of them are positively beneficial. Several others are neither harmful nor beneficial. Only a very small percentage are harmful.

Beneficial Activities. There are many kinds of bacteria without which we could not live. They are absolutely essential to the presence of life on earth. They make possible the continued existence of green plants and therefore of animals because the plants are the only source of food for animals. Following is a brief account of the more important activities of the bacteria :—

1. Role in Agriculture. (*i*) Decay and decomposition. Soil bacteria play an important role in brining about decomposition of organic matter. They serve a double purpose. In the first instance they act as scavengers removing harmful waste from the earth. Secondly, they return it to the soil as plant food. The dead bodies and wastes of organisms (both plants and animals) are decomposed by the activities of the saprophytic bacteria. In consequence a variety of elements of minerals of the earth such as carbon, oxygen, hydrogen, sulphur and phosphorus which make up their bodies are reduced to simple compounds such as carbon monoxide, water, nitrates, sulphates and phosphates. Some of these go back to the soil and the rest to the air. From the soil they can be absorbed as plant food. This activity of the bacteria is used in sewage disposal system of cities. The bacterial action on the city's sewage promotes decay. As a result water is finally purified and changed into an odourless and valuable fertiliser instead of a dangerous and expensive waste product.

(ii) **Soil fertility.** Some bacteria play an important role in maintaining and others in increasing soil fertility. The fertility of soil is proportional to its nitrogen content. Nitrogen is an essential ingredient of all living protoplasm. All growing plants, therefore, require it in their metabolism. Atmosphere, no doubt, is four-fifths (80%) nitrogen, green plants generally are unable to use it. They mostly absorb it as nitrates and to some extent as ammonia from the soil. Continuous absorption of these salts results in their exhaustion in the soil. Nearly all fertilisers for the soil include a large proportion of such soluble nitrogen compounds to promote plant growth. In nature the presence of a regular supply of these salts is ensured by bacteria of certain types. These bacteria which function as Nature's farmers belong to three categories, namely, **ammonifying bacteria, nitrifying bacteria** and **nitrogen-fixing bacteria.** They are the agents of maintaining a continual circulation of nitrogen in nature between the plant world, in soil and the atmosphere. The series of changes through which the nitrogen passes due to the activities of these organisms constitute the **nitrogen cycle** (Fig. 18.18).

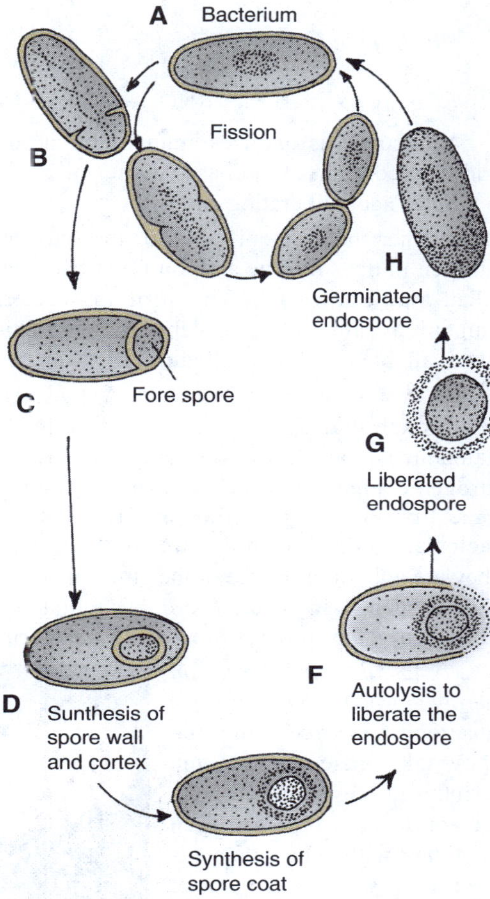

Fig. 18.17. *Bacteria.* Diagrammatic life cycle of an endospore forming bacterium.

(a) **Ammonifying Bacteria.** The saprophytic bacteria break down the proteins and other nitrogen containing remains of the plant and animal origin in the soil to amino acids by secreting enzymes. The amino acids are then converted into ammonia by a group of bacteria called the **ammonifying bacteria.** The liberated ammonia may combine with carbon dioxide and water in the soil to form ammonium carbonate. A few plants such as the common cereals can make use of ammonium compounds as a source of nitrogen. The majority of plants, however, cannot absorb ammonium compounds as a source of nitrogen.

(b) **Nitrifying Bacteria.** Ammonia is very soluble. It moves in the soil rapidly and is acted upon by microorganisms of the category of chemosynthetic autotrophs in the soil. They are the nitrifying bacteria such as *Nitrosomonas* and *Nitrobacter.* They form nitrates from ammonium compounds. The reaction takes place in the following two steps :

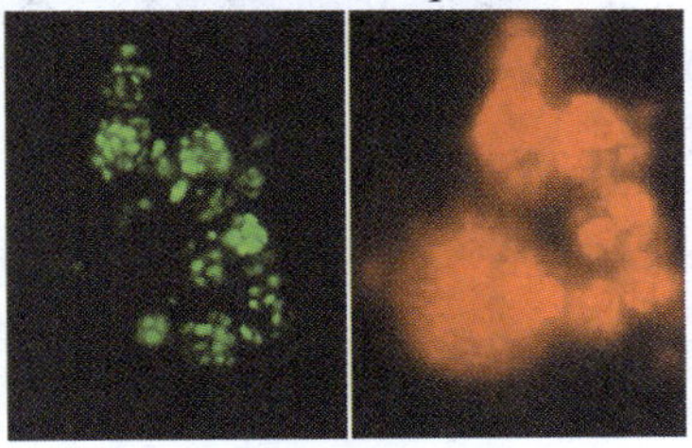

Nitrobacter

Nitrosomonas

NITRIFYING BACTERIA

Nitrosomonas	$NH_4^+ + 2O_2 \xrightarrow{\text{energy}} 2H_2O + NO_2 \uparrow$
Nitrobacter	$NO_2^- + \frac{1}{2}O_2 \xrightarrow{\text{energy}} NO_3^- \uparrow$

Nitrosomonas oxidizes ammonium carbonate to nitrous acid liberating energy. The nitrous acid then combines with bases in the soil forming potassium nitrite. Nitrobacter oxidizes nitrites to nitrates again liberating energy.

Neither the ammonifying nor the nitrifying bacteria add to the total quantity of combined nitrogen in the soil. The ammonifying bacteria convert aminoacids into ammonia. The process is called **ammonification.** The nitrifying bacteria convert nitrogen from the unavailable form of ammonium salts to the available nitrates. This process converting unavailable ammonium salts into available nitrates is called **nitrification.**

(*c*) Nitrogen-fixing Bacteria. A considerable amount of nitrogen is lost by denitrification and through drainage. The loss must be made good by equal gains if the soil fertility is to be maintained. The electric discharges in the atmosphere bring about the formation of traces of nitrogen compounds which are washed to the soil by rain water. The largest additions, however, come from a biological fixation process through the activity of two types of **nitrogen-fixing bacteria.** Some of them live free in the soil and others in the root nodules of leguminous plants. They are able to make use of the atmospheric nitrogen and change it into nitrogenous compounds. The nitrogen-fixing bacteria are thus unique in tapping a source of nitrogen not available to most other plants. This process of nitrogen transformation is called **nitrogen fixation.**

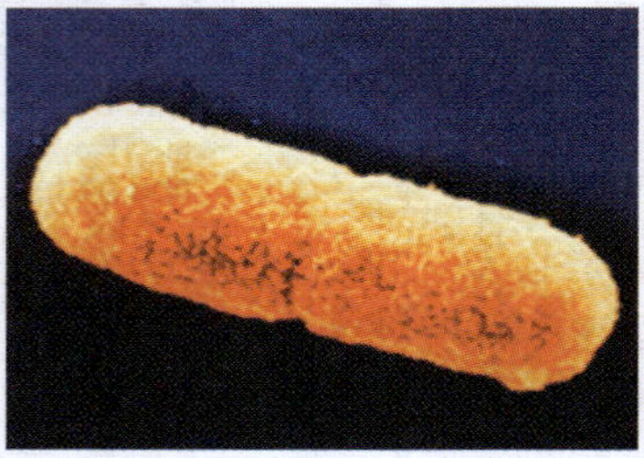

Azotobacter beijerinckia

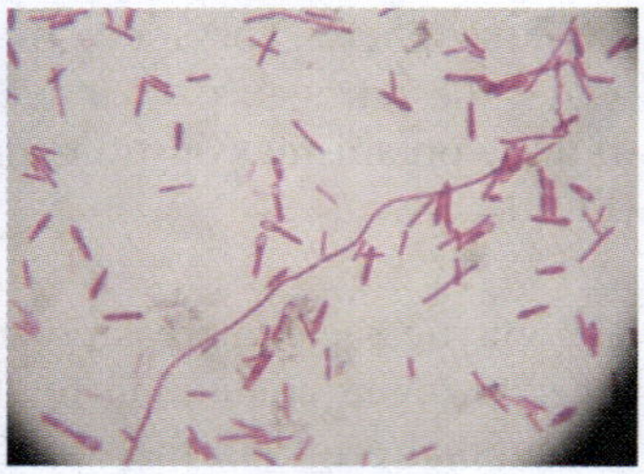

Clostridium

NITROGEN – FIXING BACTERIA

(*i*) *Azotobacter beijerinckia* (*aerobic forms*) and *Clostridium* (anaerobic) live free in the soil. They take gaseous nitrogen from the air present between the soil particles. The nitrogen combines with other elements forming organic nitrogenous compounds. These compounds are assimilated by the bacteria. In due course these bacteria die and their dead bodies containing nitrogenous compounds are decomposed by another type of bacteria called the **bacteria of decomposition.** During decomposition ammonia is produced. The nitrifying bacteria convert this ammonia first into nitrites and finally into nitrates. Nitrates constitute the form of nitrogen needed by the green plants.

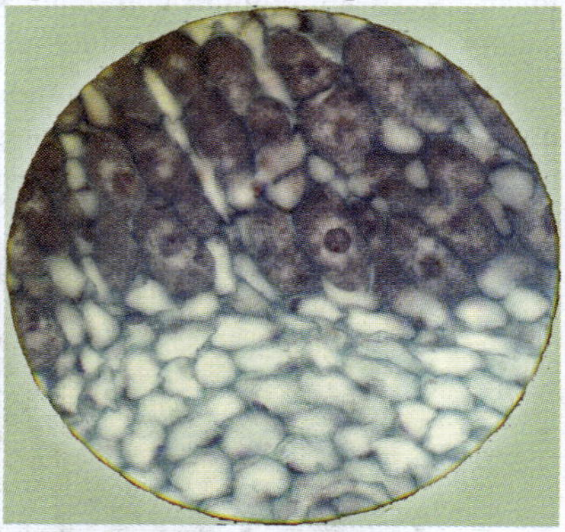

Rhizobium leguminosarum (N2-fixing; lives in root nodules of members of Leguminosae

(*ii*) *Rhizobium leguminosarum* (syn. *Bacillus radicicola*) is another nitrogen-fixing bacterium. It lives in the roots of such plants as Pea (Fig. 18.19A), Bean, *Medicago* and others. All these belong to the Pea family (Leguminoseae). Besides the legumes, the nodules are found on the roots of *Alnus glutinosa, Casuarina,* species of *Coriaria* and a few others. The symbiont in non-leguminous plants is a member of Plasmodiophorales. The presence of bacteria in the roots causes the formation of little nodules (Fig. 18.19A). In *Pavetta indica* the bacterial nodules are formed on the leaves. These nodules or the

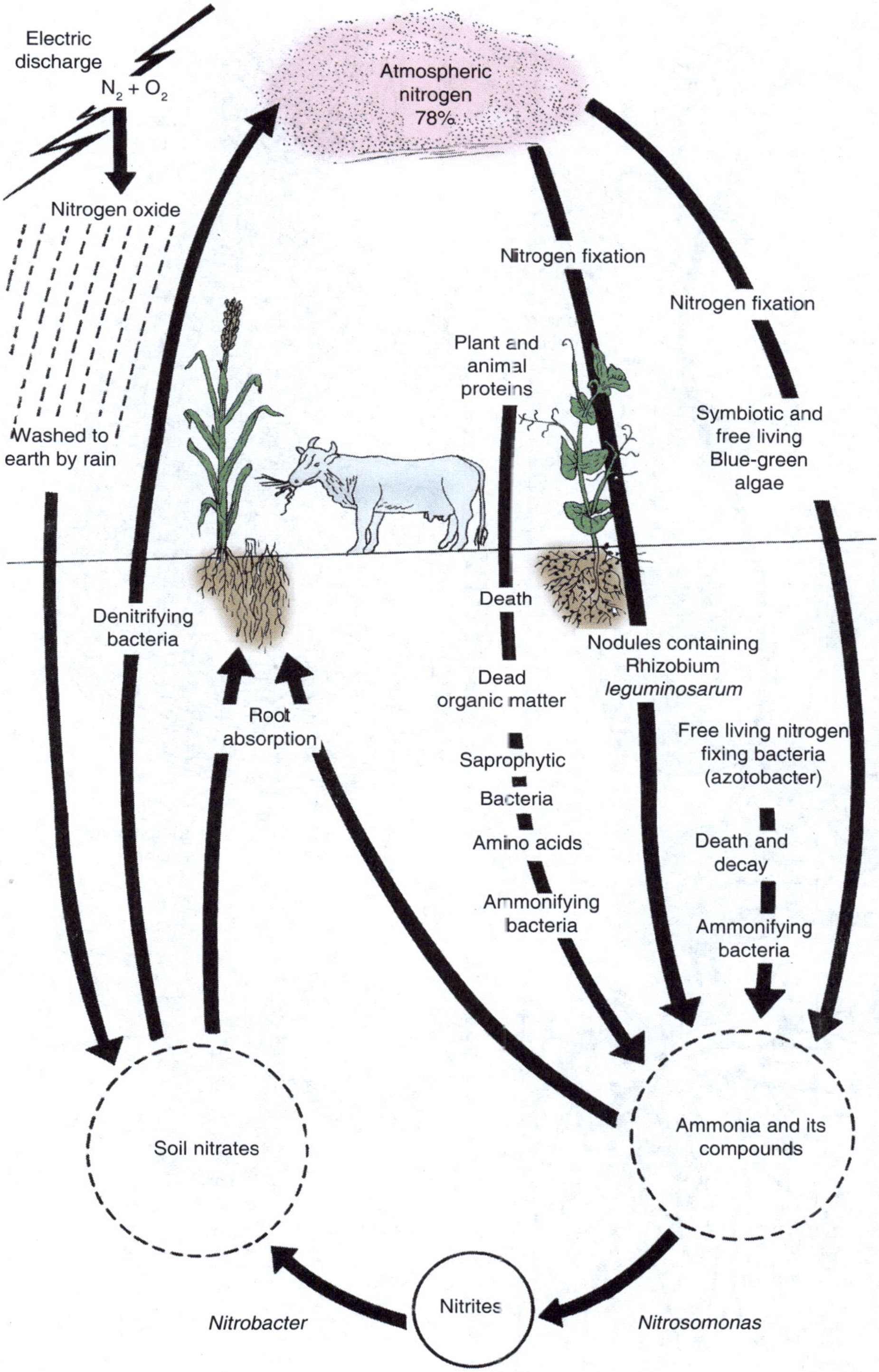

Fig. 18.18. Nitrogen cycle in nature and the role that bacteria play in it.

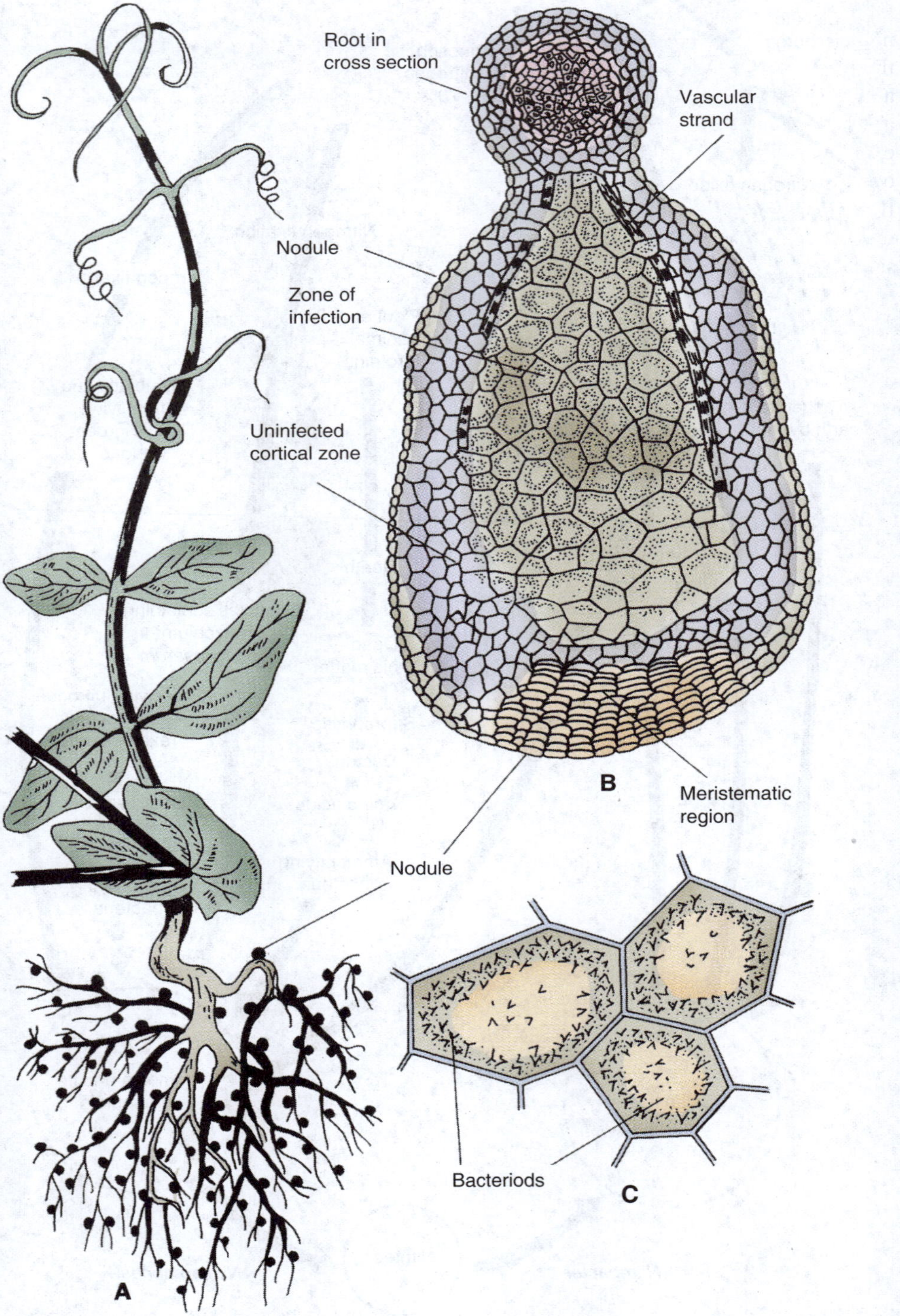

Fig. 18.19 (A-C). *Nitrogen fixing Bacteria*. A, Nodulated root of Pea; B, section of nodule; C, three host cells from the infected cortical zone containing bacteriods of *Rhizobium leguminosarum*.

tubercles are the homes of millions of these bacteria. They have the ability to take up free nitrogen of the air and convert it into nitrogen compounds. A part of the fixed nitrogen passes into the tissues of legume plant and a part diffuses into the surrounding soil. These bacteria enter into a mutually beneficial partnership with the host plant. They give to the host the nitrogen compounds and receive in return carbohydrates manufactured by the host plant. This association is an excellent example of **symbiosis.** The legumes are very rich in nitrogen because of this association. Some of our best protein plant foods come from the legume family of plants. The legume plants can flourish on land that has been depleted of its nitrogen by other plants. They are sometimes grown and ploughed under by the farmer when a foot or so in height. The decaying bodies of these legume plants enrich the soil. They furnish a rich supply of nitrogen to the future crop. This is called **green manuring.** The soil on which repeated crops of cereals are grown becomes impoverished. It can be enriched again by growing on it a crop of some plants of Pea family. This practice of alternating cereals with leguminous crop is known as rotation of crops. The leguminous plants contain more of nitrogen than they get from the soil salts. The additional quantity is obtained from the air by *Rhizobium.* Neither *Rhizobium* nor the legume root alone can fix nitrogen.

Green Manuring of Subclover

Rhizobium lives in the soil where it has the form of *coccus.* The legume roots secrete substances which attract bacteria on to their surface. The bacteria, in turn, secrete a growth hormone which causes the root hairs to curl. The cocci enter the curled root hairs. They grow in the root hair in the form of a continuous thread-like mass which finally reaches the root cortex. In this way many of the cortical cells become filled with a dense mass of these bacteroids. Their presence in the cortical cell serves as a stimulus causing abnormal growth. The cortical cells around the infection divide and redivide and grow to form a nodule. A nodule comprises a central mass of cells full of bacteroids. Around this zone of infection are a few cell layers thick of bacterial free cortical zone. At the apex of the nodule is the meristematic region and a vascular strand at its base (Fig. 18.19 B). Within the host cells the cocci feed on the carbohydrates and other foods and undergo a change in their form (Fig. 18.19 C). They become V, T or Y shaped. The core of the nodule is red owing to the formation of red respiratory pigment **haemoglobin.**

2. Role of Bacteria in Industry. Man has utilised the activities of bacteria for various industrial processes. The butter and cheese industries entirely depend upon the activities of the lactic acid bacteria. The souring and curding of milk by lactic acid bacteria is another common example of application in everyday life. It takes place in two steps. In the first step the lactose sugar of milk is fermented into glucose by an enzyme lactose secreted by the lactic acid bacteria. In the second step there is transformation of

Samples Produced from Butter and Cheese Industry

glucose into lactic acid. The latter sours the milk and coagulates the milk proteins (casein) forming curds and whey. Oxidation of alcohol into vinegar (acetic acid) is brought about by the acetic acid bacteria. The curing of tea, tobacco and manufacture of indigo are other examples of useful chemical activities of bacteria which have been controlled for the benefit of mankind. The process of tanning hides in leather making and preparing sponges also involve the use of bacteria. The production of linen is impossible without bacterial activity. The tough fibres, which are left behind, are separated. These fibres are spun and woven into linen cloth, ropes, etc. The preparation of coffee and cocoa is also dependent upon bacterial action. The cocoa beans are white in colour and quite bitter in taste. The bacteria digest the bitter coverings of seeds and give the characteristic colour, flavour and aroma. Many saprophytic bacteria in their metabolic activities excrete waste products of great commercial importance. Some of these are :

Preparing Sponges

(*i*) *Lactic acid.* It is useful in tanning industries.

(*ii*) *Citric acid.* It is used to give aroma and flavour to beverages, sweets and other foodstuffs.

(*iii*) *Vitamins.* Vitamin B is the product of fermentation of sugars and starch by *Clostridium acetobutilicum.* The vitamins are used in medicinal preparations.

(*iv*) Butyl alcohol, acetone and ethyl alcohol are produced in one fermentation operation when a certain bacterium is allowed to act on cooked corn starch. These products are important commercial solvents.

(*v*) *Acetone.* It is an important ingredient of explosives and is also used in the manufacture of photographic films.

Chemical formula of Lactic Acid - (Diagrammatic)

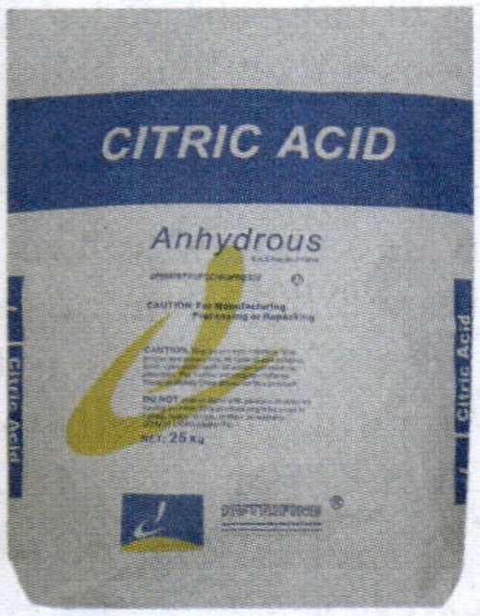

Citric Acid - Sample

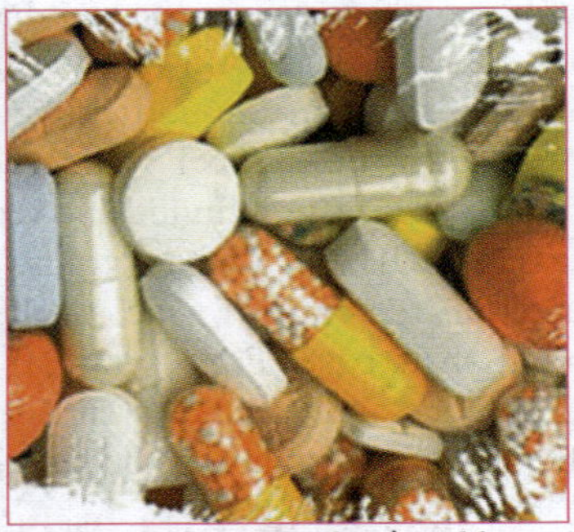

Vitamins – Capsules & Tablets

3. Role of Bacteria in Medicine

(*i*) *Source of Antibiotics.* The milder antibiotics of bacterial origin are tyrothricin, subtilin, polymyxin B, and bacitracin. *Bacillius subtilis* is the source of subtilin. Bacitracin is obtained from a stain very much like *B. subtilis.* The actionomycetes which are filamentous, bacteria-like organisms produce more powerful antibiotics such as streptomycin, aureomycin and terramycin.

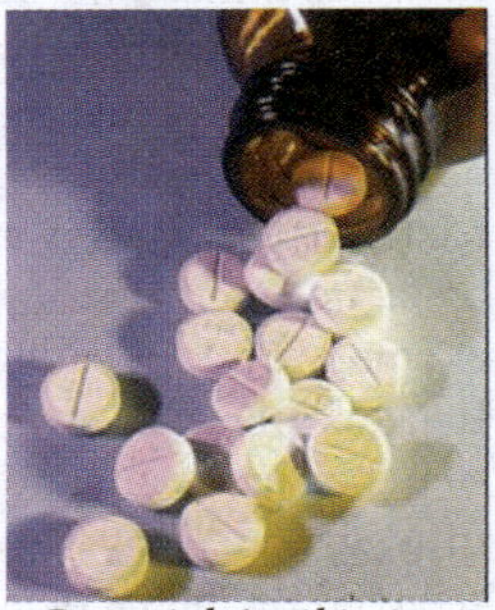

Bacterial Antibiotics - Tablets

(*ii*) *Preparation of serums and vaccines.* These are substances which are used to develop immunity to various diseases in man. Serums are used in advance as a therapeutic measure. They are also used when a person actually suffers from a disease. Diphtheria, lockjaw, pneumonia, etc. are the diseases in which the serums are effective. Vaccines are commonly used to make people immune to diseases like typhoid, small-pox, cholera, scarlet fever, etc.

In the preparation of serums, small doses of bacterial toxins (poisons) are injected into the blood of animals such as horses. To combat or neutralize the bacterial poisons, the body of the animal produces antiboidies. The blood of the animal is then withdrawn. Impurities such as blood corpuscles and other solid matter are removed from the blood. The clear blood liquid containing the anitbodies is the **serum.** It is used as weapon to combat diseases caused by these bacteria.

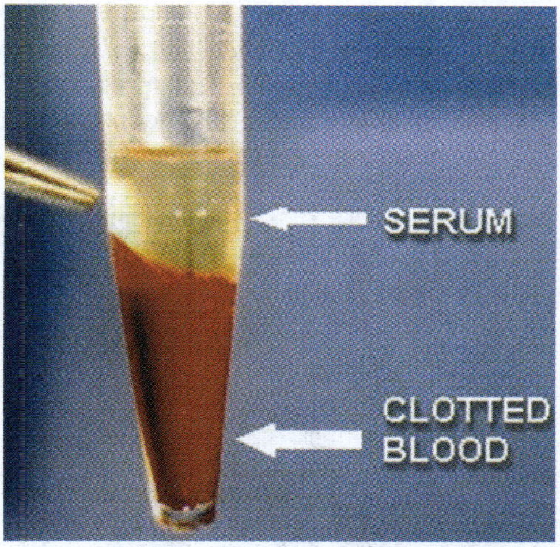

Serum (from which vaccines are prepared)

To produce vaccines dead or weakened disease producing bacteria or their diluted poisons (antigens) are directly injected into a man to cause a disease in a mild form. As a reaction the host is stimulated to form antibodies. The latter may remain for years in the body of the host imparting immunity against the same type of bacteria which may later enter his body.

This is an incomplete account of the beneficial activities of bacteria. Anyhow it indicated their extreme importance in everyday life. Because of these beneficial activities the bacteria are called the friends of mankind.

Harmful activities of bacteria.

1. Food poisoning. Of course, all activities of bacteria are not beneficial. Some saprophytic bacteria cause decay of our food and make it unpalatable. The activities of certain bacteria produce powerful toxins such as ptomins in the food. These toxins are powerful enough to cause **food poisoning** which results in serious illness and even death. Some species of *Staphylococcus* are the common offenders. There is another dangerous food poisoning bacterium known as *Clostridium botulinium.* It causes **botulism**—a fatal form of food-poisoning.

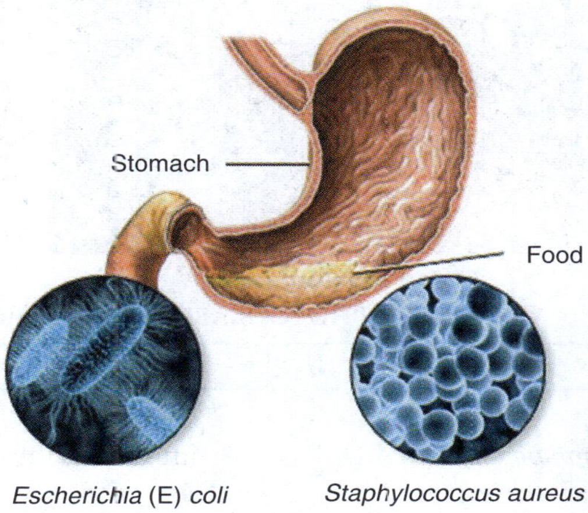

Process of Food Poisoning

2. Disease. Many parasitic bacteria are the causative agents of bacterial diseases. They cause diseases of our economic plants, domesticated animals and man. T.J. Burrill in 1878 first gave the information that bacteria cause plant diseases. There are more than 170 species of bacteria which cause plant diseases. Usually they are rod-like and non-spore forming. Many of them have flagella. The bacteria gain entry into the host through wounds or natural openings such as stomata, lenticels, hydathodes or through the thin epidermis. The bacterial diseases of plants belong to the following categories :

(i) *Wilt diseases* caused by blocking of the vessels of host plant by masses of bacteria. The common example of this category are the *wilt diseases* of potato, cucumber, water melon and egg plant.

(ii) *Crown gall and Hairy root diseases.* These are due to overgrowth or hyperplasia. The *crown* gall of beets and *hairy root* of apple are the examples.

(iii) *Narcotic blights,* leaf spots and rots caused by killing of parenchyma cells. *Fire blight* of apple, and pear and *soft rot* of carrot and turnip are the common examples. The following Table 3 gives a list of some important disease-causing bacteria, host plants and diseases.

Fire Blight of Aple

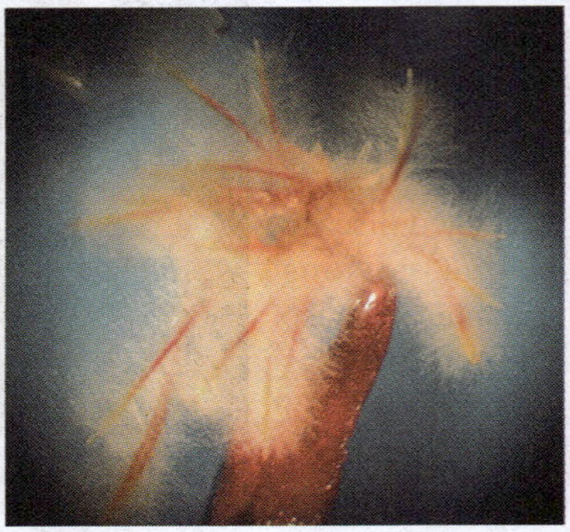
Hairy Root

Crown Gall

Bacterial Wilt of Eggplant

Bacterial Wilt of Potato

FIELD PHOTOGRAPHS OF BACTERIAL DISEASES OF PLANTS

TABLE 3

Bacteria	Host Plant	Disease
Xanthomonas citri	Citrus plants (all species)	*Citrus canker*
X. malvacearum	*Gossypium* (cotton)	Angular leaf spot or Blackarm of cotton (soft rot)
Pseudomonas solanacearum	*Solanum tuberosum* (Potato)	Bacterial brown rot or wilt disease
Eriwinia amylovora	Apple, pear and other members of Rose family	Fire blight
Erwinia carotovora	Carrots and Turnips	Soft rot

A list of principal bacterial diseases of man and animals have been given in the Table 4 given below. The most interesting thing is that these diseases are cured by the use of **antibiotics.** These antibiotics are prepared from the bacteria themselves which cause diseases.

TABLE 4

List of diseases in Man and Animals caused by bacteria.

Name of the disease	Causal organism
Abscesses	*Staphylococcus*
Anthrax	*Bacillus anthracis*
Bacillary dysentry	*Shigella sp.*
Cholera	*Vibrio cholerae*
Diphtheria	*Corynebacterium diphtheriae*
Enteritis in intestine	*Escherichia coli*
Enteric fever	*Salmonella sp.*
Gonorrhoea	*Neisseria gonorrhoeae*
Influenza	*Haemophilus influenzae*
Meningitis	*Neisseria meningitides*
Pnemonia	*Diplococcus pneumoniae*
Plague	*Pasteurella pestis*
Rat bite fever	*Spirillum minus*
Tetanus	*Clostridium tetani*
T.B.	*Mycobacterium tuberculosae*
Undulant fever	*Brucella sp.*
Urinogenital infections	*Klebsiella sp.*
Whooping cough	*Bordetella sp.*
Wound infections	*Proteus sp.*

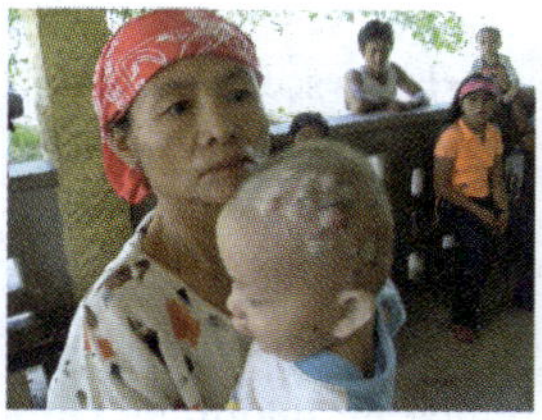
Abscesses

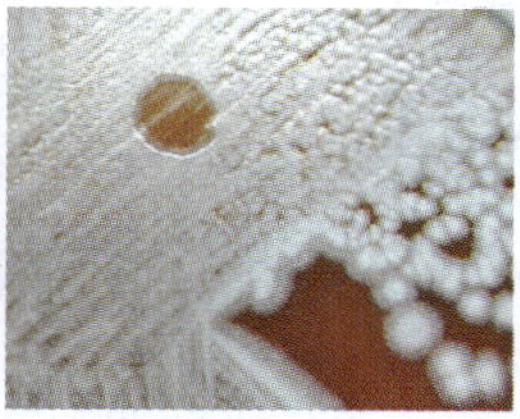
Anthrax

Dysentry

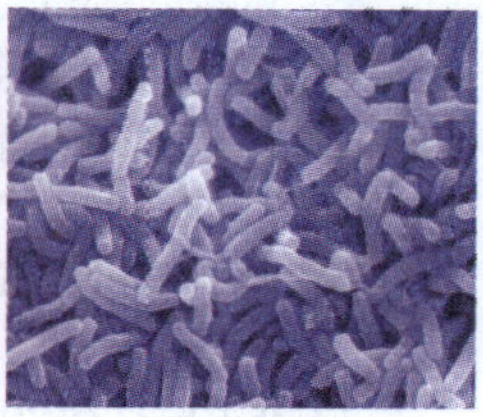
Cholera

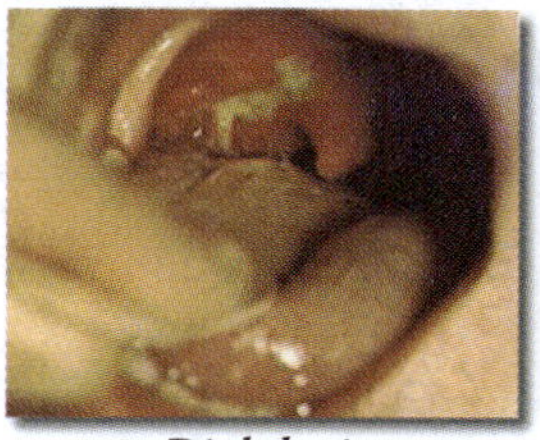
Diphtheria

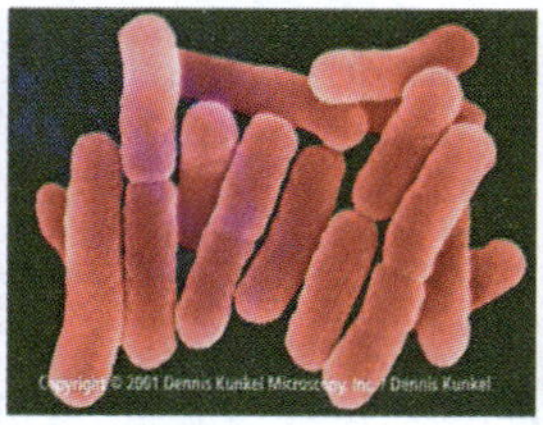

Enteric Fever

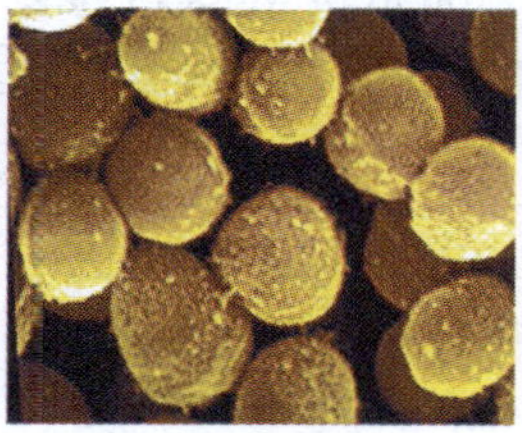
Gonorrhoea

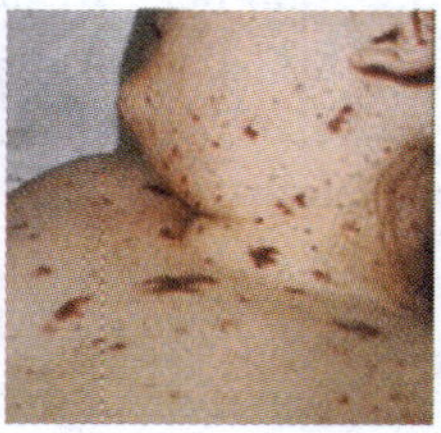
Meningitis

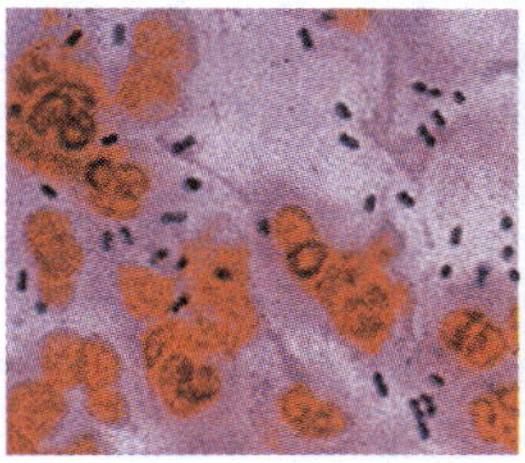
Pneumonia

Tuberculosis

Rat bite fever

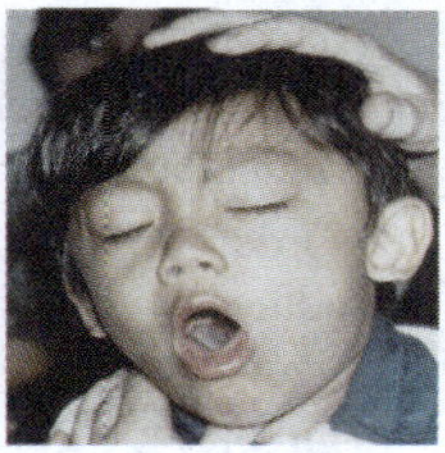
Whooping Cough

PHOTOGRAPHS OF SYMPTOMS OF HUMAN DISEASES CAUSED BY BACTERIA

3. Denitrification. There are, sometimes, a group of bacteria in the soil which reverse the nitrifying process. They injure the soil by causing the loss of a part of its combined nitrogen. This they do by breaking down nitrates into nitrites and nitrites into ammonia compounds or to free nitrogen

$$NO_3 \longrightarrow NO_2 \longrightarrow NH_3 \longrightarrow N \text{ gas}\uparrow$$

The free nitrogen passes into the atmosphere and is lost to the soil. The result is the lowering of soil fertility. This process is called **denitrification.** The bacteria which bring about denitrification are called the **denitrifying bacteria.** They are most active in the soil containing excess of nitrogen compounds such as the heavily manured soils. Soils deficient in oxygen are also favourable for the activity of this type of bacteria. Denitrification is checked if the soil is well aerated by ploughing or digging and well drained. It is uneconomic to use natural and artificial nitrate manures simultaneously. The denitrifying bacteria found in the faeces contained in the manures tend to destroy the nitrates.

Classification of Bacteria. No single system of classification of bacteria meets universal acceptance. Most of the bacteriologists, however, follow the classification proposed in "Bergey's Manual of Determinative Bacteriology". It is discussed here in brief. He divides class **Schizomycetes** into the following 10 orders :—

Order I. **Pseudomonadales.** It includes some photosynthetic, some chemosynthetic and many heterotrophic bacteria with straight, curved or spiral rods usually with polar flagellation. They are gram-negative and reproduce always by fission. The order includes 7 families. Of these the two important ones are :—

Family 1. *Nitrobacteriaceae.* It includes nitrogen-fixing bacteria and chemosynthetic species which convert ammonia to nitrites or nitrites to nitrates.

Family 2. *Pseudomonadaceae.* It includes common heterotrophic bacteria of soil and water. Some of these cause plant diseases and others fermentation. The important genera are *Pseudomonas, Xanthomonas,* some *Vibrios* and *Spirilla.*

Order II. **Chlamydobacteriales.** They form trichomes usually with sheaths and are found in water.

Order III. **Hyphomicrobiales.** Cells are ovoid or spherical and reproduce by budding. They are gram-negative and found in water or mud.

Order IV. **Eubacteriales.** Includes straight rods or spherical, simple, motile or non-motile, flagellation non-polar and peritrichous, reproduce usually by fission. The order includes a number of families :—

Family 1. *Azotobacteriaceae.* It includes soil forms which are non-symbiotic nitrogen fixers, cocci or large rods. Example *Azotobacter* sp.

Family 2. *Rhizobiaceae.* It includes gram-negative rods. Some are plant pathogens and some nitrogen-fixing symbionts in leguminous roots (*Rhizobium* sp.).

The other families included in this order are *Achromobacteriaceae, Enterobacteriaceae, Brucellaceae, Bacteriodaceae, Micrococcaceae, Neisseriaceae, Bacillaceae, Lactobacillaceae* and *Corynebacteriaceae.*

Order V. **Actinomycetales.** The order includes filamentous forms, filaments branched, reproduce by conidia, oidiospores and the like. The order comprises three families, namely, *Mycobacteriaceae, Actinomycetaceae* (*Actinomyces*) and *Streptomycetaceae* (*Streptomyces*).

Order VI. **Caryophanales.** Cells in filaments, or shorter structures, reproduce by conidia.

Order VII. **Beggiatoales.** Cells solitary or in filaments, aflagellate, reproduce by fission.

Order VIII. **Myxobacteriales.** It includes rods with slime, reproduce by fission and are found in soil or dung.

Order IX. **Spirochaetales.** Free living in water; body slender, spiral, aflagellate; reproduce by fission. The order comprises two familes, namely, *Spirochaetaceae* and *Treponemataceae.*

Order X. **Mycoplasmatales.** It includes mostly parasitic pleomorphic forms. The order comprise one family *Mycoplasmataceae.*

Salient Features. The chief characterisitics of bacteria which distinguish them from most of the other organisms are :

1. They are the smallest, simplest and chiefly unicellular organisms.
2. Their simple protoplast lacks a definite nucleus, plastids and the sap cavity.
3. Their heterotrophic mode of nutrition. A few, however, are unique in the fact that they are able to synthesize carbohydrate food without true chlorophyll.
4. Simple reproductive methods that are entirely asexual in character. These are fission, fragmentation, and in some endospore formation.
5. Entire absence of sexual reproduction. Recent investigations, however, have indicated the existence of primitive type of sexual reproduction called **genetic recombination.**
6. Their worldwide distribution.
7. They are the agents of maintaining a continual circulation of nitrogen and carbon in nature which make possible the existence of life on earth.

The above-mentioned features indicate that the bacteria are the most ancient, lowliest and the most primitive members of the plant kingdom. We call them ancient because they have been preserved as fossils in Pre-cambrian strata which are at least 620 million years old.

Evolutionary Relationships and Taxonomic Position of Bacteria. The early investigators considered bacteria as animalcules and place them with the Protozoa. This viewpoint was held up to the middle of the last (nineteenth) century. Some assigned them a position of the border line between plants and animals. They place this group of structurally simple organisms, which cannot readily be classified as either plants or animals, in the **Protista.** Even today the affinities of bacteria are uncertain. However, it has now been established beyond doubt that bacteria are members of the plant world because in most of their structural features particularly in their mode of nutrition and presence of a rigid cell wall they resemble plants more than animals. For other plant characteristics of bacteria refer to page 537.

The nearest relatives of bacteria in the kingdom Plantae are the fungi and algae. Many workers hold that the bacteria are related to fungi. In the chitinous nature of cell wall, the absence of chlorophyll and mode of nutrition as parasities or saprophytes, the bacteria show a nearer approach to the fungi. In their unicellular plant-body they resemble the yeasts. But here the resemblance ends. Bacteria as a rule reproduce by **fission.** Because of this feature and the above-mentioned resemblances with the fungi, the supporters of this viewpoint referred to bacteria as **fission fungi** and placed them in the class **Schizomycetes** coordinate with the other classes of true fungi. The opponents of this thesis point out that bacteria differ from the fungi in their small size, unicellular nature, lack of a definite nucleus, mitochondria and endoplasmic reticulum in the cell and lastly reproduction, as a rule, by fission. Furthermore, there is no group of fungi with which the bacteria share any relationships.

The relationship of bacteria with other classes of plants is totally obscure. The modern trend, however, is to club them with the Cyanopyceae (blue-green algae). In the opinion of the supporters of this hypothesis, the two groups are closely related. This assumption is based on the following common features which the two share between themselves :—

1. Both groups include members which are one-celled.

2. Obscure nuclear structure and lack of mitochondria in the cells of both.
3. Cell wall structure and cell division similar in both.
4. Frequent development of gelatinous sheaths in both.
5. Formation of resting cells or spores in both.
6. Presence of similar cell forms, such as cylindrical, spherical and spiral in both.
7. Ability of some members of both the groups to withstand extreme desiccation and endure temperatures approaching boiling point of water.
8. Ability of few blue-green algae to live on dead organic matter like saprophytic bacteria when light is not available for photosynthesis.
9. Ability of certain Cyanophyceae to fix free atmospheric nitrogen like the nitrogen-fixing bacteria.
10. Multiplication in both by fission.
11. Lack of motile asexual spores in both.
12. Entire absence of sexual reproduction in both. Recent experiments, however, indicate that some bacteria are capable of a primitive type of sexual reproduction.

The sharing of above-mentioned common features particularly the striking similarity of the nuclear apparatus in the two groups tempted some students of blue-green algae to place bacteria and blue-green algae in a single phylum "**Schizophyta**". Some even go to the length of placing them in a separate kingdom, the **Monera** which includes the **prokaryotic** organisms (lacking distinct nuclei). It is subdivied into the following two phyla :—

1. **Phylum Schizophyta** which includes the bacteria.
2. **Phylum Cyanophyta** which comprises the blue-green algae.

The International Code of Botanical Nomenclature, however, does not recognise, as yet, more than two Kingdom systems.

The evolutionary relationships of the two phyla with each other is still not clear. One view is that the bacteria and the blue-greens arose from the common prokaryotic ancestor. This hypothesis is supported by the possession of so many features in common between them. This evolutionary scheme can be represented as follows :—

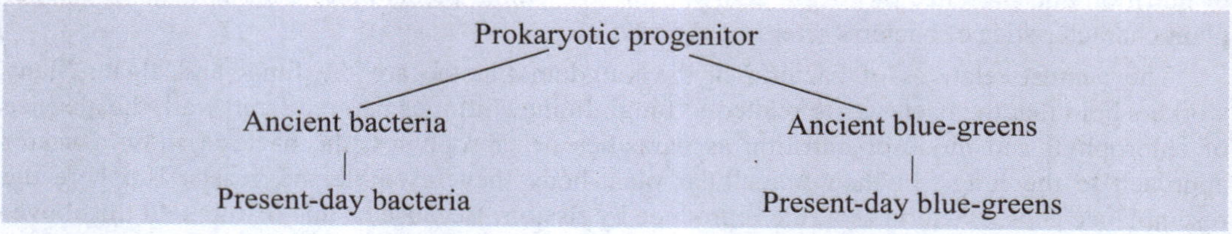

The other hypothesis is that the two groups do not share a common ancestry. They have evolved independently of each other. The proponents of this hypothesis point out the great differences between the two groups which preclude any close affinities between them. These are :-

1. Blue-greens are aerobic whereas many bacteria are anaerobic.
2. Many bacteria are flagellate whereas blue-greens are aflagellate.
3. All blue-greens are photosynthetic and use chlorophyll *a* to capture light energy. Only a few bacteria are photosynthetic but lack chlorphyll *a*.
4. In photosynthesis the blue-greens use hydroxzyl ions derived from water as electric donors and release oxygen as a by-product. In bacterial photosynthesis hydrogen sulphide is the source of hydrogen and the by-product is sulphur and not oxygen.

The supporters of the second hypothesis thus advance the following evolutionary scheme :—

Prokaryotic progenitor	Prokaryotic progenitor
|	|
Ancient bacteria	Ancient blue-greens
|	|
Present-day bacteria	Present-day blue-greens

There is a third suggestion which advocates that the bacteria have evolved from the blue-green algae by the loss of chlorophyll and adaptation to a saprophytic or parasitic mode of existence. The fossil record, at present, is of little help in deciding the issue.

The fact that some bacteria possess flagella suggests the possibility of their origin from simple, flagellated progenitors which also gave rise to the green algae. There is also a possibility that different groups of bacteria have evolved independently from different ancestors. However, our knowledge of the evolutionary relationships of bacteria is unsatisfactory.

The modern bacteriologists, however, agree that the simple flagellar structure of bacteria which differs from 9+2 pattern flagellar structure of all other organisms alone is a sufficient feature to indicate isolated position of bacteria in the plant world. In this text they have been assigned to the division Schizomycophyta or Schizomycota coordinate with the true fungi.

Taxonomic Position :

Division	:	**Schizomycota**
Class	:	**Schizomycetes**
Order	:	**Eubacteriales**
Family	:	**Bacillaceae**
Genus	:	*Bacillus*
Species	:	*subtilis*

QUESTIONS

ESSAY TYPE

1. Describe the mode of nutrition and methods of reproduction in bacteria.
2. (*a*) Give the structure of the typical bacterium cell and state the economic importance of bacteria to mankind.
   (*b*) Describe prokaryotic cell structure and its important functions. *(Lucknow, 2005)*
3. Give the account of bacteria in general and point out their importance in nature to man.
4. Give an account of the beneficial and harmful activities of bacteria to man.
5. Describe the life history of *Bacillus subtilis* and also list the distinctive features of bacteria.
6. Discuss the systematic position of bacteria.
7. (*a*) "Bacteria are our greatest friends and foes." Comment upon this statement. *(Agra, 1957; Gorakhpur, 1990)*
   (*b*) Describe the role of bateria in human welfare. *(Kanpur M.Sc.,2005)*
8. "The beneficial effects of bacteria outweigh the harm which they do." Justify the above statement by giving suitable examples. *(Agra, 1951)*
9. (*a*) Write an account of the classification, structure and mode of nutrition of bacteria and their importance in relation to the circulation and fixation of nitrogen. *(Agra. 1953)*
   (*b*) Describe the criteria used for the classification of bacteria. *(Lucknow, 2005)*
10. Give an account of the role of bacteria in relation to the fertility of the soil. *(Agra. 1958)*
11. Name five persons who have made important contributions to early bacteriology. What were

their most outstanding contributions?

(Lucknow, 1992, 1996)

12. (*a*) In relation to genetic recombination, discuss in detail the following :

(*a*) Transmission (*b*) Transduction

(*c*) Conjugation

(Bundelkhand, 1999; Lucknow, 1993, 2000, 2005; Awadh, 1997; Gorakhpur, 1991, 1996; Himachal Pradesh, 1996)

(*b*) Analyse the view that Bacteria are omnipresent. *(Kanpur M.Sc., 2006)*

13. Describe the structure of bacterial cell and importance of bacteria in human life.

(Lucknow, 1995, 1997)

14. What are bacteria? Describe their different morphological forms. What are the major differences between Gram-positive and Gram-negative bacteria?

(Lucknow, 1991, 1994; Gorakhpur, 1991; Rohilkhand, 1998)

15. Describe the structure of bacterial cell and write a note on the nutrition of bacteria.

(Agra, 1991; Kanpur, 1995; Gorakhpur, 1996; Kanpur M.Sc., 2006)

16. Write an essay on role of microorganisms in human health. *(Agra, 1993; Kanpur, 1999)*

17. What are cell type of bacteria and blue green algae (Cynobacteria) called ? Give an account of morphological features, physiological functions and methods of reproduction in any of the type studied by you. *(Agra, 1995)*

18. Give an account of sexual reproduction in bacteria with diagram.

(Kanpur, 1995, 1999; Awadh, 1999)

19. Write an essay on the beneficial and harmful activities of bacteria.

(Kanpur, 1999; Allahabad, 1996)

20. Describe in detail the mode of nutrition and method of reproduction in bacteria.

(Awadh, 1992, 1993, 1997; Gorakhpur, 1994)

21. Describe in short the various methods of reproduction in bacteria and describe the mechanism of transformation in bacteria.

(Allahabad, 1991, 1995, 1997, 2001)

22. With the help of suitable diagrams, describe the various morphological forms of bacteria. How do bacteria reproduce by asexual methods? *(Gorakhpur, 1995)*

23. Write an essay on industrial importance of bacteria.

(Magadh, 1990, 1991; Himachal Pradesh, 1996)

24. Describe briefly structure, reproduction and economic importance of bacteria.

(Kumaon, 1997, 1999; Rohilkhand, 1997; Bundelkhand, 1996, 1998; Poorvanchal, 2006)

25. Describe the cell wall, nucleoid and mesosomes in Bacteria.

(Himachal Pradesh, 1996)

26. Give an account of various modes of nutrition in bacteria. *(Lucknow, 2001; Kumaon, 2003)*

27. Give an account of economic importance of bacteria.

(Lucknow, 2001; Allahabad, 2000, 2002; Kanpur, 2005)

28. What is genetic recombination ? Discuss it in the light of bacterial cell.

(Lucknow, 2004; Poorvanchal, 2001; Meerut, 1997)

29. Describe genetic recombination in bacteria. Why does alternation of generation not occur in bacteria. *(Poorvanchal, 2003)*

30. What are bacteria ? Describe the structure and reproduction in bacteria. *(Kumaon, 2001)*

31. Give a brief account of reproduction in Bacteria. *(Allahabad, 2003)*

32. Give an account of ultrastructure of bacterial cell. *(Allahabad, 2004)*

33. Write an account on the importance of bacteria in industries. *(Rohilkhand, 2003)*

34. (*a*) Describe various forms of bacteria and explain their mode of nutrition.

(Rohilkhand, 2001)

(*b*) Describe different methods of sexual reproduction in bacteria with the help of diagrams. *(Kanpur, 2006)*

(*c*) Describe the fine structure and sexual reproduction in Bacteria.

(Poorvanchal, 2005; Kanpur, 2006)

SHORT ANSWER TYPE

35. Write short notes on the following :
    (*a*) Photosynthetic bacteria *(Lucknow, 1994, 1999)*
    (*b*) Anaerobic respiration in Bacteria *(Lucknow, 2001)*
    (*c*) Transformation *(Lucknow, 1994, 1997)*
36. Write the distinguishing features of the following :
    (*a*) Archaebacteria and Eubacteria
    (*b*) Anaerobic respiration and fermentation
    (*c*) Gram-positive and Gram-negative bacteria *(Lucknow, 1994)*
    (*d*) Gram-negative bacteria *(Allahabad, 2002)*
37. Describe various forms of bacteria and their nutrition. *(Lucknow, 1993, 1997, 1999)*
38. Describe briefly the following :
    (*a*) Culture of bacteria *(Lucknow, 1992, 1996, 1999)*
    (*b*) Motility of bacteria *(Lucknow, 1992, 1996)*
    (*c*) Bacterial endospores *(Lucknow, 1992, 1996, 1999, 2001; Allahabad, 2001 ; Gorakhpur, 1997; Poorvanchal M.Sc., 2003 ; Himachal Pradesh, 1992)*
    (*d*) Conjugation between bacterial cells. *(Lucknow, 2001, 2003; Kanpur M.Sc., 2006; Kumaon, 2002; Rohilkhand M.Sc., 2000)*
    (*e*) Reproduction in bacteria *(Guru nanak Deo, 1991, 1992; Poorvanchal, 2006)*
    (*f*) Characters of Bacteria *(Lucknow, 2004)*
39. Name the species of the lower plants which are considered to be most primitive and why?
40. Give a brief account of the following :
    (*a*) Economic importance of bacteria *(Lucknow, 1996; Kanpur 1997; Awadh, 1994, 1996; Kanpur M.Sc., 2002; Poorvanchal, 2001 ; Rohilkhand, 2002; Allahabad, 1991, 1998; Gorakhpur, 1996; Kumaon, 1996, 1998; 2000 Himachal Pradesh, 1996; Bundelkhand, 1994, 1995)*
    (*b*) Nitrogen fixation by bacteria *(Allahabad, 1995, 2001; Lucknow, 1996, 1997)*
    (*c*) Nutrition in bacteria *(Poorvanchal 2003, 2005; Lucknow, 2000, 2002, 2004; Kanpur, 1999, 2005; Awadh, 1996; Gorakhpur, 1997)*
41. Describe the various criteria for classification of bacteria. *(Lucknow, 1997, 1998, 2003; Awadh, 1991)*
42. Giving suitable examples, write an account of contribution of bacteria in industries. *(Agra, 1992)*
43. Give the names of two bacteria of each of the following group :
    (*a*) Baking Industry (*b*) Saprophytic
    (*c*) Chemosynthetic
    (*d*) Photosynthetic *(Kanpur, 1994)*
    (*e*) Structure and composition of bacterial cell wall *(Rohilkhand, 1993)*
44. Write explanatory notes on the following :
    (*i*) Use of bacteria in Agriculture/useful activities of Bacteria. *(Awadh, 1991, 1998; Rohilkhand, 1991; Himachal Pradesh, 1992)*
    (*ii*) Characteristics of bacterial nucleus. *(Awadh, 1991)*
    (*iii*) Sexual reproduction in Bacteria. *(Allahabad, 2004; Awadh, 1992, 1995)*
    (*iv*) Classification of bacteria. *(Awadh, 1996)*
    (*v*) Fine structure of bacterial cell. *(Kumaon, 2000, Poorvanchal, 2003; Awadh, 1996; Gorakhpur, 1998; Bihar, 1991; Rohilkhand, M.Sc., 1999)*
    (*vi*) Asexual Reproduction in Bacteria. *(Allahabad, 2004; Kanpur 2001, 2006; Awadh, 1996; Gorakhpur, 1990; Poorvanchal, 2004)*
    (*vii*) Heterotrophic bacteria. *(Awadh, 1998; Himachal Pradesh, 1996)*
    (*viii*) Bacteria in pharmaceutical industry. *(Awadh, 1998)*
    (*ix*) High frequency recombination in Bacteria. *(Lucknow, 2004)*
    (*x*) Contributions of Antonie Leeuwenhoek *(Lucknow, 2004)*
45. Describe the structure of bacterial cell and narrate the differences between Gram-positive and Gram-negative bacteria *(Awadh, 1992, 1995, 1998; Allahabad, 1995; Gorakhpur, 1993, 1998)*
46. Comment upon the following :
    (*a*) Flagellation in bacteria *(Allahabad, 1993, 1994, 1995, 1997, 1998, 2003; Gorakhpur, 1990, 1992; Rohilkhand, 1997)*

(*b*) Transduction
(*Rohilkhand, 2001, 2003; Allahabad, 1994, 1995, 1998; Gorakhpur, 1994, 1995; Magadh, 1991; Rohilkhand, M.Sc., 1999*)

(*c*) Gram-positive and Gram-negative bacteria
(*Allahabad, 1994, 1995, 1996, 1998; Punjab, 1998*)

(*d*) Transformation
(*Lucknow, 2002; Poorvanchal, 2002; Allahabad, 1995, 1996, 2002; Gorakhpur, 1994; Rohilkhand, 1992*)

(*e*) Gram staining (*Allahabad, 1997*)

(*f*) Classification of bacteria
(*Lucknow, 2003; Gorakhpur, 1992, 1995*)

(*g*) Genetic recombination in Bacteria
(*Gorakhpur, 1992, 1999; Punjab, 1992; Meerut, 1997*)

(*h*) Non-symbiotic nitrogen fixing bacteria
(*Gorakhpur, 1995*)

(*i*) Nitrogen Cycle (*Lucknow, 2002*)

(*j*) N2- Fixing Bacteria
(*Rohilkhand M.Sc., 2000*)

47. Give an account of the role of bacteria in increasing the fertility of soil.
(*Gorakhpur, 1999; Punjab, 1999; Rohilkhand M.Sc., 1999*)

48. Draw an ultrastructure of bacterial cell as seen in Electron Microscope.
(*Poorvanchal, 2001, Kanpur, 2005 ; Guru Nanak dev 1991; Punjab/HP, 1990, 1999, Bundelkhand, 1993, 1997*)

49. Describe

(*a*) type of bacteria on the basis of shape

(*b*) type of bacteria on the basis of flagellation

(*c*) type of bacteria on the basis nutrition

(*d*) type of bacteria on the basis of Gram staining.
(*Rohilkhand, 1993; Himachal Pradesh, 1996*)

(*e*) Bacterial Cell Structure.
(*Lucknow, 2000*)

(*f*) Useful and harmful bacteria.
(*Lucknow, 2000*)

(*g*) Nutrition in bacteria based on carbon source. (*Lucknow, 2004*)

(*h*) Useful and harmful bacteria.
(*Lucknow, 2004*)

50. (a) What is Gram stain? Differentiate between Gram-positive and Gram-negative bacteria.
(*Punjab, 1999*)

(*b*) Compare Gram positive and gram negative bacteria. (*Kanpur M.Sc., 2006*)

51. Enumerate four differences in cell wall structure of Gram positive and Gram negative bacteria. (*Lucknow, 2002*)

52. Give two examples of the following :

(*a*) Asymbiotic nitrogen fixing bacteria

(*b*) Antibiotic producing bacteria and their products. (*Lucknow, 2002*)

(*c*) Nitrifying bacteria. (*Poorvanchal, 2001*)

53. Explain High Frequency (Hfr) recombination in bacteria. (*Lucknow, 2002*)

54. Mention briefly the major contributions of Louis Pasteur. (*Lucknow, 2002*)

55. Write an explanatory note on culture of bacteria.

56. Describe the role of bacteria in soil fertility.
(*Lucknow, 2003*)

57. Write the name of the bacterium from which steptomycin is obtained ?
(*Poorvanchal, 2001*)

58. What is the nature of food reserve in Bacteria?
(*Poorvanchal, 2003*)

59. Write briefly the salient features of cytology of bacteria. (*Kanpur M.Sc., 2002*)

60. Describe the formation and nature of endospores in bacteria.
(*Kanpur M.Sc., 2003*)

61. Differentiate between/compare

(a) Gram positive and Gram negative bacteria.
(*Kanpur, 2005; Allahabad, 2003; Kanpur M.Sc., 2001, 2004, 2006*)

(*b*) Monotrichous and amphitrichous flagella.
(*Allahabad, 2004*)

(*c*) Bacterial transformation and transduction.
(*Lucknow, 2005*)

62. Draw well labelled diagram of

(*a*) Flagellar apparatus in bacteria.
(*Kanpur M.Sc., 2001*)

(*b*) Bacterial cell under Electron microscope.
(*Kanpur M.Sc., 2001*)

(*c*) Differences between process of transduction and transformation.
(*Kanpur M.Sc., 2003*)

(*d*) Different shapes of bacteria.
(*Kanpur M.Sc., 2004*)

OBJECTIVE TYPE

63. Fill in the blanks :

(*i*) Muramic acid to found in the cell wall of ....

(*ii*) Nitrifying bacterium is ....

(*iii*) The science of study of bacteria is known as ....

(*iv*) Capsule of bacteria is made up of ....

(*v*) While fungi are mostly eukaryots, the bacteria are ....

(*vi*) The bacteria have been included in the plant kingdom due to the presence of ....

(*vii*) Bacteria which do not bear flagella are known as ....

(*viii*) The bacterium which assists digestion in human beings is ....

(*ix*) The bacterium which is found in root nodules of leguminous plants is ....

(*x*) The name bacterium was first proposed by ....

(*xi*) In bacteria resperatory enzymes are located in ....

(*xii*) In bacteria the small granules containing reserve phosphates are called ....

(*xiii*) Bacteria generally reproduce by ....

(*xiv*) The bacteria are ....

(*xv*) The distinctive polymer which forms the bacterial cell wall is called ....

(*xvi*) Bacilli having a tuft of flagella at one end are called ....

(*xvii*) After gram staining the bacteria retaining crystal violet colour are called ....

(*xviii*) An antibiotic streptomycin is obtained from ....

(*xix*) Non symbiotic fixation of nitrogen takes place by ....

(*xx*) Bacterial transduction was discovered by ....

(*xxi*) Bacterial flagellum originates from ....

(*xxii*) Bacterium that causes Botulism is ....

(*xxiii*) Anthrax is caused by ....

(*xxiv*) Bacterial cell wall is made up of ....

64. Select the correct answer :

(*i*) Nitrifying bacterium is
(*a*) *Rhizobium* (*b*) *Nitrosomonas*
(*c*) *Nitrobacter* (*d*) *Bacillus.*

(*ii*) The branch of science which deals with the study of bacteria is called
(*a*) Microbiology (*b*) Mycology
(*c*) Bacteriology (*d*) Virology.

(*iii*) Capsule of bacteria is made up of
(*a*) Polysaccharides
(*b*) Amino sugars
(*c*) Polypeptides
(*d*) All of the above

(*iv*) Citrus canker is causued by
(*a*) *Rhizobium* (*b*) TMV
(*c*) *Xanthomonas* (*d*) *Aspergillus.*

(*v*) Bacteria capable of synthesizing their own food by themselves from organic and inorganic substances are called
(*a*) Autotrophic
(*b*) Chemoautotrophic
(*c*) Heterotrophic
(*d*) Nitrifying.

(*vi*) Muramic acid is found in the cell walls of
(*a*) Algae (*b*) Fungi
(*c*) Bacteria (*d*) Viruses.

(*vii*) Bacteria are considered to be plants because
(*a*) Some of them are green
(*b*) They are present everywhere
(*c*) Some of them cannot move
(*d*) They have rigid cell walls.

(*viii*) The name of the bacterium which assists digestion in human body is
(*a*) *Rhizobium leguminosarum*
(*b*) *Xanthomonas citri*
(*c*) *Escherichia coli*
(*d*) *Vibrio cholerae*

(*ix*) The bacterium cell wall is made up of
(*a*) chitin (*b*) lignin
(*c*) cellulose (*d*) pectose.

(*x*) Bacteria were first discovered by
(*a*) Louis Pasteur (*b*) Robert Koch
(*c*) T.J. Burril
(*d*) Anton von Leeuwenhoek

(*xi*) The bacteria are
(*a*) unicellular (*b*) bicellular
(*c*) multicellular (*d*) flamentous

(*xii*) Symbiotic nitrogen fixation is carried out by
(*a*) *Bacillus subtilis*
(*b*) *Rhizobium leguminosarum.*
(*c*) *Xanthomonas citri*
(*d*) *Azotobacter.*

(*xiii*) The common method of reproduction in

bacteria is
(*a*) budding
(*b*) fragmentation
(*c*) fission
(*d*) hormogone formation.

(*xiv*) Which of the following is a bacterial disease?
(*a*) typhoid (*b*) cholera
(*c*) tuberculosis (*d*) All of the above.

(*xv*) Which of the following is not found in bacterial cell?
(*a*) endoplasmic reticulum
(*b*) glycogen
(*c*) protein
(*d*) fats.

(*xvi*) The bacteria having a tuft of flagella at both the ends are known as
(*a*) lophotrichous
(*b*) peritrichous
(*c*) amphitrichous
(*d*) atrichous.

(*xvii*) The gram negative bacteria differ from gram positive in cell wall composition in having
(*a*) Lipids (*b*) Carbohydrates
(*c*) Protein (*d*) None

(*xviii*) Father of medical microbiology is :
(*a*) Robert Koch
(*b*) A.M. Campbell
(*c*) M.K. Allen
(*d*) J.Leister

(*xix*) F-factor of bacterium is made up of
(*a*) DNA only
(*b*) DNA and Protein
(*c*) RNA and Protein
(*d*) DNA and RNA

(*xx*) Process of making the milk germ free by boiling it to 70°C for 30 minutes is called
(*a*) Dehydration (*b*) Immunization
(*c*) Pasteurization (*d*) Sterilisation

(*xxi*) Transformation was discovered by
(*a*) Lederburg and Tatum
(*b*) Griffith
(*c*) Watson and Crick
(*d*) Zinder and Lederberg

(*xxii*) Which of the following is correct ?
(*a*) Bacterial cell is made up of complex chemicals
(*b*) Bacteria have definite organised nuclei
(*c*) Bacteria have chloroplast
(*d*) All of the above

(*xxiii*) Bacterial Plasmids contains
(*a*) RNA (*b*) RNA of protein
(*c*) DNA
(*d*) Photosynthetic structures

(*xxiv*) Transfer of DNA fragments from a Donor to receptor bacterial cell through the medium is
(*a*) Translation (*b*) Transcription
(*c*) Transduction (*d*) Transformation

(*xxv*) Diaminopimelic acid and muramic acid are constitutents of the cell walls of
(*a*) Fungi (*b*) Bacteria
(*c*) Mycoplasma (*d*) Archaebacteria

(*xxvi*) Jam and Jellies are free from bacteria and molds because
(*a*) they have complex substances not degraded by bacteria
(*b*) they have poisonous substances
(*c*) bacterial cells get plasmolysed
(*d*) none of these.

Cadang Coconut Viroid

CHAPTER 19

VIRUSES, VIROIDS AND MYCOPLASMA

- Viruses
- Classification and Nomenclature of Viruses
- Shape, Size and Structure of Viruses
- Economic Importance of Viruses
- Viroids
- Mycoplasma (General Features, Morphology, Cell Shape, Cell Structure, reproduction and transmission)
- Spiroplasma

I. VIRUSES

Historical. Some of the plant diseases, although not attributed to viruses at that time, were observed quite early. Breaking of Tulip was recorded as early as 1576 while potato curl disease and peach yellows were recorded in 1765 and 1791, respectively. Around 1868, Variegation in *Abutilon* gained importance in Europe. It was sometime in 1870 that tobacco crop was affected by a severe disease in Holland. Adolf Mayer, Director of Agriculture experiment station, Wageningen began his studies on this disease in 1880 and published his results in 1886. Mayer named the diease as **Mosaik krankhet** which means **mosaic**. He was able to demonstrate the infectious nature of the sap of diseased leaves which could pass through double filters. It is, therefore, considered that the science of viruses or **Virology** was born in 1886.

However, a Russian scientist named Iwanowsky in 1892 demonstrated that Tobacco Mosaic could be transmitted to healthy plants through the sap from diseased plants even though the sap had passed through double filters, fine enough to remove all bacteria. In addition, the sap, when inoculated on culture medium, could not produce any living thing capable of producing the tobacco mosaic disease. Beijerinck (1898) from Holland also found that the infectious juice remained infective even after passing through bacterial filters and remained sterile. He recognised the existance of a new type of infectious agent and he called it as "Contagium vivum fluidum" (infectious living fluid). Subsequently he referred this infective agent of tobacco mosaic disease as **Virus**.

By 1892 viral diseases of vertebrates were well known since the noted scientist Louis Pasteur had been studying cannine rabies for quite some time. The first viral disease of human beings—the yellow fever—was discovered by Walter Reed and his associates in 1900. At that time the term virus included toxins, bacteria and viruses.

However, more important events in the development of plant virology have been listed below in chronological order :

TABLE

Year	Name of the Scientist	Contribution
1906	Ball	Transmission of a vims disease by leaf-hopper.
1915	Twort	Discovered the Virus attacking bacteria.
1917	d'Herelle	Fully studied and described the Virus attacking bacteria and named it as **Bacteriophage.**
1918	Nishimura	Discovered symptomless plant carriers of Virus diseases
1923	Schultz and Folsom Murphy	Identified Aphids as vectors.
1925	Carsner	Discovered mutabitiy of virus leading to development of virus strains.
1926	Johnson	
1925	Dickson	Discovered synergism in Viruses.
1928	Beale	Described that plants infected with Tobacco Mosaic disease (TMV) contain a specific antigen.
1929	Mc Kinney	Reported cross protection
1931	K.M. Smith	Described use of indicator plants for the separation and identification of different Viruses.
1932	Knoll and Ruska	Invented Electron Microscope.
1935	Stanley	Crystallised TMV and showed that crystals were proteins.
1936	Bawden *et al.*	Reported that TMV is a Nucleo-protein and contains a nucleic acid of pentose type.
1937	Bawden and Pirie	Crystallised and purified Tomato bushy stunt Virus in three dimensional crystals.
1937	Kunkel	Reported multiplication of Aster yellows agent in leaf hopper Vector.
1939	Kaushe *et al*	Electron micrography of TMV.
1941	Barnal and Fankuchen	Conducted x-ray diffraction studies on organisation of TMV.
1952	Black and Brakke	Reported that wound tumour Virus multiplies in its leaf-hopper vector.

Year	Name of the Scientist	Contribution
1956	Gierer and Schramm Frankel-Conrat	Independently reported that infectivity of TMV lies in its RNA
1957	Crick and Watson Crick and Kendrew	Formulated general theory about structure of Viral capsids.
1958	Bancoft and Kaesberg	Reported that alfa-alfa mosaic virus is a multicomponent virus.
1960	Anderer *et al* Tsugita *et al.*	Worked out the sequence of Amino acid arrangement in protein molecule of TMV coat protein.
1961	Kassains and Nixon	Reported satellite virus of Tobacco—necrosis virus.
1962	Kassanis	
1963	Black and Markham	Reported that wound tumour virus contains double stranded RNA.
1964	Mendel *et al.* Shipp and Haselkorn Burdon *et al.*	Isolated double strand viral RNA from leaves infected with TMV and Turnip yellow Mosaic Virus.
1967	Doi *et al.*	Reported the existence of Mycoplasma like organisms as causal agents of plant diseases.
1967	Diener and Rayener	Reported that Potato Spindle tuber disease is caused by a free nucleic acid (viroid).
1968	Shephard *et al.*	Cauliflower mosaic Virus contains DNA as the nucleic acid.
1969	Schneider	Reported the existence of second satellite Virus-Tobacco ringspot virus.
1969	Hirai and Wildman	Reported that TMV inhibits transcription of chloroplast DNA leading to chloroplast destruction.
1970	Harrison *et al,*	Reported that mitochondria is concerned with replication of tobacco rattle virus.
1971	Bradley and Zaitlin	TMV tissues contain a viral-specific replicase.
1971	Harrison *et al.*	Classified plant viruses into sixteen groups.
1972	Jackson *et al.*	Conducted identification of RI and RF RNA during TMV RNA synthesis.
1972	Davis and Worley	Reported association of spiroplasma organism with corn stunt disease
1973	Francki	Created Group of plant Rhabdoviruses.
1973	Goheen *et al.*	Reported association of rickettsia like organism with Pierci's disease of grapevine and alfa-alfa dwarf.
1974	Maramorsch	Coined the term mollicute-like organism.
1974	Reddy and Black	Deletion mutations of wound tumour virus detected.
1974	Yang and Hamilton	Reported that seed transmission of tobacco ringspot virus is by infection of Megagametophytes.
1975	Semancik *et al.*	Purification of pathogenic RNA (Viroid) from exocortis disease of citrus reported.
1975	Jonard *et al.* Guilley *et al.*	Reported that replication initiating site of TMV RNA appears to be constituted by a sequence of 103 nucleotides that represents a part of the coat protein cistron.

General Characteristics

Viruses are non-cellular, ultra-microscopic particles (or structures) of protein and nucleic acid which grow and multiply only in living cells. They cause many highly infectious diseases of animals including man, angiosperms, bacteria, fungi and algae including the blue-greens. All known viruses are pathogens. No saprophytic virus has been discovered so far. The viruses can exist outside the host free in air or water for long periods but remain quite inert in the free state. They become activated only when they enter a living cell of a host organism. They can neither multiply nor metabolize even in complex mixtures of nutrients including the extracts of cells which they normally parasitize. It is possible to cultivate them only on intact living cells. The viruses thus are **obligate parasites** and are the smallest of all infective agents, much smaller than the smallest bacteria. Some are of the dimensions of large protein molecules. They are not visible with the light microscope. Even the largest ones are seen only as tiny dots at the highest possible magnification. The viruses can, however, be observed and photographed with the electron microscope. In electron micrographs some appear rod-shaped. Others have spherical, cubical or tadpole forms. Most of them have been crystallized. The virus crystals resemble simple chemical compounds and can be treated like them.

Within the host cells, the viruses possess some of the properties of life but not all. They grow, reproduce and undergo mutation but no virus has any respiration of its own. They depend on the enzymes of the host cells to do their work. The viruses cannot be placed in either the plant or animal kingdom. The branch of science which deals with them is called **virology.** One who specializes in this branch of science is called a **virologist.**

Nature of Viruses. It has long been a vexed question. In fact it is still a mystery as to what is the *nature of viruses*. The biologists hold that the viruses are **animate objects.** Biochemists vehemently oppose the living hypothesis. They say that the viruses are **inanimate objects.**

1. Organism Theory. The supporters of the living hypothesis argue that growth and reproduction are chief distinctive characteristics of living things (organisms). Viruses possess both. They grow, reproduce and undergo mutations and become adapted to the environment through natural selection. They possess genetic continuity and occur in definite races, or strains, each with its specific character. In all these respects, they show a similarity to a typical organism. A Russian virologist, Iwanowski (1890) obtained a juice from the leaves of tobacco plant suffering from *Tobacco mosaic* virus disease. The juice was filtered through an extremely fine porcelain filter that held back even the smallest bacteria. It was then inoculated in an infinitesimal dose (in dilutions up to 1—1,000,000) in the fresh, healthy leaves of tobacco. The latter showed signs of *Virus Tobacco Mosaic*. Moreover, the virus spread throughout the inoculated plant. It increased enormously in amount. Being obligate parasites, the viruses reproduce and grow only within the living tissues of the host. They do not multiply in the dead tissues of the host. Thus attempts to grow them on artificial media failed. These smallest living organisms contain all the informtion for their reproduction.

2. Molecular Theory. The proponents of the living hypothesis argue that viruses have no cellular organisation. They do not have anything resembling cytoplasm, nucleus or cell membrane. They show no physiological activity except self-duplication. They neither take food nor respire. In fact they show no signs of metabolism. All these are characteristics of inanimate objects. The supporters of this hypothesis were further encouraged in their belief by the extraction of viruses in the form of crystals from the diseased host tissues. Stanley (1935) isolated *Tobacco Mosaic Virus* in the form of rod-shaped cyrstals (Fig. 19.1 A). Since then many other viruses have been isolated. The virus crystals have a definite shape and size which varies with each virus. The virus crystals or particles which are termed virions are apparently lifeless. They can exist in the free state indefinitely like simple chemical compounds and are inert in the free state. In their chemical and physical homogeneity these are similar to the protein molecules. They seem and behave like the non-living chemicals. They are not infective by themselves. Upon contact with a suitable host they are, however, capable of infection. In view of these characteristics, the supporters of non-living hypothesis consider viruses as non-living *toxic substances* that have acquired the ability to leave the host cell and infect other cells.

The consensus of opinion at present supports the view that the viruses are neither organisms nor cells. They are considerably less than both. They represent the simplest forms of life known at present and are subcellular in organisation. In the living world they stand at the threshold of life.

Origin of Viruses. The supporters of the living hypothesis are divided into three camps about the *origin of viruses.* Some consider viruses as simple or primitive forms of living matter which represent a transition stage between the living and the non-living world. They hold them to be living forms with certain characteristics of the non-living materials. According to this view, the viruses suggest the way how life might have arisen from lifeless matter. It is generally held that life originated from simple, inorganic compounds by slow chemical reactions spread over millions and millions of years. In this way the pre-cellular first inhabitants of the earth known as the *protoviruses* may have originated. The present-day viruses are survivors of these inhabitants of the earth.

The supporters of the second hypothesis look the other way. They consider viruses as degenerate microorganism degraded from primitive, free living microorganisms. The supporters of this *regressive theory* hold that the viruses have evolved from intracellular bacterial and rickettsial forms by losing most of the functions and structures characteristic of free living organisms. Mathew (1970) derives viruses from mycoplasma-like organisms which are the simplest of the cellular microorganisms by loss of plasma membrane and change of genetic material from **DNA** to **RNA.**

The members of the third camp have put forth the *cell component theory.* They derive viruses from normal cell constituents such as mitochondrium or chloroplast. They hold that virus is nothing but a part of the cell, the *nucleic acids* or *genetic elements,* which have developed the ability to leave the cell and be transferred to other cells. Such material is independent enough to pass from cell to cell and impose its patterns on the infected host cells.

Chemical Composition of Viruses. Chemically viruses are **nucleoproteins** which are made up of two components, **protein** and **nucleic acid.** The nucleic acid forms the core of the virus particle or **virion.** It is covered by a tightly fitting **protein coat shell** or **sheath.** Each virion consists of one nucleic acid molecule but the protein coat or shell is formed of many similar protein molecules.

Cauliflower Mosaic Virus

Dahlia Mosaic Virus

CONTAIN DOUBLE - STRANDED DNA

The nucleic acid is the infective agent. The protein coat is protective in function. It is inert and non-genetic. The proportion of nucleic acid varies from about 1-40%. The nucleic acid in virus particle is of a particular type. it is either **ribonucleic acid** (RNA) or **deoxyribonucleic acid** (DNA) but not both. Plant viruses generally contain RNA. The only exceptions are *Cauliflower Mosaic Virus* and *Dahlia Mosaic Virus* which contain double-stranded DNA. The RNA in plant viruses is usually

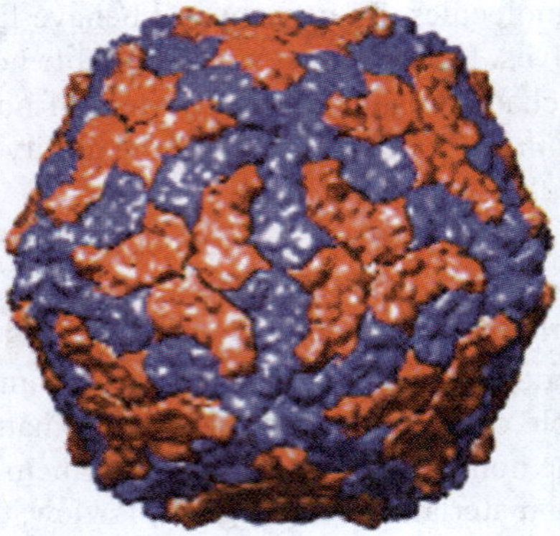

Rice Dwarf Virus

Wound Tumor Virus

DOUBLE - STRANDED RNA

single-stranded. Double-stranded RNA is, however, known to occur in two namely, *Wound Tumor Virus* and *Rice Dwarf Virus*. The DNA almost always occurs as a double strand. The phages contain DNA. About half of the animal viruses contain **DNA** and half **RNA**. Lipids and carbohydrates so common in cellular organisms are usually absent in the viruses. The large cowpox virus, however, contains lipid, carbohydrates, copper and one or two vitamin-like substances. Even such large and complex viruses exhibit no independent metabolic activity outside the host. On entering the host cell, the virus particle sheds its protein coat. The nucleic acid is released. It passes into the healthy cell and brings about multiplication of the virus. The viruses, thus represent the simplest forms of life known at present. They are below the level of cellular organisation and do not appear to have anything resembling cytoplasm, nucleus or chromosomes. The only thing they have in common with other living forms is the presence of nucleic acid, RNA or DNA but never both. Thus they have definite and stable, hereditary mechanisms. The only physiological activity of viruses is self-duplication. They exhibit no other physiological function and no metabolism. They are strict parasities alive at least while in the living cells of their hosts. They do not possess energy yielding or synthetic enzymes systems which are essential for an independent existence. Owing to the absence of the enzyme systems they cannot themselves convert energy. Energy is essential for duplication. Entering the host the virus can make use of the enzyme systems present in the cells of the host and shunt their energy to themselves. It is used for rapid multiplication at the expense of the raw materials of the host cell.

Properties Of Viruses

The viruses possess certain physical and chemical properties. On the basis of these and other characters an attempt has been made to classify them into families, genera and species. The most important properties of viruses are :—

1. They are transmissible from unhealthy to healthy plants. Some can be transmitted by juice inoculations. Others lose their activity in plant juices.
2. They are so small that they can pass through the minutest pores of the filter paper—the pores that can hold back the smallest bacteria. Now it is possible to construct filters fine enough to hold the smallest virus particles.
3. All viruses are parasitic but differ from the parasitic bacteria in that they can grow and multiply only within the living host cells. No saprophytic viruses are known. All are obligate parasites.
4. They vary greatly in their reactions towards a number of environmental conditions. Some viruses are easily killed by drying, heat or poisons. Others are highly resistant. A

few remain active for years. In dry state viruses, as a rule, can resist the action of common germicides such as alcohol, formaldehyde, carbolic acid and the like.

5. Some viruses are effective even if inoculated in infinitestimal doses. They are thus highly infectious.
6. They are unaffected by exposure to direct sunlight and many can withstand heat, cold or drought.
7. They can be precipitated from suspensions and redissolved in water.
8. Viruses have *host specificity*. It means each virus can infect only a single species and requires definite cells within host organisms for reproduction.
9. Antibiotics have no effect on viruses. Most of the antibiotics usually block a specific reaction in bacterial metabolism. The viruses have no metabolism of their own. Since they use the metabolism of the host cell for their reproduction, the antibiotics have no effect on them.

Transmission of plant viruses. The viruses are infectious and may be transmitted through several generations inside a host. They can also be taken out of the host and introduced into other living cells. The transmitted units are invisible with light microscope. They are thus recognisable only by their effects on the host organism. The chief methods of transmission of plant viruses are :—

1. *Seed Transmission of Virus.* Transmission through the seeds of the host plant was earlier considered to play a minor part in the spread of virus diseases. Recently Bennett (1969) listed 53 viruses which are transmitted by seeds of about 124 plant species. The seeds are important in the spread of a few viruses of legumes, wild cucumber, tomatoes, and curly top virus of beet sugar. In the latter case the seeds carry a high percentage of the virus. The virus, however, does not enter the embryo. It is carried in a portion of the seed of the diseased plants.

2. *Transmission by Vegetative Propagation.* It is one of the chief methods of transmission of virus diseases especially of Potato, Rose, Sugarcane, Raspberry, Strawberry, Turnips, Bulb plants, fruit trees and many ornamentals. The vegetative parts, the infected plants such as the tubers, bulbs, roots, offshoots, buds and scions which are used for propagation, will contain the virus present in the parent. The new plants raised by the above-mentioned vegetative methods are nearly always infected.

3. *Transmission by Mechanical Means.* Many mosaic viruses are transmitted mechanically from diseased plants to healthy ones by the following methods :—

(*i*) By contact of infected and healthy leaves brought about by wind.

(*ii*) By rubbing the juice of the diseased plants over the surface of the leaves of healthy plants.

(*iii*) By grafting infected buds on to healthy plants.

(*iv*) Agricultural implements also play quite an important part. The knife used for cutting the seed pieces and the pruning shears will spread the disease.

(*v*) Some viruses spread below ground by contact between the roots of diseased and healthy plants.

(*vi*) Handling plants at planting time and in cultural operation will also help in the spread of viruses such as *Sugar beet, Curly top virus and Cucumber mosaic virus.*

4. *Transmission by Cuscuta.* In many cases Dodder (*Cuscuta*) serves as a transmitting agent and an effective bridge between the infected host and the healthy plants by establishing intimate biological contact through its haustoria.

5. *Soil Transmission.* Quite a number of viruses are transmitted through the soil. Common examples of soil borne viruses are *Potato mosaic virus, Oat mosaic, Wheat mosaic,* etc. In all these cases the disease is contracted from the soil.

Sugarbeet Virus

Curly Top Virus

Cucumber Mosaic Virus

EXAMPLE OF MECHANICAL MODE OF TRANSMISSION

6. *Insect Transmission.* Some plant and animal viruses are spread and complete particles introduced into host cells by arthopod vectors and even by dog-bite as in rabies. Among the arthropods most important agents of spread of virus diseases are the insects. The insect which carries the disease is called a **vector.** The insect vectors which play a major role in the dissemination of plant viruses are the *Aphids, Leafhoppers, Flee beetles, Scale insects, thirps* and *White flies*. Most of the insect vectors are sucking insects. Aphids transmit more plant viruses than any other insects. Leafhoppers come next in the list. About three hundred plant virus diseases are known to have insect vectors. The insect obtains virus through its mouth parts at the time of feeding on the diseased plant. It is then inoculated in the healthy plant by means of the mouth parts. Inoculation in many cases must be in a certain tissue or upon young leaves. The virus may remain active in the body of the vector

Potato Mosaic Virus

Oat Mosaic Virus

EXAMPLES OF SOIL TRANSMISSION

Aphids

Leafhoppers

Fleabeetles

Whiteflies

Scale insects

Thrips

EXAMPLES OF INSECTS WHICH TRANSMIT VIRUSES

for many days. Instances are, however, known when infectivity is soon lost. There are also cases where a vector cannot infect a healthy plant immediately after it has fed on a diseased plant. There is delay in the development of infective power within the vector. This period of development of infectivity for the virus within the vector is called the **incubation period.** The duration of the incubation period varies with different viruses from a few hours to days.

There also appears to be some relationship between the plant viruses and the insect vectors which transmit them. The precise nature of this relationship is still unknown. The virus disease of sugar beet known as *curly leaf* or *curly top* is spread by the leaf-hopper *Circulifer tenellus.* Other sucking insects which feed on sugar beet are unable to transmit this virus. On the other hand peach aphid is the vector of *Sugar beet mosaic* virus. The leafhopper does not transmit this virus. Thirps transmit the spotted-wilt virus. All vectors of yellow group of viruses are leafhoppers and of mosaic group are aphids.

Circulifer (*Spreads sugarbeet curly top virus*)

7. *Transmission by Fungi.* The first proof of the fungus as a vector of plant viruses was found by Gorgon in 1958. He found that the diseased lettuce was invariably infected by a soil chytrid, *Olpidium.* Later he discovered that the fungus acts as a reservoir and vector of the big vein virus. The virus acquired by the fungus remains in the oospore. The latter germinates and produces the zoospores which function as infective agents and penetrate lettuce roots. Similarly tobacco necrosis virus has been reported by Teakle (1960) to enter roots of its host by the zoospores of *O. brassicae.*

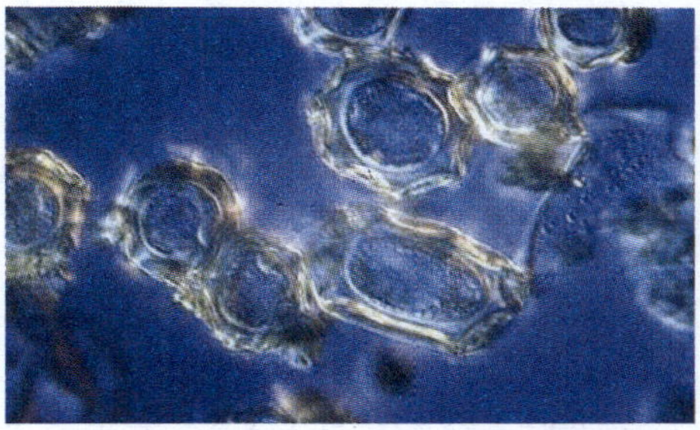

Olpidium (*Fungal vector of plant viruses*)

8. Some soil inhabiting viruses have nematode vectors.

Animal viruses may gain access to the higher animals through the mouth and nose from dust or contaminated food. Besides infection from outside, virus may also be transmitted from cell to cell but the internal transmission need not be in the form of virus particles.

Symptoms of Viral Diseases. Symptoms are the visible effects which the pathogens cause in the host plants. Viruses produce a wide range of symptoms on their hosts. They are grouped under two main categories *Primary* and *Systemic symptoms.*

(*a*) *Primary symptoms.* They relate to viruses dispersed by mechanical means. They are the result of local reaction at the spot of inoculation. These are usually of two kinds, *local lesions* and *clearing of veins.* The local lesions appear in the form of spots or rings. They are the result of death of cells in these areas. In vein clearing there is clearing or chlorosis of tissue in or adjacent to the veins, consequently the veins of young leaves become conspicuous. Frequently a broad band of chlorotic tissue appears along the veins. These two symptoms, vein clearing and vein banding may be of short duration, disappearing as the disease spreads or may remain as major symptoms.

(*b*) *Systemic symptoms.* The chief systemic symptoms are :-

1. *Mosaic Spotting.* In this case usually the leaves, sometimes the blossoms and other parts of

the host plants show spots or patches of light green, yellow or even white. These patches or chlorotic areas may be circular or irregular. They also vary in size.

2. *Ring Spotting.* This is a symptom which appears in localized spots. Sometimes circular chlorotic areas appear on the surface of leaves. These are called the **chlorotic ring spots.** Usually necrosis appears in rings alternating with normal green. The rings are concentric to a central spot. This symptom is called **necrotic ring spot.**

3. *Chlorosis.* It is uniform yellowing of leaves subsequent to infection. It occurs when young growing leaves are infected. It is not a very common symptom.

4. *Distortion.* It is a common sysmpton of virus diseases. It may be in the form of rolling and crinkling of leaves caused by retarded growth of veins. Sometimes there is production of a cluster of thin wiry shoots. It is called *witches broom.*

5. *Necrosis.* It is the death of cells in localized regions. It appears in various forms. Usually it is restricted to small areas on the leaves or streaks on the stem. In the former case the chlorotic tissue may break down and necrotic spots appear. The necrotic or dead tissue is usually demarcated from the living tissue by a distinct dark brown border. In top necrosis there is rapid killing of a terminal bud or a branch or even entire top of the plant. Sometimes it extends over large areas finally resulting in the death of the whole plant.

6. *Breaking of blossoms.* It is another sympom of virus diseases. Attractively variegated leaves of some ornamental shrubs such as *Abutilon striatum* are the result of virus infection.

Stunting and Premature Defoliation. The virus diseases in general are hypoplastic and cause under-development of the host. The host plant remains stunted with shorter internodes, smaller leaves and fruits and general reduction in size. The virus infected plants are more often stunted than killed. Cabbage mosaic causes premature defoliation.

Ring Spots

Chlorosis

Distortion of Leaves

Necrosis of Leaves

VIRAL SYMPTOMS

CLASSIFICATION AND NOMENCLATURE OF VIRUSES

(a) Classification of Viruses

It is still in a fluid state. Whether viruses are plants or animals is purely an academic question and difficult to answer with any assurance. Perhaps they are neither. According to the commonly accepted view, the viruses represent the transition stage between the living and non-living world. In other words, they stand at the very threshold of life and thus may be considered as a hypothetical model of an early stage in the evolution of living organisms (plants and animals). All biologists, at present agree to place viruses in the phylum **Vira.** Holmes (1948) placed them in the order Virales. He divided the order virales on the basis of host specificity into the following three sub-orders.

Suborder 1. **Phytophagineae.** It includes the viruses causing plant diseases.

Suborder 2. **Phagineae.** It comprises the viruses which attack bacteria and are known as the bacteriophages.

Suborder 3. **Zoophagineae.** It includes viruses causing diseases to men and animals.

On the basis of the genetic material (RNA or DNA), Dwoff and Tausnier (1966) divided the phylum Vira into two subphyla namely, (*i*) **Ribovira** containing RNA and (*ii*) **Deoxyvira** containing DNA. On the basis of the symmetry they further sub-divided subphylum Ribovira into two classes, *Ribohelica* and *Ribocubica* whereas deoxyvira was split into 3 classes, *Deoxyhelica, Deoxycubica* and *Deoxybinala* (with a combination of both helical and cuboidal symmetry). The classes are divided into orders, as detailed below :

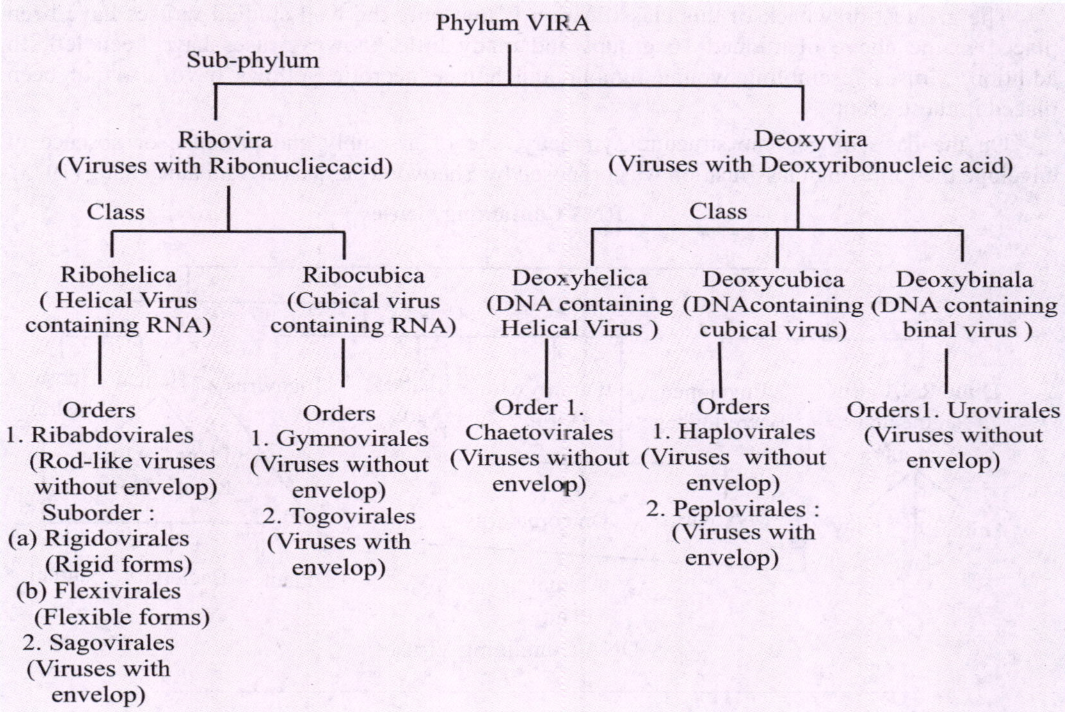

In 1968 Thornberry proposed a system of classification in which all the viruses were placed under the phylum Thallophyta, class Schizomycota, and order Biovirales. The order was further classified into three families Phytoviraceae, Zooviraceae and Rickettsiaceae. The family Phytoviraceae contained five genera on the basis of the type of hosts attacked viz. Phytovirus, Pteridovirus, Bryovirus, Thallovirus and Bacteriophagus.

Gibbs (1969) presented a computerized hierarchial classification of about 140 viruses and grouped them into two major and five minor clusters or groupings.

On the basis of the hehaviour of virus in the hosts, vector relations, particle properties and particle composition, Harrison *et al.* (1971) divided plant Viruses into 16 groups, as detailed below :

1. **Tobravirus group.** Type member : Tobacco rattle virus.
2. **Tobamovirus group.** Type member : Tobacco mosaic virus
3. **Potexvirus group.** Type member : Potato virus X
4. **Carla virus group.** Type member : Carnation latent virus
5. **Poty virus group.** Type member : Potato virus Y
6. **Cucumovirus group.** Type member : Cucumber mosaic virus.
7. **Tymovirus group.** Type member : Turnip yellow mosaic virus.
8. **Comovirus group.** Type member : Cowpea mosaic virus
9. **Nepovirus group.** Type member : Tobacco ringspot virus.
10. **Bromovirus group.** Type member : Brome mosaic virus.

11. **Tombus virus group.** Type member : Tomato bushy stunt virus
12. **Caulimovirus group.** Type member : Cauliflower mosaic virus.
13. **Alfalfa mosaic virus.** Type member : Alfalfa mosaic virus.
14. **Pea Enation Mosaic virus.** Type member : Pea enation mosaic virus.
15. **Tobacco necrosis virus.** Type member : Tobacco necrosis virus A strain.
16. **Tomato spotted wilt virus.** Type member : Tomato spotted wilt virus.

The greatest drawback of this classification is that only the well studied viruses have been placed in the above mentioned 16 groups and many little known viruses have been left. In addition, viruses resembling wound tumour and lettuce necrotic yellows have also not been placed in these groups.

On the basis of genome structure symmetry, site of assembly and presence or absence of envelope the following classification was proposed by Sherwood Gasjens and Jonathan King (1975).

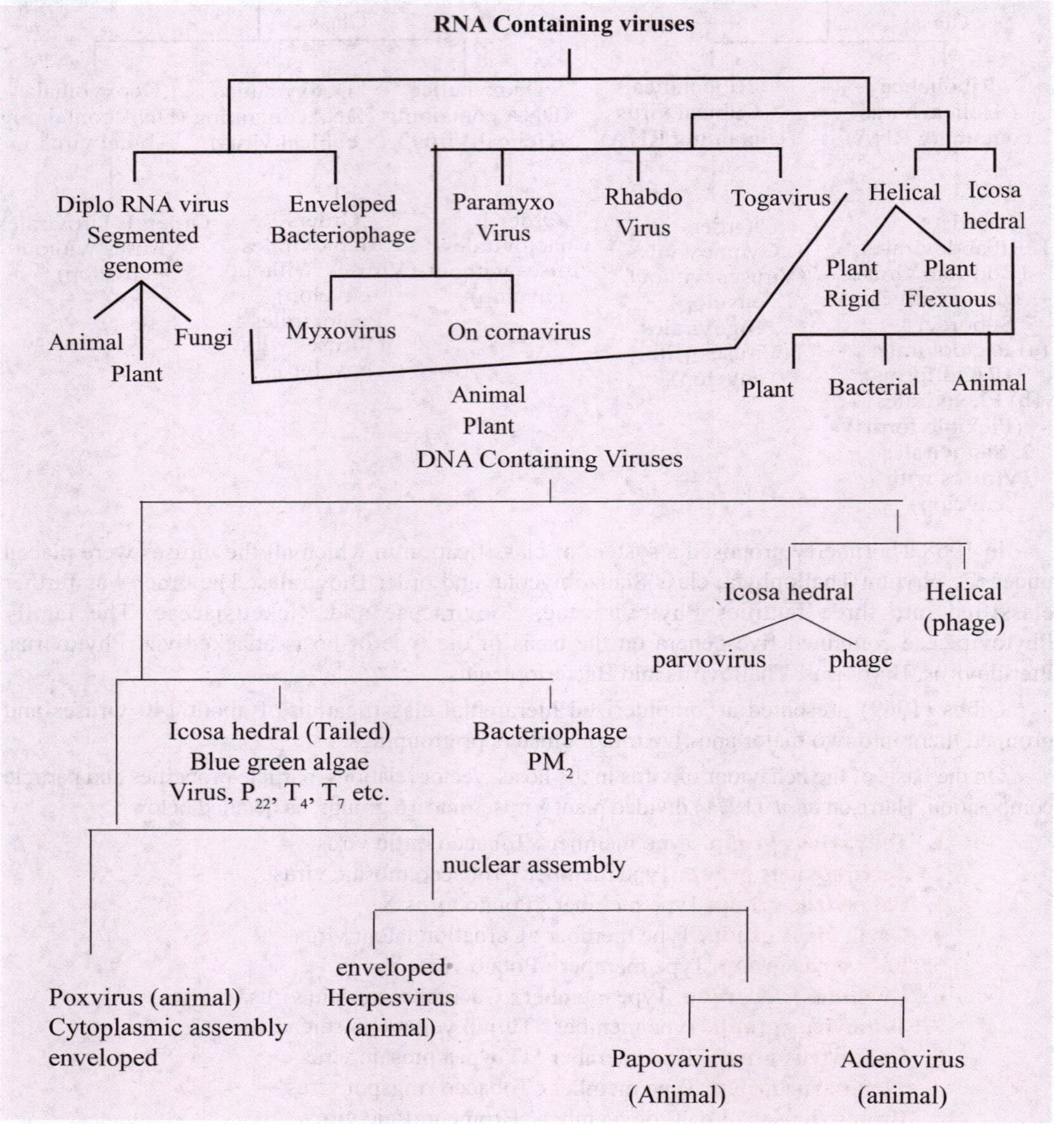

(B) Nomenclature of Viruses

It was realised by the virologists that binomial system of nomenclature proposed by Carolus Linnaeus and successfully used for both animals and plants can not be used for viruses. Johnson (1927) was the first to attempt to naming the viruses. According to him, the name of a virus consists of at least three parts : Common name of the host on which it was first discovered and described, the term 'virus' and an Arabic number indicating chronologically the order in which it was described on that host. As per this system Tobacco Mosaic virus may be named a Tobacco virus 1. A decade later, Smith (1937) modified the Johnson's system by replacing the common name of the host with the latin generic name. Thus, Tobacco virus 1 became **Nicotiana virus 1.** Holmes (1948) proposed new generic names to virus on the basis of symptoms induced while virus species was derived from the names of host plants. Accordingly, Tobacco mosaic virus was named as **Marmor tabacci.**

Hansen (1968) developed a complicated system of nomenclature in which name of the genus is constructed by putting together different syllable, each syllable being based on different property of the virus. Thus, Tobacco Mosaic virus became **Minochorda,** wherein **M** means that virus is mechanically transmitted—**in**—signifies that specific vector is unknown, and —**Chorda** signifies that the virus particles are rod shaped and stiff.

According to Thornberry (1968), Viruses of higher plants were named as **Phytovirus**, those of pteridophytes as **Pteridovirus**, viruses of Bryophytes as **Bryovirus** and those of Thallophytes as **Thalloviruses.** Specific name suffixed to the generic names, contained the name of host plant and symptoms produced. Thus, Tobacco Mosaic Virus is named as **Phytovirus nicomosaicum.**

Gibbs (1969) opinded that introduction of binomials to viruses would create more problems than solving and result in confusion. Earlier in 1953, the Sixth International Congress of Microbiology, at Rome took the decision that application of binomials to the viruses is undesirable and should be discouraged. Therefore, the International Committee of Virus nomenclature at its meeting held in Rio de Janerio also concluded that the viruses be designated by non-lineal binomials and proposed a new system of nomenclature. According to this system of nomenclature, the name of the virus consists of two parts : the first part represents the common name of virus while the second part contains coded information about the virus. The second part is known as **Cryptogram.** Each cryptogram has four pairs of symbols which give definite infomation about the virus. Meaning and details of each of the four pairs are as follows :

First pair : Gives information about the nucleic acid and it strandedness

Nucleic acid : DNA = D, RNA = R

Strandedness : Single stranded = 1

Double stranded = 2

Second pair : indicates molecular weight of nucleic acid in millions/percentage of nucleic acid in infective particles.

Third pair : gives information about outline of particle/outline of nucleocapsid.

Essentially spherical = S

Elongated with parallel sides, ends not rounded = E

Elongated with parallel sides, ends rounded = U

Complex or none of the above = x

Fourth pair : signifies the kinds of host infected and kinds of vector.

Kinds of Hosts :

Actinomycetes	=	A
Bacteria	=	B
Fungus	=	F
Invertebrates	=	I
Seed plants	=	S
Vertebrates	=	V

Kinds of Vectors

Mite and Tick	=	Ac
White fly	=	Al
Aphids	=	Ap
Leafhoppers, Plant hoppers or tree hoppers	=	Au
Mealy bug	=	Cc
Beetles	=	Cl
Fungus	=	Fu
Mind or tingid bug	=	Gy
Nematodes	=	Ne
Thrips	=	Th
Vectors known but none of the above	=	Ve
Spreads without a vector but with contaminated environment	=	O

Following two symbols can be placed at any place among the four pairs :

Ç = specific property signified by the symbol in place of asterik put is not known

( ) = Enclosed information is doubtful or unknown.

Thus, cryptogram of **Tobacco Mosaic Virus** will be : R/1 : 2/5 : E/E : S/O

Cryptograms of some other important viruses is as follows :

Tobacco Rattle virus = R/1 : 2.3/5 + 0.9/5 : E/E : S/Ne

Potato virus X = R/1 : 2.1/6 : E/E : S/(Fu) or 0

Potato virus Y = Ç/Ç : Ç/Ç : E/E : S/Ap

Turnip yellow mosaic virus = R/1 : 1.9/34 : S/S : S/Cl

Cow pea mosaic virus = R/1 : 1.5/24+2.6/33 : S/S : S/Cl

Tobacco Ringspot virus = R/1 : 2.2/40 : S/S : S/Ne

Tobacco Necrosis virus = R/1 : 1.5./19 : S/S : S/FU

Cauliflower mosaic virus = D/2 : 4.5/16 : S/S : S/Ap

Tomato spotted wilt virus = R/Ç : Ç/Ç : S/S : S/Th

SHAPE, SIZE AND STRUCTURE OF VIRUSES

The viruses, as mentioned above, are ultramicroscopic. They cannot be seen with the best light microscope. They are recognised by their effects on the host plant. The purified extracts of some viruses, when viewed in the electron mocroscope, show characteristic particles called the **virions.** The virions vary in shape (Fig. 19.1 A-D) and size depending upon the species.

(a) Shape (Fig. 19.1). The shapes of virus particles or virions are known chiefly from electron micrographs. The virions may be spherical, oval, rodlike, brick-shaped or shaped like a tadpole with a head and slender tail (D). On the basis of the shapes of the virions, the plant viruses chiefly fall under the following three categories :—

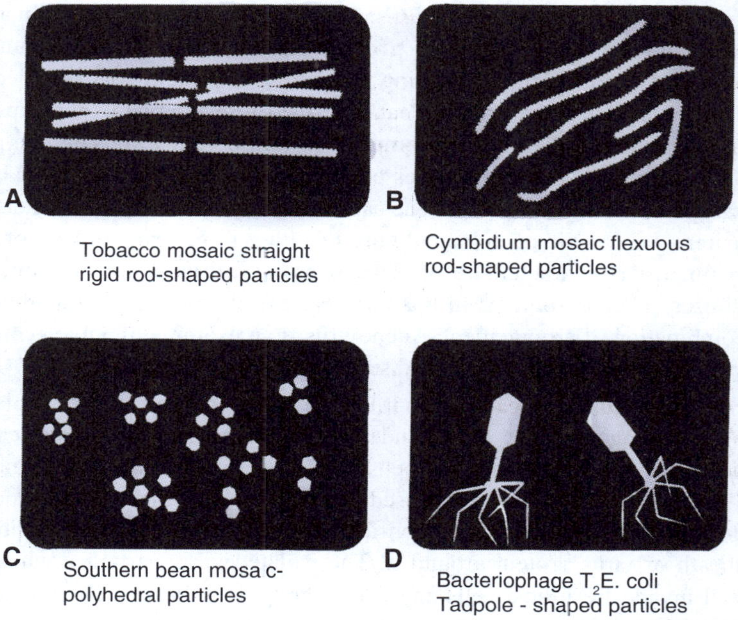

Fig. 19.1 (A-D). *Viruses.* Principal shapes of virus particles.

(*i*) *Straight tubular, long, rigid rods with a helical architecture* (A). To this category belong the viruses of Tobacco mosaic, Barley stripe mosaic and Tobacco rattle.

(*ii*) *Long flexuous thread-like rods* (D). The common examples of this type are potato latent mosaic, common Bean mosaic, Lettuce mosaic and wheat streak mosaic virus.

(*iii*) *Polyhedral virions* (C). Their gross shape has often been described as spherical. The common examples of this category are Turnip yellow mosaic, Cucumber mosaic, Squash mosaic, Tobacco ring spot. The polyhedral virus particles have numerous facets. Some of the polyhedron virions of Tomato spotted wilt have been reported to have a tail.

The Alfalfa mosaic virus particles which were formerly described as spherical have been shown to be short, rigid rods. The virions of Lettuce necrotic yellow virus, *Gomphrena* virus and potato yellow dwarf virus are bullet-shaped. The viruses in gross appearance are thus as heterogeneous as bacteria.

(b) Size. The virus particles or virions are measured by the metric system known as *millimicron,* the abbreviation for which is m μ. One millimicron is 1/1,000 or = 0.001 micron or one millionth of a millimetre.

The plant virions range in size from 10 μ to more than 300 mμ. The straight rigid rods of Tobacco mosaic virus are about 18 m μ in diameter and 300 m μ in length. The polyhedral plant virus particles vary in the range of 16 m μ to 22 m μ. Alfalfa mosaic virus is about 17 m μ in diameter. Virus of Turnip yellow mosaic is 20—30 m μ in diameter. The Hydrangia spot and Maize stunt which are short rod viruses are 44 × 16 and 240 × 50 m μ respectively.

The animal viruses have a wider range of size. Those based on the polyhedral shape vary from 8 m μ to about 100 m μ in diameter. There are larger ones which may be spherical, oval or brick-shaped. They go up to 450 m μ in smallest dimensions. The animal virus causing foot and mouth disease of cattle is 10 m μ in diameter. The spherical Mouse encephalitis virus is about 28 m μ. The small-pox virus is 150 m μ in length. The spherical virus particles causing *Polimycelitis* are between 20-25 m μ in diameter. The Psittacosis virus is about 275 m μ in diameter. There are still larger ones which may go up to 450 m μ in smallest dimenstion. To cut short the largest viruses are smaller than the smallest bacteria. The virus particles capable of infection range in size from a quarter to a tenth of that of the smallest bacteria.

(c) Microstructure of viruses. The mature virus particle (virion as mentioned before) consists of an outer, inert, non-genetic **protein coat** or **shell** which surrounds and protects the **genetic material** or **nucleic acid** forming the core. The protein coat is called the **capsid** and the nucleic acid molecules in its protein coat is termed **nucleocapsid.** The capsid is made up of numerous small subunits called the **capsomeres.** In turnip yellow mosaic, the protein shell is composed of 32 capsomeres. The capsomeres consituting the capsid are similar but those of different strains vary in kind and aminoacids. The capsomeres are closely packed and are arranged in a geometrical pattern around the nucleic acid core resulting in helical, cubical, or complex symmetry. Helical symmetry is characteristic of Tobacco mosaic virus. Cubical symmetry is illustrated by many viruses. The capsomeres in this case are closely packed in triangular, pentagonal, hexagonal or combination of arrangement. Adenovirus is an example of cubic symmetry. Examples of complex symmetry are Vaccinia virus and some large bacterial viruses.

1. Phytophagineae. The viruses of this suborder attacking plants are the most important plant pathogens. They are included in the suborder **Phytophagineae.** Some plant viruses have been separated from the host cells and crystallized. Prof. W.M.Stanley shared a Nobel Prize in 1946 in isolating from the diseased plants of tobacco the needle or rod-shaped crystals of *Tobacco mosaic virus.* On analysis, the virus particles consist of a core of ribonucleic acid (RNA) with a sheath of virus protein around it. The isolated virus particles, which apparently are non-living, will invade fresh host cells and cause the typical disease. RNA is the infective part. Within the host, it will produce many new virus particles. The virus can exist in the crystallized form indefinitely.

The plant viruses are vastly destructive. In the past few years several hundred new plant viruses have been discovered. The number of diseases they cause is not yet definitely known. Majority of these have been recognised in the flowering plants. Many of the plant viruses attack a large number of our annual crops beans, potatoes, cucumbers, sugarbeets, tobacco and tomatoes and cause considerable damage. Fortunately none of our major cereal crop plants is a victim to them. Losses are far greater in horicultural plants, plantation crops, vegetables and tuber crops. The most important plant diseases caused by the viruses are a group of *mosaics.* The common virus mosaic diseases are on tobacco, sugarcane, cucumber, tomato, peach, potato and sugar beets. The *mosaic diseases* (Fig. 19.2 A) cause mottling or variegation of leaves. They are not fatal. They cause lowering of vigour accompanied by stunted growth. There is, however, enormous loss in the crop both in quality as well as quantity of the yield. Much work in viruses has not been done in India so far. The important virus diseases

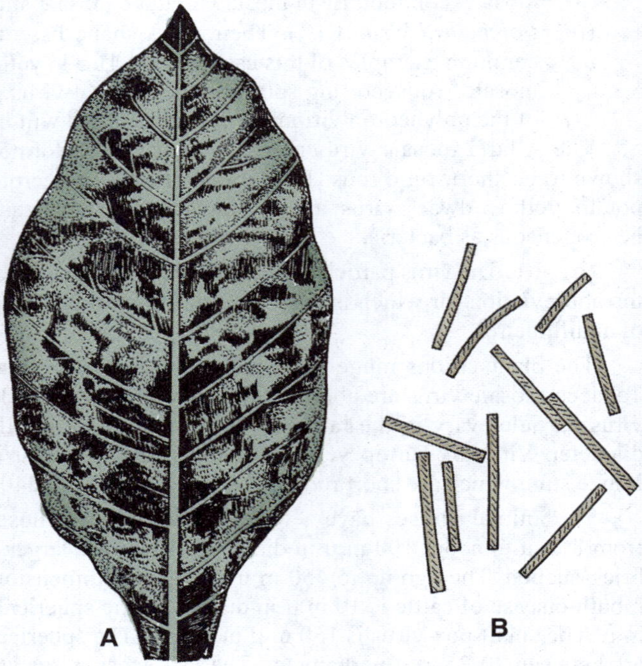

Fig. 19.2 (A-B). *Viruses.* A Mottled Tobacco leaf infected by *Tobacco mosaic virus* (TMV); B. Rod-shaped particles of TMV.

reported in India are *Rugose masaic* and *Leaf roll* of potato, *yellow-vein mosaic* of Bhindi, *leaf curl* of tomato, *mosaic* and *streak* of sugarcane, *leaf curl* of tobacco, *mosaic* of cucurbita and *groundnut rosette. Blossom malformation disease* of mango, which is the well-known national fruit of India, is also a virus disease. It is a great menace to our mango plantations. The *bunchy top disease* of banana is also equally devastating.

Leaf Curl of Tobacco

Leaf Curl of Tomato

Leaf Roll of Potato

Groundnut Rosette

Bunchy Top of Banana

Cucurbit Mosaic

Yellow Vein Mosaic of Bhindi

Control of Viral Diseases of Plants. Various practices and control measures are employed to check the spread of virus diseases. These are :

1. *Destruction of Weed Hosts of Viruses.* Many weeds harbour soil born viruses. They serve as the source of virus infection for cultivated crops grown in the same soil. Destruction of these will be of value to check the spread of a number of virus diseases.

2. *Insect Control.* To reduce the number of insect vectors by sprays of insecticides and antibiotic techniques may prove to be of value in the control of some virus diseases.

3. *Growing Resistant varieties of Crops.* The best method of control is the use of resistance varieties. Some success in the production of virus resistant varieties of plants has been achieved.

4. Careful rogueing of diseased or suspected plants and control of the vectors with effective insecticides are also helpful to a considerable extent. The plants showing suspicious symptoms shoud be removed from the field and burnt.

5. *Inactivation of Virus by Chemicals.* Fumigation by insecticides, sprays or dusts under glass house conditions are quite effective. 8-a Zaquanine, Catechol, Quinol, Phenol, Thiouracil, Nicotine are some of the chemicals with which success has been obtained.

6. *Thermal Inactivation.* Many viruses can be destroyed simply by heating. The thermal inactivation point varies with different species.

7. *Desiccation.* A few viruses can be destroyed by drying.

Ultrastructure of Plant Virus. Tobacco Mosaic Virus (TMV., Fig. 19.3 A-C), the fine structure of which has been extensively studied, is taken as a type to illustrate the ultrastructure of plant viruses.

(*a*) Structure. The virus particles or virions of TMV are straight, rigid cylindrical rods (A) with a helical architecture and remarkably uniform outer measurements of 18 nm dia and 300 nm length. The rods are hollow. The hollow core or cavity, which is 4 nm wide, extends the entire length of the virus particle (virion) and is bounded by 14 nm thick walls (C). Thus in a cross section the virion consists of two parts namely, (*i*) an outer protective protein coat or shell called the **capsid** enclosing; (*ii*) the inner infective agent, the **genetic material** or **nucleic acid.** The latter in plant viruses is, generally, a single-stranded RNA which in TMV is in the form of a long helix. It extends the entire length of virion. The capsid is built of about 2200 small, similar protein subunits called the **capsomeres.** They are closely packed (B) and arranged in a regular left handed helical around the RNA helix.

(*b*) Composition. Viruses protein comprises 95 percent of the total substance of the virus particle. The remaining 5 percent is RNA. Both protein and RNA are coaxially intertwined to form the rod-shaped virions with a helical architecture. The molecular weights of RNA molecule is about two millions. It provides a code which controls the amino acid sequence in the capsid.

The virus protein which forms the capsid is similar to other plant proteins and is made up of usual amino acids joined by the peptide links. The virus proteins, however, are unusual in their size and are of high molecular weight. The molecular weight of tobacco mosaic virus protein is nearly 40 million. Each capsomere has a molecular weight of 17,400 and is formed by the condensation of about 158 amono acid residues. The capsid forms a tightly fitting layer. The virus particles remain ineffective for about 50 years in dried tobacco leaves. They can stand boiling for 10 minutes.

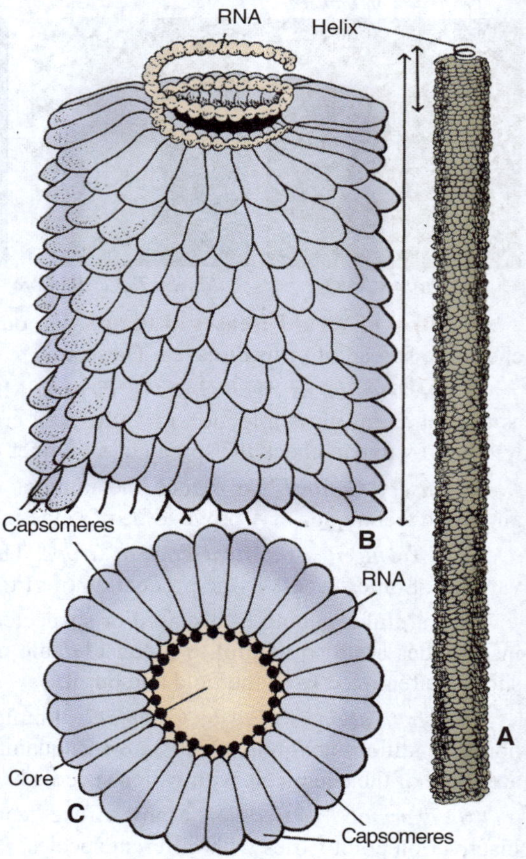

Fig. 19.3 (A-C). *Viruses.* Ultrastructure of tobacco mosaic virus (TMV). A, Diagrammatic representation of particle of TMV in surface view in electron microscope; B, A portion of the above showing structure; C, A part of the rod in section. (Adapted from Stanley and Valens)

Life Cycle of Plant Viruses. The life cycle of a typical plant virus such as TMV (*Tobacco Mosaic Virus*) comprises four stages :—

Stage 1. *Transmission of virus particles* (virions). The rod-shaped Tobacco Mosaic Virus (TMV) infects only the leaves of tobacco plants. It is transmitted and complete particles introduced into the host cells by insect vectors which are many such as *Myzus pseudosolani* (false potato aphid), *M. circumflexus, Macrosiphum solanifolii* and grasshopper. The virus is also transmitted by mechanical means such as rubbing of virus on the leaves and through handling of tobacco plants at transplanting. The virus is extremely contagious. A person who has handled cigarettes containing infected tobacco leaves may transmit the virus to the healthy plants.

Stage 2. *Infection.* Infection takes place with the whole virus particle. Separately the virus components are not infectious. Fraenkel-Conrat (1956) seprated the protein coat from RNA core by delicate chemical procedures. He found that the protein coat alone is not capable of producing infection. The **RNA** alone is infectious but much less than that of the intact virus particle. He further placed the protein shells in a solution of viral RNA. The viral particles of TMV were reconstituted. They were able to reinfect tobacco leaves. It shows that the nucleic acid alone carries full instructions to form new virus particles. Highly interesting and significant were the results of Fraenkel-Conrat and his associates in this connection. They separated the nucleic acid and protein coats of TMV and another strain of rod-shaped virus **HR.** They then mixed the protein coats of TMV strain with the **RNA** cores of **HR** strain. The result was the reconsitution of virus particles with **TMV** protein coats and **HR** core of **RNA.** Infections with this new type hybrid virus always yielded the new virus particles in which both the protein coat and RNA belonged to the **HR** strain which had furnished the infective **RNA.** The results of this experiment conclusively prove that the nucleic acid alone is coded with the specific information needed to the host cell for building complete virus particles. Infection may be with viral **RNA** or with virus particle intact. Separately both viral **RNA** and protein are readily attacked by enzymes in host cells and decomposed. Solitary **RNA** is destroyed quickly by ribonuclease enzyme present in the host cells. Virus particles intact are very resistant to decomposition and thus are stable and viable. For infection, wounding of the host cell wall is necessary to produce an infectible site. It may be caused by the mouth parts of the insect vector or by wounding epidermal cells or breaking epidermal hairs by abrasion. Through these infectible sites, the virus particles (virions) intact gain entry into the host and come in contact with the host cell membrane. By the process of pinocytosis, the virions enter the host cell protoplast.

Stage 3. *Virus multiplication.* After infection, the virus particle disappears from view. It ceases to exist as an organized unit and breaks up into its two component parts in the host cytoplasm. The viral RNA is thus released from the protein coat. Uncoating of viral RNA is the initial step towards virus multiplication. It takes place within minutes after the virion gains entry into the host cell. This is known as the **eclipse stage.** Thereafter starts the multiplication process. Viral protein coat remains in the host cell cytoplasm. The **RNA** component moves from cell to cell causing the production of new virus components which assemble to form more virus particles. Esau *et al.* (1966) reported that the virions of beet-yellow virus move intact from cell to cell through plasmodesmata.

Stage 4. *Sites of Viral RNA and Viral Protein Synthesis.* The widely held view is that the virus protein and virus RNA are formed by two separate systems. Entering the host cell the viral RNA takes control of the cell machinery and uses it to produce the viral components instead of the host cell parts. Thr viral components are then assembled into new virus particles. The actual mechanism is a mystery. The account of virus multiplication given in books is highly speculative. One view

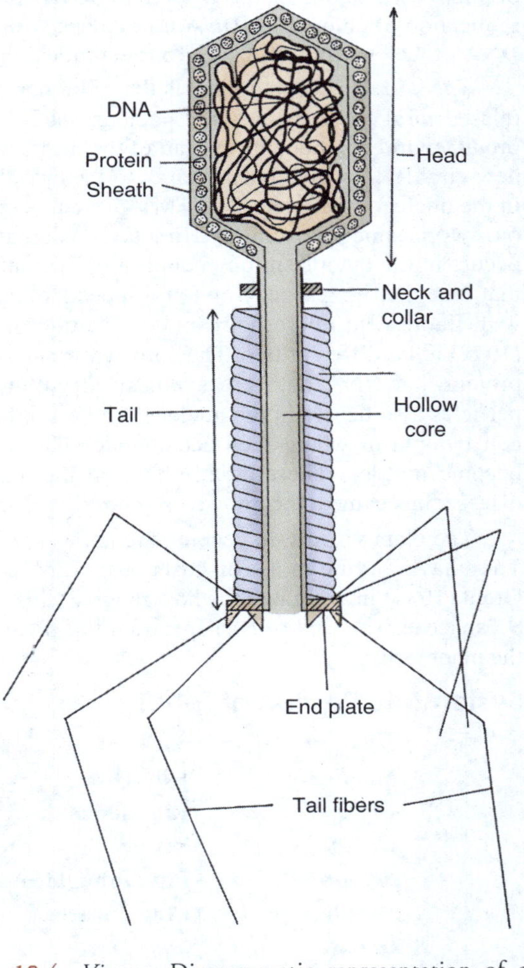

Fig. 19.4. *Viruses.* Diagrammatic representation of a longitudinal section of a bacteriophage showing structure.

is that the viral nucleic acid combines with or replaces part of that of the host cell. Thereafter the host cell produces the viral components at the expense of its own.

The other commonly held view is that the **DNA** of the host cell uses viral **RNA** to elaborate more viral **RNA** and protein. They are elaborated in the cytoplasm of the host cell, first the viral **RNA** and then the viral protein. It is difficult to understand why the host plant **DNA** which is coded with specific information to build plant **RNA** and plant protein, starts building the viral components.

There is also a suggestion that the ribosomes in the cytoplasm of the host cell are associated with the elaboration of viral components. This assumption is based on the fact that viral **RNA** is related to ribosome **RNA.** Entering the host cell viral **RNA** overrides that of the ribosomes and uses the latter to produce viral components. The ribosomes, however, are considered to have no genetic continuity. It is, therefore, difficult to associate them with the replication of specific viral **RNA.**

Some virologists associate the replication of viral RNA with the chloroplasts in the host cell. The biochemical, cytological and genetic data have proved beyond doubt that chloroplasts are semi-autonomous, self-duplicating organelles of eukaryotic cells. They thus have a genetic continuity of their own and contain **DNA** and **RNA** distinct from analogous nucelar constituents. The replication of chloroplast **DNA** takes place within the organelle. It is probable, therefore, that the **DNA** of the chloroplasts codes for the replication of viral **RNA** and protein.

The widely accepted view is that viral components are formed by two separate systems. The relased viral RNA reaches the nucleus, takes control of the host cell machinery and uses it to produce viral components instead of the host cell parts. The nucleolus in the nucleus produces the new viral RNA which later escapes to the cytoplasm around the nucleus through channels present in the nuclear membrane. The TMV protein is synthesized in the cytoplasm very near the nucleus on cytoplasmic ribosomes. (Hira and Wildman, 1969). The assembly of new TMV particles occurs in the cytoplasm on endoplasmic reticulum. The eclipse stage lasts for about 6 hours. By that time the first progeny of TMV appears in the cytoplasm. However, the results of experiments with fluorescent microscopy and/or location of labelled anti-bodies and investigations of Reddi (1972) tell a different tale. They are at variance with the majority view. These indicate that TMV protein and RNA both are synthesized within the host cell nucleus. The assembly of TMV particles also occurs in the nucleus. The assembled TMV virions or particles escape into the host cell cytoplasm where they accumulate. The attack of TMV causes the leaves to wrinkle and become mottled. The mottled effect on the leaves gives like the appearance of a *mosaic.* The disease thus came to be known as *Tobacco Mosaic* disease (host TMV).

The plant viruses, in general, are not as specific as the bacteriophages and the animal viruses. They have a wide range of hosts. The Tobacco mosaic virus for example also infects tomato. Grant (1934) investigated the host range of this virus and found 29 susceptible species outside the Solanaceae. The lack of host specificity of plant viruses is attributed to the absence of **DNA** from the plant viruses.

Systematic Postion of TMV :

Division	:	**Vira**
Subdivision	:	**Ribovira**
Class	:	**Ribohelica**
Order	:	**Virales**
Sub-order	:	**Phytophagineae**
Family	:	**Marmoraceae**
Genus	:	*Marmor*

2. Phagineae. This sub-order includes viruses capable of reproducing in and lysing their hosts. These viruses are termed the **phages.** The phage has a special structure different from a plant virus. The difference is associated with the need to penetrate rigid wall of its host. The phage particles (virions) are, in general of a tadpole shape or sperm-like with differentiation into a head and conspicuous tail. The head may be spherical, ellipsoid or hexagonal in outline. On the basis of the host, which they infect the phages are divided into two categories namely, (*i*) *Bacteriophages* and (*ii*) *Cyanophages*.

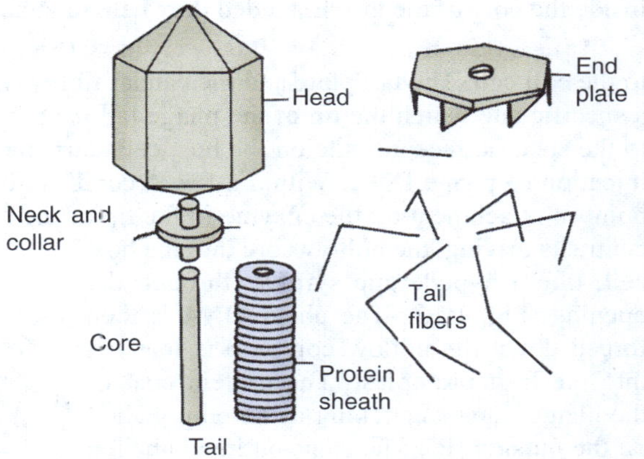

Fig. 19.5. *Viruses.* Various components of the bacteriophage shown separately.

(*i*) Bacteriophages (Bacterial Viruses). The viruses which attack, reproduce and lyse bacteria are called **bacteriophages.** The term bacteriophage literally means, "eater of bacteria". The bacteriophages do not eat bacteria but infect and kill them. Twort discovered in 1915 and d' Herelle studied them more fully two years later in 1917.

The bacteriophage which infects the colon bacterium (*Escherichia coli*) is called a **coliphage.** Several strains of coliphage are known. The best known have been designated T_2, T_4, T_6 or collectively **T-evenphages.** They are perhaps the best known structurally.

Ultrastructure of Bacteriophages (Fig. 19.4). A diagrammatic sectional view of T-4 phage is shown in Fig. 19.4.

The bacteriophage particle resembles a tiny sperm. It has a head and a tail. In addition there is some sort of attachment region adapted to stick to the surface of the host cell. The virus particle of T-even phage is about 200 to 280 m μ in lenght. The virus particle is thus too small to be seen with the best of light microscopes. Nevertheless, pictures of virus particles are obtained by the use of electron microscope.

The head of T-even phage is hexagonal in outline and bears numerous facets. It consists of protein coat surrounding a core of genetic material which in this case is **DNA** molecule. The coiled coil of a single thread-like double-stranded macro-molecule of DNA is packed tightly in the head. It is about 50 microns long. The phage DNA is said to lack the base **cytosine.** Instead it has another base **hydroxymethylcytosine** (HMC). The protein coat is built of globular proteins.

The cylindrical tail entirely consists of a protein sheath surrounding an empty core. The tail sheath can contract longitudinally. The attachment apparatus of the phage consists of six long, slender protein fibres known as the **tail** or **caudal fibres.** They arise from a plate at the basal end of the tail. The tail fibres normally remain twined

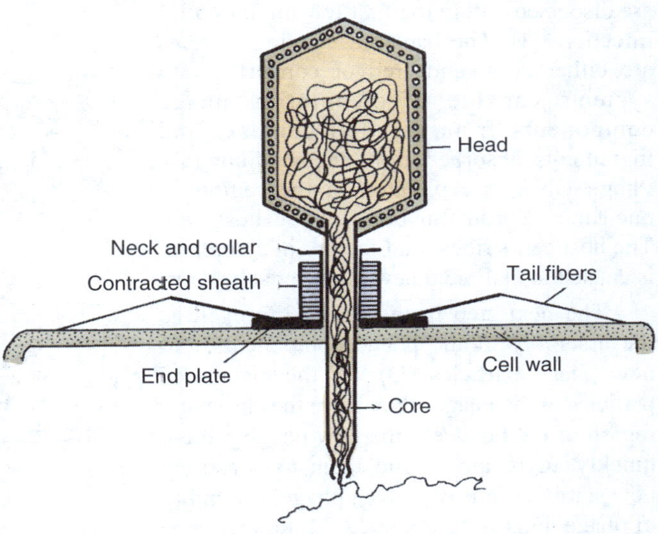

Fig. 19.6. *Viruses.* Bacteriophage in action.

inside the core of the tail. Extended they help to attach the phage particle to the coli cell.

Life Cycle of a Virulent Bacteriophage (Fig. 19.7). The coliphage particle is carried close to the coli cell. The tail plate and the caudal fibres anchor the phage virion to the host cell wall at a specific site which the tip of the phage tail happens to touch. With the attachment of the phage to the specific receptor site on the host cell starts the life cycle of the phage. The next step is the injection of phage **DNA.** Within a few seconds an opening is formed in the host cell wall at the point of attachment by the enzyme in the tip of the tail. The elastic protein sheath of the tail now contracts driving, the hollow core into the host cell, like a hypodermic syringe, through the opening (Fig. 19.6). The phage **DNA** is then forced down the hollow core and is injected into the host protoplast. The protein coat of the phage remains sticking as an empty shell on the outside (B). The steps outlined aboved constitute the **adsorption** and **penetration phase** or **infection phase** in the life cycle of the bacteriophage.

Next follows the **latent period phase** or the **period of eclipse.** During the eclipse period which is usually about 12-22 minutes no virus **DNA** as such can be seen in the host cell. It disappears as an independent infective agent. It seems that the phage **DNA** within the host cell enters the nuclear or chromatin body of the host cell and takes control of the cell's genetic machinery in a manner unknown at present. The replication of bacterial DNA and synthesis of bacterial enzymes ceases immediately. Under influence of phage DNA, the chromatin body of the host which also contains **DNA** is broken down and its contents are dispersed within the first few minutes after infection (B). The bacterial enzyme systems are either commandeered or converted into systems capable of eleborating phage components from the cell contents and ingredients absorbed from the medium (C). Phage DNA is replicated in the region of nuclear or chromatin body of the host cell. The host cell's ribosomal factory in cytoplasm is caused to elaborate new phage coat protein.

The next step in the phage life cycle is the assembly of the phage components into new phage particles (D). As the elaborated particles of the phage **DNA** leave the chromatin region of the host cell, the new phage protein quickly aggregates round them to constitute new stable virions of T-even phage. Assembly of phage units starts about 12-22 minutes after infection. The tail is finally added to the phage

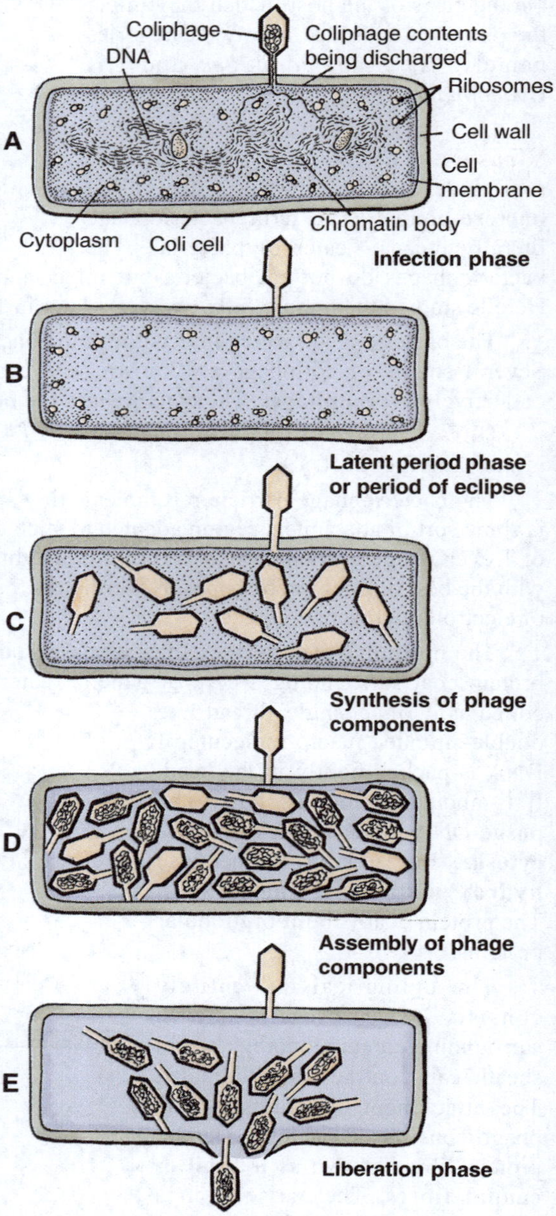

Fig. 19.7 (A-B). *Viruses.* Diagrammatic representation of life cycle of a bacteriophage.

head. Within about half an hour after infection the host cell contains two to three hundred or even more particles of T-even phage. The period of production of new phage particles is called **vegetative phase.** It comprises the replication of *DNA,* synthesis of phage protein and assembly of phage components into new phage particles.

The final step in the phage life cycle is the **liberation phase** (E). The host cell or whatever is left of it suddenly bursts to release the phage progeny. The process of disintegration or destruction of coli-phage cell is called **lysis.** The phage which brings about lysis or destruction of the host bacterium is called a **virulent phage.**

Systematic Postion of a Bacteriophage

Divison	:	**Vira**
Subdivision	:	**Deoxyvira**
Class	:	**Deoxybinala**
Order	:	**Virales**
Sub-order	:	**Phagineae**
Common name	:	*Bacteriophage*

(*ii*) Cyanophages. Sufferman and Morris (1963) were the first to report the discovery of viral agent which attacks and lysis blue-green algae. Since then many more viruses have been found which attack a wide range of blue-green algae. These viral agents attacking algae have been differently designated as *phycoviruses, algal phages, blue-green algal viruses* and *cyanophages.* To the majority of the investigators, the latter term (Cyanophages) appears more appropriate for two reasons namely, (*i*) close relationship between the blue-green algae and bacteria (Echlin and Morris, 1965) and (*ii*) similarity to a considerable extent in structure, infection and growth cycle to a bacteripohage. The term Cyanophage thus denotes a viral agent which is capable of infecting, reproducing in and lysing a blue-green alga.

Distribution. Cyanophages have been reported from many countries including India and thus are considered to be worldwide in their distribution. They occur almost throughout the year in fresh-water bodies such as streams, rivers, lakes, industrial storage tanks, farm and residential ponds. Singh, R.N., and P.K.Singh (1967) reported cyanophages from India in sewage, ponds and Singh P.K. reported cyanophages (1973) from rice fields.

Code Name	Host Range	Characteristic Features
LPP$_{1\text{-}2}$	*Lyngbya, Plectonema* and *Phormidium*	(*a*) **Head :** Polyhedron six-sided outline with 20 facets, edge to edge dia. 60±2 nm (*b*) Tail capital. (*c*) Tail short, hollow, noncontractile, length 20 nm, width 15 nm.
SM-1	1. *Synechococcus elongatus* 2. *Microcystis aeroginosa*	(*a*) Head Icosahedron, edge to edge dia. 67±1.8 nm (*b*) Tail very short, collar with thin appendage.
AS-1	*Anacystis nidulans* and *Synechococcus cadrorium*	(*a*) Head polyhedron, outline six-sided, edge to edge dia. 90 nm. (*b*) Tail long, contractile, length 243.5 nm, width 22.5 nm.
N-1	*Nostoc muscorum*	(*a*) Head Polyhedron, six-sided outline, edge to edge dia. 55 nm. (*b*) Tail long contractile, length 110 nm, width 16 nm.

Nomenclature. The cyanophages, known at present are placed under six groups namely, LPP_{1-2}, SM-1, C-1, AR-1, AS-1 and N-1. Each group is given a code name according to its host range. The code name is composed of the initials of the generic names to the known hosts. The Arabic numerals are added to the code names to denote seriological subgroups. For instance, LPP group of Cyanophages is subdivided into two sub-groups LPP_1 and LPP_2. It means another group of viruses with the same host range has since been discovered and designated LPP_2. The code names of the main cyanophage groups, with their host range and important morphological features, are given in the above Table.

Ultrastructure of Cyanophage (Fig. 19.9)

The fine structure of the intact particles (virions) of LPP_1 group has been thorougly studied. They are similar in structure to one another. Like the bacteriophage they are of a head-tail type. The cyanophage virion is differentiated into 3 regions namely, *head, tail capital* and *tail*.

(*i*) Head. The LPP_1 cyanophage head is polyhedron. It appears hexagonal in outline and has many (20) facets and thus is described as icosahedral in form. The short dia. of the viral head is 50.6 ± 2.0 nm (586 ± 20A°). It is differentiated into an outer protein coat or sheath which encloses a cavity filled with the genetic material called the nucleic acid. The latter which forms the core of the cyanophage head is the infective agent. The protein sheath which encloses it is protective in function. It is inert and nongenetic. Technically the protein sheath constitutes the head capsid. It is 8-10 nm thick. The major proteins of the head capsid are the 39,000 and 13,000 species. The nucleic acid in the cyanophages is always a DNA molecule. It is a single thread-like, double stranded macromolecule of DNA with a contour length of 13.2 ± 0.5 nm. It just fills the cavity of the head capsid.

(*ii*) Tail. The short, hollow cylindrical tail is 20-25 nm long and 15 nm wide. The upper end of the tail is attached to one of the vertices of the head capsid and from the lower end probably protrude the tail fibres (Shilo, 1967). The tail is non-contractile. It is differentiated into an outer protein sheath enclosing a cavity. The former is designated *tail sheath* and the latter *tail cavity.* The tail sheath is made up of 2-4 rings of six protein subunits each.

(*iii*) Tail Capital. Located at the junction of the tail with the head capsid is the *tail capital* about 10 nm long and 25 nm in dia. It protrudes into the head capsid. The function of the tail capital is uncertain. Probably it serves as a joint to fix the tail to the head capsid.

Infection and Growth Cycle of a Cyanophage. (*i*) *Infection process*. The LPP-1 Cyanophage particle in the aquatic environment and close to its blue-green algal host *Plectonema* affixes (adsorbs) itself to the prokaryont host cell probably at a specific receptor site which the tail end of the phage tail happens to touch on the host cell wall. According to

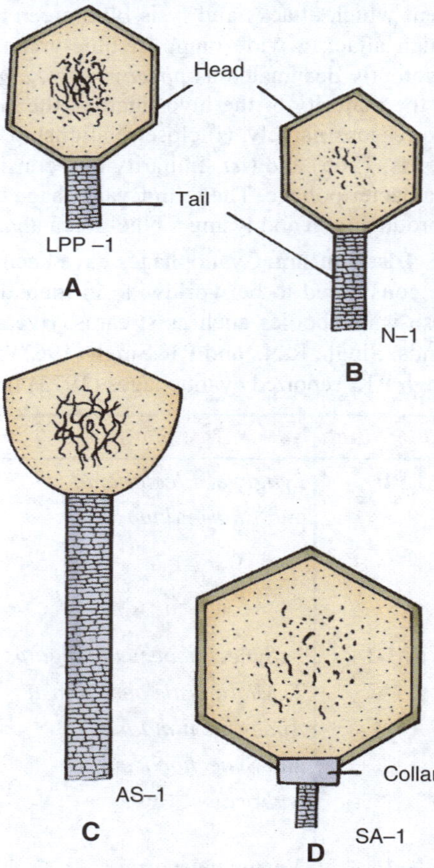

Fig. 19.8. (A-D). *Cyanophages.* Diagrammatic representation of the important groups of Cyanophages.

Smith *et al.* (1967), the tip of the phage tail appears to penetrate the host cell wall at the point of attachment. It is presumed that the cyanophage then injects its DNA into the host cell protoplast through the hollow tail. The empty protein sheath or coat of the phage called the **ghost** remains sticking to the host cell surface. The steps outlined above describing the infection process also constitute the *adsorption* and *penetration phase.*

(*ii*) *Latent Period Phase.* After completion of the infection process which takes about 1 hr. (Goldstein *et al.*, 1967) starts the *latent period phase.* The latter lasts for about 6 to 6.50 or even 7 hrs. The first three hours of this phase constitue *eclipse period* during which the injected virus DNA, as such disappears. It is not seen in the host cell. Three hours after infection, the thylakoids arranged in the form of a sheath at the periphery of the host cell of *Plectonema* begin to recede from the plasma membrane. The space thus formed between the replaced thylackoids and plasma membrane is termed the *virogenic stroma.* The virogenic stroma becomes distinct by the 4th hour corresponding to the end of the *eclipse period.* The formation of virogenic stroma is the first visible sign of infection. The cyanophage DNA, after infection, seems to reach nucleoplasm of the host cell and takes control of the host cell genetic machinery. Synthesis of host cell proteins and replication of algal DNA ceases within 5 hours after infection. Sherman and Haselkorn (1970) reported that about half of the host DNA is broken down to provide acid soluble precursors of the viral DNA. The replicated viral DNA first migrates into the spaces between the thylakoids and then into the virogenic stroma in the form of long helices. Viral heads have been seen in the virogenic stroma by the end of the latent period (7th hr.) where assembly of cyanophage virions begins by the 10th hour and is complete within 14 hours after infection. The release of mature cyanophage particles is achieved by cell lysis. The average burst size is 100 virions per cell. The cyanophage reproductive cycle is of longer duration as compared with that of the bacteriophage.

Economic Importance of Viruses. The viruses are generally considered of great negative economic importance as they cause serious human and animal diseases and vastly destructive plant diseases. However, not all viruses cause disease resulting in death or malformations of the host. In some of them, for instance, those of fig trees and many ornamental plants, bring about only motling of leaves and have no harmful effect. In some cases the viruses may destroy some chlorophyll but what usually remains is more than enough for the leaves to carry on photosynthesis. The petals of flowers of tulips develop beautiful striations of various shades due to viral infection.

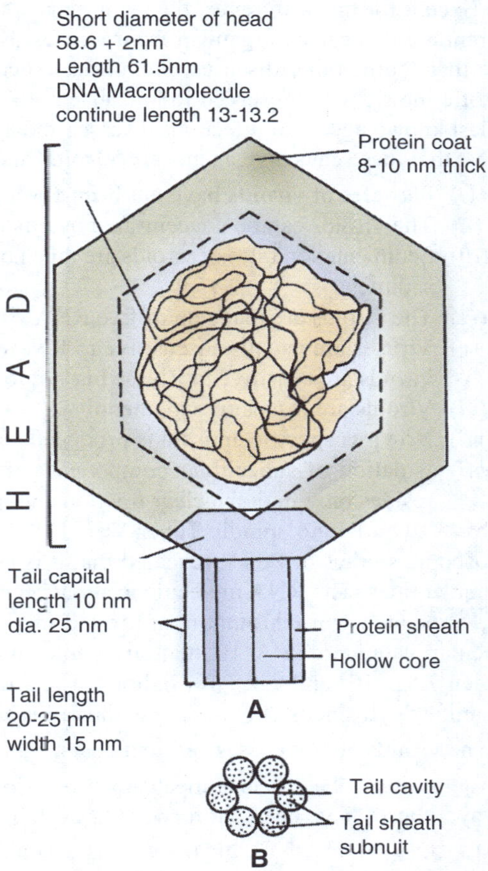

Fig. 19.9 (A-B). *Cyanophage.* Sectional view of LPP_1 virus (A) and end view of tail (B) (Diagrammatic).

The viruses also interfere in the working of many industrial processes dependent on bacterial action. They attack the bacteria involved in the industrial process and cause loss in economy. The curative properties of waters of certain springs and Ganga at Haridwar are attributed to the presence of viruses in the water which attack and lyse the pathogenic bacteria (bacteriophages) and the bacteria of decay and decomposition. From the evolutionary point of view, the viruses are important as they stand at the very threshold of life representing the starting point in the evolution of other organisms.

Affinities of Viruses. Another group of organisms with possible affinities with viruses are the bacteria. The former are entirely heterotrophic in their mode of nutrition and the latter are largely so. There is a group of organisms closely related to bacteria known as the *Rickettisiae.* The latter resemble bacteria in certain respects and viruses in others. In their spherical or rod-like form, they are bacteria but are in the range of large viruses. In fact it is doubted whether rickettsiae are organisms at all. Furthermore, like viruses they are obligate parasites. Outside the specific host cells they are quite inert like viruses and thus unable to carry out living functions.

II. VIROIDS

Diener (1971), during his studies on Potato spindle Tuber disease assumed to be caused by a virus, discovered that the causal organism was made up of naked free RNA devoid of protein coat. Except for this difference, the symptoms appeared to be similar to viral diseases. In order to differentiate this causal organism from viruses, Diener (1971, 1972, 1974) termed it as **'Viroid'.** Since then, some other diseases like, citrus exocortis, chrysanthemum stunt and chrysanthemum chlorotic mottle were observed to be caused by **Viroids.** According to Diener (1973) Viroids are smallest known agents of infectious diseases and are not related to viruses. The following evidences suggest that these causative agents are **Viroids** and not **viruses** :

(*i*) Particles of **viroids** have not been observed in infected tissues.
(*ii*) The viroids can be concentrated by ethanol precipitation.
(*iii*) Sedimentation rates of viroids are much below the range at which nucleo-proteins normally sediment.
(*iv*) The viroids are made up of free RNA without protein coat.
(*v*) Viruses are usually insensitive to RNase because of the protein coat around RNA while viroids are sensitive to RNase but not to DNase.
(*vi*) Viroids are insensitive to phenol
(*vii*) No virus specific antisera is produced by viroid-infected tissues.
(*viii*) Isolation of subcellular components of viroid-infected tissues indicate that infectivity resides only in the nuclear fraction and in the tissue debris.

Structure. Potato spindle Tuber Viroid (PSTV) disease has been studied in detail. Electron microscopic studies of PSTV revealed that it is a single stranded RNA molecule containing 250-350 nucleotides. The RNA molecule is about 50 nm long, confined within the nucleus of infected cells, associated with chromatin and found in exceedingly small amounts. Major fraction has a molecular weight of 5×10^4 daltons while minor fractions have molecular weights ranging between 7.5×10^4 and 9.4×10^4 daltons. The adenine : Uracil ratio is 21.7 : 20.9 while tha ratio of guanine : cytosine is 28.9 : 28.3 i.e. the ratios are close to unity.

The viroids do not possess protein coat around RNA molecule.

Replication. The precise mechanism of viroid replication has been a matter of controversy. Two hypotheses have been put forward to explain viroid replication :

(*a*) DNA-directed Replication. It is assumed that replication of viroids takes place at the direction of DNA which produces DNA templates. These DNA templates are either present in repressed form in uninfected tissues or are synthesised as a result of viroid infection.

(*b*) RNA-directed Replication. This hypothesis sugests that a replicase enzyme exists in uninfected tissues of the host. This enzyme accepts a wide variety of RNA species as templates. RNA-directed RNA polymerase has been observed to be present in healthy chinese cabbage and tobacco plants.

Symptoms. Symptoms of diseases caused by viroids are similar to those caused by viruses. Symptoms associated with viroid diseases include chlorosis, stunting, veinal discolouration, epinasty, vein clearing, localised necrotic spots, mottling of leaves and death of the host. Generally, the viroid-caused diseases are persistent infections, meaning thereby that the diseased plants do not recover and the viroids can be isolated from the diseased plants as long as it lives.

Transmission. The viroids known so far, are generally transmitted through mechanical means. In PSTV, the viroid is also transmitted through seed and pollen of infected plants. Insect transmission has not been reported.

Origin of Viroids. There are four possibilities of the origin of viroids :

(*i*) Viroids are derived from viruses in which protein coat is not required or is lost. Therefore, the viroids are sub-viral agents.

(*ii*) In many cases such as leaves infected with TMV, leaves infected with broad bean mottle virus and in *Escherichia coli* infected with phase Qβ low molecular weight RNAs have been discovered. Ultimately, these might have given rise to viroids.

(*iii*) Normal cells and particularly nuclei and nucleoli are known to contain RNA of low molecular weight. It has been assumed that some of these may have some regulatory function during RNA transcription from genome and may be essential for normal development of the organism. It is possible that viroids may have originated from these regulatory RNAs which have become abnormal by mutation.

(*iv*) Self-replicating RNAs normally occur in healthy animal cells as well as in tomato leaves. The viroids might have originated when one of these self-replicating RNAs becomes pathogenic and induces diseases. Potato spindle tuber viroid could be one such self-replicating RNA and PSTV, one such disease.

III. MYCOPLASMA

Mycloplasmas are the smallest, wall-less free living prokayotes belonging to class-Mollicutes. There were first discovered by Pasteur in 1843 when he was studying the causal organisms of pleuropneumonia in cattles. He named it as *Pleuropneumonia like organisms* (*PPLO*) but failed to isolate them in pure cultures. These were first isolated by two French bacteriologists E. Nocard and E.R. Roux in 1898 from pleural fluids of cattles affected with pleuropneumonia and these organisms were named as *Mycoplasma* in 1929 by Nowar. It may be mentioned that Pleuropneumonia is a highly contagious disease of cattle and it first appeared in Germany and Switzerland in 1713 and it spread throughout Europe in the 18th Century. *Mycoplasma mycoides* was found to be the causal organism of Bovine and Cattle pleuropneumonia. Mycoplasmas are highly pleomorphic, reproduce by budding and/or by fission and by producing small bodies. They lack *sterols*, a class of lipid invariably present as component of cell membrane of Eucaryotes, but are capable of incorporating exogenous sterols obtained from growth medium. Therefore, most mycoplasma require a rich nutrient medium containing *sterols* and serum protein for growth. On solid (Agar) medium mycoplasmas form minute, transparent colonies that typically have a characteristic "fried egg" appearance. Since their discovery it was found that members of mycoplasma are obligate parasites and pathogens of many mammalian and Avian hosts.

Plant Pathologists had not considered or even suspected mycoplasmas as causal agents of plant diseases before the discovery of Japanese workers Doi *etal*., (1967) and Ishii *etal*., (1967).

They found that wall-less microorganisms resembling mycoplasmas were seen with electron microscope in the phloem of plants infected with yellow type and Witche's broom of alfa-alfa diseases. Such diseases were earlier thought to be caused by viruses. Currently more than 75 distinct plant diseases are known to occur due to mycoplasma like organisms. Now the name of phytopathological mycoplasmas was changed to *Phytoplasma* (Seemuller *et al.*, 1994). Phytoplasmas are wall-less prokayotes which are phytopathogenic and belong to class Mollicutes. Phytoplasmas basically survive in Phloem tissues of plants and produce symptoms like yellowing, phyllody and witche's brooms. With the discovery of several groups of mycoplasma like organism, it has been categorised as *Mycoplasmas* (infect animals), *Phytoplasmas* (infect plants), Spiroplasmas (infect plants and insects), *Archeoplasmas* (infect animals, plants and insects) and *Entomoplasmas* (infect insects and plants).

Morphology of Mycoplasma. Since mycoplasmas pass through many filters and grow on media without living tissues, these are considered to be intermediate between bacteria and viruses.

1. Mycoplasma are very small unicellular, usually non-motile, procaryotic organisms.
2. They can grow in cell free media forming. typical "fried egg" shaped colony.
3. They are highly pleomorphic (variable in shape) showing small coccoid bodies, ring forms and five filamentous forms which may be branched.
4. Mycoplasma cells are bounded by triple layered unit membrane without a rigid cell wall. They lack ability to synthesize cell wall material.
5. They are filterable through bacterial filter.
6. They are highly resistant to pencillium but inhibited by tetracyclines.
7. They are inhibited by specific antibody.
8. They require sterols for growth.
9. Mycoplasma have no history of reversion to or derivation from a bacterial parent.
10. Mycoplasma cells contain both DNA & RNA.
11. Reproduction is controversial - perhaps develop within filaments tiny coccoid structures called elementary bodies, released by fragmentation of filaments or by binary fission or by budding.

Cell-shape : Mycoplasmas vary in shape (Fig. 19.10). These may be entire, spherical, polymorphic or irregular filamentous in form. The filaments may be branched or unbranched.

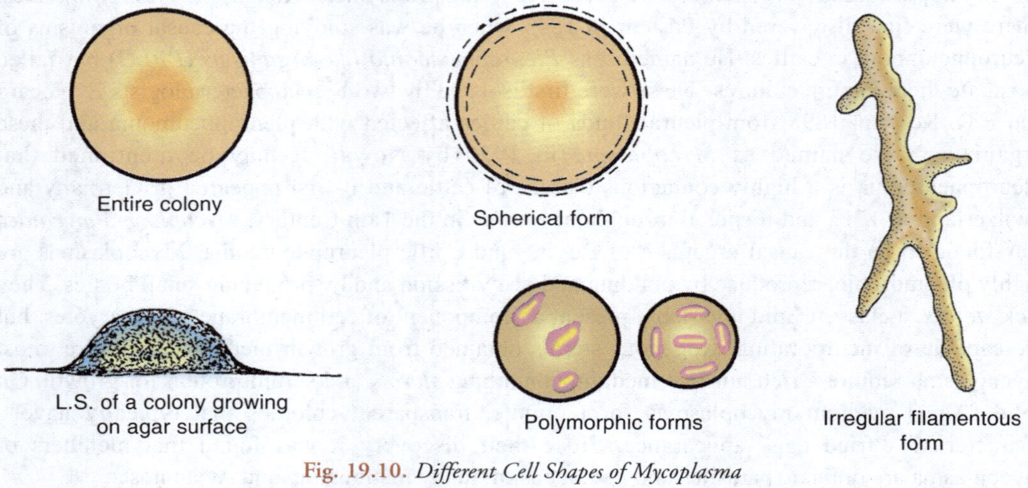

Fig. 19.10. *Different Cell Shapes of Mycoplasma*

Cell structure. The cells are small, and the diameter of the cells range between 0.3 and 0.9 μm. These are prokaryotic and do not contain a true nucleus.

Each cell is covered with a unit lipoprotein cytoplasmic membrane which is 7.5 – 10 μm thick. The cytoplasm contains genetic material (nucleoplasm like material) and ribosomes (Fig. 19.11). Both DNA and RNA act as genetic materials. It is usually half that usually occurs in other prokayotes. The DNA amounts to 4 per cent and RNA about 8 per cent. Though gram negative, mycoplasmas stain slowly on long exposure to dyes. Mycoplasmas are non-motile but show gliding slow movements. Mycoplasmas are sensitive to streptomycin, erythomycin and chloram phenicol but are insensitive to pencillin.

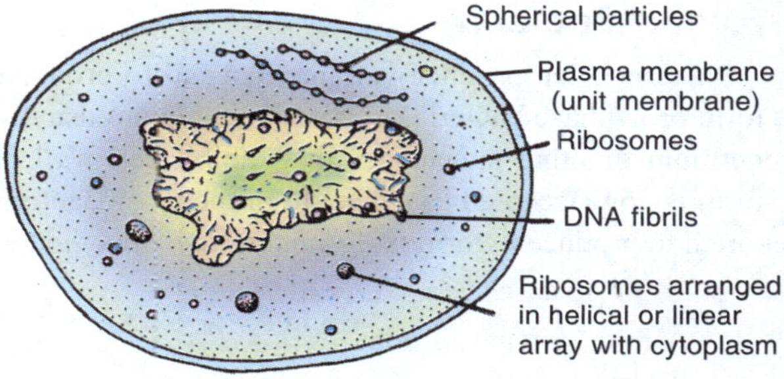

Fig. 19.11. *Cell Structure.*

Reproduction. Mycoplasmas reproduce by budding and/or binary fission (Fig. 19.12). Cells of mycoplasma divide unevenly into very minute bodies called the elementary bodies or minimal reproductive units. These are formed inside the large bodies or mature cells. Their size varies from 330 *nm* to 450 *nm*. These bodies are the smallest independent living entities so far known.

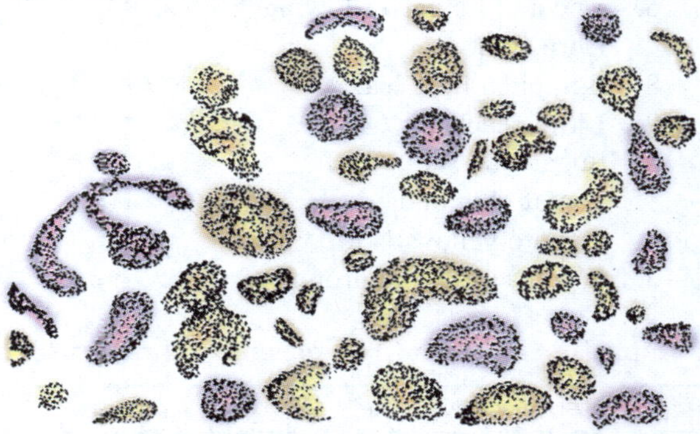

Fig. 19.12. *Several polymorphic mycoplasmas showing binary fission or budding.*

Transmission. Mycoplasma like organisms (MLO) or phytoplasmas are usually present in phloem of the host plants and are transmitted from host to another host by leaf hoppers but some are transmitted by psyllids, treehoppers, plant hoppers and some possibly by aphids and mites. Some of the pathogens are known to infect various organs of their leaf hopper or psyllid vectors and to multiply in their cells. The vectors cannot transmit the phytoplasma immediately after feeding on the infected plant but it begins to transmit if after an incubation period of 10 to 45 days depending upon the temperature.

IV. SPIROPLASMAS

Mycoplasma were recognised in 1966 by International Commission on Nomenclature of Bacteria, sub-committee on Taxonomy of Mycoplasmatales-class-Mollicutes. That time there was only one genus known-Mycoplasma. But now there has been more than one genera known. American Scientists working on corn stunt disease observed some helical form filaments from sap of infected plants in 1972. These structures had no cell wall but were surrounded by a unit membrane. The filaments had flexible bending and curling motion like spirochetes. In 1973, Davis and Worley named these organisms as *Spiroplasma*. *Spiroplasma* have so far been found is citrus plants infected with stubborn disease (*S. citri*), in corn stunt disease and in bermuda grass showing witche's broom symptoms. Like mycoplasma, *Spiroplasma* have a wide range of hosts extending from fungi to plants and arthopods to vertibrates. *Spiroplasma,* unlike mycoplasma like organism satisfy Koch's postulates.

L-Form of Bacteria

Bacteria often produce variants that fail to produce cell wall – the wall less bacteria, called L-form or L-phase bacteria. L- form bacteria are usually produced in laboratory conditions when Pencillium or other substances that inhibit cell wall production are added to culture medium. L-form is now accepted as a general property of Bacterium. The only Plant pathogenic bacterium reported to produce L-form is *Agrobacterium tumefaciens* – the causal agent of crown gall disease.

List of Plant diseases caused by MLOs

1. Alfalfa Witche's broom
2. Aster Yellows
3. Citus greening
4. Citus stubborn
5. Clover phyllody
6. Coconut Lethal Yellowing
7. Corn stunt
8. Egg plant little leaf
9. Little peach
10. Sugarcane White leaf
11. Tomato big bud
12. Tomato yellow dwaf
13. Witch's broom of groundnut
14. Papaya bunchy top
15. Peach yellows
16. Pear decline
17. Potato Purple top
18. Potato marginal flavescence
19. Potato Witche's broom
20. Rice yellow dwarf
21. Sandalwood spike
22. Sugarcane grassy shoot
23. Sunflower phyllody
24. Tomato purple top
25. Yellow leaf of areca palm.
26. Yellow wilt of sugarcane

Comparison between Mycoplasma, Plant Viruses and Bacteria

Details	Mycoplasma	Plant Viruses	Bacteria
1. Shape	Polymorphic	Rigid	Rigid
2. Size	0.1 to 7.0 microns	Variable, Spheres 18 - 80 *nm*, rods 180 - 1250 *nm*.	1 – 3 × 0.5 microns
3. Cell Wall	Mucopolysaccharide absent, instead a 10*nm* thick triple layered membrane is present.	Cell wall with mucopolysaccharides is not present	Cell wall with mucopolysaccharides is present.
4. DNA & RNA	Both present	Only RNA or in some DNA present	Both present
5. For multiplication, dependent on host nucleic acid	No	Yes	No
6. Ability to synthesize their own nucleic acid	Yes	No	Yes
7. Obligate parasitism	No	No, but considered obligate plant pathogens	No
8. Can be cultured on cell free medium	Yes	No.	Yes
9. Discovery	Recent	Early	Early

QUESTIONS

(A) VIRUSES & VIROIDS

ESSAY TYPE

1. Discuss the various systems of classification and nomenclature of viruses.
2. List general symptoms of viral diseases of plants. *(Himachal Pradesh, 1999)*
3. Describe briefly the nature of plant viruses and the method of their transmission.
4. What are viruses? Write a detailed account of transmission/ multiplication of plant viruses. *(Kanpur, 1992, 2005; Allahabd, 1997, 2003)*
5. Describe the symptoms on the plants caused due to infection by virus. *(Kanpur, 1993; Lucknow, 1995, 1997; Gorakhpur, 1992)*
6. How will you define virus? Discuss its nature and origin. *(Kanpur, 1994; Lucknow, 1997)*
7. Give an illustrated account of the structure of viruses. *(Kanpur, 1996)*
8. How are the plant viruses transmitted? Give the control measures of any two plant virus diseases. *(Kanpur, 1998)*
9. Describe the nature of plant viruses and give the methods of their transmission. *(Kanpur, 1999; Awadh, 1999; Allahabad, 2002)*
10. What are viruses? Discuss their living and non-living properties. *(Agra, 1992; Gorakhpur, 1993; Awadh, 1993)*
11. Give an account of different modes of transmission of plant viruses. *(Lucknow, 1994, 1996; Gorakhpur, 1993; Allahabad, 2004)*
12. Describe the morphological variations, internal structure and mode of replication of plant viruses. *(Lucknow, 1995, 1998, 1999)*
13. Describe the nature of TMV Given illustrated account of symptoms of virus infection in plants. *(Gorakhpur, 1990, 1994, 1996, 1998)*
14. Describe the structure of TMV. How are viruses transmitted by insect vectors? *(Gorakhpur, 1996; Allahabad, 1996, 1999)*
15. Give an illustrated account of multiplication of plant viruses. *(Allahabad, 1991, 1994)*
16. Give a detailed account of economic importance of viruses. *(Allahabad, 1995, 1998)*
17. Describe in detail the nature and structure of Tobacco mosaic virus. *(Awadh, 1994, 1996, 1997)*
18. What are viruses? Describe various morphological symptoms produced by viruses on the host. *(Awadh, 1998)*
19. Give an illustrated account of structure and life cycle of a Bacteriophage. *(Kumaon, 1997; Lucknow, 2001)*
20. What are viruses? Describe their nature, mode of transmission and multiplication. *(Kumaon, 1998; Bundelkhand, 1995)*
21. "Virus is a link between living and non-living". Justify this statement. Describe the viral transmission and diseases. *(Rohilkhand, 1997)*
22. (*a*) Describe the ultrastructure of TMV. *(Punjab, 1995, 1999; Gorakhpur, 2003)*

    (*b*) Write an essay on TMV. *(Bharathidasan, 1995)*

    (c) Write a brief account of replication of TMV. *(Gorakhpur, 2003)*
23. What are viruses? Describe the structure various methods of transmission of viruses. *(Lucknow, 2000; Kumaon, 2000)*
24. Give a detailed account of the economic importance of viruses. *(Allahabad, 2000)*
25. Give an account of chemical composition and ultrastructure of viruses. *(Allahabad, 2001)*
26. Give an account of nature, transmission and multiplication in viruses. *(Kanpur, 2001, 2003)*
27. What are viruses ? Describe the structure and replication of TMV. *(Poorvanchal, 2001, 2003; Poorvanchal M.Sc. 2003)*
28. Describe the structure and multiplication of Bacteriophage. *(Poorvanchal, 2002)*
29. How are viruses transmitted ? Give the control measures of a typical plant viral disease. *(Rohilkhand, 2003)*

SHORT ANSWER TYPE

30. Write short notes on the following :

(*i*) Bacteriophage
(Meerut, 1999; Kanpur, 1992, 1995, 1996; Agra, 1994, 1996; Meerut, 1999; Lucknow, 1997, 2000; Gorakhpur, 1999; Allahabad, 1992, 1996, 2000, 2003, 2004; Awadh, 1995, 1997, 2000, 2003, 2004; Magadh, 1991; Rohilkhand, 1991, 2003; Poorvanchal, 2006)

(*ii*) Isolation of plant viruses *(Lucknow, 1993)*

(*iii*) TMV
(Kanpur, 1992, 1994; Lucknow, 1999, 2000, 2004; Allahabad, 1992, 1998, 2000; Awadh, 1997, 1999; Magadh, 1990, 1991; Rohilkhand, 1994, 1996, 2001, 2003; Meerut, 1998; Bundelkhand, 1999; Poorvanchal, 2005)

(*iv*) Structure of Tobacco mosaic virus
(Kanpur, 1997; Agra, 1991; Gorakhpur, 1997, Lucknow, 2002; Kumaon, 2001; Kanpur M.Sc., 2003; Awadh, 1992, 1995; Rohilkhand, 1991)

(*v*) Plant viruses *(Poorvanchal, 2001; Kanpur 1999)*

(*vi*) Transmission of plant viruses
(Poorvanchal, 2001; Gorakhpur, 1992; Allahabad, 1991; Awadh, 1994; Rohilkhand, 1991, 1994, 2002)

(*vii*) Cyanobacteria *(Agra, 1991, 1994, 1996)*

(*viii*) Biological importance of Bacteriophages. *(Agra, 1993)*

(*ix*) Life history of bacteriophages *(Lucknow, 1993; Awadh, 1993)*

(*x*) Satellite virus *(Lucknow, 2000)*

(*xi*) History of viruses *(Lucknow, 2001)*

(*xii*) Internal symptoms of virus infected plants *(Lucknow, 2002)*

(*xiii*) Capsomose *(Allahabad, 2002)*

(*xiv*) Structure of bacteriophage *(Lucknow, 2005)*

31. Answer the following :

(*a*) Symptoms of virus diseases *(Gorakhpur, 1999, Awadh, 1994, 1995)*

(*b*) Transmission of viruses *(Kumaon, 2001, 2003; Gorakhpur, 1999; Awadh, 1993)*

(*c*) Methods of control of viruses. *(Agra, 1995)*

(*d*) Symptoms and control of viruses.

32. Write explanatory notes on the following :

(*a*) Insect transmission of viruses *(Kanpur M.Sc., 2004; Awadh, 1998)*

(*b*) Cryptogram of TMV and an explanation of the terms used. *(Lucknow, 1992, 1998, 2005)*

(*c*) Different morphological forms of bacteriophages.

(*d*) Mechanism of multiplication of TMV *(Kumaon, 1999)*

(*e*) Mechanical transmission of plant viruses and factors affecting it.

(*f*) Different morphological forms of plant viruses.

(*g*) Properties of virus in Crude Sap *(Lucknow, 1992)*

(*h*) Transmission of viruses *(Lucknow, 2003; Gorakhpur, 1997; Awadh, 1995, 1996)*

(*i*) External symptoms of virus infected plants *(Lucknow, 2000, 2003, 2004)*

(*j*) Replication of TMV *(Lucknow, 2003, 2005)*

33. Write not more than a paragraph on the following :

(*a*) Viroid *(Lucknow, 1992, 1998, 1999, 2003; Allahabad, 1993, 1994, 1997, 1998)*

(*b*) Cryptogram of TMV*(Lucknow, 1993)*

(*c*) Internal symptoms of virus infected plants *(Lucknow, 1994, 1998; Gorakhpur, 1991)*

(*d*) Major morphological foms of plant viruses. *(Lucknow, 1994, 1998)*

(*e*) Nature and structure of plant virus *(Rohilkhand, 1992, 1993; Bundelkhand, 1996)*

(*f*) Viral diseases *(Bundelkhand, 1993)*

34. With the help of suitable diagrams, describe the following :

(*i*) Structure and composition of TMV *(Kanpur, 2005)*

(*ii*) Life history of bacteriophages. *(Lucknow, 1994, 2004)*

(*iii*) T_4 viruses *(Lucknow, 2002)*

35. Describe briefly :

(*a*) Nature of viruses *(Allahabad, 2003, Rohilkhand, 2001; Kanpur M.Sc., 2001, 2002, 2003, 2004, 2005; Lucknow, 1995, 1999)*

(*b*) Isolation and purification of plant viruses *(Lucknow, 1995)* (*c*) Different morphological forms of bacteriophages (*Lucknow, 1996*)

(*d*) Properties of virus in Crude Sap (*Lucknow, 1996*)

(*e*) Mechanism of multiplication of TMV (*Lucknow, 1996; Gorakhpur, 1991, 1994, 1996, 1998; Allahabad, 1992*)

(*f*) Insect transmission of plant viruses (*Lucknow, 1999; Gorakhpur, 1991, 1994, 1996*)

(*g*) Symptoms produced by viruses (*Awadh, 1991*)

(*h*) Classification of viruses (*Lucknow, 2001, 2002, 2004; Meerut, 1997*)

(*i*) Structure & multiplication of Bacteriophages (*Poorvanchal, 2005*)

(*j*) Structure and Reproduction of viroids (*Lucknow, 2004*)

(*k*) Morphology and chemical nature of viruses (*Lucknow, 2004*)

(*l*) Economic importance of phage particles (*Lucknow, 2004*)

(*m*) Structure of Bacteriophage (*Rohilkhand, 2002*)

36. (a) Give the general characteristics of cyanophages.

(*b*) Discuss the structure of T_2 viruses.

37. Differentiate between:

(*a*) Virus and bacteria and explain the phenomenon of multiplication of virus. (*Bundelkhand, 1994*)

(*b*) Viruses and Viroids (*Allahabad, 2003*)

(*c*) Prophage and vegetative phage (*Allahabad, 2004*)

(*d*) Viruses and Mycoplasma (*Rohilkhand, 2002*)

(e) Virus, bacteria and bacteriophage (*Poorvanchal M.Sc., 2003*)

(*f*) Viroids and Prions (*Kanpur M.Sc., 2004*)

(*g*) Difference between Plant and Animal viruses. (*Lucknow, 2004*)

38. Give diagnostic features of Viroids. (*Lucknow, 2002*)

39. Define a virus. (*Poorvanchal, 2002*)

40. Give a brief account of

(*a*) Mechanical Transmission of a plant virus on healthy host

(*b*) Replication of TMV

(*c*) Multiplication of Bacteriophage (*Lucknow, 2002*)

(*d*) Different Modes of transmission of Viruses (*Kanpur, 2006*)

41. What is a Cryptogram ? Give the Cryptogram of TMV. (*Lucknow, 2003*)

42. Explain the structure of bacteriophage. (*Kanpur, 2002, 2006*)

43. (*a*) Define Bacteriophage. (*Poorvanchal, 2003*)

(*b*) Explain the structure of Bacteriophage

44. What are bacteriophages. (*Gorakhpur, 2003*)

45. What properties of living organisms do viruses show ? (*Poorvanchal M.Sc, 2003*)

46. What are bacteriophages ? Describe their replication. (*Kanpur M.Sc., 2006*)

47. Describe in brief different methods of transmission of viruses. (*Kanpur, 2006*)

48. Describe the main characteristics of viruses. (*Lucknow, 2003, 2004*)

OBJECTIVE TYPE

49. **Fill in the blanks :**

(*i*) The viruses were discovered by ....

(*ii*) It was .... who crystallised TMV.

(*iii*) Yellow vein mosaic disease of Bhindi is caused by ....

(*iv*) The particle shape of TMV is ....

(*v*) .... is the genetic material of TMV

(*vi*) In TMV, the amount of nucleic acid is .... per cent and amount of protein is .... per cent.

(*vii*) Viroids are made up of RNA without ....

(*viii*) Morphological units of capsid of virus are known as ....

(*ix*) Sub-microscopic infectious agents composed of nucleic acids, protein, obligately parasitic and hence replicates only in living hosts are called ....

(*x*) DNA is absent in ....

(*xi*) Propagative virus is that which ....

(*xii*) Viruses were crystallised for the first time by ....

(*xiii*) W.M.Stanley is famous for ....

(*xiv*) Bacteriophage is made up of ....

(*xv*) A viroid differs from virus in ....

(*xvi*) A substance which stimulates the formation of antibodies known as ....

(*xvii*) Viroids are made up of ....

50. Select the correct answer :

(*i*) Virology is the study of
(*a*) Algae (*b*) Fungi
(*c*) Bacteria (*d*) Virus

(*ii*) Bacteriophage is made up of
(*a*) Protein (*b*) DNA
(*c*) Nucleo protein (*d*) Lipid and protein

(*iii*) The term viroids are used for organisms which are made up of
(*a*) DNA only (*b*) RNA only
(*c*) Nucleo proteins (*d*) Lipid and proteins

(*iv*) Cryptogram of TMV is
(*a*) R/1 : 2/5 : E/E : S/0
(*b*) R/1 : 2.3/5 : E/E : S/Ne
(*c*) D/2 : 4.5/16 : S/S : S/Ap
(*d*) Ç/Ç : Ç/Ç : E/E : S/Ap

(*v*) TMV is made up of
(*a*) DNA and RNA
(*b*) DNA and protein
(*c*) RNA and protein
(*d*) DNA, RNA and proteins.

(*vi*) Size of TMV viruses is approximately
(*a*) 300 m/m × 15mm
(*b*) 500 m/m × 15mm
(*c*) 250 m/m × 150mm
(*d*) 10,200 m/m × 15mm

(*vii*) Genetic factor in bacteriophage is
(*a*) RNA (*b*) DNA
(*c*) Both (*d*) None

(*viii*) Plant viruses do not have
(*a*) Nucleic Acid
(*b*) Protein
(*c*) Cytoplasm organelle
(*d*) Amino acids

(*ix*) Who discovered Bacteriophage ?
(*a*) Iwanowsky (*b*) Twort
(*c*) Stanley (*d*) Morris

(*x*) Viruses are considered as non-living because
(*a*) they can undergo mutation
(*b*) they can be crystallised
(*c*) they can multiply
(*d*) they can grow

(*xi*) Which one of the following is neither a prokayote nor an eukaryote ?
(*a*) *Chlamydomonas*
(*b*) *Synchytrium*
(*c*) *E.coli*
(*d*) TMV

(*xii*) A bacteriophage is
(*a*) An organelle of the bacterium
(*b*) Bacterium which infects a higher plant
(*c*) Bacterium which infects an animal cell
(*d*) Virus which infects a bacterium

(B) MYCOPLASMA

ESSAY TYPE

1. What are mycoplasma? Write the name of three diseases which are caused by mycoplasma. Describe the morphological characters. (*Kanpur, 2005*)

2. Describe the characteristic features of mycoplasmas and write names of any two diseases caused by mycoplasma in plants. (*Kanpur, M.Sc., 2004*)

3. Write structure and reproduction in Mycoplasma. (*Poorvanchal, 2003*)

4. What is Mycoplasma? How does it differ from bacteria and Viruses? Name five important diseases caused by Mycoplasma. (*Rohilkhand, M.Sc., 1996*)

SHORT ANSWER TYPE

5. What is mycoplasma? Explain in brief. *(Poorvanchal, M.Sc., 2003)*
6. Write the name of two mycoplasma and two bacterial plant diseases. *(Kanpur, 2003)*
7. Describe structure of mycoplasma. *(Kanpur, M.Sc., 2006)*
8. Write notes on
   (a) Mycoplasma *(Kanpur, 2002, 2006)*
   (b) Structure of Phytoplasma *(Poorvanchal, 2002)*

OBJECTIVE TYPE

9. **Select the correct answer:**
   (*i*) The main function of primary bodies in mycoplasma is
   (*a*) Reproduction (*b*) Respiration
   (*c*) Secretion (*d*) Food storage
   (*ii*) Mycoplasmas are
   (*a*) aerobic eukaryotes with a cell wall
   (*b*) aerobic eukaryotes with out a cell wall
   (*c*) aerobic prokaryotes without cell wall
   (*d*) anaerobic prokaryotes with a cell wall
   (*iii*) Mycoplasma was discovered by
   (a) Nocard and Roux
   (b) Nowak
   (c) Lederberg
   (d) None of these
   (iv) Mycoplasma differs from bacteria in not possessing
   (a) Cellwall (b) Cell membrane
   (c) Ribosome (d) DNA

SECTION III

LICHENS

Lichens – composite organisms

CHAPTER 20

LICHENS

- Dual Nature of Lichens
- Classification
- Lichen thallus and associated structures
- Nutrition
- Reproduction
- Ecologic and Economic Importance

Dual Nature of Lichens. The lichens are a small group of curious plants with about 18,000 species. We call them curious because they are **composite** or **dual organisms**. For a long time the lichens were considered as individual plants and were lumped together with the mosses. Their true nature was discovered in the middle of the nineteenth century. During the 1860s it was found that the thallus of every lichen is composed of two quite different organisms, a **fungus** and an **alga**, forming a self-supporting combination. The fungal component is called the **mycobiont** and the algal component is known as the **phycobiont**. The two live in intimate association. So much so that they appear to be a single plant. The fungal component forms the bulk of the lichen thallus. The algal component occupies only a small part comprising about 5-10 percent of the mass of the thallus. The lichen thallus thus is essentially fungal in nature. The fungal hyphae form a close network resembling a tissue-like mass with the algal cells often embedded in it. Another extraordinary thing about the lichens is that the new individual resulting from this association differs considerably from either of the partners in form and habit. It also has a physiology of its own. The fungi obtain food either saprophytically from dead organic matter or else parasitically from the living bodies of host organisms. Algae synthesize their own food from carbon dioxide and water. In a lichen, both the components live on the food manufactured by the algal component. Lichens thus are **biotrophic** in their mode of nutrition as they obtain their food from the living algal associate. Probably food materials from the alga diffuse out and are absorbed by the fungus. The reproductive organs, however, are entirely fungal in character.

Every individual of a given lichen species contains the same alga and the same fungus. This dual nature of the lichen thallus has been confirmed experimentally. People have been successful in building up a new lichen individual out of a fungus and its appropriate alga. The algal component in the association may belong to the Cyanophyceae or to the simple Chlorophyceae. It may be filamentous or non-filamentous. In the majority of the lichens, the alga is unicellular. About 26 genera of algae are involved. Of these, eight are blue-green, seventeen green algae and one yellow green. The common blue-greens are *Nostoc, Stigonema, Rivularia* and *Gloeocapsa*. About 80 percent of all lichen thalli contain green algae. Of these, unicellular green alga *Trebouxia* is the commonest. On the basis of the algal component, the lichens may thus be divided into two groups namely, (*i*) cyanophycophilous with blue-green alga phycobionts and (*ii*) chlorophyco philous with green algal phycobionts. The fungal partner is generally an Ascomycete. Only in 2 or 3 genera out of 400, the fungal component is a Basidiomycete. No lichenized Phycomycetes have been reported so far.

Classification. Many scientists classify the lichens on the basis of, (*i*) the nature of the fungal element and (*ii*) the kind of the fructification. On this basis, the lichens are divided into the following two groups :–

1. *Ascolichens* in which the fungal component is an *Ascomycete*. The *Ascolichens* are subdivided into two sub-groups :–

(*a*) *Gymnocarpeae* in which the ascocarp is of an apothecium type.

(*b*) *Pyrenocarpeae* in which the ascocarp is perithecium type.

G.L.Chopra (1934) in his monograph on the "Lichens of Darjeeling and the Sikkim Himalayas" described 80 species belonging to 38 genera of Ascolichens. The list includes one new species, three new varieties and a new genus *Chaudhuria*, with one species *C. indica*.

2. *Basidiolichens*. In this group are placed the *Lichens* in which the fungal partner is a *Basidiomycete*.

The branch of Botany which is concerned with the study of *lichens* is called **Lichenology**. A scientist who has specialized in the study of *lichens* is called a **Lichenologist**.

Basidiolichens

Lichens – Best example of Symbiosis

Nature of the Association (Physiology). Two different hypotheses have been put forth to explain the nature of the association in this plant complex. According to one hypothesis, the alga is considered a mere victim of the fungus. In other words, the fungus lives as a parasite on the algal partner. The parasitism, of course, is of a mild nature as it permits most of the algal cells to live. The adherents of this hypothesis advance the following two facts in support of their view :–

1. In some lichens the fungal hyphae give out **haustoria** or **appressoria**. These penetrate the algal cells to obtain nutrition for the fungus mycelium.

2. When the two components of the *lichen* thallus are artificially separated the alga is able to live as an independent individual. It grows and multiplies like an independent organism. But the fungus is unable to do so. It invariably dies.

This experimental evidence is cited as a positive proof of the fact that the *fungus* lives as a **parasite** on *alga* in the *lichen* thallus. The supporters of this hypothesis do not see any real benefit which the alga derives from the fungal partner. They argue that the water and minerals in solution, the alga could get in the free living state as well by itself as in the association with the fungus.

The second hypothesis is that the two partners in the lichen thallus derive mutual benefit from their partnership. There are advantages on both sides. They hold that the fungus with its rhizoids absorbs water and minerals in solution from the substratum. It also absorbs moisture from the fog and moist air. The water and minerals thus absorbed are passed on to the alga. The fungal hyphae are also somewhat gelatinous. They absorb water readily and hold it tenaciously. In this way, they afford protection to the algal partner from fatal drying when the air humidity is very low. The alga is thus able to live in exposed places where it would have been impossible for it to grow in the free state. In addition the fungal hyphae, which form the body of the lichen, provide shelter to the alga. They thus protect it from intense light, drought, and other adverse weather conditions. The duty of alga is to synthesize the necessary carbohydrates with the help of its green chloroplasts and in the case of cyanophycophilous lichens, nitrogen as well. This it does both for itself and also for the fungus. The alga, therefore, provides food to the fungus. In the course of providing food to the fungal partner, a few of the algal cells may be exhausted. They may perish. The bulk of them, however, live. They grow and multiply within the thallus. According to this view, the nature of association between the two components of the lichen thallus appears to be a mutualistic one. The adherents of this hypothesis cite lichens as the best example of **symbiosis** (mutualism) in the plant world. The high resolution radio autography experiment of Jacob and Ahmadjean (1971) in *Cladonia* lends support to this hypothesis. In this experiment, C^{14}–labelled sodium bicarbonate was provided as a source of carbon dioxide for photosynthesis to a segment of the lichen thallus. It appeared in the organic compounds first in the algal component and 15 minutes later in the hyphae of the fungal component. This evidence supports the movement of materials from the alga to the fungus in the lichen thallus. The consensus of opinion now favours second hypothesis. The adherents of this view place the lichens in a class distinct from both the fungi and algae.

There is third view which suggests the via media between the two extremes. According to this view, the relationship between the fungus and the algal partner in the lichen thallus is an example of symbiosis but the fungus in this partnership has the upper hand. The algal partner lives as a prisoner or as a subordinate partner. The association between the two partners is thus described as beneficial slavery for the alga. The supporters of this view suggest the term **helotism** to this kind of association.

Habitat. The lichens grow in a wide variety of habitats. They are commonly found growing on the walls and roofs of houses, leaves, tree bark, bare earth and even barren, unpromising rocky surfaces. Generally they are xerophytic in nature and thus can withstand long periods of drought.

Consequently, they thrive and multiply in habitats where other vegetation is non-existent such as sand dunes, deserts and bare rocks. Here they live happily under conditions of drought and apparent starvation. Usually they are attached firmly to the substratum. However, they derive little of their inorganic requirements from the substratum. Rather they depend on the rain or spray. They absorb, whatever substances they require through the whole surface of the thallus from those dissolved in the rain water. Many species are able to withstand low temperature. They are abundant on high mountain elevations, and grow in extremely cold regions such as the arctic tundras where conditions are unfavourable for the growth of other plants. Many lichens grow in the tropics and subtropics. These mesophytic lichen species grow in the abundance of moisture. Some aquatic species occur only on the sea shore. According to their habitat, Lichenologists divide the lichens into three categories, **saxicolous, corticolous** and **terricolous**. The **saxicoles** are predominantly stone or rock lovers. They grow on firm substratum in cold regions. The **corticoles** are generally bark lovers. They are mostly confined to the tropics and subtropics where there is abundance of moisture. Here they grow on the leaves and bark of trees as epiphytes depending on their hosts for anchorage only. The **terricolous** species are terrestrial and thus inhabit the soil. The factors favouring lichen growth are direct light, moderate or cold temperature, pure atmosphere, firm substratum and sufficient atmospheric moisture. The conditions unsuited for the growth of lichens are scanty precipitation, hot and dry summer. The growth in lichens is very slow.

Corticolous Lichens

Saxicolous Lichens

Terricolous Lichens

Distribution. Lichens are one of the most widely distributed groups of plants. They are found all over the globe from the Arctic to the Antarctic and all regions in between and in diverse habitats. The factors which help in their worldwide distribution are :–

1. Their symbiotic life.
2. Their prolific methods of vegetative propagation and efficient means of dispersal.
3. Resistance to extremes of temperature and moisture which enables them to grow at all places where life can be supported at all.

They are found far north and far south than any other plants of the arctic region. Here they along with mosses comprise the entire vegetation of the place. Far north lie the tundras of Iceland, Lapland, Greenland and Canada and other subarctic regions. The lichens are also met with at higher altitudes on land surfaces than any other plants. Here they colonize the fresh rocky surfaces. The lichens are equally at home in the equatorial jungles. In India they are very common all over the Himalayas particularly the Eastern side. They are also found in the higher hills of Peninsular India.

Lichen thallus. The plant body in the lichens is a **thallus**. It is a grey or greyish green irregularly shaped structure. Some species are strongly pigmented and may be yellow, orange, brown or red in colour. In form and habit, the lichen thalli are highly diverse. Some species simply form a powdery layer over the substratum (tree bark and rocks) and others form a thin crust closely adherent to the substratum. Still others take on leafy on leafy forms and the complex ones resemble tiny shrubs. A few species present intermediate forms. Basically, however, there are three main types of lichen thalli recognisable on the basis of their general habit of growth, form and manner of attachment to the substratum :–

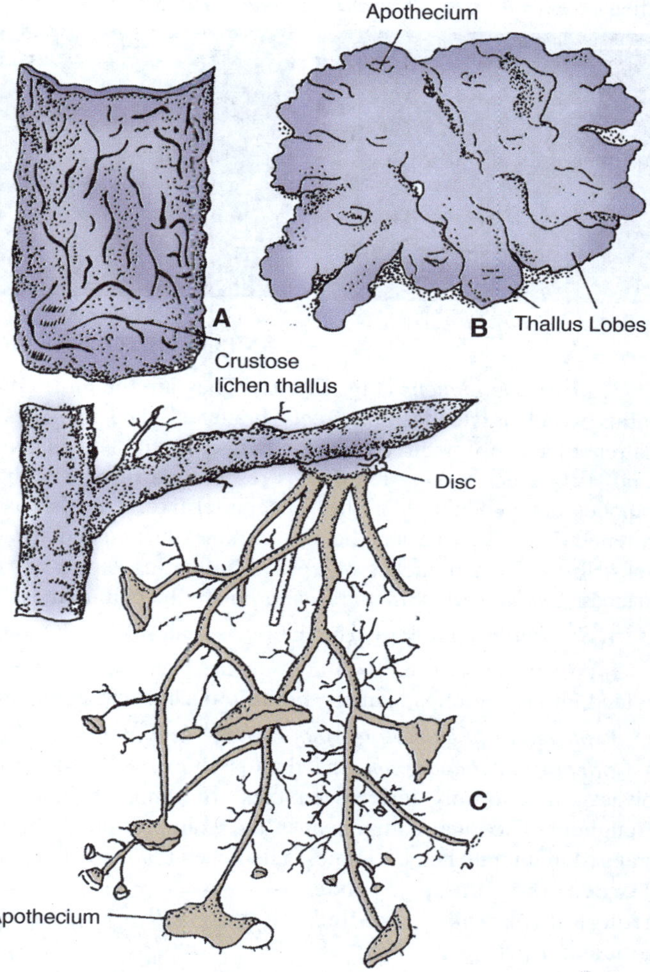

Fig. 20.1 (A-C). *Lichens.* Three fundamental types of lichen thalli. A, Crustose lichen; B, Foliose lichen; C, Fruticose lichen.

1. *Crustose* (*Crustaceous*) *Lichens* (Fig. 20.1 A). The thallus is of insignificant size. It is flat, thin usually without any distinct lobes. It is just like a thin layer or crust closely attached by the whole of its lower surface to stones or rocks, bark and similar hard substrata that the crustose lichen appears to be painted on. Normally it is impossible to dislodge it without breaking it. The surface of the thallus is usually divided into more or less hexagonal areas called the **areolae**. In many species, the thallus is partly buried in the substratum. *Graphic scripta* and *Haematomma puniceum* are the best examples of crustaceous lichens.

2. *Foliose* (*Foliaceous*) *Lichens* (Fig. 20.1 B). The foliose thallus is more striking. It is flat, broad, much lobed and leaf-like. In form it often resembles crinkled and twisted leaves. It grows more or less free of the substratum but close to it. It has a distinct upper and a lower surface. The lower surface may be white or sooty. The edges are usually curled up. The foliose thallus is attached to rocks and twigs by rhizoid-like outgrowths called the **rhizinae**. The rhizinae arise from its lower surface. In some foliose lichens, the rhizine consists of a single, simple to branched hypha. In others it is made up of a number of hyphae lying parallel and closely applied to one another. The free end of the rhizine broadens to form a flat disc. The disc secretes mucilage and attaches itself firmly to the substratum. The rhizines thus function as anchorage and absorptive organs. They are dark or dark brown in colour. The thallus, in some species, is attached to the substratum by means of a single rhizine. It grows from the centre of the lower surface of the thallus. In others, the thallus is attached by several rhizines. The common examples of foliose

lichens are *Xanthoria,, Physcia, Peltigera, Parmelia, Cetraria* and *Chaudhuria.*

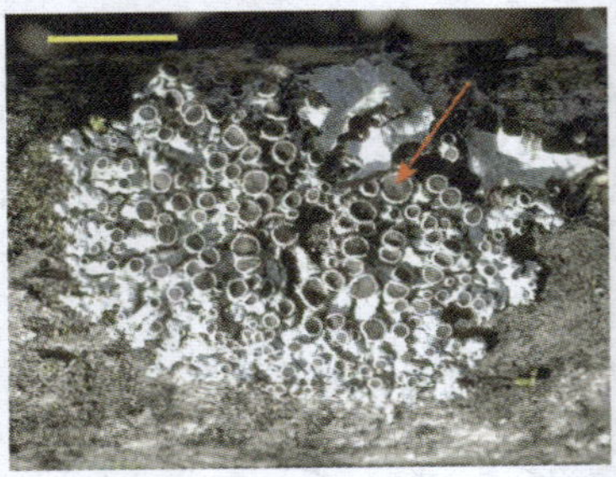

Physcia

Peltigera

Parmelia

EXAMPLES OF FOLIOSE LICHENS

3. *Fruticose Lichens* (Fig. 20.1 C). They are the most conspicuous and have a most complex thallus which is slender and freely branched. The branches may be cylindrical or ribbon-like (flattened) and form thread-like or twig-like tufts. In many species of *Cladonia,* the branches stand stiffy erect making the thallus resemble a tiny bud. There are others (*Usnea*) in which the branches are pendant (hanging) and tassel-like. The thallus is attached only at the base by a flattened **disc**. The fruticose lichens **make** extensive and attractive growths standing out from the rocks, foliage and branches of trees. **The** thallus shows no differentiation into upper and lower surfaces. *Usnea, Cladonia* and *Ramalina* are the common examples of fruticose lichens.

4. *Squamulose Lichens*. The thallus is scale-like and composed of small lobes.

Structure of Lichen Thallus. On the basis of internal structure of thallus, the lichens are divided into two groups, namely, **homoiomerous** and **heteromerous** lichens.

1. *Structure of Homoiomerous Lichen Thallus* (Fig. 20.2). In the gelatinous lichen thalli such as *Collema* and *Leptogium,* the thallus shows a simple structure with little differentiation. It consists of a loosely interwoven mass of fungal hyphae with algal cells equally distributed throughout. The algal component in the examples cited above is a blue-green one with the cells arranged in unbranched trichomes. The lichen thallus with the algal component scattered uniformly between the fungal hyphae throughout is called **homoiomerous**.

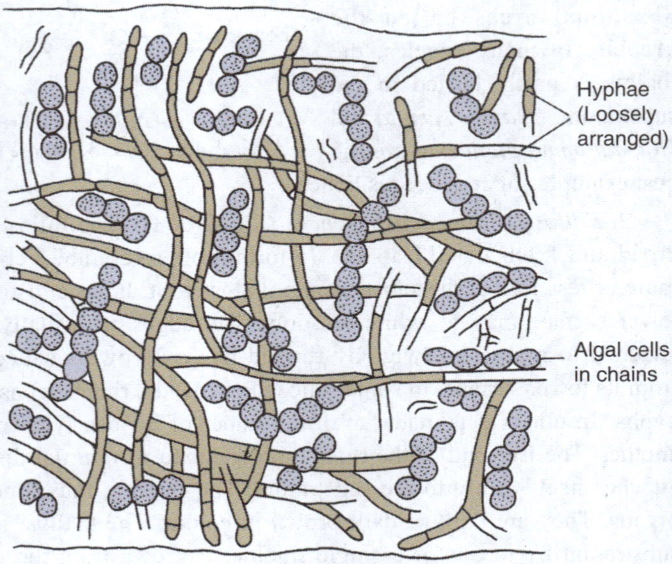

Fig. 20.2. *Lichens*. V. S. Thallus of *Collema* sp.

2. *Structure of Heteromerous Lichen Thallus* (Fig. 20.3). Most of the lichens belong to this category. They exhibit considerable differentiation and layered structure. The algal component in a **heteromerous** thallus is restricted to a specific zone or layer. A vertical section through the foliose thallus such as that of *Parmelia* or *Xanthoria* reveals the following four distinct zones :–

(*a*) *Upper Cortex*. It forms the upper surface which is generally thick and protective. The fungal hyphae in this region grow more or less vertically and are compactly interwoven to produce a tissue-like layer (**Plectenchyma** or **pseudoparenchyma**) called the **upper cortex**. The fungal cells in the upper cortex are either closely packed without intercellular spaces between them or with intercellular spaces filled with gelatinous material. Generally the upper cortex has epidermis-like configuration on the surface.

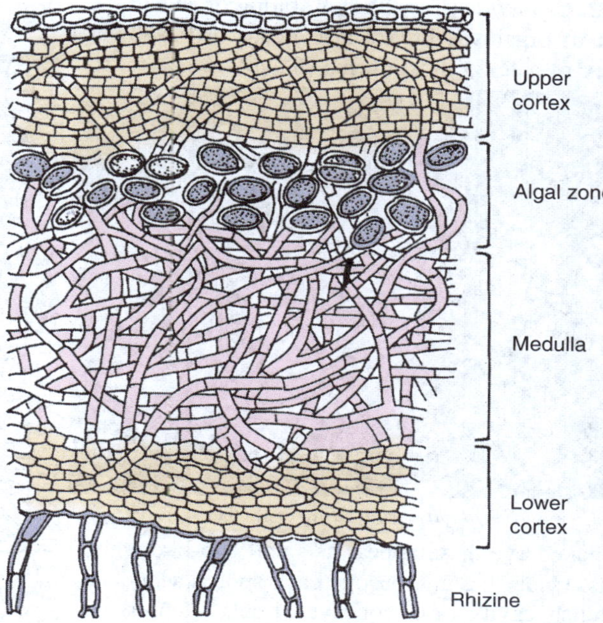

Fig. 20.3. *Lichens*. V. S. foliose lichen thallus.

(*b*) *Algal Zone*. It is the blue-green or the green zone which lies immediately beneath the upper cortex. It consists of a tangled network of loosely interwoven fungal hyphae with the algal cells of a green alga (in *Xanthoria*) or of a blue-green alga (in *Peltigera canina*) intermixed with the fungal hyphae. Common among the unicellular green algae present in this layer are *Chlorella, Pleurococcus* and *Cystococcus*. *Gloeocapsa* is a common example of a unicellular blue-green alga present in this region. The filamentous blue-green algae found in the algal zone are *Nostoc* and *Rivularia*. The algal region is the photosynthetic region of the lichen thallus. Formerly it was called the **gonidial layer**–a misnomer. The algal cells multiply by cell division or aplanospore formation. The enveloping fungal hyphae, in some species, send haustoria into the algal cells. The haustoria absorb nutrition for the fungus.

(*c*) *Medulla*. It forms the central core of the thallus. It is less compact and consists of loosely interwoven hyphae with large spaces between them in certain regions. The fungal hyphae in this region are scattered and usually have thick walls. They run in all directions. The central hyphae of the medullary region usually run longitudinally. They become very thick in the region of the vein and thin at the margin. Here and there they give out anastomosing strands.

(*d*) *Lower Cortex*. It forms the lower surface of the thallus and is composed of densely compacted hyphae. They may run perpendicular to the surface of the thallus or parallel to it. Bundles of hyphae (rhizinae) often arise from the surface of the, lower cortex and penetrate the substratum to function as anchoring organs. In some lichen species the lower cortex is absent. Its place is taken up by a thin sheet of hyphae constituting the hypothallus. It persists chiefly at the margins of the thallus. The rhizines in these species arise directly from the thicker part of the medullary region.

Structures associated with the lichen thallus. Associated with the lichen thallus are certain other vegetative structures peculiar to lichens only. The most important or these are :–

1. *Breathing Pores*. In certain species of lichens particularly the foliose forms the compact nature of the upper cortex is interrupted at inervals. In these localised areas, which are called the **breathing pores,** the fungal hyphae are loosely interwoven. The tissue beneath the breathing pore is more of less medullary in nature. The breathing pores serve for aeration and may be in level with the surface or raised on cone-like elevations on the thallus.

2. *Cyphellae* (Fig. 20.4). Aerating organs in the form of organised breaks also occur in the lower cortex of a few foliose forms (*Sticta sylvatica*). To

Sticta sylvatica bearing cyphellae

the naked eye these appear as small cup-like white spots. Under the microscope each spot is seen as a roundish cavity or a concave circular depression where white medulla is exposed. Here the hyphae grow directly from the medulla and abstrict empty rounded cells in a spore-like manner at their tips. Such aerating or breathing pores in the lower cortex may or may not have a definite border formed by the edge of the cortex. In the former case they are called the **cyphellae** and in the latter **pseudocyphellae**.

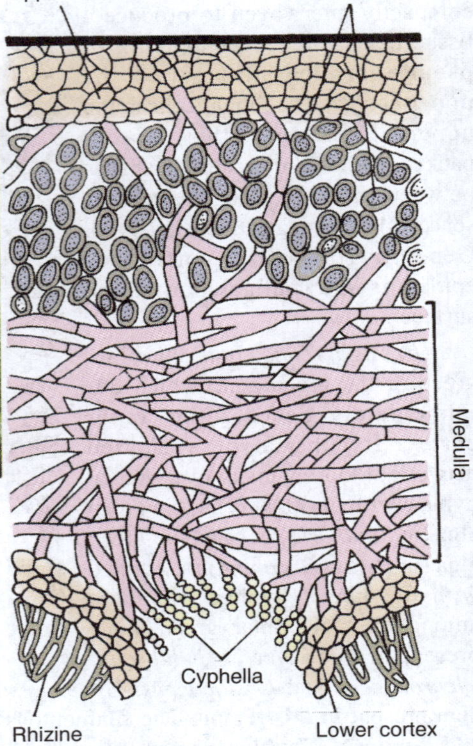

Fig. 20.4. *Lichens.* V. S. thallus passing through a *Cyphella.*

3. *Cephalodia* (Fig. 20.5 A-B). These appear as small, hard, dark-coloured, gall-like swellings on the free surface of some lichen thalli such as *Peltigera aphthosa*. The cephalodium contains the same fungal hyphae as in the thallus but the algal component is always different. For example, in *Peltigera aphthosa* the cephalodium contains a blue-green alga but the algal component in the thallus is of a bright, green kind.

4. *Isidia* (Fig. 20.6). These are small outgrowths on the upper surface of the lichen thallus each consisting of an outer cortical layer followed by an algal layer of the same kind as in the thallus. The chief function of isidia appears to increase the photosynthetic surface of the lichen thallus. The isidia vary in form in different lichen species. In *Parmelia sexuatilis* they are rod-shaped, but coralloid in *Umblicaria postulata*, cigar-shaped in *Usnea comosa*, tiny coral-like buds in *Peltigora praetexta* and scale-shaped in *Collema crispum*.

Collema

Usnea

EXAMPLES OF LICHENS HAVING ISIDIA

Nutrition. In a lichen thallus the body of the organism is provided by the fungal partner. In spite of this, the lichen in its mode of nutrition is totally different from a typical fungus. As a whole it is neither a parasite nor a saprophyte. Owing to symbiosis it obtains its nutrition like a green plant. The algal cells obtain water and minerals from the thallus of which they form a part. They provide themselves with carbon from the carbon dioxide of the air. From these raw materials they synthesize the necessary carbohydrates with the help of their chloroplasts. A part of these carbohydates the alga uses in its nutrition, the rest is supplied to the fungal partner. The actual mechanism of interchange of food between the two symbionts is still obscure. The movement of food materials probably occurs by diffusion. The soluble food material diffuses out from the alga and is absorbed by the fungal partner. In addition, the fungus may augment its food supply by annexing the living algal cells with its haustoria. It may also consume the dead bodies of algal cells in due course of time.

Reproduction. Lichen thallus reproduces itself asexually by various vegetative methods. In addition both the components reproduce independently each in its own characteristic manner.

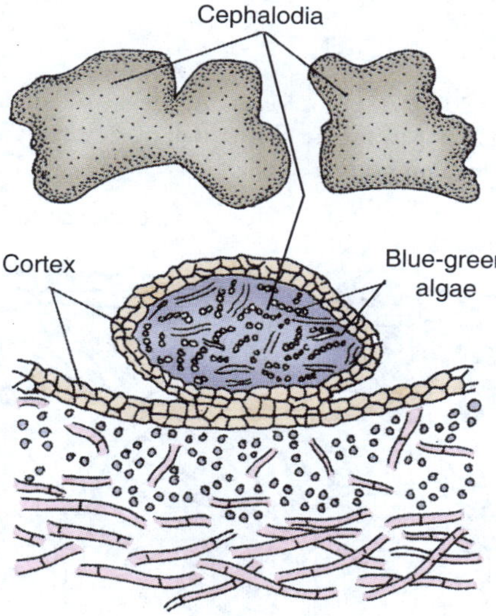

Fig. 20.5 (A–B). *Lichens.* A, Cephalodia in surface view; B, V. S. Thallus passing through the Cephalolium.

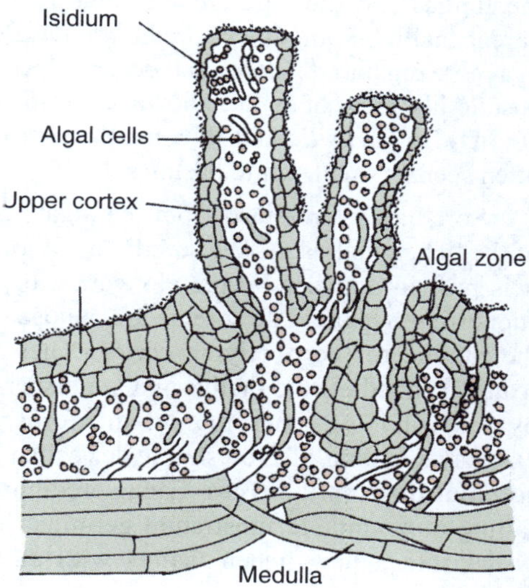

Fig. 20.6. *Lichens.* V. S. Lichen thallus passing through isidia which are cut longitudinally.

A. Reproduction of Lichen Thallus. It multiplies by vegetative methods of **fragmentation** and also by the formation of highly differentiated propagules as follows :–

(*i*) *Fragmentation.* It consists in the breaking up of the established thalli into segments which are distributed to start new growths. Fragmentation is asscomplished by *ageing* and *accidental severing*. In the first case, the older cells in the posterior or basal parts of the thallus die and separate the branches or lobes. Each of these continues to grow independently by apical growth. In this way a number of separate plants, all derived from the parent thallus, are formed. Sometimes larger portions of the established thalli are accidentally broken off the parent plant. Each of these, provided it contains both the symbionts, develops into a new lichen individual. In some fruticose lichens, these detached portions are carried by the wind to other trees. There they develop into new lichen plants. Vegetative reproduction by fragmentation thus is an efficient means of propagation. It leads to increase in the number of individuals of the species as well as constant invasion of new territories.

(*ii*) *Propagules*. Lichens also reproduce by vegetative structures peculiar to themselves. These are the **soredia** and **isidia**. Both these reproductive bodies or propagules contain elements of the lichen thallus (fungus and alga).

(a) Soredia (Fig. 20.7). These are small, rounded granules or bud-like outgrowths which develop in the form of a greyish white or greyish green powder in extensive patches usually over the upper surface or edges of the thalli of many species of lichens. Each soredium contains one or a few algal cells closely surrounded, by a little weft of fungal hyphae produced by branching of a hypha from the algal region (Fig. 20.7 A). Both the fungus and the alga are the same as in the parent thallus. Sometimes soredia are developed in a more organised manner and occur in localised, pustule-like areas of thallus known as **soralia** (Fig. 20.7 D). In *Parmelia* and *Physcia*, the soralia are often seen as small, white pustules.

Soredia are produced from the algal layer of the lichen thallus. As the soredium is formed it is pushed outward by the elongation of the supporting hyphae. The soredia are exposed by the rupture of the overlying cortex. They are readily rubbed off the thallus and are dispersed by the wind. At times the soredia forms an extensive coating of greyish powder called the **soredial dust** on the **trees**. Each soredium on falling on a suitable substratum germinates. It grows into a new lichen thallus with all the characteristics of the parent.

(*b*) Isidia (Fig. 20.6). As mentioned before these are small, conical warts developed on the thallus of many lichens. Isidia are usually constricted at the base and thus can easily be broken off. Under favourable conditions each isidium grows into a new lichen thallus. The isidium differs from a soredium in being covered by cortex.

B. Reproduction of Algal Component (Phycobiont). The green algal component of the lichen thallus multiplies by cell division. Ahmadjean (1967) reported the formation of **aplanospores**. Slocum *et al.* (1980) observed zoospores in the thallus of *Parmelia* with *Trebauxia* as the algal component. The blue-green algal components in the algal thallus have

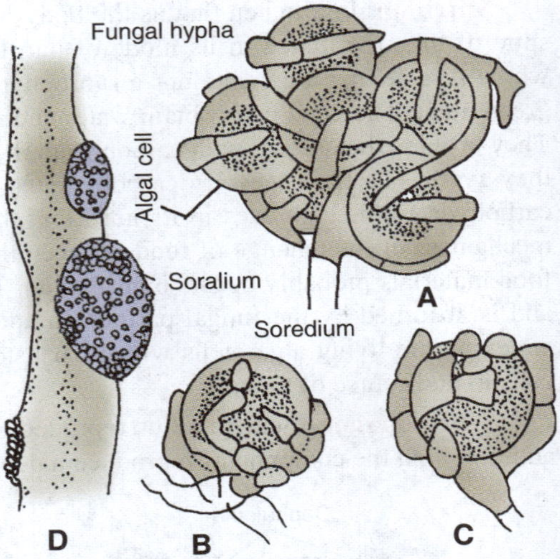

Fig. 20.7 (A–D). *Lichens.* A, single soredium; B–C, stages in the formation of a soredium; D, a thallus bearing soralia.

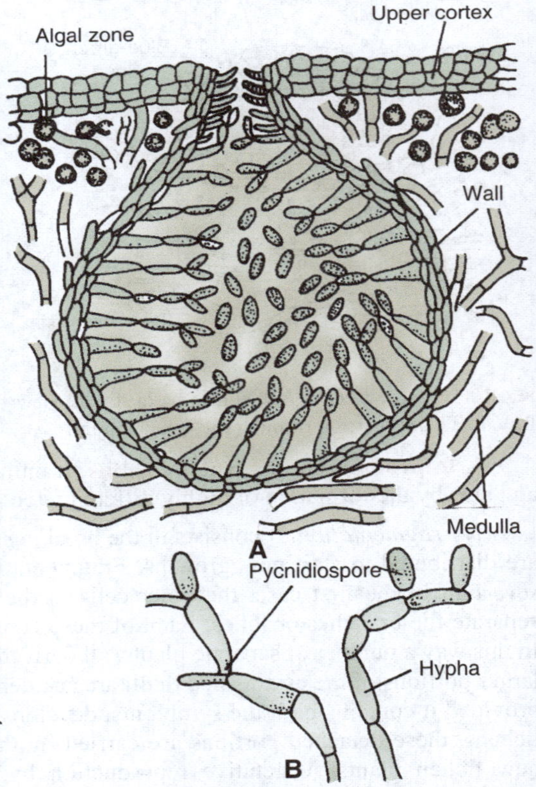

Fig. 20.8 (A-B). *Lichens.* V. S. thallus passing through a pycnidium (A); B, pycnidial hyphae bearing pycnidiospores (highly magnified).

been reported to multiply by cell division, hormogonia, akinetes and heterocysts.

C. Reproduction of Fungal Component (Mycobiont). The fungal component of the lichen thallus reproduces by spore formation (Sporulation) and sexual reproduction.

1. Sporulation (Asexual Reproduction) (Fig. 20.8). The fungal partner produces small, non-motile asexual spores known as the **pycnidiospores**. They are produced in large numbers in special, conical, flask-shaped cavities called the **pycnidia** (A). The pycnidia are found sunk on the upper surface of the lichen thallus in certain species. Each pycnidium opens to the surface through a small pore called an **ostiole**. The wall of the pycnidium consists of sterile fungal hyphae. From the wall arise fertile hyphae. The latter abstrict asexual spores (pycnidiospores) at their tips (B). The pycnidiospores in certain species of lichens are capable for germination. Each produces a fungal hypha which coming in contact with an appropriate alga develops further into a new lichen thallus.

2. Sexual Reproduction (Fig. 20.9). It is entirely the concern of the fungal partner on which the sexual organs are developed. The male reproductive organ is called the **spermogonium** and the female **carpogonium**.

(*a*) *Spermogonia*. In certain species of lichens the pycnidia-like structures (Fig. 20.8 A) are reported to function as **spermogonia**. Each spermogonium is a flask-shaped receptacle immersed in a small elevation on the upper surface of the thallus. It opens by a small pore, an **ostiole**, at the surface. The cavity of the spermogonium is filled with the fertile and sterile hyphae. The fertile hyphae abstrict minute, rounded cells at their tips. These are the male cells and are called the **spermatia**. They are non-motile and are produced in large numbers in each receptacle. Each spermatium has a cell wall around it. The spermatia are set free in a slimy mass which oozes out through the ostiole.

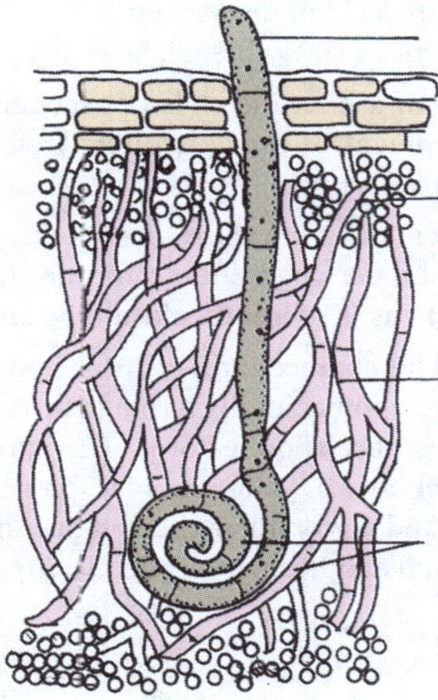

Fig. 20.9. *Lichens*. V. S. Thallus showing coiled ascogonium with a long and multicellular trichogyne.

(*b*) *Carpogonia* (Fig. 20.9). The carpogonium (Fig. 20.9) is a special cellular filament. It consists of two portions, the lower coiled portion and the upper straight portion. The coiled portion constitutes the **ascogonium**. It is multicellular. The cells are uninucleate. In certain species they are multinucleate. The ascogonium lies deep in the medullary region of the thallus. The carpogonia either develop from the hyphae of the medullarly region or from the hyphae deep in the algal layer of the lichen thallus. The straight upper portion of the carpogonium is called the **trichogyne**. It is also multicellular. The component cells are elongated, the septa between the cells have minute holes in the centre, one each. The terminal portion of the trichogyne ends in a long cell which projects beyond the surface of the thallus and has a gelatinous cell wall. From the account given above it is evident that the sexual apparatus of the lichens differs from that of the Red Algae in the following respects.

(*i*) Coiled ascogonium sunken in the medulla.

(*ii*) Long multicellular trichogyne with the terminal cell only projecting above the sur-face of the thallus.

(*iii*) Absence of supporting and auxiliary cells characteristic of red algae.

(*iv*) Antheridia in the form of flask-shaped cavities or receptacles sunk in the thallus, each opening through a pore on the surface of the thallus.

(*v*) Male cells or spermatia with cell walls.

(*c*) *Fertilisation*. Spermatia or the male cells have been found adhering to the gelatinous wall of the projecting terminal cell of the trichogyne. From this, some Lichenologists conclude that the spermatia are functional. At the point of contact the intervening walls between the spermatium and the trichogyne dissolve. The contents of the spermatium migrate through the spore into the trichogyne. The supporters of this view hold that only the carpogonia of which the trichogynes have been spermatised develop into ascus fruits. The presence of empty spermatia in contact with the trichogyne supports this view. However, it is not accepted by all. The opponents hold that the actual migration of the male nuclei down the trichogyne has never been observed.

(*d*) *Post-plasmogamy changes*. Fertilisation is followed by the following changes in the carpogonium :

(*i*) Gradual withering of the trichogyne.

(*ii*) Development of freely branched ascogenous hyphae from the ascogonium.

(*iii*) Formation of asci at the ends of ascogenous hyphae or their branches. In some lichens the development of ascus is direct from the terminal cell, whereas in others from the typical croziers.

(*iv*) Development of an envelope of sterile hyphae around the developing asci, the ascogenous hyphae and the ascogonium to form a fructification or the spore fruit which in many of the ascolichens is of **apothecium** type and in a few others of **perithecium** type.

1. Apothecia. The apothecium is a fruit body of an ascomycete fungus borne in the lichen thallus. Generally the apothecium is a rounded, cup-shaped structure. Sometimes it may be plate-like and rarely it has an elaborate form. The apothecia of lichens vary in colour which ranges from reddish, reddish brown, yellow to black depending upon the species. In many lichen species (*Lecidea, Cladonia* and *Gyrophora*) the apothecium consists of fungal elements only. There is no algal component. Such apothecia are called **lecideine** type. Many species (*Parmelia, Physcia,* and

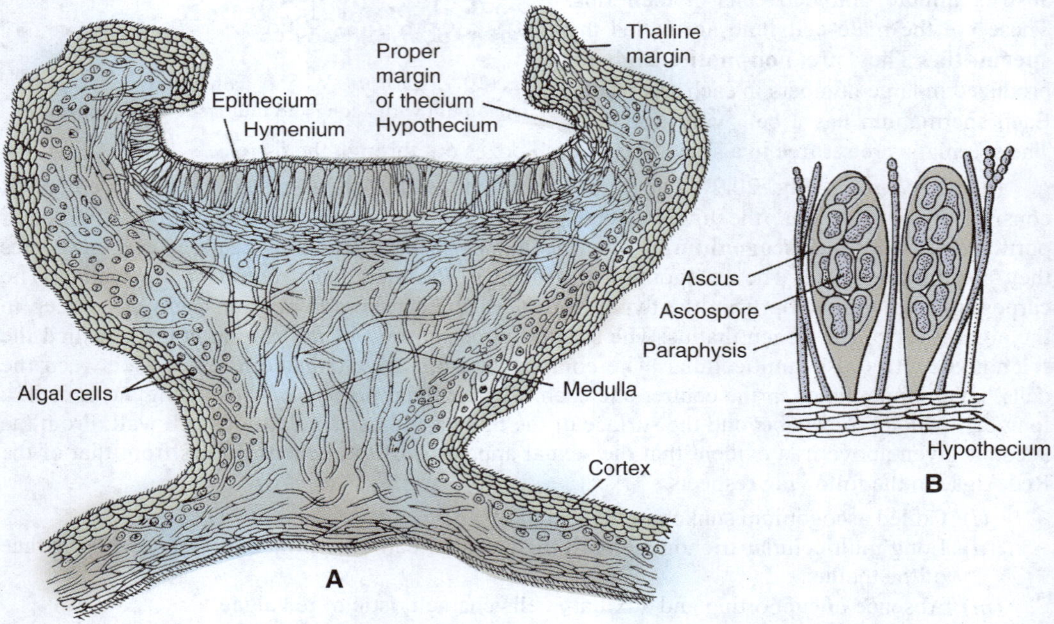

Fig. 20.10 (A-B). *Lichens*. A, V. S. Apothecium showing structure; B, A portion of hymenium (highly magnified).

Lecanora) have more highly developed apothecia in which algal component of the thallus also takes part in the formation of the apothecial margin. Such apothecia are known as **lecanorine** type.

(*a*) *Structure of the Apothecium* (Fig. 20.10 A). Whatever its form, the apothecium from outside is usually seen to consist of two parts, namely, the **disc** and the **margin**.

(*i*) *Disc of the apothecium*. It is the fertile portion of the apothecium. In many lichens it is soft and bright in colour and in others hard and black. In a vertical section of the apothecium, the disc consists of a closely packed, palisade-like layer of sac-like **asci** and sterile, hair-like fungal **hyphae** known as the **paraphyses**. The stout club-shaped asci grow upwards between the paraphyses. They do not project beyond them. Each ascus contains eight ascospores. The asci and the paraphyses are closely packed to form a continous fertile layer. It is called the **hymenial layer** or **hymenium**. It is also known as **thecium**. The projecting ends of paraphyses form a smooth surface, the **epithecium**. Beneath the hymenium is a region consisting of a dense mass of sterile hyphae. It is the **subhymenium**. It consists of closely interwoven hyphae from which arise the vertically growing paraphyses.

(*ii*) *Margin of the apothecium*. Surrounding the disc and forming the edge of the apothecium is the **proper margin**. The lecideine type of apothecium has only a **proper margin** consisting of fungal hyphae alone. The lecanorine type of apothecium (*Xanthoria* and *Physcia*), in addition to the proper margin has a second appreciably thickened margin surrounding the **proper margin**. The second margin is an upgrowth of the thallus and thus in colour and appearance it resembles the thallus itself. It is called the **thalline margin**. The apothecia with a thalline margin (lecanorine type) have both the fungal and the algal elements.

(*b*) *Development of Ascospores*. The young ascus mother cell contains a single diploid nucleus formed by the union of two haploid nuclei. As it advances towards maturity, the nucleus undergoes three divisions to form eight daughter nuclei. The first two divisions of the ascus nucleus constitute **meiosis**. The resultant eight nuclei are thus all haploid. By the accumulation of small amount of cytoplasm around each nucleus, eight ascospores are formed. The residual cytoplasm in the ascus becomes gelatinous. It takes up water during wet weather and swells. Owing to the pressure from within, the mature asci burst open at their tips. The ascospores are liberated with considerable force. Before liberation, the ascospores in many lichen species may undergo division to form two-celled to many-celled structures. The asci in the apothecium come to maturity at different times. As they mature the dehiscence takes place whenever wet weather occurs. The liberated ascospores are very light. They remain suspended in the air after their forcible discharge and thus may be carried by wind or air over long distances. On falling on suitable soil, the ascospore germinates. It produces a fungal hypha. The latter grows into a new lichen thallus if it comes in contact with an appropriate alga.

2. Perithecia. In the second sub-group *Pyrenocarpeae*, the ascus fruit is of **perithecium type**. It is much smaller in size than the apothecium and appears as a dot on the surface of the thallus. In a V. S., the perithecium is seen as a flask-shaped structure. It is bounded by a dark-coloured wall. The cavity is lined with asci. The paraphyses are either few in number or lacking. The perithecium is embedded in the elevations of the thallus on its upper side. It opens at the surface through a small pore called an **ostiole**. For structure of the perithecium refer to Chapter 8. The parthenogenetic development of the ascus fruit from the ascogonium has also been reported.

The basidiolichens produce basidiospores exogenously on basidia in the same manner as in *Agaricus* :

Features of Special Interest in Lichens

1. The Lichens are dual or composite organism.
2. The thallus-like plant body is made up of a Fungus and an Alga living in closest association. The nature of association between the two partners is the best example of symbiosis in the plant kingdom.
3. Fungus provides the body of the organism whereas Alga synthesizes carbohydrate food for itself and the Fungus.
4. The Lichen thalli are generally of three kinds, **crustose, foliose** and **fruticose**.
5. Internally the thallus particularly in the foliose lichens consists of four regions namely, the upper cortex, the algal layer, the medulla and the lower cortex.
6. Asexual reproduction by asexual spores and sexual reproduction are entirely the functions of the fungal partner.
7. The carpogonium is a coiled, multicellular filament.
8. It consists of a coiled, multicellular ascogonium and **a straight**, multicellular trichogyne.
9. There is a pore in the centre of each septum **between** the **cells** of the trichogyne.
10. The antheridia which are flask-shaped receptacles are sunk in the upper surface of the thallus.
11. Male cells or spermatia are non-motile. Each has a cell **wall around it**.
12. The ascus fruit in many species is of apothecium type **and in others** of perithecium **type**.
13. The contents of each ascus are fashioned into eight haploid ascospores. The ascospores may be simple or septate.
14. Each ascospore under suitable conditions germinates to produce a fungal hyphae which if it comes in contact with a proper alga, develops into a new lichen. If the fungal hypha fails to find an appropriate alga, it perishes.

It has been mentioned in the first paragraph of this chapter that the fungal component constitutes the major part of the body of the lichen thallus. As a matter of fact it provides the body of the lichen. In spite of this the lichens differ strikingly from the fungi in habit, in form and in their physiology. They agree in their methods of sexual and asexual reproduction which are all fungal in character. The important points in which they differ from each other are listed below in a tabulated form :–

TABLE I
Difference between Lichens and Fungi

Lichens	Fungi
1. They flourish the most in the temperate and colder regions of the globe.	They thrive best in the tropical and subtropical regions of the world which are warm.
2. Grow freely exposed to air and light and dislike smoky atmosphere of towns.	Prefer moist, shady, dark places and vitiated atmosphere of towns.
3. Can grow on barren substratum under drought and starvation conditions.	Require organic matter dead or living on which they feed and demand much moisture.
4. Grow generally as epiphytes or as terrestrial autophytes.	Grow either as parasites or as saprophytes.
5. The thallus is generally conspicuously coloured due to the presence of various organic acids.	The plant body is generally colourless.
6. The thallus is generally tough, leathery in texture and in many lichens it is soft as a rubber sponge.	The thallus is usually a filamentous mycelium, delicate in texture, immersed in substratum. In many it is slimy and gelatinous.

Ecological Importance. The lichens are of considerable ecological importance. They are slow but efficient soil farmers. They are the pioneer plants to grow on barren, naked, rocky surfaces where no other plants can grow. Crustaceous forms are usually the first to appear. They are followed by the foliaceous and eventually the fruticose types. They live there happily for years under conditions of drought and apparent starvation. When there is occasional shower of rain the lichen thalli soak it up like a blotting paper and retain it to support a more active life over long dry spells. Their other requirements are met with by the substances dissolved in rain water which is absorbed through the entire surface of the thallus. The continuous growth of the lichen thalli brings about gradual disintegration of the rocks immediately beneath. The lichen thalli secrete certain organic acids which gradually dissolve and disintegrate the rocks to which they cling. The disintegration is also helped by the stresses and the strains caused by the expansion and the contraction of the gelatinous lichen thalli. The rock particles together with the decaying and dead lichen thalli form a soil fertile enough for other plants to make appearance. The successors are the mosses. Sooner or later flowering plants begin to grow in this soil.

Economic Importance

The lichens, in spite of their beauty, variety and abundance include only a few species which are useful to man.

1. Certain species of lichens are valuable **sources of food**. Most important of these are the fruticose lichen *Cladonia rangiferina* (Reindeer Moss) and other tundra lichens which serve as food for reindeer, caribou, musk ox and other wild animals of the Arctic tundra zone. A few species have been used as food by man. One of the rocky lichens is still regarded a delicacy in China and Japan. A species of *Parmelia* is prized as food in Southern India. *Cetraria islandica* (the Iceland Moss) and *Lecanora esculenta* are the other edible species. The edible lichens are harvested and dried for human consumption or as fodder for cattle, swine and horses. The lichens, however, have little nutritive value.

Cladonia rangiferina

Parmelia

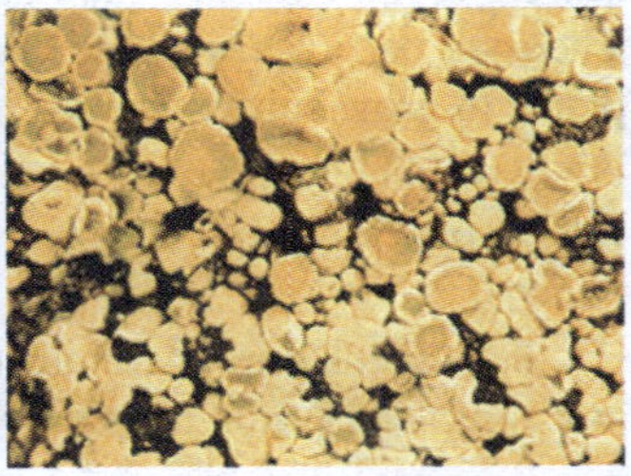

Lecanora esculenta

EXAMPLES OF LICHENS USED AS SOURCE OF FOOD

2. The traditional use of lichens in the **preparation of dyes** deserves a special consideration. The fungal components of certain species of lichens produce coloured pigments that have been used for centuries as **dyes** in colouring fabrics and paints. One of these is **orchil**. It is a blue dye formerly used to dye woollens. **Litmus** is another important and widely used dye in chemical laboratories as an acid-base indicator. It is obtained from *Rocella montaignei*. Besides these, there are many other brown and yellow dyes. The use of lichens as a commercial source of dyes has, however, been abandoned in favour of synthetic dyes.

Rocella (*Source of litmus*)

3. Medicinal use. A few species of lichens have been used in the medieval times as a cure for jaundice, fevers, diarrhoea, epilepsy and various skin diseases. The chief among them were *Peltigera camana, Lobaria pulmonaria* and *Everina.*The supposed medicinal benefits of lichens are now known to have no scientific basis. Recently a yellow substance **usnic acid** has been found in a species of *Usnea, Cladonia* and certain other lichens. It is a broad spectrum antibiotic useful in the treatment of various infections. It is used in the preparation of ointments for wounds and burns. A mucilaginous substance obtained from *Cetraria islandica* (Iceland Moss) is used as a laxative.

Lobaria Pulmonaria

Usnea

Cladonia

EXAMPLES OF LICHENS USED AS MEDICINES

4. The sweet-scented thalli of some lichens are used in the manufacture of dhup, hawan samagris and other perfumeries. A lichen popularly known as an Oak Moss is still used as a fixative for perfumes in Southern Europe.

5. Some lichen species synthesize and yield specific organic compounds commonly called *lichen acids*. These are useful in identification and classification of lichens.

6. Lastly the lichens play an important role in nature in initiating **soil formation**. They secrete organic acids which gradually dissolve and disintegrate the rocks over which they grow. The rock particles with the dead organic remains of thalli form substrata fertile enough for other plants to grow subsequently.

7. Lichens with blue-green algal phycobionts (Cyanophycophilous lichens) play an important role in providing fixed nitrogen in certain ecosystems such as deserts covered with crusts of soil lichens.

8. Lichens serve as indicators of air pollution. Lichens grow merrily exposed to air and light and completely shun the smoky polluted atmosphere of centres of heavy industry.

Two Common Examples of Lichens : 1. Parmelia (Fig. 20.1 B). It is a common example of a **foliose lichen** which is found growing on the bark as well as the rock intermixed with the mosses at higher altitude (above 4,000 ft.). The heteromerous thallus is lobed to deeply incised. It is attached to the substratum by **rhizinae** which arise from its darker lower surface. Internally the thallus (Fig. 20.3) is differentiated into four tissues, namely, **upper cortex, algal layer, medulla** and **lower cortex**. The phycobiont is a green alga *Pleurococcus*.

The thallus reproduces vegetatively by fragmentation and also by **isidia** (Fig. 20.6) and **soredia** (Fig. 20.7). The spores are produced in typically cup-shaped fructifications called **apothecia** which may be sessile or shortly stalked. The apothecia are scattered on the upper surface of the thallus lobes. They are large, round and have a **thalline margin** (Fig. 20.10). The asci are usually 8-spored. The ascospores are typically simple and colourless. The paraphyses are septate and may be branched. They arise from the hypothecium which is colourless. Underneath the hypothecium are present the algal cells.

Taxonomic Position :

Division	:	**Mycota**
Subdivision	:	**Eumycotina**
Class	:	**Lichens**
Sub-class	:	**Ascolichens**
Series	:	**Gymnocarpeae**
Order	:	**Parmeliales**
Family	:	**Parmeliaceae**
Genus	:	*Parmelia*

2. Usnea (Fig. 20.1 C). It is an example of a **fruticose lichen** with a cylindrical to ribbon-like, much branched thallus. It grows erect or pendant attached by a holdfast at the base only to the substratum (branches of trees). The thallus is usually grey-green and consists of a single main stalk or many. The stalks arise from the base and fork repeatedly. The main stalks and the branches are covered with conspicuous, branchlet-like bristles or fibrils.

A cross section of the thallus reveals a central medullary tissue consisting of loosely arranged hyphae with many interspaces between. It is surrounded by an algal layer with a thin cortex external to it. The phycobiont is a green alga, *Protococcus*.

The thallus reproduces vegetatively by fragmentation. Some species (*U. comosa*) bear cigar-shaped soredia in whitish soralia.

The apothecia are large, plate-like and terminal in position. The apothecium has a thalline margin fringed with bristle-like outgrowths which may be simple or branched. The disc of the apothecium is fawn in colour and about 1 cm in diameter. Each ascus contains 8 ascospores which are colourless and simple. Beneath the colourless hypothecium is the algal layer.

Systematic Position :

Division	:	**Mycota**
Subdivision	:	**Eumycotina**
Class	:	**Lichens**
Sub-class	:	**Ascolichens**
Series	:	**Gymnocarpeae**
Order	:	**Parmeliales**
Family	:	**Usneaceae**
Genus	:	*Usnea*

Some Unique Features of Lichens.

1. The lichens are dual organisms, each consisting of two different individuals, an **alga** and a **fungus**.
2. The new individual (lichen), resulting from this association, differs from either of the partners in its form, habit and physiology.
3. The lichens are the slowest growing of all the plants. The average rate of growth of lichen thallus ranges from 1 mm to 10 mm in radius a year.
4. The lichens have an extraordinary ability to withstand complete drying and for a very long period. The completely dried up thalli, on wetting absorb moisture and resume growth.
5. They are highly sensitive to atmospheric pollution and thus will not grow near large industrial towns.

6. The lichens can grow on barren substratum under conditions of drought and starvation and in situations where no other vegetation can exist.
7. The thallus of certain species of lichens contains characteristic kinds of **lichen acids** found in no other plants. The lichen acids are important from the taxonomic point of view.
8. The lichens play a very important role in soil formation.

QUESTIONS

ESSAY TYPE

1. What are Lichens? Give the various views put forth to explain the nature of the association between the two components in the Lichen thallus.
2. What are Lichens? Give an account of the habit, habitat and external morphological features of Lichens. *(Gorakhpur, 2003)*
3. Describe the structure of the Lichen thallus and the various vegetative structures associated with it. *(Allahabad, 1999)*
4. Give an account of ecological and economic importance of the Lichens.
5. How do Lichens get their nutrition? Give the diagnostic characters of Lichens. *(Allahabad, 2004)*
6. Give an illustrated account of methods of reproduction met in the Lichens. *(Gorakhpur, 1998)*
7. (*a*) Describe the structure of the ascus fruit of any Lichen.
   (*b*) Describe structure and economic importance of lichens. *(Poorvanchal, 2006)*
8. Give an account of sexual reproduction and development of fructification in the Lichens. *(Kanpur M.Sc., 2006)*
9. In what respects do Lichens differ from and agree with the Fungi? *(Himachal Pradesh, 1993)*
10. Give an economic importance of Lichens and describe the life history of any one of them.
11. Give an illustrated account of the internal structure of typical Lichen thallus. *(Lucknow, 2004; Rohilkhand,2001)*
12. Give detailed account of morphology, classification and reproduction in lichens. *(Kanpur, 2001, 2003)*
13. Describe the structure and asexual reproduction of Lichens. *(Poorvanchal, 2002)*
14. (*a*) Define the name of vegetation made by the association of algae and fungi. Describe their main characters. *(Kanpur, 2005)*
    (*b*) Describe sexual reproduction in Lichens. *(Kanpur, 2006)*

SHORT ANSWER TYPE

15. Discuss the following :
    (*a*) Foliose lichens *(Lucknow, 1999, 2001)*
    (*b*) Internal structure of Lichens *(Lucknow, 1999)*
    (*c*) Economic Importance of Lichens *(Rohilkhand, M.Sc., 1990; Lucknow, 1999)*
    (*d*) Crustose Lichens *(Lucknow, 2000, 2003; Allahabad, 2004)*
    (*e*) Lichens and Pollution *(Lucknow, 2004)*
    (*f*) Physiology of Lichens *(Lucknow, 2002)*
    (*g*) Homoiomerous Lichens *(Allahabad, 2000, 2002)*
    (*h*) Mycobiont *(Allahabad, 2001, 2002)*
    (*i*) Kinds of thalli in lichens *(Allahabad, 2001)*
    (*j*) Heteromerous Lichen *(Allahabad, 2001)*
    (*k*) Dual hypothesis of lichen nature *(Meerut, 1998)*
    (*l*) Lichens as bioindicators of pollution *(Rhoilkhand M.Sc., 1990; Poorvanchal M.Sc., 2003)*
    (*m*) Lichens as plant indicators *(Kanpur M.Sc., 2004)*
16. What are foliose Lichens? *(Lucknow, 2001)*
17. What are crustose lichens ? Mention their habitat with two examples. *(Lucknow, 2004)*

18. Differentiate between :

(*a*) Crustose Lichens and Foliose Lichens *(Allahabad, 2003)*

(*b*) Fruticose and foliose Lichens *(Allahabad, 2004)*

19. What are Lichens? Describe different methods of reproduction in Lichens. *(Allahabad, 1991; Gorakhpur, 1992, 1994, 1999; Poorvanchal, 2004; Rohilkhand, 2002)*

20. Give an illustrated account of external and internal structures of Lichens. *(Allahabad, 1992, 1996)*

21. What is a heteromerous Lichen? How does it differ from a homiomerous one? Give an illustrated account of external and internal structure of homiomerous lichen studied by you. *(Allahabad, 1993)*

22. Give an account of structure and economic importance of lichens. *(Allahabad, 1998, 2003)*

23. Give an illustrated account of structure and sexual reproduction in *Parmelia*. Comment upon the importance of lichens in medicines, cosmetics and soil formation. *(Allahabad, 1999)*

24. Give an account of range of thallus structure in Lichens. *(Allahabad, 1999; Awadh, 1999)*

25. Give a brief account of reproduction in lichens and their importance in relation to environment pollution. *(Lucknow, 1992)*

26. Describe structure and classification of lichens. *(Lucknow, 1998)*

27. What are lichens? Write an essay on their economic importance. *(Agra, 1991; Kanpur, 1998)*

28. How lichens reproduce in nature? Give the details. *(Awadh, 1991)*

29. What are lichens? Detail the biotic relationship between the components of the thallus of lichens. *(Bundelkhand, 1998)*

30. Define the term lichen. How many types of lichens have you studied? Give external and internal morphology of lichens. *(Rohilkhand, 1993)*

31. Write notes on

(*i*) The classification of Lichens.

(*ii*) Structure of thallus of Lichens. *(Gorakhpur, 1998, 1999; Lucknow; 1992, 2001, 2002; Awadh, 1998)*

(*iii*) Economic importance of Lichens mentioning the names of genera concerned. *(Allahabad, 1991)*

(*iv*) Physiology of symbiotic association of Lichens. *(Allahabad, 1991, 1994)*

(*v*) *Parmelia* *(Allahabad, 1992)*

(*vi*) Phycobiont in Lichens. *(Allahabad, 1992, 1995, 2000, 2002)*

(*vii*) Medicinal importance of Lichens. *(Allahabad, 1993)*

(*viii*) Sexual reproduction in Lichens *(Allahabad, 1994, 2002; Kanpur, 2006, 2005)*

(*ix*) Reproduction in Lichens *(Allahabad, 1997, 2000, 2003; Lucknow, 1995, 1997, 2001, 2003; Awadh, 1997; Kumaon, 2000, 2001, 2004)*

(*x*) Soredia *(Allahabad, 1997, 2001, 2002; Lucknow, 1993, 2002; Awadh, 1997; Allahabad M.Sc., 1999)*

(*xi*) Isidia *(Allahabad, M.Sc., 1999; Allahabad, 2001)*

(*xii*) Crustose Lichen *(Gorakhpur, 1991)*

(*xiii*) Asexual Reproduction in Lichen *(Gorakhpur, 1991, 1996, 1997; Rohilkhand, 1991; Kanpur, 2003, 2005; Kanpur M.Sc., 2003)*

(*xiv*) Sexual Reproduction in Lichen *(Gorakhpur, 1992, Awadh, 1998)*

(*xv*) Structure of foliose Lichen *(Gorakhpur, 1994; Lucknow, 1997)*

(*xvi*) Morphological forms and internal structure of Lichen thallus *(Gorakhpur, 1995; Lucknow, 1993, 1997)*

(*xvii*) Nutrition in Lichens *(Gorakhpur, 1996; Lucknow, 1996)*

(*xviii*) V. S. of lichen fruiting body *(Gorakhpur, 1997; Awadh, 1997)*

(*xix*) Economic importance of Lichens. *(Allahabad, 1994, 1997, 2001; Gorakhpur, 1991, 1995, 2003; Lucknow, 1992, 1996, Kanpur, 1996; Awadh, 1996; Bundelkhand, 1994, 1995, 1998, 1999; Kumaon, 1999,2000, 2001, 2002, 2003; Poorvanchal 2002; Rohilkhand, 2003; Poorvanchal M.Sc., 2003)*

(*xx*) Fruticose lichens *(Allahabad, 2000;Lucknow, 1993)*

(*xxi*) Classification of Lichens. *(Awadh, 1996, 1997, 1998; Rohilkhand M.Sc., 2006)*

(*xxii*) Types of Lichens *(Kanpur, 1999; Rohilkhand, 1992; Kumaon, 2004)*

(*xxiii*) *Parmelia* lichen *(Punjab, 1999, 1998)*

(*xxiv*) T. S. of lichen thallus *(Bundelkhand, 1995, 1997)*

(*xxv*) Symbiotic relationship in Lichens *(Rohilkhand M.Sc., 1990; Kumaon, 2002; Allahabad, 2002)*

(*xxvi*) Different structures of Lichens *(Poorvanchal, 2005)*

32. Lichens are rare or absent in cities. What conclusion can be drawn from this or their habitat. Why? *(Gorakhpur, 1993)*

33. (*a*) Describe the characteristic features of *Usnea* and its importance. *(Guru Nanak Deo, 1991)*

(*b*) Describe the sex organs of lichens. *(Punjab, 1991)*

34. With examples, explain ecological importance of lichens. *(Kanpur, 1995; Awadh, 1998)*

35. Describe the economic importance of lichens. *(Kanpur, 2001, 2002)*

36. Why are lichens called pollution indicators ? *(Poorvanchal, 2002)*

37. Draw V.S. of heteromerous lichen thallus. *(Poorvanchal, 2003)*

38. Describe the nature of association between the two components of a lichen thallus. *(Poorvanchal, 2003)*

39. Draw neat and labelled diagrams of heteromerous and homosporous lichens. *(Kanpur, 2005)*

40. What are lichens ? Differentiate between crustose and foliose lichens. *(Kanpur, 2006)*

41. Make neat and labelled diagrams of

(*a*) morphological structures of different kinds of lichen thalli. *(Kanpur M.Sc., 2004)*

(*b*) V.S. apothecium of Lichens. *(Kanpur M.Sc., 2003)*

(*c*) V.S. cephalodium.

(*d*) V.S. lichen through isidium.

OBJECTIVE TYPE

42. Fill in the blanks :

(*i*) The algal component of a lichen thallus is known as ....

(*ii*) The lichens consisting of a member of Basidiomycetes are known as ....

(*iii*) Fructification bearing lichens are called .... lichen.

(*iv*) Lichens are composite thallophytes containing an alga and a ....

(*v*) In majority of lichens, the fungal component is a member of class ....

(*vi*) The mode of life of a lichen is ....

(*vii*) A group of algal cells imprisoned in a fungal mycelium, which is used for vegetative reproduction is ....

(*viii*) The first colonisers of rocks in mountains are ....

(*ix*) Saxicolous lichens grow on ....

(*x*) A combination of an alga and a fungus in which the two components are so interwoven as to form a single plant is known as ....

(*xi*) .... constitute the phycobiont of a lichen.

(*xii*) The name of the lichen which produces antibiotics is ....

(*xiii*) Lichens are bioindicators of ....

(*xiv*) Lichens are the association of an .... and a ....

(*xv*) .... is a fertile layer of fruit body bearing asci along with paraphyses.

(*xvi*) Soredia are found in ....

(*xvii*) Litmus dye is obtained from Lichen ....

43. Select the correct answer :

(*i*) The fertile layer of a fruit body bearing asci along with paraphyses is known as
(*a*) Hymenium
(*b*) Sub-hymenium
(*c*) Medulla
(*d*) Cortex.

(*ii*) Lichen is the association of an alga and a
(*a*) fungus
(*b*) bacterium
(*c*) virus
(*d*) bryophyte.

(*iii*) The mode of life in lichen thallus is
(*a*) saprophytic
(*b*) parasitic
(*c*) symbiotic
(*d*) epiphytic.

(*iv*) In most of the lichens the fungal component is generally a member of class
(*a*) Phycomycetes
(*b*) Ascomycetes
(*c*) Basidiomycetes
(*d*) Deuteromycetes.

(*v*) When the fungal components are Basidiomycetes the lichen is known as
(*a*) Phycolichen
(*b*) Ascolichen
(*c*) Basidiolichen
(*d*) Deuterolichen.

(*vi*) Lichens are bioindicators of
(*a*) air pollution
(*b*) water pollution
(*c*) soil pollution
(*d*) Noise pollution.

(*vii*) The vegetative reproduction generally takes place by
(*a*) Oospores
(*b*) Ascospores
(*c*) Basidiospores
(*d*) Soredia.

(*viii*) Saxicolous lichens grow on
(*a*) Rocks (*b*) Soil
(*c*) Barks of trees (*d*) Water.

(*ix*) Lichens are composite organs consisting of algae and
(*a*) mosses (*b*) fungi
(*c*) protozoa (*d*) bacteria.

(*x*) Which of the following is of medicinal importance?
(*a*) *Lecanora*
(*b*) *Rocella*
(*c*) *Lobaria pulmonaria*
(*d*) *Cladonia rangifera.*

(*xi*) Foliose lichens are internally differentiated into
(*a*) Root, stem and leaves
(*b*) Epidermis, Cortex and Medulla
(*c*) Upper and lower cortex, algal zone andmedulla
(*d*) Epidermis ,cortex, algal zone and stele.

(*xii*) Lichens reproduce by
(*a*) Soredia (*b*) Conidia
(*c*) Conidia (*d*) Oidia.

(*xiii*) Which of the following is a crustose lichen ?
(*a*) *Usnia* (*b*) *Cladonia*
(*c*) *Eversia* (*d*) *Graphis.*

(*xiv*) In India
(*a*) lichens are common in
(*b*) Eastern Himalayas
(*c*) Western Himalayas
(*d*) South India
(*e*) Nilgiris Hills.

(*xv*) Female reproductive structure in many lichensis called as
(*a*) Carpogonium (*c*) Oogonium
(*b*) Ascogonium (*d*) Ascostomata.

SECTION IV

ELEMENTARY PLANT PATHOLOGY

Downy Mildew (*caused by Peronospora destructor*)

CHAPTER 21

PRINCIPLES OF PLANT PATHOLOGY

- Introduction
- Definition of Disease
- Classification of plant diseases
- Disease Pyramid
- Definitions and terms
- General symptoms
- Defence Mechanisms in Plants
- Disease Control
- Plant Quarantine
- IDM

Introduction

In the preceding pages of this book we have made the reader acquainted with the structure, life histories, modes of nutrition and food requirements of the non-vascular plants lacking chlorophyll. The following pages will deal briefly with the economic aspect *i.e.*, the diseases caused by these organisms in plants. The science which deals with the plant diseases is known as **Plant Pathology** or **Phytopathology**. It is one of the botany's youngest children. One who specializes in the study of plant diseases is called a **plant pathologist** or **phytopathologist**. He is chiefly concerned with the life histories of disease-causing organisms. His final goal is to prevent or control plant diseases.

Definition of disease

Any deviation from the healthy condition which interferes with the normal structure and performance of vital functions of a plant is called

a disease. The diseased plants can thus be spotted by changes in their structure or physiological processes. The deviations from the normal structure and function often develop to such an extent that signs of such effects become obvious. They are characteristic of a given disease. The external signs on the plant which indicate disease are called **symptoms**. The injurious effects of the disease may lower its value or even threaten its existence. The disease may effect the state of the whole plant. It is called a **systemic disease**. Some diseases, however, effect certain organs of the host or parts of those organs. These are called the **localised diseases**.

Classification of Plant Diseases

The plant diseases are classified differently by different scientists. The scheme followed in the present text is based on the nature of the causal agent, *living* or *non-living*. This scheme is considered more logical, convenient, simple and accurate. The only drawback of this system is that it simply indicates the taxonomic position of the pathogen causing the disease. It throws no light on the nature of the disease itself. On the basis of the nature of the causal agent, the plant diseases may be classified into two main categories as follows:

I. *Non-parasitic or abiotic diseases*. These are induced by unfavourable environmental conditions of soil or air such as mineral deficiencies or excesses in the soil, low or high temperatures, improper water, oxygen and light relations. The non-parasitic diseases may also be caused by mechanical injuries. Common examples of the non-parasitic diseases are blossom and rot of potatoes, black heart of potatoes, scald of apples, tip burn of lettuce and heat canker of flax.

II. *Parasitic diseases*. These are caused by the attack of some other living agents called the **causal agents**. The causal agent may be a plant or an animal or a virus. It is called the **parasite**. In the case of a disease caused by a parasitic organism the diseased plant is called a **host** and the

Scald of Apples (Abiotic Disease)

Alga Cephaleuros virescens

Peronospora destructor (Fungus)

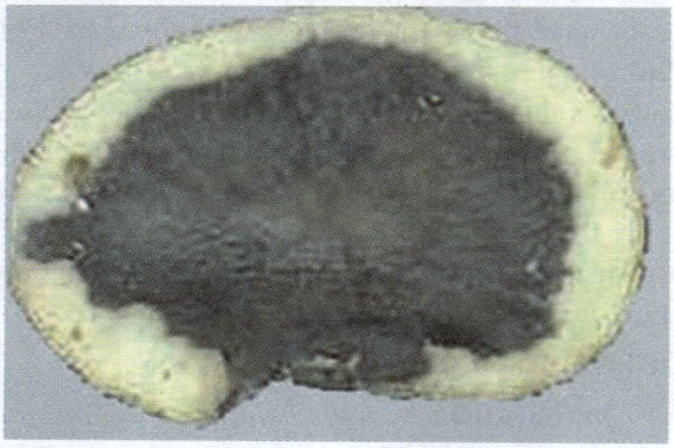

Black Heart of Potatoes

Tip Burn of Lettuce

PHOTOGRAPHS OF SOME PARASITIC AND NON PARASITIC DISEASES

causal agent is called pathogen. The parasite may subsist in whole or in part upon the living tissue of the host. The host is more appropriately called the **suscept**. It is the term applied to the plant which is prone to a given disease caused by a given causal agency. The parasitic diseases of plants may also be caused by the animal parasites such as the nematodes, molluscs (snails and snugs) and insects. Mostly the plant diseases are caused by the following pathogens:–

1. *Parasitic fungi*. They produce a great majority of the parasitic plant diseases called the **fungal diseases**. The fungal diseases are of first importance because they sometimes cause much damage and destruction.

2. *Parasitic bacteria*. They are also the causative agents of many plant diseases. The **bacterial diseases** of plants are second in importance to the fungal diseases.

3. *Parasitic seed plants*. There are a very few phanerogamic parasites which cause plant diseases. The most common parasitic forms are the mistletoes (*Viscum*), Dodders (*Cuscuta*) Cystanche, *Orobanche*, and a few others.

4. *Parasitic algae*. There are a very few parasitic algae that cause plant diseases. The most important ones are:–

(*i*) *Cephaleuros virescens* which causes red rust of tea.
(*ii*) *C. coffeae* is parasitic on coffee.
(*iii*) *C. parasitica* incites 'Orange rust of tea'.

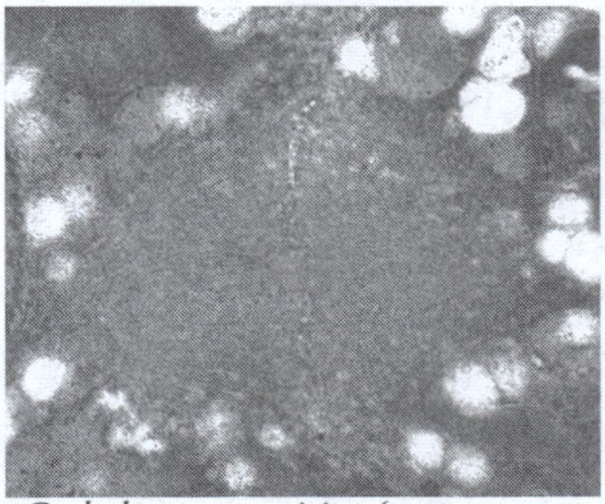

Cephaleuros parasitica (causes orange rust of tea)

Cephaleuros coffeae

Red Rust of Tea (caused by Cephaleuros virescens)

DISEASES CAUSED BY ALGAE

5. *Viruses*. The viruses are pathogens all of which are the most extreme parasites being almost completely dependent upon their host for their growth and replication. They cannot be seen with the light microscope and are best known from their effects on the host plants or animals. None of them has ever been cultivated outside the bodies of plants or animals. They cause serious plant and animal (including man) diseases. All diseases caused by the viruses are of **systemic type**. The common examples of virus diseases of plants are tomato and tobacco mosaic, potato and cucumber mosaic, curly top of sugar beet, spotted wilt of various economic plants such as pea, potato, tomato, tobacco, and spinach, and yellow diseases of peach and aster.

Curly Top of Sugarbeet

Cucumber Mosaic

Yellow Disease of Peach

VIRAL DISEASES

The parasitic diseases are often subdivided in relation to their occurrence into three categories, namely, endemic, epiphytotic (or epidemic) and sporadic.

(*a*) Endemic diseases. A disease which is regularly present (from year to year) in a certain region (country) or a part of a region (district) in a moderate to severe form is referred to as **endemic**. This is possible only when the causal agent is well established in that locality and can persist by virtue of its ability to survive from one crop season to the next in soil, crop debris or in wild hosts and the environmental conditions are favourable for its development.

(*b*) Epiphytotic (epidemic) diseases. Epiphytotic is a term applied to the infectious plant diseases which spread widely but occur periodically. The causal agent may be regularly present in the locality but the environment favourable for its rapid development occurs only periodically. The epiphytotic diseases are thus very responsive to variations in the environment.

(c) Sporadic diseases. These are plant diseases which occur only here or there at irregular intervals or occasionally.

On the basis of method of perpituation and mode of primary infection plant diseases are classified as :

1. *Soil Borne Diseases*: There are several plant pathogens whose infectious agents are present in soil and primary infection takes place through soil. They all attack roots of host plants. e.g. *Phytophthora, Fusarium, Pythium*, etc.
2. *Seed Borne Diseases*: Same pathogens survive as dormant mycelium in the seeds or other propagating materials of host plants. e.q. Smut (*Ustilago*)

   The mycelium grow become active at the time of seed germination and mycelium grows with the growing apex of the seedlings.
3. *Air Borne Diseases*: There are pathogens which infect host plant through air and bring primary as well as secondary infection. The spores of such pathogens are disseminated by wind and infect the host plant. e.g., *Ustilago*, *Pucuinia*, *Rhizopus*, *Colletotrichum*, *Erysiphe* etc.

There are few pathogens which are soil borne as well as air borne e.g. *Phytophthora* sps.

Definitions and Terms. There are certain terms which will be used repeatedly in the following chapters. It will be advisable to explain them at the outset:

(*i*) **Incitant.** The organism which causes the disease is termed **incitant** or **causal agent or organism.**

(*ii*) **Parasite.** It is an organism which lives upon living organic matter such as tissues of plants and animals. A parasite may not always cause a disease.

(*iii*) **Pathogen.** It is a living entity which causes disease. The student is advised to note the difference between a pathogen and a parasite.

(*iv*) **Inoculum.** The disease-causing organism usually survives from one growing season to the other (over-winters or over-summers) in some form such as spores, sclerotia, dormant mycelium or in various other forms in or on the host seed, wild herbs growing in the vicinity, and remains of organic matter of diseased plants of the previous season. The form of the infective agent whether a dormant spore, a sclerotium or a dormant mycelium constitutes the **primary inoculum** and source becomes the **source of primary inoculum**. On the host the causal organism may produce a fresh crop of infective bodies which serve as **secondary inoculum** and bring about secondary infection.

(*v*) **Penetration.** Prior to infection the causal organism lands on the host, resumes activity or germinates and gains entry into it directly or indirectly. This invasion of the host tissue by the living microorganism is called **penetration**.

(*vi*) **Infection.** Following penetration, the causal organism grows vigorously in the host tissues and establishes itself firmly. The establishment of the causal organism in the body tissue after penetration is termed **infection**.

(*vii*) **Incubation Period.** The interval or intervening period between infection of the host by an infective agent and the appearance of first symptoms of the disease is referred to as the **incubation period** of disease. It may be long or follows closely upon infection.

(*viii*) **Disease cycle.** A series of events between infection and complete development of a given disease constitutes its **disease cycle**.

DISEASE TRIANGLE DISEASE PYRAMID

A plant is said to be diseased one only if it is attacked by a living or non living entity. During pathogenesis environment plays an important role in the development of disease. Environment affects both the host and the pathogen simultaneously thereby affecting the host pathogen interaction as well. Earliest plant pathologists must have thought about this. For example Duggar wrote in 1909: "... the abundance of a very large number of fungous [sic] diseases is directly connected with or conditioned by climatological factors. Factors may affect independently host and parasite, and they may affect the interrelations of these organisms." Gaumann, who analyzed *examples of* crucial environmental, host, and pathogen determinants and their effects on disease development while discussing three way treatments of host pathogen interaction Perhaps these findings led Stevens in 1960 to draw a disease triangle, representing the three component: host, pathogen and environment.

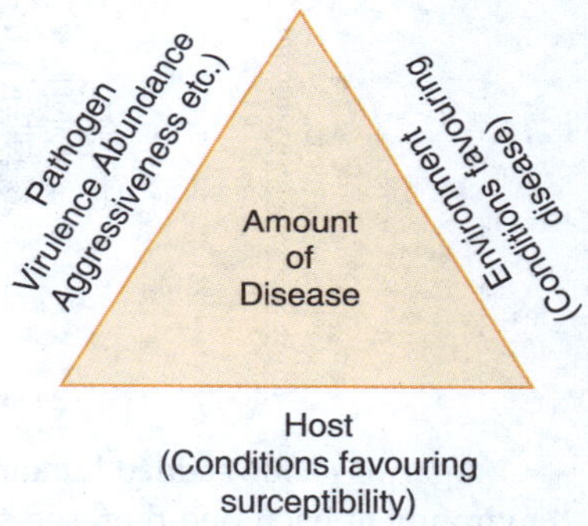

Fig. 21.1 The Disease Triangle

In a qualitative sense, the disease triangle concisely illustrates the phenomenon of plant disease as occupying the interior space of a triangle with the thee essential factors at the vertices (Figure 21.1) Alternatively, the three factors may be associated with the line segments (i.e., triangle sides); then, line length and interior volume can show variation in the strength of the relationship in a quantitative sense (Figure 21.2)

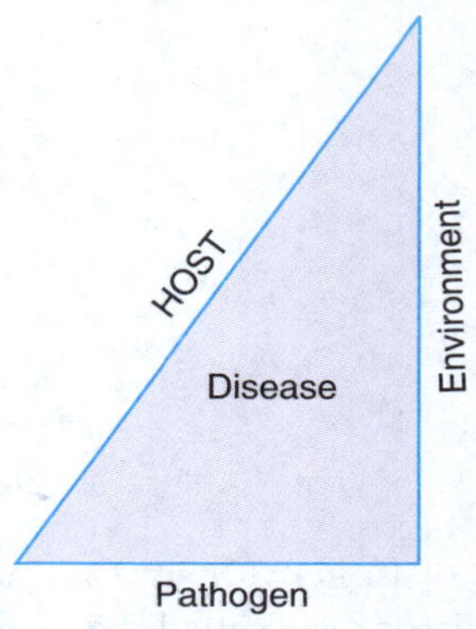

Fig 21.2 A Disease Triangle Showing an unequal relationship between Host, Pathogen and Environment

For example, a host with some degree of resistance, but not immune, will result in an overall lower level of disease. Used in this sense, the disease triangle illustrates the continuum of host reaction from complete susceptibility to immunity. So too, the degree of pathogen virulence and environmental conduciveness may be conveyed equally well. If any one element is reduced to a null variable. The geometric figure transforms in to a line and the area occupied by disease collapses to zero. Aside from this null case, the alternative quantitative representation treats disease as a degree of intensity (i.e., incidence or severity) rather than as a phenomenon (Franci, 2001).

Some plant pathologists have elaborated on the disease triangle by adding one or more parameters (1). Suggested additional parameters have included humans, vectors and time. Of these, only time is absolutely required so other elements represent special case applications. Indeed, the amount of each of three components of plant disease and their interaction in the development of disease are affected by fourth component; time. The interaction of fourth components can be visualized only in a tetrahedron: a three-dimensional disease pyramid (Figure 21.3) which has been the most common figure drawn after addition of a single parameter.

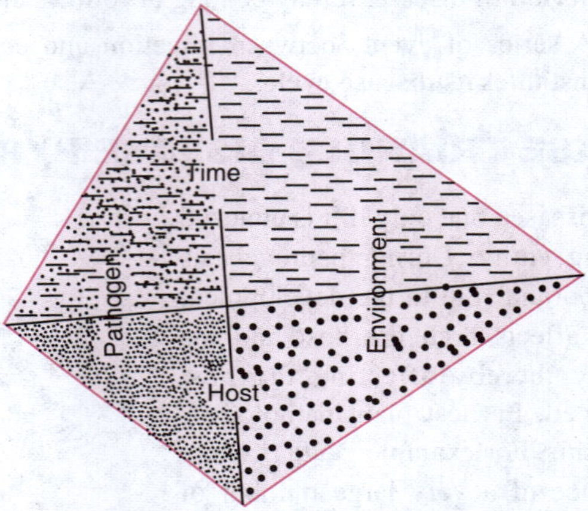

Fig. 21.3 Disease Pyramid with human as Fifth Factor.

Agarios (2006) added human a fifth factor at the peak of disease triangle. Host, pathogen and environment each one represents three sides of triangle and time is represented as perpendicular line arising from centre of the triangle (Fig. 21.4). This model shows that humans interact and influence each component of the disease pyramid. Sometimes even humans themselves can be affected to a greater or lesser extent by plant disease epidemics.

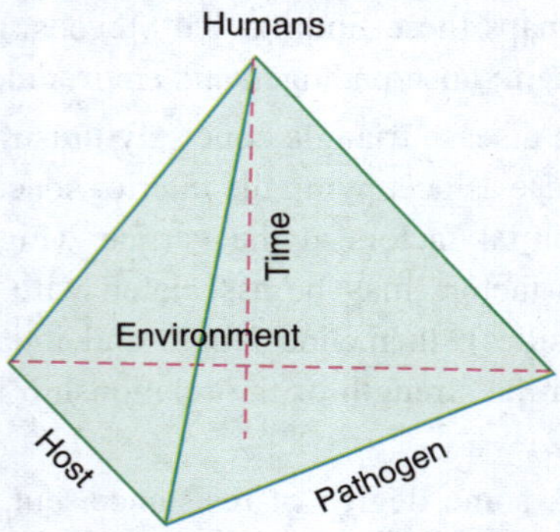

Fig. 21.4 The disease tetrahedron.

General Symptoms of Diseases. The changes in the host plant which serve to recognise the disease are called the **signs** and **symptoms** of the disease. The **sign** of a disease is the external appearance of some portion of the pathogen of the host. It may be some portion of the mycelium of the parasite or some spore stage. The best examples are the rusty uredosori and black teleutosori of the *Stem Rust* of wheat, the smut spore stage of *Ustilago* and white blisters of white rusts (*Albugo*). These stages of the pathogens are at first covered but become exposed as the spores mature.

Stem Rust of Wheat

White Rusts

FUNGAL DISEASES

The **symptoms** are the visible effects which the parasite induces on the host plant. Any visible deviation on the host plant (both physiological and morphological) from the normal in structure and function is called a **symptom**. Generally the symptoms are growth responses. These are induced by the causal agent operating on the host. They furnish clues to find out the nature of the disease. Symptoms may affect the whole plant or be restricted to a particular organ or parts of an organ. They are, as a matter of fact, the danger signals of a disease. The modern plant pathologists do not make any distinction between the two terms (signs and symptoms). They club them together as symptoms of a disease and define the term as external signs on the host plant which are characteristic of a given disease. These are the result of interaction between the host and the pathogen. They divide the symptoms of a disease into two categories:–

Masked symptoms and symptomless carrier

There are many viruses which infect certain hosts without producing visible symptoms on them. Such viruses are called **latent viruses** and host is considered **symptomless carrier.** In many cases plants usually develop symptoms following viral infection but become temporarily symptomless under certain environmental conditions (eg: high or low temperature and such symptoms are called masked **symptom.**

A. *Symptoms due to the external appearance of the pathogen or some portion of the pathogen on the host* (signs):–

The somatic phase in most of the pathogens is usually invisible. It lies within the tissues of the host plant. During the disease cycle some portion of the parasite such as the reproductive or resting structures become visible by rupturing the overlying tissues of the host. Some pathogens, however, are ectoparasites. In their case both the somatic and reproductive structures are visible externally and furnish a clue to the diagnosis of the disease. Common examples of such symptoms (signs) are:–

1. Mildews. These are a group of important fungal diseases of seed plants in which the parasite is seen as a superficial growth on the host surface (leaves, green stems and fruits) in the form of patches of varying sizes and colours. The mildews are of two kinds, **downy mildews** and **powdery mildews**. The downy mildews are all internal obligate parasites. They are characterised by superficial downy growth consisting of conidiophores and conidia on the host lesions in damp and warm weather. The powdery mildews are external parasites in which the mycelium forms whitish patches on the surface of the leaves of the host plant. The patches appear dusty or powdery with the formation of numerous white conidia which form a coating on the host surface.

Downy Mildews

Powdery Mildew

FUNGAL DISEASES

2. Smuts. These are the fungal diseases of cereals and other members of grass family that cause the ears (particularly the ovaries) to turn black or sooty. The smut symptoms may as well appear on other parts (leaves, stem or roots) of the host plant.

3. Rusts. These are fungal diseases of grasses and other plants which appear on the host surface as small, coloured pustules–red, brown, yellow, orange, or black in colour.

4. Sclerotia. The conspicuous phase of some fungal diseases such as Ergot of rye is the formation of sclerotia in the position of kernels in the spike. The sclerotium is a compact, hard mass of dormant fungal hyphae. It may be black, greyish violet, dark brown or purplish in colour.

5. White blisters. These are white, shining, blister-like pustules found on the leaves of cruciferous plants caused by *Albugo candida*, the white rust. These pustules expose the white, powdery masses of spores.

Corn Smuts

Rusts

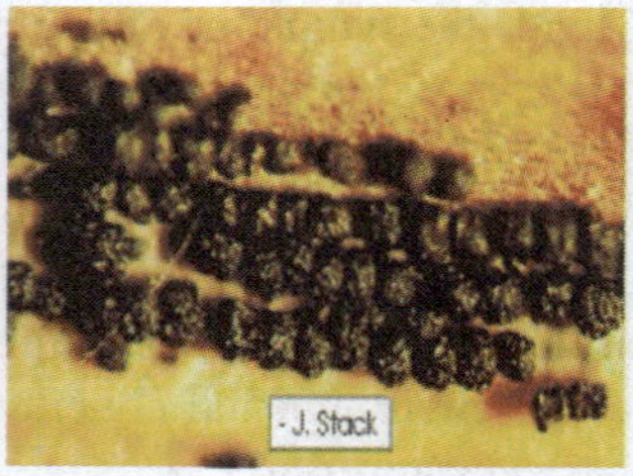

Sclerotia

White Blisters

VARIOUS SYMPTOMS

B. *Symptoms which are the visible effects induced by the parasite on the host plant:–*

These are grouped under three categories **necrosis, hypoplasia** and **hypertrophy**. The other important symptoms are wilts, and damping off.

1. Necrosis. Death or killing of the host tissues induced by the attack of a pathogen is called **necrosis**. It may be caused by rots, blights, wilts, die back and cankers and may be general or local, gradual or rapid. When necrosis is general it is called **decay** and **rotting**. The causal organism may be a fungus or a bacterium. Rotting is universal in plants. It takes a heavy toll of useful plants. The rots are of many kinds. The common ones are dry rots, wet rots, soft rots, hard rots, brown rots, and black rots. These rots cause general necrosis and progress slowly or rapidly.

(a) Lesions. In many cases necrosis is limited in extent. It is confined to small areas and is called **local necrosis**. It results in localised spots of diseased tissues. These are called the **lesions**. The tissue in the **spot** or the lesion area is dead. Examples of local necrosis are leaf and fruit spots. They are local lesions caused by fungi, bacteria and viruses. The **spots** are minute, circular or sub-circular and sometimes angular in outline. They are of different colours such as brown, dark, orange, red or white. The colour of the spot depends upon the nature of fungal spores present in the area. The leaf spot causes reduction in leaf surface and sometimes defoliation. The dead tissue of the leaf spot may fall out leaving circular or irregular perforations. These are called **shot holes**. The leaf shot holes are sharply bordered.

Necrosis

Lesions

(b) Cankers. Local necrosis also results in open wounds often sunken in stems and surrounded by living tissue. These are called the **cankers**. Canker is a dead lesion. Sometimes it is surrounded by a raised margin. Cankers may be due to slow rots of the outer parts of herbaceous and woody stems. They are thus usually limited in size. They may also be restricted in extent due to the formation of cork around and edges of the wound in the woody plants. When the canker partly encircles the branch its growth is subnormal. The branch is underdeveloped. If the canker completely encircles the branch the part of the branch beyond the canker dies.

(c) Blight. In many cases the leaves, stems or twigs, in response to the attack of the pathogen undergo rapid discolouration followed by death. The dead parts become dark or brown in colour. This condition of the affected organ is called the **blight**.

(d) Rot. It is a disease in which the affected tissues die and undergo decay. The rot diseases may be classified on the basis of the organ attacked such as leaf, stem or root rots, bud rot and fruit rot. On the basis of type of dissolution brought about by the causal agent, the rots may be classified as soft rot, dry rot, black rot and wet rot.

Cankers

Blight

Soft Rot

DIFFERENT SYMPTOMS

2. Hypoplasia. It is a subnormal cell production in response to the attack of the pathogen. It results in the subnormal growth of the parts attacked resulting in stunted growth and dwarfing of the host plant.

3. Hypertrophy (overgrowths). It is an abnormal increase in size of one or more organs of a plant in response to the attack of a pathogen. It results in abnormal growth causing *distortions, swellings, leaf curls* and *galls*. The excessive growth may be due to two processes, **hyperplasia** and **hypertrophy**. The former consists in rapid cell division and thus increase in the number of cells of which the organ is composed. Hypertrophy leads to abnormal increase in the size of cells only. The attack of Corn smut stimulates abnormal and rapid cell division (hyperplasia) and

Hypertrophy *Leaf Curl* *Witches Broom*

DIFFERENT SYMPTOMS

enlargement of individual cells (hypertrophy) of the infected organs resulting in the formation of overgrowths called the **corn galls** or **tumours**. *Witches broom* is another result of hypertrophy. It is a compact cluster of fine slender branches generally developing from an enlarged axis and the whole looking like a broom (Witches broom of Cherry). The hypertrophied parts become inedible.

Symptoms of damping off

4. Wilts. The drooping of the entire plant due to loss of turgidity is a common symptom of disease. The causes may be varied. The wilting due to a disease is permanent and eventually leads to death of the plant.

5. Damping off. Species of *Pythium* and *Rhizoctonia* are most important among the fungi which cause damping off disease of seedlings. Either the stem is attacked near the soil level or the crown of roots. The attacked region becomes weak and thus is not able to bear the load of the upper portion of the seedling. Consequently the seedling collapses. It topples down and dies. The seedling of many plants such as chilli, tobacco, tomato and mustard are prone to the damping-off disease.

DEFENCE MECHANISM IN PLANTS

It is natural for every living organism to defend the attack of the pathogen causing disease. Plants are no exception and the plants too have several mechanisms to defend themselves from the attack. These develop several features for this purpose. The defence may be structural or biochemical. In the structural defence, the plants develop or have features which makes them to withstand the attack. In the biochemical defence the plant releases chemical or chemicals which inhibits the attack. In certain plants, certain phytochemical compounds are formed in response to the attack that checks the further loss of host. Both the structural and biochemical defence mechanisms will be discussed briefly.

I. Structural defence : In plants some structures are already present to defend the attack while in others, the structures to defend the host develops after the infection. In this way, structural defence can be characterised as (A) Preexisting defence structures and (B) Defence structures developed after the attack of the pathogen.

(A) Preexisting defence structures

(*i*) *Cuticular Wax*. Wax-mixtures of long chain aliphatic compounds get deposited on the cuticular surface of some plants. Deposition of wax on the cuticular surface is thought to play a defensive role by forming a hydrophobic surface where water is repelled. As a result, the pathogen does not get sufficient water to germinate or multiply. In addition, a negative charge usually develops on the leaf surface due to the presence of fatty acids – the main component of cuticle. The negative charge prevents/reduces the chance of infection by many pathogens.

(*ii*) *Cuticle Thickness*. The thickness of cuticle is most important for those which try to enter the host through the leaf surface. The cuticle thickness obstructs the path of pathogen. In addition, a thick cuticle checks the exit of the pathogen from inside the host, thus reducing the secondary infection.

(*iii*) *Structure of Epidermal cell walls*. Tough and thick outerwalls of epidermal cells may directly prevent the entry of the pathogen completely or make the entry difficult. The presence or absence of lignin and silicic acid in the cell walls may show variation in resistance to penetration of the pathogen. Most outer walls of epidermal cells of rice plants are lignified and are seldom penetrated by blast disease of rice pathogen. In resistant varieties of potato tubers (resistant to *Pythium debaryanum*) the epidermal cells contain higher fibre content than the susceptible ones.

(*iv*) *Structure of Natural openings*. Structure of natural openings like stomata lenticels etc. also decide the fate of the entry of the pathogen. In *Szincum* variety of citrus, the stomata are small and possess very narrow openings surrounded by broad lipped raised structures which prevent entry of water drops containing citrus canker bacterium.

In the same way, the size and internal structures of lenticels may play a defensive role against the pathogens. Varieties having small lenticels in the apple fruits prevent the entry of the pathogen while those having large openings easily allow the pathogen to enter. Nectaries provide openings in the epidermis and may play a defensive role due to high osmotic concentration of the nectar. In resistant varieties of apple, presence of abundant hairs in the nectaries acts as a defence mechanism while susceptible varieties are devoid of abundant hairs.

Internal defence structures. There are many preexisting internal defence structures inside the plant that prevent the entry of pathogen beyond these structures. In some plants, cell walls of certain tissues become thick and tough due to environmental conditions and this makes the advance of the pathogen quite difficult. In case of stems of cereal crops, vascular bundles or extended areas of sclerenchyma cells checks the progress of rust pathogen. Leaf veins effectively obstruct the spread of pathogen like the angular leaf spot pathogen.

(B) Defence structures developed after the attack of the pathogen

After the pathogen has successfully managed to overcome the preexisting defence mechanisms of the host, it invades the cells and tissues of the host. In order to check the further invasion by the pathogen, the host plants develop some structures/mechanisms which may be defence reactions in the cytoplasm, cell wall defence structures, defence structures developed by the tissues and ultimately the death of the invaded cell i.e. necrosis. These will be briefly discussed here.

(*i*) *Defence Reactions in the Cytoplasm.* The cytoplasm of the invaded cell surrounds the hyphae of the pathogen and the nucleus of the host cell gets stretched to break into two. In some host cells, the cytoplasm and the nucleus of the infected cells enlarge. The cytoplasm becomes granular and dense and develops granular particles. These result in the disintegration of the pathogen mycelium and thus the invasion stops. Such cytoplasmic defence mechanisms can be seen in weak pathogens like *Armillaria* and some mycorrhizal fungi.

(*ii*) *Cell wall defence structures*. Cell wall defence structures are of limited help to the host. These include morphological changes in the cell wall of the host. Three types of cell wall defence structures are generally observed : (*i*) cell walls thicken in response to the pathogen by producing a cellulose material, thus preventing the entry of the pathogen (*ii*) the outer layer of cell walls of the parenchyma cells in contact with invading bacterial cells produce an amorphous fibrillar material that traps the bacteria thus preventing them to multiply and (*iii*) Callose papillae get deposited on the inner layers of the cell walls due to invasion by fungal pathogens. In raw cases, the hyphal tips of the infecting fungal pathogen penetrating the cell wall and thereafter growing into the cell lumen get enveloped by callose material that, later become infused with phenolics forming a sheath around the hyphae.

(*iii*) *Defence structures developed by the tissues*. The following four developments take place in the tissues after penetration:

(*a*) *Gum deposition*. Plants produce a variety of gummy substances around lesions or spots as a result of infection. These gummy substances inhibit the progress of the pathogen. The gummy substances are commonly produced in stone fruits. (Fig. 21.5A)

(*b*) *Abcission layers*. Abscission layers are usually formed to separate the ripe fruits and old leaves from the plant. But in some stone fruit trees, these layers develop in their young leaves in response to infection by several fungi, bacteria or viruses. An abscission layer is a gap formed between two circular layers of cells surrounding the point of infection. This gap is created by the dissolution of one or two layers of the middle lamella, one or two layers of cells surrounding the infected loci resulting in the infected locus becoming unsupported, shrivels, dies and falls down along with the pathogen. Abscission layer formation protects the healthy leaf tissue from the attack of the pathogen. (Fig. 21.5B)

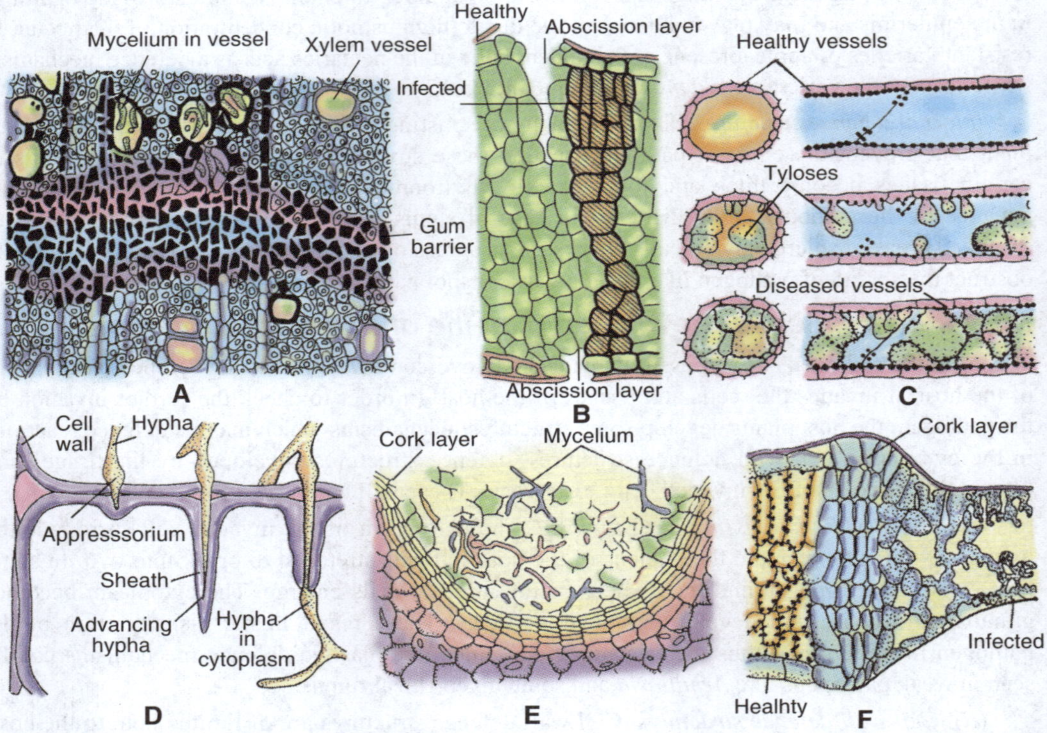

Fig. 21.5. (A-F). Different defence structures developed after infection : (A) Gum deposition in cells ; (B) Abscission layer ; (C) Tyloses ; (D) Formation of sheath around developing hyphae (E, F)./ Development of cork layer in tuber (E) and leaf (F).

(C) *Tyloses*. Tyloses are out growths of protoplasts of adjacent live parenchyma cells protruding into xylem vessels through pits under stress or in response to attack by the vascular pathogens. Their development blocks the Xylem vessels, obstructing the flow of water and resulting in the development of wilt symptoms. However, tyloses are formed in some resistant plants ahead of infection and the prevent the plant from being attacked. (Fig. 21.5C)

(D) *Formation of layers*. Some pathogens like certain bacteria, some fungi and even some viruses and nematodes stimulate the host to form multilayered cork cells in response to infection. These develop as a result of stimulation of host cells by substances secreted by the, pathogen. These layers inhibit the further invasion by the pathogen and also block the flow of toxic substances secreted by the pathogen. Cork layers also stop the flow of nutrients of the host thus also depriving the pathogen of the nutrients. Examples of cork layer formation as a result of infection are : soft rot of potato caused by *Rhizopus* sp., potato tuber disease caused by *Rhizoctonia* sp., Scab of potato caused by *Streptomyces scabies* and necrotic lesions on tobacco caused by tobacco mosaic virus. (Fig. 21.5 D-F)

IV. *Necrosis or Hypersensitive type of defence*. Necrosis or hypersensitive type of defence is another defence mechanism adopted by some pathogens like *Synchytrium endobioticum* causing wart disease of potato, *Phytophthora infestans* causing late blight disease of potato and *Pyricularia oryzae* causing blast of rice etc. In such diseases, the host nucleus moves toward the pathogen when the latter comes in contact with the protoplasm of the host. The nucleus soon disintegrates into brown granules which first accumulate around the pathogen, later dispersing throughout the host cytoplasm. Soon the cell membrane swells and finally the cell bursts and dies. These cause the pathogen nucleus to disintegrate into a homogenous mass and its cytoplasm dense. As a result, the pathogen fails to grow beyond the necrotic or dead cells and the further growth of the pathogen is stopped.

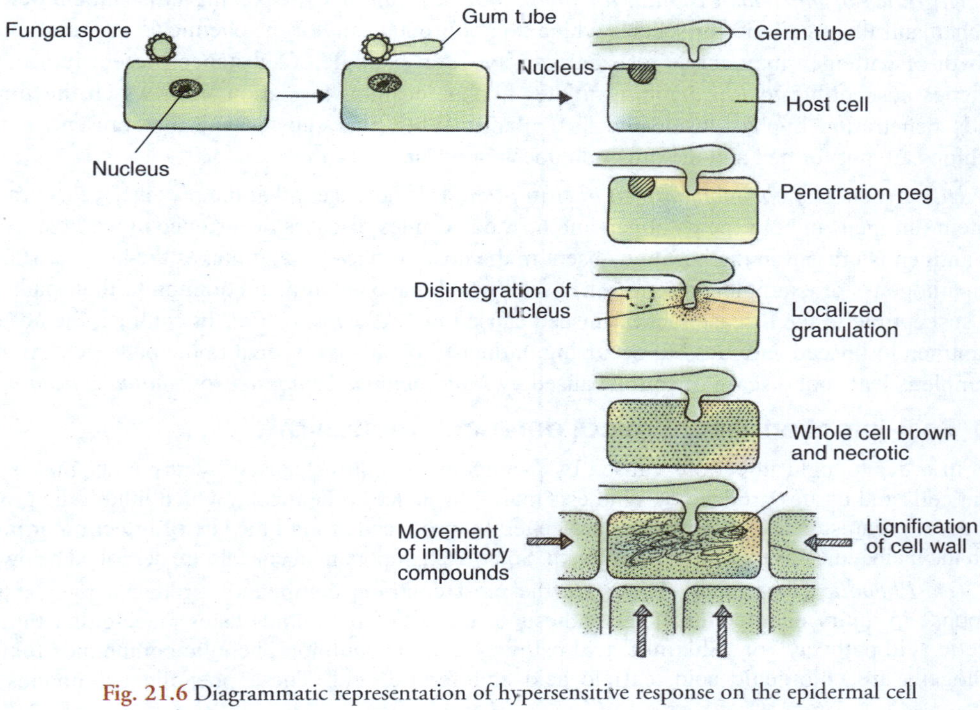

Fig. 21.6 Diagrammatic representation of hypersensitive response on the epidermal cell

II *Biochemical Defence.* Although structural defence mechanisms do prevent the attack of the pathogen, the defence mechanism also includes the chemical substances produced in the plant cells before or after the infection. It has now been established that biochemical defence mechanisms play more important role than the structural defence mechanisms. This has been supplemented by the fact that many pathogens entering non host plants naturally or artificially inoculated fail to cause infections in absence of any structural barriers. This does suggest that chemical defence mechanisms rather than structural mechanisms are responsible for resistance in plants against certain pathogens.

(A) Pre-existing Biochemical defence

(*i*) *Inhibitors released in the prepenetration stage* . Plant generally exudes organic substance through above ground parts (phyllosphere) and roots (rhizosphere). Some of the compounds released by some plants are known to have an inhibitory effect on certain pathogens during the prepenetration stage. For example fungistatic chemicals released by tomato and sugar beet prevent the germination of *Botrytis* and *Cercospora*. Presence of phenolics like protocatechuic acid and catechol in scales of red onion variety inhibit the germination of conidia of *Colletotrichum circinans* on the surface of red onion.

Inhibitors present in high concentrations in the plant cells also play an important role in defence of plants. Presence of several phenolics, tannins and some fatty acid like compounds such as dienes in cells of young fruits, leaves or seeds afford them resistance to *Botrytis*. The tubers of resistance vars of potato against potato scab disease contain higher concentrations of chlorogenic acid around the lenticels and tubers than the susceptible vars. Several other compounds like saponin, tomatin in tomato and avinacin in oats have antifungal activity. Some enzymes like glucanases and chitinases present in cells of some plants may break down the cell wall components of pathogens.

(*ii*) *Lack of nutrients essential for the pathogen* is another preexisting biochemical defence mechanism. Plant varieties or species which do not produce any of the chemicals essential for the growth of pathogen may act as resistant variety. For example, a substance present in seedling varieties susceptible to *Rhizoctonia* initiates hyphae cushion formation from which the fungus sends penetration hyphae inside the host plants. When this substance is not present, hyphal cushions are not formed and the infection does not occur.

(*iii*) *Absence of common antigen in Host plant*. It is now clear that the presence of a common protein (antigen) in both the pathogen and host determines diseases occurrence in the host. But if the antigen is present in the host and absent in the host or vice-versa, it makes the host resistant to the pathogen. For example, varieties of linseed which have an antigen common to their pathogen are susceptible to the disease rust of linseed caused by *Melampsora lini*. In contrast, the absence of antigen in linseed varieties but occurring in the pathogen are resistant to the pathogen. Another example is leaf spot disease of cotton caused by *Xanthomonas campestris* pv. *malvacearum*.

(B) Post-infection-Biochemical defence mechanism

In order to sight infections caused by pathogens or injuries caused by any other means, the plant cells and tissues produce by synthesis many substances (chemicals) which inhibit the growth of causal organism. These substances are generally produced around the site of infection or injury with the main aim at overcoming the problem. Some such important chemicals are described below:

(*i*) *Phenolic compounds*. These are the most common compounds produced by plants in response to injury or infection. The synthesis of phenolic compounds takes place either through "acetic acid pathway" or "Shikimic acid pathway". Some common phenolic compounds toxic to pathogens are chlorgenic acid, caffeic acid and ferulic acid. These phenolic compounds are produced at a much faster rate in resistant varieties than in susceptible varieties. Probably that the combined effect of all phenolics present is responsible for inhibiting the growth of the infection.

(*ii*) *Phytoalexins*. Phytoalexins are toxic antimicrobial substances synthesized '*de novo*' in the plants in response to injury, infectious agents or their products and physiological stimuli. The term phytoalexin was first used by the two phytopathologists Muller and Borger (1940) for fungi static compounds produced by plants in response to mechanical or chemical injury or infection. All phytoalexins are lipophilic compounds and were first detected after a study of late blight of potato caused by *Phytophthora infestans*. Phytoalexins are believed to be synthesized in living cells but surprisingly necrosis follows very quickly. According to Bill (1981), peak concentration of phytoalexins almost always coincides with necrosis Although the exact mechanism of production of phytoalexin has not been properly understood, it is considered that a metabolite of the host plant interacts with specific receptor on the pathogen's membrane resulting in the secretion of "phytoalexin elicitor" which enters the host plant cells and stimulates the phytoalexin synthesis. Phytoalexins are considered to stop the growth of pathogens by altering the plasma membrane and inhibiting the oxidative phosphorylation.

Phytoalexins have been identified in a wide variety of species of plants such as Soyabean, Potato, sweet potato, barley, carrot, cotton etc. are being investigated. Some common phytoalexins are Ipomeamarone, Orchinol, Pistatin, Phaseolin, Medicarpin, Rishitin, Isocoumarin, 'Gossypol' Cicerin, Glyceolin, Capisidiol etc.

The following Table gives a list of phytoalexins, chemical nature the host and the pathogens in response to which these are produced :

Phytoalexin	*Host*	*Pathogen*	*Chemical Nature of the phytoalexin*
1. Ipomeamarone	*Ipomoea batata*	*Ceratocystis fimbriata*	Furanosesquiterepene ketone.
2. Orchinol	*Orchid malitaris*	*Rhizoctonia repens*	Phenanthrene
3. Pistatin	*Pisum sativum*	*Ascochyta pisi*	Kievitone
4. Phaseolin	*Phaseolus* sp.	*Monilia fruticola*	
5. Medicaripin	*Medicago lupulina*	*Colletotrichum phomoides*	Dimethyl-homoptero-carpin
6. Rishitin	*Solanum tuberosum* var. *rishiri*	*Phytophthora infestans*	Nonsequiter-pene
7. Isocoumarin	*Daucus carrota*	*Fusarium oxysporum f. lycopersici* *Rhizopus solonifer*	–
8. *Cicerin*	*Cicer arietinum*	*Ascochyta rabie*	
9. Gossypol	*Gossypium* sp.	*Verticillium alboatrum*	
10. Capisidiol	*Capsicum fruitescens*	*Peronospora trifolium*	
11. Wyerone acid	Broad bean	*Botrytis fabae*	

(*iii*) *Substances produced in Host to Resist enzymes produced by pathogen.*

Some hosts produce chemicals which neutralise the enzymes produced by pathogen, thus defending the host. Therefore these substances help plants to defend themselves from the attack of the pathogen. In bean plants, infection with *Rhizoctonia solani* causes necrosis. In resistant bean varieties, the entry of pathogen causes the separation of methyl group from methylated pectic substances and forms polyvalent cations of pectic salts which contain calcium. The calcium ions accumulate in infected as well as neighbouring healthy tissues and because of the calcium accumulation, the pathogen fails to disintegrate middle lamella by its polygalacturonase enzymes.

These are known to dissolve the middle lamella of healthy tissue in susceptible varieties.

(*iv*) *Detoxification of pathogen Toxins and Enzymes*. In some cases, the plants produce chemicals which deactivate the toxins produced by the pathogens. For example, *Pyricularia oryzae* which causes blast disease of rice produces Picolinic acid and pyricularin as toxins. Although resistant varieties convert these toxins into N-methyl picolininic acid pyrecularin into other compounds, the susceptible varieties do get affected by these toxins. Similarly in case of cotton and tomato wilts, the toxin fusaric acid produced by the pathogen gets converted into non-toxic N-methyl-fusaric acid amide in resistant varieties.

As in case of detoxification of toxins, the toxic enzymes produced by the pathogen is deactivated by phenolic compounds or their oxidation products. Some varieties of cider apple are resistant to brown not disease caused by *Sclereotinia fructigena*. It may be because of the resistant varieties producing pheolic oxidation products which inactivate the pectinolytic enzymes produced by the pathogen.

(*v*) *Biochemical Alterations*. It has been observed that infection of the host by the pathogen brings about biochemical changes in the host which may prove toxic to the pathogenic microorganisms and cause resistance to the pathogen. Production of certain new enzymes and other compounds are synthesized and accumulated in higher concentration. This may also add to the resistance of the plant by being toxic to pathogenic microorganisms.

DISEASE CONTROL

There are various methods for the control of plant diseases. The chief aim of all these is at first, to check economic loss and secondly to increase the value of crop. Disease control follows two main principles, **prevention** and **cure**.

Prevention or Prophylaxis. It is the protection of the host from infection and exposure to pathogen. Prevention includes a wide variety of control measures. Thses are:–

1. *Immunity*. It means the resistance of the host to infection and disease development. It is most effective and promising method of protecting a crop against disease. It consists in developing and growing disease resistant varieties of crop plants. It is an ideal, the cheapest and the best method of controlling all plant diseases. In nature it is a very slow and gradual process. Plant breeder can, however, accelerate the process by continued selection and hybridisation. Some success in this direction has already been achieved. Several disease resistant varieties of crop plants have been produced. Much more, however, has still to be done. Breeding for disease resistance, however, is a very complicated and arduous task. It is made still more difficult by the occurrence of new physiologic races of pathogens from time to time.

2. *Protection of non-immune varieties*. In majority of the crop, plants immune varieties are either not available or have not been so far developed. In such cases the development of disease-free crops requires much attention and forethought. They are protected by more positive measures. All these methods involve eradication of the pathogen. It is carried out in a wide variety of ways. These methods include sanitation and eradication.

(*a*) *Sanitation*. Many plant pathogens over-winter or over-summer on the remnants of previous year's crop in the soil, wild perennial hosts or weeds and seeds. They serve as a source of primary inoculum. The inoculum is transmitted to the hosts by various agencies such as wind, water, insects, manure of livestock and agricultural implements. To reduce infection the following sanitation measures are put into practice:–

1. Destruction of crop debris by burning or burying.
2. Removal of plant parts from the field as they become diseased and their destruction.
3. Soil sterilisation by physical treatment such as flooding, application of heat or steam or hot water.

4. Chemical treatment of soils for controlling soil borne diseases is very effective.

5. Disinfection of agricultural implements and pruning tools by the use of fungicides.

(*b*) *Eradication of the Pathogen.* It is carried on in a wide variety of ways. The most important ones are:–

1. *Rotation of Crops.* The chief object of rotation of crops is to prevent the increase of inoculum in the soil and consequently to reduce losses from diseases. When the same crop is grown year after year in the same soil the fungus may increase enormously. Economic production is no longer possible in such soils unless disease resistant varieties are grown or rotation of crops is resorted to. The rotation of crops allows the disease producing organism in the soil to die before the suitable host is again sown there.

2. *Elimination of Weed hosts.* Many weeds particularly biennial and perennial ones are potential sources of infection. They harbour the pathogens which later infect the cultivated crops seeded in the same soil. Elimination of these wild and alternate hosts is a very effective measure of disease control.

3. *Prevention of the disease carried by the Plant Parts.* Normally the seeds, tubers, rhizomes, bulbs, etc., carry heavy loads of disease causing organisms or their spores. Under favourable conditions they cause disease. In such cases the disinfection of these infected plant organs by using sterilising agents, before they are sown, is quire effective. The choice of the agent to be employed depends upon whether the causal organism is borne on the surface or has invaded the deeper tissue.

Against surface borne causal organisms, surface disinfection of propagative plant parts by chemical disinfectants is very effective. It will destroy the pathogen. These external plant protectants may be used in the form of gases in fumigation or as solids in the form of dusts. These plant parts are also disinfected through the use of soaks, dips, washes, etc., in liquids containing toxic compounds. The chief character of these disinfectants is that they must be cheap, effective, easy to apply, persistent and safe. They should be toxic to the pathogen but not to the host. Chemical compounds in dust form are now considered much more popular as external protectants.

In case the disease is carried inside the propagative plant part the application of moist heat appears to be the only effective control.

In the case of crop plants growing in the field, external protection is provided for foliage and other growing parts by protective sprays and dusts of chemicals.

Cure. It involves the destruction of the pathogen after it has gained entry into the host. In fact it is the cure or control of disease in the growing crops. It is done by the application of strong chemicals in the form of sprays. These actually destroy the pathogen within the diseased tissues of the host. A thorough spray of lime sulphur, Bordeaux mixture and many inorganic mercury compounds will destroy the spores or the exposed fungus hyphae. The spray is delivered with considerable force and in the form of fine mist. This conventional method of plant disease control by fungicidal sprays or dusts is cheap and effective against a wide range of pathogens. However, it suffers from the following disadvantages:

Bordeaux Mixture (Spray destroys Fungal Hyphae)

(*i*) They are bulky and corrosive.

(*ii*) Bordeaux mixture is somewhat difficult to prepare.

(*iii*) Lime sulphur is unpleasant to use.

(*iv*) Spraying and dusting is too expensive against diseases of low cost crops.

Because of the limitations of the above-mentioned conventional method, pathologists were induced to make a search for new compounds having fungicidal potentialities. Consequently various chemicals, organic and inorganic and antibiotics have been tested. Some of them have shown encouraging results.

Antibiotics. The use of antibiotics is the most recent development in plant chemotherapy. It is receiving increasing attention. The antibiotics are most effective in the control of bacterial plant pathogens. However, their use as disease controlling agents in gereral is still in the experimental stage. Brown and Boyle (1945) were the first to use antibiotics for the control of bacterial diseases. They worked on the crown galls of various plants caused by *Bacterium tumefaciens*. For this they used filtrate of *Penicillium notatum*. The fluid was either injected into the galls or applied with cotton wool soaked in it. The result was encouraging. The expansion of galls was inhibited. Identical results were obtained by Hampton (1948) with penicillin and streptomycin. The mixture of streptomycin with terramycin or the former alone when sprayed on pear blossom suffering from fireblight gave good results (Ark and Alcern, 1956). Agrimycin which is a mixture of streptomycin and terramycin proved effective in obtaining an economic control of bacterial spot of tomatoes. The causative agent in this case is *Xanthomonas vasicatoria*.

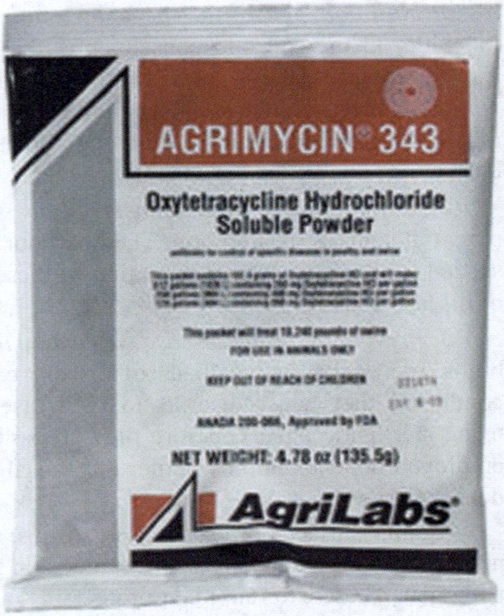

Agrimycin (controls bacterial spot of tomatoes)

The results of the above-mentioned experiments and others have shown that the use of antibiotics to combat plant diseases is of more than academic interest. Whether they can be used economically as plant protectants is difficult to say. An antibiotic which is antifungal has been discovered at the Indian Agricultural Research Institute. It is named as "Bulbiformin". Experiments are being carried out to test its efficacy in controlling disease. The antibiotics are considered to yield promising results as:–

1. Soil disinfectants for the control of pathogens in the soil.
2. Seed treatment both when the pathogen is surface borne and when it has invaded the deeper tissues.
3. External protectants against many diseases such as rots, mildews, leaf spots, etc.
4. Systemic plant protectants against pathogens causing diseases of vascular tissues.

PLANT QUARANTINE

It is a legal remedy to control the spread of disease from infected area to non-infected area. "as the restriction imposed by duly constitute authorities on the production, movement and existence of plants or plant materials, animals or animal products, or any other article or material or normal activity of persons." This is brought under regulation in order to see that the introduction or spread of a pest may be prevented or limited so that the pest already introduced may be controlled and also to avoid losses that would otherwise occur through the damage done by the pest or through the continuing cost of their control. When plant pathogens are introduced into an area in which they did not previously exist, they are likely to cause much more catastrophic epidemics than do the existing pathogens.

History

A quarantine law was enacted in Rouen, France in 1660, ordering the eradication of barberry plants from the vicinity of grain fields. The first British legislation against a disease in animal or plants was an Act of 1866 granting emergency powers for the destruction of all cattle affected by rinder pest which has been introduced into Britain by imported Russian cattle the previous year. In US, the first plant quarantine measure was initiated in 1891 by setting up a seaport inspection station at San Padro, California. In 1912, the US Congress enacted the Federal Plant Quarantine Act, which among other things, prohibits the entry of plants into the United States. It was phylloxera gall louse (*Phylloxera vastatrix*) that provided the initial stimulus for the establishments of effective international cooperation on plant disease legislation. Experienced quarantine inspectors stationed at all points of entry into the country enforce quarantine of produce likely to introduce new pathogens. Now quarantine regulations are enforced in many countries and through these regulations several plant disease have been successfully restricted to their respective countries.

Plant quarantine regulation in India

In India Plant Quarantine regulatory measures are operative through the *'Destructive Insects & Pests Act, 1914 (Act 2 of 1914*). Commonly known as DIP Act. The purpose and intent of this Act is to prevent the introduction of any insect, fungus or other pest, which is or may be destructive to crops. The import of agricultural commodities is presently regulated through the *plant Quarantine (Regulation of Import into India) Order, 2003* the provisions of New Policy on Seed Development, 1988. Further, the significance of Plant Quarantine has increased in view of Globalization and liberalization in International trade of plants and plant material in the wake of *Sanitary* and Phytosanitary (SPS) Agreement under WTO. The phytosanitary certification of agricultural commodities being exported is also undertaken as per International Plant Protection Convention (IPPC), 1951. In line with the New Policy on Seed Development, 1988 and the provisions of PQ Order, 2003, the specified planting material for propagation (viz., cuttings, saplings, bud woods, etc.) require growing under Post Entry Quarantine (PEQ) for a specified period. The import permit for such planting material is granted based on a certificate from Designate authorities.

There are 35 Plant quarantine Stations at different Airports, Seaports and Land frontiers implementing the Plant Quarantine regulations, The Directorate of Plant Protection and Plant Quarantine, Ministry of Agriculture, Government of India is overall in-charge of plant quarantine regulations. The National Plant Quarantine Station (NPQS), New Delhi and Regional Plant Quarantine Stations (RQQSs) at Chennai, Kolkata, Amritsar and Mumbai have been strengthened with modern equipment for plant quarantine testing. etc., to facilitate speedy clearance of imports and exports under the FAO-UNDP Project.

Plant quarantine regulations exist at two levels. The one at National level and the other at domestic level to prevent international and intra-national movement of infected press and diseases, respectively. Under domestic quarantine regulations transportation of potato tubers of Darjeeling district are completely banned to other parts of the country because Wart disease of Potato (caused by *Synchytrium endobioticum*) is very common in Darjeeling and is recurring there. Similarly, potato tubers are not allowed into India unless it is certified by Ministry of Agriculture of the exporting country that the tubers are free from wart disease and there was no wart in the vicinity of 8 kms of potato crops.

National Bureau of Plant Genetic Resources, New Delhi which is the nodal institution for exchange of plant genetic resources has been empowered under PQ order to handle quarantine processing of germplasm and transgenic planting material being imported for research purposes in the country. As per the recent amendments made under the PQ order, the Advanced Centre for Plant Virology at IARI, New Delhi, Indian Institute of Horticultural Research, Bangalore and Institute of Himalayan Bioresource Technology, Palampur have been identified for ensuring virus-free status in the imported *in vitro* material.

Under the DIP Act, there is a provision of Domestic Quarantine to restrict the inter-state movement of nine invasive pest viz flute scale, sanjose scale, coffee berry borer, codling moth, banana bunchy top and mosaic viruses, potato cyst nematode, potato wart and apple scab. Keeping in view the biosafety concerns associated with the growing of imported transgenic material, a containment facility has been established at NBPGR, New Delhi for their quarantine processing.

Seed inspection and certification

Seed certification or registry is a process of certifying seeds and other plant propagating materials which have the potential to carry and spread plant pathogens. It is a voluntary process where the farmers of the crop if want to get their seeds or plant propagating material certified approach the qurantine regulation agencies. The quarantine inspectors or plant Pathologist inspect the seed or plant propagating materials and if they find it free from the pathogen inoculums, the material can be certified or registered as free from plant pathogens. Usually certification is needed for commercial activity of plant seeds and plant propagating materials. Many viral, bacterial, fungal and nematode induced diseases of plant have been controlled through seed inspection and certification. The Plant Protection Advisor heads the seed *inspection and certification* process and is the one who issue the certificates and import permits.

INTEGRATED DISEASE / PEST MANAGEMENT

"a sustainable approach to managing diseases by combining biological, cultural, physical and chemical tools in a way that minimizes economic, health and environmental risk". This concept evolved from the original IPM definition after responding to today's call for ecologically based pest management (Overton, 1996).

In an effort to increase crop production and reduce crop losses from plant diseases indiscriminate use of chemicals and pesticides in agricultural systems dominated during the past few decades. Though, these efforts have significantly contributed in elevating food problem all over the world but simultaneously it has resulted in some environmental and health concerns. Chemical fertilizers at one side weaken the natural health of soil; the residual effect of pesticides, fungicides, bactericides, and nematicides etc. becomes fatal for many animals and birds even for human beings. In view of these finding concept of IDM evolves to reduce and rationalize the used chemicals and pesticides in such a way that not only take care of crop productivity but also deal with health concerns and environmental hazards. Concept of Integrated disease management (IDM) is "*A disease management system that, in the context of the associated environment and the population dynamics of the pest / pathogen species, utilises all suitable techniques and method in a manner as compatible as possible and maintains the pest/pathogen population at levels below those causing economic injury*" In the broad sense has been defined as "*the optimization of pest/pathogen control in an economically and ecologically sound manner, accomplished by the coordinated use of multiple tactics to assure stable crop production and to maintain pathogen pest damage below the economic injury level while minimizing hazards to humans, animals, plants and the environment*".

IDM is strategy is basically managing the three components of disease triangle simultaneously viz: management of crop, management of pathogen and management of environment. While manipulating these components considering the time frame of disease development is necessary. Effective disease management must be integrated with management of the whole farm (Kumar and Shivay, 2008). Basic strategies should be implemented regardless of whether or not a significant disease problem exists. These basic strategies include the following:

(*i*) conduct disease survey in each season

(*ii*) practice farm hygiene principles

(*iii*) use resistant varieties where available

(*iv*) provide a balanced crop nutrition

(*v*) manage crop residues to minimize carryover of pathogens into subsequent crops

(*vi*) develop a sound crop rotation strategy

(*vii*) use chemical and biological means to control diseases, if required

In short the integrated disease management practice involves more of preventive measures less stress on curative measures for plant pathogens. Right selection of seed, proper and healthy soil, healthy crop cultural practices, biological control, use of chemicals and pesticides as last resort only if it is very necessary, are some of the key points of IDM. IDM strategy can produce several desirable outocomes like: prevention or postponement of the introduction of a disease (s) that is not already present in the area, reduction of the rate of build-up of a disease that has been introduced to the new area elimination of a disease that may be affecting profitability or minimization of the impacts of a disease on a crop growth and/or productivity.

QUESTIONS

ESSAY TYPE

1. How will you define the term disease? Classify the diseases in plants.
2. What are the signs and symptoms of diseases? Give a brief account of the general symptoms of plant diseases. *(Meerut, 1997; Allahabad M.Sc., 1999)*
3. What are chief aims of the disease control measures? Give a brief account of the various control measures used to check the spread of a disease in plants.
4. What are antibiotics? What role they are expected to play in plant chemotherapy?
5. Define Disease. Describe the classification and symptoms of various diseases.
6. Write the prophylactic methods of disease control. *(Punjab, 1998; Himachal Pradesh, 1999)*
7. Describe in detail the various symptoms of plants attacked by fungi. *(Meerut, 1998, 1999; Kumaon, 2004)*
8. Describe the defence mechanisms adopted by plants to resist pathogen and disease. *(Kanpur, 2002)*
9. Describe the symptoms produced by fungal pathogen. *(Poorvanchal, 2003)*
10. Justify the necessity of quarantine. *(Kanpur M.Sc. 2006)*
11. Give an account of plant disease control by application of chemicals. *(Kanpur M.Sc. 2006)*
12. Define fungicide and describe its important characteristic features. *(Kanpur, 2006)*
13. Describe the different kinds of symptoms of plant diseases with the help of examples and diagrams. *(Poorvanchal, 2005; Kanpur, 2006)*

SHORT ANSWER TYPE

14. Differentiate between
  (*i*) Communicable and non-communicable diseases *(Kanpur, 1998)*
  (*ii*) Epidemic and endemic diseases
  (*iii*) Parasitic and non-parasitic diseases.
  (*iv*) Smuts and Rusts
  (*v*) Black Rust and White Rust *(Agra, 1993)*
  (*vi*) Hypertrophy and Hyperplasia
  (*vii*) Parasitism and symbiosis *(Allahabad, 1991)*
  (*viii*) Chlorosis and Necrosis

15. (*a*) What is a fungicide ? Name two fungicides. *(Lucknow, 2003)*
(*b*) Write definition of fungi cide and describe its important characteristic features. *(Kanpur, 2006)*

16. What is hypertrophy ? *(Poorvanchal, 2002)*

17. Define pathogenicity. *(Poorvanchal, 2003)*

18. What are the main principles of disease control?

19. How will you classify diseases on the basis of degree of their spreading?

20. Write not more than five lines on the following :
  (*i*) Epiphytotic *(Lucknow, 1992)*
  (*ii*) Hypertophy *(Lucknow, 2004; Allahabad, 2002; Kanpur M.Sc., 2006)*
  (*iii*) Hyperplasia *(Lucknow, 2004; Kanpur M.Sc., 2006)*
  (*iv*) Endemic
  (*v*) Necrosis *(Allahabad, 2001)*
  (*vi*) Rusts
  (*vii*) Smuts

21. Describe various inanimate causes of plant diseases. *(Punjab, 1999; Himachal Pradesh, 1999)*

22. Write short notes on :
  (*a*) Diseases
  (*b*) Control measures or control of plant diseases *(Kumaon, 2004)*
  (*c*) Fungicides *(Allahabad M.Sc., 1999)*
  (*d*) Classification of diseases
  (*e*) Symptoms of plant diseases.
  (*f*) Bordeaux mixture. *(Meerut, 1997)*
  (*g*) Phytoalexins *(Lucknow, 2005)*
  (*h*) Pathogenicity *(Kanpur M.Sc., 2006)*
  (*i*) Quarnatine. *(Rohilkhand M.Sc., 2006, Kanpur, 2000)*

OBJECTIVE TYPE

23. (A) Fill in the blanks :
  (*i*) A plant disease occuring on a large scale is called as .... *(Lucknow, 1996)*
  (*ii*) Necrosis means .... of the ....
  (*iii*) The study of plant disease is known as ....
  (*iv*) The chemicals used for killing fungal pathogen are called ....
  (*v*) Chlorosis means .... of .... in the green parts of the plants.

(B) Select the correct answer :
  (*i*) Necrotic symptoms mean
    (*a*) increase in the number of cells
    (*b*) decrease in the number of cells
    (*c*) death of the cells
    (*d*) loss of chlorophyll.
  (*ii*) The chemical used for killing fungal pathogens are known as
    (*a*) pesticides (*b*) bactericides
    (*c*) nematicides (*d*) fungicides.
  (*iii*) In most of the leaf spot diseases, before the appearance of the lesions, there is loss of chlorophyll in the leaves. This phenomenon is referred to as
    (*a*) mottling (*b*) chlorosis
    (*c*) necrosis (*d*) distortion.
  (*iv*) A disease which is more or less constantly present from year to year in a locality is known as
    (*a*) epidemic (*b*) endemic
    (*c*) pandemic (*d*) sporadic.
  (*v*) The pathogens which can not live

without the presence of the host and which does not allow the host to die is called.

(*a*) obligate saprophyte

(*b*) obligate parasite

(*c*) facultative saprophyte

(*d*) facultative parasite.

(*vi*) Father of Indian Plant Pathology is

(*a*) K. C. Mehta

(*b*) B. B. Mundkur

(*c*) E. J. Butler

(*d*) S. N. Das Gupta.

(*vii*) Excessive enlargement of cells in response to infection by a pathogen is called as

(*a*) hyperplasia

(*b*) hypoplasia

(*c*) hypertrophy

(*d*) atrophy.

(*viii*) Antibiotics have no effect on

(*a*) algae

(*b*) fungi

(*c*) bacteria

(d) viruses.

(*ix*) The scientist who is often credited universally as 'Father of Modern Plant Pathology' is

(*a*) E. J. Butler

(*b*) Anton de Bary

(*c*) Robert Koch

(*d*) J. G. Kuhn.

(*x*) A chemical which does not kill but inhibits fungal growth temporarily is called

(*a*) fungistat

(*b*) fungicide

(*c*) antisporulant

(*d*) adjuvant.

(*xi*) Bordeaux mixture consists of

(*a*) lime and Copper Sulphate

(*b*) lime and Benzene

(*c*) lime and Zinc Sulphate

(*d*) cooper Sulphate and Sulphur.

(*xii*) Crop rotation eradicates

(*a*) soil inhabitants

(*b*) soil invaders

(*c*) fertilisers

(*d*) none of the above

Powdery Mildew Caused by *Erysiphe*

CHAPTER 22

EXAMPLES OF IMPORTANT PLANT DISEASES

- Diseases caused by Fungi
- Diseases caused by Bacteria
- Diseases causes by Viruses
- Diseases caused by Mycoplasma

I. DISEASES CAUSED BY FUNGI

(a) DISEASES CAUSED BY PHYCOMYCETES

1. WART DISEASE OF POTATO

Host : *Solanum tuberosum*

Pathogen : *Synchytrium endobioticum*

General. The disease, wart disease of potato also known as Black wart of potato, was first described in 1895 from Hungary. Since then it has been reported from the regions of temperate climate like Europe, North America, mountainous areas of South America as well as South Africa. In India, the disease was first reported from Darjeeling by Ganguly and Paul (1953) and continues to be restricted to that area.

Effect. The damage caused by the disease is frequently high since the potato tubers get de- shaped, become tasteless loosing flavour and therefore unfit for human consumption. Young tubers get deformed due to infection and have to be thrown away.

Symptoms. The symptoms of the disease appear only on underground parts except roots of the plant i.e. tubers, buds of stems, and stolons. These are characterised by warty, tuberous and dirty Cauliflower like outgrowths on infected parts. The warty outgrowths vary in size from small protuberances to large intricately branched systems. Early in the growing season they are green or greenish-white in colour if exposed to light but are cream coloured or black on underground parts. On the tubers, mostly the warts are larger than the tuber itself covering the whole tuber. In advanced stages, the warts become dark black in colour and may sometimes be attacked by saprophytic fungi. The wart usually consists of distorted proliferated branched structures grown together into a mass of hypertrophied tissue (Fig. 22.1)

Wart Disease of Potato

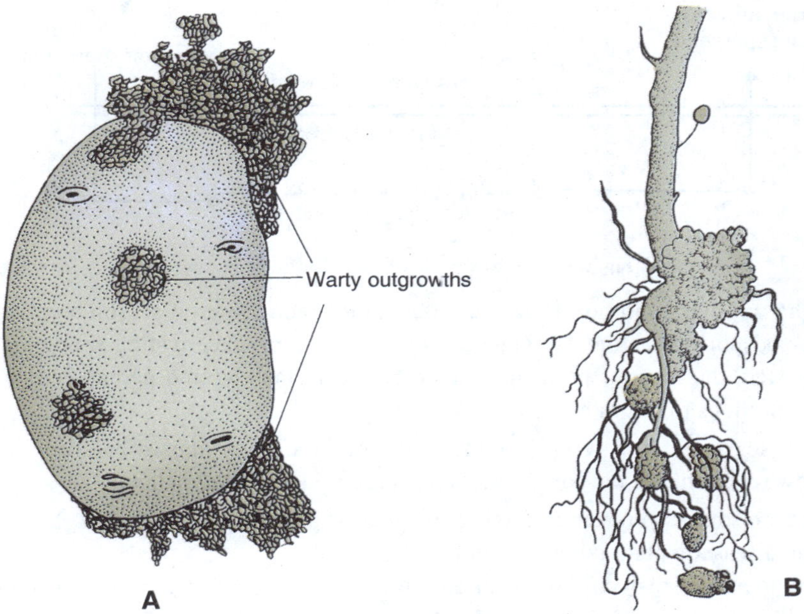

Fig. 22.1. Symptoms of wart disease of potato on a potato tuber (A) and underground parts of a plant (B).

Causal Organism. The causal organism is **Synchytrium endobioticum** (Schilb.) Percs.—a holocarpic endoparasite.

Disease cycle. The primary infection of the young potato crops available in the field takes place through the germination of resting spores present in the soil into Zoospores. These zoospores after swimming for a while land on the underground shoot parts, withdraw their flagella and enter the host cells where they settle down at the bottom of the host cells. Each of the neighbouring cells start dividing repeatedly followed by swelling, ultimately resulting in the formation of wart. There warts are the main symptoms and contain the material for secondary infection. The sporangia contained in the warts produce zoospores upon disintegration of the hosts. These zoospores are responsible for secondary infection. If environmental conditions are favourable, the zoospores are formed repeatedly and secondary infections may take place as well causing serious damage to the potato crop.

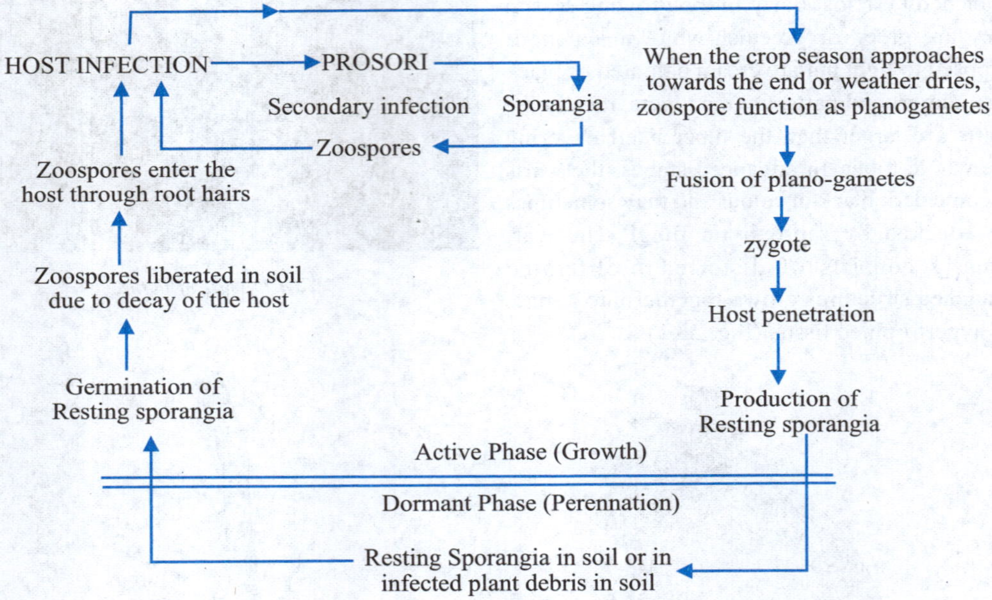

Fig. 22.2. Disease cycle of wart disease of Potato.

When the weather conditions become unfavourable and the crop season approches towards the end, the zoospores function as planogametes, fuse and produce resting sporangia. While the crop is harvested, the resting sporangia remain buried in the soil and perennate waiting for the favourable conditions to return back.

Control Measures. It is very difficult to control the disease once it has been introduced in the field. However, some of the control measures practised are listed below :

(*i*) Entry of diseased material into healthy areas should be prevented.

(*ii*) The diseased potato tubers should be discarded.

(*iii*) Soil treatment may control the disease to a large extent. These include steam sterilisation and application of mercuric chloride—copper sulphate and 5 percent formaline. But these are very costly.

(*iv*) Cultivation of disease resistant varieties continuously for 8-10 years is the only effective control measure.

2. WHITE RUST OF CRUCIFERS

Host : Members of family Cruciferae like Brassica sp., Raphanus sativus, Eruca sativa etc.

Pathogen : Albugo candida.

General. White rust or white blisters disease is one of the common diseases of crucifer crops. It is worldwide in distribution occurring in all the areas wherever crop is cultivated. Both wild and cultivated varieties are attacked. The disease affects a large number of crucifer crops of economic importance like Mustard, Cress, Rape, Radish, Cabbage, Cauliflower, turnip etc. In India the disease is reported on Mustard, Rape, *Eruca sativa,* turnip, Cauliflower and *Cleome viscosa.*

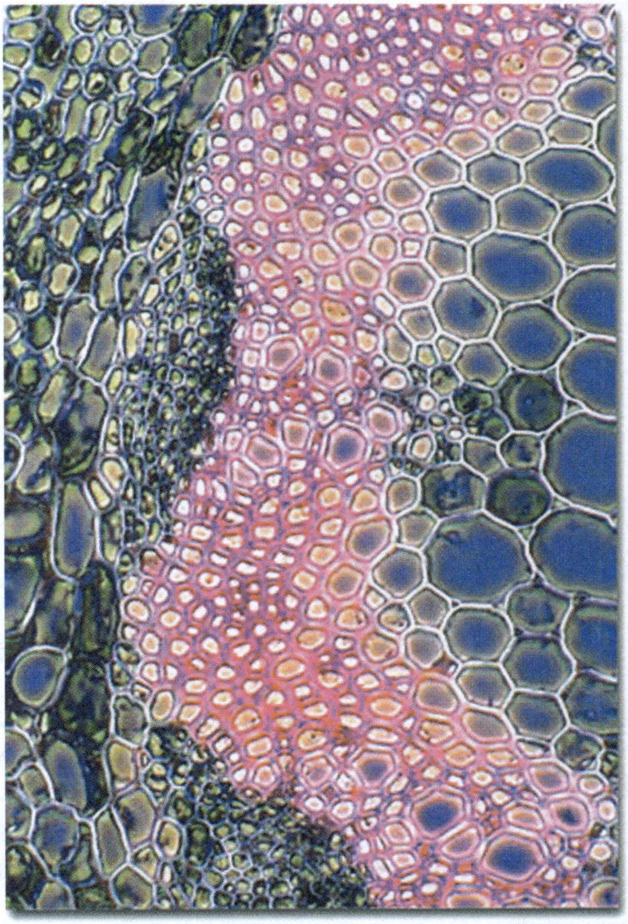

White Rust of Crucifers Albugo candida in hypertrophied cells.

Effect. Although considered unimportant in proportion to its widespread occurrence, the disease may cause serious damage in certain areas and the losses may be upto 25 percent when the floral parts get malformed and seeds are not formed at all. The disease in association with downy mildew disease of crucifers caused by *Peronospora parasitica* cause severe damage to the crop.

Symptoms. The disease affects all the aerial parts of the plant, the roots are not attacked. Symptoms may appear as a result of two types of infection : Local and Systemic.

In case of **local** infection, isolated spots or pustules appear on leaves or stems or inflorescence. The pustules are of variable size, measuring 1-2 mm in diameter and are raised shiny white areas. These may arise in close proximity and coalesce to form large irregular patches. Usually, the pustules appear in circular or concentric arrangement with one or two central areas. The host epidermis ruptures exposing white powdery mass consisting of spores of the fungus. Pustules occurring on leaves are usually confined to the lower surface only.

In **systemic** infections, young stems and inflorescence are infected. The fungus becomes systemic in these parts and the affected tissues are stimulated to various types of deformities. The most prominent is Hypertrophy of the affected parts. Due to Hypertrophy and Hyperplasia of floral parts, these show swellings and distortion. The peduncle and pedicel may become enormously thickened upto 12-15 times, the normal diameter. Floral parts become fleshy, swollen, green or violet in colour, the stamens falling off early. The petal may turn green sepal like and stamens and carpels are also converted to swollen leaf like structures. The ovules are usually atrophied as also the pollen grains resulting in total sterility. Pustules may also appear on these parts. However, the affected parts are full of oospores and starch.

When the systemic infection has taken early, the growth of the entire plant is checked, stunted and only small leaves may be formed.

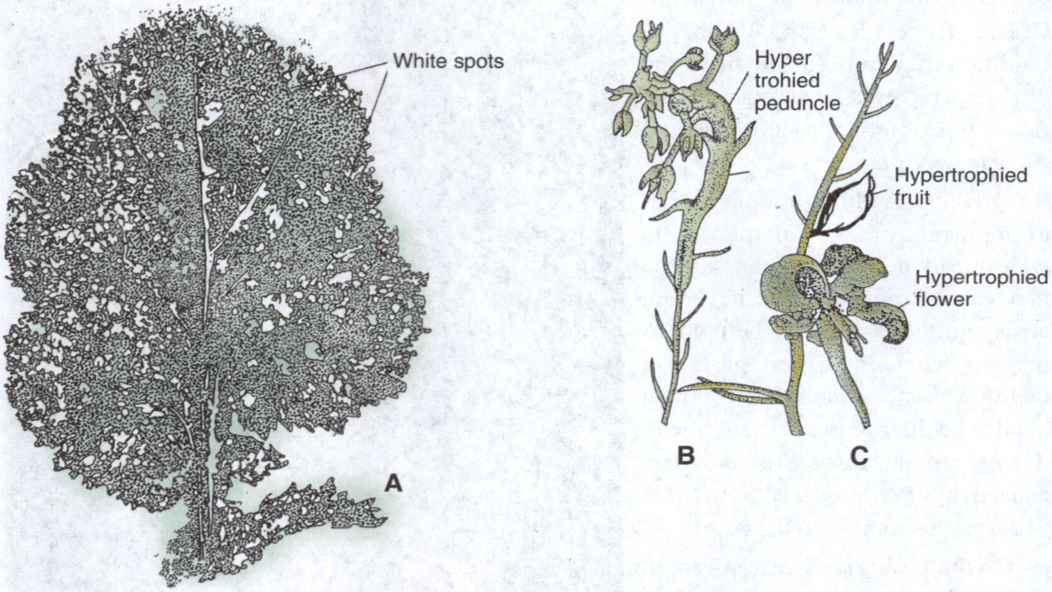

Fig. 22.3. Symptoms of white rust disease on leaves (A) and inflorescence (B and C).

The stem and the axis of the inflorescence may get twisted appearing in a zigzag sequence. Normal dormant buds are stimulated and grow into lateral shoots.

Causal Organism. The causal organism *Albugo candida* (Lev.) Kunze or *Cystopus candidus* Lev. is an obligate parasite.

Disease Cycle. The primary infection occurs due to oospores perennating in the soil or due to mycelium perennating on perennial hosts. These serve as primary inoculum when the environmental conditions are favourable. Oospores germinate in presence of water to form a vesicle in which a large number of zoospores are formed. These zoospores swim in a film of water and land on the suitable host, germinate by germ tubes, enter the host and establish infection. The mycelium in the host is intercellular with globose haustoria. Soon the mycelium after absorbing nutrients and food materials from the host, accumulates below the lower epidermis. Conidiophores, which are clavate, and formed at the tip of hyphae, begin to produce conidiosporangia in basipetal succession. The pressure of these breaks open the lower epidermis and white rust symptoms become apparent on the leaves.

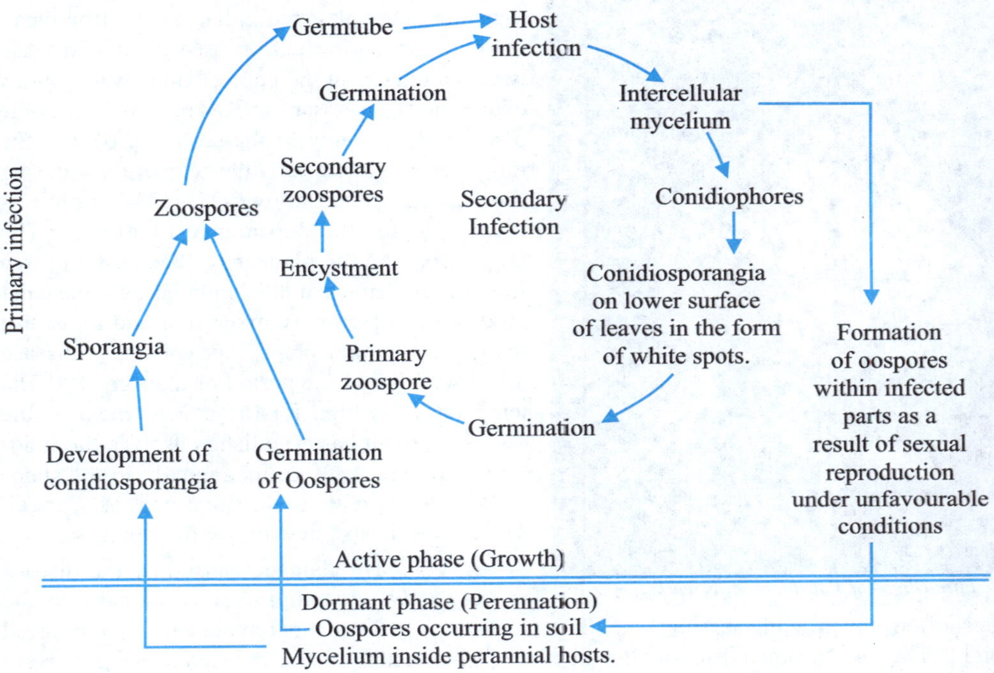

Fig. 22.4. White rust of crucifers–Disease cycle.

The conidiosporangia produced during early phase of the growing season cause secondary infection in the host. These are blown away by wind or any other agency, land on the host surface and germinate to form zoospores. The zoospores germinate by formation of germ tubes which enter the host and cause secondary infection. If the conditions are favourable, this is repeated.

When the conditions become unfavourable or during the later phase of the growing season, the fungus begins sexual reproduction producing oospores. These oospores, being thick-walled, can withstand the unfavourable conditions. During harvesting of the crop, the diseased hypertrophied portions of the plant are generally left in the field where they perennate waiting for the favourable conditions to return back.

Control Measures. The disease may be controlled by the following methods :

(*i*) Clean cultivation and destruction of weed should be practised.
(*ii*) Crop rotation will avoid the soil borne primary inoculum.
(*iii*) Spraying with 0.8 percent Bordeaux mixture or Dithane M-45 (0.2%) may be undertaken to check the spread of the disease.
(*iv*) Disease resistant varieties be preferred.

3. LATE BLIGHT OF POTATO

Host : Solanum tuberosum

Pathogen : Phytophthora infestans

General. Late blight is a serious fungal disease of potatoes. It is worldwide in its distribution. It occurs in potato growing areas of the world. Winter is the main potato growing season in India. It is followed by hot summer months in the plains. The drought and high temperature kill the

Late Blight of Potatoes - infected leaf

fungus in the soil. The late blight epidemics are thus rare in the plains in India. It is destructive to the crop grown in the rainy season. The disease occurs annually in the cooler Himalayan regions extending from Assam to Kashmir at an altitude of 6,000 ft. or more as the crop is grown in the rainy season. Moreover, the temperature during the day is never above 22°-23°C which is favourable for the appearance of disease. The crops grown in the plains have been usually free from the epidemics of late blight because the chief predisposing factors (temperature and moisture) that render potato plants susceptible to disease are absent during the period of their growth. The temperature is high for the development of the disease. Now it has established itself in the Indo-Gangetic plain and occurs annually in the states of Punjab, Uttar Pradesh, Bihar, and W. Bengal. The disease is also destructive to tomatoes.

Effects. The damage caused by the disease is frequently very high. Severe damage to the foliage shortens the growing season (Fig. 22.5). Consequently the tubers remain small and reduced in weight. They are produced in smaller numbers. This results in the reduced yield. In severe cases of infection there is complete loss of the crop. Infection also results in the decay of tubers in the field and storage.

Symptoms. The disease first appears on the tops of the plants generally after the blossoming period but mostly in the month of January. It may appear as well at any time during the growth period of the plant. The conditioning factor is the favourable environment. The disease makes its appearance as small, dead, brownish to purplish black areas or lesions. These appear on the tips and margins of the leaflets, rachis, petiole and stem. Under favourable conditions (low temperature and high humidity) the lesions rapidly increase in size involving the whole surface of the leaf. The disease generally first attacks the leaves, and petioles near the ground and the lesions appear on the lower surface of the leaflets on individual plants and then spreads upwards. Finally, a rapid and general blighting of foliage occurs. The blighted leaves curl and shrivel in dry weather. Under moist conditions they decay and emit a characteristic offensive odour.

Examination of the lesions on the lower surface of the leaf on a dew morning reveals a delicate growth of the fungus parasite in the form of whitish powdery bloom. It consists of sporangiophores and sporangia (Fig. 22.7 E) of the pathogen pushing out through the stomata. The sporangia serve to spread the disease in the growing season.

Fig. 22.5 Late blight of Potato. Infected leaf.

Potato tubers are often infected in the field after the tops have been blighted. They get separate infections while in the hill. There is brownish discoloration of the skin of those parts of the tubers which lie nearest the surface of the soil. These dry rot spots remain firm and extend to about half an inch below the surface. During storage, the bacteria assist to set in the wet rot phase. In cool and dry conditions the progress of the disease is slower and the wet rot phase is generally checked. Under moist conditions hyaline mycelial hyphae and sporangiophores push out through the lenticels and appear on the surface of infected tubers.

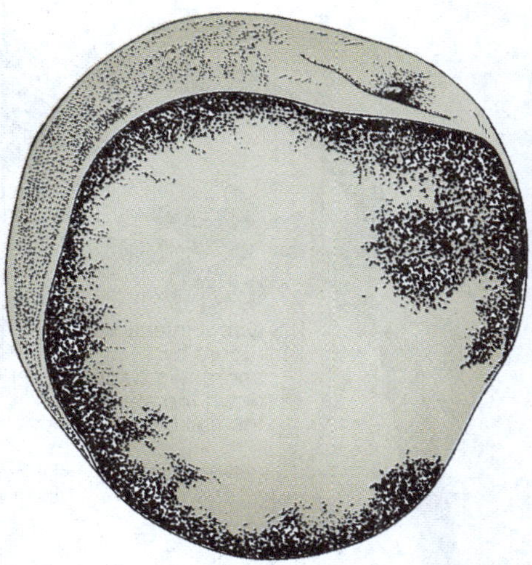

Fig. 22.6 Late blight of Potato. Infected potato tuber.

Causal organism and over-wintering. The causal organism is *Phytophthora infestans* (Mont.) De Bary. The mycelium is aseptate conenocytic, hyaline and branched. The hyphae are both intercellular and intracellular (Walker). They form rudimentary haustoria in the host leaf cells but in the tubers the haustoria are more common and elaborate (club-shaped, hooked or spirally twisted). According to De Bary (1876), the mycelium overwinters in the infected tubers. Melhus (1915) confirmed De Bary's observation. De Bruyn (1926) opined that the fungus overwinters in the soil but this remains unconfirmed. Kaung (1956) stated that in the temperate regions the fungus perannates in soil in the form of sporangia and germ tubes.

Disease Cycle (Fig. 22.7). The infected tubers (A) are generally considered as the main source of primary infection in India. The survival of the fungus in the soil in the Indian climatic conditions in any form appears remote. According to the widely held view, the fungal parasite overwinters as a dormant mycelium in the infected tubers. It becomes activated at the time of germination of the diseased seed tubers among the planting stock or waste tubers in dump heaps or infected tubers remaining in the ground after a previous crop. The activated mycelium invades the healthy sprouts (B). The second view is that the thick-walled resting oospores which are found in abundance in the infected tubers are the important overwintering structures. They play a significant role as the source of primary infection. At the planting time, the resting oospore germinates. The germ tube after emergence usually ends in a terminal sporangium. The contents of the latter divide to form zoospores. The released zoospores invade the healthy sprouts and bring about infection. According to some, the sexual phase seems to paly no role in the life history of the pathogen. The infected sprouts emerge above ground and produce shoots which contain the mycelium (C). It grows and ramifies in the intercellular spaces absorbing nutrition by putting haustoria into the host cells (D). Under suitable conditions of temperature and humidity, the mycelium pushes out hyaline, branched, indeterminate sporangiophores through the stomata of the host leaves (E). The thin-walled, ovoid or lemon-shaped sporangia, each with an apiculate tip, are borne singly at the tips of sporangiophores or their branches. As the sporangium reaches maturity, the supporting hyphal branch immediately below it swells slightly and continues to grow turning the attached sporangium to the side. The elongation of the branch proceeds and a new sporangium is formed. The process is repeated. A fertile branch or sporangiophore is thus characterised by 9 or 10 such swellings occurring at intervals. Each nodular swelling marks the point where the sporangium was borne. The mature sporangia are readily detached and spread by splashing rain or air currents to new potato plants (F_1 and *a*). Grosier (1934) who studied the biology of *Phytophthora infestans* reported that the sporangia are formed within a temperature range of 3° to 26°C with an optimum of 18° to 22°C. The minimum relative humidity required is

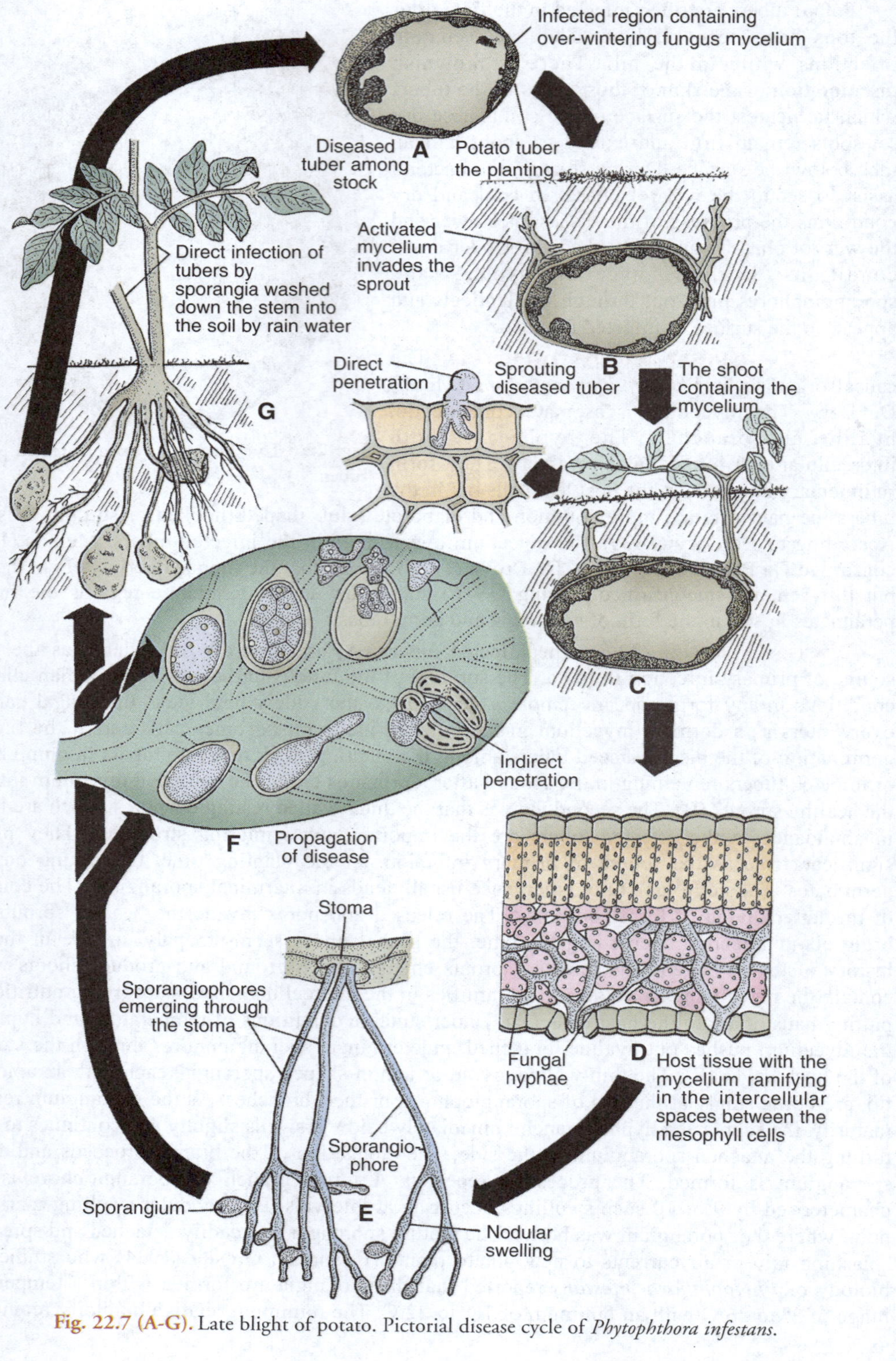

Fig. 22.7 (A-G). Late blight of potato. Pictorial disease cycle of *Phytophthora infestans*.

91 per cent with an optimum of 100 per cent. On reaching a suitable host (potato), the sporangia germinate on the leaves (F). Germination is influenced by moisture and temperature conditions.

(*a*) *Indirect Germination.* In cool moist weather, the sporangia function as zoosporangia (F_{1-3}). The optimum temperature for the formation of zoospore is 12°C (54°F). In the indirect germination the protoplasmic contents of sporangium divide to form a number of (usually 8) biflagellate zoospores (F_3). They are liberated in a group through terminal pore formed by rupture of the apical papilla. The released zoospores, after a brief period of activity in rain water or dew come to rest. Each retracts its flagella and secretes a wall around it. The clothed zoospores (cyst) then germinates by pushing out a germ tube or infection thread (F_4). The zoospores germinate rapidly at 12° to 15°C. Cool and moist nights are thus favourable for the formation and germination of zoospores. The germ tubes show rapid growth at 21°C. After infection they grow best at a slightly higher temperature.

(*b*) *Direct germination.* Under dry and warmer conditions no zoospores are formed. The sporangium functions as a conidium (F_a). It directly puts out a germ tube or infection thread (F_b). The optimum temperature for this direct germination of sporangia is about 24° or 25°C. The indirect method of germination of sporangia by the formation of zoospores in a terrestrial late blight fungus is an instance of retention of an ancestral primitive character which was normally used by its aquatic ancestor.

(*c*) *Spread of the disease* (*Secondary Infection*). The infection thread produced on the surface of the host leaf in either of the two above-mentioned methods enters the host tissue (leaves or stem). It makes its entry occasionally through the stoma but more often it penetrates directly through the cuticle (Pristou and Gallegly, 1954) by a penetration hypha arising from an appresorium (F_4). The lower surface of the leaf is more susceptible than the upper. The infected leaves produce another crop of sporangia. These are carried by wind to the healthy plants which are thus infected. This constitutes secondary infection. The process is repeated. As a result the disease spreads during the growing season over large tracts under potato cultivation. The disease spreads quickly when cool and wet nights alternate with warm moist days. Low temperature and high humidity favour the spread of the disease.

Field infection of potato tubers. The tubers get separate infections (G). It is caused by zoospores produced in foliage lesions (blighted tops) or present in the contaminated soil. Sporangia and zoospores come in contact with the tubers in two ways. Firstly, by contact freshly lifted healthy and wounded tubers with diseased haulms and contaminated soil. Secondly, during crop growth, the zoospores and sporangia washed down the stems into the soil by rain come in contact with the tubers (Lapwood, 1977). Tuber infection is dependent on the germination of sporangia, release and motility of zoospores. The released zoospores have to move through soil to the infection sites. The longer the zoospores continue to swim and greater their number, the greater are the chances of infection. The germ tubes gain entrance through the eyes, wounds and lenticels. According to Sato (1979), wet cool soil promotes infection but wet warm soil lowers it because cool water at 16°C or below 12-14°C favours indirect germination of sporangia and prolongs motility of zoospores. The sexual phase seems to play no significant role in the life history of the pathogen.

The severity of late blight infection is governed by environmental conditions. The chief among them are:–

(*i*) Night temperature below dew point for 4 or more hours,
(*ii*) Minimum temperature 10°C or slightly above,
(*iii*) Mean cloudiness not below 0.8°C on the next day,
(*iv*) Rainfall during next 24 hours, at least 0.1 mm.

Control Measures. Various methods of control of the disease are known. These are:–

1. *Selection of Seed (Planting) Tubers*. The seed tubers should be free from the disease. This requires strict seed tuber inspection at the cutting time. This measure will eliminate direct infection.

2. *Storage of Tubers at 40° F or below*. Storage of potato tubers in cold storage rooms reduces or even checks the progress of the rot.

3. *Growing Disease Resistant Varieties*. Considerable success has been achieved in the perfection of resistant varieties of potato at the potato breeding station, Simla. Growing these will provide an increasing opportunity to combat the disease.

4. *Use of Fungicides*. Resistance alone has not effectively checked the disease. Therefore, the complete control of blight is accomplished by the application of protectant fungicides. These are applied before infection for effective control in two ways namely by spraying or dusting as follows:–

4*a*. *Spraying*. The best method of control is the timely and repeated foliage spray schedule with copper fungicides such as Perenox, Blitox-50 and Fytolan. Dithane Z-78, and Dithane M-22 have proved more effective than the copper fungicides. The spraying should start when the plants are 8 inches tall. It should continue until the harvest time at 10 days' interval. Both the surfaces of foliage should be properly protected by adequate spraying delivered with a considerable force in the form of fine mist. Roy and Das (1968) found Brestan 60, Dithane M-45 and Zineb useful for the control of late blight in Assam. Mistiming of sprays may have serious consequences for late blight control in N.E.U.S.A.

Mancozeb and Chlorothalonil are the major fungicides which are presently used (Buck *et al.* 1981). Both fungicides inhibit sporangial and spore germination but has little effect on the mycelium in the leaf tissue.

4*b*. *Dusting*. Some people claim that dusting the foliage with copper-lime dust is a more effective control measure. Dusting is done in the morning when the plants are wet with dew.

5. *Sanitation*. Destruction or proper disposal of potato tuber refuge from pits and store houses is another practical measure to reduce the incidence of disease.

6. *Tuber treatment before storage*. The tubers should be dipped in 1 : 1,000 mercuric chloride solution for 90 minutes before storage. Before use they should be washed.

7. Avoidance of injuries to tubers at harvest is also important.

8. In cool humid areas killing of foliage a few days before harvest proves beneficial. This is accomplished by spraying with herbicides or flame throwers or by the use of mechanical vine beaters.

4. GREEN EAR DISEASE OF BAJRA

Host : Pennisetum typhoides

Pathogen : Sclerospora graminicola

General. The green ear disease of Bajra is a common disease and has been reported from several countries including India, Iran, Israel, China, Fji, Japan and the countries wherever Bajra crop is grown. In India, the disease was first reported and studied by Butler (1907) who considered the disease to be sporadic in nature not causing much damage to the crop. Mitter and Tandon (1930) reported the disease from Allahabad, Uttar Pradesh and confirmed the observations of Butler (1907). Since then, the disease on Bajra has been reported from all the states wherever Bajra is cultivated as one of the crops of 'Kharif' season.

Effect. Initially the disease was regarded as sporadic in nature causing not much damage to the crop except in lowlying fields where the loss could be quite significant. However, Rai and

Sinha (1965), Mathur and Dalela (1971) and Nene and Singh (1976) have estimated the loss to be as high as 27-30 percent. The disease appeared in an epidemic form in Karnataka and Maharashtra during 1975 causing almost total loss of the crop.

Symptoms. Two stages of the symptoms of the disease have been recorded : the downy mildew stage prominent on the leaves and the green ear stage affecting the inflorescence (ears). The latter is more prominent than the former symptom since it has been reported that the strain of the pathogen occurring in India produces more oospores than the sporangia.

The growth of the plants affected is checked and the plants appear stunted and sick, pale yellow in colour. Even young plants show this symptom. Soon, the leaves begin to show chlorosis in streaks on the upper surface while the lower surface shows downy growth of the fungus. After some time chlorotic areas turn brown and in advanced stages shredding of the leaves has also been observed. Since the plant is unbranched, the disease causes stimulation of nodal bud's and lateral shoots develop which is an abnormal character. Associated with it, the leaves also show additional symptoms. The usually green colour of the leaves may change wholly or partly, to whitish or brown colour. Whitening of the younger leaves is evident along the streaks while in older leaves, the colour changes to brown. In older plants, this occurs especially in those leaves from the axil of which the inflorescence develops. The change of the colour of leaves is associated with twisting, folding and shredding towards the tip. In older plants where green ears have been formed the upper leaves are mostly brown and split. A large number of leaf buds develop into a large mass of white or twisted brown leaves. The axil of these buds produce green leafy human ear like structures.

The disease becomes prominent when the inflorescence gets converted into green leafy ears.

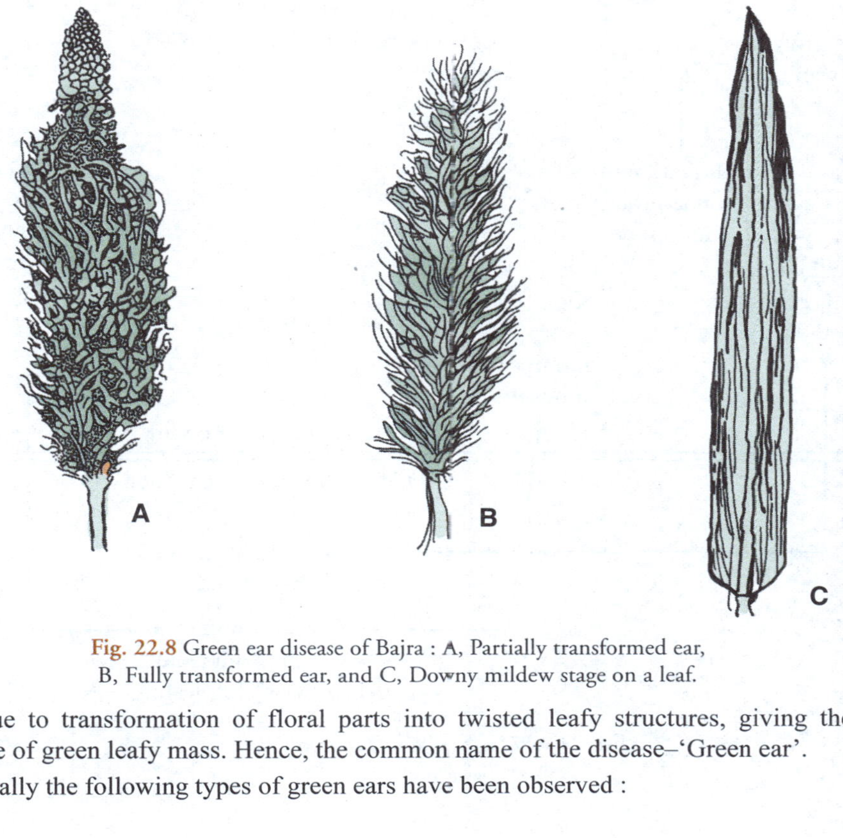

Fig. 22.8 Green ear disease of Bajra : A, Partially transformed ear, B, Fully transformed ear, and C, Downy mildew stage on a leaf.

This is due to transformation of floral parts into twisted leafy structures, giving the ear an appearance of green leafy mass. Hence, the common name of the disease–'Green ear'.

Generally the following types of green ears have been observed :

(*i*) Entire ear (inflorescence) is transformed into green leafy mass.

(*ii*) The lower part of the inflorescence is converted to green leafy mass but the upper part bears seeds.

(*iii*) The development of inflorescence is completely suppressed and in its place is formed a small bunch of leafy structures.

The disease causes changes in the florets as well as floral parts. The bristles of spikelets become contorted and hypertrophied. The number of spikelets may change from one to two in the same pedicel. A very common feature is the increase in the number of florets in each spikelet.

The fertile parts especially the stamens and carpels usually become sterile, leaf like and suppressed to rudimentary peglike structures. Sometimes the stamens appear as brown leathery, pointed bodies without any differentiation into filament and anther. In severely attacked plants, the carpels rarely develop and is replaced by small, leafy shoots or horn like outgrowths. This may be attributed to hypertrophy and hyperplasia of the affected tissues.

Causal Organism. The causal organism is *Sclerospora graminicola* (Sacc.) Schroet. It is an obligate parasite.

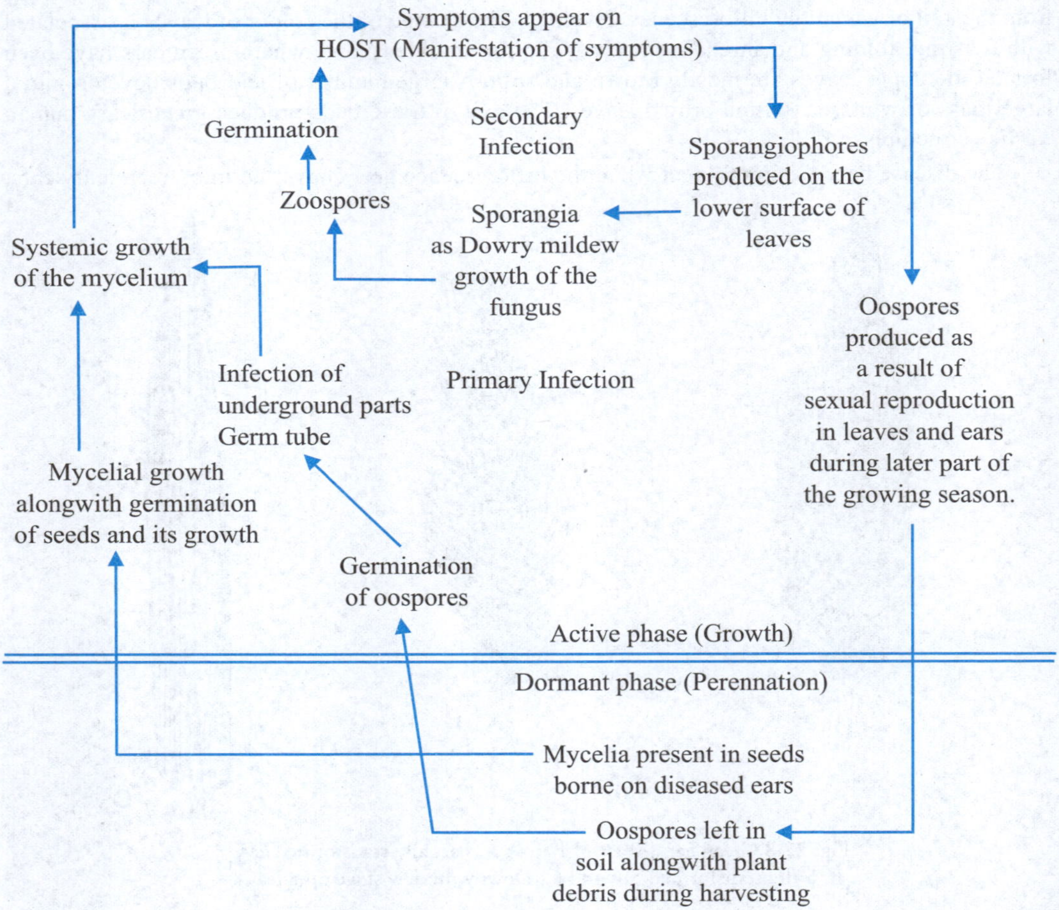

Fig. 22.9 Disease cycle of 'Green ear' disease.

Disease cycle. The primary infection occurs through oospores or mycelium present in the seeds produced on diseased ears. After the return of favourable conditions, the oospores present in the soil, germinate and enter the young seedlings of host through the underground parts and cause primary infection. When seeds containing mycelium are sown in the fields, these germinate and grow and along with it the mycelium also grows and causes primary infection as well. The infection spreads systemically alongwith the growth and development of seedlings into mature plants. In early stages, symptoms appear on leaves as downy growth of the fungus on the lower surface of leaves. After some times the characteristic green ears appear on the diseased plants.

The secondary infection takes place through sporangiophores and sporangia produced on the host as a result of primary infection. These sporangia are disseminated by wind, water and insects, land on the susceptible parts of the host, and germinate producing zoospores. These zoospores germinate by germ tube and cause secondary infection. Since the sporangia are reported to germinate only in the presence of enough dew on leaf surface, the chances of secondary infection are remote.

Towards the end of the growing season of the crop, the pathogen produces oospores through sexual reproduction. The oospores are thick walled and serve as resting spores, keeping the pathogen alive during the unfavourable conditions. During harvesting of the crop, the oospores alongwith plant debris are left in the soil of the fields where the oospores perennate waiting for the favourable conditions to return back so that these can serve as the source of primary inoculum.

Control Measures. Green ear disease is primarily soil borne and the following control measures have been suggested :

(*i*) Rotation of crop, with non-host types may check the disease.

(*ii*) Removal of diseased plants and their burning within a month of the disease detection checks the disease to a large extent.

(*iii*) Spraying with Dithane M-45 also helps in controlling the disease.

(*iv*) Seed treatment with 0.1 pecent Agrosan GN and 0.4 percent thiram has been reported to control the disease to 50 percent.

(*v*) Use of disease resistant varieties like HB-15, PHB-10 and PHB-14 has been recommended.

(B) DISEASES CAUSED BY ASCOMYCETES

5. POWDERY MILDEW OF PEAS

Host : Pisum sativum

Pathogen : Erysiphe polgoni.

General. The powdery mildew of peas is worldwide in occurrence causing serious losses to the crop. The disease is considered to be much more serious than the downy mildew of peas. Physiologic races of the fungus attack beans, urd, lucerne, coriander, turnip, cabbage and many other plants. In India, the disease generally appears during November-December.

Effect. The disease appears in epiphytotic form in the plains almost every year. The disease is worst in dry

Powdery Mildew of Pea (caused by Erystiphe polygoni)

weather causing great loss to the crop, although varieties maturing in January escape the disease. The loss to the crop is proportional to the intensity of the disease. In a 100 percent infected crop, the loss is estimated to range between 20-25 perecnt. Munjal *et al.* (1963) have estemated the loss in pod number to 21-30 percent and reduction in pod-weight by about 26-50 percent.

Symptoms. First symptoms appear on the leaves in the form of white floury patches on both sides of leaves. The disease then spreads to other green parts of the plant such as tendrils, pods, stems etc. The patches on the leaves originate in the form of minute discoloured specks from which powdery mass radiates on all sides. In the advanced stages of the disease, large areas of the host get covered with white floury patches. Infected plants impart dirty appearance. In extreme severe infections the infected leaves are shed leaving stem devoid of the leaves.

Causal organism. The causal organism is *Erysiphe polygoni* DC, although *E. pisi* and *E. cichoracearum* have also been reported on peas, but *E. polygoni* is more destructive. The fungus is an obligate parasite having septate, hyaline, profusely branched superficial mycelium sending finger shaped haustoria to the host cells.

Disease Cycle. The primary infection takes place through ascospores released in soil as a result of disintegration of the wall of cleistothecium. Under favourable conditions, the ascospores come in contact with the lower most leaves of the host, germinate by germ tube and cause primary 'infection'. From the lower most leaf, the infection spreads to other leaves. Later the symptoms in the form of white powdery mass appear. These patches contain conidia and conidiophores. The primary infection may also come from the conidia produced on other hosts as the causal organism *E. polygoni* has a large host range.

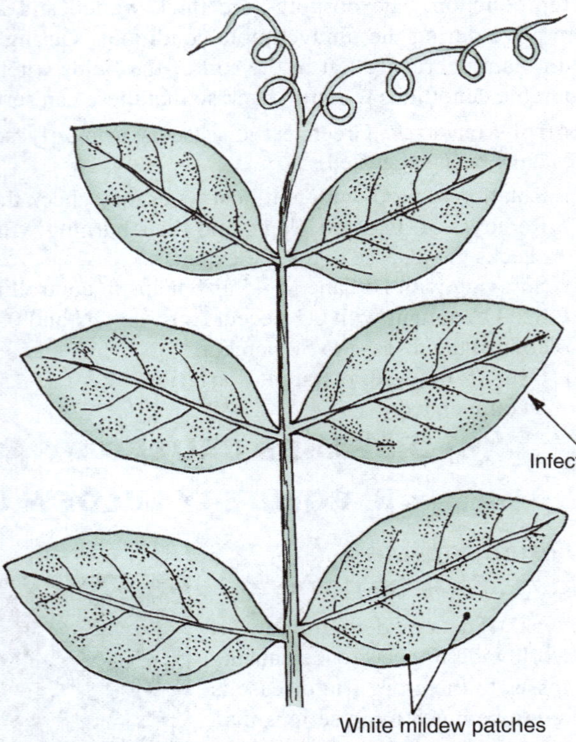

Fig. 22.10. Symptoms of powdery mildew of peas.

The secondary infection of the host takes place by conidia which are produced in large numbers and are disseminated by wind. Upon reaching the host the conidia germinate and cause infection. During the growing season, more than one secondary infection may occur because the establishment of infection and conidia production is a quick process. This results in severe large scale infection on the crops.

The cleistothecia–the perennating structures are produced on fallen infected debris in soil. These remain buried in the soil waiting for the favourable conditions. These contain asci and ascospores which serves as primary inoculum of the disease infection.

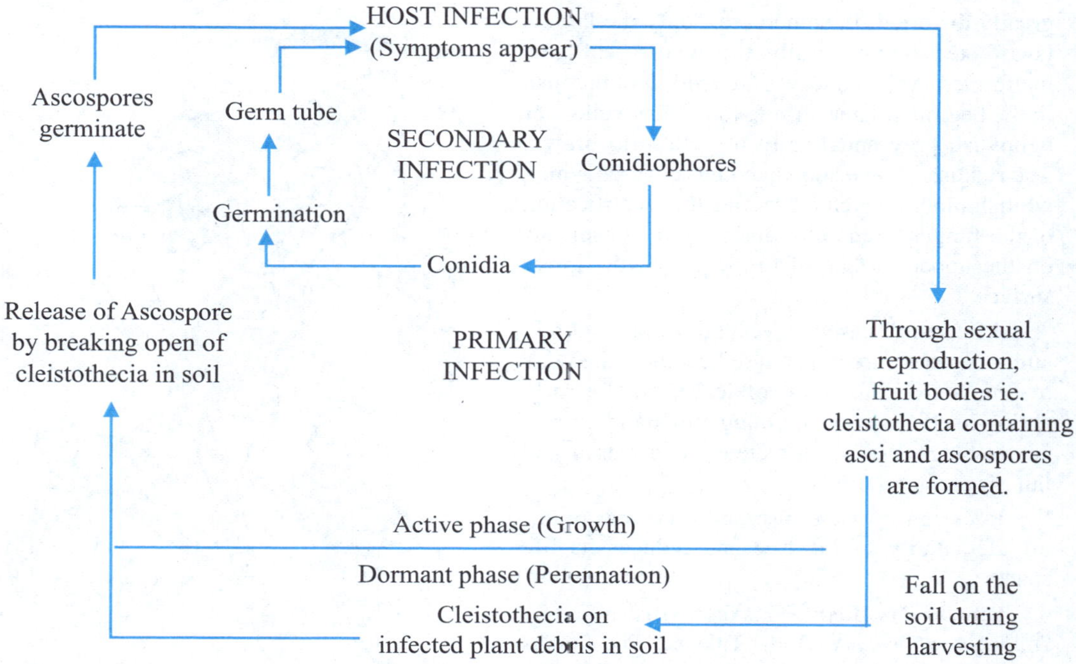

Fig. 22.11 Disease cycle of powdery mildew of peas.

Control Measures :

(*i*) Field sanitation, rotation of crop and destruction of diseased plant debris may effectively control the disease.

(*ii*) Sulphur dusting @ 25-30 Kg/hectare gives effective results. Only one dusting is sufficient.

(*iii*) Spraying with Karathane (0.2%), Elosal (0.5%) and Morocide (0.1%) have been found effective in controlling the disease.

(*iv*) Early maturing and resistant varieties like P185, 388, 61113, 1683, 6588, Freazer-656, T-10, T-163 and IC775 should be cultivated.

6. LEAF CURL OF PEACHES

Host : Prunus persica

Pathogen : Taphrina deformans

Peach Leaf Curl (Taphrina deformans)

General. Although Peach leaf curl disease is worldwide in distribution, it occurs more predominantly in Europe and North America. In India, the disease is prevalent in peach orchards of Kumaon, Kulu and other parts of the Sub-Himalayan range.

Effect. The disease causes much defoliation and distortion of leaves. These generally result in the formation of small fruits or fruit drops, thus reducing life span of the plants.

Symptoms. Symptoms first appear in early spring when the leaves come out of the bud. Some of the leaves appear twisted, puckered, thickened, curled downwards and often greatly distorted. In some cases, only the lamina is affected but generally the whole leaf gets malformed. Affected leaves are reddish or purplish in the begining, later turning to reddish yellow or yellowish grey and finally to yellow to brown. The reddish lamina soon becomes covered with a whitish bloom which represents the fructification of the fungus. Generally this is more conspicous on the upper surface of the leaf than the lower surface.

Affected diseased leaves fall off prematurely and in serious infections, tree become naked due to premature defoliation of leaves in the early spring itself. Blossoms, young fruits and current year's twigs also get affected and diseased and fall off prematurely.

Recurrence of the disease every year drains off the vitatily of the tree and reduces its life span.

Fig. 22.12. A twig of peach showing symptoms of the leaf curl disease.

Causal Organism. This disease is caused by *Taphrina deformans* (Berk) Tulasne. The fungus is characterised by intercellular sub-cuticular mycelium, absence of fruit body and the asci produced naked on the host surface.

Disease Cycle. Primary infection takes place through ascospores or blastospores or conidia which are present in the soil of the orchards. Due to rains, these ascospores or blastospores or conidia are washed, splashed or blown onto young tissues where they germinate and cause

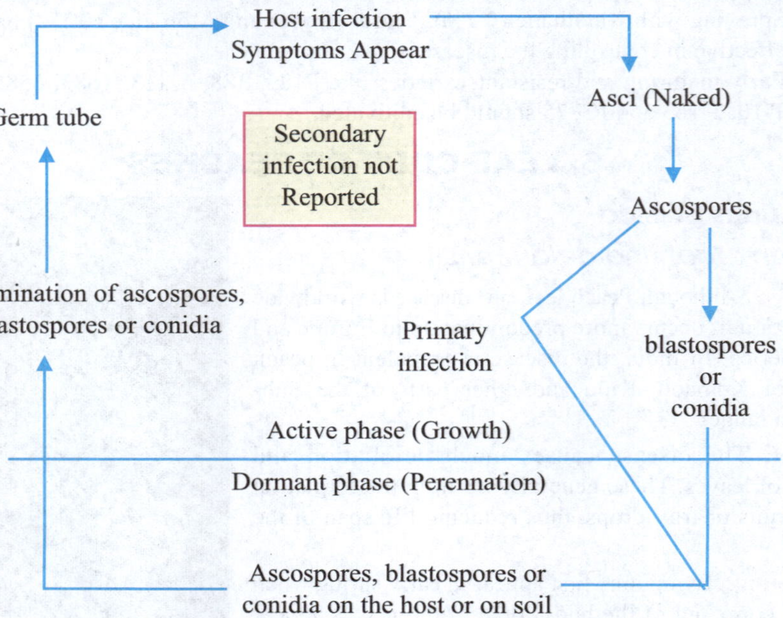

Fig. 22.13. Disease cycle of Peach leaf curl.

infection. At the time of penetration the, germ tube becomes binucleate due to mitotic nuclear division. The binucleate nucleus upon entering the host tissue forms the binucleate mycelium. After the establishment of the infection inside the host, disease develops and symptoms become apparent. Soon, the asci and ascospores are formed on host surface. Ascospores upon release usually germinate to produce blastospores or conidia.

Secondary infection has not been reported as the fungus can infect only the younger tissues. Older tissues generally become resistant to infection.

Ascospores perennate usually in the fallen infected plant debris in soil. They are also reported to perennate in the bud scales.

Control Measures

(*i*) Sanitation of the trees and orchards is important in controlling the disease.

(*ii*) The tree must not have diseased leaves. Therefore, as soon as the disease leaf is noticed, it should be plucked and burnt.

(*iii*) Fallen infected leaves should be collected and burnt.

(*iv*) Spraying with lime sulphur or copper fungicides helps in controlling the disease. Among the fungicides, Bordeaux mixture (4 : 4: 50), Perenox or Fytolan (0.3%) Blitox-50 (0.25%) are commonly used. One spraying in autumn and another just before openning of the buds considerably reduce the incidence of the disease.

(*v*) Spraying with Cuman five times a year brings about total control of the disease.

(C) DISEASES CAUSED BY BASIDIOMYCETES

7. LOOSE SMUT OF WHEAT

Host : Triticum vulgare
Pathogen : Ustilago tritici

General. This disease is very common and widespread. It causes great damage in the wheat growing tracts of India, particularly in the Punjab, Uttar Pradesh and certain districts of Madhya Pradesh. Luthra (1953) reported that in India, the disease causes loss of over 50 million rupees annually. In Punjab the disease is called *Kangiari*.

Loose Smut of Wheat (caused by Ustilago tritici)

Symptoms. The smutted ears (Fig. 22.14 E) emerge from the boot leaves a little earlier than the healthy ones. They bear loose, black, powdery masses of smut spores instead of flowers. All the ovaries and other floral parts except the awns and rachics are converted into masses of smut spores. In the young spikelets before emergence each ovary has become a spore sac. It is the home of millions of spores. The spores of each spikelet are covered by a thin greyish or silvery membrane. By the time the ear emerges from the boot leaf the membrane ruptures to expose the black powdery mass of spores. The ear is generally completely destroyed except the awns and the rachis. When the wind blows the spores are blown off and the bare rachis and central axis is left behind. To it may cling a few spores that have not been blown off by the

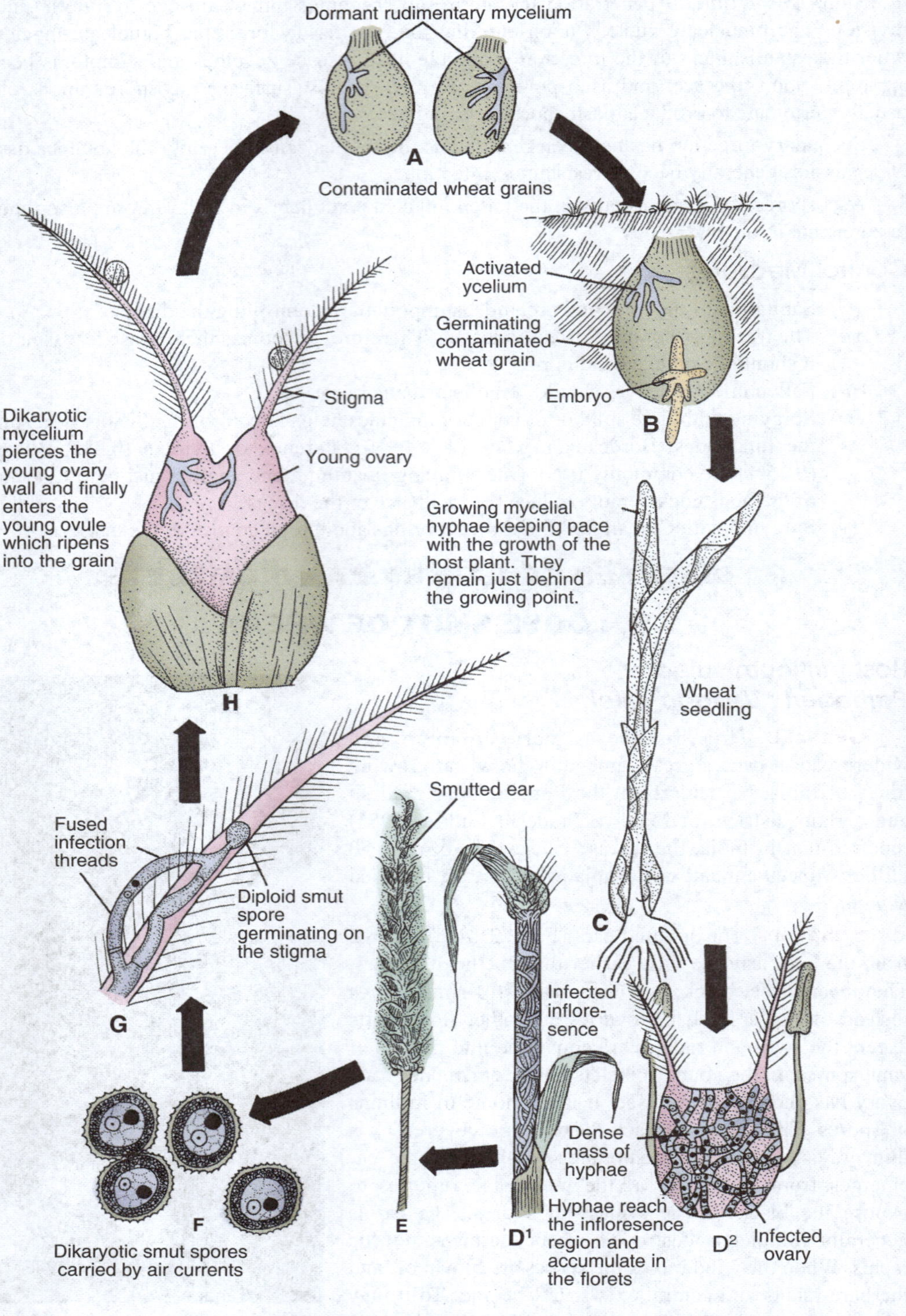

Fig. 22.14 Loose smut of wheat. A pictorial disease cycle of *Ustilago tritici*, a blossom-infecting smut.

wind. It is not necessary that all the ears of a wheat plant may be smutted. Some may be found to be healthy and others diseased.

Causal Organism. The causal organism of this disease is *Ustilago tritici* (Pers.) Rostr. and the host is *Triticum vulgare*. Fisher transferred *U. tritici* to *U. nuda*. Popp holds that since the two species differ in the type of teliospore germination, they should be considered as distinct species. The disease is internally seed borne. The mycelium of the fungus lies dormant in the grain (Fig. 22.14 A).

Mycelium. Ustilago tritici is an internal parasite. It has a dikaryotic mycelium. The hyphae ramify the intercellular in spaces of the host tissue. They absorb nutrition from the host cells by diffusion. The hyphae do not produce haustoria.

Spore Formation. The mycelium grows keeping pace with the growth of the host plant. It is chiefly confined to the stem (Fig. 22.14 C). At the time of flowering and when the inflorescence is still enclosed by the boot leaf, the mycelial hyphae enter into the ovaries of flowers. Within the ovary each hypha grows vigorously and branches repeatedly to form a dense mass of hyphae (Fig. 22.14 D^2). The latter destroy the host tissue in the ovaries and surrounding floral parts. The cells of these hyphae are binucleate. The hyphae undergo additional septation to form short binucleate cells. These cells swell and round off to form binucleate smut spores (Fig. 22.14 F). The smut spores are called the **brand spores**. Some mycologists prefer to call them **teliospores**. They are spherical to oval and measure 5.9 μ in diameter. They have a finely echinulate thick spore wall which is olivaceous brown but slightly lighter on one side. The teliospores are produced in enormous numbers.

Disease cycle (Fig. 22.14). The loose smut of wheat is a systemic disease. It is seed borne (A). As the infected grain is sown and germinates (B), the dormant fungus mycelium within the grain resumes activity. It grows best in or near meristematic tissues keeping pace with the growth of the host plant (C). The hyphae thus grow just behind the growing point. The presence of the fungus in the meristematic tissue exercises accelerating influence on the growth of the host which matures early producing flower heads. At the flowering time, the hyphae reach the inflorescence region (D^1) and accumulate in the floral parts chiefly the florets (D^2) which are subsequently completely destroyed. The hyphae become swollen and additionally septate. The segments, which are binucleate, round off, separate and secrete thick walls to become smut spores which are frequently called **teliospores**. The teliospores serve as a means of propagating the disease during the growing season. They are readily carried from the smutted ears (E) by air current (F) at a time when the healthy plants are in the flowering stage. There are clouds of spores in the atmosphere over the wheat fields. The glumes of healthy flowers are wide open and the stigmas sufficiently exposed and spread in the dry weather to the spore dust. Thus the teliospores fall on the feathery stigmas of healthy wheat flowers. Under suitable conditions (warmth and moisture) the spores germinate on the stigma (G). Before germination the two nuclei of a dikaryon fuse in the smut spore to form a single fusion nucleus or the synkaryon. The diploid teliospore represents the hypobasidium or probasidium stage. At the time of germination the exosporium ruptures. The endosporium grows out in the form of a short tubular hypha called the **promycelium** or **epibasidium**. The diploid synkaryon migrates into the epibasidium and undergoes meiosis, which consists of two nuclear divisions. As a result four haploid nuclei are formed. They are arranged in a row. Two of these are of *plus* and two of *minus* strain. Walls are laid between the daughter nuclei. The epibasidium now consists of four haploid cells arranged in a row. Each cell of the epibasidium produces a slender hypha called the **infection thread**. It contains a single haploid nucleus. Out of the four infection threads two contain haploid nuclei (one each) of plus strain and two of minus strain. Basidiospores or sporidia are not produced by the basidia of *Ustilago tritici*. Their place is taken up by the infection threads. The neighbouring infection threads of opposite strains fuse (Fig. 22.14 G). Consequently one of the infection threads becomes binucleate or dikaryotic. The dikaryotic infection threads become binucleate or dikaryotic. The dikaryotic infection thread grows by elongation and clamp connections. It enters the style, grows forwards through the intercellular spaces and the channels left by the pollen tubes and reaches the ovary. Recently Pedersen (1956) and Malik and Batts (1960 *a*, 1960 *b*) opposed this view. They hold that the infection thread pierces the young ovary wall in a week or so. From there it makes its way into the developing ovule through the integuments. Penetration into the immature ovule occurs between the 7th and 10th day. About 10 days after infection the integuments become cutinised and resistant to infection. In the ovule the dikaryotic hypha passes into the space between the endosperm and nucellus. Here it branches freely. The branch hyphae take about three weeks to reach the base of the raphe. Some of these pass round the bottom of the endosperm to penetrate the embryo through the scutellum. The hyphae in the scutellum are irregularly swollen and have thick and oily walls. As the ovary ripens into the grain the fungus mycelium becomes inactive (Fig. 22.14 A). It remains small and lies dormant in the embryo chiefly in the scutellum. The dormant mycelium within the embryo carries the fungal pathogen over seasons unfavourable to growth. It becomes activated again at the time of germination of the grain (Fig. 22.14 B). The presence of the dormant fungus mycelium in the grain shows no external signs of its infection. The infected grains apparently look like the healthy ones. *Ustilago tritici* thus provides an excellent example of infection through the flower.

Host–Parasite relations in Loose Smut of Wheat. Kourssanow (1928) reported that the infected plants were generally smaller and had higher respiration rate than the healthy ones. Mather and Hausing (1960) found that the total dry weight was reduced by 33 percent, root dry weight by 32 percent, height by 11 percent in infected plants. Gaunt and Manners (1971) reported reduction in the photosynthetic area. Expansion of successive leaves of the host plant was delayed. This reduced the availability of assimilates for the development of the infected plants which consequently showed restricted growth of the roots and the tillers. Such effects have been attributed to, (*i*) increased respiration and decreased photosynthesis and (*ii*) to the deprivation to the host by the pathogen (*U. tritici*) of necessary metabolates for further development.

Effect. The disease results in the reduction of yield from 20 to 50 percent. Quality of the grain is, however, not affected.

Control Measures. Since the mycelium of the parasite is lodged inside the grain, external application of disinfectants is ineffective. Direct attack on the fungus living deep in the tissues is very difficult. In the first instance most of the chemicals do not reach the seat of trouble. Some which do may injure the embryo as well. The dormant mycelium in the grain is very resistant to heat. Hence in all methods of treatment the first step is to make the dormant mycelium active. In the activated condition it is vulnerable. It is killed by the application of moist heat. The following methods are generally employed to kill the mycelium in the embryo of the grain:–

1. *Hot Water Treatment*. The wheat grains are at first soaked in water kept within a range of temperature between 26ºC-30ºC. They are allowed to remain there for about 4-5 hours. In the softened grains the dormant mycelium becomes active. The temperature of water is then raised and kept constant at 54ºC for about 10 minutes. At this temperature the activated mycelium is killed. This method requires strict care and supervision. The temperature should be carefully controlled. At a range a little too low, it will fail to kill the mycelium and at a degree or so too high it will kill the embryo. In this case the embryo is killed at 56ºC. The margin or error either way is thus very little. After the treatment the water is drained off. The grains should then be dried and sown. The hot water treatment was first evolved by Jensen (1888-89) against the late blight of potato. Swingle (1892) applied it against the loose smut of wheat.

2. *Sun Heating*. This method is in vogue in the Punjab and U.P. Here the sun in the months of May and June is very hot. The atmospheric temperature is very high. The suspected grains are soaked in water in flat, shallow bottomed basins with water level about two inches above the level

of grain. The basins are placed in the direct rays of the summer sun for about 4 to 6 hours, say from 8 a.m. to 12 noon. During this period the dormant fungus mycelium becomes active. The water is then drained off. The softened grains are spread in thin layers on the brick floor in the midday sun to dry. In the cooler regions the use of galvanised iron sheet to spread and dry the grain in the sun has been recommended. This treatment kills the activated mycelium. Mitra and Taslim (1936) recommended the sun heating method for controlling the disease in North Bihar. Luthra found the solar treatment quite suitable in the Punjab where the day temperature in summer goes very high. Bedi (1957) suggested a modification in the method. He found 4 hours presoaking period followed by one hour exposure to sun under Punjab conditions quite sufficient to kill the activated mycelium. Extra exposure to sun heat serves to dry the grain.

3. *Growing Resistant Varieties*. Sowing grains of varieties of wheat which are immune from or resistant to this disease is the best method of controlling the disease. Some of the wheat resistant varieties are Np 710, Np 120, and Pb 90. The other equally effective methods are:–

4. The wheat plants with infected ears, which emerge out of the boot leaves earlier than the healthy ones, may be uprooted at once and burnt. This practice is called *rogueing*.

5. The grains for sowing purposes should be thrashed from uninfected wheat ears.

6. *Use of systemic fungicides*. The use of fungicides, which till recently was considered impraticable to control the seed borne loose smut disease of wheat, has received much attention. Chatrath *et al.* (1969) found that two systemic fungicides D735 (Vitavax) and F 461 (Plantavax) give quite encouraging results when applied as seed dressing fungicides at the rate of 2.50 gm per Kg. The use of Benomyl and Carboxin to control loose smut of wheat (*U. nuda* var. *tritici*) has been recommended by many workers. Joshi *et al.* (1975) reported that seed dressing with 0.25 per cent Benomyl can effectively control the disease. Thomas and Chatrath (1975) found that a systemic fungicide thiabendazole employed as a seed treatment at the rate of 0.1 to 0.2 per cent is highly effective to control the disease without affecting germination.

8. COVERED SMUT OF BARLEY

Host : Hordeum vulgare

Pathogen : Ustilago hordei

Covered Smut of Barley (caused by Ustilago hordei)

General. The disease is of common occurrence in the barley growing tracts of India. It causes considerable damage in the Punjab, Uttar Pradesh, Madhya Pradesh and certain parts of Bihar.

Symptoms. The first symptom of the disease makes its visible appearance at the flowering stage. From the flowering time onwards the ears of the infected barley plants in the fields appear blackened and shrivelled as soon as they emerge out of the boot leaves. They can be seen from a distance. The smutted ears emerge from the boots at the same time as the healthy ones. They usually remain shorter and are retained within the boot leaves for a longer time. Sometimes they fail to come out. Every ear in a diseased plant and every ovary of the smutted ear contains a compact spore mass called a **smut sorus**. The grains are thus replaced by the smut balls or sori. The spores in the sorus do not fall apart due to the deposition of a fatty substance. They are held together by an investment provided by the walls of their respective ovaries and the basal parts of glumes of the spikelets. The smut spores in each sorus being firmly enclosed within a white, shining, silvery membrane are not blown by the wind. Flower infection is thus precluded. The smut spores do not serve as means of propagation of the disease during the growing season. They are liberated by the rupture of the surrounding membrane as the ears get crushed under heavy pressure during threshing. The spores thus released adhere to the rough skinned healthy grains (Fig. 22.15 A).

Causal Organism. The causal organism of this disease is *Ustilago hordei* (Pers.) Lagerheim.

Mycelium. The mycelium is internal and dikaryotic. The hyphae are septate and branched. They ramify in the intercellular spaces between the cells in the host tissue. The mycelium grows in the stem as well as in the leaves but generally it is confined to the stem. By the flowering time it reaches the ear where it grows more vigorously. The fungal hyphae enter the ovaries of the flowers. There they branch repeatedly and become additionally septate. The short, binucleate segments thus formed round off to form the binucleated smut spores. The latter remain compacted together into a ball-like mass in each ovary. They are liberated at the threshing time by the breaking of the walls of the ovary. The spores thus set free remain attached to the surface of the healthy grains.

The spores are brown in colour but lighter on one side and globose to ellipsoidal in shape. In a mass they appear black. Each measures 9 μ in diameter. The spore wall is smooth.

Disease Cycle (Fig. 22.15). The disease is externally seed borne and perennates through the smut spores which adhere to the healthy grains during threshing-time (A).

(*a*) *Germination of Teliospores*. The binucleate smut or teliospores present in the soil or sticking to the grains (B) germinate in the presence of warmth and moisture when the contaminated barley grains are sown in the field. Prior to germination, the two nuclei of dikaryon in the spore fuse to form a single fusion nucleus or the synkaryon. The diploid teliospore or probasidium as it is now called, germinates to produce a germ tube, the **promycelium** or the **epibasidium**. In the epibasidium, the synkaryon undergoes meiosis to form a row of four haploid nuclei. Segregation of plus and minus strains takes place at the time of division. Two of the nuclei are of plus and two of minus strain. Septa are laid between the nuclei. The epibasidium now consists of four haploid segments or cells (G). Each segment of the epibasidium produces one uninucleate sporidium or basidiospore. Two of these basidiospores are of plus and two of minus strain. They are ovate to oblong in form. The basidiospores of covered smut of barley are capable of multiplying by budding in the presence of nutrition like yeast. The secondary spores produced by budding germinate like the basidiospores and are capable of infecting fresh barley plants.

(*b*) *Infection*. The basidiospores or secondary spores germinate on the surface of the grain. Each produces a slender haploid germ tube. The haploid germ tubes by themselves are incapable of infecting the barley seedlings. The germ tubes from two different basidiospores of opposite strains growing in the vicinity fuse (C). As a result one of them becomes dikaryotic. The dikaryotic germ tube or the infection thread infects the barley seedling at a very early stage. The covered smut of barley therefore provides an example of seedling infection. The dikaryotic hypha penetrates the young coleoptile and enters the growing tip of the embryo. There it grows upwards keeping pace with the growing point (D). It forms a dikaryotic mycelium. The mycelium grows upwards in the leaves and the stem. It grows more vigorously in the stem than in the leaves. By the flowering time the mycelial hyphae reach the inflorescence region (E_1) and enter the ovaries to produce the smut spores (E_2). Infection fails to take place if the primary shoot of the host has come above the soil surface.

Effect. The host is not killed outright. Its vitality is no doubt lowered. Thus the infected plants show stunted growth. In severe cases of infection the yield may be lowered by 50 per cent of the crop.

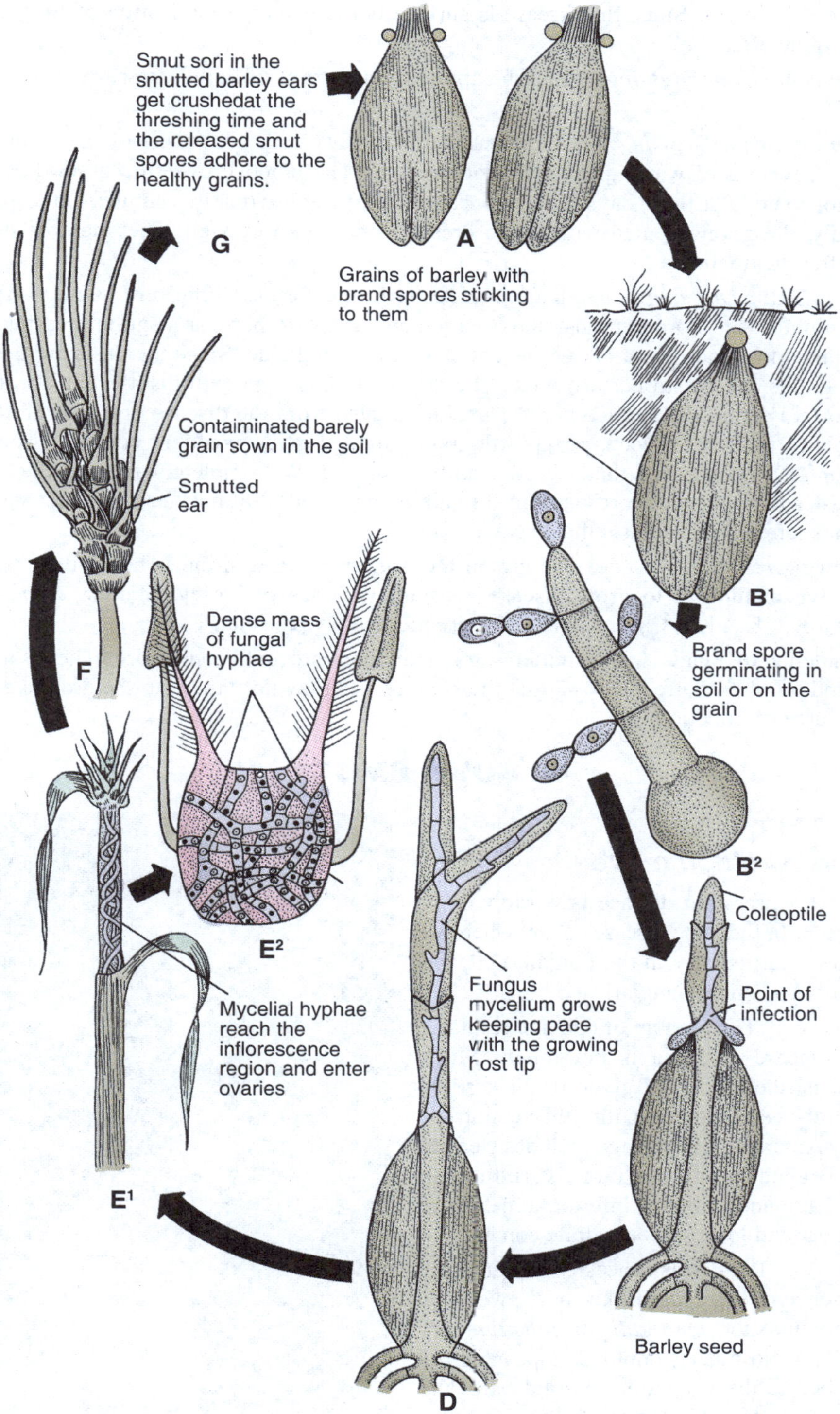

Fig. 22.15 (A-G). Covered Smut of barley—a pictorial disease cycle of *Ustilago hordei*, a seedling-infecting smut.

Control Measures. Since the disease is surface borne external application of fungicides to the grain is quite effective.

1. *Chemical Grain Dressing*. It may be applied in the form of liquid dressing or in the form of dust.

(*a*) *Liquid Grain Dressing*. Dressing the barley grains with a solution of one fluid oz. of formalin to 2 pounds of water gives very good results. The practice of liquid grain dressing is, however, not so popular these days. Firstly, because the procedure requires cumbersome apparatus and secondly, the grains treated with liquid dressings do not store well. They have to be sown promptly after the treatment.

(*b*) *Dry Grain Dressing*. Chemical compounds in the form of lime dust such as Agrosan, Ceresan, Sulphur dust, Formalin dust and Spergon etc., are very popular as grain dressings. Care should be taken that each grain is lightly coated with the fungicide, Since the dusts are persistent the dusted grains can be stored. Moreover, the dry grain dressings with disinfectants protect the seedling not only from the surface borne parasite but also from the disease organism in the soil. Agrosan GN at the rate of 4 oz. per 37 kilograms of seed gives excellent control. Grewal and Dharam Vir (1964) reported that Ceresan and Agrosan 5 W completely control the disease. Sulphur dust which is capable of passing through a sieve with 300 meshes per linear inch (300 mesh) is most effective to control the disease.

2. *Growing resistant varieties*. Acting on the principle, "Prevention is better than cure" the most effective method is to grow disease resistant varieties of barley. Mathur, *et al.* (1960) reported that K_{12}, K_{18}, K_{19}, C_{50}, C_{84}, and CN_{249} are moderately resistant.

3. Rogueing of plants with smutted ears is another preventive measure. They should be uprooted and burnt with great care so that the smut spores may not fall on to the ground and be a potential source of infection.

9. CORN SMUT

Host : Zea mays

Pathogen : Ustilago maydis

General. Corn smut disease is worldwide in its distribution. In India the disease occurs wherever corn is grown particularly in the Punjab, U.P. and Kashmir. It is known as Kangiari.

Importance. The amount of damage resulting from this disease in India is not much. Much importance is, therefore, not given to the disease in this country as compared with other countries, particularly America. The damage includes sterility of the diseased plants. There is reduction in yield and amount of fodder, inferior kernel lustre of smutted ear and lower carbohydrate content.

Corn Smut caused by Ustilago maydis

Symptoms. The disease is easily recognisable by the presence of large knob-like sooty swellings or boils known as the *smut galls* or *balls tumours* (Fig. 22.16 E). No other symptoms appear on the host plant before the formation of smut ball. The latter appear on ears or stems (stalks). Tumours are also formed on the leaves and tassels (male

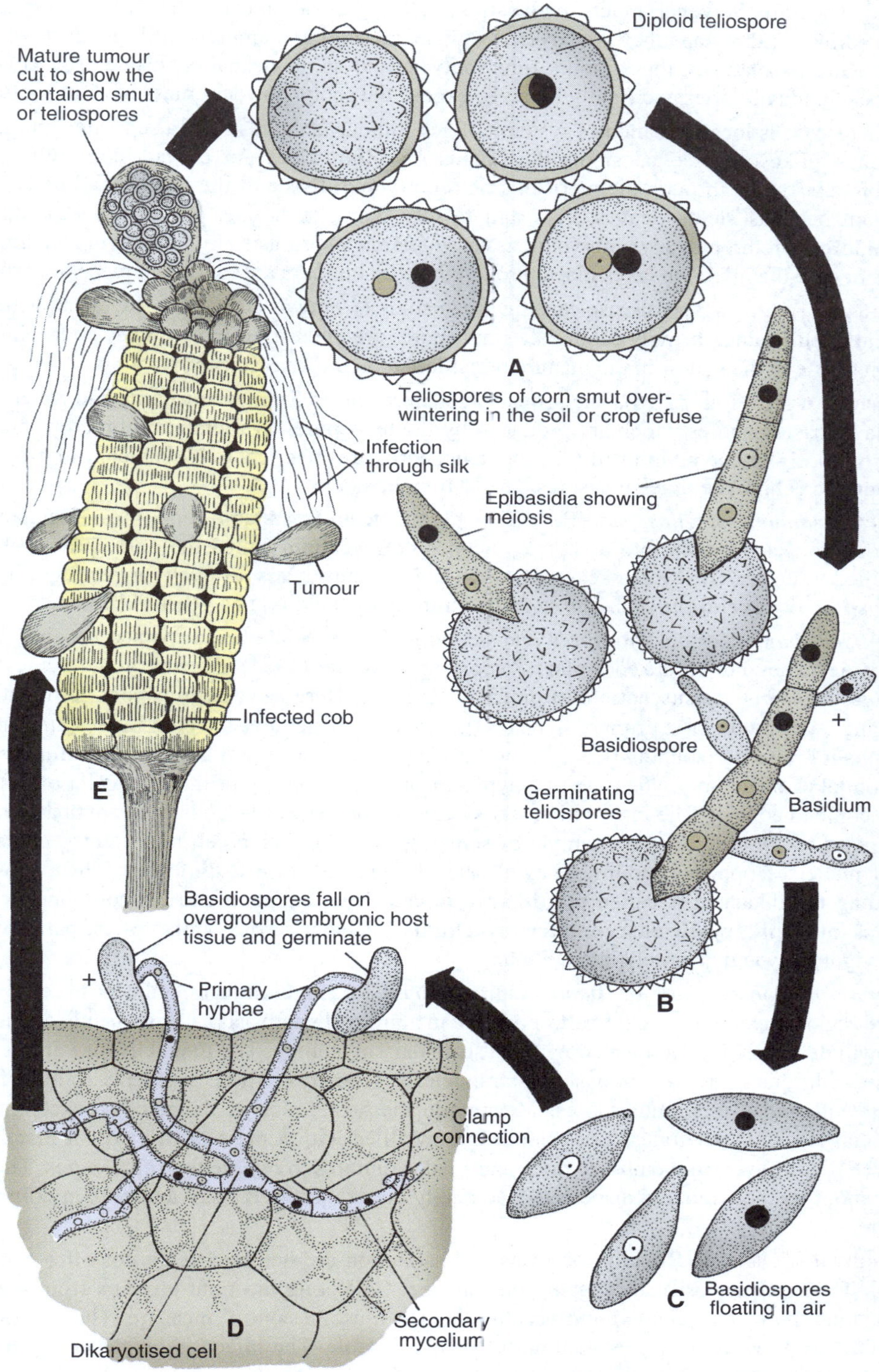

Fig. 22.16. (A-E) Corn Smut. Pictorial disease cycle of *Ustilago maydis*, an example of general infection through any overground embryonic tissue.

flowers). They are, however, much smaller in size. The galls at an early stage are light coloured and are edible. At this stage they are covered with a firm shining membrane. It is greenish white in colour. Towards maturity the tumours turn sooty due to spore formation inside. The covering membrane gradually dries and finally bursts to expose the sooty, powdery mass of smut spores.

The disease is localised and not systemic. Each tumour is the result of a separate and distinct infection. Unlike other cereal smuts the fungus mycelium does not pervade the entire plant. Infection takes place through the meristematic (embryonic) tissue of the overground parts of the host plant such as stems, leaves, ears and tassels. Corn smut thus provides an example of **general infection** through any embryonic tissue. Infection takes place any time during the growing season. In case the infection takes place through the silk, tumours are formed in place of kernels.

Causal organism. The disease causing organism is *U. maydis* (Dc) Cda. The black, generally spherical to ellipsoidal, heavily echinulate smut spores are the resting spores. They over-winter on the crop refuse in the soil or in the manure of cattle fed on the diseased corn.

Disease Cycle (Fig. 22.16). The primary inoculum comes from the air borne and water borne sporidia or basidiospores which are produced by the teliospores overwintering on crop refuse or in the soil rarely on the grain itself (A). The teliospores are spherical to ellipsoidal in form, black in colour and echinulate, Each measures 8 to 11 μ in diameter.

(*a*) *Germination of teliospores* (B_{1-3}). They germinate under suitable conditions of temperature and moisture. Each grows into a septate, slender promycelium called the **epibasidium**. It is a four-celled structure (B_2). Each segment of the epibasidium bears an ovate, unicellular, hyaline **basidiospore** or **sporidium** (B_3). The sporidia multiply by budding in a yeast-like manner.

(*b*) *Germination of basidiospores and infection*. The sporidia are dispersed by air currents (C) or water. They fall on the corn plant. According to Walter (1934), the embryonic tissues of all above ground parts are susceptible to direct penetration. Here each basidiospore germinates by producing a germ tube or a primary hypha (D). It directly penetrates the corn tissue. Entering the epidermis it grows horizontally beneath it to form a **primary** or **monokaryotic mycelium**. Active infection takes place only when the two hyphae from neighbouring primary mycelia of opposite strains come in contact. The intervening walls between the vegetative cells of the hyphae dissolve at the point of contact. It is an example of somatogamy. The fusion cell thus formed contains a pair of nuclei of opposite strains. They lie side by side to form a **dikaryon**. The fusion cell containing the dikaryon is called the **dikaryotic cell**. By elongation and clamp connections, it develops into a **dikaryotic** or **secondary mycelium**. The latter carries on the role of parasitism. It grows vigorously and is largely intercellular.

(*c*) *Formation of Galls or Tumours*. In response to fungus infection the host cells in the infected region increase in size (hypertrophy) and in number (hyperplasia). Consequently a tumorous outgrowth develops. It is permeated with the secondary mycelium. The fungal hyphae at this stage become additionally septate into smaller binucleate segments. These segments round off and separate into binucleate smut spores. Meanwhile the host tissue in the tumour, excepting the superficial layer, disintegrates. The tumour is now filled with a mass of myriads of black smut spores (E). The covering membrane dries and finally ruptures to release the smut spores. They are blown about by the wind and dispersed. Eventually they return to the soil and remain dormant in winter.

Control Measures. Rotation of crops and sanitation are helpful but do not offer complete control. They simply assist in reducing the damage. Collecting the smut tumours from the field before smut spores are released and burning them is a useful control measure. The best control measure is to grow the most resistant corn hybrid available. The infection incidence can also be reduced by the application of insecticides such as DDT, Nicotine etc.

10. STEM OR BLACK RUST OF WHEAT

Host : Triticum vulgare

Pathogen : Puccinia graminis tritici.

General. The disease occurs in all wheat growing countries of the world. In India it appears at different times of the year in different parts of the country. It appears in the month of March in Northern India. In Southern and Peninsular India it appears very early in the 4th week of November.

Stem Rust of Wheat (caused by Puccinia graminis tritici)

Effects and Importance. The stem rust is a serious threat to wheat in India. It causes serious damage in moist areas and moist season. The nature and extent of damage varies from slight to almost complete failure of the crop. The annual loss from rusts in India was estimated to be about Rs. 40 million (Butler and Hayman, 1906). Mehta (1942) put this loss at about Rs. 60 million annually both in wheat and barley. Prasad (1960) reported that nearly one million tons wheat worth about Rs. 392 million was lost during 1958-59. The losses include reduction in the quantity of the thrashed grain (low yield), its poor quality and shrivelled lighter grains. These effects are due to two factors:–

1. Robbing off of the food by the pathogen for its own use.

2. Excessive transpiration due to numerous slits in the host tissue (epidermis). These factors result in lessened metabolism, in partial sterility, reduction in the number of grain in the head and shrivelled grain.

Samborski and Shaw (1956) reported increase in the respiratory activity of rusted plants especially at the site of infection. Shaw and Colotelo (1961) found increase in total nitrogen, protein nitrogen, soluble nitrogen and the ratio of soluble to insoluble nitrogen as the rust develops. Prasad (1967), opined that aminoacids and proteins are synthesized by the rust fungus itself.

Symptoms on the Grain host. Elongated pustules or streaks appear chiefly on the stem and leaf bases, leaf sheaths and even on the glumes. These are the **uredia** or **uredosori**. Each uredosorus contains a mass of reddish brown or rusty red, one-celled, binucleate **uredospores.** They are exposed by the rupture of the leaf epidermis. The torn epidermis forms a collar-like structure around the oblong sorus. The uredospores are produced by the dikaryotic or secondary mycelium parasitizing the tissue of the host. This spore stage of the parasite is called the **red rust** or **summer stage**. Later in the season the sori turn black. This is due to the appearance of two-celled, black teleutospores or teliospores in the place of uredospores in the old uredia. When the grains are almost ripe, new and independent oblong to linear teleutosori make their appearance on the stem of the host. They are smooth and dark brown or black in colour. The overlying epidermis is ruptured and the spores are exposed. The stems at this stage are dry and cracked. The teleutosori contain masses of black two-celled spores called the **teleutospores**. There is no other marked abnormality in the appearance of the plants. However, the plants look sickly and fail to develop normal ears in the case of a severe attack.

Symptoms on Barberry Host. The infected barberry leaves show small circular, slightly elevated, yellowish or orange yellow spots on the upper surface. Each spot is dotted with an indistinct pimple, an ostiole of the **spermagonium**. A sticky liquid begins to exude through the ostiole. On the lower side of the leaf there appear cupshaped **aecidia** or **aecial cups**. They usually appear opposite the spermogonial or pycnidial spots. The aecidial cups are orange yellow in colour. Each one of them contains a mass of orange coloured, binucleate spores called the **aecidiospores** or **aeciospores**. The infection causes no serious damage to the Barberry host.

Causal Organism. The causal agent of this disease is *Puccinia graminis tritici* Erikss. and Henn. (*P. graminis Pers.,*). It is a heteroecious parasite which completes its disease cycle in two hosts namely wheat and barberry (*Berberis*) or *Mahonia*.

Disease Cycle (Fig. 22.17). Eriksson (1896, 1910, 1922) on the basis of his observations in cereal rusts, especially *Puccinia glumarum* proposed the "mycoplasm theory". He held that on the onset of winter, the fungal hyphae degenerated in the host plant. The fungus cytoplasm mingled with that of the host protoplasm in the cell. As soon as the winter was over, the mycoplasm (fungal cytoplasm) migrated into the intercellular spaces of the host. There the mycoplasm reforms the hyphae and haustoria. The mycoplasm theory was vehemently opposed by (Bolley, 1898; Ward, 1903 a, b; Kiebahn, 1904 a, b). It was not widely accepted because of opposition and thus fell by the wayside. The commonly held view is that normally teleutospores are the overwintering structures (A). But in the plains in India uredospores constitute the primary inoculum. They are carried from the distant high altitude hills by wind. After the usual resting period the teleutospores germinate *in situ* (on wheat stem and stubbles in the field). Each cell produces a short promycelium (epibasidium) into which the synkaryon migrates (B). Each synkaryon undergoes meiosis in the promycelium or the epibasidium. Segregation of the sexual strains takes place. Walls are laid between the haploid nuclei so that each promycelium or epibasidium becomes septate and four-celled (B). Each cell of the basidium produces a single, uninucleate haploid basidiospores at the end of a sterigma. Of the four basidiospores thus produced two are of plus strain and two minus strain.

The basidiospores are disseminated by air currents. While floating in the air they may chance to fall on young barberry leaves. If temperature and moisture conditions are favourable the basidiospores germinate. Each basidiospore develops a germ tube or a primary hypha (C). It penetrates the cuticle directly and brings about infection of the new host (Waterhouse, 1921). Within the host tissue the primary hypha branches freely to form a monokaryotic or haplomycelium. The hyphae constituting it ramify in the intercellular spaces between the mesophyll cells. The cells are uninucleate. The nuclei in the mycelium are either of plus strain or minus strain depending upon the nature of the germinating spore. The mycelium feeds and grows vigorously. Eventually it enters the reproductive phase and forms thick mats of hyphae here and there beneath the upper and lower epidermis. The hyphal mats beneath the upper eipdermis function as primordia of spermogonia. Those beneath the lower epidermis function as primordia of aecidia or aecia. In about a week's time the primordia beneath the upper epidermis produce small flask-shaped fruiting bodies called the **spermagonia** (Allen, 1930). They are embedded in the tissue in orange yellow spots on the upper surface of the leaves of barberry bush (E). Each spermagonium opens to the outside through a small aperture called an **ostiole** which projects above the surface of the leaf. The spermagonium contains three types of hyphal threads:

1. *Periphyses*. These are slender sterile, hyphae guarding the ostiole and projecting through it.

2. *Spermatiophores*. These are numerous, fine, elongated hyphae arising from the interior of the swollen portion of the spermagonium. They abstrict small, hyaline spermatia at their tips in succession. The abstricted spermatia lie free in the cavity of the spermagonium.

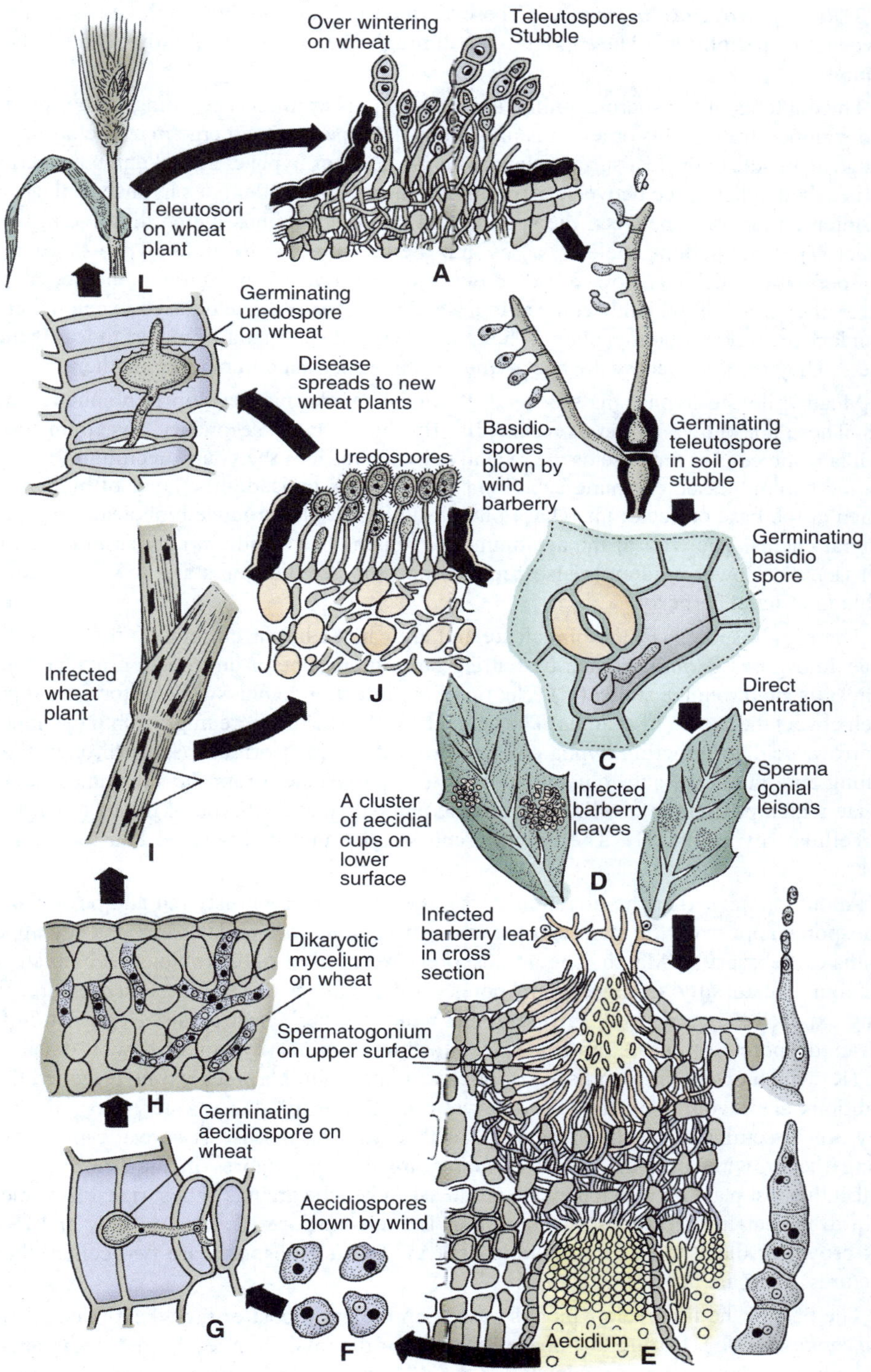

Fig. 22.17. Black or Stem Rust of Wheat. A pictorial disease cycle of *Puccinia graminis tritici*.

3. *Receptive or Flexuous hyphae*. These are the fine, hair-like hyphal threads seen interspersed between the periphyses. They extend out through the ostiole and project much beyond the periphyses.

The contents of the spermagonium are entirely plus or minus according as the spermagonium has developed from a plus or a minus mycelium. The spermatia emerge in a viscous sugary liquid through an ostiole to the leaf surface along with the flexuous hyphae. Sexual union or spermatisation, as it is called, takes place between spermatia of one strain and flexuous hyphae of the other strain. The intervening walls between the spermatium and the flexuous hypha dissolve at the point of contact. The spermatium nucleus passes into the receptive or flexuous hypha through the pore. The spermatium nucleus now passes down the receptive hypha through the septal pores and reaches the basal cell which becomes binucleate or dikaryotic. The dikaryotic cell develops into a secondary or a dikaryotic mycelium. The transference of spermatia from leaf to leaf is the work of insects. They are attracted by the nectar and visit one spermagonium after another.

Meanwhile the hyphal mats beneath the lower epidermis develop into spherical masses of cells. These are known as the **protoaecidia.** By this time the secondary mycelium formed from the dikaryotic cells at the base of the receptive hypha reaches the young aecidium. Its cells mingle with the haploid tissue of young aecidium. As a result a palisade-like layer of binucleate cells is formed at the base of aecidium. These binucleate basal cells produce binucleate aecidiospores in terminal chains. The wall of the aecidium splits open. The aecidium now assumes a cup-shaped form (E). The lower epidermis also ruptures and the aecidiospores are now exposed. They are unable to reinfect barberry.

The aecidiospores (F) are binucleate and are carried by air currents to the wheat host. Here the aecidiospores germinate each by putting out a germ tube or an infection hypha which enters the host tissue through a stoma (G). The tip of the infection hypha swells to form an appressorium which covers the mouth of a stoma. The contents of the aecidiospore migrate into the appressorium. A narrow, peglike infection hypha emerges from the appressorium, passes through the stomatal opening and enters the substomatal cavity to form a vesicle. From the substomatal vesicle arise hyphae which proceed intercellularly into the parenchymatous tissue of the host leaf to form the intercellular mycelium (H). The hyphae send small round or branched haustoria into the host cells.

About ten days after the infection of the grain host (wheat) rusty red and powdery masses of uredospores appear on the stem and the leaves in oblong to circular sori (I). They appear in the months of February to March. The uredospores are shortly stalked, oblong, echinulate structures with four equatorially arranged germ pores on the outer wall (J). These spores are binucleate. Being exposed they are easily carried by the wind to other wheat plants. Here, within few hours each uredospore germinates in the surface moisture provided by rain or dew. It develops a germ tube (K) which enters the host tissue through a stoma. Within a week's time provided the weather conditions are favourable, the new dikaryotic mycelium produces a new crop of uredospores. They are disseminated in the same way. In this way the disease is spread rapidly and widely during the growing period. The uredospores are the only spores in the life cycle which can reinfect the host plant on which they are produced. When the grain is almost ripe, black teleutospores begin to appear in the uredosori. Soon after the teleutospores develop in new and independent dark brown or black teleutosori or teliosori (A). The teleutospores are two-celled, thick-walled structures. They are the resting spores.

The part of the life cycle which is passed on the grain host or the wheat plant represents the *dikaryophase* (H-L). During this phase each cell of the mycelium, each uredospore and each cell of teleutospore has a pair of nuclei called the **dikaryon**. One of these nuclei is of plus strain and the other of minus strain. Towards maturity plus and minus strain nuclei in each cell of the teleutospore fuse to form a *synkaryon* or the fusion nucleus. The mature teleutospores thus represent the reduced diplophase (A).

Puccinia graminis tritici is a **polymorphic species** as it produces a succession of different types of spores. However, investigations carried out by late Dr. K.C. Mehta (1923-40) in India revealed that there is no local source of infection in the plains. The uredospores perish in the high summer temperature of the plains. There are no barberry bushes in the plains and thus the teleutospores which are unable to survive the high summer temperatures following the harvest remain ineffective. Hence, the sexual stage comprising the spermagonium and aecidia is cut out from the life cycle. The actual disease cycle is completed with uredospores alone. The teleutospores are produced but they are non-functional. This eliminates the chances of perennation of the rust on the alternate barberry host in the hills. This has been confirmed by the fact that the races of *P. graminis* found on the barberry bushes on the hills are different from those occurring on the cereals in the plains.

In spite of the above-mentioned facts, there is annual recurrence of the rusts in the plains in India. It is due to the fact that the uredospores remain viable on the hills. The summer temperature at an altitude of 5,000 ft. and above is quite congenial for their survival. There at different altitudes, the uredospores oversummer on the out of season wheat plants, stubbles, and grass hosts in the uredial stage. These serve as primary inoculum for the wheat crops when sown. The infection thus starts from the hills. The uredospores are wafted down from the higher altitudes to the foot hills. From there the infection is carried to the crops in the plains.

Control Measures

1. *Cultivation of Rust Resistant Varieties*. The cultivation of varieties immune to the rust disease is an important means of combating the disease. Some rust resistant varieties of wheat are available in India. Np 710, Np 718 and Np 770 find favour with the farmers. The newly bred hybrids Np 822, Np 823 and Np 825 have given good results. They possess high degree of rust resistance. In addition they are high yielders. The recently introduced dwarf Mexican wheat varieties such as Sonora 64 and Lerma Rojo are almost completely resistant to black rust.

2. *Eradication of Barberry*. Formerly it was believed that the eradication of the less important alternate host barberry is a possible means of eliminating the disease. We now definitely know that the control of stem rust by this method is not possible in India. The uredospores which are able to survive on stray and self-sown wheat plants on the hills serve as an inoculum.

3. *Use of fungicides including antibiotics*. Despite ceaseless efforts to control wheat rusts through resistant varieties, much success has not been achieved. Practically no variety is resistant for a long period due to emergence of new physiological races. Hence the use of fungicides to control rust diseases has received much more attention in the recent years than in the past. Grewal and Dharam Vir (1959), Mathur *et al* (1961) and Tandon *et al* (1968) recommended Zineb and Maneb to control wheat rusts effectively. Srivastava, Rai and Aggarwal (1972) recommended the use of Dithane Z-78 and Dithane M-45. The latter controls the disease even at lower conc. (1.5 lbs/acre) when applied with sticker. The yield increased by 33.3% over the control. A number of other chemicals like RH-124 and Plantavax have as well given quite encouraging results. Heagle and Key (1973) reported that Ozone (O_3) inhibits infection, hyphal growth and uredospore formation of wheat stem rust fungus. Actidione has been recommended to be a useful antibiotic as a fungicide for the control of rusts plus zinc sulphate at fifteen days interval from the first week of February are quite effective.

11. RUST OF LINSEED (FLAX)

Host : Linum usitatissimum.

Pathogen : Melampsora lini.

General. Rust of linseed is a common disease of flax growing areas of the world. In India flax is a major oilseed crop cultivated in almost all the states. The crop is generally cultivated during October–April. The disease generally appears in February or later but Butler (1918) has observed the disease in central India in early November. At the time of crop harvesting the affected plants get a fired appearance due to the presence of telial sori.

Linsed Rust (Linum usitatisimmum)

Effects. The disease causes disastrous effects on the plants. Once the disease sets in a field, most of the linseed fields in the locality get affected within no time. The disease generally appears in epidemic form. Severly infected plants get mostly defoliated due to which starch formation is also reduced. Most of the food reserves of the host is consumed by the pathogen resulting in decrease in the yield of seeds and quality of fibres. Oil content of the seeds is also reduced.

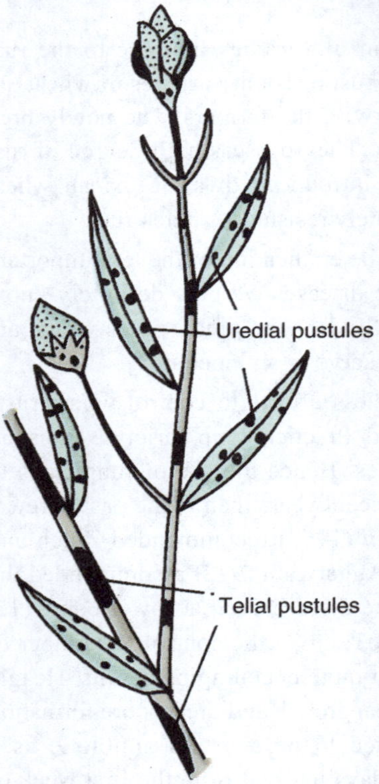

Fig. 22.18. Symptoms of Rust of Linseed.

Symptoms. The leaves are the first to show the symptoms and gradually all the aerial parts of the plant get infected. Large, orange coloured pustules generally appear on the leaves. Small pustules are initially surrounded by chlorotic areas. Little necrosis of the leaves is at first observed but it grows, becomes more general and the leaves prematurely die. The pustules on the leaves are uredopustules containing uredospores. Uredopustules may also appear on stems. While the uredopustules on leaves are round and small, those on stems are elongated and irregular.

Teleutopustules are generally produced on stems and rarely on leaves, if they have not been shed prematurely. Orange yellow uredopustules are often surrounded by reddish-brown teleutopustules. The contents of telial pustules do not break the epidermis of the host and remain buried subepidermally appearing glossy.

Causal organism. The causal organism is *Melampsora lini* (Pers.) Lev.–an autoecious rust.

Disease cycle. In temperate countries, primary infection takes place through basidiospores which are produced as a result of germination of teleutospores perennating in the soil. But in tropical countries, the teleutospores, produced at the end of growing season i.e. April-May, loose their viability due to excessive heat of summers. Uredospores too are killed due to excessive temperatures. It is presumed that the uredospores produced on linseed at hills come down to plains to cause infection. Thus the primary inoculum, wind blown, fall on the host, germinate and cause infection. Because the transportation of uredospores from hills takes some time, the out break of the disease in plains takes place only after 2-3 months of the sowing of the crop.

Uredospores produced as a result of primary infection cause secondary infection. Since the uredospores take very little time to germinate the secondary infection is caused within a short time. Secondary infection alongwith primary infection cause heavy damage to the seed crop. Teleutospores produced at the end of the growing season appear to have no role to play in the continuance of the life cycle of the rust.

Control Measures

(*i*) Use of disease resistant vars. like NP (RR) 9, 10, 56, 95, 218, 279B 279K3, 368, 381, 389, 415 and 501 is the only effective method to control the disease.

(*ii*) Seed treatment to kill the teleutospores in hills has also been suggested.

(*iii*) Avoidance of excessive nitrogenous manures is recommended.

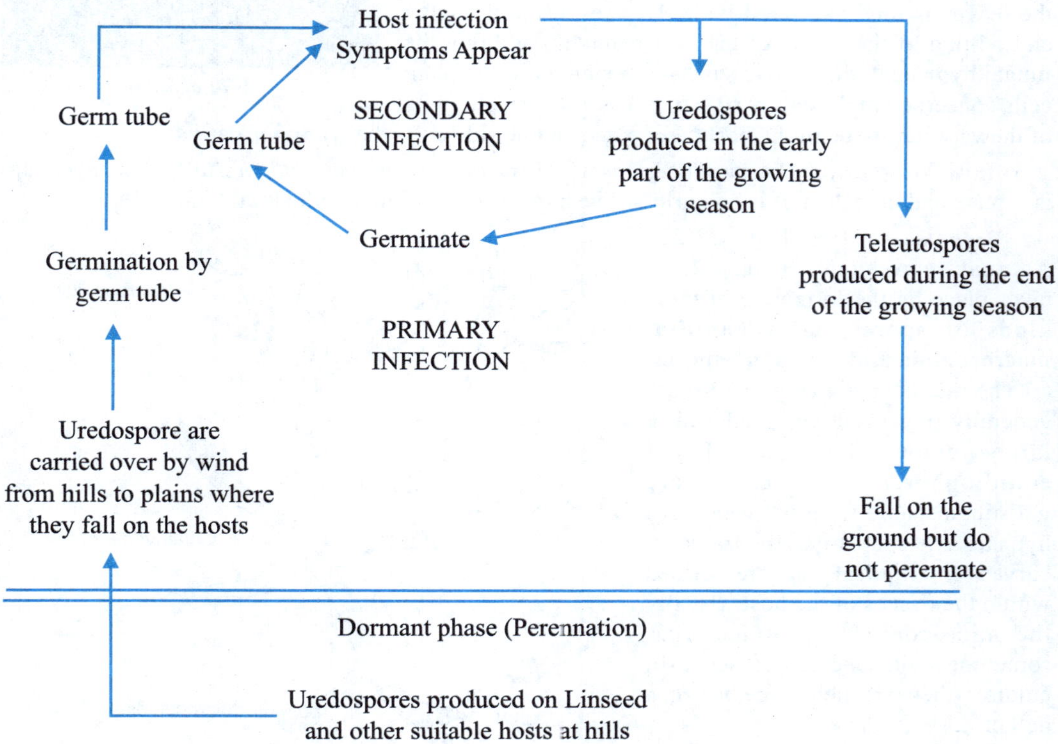

Fig. 22.19. Disease cycle of Rust of Linseed.

(d) DISEASES CAUSED BY DEUTEROMYCETES

12. WILT OF ARHAR

Host : Cajanus cajan
Pathogen : Fusarium oxysporum f.udum

General. The wilt disease of arhar is a most severe threat to the arhar producing states in the country such as U.P., M.P., Bihar and in parts of Bombay.

Symptoms. The susceptible plants are attacked when young about five to six weeks old. The symptoms are variable. Typically the first symptom is the premature yellowing of the leaves. The next symptom is the wilting or withering of the leaves of the diseased plants. The plant is in a blighted condition. The leaves appear to be affected progressively from the bottom towards the top. Sudden wilting is rare. Finally the entire plant completely dries up. In severe cases of the infection more than half the plants in the field may fall victim to this disease.

Wilt of arhar

The wilting is brought about by the plugging of the vascular tissue of the stem and roots of the host plant by dense masses of mycelial hyphae. The free flow of water to the leaves is thus interfered with. This results in drooping and wilting of the leaves of the host plants. In addition the fungal hyphae produce toxic substances which kill the plant cells concerned in the ascent of sap. In cases where plugging of the vascular tissue of the host takes place on one side only, the wilting is partial.

Causal organism. The causal organism is *Fusarium oxysporum f. udum* Butler. The mycelium is septate and branched. It is colourless. The hyphae are both intercellular and intracellular.

Disease cycle. (Fig. 22.22) Asexual reproduction (Fig. 22.20) takes place by the formation of three kinds of spores, **microconidia, macroconidia and chlamydo-spores.** (*i*) The microconidia (Fig. 22.20) are generally tiny structures produced in large numbers from the tips of conidiophores which are indistinguishable from the vegetative hyphae. The spores may be rounded, curved or elliptical and are formed within the tissues of the host. Usually the microconidia are unicellular, sometimes with one or two septa. In cultures they are held together in a ball or false head.

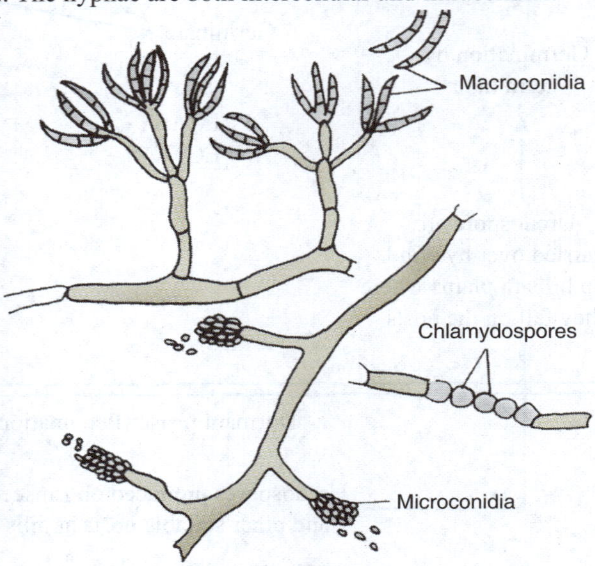

Fig. 22.20. *Fusarium.* Asexual Reproduction.

(*ii*) Macroconidia (Fig. 22.20) are long, sickle-shaped or crescent-shaped spores. They are septate (with 3-4 septa) and have pointed tapering ends. The macroconidia are produced at the tips of short conidiophores which arise from the upper surface of cushion-shaped stromatic mass of hyphae. The whole fruit body is called the **sporodochium** (Fig. 22.21). The macroconidia are shed as soon as they reach maturity.

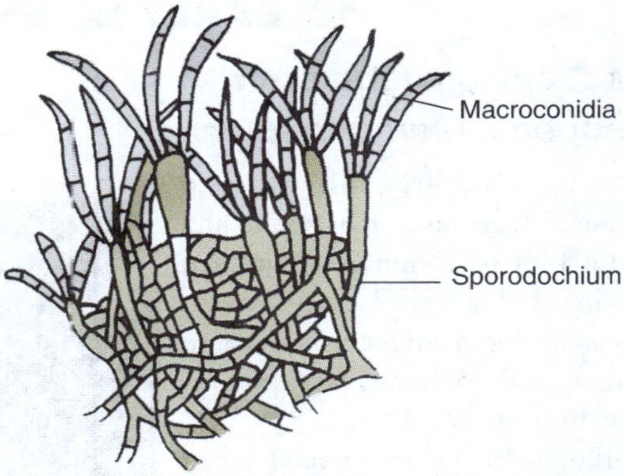

Fig. 22.21. Sporodochium of *Fusarium*

(*iii*) The chlamydospores (Fig. 22.20) are thick-walled, rounded or oval cells formed in the hyphae. They occur singly or in chains and may be terminal or intercalary. The chlamydospores become separated from the parent hypha after maturing. They function as resting spores.

Infection. Wilt of arhar is a soil borne disease. The microconidia, macroconidia and chlamydospores remain viable in the soil. At the sowing time the spores germinate and initiate new infections. Infection takes place through the young roots or rootlets possibly through the root hairs. Infection of the aerial parts of the plant through air-borne spores never takes place. The fungus mycelium also exists in the soil as an active facultative saprophyte so long the dead host roots and other parts are present in the soil. It gains entry into the host through the rootlets and finally passes into the larger roots.

Control Measures. The best method to control this disease is to produce wilt resistant varieties. Varieties N.P. 15 and N.P. 38 are fairly resistant to the disease. Crop rotation, field sanitation and hot weather cultivation are some of the methods to eliminate soil borne infection. The other recommendation is to protect the plants from the attacks of the wilt organism by the use of antibiotics and chemicals. Nothing positive has been done in this direction so far. Vasudeva, *et al.* (1952, 1954) reported that *Bacillus subtilis* produces an antibiotic called **bulbuformin** which inhibits the growth of the pathogen.

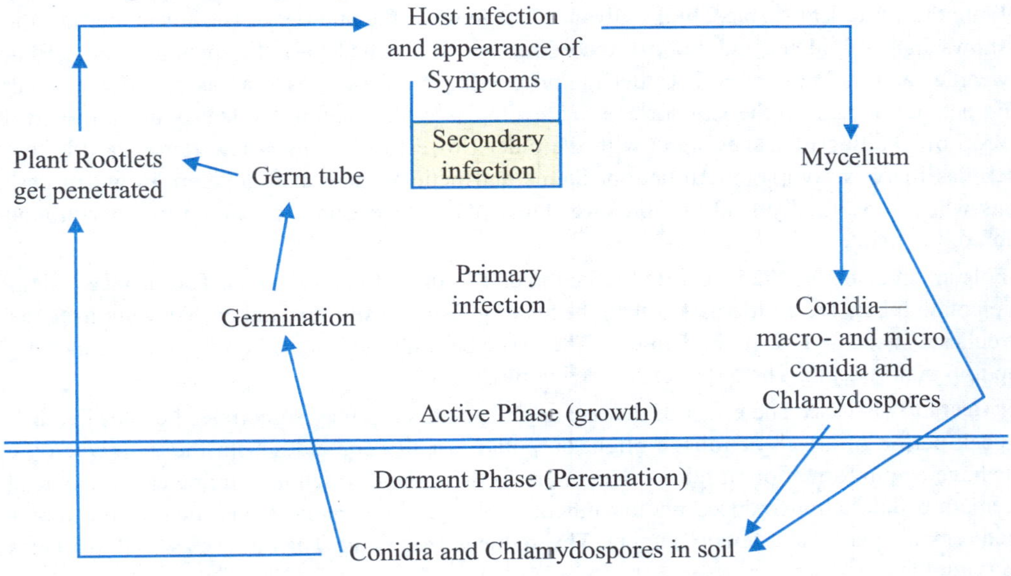

Fig. 22.22. Disease cycle of wilt of Arhar.

13. EARLY BLIGHT OF POTATO

Host : *Solanum tuberosum*
Pathogen : *Alternaria solani*

General. Early blight is a common foliage disease of potato and tomato. It is of common occurrence both in cold as well as in warm regions in India and abroad where potatoes and tomatoes are grown. According to Mathur, Singh and Nagarkote (1971), there have been outbreaks of early blight of potato during the past 5 or 6 years in the plains of Uttar Pradesh. The pathogen causes injury to the leaves as a **leaf spot disease** and instigates permature defoliation. Finally it incites **tuber rot** of potato and **fruit rot** of tomato.

Early Blight of Potato (caused by Alternaria solani)

Symptoms. The disease appears on the leaflets, 3-4 weeks after the crop is sown as small, isolated, scattered pale brown to dark spots, oval or angular in shape mostly up to 3 or 4 mm in diameter. Each spot is usually delimited by a narrow chlorotic marginal zone which fades into the normal green. The chlorotic zone increases with the increase in size of the spot. As a rule the oldest (lowest) leaves are affected first and the disease progresses upwards. The necrotic tissue of the spot often shows a series of concentric ridges which produce a target-board effect, a symptom characteristic of this disease (Fig. 22.23 D). The number of spots on the leaflets may be a few but if the conditions are favourable the spots increase in number and size involving the entire leaf surface. In the advanced stage when the number of spots is numerous, the leaf shows signs of old age and droops. According to R.S. Singh (1968), the spots become hard in dry weather and the leaflets curl. Under humid conditions, the diseased areas coalesce and big rotting patches appear on the leaf surface. In severe cases of infection the leaves dry up, shrivel and drop off. Falling of leaves starts with the older (lower) ones until a few remain at the top. Sometimes there is complete defoliation. Stems and petioles may also develop brown to dark lesions which may finally lead to either worthless plants or collapse of the entire over-ground portion of the plant.

Folsom and Bonds (1925) reported infection and rotting of potato tubers. The surface lesions on the potato tubers are a little darker than the healthy skin and slightly sunken. They are irregular or circular in shape up to 2 cm in diameter. The tissue beneath the lesion shows a brown corky dry rot up to 6 mm in depth. The older lesions develop fissures.

Effects of Disease. The effect of disease on the potato crop may sometimes be considerable. The peak period of attack of foliage blight on potato usually coincides with the time when the plants have begun formation of tubers. The injury to the foliage which in extreme cases may lead to premature defoliation reduces photosynthetic activity of the plant. Consequently the tubers formed remain small and a few in number. This reduces crop yield. The disease as well instigates rotting of tubers.

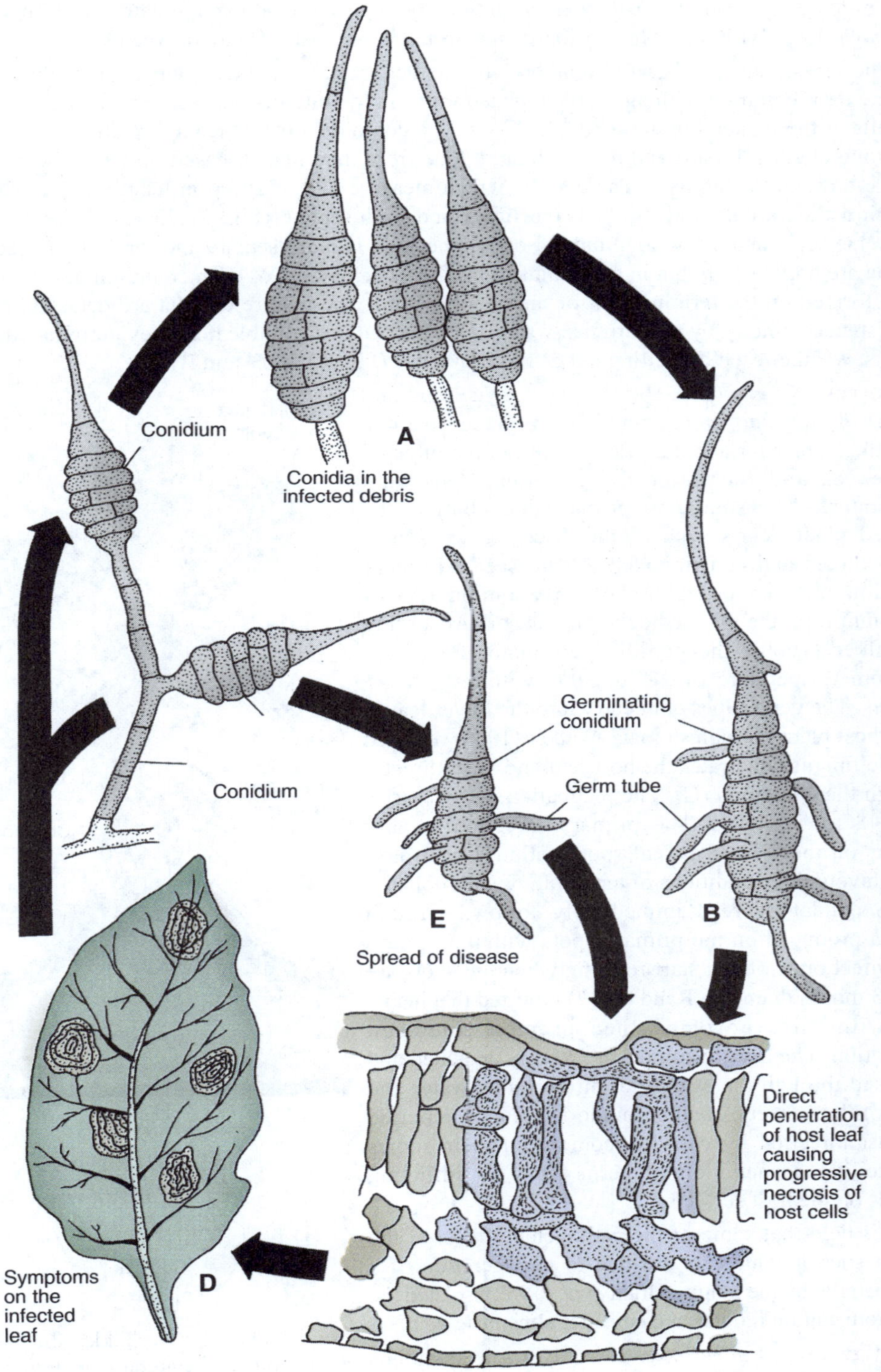

Fig. 22.23 (A–B). Early blight of potato—Pictorial disease cycle of *Alternaria solani*.

Causal Organism. The pathogen causing early blight of potato and tomato is a form-species *Alternaria solani* (Ell. and Martin) Jones and Grout of form class Deuteromycetes.

The mycelium consists of light brown, slender, septate sparsely branched hyphae which become dark-coloured with age. The hyphae ramify in the intercellular spaces but later penetrate the cells of the invaded tissues (Fig. 22.23 C). The conidiophores which are relatively short (50–90 μ long and 9 μ broad) and dark-coloured arise from the older diseased tissue of the host and emerge through the stomata. The conidia which measure 120 to 296 μ in length and 12–20 μ in breadth are dark-coloured, beaked, muriform and multiseptate (Fig. 22.23 E). There are 5–10 transverse septa and a few longitudinal ones. The beak is long, septate and rarely branched. The conidia are borne singly but in pure cultures, in short chains of two. Each conidium develops from a bud formed on the terminal cell of the conidiophore. The mature conidia are detached readily and dispersed chiefly by air currents, water and insects. On a suitable host they germinate readily in moist weather each by putting out 5–10 germ tubes (Fig. 22.23 E^1 and B).

Disease Cycle (Fig. 22.23). The mycelium and conidia of the pathogen remain viable for a considerable time, the former for about a year or more in the infected dry leaves and the latter for 17 months at room temperature. The source of primary inoculum is the infected plant debris such as the dried leaves, stems, potato tubers and contaminated tomato seeds. Primary infection may be brought about by conidia (A) or mycelium from the infected debris in the soil. According to Walker (1969), the conidia germinate (B) at the optimum temperature of 28° to 30°C within 35 to 45 minutes. The germ tubes gain entry into the lower leaves of the host plant through stomata. Walker (1969) reported that the fungus penetrates the host leaf and stem directly through the epidermis (C). The incubation period varies from 48 to 72 hours. The primary infection usually appears on the foliage as leaf spots within 2 or 3 days under favourable conditions of temperature and moisture (D). Secondary spread immediately follows through conidia produced on the primary spots within 5–7 days after infection. Conidia start forming when the spots are about 3 mm in diameter. Rand (1917) reported that heavy dew with rains now and then promote abundant sporulation. The mature conidia are readily detached and dispersed through the agency of air currents, water and insects. Flees, beetles and colorado bettles are reported to assist infection by carrying conidia on their bodies and facilitate penetration of germ tubes by inflicting wounds on the host surface.

It is held that some of the pathogenic effects of the disease such as chlorosis are due to the secretion of a toxic material by the fungus. Brain *et al.* (1952) isolated the toxin from culture filtrates and named it **alternaric acid**.

Epidemiology of Disease. Climatic or soil conditions which are unfavourable to the host and thus reduce its physiological vigour tend to promote the disease.

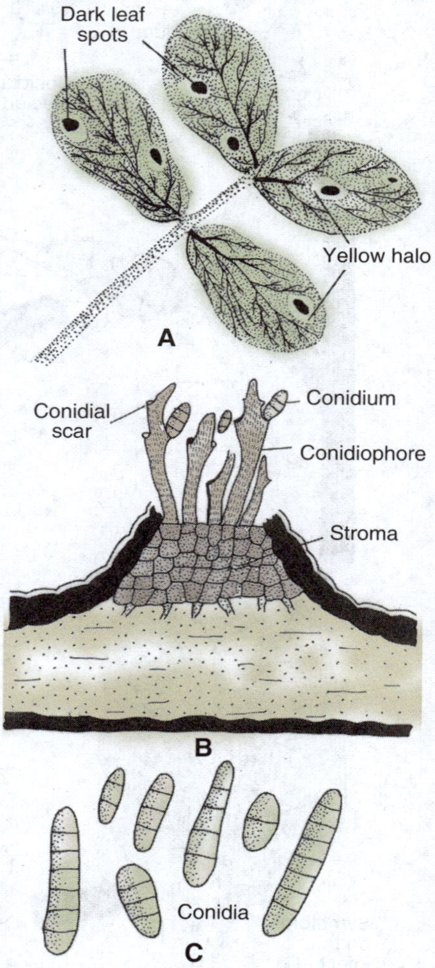

Fig. 22.24 (A–C). Tikka disease of groundnut. A, symptoms on leaflets; B, geniculate conidiophores arising from the stroma; C, septate conidia.

The disease spreads and becomes serious when the season begins with abundant moisture followed by high temperatures because these conditions are unfavourable to the host. They reduce its physiological vigour. Weaker plants are more susceptible to disease than the healthier ones. On the other hand, high soil fertility tends to reduce the severity of disease. Periods of continued drought also check its spread.

Control Measures. As in the case of other soil borne diseases, crop rotation and field sanitation are the two rational measures which provide effective check against primary infection from spores in the infected plant debris. Infected dry leaves and dead haulms should be raked together and burned immediately after harvest.

Application of regular spray schedule is another effective measure. When the plants are 15–20 cm high, weekly sprays (with Bordeaux mixture or other suitable fungicides) throughout the period of plant growth effectively control the early blight. Mathur, Singh and Nagarkote (1971) found that spraying potato crop with Dithane M–45 was most effective against early blight disease. It also resulted in significant increase in the yield. Vorster (1962), Bebrchen and Brien (1965) and Harison *et al.* (1965) reported similar results with manganese carbamate from other parts of the world. Compacci and Santo (1959), Paharia (1961) and Kadyroa (1964) recommended repeated application of Zineb. The use of Brestan 60 has also proved profitable.

14. LEAF SPOT (*TIKKA*) DISEASE OF GROUNDNUT

Host : Arachis hypogea L

Pathogen : Cercospora sp

Leaf Spot of Groundnut (caused by Cercospora sp.)

General. One of the best known leaf spot diseases is that of *Arachis hypogea L.* (groundnut). Popularly it is called the **tikka disease**. The tikka disease is a serious disease occurring in areas where the groundnut crop is grown in India. The spots appear on the host leaves when the plants are one or two months old. Later necrotic lesions appear on the stem as well. In fact there are two leaf spot diseases of groundnut caused by two different species of form genus *Cercospora* namely, *C. personata* and *C. arachidicola*. The leaf spot disease caused by the former is more common, dangerous and does greater damage than the latter. The spots produced by *C. personata* are numerous. These weaken the host plant and lead to defoliation which adversely affects size and quality of the fruit. The latter remains small in size and fewer in number. Sundaram (1965) reported that in the case of severe infection the crop yield may be reduced to 22%.

Causal organism. Tikka disease of groundnuts is caused by two species of *Cercospora : Cercospora personata* (Berk. & Curt.) Elle. and Eve., now known as **Cercosporidium personatum** (Berk. & Curt.) Deighton and **Cercospora arachidicola** Hori. The two form-species differ from each other with respect to the size, shape and colour of necrotic lesions they produce, conidia formation and the nature of the mycelium. These differences are summed up in the Table given below:

TABLE

C. personata (Fig. 22.24)	*C. arachidicola*
Symptoms :	
1. Leaf spots are small, more circular and about 1–6 mm in diameter (A).	The leaf spots are comparatively larger in size, irregularly circular in outline and 4–10 mm in diameter.
2. The necrotic lesions appear on both the leaf surfaces and change from dark brown to dark in colour.	The leaf spots are often confluent and necrotic lesions occur on both the surfaces. Those on the upper surface are reddish brown to black and on the lower surface are light brown.
3. Young spots lack bright yellow halo which, however, develops around the older ones.	There is a yellow halo around each spot but halos on the lower surface are less distinct.
Mycelium :	
4. The mycelium consists of hyphae which are entirely internal.	It consists of both external and internal hyphae.
5. The septate hyphae ramify in the intercellular spaces and obtain nutrition by sending branched haustoria into the mesophyll (both spongy and palisade) cells.	The internal hyphae are both intercellular and intracellular. The haustoria are absent.
6. The unbranched conidiophores arise in tufts from a dense, globular, brown to black stroma (23–30 μ in dia.) and emerge by rupturing the host epidermis (B).	The unbranched conidiophores arise scattered from a deep brown stroma (25–100 μ in dia.).
7. They are geniculate, olivaceous brown, 24–54 μ long and 5–8 μ broad and are either aseptate or septate with 1–2 septa.	They are geniculate, yellowish brown and 15 to 45 μ long and 3-5 μ broad and are either aseptate or septate with 1 to 2 septa.
8. The conidiophores are confined to the lower surface of the host leaf and are seen arranged in concentric circles in the tuft.	The conidiophores usually occur on the upper surface of the host leaf but occasionally are found on the lower surface as well and thus described as **amphigenous**. They are sparse and do not occur.
9. The conidia (C) are obclavate or cylindrical, light coloured, 18–60 μ in length and 6–11 μ in breadth. They are septate with 1–7 septa and have bluntly rounded ends.	The conidia are obclavate hyaline or pale yellow to slightly olivaceous, 38–108 μ long and 8–6 μ broad. They are septate with 4–12 septa and have rounded to distinctly truncate base and sub-acute apex.

Disease Cycle. Primary infection takes place by soil borne conidia, which survive in the infected plant debris in the soil and on seeds in the shells. The spread of the disease during the season is by means of wind disseminated conidia. Penetration is direct through the epidermis and also by way of stomata. Chiefly it occurs through the upper epidermis. However, there is possibility of infection through both the leaf surfaces.

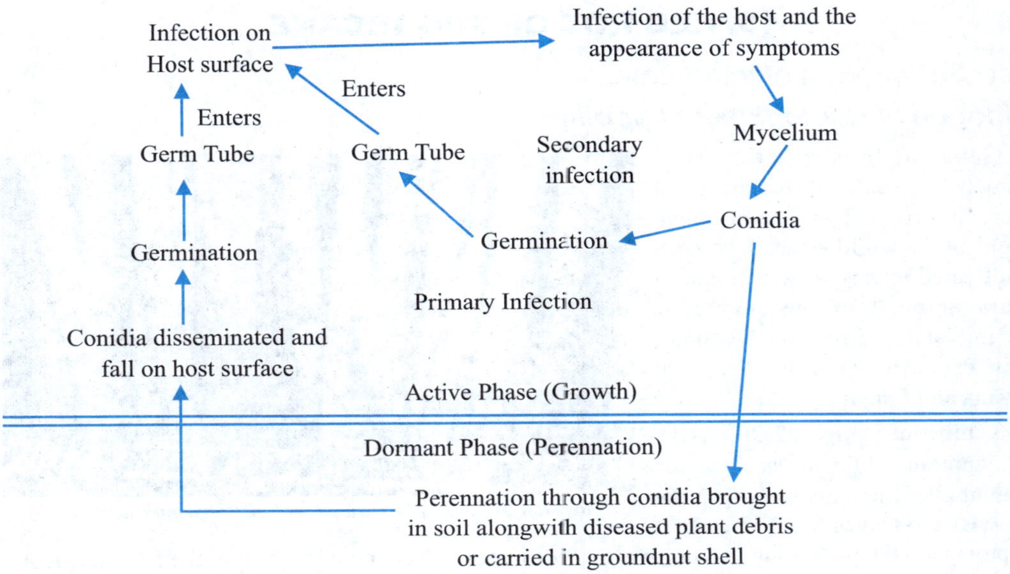

Fig. 22.25. Disease cycle of Tikka disease of Groundnuts.

Predisposing factors. According to Sulaiman and Agashe (1965), maximum predisposing factors for the development of disease prevail in the month of September. Prolonged low temperature and heavy dew favour severe infection. According to Shanta (1960) and Jensen and Boyle (1965), relative humidity plays an important role in infection. Ramakrishnan and Apparao (1965) reported that a period of 3 days of high humidity is essential for maximum infection.

Shanta (1960 a) studied the effect of **NPK** fertilizers on the incidence of disease and found that the application of nitrogen and phosphatic fertilizers increase disease incidence whereas potash decreases it slightly. Ravindar Nath, Kulkarni and Appaji Rao (1965) who carried on preliminary biochemical observations of Tikka disease, have suggested that ascorbic acid and riboflavin content of host leaves appear to play some part in susceptibility and resistance to the disease. According to them, increase in riboflavin content of the leaves is correlated with resistance whereas increase in ascorbic acid is directly related to susceptibility.

Control Measures. Rotation of crops, seed treatment and disposal of infected host debris by burning or burying in deep pits eliminate chances of primary infection from the soil borne inoculum. The seeds within the shells are disinfected with sulphuric acid. Without shells, they are soaked for half an hour in 0.5% copper sulphate solution. Agrosan GN dressing of naked seeds is equally effective.

To check secondary spread of disease in the field, spraying with suitable fungicides is the only remedy. Among the fungicides the use of Bordeaux mixture has given good results. Addy and Dash (1967) recommended Dithane while Shanta (1960) considered red oxide of copper plus sulphur as highly effective. Sulaiman (1964) preferred copper dust (300 mesh) plus sulphur in the ratio of 1 : 1. Harrison (1973) found that Daconil and Benlate are the two more effective fungicides than Dithane M45 for pea nut leaf spot disease control. The sprays of these fungicides increased yield as much as 100%. Early maturing varieties are reported to possess resistance to the disease.

15. RED ROT OF SUGARCANE

Host : Saccharum officinarum L.

Pathogen : Colletotrichum falcatum

General. It is a serious and destructive disease of sugarcane. It occurs in tropical and subtropical regions of the world where sugarcane is cultivated extensively. In India, it occurs in most of the sugarcane growing states particularly Bihar, Uttar Pradesh, Madhya Pradesh, Haryana and Punjab.

Red Rot of Sugarcane (caused by Colletotrichum falcatum)

Symptoms (Fig. 22.26, A-B). The symptoms of the disease become apparent after the rainy season on the stems (B) and leaves (C). The earliest symptoms are the yellowing and drooping of the upper leaves (near the tip). In the later stages the stems shrivel, the rind loses its **bright** colour, becomes dull in appearance and longitudinally wrinkled. If the diseased stems are **split** open lengthwise the white pith will be found reddened with patches of white running transversely across. At the advanced stage of disease blood red lesions with dark margins develop on the mid ribs of the host leaves (B). With the formation of asexual fruit bodies (acervuli), these lesions become covered with powdery masses of conidia which serve to spread the disease. Late in the season, acervuli appear as dark, velvety dotlike structures in the shrivelled areas on the internodes and near the nodes of diseased canes.

Effects of disease. The red rot disease attacks the standing sugarcane crop and thus causes enormous losses. In an epidemic form, it causes heavy reduction in stands and even threatens to wipe out the entire crop in the field. Due to enzymatic action of the pathogen, there is conversion of sucrose in the cane juice into glucose and alcohol. This loss in sucrose may be as high as 33%. Consequently there is reduction in the recoverable sugar in the factory.

Causal Organism. The causal agent of the red rot disease of sugarcane is the form-species *Colletotrichum falcatum* Went, of form-class Fungi Imperfecti. The perfect or sexual stage of the fungus has also been found by Arx and Muller. It is described in the Ascomycetes as *Glomerella tucumanensis*.

Mycelium. The fungus mycelium is found within and in the intercellular spaces between the pith cells of the host. The hyphae constituting the mycelium are thus both inter- and intracellular. They are slender, colourless, branched and septate. The cells contain droplets of oil. The hyphae produce terminal or intercalary, thick-walled greenish black chlamydospores in the pith cells. The spherical chlamydospores remain dormant in the soil for a long period.

Asexual Reproduction (Fig. 22.26, C-D). Finally the mycelial cells begin to collect beneath the epidermis to form stromata consisting of densely packed cells. From the upper surface of each stroma arise the **conidiophores** and bristles called **setae** (C). The setae are rigid, long hair-like structures measuring 100-200 μ long and 4μ broad. They are septate, each with 4 septa. They either form a ring around the stroma or are intermixed with the conidiophores in the acervulus. As the setae grow and push their way up, the overlying epidermis ruptures and exposes the conidia borne on small conidiophores in the acervulus (C).

The conidiophores are small, unseptate structures measuring 20 μ long and 8 μ broad (C). The conidia are one-celled and usually sickle-shaped (falcate) in form (D). They are hyaline, measure 16 to 48 μ long and 4 to 8μ broad. Each conidium has a large oil globule in the centre. The conidia are readily detached and dispersed by air currents, rain, rain drop splashes and insects. They are short-lived and thus germinate immediately in the presence of moisture. On germination, each conidium produces a germ tube which develops either into a mycelium or produces an appressorium on coming in contact with a hard surface (E). The appressorium is a thick-walled structure which functions like a chlamydospore.

Chona and Srivastava (1952) found the perithecial stage of the pathogen for the first time in India in cultures at the I.A.R.I. New Delhi. A year later, Chona and Bajaj (1953) reported its occurrence in nature on sugarcane leaves.

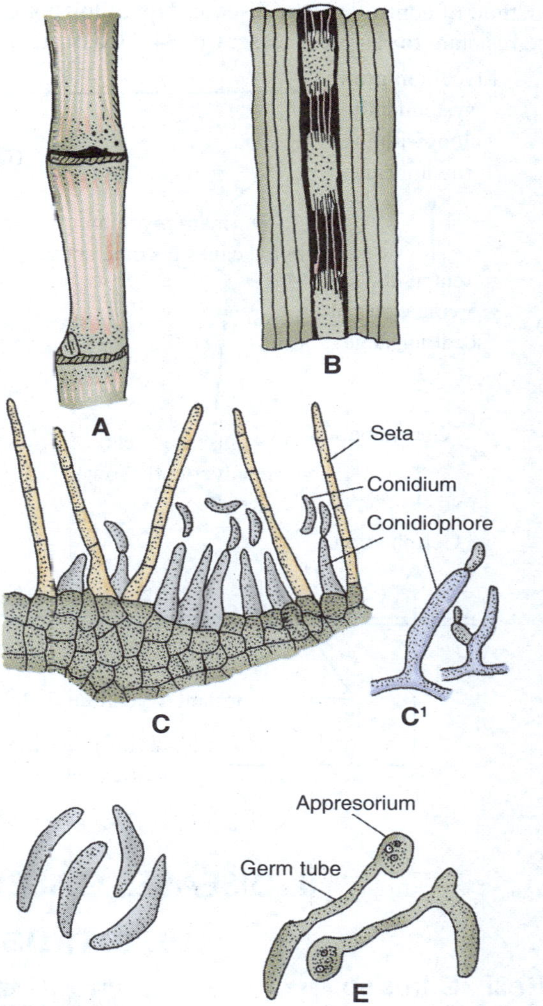

Fig. 22.26 (A–F) *Red Rot Disease of Sugarcane caused by Colletotrichum falcatum.* A, symptoms on the stem; B, symptoms on the leaf midrib; C, acervulus; C[1] coniciophores bearing conidia; D, conidia; E, germinating conidia. (After Buller)

Disease Cycle. The incitant lives from one growing season to the next on the debris of the diseased plants. The conidia are short-lived and thus play no role in the perennation of the pathogen. The thick-walled chlamydospores and perithecia are considered as probable means of survival.

Chona (1950) and later Chona and Nariani (1952) reported that the fungus is capable of growing and producing acervuli in the soil. It thus survives in the active stage for 3 or 4 months. The survival of the mycelium for this limited period is sufficient to provide easy catching of the succeeding crop because sugarcane practically has no dead season.

There is evidence to show that the disease is borne in the seed sets which serve as chief means of survival and spread of the disease. The diseased sets sown in the soil, sprout into infected shoots which soon produce conidia in the acervuli. The conidia serve as a secondary means of infection and spread of the disease. They get detached and are dispersed through the agency of wind, water and insects. On reaching the surface of the healthy sugarcane plants, they germinate immediately in the moisture retained in the enclosing sheaths. Penetration into the host takes place at the nodes through the wounds left by the tearing of leaves or caused by insects.

Control Measures. Field sanitation is an important measure to prevent the build up of source of primary inoculum. It consists in the collection and burning of sugarcane trash in the field. The other equally important preventive measure is the use of sound and healthy seed sets. Long rotation of crops minimizes soil borne infection. The use of resistant varieties is the most effective method of controlling the disease. The following *CO* and *BO* varieties have been recommended to be resistant to red rot disease, Co. 846, Co. 951, Co. 1148, Co. 561, B.O.3, B.O. 7 and B.O. 32.

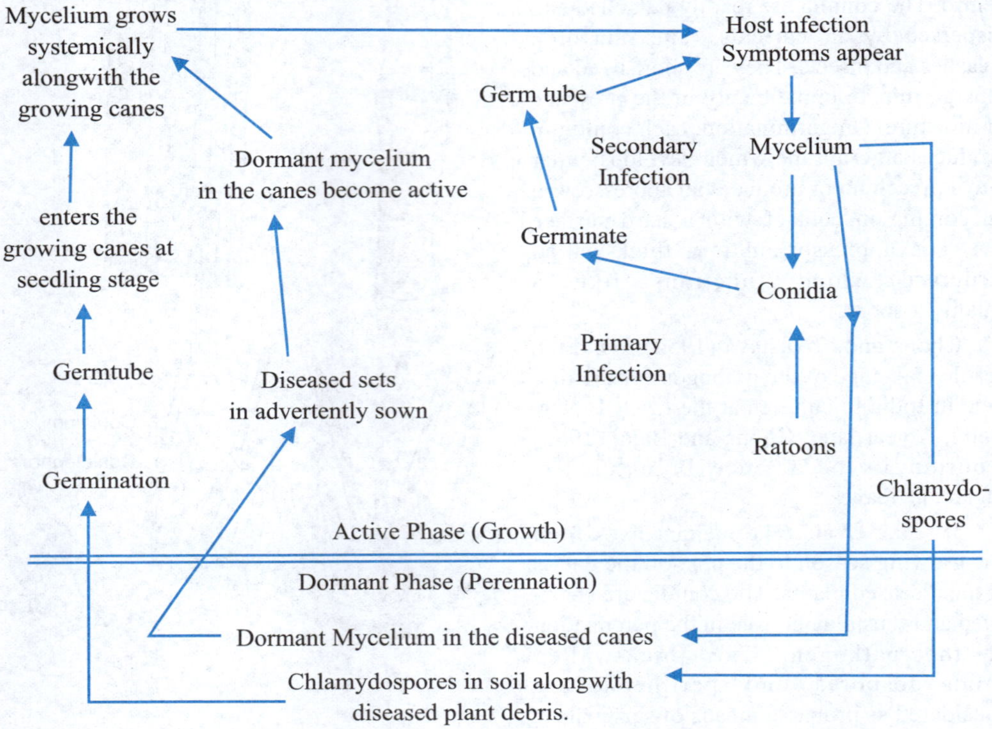

Fig. 22.27. Disease cycle of Red Rot of Sugarcane.

II. DISEASES CAUSED BY BACTERIA

16. CITRUS CANKER

Host : Citrus sp.

Pathogen : Xanthomonas sp.

General. Citrus Canker is a bacterial disease of worldwide distribution occurring wherever citrus is grown. It is a serious menace to our most valued citrus orchards causing objectionable blemishes on the fruit. The disease causes serious damage in India, China, Japan and Java. The pathogen incites severe canker disease in a number of citrus species on stems, leaves and fruits. The disease attacks most of the species/varieties of citrus. The most susceptible species are the acid lime plants, the sweet orange and the grape fruit.

Citrus Canker (caused by Xanthomonas sp.)

Symptoms. Crust-like disease lesions or scabby spots and small cankers (open wounds or dead tissue surrounded by living tissue) appear on all overground parts of the plant such as leaves, young branches and fruits. The trees are, however, not commonly killed. The lesions on the foliage, at first, appear on the lower surface as small round raised spots. These are translucent and of yellowish brown colour. Later the spots turn white or greyish and finally rupture. The older lesions are corky and brown, sometimes purplish. The necrotic brownish canker regions are surrounded by a yellowish brown to green raised margin and distinct watery yellow halo region. The yellow halo ragion is free from the pathogen. The cankerous lesions contain the pathogen in millions. Mairie suggested that the halo regions are formed due to the response of the host tissue to a diffusible metabolite of the pathogen. Padmanabhan *et al.* (1975) reported accumulation of malic acid in the halo region due to increased respiration in this region. The lesions on the **twigs** are usually irregular in form. The lesions on the fruit are similar to those on the leaves but lack the yellow halo.

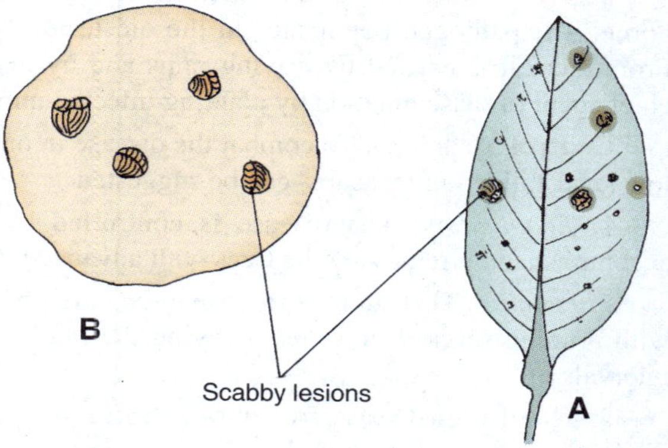

Fig. 22.28. Symptoms of Citrus Canker on leaves (A) and fruits (B).

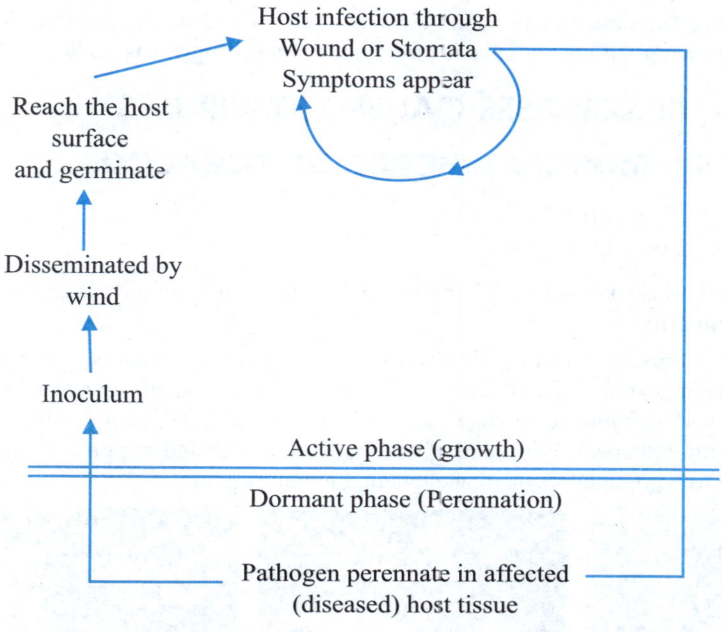

Fig. 22.29. Disease cycle of Citrus Canker.

Causal organism. The causal organism is the bacterial pathogen *Xanthomonas citri*, now called *X. campestris* pv. citri (Hasse) Dowson. It consists of a short, motile rod (1.5–2.0 × 0.5–0.75 μ) furnished with a single polar flagellum (monotrichous). It lacks endospore formation. It is a gram negative, aerobic form surrounded by a mucilaginous capsule. It forms chains.

The climate factors which favour the disease are the mild temperature and wet weather. The most suitable range of temperature appears to be 20°C to 30°C.

Disease Cycle. Infection takes place through the stomata and wounds. The disease is not soil borne. The pathogen perennates in the old lesions on the twigs still attached to the host plant. From there it is carried by driving rains and by insects to new localities. Man functions as the chief agent of dissemination by planting infected nursery stock in new localities.

Control Measures. To combat the disease in order to prevent economic loss or to reduce to a low level following measures can be suggested:–

1. *Eradication*. The disease is controlled by the eradication of diseased trees. This is accomplished by removing the trees with advanced infection and burning them.

2. *Pruning*. The infected trees may be cured by removing the diseased foliage and branches with pruning scissors and then spraying the trees with one percent Bordeaux mixture at regular intervals.

3. The use of disease free nursery stock for planting is the best method of controlling the disease.

4. The fallen infected leaves and twigs should be collected and burnt.

5. *Spraying*. Spraying with Bordeaux mixture and lime sulphur is a useful measure to protect the fruit. It should be done during the first three months after the beginning of fruit formation. Spraying should commence before the onset of rains and repeated during the rainy season.

6. Citrus nurseries should be raised in places away from the regions of heavy and protracted rainfall. There should be no "khatti" hedge around the nurseries.

7. Rangaswamy (1957) reported that the use of antibiotic sprays is useful in controlling the disease. Streptomycin sulphate and Phonomycin have been found to be effective. Vaheeddudin (1959) found that spraying with neem-cake is effective in controlling citrus canker.

III. DISEASES CAUSED BY VIRUSES

17. MOSAIC DISEASE OF TOBACCO.

Host : Nicotiana tabacum

Pathogen : Nicotiana Virus

General. It is the best known of all the virus diseases and is found wherever susceptible plants of tobacco will grow.

Effects. Tobacco-mosaic virus (TMV) also known as *Nicotiana Virus,* is a serious pathogen. It is the most damaging among the virus infection of plants. It causes enormous losses in the tobacco crop. The losses from the disease include reduced yield and poor quality. The mosaic virus can attack about 200 species of plants. The most susceptible and important economic hosts besides tobacco are tomato, potato, sugarcane, cucumber and pepper.

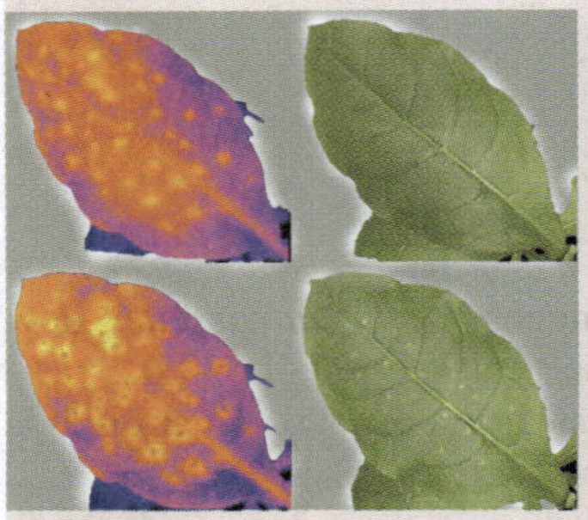

Tobacco Mosaic Disease

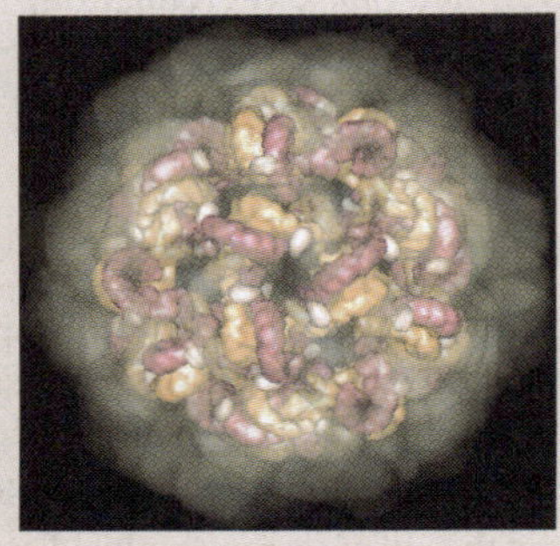

Satellite Tobacco Mosaic Virus

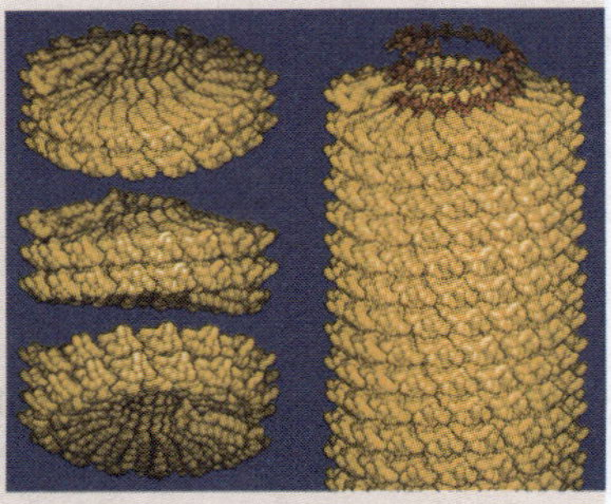

Tobacco Mosaic Virus

Symptoms. The first common symptom is the downward curling and distortion of the youngest leaves accompanied by chlorosis in the young tobacco plants. A number of dark green spots appear as the leaves enlarge. These spots develop into irregular blister-like areas whereas the other parts become chlorotic. These light and dark green patches give the leaf a mosaic pattern. The other symptoms which follow are various types of leaf malformations. The old leaves may develop large yellowish blotches without any sign of mottling. The plants remain dwarf or stunted. The disease is seldom fatal to the host.

Causal organism. The causal organism is Tobacco Mosaic Virus of tobamo Virus group. The tobacco mosaic virus particles are *rod-shaped* (Fig. 22.30, B). They are approximately 280 $m\mu$ long by 15 $m\mu$ in diameter. The virus is highly resistant. It survives immersions in absolute alcohol and other germicides. It is not easily killed by drying, heat or poisons. As a matter of fact it remains active for years in the dry state. It can withstand heating up to 90ºC for 10 minutes.

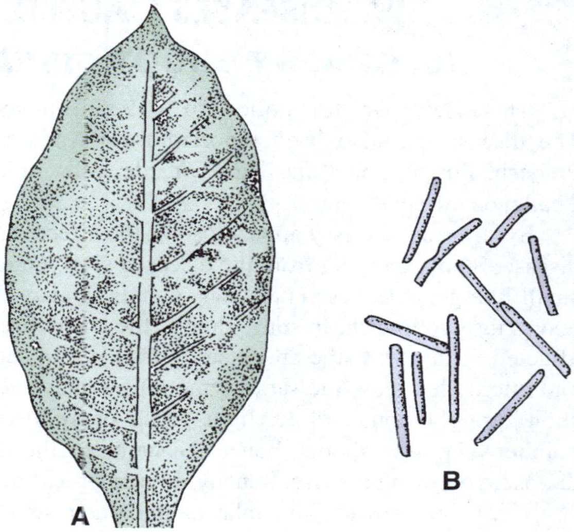

Fig. 22.30 (A-B). *Tobacco Mosaic virus.* A, leaf of Tobacco suffering from Tobacco mosaic; B, Rod-shaped particles of Tobacco mosaic virus. (highly magnified)

Methods of Transmission (Disease Cycle). It is one of the most infectious of the plant viruses. It is sap transmissible. It can stand a high degree of dilution. It gives good infection up to a dilution of 1 : 10,00,000. The virus enters the host through the minute wounds, scratches or abrasions, rubbing leaves and also through broken hairs. Breaking of a hair with contaminated instrument serves as a means of transmitting the virus. The virus remains active in cigarettes, biris and pipe tobacco. The workers in the tobacco growing industry by the use of contaminated tobacco carry the virus on their hands to susceptible crops handled in operations of transplanting, pruning and harvesting. Once within the host the virus multiplies rapidly. This virus is not insect borne. The primary source of inoculum is the infected organic debris of tobacco in the soil. Taylor (1961-1962) reported transmission of virus in tobacco and tomato seed. He located the virus in and on the testa and sometimes in the endosperm. Broadbent *et al.* (1962-1966) found seeds and root debris as the chief sources of primary inoculum.

Control Measures. Following are some important recommendations for the control of the disease:–

(*i*) Weeding out and destroying diseased plants, their immediate neighbours and the plants showing suspicious mottling as soon as they appear.

(*ii*) Destruction by burning of the mosaic infected tobacco and tomato crop refuse on land.

(*iii*) Thorough washing of hands with 3 per cent trisodium phosphate followed by scrubbing with soapy water before handling the plants.

(*iv*) Since pipe tobacco, cigarettes and chewing tobacco are all potential sources of **virus** infection, the workers in the tobacco fields should be furnished with heat sterilized or mosaic resistant smoking and chewing tobacco.

(*v*) Removal and destruction of solanaceous weeds, which may be suspected of harbouring the virus from the vicinity of the tobacco fields.

(*vi*) Sterilisation of instruments and equipment used in tobacco crop plantations in trisodium phosphate without subsequent rinsing.

(*vii*) John and Sova (1955) found seed treatment with 10 per cent trisodium phosphate for 10 minutes effective in removing TMV.

(*viii*) *Growing Resistant Varieties*. A variety of tobacco known as Ambalema resistant to the mosaic virus has been found. Other mosaic resistant commercial tobaccos have been produced and are being grown.

IV. DISEASES CAUSED BY MYCOPLASMA

18. GRASSY SHOOT DISEASE OF SUGARCANE :

This disease was first noticed in 1919 in Bombay and is still a serious problem in Maharashtra. The disease has also been reported from Andhra Pradesh, Tamil Nadu, Orissa, Bihar, Uttar Pradesh, Punjab and Rajasthan. Outside India, it is reported from Burma. Sri Lanka, Taiwan, Thailand and Sudan etc.

Symptoms : Grassy appearance of the affected stools is the most pronounced symptom of the disease. Stools growing from the diseased setts remain dwaf or stunted. The leaves are narrow and small like grass leaves. Thickness of stalk is also like grasses. The internodes become short and leaves turn yellowish. In some shoots, the leaves are completely yellow. If many of the shoots are affected in this way, the entire stool dries up. In secondary infections, ordinarily the new leaves and internodes show the same symptoms as has been observed in primary infections. The leaves show straight, long, white, light green or yellowish streaks. The lower nodes produce large number of grassy shoots, hence the name of the disease. The disease is also known as albino disease, yellowing disease, bunchy disease or leaf tuft.

Causal organism. Mycoplasma like organism (MLO) Disease Cycle. Ratooning of a diseased crop is an important source of perennation. Aphids like *Rhopalosiphum maidis* (*Aphis maidis*), *Aphis sacchari* and *Aphis indosacchari* besides leaf-hopper, transmit the disease. *Aphis sacchari* transmits the disease from sugarcane to sorghum and again from sorghum to sugarcane. *Sorghum* is also a natural host of mycoplasma like organisms (MLOs).

Control Measures

1. Use of healthy setts, seed material should not be taken from that field where the disease was detected.
2. Diseased plant should be dug out and burnt.
3. Heat therapy has also been recommended. Setts to be kept in hot water at 50°C for 2 hours and/or hot air at 54°C for 8 hours.
4. During early stage of crop growth insects, vectors can be controlled by using malathion spray (0.16%) on crops.

19. LITTLE LEAF OF BRINJAL (EGGPLANT)

The disease is found throughout India and Sri Lanka and probably occurs in other neighbouring countries. In India it was first reported from Coimbatore (Tamilnadu). When young plants are infected, they do not produce flowers and fruits and therefore quite adversely affect the profitability of farmers.

Symptoms

The main symptom of the disease is the production of very short leaves by affected plant. The petioles are so much reduced in size that leaves appear sticking to the stem. Such leaves are narrow, soft, smooth and yellowish in colour. Newly formed leaves are further reduced in size. The internodes are shortened and at the same time large number of axillary buds are stimulated to grow into short branches with small leaves. This gives whole plant a bushy appearance. Usually such plant unable to form flowers. Fruiting is very rare.

Causal Organism: Mycoplasma like organism (MLO).

Disease cycle : The disease is transmitted through by the vector *Cestius phycitis*. Artificially the disease has been transmitted successfully to tomato, potato and tobacco. Probably during the season of Brinjal crop, the causal agent survives on weed hosts and from there it is transmitted to main crop by its insect vector.

Control Measures :

Since no effective control measure is found it is better to eradicate the weed host and remove the diseased Brinjal plants. Tetra-cycline has been reported to control the disease.

20. MYCOPLASMA DISEASES OF POTATO

Three mycoplasmal diseases of potato have been reported by Nagaich et. al. in 1974.

(*a*) Purple top roll

(*b*) Marginal flavescence

(*c*) Witche's Broom

(a) Purple Top Roll

The disease occurs particularly in North America but reported for the first time in India which can cause 15 to 20% of loss of Potato yield.

Symptoms: The disease is characterised by rolling and purple or pink coloration of the basal part of leaflets of top leaves. Certain varieties develop only rolling and no coloration. Stunting, chlorosis, profuse axillary shoots with aerial tubers and swelling of nodes are other symptoms. No wilting or necrosis of tuber takes place.

Causal Organism : Mycoplasma like organisms belonging to "Aster Yellows".

Disease Cycle : Vector *Orosius albicinctus* transmits the disease. Infected tubers are the most important source of transmission. The disease has been transmitted to clove, tomato, datura and tobacco also.

Control Measures. No effective control measure has been reported.

(b) Marginal Flavescence of Potato

It is another mycoplasmal disease reported from India which is prevalent in hills and plateau areas where cool weather prevails for a longer duration during the crop season. The crop loss due to this disease varies between 40-75% (Nagaich, *et. cl*. 1974)

Symptoms : Chlorosis on the margins of upper young leaves is the main symptom of the disease. In rare cases, the severely affected plants after exhibiting flavescence of leaf margins for quite sometime develop pinkish coloration on leaflets bases and start rolling. Hairy root symptoms on tubers of 60-65% plants affected with marginal flavescence has been observed.

Disease cycle : The disease is transmitted by leaf hopper and grafting. 5-35% tubers carry the pathogen and serve as source of primary inoculum.

Control Measures : Hot water treatment of potato tubers at 50°C for 10-15 minutes eliminates the pathogen without adversely affecting the tubers (Sangav & Singh, 1985).

(c) Witches' Broom

The disease results in extreme stunting of plants and numerous filaments stems with simply leaves. Very small tubers are produced by the affected plant resulting in total loss of the plant.

Disease Cycle : Mode of transmission is similar to other two disease control measures.

Control Measure. No effective control measure has so far been suggested.

21. SANDAL SPIKE DISEASE

The disease was first reported from Coorg distinct of Karnataka State in 1903 by Barban. The disease was first thought to be caused by Viruses but in 1969, Varma et al reported that Mycoplasma like organism is the causal organism. The disease is one of the most destructive diseases of Sandal tree – an economically important tree and is prevalent in the areas whereever Sandal trees are grown. Significantly the disease is confined to India and Indonesia. In India, Sandal tree to common in Southern States including Karnataka, Kerala and Tamilnadu. According to Iyengar (1969), an annual loss amounting to Rs. 3 million has been estimated due to this disease.

Symptoms : Two types of symptoms are produced. The common symptom called the "rosette spike' is characterised by severe reduction in leaf size and reduction of internodes. This results in crowding of leaves on leaf bearing branches. The new leaves are further reduced in size. Such leaves become stiff and stand out as spikes. In advanced stages of the disease, the leaves become reddish or yellowish. This symptom appears just before death of the tree. The flowers of the infected trees become leaf like and the infected plants do not bear any fruits. the ends of roots of infected plants die out resulting in the loss of the haustorial connection with the host plants. Ultimately the infected plants drie up and die.

The other symptom called 'pendulous spike' is due to continuous apical growth of individual shoots without proper thickening and results in drooping of shoots. The dormant buds do not grow and no rosettes are produced. In this symptom, the roots and haustorial connection are not damaged.

Causal Organism. MLO (Mycoplasma like organisms) is the causal organism. MLO size in the host ranges from 60-750 nm. The common form is ellipsoid , measuring 180-220 × 250-300 nm.

The MLO is reported to infect a large number of plant species such as *Eucalyptus grandis*, *Vinca rosea*, *Zizyphus oenoplea*, *Dodonea viscosa*.

Disease Cycle. The disease is transmitted through root contacts and insect vectors. The disease can also be transmitted through *Vinca rosea* and *vice versa* through dodder. The most common insect Vectors are *Moonia albimaculata* and *Nephotettix virescence*.

Control Measures. No specific method to control the disease has so far been developed. However, it is claimed that planting of Mysore gum trees (ahybriod of *Eucalyptus tereticornis*) at a distance of 10-20 metres from the Sandal trees keeps the latter free of infection. As with other mycoplasmal diseases, treatment with tetracycline checks the disease. Attempts are also being made to develop disease resistant trees via genetic engineering and tissue culture technique.

QUESTIONS

ESSAY TYPE

1. Give an account of loose smut of wheat and mention clearly the method by which it attacks the host.
2. Give an account of any smut disease you have studied and list the important methods of controlling this disease.
3. Describe 'late blight' of potatoes laying emphasis on the symptoms of the disease, modes of infection and methods of control.
4. Give the etiology of stem or black rust of wheat.
5. Write the name of one disease each caused by fungus, virus and bacteria and methods of their control. *(Kanpur, 1998)*
6. Describe the mode of infection, symptoms, name of the disease and its methods of control caused by the only one holocarpic fungus studied by you. *(Kanpur, 1999; Agra, 1993)*
7. Describe the causal organism, symptoms, disease cycle and methods of control of the following diseases :
   (*i*) Wart disease of potato *(Bundelkhand, 1999; Poorvanchal, 2002)*
   (*ii*) Loose Smut of wheat(*Bundelkhand, 1995; Kanpur, 1998, 1999*)
   (*iii*) White rust of Crucifers *(Agra, 1991; Gorakhpur, 1991; Meerut, 1998; Bundelkhand, 1995; Kanpur M.Sc., 2004)*

(*iv*) Covered Smut of Barley (*Agra, 1991*)
(*v*) Early blight of potato
(*Agra, 1991; Gorakhpur, 1991; Meerut, 1997, Bundelkhand, 1995; Magadh, 1991; Rohilkhand, 2001; Rohilkhand M.Sc. 2006; Kanpur M.Sc,. 2003*)
(*vi*) Citrus Canker
(*Gorakhpur, 1991; Meerut, 1997, 1999*)
(*vii*) Wilt of Arhar / Cotton
(*Kanpur M.Sc., 2003*)
(*viii*) Late blight of potato
(*Bihar, 1991; Magadh, 1990*)
(*ix*) Red rot of Sugarcane
(*Meerut, 1999; Kanpur M.Sc., 2004*)
(*x*) Rust of Linseed
(*Bihar, 1991; Kanpur M.Sc., 2003*)
(*xi*) Stem Rust of wheat (*Punjab, 1999*)
(*xii*) Green ear disease of Bajra
(*xiii*) Powdery mildew of peas
(*xiv*) Peach leaf curl
(*Kanpur M.Sc., 2001, 2004*)
(*xv*) Mosaic disease of Tobacco
(*xvi*) Damping off of seedlings
(*Lucknow, 2004*)
(*xvii*) Tikka disease of ground nuts
(*Kanpur, 2005; Kanpur M.Sc., 2004*)
(*xviii*) Gall disease of Coriander
(*Kanpur M.Sc., 2004*)

8. Describe the symptoms and general methods of control of the following diseases of wheat
(*a*) Black stem rust
(*b*) Loose smut (*Lucknow, 1991, 1996, 1998*)
(*c*) Early blight of potato (*Rohilkhand, 1993*)

9. (a) Name the disease caused by *Phytophthora infestans*. Describe the symptoms, disease cycle and control measures.
(*Poorvanchal 2004, 2006; Himachal Pradesh, 1994*)
(*b*) Describe annual recurrence of rust in India
(*Kanpur, 2006*)

10. (*a*) Describe the disease cycle and control measures of loose smut of wheat.
(*b*) Write the name of any one disease caused by bacteria and methods of its control
(*Kanpur, 2006; Poorvanchal, 2004*)

11. Describe in detail about the mode of infection by *Colletotrichum* and its control measures.

12. What do you understand by smuts? Describe the loose smut of wheat. (*Bundelkhand, 1997*)

13. Give a detailed account of rust of Linseed. What measures should be taken for its control? (*Magadh, 1990*)

14. Give an account of symptoms, causal organism, disease cycle and control measures of peach leaf curl.

15. Describe in detail the symptoms, causal organism and disease cycle of powdery mildew of peas.

16. Describe in detail the symptoms, causal organism, disease cycle and control measures of Green ear disease of Bajra.

17. Describe the disease cycle and symptoms of black rust of wheat.
(*Poorvanchal, 2003, 2005*)

18. Describe the symptoms, causal organism and disease cycle of Loose smut of wheat or citrus canker. (*Poorvanchal, 2002*)

19. What is smut disease ? Describe life cycle of a Smut fungus.
(*Rohilkhand, 2001; Meerut, 1994*)

20. (a) Describe the method of a sexual reproduction of the fungus causing, white rust'. What are its symptoms?
(*Gorakhpur, 2003*)
(b) Describe disease cycle of white rust of mustard. (*Poorvanchal, 2005*)

SHORT ANSWER TYPE

21. **Differentiate between :**
(*i*) Rust and Smut
(*Kanpur, 1995; Guru Nanak Deo, 1991; Rohilkhand, 1996*)
(*ii*) White rust and rust
(*Kanpur, 1995; Agra, 1993, 1994*)
(*iii*) Loose smut and covered smut
(*Kanpur, 1998, 2001; Agra, 1993, 1994; Bundelkhand, 1996*)
(*iv*) Black rust and White rust
(*Kanpur, 1998*)
(*v*) Autoecious and Hetroecious Rust
(*Bundelkhand, 1997*)
(*vi*) Primary and Secondary Host
(*Rohilkhand, 1992*)
(*vii*) Early blight and late blight of potato
(*Rohilkhand, 1996*)

22. **Write short notes on :**
(*a*) Smut diseases and their control
(*Kumaon, 2003*)
(*b*) Mechanism of pathological wilting in plants. (*Lucknow, 1995*)
(*c*) Symptoms of loose smut of wheat
(*Gorakhpur, 1994; Kumaon, 2002*)
(*d*) Epidemic and sporadic diseases
(*Himachal Pradesh, 1995*)

(*e*) Necrosis.
(*f*) Hyperplasia
(*g*) Tobacco Mosaic Virus disease *(Bundelkhand, 1999)*
(*h*) Black Wart disease of potato *(Bundelkhand, 1993, 1995; Rohilkhand, 2003; Kumaon, 2004)*
(*i*) White rust of Crucifers *(Bundelkhand, 1993, 1995)*
(*j*) Control of wheat rust *(Bundelkhand, 1993, 1995;)*
(*k*) Black rust of wheat *(Bundelkhand, 1998; Rohilkhand, 1997)*
(*l*) Tikka disease of groundnut. *(Rohilkhand, 1997; Allahabad, 2001; Poorvanchal, 2001)*
(*m*) Loose and covered smut *(Kumaon, 2000)*
(*n*) Earlyblight of potato *(Meerut, 1997; Rohilkhand, M.Sc, 2006)*

23. Write the name of the pathogen and their taxonomic position of the following diseases:
(*a*) White rust of crucifers *(Kanpur M.Sc., 2006)*
(*b*) Stem rust of wheat
(*c*) Citrus Canker *(Lucknow, 2004; Kanpur M.Sc., 2006)*
(*d*) Late blight of potato
(*e*) Loose smut

24. (*a*) Explain the Annual recurrence of stem rust in plains of India *(Punjab, 1999)*
(*b*) How will you classify the diseases on the basis of degree of their spreading.

25. Which fungus causes white rust of crucifers? *(Poorvanchal, 2001)*

26. Write a note on
(*i*) any four historical symptoms of fungal diseases. *(Poorvanchal, 2002)*

27. Write a note on historical significance and causal organisms of late blight of Potato. *(Poorvanchal, 2003)*

28. What is chlorosis? *(Poorvanchal, 2003)*

29. (a) Discuss symptoms of bacterial diseases. *(Kanpur, 2005)*
(b) Write the name of anyone disease caused by bacteria and methods of its control. *(Kanpur, 2006)*

30. Write the causal organism of any three of the following :
(*a*) Rust of linseed
(*b*) *Ergot of rye*
(*c*) Early blight of potato
(*d*) Smut of sugarcane *(Kanpur M.Sc.,2003)*

31. Mention the names of causal organism of any four of the following :
(*a*) Late Blight of potato
(*b*) rust of pea
(*c*) White rust of crucafer
(*d*) Gall disease of coriander
(*e*) Tikka disease ground nut
(*f*) Peach leaf curl *(Kanpur, M.Sc., 2004)*

32. Describe disease symptoms of *Alternaria*. *(Kanpur, 2005)*

33. Describe annual recurrence of rust of wheat in India. *(Kanpur, 2005, 2006)*

34. Describe the disease cycle of
(*a*) Loose smut of wheat *(Poorvanchal, 2003)*
(*b*) Late blight of potato *(Poorvanchal, 2003)*

OBJECTIVE TYPE

35. Fill in the blanks :
(*i*) Red rot of Sugarcane is caused by ....
(*ii*) In acervuli of .... besides conidiophores, chlamydospores are also present.
(*iii*) *Phytophthora infestans* causes the disease known as ....
(*iv*) The name of the disease caused by *Albugo candida* is ....
(*v*) The name of the disease caused by *Ustilago hordei* is ....
(*vi*) *Colletotrichum falcatum* causes the disease ....
(*vii*) Loose smut of wheat is caused by ....
(*viii*) The name of the causal organism of early blight of potato is ....
(*ix*) Green ear disease of Bajra is caused by ....
(*x*) The inflorescence of wheat or barley is converted into a black mass of spores by ....
(*xi*) Covered smut of Barley is caused by ....
(*xii*) Naked asci are produced in .... which causes peach leaf curl.
(*xiii*) The ascocarp of the causal organism of powdery mildew of peas is ....

(*xiv*) Repeatedly produced spores on wheat in the life cycle of *Puccinia* is ....

(*xv*) Citrus canker diseases caused by bacterium ....

(*xvi*) Black stem rust of wheat is caused by ....

36. Select the correct answer :

(*i*) Aecial stage of *Puccinia* develops on
(*a*) Wheat leaf
(*b*) Wheat stem
(*c*) Lower surface of barberry leaf
(*d*) Upper surface of barberry leaf.

(*ii*) Loose smut of wheat is caused by
(*a*) *Ustilago avenae*
(*b*) *Ustilago nuda*
(*c*) *Ustilago tritici*
(*d*) *Ustilago hordei.*

(*iii*) Citrus Canker is caused by
(*a*) *Rhizobium*
(*b*) TMV
(*c*) *Xanthomonas*
(*d*) *Aspergillus.*

(*iv*) The genus in which naked asci are produced is
(*a*) *Puccinia*
(*b*) *Taphrina*
(*c*) *Ustilago*
(*d*) *Peronospora*

(*v*) The ascocarp of the genus *Erysiphe* is known as
(*a*) Apothecium
(*b*) Perithecium
(*c*) Cleistothecium
(*d*) Pycnidium.

(*vi*) Repeatedly produced spores in the life cycle of the fungus causing rust of wheat are
(*a*) Uredospores (*b*) Teleutospores
(*c*) Basidiospores (*d*) Aeciospores.

(*vii*) *Sclerospora graminicola* is a
(*a*) Facultative sporophyte
(*b*) Facultative parasite
(*c*) Obligate parasite
(*d*) Saprophyte.

(*viii*) Early blight disease of potato is caused by
(*a*) *Alternaria solani*
(*b*) *Phytophthora infestans*
(*c*) *Colletotrichum falcatum*
(*d*) *Puccinia graminis*

(*ix*) When a fungus requires two different hosts to complete its life cycle the fungus is referred to as
(*a*) Heterothallic
(*b*) Homothallic
(*c*) Autoecious
(*d*) Heteroecious

(*x*) The infection of *Ustilago tritici* is caused by
(*a*) Basidiospore
(*b*) Infection Hyphae
(*c*) Germ tube
(*d*) Teleutospores

(*xi*) Which of the following is **not** dikaryotic?
(*a*) Basidiospores
(*b*) Uredospores
(*c*) Teleutospores
(*d*) Aeciospores

(*xii*) .... are the sexual spores in rust fungi.
(*a*) Pycniospores
(*b*) Basidiospores
(*c*) Teleutospores
(*d*) Uredospores

(*xiii*) The alternate host of *Puccinia graminis tritici* is
(*a*) *Triticum vulgare*
(*b*) *Barberis vulgaris*
(*c*) *Briza minor*
(*d*) *Pennisetum typhoides.*

(*xiv*) Emittance of mouldy odour from diseased potato fields is an indication of the incidence of
(*a*) Early blight
(*b*) Late blight
(*c*) Wart disease
(*d*) Black rot.

(*xv*) In problem areas, red rot incidence in sugarcane can be minimised by
(*a*) Roguing
(*b*) Fungicidal sprays
(*c*) Avoiding ratooning
(*d*) Following crop rotation with non-hosts.

(*xvi*) White rust of crucifers is a pseudorust because
(*a*) The causal organism does not belong to rust family
(*b*) The colour of the spots is not red
(*c*) The disease is not produced on wheat
(*d*) The disease is restricted to crucifers.

(*xvii*) Tikka disease of groundnuts is caused by
(*a*) *Cercospora*
(*b*) *Alternaria*
(*c*) *Phytophthora*
(*d*) *Rhizoctonia*

(*xviii*) Irish famine of potato occurred due to
(*a*) Early blight disease of potato
(*b*) Late blight of potato
(*c*) Wart disease of potato
(*d*) none of the above

(*xix*) Little leaf of Brinjal is caused by
(*a*) *Bacteria*
(*b*) Mycoplasma
(*c*) Viruses
(*d*) *Fungi*

(*xx*) Later Blight of potato is caused by
(*a*) *Albugo candida*
(*b*) *Melamspora line*
(*c*) *Phytophthora infestans*
(*d*) *Puccinia*

(*xxi*) Covered smut of Barley is caused by
(*a*) *Ustilago nuda graminis*
(*b*) *U. hordei*
(*c*) *U. tritici*
(*d*) *U. graminis*

(*xxii*) Mosaic symptom indicates
(*a*) Reduction of protein
(*b*) Reduction of chlorophyll
(*c*) Increase of chlorophyll
(*d*) All of the above

(*xxiii*) Green ear disease of Bajra is caused by
(*a*) *Sphacelotheca sorghi*
(*b*) *Sclerospora graminicola*
(*c*) *Sclereospora sorghi*
(*d*) *Erysiphe graminis tritici*

(*xxiv*) Red Rust cisease of Tea is caused by
(*a*) *Noctiluca*
(*b*) *Cephaleuros*
(*c*) *Polysiphonia*
(*d*) *Bactrachospermum*

(*xxv*) *Agrobacterium tumefasciens* causes
(*a*) Wilt of crop plants
(*b*) Rust of cereals
(*c*) Damping off of seedlings
(*d*) Crown gall of horticultural plants.

CHAPTER 23

PLANT PARASITIC NEMATODES

- Introduction
- Morphology, Anatomy & Life Cycle
- Symptoms
- Root-knot nematode (Meloidogyne)
- Cyst Nematodes (Heterodera, Globodera)
- Seed Gall Nematode (Anguina sp.)

Introduction

Plant-parasitic nematodes are recognized as one of the greatest threat to crops throughout the world especially in the Asian and African countries. Several hundreds of species of nematodes are known to cause a number of plant diseases worldwide affecting severely the crops belong to cereals, fruits, vegetables, ornamental and cash crops like Sugarcane, Banana, Coconut, Potato etc. Crop losses reported from various part of the world vary between 5 to 75% in case of heavy infections.

Nematodes belong to kingdom Animalia, and phylum Nematoda (round worm). The rhizosphere soil around small plant roots and root hairs is particularly rich habitat for various kinds of nematodes. Most of nematodes do not parasitize on plants but they help in decomposition of organic matter, often referred as free living nematodes. Some of the soil inhibiting nematodes cause severe plant diseases affecting adversely the plant yield. Soil-inhabiting nematodes are also classified according to their feeding habits. They are grouped as Herbivores, Bacterivores, Fungivores, Predators and Omnivores depending upon their feeding preferences.

Herbivorous plant parasitic nematodes are relatively well known. This group includes many members of the order Tylenchida (*Anguina,Haplolaimus, Heterodera, Globodera, Meloidogyne*) as well as a few genera in the sub-orders Aphelenchina (*Aphelenchoides*) and order Dorylaimida (*Longidorus, Trichodorus*). Important difference with other free living nematode is in their mouthpart. Mouthpart has a needlelike stylet which is used to puncture cells during feeding. Plant parasitic nematode are either freely moving (migratory) around the plant and feeding on plant roots as ectoparasite or as endoparasite nematodes enter the root system once inside the root they do not move (sedentary) and can live and feed within the root.

Plant-parasitic nematodes are recognized as one of the greatest threat to crops throughout the world. Several hundreds of species of nematodes are known to cause a number of plant diseases worldwide affecting severely the crops belong to cereals, fruits, vegetables and cash crops like Sugarcane, Banana, Coconut , Potato etc.

Some important Plant Parasitic Nematodes are:

Root knot nematodes : *Meloidogyne spp*.
Cyst nematodes: *Heterodera* (infecting cereals *H.avenae;* infecting soyabean *H.glycines) and Globodera* (golden cyst nematode infecting potato *G.pallida and G.rostochiensis*)
Stem Gall nematode) : (infecting wheat *Anguina tritici)*
Burrowing nematodes: Radopholus (infecting banana *R.similis*)
Foliar nematodes : Aphelenchoides (infecting Chrysanthemum *A.fragariae*)

MORPHOLOGY AND ANATOMY (FIG. 23.1)

Plant-parasitic nematodes are microscopic mostly invisible to naked eye but can be seen under compound microscope. Their shapes and size varies according to the species. The typical nematode shape is a long and slender worm-like animal (commonly known as **eel worm**), but often the adult animals are swollen and no longer even resemble worms (*Heterodera* spp.). Plant-parasitic nematodes usually range from 250 μm to 1000 μm with some upto 04 μm in length, and about 15-35 μm in width. While nematodes may look quite different but they all share some common features. Nematodes often look segmented because of the numerous annulations actually are unsegmented and have no replication of body parts throughout the worm. Nematodes possess bilateral symmetry and have pseudocoelomic body cavity (body cavity that is not totally surrounded by mesoderm). Digestive system has all important part like esophagus, intestine rectum and anus. Mouth usually surrounded by six lips and has a hollow stylet or a solid spear in all plant parasitic nematodes. The stylet/solid spear used to puncture holes in the plant cells and then cell nutrients are withdrawn to meet their own nutritional requirements. Nematodes are dioecious. The male and female reproductive parts are similar except that the females have one or two ovaries followed by an oviduct and uterus terminating into vulva whereas; males have a testis, seminal vesicle terminating in a common opening with intestine. Males also have a pair of protrusible and copulatory spicules. Reproduction is through eggs which may be sexual or parthogenetic as many species lack males.

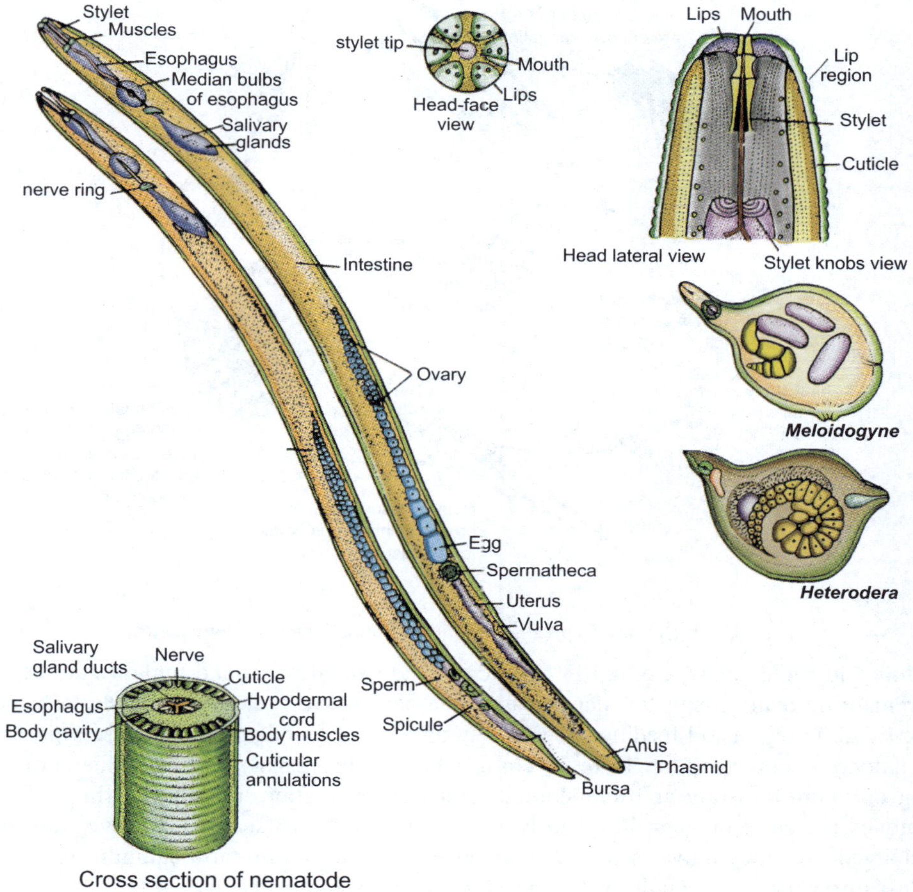

Fig. 23.1. Morphology and main characteristics of typical male and female plant parasitic nematodes. Females of the two most common genera, *Meloidogyne* and *Heterodera* are also shown.

LIFE CYCLE (FIG. 23.2)

Almost similar life cycle occur in all plant parasitic nematodes which have simple life cycle of six stages. First stage represents egg, followed by four juvenile stages, and finally adult stage. The embryo develops inside the egg to become the first-stage juvenile. The first-stage juvenile molts inside the eggshell to become a second-stage juvenile, which hatches from the egg, and in most species must feed before continuing to develop. The nematode molts three more times to become a fully developed adult. Final molting results into differentiation of male and female adult. Adult females used to mate with males to produce fertile eggs or in absence of males parthenogenetically. The whole life cycle completes within 2 to 4 weeks under favourable environmental conditions. Usually first two stages of juveniles are not infectious, when infective stage comes they must feed on susceptible host or else they die in few months. In many species of

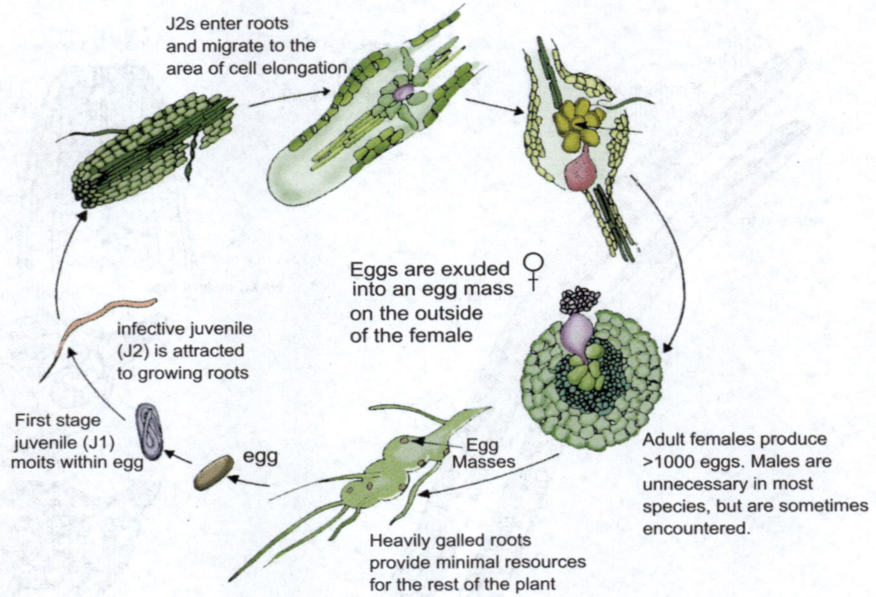

Fig.23.2. Typical life cycle of Plant Parasitic Nematodes.

nematodes juveniles may dried up in absence of the suitable host and remain quiescent or egg may remain dormant in soil for many years. So a part of life cycle of nematodes also occurs in soil or living freely in soil feeding superficially on roots and underground stems. Highest density of nematodes occurs is in top 15 to 20 cm of soil. Greater density will be in the cultivated soil having optimum moisture and temperature with proper aeration, usually the rhizosphere zone. Nematodes are slow movers they hardly move few meter per season. However, in moist and waterlogged soils they move faster. They are also disseminated by farm equipments, bird animal feet dust storms besides though irrigation water, flood and rains drainage.

COMMON SYMPTOMS OF NEMATODES CAUSING DISEASES (FIG. 23.3)

Nematodes usually infect the underground part of the plant thus most of the symptoms produced on roots however above ground parts are also affected thus showing many symptoms in the leaves and stems also. Root symptoms are in the form of root knot, root galls excessive branching of roots, root tip injury and even root rot. Nematode infections are usually accompanied by many fungal and bacterial infections showing various non-characteristic symptoms on the aboveground part of the plant. Reduced growth, chlorosis, yellowing of leaves, various nutrient deficiency syndromes, reduced yield and poor quality of yield are manifestations of nematode infections.

Control Measures

Modification in cultural practices, Biological control measures as well as chemical control measures are recommended for nematode infections. Adaptation of cultural

(A)

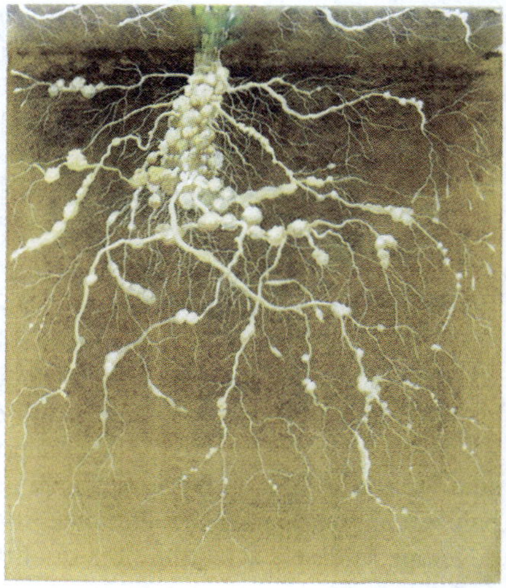

(B)

(C)

(D)

Fig.23.3.

practices like crop rotation, fallowing, crop cover, use of disease resistant varieties, organic amendments of the soil are some of the strategies can be used to restrict nematode infections. Proper tillage, soil solarization, heating, flooding, use of chemicals, suitable nematicides are also recommended to control nematode infections. Many biological agents like *Trichoderma sp.*are used in controlling and checking the spread of nematode infections. Integrated disease/pest management is the best way to control the nematode infections.

ROOT-KNOT NEMATODES (MELOIDOGYNE SP.)

Root knot nematode disease was first reported by Berkeley in 1855 in England on cucumber. Now the nematode has been reported from throughout the world where climate is hot with short winters. Among all plant parasitic nematodes attacking vegetables, Root knot nematode (*Meloidogyne sp.*) has got maximum significance due to its wide range of hosts and the damage they cause to the plants. More than 2000 species of plants, including almost all cultivated plants that are attacked by root knot nematodes. If the susceptible plants are infected by nematodes at early or juvenile stage the chances of crop loss is more even complete destruction of the crop may also occur. Estimated crop loss due to root knot may vary between 5 to 60% depending upon the severity of infection.

Symptoms

Most characteristic symptom appears in the form of root-knot galls which are two to three times larger in diameter than the healthy roots. These gall formations are induced by increased activity of growth regulators introduced by the second stage juveniles of nematodes. Infected roots remain smaller give stunted and cracked appearance. In case of severe infection roots may rot at later stages. In some other nematode infections roots may be bushy in appearance with smaller with root knots. As for above ground symptoms are concern these are small malnourished pale green, yellowish leaves, excessive wilting in hot and dry conditions, smaller and poorly developed flowers affecting both quality and quantity of yield of crop.

Pathogen : *Meloidogyne sp.*(eel worm)

Host : There are more than 2000 species of plants includes many vegetable crops like cucurbits, potato, tomato, onion, cabbage, cow pea, carrot, beans, egg plant, colocasia, peanuts etc.

Disease Cycle (Fig. 23.4)

There are four common species of *Meloidogyne* which infect most of the vegetable crops these are *M. incognita, M. javanica, M. arenaria and M. hapla.* Of these species *M. incognita* is most common. These all species are found in India also. These nematodes are sedentary parasites affecting the underground part of the host plant. Life cycle starts with egg laying by the females inside the host tissues. Each adult female lays about 500 eggs in a gelatinous substance. These eggs start molting, the first two stage juveniles are worm like the second stage juvenile comes out from egg into soil which is the infective stage of nematode. These infective worms move slowly towards roots of susceptible host. Once they come in contact with the susceptible host they enter their roots by piercing the root surface near root tip just behind the root cap with the help of their stylet and secreting the saliva into cell. The saliva stimulates infected cell enlargement at the same time cell contents also gets liquefied following infections. These cell contents are then withdrawn by the nematodes through its stylet. Further molting occur inside the cell only and after final molting (fourth) adult nematodes emerges out from the root either as worm like adult male or pear shaped female. Adult males are free living in the soil. Females continue to swell with or without fertilization to produce eggs outside root tissues. Adult males and females are easily recognizable by their shape and size. Adult males are worm like about 1.2 to 1.5 mm long and 30 to 36 μm in diameter. Adult females are white pear shaped of about 0.4 to 1.3 mm long and 0.2 to 0.75 mm wide. It takes around 25 days to complete the life cycle at optimum environmental conditions. Life cycle may be longer in case of adverse environmental conditions.

Gall Formation (Fig. 23.5)

Two to three days after infection the juvenile settles in the vascular region of the host. Some of the cells around the juvenile head starts enlarging, their nuclei divides, existing wall between the cells dissolves giving rise to giant cell. Enlargement coalescence of the giant cell continues for 2 to 3 weeks affecting other surrounding cells of the host root. The giant cell contains the chemical substances released by the nematode during feeding. Giant cell also absorb nutrients from the surrounding cells to meet the nutritional requirement of the nematode. Along with giant cells there is repeated division of other surrounding cells, cells become hyperplasic eventually aid in gall (Knot) formation.

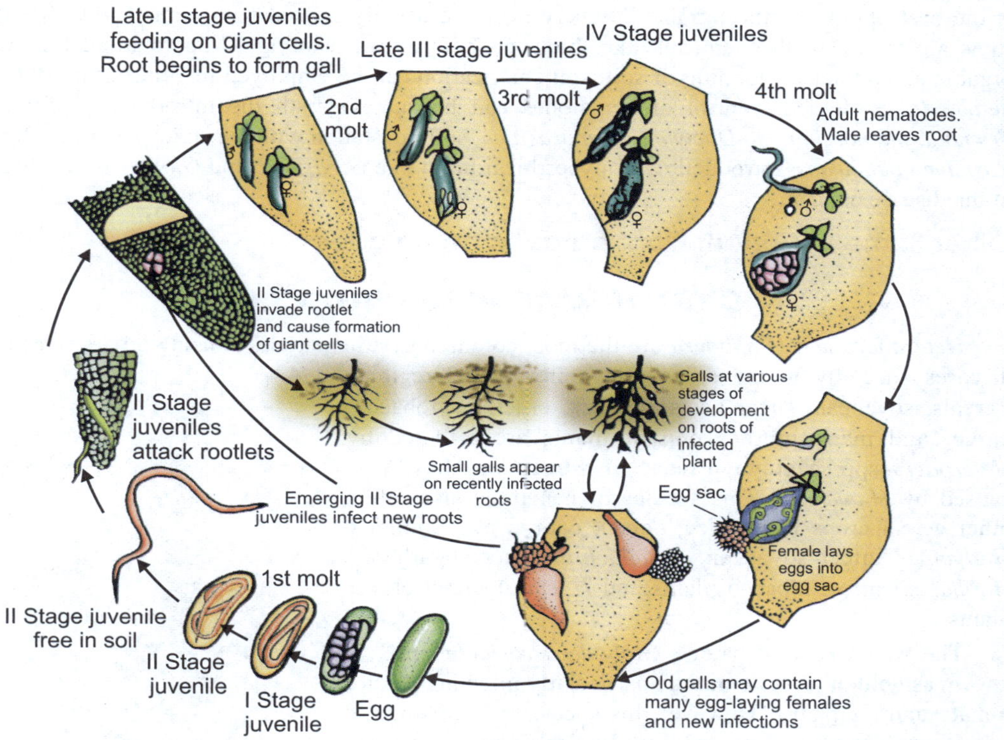

Fig.23.4. Disease cycle of root knot caused by nematodes of the genus Meloidogyne.

Control Measures

Integrated disease management is the best way to control the root knot nematodes. Cultural, Biological and Chemical methods are used to control the disease. Use of disease resistant variety

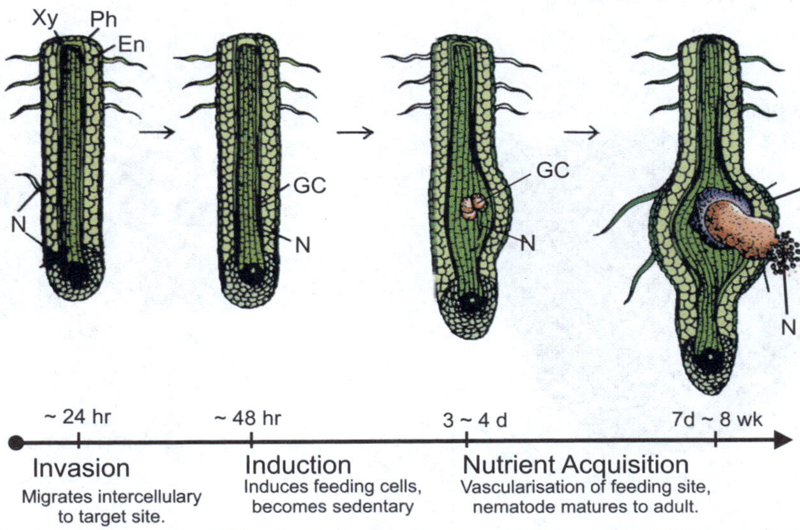

Fig.23.5. Root Gall formation following infection by juvenile nematode. (N-Nematode, Xy-Xylem, Ph-Ploem, En-Endodermis, GC - Giant Cell

is the best option for the farmers. Nursery beds are usually treated with nematicides. Seedling roots are treated with chemicals like thionazin before transplanting. Crop rotation, fallowing, organic amendments, flooding of soil, soil solarization can be employed to control the root knot nematode. Biological control of nematodes can be achieve with the introduction of fungus *Trichoderma harzianum*, *Dactylella oviparasitica*, soil infestation with the endospore of bacterium *Pasteuria penetrans*. Introduction of mycorrhizal fungi like *Gigaspora* and *Glomus* is also helpful in the disease control.

Other Some Important Plant parasitic nematodes

CYST NEMATODES (FIG. 23.6)

Heterodera and *Globodera* are the most common cyst nematodes causing a variety of plant diseases specially in the crops grown in tropical regions. Cereals, soyabean, sugar beets, spinach, crucifers, tobacco, clover and many more group of plants are affected by *Heteroders* spp. **Molaya disease of wheat and barley** is caused by *H.avenae*. This species of nematode also infect other cereal crops like maize, bajara, oat,rye etc. Similarly *H.glycines* infect soyabean, *H.schachtii* infect sugarbeets, *H.tabacum* infect tobacco plants and *H.trifolii* infect clover plants.

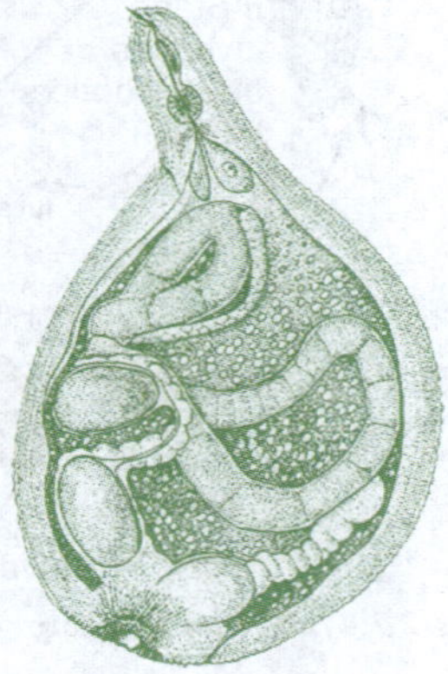

(A)

The round cyst nematode *Globodera rostochinensis* is known as **golden cyst nematode** particularly infecting potato, tomato and eggplants. However this species of nematode is not common in India, reported from British Columbia, United States and New York. The adults are almost spherical of about 450 um in diameter with the projecting neck and head. The infection cause reduced size of tubers with no symptoms except that the infected plants are shorter and yellowish. The nematodes are difficult to eradicate as the eggs in the cyst in soil survive for more than 20 years.

Globodera pallida cyst 63 X

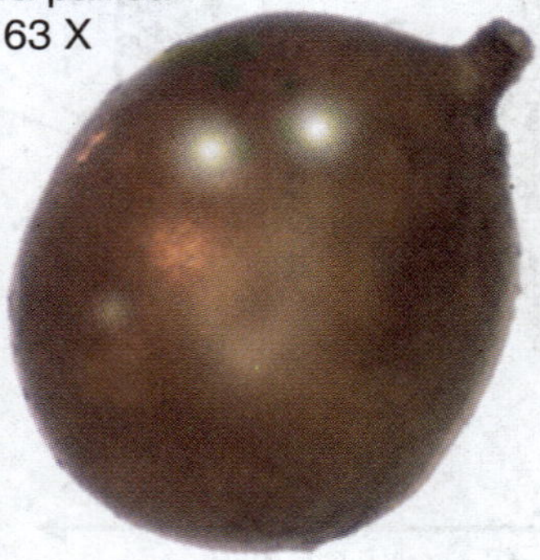

(B)

Golden nematode cysts

(C)

Fig.23.6.

The diagnostic feature of *Heterodera* cyst nematode is the presence of cysts on the host roots and usually the proliferation of roots and production of shallow and bushy root system. Adult females are lemon shaped .6 to .8 mm long and .3 to .5 mm in diameter. Each female cyst contains about 250 to 300 eggs. The adult males are worm like around 1.3 mm long and 30 to 40 μm in diameter. Second stage juveniles are the infective stage of nematode which penetrates the primary and secondary roots are the apical soft portion through its styletes. The juvenile pierced cells of cortex, endodermis and pericycle get enlarged. The group of enlarged cells are known as **syncytia** which serve as feeder cell for nematodes. Rest of the life cycle follow the similar pattern as of a typical nematode.

SEED GALL NEMATODES: *ANGUINA SP.* (FIG. 23.7 & 23.8)

These are the first recorded plant parasitic nematodes, discovered in 1743 when the infected wheat seed (seed gall) are crushed in water and observed under microscope. Now several species of *Anguina* are known which are responsible for gall formation on seeds. The **ear cockle disease** of wheat commonly known as **Sehun** is caused by *Anguina triticii*. The disease is present in all major wheat growing countries of the world. In India the disease was first reported in 1919 from Punjab, now from all part of the country where ever wheat is grown. The symptoms caused by seed gall nematodes are found in all growth stages. The infected seedlings are stunted show characteristic rolling or twisting of leaves. In the diseased head the seeds are replaced by nematode galls (cockles). Certain species of *Anguina* are vectors into seeds of certain pasture grasses of plant phytopathogenic bacterium *Clavibacter toxicus* (cause **Tundu disease of wheat**). Adult females are about 3.2 mm long and 120 μm in diameter. The seed gall nematodes overwinters as second stage juveniles in seed

Fig. 23.7. Damaged wheat kernels cuased by seed Gall nematode *Anguina Sp.*

galls or in plants infected in the fall. The second stage juveniles are very resistant to desiccation and can survive in galls upto 30 years.The seed gall nematode produce only one generation per year.The nematode is spread in the infected seed.

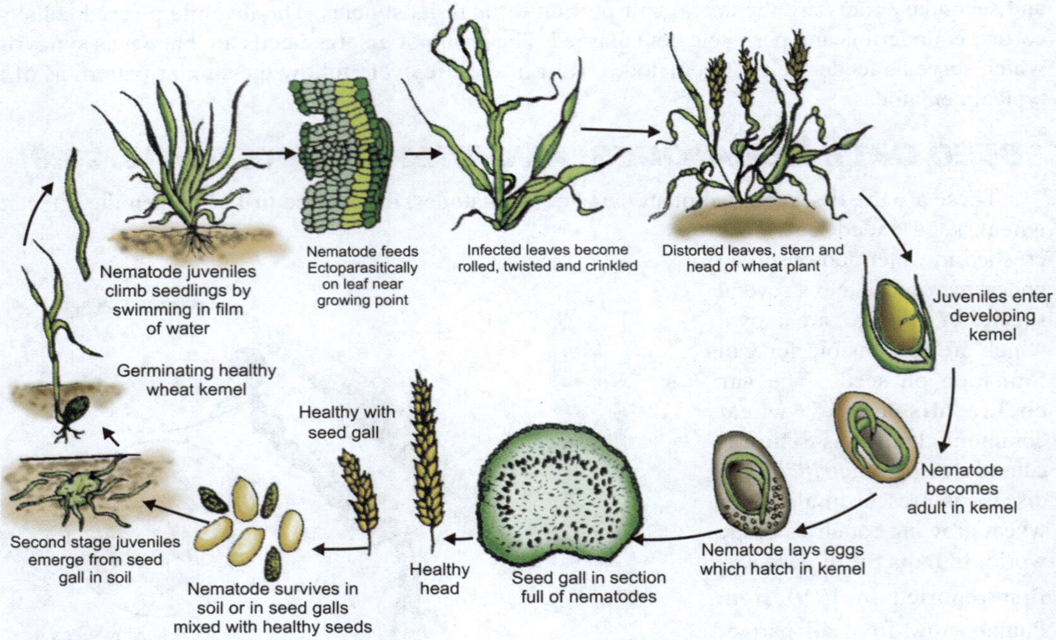

Fig.23.8. Disease cycle of ear cockel (seed-gall) disease of wheat \caused by *Anguina tritici*.

SELECTED BIBLIOGRAPHY

1. Abbot, E.V. 1938. Red rot of sugarcane. *Tech. Bull. U.S. Dept. Agri.,* 641.
2. Adams, M.H. 1959. *Bacteriophages.* Interscience Publisher, New York.
3. Agar, H.D., and H.C. Douglas. 1957. Studies in the Cytological structure of yeast; electron microscopy of thin sections. *Jr. Bact.,* **73** : 365-375.
4. Agnos, G.N. 1969. *Plant Pathology,* Academic Press, New York.
5. Agrios (2006): *Plant Pathology* 5th Edition. Elsevier academic press
6. Ahmad M. 1953. The mating system in *Saccharomyces. Ann. Bot.* n.s., **17** : 329-342.
7. ______, and A.A. Khan. 1955. Studies on baker's yeast of East Pakistan. *Mycologia,* **47** : 329-338.
8. Ahmadjian, V. 1959. A contribution toward lichen synthesis. *Mycologia*, **51** : 56-60.
9. ______, 1960 (1961). The lichen association. *Bryologist*, **63** : 250-254.
10. ______, 1962. Investigations on lichen synthesis. *Amer. Jr. Bot.,* **49** : 277-285.
11. ______, and E. Henriksson. 1959. Parasitic relationship between two culturally isolated and unrelated lichen components. *Science*, **130** : 1251.
12. Ainsworth, G.C. 1973. Introduction and Keys to Higher Taxa. In G.S. Ainsworth *etal*. *The Fungi – An Advanced Treatise*. Vol. IV A : 1-7. Academic Press, New York.
13. Ainsworth, G.C., and G.R. Bisby. 1961. *A dictionary of the fungi*. 5th edition. C.M.I. Kew, Surrey.
14. Ainsworth, G.C., and A.S. Sussman (Eds.) 1973. *The Fungi – An Advanced Treatise.* Vols. I-IV. Academic Press, New York.
15. Aist, J.R., and P.H. Williams, 1971. The cytology and kinetics of cabbage root hair penetration by *Plasmodiophora brassicae. Can J. Bot.,* **49** : 2023-2034.
16. Aist., J.R. 1974. A free—etch study of membranes in *Plasmodiophora* infected and non-infected cabbage root hairs *Can. J. Bot*., **52** (7) : 1441.
17. Aldrich, H.C. 1967. Ultrastructure of meiosis in three species of *Physarum. Mycologia,* **59** : 127-148.
18. ______, and C.W. Minis. 1970. Synaptonemal Complexes and meiosis in Myxomycetes. *Am. J. Bot.* **57** : 935-941.
19. Akai, S., M. Fukutomi, H. Kunoh, and M. Shiraishi. 1976. Fine structure of the spore wall and germ tube change during germination. *In the Fungal Spore, Form and Function*. Edited by D.J. Weber and W.H. Hess. Wiley and Sons, N. York, Lond., Sydney, Trouts pp. 355-410.
20. Alexopoulos, C.J. 1960. Gross morphology of the plasmodium and its significance in the relationship among the Myxomycetes. *Mycologia*, **52** : 1-20.
21. ______, 1961. Myxomycetes. II. *Bot. Rev.*
22. ______, 1962. Introductory Mycology. 2nd edition. John Wiley and Sons, New York.

23. Alexopoulos, C.J. and C.W. Minis. 1979. *Introductory Mycology*. Third edition. Wiley Eastern Limited, New Delhi.

24. Alexopolous C.J; Mims C.W. & Blackwell, M (1996): *Introductory Mycology*, 4th edi; Wiley.

25. Allen, R.F. 1933. Further cytological studies of heterothallism in *Puccinia graminis. Jr. Agr. Res.* **47** : 1-16.

26. Alvin, K.L. and K.A. Kershaw. 1966. *The Observer's Book of Lichens*. London.

27. Atkinson, G.R. 1906. The development of *Agaricus campestris Bot-Gaz.,* **42** : 241-264.

28. ______, 1909. Some problems in the evolution of the lower fungi. *Ann. Mycol.* **7** : 441-472.

39. ______, 1914. Phylogeny and relationship in the Ascomycetes. *Ann. Missouri Botan. Garden,* **2** : 315-376.

30. Aswatha Narayana, N.V. 1956. The tonoplast in yeast. *Curr. Sci.* **25** : 88-89.

31. Awasthe, D.D. 1953. Studies on lichens of India. I. A note on lichens from Simla. *Proc. 40th Indian Sci. Cong.* Pt. III, 72-73.

32. Ayers, G.W. 1944. Studies on the life history of the club root organism, *Plasmodiophora brassicae. Can. Jr. Res.,* C **22** : 143-149.

33. Banzieger, P.S. and R.A. Kilpatrik, 1979. Reduced root and shoot growth caused by Erysiphe graminis tritici in related wheat grown in nutrient culture solution. *Can. J. Bot.* **57** (12) : 1345-1348.

34. Bakerspigel, A. 1960. Nuclear structure and division in the vegetative mycelium of Saproleginaceae. *Am. J. Bot.* **47** : 94-100.

35. ______, 1959. **a**. The structure and manner of division of nuclei in the vegetative mycelium of *Neurospora crassa. Am. J. Bot.*, **46** : 180-190.

36. Bakshi, B.K. 1956. Occurrence of *Polyporus squamosa* (Huds). Fr. in India. *Indian Phytopathol.*, **9** : 191-194.

37. Banbury, G.H. 1954. Processes controlling zygospore formation and zygotropism in *Mucor mucedo* Brefeld. *Nature* (Lond.), **173** : 499-500.

38. Banbury, G.H. 1955. Physiological studies in Mucorales. III. The zygotropism of zygospores of *Mucor mucedo* Brefeld. *J. Exp. Bot.*, **6** :235-244.

39. Barksdale, A.W. 1960. Inter-thallic sexual reactions in Achlya, a genus of the aquatic fungi. *Am. J. Bot.*, **47** : 14-23.

40. Bartinicki Garcia, S.N. Nelson, and E. Cota-Robbs. 1968. A novel apical corpuscle in hyphae of Mucor rouxii. *J. Bacteriol.* **95** : 2399-2402.

41. Bawden. F.C. 1958. *Plant viruses and virus diseases*. Waltham Mass, *Chronica Botanica*.

42. ______ 1959. Viruses in *Vistas in Botany*. 291-312. Pergamon Press, London.

43. Beakes, G.W. 1980. Electron microscopic study of oospore maturation and germination in an emasculate isolate of *Saprolegnia ferax*. 1. Gross changes. *Can. J. Bot.*, **58** : 182-194.

44. Beakes, G.W. 1980 a. Electron microscopic study of oospore maturation and germination in an emasculate isolate of *Saprolegnia ferax*. 2. Wall formation.. *Can. J. Bot.*, **58** : 195-208.

45. Beakes, G.W., and J.L. Gay. 1977. Gametangial nuclear division and fertilization in *Saprolegnia furcota* as observed by light and electron microscopy. *Trans. Br. Mycol. Soc.* **69** : 459-471.

46. Bedi, K.S. 1967. Further studies on the control of loose smut of wheat in the Punjab. *Indian Phytopathol.*, **10** : 138-141.

47. Benjamin, C.R. 1955. Ascocarps of *Aspergillus* and *Penicillium. Mycologia*, **47** : 669-687.

48. Berlin, J.D., and C.C. Bowen. 1964. The host-parasite interface of *Albugo candida* on *Raphanus sativus. Am. J. Bot.* **51** : 445-452.

49. ______ 1964. Centrioles in the fungus *Albugo candida. Am. J. Bot.* **51** : 650-652.

50. Bessey, E.A. 1942. Some problems in fungus phylogeny. *Mycologia* **34** : 355-379.

51. ______ 1950. *Morphology and Taxonomy of Fungi*. The Blakistan Co., Philadelphia.

52. Bhatnagar, G.M., and P.S. Krishnan. 1959. Studies on the germination of spores of *Aspergillus niger*. Arch. Mikrobiol., **33** : 395-405.

53. Bisset, K.A. 1955. *The cytology and life history of bacteria.* Williams and Wilkins Co., Baltimore.

54. ______ 1959. Bacteria in *Vistas in Botany*. 313-327. Pergamon Press, London.

55. Bistis, G.N. 1956 b. Sexuality in *Ascobolus stercorarius* I. Morphology of ascogonium, Plasmogamy, evidence for sexual hormone mechanism. *Am. J. Bot.*, **43** : 389-394.

56. ______, 1957. Sexuality in *Ascobolus stercorarius* II. Preliminary aspects of the sexual process. *Am. J. Bot.*, **44** : 436-443.

57. ______, and J.R. Raper. 1963. Heterothallism and sexuality in *Ascobolus stercorarius. Am. J. Bot.*, **50** : 880-891.

58. ______, and L.S. Olive. 1968. Induction of antheridia in *Ascobolus stercorarius. Am. J. Bot.*, **55** : 829-834.

59. Biswas, K., and D.D. Awasthe. 1948. Distribution of Indian lichens. Proc. 35th *Indian Sci. Cong*. part III, 216.

60. Blackwell, E. 1943, the life history of *Phytophthora cactorum. Trans. Brit. Mycol. Soc.,* **26** : 71-89.

61. ______, and G.M. Waterhouse, 1931. Spores and spore germination in the genus *Phytophthora. Trans. Brit. Mycol. Soc.,* **15** : 294-310.

62. Blackslee, A.F. 1904. Sexual reproduction in Mucorineae. *Proc. Am. Acad. Arts and Sci*., **40** : 205-319.

63. ______, 1920. Sexuality in *Mucor. Science*, **51** : 375-409.

64. Bonner, J.W., K.K. Kane, and H.L. Raphael. 1956. Studies in the Mechanism of growth in the common mushroom *Agaricus campestris, Mycologia*, **48** : 13-19.

65. Bose, S.R. 1937 b Cytological study of basidia of Polyporaceae. *Jour. Indian Bot. Soc.*, **16** : 119-128.

66. Bracker, C.E. 1967. Ultrastructure of Fungi, *Annu. Rev. Phytopathol.*, **5** : 343-373.

67. ______, 1968. Ultrastructure of the haustorial apparatus of Erysiphe graminis and its relationship to epidermal cell of barley. *Phytopathol*. **58** : 12-30.

68. Bracker, C.E., and E.E. Butler, 1963. The ultrastructure and development of septa in the hyphae of *Rhizoctonia solani. Mycologia*, **55** : 35-58.

69. Bracker, C.E. and L.J. Littlefield. 1973. Structural concepts of host-pathogen interfaces. In fungal Pathogenicity and the Plants Response. Edited by R.J.W. Byrde and C.V. Cutting. Academic Press, London and New York, pp. 159-318.

70. Brock, T.D. 1951. Studies on the nutrition of *Morchella esculenta*. Frieds, *Mycologia*, **43** :

72. ______, 1951 C. The function of the cups of *Polyporus conchifer. Science*, **144** : 634.
73. Buckley, P.M., N.F. Soneur, and T.T. Matsunoto. 1968. Ultrastructural details in germinating sporangiospores of *Rhizopus stolonifer* and R. Arrhixus. *J. Bacteriol.* **95** : 2365-2373.
74. Buck R.J.W.F. Fry, A.E. Apple, and C.C. Mundt. 1981. Effect of protectant fungicides on the developmental stages *Phytophthora infestans* in potato foliage *Phytopathol.* **71** (2) : 164-166.
75. Buczacki, S.T. and S.E., Moxham. 1983. Structure of the resting spore wall of *Plasmodiophora brassicae* revealed by electron microscopy and chemical digestion. Trans. *Br. Mycol. Soc.*, **81** (2) : 221-231.
76. Buller, A.H.R. 1933, *Researches in Fungi*. Longmans, Green and Co. London.
77. ______, 1934. The biology and taxonomy of *Pilobolus* in *Researches of Fungi*. Vol. **6** : 1-224. Longmans, Green and Co. London.
78. ______, 1938. Fusion between flexuous hyphae and pycnidiospores in *Puccinia graminis. Nature* (Lond.), **141** : 33-34.
79. ______, 1950. *Research on Fungi*. Vol. VII. University of Toronto Press, Toronto.
80. Burges A (1939): Soil Fungi and root infection. *Broteria8(fasc.2).*
81. Burnett J H (1976): *Fundamentals of Mycology*. 2nd edi, London,
82. Burnett, J.H. 1968. *Fundamentals of Mycology*. Edward Arnold (Publishers) Ltd., London.
83. Burnet, F.M., and W.M. Stanley. 1959. *The Viruses*. 3 Volumes, Academic Press, New York.
84. Bushnell, W.R. 1972. Physiology of fungal haustoria. *Annu. Rev. Phytopathol.*, **110** : 151-176.
85. ______, 1976. Growth of races 38 and 17, *Puccinia graminis* f. sp. *tritici* on artificial media *Can. J. Bot.,* **54** : 1490-1498.
86. Butler, E.J. 1918. *Fungi and Diseases in Plants*. Thacker Spink & Co. Calcutta.
87. ______, and G.R. Bisby, 1931. *Fungi of India*., ICAR.
88. Butler, E.J., and Hafiz Khan. 1913. Red rot of sugarcane. *Mem. Dept. Agri., India Bot. Ser.*, **6** : 151-178.
89. ______, and S.G. Jones. 1955. *Plant Pathology*. MacMillan and Co., New York.
90. Callen, E.O. 1940. The morphology, cytology and sexuality of the homothallic *Rhizopus sexualis* (Smith) Calen. *Ann. Bot.,* **4** : 791-818.
91. Camp, W.G. 1937. The structure and activities of myxomycete plasmodia. *Bull. Torrey Bot. Club.* **64** : 307-335.
92. Campbell, C.K. 1970. Fine structure of the vegetative hyphae of *Aspergillus fumigatus. J. Gen. Microbiol*. **64** : 373-376.
93. Campbell, T.H.M.P. Backus and J.F. Stauffer. 1956. Cytological studies on *Penicillium chrysogenum*. Thom. Bull. *Torrey Bot. Club*. **83** : 93-106.
94. Campbell, W.E. 1957. Studies in ergot infection in graminaceous hosts. *Can. J. Bot.* **35** : 315-320.
95. Cantino, E.C., and G.F. Turian. 1959. Physiology and development of lower fungi (Phycomycetes). *Ann. Rev. Microbiol*., **13** : 97-124.
96. Carpenter, Phillip L. 1967. *Microbiology*. 2nd edi. W.B. Saunders Co., Philadelphia—London.
97. Carvajal, F., and C.W. Edgerton. 1944. The perfect stage of *Colletotrichum falcatum. Indian Phytopathol.* **34** : 206-213.

99. Cayley, D.M. 1935. Spores and spore germination in wild and cultivated mushrooms *Trans. Br. Mycol. Soc*. **20** : 225-241.

100. Chandra, A., and R.P. Purkayastha. 1977. Physiological studies on Indian edible mushrooms. *Trans. Br. Mycol. Soc*. **69**(1) : 63-70.

101. Chattopadhyay, S.B., and C. Das Gupta. 1953. Contribution to the knowledge of Indian Aspergilli. *Indian Phytopathol*., **6** : 92-95.

102. Chaudhary, H. 1932. *Sclerospora gramminicola* on Bajra (*Penninsetum typhoides*) *Phytopathology*, **22** : 270-274.

103. Chet, I., and N. Kislev. 1973. Scanning electron microscopy of spherules of *Physarum polycephalum, Tissue and Cell* **5**(4) : 545-551.

104. ______, and H.P. Rusch. 1969. Induction of spherule formation in *Physarum polycephalum* by polypols. *J. Bacteriol* **100** : 673-678.

105. Chetrath, M.S., B.L. Renfro, Y.L. Nene, R.K. Grover, N.K. Ray, D.V. Singh and S.H. Gandhi. 1969. Control of loose smut of wheat with fungicides. *Indian Phytopathol* **XXII**(2) : 184-187.

106. Chona, B.L. 1943. Red rot sugarcane and its control. *Indian Farming*, **41** : 27-32.

107. ______, 1943. Sugarcane smut and its control. *Indian Farming*. **41** : 401-404.

108. ______, and D.N. Srivastava. 1952. The perithecial stage of *Colletotrichum falcatum*. Went in India. *Indian Phytopathol*., **5** : 158-160.

109. Chopra, G.L. 1934. *Lichens of Darjeeling and Sikkim Himalayas*. Punjab University, Lahore.

110. Chou., C.K. 1970. An electron microscope study of host penetration and early stages of haustoria formation of *Peronospora parasitica* (Fr.) Tul. on cabbage cotyledons. *Ann. Bot*. (Lond.) **34** : 189-204.

111. Christensen, C.M. 1943. *Common edible mushrooms*. Univ. of Minnesota Press, Minneapolis.

112. Chung, J. Know, and J. Kim Sang. 1974. A second heterothallic Aspergillus. *Mycologia*. **66**: 628-638.

113. Clark, J., and O'Neil, Ray Collins. 1976. Studies on the mating system of eleven species of Myxomycetes. *Am. J. Bot*., **63** (6) : 783-789.

114. Claussen, P. 1912. Zur entwicklungsgeschichte der Ascomyceton *Pyronema confluens. Zeitschr. Botanik*, **4** : 1-64.

115. Cochrane, V.W. 1958. The *physiology of fungi*. Johnson Wiley and Sons, New York.

116. Coffey, M.D. 1975. Ultrastructural features of the haustorial apparatus of the white blister fungus *Albugo candida. Can. J. Bot*. **53** : 1285-1299.

117. Coffey, M.D. 1983. Cotychemical specialization of the haustorial interface of a biotrophic fungal parasite *Albugo candida*. *Can. J. Botany* **61**(7) : 2004-2014.

118. ______, and U.E. Wilson. 1983. An ultrastructure study of the late blight fungus *Phytophthora infestans* and its interaction with the foliage of two potato cultivars possessing different levels of general (field) resistance. *Can. J. Bot*. **61**(10) : 2669-2685.

119. Cohen, A.L. 1939. Nutrition of myxomycetes. Bot. *Gaz.,* **101** : 243-275.—1959. An electron microscope study of flagellation in mycomycete swarm cells (Abst.) *Proc. IX. Int. Bot. Congr.* **2** : 77.

120. Colins, O.R. 1961. Heterothallism and homothallism in two myxomycetes. *Ame. Jr. Bot*., **48** : 674-683.

121. Colson, B. 1935. The cytology of the mushroom. *Psalliota campestris. Ann. Bot*.. **49** : 1-18.

123. Cook, A.H. (Editor). 1958. *The chemistry and biology of yeasts.* Illustr. Academic Press, New York.

124. Cook, W.R.I., and E.J. Schwartz 1930. The life history, cytology and method of infection of *Plasmodiophora brassicae* Woron., the cause of finger and toe disease of cabbages and other crucifers, *Phil. Trans. Roy. Soc.* London, **B 218** : 283-314.

125. Copeland H.F (1956): *Classification of lower organisms*, 302pp.

126. Corner, E.J.H. 1929. Studies in the morphology of Discomycetes II. The structure and development of the ascocarp. *Trans. Brit. Mycol. Soc.* **14** : 275-291.

127. Cother, E.J., and D.M. Griffin. 1973 b. Formation of Chlamydospores by *Phytophthora drechsleri, Trans. Brit. Mycol. Soc.,* **61** : 379-383.

128. Couch, I.N. 1927. Discovery of the function of the Pycnia of rust fungi. *Nature* (London), **120** :765-767.

129. Craig, G.D., R.J. Newsam, K. Gull and D.A. Wood. 1979. An ultrastructural and auto-radiographic study of stipe elongation in *Agaricus bisporus. Protoplasma* **98** : 15-29.

130. Craigie, J.H. 1931. An experimental investigation of sex in the rust fungi. *Phytopathol*, **21** : 1001-1040.

131. ______, 1933. Union of pycniospores and flexuous hyphae in *Puccinia helianthi* Schiv. *Nature* (London), **131** : 23.

132. Crump, E., and V.D. Branton. 1966. Behaviour of primary and secondary zoospores of *Saprolegnia* sp. *Can. J. Bot.*, **44** : 1393-1400.

133. Cunfer, B.M. 1976. Inhibition of conidial germination by *Claviceps purpurea* honey dew fluid. *Can. J. Bot.*, **54** : 2541-2545.

134. ______, and A. Sectinger. 1977. Survival of *Claviceps purpurea* sclerotia. *Mycologia* **69** : 1147-1148.

135. Curtis, K.M. 1921. The life history and cytology of *Synchytrium endobioticum. Trans. Roy. Soc. London*, B. **210** : 409-478.

136. Cutter, V.M. Jr. 1942 a. Nuclear behaviour in the Mucorales I. The Mucor pattern. *Bull. Torrey Bot. Club*, **69** : 480-508.

137. Cutter, V.M. Jr. 1942 b. Nuclear behaviour in the Mucorales II. The Rhizopus, Phycomyces and Sporodinia patterns. *Bull. Torrey Bot. Club*, **69** : 592-616.

138. Daniel, J.W., and U. Jarlfors. 1972. Plasmodial ultrastructure of the myxomycete *Physarum polycephalum. Tissue and Cell* 4(1) : 15-36.

139. Dastur, J.F. 1915. The potato blight in India. *Ibid.*, **7** :163-176.

140. Davis, B.M. 1900. The fertilization of *Albugo candida. Bot. Gaz.,* **29** : 297-311.

141. Davison, E.M. 1968. Cytochemistry and ultrastructure of haustoria and hyphae of *Peronospora parasitica..Ann. Bot.* (*Lond.*), **32**: 613-621.

142. De Bary, A. 1887. *Comparative morphology and biology of the fungi, mycetozoa and bacteria*. Clarendon Press, Oxford.

143. Delamater, E.D. 1950. The nuclear cytology of the vegetative diplophase of *Sachharomyces cerevisiae. Jou. Bact.*, **60** : 321-332.

144. Denil, G.M. Nebel, and F. Overwinkler 1981. Light and Scanning electron microscopic studies of spore formation in *Ustilago pustulata* and *U. scabiosae. Can. J. Bot.* **59**(2) : 122-128.

145. Dennett, C.W., and M.E. Stanghellini. 1977. Genetic and cytological evidence for diploid life

147. Dickinson, S. 1949. Studies on the morphology of obligate parasitism IV. The formation on membranes of haustoria by rust hyphae and powdery mildew germ tubes. *Ann. Bot.*, n.s., **13** : 345-353.

148. Doge, B.O. 1914. The morphological relationships of the Florideae and the Ascomycetes. *Bull. Torrey. Bot. Club*, **41** : 157-202.

149. ______, 1920. Life history of *Ascobolus magnificus. Mycologia*. **12** : 115-134.

150. ______, 1928. Production of fertile hybrids in the Ascomycete *Neurospora. Jr. Agri. Res.*, **36** : 1-14.

151. ______, 1933. The perithecium and ascus of *Penicillium. Mycologia*. **25** : 90-104.

152. ______, 1935. The mechanics of sexual reproduction in *Neurospora Mycologia*. **27** : 418-438.

153. Donk, M.A. 1954. A note on sterigmata in general. *Bothalia*, **6** : 301-302.

154. Drechsler, C. 1947. Germination of oospores of *Pythium butleri* and *P. tardicrescens. Phytopath*., **37** : 438-439.

155. ______, 1952. Production of zoospores from germinating oospores of *Pythium ultimum* and *Pythium debaryanum. Bull. Torrey Bot. Club*, **79** : 431-450.

156. ______, 1956. Production of zoospores from germinating oospores of *Pythium mamillatum. Bull. Torrey. Bot. Club*. **83** : 196-206.

157. Dube, H.C. 1993. *An introduction to Fungi*. Vikas, New Delhi.

158. Duggar, B.M. (1909). *Fungous Diseases of plants*. Ginn and co., N.Y.

159. Ehrlich, H.G., and M.A. Ehrlich. 1963. Electron microscopy of the host parasite relationship in stem rust of wheat. *Am. J. Bot.*, **50** : 123-130.

160. Ehrlich, M.A., and H.G. Ehrlich. 1965. Ultrastructure of *Phytophthora infestans. Phytopathology*. **55**: 1056.

161. ______, 1971. Fine structure of the host parasite interfaces in mycoparasitism. *Annu. Rev. Phytopathol*., **9** : 155-184.

162. ______ and ______, 1963. Electron microscopy of the sheath surrounding the haustorium of *Erysiphe graminis. Phytopathol*. **53** : 1378-1380.

163. Ekundayo, J.A and M.J. 1964. The germination of the spores of *Rhizopus arrhizus*, spore swelling and germ tube emergence. *J. Gen. Microbiol*. **35** : 261-269.

164. Ekundayo, J.A. 1966. Further studies on germination of sporangiospores of *Rhizopus arrhizhus. Jr. Gen. Microbiol.*, **42** : 283-291.

165. Elliot, E.W. 1948. The swarm cells of Myxomycetes *Jr. Wash. Acad. Sci.* **38** : 133-137.

166. ______, 1949. The swarm cells of Myxomycetes. *Mycologia*, **41** : 141-170.

167. Elliot, M.E. and K. Kraufert. 1974. *Peziza badia* and *Peziza badioconfusa Can. J. Bot.*, **52** : 467-472.

168. Elison, B.R. 1945. Flagellar studies on zoospores of some members of the Mycetozoa. Plasmodiophorales and Chytridiales. *Mycologia*, **37** : 449-459.

169. Ellzey, J.T. 1974. Ultrastructural observations of meiosis within antherogdia of *Achlya ambisexualis. Mycologia* **66** : 32-47.

170. Eloriza, M.V., G. Larriba, J.R. Villanueva and R. Sentandreu. 1977. Budding *Sachharomyces cerevisae* : formation of cross bar. *Trans. Br. Mycol. Soc.* **69** : 451-457.

171. Elsner, P.G., G.E. Vander Molen, J.C. Horton, and C.C. Bowen 1970. Fine structure of *Phytophthora infestans* during sporangial differentiation and germination. *Phytopathology*. **60** : 1765-1772.

172. Fischer, and C.S. Holton. 1957. *Biology and control of the smut fungi*. The Ronald Press, Co., New York.
173. Fitzpatrick, H.M. 1930.*The lower fungi—Phycomycetes*. McGraw Hill Book Co., New York.
174. Fletcher, J. 1971. Fine structural changes during germination of conidia of *Penicillium griseofulvum* Dierkx *Ann. Bot.,* **35** (140) : 441-449.
175. ______, 1971. Conidium ontogeny in *Penicillium. J. Gen. Microbiol*., **67** : 207-214.
176. ______, 1969. Morphology and nuclear behaviour of germinating conidia of Penicillium grisefulvum. *Trans. Br. Mycol. Soc.* **53** : 425-432.
177. Forbischer. M, 1957. *Fundamentas of Microbiology*. W.B. Saunders Co., Philadelphia.
178. Foudin, L.L., K.E. Papra and R.T. Hanlin 1985. Nuclear behaviour during conidiogenesis in *Aspergillus flavus. Can. J. Bot*., **59**(11) : 2116-2120.
179. Frank. H. Gleason, 1973. Nuclear tubules in *Saprolegnia. Cytobios.,* **8**(31) : 185-187.
180. Franci L.J. (2001): The Disease Triangle: a plant pathological paradigm revisited *The Plant health Instructor.* DOI: 10.1094/PHI-T-2001-0517-01.
181. Fraymouth, J. 1956. Haustoria of the Peronosporales. *Trans. Br. Mycol. Soc.,* **39** : 79-107.
182. Fuller, M.S. and R.E. Reichle. 1968. The fine structure of *Monoblepharis* sp. zoospores, *Can. J. Bot.,* **46** : 279-288.
183. ______, and L. W. Olson. 1971. The zoospore of allomyces. *J. Gen. Microbiol*. **66** : 171-183.
184. Gaither, W. Thomas. 1974. Ultrastructure of the capillitium and spores of *Physarum polycephalum. Bull. Torrey Bot. Club,* **101** (4) : 186-190.
185. Galindo, J., and M.E. Gallegly. 1960. The nature of sexuality in *Phytophthora infestans. Phytopathol.,* **50** : 123-128.
186. Gallegly, M.E., and J.A. Gallindo. 1960. Mating types and oospores of *Phytophthora infestans* in nature in Mexico. *Phytopathol*., **48** : 274-278.
187. Gamundi, I.J., and M.E. Ranalli. 1963. Apothecial development in *Ascobolus stercorarius. Trans. Br. Mycol. Soc*., **46** : 393-400.
188. Ganesan, A.T. 1956. The nucleus of yeast cell. A study of five-day-old fermenting culture. *Cytologia*, **21**: 124-134.
189. ______, 1959. The cytology of Sachharomyces. *Compt. rend. lab. Carlsberg.* **31** : 149-174.
190. Garber, R.C. and J.R. Aist. 1979. The ultrastructure of Meiosis in Plasmodiophora brassicae (Plasmodiophorales), *Can. J. Bot.* **57** : 2509-2518.
191. Garrett S.D. (1951): Ecological groups of soil Fungi. *New Phytologist*. **50** (2):149-166
192. Gauman, E.A. 1952. *The fungi*. Hafner Publishing Co., New York.
193. Gaumann, .E. (1950). *Principles of Plant infection. Hafner publi*., NY.
194. Ghorse, W.C. and M.R. Edwards. 1973. Ultrastructure of *Aspergillus fumigatus* conidia, Development and maturation. *Protoplasma*. **76** :49-59.
195. Gibbs, A.J. 1968. Cryptograms in Plant virus names. In Martyn E.B. (Ed.) Plant Virus names phytopathological paper No. 9, C.M.I. Kew. Surrey.
196. ______, 1969. Plant Virus classification. Adv. Virus Res. **14** : 263-328.
197. Gilbert, F.A. 1928 b. Observations on the feeding habits of swarm cells of Myxomycetes. *Ann. Jr. Bot*., **15** : 473-484.
198. ______, 1929 a. Factors influencing the germination of myxomycetes spores. *Am. Jr. Bot.,* **16** : 280-286.
199. Gilbert, F.A. 1929 b. Spore germination in the Myxomycetes, a comparative study of spore germination by families. *Am. Jr. Bot.,* **16** : 421-432.

200. Gokhale, V.P., and M.K. Patel, 1952. Effect of stem rust particularly on grain weight of some improved Indian wheats. *Indian Phytopathol*., **5** : 89-109.

201. Gray, W.D. 1938. The effect of light on the fruiting of Myxomycetes. *Am. Jr. Bot*., **25** : 511-522.

202. ______, and C.J. Alexopoulos. 1968. *Biology of Myxomycetes*. Ronald Press, New York.

203. Gregory, C.T. 1912. Spore germination and infection with *Plasmopara viticola. Phytopath*. **2** : 235-249.

204. Greuter, B., and D. Rast. 1975. Ultrastructure of the dormant *Agaricus bisporus* spore. *Can. J. Bot*., **53** : 2096-2101.

205. Guha, A.K. and A.B. Banerji. 1971. Effect of different carbon compounds on the submerged production of *Agaricus campestris. J. Food Science and Technolocy* **8** : 82-83.

206. Guilliermond, A., 1940. Sexuality, development cycle and phycology of yeasts. *Bot. Rev.,* **6** : 1 : 24.

207. Gupta, D.D. 1951. Sporulation and aversion in *Eurotium herbariorum (wigg.), Indian Phytopathol*, **4** : 106-115.

208. Guttes, E., Sophia Guttes, and H.P. Rusch. 1961. Morphological observations on growth and differentiation of *Physarum polycephalum* growth in pure culture. *Devel. Biol*., **3** : 588-614.

209. Gwynne Vaughan, H.C.I., and B. Barnes. 1927. *The structure and development of the fungi.* University Press, Cambridge.

210. ______, and H.S. Williamson. 1930. Contributions to the study of *Humaria granulata. Ann. Bot*., **44** : 127.

211. ______, 1931. Contributions to the study of *Pyronema confluens. Ann. Bot*., **45** : 355-371.

212. ______, 1932. Cytology and development *Ascobolus magnificus. Ann. Bot*., **46** : 653-670.

213. ______, 1933. Notes on the Ascoboleae. *Trans. Br. Mycol. Soc*. **18** : 127-134.

214. Hammond, J.B.W., and R. Nichols. 1976. Glycogen in *Agaricus bisporus. Trans. Br. Mycol. Soc*. **66** : 325-369.

215. Hanlin, R.T. 1976. Phialide and conidium development in *Aspergilus clavatus. Am. J. Bot.,* **63** : 144-155.

216. Harder, D.E., J. Chong, R. Rohrenger and W.K. Kim. 1986. Structure and Cytochemistry of the walls of uredospores germ tubes and appresoria of *Puccinia graminis tritici. Can. J. Bot*., **64**(3) : 476-485.

217. Hardison, J.R. 1972. Prevention of ascocarp formation in *Claviceps purpurea* by fungicides applied over sclerotia at soil surfaces. *Phytopathol*. **62** : 609-611.

218. Harper, R.A. 1900. Sexual reproduction in *Pyronema confluens* and morphology of ascocarp. *Ann. Bot.,* **14** : 321-400.

219. ______, R.A. and B.O. Dodge. 1914. The formation of capillitium in Myxomycetes. *Ann. Bot*., **28** : 1-18.

220. Hartwell, L.H. 1974. *Saccharomyces cerevisiae*, Cell cycle. *Bact. Rev.* **38** : 164-198.

221. Hashimoto, T.S., S.F. Conti, and H.B. Naylor. 1958. Fine structure of microorganisms, III. Electron microscopy of resting and germinating ascospores of *Sachharomyces cerevisiae Jr. Bact*. **76** : 406-416.

222. Hashimoto, T., S.F. Conti, and H.B. Naylor. 1959. Studies of fine structure of microorganisms, IV. Observations on budding *Sachharomyces cerevisiae* by light and electron microscopy. *Jr. Bact.* **77** : 344-354.

223. Haskins, E.F., and A.A. Hinchee. 1974. Light and ultramicroscopical observations on the surface structure of the protoplasmodium, aphanoplasmodiuma nd phaneroplasmodium (Myxomycetes). *Can. jr. Bot.*, **52** : 1835-1839.

224. Haskins, R.H., J.A. Brushaber, J.J. Child and L.B. Holtby. 1976. The Ultrastructure of sexual reproduction in *Pythium acanthecum. Can. J. Bot.*, **54** : 2193-2203.

225. Hatch, W.R. 1938. Conjugation and zygote germination in *Allomyces arbuscula. Ann. Bot.,* n.s., **2** : 583-614.

226. Hawker, L.E. 1957. *The physiology of reproduction in fungi*. Cambridge University Press, Cambridge.

227. ______, 1965. Fine structure of fungi as revealed by electron microscopy. *Biol. Rev.,* **40** : 52-92.

228. Hawker, L.E. 1963. Fine strucrure of *Pythium debaryanum* and its probable significance. *Nature* **197** : 618.

229. Hawker, L.E. 1967. *An Introduction to fungi*. Hutchinson and Co. Ltd., London

230. Hawker, L.E., and Albot. 1963. Fine structure of young vegetative hyphae of *Pythium debaryanum. J. Gen. Microbiol.* **31** : 491-494.

231. ______, 1963. An electron microscope study of maturation and germination of sporangiospores of two species of *Rhizopus. J. Gen. Microbiol*. **32** : 295-303.

232. ______, and C.E. Bracker. 1966. Plasmodesmata in fungal cell walls *Nature*, **212** : 636.

233. Hawsworth D.L. Kirk P.M. Pelger D.N. & Sutton B.C. (1995): *Ainsworth & Bisby's Dictionary of Fungi*, 8th edition, CAB international, Wallingford.

234. Hein, I. 1930. Studies on the mycelium of *Psalliota campestris. Am. J. Bot.* **17** : 197-211.

235. Hemmes, Don. E., and S. Bartricki-Garcia. 1974. Electron microscopy of gametangial interaction and oospore development in *Phytophthora capsici. Am. J. Bot*., **61** (5) Supp. Abstract of papers presented at the meeting of Bot. Soc., America, 1974.

236. Henney, R.H. jr. and M. Asgari. 1975. The function of slime from *Physarum flavicomune* in the control of cell division *Can. J. Microbiol*. **21** : 1866 : 1876.

237. Hess, W.M., D.J. Webex and D. Hohnson. 1971. Ultrastructure and lipid analysis of Erysiphe graminis tritici Dc conidia. *Am. J. Bot.*, **61** (5) Supp, 1974.

238. Hibbett, D.S. *et al*. (2007) A higher-level phylogenetic *classification of the Fungi*. *Mycol. Res*. 111, 509-547

239. Hickey, E.L., and M.D. Coffey. 1977. A fine structural study of Pea downy mildew fungus *peronospora pisi* on its host *Pisum sativum. Can. J. Bot*. **55** : 2845-2858.

240. Hill, E.P. 1969. The fine structure of the zoospores and cysts of *Allomyces* macrogynous. *J. Gen. microbiol*. **56** : 125-130.

241. Hirsch, H.E. 1950. No brachymeiosis in *Pyronema confluens. Mycologia*, **42** : 301-305.

242. Hirst, J.M., and O.J.S. Stedman. 1960. The epidemeology of *Phytophthora infestans*. I. II. *Ann. Appl. Biol.,* **48** : 471-517.

243. Ho, H.H., K. Zachariah and C.J. Hickman. 1968. The ultrastructure of zoospores of *Phytophthora megasperma* Var. Sojae. *Can. J. Bot.*, **46** : 37-41.

244. Ho, H.H., G.A. Zentmyer and D.C. Erwin. 1977. Morphology of the sex organs of *Phytophthora cambivora. Mycologia* **69** (3) : 641-646.

245. ______, 1977. Morphology of *Phytophthora cinnamomi. Mycologia* **69** (4) : 701-713.

246. Ho, H.H. 1975. Observations on the behaviour of zoospores of *Saprolegnia* sp. *Mycologia* **67** : 425-428.

247. Ho, A.H. 1979. The nature of amiphigyny in *Phytophthora. Mycologia* **LXXI** (5) : 1057-1064.

248. Holloway, S.A., and I.B. Heath. 1974. Observations on the mechanism of flagellar retraction in *Saprolegnia terrestris. Can. J. Bot.*, **52** : 939-942.

249. ______ and ______, 1977 An ultrastructural analysis of the changes in the organelle arrangement and structure between the various spore types of *Saprolegnia. Can. J. Bot.* **55** : 1328-1339.

250. Hohl, H.R., and S.T. Hamamoto. 1967. Ultrastructural changes during zoospore formation in *Phytophthora parasitica. Am. J. Bot.*, **54** (9) : 1131-1139.

251. ______ and Stossel. 1976. Host parasite interfaces in a resistant and susceptible cultivar of *Solanum tuberosum* inoculated with *Phytophthora infestans* : tuber tissue. *Can. J. Bot.*, **54** : 900-912.

252. ______ and E. Suter. 1976. Host-parasite interfaces in a resistant and a susceptible cultivar of *Solanum tuberosum* inoculated with *Phytophthora infestans* : Leaf tissue. *Can. J. Bot.*, **54** : 1956-1970.

253. Hohnk, W. 1933. Polyplanetism and zoospore formation in Saprolegniaceae and *Pythium. Am. J. Bot.*, **20** : 45-62.

254. Horne, A.S. 1930. The nuclear division in the Plasmodiophorales. *Ann. Bot.*, **44** : 199-231.

255. Hosoda, E. 1980. Culture methods and sporulation of *Physarum polycephylum. Mycologia* **72** (3) : 500-504.

256. Howard, F.L. 1931. The life history of *Physarum polycephalum. Am. J. Bot.*, **18** : 116-133.

257. Howard, K.L., and R.T. Moore. 1970. Ultrastructure of oogenesis in *Saprolegnia terrestrics. Bot. Gaz.* **131** : 311-336.

258. Hughes, S.J. 1971. Percurrent proliferation of sporangia in fungi, algae and mosses. *Can. J. Bot.* **49** : 215-231.

259. Hughes, S.J. 1953. Conidiophores, conidia and classification. *Can. J. Bot.* **31** : 577-659.

260. Hung, Ching Yuan. 1977. Ultrastructural studies of ascospore liberation in *Pyronema domesticum. Can. J. Bot.* **55** (19) : 2544-2549.

261. Ingold, C.T. 1963. *Dispersal in Fungi.* Clarendon Press. Oxford.

262. ______, 1954. Aquatic Ascomycetes. Discomycetes from lakes. *Trans. Br. Mycol. Soc.*, **37** : 1-18.

263. ______, 1956. Dispersal in cup fungi. *Essex Naturalist*, **29**, pt. 5.

264. ______, 1960. Spore discharge in Pyrenomycetes. *Friesia*, **6** : 148-163.

265. ______, 1961. *The biology of fungi*. Hutchinson Educational Ltd., London.

266. Ingram, D.S. 1969. Abnormal growth of tissues infected with *Plasmodiophora brassicae. J. Gen. Microbiol*. **56**: 55-67.

267. Ingram, D.S., J.A. Argent and I.C. Tommercup. 1976. Structural aspects of infection by biotrophic fungi. *In* Biochemical aspects of plant-parasite relationships. Edited by J. Friend and D.R. Threefall. Academic press, hondose, New York, pp. 43-78.

268. Jackson, H.S. 1944. Life cycles and phylogeny in the higher fungi. Presidential address. *Trans. Roy. Soc. Can.*, **38** : 1-32.

269. James, T.Y. et al. (2006) Reconstruction of the early evolution of Fungi using a six-gene phylogeny. *Nature* **443** : 818-822.

270. James et al (2007): A molecular phylogeny of the flagellated Fungi (Chytridiomycota) and proposal for a new phylum (Blastocladiomycota). *Mycologia* **98** : 860-871.

271. Jones et al (2011): Discovery of novel intermediate forms redefines the fungal tree of life. *Nature*, **474** : 200-203.

272. Johnson, L.E.B., W.R. Bushnell and R.J. Zeyen, 1979. Binary pathways for analysis of primary infection and host response in populations of powdery mildew fungi. *Can. J. Bot.* **57** : 497-511.

273. Joshi, L.M. 1976. Recent contributions towards epidemiology of wheat rusts in India. *Indian Phytopathol.* **29** : 1-16.

274. Jump, J.A. 1954. Studies on sclerotization in *Physarum polycephalum. Am. J. Bot.*, **41** : 561-567.

275. Kajiwara, T. 1971. Structure and physiology of haustoria of various parasites. In Morphological and chemical events in plant parasite interaction. Edited by S. Akai and S. Ouchi. *The Phytopathological Society of Japan*, Tokyo pp. 225-277.

276. Kamat, M.N. 1959., 1961. *Handbook of mycology*. Prakash Publishing House, Poona.

277. Kamiya, N. 1950 b. The rate of protoplasmic flow in the myxomycete plasmodium. I. *Cytologia*, **15** : 183-193. II. 195-204.

278. ______, 1959. Protoplasmic streaming. *Protoplasmatologia*, Vol. 8, pt. 3a. 199 pp.

279. Karling, J.S. 1932. Studies in the Chytridiales VII. The organisation of the Chytrid thallus. *Am. J. Bot.*, **19** : 41-47.

280. ______, 1942. *The plasmodiophorales*. New york.

281. ______, 1954. The cytology of host reaction to infection by *Synchytrium australe. Am. J. Bot.*, **41** : 651-663.

282. ______, 1955 b. Prosori in *Synchytrium. Bull. Torrey. Bot. Club*, **82** : 218-236.

283. ______, 1955 c. Resting scpore germination in *Synchytrium australe* in relation to its classification. *Mycologia*, **47** : 185-192.

284. ______, 1955 d. The cytology of prosoral, soral and sporangial development in *Synchytrium australe. Am. J. Bot.*, **42** : 37-41.

285. Kershaw, K.A., and K. Alvin. 1966. *The observer's book of Lichens.* Frederick Warne and Co. Ltd. London.

286. Khan, S.R. 1977. Light and electron microscope observations of sporangium formation in *Albugo candida. Can. J. Bot.* **55** : 730-739.

287. ______, 1976. Ultrastructural changes in maturing sporangia of *Albugo candida. Am. Bot.* **40** : 1285-1292.

288. Kimbrough, J.W. 1970. Current trends in the classification of Discomycetes. *Bot. Review* **36** : 91-161.

289. Kislev, N., and I. Chet. 1973. Scanning electron microscopy of sporulating clusters of the myxomycete *Physarum polycephalum. Tissue and Cell* **5** : 349-357.

290. ______ ______, 1974. Scanning electron microscopy of Freeze-Fractured sclerotia of *Physarum polycephalum Tissue and Cell* **6** : 209-214.

291. Kirk P M, Cannon PF, David J C, Stalpers. J (2001): In: *Ainsworth & Bisby's dictionary of fungi*. 9th. Di, Wallingford, CAB international.

292. Kirk et al (2008): *Dictionary of the fungi* , (10th edn), CAB international.

293. Knoll M & Ruska E (1939): Das elektronenmikroskop. Zeitschrift fur physic, *Springler*, **78 (5-6)**: 318-339.

294. Koch, W.J. 1956, 1958, 1961. Studies of the motile cells of Chytrids I, II, III. *Am. J. Bot.* **43** : 811-819; **45** : 59-72; **48** : 786-788.

295. Koch, W.J. 1968. Studies of motile cells of Chytrids. V. Flagellar retraction in posteriorly uniflagellate fungi. *Am. J. Bot*., **55**: 841-859.

296. Koga, H.S. Mayama, and J. Shishiyama. 1980. Correlation between the deposition of fluorscent compounds in papillae and resistance in barley against *Erysiphe graminis hordei Can. J. Bot*. **58** (5) : 536-541.

297. Kole, A.P. 1957. Electron microscope observations on the soral zoospores of *Synchtrium endobioticum. Tijd. Plantenzeikt*. **63** : 361-364.

298. ______, and K. Hostra. 1959. Electron microscope observations on the flagella of the zoospores of *Phytophthora infestans.* Konikl, Nederl, *Akad. Wetensch. proc. ser. C. Biol. Med. Sci.* **62**: 404-408.

299. Korf. R.P. 1973. Discomycetes and Tuberales pp. 249-319. In G.C. Aisnworth, F.K. Sparrow and A.S. Sussman (eds.). *The Fungi*. Vol IV A. Academic Press, New York.

300. Kreiger, I.C.C. 1930. *The Mushroom Handbook.* McMillan Co., New York.

301. Kulkarni, G.S. 1942. Ergot in India. *Curr. Sci*., **11** : 246.

302. Kukkonen, I., and T. Vaissalo, 1964. An electron microscope study of spore formation in a smut. *Ann. Bot. Fenn.,* **1** : 236-249.

303. Lamb, I.M. 1959. Lichens. *Sci Am.* **201** : 144-156.

304. Langdon, R.F.N., and R.A. Fullerton. 1975. Sorus ontogeny and sporogenesis in some Smut Fungi. *Aus. J. Bot.*, **23** (6) : 915-930.

305. Lange, L., and L.W. Olson. 1978. The zoospore of *Synchytrium endobioticum Can. J. Bot*. **56** (10) : 1229-1239.

306. Laviola, C. 1974. on the fertilization tube of *Phytophthora infestans. Phytopathol*. **13** : 55-58.

307. Leonian, L.H. 1925. Physiological studies on the genus *Phytophthora. Am. J. Bot*., **12** : 444-498.

308. Leppik, E.E. 1953-1959. Some viewpoints on the phylogeny of rust fungi. I, II, III. Mycologia, **45**: 46-74; **48** : 637-654; **51** : 512-528.

309. Lessie, P.E., and J.S. Lovertt. 1968. Ultrastructural changes during sporangium formation and zoospore differentiation in *Blastocladiella emersonii. Am. J. Bot*., **55** : 220-236.

310. Lilly, V.G., and H.L. Barnett. 1951. *Physiology of the fungi*. McGraw Hill Book Co., New York.

311. Lindengren, C.C. 1945. An analysis of the method of budding in yeasts and some observations on the structure of the yeast cell. *Mycologia*, **37** : 767-780.

312. ______, 1949 b. *The yeast cell*, *its genetics and cytology*. Educational Publishers, Saint Louis.

313. ______, 1944. Sporulation in *Saccharomyces cerevisiae. Bot. Gaz.,* **105** : 304-316.

314. Lindier, D.H. 1940. Evolution of Basidiomycetes and its relation to terminology of the basidium *Mycologia*, **32** : 419-447.

315. Lodder, J. and N.J.W. Kreger-van Rij. 1952. *The Yeasts.* North Holland Publishing Co., Amsterdam.

316. Lodder, J., *et al*. 1958. The classification of yeasts. In *The chemistry and biology of yeasts.* A.H. Cook (Editor), Academic Press, New York.

317. Losel, D.M. 1964. The stimulation of spore germination in *Agaricus bisporus* by living mycelium. *Ann. Bot*. **28** : 541-554.

318. ______, 1967. The stimulation of spore germination in *Agaricus bisporus* by organic acids. *Ann. Bot*. **31** : 417-425.

319. Lunney, C.Z., Ch. E. Bland. 1976. Ultrastructural observations of mature and encysting zoospores of *Pythium proliferum* de bary. *Protoplasma* **90** : 119-137.

320. Luthra, J.C. and A. Sattar. 1934. Some experiments on the control of the loose smut of *Ustilago tritici* of wheat. *Indian Jour. Agri Sci.*, **4** : 177-189.

321. ______, 1953. Solar energy treatment of wheat loose smut (*Ustilago tritici*). *Indian Phytopathol*, **6** : 49-56.

322. Luttrell, E.S. 1980. Host-parasite relationship and the development of ergot sclerotium in *Claviceps purpurea. Can. J. Bot.* **58** (8) : 942-958.

323. Lynn, R.R., and P.T. Magee. 1970. Development of spore wall during ascospore formation in *Sachharomyces cerevisiae. J. Cell Biol.*, **44** : 688-692.

324. Macbride, Th. H., and G. W. Martin. 1934. *The Myxomycetes*. The Mc Millan Co., New York.

325. Macdonold, T.A. 1951. *Introduction to mycology*. Butterworths Scientific Publication, London.

326. Macfarlane, I. 1970. Germination of resting spores of *Plasmodiophora brassicae. Trans. Br. Mycol. Soc.* **55** : 97-112.

327. Machlis, L. 1958 a. Evidence for a sexual harmone in *Allomyces. Physiol. Plant.*, **11** : 181-192.

328. Maheshwari, P. and R.N. Kapil. 1963. *Progress of Botany*. Indian Science Congress Association, Calcutta.

329. Malik, M.M., and C.C. Batts. 1960. The infection of barley by loose smut (*Ustilago nuda*). *Trans. Brit. Mycol. Soc.*, **43** : 117-125.

330. ______, 1960. The development of loose smut (*Ustilago nuda*) in the barley plants and observation on spore formation in nature and culture. *Trans. Brit. Mycol. Soc.*, **43** : 126-131.

331. Malloch, D. 1976. *Agaricus brunnescens* : The cultivated mushroom. *Mycologia*. **68** (4) : 910-919.

332. Mandahar, C.L. 1978. *Introduction to plant viruses*. S. Chand & Co. Ltd., New Delhi.

333. Manocha, M.S. 1965. Fine structure of *Agaricus* carpophore. *Can. J. Bot.* **43** :1329-1333. ______ and J.R. Calvin, 1968. Structure of cell wall of *Pythium debaryanum. Jour Bacteriol.* **95** : 1140-1152.

334. Mantle, P.G., and I. Shaw. 1976. Role of ascospore reproduction by *Claviceps purpureae* in etiology of ergot disease in male sterile wheat. *Trans. Br. Mycol. Soc*. **67** : 17-22.

335. Manton, I., B. Clarke, and A.D. Greenwood. 1951 observations with the electron microscope on a species of *Saprolegnia. Jr. Exp. Bot.*, **2** : 321-331.

336. Marchant, R. 1968. An ultrastructural study of sexual reproduction in *Pythium ultimum. New Phytologist* **67** : 167-171.

337. Marchant, R., A. Peat, and G.T. Banbury. 1967. The ultrastructural basis of hyphal growth. *New Phytol.*, **66** : 623-629.

338. Martin, G.W. 1932. Systematic position of the slime molds and its bearing on the clssification of fungi. *Bot. Gaz.,* **93**: 421-435.

339. ______, 1933. The morphology of basidium. *Am. J. Bot.*, **25** : 682-685.

340. ______, 1940. The Myxomycetes. *Bot. Rev.,* **6** : 356-388.

341. ______, 1946. The genus *Aspergillus Science*, **193** : 116-117.

342. ______, 1950. *Outline of the Fungi.* W.C. Brown Co., Dubuque.

343. ______, 1955. Are fungi plants ? *Mycologia*, **47** : 779-792.

344. ______, 1961. The systematic position of the Myxomycetes. *Mycologia*, **52** : 119-129.

345. Martin, J.F., F. Uruburu, and J.R. Villaneva 1973. Ultrastructural changes in the conidia of *Penicillium notatum* during germination. *Can. J. Microbiol* **19** : 797-801.

346. Martin, M. and J.L. Gay 1983. Ultrastructure cf Conidium development in *Erysiphipisi. Can. J. Bot.* **6** (9) : 2472-95.

347. McIntosh, D.L. 1954. A cytological study of ascus development in *Pyronema confluens Can. J. Bot.* **32** : 440-446.

348. McKeen, W.E. 1962. The flagellation, movement and encystment of some phycomycetous zoospores. *Can. J. Microbiol.*, **8** : 897-903.

349. McLean, R.C. and W.R.I. Cook. 1951. *A text book of theoretical Botany*. Vol. I. Longmans, Green and Co., London.

350. McKeen, W.E. 1975. Electron microscope studies of a developing Pythium oogonium. *Can. J. Bot.* **53** : 2354-2360.

351. McLeague, D.M. Hutchinson, S., and Reeds, R.I. 1959. Spore germination in *Agaricus campesiris. Nature*, Lond. **183** : 1736.

352. McNeiil J, Barrie F R, Burdet H.M, Demoulin V, Hawks worth DL, Marholdk, Nicolson Dh, Prado J, Silva PC, Skog J E et al (2006): *International code of Botanical Nomenclature* (Vienna Code) adopted by 17th International Botanical Congress, Vienna, Austria, July,2005.

353. Mehrotra, B.S. 1976. *The Fungi.* Oxford-IBH R.S. 1980. *Plant Pathology*. Tata McGraw Hill, Calcutta.

354. Mehta, K.C. 1931. Annual recurrence of rust on wheats and barley in plains of India. *Indian Jour. Agri. Sci.*, **1** : 297-391.

355. ______, 1933. Rusts of wheat and barley in India. *Indian Jour. Sci.*, **13** : 939-966.

356. ______, 1940. *Further studies on cereal rusts in India*, Vols. I, and II. I. C.A.R., New Delhi.

357. ______, and B.P. Pal. 1954. Rust resistant wheat for India, *Nature*. **146** : 98.

358. Meier, H., and J. Webster. 1954. An electron microscope study of cyst in the Saprolegniaceae. *J. Exp. Bot.,* **5** : 401-409.

359. Meyer, R.J., and M.S. Fuller. 1985. Structure and development of hyphal septa in *Alloymyces. Am. J. Bot.* **72** (10) : 1458-1465.

360. Middleton, J.T. 1943. The taxonomy, host range and geographic distribution of the genus *Pythium. Mem. Torrey. Bot. Club.* **20** : 1-171.

361. Miller, J.H. 1949. A revision of the classification of the Ascomycetes with special emphasis on the Pyrenomycetes. *Mycologia*, **41** : 99-127.

362. Mitchell, D.T., and R.C. Cooke. 1968. Some effects of temperature on germination longevity of sclerotia in *claviceps purpurea. Trans. Br. Mycol. Soc.* **51** :721-729.

363. Mitter, J.H. 1939. Some recent contributions to our knowledge of heterothallism in fungi. *Jour. Indian. Bot. Soc.* , **15** : 183-192.

364. Miyake, K. 1901. The fertilization of *Pythium debaryanum. Ann. Bot.,* **15** : 653-667.

365. Moens, P.B. 1971. Fine structure of ascospore development in the yeast. *Sachharomyces cerevisiae. Can. J. Microbiol.* **17** : 507-510.

366. Mohanty, U. 1948. Studies on Indian Aspergilli. *Indian Phytopathol.*, **1** : 55-65.

367. Moore, H., and J. Muhlethaler. 1963. Fine structure of frozen-etched yeast cells. *J. Cell Biology* **17** : 609-628.

368. Moore, E.J. 1963. The ontogeny of apothecium of *Pyronema domesticum. Am. Jr. Bot.*, **50** : 37-44.

369. Moor, R.T., and J.H. McAlear. 1962. Fine structure of Mycota 7. Observations on septa of Ascomycetes and Basidiomycetes. *Am. Jr. Bot.*, **49** : 86-94.

370. Moore, R.T. 1965. The ultrastructure of fungal cells. In *The Fungi*; Vol. I, edited by Editors G.C. Ainsworth and A.S. Sussman; pp. 95-118. Academic Press, New York.

371. Moore, R.T. *et* R. Marchant. 1972. Ultrastructural characterisation of the basidiomycete septum of *Polyporus biennis. Can. J. Bot.*, **50** : 2463-2469.

372. Mukerjee, K.G. 1974. *Fungi of India* (1962-1973) Emkay Pub., Delhi.

373. Mullins, Thomas J. and Charles O. Warren. 1975. Nutrition and sexual reproduction in the water mold *Achlya. Am. Jour. Bot.*, **62** (7) : 770-774.

374. Mundkur, B.B. 1945. Studies in Indian cereal smuts VIII. Nomenclature of Indian smut fungi and their probable mode of transmission. *Ind. Jour. Agric. Sci.*, **15** : 108-110.

375. ______, and M.J., Thirumalachar, 1952. *Ustillaginales of India*. C.M.I. Kew, Surrey, England.

376. Mundkur, B.B. 1967. *Fungi and plant diseases*. Second edition. MacMillan and Co. Ltd., London.

377. Munk. A. 1954 b. The basis of the systematic relationships of Pyrenomycetes. *8th Int. Bot. Congr., Sect.* 18, 19, 20. pp. 35-44.

378. Mussak R and BEchtold T (2009) : *Handbook of natural colorants*. John Wiley & sons. New York.

379. Nickerson, W.J. and K.V. Thiman. 1943. The chemical control of conjugation in *Zygosaccharomyces*, II. *Am. J. Bot.*, **30** : 94-101.

380. Niederhauser, J.S. 1956. The blight, the blighest and the blighted *Trans. N.Y. Acad. Sci.*, **19** : 55-63.

381. Northcote, D.M., and R.W. Horne. 1952. The chemical composition and structure of the yeast cell wall. *Biochem. Jour*., **51** : 232-236.

382. Nygaard, O.F., Sophia Guttes, and H.P. Rusch. 1959. Nucleic acid metabolism in a mold with synchronous mitosis. *Biochem et Biophys. Acta*, **38** : 298-306.

383. O'Donnel, K.L. and E.S. Beneke. 1973. Apothecial formation by *Peziza quelepidotia* in pure culture. *Mycologia*, **65** : 913-915.

384. ______, and G.R. Hooper. 1974. Scanning and ultrastructural ontogeny of paragymnophymenial apothecia in the operculate discomycete *Peziza quelepidotia. Can. J. Bot.*, **52** : 873-876.

385. Olive, L.S. 1944. Development of the perithecium in *Aspergillus fisheri* with a description of crozier formation. *Mycologia*, **36** : 266-275.

386. ______, 1953. The structure and behaviour of fungus nuclei. *Bot. Rev.*, **19** : 439-586.

387. ______, 1954. Heterothallic behaviour in Aspergillaceae. *Mycologia*, **46** : 254-255.

388. ______, 1958. On the evolution of heterothallism in fungi. *Am. Natural*., **92** : 233-251.

389. Oliver, P.T.P. 1972. Conidiophore and spore development in *Aspergillus nidulans. J. Gen. Microbiol.*, **73** : 45-54.

390. Olson, L.W. 1973. The meiospore of *Allomyces. Protoplasma* **78** : 113-127.

391. Oso, B.A. 1969. Electromicroscopy of ascus development in *Ascobolus. Ann. Bot.* (Lond.), **33** : 205-209.

392. O'Sullivan, J., and D.M. Losel. 1971. Spore lipids and germination in *Agaricus bisporus, Arch. Microbiol* **80** : 277-283.

393. Ower, R. 1982. Notes on the development of morelascocarp, *Morchella esculenta. Mycologia* **74** : 142-144.

394. Padan, F., and Shilo, M. 1973. Cyanophages—viruses attacking blue-green algae. *Bacteriol. Review*, **37** : 343-370.

395. Paden, J.W., and E.A. Stanlake. 1973. Ascocarp development in *Ascobolus michandii. Can. J. Bot.* **51** : 1271-1273.

396. Padmanabhan, D., P. Vidyasekaran and C.K. Saumini Rajagopalan. 1974. Mineral composition of canker. halo and green regions of Citrus leaves infected by *Xanthomonas citri, Indian Phytopathology*. **XXVII** (4) 640-641.

397. ______, ______, ______, 1975. Accumulation of malic acid in halo region of citrus leaves infected by *Xanthomonas citri, Indian Phytopathol*. **28**: 276-277.

398. Page, R.M. 1952. The effect of nutrition on the growth and sporulation of *Pilobolus. Am. J. Bot*., **39** : 731-738.

399. ______, 1956. Studies on the development of asexual reproductive structures in *Pilobolus. Mycologia,* **48** : 206-224.

400. Pallotia, D.J., P.Y. Youngman, T.M. Shinnick and C.E. Holt. 1979. Kinetics of mating in *Physarum polycephalum. Mycologia* **LXXI** : 68-74.

401. Palm, B.T. 1932. Biological notes on *Albugo. Ann. Mycol.* **30** : 421-426.

402. Patel, M.K., G.W. Dhande, and Y.S. Kulkarni. 1950. A modified treatment of loose smut of wheat. *Curr. Sci.*, **19** : 324-325.

403. Patterson, M.A. and R.W. Sheetz.1977. Spherule formation in *Physarum polycephalum. Protoplasma*. **92** : 253-261.

404. Person, L.H., and G.B. Lucas. Oospore germination in *Peronospora tabacina. Phytopath*., **43** : 701-702.

405. Peterson, R.H. 1974. The Rust Fungus Life Cycle. *The Botanical Review*. Vol. **40** (4) : 453-513.

406. Pommerville, J. 1978. Analysis of gametic and Zygote motility in Allomyces. *Exp. Cell Research* **113** : 161-172.

407. Pontecorvo, G. and J.A. Roper. 1952. Genetic analysis without sexual reproduction in *Aspergillus niger*. J. Gen. Microbiol. **6** : VII (Abstr.).

408. Popp, W. 1959. A new approach to the embryo test for predicting loose smut of wheat in adult plantsd. *Phyto pathol*. **49** : 75-77.

409. Prahlad Rao, L.S., and M.K. Subramaniam. 1953. Studies on the cytology of yeasts. Part VII. Nuclear phenomenon in cells from 24 hour slants. *Proc. Indian Acad. Sci.* **37** : 72-81.

410. Prasada, R. 1948. The formation and germination of teleospores of rusts. I. *Indian Phytopathol*. **1** : 119-126.

411. ______, and V.C. Lele. 1952. New Physiologic races of wheat rusts in India. *Ind. Phytopath*. **5** : 128-129.

412. Pristou, R., and M.E. Gallegly. 1954. Leaf penetration by *Phytophthora infestans. Phytopath*., **44** : 81-86.

413. Purkayastha, R.P., and A. Chandra. 1974. A new species of edible mushrooms from India. *Trans. Br. Mycol. Soc*. **62** : 416-418.

414. ______, and ______, 1975. *Termytomyces eurhizus*, a new Indian edible mushroom. *Trnas. Br. Mycol. Soc*. **64** : 168-170.

415. Pushkar Nath, and M.W. Padwick. 1941. Ergot in India. *Curr. Sci*., **10** : 488-489.

416. Rai, J.N. 1970. Ecological specialisation in microfungi of Usar (alkaline) soils of India. U.S. PL 480 report, Lucknow University.

417. Ramakrishnan, T..S. 1941. Studies in the genus *Colletotrichum* I. *Proc. Ind. Acad. Sci*., **3** : 60-70.

418. ______, 1950. Some interesting rusts of South India. *Ind. Phytopath*., **3** : 43-50.
419. Ramsbottom, J. 1953. *Mushrooms and toadstools.* Collins, London.
420. Randall, L.P., and Lynch. D.L. 1974. Spore ultrastructure of the Myxomycete *Physarum polycephallum. Am. J. Bot*. **61** : 513-26.
421. Rane, M.S. and M.K. Patel. 1959. Diseases of cotton in Bombay, i. Altemaria leaf spot. *Ind. Phytopath.,* **9** : 106-113.
422. Raper, J.R. 1939. Sexual hormones in *Achlya.* I. Indicative evidence for a hormonal coordinating mechanism. *Am. Jr. Bot.* **26** : 639-650.
423. Raper J R (1954): Life Cycles, sexuality and mechanism in the fungi. In *Sex in Microorganisms*. Amer. Assoc. Adv. Sci. Washington.pp 42-81.
424. ______, 1940. Sexuality in *Achlya ambisexualis. Mycologia.,* **32** : 710-727.
425. ______, 1951. Sexual hormones in *Achlya. Am. Scientist*., Vol. 39. no. 1, pp. 110-120.
426. ______, 1957. Hormones and sexuality in lower plants. *Symp. Soc. Exp. Biol.*, **XI**, pp. 143-165.
427. ______, 1959. Sexual versatility and evolutionary processes in fungi,. *Mycologia*., **51** : 107-124.
428. ______, 1959 a. Parasexual phenomenon in Basidiomycetes. *Proc. IX Int. Bot. Cong*., **2** : 318.
429. ______, 1960. The control of sex in fungi. *Am. J. Bot.* **47** : 794-808.
430. Rast, D., and G.O. Hollenstein. 1977. Architecture of the *Agaricus bisporus* spore wall. *Can. J. Bot.* **55** : 2251-2262.
431. Reeves, F., Jr. 1967. The fine structure of ascospore formation in *Pyronema domesticum. Mycologia*. **59** : 1018-1033.
432. Reichle, R.E., and M.S. Fuller. 1967. The fine structure of *Blastocladiella emersonii* Zoospore. *Am. Jr. Bot.*, **54** : 81-92.
433. Remy W, Taylor T.N. , Hass H, & Kerp H (1994): Four hundred million year old vesicular arbuscular mycorrhizae. *Proc. Natl. Acad. Sci*., USA **91** : 11841-11843.
434. Rice, M.A. 1927. The haustoria of certain rusts and the relation between the host and the pathogen. *Bull. Torrey. Bot. Club*., **79** : 126-138.
435. Rijkenberg, F.H.J., and S.J. Trutter. 1973. Haustoria and intercellular hyphae in Rusts. *Phytopathol*, **63** : 281-286.
436. Robards, A.W. 1968. A new interpretation of plasmodesmatal ultrastructure *Planta*, **82** : 200-210.
437. ______, 1971. The ultrastructure of plasmodesmata. *Protoplasma*, **72** : 315-326.
438. Robinow, C.F. 1957. The structure and behaviour of the nuclei in spores and growing hyphae of Mucorales. I, II. *Can. Jr. Microbiol*., **3** : 771-789.
439. Rogers, D.P., 1934. The basidium. Univ. *Jawa Stud. Nat. Hist*., **16** : 160-183.
440. Roland, J.F., *et al*. 1960. Calvacin, a new antitumor agent. *Science*, **132** : 1897.
441. Romero, S., and M.E. Gallelgy. 1963. Oogonium germination in *Phytophthora infestans Phytopathol.* **53** : 899-903.
442. Ross, J.K. 1961. Further studies in meiosis in the Myxomycetes. *Am. Jr. Bot*., **48** : 244-248.
443. Rovira A.D. (1965): Plant root exudates and their influence upon soil micro-organisms. In: *Ecology of soil borne Plant pathogens*- prelude to Biological control (Eds: K.F. Baker 7&W.C.Synder), 170 -186pp. John Murry, London.

444. Rovira A.D., Foster R.C., Martin J.K. (1979): Note on terminology: origin, nature and nomenclature of the organic materials in the rhizosphere. In: *The root-soil Interface* (Eds; J.L. Harley and R.S. Russell) 1-4pp. acad. press, London.

445. Royan, S. 1956. Relation between the nucleus and the vacuole in yeast. *Curr. Sci.*, **25** : 397-398.

446. Ruber, D.M., and M.E. Stanghellini. 1978. Ultrastructure of oospore germination in *Pythium aphanidermatum Am. J. Bot.* **65** (5) : 491-501.

447. Ruinen J(1961): Phyllosphere. I.An ecologically neglected milieu. *Plant Soil* **15** : 18-109.

448. Saccardo, P.A. 1882-1931. *Syllage Fungorum Hucusque Cognitonum*. 25 Vols. Published by the author himself Pavia.

449. Saksena, R.K. 1953. Some aspects of physiology and cytology of Pythiaceae. Presidential address. *Proc. 4th Indian Sci. Congress*. Pt. III : 1-17.

450. ______, 1953. A new genus of Mucorales. *Mycologia*. **45** : 426-436.

451. ______, 1957. On some affinities of Mucorales. *Bull. Bot. Soc. Univ. Saugor*. **9** : 1-6.

452. Salvin, S.B. 1940. The occurrence of five successive swarming stages in a nonsexual *Achyla. Mycologia*. **32** : 148-154.

453. Sansome, E. 1963. Meiosis in *Pythiyum debaryanum* and its significance in the life history of the Biflagellate. *Trans. Br. Mycol. Soc*. **46** : 63-72.

454. ______, 1965. Meiosis in the diploid and polyploid sex organs of *Phytophthora* and *Achlya. Cytologia* (Tokyo), **30** : 103-117.

455. ______, and C.M. Braisier. 1973. Diploidy and chromosomal structural hybridity in *Phytophthora infestans. Nature* (Lond.). **241** : 344-345.

456. ______, and B. J. Harris. 1962. Use of Camphor-induced polyploidy to determine the place of meiosis in the fungi. *Nature* (Lond.), **196** : 291-292.

457. ______, and F.W. Sansome. 1974. Cytology and life history of *Peronospora parasitica* on *Capsella bursa-pastoris* and Albugo candida on *C. bursa pastoris. Trans. Br. Mycol. Soc.* **62** : 323-332.

458. Sansome, E.R. 1976. Gametangial meiosis in *Phytophthora capsica. Can. J. Bot*. **54** : 1535-45.

459. Santelli, V. 1958. Zoospore morphology of the Californian strain of *Plasmopara viticola. Nature* (London), **181** : 924-925.

460. Sarbhoy, A.K. 1997. Fungi of India (1982-1992). Daya Pub. House, Delhi.

461. Sauer, H.W., K.L. Babcock, and H.P. Rusch 1969. Sporulation in *Physarum polycephalum Exp. cell Res.* **57** : 319-327.

462. Savage, E.J., C.W. Clayton, J.H. Hunter, J.A. Brenneman, C. Laviola and M.E. Gallegly 1968. Homothallism, heterothallism and interspecific hybridization in the genus *Phytophthora Phytopathol* **58** : 1004-1021.

463. Schmidt, E.C. 1983. Spore germination and carbohydrate colonization by *Marchella esculenta* at different soil temperatures. *Mycologia* **75** (5) : 870-875.

464. Schuttle, K.H. 1956. Translocation in the fungi. *New Phytol*, **55** : 164-182.

465. Scott, G.D. 1960. Studies of the lichen symbiosis. 1. The relationship between nutrition and moisture content in the maintenance of the symbiotic state. *New Phytol*., **59** : 374-381.

466. Scale, T. 1973. *Life cycle of Neurospora crassa* viewed by scanning electron microscopy. *J. Bacteril.*, **113** : 1015-1025.

467. Seaman, W.L.R., H. Larson and J.C. Walker. 1961. Tinsel type structure in the flagella of zoospores associated with cabbage club root tissue. *Nature* (Lond.). **190** : 186-187.

468. Sentandreu, R.D.H. Northcote, 1969. The formation of buds in yeast. *J. Gen. Microbiol*. **55** : 393-398.
469. Sharma, P.D. 1998. *The Fungi*, Rastogi Publications, Meerut.
470. Sharma, R. and the Late R.H. Cammack. 1976. Spore germination and taxonomy of *Synchytrium endobioticum* and *S. succisae Trans. Br. Mycol. Soc*., **66** : 137-147.
471. Sharma D, C. Gupta, S. Aggarwal & N.Nagpal (2012): Pigment extraction from fungus for textile dyeing. *Ind .J. of Fiber and textile Res*, **37** : 68-73.
472. Shaw, B.I., and P.G. Mantle, 1980. Host infection by *Claviceps purpurea. Trans. Br. Mycol. Soc*. **75** (1) : 77-90.
473. Shaw, M. *et* M.S. Manocha. 1965. The physiology of host-parasite relations, 15. Fine structure in rust infected wheat leaves. *Can. J. Bot*., **43** : 1285-1292.
474. Shepherd, C.J. 1978. Mating behaviour of Australian isolates of *Phytophthora* species I. Intra-and interspecific mating. *Aust. J. Bot*. **26** (2) 123-38.
475. Shimony, C., and J. Friend. 1973. Ultrastructure of the interaction between *Phytophothora infestans* and the leaves of two cultures of Potato (*Solanum tuberosum*) Orion and Majestic. *New Phytologist* **74** : 59-65.
476. Singer, R. 1949. The "Agaricales" (Mushrooms) in modern Taxonomy. *Lilloa*, **22** : 5-832.
477. Singh, R.N., and P.K. Singh. 1967. Isolation of cyanophages from India. *Nature* (Lond.) **216** : 1020-1021.
478. Singh, P.K. 1973. Occurrence, distribution of Cyanophages in ponds, sewage and rice fields. *Arch. Microbiol*. **89**: 169-172.
479. Singh, R.B. 1997. *Plant Pathology*. Central Book Depot., Allahabad.
480. Singh, R.S. 1969. *Plant diseases*, Second edition. Oxford and I.B.H. Publishing Co., Calcutta.
481. Singleton, J.R. 1953. Chromosome morphology and the Chromosome cycle in the ascus of *Neurospora crassa. Am. Jr. Bot*., **40** : 124-144.
482. Sinha, A.K. 1965. Studies in Physiology of Fungi. Ph. D. thesis, Lucknow University.
483. Smith, A.L. 1931. Lichens from Northern India. *Trans. Brit. Mycol. Soc*., **16** : 121-132.
484. Smith, D.C. 1960. Studies in the physiology of lichens, 3. Experiments with directed discs of *Peltigera polydactyla. Ann. Bot*., **24** : 186-199.
485. Smith, G. 1900. The haustoria of Erysipheae. *Bot. Gaz*., **29** :153-184.
486. Smith, G.M. 1955. *Cryptogamic Botany*. Vol. I, McGraw Hill Book Co. New York.
487. Smoot, J.J., *et al*. 1958. Production and germination of oospore of *Phytophthora infestans. Phytopath*., **48** : 165-171.
488. Sparrow, F.K. 1960. *Aquatic Phycomycetes*. Second rev. ed. Univ. Michigan Press, Ann. Arbor.
489. ______, 1958 (1959). Inter-relationship and phylogeny of the aquatic Phycomycetes. *Mycologia*. **50** : 797-813.
490. Sporne, K.R. 1959. On the phylogenetic classification of plants. *Am. J. Bot.*, **46** : 385-394.
491. Srinath, K.V. 1946. The cytology of yeast. *Curr. Sci*., **15** : 25.
492. Srivastava, H.C., and P.R. Mehta. 1951. A new species of *Cerospora* on *Grewia asiatica Linn. Indian Phytopath*. **4** : 67-70.
493. Srivastava, J.P. 1966. *An Introduction to Fungi*. Central Book Depot. Allahabad.
494. Srivastava, U.S., R.A. Rai, and J.M. Aggarwal. 1972. Chemical control of wheat rusts in Kota region of Rajasthan. *Indian Phytopathol*. **25** (2) : 295-297.
495. Stakman, E.C., and J.J. Christensen. 1927. Heterothallism in *Ustilage Zeae. Phytopathol*. **17** : 827-834.

496. ______, 1947. Variations in *Ustilago Zeae*. Science, **85** : 58-59.

497. Stephenson, L.W., and D.C. Erwin. 1972. Encirclement of the oogonial stalk by the amphigynous antheridium in *Phytophthora capsici Can. J. Bot.*, **50** : 2439-2441.

498. ______, and J.V. Leary. 1974. Meiotic configurations in the oospore of *Phytophthora capsici. Can. J. Bot.*, **52** : 2141-2143.

499. ______, 1974. Cytology of somatic and gametangial nuclei in *Phytophthora capsici* and *Phytophthora megasperma. Can. J. Bot.,* **52** : 2055-2060.

500. Stevens, F.L. 190. Gametogenesis and fertilization in *Albugo. Bot. Gaz.,* **32** : 77-98; 157-168; 238-261.

501. Storck, R., and R.C. Morril. 1977. Nuclei, nucleic acids and protein in the sporangiospores of *Mucor bacilliformis* and other *Mucor* species. *Mycologia* **69** : 1031-1041.

502. Subramanian, C.V. 1973. Reflections on the conidium. *J. Ind. Bot. Soc*. **52** 3 -16.

503. Subramaniam, M.K. 1946. Studies in the cytology of yeasts. Mitosis in *Saccharomyces cerevisiae. Proc. Nat. Inst. Sci. India*. **12** : 143-149.

504. Sullivan, T.P., W.R. Bushnell, and J.B. Rowell, 1974. Relations between *Erysiphe graminis* and host cytoplasm in cells opened by microsurgery. *Can. J. Bot*. **52** : 987-998.

505. Suryanarayana, D. 1965. Studies on the Downy Mildew disease of Millets in India. *Indian Phytopath. Soc. Bull*. **3** : 72-78.

506. Talbot, P.H.B. 1971. *Principles of Fungal Taxonomy.* The MacMillan Press, London.

507. Tanaka, K., and T. Yanagita. 1963. Electron Microscopy on ultrathin sections of *Aspergillus niger* 1. Fine Study of hyphal cells. *J. Gen. Applied Microbiol*. (Tokyo). **9** : 101-118.

508. Tandon, R.N. 1959. *Mucorales of India*,. I.C.A.R., New Delhi.

509. Tandon, R.N., and K.S. Bilgrami. 1954. *Phyllosticta cycadina* (Passer) on *Cycas revoluta Curr. Sci.* **23** : 370-371.

510. ______, 1957. Leaf spot disease of *Artocarpus heterophyllous* caused by *Phyllosticta artocarpina. Proc. Nat. Acad. Sci*., **27** : 204-209.

511. Tandon, R.N., and J.P. Srivastava, 1958. Studies on spore characters of *Alternaria tenuis. Proc. Nat. Acad. Sci*. **28** : 155-171.

512. Tatum, F.L., and J. Lederberg. 1947. Gene recombination in the bacterium *Eischerichia coli. Jour. Bact.*, **53** : 673-684.

513. Taylor T.N, WinfriedR, Hass H (1992) Fungi from lower devonian Rhynie chert: Chytridiomycetes. *Amm. J. Bot.* **79 (11)**: 1233-1241.

514. Tewari, J.P., and W.P. Skoropad. 1977. Ultrastructure of oospore development in *Albugo candida* on rape seed (*Brassica campestris*). *Can. J. Bot*. **55** : 2348-2357.

515. Tewari, M.M. and H.C. Arya. 1966. Studies on green ear disease of Bajra (*Pennisetum typhoides*) caused by *Sclerospora graminicola*. *Indian Phytopath*. **19** : 125 (Abs.).

516. Thakur, S.B. 1977. Percurrent proliferation of sporangiospores in the genus *Albugo Mycologia* **69** : 637-641.

517. Thiman, K.V. 1955. *The life of bacteria*. MacMillan Co., New York.

518. Thirumalacher, M.J., M.D. Whitehead and J.S. Boyle. 1949. Gametogenesis and oospore formation in *Cystopus* (*Albugo evolvuli*). *Bot. Gaz.,* **110** : 487-491.

519. Timmer, L.W., J. Castro, D.C. Erwin, W.L. Belser and G.A. Zentmyer. 1970. Genetic evidence for zygotic meiosis in *Phytophthora capsici. Am. Jr. Bot.*, **57** : 1211-1218.

520. Thom. C. and K.B. Raper. 1945. *A Manual of the Aspergilli*. Williams and Wilkins Co., Baltimore.

521. Thomas, K.M. and T.S. Ramakrishnan, 1947. Studies in the genus *Phytophthora*. Oospore formation and taxonomy of *Phytophthora palmivora* Butler. *Proc. Ind. Acad. Sci.*, **26** : 147-163.

522. Thomas, N.T. and M.S. Chatrath. 1974. Influence of various factors on the efficacy of two systemic fungicides for control of loose smut of wheat. *Indian Pathology*. **XXVII** (4) : 493-498.

523. ______, 1975. Control of loose smut of wheat by seed treatment with thiabendazol. *Indian Phytopathol*. **28** : 300-301.

524. Tkacz, J.S. and V.L. Mackay. 1979. Sexual conjugation in yeast, cell surface changes in response to action of mating hormones. *J. cell biology* **80** : 326-333.

525. Tommercup, I.C., and D.S. Ingram. 1971. The life cycle of *Plasmodiophora brassicae* in *Brassica* tissue cultures and intact roots. *New Phytol*. **70** : 327-332.

526. Thurston, H.D. 1971. Relationship of general resistance to *Phytophthora infestans*. Late blight of potato. *Phyto pathol*. **61** : 620-626.

527. Vanterpool, T.C. 1939. Homothallism in *Pythium. Mycologia*. **31** : 124-127.

528. ______, 1959. Oospore germination in *Albugo candida. Can. Jr. Bot.*, **37**: 169-172.

529. Vasudeva, R.S. 1960. *The Fungi of India*, ICAR, New Delhi.

530. Vasudeva, R.S., and M.R.S. Iyengar. 1950. Control of loose smut of barely. *Curr. Sci.*, **19** : 218-219.

531. ______, *et al*. 1961. Distribution and prevalence of physiologic races of wheat and barley in India during 1952-57. *Ind. Phytopath*. **14** : 61-76.

532. Velisek J & Cejpek K. (2008); *Biosynthesis of food components*. Ossis, Tabor.

533. Velisek J and Cejpek K (2011): Pigments of higher fungi: A Review. *Czech J. food sci*., **29** (2): 87-102

534. Verma, J.P. 1986. *The Bacteria*. MPH, New Delhi.

535. Verma, P.R., and G.A. Petrie. 1975. Germination of oospores of *Albugo candida. Can. J. Bot.*, **53** (9) : 836-842.

536. ______, H. Harding, G.A., Petrie, M. William 1975. Infection and temporal development of mycelium of *Albugo candida* in cotyledons of 4 Brassica species. *Can. J. Bot.* **53** : 1016-1020.

537. Vining, L.C., and W.A. Taber. 1959. Estimation of ergot alkaloids in cultures of *Claviceps purpurea. Can. Jr. Microbiol*. **5** : 441-451.

538. Vogel, F.S., and R.F. Weaver. 1972. Concerning the dormancy in spores of *Agaricus bisporus Exp. cell Res.* **75** : 95-104.

539. Vujicic. 1971. Ultrastructural study of sexual reproduction in *Phytophthora palmivora. Trans. Br. Mycol. Soci.* **57** : 525-530.

540. Wager, H. 1896. On the structure and reproduction of *Cystopus candidus. Ann. Bot.*, **10** : 295-302.

541. Walker, J.C. 1969. *Plant Pathology*. 3rd edition. McGraw Hill Co., New York.

542. Ward, E.W.B., and K.W. Ciurysek. 1961. Somatic mitosis in a basidiomycete. *Can. Jr. Bot.,* **30** : 1497-1504.

543. Waterhouse, G.M. 1956. *The genus Phytophthora*. C.M.I., Kew, Surrey, England.

544. Waterman, M.A., J.R. Aist and H.W. Israel. 1978. Centrifugation studies to clarify the role of papilla formation in compatible barley powdery mildew interactions. *Phytopathol*., **68** : 797-802.

545. Walkins, J.E., and L.J. Littlefiled. 1976. Relationship in Waldron wheat to infection by *Claviceps purpurea. Trans. Br. Mycol. Soc*. **66** : 362-363.

546. Webster, J. 1970. *Introduction to Fungi.* Cambridge Univ. Press.

547. Webster J & Weber R (2007): *Introduction to Fungi*. 841p, Cambridge Univ.Press.

548. Weber R.W.S. (2009): Recent developements in molecular taxonomy of Fungi. (In *Physiology & Genetics*, Eds. T. Anke and Weber D.) Springler Verlag.

549. Whitehouse, H.L.K. 1949. Heterothallism and sex in fungi. *Biol. Rev. Cambridge Phil. Soc.*, **24** : 411-447.

550. Whittaker (1996): New Concepts of kingdoms of organism. *Science*, **163** : 150-160.

551. ______, 1954. Incompatibility in fungi. *Proc. 8th Int. Bot. Congr.* pp. 151-160.

552. Widra, A., and E.E. Delamater. 1955. The cytology of meiosis in *Schizosaccharomyces octosporus. Am. Jr. Bot.*, **42** : 423-435.

553. Wilkins, A.S., and G. Reynolds. 1979. The development of sporulation competence in *Physarum polycephalum. Develop. Biol.* **72** : 175-181.

554. Williams, P.H., S.J. Aist and J.R. Aist. 1971. Response of Cabbage root hairs to infection by *plasmodiophora brassicae. Can. J. Bot*. **49** : 41-47.

555. ______, and S.S. McNabola 1970. Fine structure of the host parasite interface of *Plasmodiophora brassicae*. in cabbage *Phytopath*., **60** : 1557-1561.

556. ______, 1967. Fine structure of *Plasmodipohora brassicae* in sporogenesis *Can. J. Bot*. **45** : 1665-1669.

557. ______, and Y.B. Yukawa. 1967. Ultrastructural studies on the host parasite relations of *Plasmodiophora brassicae. Phytopathol*. **57** : 682-687.

558. Williams, W.T., and R.K. Webster. 1970. Electron microscopy of the sporangium of *Phytophthora capsici. Can. Jr. Bot.*, **48** : 221-227.

559. Wilson, C.L., and I.K. Ross. 1957. Meiosis in the Myxomycetes. *Am. Jr. Bot.*, **42** : 743-749.

560. Wilson, I.M. 1952. The ascogenous hyphae of *Pyronema confluens. Ann. Bot.,* **16**: 321-339.

561. Wilson, U.E., and M.D. Coffey. 1980. Cytological evaluation of general resistance to *Phytophthora enfestans. Ann. Bot. (*Lond.) **45** : 81-90.

562. Winge, O., and C. Roberts. 1958. Life history and cytology of yeasts. In *Chemistry and biology of yeasts.* A.H. Cook (Editor). Academic Press, New York.

563. Wolf, F.A., and F.T. Wolf. 1947. *The Fungi*. Vols. I and II. John Wiley and Sons, New York.

564. Wolf, F.T. 1959. Chemical nature of the photoreceptor pigment inducing fruiting of plasmodia of *Physarum polycephalum*. In *Photoperiodism and related phenomenon in plants and animals.* pp. 321-326. American Association for the Advancement of Science, Washington, D.C.

565. ______, 1974. The cultivation of rust fungi upon artificial media. *Can. Jr. Bot.*, **52** : 767-772.

565. Yadav, A.S. 1953. A contribution to the Uredineae of Bihar. *Ind. Phytopath*., **6** : 86-91.

567. Yang, S.L., and A.H. Ellingboe. 1972. Cuticle layer as a determining factor for the formation of mature appressoria of *Erysiphe graminis* on wheat and barley. *Phytopathol*. **62** : 708-714.

568. Yoon, K.S., and D.J. Mclaughlin. 1979. Formation of a hilar appendix in basidiospores of *Boletus rubinellus. Am. J. Bot.*, **66** : 870-873.

569. Yu, C.C. 1954. The culture and spore germination of Ascobolus with special emphasis on *Ascobolus magnificus. Am. J. Bot.,* **41** : 21-30.

570. Yuill, E. 1950. The numbers of nuclei in conidia of aspergilli. *Trans. Br. Mycol. Soc*., **33** : 324-331.

571. Yukawa, Y.S., and S. Tanaka. 1979. Scanning electron microscope observations on the resting sporangia of *Plasmodiophora brassicae* in the clubroot tissue of turnip *(Brassica rapa* var. *globra*) after alcohol cracking. *Can. J. Bot.*, **57** : 2528-2532.

572. Zeyen, R.J., and W.R. Bushnell 1979. Papilla response of barley epidermal cells caused by *Erysiphe graminis* : rate and method of deposition determined by micro-cinematography and transmission electron micro-microscopy. *Can. Jr. Bot*. **57** : 898-913.

573. Ziegler, A.W. 1948. A comparative study of zygote germination in the Saprolegniaceae. *Jr. El. Mitchell Sci. Soc.*, **64** : 13-40.

574. _______, 1953. Meiosis in Saprolegniaceae. *Am. J. Bot.*, **40** : 60-66.

575. Zentmyer, G.A., and D.C. Erwin. 1970. Development and reproduction of *Phytophthora. Phytophathol*. **60** : 1120-1127.

INDEX

A

C

D

G

H

I

K

L

M

N

O

P

R

S

W

X

Y

Z